Handbook of Systems and Complexity in Health

Joachim P. Sturmberg • Carmel M. Martin
Editors

Handbook of Systems and Complexity in Health

Springer

Editors
Joachim P. Sturmberg
Departments of General Practice
Monash University
Melbourne, VIC, Australia

The Newcastle University
Newcastle, NSW, Australia

Wamberal
NSW, Australia

Carmel M. Martin
Northern Ontario School of Medicine
Greater Sudbury, Canada

Department of Public Health
and Primary Care
Trinity College
Dublin, Ireland

Merchandise Mart Plaza
Chicago, IL, USA

ISBN 978-1-4614-4997-3 (hardcover) ISBN 978-1-4614-4998-0 (eBook)
ISBN 978-1-4939-2255-0 (softcover)
DOI 10.1007/978-1-4614-4998-0
Springer New York Heidelberg Dordrecht London

Library of Congress Control Number: 2012953607

Printed on acid-free paper

Springer is part of Springer Science+Business Media (www.springer.com)

We dedicate this book to

Our teachers, in particular

Aristotle and Plato
Paul Cilliers
Edgar Morin
Ilya Prigogine
Stuart Kauffman
Fritjof Capra
Thure von Uexküll
Hannes Pauli
Ian McWhinney
David Snowden
and
Reuben McDaniel

Our colleagues,

Those who will take up the challenge to carry the agenda forward,

and last but not least

Our families for their support during this endeavour.

Prologue

Today the network of relationships linking the human race to itself and to the rest of the biosphere is so complex that all aspects affect all others to an extraordinary degree. Someone should be studying the whole system, however crudely that has to be done, because no gluing together of partial studies of a complex nonlinear system can give a good idea of the behavior of the whole.

Murray Gell-Mann

The *Handbook of Systems and Complexity in Health* is the first of its kind, spanning the theoretical and pragmatic dimensions of the nonlinear and complex systems sciences for all involved in healthcare. The editors are particularly thankful to the early thinkers in the field for their passion, innovative minds and their generosity. We really appreciate the time and effort contributed to writing for an audience still largely unaware of this field of inquiry. We also thank our publisher, Khristine Queja, for her patience and ongoing support in putting this book together. It is hoped that this *Handbook* will stimulate students, researchers and practitioners to take up the challenge of other ways of taking the emergent field into the mainstream of health-related knowledge generation.

Two historical notes: The earliest paper we were able to identify goes back to 1935, published in the New England Journal of Medicine, entitled *Physician and Patient as a Social System.* Its author, Henderson, alludes to the consequences of a simple statement like *"This is a carcinoma."* can have on the consultation. He writes:

> … [this statement] *will produce a response and the response, together with the mechanism that is involved in its production, is an extremely complex one, at least in those cases where a not too vague cognition of the meaning of the four words is involved in the process. For instance, there are likely to be circulatory and respiratory changes accompanying many complex changes in the central and peripheral nervous system. With the cognition there is a correlated fear. There will probably be concern for the economic interests of others, for example, of wife and children. All these intricate processes constitute the response to the stimulus made up of the four words, "This is a carcinoma", in case the statement is addressed by the physician to the patient, and it is obviously impossible to produce in the patient cognition without the accompanying affective phenomena …* (NEJM 1935, 212(18): 819–822)

In a second paper, published in 1949 in The Journal of the American Medical Association, Reimann for the first time describes the dynamic behaviour of disease: periodic fever, periodic abdominalgia, cyclic neutropenia, intermittent arthralgia, angioneurotic edema, anaphylactoid purpure and periodic paralysis (JAMA 1949; 141(3): 175–183).

The first paper precedes, and the second coincides with the "formal" emergence of the systems and complexity literature and its seminal papers by Ludwig von Bertalanffy and Norbert Weiner. However, these early "revolutionary" papers had no impact on the practice or the research agendas in the health professional domains. It took another 30 years before the next systems and complexity papers were published.

Systems and complexity thinking, today well established in many science areas, is still in its infancy in the health disciplines. We hope that this *Handbook*, by providing an overview of the philosophical and scientific background, and by demonstrating its applicability across all areas of the health professional work, will propel the field. There are significant benefits to be gained and shared for health professionals and patients alike from this different way of looking and understanding health, illness and disease, healthcare delivery, from teaching health professionals and from organising health services. Let us meet the challenge, and challenge ourselves.

> *I know that most men, including those at ease with problems of the greatest complexity, can seldom accept even the simplest and most obvious truth if it be such as would oblige them to admit the falsity of conclusions which they have delighted in explaining to colleagues, which they have proudly taught to others, and which they have woven, thread by thread, into the fabric of their lives.*
>
> Count Leo Tolstoy

About the Editors

Joachim P Sturmberg MBBS, DORACOG, MFM, PhD, FRACGP I am Adjunct A/Prof of Monash University, Melbourne, and Conjoint A/Prof of General Practice in the Department of General Practice in the Department of General Practice, General Practice, The Newcastle University, Newcastle, Australia. I am a graduate from Lübeck Medical School, Germany, where I also completed my PhD. Since 1989 I work in an urban group practice on the NSW Central Coast, with a particular interest in the ongoing patient-centred care of patients with chronic disease and the elderly. In 1994 I started to pursue systems and complexity research with an inquiry into the effects of continuity of care on the care processes and outcomes. Since then my research has expanded and includes the areas of understanding the complex notion of health, health care and healthcare reform, showing that health is an interconnected multi-dimensional construct encompassing somatic, psychological, social and semiotic or sense-making domains, that health care has to embrace the patient's understanding of her health as the basis for effective and efficient care, and that an effective and efficient healthcare system ought to put the patient at the centre. I am joint chief editor of the Forum on Systems and Complexity in Medicine and Healthcare in the Journal of Evaluation in Clinical Practice. Together with Carmel M. Martin and Jim Price I chair the Complexity SIG in WONCA.

Carmel M. Martin MBBS, MSc, PhD, MRCGP, FRACGP, FAFPHM I am Visiting Academic Department of Public Health and Primary Care, Trinity College, Dublin, and an Associate Professor of Family Medicine at the Northern Ontario School of Medicine. I am active in clinical general practice with a particular interest in chronic disease and illness and patient-centered care and integrated care using information technology, informed by complex adaptive systems theories. I am an Australian medical graduate from the University of Queensland. I completed my Masters in Community Medicine at the London School of Hygiene, University of London, and my PhD at the Australian National University. My research in Australia, Canada and Ireland has focused on reforms related to primary health care and chronic care, the nature of health in body, mind, society and the environment and meaning and sense-making about personal health. My PhD on the care of chronic illness in general practice explored the nature of the experience of illness and care associated with multi-morbidity from the perspectives of those afflicted and their general practitioner/primary care physician as the key ‘users’ of care. This PhD led to a wide range of systems-based interventions, underpinned by complex adaptive systems theory and social constructionist perspectives in Australia, Canada and Ireland. I am joint chief editor of the Forum on Systems and Complexity in Medicine and Healthcare in the Journal of Evaluation in Clinical Practice.

Contents

Contributors

Khaled Ali School of Electrical Engineering and Computer Science, University of Ottawa, Ottawa, ON, Canada

Polemnia Amazeen Department of Psychology, Arizona State University, Tempe, AZ, USA

Carlos Arce Billings Clinic, Billings, MO, USA

Aladdin Baarah School of Electrical Engineering and Computer Science, University of Ottawa, Ottawa, ON, Canada

Yaneer Bar-Yam New England Complex Systems Institute, Cambridge, MA, USA

Patrick Beautement The abaci Partnership LLP, Tewkesbury, UK

Jeanette M. Bennett Psychology Department, University of North Carolina at Charlotte, Charlotte, NC, USA

Leonard L. Berry Marketing Department, Mays School of Business, Texas A&M University, College Station, TX, USA

George E. Billman Department of Physiology and Cell Biology, The Ohio State University, Columbus, OH, USA

Rakesh Biswas Department of Medicine, People's College of Medical Sciences and Research Center, Bhopal, Madhya Pradesh, India

Rick Botelho Rochester, NY, USA

Christine Broenner The abaci Partnership LLP, Tewkesbury, UK

Mathew R. Broome Warwick Medical School, University of Warwick, Coventry, UK

Cary A. Brown University of Alberta, Edmonton, AB, Canada

Brian Castellani Kent State University, Ashtabula, OH, USA

Pierpaolo Cavallo University of Salerno, Fisciano, SA, Italy

Pranab Chatterjee University College of Medical Sciences and Guru Teg Bahadur Hospital, New Delhi, India

Sarah Cherng University of Michigan, Ann Arbor, Michigan, USA

Leonid Churilov Melbourne Brain Centre, Heidelberg, VIC, Australia

Elizabeth Ciemins Billings Clinic, Billings, MO, USA

Paul Cilliers c/o Jan-Hendrik Hofmeyr, Department of Philosophy, Centre for Studies in Complexity, University of Stellenbosch, Stellenbosch, South Africa

Renee Crichlow Department of Family Medicine, MMC 381, North Memorial Family Medicine Program, School of Medicine, University of Minnesota, Minneapolis, MN, USA

Ana Diez Roux Center of Social Epidemiology and Population Health, University of Michigan, School of Public Health, USA

Beverley Ellis School of Health, University of Central Lancashire (UCLan), Preston, UK

Christopher P. Fagundes Institute for Behavioral Medicine Research, Ohio State University College of Medicine, Columbus, OH, USA

Howard J. Federoff Georgetown University Medical Center, Washington, DC, USA

Margot Félix-Bortolotti Independent Researcher, Ottawa, ON, Canada

Robert Ferrer Department of Family & Community Medicine, University of Texas Health Science Center at San Antonio, San Antonio, TX, USA

Patrick H. Finan Department of Psychiatry and Behavioral Sciences, Johns Hopkins University School of Medicine, Baltimore, MD, USA

Winston Garira Modelling Biomedical Systems Research Group, Department of Mathematics and Applied Mathematics, University of Venda, Thohoyandou, South Africa

Brandon L. Gillie Institute for Behavioral Medicine Research, Ohio State University College of Medicine, Columbus, OH, USA

Deirdre Grady Patient Journey Record Campus Company, Trinity College Dublin, Dublin D2, Ireland

Geoffrey C. Green Therapeutic Monitoring Systems, Ottawa, ON, Canada

Stephen J. Guatello Marquette University, Milwaukee, WI, USA

Frederic Hafferty Program in Professionalism and Ethics, Mayo Clinic, Minneapolis, MN, USA

Dion Harmon Two Sigma Investments, LLC, New York, NY, USA

Margaret Hatch Billings Clinic, Billings, MO, USA

Iona Heath Past President - Royal College of General Practitioners, London, UK

Lucy Hederman School of Computer Science and Statistics, Trinity College, Dublin, Dublin 2, Ireland

Henry H.Q. Heng Center for Molecular Medicine & Genetics, Wayne State University School of Medicine, Detroit, MI, USA

Christophe L. Herry Ottawa Hospital Research Institute (OHRI), Ottawa, ON, Canada

Eric E. Hessler Department of Psychology, University of Minnesota, Duluth, MN, USA

Tim A. Holt Department of Primary Care Health Sciences, University of Oxford, Oxford, UK

S. Lee Hong Ohio University, Athens, OH, USA

Peter Hovmand Director, Social System Design Lab, Brown School

Timothy Hower Associate Director, Social System Design Lab, Brown School, USA

Carlos R. Jaén Department of Family & Community Medicine, University of Texas Health Science Center at San Antonio, San Antonio, TX, USA

Rajkumar Kamatchi The Caludon Centre, Coventry, UK

David A. Katerndahl Department of Family & Community Medicine, University of Texas Health Science Center at San Antonio, San Antonio, TX, USA

Leslie I. Katzel Baltimore VA Medical Center, Baltimore, MD, USA

Janice K. Kiecolt-Glaser Institute for Behavioral Medicine Research, Ohio State University College of Medicine, Columbus, OH, USA

Holly J. Lanham The University of Texas Health Science Center at San Antonio, San Antonio, TX, USA

Stephen Lewis Department of Biological Sciences, University of Chester, Chester, UK

May Lim University of the Philippines Diliman, Quezon City, Philippines

Curt Lindberg Billings Clinic, Billings, MO, USA

Monica E. Lindgren Institute for Behavioral Medicine Research, Ohio State University College of Medicine, Columbus, OH, USA

M.A. Little Oxford Centre for Industrial and Applied Mathematics (OCIAM), University of Oxford, Oxford, UK

Media Lab, Massachusetts Institute of Technology, Cambridge, MA, USA

Carmel M. Martin Department of Public Health and Primary Care, Trinity College, Dublin, Dublin D2, Ireland

Geoffrey D. McDonnell Centre for Health Informatics, University of New South Wales, Waverley, NSW, Australia

P.E. McSharry Smith School of Enterprise & the Environment, University of Oxford, Oxford, UK

Neil Mehta Cleveland Clinic, Lerner College of Medicine of Case Western Reserve University, Cleveland, OH, USA

Stewart P. Mennin Sao Paulo, Brazil

Andrew Miles WHO Center for Public Health Education and Training, Imperial College, London, London, UK

Virginia Mohl Billings Clinic, Billings, MO, USA

Hussein T. Mouftah School of Electrical Engineering and Computer Science, University of Ottawa, Ottawa, ON, Canada

Alain Mouttham School of Electrical Engineering and Computer Science, University of Ottawa, Ottawa, ON, Canada

Keith Nesbitt School of Design, Communication & Information Technology, University of Newcastle, Callaghan, NSW, Australia

Cameron D. Norman Dalla Lana School of Public Health, Toronto, ON, Canada

Di M. O'Halloran Glenorie, NSW, Australia

Sergio Pagano University of Salerno, Fisciano, SA, Italy

Bradley Perkins Vanguard Health Systems, Nashville, TN, USA

Liam Peyton School of Electrical Engineering and Computer Science, University of Ottawa, Ottawa, ON, Canada

David Pincus Crean School of Health and Life Sciences, Chapman University, Orange, CA, USA

Rick Riolo Center for the Study of Complex Systems, University of Michigan Ann Arbor, MI

Johnie Rose Division of Research, Department of Family Medicine, School of Medicine, Case Western Reserve University, Cleveland, OH, USA

Corey Schimpf School of Engineering Education, Purdue University, West Lafayette, IN, USA

John G. Scott Northeastern Vermont Regional Hospital, Lyndonville, VT, USA

Andrew J.E. Seely Ottawa Hospital, Ottawa, ON, Canada

Kent D. Seltman Mayo Clinic, Rochester, MN, USA

Kevin Smith Department of Medicine, University College Cork, Cork, Ireland

Suzanne Smith Speech Technology and Applied Research (STAR) Corp., Bedford, MA, USA

Kurt C. Stange Case Western Reserve University, Cleveland, OH, USA

Joachim P. Sturmberg Department of General Practice, Monash University, Wamberal, NSW, Australia

The Newcastle University, Newcastle, NSW, Australia

Jing Su School of Computer Science and Statistics, Trinity College, Dublin, Dublin D2, Ireland

N. Marcus Thygeson Blue Shield of California, San Francisco, CA, USA

Stefan Topolski University of Massachusetts Medical School, Shelburne Falls, MA, USA

A. Tsanas Department of Engineering Science, Systems Analysis, Modelling and Prediction (SAMP), University of Oxford, Oxford, UK

Oxford Centre for Industrial and Applied Mathematics (OCIAM), University of Oxford, Oxford, UK

Carl Vogel School of Computer Science and Statistics, Trinity College, Dublin, Dublin D2, Ireland

Shari R. Waldstein Department of Psychology, University of Maryland, Baltimore County, Baltimore, MD, USA

Division of Gerontology & Geriatric Medicine, Department of Medicine, University of Maryland School, of Medicine, Baltimore, MD, USA

Geriatric Research Education & Clinical Center, Baltimore VA Medical Center, Baltimore, MD, USA

Carrington R. Wendell Department of Psychiatry, Johns Hopkins University School of Medicine, Baltimore, MD, USA

Bruce J. West Army Research Office, Research Triangle Park, NC, USA

Ralph Wittenberg Georgetown University, Washington, DC, USA

Andrea L. Yip Dalla Lana School of Public Health, Toronto, ON, Canada

Atieh Zarabzadeh National University of Ireland, Galway, Ireland

1 Complexity in Health: An Introduction

Joachim P. Sturmberg and Carmel M. Martin

The world will not evolve past its current state of crisis by using the same thinking that created the situation.

Albert Einstein

More of the same, paraphrasing Einstein, can only lead to more of the same, or using Lakoff's terminology [1], the way we *talk* about things is the way we *think* about them. Currently, and in contrast to most other disciplines, medicine remains largely stuck in the simplistic "reductionist" scientific world view and is resisting the move to the complex dynamic "holistic" scientific world view (Table 1.1).

1.1 Complexity

Complexity arises from the Latin word *complexus; com-* meaning "together" and *plectere* "to wave" or "braid". Thus complexity study aims to understand how things are connected with each other, and how these interactions work together. Something is *complex* if it is made up of usually several closely connected parts; the more parts and the more connections are entwined within a system, the more complex it will be, and the more difficult it will be to analyse such a system.

Complexity science and complexity theories represent a convergence of different types of ideas and theories to address the nonlinearity and dynamics of the real world systems, often known as complex adaptive systems (CAS).

Complexity thinking is a change in mindset—away from understanding the whole arising from an understanding of its individual parts (the Newtonian approach) towards an appreciation that the whole is different and less than the sum of its parts; viewed in isolation the parts exhibit different properties to those seen in the context of the whole. In addition, the behaviour of system components varies depending on context; changing context may result in "unexpected" changes in the component's and therefore the system's behaviour.

1.1.1 Be Aware

It is important to distinguish between *complicatedness* and *complexity* (Fig. 1.1). Complicated objects, like a plane, have many parts that act together in a perfectly predictable way—who would otherwise trust to travel on a plane. A children's birthday party, on the other hand, has many different actors who behave in rather unpredictable ways, and the behaviour of a party can change abruptly—unforeseen or unpredictably—with only minor changes in its environment.

J.P. Sturmberg (✉)
Department of General Practice, Monash University, Melbourne, VIC, Australia

The Newcastle University, Newcastle, PO Box 3010, Wamberal, NSW 2260, Australia
e-mail: jp.sturmberg@gmail.com

C.M. Martin
Northern Ontario School of Medicine, Greater Sudbury P3E 2C6, ON, Canada

Department of Public Health and Primary Care, Trinity College, Dublin, College Green, Dublin D2, Ireland
e-mail: martinc7@tcd.ie

J.P. Sturmberg and C.M. Martin (eds.), *Handbook of Systems and Complexity in Health*, DOI 10.1007/978-1-4614-4998-0_1, © Springer Science+Business Media New York 2013

Table 1.1 The differences between the simple and complex scientific world views

Simple scientific world view	Complex scientific world view
• Linear, output is proportional to input	• Nonlinear, small changes may diverge
• Additive	• Multiplicative
• Simple rules yield simple results	• Simple rules yield complex results
• Stable	• Unstable
• Predictable	• Limited predictability
• Quantitative	• Qualitative plus quantitative
• Normal distribution	• Inverse power–law distribution

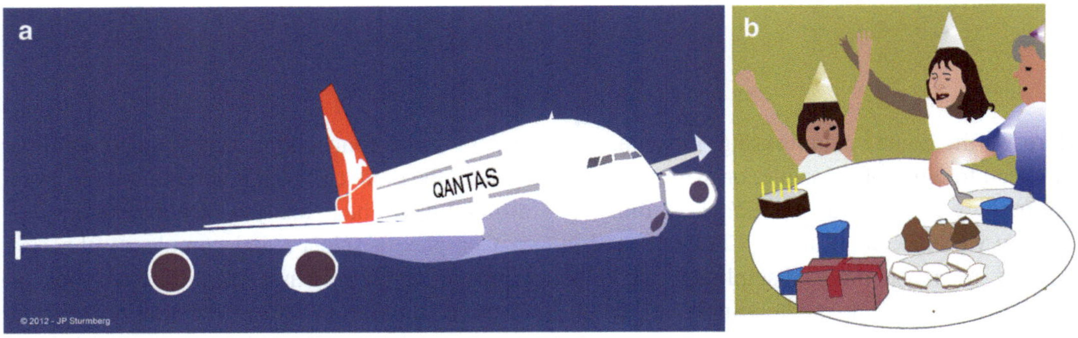

Fig. 1.1 The difference between a complicated and a complexity phenomenon

1.1.2 Coping with Complexity

As Dörner [2] has shown, the difficulties we experience when confronted with complex problems arise for psychological reasons (Tables 1.2 and 1.3); humans cannot keep more than a few things (on average 7 ± 2) in mind at any one time, they cannot easily detect connections between seemingly unconnected objects or facts, and they cannot easily anticipate—especially nonlinear—behaviours more than a step or two ahead.

Table 1.2 Observations about unsuccessful decision makers (Dörner [2], p. 18)

• Acted without prior analysis of the situation
• Failed to anticipate side effects and long-term repercussions
• Assumed that the absence of immediately obvious negative effects meant that correct measures had been taken
• Let over-involvement in "projects" blind them to emerging needs and changes in the situation
• Were prone to cynical reactions

1.1.3 Linear Versus Nonlinear Distributions

The common understanding of "normal distribution" goes back to the German mathematician Karl Friederich Gauss (1777-1855). Normal "Gaussian" distribution refers to a continuous probability distribution with all variables distributing symmetrical around the mean, resulting in the characteristic bell-curve.

Vilfredo Pareto, an Italian engineer, sociologist, economist, political scientist and philosopher (1848–1923), however, observed that most natural phenomena are not linearly distributed; they follow a nonlinear power law (or "Pareto" probability) distribution. The Pareto distribution is also known as the "80–20 rule" resulting from Pareto's initial observation of the distribution of wealth in his community—20% of the population owned 80% of the wealth (Fig. 1.2).

The implications of Pareto's insights so far have largely failed to be taken into account in most medical research. The Gaussian definition of normality turns the life of many *healthy people* to being *patients*—meaning suffers—to interventions which have no benefit but cause a lot of

Table 1.3 Differences in approaches to solving complex problems between successful and unsuccessful volunteers (adapted from Dörner [2], Chap. 1)

Characteristic approaches of successful problem solvers	Characteristic approaches of unsuccessful problem solvers
• Made more decisions	
• Considered not just the primary goal of any given measure but also its potential effect on other sectors of the system	
• Acted "more complexly". Their decisions took different aspects of the entire system into account, not just one aspect	
• Tested hypotheses frequently	• A proposed hypothesis equals reality; testing the hypothesis was unnecessary
• Asked more *why* questions (as opposed to *what* questions)	
• Were more interested in the causal links behind events, in the causal network that made up …, dug deeper in their analyses	
• Uses similar decision strategies over time	• High degree of "ad hocism"
• Focuses on the same topics within the problem area	
• Reflects more on own behaviour, comments critically on it, and made efforts to modify it	• Recapitulates behaviours
• More structured behaviour, thinking out loud more frequently displaying sequencing like "First I have to deal with A, then with B, but I shouldn't forget to think about C as well"	

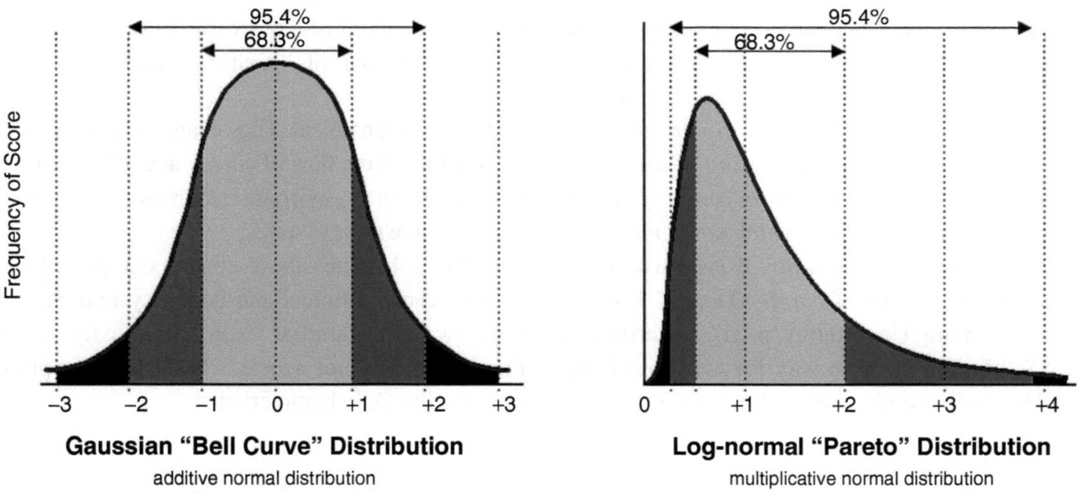

	Characterisation	
$\bar{x}$, Arithmetic	Mean	$\bar{x}$ *, Geometric
s, Additive	Standard Deviation	s*, Multiplicative
cv=s/$\bar{x}$	Measure of Dispersion	s*
	Interval of Cofidence	
$\bar{x} \pm$ s	68.3%	$\bar{x}$ *x/s*
$\bar{x} \pm$ 2s	95.5%	$\bar{x}$ *x/(s*)^2
$\bar{x} \pm$ 3s	99.7%	$\bar{x}$ *x/(s*)^3

Fig. 1.2 Comparing Gaussian and Pareto distributions

harm. The age-old doctrine of *primum non nocere* is jeopardised by ignoring the nonlinear distribution of living systems.

1.1.4 Certainty Versus Uncertainty

Scientific enquiry is driven by a desire to find certainty to the many confusing observations and experiences in daily life. Certainty—defined as either perfect knowledge or the mental state of being without doubt—reflects a deeply human desire. Its limitation though have already been described by Plato who said: "*I am wiser than the average man in that I know that I know nothing*".

Uncertainty not only reflects on the limited state of knowledge one has, it is a key characteristic of all CAS—the future state, or the outcome of a system's dynamics, are impossible to predict.

The conundrum of certainty and uncertainty has been poignantly summarised by Dennis Lindley[1] [3]: *There are some things ... that you know to be true, and others that you know to be false; yet, despite this extensive knowledge that you have, there remain many things whose truth or falsity is not known to you. We say that you are uncertain about them. You are uncertain, to varying degrees, about everything in the future; much of the past is hidden from you; and there is a lot of the present about which you do not have full information. Uncertainty is everywhere and you cannot escape from it* (Dennis Lindley, Understanding Uncertainty, p. xi). Nevertheless, CAS thinking offers a way forward to a better understanding and handling these uncertainties.

1.2 Characteristics of Complex Adaptive Systems

CAS are dynamic networks of many agents acting in parallel; they constantly act and react to the other agents' behaviours. The control of a CAS is highly dispersed and decentralised and its coherent behaviour arises from competition and cooperation among its agents. The overall behaviour of a system is the result of a huge number of decisions being made at every moment by interacting individual agents.

Cilliers[2] [4] described the key characteristics of CAS as follows:

- Complex systems consist of many different components that interact in nonlinear ways.
- They are open to their environment.
- Interactions occur at many different levels and influence each other through recursive feedback loops—they are self-organising.
- Pattern and organisation develop iteratively through interactions among the system's components in the absence of any external supervisory influence.
- Some simple rules for self-organisation in human systems include shared values and principles, connectivity and feedback, dialogue, memory and interdependency.
- A complex system is defined by its relationships or patterns of interaction, not its constituent components.
- The behaviour of a CAS cannot be reduced to the behaviour of specific components, it is emergent.
- CAS are dynamical. They change over time as a function of the flow of energy and information.
- CAS adapt to environmental pressures, agents co-evolve to new states.

Table 1.4 relates these complexity principles to well-known clinical and health system examples—we are familiar with complexity even though we may not necessarily relate these phenomena to CAS characteristics.

1.3 Clarifying Some Common Concepts from a Complexity Perspective

Before proceeding it is necessary to clarify the meaning of some commonly used concepts—knowledge and health—illness—disease—from a complexity perspective.

[1] British statistician, decision theorist and leading advocate of Bayesian statistics.

[2] For a detailed discussion, see Chap. 3.

Table 1.4 System properties are abundant in everyday clinical life

Properties	Clinical examples	Health system examples
Nonlinearity		
• Results not proportional to stimulus • Can lead to sudden massive and stochastic changes of the system • Sensitive to initial conditions	• Allergic responses and anaphylaxis • More intensive glucose control increase mortality [5] • Response to coumadin therapy • Increasing the dose of chemotherapy does not improve therapeutic response or survival [6] • Chemotherapy initially reduces tumour size but also includes the promotion of secondary tumours [7]	• Large investment in health services has not been matched by a similar magnitude of improvement in inequity between social classes [8] • The introduction of electronic prescribing systems had mixed impacts on appropriateness and safety of prescribing and patient health outcomes [9, 10]
Open to environment		
• A system continuously interacts with its environment, e.g. Exchanging material, energy, people, capital and information • Nonlinear responses to the external environment can lead to sudden massive and stochastic changes	• Physiological function – Immune system – Respiratory tract – Gastrointestinal tract – Skin – Semi-permeable membranes • Pathological function – HIV/AIDS – Asbestosis – Food poisoning – Burns	• Strategies to train and maintain more health professionals need to account for competing individual, organisational and social factors in motivation and other markets [11] • An epidemic like SARS arises from the global openness to fluidity, flows, mobility and networks [12]
Self-organisation		
• Relies on four basic principles – Recursive feedback (positive and negative) – Balance of exploitation and exploration – Multiple interactions	• "Homeostasis" in health, e.g. – Blood glucose levels – Thyroxin levels – Water balance and creatinine levels • Disease, e.g. – Stable heart failure – Intermittent claudication – Hypogonadism	• DRG (Diagnostic Related Group) payment mechanisms leads to – Gaming – Category creep – Shift of emphasis [13] • The natural formation of viable high performing teams is based on multiple interactions and feedback [14]

(continued)

Table 1.4 (continued)

Properties	Clinical examples	Health system examples
Emergence		
• Occurs when a number of simple entities (agents) operate in an environment, forming more complex behaviours as a collective • Arises from intricate causal relations across different scales and feedback—interconnectivity • The emergent behaviour or properties are not a property of any single such entity, nor can they easily be predicted or deduced from behaviour in the lower-level entities: they are irreducible	• Appearance of superbugs in response to antibiotic therapies • Appearance of previously unknown infectious disease epidemics like SARS [15] • Emergence of drug side effects in particular individuals • Emergence of new patterns of morbidity, gene expression, as the population ages • Brain function from complex cellular self-organisation	• Prevention paradox—inequities emerge when "innovative" health promotion guidelines are put into place without considering social and cultural assumptions between public health practitioners and target groups as is seen in – Screening programmes – Well baby checks – Teenage pregnancy education – Smoking cessation programmes [16] • The addition of nurse practitioners to primary care – Did not alter costs or efficiencies – Did address considerable other unmet needs [17]
Pattern of interaction		
• Different combinations of agents lead to the same outcome, or • The same combination of agents leads to different outcomes	• Sinus-rhythm heart-rate variability diminished in patients with severe congestive heart failure [18] • Loss of beat-to-beat variability in autonomic neuropathy [19] • Cheyne–Stokes breathing [18] • Most patients with cancer display drastically different patterns of genetic aberrations [20] • Many biological factors (genetic and epigenetic variations, metabolic processes) and environmental influences can increase the probability of cancer formation, depending on the given circumstances [21]	• Patterns of maternity provider interaction appropriate for the local context influence the emotional well-being of rural mothers [22] • International comparison shows that many diverse multifaceted health services lead to remarkably similar outcomes – Smoking cessation successes [23] – Obesity challenges exist across diverse cultures and levels of development despite evidence-based national dietary guidelines [24]

Adaptation and evolution		
• In the clinical context, numerous diseases develop over many years, during which time the "'whole body system" has adapted to function in the altered environment • Changes involve the whole system and are not restricted to a few clinically measurable factors • Adaption leads to a new homeostasis with new dynamic interactions [25]	• Hypothyroidism • Coronary artery disease due to stable plaques • "Burnt-out" rheumatoid arthritis • Stable chronic obstructive airways disease • Celiac disease • Cataract • Hearing impairment	• Adjustments to the health care system are needed due to changing patterns in – Health care delivery – Health financing – The rate of development of new health technologies – Rising community expectations [26] • Stable ritual of clinical care delivery despite ongoing reforms, research and interventions [27] • Healing tradition moves from mainstream health care to alternative health care [28]
Co-evolution		
• Each agent in the exchange is changing in response • Parallel development of a subsystem with new characteristics and dynamics	• The physician learns from the patient and the patient learns from the physician [29] • A person becomes blind and develops superb hearing • Microorganisms succumb to antibiotic therapies and some develop drug resistance	• Local systems function well in response to local need in spite of or in parallel to top down health initiatives – User driven health care [30] – Self-help groups [31] – Health 2.0 [32]

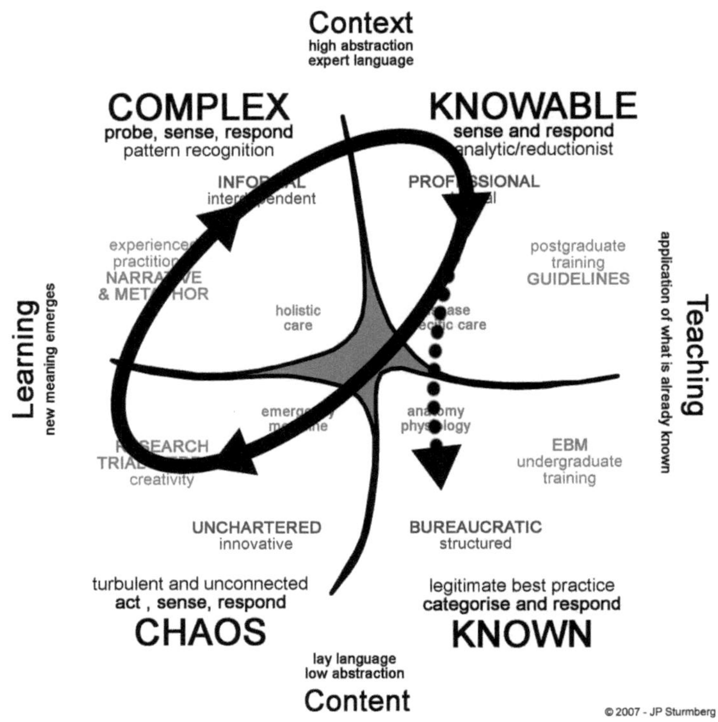

Fig. 1.3 Cynefin framework of knowledge

1.3.1 Knowledge

Knowledge[3] is often seen as objective and equated to truth; science regarding observation as the means to deriving truth that can be expressed as "natural laws". Some important limits to this notion have been outlined by Popper [33]—observations are always subjective and context bound, and Polanyi [34]—knowledge is always personal: *I know*.

Knowledge, as defined by the *Oxford English Dictionary*, variably refers to:

1. *Expertise, and skills acquired by a person through experience or education; the theoretical or practical understanding of a subject*;
2. *What is known in a particular field or in total; facts and information*, or
3. *Awareness or familiarity gained by experience of a fact or situation.*

These definitions imply that knowledge is a multidimensional construct. Its acquisition involves multiple interconnected processes, including perception, learning, communication, association and reasoning. The most commonly used philosophical approach to understanding knowledge is to distinguish the notions of propositional knowledge, that is, "knowing-that", from that of "knowing-how". However, as Polanyi pointed out, these two forms of knowledge coexist. He rejected the notion that knowledge can be completely objective and, instead, elaborated on the personal nature of knowing, particularly emphasising the tacit aspects of knowing, and its implications for knowledge transfer and learning [34].

Knowledge has multiple dimensions—it can be ordered and predictable, or complex and unpredictable—and thus can be simultaneously perceived in different, but mutually agreeable ways. Knowledge is simultaneously a thing and a flow; its complex adaptive nature has been visualised by Kurtz and Snowden through the Cynefin framework [35]. A Cynefin view of medical knowledge is illustrated in Fig. 1.3 [36].

Using this framework, the focus of knowledge generation is dynamic and fluid. It shifts between context and narrative, rather than being fixed on

[3] For a more detailed discussion, see Chap. 4.

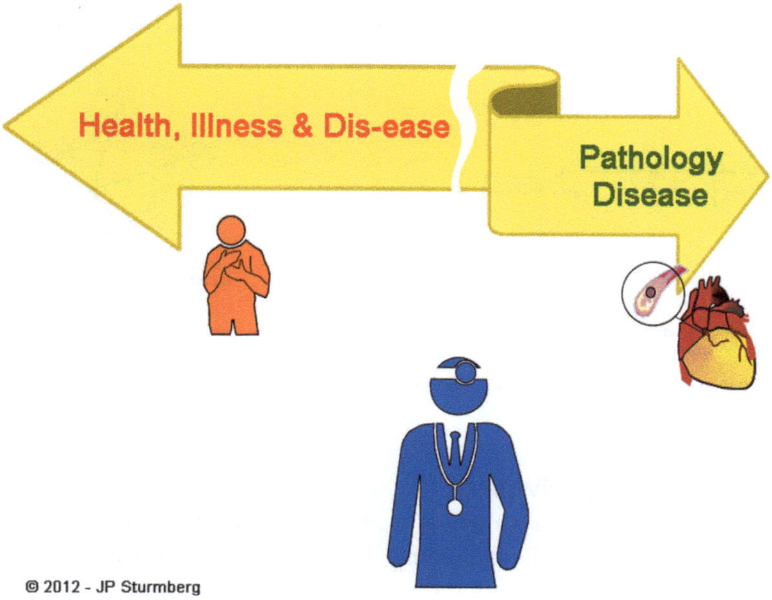

Fig. 1.4 Health, illness and dis-ease versus pathologies and disease classifications. The ***clinical encounter*** is the meeting place of the subjective experience of the patient and the objective world of the pathologist and the medical professional classification system based on a Gestalt of aetiology, function and genetics [37]. In fact, with increasing refinements and changing taxonomies of disease, there are major issues which need to be addressed to deal with increasing embedding of these systems into electronic financial and clinical systems [38]

content alone, and between inductive and deductive approaches. Understanding knowledge as complex and fluid overcomes the divides created by specific viewpoints and ways of thinking, making visible and understandable the dynamic nature of the different sources of knowledge we use in specific instances. This approach highlights that *our*, i.e. personal perspective, of knowing "will always contain uncertainty".

1.3.2 Health—Illness—Disease

Commonly health, illness, disease[4] and sickness are used as if being mutually interchangeable. This confusion unfortunately has been perpetuated by the WHO's definition of health through its inverse—absence of disease, and the preceding "not merely" has largely been forgotten. Health, illness and disease are points on the same *subjective* scale as experienced by the patient, and needs to be distinguished from the *objective* findings of disease at the organ, cell or sub-cellular changes as seen by the pathologist, and the classification of disease by the health professions in the ICD (Fig. 1.4) [39].

The doctor's function is that of a translator, between the subjective experience of the patient and the potentially objective bodily changes in the patient. The consultation provides legitimacy to the person's experience, having been validated, society provides certain privileges to its members who are sick [40].

1.3.2.1 Health: A Dynamic State

The experience of health, illness and dis-ease are therefore dynamic and adaptive states. They can be experienced as much in the *absence* as *presence* of identifiable pathologies, and clinical experience suggests that *the length of a patient's problem list is inversely related to his subjective health experience*. We have previously suggested that health should be seen as a dynamically balanced state, its utility being demonstrated by the two patients in Fig. 1.5, both having suffered an "acute coronary event" with markedly different outcomes in terms of objective and subjective adaptation.

[4] For a detailed discussion, see Chaps. 14–18.

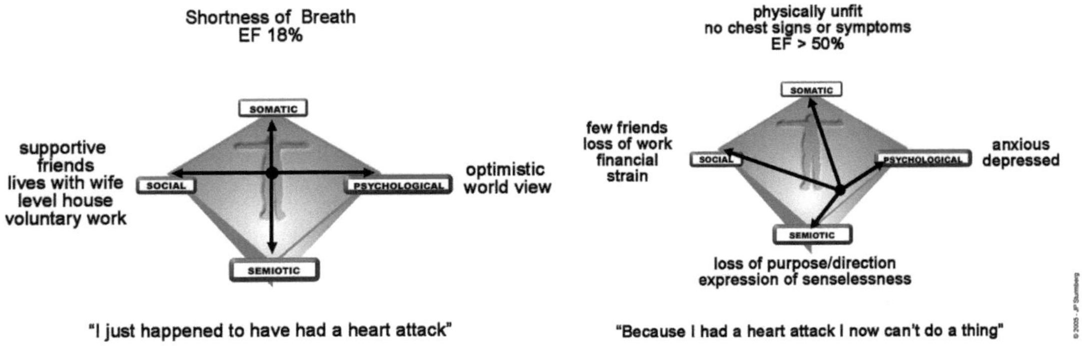

Fig. 1.5 Patient experience of health and illness following myocardial infarction

1.3.2.2 Disease: Not an Objective State

As outlined above, dis-ease is a subjective state and disease is a medical classification that has been objectified to mean pathology; the cross-over of the subjective meaning the objective, which however is only true in a small number of patients presenting to a doctor, has become the preoccupation of the "medical industry". This objectification of disease as a specific entity is a fundamental aspect of Western culture. Suffering without the objective identification of a disease has no legitimacy, and in many parts of the world reimbursement for medical services has been linked to disease activities [41].

The objectification of disease as an objective state is a great fallacy. Disease, to quote Per Fugelli [42], *does not exist, only the experience of disease* [does] (p. 185). Disease, however, is the currency of the medical industrial complex.

Dispelling this fallacy is of obvious importance as it distorts the purpose and the function of *health care* delivery. The negative impacts of the objectified disease focus are summarised by Barbara Starfield [43]: diseases (1) are professional constructs, (2) can be and are artificially created to suit special interests with the peculiar outcome that the sum of deaths attributed to diseases exceeds the number of deaths, (3) do not exist in isolation from other diseases and are, therefore, not an independent representation of illness, and (4) are but one manifestation of ill health.

1.4 Examples of Nonlinearity in Health and Health Care

Three examples show the nonlinear distribution of variables and illustrate the implications on clinical and health service thinking, planning and implementation. The first example illustrates that very few in the community require tertiary level health care, the second examples demonstrates the threshold behaviour of blood pressure and mortality, and the third example the exponential rise in life expectancy with small changes of rise in income for the poor and virtually no change for the rich.

1.4.1 Utilisation of Health Services

The community experience of health and illness and its consequences on health service utilisation was first examined by White et al. [44] in 1961, and re-examined by Green et al. [45] in 2001, showing that people are healthy most of the time.

20% of patients report no illness symptoms at all. Of the 80% with illness symptoms 80% have no immediate health care needs, and of the 20% seeking health care, 80% only require care from their trusted primary care physician (i.e. 16% of the community). Some 80% of the remaining 20% need care only from secondary services (i.e. 3.2% of the community), leaving a mere 20% of this already small group requiring tertiary care (i.e. 0.8%) (Fig. 1.6).

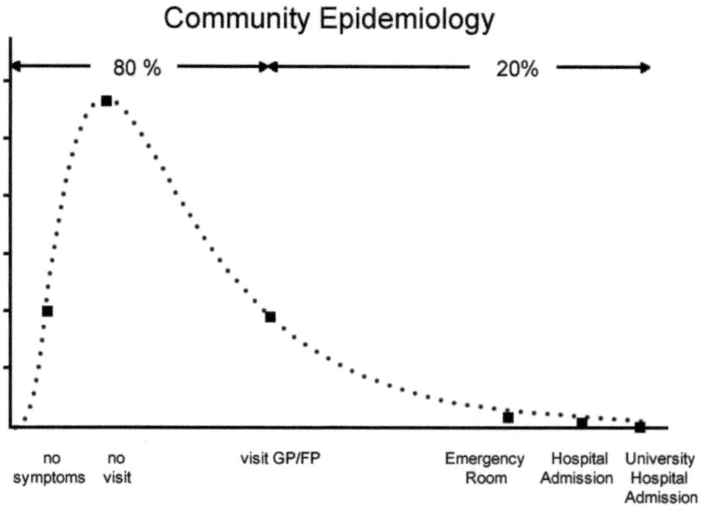

Fig. 1.6 Community epidemiology of health, illness, and health service utilisation

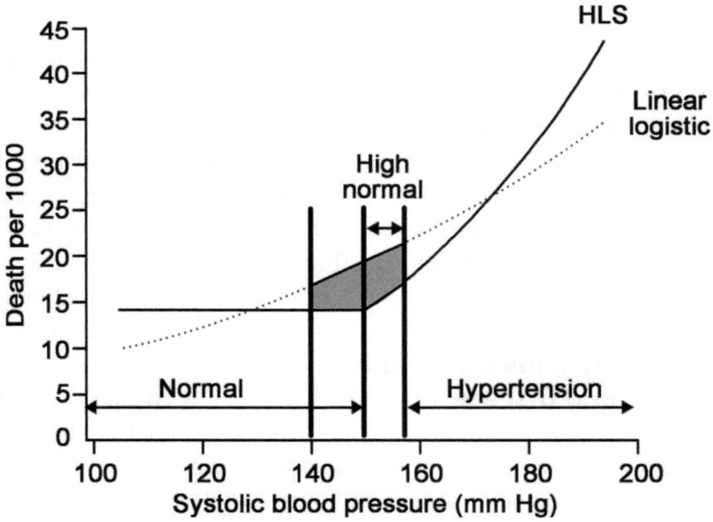

Fig. 1.7 Blood pressure related ABSOLUTE mortality for 50-year-old males. Superimposed is the relative mortality derived from linear logistic regression analysis (HLS: horizontal logistic spline) (from [46], with permission)

1.4.2 Blood Pressure Levels and Mortality

Port and colleagues [46] re-examined the Framingham data in relation to blood pressure related mortality. Plotting the absolute number of death for age and gender groups showed threshold behaviour of blood pressure mortality: mortality rates are unrelated to blood pressure readings up to approximately 100+age, before slowly rising for the next 20 mmHg, and only after that point mortality rises exponentially (Fig. 1.7).

1.4.3 Life Expectancy and Income

Income per capita and income inequality studies have not shown any direct causal effect on health as such; however, they have shown a strong link of small rises in income for the most disadvantaged

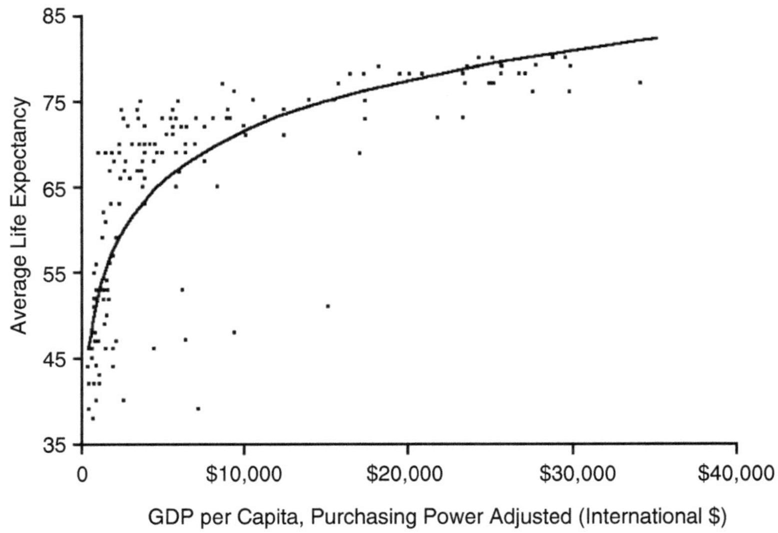

Fig. 1.8 Life expectancy and income—(reprinted by permission of the publisher: World Bank. 2002. *The 2002 World Development Indicators CD-ROM*. Ver. 4.2.Washington, D.C.: The World Bank.)

on health and life expectancy [47]. This should not be at all surprising as income reflects a variety of environmental inputs, and allows for a variety of health enabling outputs—all of which feedback on each other and where a small change in a key variable may be responsible for a disproportionate effect on the gain seen (Fig. 1.8).

1.5 Dynamics in Health and Disease

Health and disease are not a static equilibrium states. Physiological parameters vary within ranges day by day, diseases show "characteristic alterations" in their disease-specific variables that return back to pre-disease levels in self-limiting, or to a new level in chronic diseases. Variables show a great deal of variability within a patient over time, and between people at any one time (Table 1.5).

Variability is a normal phenomenon[5] reflecting a high degree of complexity in the interaction of a well-functioning body—variability is a sign of health. Loss of variability, whether too little or too much, is a sign of loss of complexity, and a sign of disease, a finding first shown by Goldberger in relation to heart beat variability changes in cardiac disease [18]. Too little beat-to-beat variability is associated with cardiac failure, whereas too much variability is resulting in atrial fibrillation (Fig. 1.9).

Aging is another example of progressive loss of complexity in physiologic dynamics and can be caused by loss or impairment of the system's functional components, and/or an impairment of the coordinated interactions between these components. Such loss can be seen in the aging characteristics of the heart; though mean heart beat in a young and old person may be very similar, the variability over time does change significantly. Table 1.6 summarises some of the other dynamic changes of aging [74].

1.6 Understanding Systems: Causal Loop Diagrams

Causal loop diagrams are a common tool to visualise systems and system behaviour.[6] The regulation of thyroid function is an example of a

[5] For more detail, see Chaps. 5, 11 and 12.

[6] For more detail on system dynamics and modelling, see Chap. 6; applications of modelling are illustrated in Chaps. 33, 44 and 45.

Table 1.5 Examples of regular and irregular dynamics in health and disease (from [70], with permission)

Field	Regularly recurring dynamics	Irregularly recurring dynamics	References
Behaviour	Affective disorders "rapid cyclers"	Affective disorders "rapid cyclers"	[49–51]
Cardiology	Sinus rhythm	Atrial fibrillation	[52]
	Wenckebach phenomenon	Ventricular fibrillation	
	Ventricular bigeminy		
Electrophysiology			
Beta cells / Molluscan neuron / Thalamus	Rhythms and burst	Irregular spiking	[53, 56] [57] [58]
EEG			
Recurrent inhibition	Spike and wave	Background activity	[59]
	Hippocampal activity	Penicillin epilepsy model	[60]
Haematology	Periodic haematopoiesis	Periodic CML	[61]
	Autoimmune haemolytic anaemia	Cyclical thrombocytopenia	
Movement			
Locomotion	Gait	Cerebellar gait	[61, 62]
Coordinated activity	Tremors	Choreo-athetosis	
	Hiccups		
Nerve-muscle	Fibrillations	Myoclonus	[63, 64]
	Myotonic discharges	Fasciculations	
	Myokimia		
Neuro-ophthalmology			
Pupil diameter	Pupic cycle time	Hippus	[65–68]
Eye movements	Nystagmus	Opsoclonus	[69, 70]
Respiration	Periodic breathing	Ataxic breathing	
	Cheyne–Stokes	Cluster breathing	

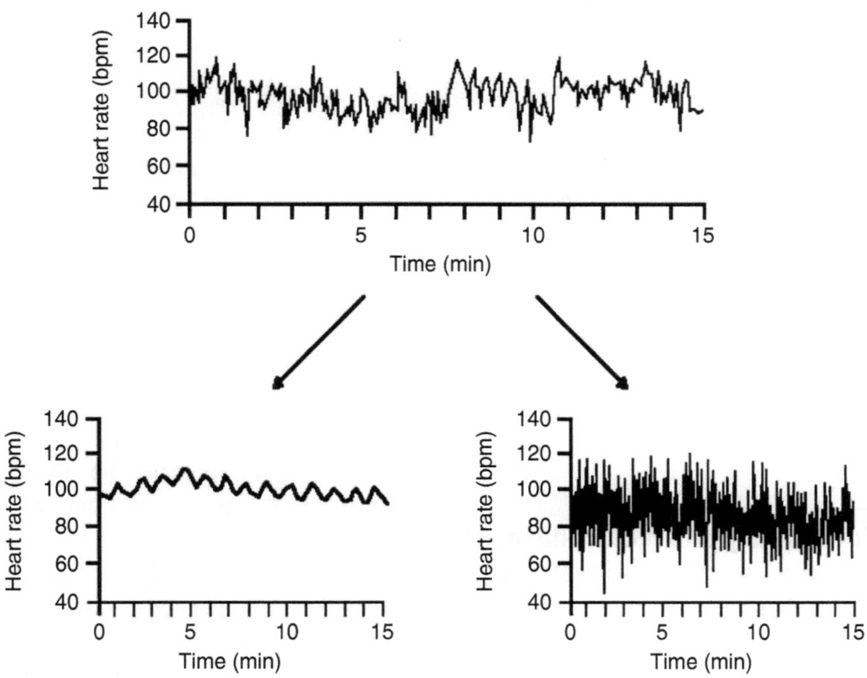

Fig. 1.9 Healthy dynamics (top), showing multiscale, long-range order; pathological breakdown of fractal dynamics, leading to single-scale (bottom left) or uncorrelated randomness (bottom right) (from [18], with permission)

Table 1.6 Examples of decreased structural (anatomic) and functional (physiologic) "complexity" in advanced age (integrity)

	Measure of complexity	Age effect	References
Anatomic structures			
Neuronal dendrites	Branching arbour	Dendrite loss and reduced branching	[71, 72]
Bone trabeculae	Meshwork	Trabecular loss, disconnection	[73]
Physiologic systems			
Heart rate variability	Dimension, entropy	Decrease	[74–76]
BP variability	Dimension, entropy	Decrease	[76]
Pulsatile TRH release	SD of interpulse interval	Decrease	[77]
EEG evoked potentials	Range of frequencies evoked	Decrease	[78]
Auditory	Range of audible frequencies	High-frequency loss	[79]

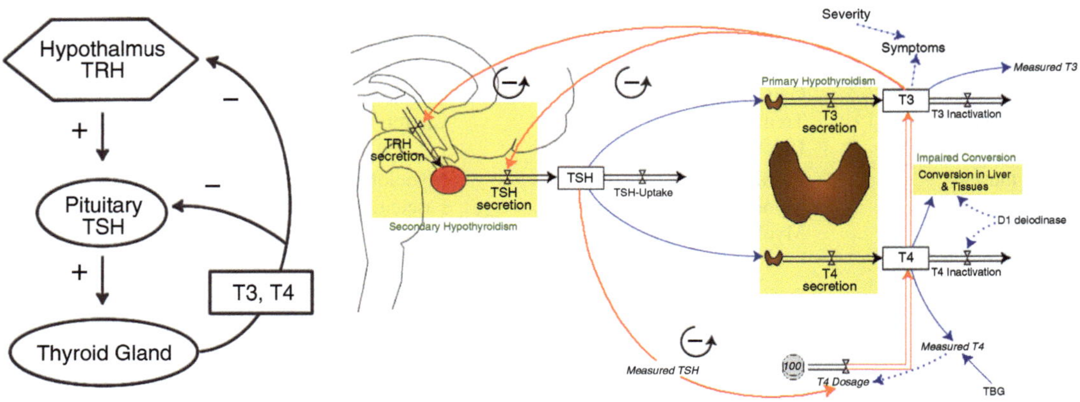

Fig. 1.10 Feedback loops regulating thyroid function

self-stabilising feedback loop. Figure 1.10 (left) depicts a simplified version; Fig. 1.10 (right) an extensive version of the regulatory cycles controlled by the thyroid gland.

This technique can be applied to model more complex systems as a starting point to explore the interactions and interdependencies within it. The example in Fig. 1.11 models *continuity of care* in primary care. The theory and technique of modelling is described in detail in Chap. 6.

1.7 Complexity and Nonlinearity: A Way Forward to Understanding Our World

VUCA—volatility, uncertainty, complexity and ambiguity—is an aphorism to describe the reality of the world we live in. The acronym has been coined by the military in the late 1990s to help them better understand the challenges for their missions [80]. VUCA—vision, understanding, clarity and agility—provides guides for actions in a complex world [81]. VUCA reminds us that to be successful we constantly have to make sense of our environment before acting, and to re-evaluate the outcome of our actions to remain successful.

We hope that this short introduction has helped to dispel some of the mysteries about systems and complexity and enticed your—the reader's—curiosity to further explore the "real world of healthcare". The remainder of this book will explore the complexity view of health and healthcare in great detail, and it will provide guidance for readers to further their personal interests and developments within a complex systems framework.

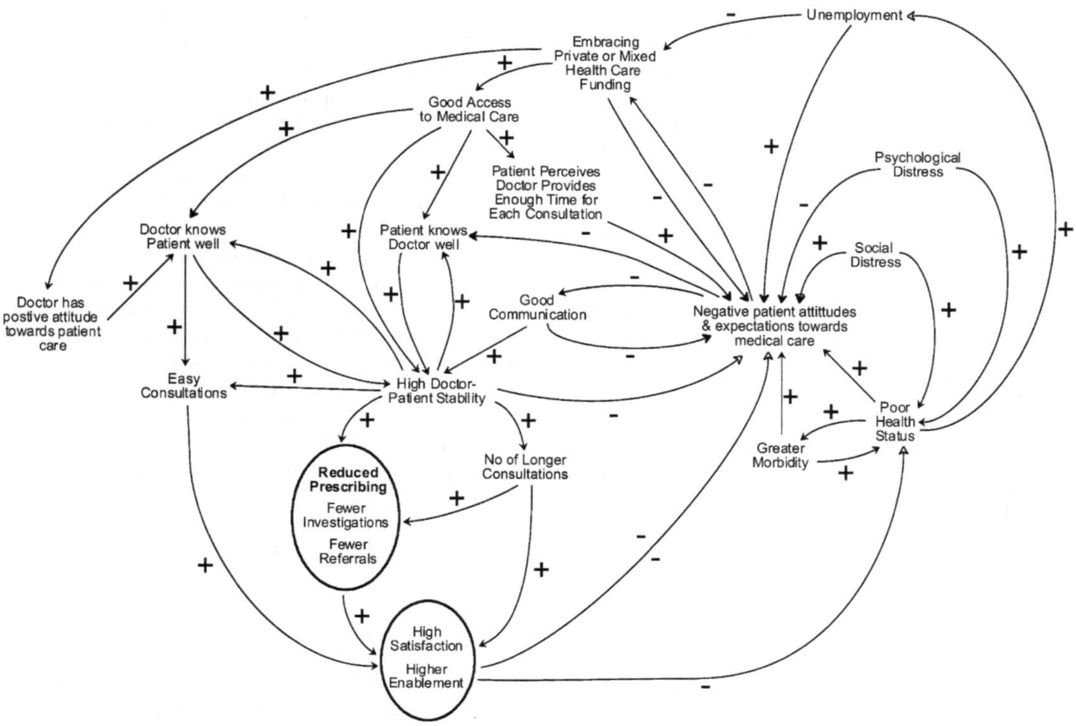

Fig. 1.11 Continuity of care model

References

1. Lakoff G, Johnsen M. Metaphors we live by. London: The University of Chicago Press; 2003.
2. Dörner D. The logic of failure: recognizing and avoiding error in complex situations. New York: Metropolitan Books; 1996.
3. Lindley DV. Understanding uncertainty. Hoboken, New Jersey: Wiley; 2006.
4. Cilliers P. Complexity and postmodernism. Understanding Complex Systems. London: Routledge; 1998.
5. The Action to Control Cardiovascular Risk in Diabetes Study Group. Effects of intensive glucose lowering in type 2 diabetes. N Engl J Med. 2008;358(24):2545–59.
6. Leyvraz S, Pampallona S, Martinelli G, Ploner F, Perey L, Aversa S, et al. A threefold dose intensity treatment with ifosfamide, carboplatin, and etoposide for patients with small cell lung cancer: a randomized trial. J Natl Cancer Inst. 2008;100(8):533–41.
7. Mittra I. The disconnection between tumor response and survival. Nat Clin Pract Oncol. 2007;4(4):203.
8. Blas E, Gilson L, Kelly MP, Labonté R, Lapitan J, Muntaner C, et al. Addressing social determinants of health inequities: what can the state and civil society do? Lancet. 2008;372(9650):1684–9.
9. Hider P. Electronic prescribing: a critical appraisal of the literature. New Zealand Health Technology Assessment (NZHTA), Vol. 5, No. 2. 2002. http://nzhta.chmeds.ac.nz/publications/elpresc.pdf.
10. Ammenwerth E, Schnell-Inderst P, Machan C, Siebert U. The effect of electronic prescribing on medication errors and adverse drug events: a systematic review. J Am Med Inform Assoc. 2008;15(5):585–600.
11. Zurn P, Dolea C, Stilwell B. Nurse retention and recruitment: developing a motivated workforce. Geneva: International Council of Nurses; 2005.
12. Ali S. Infectious disease, global cities and complexity. Paper presented at the annual meeting of the American Sociological Association Annual Meeting, Sheraton Boston and the Boston Marriott Copley Place, Boston, MA. http://www.allacademic.com/meta/p_mla_apa_research_citation/2/3/7/3/0/p237304_index.html. Accessed 31 Jul 31 2008.
13. Kuhn M, Siciliani L. Upcoding and optimal auditing in health care (or The economics of DRG creep). SSRN, CEPR Discussion Paper No. DP6689. 2008.
14. Grumbach K, Bodenheimer T. Can health care teams improve primary care practice? JAMA. 2004;291(10):1246–51.
15. Smith RD. Responding to global infectious disease outbreaks: lessons from SARS on the role of risk perception, communication and management. Soc Sci Med. 2006;63(12):3113–23.

16. Frohlich KL, Potvin L. Transcending the known in public health practice: the inequality paradox: the population approach and vulnerable populations. Am J Public Health. 2008;98(2):216–21.
17. Laurant M, Reeves D, Hermens R, Braspenning J, Grol R, Sibbald B. Substitution of doctors by nurses in primary care. Cochrane Database Syst Rev. 2005;(2):CD001271.
18. Goldberger A. Non-linear dynamics for clinicians: chaos theory, fractals, and complexity at the bedside. Lancet. 1996;347:1312–4.
19. Stella P, Ellis D, Maser RE, Orchard TJ. Cardiovascular autonomic neuropathy (expiration and inspiration ratio) in type 1 diabetes. Incidence and predictors. J Diabetes Complications. 2000;14(1):1–6.
20. Wood LD, Parsons DW, Jones S, Lin J, Sjoblom T, Leary RJ, et al. The genomic landscapes of human breast and colorectal cancers. Science. 2007;318(5853):1108–13.
21. Heng HQH. Cancer genome sequencing: the challenges ahead. Bioessays. 2007;29(8):783–94.
22. Sutherns R. Adding women's voices to the call for sustainable rural maternity care. Can J Rural Med. 2004;9(4):239–44.
23. Agency for Healthcare Research and Quality. Treating tobacco use and dependence—a systems approach. A guide for health care administrators, insurers, managed care organizations, and purchasers. In: 2000 UPHS, editor. Washington DC: Agency for Healthcare Research and Quality. 2000. http://www.surgeongeneral.gov/tobacco/systems.htm.
24. Nishida C, Uauy R, Kumanyika S, Shetty P. The joint WHO/FAO expert consultation on diet, nutrition and the prevention of chronic diseases: process, product and policy implications. Public Health Nutr. 2004;007(01A):245–50.
25. Heng HHQ. The conflict between complex systems and reductionism. JAMA. 2008;300(13):1580–1.
26. House of Representatives Standing Committee on Health and Ageing. The blame game. Report on the inquiry into health funding. Canberra: Commonwealth of Australia; 2006.
27. Plamping D. Looking forward: change and resistance to change in the NHS. Br Med J. 1998;317(7150):69–71.
28. Snyderman R, Weil AT. Integrative medicine: bringing medicine back to its roots. Arch Intern Med. 2002;162(4):395–7.
29. Suchman AL. A new theoretical foundation for relationship-centered care. Complex responsive processes of relating. J Gen Intern Med. 2006;21 Suppl 1:S40–4.
30. Biswas R, Martin CM, Sturmberg JS, Mukherji KJ, Lee EWH, Umakanth S, et al. Social cognitive ontology and user driven health care. In: Hatzipanagos S, Warburton S, editors. Handbook of research on social software and developing community ontologies. London, UK: IGI Global; 2009.
31. Martin CM, Peterson C, Robinson R, Sturmberg JP. Care for chronic illness in Australian general practice—focus groups of chronic disease self-help groups over 10 years. implications for chronic care systems reforms. Asia Pac J Fam Med. 2009;8(1):1.
32. Eysenbach G. Medicine 2.0: social networking, collaboration, participation, apomediation, and openness. J Med Internet Res. 2008;10(3):e22. http://www.jmir.org/2008/3/e22/).
33. Popper K. Conjectures and refutations: the growth of scientific knowledge. London: Routledge and Kegan Paul; 1972.
34. Polanyi M. Personal knowledge. Towards a Post-Critical Philosophy. London: Routledge; 1958.
35. Kurtz CF, Snowden DJ. The new dynamics of strategy: sense-making in a complex and complicated world. IBM Syst J. 2003;42(3):462–83.
36. Sturmberg JP, Martin CM. Knowing—in medicine. J Eval Clin Pract. 2008;14(5):767–70.
37. Manchikanti L, Falco FJ, Hirsch JA. Necessity and implications of ICD-10: facts and fallacies. Pain Phys. 2011;14(5):E405–25.
38. Cookson WO. Disease taxonomy—polygenic. Br Med Bull. 1999;55(2):358–65.
39. Lewis C. Exploring the biological meaning of disease and health. 2003. http://sites.google.com/site/sjlewis55/presentations/vienna2003. Accessed 14 May 2011.
40. Sturmberg JP. The foundations of primary care. Daring to be different. Oxford, San Francisco: Radcliffe Medical Press; 2007.
41. Rosenberg CE. Disease in history: frames and framers. Milbank Q. 1989;67 Suppl 1:1–15.
42. Fugelli P. Clinical practice: between Aristotle and Cochrane. Schweiz Med Wochenschr. 1998;128:184–8.
43. Starfield B. Diseases or morbidity burden? The John Hopkins University's 2008 ACG International Risk Adjustment Conference, Las Vegas. 2007. http://www.acg.jhsph.edu.
44. White K, Williams F, Greenberg B. The ecology of medical care. N Engl J Med. 1961;265(18):885–92.
45. Green L, Fryer G, Yawn B, Lanier D, Dovey S. The ecology of medical care revisited. N Engl J Med. 2001;344(26):2021–5.
46. Port S, Demer L, Jennrich R, Walter D, Garfinkel A. Systolic blood pressure and mortality. Lancet. 2000;355(3):175–80.
47. Lynch J, Smith GD, Harper S, Hillemeier M, Ross N, Kaplan GA, et al. Is income inequality a determinant of population health? Part 1. A systematic review. Milbank Q. 2004;82(1):5–99.
48. Wehr TA, Goodwin FK. Rapid cycling in manic-depressives induced by tricyclic antidepressants. Arch Gen Psychiatry. 1979;36(5):555–9.
49. Wehr TA, Goodwin FK, Wirz-Justice A, Breitmaier J, Craig C. 48-Hour sleep-wake cycles in manic-depressive illness: naturalistic observations and sleep deprivation experiments. Arch Gen Psychiatry. 1982;39(5):559–65.
50. Wehr TA, Goodwin FK. Circadian rhythms in psychiatry. Pacific Grove, CA: Boxwood Press; 1983.
51. Marriott HJL. Practical electrocardiography. Baltimore: Williams & Wilkins; 1980.

52. Chay TR, Rinzel J. Bursting, beating, and chaos in an excitable membrane model. Biophys J. 1985;47(3):357–66.
53. Lebrun P, Atwater I. Chaotic and irregular bursting electrical activity in mouse pancreatic B-cells. Biophys J. 1985;48(3):529–31.
54. Holden A, Winlow W, Haydon P. The induction of periodic and chaotic activity in a molluscan neurone. Biol Cybern. 1982;43(3):169–73.
55. Holden AV, Muhamad MA. The identification of deterministic chaos in the activity of single neurons. J Electrophysiol Tech. 1984;11(3):135–47.
56. Roy JP, Clercq M, Steriade M, Deschenes M. Electrophysiology of neurons of lateral thalamic nuclei in cat: mechanisms of long-lasting hyperpolarizations. J Neurophysiol. 1984;51(6):1220–35.
57. Spehlmann R. EEG primer. New York: Elsevier Biomedical Press; 1985.
58. Prince DA. Microelectrode studies of penicillin foci. In: Jasper HH, Ward AA, Pope A, editors. Basic mechanisms of the epilepsies. Boston: Little Brown; 1969. p. 320–8.
59. Wintrobe MM. Clinical hematology. 7th ed. Philadelphia: Lea & Febiger; 1976.
60. Adams RD, Victor M. Principles of neurology. Toronto: McGraw-Hill; 1985.
61. Marsden CD. The pathophysiology of movement disorders. Neurol Clin. 1984;2(3):435–59.
62. Purves D, Sakmann B. Membrane properties underlying spontaneous activity of denervated muscle fibres. J Physiol. 1974;239(1):125–53.
63. Kimura J. Electrodiagnosis in diseases of nerve and muscle: principle and practice. Philadelphia: Davis; 1983.
64. Miller NR. Walsh and Hoyt's clinical neuro-opthalmology. Baltimore: Williams & Wilkins; 1985.
65. Sharpe J, Fletcher W. Saccadic intrusions and oscillations. Can J Neurol Sci. 1984;11(4):426–33.
66. Yoss R, Moyer N, Hollenhorst R. Pupil size and spontaneous pupillary waves associated with alertness, drowsiness, and sleep. Neurology. 1970;20(6):545–54.
67. Yoss RE, Mover NJ, Hollenhorst RW. Hippus and other spontaneous rhythmic pupillary waves. Am J Ophthalmol. 1970;70(6):935–41.
68. Plum F, Posner B. The diagnosis of stupor and coma. Philadelphia: Davies; 1980.
69. Specht H, Fruhman G. Incidence of periodic breathing in 2000 subjects without pulmonary or neurological disease. Bull Physiopathol Respir (Nancy). 1972;8(5):1075–82.
70. Mackey MC, Milton JG. Dynamical diseases. Ann N Y Acad Sci. 1987;504:16–32.
71. Vaillancourt DE, Newell KM. Changing complexity in human behavior and physiology through aging and disease. Neurobiol Aging. 2002;23(1):1–11.
72. Scheibel AB. Falls, motor dysfunction, and correlative neurohistologic changes in the elderly. Clin Geriatr Med. 1985;1(3):671–6.
73. Mosekilde L. Age-related changes in vertebral trabecular bone architecture—assessed by a new method. Bone. 1988;9(4):247–50.
74. Lipsitz LA, Goldberger AL. Loss of 'complexity' and aging. Potential applications of fractals and chaos theory to senescence. JAMA. 1992;267(13):1806–9.
75. Iyengar N, Peng CK, Morin R, Goldberger AL, Lipsitz LA. Age-related alterations in the fractal scaling of cardiac interbeat interval dynamics. Am J Physiol Regul Integr Comp Physiol. 1996;271(4):R1078–84.
76. Kaplan DT, Furman MI, Pincus SM, Ryan SM, Lipsitz LA, Goldberger AL. Aging and the complexity of cardiovascular dynamics. Biophys J. 1991;59(4):915–49.
77. Greenspan SL, Klibanski A, Rowe JW, Elahi D. Age-related alterations in pulsatile secretion of TSH: role of dopaminergic regulation. Am J Physiol Endocrinol Metab. 1991;260(3):E486–91.
78. Frolkis VV, Bezrukov VV. Aging of the central nervous system. Interdiscip Top Gerontol. 1979;16: 87–9.
79. Mader S. Hearing impairment in elderly persons. J Am Geriatr Soc. 1984;32(7):548–53.
80. Stiehm JH, Townsend NW. The U.S. Army War College: military education in a democracy. Philadelphia: Temple University Press; 2002.
81. Caron D. It's a VUCA world. 2009. http://www.slideshare.net/dcaron/its-a-vuca-world-cips-cio-march-5-2009-draft. last accessed: 22-Nov-21012.

2 Complexity, Uncertainty and Mess as the Links Between Science and the Humanities in Health care

Iona Heath

2.1 Introduction

In medical school I was taught the mantra of history, examination, diagnosis and treatment. Each was supposed to follow the other in a rational and linear sequence with a single diagnosis suggesting appropriate treatment. Yet in practice, it is almost never like this. A single patient has hypertension—so far so good—but she also has bronchiectasis after many years of smoking. She has just been diagnosed with cancer of the oesophagus. She has a daughter with severe learning difficulties and she is fearful, not only for herself but also about what will happen to her child. She is married and the relationship is difficult. Her sister has lung cancer and is already very ill. The sister's children have problems and children of their own and this is only the beginning of a story which has become richer and richer the longer I have known this patient and her family. It is a story with multiple components each of which interacts with the others unpredictably. Each of the components has a history which affects the interaction and each has the capacity to affect my patient's blood pressure and to support or undermine the treatment that I am prescribing for her. Seeing this situation as a complex adaptive system is much more useful than trying to stick to the medical school mantra because sensitivity, intuition, commitment and a pragmatic preparedness to muddle through then become as important as a sound grasp of biomedical knowledge.

There appear to be two quite distinct groups of health care professionals interested in complexity theory. The first group is formed by those with an understanding and talent for mathematics who are excited by the possibilities of using mathematical modelling of chaos and complexity [1] to increase our understanding of what happens within organisations and within human bodies [2]. The other group is motivated by the new insights which can be gained from using a much more limited understanding of chaos and complexity as a metaphor to make sense of the experience of caring for patients and to reunite the apparent polarities described as the art and science of medicine. This use of complexity as metaphor was roundly dismissed by a vituperative correspondent to the *British Medical Journal* as "intellectual snake oil" [3], but I think he was wrong and I confess to belonging firmly in this second group.

Metaphors provide us with ways of imagining and understanding the world and events within it. We all use metaphors: poets use them explicitly; some scientists aspiring to an objective rationality regard them as slightly reprehensible and have less insight into the pervasiveness of metaphor at the root of all thinking.

The core of this chapter was first written as a keynote presentation to the Annual Conference of the Royal New Zealand College of General Practitioners in Wellington in 2004 (with permission from NZFP)

I. Heath (✉)
e-mail: iona.heath22@yahoo.co.uk

J.P. Sturmberg and C.M. Martin (eds.), *Handbook of Systems and Complexity in Health*,
DOI 10.1007/978-1-4614-4998-0_2, © Springer Science+Business Media New York 2013

> Our visions—our ways of imagining the world—determine the direction of our thoughts, as well as being the source of our poetry. Poetry exists to express those visions directly, in concentrated form. But they are also expressed less directly in all our thoughts and actions, including scientific ones, where they often pass unnoticed and uncriticised [4].

Most biomedicine is based on the principles of Newtonian science using a metaphor of the body as a machine and every effect having an identifiable cause. Anyone who has worked with the uncertainties of clinical practice is painfully aware of the limitations of this model, and the capacity of complexity science to offer different metaphors is enormously welcome. We are enabled to look at old and intractable situations in a new way and provided with new defences against the simplistic and deterministic reductionism which informs much biomedical science and much of the organisation and evaluation of health care [5].

2.2 Time's Arrow

Complexity theory acknowledges the power of time much more than traditional Newtonian science. The classical laws of physics work either backwards or forwards in time, but complexity science demonstrates that changes over time within complex systems are irreversible, making such systems inherently evolutionary and historical [6]. This single difference makes the complex adaptive system much more effective than the machine as a metaphor for the human body. Human bodies decay over time and many of the changes brought by disease may be controllable but are essentially irreversible. Everything is evolving and has a unique history and we can never extrapolate either backwards or forwards in time with any degree of certainty and time flows only one way. The lesson that systems are essentially historical has been insufficiently learnt. It is the key to the uncertainty in medical practice—the explanation of why two individuals with the same diagnostic label and given the same recommended treatment can have quite different outcomes. It explains why there can never be one rule for everyone and why the much vaunted rolling out of good practice from one place to another so often fails. Every interaction between people, including between doctor and patient, contains the possibility of regret and lost opportunity. Information or pharmaceuticals or fear, once given, cannot be taken back, which seems to mean that we should be less certain and less arrogant in what we do. Consultations and a sequence of consultations are irreversible. A tenth consultation between a particular doctor and patient is inherently different from the first one.

Michael Frayn's magnificent play *Copenhagen* has three protagonists: Niels Bohr, Bohr's wife Margrethe and Werner Heisenberg. Between 1924 and 1927, in Copenhagen, Bohr (a Dane) and Heisenberg (a German) revolutionised atomic physics, and indeed the whole foundation of science, with the Copenhagen Interpretation which incorporated the twin principles of uncertainty and complementarity. In the play [7], these principles are summarised by Bohr:

> **Bohr**... *Particles are things complete in themselves. Waves are disturbances in something else. ... They're either one thing or the other. They can't be both. We have to choose one way of seeing them or the other. But as soon as we do we can't know everything about them.*

The play is also about the inexorability of time and the inevitability of regret. One of the Bohr's six sons died in a sailing accident:

> **Heisenberg**...*Those short moments on the boat, when the tiller slams over in the heavy sea, and Christian is falling.*
> **Bohr**...*If I hadn't let him take the helm ...*
> **Heisenberg**...*Again and again the tiller slams over. Again and again ...*

Wendy Lesser has written about the experience of reading the same book twice in a lifetime, as a younger and an older person [8]. She describes the dizzying effect of looking at a stationary work of art from two different points in time. The irreversibility of time means that the older reader is conscious of the younger one but the younger completely oblivious to the older and this asymmetry adds to the sense of vertigo. In the continuing relationships between doctors and patients, both parties are moving through time but both carry a perception of their younger selves interacting with the other and I find echoes of the same dizziness.

2.3 Uncertainty and Freedom

Each generation looks back on the knowledge of earlier generations and sees the extent of ignorance, yet each, in turn, seems to believe itself immune to that ignorance. Ours is no exception. Coleridge, like Werner Heisenberg several generations later, recognised that it is the unknown that offers us the possibility of new discoveries and new understanding.

> I do not like that presumptuous philosophy which in its rage of explanation allows no XYZ, no symbol representative of the vast Terra Incognita of Knowledge, for the Facts and Agencies of Mind and Matter reserved for future Explorers, while the ultimate grounds of all must remain inexplorable or Man ceases to be progressive. Our ignorance, with all the intermediates of obscurity, is the condition of our ever-increasing Knowledge [9].
>
> One may say that the human ability to understand may be in a certain sense unlimited. But the existing scientific concepts cover always only a very limited part of reality, and the other part that has not yet been understood is infinite. Whenever we proceed from the known to the unknown we may hope to understand, but we may have to learn at the same time a new meaning of the word understanding [10].

A precious freedom exists in the uncertainty and doubt that surrounds us and which alone gives us room to manoeuvre, to explore our possibilities and to be fully human. Here is the key. The hope of both researchers and practitioners in health care is to make the world a better place by easing suffering, but the hubris is to exaggerate the achievement by extrapolating far beyond the findings and constructing an over-certain future. Researchers have a grave responsibility always to explain and indeed to emphasise the limitations of their findings. Changes in the levels of risk to health should always be explained by researchers, practitioners and journalists in absolute rather than relative terms [11].

The outcomes of proposed treatments should always be discussed in terms of "Numbers Needed to Treat" and "Numbers Needed to Harm" [12]. Only then can patients make properly informed choices about whether to change their behaviour or to accept medication and only in this way can human autonomy be enhanced rather than eroded.

This concerns the old division between utilitarianism which asserts that decisions should be judged by their consequences and liberalism which is focussed on rights and opportunities, on where people start rather than on where they end up. Today, utilitarianism has a new underpinning in modern epidemiology and systems of health care supported by information technology and sponsored by pharmaceutical companies.

We have embarked on a dangerous road.

> The human being, who appears to be thrilling and wonderful, may turn out at the same time to be monstrous in its ambition to simplify and control the world. Contingency, an object of terror and loathing, may turn out to be at the same time wonderful, constitutive of what makes a human life beautiful and thrilling [13].

It is contingency, chance, fate and uncertainty that makes life beautiful; the enduring truth that we can never know what will happen tomorrow, whether or not we have taken our aspirin and statin and declined to have butter on our bread, that makes life thrilling.

2.4 The First-Person View of Agency

The brilliance of Michael Frayn's *Copenhagen* is the parallel he draws between the uncertainty principle of the wave and the particle and the uncertainty of human thought and action which is due to the subjective sense of agency.

> **Heisenberg**... *Exactly where you go as you ramble around is of course completely determined by your genes and your upbringing and the earth's magnetic field and the gravitational pull of the moon. But it's also completely determined by your own entirely inscrutable whims from one moment to the next. So we can't completely understand your behaviour without seeing it both ways at once, and that's impossible, because the two ways are mutually incompatible.*

A similar point is made rather less elegantly in the Harvard Law of Animal Behavior:

> Under carefully controlled experimental circumstances, an animal will behave as it damned well pleases.

And if this is true for animals, how much more so is it true for humans. Part of the impossibility of predicting human behaviour is to do with conscious decisions made by self-determining autonomous individuals, but another part is to do with the randomness of thoughts and ideas.

> **Heisenberg**... *There's no reason at all. I didn't tell Speer simply because I didn't think of it. I came to Copenhagen simply because I did think of it. A million things we might do or might not do every day. A million decisions that make themselves.*

For doctors and patients there are also a million decisions that make themselves—whether to consult the doctor, whether to consider a particular diagnosis, whether to think of prescribing a particular medication, whether to mention a particular worry. No amount of advice or guidance will change the shifting and elusive nature of thoughts and intentions.

In his postscript to *Copenhagen*, Michael Frayn puts it like this:

> Thoughts and intentions, even one's own—perhaps one's own most of all—remain shifting and elusive. There is not one single thought or intention of any sort that can ever be precisely established. What the uncertainty of thoughts does have in common with the uncertainty of particles is that the difficulty is not just a practical one, but a systematic limitation which cannot even in theory be circumvented.

2.5 Meaning as Emergent Property

An emergent behaviour or emergent property is shown when a number of simple *entities* operate in an environment, forming more complex behaviours as a collective. A *system* made of several things can host properties which the things themselves do not have. George Henry Lewes, the nineteenth century English philosopher and common-law husband of the novelist George Eliot, distinguished between resultants and emergents: phenomena that are predictable from their constituent parts and those that are not.

Poetic meaning emerges from combinations of words and thought. Meaning, ideas and language reinforce and change each other in a continuously evolving process described by George Steiner as "the informing reciprocities between grammar and vision" [14]. The London Language Line provides telephone translation for more than 100 different languages and each of these languages sees the world slightly differently. William Carlos Williams was both a family doctor and a poet and his view was that what patients said to their doctors in the privacy of the consultation room was the closest many of them would come to the creation of poetry—choosing words to express the deepest of feelings and fears [15]. The patient's presentation of his or her symptoms emerges rather than results from their experience of the symptoms themselves. Each individual chooses and uses words differently and gives different expression to his or her symptoms. And, similarly, each dyad of doctor and patient will generate a different response to the particular patient's predicament.

Science has its own "informing reciprocities" between grammar and vision and Weiner Heisenberg argued that "natural language" using ordinary words is much more useful in the attempt to communicate innovative ideas and new knowledge than a specialist vocabulary which has much less flexibility [16].

Ralph Stacey argues that the creativity and capacity for learning of an organisation is dependent on the number and the quality of the conversations within it [17]. The crucial conversations within health care occur between patients and clinicians, and complexity science suggests that health services should be organised to maximise and to sustain these conversations. The outcome of these conversations is always uncertain and if an authentic interaction is achieved both parties become caught up in it and neither controls it [18].

2.6 Ethical Implications

It is important to distinguish between messes and difficulties. A difficulty may be complicated but it is capable of solution whereas a mess presents challenges which cannot even be clearly defined and on which there are many different perspectives offering a variety of competing approaches. Repairing a car is a difficulty, whereas devising a policy to improve the quality of care in a health service is a mess [19].

The problem of the mess permeates health care and is often tackled by pretending that things are simpler and more straightforward than they are and that situations will respond to a linear, rational model of decision or policy making. Complexity science tells us that the delivery of health care to a unique individual can never be simple. Standardised diagnostic categories and treatment guidelines can be used to predict probable outcomes at a population level but can tell us nothing about the likely outcome for a particular individual. This situation is problematic in curative or therapeutic health care but at least in this situation the patient has sought care. The problems posed by preventive health care in this uncertain situation are much more ethically dubious and have led to a situation where the populations of the richest countries in the world are healthier by objective standards than they have ever been before and yet have higher rates of self-reported illness than those living in poorer countries. People live longer lives than ever before but live them feeling fearful and ill, exposed to one health scare after another.

2.7 Multiple and Compounding Health and Social Problems

Linear reasoning has some usefulness in the management of patients with single diseases and this usefulness has been expressed in the explosion of management guidelines that have been made available to clinicians over recent years. However, comparatively few people have only one diagnosis and many, particularly those who are poor and/or old, have multiple and compounding problems. There is a socio-economic gradient in the incidence and prevalence of almost all major disease categories, meaning that individuals and families who are socio-economically disadvantaged are at risk of a compounding multiplicity of health and social problems. Multiple morbidity is a major component of health inequalities and can be seen in part as a direct consequence of the wider societal determinants of ill-health. Health care that is both driven and evaluated increasingly by protocols derived from studies of single disease conditions seems likely to disadvantage systematically those with complex and overlapping health problems. There is an urgent need to know much more about the optimal treatment of multiple morbidity. How should the care of different diseases be prioritised in situations where treatments are incompatible or the burden of treatment becomes too great? Primary care clinicians have considerable experience and skills in managing multiple health problems to achieve optimal outcomes for each particular individual [20]. This is done by the careful negotiation of an individual care plan that makes sense to the patient in the context of his life story and the full diversity of his health and social problems, and which accommodates his values and aspirations. The necessary skills have been mostly unrecognised and undervalued and are systematically concealed by routine audits of the management of single disease states.

2.8 The Urge to Simplify

There seems to be an insatiable human need to make sense of the world through a process of simplification that denies the complexity and details of experience. The reductive nature of biomedical science is one response to this need and it has produced huge benefits. Biomedical science is based on the relatively crude generalisations which we recognise as diseases [21]. If we group people together according to these disease categories, we can extend our knowledge about the phenomenon they have in common—be it diabetes or epilepsy. As a direct result there has been enormous progress in clinical medicine—but the process of generalisation devalues individual experience and can leave individuals feeling unrecognised and the reality of their symptoms unheard. It is a means of making judgements about people that are both constructive and destructive. The general practitioner, while actively using the generalisations of biomedical science, has a constant responsibility to re-focus on the individual, the detail of their experience and the meaning they attach to that experience.

Philip Roth argues that the task of the writer is "keeping the particular alive in a simplifying, generalizing world" [22]. Novels are the literary form most committed to this task and the writers of novels achieve this by evoking the detail of individual thought and the precise words of dialogue and investing each life with dignity and value and a sense of the infinite importance of what happens in the life of each individual [23].

> —this great and moving novel, which looks so quiet and provincial, opens out through its small frame to our most troubling and essential questions. How well do we remember? How do we make our choices in life? Why do we need repetition? What is to remain of us? Above all, what can happiness consist in? [24]

This commitment of the novelist to the complexity and authentic nature of his or her characters has the same quality as the commitment of the general practitioner to the ongoing care of his or her patients. Ian McWhinney has described this as commitment to a person whatever may befall them [25]. Both novelist and practitioner are committed to keeping the particular alive, resisting simplification and embracing the complex reality of lived experience.

References

1. Stewart I. Sources of uncertainty in deterministic dynamics: an informal overview. Phil Trans A Math Phys Eng Sci. 2011;369(1956):4705–29.
2. Holt TA. A chaotic model for tight diabetes control. Diabet Med. 2002;19(4):274–8.
3. Reid I. Let them eat complexity: the emperor's new toolkit. Br Med J. 2002;324:171.
4. Midgley M. Science and poetry. London: Routledge; 2001.
5. Goodwin JS. Chaos, and the limits of modern medicine. JAMA. 1997;278(17):1400.
6. Byrne D. Complexity theory and social research. Social Research Update, Issue 18, 1997.
7. Frayn M. Copenhagen. London: Methuen Drama; 1998.
8. Lesser W. Sauntering through the commas: the pleasures of a book revisited. The Guardian, Saturday 22 Dec 2001.
9. Coleridge ST. Notebooks, Volume 3, 1809-1818. In: Holmes R, editor. Coleridge: darker reflections. London: Harper Collins; 1998.
10. Heisenberg W. In: Berger J. The moment of cubism and other essays. London: Weidenfield & Nicolson; 1969.
11. Gigerenzer G, JA Muir Gray JA, editors. Better doctors, better patients, better decisions: Strüngmann forum report. Boston: MIT Press; 2011.
12. http://www.medicine.ox.ac.uk/bandolier/booth/glossary/NNH.html. Accessed 28 Dec 2011.
13. Nussbaum MC. The fragility of goodness. Cambridge: Cambridge University Press; 1986.
14. Steiner G. Errata: an examined life. London: Phoenix; 1998.
15. Williams WC. The doctor stories. New York: New Directions Books; 1984.
16. Heisenberg W. Physics and philosophy: the revolution of modern science. New York: Harper; 1958.
17. Stacey RD. Complexity and group processes: a radically social understanding of individuals. London: Brunner-Routledge; 2003.
18. Nessa J, Malterud K. Tell me what is wrong with me: a discourse analysis approach to the concept of patient autonomy. J Med Ethics. 1998;24:394–400.
19. Chapman J. System failure. Why governments must learn to think differently. London: Demos; 2002.
20. Schellevis FG, van der Velden J, van de Lisdonk EH, van Eijk JTM, van Weel C. Comorbidity of chronic disease in general practice. J Clin Epidemiol. 1993;46:469–73.
21. Rosenberg CE. The tyranny of diagnosis: specific entities and individual experience. Milbank Q. 2002;80(2):237–60.
22. Roth P. I married a communist. London: Jonathan Cape; 1998.
23. Nussbaum MC. Poetic justice: the literary imagination and public life. Boston: Beacon Press; 1995.
24. Lee H. Review of John McGahern's 'That they may face the rising sun'. The Observer, 6 Jan 2002.
25. McWhinney IR. Being a general practitioner: what it means. Eur J Gen Pract. 2001;6:135–9.

Part I

Complexity Theory and Methods

Systems and complexity thinking as we know it today emerged in the latter half of the nineteenth century. Physicists, mathematicians, chemists and others observed that the prevailing reductionist framework of inquiry and explanation no longer sufficiently described the behaviour of the phenomena under study. This resulted in the new propositions of quantum mechanics (Heisenberg) and relativity theory (Einstein) in the early parts of the twentieth century. The second half of the twentieth century saw an expansion of the field—cybernetics (Wiener), general systems theory (von Bertalanffy), concepts of self-organization (Prigogine), chaos theory (Mandelbrot, Lorenz), concepts of autopoiesis and adaptation (Verala and Maturana), emergence and dynamic systems thinking (Kauffman and Bak), complex adaptive systems models (Holland, Gell-Mann) and the sciences of networks (Barabasi).

Systems and complexity sciences, as a scientific approach, have their own distinct philosophy and methodologies. This section introduces the philosophical underpinnings of systems and complexity sciences, and highlights its impact on knowledge and knowledge generation, before introducing some methodological approaches to systems and complexity-based research most amenable to the medical environment. These chapters aim to provide an overview, and detailed instructions to the background and the application of these research methodologies have to be gained from specific publications.

The apparent tensions, contradictions and synergies between reductionism and systems thinking are just that—perceptions. Systems and complexity thinkers see the world not as *dichotomous either-ors* but *mutually agreeable ands* that always have and will continue to shape human endeavours.

Understanding Complex Systems 3

Paul Cilliers

3.1 Introduction

Health is a complex notion and it, or its absence, manifests itself under complex conditions. The acknowledgement that a certain situation is complex has a number of implications for the way in which we deal with that situation. First and foremost among these is the fact that traditional "reductive" approaches lead to problematic results. The problem is exacerbated by the fact that some vague "holistic" approach is not an alternative. In order to deal with complex situations as best we can, we need to understand something of the dynamics of complex systems and then reflect on the implications of this understanding.

The argument in this chapter[1] will be built up by first showing what the problems are with a reductionist approach. Complexity theory will be introduced as a critical alternative, with the qualification that there are also different types of "complexity theory". After characterising complexity in a general way, some specific issues will be addressed, including the problem of modelling complexity and the issue of boundaries. In conclusion, some of the normative issues which follow from acknowledging complexity will be discussed.

3.2 Why Reductionism Fails

To a large extent we still live in a world where "scientific" knowledge trumps all other forms of knowledge. This state of affairs is a legacy of a certain interpretation of Enlightenment thinking. In this interpretation, the quest for verifiable knowledge, at least since the Renaissance, presupposes the need for objectivity. Novotny et al. ([2]: 50–51) describe this process in the following way: *"In its historical contest with religion, a triumphant science acquired a monopoly of describing and explaining 'reality', which both resisted and also validated human wishes, fancies and follies. Because the physical world, including its chemical and biological processes, came to be regarded as the most substantial component of the 'real world', a scientific definition of reality became ever more plausible. As a result the authority, values and practices of science permeated many other dimensions of society. The everyday world shrank to what scientists had 'discovered' and were able to exploit"*.

This traditional or, as it is often called, modernist style of scientific thinking is no longer adequate—to the extent that it ever was. The reason for this is not because of a frivolous postmodern

[1] This chapter is a slightly altered version of a chapter called "Complexity Theory as General Framework for Sustainability Science" which appeared in Burns and Weaver [1]. Permission to reuse the material is acknowledged.

P. Cilliers (✉)
c/o Jan-Hendrik Hofmeyr, Department of Philosophy, Centre for Studies in Complexity, University of Stellenbosch, Stellenbosch, South Africa
e-mail: jhsh@sun.ac.za

J.P. Sturmberg and C.M. Martin (eds.), *Handbook of Systems and Complexity in Health*,
DOI 10.1007/978-1-4614-4998-0_3, © Springer Science+Business Media New York 2013

reaction to modernity,[2] nor is it merely because of some logical problem with the verification of experimental processes (in the tradition of Popper, Kuhn or Feyerabend), it is a result of the complexity of the phenomena we deal with. As Novotny et al. ([2]: 21) state: *"Contemporary society is characterized—irreversibly—by pluralism and diversity and also, we argue, volatility and transgressivity. It can no longer be understood either in terms of the norms and practices of scientific rationality ...".*

What is at stake when we deal with complex things is thus the appropriate style of rationality. The argument is that the traditional modernist rationality—established in the first half of the seventeenth century and based on the ideas of Galileo, Newton and Descartes in the context of a more settled Europe after the peace of Westphalia (see [3]); a style of thinking fundamental to the establishment of the Royal Society (or more precisely, The Royal Society of London for the Improvement of Natural Knowledge, with the now extremely disconcerting motto *nullius in verba*)—is *not* adequate to complexity. Edgar Morin ([4]: 5) gives specific content to the inadequacy of what he calls "classical science". For him, *"classical science rejected complexity in virtue of three fundamental explanatory principles:*

1. *The principle of universal determinism, illustrated by Laplace's Daemon, capable, thanks to his intelligence and extremely developed senses, of not only knowing all past events, but also of predicting all events in the future.*
2. *The principle of reduction, that consists in knowing any composite from only the knowledge of its basic constituting elements.*
3. *The principle of disjunction, that consists in isolating and separating cognitive difficulties from one another, leading to the separation between disciplines, which have become hermetic from each other".*

For Morin, this tradition has led to wonderful results, but only in a limited context. In order to deal with a complex world, however, we need to acknowledge the limitations of this approach. An epistemological shift is required which replaces "reduction" with "distinction" and "disjunction" with "conjunction" ([4]: 10).

In many ways one can argue that we are dealing with an epistemological crisis in contemporary theory. It is generally acknowledged that simple reductive thinking is not adequate, but there is also a fear of anything which could be a form of relativism. If this was merely a theoretical problem we could let the philosophers argue about it. Unfortunately, different strategies of thinking lead to different forms of action in the world. The disastrous effects of reductive thinking are evident in many spheres, including the social, the political, the economical, the environmental, and the healthcare. The epistemological shift Morin talks about is, therefore, not merely a theoretical issue, but one with practical and ethical implications.

In what follows, an attempt will be made to give some content to the nature of this shift. It will be argued that we do not have a consistent or complete language which can replace the reductive one, but that we can articulate some of the shortcomings and limitations. This articulation is primarily critical at this stage, but it is a critique which leads to different forms of action, and thus to real results.

[2] The notions "modern" and "postmodern" have to be used with caution. Modernism is often treated in a much too simplistic way, as if there was one coherent "movement" which simply relied on an oversimplified understanding of rationality. Modernism was, or is, a divided strategy containing different strategies not easily reducible to one another. Sophisticated attempts to clarify the role and limits of rationality, as in the work of Habermas, for example, cannot be treated as if they are simply an extension of the Cartesian/Newtonian paradigm. The notion "postmodern" is also misused frequently. For some it simply means the justification of relativism, while for others it is merely a tag of approval without much content. These misunderstandings should not get in the way of recognizing the real problem, namely the inadequacy of reductive thinking when dealing with complex things.

The notion "scientific" is similarly problematic. What is criticised in this chapter is probably described better by the notion "scientistic"; i.e. an uncritical reliance on first-order logic and verifiable observation. The critical use of complexity theory in this chapter in no way intends to dismiss science; it seeks to expand the notion, or at least, to mark its limits.

3.3 Dealing with Complexity

An interest in complexity science has blossomed in the last three decades or so. Fuelled by the work of, amongst many others, Prigogine, Maturana and Varela, Mandelbrot, Kaufman, Gell-Mann and a generation of chaos theorists, the characteristics of complex systems have been studied intensively. Numerous institutions have been founded devoted solely to the investigation of these issues. As can be seen from the present volume, the topic is also becoming important in a host of other disciplines.

The mere fact that a lot of attention is being paid to complexity is, however, no guarantee that the epistemological shift referred to above has taken place. Morin is quite explicit that, even in complexity theory, the dominant or traditional rationality has largely been retained. In order to make this explicit he makes a distinction between "restricted complexity" and "general complexity".

Restricted complexity is, for Morin, exemplified in those approaches to complexity which developed from chaos theory and fractal mathematics. These approaches focus on underlying patterns and universal principles which are still highly reductive in nature.[3] *"Restricted complexity made [...] possible important advances in formalization, in the possibilities of modelling, which themselves favor interdisciplinary potentialities. But one still remains within the epistemology of classical science. When one searches for the "laws of complexity", one still attaches complexity as a kind of wagon behind the truth locomotive, that which produces laws. A hybrid was formed between the principles of traditional science and the advances towards its hereafter. Actually, one avoids the fundamental problem of complexity which is epistemological, cognitive, paradigmatic. To some extent, one recognizes complexity, but by decomplexifying it. In this way, the breach is opened, then one tries to clog it: the paradigm of classical science remains, only fissured"* ([4]: 10).

General complexity, Morin argues, is not merely a methodology; it involves a rethink of our fundamental definitions of what knowledge is. When dealing with complexity, the traditional method of analysis does not work. What is more, the divide between subject and object cannot be maintained in any clear way. This is how Morin formulates it: *"In opposition to reduction, [general] complexity requires that one tries to comprehend the relations between the whole and the parts. The knowledge of the parts is not enough, the knowledge of the whole as a whole is not enough ... Thus, the principle of reduction is substituted by a principle that conceives the relation of whole-part mutual implication. The principle of disjunction, of separation (between objects, between disciplines, between notions, between subject and object of knowledge), should be substituted by a principle that maintains the distinction, but that tries to establish the relation"* ([4]: *10–11).*

From this formulation it should be clear that Morin is not advocating a relativistic position, nor is he arguing for a "generality" which is naively holistic or vague. One cannot say anything without making distinctions, but these distinctions are always contextualised within a set of relationships ([4]: 18–20).

To my mind, Morin is absolutely correct about the fact that much of current complexity theory remains trapped within a traditional rationality. I witnessed a prominent theorist in the field claiming that everything which happens in society is reflected in the Dow Jones index, thereby reducing all the complexities of human

[3] Byrne [5] argues in the same way as Morin. He distinguishes between "simple" complexity and "complex" complexity, and then insists that simple (restricted) complexity plays in the court of current orthodoxy: "This is why simplistic complexity is so attractive to the worst sort of evolutionary psychology and contemporary ideologues of market models. Write a few rules—the selfish gene, the territorial imperative, profit maximisation, rational choice, or, preferably, a combination of all of these, and away we go. Simplistic complexity does deal with a kind of complex emergence but it remains reductionist" ([5]: 103). McLennan [6] makes a similar argument about the way in which complexity theory has been applied to sociology. It seems, for him, as if complexity theory—and what he refers to as "restricted complexity"—is not providing a critique of outdated meta-paradigms; it is simply providing a new one.

society to a single index, and a financial one at that. The current interest which many complexity theorists have in power laws is evidence of a similar form of reduction. It rests on the problematic assumption that the world is fractal in nature. At least in the biological and social domains, scale matters very much. Our knowledge of things complex cannot be free floating and abstracted; it is contingent and historically determined. Perhaps certain natural phenomena are more amenable to descriptions in terms of a "restricted" complexity, but to develop a deeper understanding of ecological, human and social phenomena we will have to move beyond a simplified Enlightenment rationality, even if that rationality is spruced up with a bit of chaos theory. For that we need to develop an understanding of complexity which is brave enough to think the uncomfortable consequences of its insights through to the end.

3.4 Characterising Complexity

The notion "complexity" has up to now been used in a somewhat general way, as if we know what the word means. According to conventional academic practice it would now be appropriate to provide a definition of "complexity". I will nevertheless resist this convention. There is something inherently reductionist in the process of definition. This process tries to capture the precise meaning of a concept in terms of its essential properties. It would be self-defeating to start an investigation into the nature of complexity by using exactly those methods we are trying to criticise! On the other hand, we cannot leave the notion of "complexity" merely dangling in the air; we have to give it some content. This will be done by making a number of distinctions which will constrain the meaning of the notion[4] without pinning it down in a final way. The characterisation developed in this way is thus not final—in specific contexts there may be more characteristics one could add, and some of those presented here may not always be applicable—but it helps us to make substantial claims about the nature of complexity, claims that may shift our understanding in radical ways.

In the first place, one should recognise that complexity is a characteristic of a *system*. Complex behaviour arises because of the *interaction* between the components of a system. One can, therefore, not focus on individual components, but on their relationships. The properties of the system *emerge* as a result of these interactions; they are not contained within individual components.

A second important issue is to recognise that a complex system generates a new structure internally. It is not reliant on an external designer. This process is called self-organisation. In reaction to the conditions in the environment, the system has to adjust some of its internal structure. In order to survive, or even flourish, the tempo at which these changes take place is vital (see [7] for detail in this regard). A comprehensive discussion of self-organisation is beyond the scope of this chapter (see Chap. 6 in [8] for such a discussion), but some aspects of self-organisation will become clear as we proceed.

An important distinction can be made between "complex" and "complicated" systems. Certain systems may be quite intricate, say something like a jumbo jet. Nevertheless, one can take it apart and put it together again. Even if such a system cannot be understood by a single person, it is understandable in principle. Complex systems, on the other hand, come to be in the interaction of the components. If one takes it apart, the emergent properties are destroyed. If one wishes to study such systems, examples of which are the brain, living systems, social systems, ecological systems and social-ecological systems, one has to investigate the system as such. It is exactly at this point that reductionist methods fail.

One could argue, however, that emergence is a name for those properties we do not fully understand *yet*. Then complexity is merely a function of our present understanding of the system, not of the system itself. Thus one could distinguish between *epistemological* complexity—complexity as a

[4] The significance of "constraints" is discussed in the chapter.

function of our description of the system—and *ontological* complexity—complexity as an inherent characteristic of the system itself. Perhaps, the argument might go, all complexity is merely epistemological, that finally all complex systems are actually just complicated and that we will eventually be able to understand them perfectly.

If one follows an open research strategy—a strategy which is open to new insights as well as to its own limitations—one cannot dismiss the argument above in any final way. Nevertheless, until such time as the emergent properties of a system are fully understood, it is foolish to treat them as if we understand them already. Given the finitude of human understanding, some aspects of a complex system may always be beyond our grasp. This is no reason to give up on our efforts to understand as clearly as possible. It is the role of scientific enquiry to be as exact as possible. However, there are good reasons why we have to be extremely careful about the *reach* of the scientific claims we make. In order to examine these reasons in more detail, a more systematic discussion of the nature of complex systems is required. The following characteristics will help us to do this [5]:

1. Complex systems are open systems.
2. They operate under conditions not at equilibrium.
3. Complex systems consist of many components. The components themselves are often simple (or can be treated as such).
4. The output of components is a function of their inputs. At least some of these functions must be non-linear.
5. The state of the system is determined by the values of the inputs and outputs.
6. Interactions are defined by actual input–output relationships and these are dynamic (the strength of the interactions change over time).
7. Components, on average, interact with many others. There are often multiple routes possible between components, mediated in different ways.
8. Many sequences of interaction will provide feedback routes, whether long or short.
9. Complex systems display behaviour that results from the *interaction* between components and not from characteristics inherent to the components themselves. This is sometimes called emergence.
10. Asymmetrical structure (temporal, spatial and functional organisation) is developed, maintained and adapted in complex systems through internal dynamic processes. Structure is maintained even though the components themselves are exchanged or renewed.
11. Complex systems display behaviour over a divergent range of timescales. This is necessary in order for the system to cope with its environment. It must adapt to changes in the environment quickly, but it can only sustain itself if at least part of the system changes at a slower rate than changes in the environment. This part can be seen as the "memory" of the system.
12. More than one legitimate description of a complex system is possible. Different descriptions will decompose the system in different ways and are not reducible to one another. Different descriptions may also have different degrees of complexity.

If one considers the implications of these characteristics carefully a number of insights and problems arise:

- The *structure* of a complex system enables it to behave in complex ways. If there is too little structure (i.e. many degrees of freedom), the system can behave more randomly, but not more functionally. The mere "capacity" of the system (i.e. the total amount of degrees of freedom available if the system was not structured in any way) does not serve as a meaningful indicator of the complexity of the system. Complex behaviour is possible when the behaviour of the system is constrained. On the other hand, a fully constrained system has no capacity for complex behaviour either. This claim is not quite the same as saying that complexity exists somewhere on the edge between

[5] These characteristics were formulated in collaboration with Fred Boogerd and Frank Bruggemans at the Department of Molecular Cell Physiology at the Free University, Amsterdam, based on the arguments in Cilliers [8], and used in Cilliers [9].

order and chaos. A wide range of structured systems display complex behaviour.

- Since different descriptions of a complex system decompose the system in different ways, the knowledge gained by any description is always relative to the perspective from which the description was made. This does not imply that any description is as good as any other. It is merely the result of the fact that only a limited number of characteristics of the system can be taken into account by any specific description. Although there is no a priori procedure for deciding which description is correct, some descriptions will deliver more interesting results than others.
- In describing the macro-behaviour (or emergent behaviour) of the system, not all the micro-features can be taken into account. The description on the macro-level is thus a reduction of complexity, and cannot be an exact description of what the system actually does. Moreover, the emergent properties on the macro-level can influence the micro-activities, a phenomenon sometimes referred to as "top-down causation". Nevertheless, macro-behaviour is not the result of anything else but the micro-activities of the system, keeping in mind that these are not only influenced by their mutual interaction and by top-down effects, but also by the interaction of the system with its environment. When we do science, we usually work with descriptions which operate mainly on a macro-level. These descriptions will always be approximations of some kind.

These insights have important implications for the knowledge claims we make when dealing with complex systems. Since we do not have direct access to the complexity itself, our knowledge of such systems is in principle limited. The problematic status of our knowledge of complexity needs to be discussed in a little more detail. Before doing that, some attention will be paid to three problems: identifying the boundaries of complex systems, the role of hierarchical structure and the difficulties involved in modelling complexity.[6]

3.5 Boundaries

In order to be recognisable as such, a system must be bounded in some way. However, as soon as one tries to be specific about the boundaries of a system, a number of difficulties become apparent. For example, it seems uncontroversial to claim that one has to be able to recognise what belongs to a specific system, and what does not. But complex systems are open systems where the relationships amongst the components of the system are usually more important than the components themselves. Since there are also relationships with the environment, specifying clearly where a boundary could be, is not obvious.

One way of dealing with the problem of boundaries is to introduce the notion of "operational closure".[7] For a system to maintain its identity, it must reproduce itself (internally). These arguments often follow from the work by Maturana and Varela on autopoiesis. Zeleny ([12]: 123) defines an autopoietic system as: "*... a system that is generated through a closed organisation of production processes such that the same organisation of processes is regenerated through the interaction of its own products (components), and a boundary emerges as a result of the same constitutive processes*".

When dealing with complex systems in an "operational" way, there is nothing wrong with this approach. One should be careful, however, not to overemphasise the closure of the boundary. The boundary of a complex system is not clearly defined once it has "emerged". Boundaries are simultaneously a function of the activity of the system itself, and a product of the strategy of description involved. In other words, we frame the system by describing it in a certain way (for a certain reason), but we are constrained in where the frame can be drawn. The boundary of the system is, therefore, neither purely a function of our description, nor is it a purely natural thing.

[6] The following two sections are based on Cilliers [10].

[7] The work of Niklas Luhmann provides a good example of this approach. For a monograph in English, see Luhmann [11].

We can never be sure that we have "found" or "defined" it clearly, and therefore the closure of the system is not something that can be described objectively. An overemphasis on closure will also lead to an understanding of the system that may underplay the role of the environment. However, we can certainly not do away with the notion of a boundary.

Our understanding of boundaries can be given a little more content by considering the following two issues. The first concerns the "nature" of boundaries. We often fall into the trap of thinking of a boundary as something that separates one thing from another. We should rather think of a boundary as something that constitutes that which is bounded. This shift will help us to see the boundary as something enabling, rather than as confining. To quote Zeleny ([12]: 133) again: *"All social systems, and thus all living systems, create, maintain, and degrade their own boundaries. These boundaries do not separate but intimately connect the system with its environment. They do not have to be just physical or topological, but are primarily functional, behavioral, and communicational. They are not "perimeters" but functional constitutive components of a given system".*

As an example of this logic, think of the eardrum. It forms the boundary between the outer and the middle ear, but at the same time exists in order to let the sound waves through. As a matter of fact, if it was not there, the sound waves would not be able to get through at all! If the boundary is seen as an interface participating in constituting the system, we will be more concerned with the margins of the system, and perhaps less with what appears to be the evident or "central" components of the system.[8]

A second boundary issue concerns the "place" of the boundary. The propensity we have towards visual metaphors inclines us to think in spatial terms. A system is, therefore, often visualised as something contiguous in space. This tendency is reinforced by the prevalence of biological examples of complex systems. We think of systems in an "organistic" way. Social systems are obviously not limited in the same way. Parts of the system may exist in totally different spatial locations. The connections between different components could be seen as virtual, and therefore the system itself may exist in a virtual space. This much should be self-evident to most inhabitants of the global village, but two important implications should be kept in mind. The first is that non-contiguous sub-systems could be part of many different systems simultaneously. This would mean that different systems interpenetrate each other, that they share internal organs. How does one talk of the boundary of the system under these conditions? A second implication of letting go of a spatial understanding of boundaries would be that in a critically organised system we are never far away from the boundary. If the components of the system are richly interconnected, there will always be a short route from any component to the "outside" of the system. There is thus no safe "inside" of the system, the boundary is folded in, or perhaps, the system consists of boundaries only. Everything is always interacting and interfacing with others and with the environment; the notions of "inside" and "outside" are never simple or uncontested.

In accepting the complexity of the boundaries of complex systems, we are committed to be critical about how we use the notion since it affects our understanding of such systems, and influences the way in which we deal with them. The notion of "boundary critique" is not a new one (see [13]), but in this critique we have to keep the enabling nature of boundaries as well as their "displacement" in mind.

3.6 Hierarchies

An analysis of the importance of hierarchies has been part of the study of complex systems for a long time [14, 15]. In his seminal paper, Simon gives at least three reasons why hierarchies are important. In the first place, a modular structure would make it easier for new complex systems to

[8] Although it will not be elaborated on in this text, a number of the ideas presented have a close affinity to arguments from deconstruction. For more detail, see Cilliers [8], especially Chap. 3.

be generated. He uses the example of two watchmakers, one building each watch from scratch, the other first constructing basic subassemblies, and then connecting these together. The second, he argues, will be more efficient. This "hierarchical" structure would also allow the system to take better advantage of evolutionary opportunities. In the second place, hierarchies establish unambiguous routes of communication. If the system is hierarchical, an algorithm can be developed that would ensure that information would get from A to B. In the third place, Simon argues that hierarchical systems have a lot of redundancy, and that it is, therefore, possible to construct models of such systems that are simpler than the system itself (a claim which is obviously somewhat at odds with the position argued for here).

A somewhat contrary position is taken in some contemporary discussions of complex systems in the context of the theory of organisations. A lot of emphasis is placed on self-organisation and the "distributed" nature of the structure in a system. According to these arguments, complex systems do not have central control systems. They have to be dynamic and adaptable, not rigid or invariable. Consequently, the notion of hierarchy is resisted. In terms of the structure of organisations, it is often argued that to the extent that there should be hierarchies at all, they should be shallow and loose. There must be enough space for innovation.

Both these positions oversimplify the role of hierarchies in complex systems. Hierarchies are certainly necessary, but the way in which they work differs in important respects from the classical understanding. Let us examine these differences.

In the first place, it must be underscored that systems cannot do without hierarchies. Complex systems are not homogeneous things. They have structure, and moreover, this structure is asymmetrical (see [8]: 120, 124, 147–148). There are subsections with functions, and for them to exist at all there has to be some form of hierarchy. Problems arise, however, when these hierarchies are seen as either too clearly defined, or too permanent. The classical understanding of hierarchies tends to view them as being nested.[9] In reality however, hierarchies are not that well-structured. They interpenetrate each other; i.e. there are relationships which cut across different hierarchies. These interpenetrations may be fairly limited, or so extensive that it becomes difficult to typify the hierarchy accurately in terms of prime and subordinate parts. Simon [15], of course, knows this. Nevertheless, in the hope of coming up with enough hierarchical structure to enable modelling of the system, he emphasises that which falls within the hierarchies, and not the interpenetrations. He argues that many complex systems are "near decomposable", meaning that hierarchical models will provide a fair approximation. This view would see the interpenetrations as part of the messiness of complexity, whereas I would rather see them as indispensable. Similar to the notion of boundaries discussed above, the structure of a complex system cannot be described merely in terms of clearly defined hierarchies. The cross-communications between hierarchies are not accidental, but part of the adaptability of the system. Alternative routes of communication are vital in order to subvert hierarchies that may have become too dominant or obsolete. Cross connections may appear to be dormant for long, but in the right context may suddenly play a vital role.

This leads directly to the next point: part of the vitality of a system lies in its ability to transform hierarchies. Although hierarchies are necessary in order to generate frameworks of meaning in the system, they cannot remain unchanged. As the context changes, so must the hierarchies. Some hierarchies may be more long-lived than others, but it is important to perceive of hierarchies as transformable entities.

To summarise then, hierarchies are necessary, but they are not neatly nested. The hierarchies in a system have a complex structure themselves. In dealing with complex systems, irrespective of whether we are merely trying to describe them or

[9] This is perhaps again a legacy of biological models—subsystems are seen as "organs". Biological systems are subjected to constraints that may not apply to all complex systems, especially not social systems.

whether we are trying to influence or change them, we have to take the existence of hierarchies into consideration, as well as the processes which undermine them.

3.7 Models of Complex Systems

The notion of a model is central to scientific understanding. The notion will be used here in a wide sense—i.e. theories and systems of rules can also be seen as models. To the extent that novels and works of art increase our understanding, it is useful to see them as "models" as well. Models, in short, are what we use to generate understanding. In the context of complexity, the role of models is described in the following way by Csányi (in [12]: 148): *"Any kind of scientific statement, concept, law, and any description of a phenomenon is a model construction which tries to reflect phenomena of the external world. Reality is extremely complex; it consists of strongly or more weakly related events. Science makes an attempt to separate and isolate different effects and phenomena. It seeks the simplest relationships by which examined phenomena can at least be described or demonstrated. It creates simplified models which only partly reflect reality, but which allow contemplation, and what is most important, pragmatic, even if sometimes modest, predictions".*

We cannot deal with reality *in* all its complexity. Our models have to reduce this complexity in order to generate some understanding. In the process something is obviously lost. If we have a good model, we would hope that which is left out is unimportant. It should be clear already that purely quantitative models of complex systems, which abstract from a set of real properties to numerical values, will be problematic ([16]: 54). The underlying problem with models of complexity is, however, even more serious. No matter how we construct the model, it will be flawed, and what is more, we do not know in which way it is flawed.

In order to understand this claim we have to remember the non-linear nature of the interactions in complex systems. This non-linearity has two important consequences. In the first place, when there are a lot of simultaneous, non-linear interactions, it soon becomes impossible to keep track of causal relationships between components. Secondly, from the non-linear nature of complex systems we can deduce that they are incompressible ([8]: 10). If we add to this the historical nature of complex systems, the problem should become clear: Models have to reduce the complexity of the phenomena being described, they have to leave something out. However, we have no way of predicting the importance of that which is not considered. In a non-linear world where we cannot track a clear causal chain, something that may appear to be unimportant now, may turn out to be vitally important later. Or vice versa, of course. Our models have to "frame" the problem in a certain way, and this framing will inevitably introduce distortions. Different frames provide different perspectives on the system, and these differences are not reducible to each other.

This is not an argument against the construction of models. We have no choice but to make models if we want to understand the world. It is just an argument that models of complex systems will always be flawed in principle, and that we have to acknowledge these limitations.

What then of the argument that it may be possible to incorporate absolutely all the information concerning a complex system into some fancy (neural network) model? I do not wish to argue that it is impossible to repeat the complexity of a system in another medium, but one should remember that we now have a "model" that is as complex as the system being modelled. It will be as difficult to understand as the system itself, and its behaviour will be as unpredictable. If the history of the model and the history of the system is not kept identical (and I cannot see how this can be done in anything but the most trivial of cases), the two will soon become uncorrelated. My conclusion is that it is impossible to have a perfect model of a complex system. This is not because of some inadequacy in our modelling techniques, but a result of the meaning of the notions "model" and "complex". There will always be a gap between the two. This gap should serve as a creative impulse that continually challenges us to transform our models, not as a reason to give up. Models provide a particular insight which can never be interpreted independently of the empirical context in which they were constructed. Context is always crucial.

3.8 Complexity, Limits and Knowledge

In recapitulation, the argument that our understanding of complex systems is problematic in principle can be made in the following way: To *fully* understand a complex system, we need to understand it in all its complexity. Furthermore, because complex systems are open systems, we need to understand the system's complete environment before we can understand the system, and, of course, the environment is complex in itself. There is no human way of doing this. The knowledge we have of complex systems is based on the models we make of these systems, but in order to function as models—and not merely as a *repetition* of the system—they have to *reduce* the complexity of the system. This means that some aspects of the system are always left out of consideration. The problem is compounded by the fact that that which is left out, interacts with the rest of the system in a non-linear way and we can, therefore, not predict what the effects of our reduction of the complexity will be, especially not as the system and its environment develop and transform in time.[10]

We cannot have complete knowledge of complex systems; we can only have knowledge in terms of a certain framework. There is no stepping outside of complexity (we are complex but finite beings, embedded in a complex context), thus there is no framework for frameworks. We *choose* our frameworks. This choice need not be arbitrary in any way, but it does mean that the status of the framework (and the framework itself) cannot be used as the basis for objective knowledge. The generation of knowledge of complex systems is an exploratory process. As the context in which this knowledge is to be useful changes, we will have to continually revise the framework which generates this knowledge. Our knowledge of complex systems is thus always provisional. We have to be modest about the claims we make about such knowledge.

An understanding of knowledge as constituted within a complex system of interactions would, on the one hand, deny that knowledge can be seen as atomised "facts" that have objective meaning. Knowledge comes to be in a dynamic network of interactions, a network that does not have distinctive borders. On the other hand, this perspective would also deny that knowledge is something purely subjective, mainly because one cannot conceive of the subject as something *prior* to the "network of knowledge", but rather as something constituted *within* that network. The argument from complexity thus wants to move beyond the objective/subjective dichotomy, as Morin [4] also argues. The dialectical relationship between knowledge and the system within which it is constituted has to be acknowledged. The two do not exist independently, thus making it impossible to first sort out the system (or context), and then to identify the knowledge within the system. This co-determination also means that knowledge, and the system within which it is constituted, is in constant transformation. What appears to be uncontroversial at one point may not remain so for long.

One should also be careful not to interpret this state of affairs as somehow inadequate, as something to be improved upon. There is a necessary relationship between the imposition of a limiting framework and the generation of knowledge. One cannot have knowledge without a framework. Despite the fact that our knowledge is necessity limited, these limits are enabling; they allow us to make claims which are neither relativistic nor vague (see [9]). At the same time, however, such knowledge is not the result of free-floating truths; it is contextualised in time and space. Because it is not objective, and because we know that, we cannot *use* this knowledge as if it is objective. There is always a normative dimension to the claims we make, and we have to stand in for them. We cannot shift the responsibility for the

[10] These ideas are elaborated upon in Cilliers [10, 17].

effects of our claims onto some process we call "scientific".

3.9 Assuming Responsibility

To a large extent, the framework within which the natural sciences operate *has* to attempt the illusion of objectivity. It is necessary to be absolutely as objective as possible, while simultaneously acknowledging the limits of the strategy at stake and of the reach of the claims made. Most good scientists will acknowledge this. The problem is, however, severely compounded when the methods of the natural sciences are imposed upon or, even worse, embraced in a simplistic way by the social sciences and humanities. The impression is then created that a traditional understanding of truth, which is problematic even in the natural sciences, should form the criterion for proper work in social sciences. It is not possible to think what the motto *nullius in verba* could *mean* in the human context! We are not faced with a set of problems that we can solve in a piecemeal way by chipping away at it using experimental procedures and good old Enlightenment rationality. We are confronted by a complex problem which is transforming not only while we are investigating it, but *because* we are investigating it.

An understanding of "general complexity" should provide us with some insight into this process. The generation of knowledge is not a linear process, but one which is folded in on itself. Useful knowledge is making and unmaking itself continuously. For example, a Marxist perspective on the economy can be proven wrong at some stage, for example, by the failure of Stalinist communism, but can again become useful as something containing important critiques of rampant and destructive capitalism. The central thrust of a general theory of complexity should, therefore, remain a critical one. It should be constantly vigilant about the limits of our understanding instead of making brash and reductive claims about the insights gained from some mathematical model, like a power law, for example. This is not to argue that complexity theory cannot provide us with useful tools or help us with the generation of new insights. On the contrary, the insights from fractal mathematics, chaos theory and complex adaptive systems could be fascinating. The argument is about the *reach* of these theories.

If we still operate in a context where knowledge generated in the framework of the natural sciences trumps other forms of knowledge, it has become necessary to actively *resist* this tendency. This is the motivation behind seeing complexity theory as a critical position. Such a role for complexity theory would entail to simultaneously work on the generation of new ideas and to resist a simplistic assimilation of these ideas. It involves acknowledging the temporal nature of what we do and not to be intimidated by a culture of performance in which everything has to be done quickly and efficiently without giving enough consideration to the costs involved (see [7]). More importantly, it involves an acknowledgement of the *limits* of our knowledge. If our knowledge is limited and if we cannot claim a purely objective status for it, then we can also not *use* that knowledge as if it is so. Our decisions and actions cannot be justified on purely rational grounds. Of course we do all the rational calculation we can, but in the end we choose the framework within which we interpret and give content to our insights. As a result, we cannot blame the outcomes of our decisions and actions on some procedure or method, not even if we incorporate complexity theory. We have to assume responsibility for them ourselves.

Assuming this critical position is, nevertheless, not a cry of desperation or helplessness. It has a profound influence on the way in which we tackle the problems of the world, including the challenges of sustainable development. Being more careful with a certain approach does not imply that it should be abandoned, just that it should continually be reflected upon. It is *not* assuming the critical position which leads to disaster. The price we pay for clinging to a constricted modernist rationality can be seen on many levels in our globalised world. From the perspective of philosophy, the most important one is the distortion it brings to our understanding of what it is to be human. Complexity theory should help us to deal with this question, rather than to play in the park with big business and grand politics. We cannot look to scientific rationality to solve our ethical dilemmas. We will have to deal with them as contingent and unique things. Acknowledging complexity should help us to humanise science, not the other way round.

If it is argued that the view from complexity is a critical one, the notion "critical" is used in more than one sense of the word. This view is also critical in the sense that it is of vital importance. It has been shown that a reductionist approach often destroys the emergent properties of a system. If a reductionist approach is coupled with a purely instrumental rationality, one which calculates in terms of direct and immediate results, it generates a strategy which is exploitative, even destructive. In order to resist this destruction, in order to think, for example, of sustainable social-ecological systems, it has become necessary to acknowledge the complexity of understanding and managing them. Because there is not a clear-cut paradigm for complexity theory at this stage, it may take a bit of courage to follow this route; but, to my mind, we have little choice. We have already seen the disastrous results of following the more established alternatives.

References

1. Burns M, Weaver A. Exploring sustainability science: a southern African perspective. Stellenbosch: African Sun Media; 2008.
2. Novotny H, Scott P, Gibbons M, editors. Re-thinking science. Cambridge: Polity Press; 2001.
3. Toulmin S. Cosmopolis: the hidden agenda of modernity. Free Press: New York; 1990.
4. Morin E. Restricted complexity, general complexity. In: Gershenson C, Aerts D, Edmonds B, editors. Worldviews, science and us: philosophy and complexity. Singapore: World Scientific; 2007. p. 5–29.
5. Byrne D. Complexity, configurations and cases. Theor Cult Soc. 2005;22(5):95–111.
6. McLennan G. Sociology's complexity. Sociology. 2003;37(3):547–64.
7. Cilliers P. On the importance of a certain slowness. In: Gershenson C, Aerts D, Edmonds B, editors. Worldviews, science and us: philosophy and complexity. Singapore: World Scientific; 2007. p. 53–64.
8. Cilliers P. Complexity and postmodernism: understanding complex systems. Routledge: London; 1998.
9. Cilliers P. Complexity, deconstruction and relativism. Theor Cult Soc. 2005;22(5):255–67.
10. Cilliers P. Boundaries, hierarchies and networks in complex systems. Int J Innov Manage. 2001;5(2):135–47.
11. Luhmann N. Ecological communication. University of Chicago Press: Chicago; 1989.
12. Khalil EL, Boulding KE, editors. Evolution, order and complexity. Routledge: London; 1996.
13. Midgley G, Munlo I, Brown M. The theory and practice of boundary critique: developing housing services for older people. J Oper Res Soc. 1998;49:467–78.
14. Pattee HH. Hierarchy theory: the challenge of complex systems. New York: George Braziller; 1973.
15. Simon HA. The architecture of complexity: hierarchic systems. Proc Am Phil Soc. 1962;106(6):467–82.
16. Emmeche C. Aspects of complexity in life and science. Philosophica. 1997;59(1):47–68.
17. Cilliers P. Knowledge, complexity and understanding. Emergence. 2000;2(4):7–13.

The Complex Nature of Knowledge

4

Joachim P. Sturmberg and Andrew Miles

There are all kinds of sources of our knowledge but none has authority.

Karl Popper

We know more than we can tell.

Michael Polanyi

Where is the life we have lost in living? Where is the wisdom we have lost in knowledge? Where is the knowledge we have lost in information?

TS Eliot

These three aphorisms describe the conundrum of knowledge—it clearly is not a straightforward concept. Nevertheless, the medical enterprise has invented a stamp of 'unquestionable knowledge', namely that based on evidence, and in particular that one derived from the randomised controlled trial and/or meta-analysis.

In this chapter, we will outline some of the conceptual issues underlying the notion of knowledge, offer a complexity based understanding of knowledge, and highlight some of the implications of a complex understanding of knowledge for the practice of medicine.

4.1 Philosophical Notions

Knowledge, according to most thinkers, must follow three defining characteristics—it must be justified, true, and believed. However, meeting these qualifications may be difficult or impossible. Truth and belief are often founded on the stereotypical *'seeing is believing'*, i.e. once own perceptual experience. Our perceptual capacities, however, can easily be delusional as is illustrated by the Adelson's chequer shadow illusion or the Necker cube and Rubin vase, and our beliefs can be limited by our expectations, like all swans are white, except in Australia where they are black (Fig. 4.1). Not only does knowledge not equate to truth [1], but the notion of truth—or certainty—itself is one that contradicts the basic foundations of scientific endeavour [2].

In pragmatic terms, knowledge is information we are aware of. Knowledge falls into two categories, *knowing what*—naming facts and relationships—and *knowing how*—explaining procedures. Furthermore, knowledge can be *explicit* referring to knowledge which can be codified and easily communicated, and *tacit* meaning knowledge which cannot be codified and therefore cannot be easily passed on to others (Fig. 4.2).

Michael Polanyi (1891–1976), in his book *Personal Knowledge*, extensively explored the meaning and understanding of knowing and

J.P. Sturmberg (✉)
Department of General Practice, Monash University, Melbourne, VIC, Australia

The Newcastle University, Newcastle, PO Box 3010, Wamberal, NSW 2260, Australia
e-mail: jp.sturmberg@gmail.com

A. Miles
World Health Organisation Collaborating Centre for Public Health Education and Training, Imperial College, London, SE11 9AN, UK

Faculty of Medicine, University of Milan, Milan, Italy

Plovdiv Medical University, Plovdiv, Bulgaria
e-mail: Andrew.Miles@keyadvances.org.uk

J.P. Sturmberg and C.M. Martin (eds.), *Handbook of Systems and Complexity in Health*,
DOI 10.1007/978-1-4614-4998-0_4, © Springer Science+Business Media New York 2013

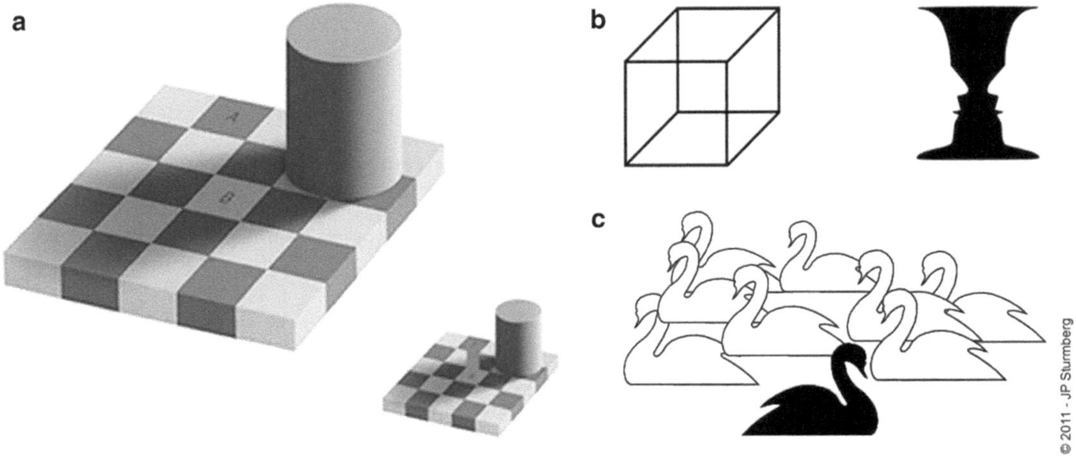

Fig. 4.1 *'Seeing is believing'*—Adelson's chequer *shadow* illusion (**a**), reproduced with permission from Wikimedia Commons; Necker cube and Rubin vase (**b**), reproduced with permission from Wikimedia Commons; (**c**) all swans are *white*, except the *black* one

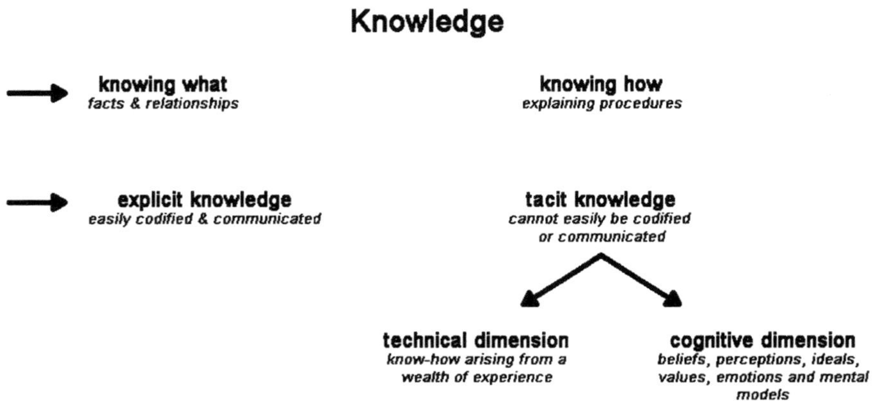

Fig. 4.2 Conceptualisation of knowledge

knowledge [3]. He defined the term tacit knowledge to describe the domain of *'personal knowing'*. Tacit knowing is further divided into a *technical dimension*—also described as know-how, and arises from a wealth of experience, though being hard to explain or transfer, and a *cognitive dimension*—consisting of beliefs, perceptions, ideals, values, emotions, and mental models. It is this latter meaning of tacit knowledge that largely shapes the way we perceive the world around us. In relation to medical sciences Polanyi stated that

> … personal knowledge in science is not made but discovered, and as such it claims to establish contact with reality beyond the clues on which it relies. It commits us, passionately and far beyond our comprehension, to a vision of reality. Of this responsibility we cannot divest ourselves by setting up objective criteria of verifiability—or falsifiability, or testability, or what you will. For we live in it as in the garment of our own skin (p. 64).

For Polanyi, tacit and explicit are different but inseparable aspects of knowledge [4]. Both can be learnt though the learning modes are different. The gaining of tacit knowledge, in contrast to the route learning of explicit facts, is experiential and typical of apprenticeships [3, 5]—the major mode of learning clinical health sciences and patient-centred care.

4.1.1 Shifting from a Static to a Dynamic Understanding of Knowledge

Kant stated that knowledge is a *thing waiting there to be discovered*. This view, however, has significant deficiencies which became apparent through the insights from the science of 'complex adaptive systems' in the late 1990s. In a complex adaptive systems framework knowledge in addition to being a *thing* simultaneously also is a *flow*, as the focus of knowledge generation shifts primarily to *context* and *narrative*, rather than content alone [4]. Snowden illustrates the dynamic nature of knowledge generation [in the context of organisational knowledge] through three heuristics:

1. ***Knowledge can only be volunteered; it cannot be conscripted*** *for the very simple reason that I can never truly know if someone is using his or her knowledge. I can know they have complied with a process or a quality standard. But, we have trained managers to manage conscripts not volunteers.*
2. ***We can always know more than we can tell, and we will always tell more than we can write down.*** *The nature of knowledge is such that we always know, or are capable of knowing more than we have the physical time or the conceptual ability to say. I can speak in 5 min what it will otherwise take me 2 weeks to get round to spend a couple of hours writing it down. The process of writing something down is reflective knowledge; it involves both adding and taking away from the actual experience or original thought. Reflective knowledge has high value, but is time consuming and involves loss of control over its subsequent use.*
3. ***We only know what we know when we need to know it,*** *human knowledge is deeply contextual, it is triggered by circumstance. In understanding what people know we have to recreate the context of their knowing if we* [are] *to ask a meaningful question or enable knowledge use. To ask someone what he or she knows is to ask a meaningless question in a meaningless context, but such approaches are at the heart of mainstream consultancy method* (p. 102).

4.1.2 Knowledge Generation and Knowledge Management: A Sense-Making Process

Kurtz and Snowden describe the dynamic understanding of knowledge generation and knowledge management as a sense-making process [6]. They used the Welsh word *Cynefin* (pronounced kun-ev'in)—a word that has no direct English translation—to describe both, the state as well as the process of knowledge generation and sense-making as follows:

> It is more properly understood as the place of our multiple affiliations, the sense that we all, individually and collectively, have many roots, cultural, religious, geographic, tribal, and so forth. We can never be fully aware of the nature of those affiliations, but they profoundly influence what we are. The name seeks to remind us that all human interactions are strongly influenced and frequently determined by the patterns of our multiple experiences, both through the direct influence of personal experience and through collective experience expressed as stories. (p. 467)

The Cynefin model of knowledge achieves a synthesis of the different notions of knowledge—being explicit and tacit, technical and cognitive, a thing and a flow—and provides a framework to understand knowledge as *a personal construct* that is achieved through a *sense-making process*.

The Cynefin framework breaks *knowing* into five domains

- ***The known***, where cause and effect relationships are generally linear, empirical, and not disputed
- ***The knowable***, where cause and effect relationships exist but may not be fully known or only known by experts, gained through systematic methodologies, and relies on trust
- ***The complex or emergent***, where cause and effect relationships can be perceived but not clearly defined or predicted and events are fully understood only in retrospect
- ***The chaotic or random***, where no apparent cause and effect relationships are evident
- ***The central space of disorder***, where conflicting views reside resulting from different perspectives on the same issue

4.2 The Multiple Dimensions of Medical Knowledge

The Cynefin framework organises knowledge in terms of *what is already known* and thus *can be taught* on the right as opposed to *what is not yet fully known* and thus *needs to be learnt* on the left hand side. The bottom half is predominantly concerned with *content*, whereas the top half deals with *context*. It is important to note that each domain within the Cynefin framework is value-free, i.e. neither is intrinsically better than any other; rather, each uses distinct thinking strategies to generate knowledge and to solve problems, which have inherent legitimacy and utility within *that* domain. Figure 4.3 illustrates our attempt to translate the 'knowledge of medicine' and the 'enacting of that knowledge' into the Cynefin framework.

The way *we know* also defines the way *we organise* and *act*. As the Cynefin framework of medical knowledge highlights even if we predominantly 'know' and 'operate' in one particular domain, we always also operate some of the time in some or all of the other domains, or move to and from one domain to another, dictated by contextual needs. For example, a primary care physician spends a lot of time helping patients to achieve the best health experience under their particular circumstances, however, in the care of a particular condition utilises condition-specific knowledge, whereas in the case of an unconscious patient applies emergency procedures.

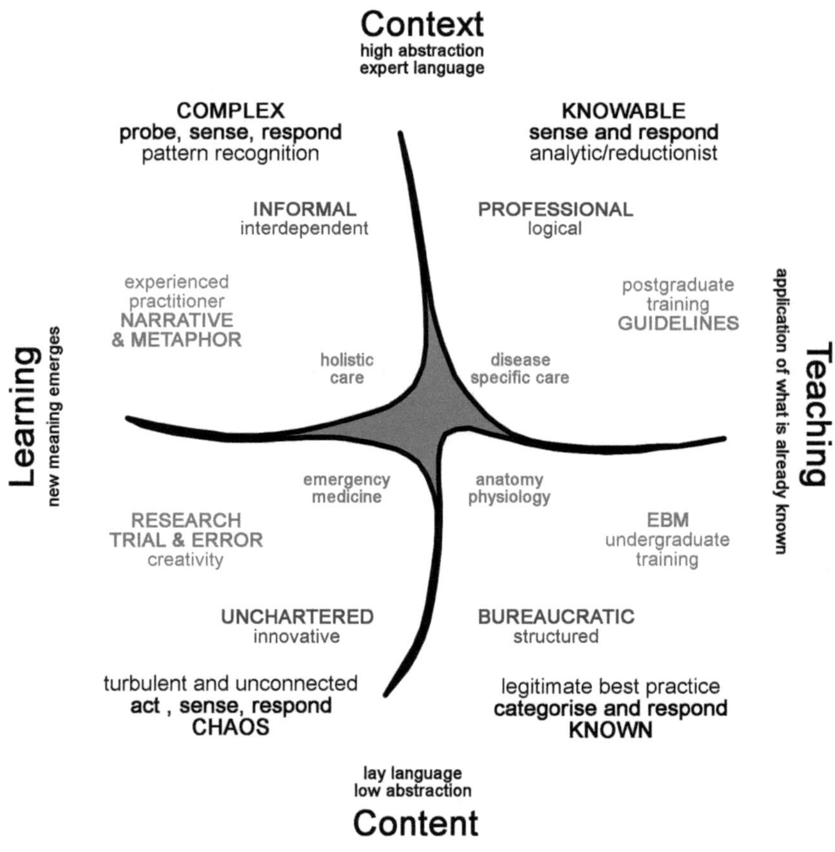

Fig. 4.3 Medical knowledge in the Cynefin framework

4.3 Knowledge Generation and Application in Each of the Four Domains

This section describes some approaches to generate and apply knowledge within each of the four domains of knowing and the approaches to uncertainty in the Cynefin framework.

4.3.1 The Known Domain

The known domain usually only applies established knowledge. Cause and effects are generally linear in nature, and can be applied repeatedly, predictably, and safely. This type of knowledge underpins legitimate best practice procedures, like 'time-out' at the beginning of an operative procedure.

4.3.2 The Knowable Domain

The knowable domain gains its knowledge from the scientific experiment and utilises well-defined methodologies. Its aim is to identify cause and effect relationships through the study of the properties of associated agents. This approach is legitimate and desirable when the patterns of the agents are relatively stable, like in patients presenting with ischaemic heart disease.

This type of knowledge is applied by analysing/interpreting the presenting data in light of the known facts. Usually, structured techniques have been developed and codified in the form of guidelines and decision trees.

However, this domain entails high risks. Since entrained pattern recognition and patterned responses are based on pre-defined assumptions, any error in any assumption can lead to false conclusions with catastrophic consequences.

4.3.3 The Complex or Emergent Domain

The complex domain gains its knowledge from sensing the emerging relationship patterns between many different agents. The relationships between the agents are not predictable, and an observed pattern is understandable only in retrospect (retrospective coherence), however, any one observed pattern is only one of many possible ones. Although patterns may be repeatable for some time, since the underlying interactions resulting in a pattern are not visible, one cannot rely on historically entrained patterns of past experiences. Understanding context, typically utilising narrative approaches, achieves new insights, e.g. though Mrs Jones has presented in the past on several occasions feeling tired secondary to her work and relationship stresses, she may well present today with the same complaint but exhibit features pointing towards hypothyroidism.

Acting in the complex/emergent domain requires one to entertain multiple perspectives on a given situation. One needs to pause and pay attention to gain new insights rather than acting automatically based on past experience.

Why is it so difficult to operate in the complex or emergent domain [7]? One part of the answer involves the structure and function of our brains. Andy Clark [8] showed that our perception involves 'predictive coding', i.e. the brain exploits prediction and anticipation to make sense of signals, which in turn guide perception, thought, and action.

He suggests that perceiving the world depends on successfully predicting our own sensory status. Our brain uses a recursive process of comparing incoming signals with previously stored information to predict what our current state is likely to be, given our previous experience—consider Mrs Jones above. Mismatches between the prediction and the incoming signals result in error messages that modify the prediction, or if the error messages are great, we actually learn something new.

We start to make sense of the world first based on general expectations, thus providing a constraining framework, before making more detailed predictions in light of continuing input. One of the consequences of 'predictive coding' is a blurring of the distinction between perception and cognition. Clark highlights that what we perceive (or think we perceive) is heavily determined by what we know, and what we know (or think we know) is constantly conditioned on what we perceive (or think we perceive). Again, Mrs Jones having hypothyroidism is a good example. This fallibility is well recognised, hypothyroidism is frequently overlooked by a patient's usual physician, but picked up 'on the spot' by the locum doctor.

Clark concludes that we are primarily in contact with the world through our expectations about the way we see and experience the world, which has been illustrated at the beginning of this chapter and can be retested by looking again at the illustrations in Fig. 4.1.

4.3.4 The Chaotic or Random Domain

The chaotic domain demands action to stabilise a critical situation, often created as a result of overdependence on rule-based actions. Following an action it is important to sense its result which in turn will provide guidance for the next response. The aim is to shift the situation into the complex domain, or to use an authoritarian approach to bring it into the known one.

4.3.5 Disordered Domain

The central disordered domain represents the space of uncertainty and conflict that arises when people have markedly opposing perception of a problem. This domain highlights important differences in perception, and it is important to appreciate the psychological dimension underlying the observation that every person involved in a conflict tends to argue for a solution that is most consistent with their 'experiential corner'.

The Cynefin model shows that all domains are interdependent and interconnected through the *flow* dimension inherent in knowledge generation; knowledge is produced through a perpetual process that at times leads to an end product, a temporarily static 'known fact', within a given framework (Fig. 4.4). As more knowledge is generated current facts become superseded by new ones.

4.4 Approaches to Knowledge Management in Health Care

Principally one can divide the approaches to knowledge generation into two categories, quantitative and qualitative. Quantitative approaches are primarily concerned with gaining mechanistic understandings whereas qualitative approaches usually focus on contextual understandings. In that sense, neither can claim to provide a *'complete picture'* about a phenomenon under study. We content that research findings are better viewed as providing *information*, and either approach in itself will always only providing *bits of information* about the whole picture, hence the colloquial notion of *'gaining some more knowledge'*.

4.4.1 The Randomised Clinical Trial

The randomised clinical trial is seen as the gold standard to establish clinical effectiveness and efficiency. The theory of such trials is based on the assumptions that only the intervention will explain the observed outcome as all other parameters are equally distributed in the control and intervention populations. Based on statistical assumptions of precision, a sufficiently large population will show even small differences in outcomes at a pre-defined level of statistical significance. The intervention of a trial is then seen as 'having caused', rather than *being associated with*, the observed outcome.

So far so good, however, there are important limitations inherent in these assumptions that question its proponents' trust in the methodology and interpretation of findings.

1. The comparative trial has its origins in agriculture to establish which farming method would result in better yields [9]. The emphasis was on a collective outcome, i.e. *pragmatic* rather than *statistically significant*.

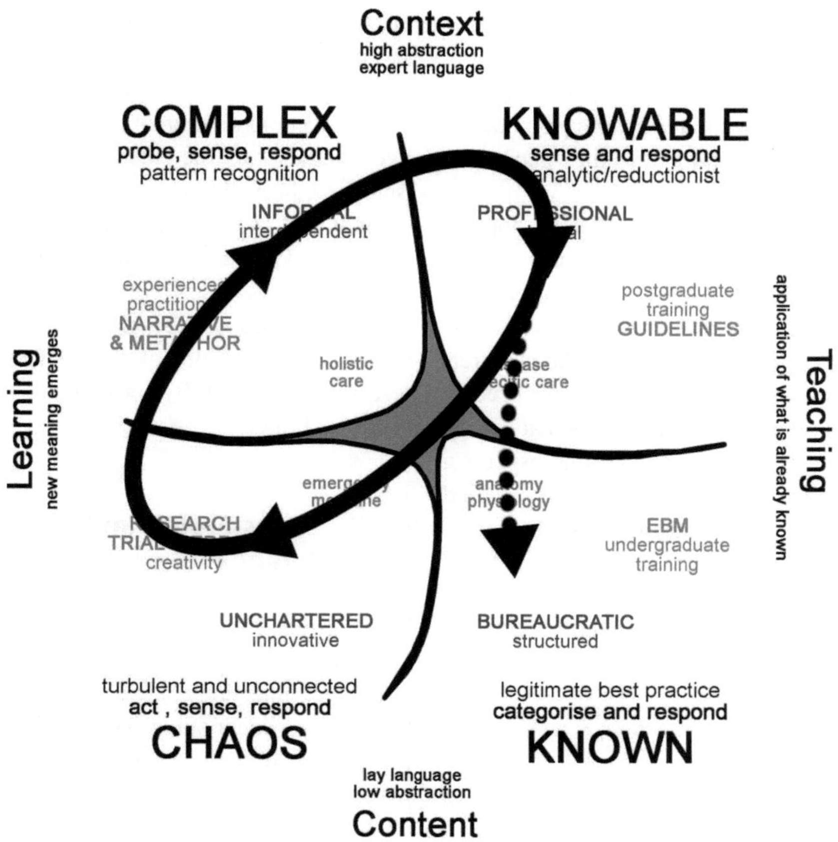

Fig. 4.4 Knowledge creation in the Cynefin framework

2. As a pragmatic tool the individual's performance to the collective outcome was of no particular importance. Thus, this type of enquiry is most useful to gain *epidemiological* insights.
3. Biological characteristics are invariably non-linearly distributed, violating the Gaussian assumptions underpinning the commonly applied statistical techniques applied to medical data sets.[1]
4. The assumption that randomisation distributes all *relevant* variables evenly between a control and intervention group is at best a limited view. This assumption presupposes that only 'those measurable' characteristics, which typically are markedly limited, are of importance in contributing to observed outcomes.
5. Statistics are a tool, statistics do not answer the question of relevance, i.e. does the observed *statistically significant* difference (expressed as *p-value*) translate into meaningful clinical outcomes?[2] And to whom are

[1] For more detail, see Chap. 5 by Bruce West.

[2] *Gradually, however, it has dawned on a few scientists that something is screwy. An obvious problem is that with so many people doing so many studies, some of them are going to run into that one-in-20 chance of believing in a mirage. The converse mistake is more subtle: scientists care about whether a result is statistically significant, but they should care much more about whether it is meaningful—whether it has, to use a technical term, oomph.*

Sadly, many scientists have started thinking that statistical significance measures oomph. If an answer meets the 95% confidence criteria, it must be important; if it doesn't, it isn't [10].

the outcomes meaningful, researchers, policy makers, doctors, or the patient?[3]

6. The interpretation of a positive statistical result is *only* showing a *probable association*, rather than *causation*, of the dependent variable.
7. The assumption that involving large populations into a trial increases the findings clinical veracity is invariably untrue. The fact is simple and straightforward, as any statistical text states: if one expects to detect a large difference in the variable studied one only needs a small sample size, the corollary is that a large sample size is required only if one expects a small difference in the variable studied between the control and intervention groups. Large trials thus do not entail greater truth, and especially do not imply *'unequivocal'* clinical veracity, as is often stated.[4]

4.4.2 Case Studies

The case study is the naturalistic approach to research [11]. Many important breakthroughs in medicine did not arrive from trials but observations, be it Jenner's cowpox vaccination, Shaw's removal of the Broad Street pump handle in the fight of a cholera outbreak, McKenzie's descriptive studies of the mechanisms underlying ischaemic heart disease, or von Behring's development and application of diphtheria toxoid in treating and preventing an inevitably lethal childhood disease.

Case studies and single observations often identify a new problem that demands further research to more fully understand the 'mechanisms' behind it.

4.4.3 Focus Groups and In-Depth Interviews

Focus group and in-depth interview studies are approaches to explore the multiple perspectives and interconnections between understandings of the problem under study. They provide a contextual account of the issue and highlight most of the perceived variables and potential interactions contributing to the problem.

[3] An NNT of 1 is the most effective and means each patient treated responds. An NNT of 2 or 3 indicates that a treatment is quite effective (with one patient in 2 or 3 responding to the treatment, **the inverse meaning** 1 or 2 patients not responding). The current belief is that an NNT of 20–40 can still be considered clinically effective; however, patients who fully appreciate that these NNTs reflect only a small likelihood of personal benefit may decline treatments if the side effects of treatments impact negatively on their health experiences and life expectancies.

[4] The implication of sample size on statistical significance is illustrated by the following hypothetical examples.

Study: In the treatment group, one out of 50 subjects died. Hence, risk of death = 1/50 = 0.02. In the control group, five out of 50 subjects died. Hence, risk of death = 5/50 = 0.1. Hence, absolute risk reduction (ARR) = 0.1 − 0.02 = 0.08. Number needed to treat (NNT) = 1/ARR = 1/0.08 = 12.5.

The table shows how by simply enlarging the study population the initial ***non-significant*** outcome becomes significant, even though the **ARR** and **NNT** remain **the same.**

Risk in experimental group	Risk in control group	ARR (95% CI)	NNT (95% CI)
1/50 = 0.02	5/50 = 0.1	0.1 − 0.02 = **0.08** (−0.012–0.172)	1/0.08 = **12.5** (5.8 to infinity)
A five times larger study group having the same risk characteristics			
5/250 = 0.02	25/250 = 0.1	0.1 − 0.02 = **0.08** (0.039–0.121)	1/0.08 = **12.5** (8.3–25.7)
A 25 times larger study group also having the same risk characteristics			
25/1,250 = 0.02	125/1,250 = 0.1	0.1 − 0.02 = **0.08** (0.062–0.100)	1/0.08 = **12.5** (10.2–16.2)

(Adopted from [10])

Large sample sizes will make small differences statistically significant without having any clinically meaningful implications. Paraphrasing David Healy, professor of psychiatry and critical commentator on drug trials at Cardiff University: the larger the study population the greater the certainty that the intervention **does not** work.

4.5 Scientific Evidence and Clinical Knowledge: *Evidence-Based Medicine*

We have already discussed the limitations of the randomised controlled trial as the so called 'gold standard' for establishing treatment effectiveness and efficiency, but no chapter on the complexities of what it is that constitutes knowledge for decision making in medicine would be complete without a detailed consideration of the evidence-based medicine (EBM) movement. The term 'evidence-based medicine' was first employed rhetorically by Gordon Guyatt in 1990 [12], with an initial attempt at its codification as a philosophy of medicine a year later within the *Journal of the American Medical Association* in 1991 [13] as a 'new approach to teaching the practice of medicine'. The assertion was predicated on Thomas Kuhn's philosophy of the structure of scientific revolutions [14] and argued that current medical practice had become untenable, being based more on tradition, opinion, and expertise, rather than evidence derived from systematic scientific study and experimentation, requiring a major redefinition of the 'base' of medicine as one resting on scientific data rather than other forms of knowing in the care of patients. The authors argued that the nature and scale of the redefinition and reorganisation of care declared necessary by their thesis was such as to constitute a 'paradigm shift' in accordance with Kuhnian thinking, explicitly recommending that medical practice should be *based* on the principles and methods of clinical epidemiology, though astonishingly acknowledging that no definitive evidence existed to demonstrate the superiority of their proposed 'new paradigm' above the form of medicine existing at the time [13]. Terminologically, EBM became rapidly described as neologistic and seductive [15, 16], exciting visceral disdain in some areas of international medicine and generating great approbation from others (cf. [17–31] and extensive bibliography referenced therein).

4.5.1 The EBM 'Hierarchy of Knowledge'

The principal characteristic of EBM's 'paradigmatically novel' approach was the ordering of sources of knowledge into a 'hierarchy of evidence' [32–34], with tensions being created by the demand of the new paradigm that the *basis* of clinical practice should be biostatistical data derived from quantitative study designs, with the randomised controlled trial reified as the ultimate and most superior form of methodological study design for this purpose and the cumulative meta-analysis of RCTs advanced as the most powerful source of knowledge for practice decisions within the whole of medicine itself [32–34]. Within this model of practice, primacy in clinical decision making was given to scientific evidence derived from epidemiological study designs conducted in rarified trial populations, these data constituting the so-called 'E' of EBM. Here, EBM traditionally advocated the searching of primary electronic database literature by individual clinicians as a means of identifying studies of relevance to the treatment of a given patient. When identified, these studies would require critical appraisal to select studies judged to be of adequate methodological rigour and as a means of rejecting those judged not to be sufficiently rigorous. Despite the organisation of EBM workshops and other such educational initiatives to train practising clinicians in the technique of literature searching and study evaluation, the practical limitations of this approach to identifying evidence for clinical practice became rapidly clear in terms of the time constraints in everyday clinical care as well as the motivations of clinicians to engage in complex scientific analyses of this type [30, 31].

4.5.2 Individual Clinician Evidence Gathering or Reliance on EBM-Style Practice Guidelines and Systematic Reviews?

These practical limitations of the EBM 'method' led to a change within the overall EBM methodological approach away from the concept of the individual clinician as harvester and appraiser of individual studies within the literature [32] to a simple user of practice guidelines developed through EBM's codification of the results of meta-analyses of randomised controlled studies and of systematic reviews of the medical literature (e.g. [35, 36]). But the meta-analysis and systematic review, while representing unquestionably powerful methods for the synthesis of the results of large numbers of individual clinical studies, are nevertheless limited by a range of methodological deficiencies, ranging from the inherent problem of heterogeneity, the minimised but nevertheless inescapable subjectivity of the reviewer, and, particularly worryingly, the systematic exclusion of studies which examine the dimensions of medical practice that by their nature do not address scientific questions, but which study issues of considerable importance in the making of good clinical decisions—so-called patient factors as opposed to disease factors. The results of systematic reviews and therefore the applicability of guidelines derived from then therefore hold only partial importance for clinical practice even when the study populations match those characteristics of the individual clinical case.

Nunn, writing in the *Journal of Evaluation in Clinical Practice*, has addressed many of these concerns. In an article entitled: 'Evidence-based medicine and the limits to the literature search' [37], the author was concerned to remind his readers of the historical obsession of the EBM enthusiasts to liberate clinicians from 'expert authority'. Following his review of the advice given to the international medical community by Sackett et al. to pursue *'aggressive strateg(ies) to tame the unruly literature (and to) read only those original articles that if valid, would be of direct importance to your own clinical practice (and) scan the methods and results of these articles to determine whether they are likely to be true before reading them in detail'* [33], Nunn continues by addressing the multiple methodological limitations of classical literature review but, importantly, demonstrates how EBM, in its preferential selection of 'checklist authority' over 'expert authority', naively privileges checklist authority over legitimate clinical authority derived from extensive experience of disease manifestation and treatment so that, for EBM, the 'competent' and 'modern' clinician is typified by a 'consumer of procedures and summaries by digesters who have (themselves) determined what is (and what is not) relevant for clinicians'. It is noteworthy that the majority of these reviewers and producers of clinical digests are remote from clinical practice, but that they have appeared content to make recommendations formulated from their activities that subsequently become codified into clinical practice guidelines intended for routine use in everyday practice—a classic example, perhaps, of power without responsibility. It is this ideological shift away from philosophically tenable understandings of clinical knowledge, towards the conclusion that only RCT-derived evidence 'counts' through reductionist and often frankly scientistic reasoning, that remains of great concern [38].

4.5.3 Scientific Data Only or Other Sources of Clinical Knowledge in Addition?

By continuing to adhere to such reasoning, EBM effectively precludes the use within decision making of the plurality of other sources of knowledge for clinical practice that are of great relevance to the care of patients. Perhaps unsurprisingly, then, EBM has become as a consequence described as dogmatic, reductionist, and scientistic [29, 39]. Feinstein, for example, was clear that EBM could not, as a direct function of its philosophy, assimilate the humanistic dimension of clinical practice which, as he discussed, included psychosocial factors, support, personal preferences, and strategies for giving comfort and reassurance [40]. The fundamental inability of the EBM philosophy to incorporate these vital factors in the care of patients when they are in conflict (as can often be so depending on the individual case) with the E of EBM has been described by prominent EBM advocates as 'vexing' [41], during a rare response by three of EBM's leaders to academic criticism. The intractable nature of this fundamental limitation of EBM was confirmed and discussed by a large number of international scholars [42–48] and even the latest *Apologia* from the EBM community [49] fails, on analysis [50], to resolve the philosophical and clinical conundrum, directly illustrating, it seems, a fundamental irreconcilability between the fundamental tenets of EBM and those of patient-centred care—a profound difficulty to which we will turn below.

4.5.4 EBM and the Inability to Integrate Patients' Preferences and Values into Decision Making

In its inability to integrate patients' values and preferences in any meaningful sense when these conflict with a clinical decision dictated by scientific evidence, EBM stands exposed as an incomplete model of medical practice. This is the seemingly intractable philosophical 'impossibility' of EBM [42–48, 50]. The fundamental irreconcilability of EBM with good clinical medicine is illustrated by its own nomenclature, which insists that the *basis* of clinical practice is scientific evidence, yet it has been convincingly argued that medicine does not have or need a 'base'. In a review of the interpretive approach in medicine alongside a consideration of phenomenological approaches to clinical practice, Upshur [51] has argued that the value of anti-foundationalism in medicine is that it acts to prevent simplistic conceptions of the relationship of science and medicine and prevents incomplete and reductionist forms of practice—such as EBM—from prevailing. With reference to this thinking, Miles and Loughlin [52], acknowledging the 'impossibility' of an evidence-*based* medicine, have argued preferentially in favour of an evidence-*informed* model of practice, where scientific evidence represents one input to the clinical decision-making process, but where it is constrained from an automatic displacement of the others. For us, such a model is far from controversial and, indeed, in full accordance with the understanding of medicine as a science—using practice, as essentially a human endeavour which employs science, but cannot be equated with science in any meaningful sense [39].

Here, the casuistic approach advanced by Tonelli is of particular relevance. Tonelli is convinced that a focus of clinical practice primarily on quantitative scientific evidence as the basis for decision making can directly result in a minimisation of the importance of the intangible, physical, emotional, and spiritual aspects of illness in the healing process, impoverishing the philosophy of medicine as a practice aimed at individuals. The casuistic approach he advances describes a range of 'warrants' for decision making which include empirical evidence, experiential evidence, pathophysiological rationale, patients' values, and preferences and system features and a consideration of the social context of the patient. As he emphasises, none of these warrants can stand on their own and certainly not hierarchically, yet they do enable the accomplished clinician to take into account a fuller appreciation of the patient as a *person* with unique needs and circumstances [53–68].

4.6 Knowledge as a Decision Tool in Clinical Practice

Using 'knowledge' as a decision tool in the first instance requires *making sense* of the available information *in the context of this particular patient*. An example shall illustrate the difficulties of applying different sources of information to *knowing* what the right thing to do is.

> A 45-year-old accountant presents on Monday morning with chest pain after having spread a truck load of wood chips in his garden over the weekend. The pain gradually increased during the weekend, and was relieved by rest. He intermittently woke throughout the night due to pain across his chest. On questioning, the pain radiated into both of his arms but worse into the left, he did not notice any palpitations or

(continued)

(continued)

breathing difficulties. He used to smoke 30 cigarettes/day for 20 years, but ceased about 4 years ago; he drinks about three to four standard drinks/day, and more on the weekend. He states that his family life is strained, and that work is very stressful.

Clinical findings. He is overweight (BMI 29.2), his blood pressure is 155/85, resting pulse rate 88/min and regular, his heart sounds are dual, there are no murmurs, his chest is clear, and he has no ankle oedema. A resting ECG shows sinus rhythm without any electrocardiographic abnormalities. A blood test shows: cholesterol 6.9 mmol/l; triglycerides 1.8 mmol/l; HDL-cholesterol 1.4 mmol/l; calculated LDL-cholesterol 4.6 mmol/l; normal creatinine and blood sugar levels.

4.6.1 What Does *Effective* Mean?

The current definition of *effective treatments* has a truly interconnected history, and illustrates how well-meaning interventions in a complex system can result in huge unintended consequences. There are four stands interacting with each other.

The patient. Before the availability of effective drugs many patients died prematurely form infectious diseases like diphtheria or endocarditis.

The Doctor. At that time doctors could only *care*, i.e. sit with the patient and provide comfort during the *crisis*,[5] if the patient struggled through he eventually would get healthy again, though many died.

The apothecary.[6] Apothecaries produced and patented medicines. These were marketed for making you look/feel good and for 'curing everything' from simple aches to martial indiscretions. The ingredients of these medicines and their safety and efficiency were usually not known.

The chemist, the modern extension of the apothecary. In the early to mid-nineteenth century the chemist developed dyes which not only stained clothes but which could also visualise cells and the bacteria that caused many of the curses of the time. Some of these dyes, like acetylsalicylic acid, could relieve fever and pain, or prontosil, which was found to be highly effective in killing the bacteria causing endocarditis.

However, patent laws caused a problem, as only the process of production of a chemical could be patented. Many chemists were able to develop different ways of producing the same compound, competition to sell them was fierce, and profits low. To gain market share, chemical industries started to trademark their particular compound, a move that proved highly successful, e.g. acetylsalicylic acid marketed as Aspirin became the first blockbuster drug.

Government/regulator. Some of these new compounds proved not only to be highly effective, but also could kill, like diacetylmorphine, also known as heroin. Governments and regulators became concerned and restricted the sale of the medicines 'on prescription only'. Now doctors rather than patients decided if a medicine was needed.

This move had two consequences, the trade in medicines became limited and patients were 'safer' but prevented from 'self-indulgence'.

Chemists were limited in developing new compounds, so their employers, now calling themselves pharmaceutical industry, had to find new ways of selling their products. This was only possible if they were able to find new diseases to treat.

Government/regulator. As most medicines became regulated as 'on prescription only', new

[5] *Crisis*—from Greek *krisis,* literally, decision; an unstable or crucial time or state of affairs in which a decisive change is impending; *especially* one with the distinct possibility of a highly undesirable outcome; the turning point for better or worse in an acute disease or fever, an emotionally significant event or radical change of status in a person's life.

[6] Historical name for a medical professional who formulates and dispenses materia medica to physicians, surgeons and patients—a role now served by a pharmacist (or a chemist or dispensing chemist).

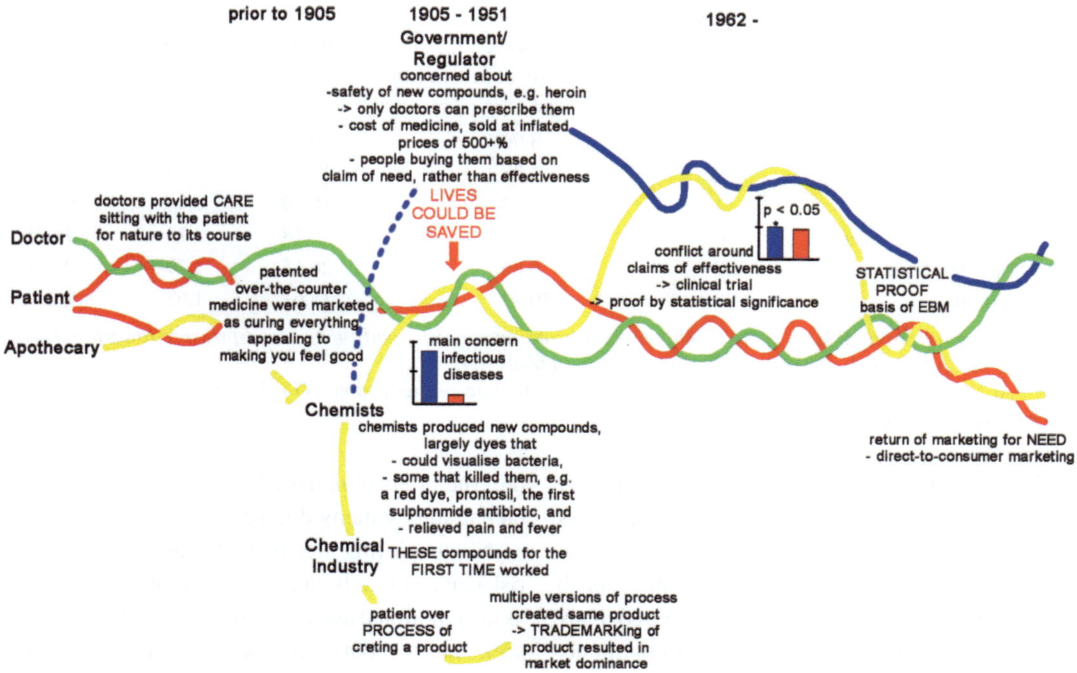

Fig. 4.5 The interlinked events that resulted in the current pharmaco-therapeutic paradigm (based on 2009 presentation by Prof David Healy, psychiatrist at Cardiff University, at Cornell University (available at http://www.youtube.com/watch?v=X53r3zTQJNk)

compounds could only be sold as medicines if they could demonstrate effectiveness and safety. The outcome—the clinical trial, now usually placebo-controlled, was born, and statistical significance, a concept the FDA insisted on, became the standard. The problems of statistical versus clinical significance have been discussed earlier in this chapter.

Three seemingly unrelated events have resulted in the predominance of drug-related treatments as we know it today: the discovery and development of highly effective compounds that 'saved lives' (namely antibiotics), the regulation of medicines 'on prescription only' to ensure effectiveness and safety, and the redefinition of effectiveness and safety based on 'statistically significant difference' in a 'randomised controlled trial' (Fig. 4.5).

The latter point especially is significant in understanding the way we perceive patients' complaints and the actions we take. The remainder of this section will illustrate some of the pragmatic problems for *good patient care*.

4.6.1.1 Information Reporting: A Biased Business

As a society we have a desire to hear 'good news stories' about the latest 'breakthroughs' in medical developments. This perceptual bias results in largely uncritical appreciation about their real impact on and their cost to individuals and society at large. In addition, journals for the most part do not publish negative research findings, compounding the perceptual bias of medical intervention always having benefits without causing substantial harm.

Of note the *research industry* is largely funded by the pharmaceutical and devices industry. This has raised substantial ethical issues; as has been shown the codependencies between researchers, their institutions, and the research sponsors have resulted not only in biased research studies but also

Table 4.1 The multiple interpretations of research findings

	ACE-I No. of events/100 patients	Placebo No. of events/100 patients	RRR (%)	ARR (%)	ITI (%)
Stroke	2.74	3.96	30.79	1.22	98.78
Coronary heart disease	8.89	11.08	19.74	2.19	97.81
Heart failure	2.54	3.02	15.79	0.48	99.52
Major cardiovascular event	13.86	17.15	19.18	3.29	96.71
Cardiovascular death	5.07	6.86	26.15	1.79	98.21
Total mortality	8.80	10.42	15.61	1.63	98.37

Figures compiled from *Effects of ACE inhibitors, calcium antagonists, and other blood-pressure-lowering drugs: results of prospectively designed overviews of randomised trials* [69]
ACE-I ACE-I-inhibitors; *RRR* relative risk reduction; *ARR* absolute risk reduction, *ITI* index of therapeutic impotence (1 – ARR) or absolute failure rate [70]

in frank harm to patients.[7] In response a counter movement involving longitudinal open-access recording by patients of their own illness trajectories has emerged which has resulted in a much greater understanding of diseases, and revised many fallacies of effectiveness and safety of commonly promoted treatments, e.g. see PatientsLikeMe.[8]

4.6.1.2 Risk and Benefit Reporting: Relative Versus Absolute

The reporting of results in *relative percentage* terms tends to hugely overstate the real or *absolute* risk and benefit of the results reported (Table 4.1). It is usually the larger figure of relative rather than the pragmatic smaller figure of absolute outcome that we 'hang' our judgement on. Reasons are conceptual as well as psychological.

Risk can be conceptualised as much in numerical as in subjective terms. The numerical concept usually refers to what is known at the group level (the epidemiologic concept), the subjective meaning is its translation to the individual level where people who are ill seek a real answer, not a probabilistic one [69]. It is common to mistake the reported epidemiological, i.e. relative risk as the true or absolute risk for the individual, a mistake made by many doctors and patients alike.

Not exploring the patient's understanding of risk can result in harm, both at the physical as well as the emotional and cognitive level. When patients seek clarity about *their risk* and doctors are uncertain, they use a mix of three approaches to create certainty in the mind of their patients: These are:

- Constructing temporary certainty and offering general reassurance
- Developing a coherent story of certainty despite implying uncertainty
- Frankly acknowledging uncertainty [70].

The latter approach would be the most honest but potentially the hardest to convey sensitively to the patient, and the most time consuming as one has to deal with the *emergent outcomes* resulting from the patient's sense-making journey.

4.6.2 Conveying Our Knowledge—or Rather Information—to this Patient

The typical response to our patient's presentation would be, loosely based on 'current promoted best practice': blood pressure is well above 120/80,[9] he is markedly overweight, his cholesterol is above 5.5 mmol/l, he is sedentary, his drinking habits are outside the safe drinking limits and he is an ex-smoker. Based on individual parameters he

(continued)

[7] For details see the SSRI studies. It had been well established that the introduction of SSRIs is associated with a substantial risk of suicide (e.g. MB *Stone* and L *Jones*. Clinical Review: Relationship between antidepressant drugs and suicidality in adults. Presented at FDA Meeting of Psychopharmacologic Drugs Advisory Committee (PDAC) 13-Dec-2006 (available at http://www.fda.gov/ohrms/dockets/ac/06/briefing/2006-4272b1-01-fda.pdf).

[8] James Allen Heywood, Founder of PatientsLikeMe, at GoldLab Symposium 2011—http://www.youtube.com/watch?v=ZAb3lIgyWmk.

[9] Heart Foundation—Guide to management of hypertension 2008, updated December 2010.

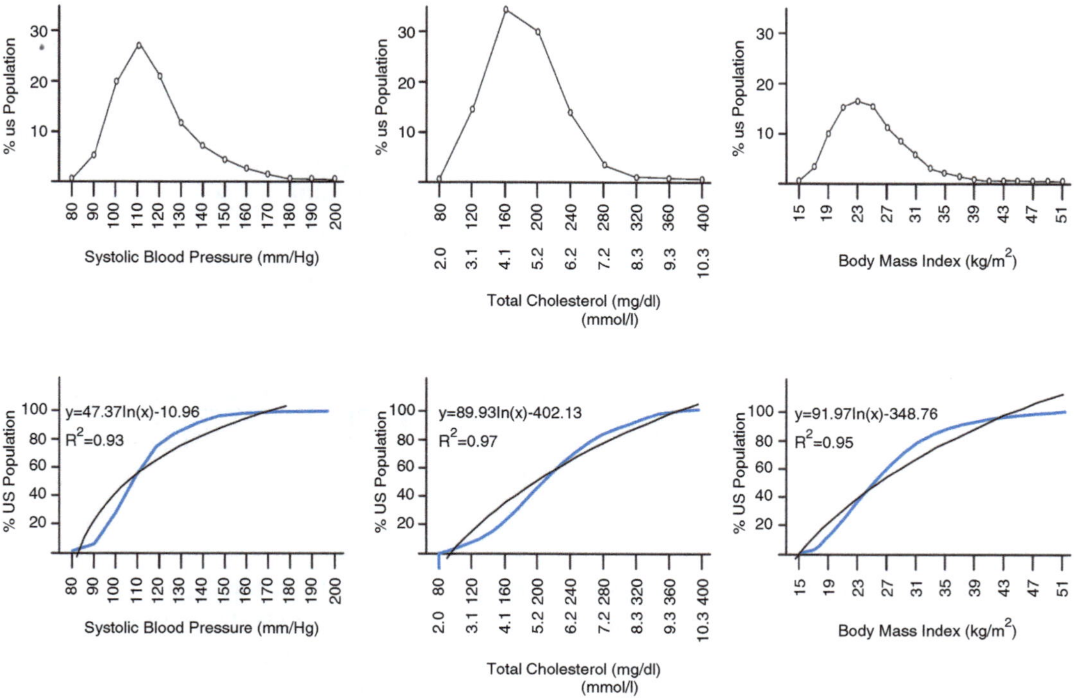

Fig. 4.6 Non-linear distribution of three physiological parameters—blood pressure, cholesterol, and body mass index—in the population. (Based on figures in [71])

(continued)

would require an antihypertensive, a lipid lowerer, an exercise programme, a dietician, and an alcohol counsellor.

A more sophisticated approach would be to check his absolute cardiovascular risk on a risk calculator, the surprise is, that he only has a low risk for a cardiovascular event at 5% over the next 5 years.[10] However, this approach to estimating risk is limited as it only takes account of a small number of risk factors—gender, age, blood pressure, smoking status, total cholesterol, and HDL-cholesterol, supplemented by two markers that indicate existing disease—diabetes and left ventricular hypertrophy (LVH).

Important other variables impacting on risk, like family history, ethnicity, stress, and socio-economic disadvantage have not been considered.

Other issues to take into account are our limited understanding of threshold behaviours of physiological parameters, the appreciation of adaptation of the *whole body* to the developing changes with the resultant risk of causing harm by over-treating a person based on data arising from single condition populations.

The further discussion highlights that the cause and effect mechanisms underpinning current clinical practice are hugely problematic in the same sense as the discussions about genes causing disease. We fully agree with the notions expressed by Heng in his chapter that we need to understand all of the interactions in the particular context, or as Heng puts it, it is the genome (the whole person), not the genes (the physiological parameters) that matters.

The following research findings highlight various observations that have been identified in relation to cardiovascular disease and mortality. These findings emphasise the complex interconnected relationships at various levels of observation; we

[10] Australian absolute cardiovascular risk calculator, http://www.cvdcheck.org.au/.

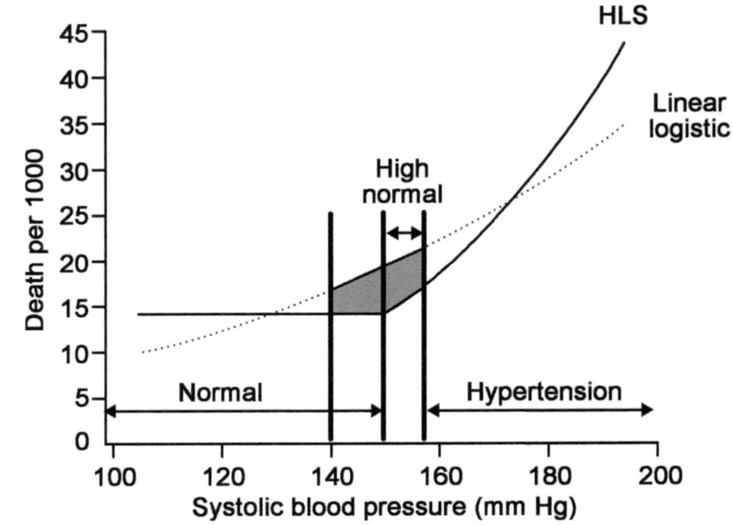

Fig. 4.7 Threshold behaviour of mortality in relation to systolic blood pressure. The figure superimposes the horizontal-logistic-spline and linear-logistic regression models. The curves are for men aged 55–64 years. The shaded area indicates individuals who may not require any treatment (about 40% of the population) (reproduced from [72] with permission)

neither proclaim that these examples provide irrefutable evidence nor that they accurately reflect all that is known in the field. However, from the perspective of physicians interacting with patients in the primary care environment, they address many of the issues that patients raise in the consultation, and that we need to make sense of with our patients.

It also needs to be stressed here that we clearly distinguish between *primary* and *secondary/tertiary* care approaches. The latter patient demonstrates the interacting consequences of one or more 'risk factors', however, these patients—according to the Pareto distribution principle—represent only about 20% of all patients with a risk factor.[11]

4.6.2.1 Physiological Parameters Are Non-linearly Distributed

Contrary to the general assumption most physiological parameters distribute in a non-linear fashion. Three common clinical measures—blood pressure, cholesterol, and body mass index—taken from a national survey in the USA [71] all show a long-tail frequency distribution; each accumulative distribution curve's trendline has a near perfect logarithmic fit (Fig. 4.6).

4.6.2.2 Threshold Behaviour of Physiological Parameters and Mortality

Physiological parameters have a range in which they show threshold characteristics, i.e. there is no strict linear increase between their levels and a secondary observation. Port reanalysed the Framingham data of blood pressure levels and age-specific all-cause mortality. He confirmed earlier findings that blood pressure levels and all-cause mortality are not linearly related. Rather, blood pressure and mortality are described by a horizontal logistic spline model, indicating threshold behaviour of blood pressure up to a certain point,[12] only after which mortality initially rises slightly before rising in an exponential fashion. Figure 4.7 shows the difference between the simple linear logistic and the horizontal logistic spline analysis. The simple linear logistic approach overestimates the risk of death for up to 70% of the population [72].

4.6.2.3 Cardiovascular Mortality Varies Widely Between Communities

Cardiovascular mortality is not solely attributable to absolute levels of a risk factor, even when adjusted for in relation to other risk

[11] For more detail see Chap. 5—West, Tail of Tales.

[12] Rules of thumb for the threshold of blood pressure-related mortality for a person of a given age and sex: Threshold = 110 + (2/3)(age) for men; Threshold = 104 + (5/6) (age) for women.

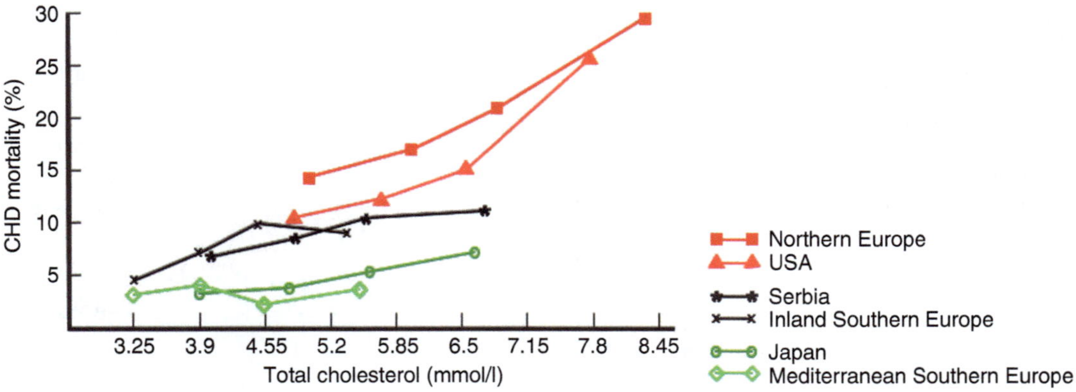

Fig. 4.8 25-year coronary heart disease mortality rates to baseline cholesterol quartile, adjusted for age, smoking, and blood pressure ([73] with permission)

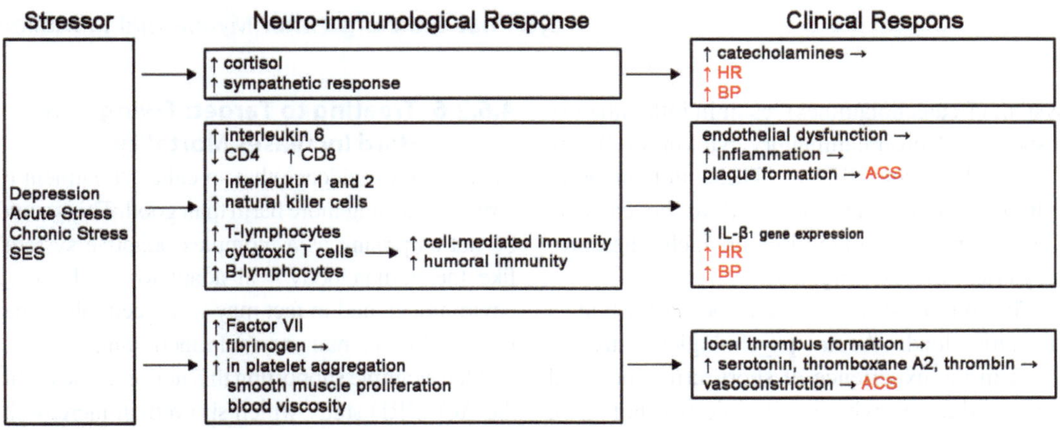

Fig. 4.9 Neuro-immunological response and clinical manifestations of stress, a schematic overview

factors. As the Seven Countries Study indicated, cholesterol has a significantly different association with cardiovascular mortality dependent on geography (and by implication ethnicity) (Fig. 4.8), leading to the conclusion not only that there is a genetic predisposition to cardiovascular disease, but also that there is an important gene-environment dimension [73], thus the need to take biogeographic (ethnic) aspects into consideration when calculating cardiovascular risk [74].

In addition, a recent comparative study of low versus high dose simvastatin has shown that there is no difference in cardio- and cerebrovascular mortality [75], but that high dose simvastatin use is associated with marked risk of rabdomyalisis [75]. It is fair to assume that this is the case with other statin drugs as well, as all affect the enzyme pathways implicated in causing this side effect. Also one should take note of the finding that even patients without objective or subjective muscle damage lose 16% push power in their legs when taking statin drugs and tend to have a greater falls risk, both of which fortunately are reversible when ceasing the drug [76].

4.6.2.4 Stress and Cardiovascular Disease/Mortality

Stress has long been known to reduce one's coping abilities and to negatively impact on neuroendocrine and immune function. The dysregulatory mechanisms between the hypothalamic–pituitary–adrenal axis (↑ CRH→↑ ACTH→↑ cortisol and glucocorticoids) and the sympathetic nervous

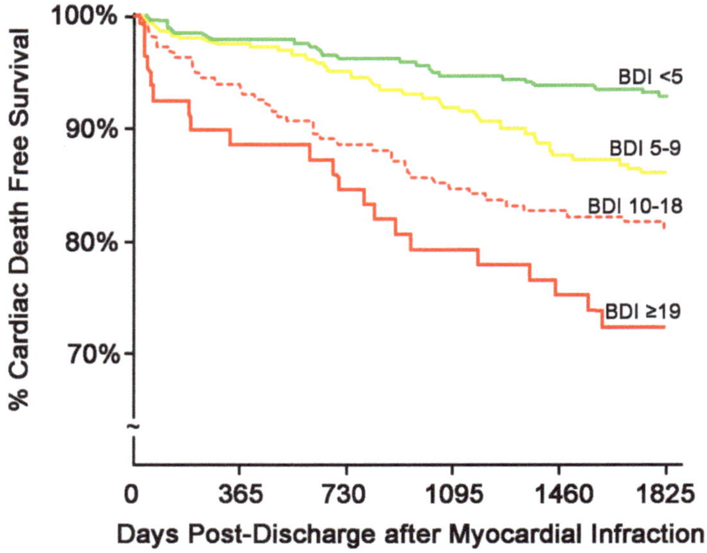

Fig. 4.10 Five-year cardiac mortality and level of depression. *BDI* = Bristol Depression Index, <5 low normal, 5–9 high normal, 10–18 mild depression, ≥19 moderate to severe depression (modified from [77])

system (↑ catecholamines) result in inflammatory responses (↑ proinflammatory cytokines—IL-1β, IL-6, TNF-α and IFN-γ). These inflammatory cytokines have been identified to be causally linked to many chronic diseases, including coronary artery disease (Fig. 4.9).[13]

The amount of stress is described as the body's allostatic load. Acute psychological stresses result in the experience of short-term emotional stress and anger, whereas chronic psychological stress results from low socio-economic status, work stress, chronic strain, social isolation, depression, anxiety, and hostility [77].

Depression [78, 79] and anxiety [79, 80] are independent risk factors for cardiovascular disease and mortality. Even mild distress outside the 'defined' depression range has a substantial effect on cardiac mortality as is illustrated in Fig. 4.10.

4.6.2.5 Socio-economic Dimensions and Cardiovascular Mortality

Like stress, socio-economic inequalities are also a significant independent predictor of all-cause and ischaemic heart disease mortality (Fig. 4.11) [81]. These data suggest that reducing socioeconomic inequality between the highest and lowest groups could have a major impact in improving cardiovascular morbidity and mortality.

[13] For more details see Chap. 19, Bennett—inflammation through a psychoneuroimmunological lens.

4.6.2.6 Treating to Target: Trying Too Hard Increases Mortality

There is now evidence that so-called 'treatment to target' is causing more harm than good. This should be not surprising as a 'complex adaptive system' like the human body will adapt to its changing environment, and in fact may well need 'abnormal readings' to maintain normal functioning.

One example supporting this notion comes from the ACCORD study which showed an increase in all-cause and cardiovascular mortality in treating diabetics to a HbA1c target of below 6.0% rather than achieving a level of between 6.9 and 7.9% [82]. A Cochrane review on the treatment of hypertension concluded that lowering blood pressure to less than 140–160/90–100 mmHg does not reduce mortality or morbidity, however at present the benefit of lower blood pressure for patients with diabetes mellitus and/or chronic renal disease remains unclear [83]. Equally, a meta-analysis of lipid lowering concluded that the effect is small on all-cause and cardiovascular mortality, and that this difference did not vary for primary and secondary prevention [84] (Fig. 4.12).

4.6.2.7 Obstructive Sleep Apnoea

Obstructive sleep apnoea has been shown to be another marker of premature cardiovascular mortality, especially for those with a family history of premature cardiovascular death [85].

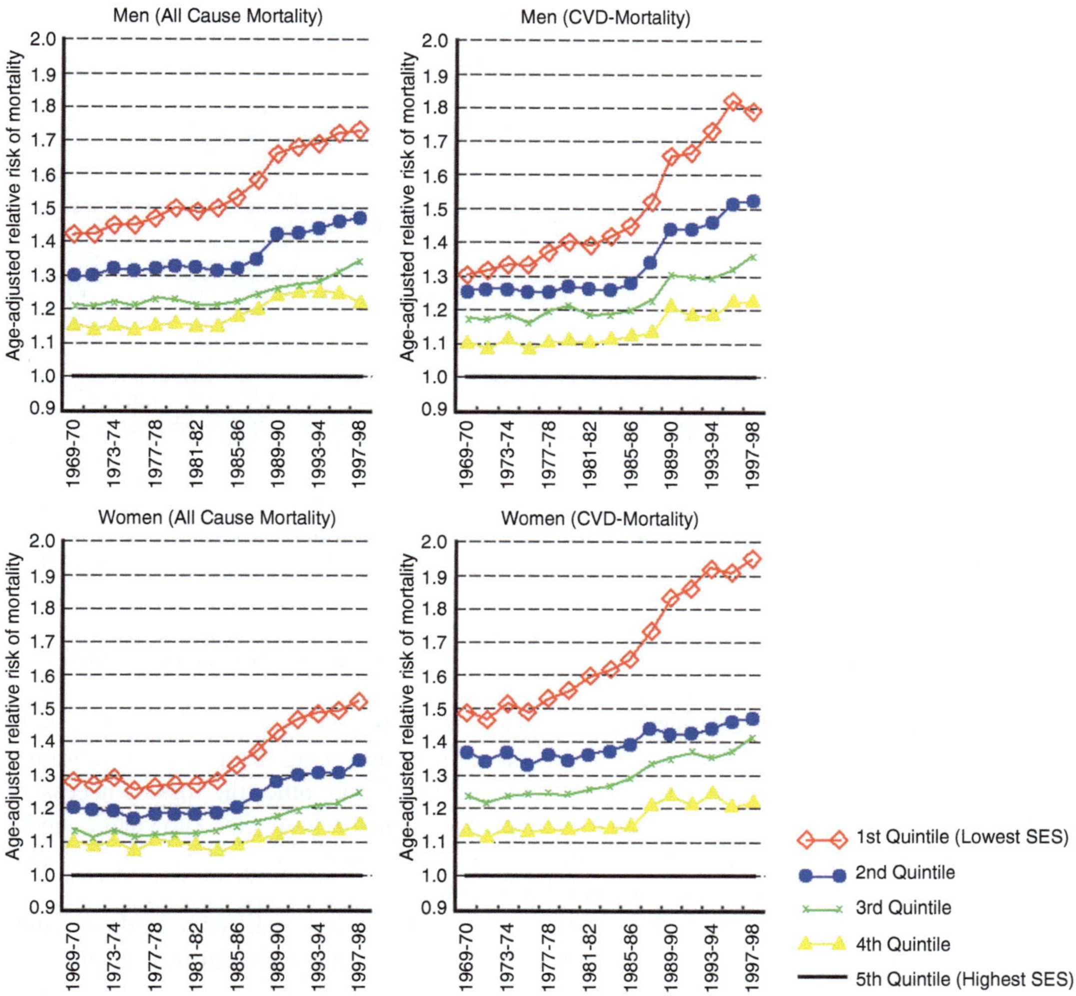

Fig. 4.11 Age-adjusted relative risks of all-cause and cardiovascular mortality among US men and women aged 25–64 years by 1990 area socio-economic status derived from Poisson regression models, 1969–1998. Notes. All relative risks of CVD mortality for men and women were statistically significant at $P < 0.0001$. All-cause relative mortality risks for men in all SES quintiles and for women in 1st and 2nd SES quintiles were significant at $P < 0.05$. All-cause relative mortality risks for women in 3rd SES quintile was significant at $P < 0.05$ for all years except during 1969–1976 and for women in 4th SES quintile significant at $P < 0.05$ only for 1997–1998. All trend tests were significant at $P < 0.0001$ ([80] with permission)

4.7 *Phronesis* or Practical Wisdom: Applying the Information from Different Perspectives to *the Particular of this Patient*

Being confronted with many pieces of information—some harder to understand than others, some of doubtful clinical significance, some outside what one has learned in medical and allied health courses, and some outside the direct reach of health professional practice—in an environment of conflicting views and vested interest, one after all has to work with one's patients' attitudes and expectations and one's health systems abilities to provide resources equitably.

As 'the evidence' as outlined above is pointing in many different directions a different approach may need to be pursued. Ian McWhinney alludes to the notion that doctors and other health

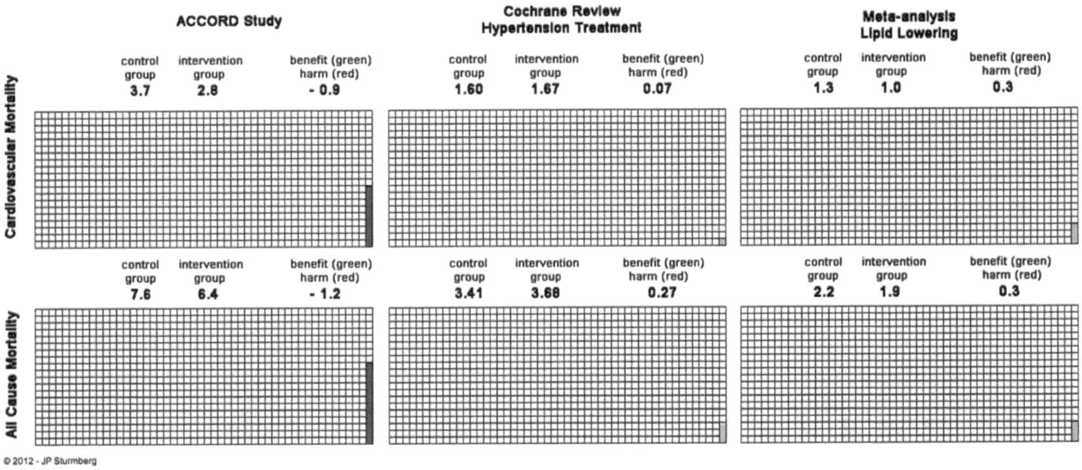

Fig. 4.12 Treating to target. The absolute benefit (*green*) and harm (*red*) of normalising physiological parameters in three recent studies on cardiovascular and all-cause mortality. The ACCORD study showed harm for 9–12 patients/1,000 treated, the hypertension and lipid lowering reviews indicate benefits for only <1–3 patients/1,000.

professionals *deal with particulars* [86], and Per Fugelli reminds us of the Aristotelian approach of *phronesis*, personalised clinical wisdom, as a guide to *'good practice'* [87].

Particulars are different from the generalised abstractions of a 'class of particulars'. Complex natural systems like a human in his/her environment are *particulars*. As each person is unique one cannot predict his/her characteristics and behaviours from that known about its class, the population. Hence, each person responds in non-linear ways and in multiple directions, and responses typically have feedback characteristics. Such circular causality defies prediction; medical practice is not about predicting events and outcomes, but rather about *care* and *understanding*, i.e. making sense of the illness [86].

Dealing with particulars is the approach to medical care, uniquely so in general practice/family medicine. And *phronesis*, the Aristotelian notion of practical wisdom, is gained by learning from individual cases. *Phronesis* reflects personal knowledge, instinct, feeling, and experience. Good practice results from the amalgamation of 'science and *phronesis*'; it requires open mindedness, critical reflection, and an ability to change and grow.

4.7.1 What to Say and What to Do?

Two flaws are readily evident in dealing with our patient, one being the de-contextualisation of knowledge, the other the loss of necessary flexibility when a particular knowledge model is not sufficient to achieve the desired outcome.

As he has a low risk for a cardiovascular event, and specific interventions in any case are of minimal benefit, a broader approach may be more appropriate. In the first instance, his risk of cardiovascular disease is determined by his genetic make-up (he is lucky, he has no family history of premature cardiovascular disease or death), he already has modified some of his health habits (stopped smoking), but could do better in relation to alcohol consumption and exercise. His high socio-economic status ameliorates his general risk; however, his stress levels (strained family relationships and work stresses) are a concern, though amenable to counselling-based interventions. Given the now well-established pathways of stress and its neuro-immunological responses, resolving these issues may well return his 'hard risk factors' (elevated blood pressure, high weight) back to 'normal'.

4.8 Knowledge is Contextual

Therefore, it is less important to focus on 'what constitutes the right kind of knowledge' but rather more important to consider all of the 'available knowledge' in context to achieve 'best practice'. It is time to 'acknowledge' that 'medical knowledge' is inherently uncertain, and to 'enact' a context-driven flexible approach to 'medical care models' congruent with the limitations and the emergence of 'knowing' [88, 89]. Doctrine, i.e. seeing only part of the whole picture arising from only one particular domain in the Cynefin framework, will fail each patient, and blinds our ability to create new insights and knowledge.

4.9 Conclusion

In this chapter we have attempted to review for the reader a subject of great complexity—the theoretical basis and practical application of knowledge for action in clinical practice. Beginning with philosophical notions, we have considered the shift from a static to a dynamic understanding of knowledge, the generation of knowledge and its multiple dimensions, its domain-specific application, the management of knowledge, and its use as a tool in clinical practice, the strengths and limitations of various study designs, including how these demonstrate the 'impossibility' of the EBM model and the vital role of *phronesis*, or practical wisdom, in the care of the patient. In concluding, we have discussed an emergent model of clinical practice that seeks to demonstrate the proper application of the various sources of knowledge for decision making and which has been termed 'person-centred medicine'.

Certainly, global health services appear to be undergoing a fundamental exercise in reflection on the nature of knowledge for practice as part of a current crisis within international health services—a crisis of knowledge, care, compassion, and costs [52]. A major concern in this context has been the observation of a growing distortion in the ethos and priorities of medicine, where a reductive focus on disease processes and organ systems alone has led to the compartmentalisation of knowledge, the fragmentation of services and to documented increases in a frank neglect of patients' concerns, needs, and values. It has been observed that scientific medicine is being replaced by scientistic medicine, with an accompanying collapse in the imperative to care as well as to cure, so that the human dimension of medicine is becoming lost and with it the fundamental purpose of medicine. The dehumanisation of medicine has been observed for some time, but in the last 20 years has been given impetus by the evidence-based medicine movement which has grown up alongside the patient-centred care movement, but rarely have the two entered into dialogue with each other [60, 90].

Miles and Mezzich [90] have argued that the two movements cannot continue to evolve—in the interests of patients and the medical profession itself—in parallel with one another, but must seek to achieve a rational form of integration, a 'giving of ground', where the more philosophically tenable components of each combine together to form a more coherent model of clinical practice for the future. This is, perhaps, the 'coalescence' for which Hartzband and Groopman have prominently called [91] and which Miles and Mezzich have endorsed, calling for a fifth reconstitution of the EBM model in order to facilitate it [90, 92]. Although much argumentation will be required to progress the associated philosophical reasoning from the status of concept to practice, we believe that the general direction for which it argues to be correct. If the crisis of knowledge, compassion, care, and costs within medicine is to be resolved, then we need to develop a model of practice which allows continuing biomedical and technological advance to be delivered to patients, but within a humanistic dimension that recognises the importance of applying science in a manner which respects the patient as a person and which takes full account of his values, preferences, aspirations, stories, cultural context, fears, worries, and hopes and which thus recognises and responds to his emotional, social, and spiritual necessities in addition to his physical needs.

In concluding, we present the reader with a stark truth: both clinicians and veterinary surgeons employ knowledge within their work—it is the *humanistic dimension of medicine* that separates the physician from the vet [93].

References

1. Popper K. Conjectures and refutations: the growth of scientific knowledge. London: Routledge and Kegan Paul; 1972.
2. Rovelli C. The uselessness of certainty 2011. http://www.edge.org/q2011/q11_4.html#rovelli.
3. Polanyi M. Personal knowledge. Towards a post-critical philosophy. London: Routledge; 1958.
4. Snowden D. Complex acts of knowing: paradox and descriptive self-awareness. J Knowl Manage. 2002; 6(2):100–11.
5. Wyatt J. Management of explicit and tacit knowledge. J R Soc Med. 2001;94(1):6–9.
6. Kurtz CF, Snowden DJ. The new dynamics of strategy: sense-making in a complex and complicated world. IBM Syst J. 2003;42(3):462–83.
7. Sturmberg JP. The illusion of certainty—a deluded perception? J Eval Clin Pract. 2011;17(3):507–10.
8. Clark A. Predictive Coding. 2011. http://www.edge.org/q2011/q11_6.html#clark.
9. Balfour EB. The living soil. Evidence of the importance to human health of soil vitality, with special reference to national planning. revised ed. London: Faber and Faber Ltd; 1948.
10. Leung W-C. Balancing statistical and clinical significance in evaluating treatment effects. Postgrad Med J. 2001;77(905):201–4.
11. McWhinney IR. The naturalist tradition in general practice. J Fam Pract. 1977;5(3):375–8.
12. Guyatt GH. Evidence-based medicine. Ann Intern Med. 1991;114(ACP (Suppl 2)):A-1612.
13. Evidence-Based Medicine Working Group. Evidence-based medicine. A new approach to teaching the practice of medicine. JAMA. 1992;268(17):2420–5.
14. Kuhn T. The structure of scientific revolutions. Chicago: University of Chicago Press; 1970.
15. Polychronis A, Miles A, Bentley P. Evidence-based medicine: reference? Dogma? Neologism? New orthodoxy? J Eval Clin Pract. 1996;2(1):1–3.
16. Polychronis A, Miles A, Bentley P. The protagonists of 'evidence-based medicine': arrogant, seductive and controversial. J Eval Clin Pract. 1996;2(1):9–12.
17. Miles A, Bentley P, Polychronis A, Grey J. Evidence-based medicine: why all the fuss? This is why. J Eval Clin Pract. 1997;3(2):83–6.
18. Miles A, Bentley P, Polychronis A, Grey J, Price N. Recent progress in health services research: on the need for evidence-based debate. J Eval Clin Pract. 1998;4(4):257–65.
19. Charlton BG, Miles A. The rise and fall of EBM. Q J Med. 1998;91(5):371–4.
20. Miles A, Bentley P, Polychronis A, Grey J, Price N. Advancing the evidence-based healthcare debate. J Eval Clin Pract. 1999;5(2):97–101.
21. Miles A, Charlton B, Bentley P, Polychronis A, Grey J, Price N. New perspectives in the evidence-based healthcare debate. J Eval Clin Pract. 2000;6(2): 77–84.
22. Miles A, Bentley P, Polychronis A, Grey J, Melchiorri C. Recent developments in the evidence-based healthcare debate. J Eval Clin Pract. 2001;7(2):85–9.
23. Miles A, Grey J, Polychronis A, Melchiorri C. Critical advances in the evaluation and development of clinical care. J Eval Clin Pract. 2002;8(2):87–102.
24. Miles A, Grey JE, Polychronis A, Price N, Melchiorri C. Current thinking in the evidence-based health care debate. J Eval Clin Pract. 2003;9(2):95–109.
25. Miles A, Grey JE, Polychronis A, Price N, Melchiorri C. Developments in the evidence-based health care debate—2004. J Eval Clin Pract. 2004;10(2):129–42.
26. Miles A, Polychronis A, Grey JE. The evidence-based health care debate—2006. Where are we now? J Eval Clin Pract. 2006;12(3):239–47.
27. Miles A, Loughlin M, Polychronis A. Medicine and evidence: knowledge and action in clinical practice. J Eval Clin Pract. 2007;13(4):481–503.
28. Miles A, Loughlin M, Polychronis A. Evidence-based healthcare, clinical knowledge and the rise of personalised medicine. J Eval Clin Pract. 2008;14(5):621–49.
29. Miles A. On a medicine of the whole person: away from scientistic reductionism and towards the embrace of the complex in clinical practice. J Eval Clin Pract. 2009;15(6):941–9.
30. Kristiansen IS, Mooney G. Evidence-based medicine—in its place. London and New York: Routledge; 2004.
31. Hickey S, Roberts H. Tarnished gold. The sickness of evidence-based medicine. Charleston SC, USA: CreateSpace; 2011.
32. Sackett DL, Richardson WS, Rosenberg WMC, Haynes RB. Evidence-based medicine: how to practice and teach EBM. New York: Churchill Livingstone; 1992.
33. Sackett DL, Straus SE, Richardson WS, Rosenberg WMC, Haynes RB. Evidence-based medicine: how to practice and teach EBM. 2nd ed. Edinburgh: Churchill Livingstone; 2000.
34. Guyatt G, Drummond R, Meade M. Users' guides to the medical literature. Essentials of evidence-based clinical practice. New York, USA: McGraw Hill & Co; 2008.
35. Clinical Evidence. Learn, teach and practise evidence-based medicine. London: BMJ Publishing Group Ltd. http://clinicalevidence.bmj.com/x/index.html. Accessed 27 Feb 2012.
36. National Institute for Health and Clinical Excellence. Published practice guidelines. London. http://www.nice.org.uk/guidance/index.jsp?action=byType&type=2&status=3. Accessed 27 Feb 2012.
37. Nunn R. Evidence-based medicine and limits to the literature search. J Eval Clin Pract. 2008;14(5):672–8.
38. Miles A. Evidence for practice and the authority of experts: there can be no former without the latter: a commentary an Nunn (2008). J Eval Clin Pract. 2008;14(5):679–81.
39. Miles A. Science: a limited source of knowledge and authority in the care of patients. A review and analysis

of: 'How Doctors Think. Clinical Judgement and the Practice of Medicine'. Montgomery, K. Oxford University Press, Oxford. ISBN 978-0-19-518712-0, pp. 256. J Eval Clin Pract. 2007;13(4):545–63.
40. Feinstein AR, Horwitz RI. Problems in the "evidence" of "evidence-based medicine". Am J Med. 1997; 103(6):529–35.
41. Djulbegovic B, Guyatt GH, Ashcroft RE. Epistemologic inquiries in evidence-based medicine. Cancer Control. 2009;16(2):158–68.
42. Miles A. Evidence-based medicine: requiescat in pace? A commentary on Djulbegovic, B., Guyatt, G. H. & Ashcroft, R. E. (2009) Cancer Control, 16, 158–168. J Eval Clin Pract. 2009;15(6):924–9.
43. Silva SA, Wyer PC. Where is the wisdom? II—Evidence-based medicine and the epistemological crisis in clinical medicine. Exposition and commentary on Djulbegovic, B., Guyatt, G. H. & Ashcroft, R. E. (2009) Cancer Control, 16, 158–168. J Eval Clin Pract. 2009;15(6):899–906.
44. Tonelli MR. A late and shifting foundation: a commentary on Djulbegovic, B., Guyatt, G. H. & Ashcroft, R. E. (2009) Cancer Control, 16, 158–168. J Eval Clin Pract. 2009;15(6):907–9.
45. Loughlin M. The search for substance: a quest for the identity-conditions of evidence-based medicine and some comments on Djulbegovic, B., Guyatt, G. H. & Ashcroft, R. E. (2009) *Cancer Control*, 16, 158–168. J Eval Clin Pract. 2009;15(6):910–4.
46. Tanenbaum SJ. More of the same: a commentary on Djulbegovic, B., Guyatt, G. H. & Ashcroft, R. E. (2009) Cancer Control, 16, 158–168. J Eval Clin Pract. 2009;15(6):915–6.
47. Sturmberg JP. EBM: a narrow and obsessive methodology that fails to meet the knowledge needs of a complex adaptive clinical world: a commentary on Djulbegovic, B., Guyatt, G. H. & Ashcroft, R. E. (2009) Cancer Control, 16, 158–168. J Eval Clin Pract. 2009;15(6):917–23.
48. Charlton BG. The Zombie science of evidence-based medicine: a personal retrospective. A commentary on Djulbegovic, B., Guyatt, G. H. & Ashcroft, R. E. (2009). Cancer Control, 16, 158–168. J Eval Clin Pract. 2009;15(6):930–4.
49. Howick J. The philosophy of evidence-based medicine. Oxford: Wiley-Blackwell, BMJ Books; 2011.
50. Tonelli MR. Not a philosophy of clinical medicine: a commentary on 'The Philosophy of Evidence-based Medicine' Howick, J. ed. (2001). J Eval Clin Pract. 2011;17(5):1013–7.
51. Upshur REG. If not evidence, then what? Or does medicine really need a base? J Eval Clin Pract. 2002;8(2):113–9.
52. Miles A, Loughlin M. Models in the balance: evidence-based medicine versus evidence-informed individualized care. J Eval Clin Pract. 2011;17(4):531–6.
53. Tonelli M. Integrating evidence into clinical practice: an alternative to evidence-based approaches. J Eval Clin Pract. 2006;12(3):248–56.
54. Djulbegovic B. Evidence and decision making. Commentary on M.R. Tonelli (2006), Integrating evidence into clinical practice: an alternative approach to evidence-based approaches. Journal of Evaluation in Clinical Practice 12, 248–256. J Eval Clin Pract. 2006;12(3):257–9.
55. Miettinen OS. Evidence-based medicine, case-based medicine; scientific medicine, quasi-scientific medicine. Commentary on Tonelli (2006), Integrating evidence into clinical practice: an alternative to evidence-based approaches. Journal of Evaluation in Clinical Practice 12, 248–256. J Eval Clin Pract. 2006;12(3):260–4.
56. Porta M. Five warrants for medical decision making: some considerations and a proposal to better integrate evidence-based medicine into everyday practice. Commentary on Tonelli (2006), Integrating evidence into clinical practice: an alternative to evidence-based approaches. Journal of Evaluation in Clinical Practice 12, 248–256. J Eval Clin Pract. 2006;12(3):265–8.
57. Lipman T. Evidence and casuistry. Commentary on Tonelli (2006), Integrating evidence into clinical practice: an alternative to evidence-based approaches. Journal of Evaluation in Clinical Practice 12, 248–256. J Eval Clin Pract. 2006;12(3):269–72.
58. Tanenbaum SJ. Evidence by any other name. Commentary on Tonelli (2006), Integrating evidence into clinical practice: an alternative to evidence-based approaches. Journal of Evaluation in Clinical Practice 12, 248–256. J Eval Clin Pract. 2006;12(3):273–6.
59. Sá Couto J. Can we forget how to treat patients? Commentary on Tonelli (2006), Integrating evidence into clinical practice: an alternative to evidence-based approaches. Journal of Evaluation in Clinical Practice 12, 248–256. J Eval Clin Pract. 2006;12(3):277–80.
60. Upshur REG. The complex, the exhausted and the personal: reflections on the relationship between evidence-based medicine and casuistry. Commentary on Tonelli (2006), Integrating evidence into clinical practice: an alternative to evidence-based approaches. Journal of Evaluation in Clinical Practice 12, 248–256. J Eval Clin Pract. 2006;12(3):281–8.
61. Loughlin M. The future for medical epistemology? Commentary on Tonelli (2006), Integrating evidence into clinical practice: an alternative to evidence-based approaches. Journal of Evaluation in Clinical Practice 12, 248–256. J Eval Clin Pract. 2006; 12(3):289–91.
62. Malterud K. The social construction of clinical knowledge—the context of culture and discourse. Commentary on Tonelli (2006), Integrating evidence into clinical practice: an alternative to evidence-based approaches. Journal of Evaluation in Clinical Practice 12, 248–256. J Eval Clin Pract. 2006;12(3):292–5.
63. Gupta M. Beyond 'evidence'. Commentary on Tonelli (2006), Integrating evidence into clinical practice: an alternative to evidence-based approaches. Journal of Evaluation in Clinical Practice 12, 248–256. J Eval Clin Pract. 2006;12(3):296–8.

64. Geanellos R, Wilson C. Building bridges: knowledge production, publication and use. Commentary on Tonelli (2006), Integrating evidence into clinical practice: an alternative to evidence-based approaches. Journal of Evaluation in Clinical Practice 12, 248–256. J Eval Clin Pract. 2006;12(3):299–305.
65. Buetow S. Opportunities to elaborate on casuistry in clinical decision making. Commentary on Tonelli (2006). Integrating evidence into clinical practice: an alternative to evidence-based approaches. Journal of Evaluation in Clinical Practice 12, 248–256. J Eval Clin Pract. 2006;12(4):427–32.
66. Tonelli MR. Advancing a casuistic model of clinical decision making: a response to commentators. J Eval Clin Pract. 2007;13(4):504–7.
67. Tonelli MR. Evidence-free medicine: forgoing evidence in clinical decision making. Perspect Biol Med. 2009;52(2):319–31.
68. Tonelli MR. The challenge of evidence in clinical medicine. J Eval Clin Pract. 2010;16(2):384–9.
69. Reventlow S, Hvas AC, Tulinius C. "In really great danger ..." The concept of risk in general practice. Scand J Prim Health Care. 2001;19(1):71–5.
70. Griffiths F, Green E, Tsouroufli M. The nature of medical evidence and its inherent uncertainty for the clinical consultation: qualitative study. Br Med J. 2005;330:511–7.
71. Schwartz L, Woloshin S. Changing disease definitions: implications for disease prevalence. Eff Clin Pract. 1999;2(2):76–85.
72. Port S, Demer L, Jennrich R, Walter D, Garfinkel A. Systolic blood pressure and mortality. Lancet. 2000; 355(3):175–80.
73. Kromhout D. On the waves of the Seven Countries Study; a public health perspective on cholesterol. Eur Heart J. 1999;20(11):796–802.
74. de Lorgeril M, Salen P, Paillard F, Laporte F, Boucher F, de Leiris J. Mediterranean diet and the French paradox. Cardiovasc Res. 2002;54(3):503–15.
75. Study of the Effectiveness of Additional Reductions in C, Homocysteine Collaborative G. Intensive lowering of LDL cholesterol with 80 mg versus 20 mg simvastatin daily in 12064 survivors of myocardial infarction: a double-blind randomised trial. Lancet. 2010;376(9753):1658–69.
76. Scott D, Blizzard L, Fell J, Jones G. Statin therapy, muscle function and falls risk in community-dwelling older adults. QJM. 2009;109(2):625–33.
77. Steptoe A, Brydon L. Psychoneuroimmunology and coronary artery disease. In: Vedhara K, Irwin M, editors. Human psychoneuroimmunology. Oxford: Oxford University Press; 2005.
78. Lespérance F, Frasure-Smith N, Talajic M, Bourassa MG. Five-year risk of cardiac mortality in relation to initial severity and one-year changes in depression symptoms after myocardial infarction. Circulation. 2002;105(9):1049–53.
79. Januzzi Jr JL, Stern TA, Pasternak RC, DeSanctis RW. The influence of anxiety and depression on outcomes of patients with coronary artery disease. Arch Intern Med. 2000;160(13):1913–21.
80. Kendall-Tackett K. Psychological trauma and physical health: a psychoneuroimmunology approach to etiology of negative health effects and possible interventions. Psychol Trauma Theor Res Pract Pol. 2009; 1(1):35–48.
81. Singh GK, Siahpush M. Increasing inequalities in all-cause and cardiovascular mortality among US adults aged 25–64 years by area socioeconomic status, 1969–1998. Int J Epidemiol. 2002;31(3):600–13.
82. The Accord Study Group. Long-term effects of intensive glucose lowering on cardiovascular outcomes. N Engl J Med. 2011;364(9):818–28.
83. Arguedas JA, Perez Marco I, Wright James M. Treatment blood pressure targets for hypertension. Cochrane Database of Systematic Reviews [serial on the Internet]. 2009; (3). Available from: http://www.mrw.interscience.wiley.com/cochrane/clsysrev/articles/CD004349/frame.html.
84. Yusuf S, Lonn E, Bosch J. Lipid lowering for primary prevention. Lancet. 2009;373(9670):1152–5.
85. Gami AS, Rader S, Svatikova A, Wolk R, Herold DL, Huyber C, et al. Familial premature coronary artery disease mortality and obstructive sleep apnea. Chest. 2007;131(1):118–21.
86. McWhinney IR. An acquaintance with particulars. .... Fam Med. 1989;21(4):296–8.
87. Fugelli P. Clinical practice: between Aristotle and Cochrane. Schweiz Med Wochenschr. 1998;128:184–8.
88. Sturmberg JP. The foundations of primary care. Daring to be different. Oxford San Francisco: Radcliffe Medical Press; 2007.
89. Martin C. Chronic disease and illness care. Additional principles of family medicine are needed to address health system redesign. Can Fam Phys. 2007;53(12): 2086–91.
90. Miles A, Mezzich JE. The care of the patient and the soul of the clinic: person-centered medicine as an emergent model of modern clinical practice. Int J Person Centered Med. 2011;1(2):207–22.
91. Hartzband P, Groopman J. Keeping the patient in the equation—humanism and health care reform. N Engl J Med. 2009;361(1):554–5.
92. Charles C, Gafni A, Freeman E. The evidence-based medicine model of clinical practice: scientific teaching or belief-based preaching? J Eval Clin Pract. 2011; 17(4):597–605.
93. Houston W. Reported and discussed within the Cambridge History of Medicine. Cambridge: Cambridge University Press; 2006.

5 Mathematical Principles: Tales of Tails

Bruce J. West

5.1 A View from Above

A mathematical principle is an overarching statement of what will always occur given a prescribed set of circumstances. Such principles are very useful because they can guide us in making decisions about the real-world even though their basis can be quite abstract. For example, in physics, we have the second law of thermodynamics, which is a phenomenological generalization of vast amounts of experimental data. The empirical law is that heat flows from hotter regions to colder regions, but the implied physical principle is that the universe is slowing down and the total entropy of the universe is increasing. The mathematical principle associated with the second law governs how entropy changes under specified conditions. The good news is I will not talk further about the second law, but I thought it essential that you realize at the outset that we will discuss the connection between data, the patterns we identify within data and the theories we construct to understand those patterns, all of which form our view of the world and how the world responds to our interventions.

My particular prejudice is that all knowledge is theoretical, and it is only through theory that we are able to make sense of the data that our world provides. Consequently, in this chapter, I underscore the theory or world view that dominates western society, particularly how that view influences medicine. It is perhaps not apparent but the dominant world view is a consequence of a specific interpretation of science and the success of the industrial revolution, all of which can be summarized in a handful of mathematical principles. We begin this particular conversation where all science begins and that is with data and the identification of patterns within the data.

5.1.1 A Little History

At the opening of the nineteenth century, the normal distribution introduced by the German polymath Gauss (1777–1855) [1] and the American mathematician Adrian (1775–1843) [2] provided a new scientific perspective of the world through their explanation as to why experimental results always vacillate from one experiment to the next. The interpretation of the normal distribution developed into the Law of Frequency of Error (LFE) in which physical measurements have a proper value and that value is determined by the underlying dynamics of the phenomenon being measured. This world view was consistent with Newtonian mechanics that determined the ballistics of cannon and rifles, the inertia of horse and carriage, and the orbits of the planets. The universe was understood as a clockwork mechanical process and therefore measured quantities ought to be predictable, quantifiable and measureable.

B.J. West (✉)
Army Research Office, 4300 S. Miami Blvd., Research Triangle Park 27709, NC, USA
e-mail: bruce.j.west@us.army.mil

J.P. Sturmberg and C.M. Martin (eds.), *Handbook of Systems and Complexity in Health*,
DOI 10.1007/978-1-4614-4998-0_5, © Springer Science+Business Media New York 2013

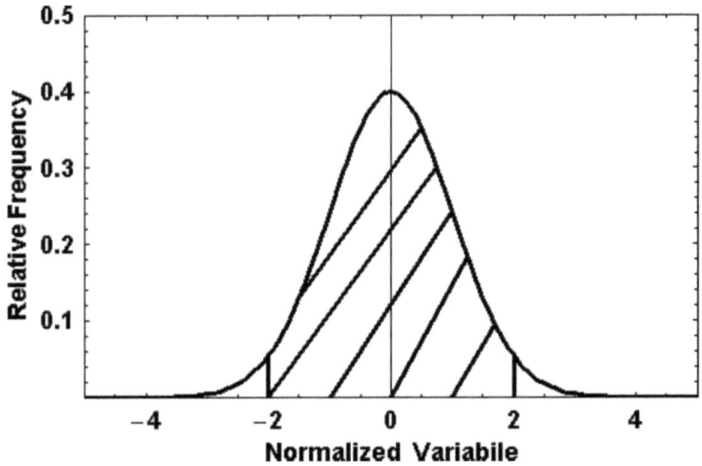

Fig. 5.1 The bell-shaped curve of Gauss, Adler and Laplace concerns errors, consequently the experimental value has the average value subtracted so the curve peaks at zero, and the new variable is divided by the standard deviation (width of the distribution) of the old variable and the new variable is therefore dimensionless. The cross-hatched region, between two standard deviations above and below the average value, contains 95% of the errors (the deviations from the average value)

However, it was observed that no matter how skilled the experimentalist, how high the quality of the instruments or how well the experiment was designed, the outcome of no two experiments is exactly the same. The measured values were rarely identical to the predicted ones and were instead scattered in the vicinity of the predictions. Gauss and Adrian maintained that most of the measured values are close to the predicted where the distribution peaks with as many data points above as there are below the predicted value. The peak of the symmetric normal (bell-shaped) distribution of Gauss and Adrian occurs at the average value and they concluded that the average value is the best representation of the ensemble of measurements. All this information is compacted into Fig. 5.1.

They also argued that the deviations from the predicted value are produced by errors in the measurements, deviations are errors because the average is the proper value of the variable—the one the experiment ought to yield each and every time. Thus, even when statistics enter the picture, where uncertainty blurs what is expected, the experimentalist still believes there ought to be a proper value that characterizes the process. In this interpretation, statistical fluctuations do not invalidate the mechanical world view; rather they are a complement to it making predictions less certain than the ticking clock.

Shortly after the introduction of the normal distribution, Laplace (1740–1847) [3] presented a proof of the central limit theorem establishing that the validity and applicability of the normal distribution was much broader than the LFE. The conditions for the proof were (1) the errors are independent, (2) the errors are additive, (3) the statistics of each error is the same and (4) the width of the distribution is finite. The mathematician Poincaré (1854–1912) commented on the LFE [4]:

> All the world believes it firmly, because the mathematicians imagine that it is a fact of observation and the observers that it is a theorem of mathematics.

Such remarks were a caution that physical phenomena do not always satisfy the mathematical conditions under which the central limit theorem is proved. However, by the time Poincaré and others were issuing such warnings the normal distribution had been accepted by a broad range of sciences most of whom adopted the intuition of normality and ignored the cautionary tales of mathematicians.

The assumptions of linearity and additively of errors necessary for the proof of the central limit theorem are fundamental to the linear world view with its reductionist basis and ambiguity produced by limitations in knowledge. This view was readily accepted by the nascent sciences of sociology, psychology and various branches of the life sciences. In this way, the nineteenth century saw the birth of sociophysics, psychophysics and was the basis of a theory of medicine whose foundation was and to a large extent remains homeostasis. The concept of homeostasis, like that of the LFE, adopted the view that there is a best value for the variable characterizing a physiological system. The heart rate, blood pressure, breathing and stride rate are all average values, all presumed to give the best measure of the physiological system being probed. But do not think that medicine had an isolated perspective. In his excellent review Norwich [5] points out that the "Le Chatelier's principle" was discovered in chemistry independently by Henri Louis Le Chatelier (1888) and Karl Ferdinand Braun and states that when a system in dynamic equilibrium is acted on by an external stress, it will adjust in such a way as to relieve the stress and establish a new equilibrium. Sound familiar?

Homeostasis assumes that measured deviations are disruptions of the network's dynamics produced by either internal or external perturbations. Control of the dynamics is provided by feedback that suppresses the fluctuations and relaxes the network back to its "best" value. Over time one expects that these perturbations produce as many deviations above as they do below the best value and that the statistics of these deviations are consequently the same as the statistics of error. The difference here is the notion of control introduced into medicine by Claude Bernard in his study of stability of the human body in his book *Introduction to Experimental Medicine* (1865). Homeostasis was hypothesized to dynamically maintain the physiological network at the average value. This feedback control was the simplest dynamics necessary to maintain "normality" in the physiological time series consistent with Le Chatelier's principle. This theoretical perspective of medicine co-evolved with the mathematical/engineering ideas of control theory that were so successful in the design engineering of the products and systems of the industrial age and subsequently the information age. As West [6] maintains, traditional control theory is not adequate for designing truly complex adaptive networks or devices and so too homeostasis lacks the dynamic flexibility to describe the richness of the non-equilibrium phenomenon in medicine.

Consequently, just as the errors in a physical network can be tested to determine if real-world data satisfy the normal distribution, the same ought to be true of physiological time series.

5.1.2 What the Data Said

Experimental data gathered throughout the nineteenth century supported the linear world view of Gauss, Adrian and Laplace. No one was more enamoured with the new perspective than the cousin of Charles Darwin (1809–1882), Sir Francis Galton (1822–1911) who compiled statistics on the properties of identical twins, the frequency of yawns, lifespans, the sterility of heiresses, and the inheritance of physical and mental characteristics [7]. He introduced the normal bell-shaped distribution into the social sciences as well as the statistics of human measure and behaviour. Sir Francis observed:

> I know of scarcely anything so apt to impress the imagination as the wonderful form of cosmic order expressed by the "law of frequency of error." The law would have been personified by the Greeks and deified if they had known of it. It reigns with serenity and in complete self-effacement amidst the wildest confusion. The larger the mob, and the greater the apparent anarchy, the more perfect is its sway. It is the supreme law of unreason. Whenever a large sample of chaotic elements are taken in hand and marshalled in the order of their magnitude, an unsuspected and most beautiful form of regularity proves to have been latent all along.

As pointed out by West [8] the normal distribution has a regularity and stability that is characteristic of the system as a whole and is quite distinct from the variability observed in the individual elements of data.

In the middle of the century, Adolphe Quetelet (1796–1874) helped erect the edifice of "social physics," which was modelled after celestial mechanics. He used the LFE to explain the

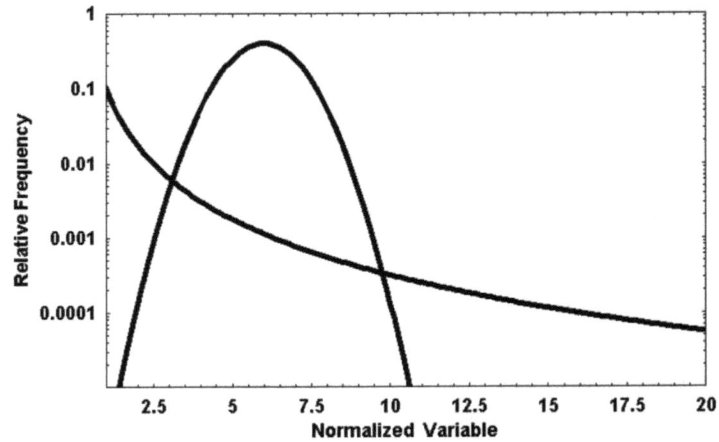

Fig. 5.2 The bell-shaped curve of Gauss, Adrian and Laplace is here compared with the inverse power law of Pareto. The much longer tail of the Pareto distribution is evident, indicating that large events are much more probable in the latter case than in the former and consequently much more important

deviations from the notion of what was believed to be the biological reality of the "average man", reality being the end result of genetics and evolution. He reasoned that human social variability was the result of error, much as the physical and mathematical scientists had argued for in the context of physical measurement. The average was stable and therefore so was the society it purported to represent as long as the errors were not too large and the network dynamics could control things through feedback.

However, even Sir Francis despite his affection for the normal distribution began to realize that data from many phenomena deviate significantly from normality. He argued that certain classes of events are better described by geometric averages rather than by the arithmetic averages. In the latter the average is determined by adding all the data together and dividing by the total number of data points N; in the former the average is determined by multiplying all the data together and taking the Nth root of the product. The arithmetic average characterizes a linear additive process whereas the geometric average characterizes a non-linear multiplicative process.

It was not until the end of nineteenth century that things began to unravel, a consequence of analysis of the distribution of income. In the linear additive world an individual makes an average income with some people making a little more and some a little less, but the distribution of income would be bell-shaped. Consequently, the distribution of income would be fair. The Italian engineer/economist/sociologist Marquis Vilfredo Frederico Damoso Pareto (1848–1923) at the end of the nineteenth century showed that this was not the case. Pareto worked as an engineer until he was middle aged and with the death of his father he left engineering and took a faculty position in Lausanne, Switzerland. Using his training as an engineer he was the first person to determine that the distribution of wealth in a society was not random but followed a consistent pattern. He described this pattern with an inverse power-law distribution which he called "The Law of the Unequal Distribution of Results" [9]. He referred to the inequality in his distribution more generally as a "predictable imbalance", which he was able to find in a variety of phenomena, including the distribution of talent among individuals. West [8, 10] noted that this imbalance is ultimately interpretable as an implicit unfairness that is invariably found in complex networks and social networks are no exception. The marked difference between the bell-shaped curve of Gauss, Adrian and Laplace and the inverse power law of Pareto is depicted in Fig. 5.2, where the inverse power law is seen to have a very long tail. The tail extends far beyond the region of the bell-shaped curve and often indicates a lack of scale resulting in the divergence of the variance.

In the absence of a variance there is no standard deviation and consequently the usual way to determine how well the average value represents the data is lost, that being, that the width of the distribution is very much less than the average. This loss of a reliable measure presents a major problem for those attempting to understand the world through the use of experimental data, particularly those complex biomedical phenomena that are often dominated by extreme events such as death.

5.2 A Conflict of Views

In this section we examine what we can learn about the world through an analysis of what we do not know but can count. The world's ambiguity is contained in the distribution of the fluctuations in the outcomes of experiments. The functional form of the distribution reveals certain general properties of the world's influence on the simple predictions we make whether it is the stopping distance when tail gaiting at 60 mph or how a patient will react to a new medicine. As human beings one of our strongest urges is to make the future knowable and through that knowledge control our destiny and the destiny of those in our charge.

5.2.1 The Linear Perspective

The linear world view has simple rules and yields simple results. The input is proportional to the output; the response is proportional to the stimulus. If I double the pull on my bow string my arrow flies twice as far. Actions and events are comfortably predictable with only a little variation that is readily determined by the bell-shaped distribution with the result that experiment gives close to the right answer.

This view of things provides an understanding of complexity that relies on the reductionist perspective. In its strong form, reductionism states that to understand complex phenomena one only needs to understand microscopic laws governing all the underlying elements. This reasoning implies that once you understand all the parts of a process, you can "add them up" to understand the total. In other words, like the LFE, the world consists of networks of linear additive processes, but unlike LFE, these parts are inherently knowable. The linear world is predictable, like Newton's mechanical clock works.

The linear model is very useful in manufacturing, where the specification for the production of an object can be given to as high a tolerance as a machine can achieve. Of course no two production pieces are exactly the same and this variability must be taken into account. Suppose we are making rulers in which case the length is very important and the variability in length coming off the production line can be measured in terms of the standard deviation denoted by sigma (σ), see Fig. 5.1. The variation around the average length is given by the bell-shaped curve and the goal of manufacturing is to reduce sigma to as small a value as possible and thereby achieve standardization. Implicit in this manufacturing model is that variability is bad and must be suppressed. Additional discussion of this in the context of Six-Sigma and health care delivery is given in West [10].

The production line reinforces the linear additive world view and the natural desire to control one's life conspires with it to make suppression of variability a human desideratum. But does this strategy for attaining uniformity capture a deep truth about the world or does it merely gloss over all the difficulties? Nearly every student believes in the linear world view because it is forced on them in the first large class they take in college. They find that the professor does not determine absolute grades, but instead doles out relative grades on a curve, the infamous bell-shaped curve, with 68% getting C's, another 27% dividing B's and D's and the remaining 5% sharing A's and F's. But does this accurately reflect the mastery of the material? Does the LFE have anything to do with the complex process of learning?

Figure 5.3 records the data and analysis of the achievement test scores of more than 65,000 students graduating high school and taking the university entrance examination of *Universidade Estadual Paulista* (UNESP) in the state of Sao Paulo, Brazil [11]. Of most significance to us here is the fact that the humanities data in Fig. 5.3a seem to support the conjecture that the normal

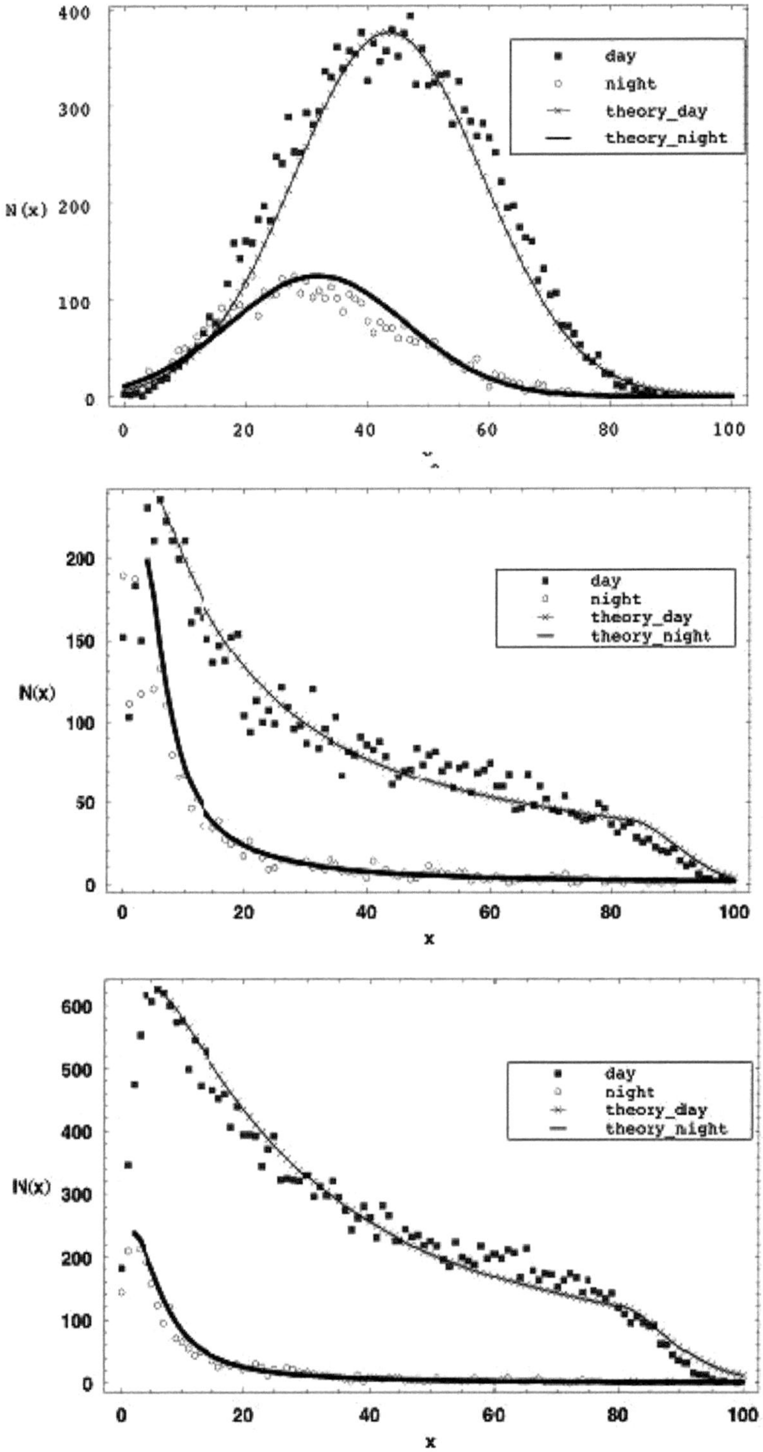

Fig. 5.3 The distribution of 65,000 student grades on the university entrance examination *Universidade Estadual Paulista* (UNESP) in state of Sao Paulo, Brazil [11]: (**a**) humanities; (**b**) physical sciences; (**c**) biological sciences

distribution is appropriate for describing the distribution of grades in a large population. The solid curves are the best fits of a normal distribution to the data in Fig. 5.3a. The data recorded in Fig. 5.3b, on the other hand, are from the physical sciences and graphed under the same groupings as that of the humanities in Fig. 5.3a. One thing that is clear is that the distribution is remarkably different from the normal. Figure 5.3c depicts the distribution under the same grouping for the biological sciences. The distribution for the biological sciences is more like those for the physical sciences than those for the humanities. In fact, the distribution of grades in the sciences is nothing like that in the humanities. The distributions in the sciences and humanities are so different that if the curves were unlabelled it would not be possible to associate them with the same general phenomenon, that is, Fig. 5.3b, c would not be identified with learning that evidently taking place in Fig. 5.3a. So can we explain why normalcy applies to the humanities but not to the sciences?

West [10] conjectures that the distribution differences are a consequence of the structural distinction between the two learning categories. The humanities collect a disjoint group of disciplines including language, philosophy, sociology, economics and a number of other relatively independent areas of study. Consequently, the grades obtained in each of these separate disciplines are essentially independent of one another, or at most weakly dependent on one another, thereby satisfying the conditions of the LFE or more generally the central limit theorem. In meeting the conditions articulated by Gauss, Adrian and Laplace the humanities take on normalcy.

Science, on the other hand, builds on previous knowledge. Elementary physics cannot be understood without algebra, and the more advanced physics cannot be understood without calculus, which also requires the understanding of algebra and arithmetic. Similarly, understanding biology requires mastery of some chemistry and some physics. The different scientific disciplines form an interconnecting web, starting from the most basic and building upward, thereby violating the independence assumption necessary to prove the central limit theorem. The empirical distribution for the sciences show extensions out into the tail region with no defined peak and consequently no characteristic value with which to characterize the data. The average values, so important in normal processes, become almost irrelevant in complex networks. A better indicator of complex processes than the average is one that quantifies how rapidly the tail decreases in value.

These data show that the normal distribution does not describe the normal situation. The bell-shaped curve is imposed through education orthodoxy and by our preconceptions and is not indicative of the process by which students master information and knowledge, at least not those in the sciences. Thus, the pursuit and achievement of intellectual goals in science is not normal. So how does this change our view of the world?

5.2.2 The Non-linear Perspective

In the linear additive world of Gauss, Adrian and Laplace, the average value dominates and that view was eventually accepted by the physical, social and life sciences. It implies that phenomena are predictable; a small change in the present state of a process produces a relatively small change in its future behaviour; the output is proportional to the input. The conclusion is that physical phenomena can be controlled in a straightforward way; the world is stable and the appearance of instability is just that, an appearance, not a reality. But is that the world of stock market crashes, staggering unemployment and the failure of medical protocols; the world in which we live?

Recall the Pareto or inverse power-law distribution of income in which a disproportionately small number of people have a disproportionately large fraction of the income. These are the individuals out in the tail of the distribution shown in Fig. 5.2. The imbalance in the distribution of income was identified by Pareto to be a fundamental inequality and he concluded that society was not fair. The Pareto world view is very different from that of Gauss, Adrian and Laplace as indicated by the two curves depicted in Fig. 5.2. It is also apparent from Fig. 5.3 that the same imbalance occurs in the sciences, where the truly gifted individuals have grades out in the tails of the distribution.

In the world of Pareto small changes can have large if not catastrophic effects; this world is not linear. The output of a network is not proportional to the input and the response is not proportional to the stimulus. In fact simple but non-linear rules lead to complicated outputs, so complicated that they are virtually unpredictable and we have the phenomenon of chaos. The circumstances attendant to the input must therefore be known to a high degree of accuracy to determine the output and very often even this information is insufficient for reliable predictions for any significant length of time.

5.2.3 The Historical Resolution

The historical strategy of scientists has been to restrict how they look at phenomena and therefore to predict over time intervals where their constrained view holds. For example a rubber band is elastic and springs back to its original shape if it is not stretched too far (the linear region). However, if pulled beyond the elastic limit it permanently deforms and eventually breaks (the non-linear region). But of course that means that the control humans have over their physical world, as they do over their machines, is only a consequence of operating in this linear region. But societies evolve, becoming more complex, less predictable and the levers of control become more non-linear. This insight into the physical and social sciences has also provided a new perspective for the life sciences. This is particularly true in medicine where the assumed linear response of homeostasis has given way to the non-linear nature of physiologic networks and our inability to control these networks with traditional devices has been increasingly acknowledged [12].

Homeostasis is part of the linear world view with its simple feedback mechanisms producing the relaxation of a disruption back to the relatively quiet equilibrium state. In the real world, health is a state of great variability not a quiescent state of equilibrium. This dynamic range is necessary in order for the human body to rapidly adapt to short-term changes in a complex environment. Variability is manifest in the time series associated with the dynamics of physiologic networks, such as the cardiovascular system, the respiratory system, the motor control system and so on. However, the importance of variability in biomedical phenomena was not pioneered in medicine, but was the result of physicists and engineers attempting to understand complexity through the analysis of data and the understanding of patterns. The medical community had long since decided that there was no useful information in the fluctuations of the data. This proclivity to ignore variability was brilliantly summarize by Stephan Jay Gould [13]:

> ...our culture encodes a strong bias either to neglect or ignore variation. We tend to focus instead on measures of central tendency, and as a result we make some terrible mistakes, often with considerable practical import.

In the real world, simple rules yield complex results; small changes may lead to divergences; predictability is limited and the inverse power law of Pareto describes natural phenomena.

So, what do we use when the linear additive view breaks down and the average is no longer the important quantity? It may not be obvious, but the slope of the Pareto distribution is the best indicator of the process. The slope of the Pareto distribution gives the Pareto index that provides a direct measure of the imbalance observed in the data. But why is the slope so important? The answer may surprise you.

A famous mathematician, Daniel Bernoulli (1700–1782), became interested in how people value what they have. He discounted such things as wealth reasoning that a single gold coin to a rich person cannot mean the same as it does to a day labourer. He introduced the notion of utility and argued that the utility of something such as a gold coin depends on how much gold one already possesses. The change in utility associated with a given quantity was therefore assumed to be given by the percentage change in that quantity.

For example, suppose you invest $1000 in the stock market and receive back $1100 and I invest $100 and receive back $200; who is happier? We both made $100 profit, but I doubled my money whereas you increased yours by 10%. Most people would agree that I am happier given that it has been determined that people respond to the percentage change and not to the absolute change. This is what the inverse power law captures; for example, the percentage change in the number of people with a given level of income is proportional to the percentage change in the level of income, with the proportionality constant being given by the power-law index. The greater the power-law index the stronger the response to changes in income levels.

A normal healthy person would have a Pareto index in a specified domain, but with age the slope changes, become steeper, that is, the power-law index becomes larger, and this change is diagnostically useful [12]. The steeper the slope the greater the loss of variability of the associated time series and therefore age and disease leads to more, not less, regularity. The non-linear view is that disease is the loss of variability and not, as is traditionally taught, the loss of regularity.

5.3 Universality of Tails

This is a handbook whose presumed purpose is to provide the reader with formulas and algorithms with which to analyse data, identify patterns, interpret those patterns, and draw conclusions regarding the phenomenon under investigation. However, you may feel that all we have done so far is alert you to the fact that the traditional methods of data analysis, based as they are on normal statistics, are not very useful for understanding complex phenomenon. If I have achieved this in some small way it should be seen as a victory. I have tried to accomplish what Leo Tolstoy (1828–1910) put so eloquently:

> I know that most men, including those at ease with the problems of the greatest complexity, can seldom accept even the simplest and most obvious truth if it be such as would oblige them to admit the falsity of conclusions which they have delighted in explaining to colleagues, which they have proudly taught to others, and which they have woven, thread by thread, into the fabric of their lives.

Recognizing the truth of Tolstoy's remarks with regard to the LFE would be a giant step towards the proper interpretation of complex data sets and subsequently changing one's world view from the linear to the non-linear.

5.3.1 Does your Data Scale?

A fundamental issue concerns measurement and how the notion of a characteristic scale determines what we think we know about the quantities we measure. Mandelbrot [14] brought to the scientific world's attention the fact that many natural objects simply do not satisfy this preconception: what is measured is not independent of how the measurement is made. Living things have structures in space and fluctuations in time that cannot be characterized by one spatial or temporal scale, but extend from the "here and now" to the "there and then". He introduced the notion of fractal geometry and fractal statistics to describe this lack of characteristic scale that was eventually incorporated into both physiological structure and dynamics.

If we magnify (use a smaller ruler) a fractal structure, new and ever finer details are revealed, whether the structure is in space or time, that is, whether we are looking at an X-ray or an electrocardiogram. Consequently, a cognizance of fractals and their mathematical elaborations provides the tools needed to describe, measure, model and understand many objects and processes in living things. However, there are straightforward ways to determine if the data set in hand is fractal or not without being master of these mathematical techniques and we concentrate on one such technique here.

The question to be answered is does your data scale? If the answer is yes, then the data may well be fractal and a number of interesting implications can be drawn.

The term scaling denotes a power-law relation between two variables X and Y

$$X = AY^{\alpha}. \quad (5.1)$$

Such scaling laws never appear by accident and they always reveal self-similarity, which is an important property of the phenomena being studied. In biology (5.1) is historically referred to as an allometric relation between two observables and was discovered experimentally at the beginning of the nineteenth century. Typically an allometric relation interrelates two properties of a given organism one of which is always the size.

Herein the allometric relation is extended to include measures of time series. In this extended view X is interpreted to be the variance and Y the average value of the quantity being measured. The fact that these two central measures of a time series satisfy an allometric relation implies that the underlying time series is a fractal random process. Consider a sequence of n independent measurements of a time series denoted by the data set $\{Z_1, Z_2, \ldots\}$. The mean $\overline{Z}$ and variance σ_z^2 are calculated in the usual way and give a single pair of points to characterize the data. Suppose we wanted to determine if the data points were correlated. One way would be to calculate the auto-correlation function, but we do not do that here. Instead, correlation within the data is determined by grouping the data into aggregates of two and more of the original data points and calculating the mean and variance at each level of aggregation.

We determine something remarkable using this aggregation approach, but before we introduce any equations let us process some data. We computer generate a million data point from a normal distribution having unit mean and unit variance. Using all the data points we obtain the first point in the lower left hand corner of Fig. 5.4 that records the standard deviation versus the average on log–log graph paper. Aggregating the nearest neighbour data points we obtain a data set half as large as the original and calculating the mean and variance for these data we obtain the second point recorded on the lower left in Fig. 5.4. The third point is obtained using the three point aggregate and so on. It is evident that the logarithm of the standard deviation (σ_z) increases linearly with the logarithm of the average, so we have our second equation

$$\log \sigma_z = \alpha \log \overline{Z} + \log A, \quad (5.2)$$

and comparing the two equations we obtain $X = \sigma_z$, $Y = \overline{Z}$, the power-law index α is the slope of the line segment that can be drawn through the points on the graph and $A = 1$ for these computer-generated data.

It is well established [15] that the exponent in a scaling equation such as (5.1) is related to the fractal dimension D of the underlying time series by

$$D = 2 - \alpha / 2. \quad (5.3)$$

A simple mono-fractal time series therefore satisfies the power-law relation of the allometric form given by (5.1). Another handy formula is the correlation index r that is related to the fractal dimension as follows:

$$r = 2^{3-2D} - 1. \quad (5.4)$$

The slope of the imaginary curve in Fig. 5.4 is $\alpha/2 = \text{slope} = 0.5$ so the fractal dimension given by (5.3) is $D = 1.5$, which is what one expects for an uncorrelated random process with normal statistics. The correlation coefficient given by (5.4) for $D = 1.5$ is $r = 0$ verifying the uncorrelated nature of the time series.

The allometric aggregation approach has been applied to a number of data sets implementing the method of linear regression analysis on the logarithm of the variance (or standard deviation) and the logarithm of the average value. Consequently, all the processed data from self-similar dynamical networks would appear as straight lines on log–log graph paper. For example, in Fig. 5.4, this process of aggregating the data is equivalent to decreasing the resolution of the time series and as the resolution is systematically decreased, the adopted measure, the relationship between the mean and standard deviation, reveals an underlying property of the time series that was present all along. The increase in the standard deviation with increasing average for increasing aggregation number shown in the figure is not an arbitrary pattern. The relationship indicates that the aggregated

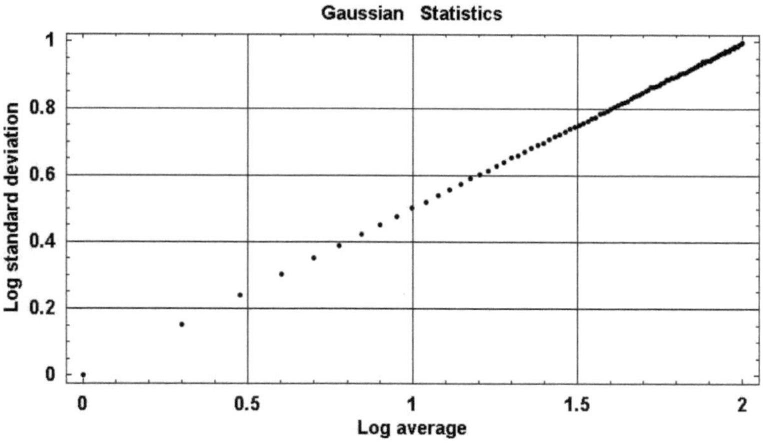

Fig. 5.4 The logarithm of the standard deviation is plotted versus the logarithm of the mean for the successive aggregation of 10^6 computer-generated random data points with normal statistics of mean one and variance one. The slope of the curve is essentially one-half so the fractal dimension of the time series given by (5.3) is 1.5

data points are interconnected. The original data points are by construction not correlated, but the addition of data in the aggregation process induces a correlation, one that is completely predictable. The induced correlation has $\alpha = 1$ if the original data are uncorrelated, but the induced correlation has $\alpha \neq 1$ if the original data are correlated.

In the same way a completely regular time series would have $\alpha = 2$, so that $D = 1$. The fractal dimension for most time series fall somewhere between the two extremes; the closer the fractal dimension is to one, the more regular the process; the closer the fractal dimension is to 1.5, the more it is like an uncorrelated random process. The data analysed in Fig. 5.4 certainly have a single fractal dimension characterizing the entire computer-generated time series. If the power-law index, the slope of the above curve, is ½ then the data are an uncorrelated random process. If the index were greater than ½ then the data cluster, indicating correlations in the random process that are not those induced by aggregating the data.

5.3.2 Physiological Data

Heart rate variability (HRV) provides a window through which we can observe the heart's ability to respond to normal disturbances that can affect its rhythm. HRV time series have become very well known over the past two decades as a quantitative indicator of autonomic activity. The medical community became interested in developing such an indicator of heart rate because observations indicate a relationship between lethal arrhythmias and such activity. The importance of HRV to medicine became widely apparent when a task force was formed by the *Board of the European Society of Cardiology and the North American Society of Pacing and Electrophysiology*, and was charged with the responsibility of developing the standards of measurement, physiological interpretation and clinical use of HRV. The task force published their findings in 1996.

There are a number of ways to analyse HRV, some 16 in all, each related to scaling in one way or another, but we do not want to go into them all here. Instead, we identify the scaling exponent for the HRV time series using the allometry aggregation approach as being among the most revealing of the nature of cardiac variability for the amount of effort involved in its determination. Consider the beat-to-beat intervals shown in Fig. 5.5, a typical HRV time series for a healthy young adult male. The data points in the figure are connected to aid in visualizing how the time intervals between heartbeats are changing. It is evident that the variation in the time intervals

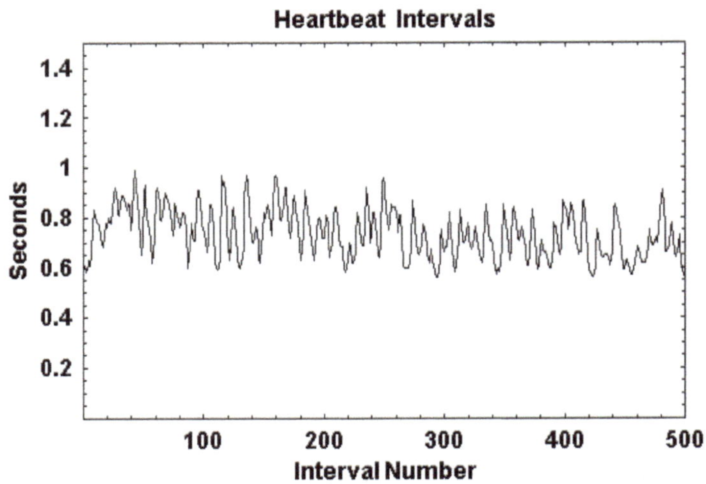

Fig. 5.5 The time series of heartbeat intervals of a healthy young adult male is shown. It is clear that the variation in the time interval between beats is relatively modest, but certainly not negligible (from West [12] with permission)

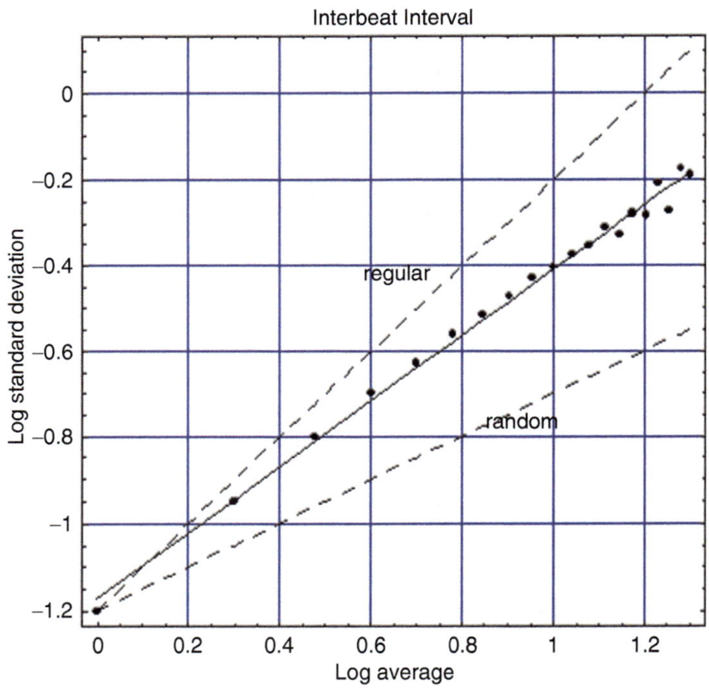

Fig. 5.6 A log–log plot of standard deviation versus the average value for HRV data from Fig. 5.5. The solid line is the best fit to the aggregated data points with fractal dimension $D = 1.24$ midway between the curve for a regular process and that for an uncorrelated random process as indicated by the *dashed curves* (from West [12] with permission)

between heartbeats is relatively small, the mean being 0.72 s and the standard deviation being 0.1 s. This relatively modest variance supports the frequently used medical term "normal sinus rhythm". Therefore, the question of what is learned by applying allometry aggregation approach to these data and plotting the standard deviation and average as a function of aggregation number becomes interesting.

In Fig. 5.6 the logarithm of the standard deviation is plotted versus the logarithm of the average value for the HRV time series depicted in Fig. 5.5. At the left-most position the data point indicates the standard deviation and average using all the data points. Moving from left to right the next data point is constructed from the time series with two nearest-neighbour data points added together and the procedure is repeated moving right until the right-most data point has 20 nearest-neighbour data points added together. The solid line is the best linear representation of the scaling and intercepts most of the data points with a positive slope of 0.76. We can see that the slope of the HRV data is midway between the dashed curves depicting an uncorrelated random process (slope = 1/2) and one that is deterministically regular (slope = 1).

We emphasize that the conclusions drawn here are not from this single figure or set of data presented; these are only representative of a much larger body of work. The conclusions are based on a large number of similar observations [16] made using a variety of data processing techniques, all of which yield results consistent with the scaling of the HRV time series indicated in Fig. 5.6. The heartbeat intervals do not form an uncorrelated random sequence; instead, the analysis suggests that the HRV time series is a statistical fractal, indicating that heartbeats have a long-time memory as indicated by their scaling.

Phenomena obeying scaling relations, such as those shown for the HRV time series data in Fig. 5.6 are said to be self-similar. The fact that the standard deviation and average values change as a function of aggregation number implies that the magnitudes of these quantities depend on the size of the ruler used to measure the time interval. Recall that this is one of the defining characteristics of a fractal curve; the length of a fractal curve becomes infinite as the size of the ruler used to measure it goes to zero. The dependence of the average and standard deviation of the ruler size, for a given time series, implies that the statistical process is fractal and consequently defines a fractal dimension for the HRV time series. These results are consistent with those first obtained by Peng et al. [17] for a group of ten healthy subjects having a mean age of 44 years, using 10,000 data points for each subject. They concluded that the scaling behaviour observed in HRV time series is adaptive for two reasons: firstly that the long-time correlations constitutes an organizing principle for highly complex, non-linear processes that generate fluctuations over a wide range of time scales; secondly, the lack of a characteristic scale helps prevent excessive mode-locking that would restrict the functional responsiveness of the organism.

The global scaling exponent determines the properties of mono-fractals, but there exists a more general class of heterogeneous signals known as multi-fractals, which are made up of many interwoven subsets with different local exponents. The local and global exponents are only equal for infinitely long time series; in general the scaling exponent and the fractal dimension are independent quantities. The statistical properties of the interwoven subsets may be characterized by the distribution of fractal dimensions. In general, time series have a local fractal exponent that varies over its course. The multi-fractal spectrum, describes how the local fractal exponents contribute to such time series. A number of investigators have used this spectrum to demonstrate that HRV time series are multi-fractal [18].

The multi-fractal character of HRV time series further emphasizes the non-homeostatic physiologic variability of heartbeats. Longer time series than the one presented here clearly show patchiness associated with the fluctuations; a patchiness that is usually ignored in favour of average values in traditional data analysis. This clustering of the fluctuations in time can be symptomatic of the scaling with aggregation observed in Fig. 5.6 or if particularly severe it can be indicative of multi-fractality. However, due to limitations of space, we do not pursue the multi-fractal properties of time series here, but refer the interested reader to the literature [15].

The second physiologic exemplar is the dynamics of breathing; the apparently regular breathing as you sit quietly reading this chapter. To understand the dynamics we first acknowledge that it is not by accident that the cascading branches of the bronchial tree become smaller and smaller, nor is it good fortune alone that ties the dynamics of our every breath to this biological structure. We argue that, like the heart, the lung is made up of fractal processes, some dynamic and others now static. However, both kinds of processes lack a characteristic scale and a simple argument establishes that such a lack of scale has evolutionary advantages [19].

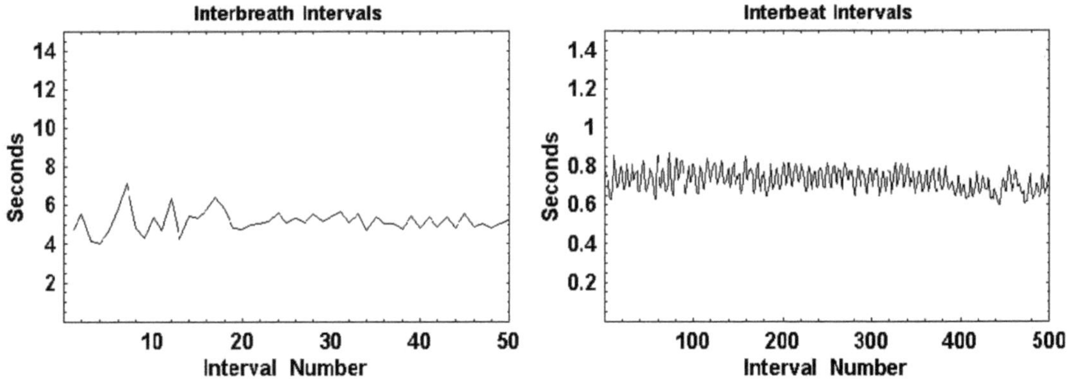

Fig. 5.7 Typical time series from one of the 18 subjects in the study conducted by West et al. [20], while at rest, is shown for the BRV and HRV time series. Not all the data are shown, just enough to indicate the relative quality of the two time series

As with the heart, the variability of breathing rate using breath-to-breath time intervals is denoted by breathing rate variability (BRV), to maintain a consistent nomenclature. Examples of HRV and BRV time series data on which scaling calculations are based are shown in Fig. 5.7. Because heart rate is higher than respiration rate, in the same measurement epoch there is a factor of five more data for HRV than there is for BRV time series. We have adjusted the scales in the two graphs to highlight their differences. The data displayed in Fig. 5.7 were collected under the supervision of Dr. Richard Moon, the Director of the Hyperbaric Laboratory at Duke Medical Center. Looking at these two time series together one is struck by how different they are. It is not apparent that both physiologic phenomena scale in essentially the same way, but they do [20].

West et al. [20] applied the allometry aggregation approach to the various HRV and BRV time series and obtained the typical results depicted in Fig. 5.8. The logarithms of the aggregated standard deviation and aggregated average were determined in the manner described above. Note that we stop the aggregation at ten points because of the small number of data in the breathing sequence. The solid curve at the right in Fig. 5.8 is the best least-square fit to the aggregated BRV data and has a slope of 0.86; the scaling index. A similar graph is constructed for the HRV data in the left curve, where we obtain a slope of 0.80 for the scaling index. The latter slope is so different from that obtained in Fig. 5.5 because of the age difference in the two subjects.

The self-similar nature of breathing time series deduced from the scaling in Fig. 5.8 has been used to produce a revolutionary way of utilizing mechanical ventilators. Historically ventilators facilitate post-operative breathing and have a built-in frequency of ventilation. Mutch et al. [21] challenged the single-frequency ventilator design by using an inverse power-law spectrum of respiratory rate to drive a variable ventilator. They demonstrated that this way of supporting breathing produces an increase in arterial oxygenation over that produced by conventional control-mode ventilators. This comparison indicates that the fractal variability in breathing is not the result of happenstance, but is an important property of respiration. A reduction in variability of breathing reduces the overall efficiency of the respiratory system.

Altmeier et al. [22] measured the fractal characteristics of ventilation and determined that not only are local ventilation and perfusion highly correlated, but they scale as well. Finally, Peng et al. [23] analysed the BRV time series for 40 healthy adults and found that under supine, resting, and spontaneous breathing conditions, the time series scale. This result implies that human BRV time series, like HRV time series, have long-time correlations across multiple time scales and therefore breathing is a fractal statistical process.

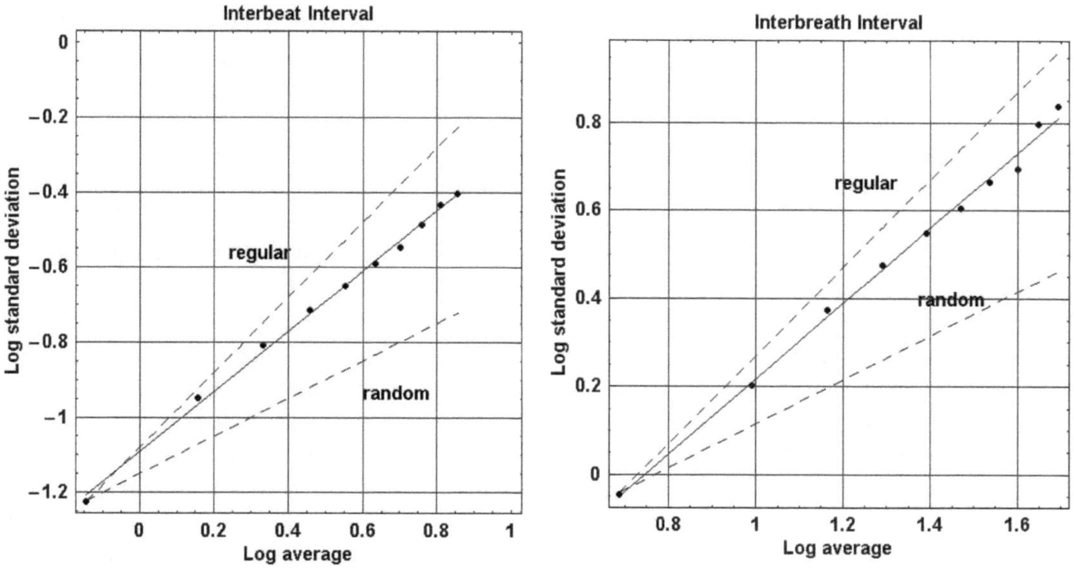

Fig. 5.8 A typical fit to the aggregated standard deviation versus the aggregated mean for BRV and HRV time series [20] and depicted in Fig. 5.7. The points are calculated from the data using (5.2) and the solid curve is the best least-square fit to the data. The left curve is the fit to the HRV data (slope = 0.80), the right curve is the best fit to the BRV data (slope = 0.86)

Of course there is a variety of other physiological phenomena whose time series have been discovered to be fractal and multi-fractal. For example, the regular gait cycle, so apparent in everyday experience, is not as regular as we once believed it to be. Gait is no more regular than the normal sinus rhythm or the breathing cycle just discussed. The stride rate variability (SRV), although small, is non-negligible and indicates an underlying complex structure that cannot be treated as uncorrelated random noise. West [12] reviews SRV and other physiological time series and discusses the nature of their scaling.

The interbeat intervals of the human heart, the interstride intervals of human gait and the interbreath intervals of human breathing are apparently random, but they in fact have long memories. This combination of randomness and order has been used as the defining characteristic of complexity. We used the allometry aggregation approach to establish that such dynamic physiologic phenomena generate time series that are statistical fractals. The scaling behaviour of such time series determines the overall properties such complex networks must have and they are unlike the older analysis of errors and noise in physical systems.

5.4 Some Closing Observations

The historical view of complexity involved having a large number of variables, each variable making its individual contribution to the operation of the network and each variable responding in direct proportion to the changes in the other network variables. Small differences in the input could be washed out in the fluctuations of the observed output. The linear additive statistics of measurement error or biological noise is not applicable to complex medical phenomena. The elements in complex physiologic networks are too tightly coupled, so instead of a linear additive process, non-linear multiplicative statistics more accurately represent the fluctuation, where what happens at the smallest scale can and often is coupled to what happens at the largest scale. This coupling is manifest through the scaling index.

West and Griffin [24] in their book on biomechanics emphasized that the signal-plus-noise paradigm used by engineers for signal processing ought to be replaced in biomechanics by the paradigm of the high wire walker. In the engineering paradigm, the slow regular behaviour of the time series is interpreted as signal and the fast erratic variations are treated as noise. But this historical separation into signal and noise is not applicable to the time series from complex phenomena. In the circus, high above the crowd and without a net, the tightrope walker carries out smooth average motions plus rapid erratic changes of position, just as in the signal-plus-noise paradigm. However, the rapid changes in position are part of walker's dynamical balance; so that far from being noise as they are for the engineer, these apparently erratic changes in position serve the same purpose as the slow graceful movements; they maintain the walker's balance. Thus, both aspects of the time series recording the wirewalker's position constitute the signal and both contain information about the dynamics. Consequently, if we are to understand how the wirewalker retains balance on the wire, we must analyse the fluctuations, that is, the wirewalker's fine-tuning to losses of balance as well as the slow movements. This picture of the wirewalker more accurately captures the view of physiological signals developed herein.

The individual mechanisms giving rise to the observed statistical properties in physiological networks are very different, so we did not even attempt to present a common source to explain the observed scaling in walking, breathing and the beating heart. On the other hand, the physiological time series in each of these phenomena scale in the same way, so that at a certain level of abstraction the separate mechanisms cease to be important and only the relations matter and not those things being related. It is the relation between blood flow and heart function, between locomotion and postural balance, between breathing and respiration, which is important. The thesis of complexity theory is that such relations often have a common form for complex phenomena. This assumption is not so dramatic as it might at first appear. Consider that traditionally such relations were assumed to be linear, in which case their control was assumed to be in direct proportion to the disturbance. Linear control theory has been the backbone of homeostasis, but it is not sufficient to describe the full range of HRV, SRV and BRV. Traditional linear control theory cannot explain how the statistics of time series become fractal, or how the fractal dimension changes over time.

Scaling suggests that the historical notion of disease, which has the loss of regularity at its core, is inadequate for the treatment of dynamical diseases. Instead of loss of regularity, we identify the loss of variability with disease [25], so that a disease not only changes an average measure, such as heart rate, which it does in late stages, but is manifest in changes in variability at very early stages, such as in HRV. Loss of variability implies a loss of physiologic control and this loss of control is reflected in the change of fractal dimension, that is, in the scaling index of the corresponding time series [25, 26].

We have concluded that the world is not a linear place, the statistics are not normal and so we are left to speculate or at least think about what that implies about daily living. For one thing it means that the world contains unexpected strange events, and these events often control our lives. The daily routine is disrupted by the plunging stock market, changes in health insurance policies, illness, and death. The inverse power laws cast shadows on how we believe the world ought to be. So let us examine a number of examples to peer into the shadows.

Pareto determined that most workers earn <u>less</u> money than average as described by his inverse power law in the level of income. Of course the income imbalance is manifestly unfair, and guilds, unions and other social organizations have worked over the years to reduce this inequity. The income imbalance in stable societies poses the question of whether the social concept of fairness is necessary or even desirable if the trade off is social instability; historically revolutions have been fought to redress such wrongs. So how much imbalance is a tolerable cost of stability?

Another example of Pareto's imbalance is in the number of scientists that have a given number of scientific publications, which is also an inverse power-law distribution. Consequently, most scientists publish <u>fewer</u> papers than average and their papers are cited by other scientists <u>fewer</u> times than average. The distribution in the number of citations is also of the Pareto form so that a scientist being cited an average number of times is in the highest 4 percentile of the distribution; recall the distribution of grades in science discussed earlier. One can therefore observe that an average scientist is truly outstanding even among other scientists. It appears that in this social context the imbalance is not only acceptable but welcomed.

These examples might seem esoteric, except perhaps to the academic, so let us look at some distributions more closely related to medicine and health care. The length of stay for patients in emergency wards (EW) is also determined to be inverse power law [27]. Consequently, most EW patients stay in hospitals less time than average.

The well being of the body's network of networks is measured by the fractal scaling properties of the various dynamic networks and such scaling determines how well the overall harmony is maintained. Once the perspective that disease is the loss of complexity has been adopted, the strategies presently used in combating disease must be critically examined. Life support equipment is one such strategy, but the tradition of such life support is to supply blood at the average rate of the beating heart, to ventilate the lungs at their average rate and so on. So how does the new perspective regarding disease influence the traditional approach to healing the body?

Alan Mutch applied the lessons of fractal physiology to point out that blood flow and ventilation are delivered in a fractal manner in both space and time in a healthy body. However, he argues, during critical illness, conventional life support devices deliver respiratory gases by mechanical ventilation or blood by cardiopulmonary bypass pump in a monotonously period fashion. This periodic driving overrides the natural á periodic operation of the body. Mutch speculates that these devices result in the loss of normal fractal transmission and consequently life support winds up doing more damage the longer they are required and become more problematic the sicker the patient [26]. From this perspective the loss of complexity is the loss of the body as a cohesive whole; the body is reduced to a disconnected set of organ systems.

One of the traditional views of disease is what Tim Buchman calls the "fix-the-number" imperative [28]. He argues that if the bicarbonate level is low then give bicarbonate; if the urine output is low then administer a diuretic; if the bleeding patient has a sinking blood pressure then make the blood pressure normal. He goes on to say, that such interventions are commonly ineffective and even harmful. For example, sepsis—which is a common predecessor of multiple organ dysfunction syndrome (MODS)—is often accompanied by hypocalcaemia; where in controlled experimental conditions, administering calcium to normalize the laboratory value increases mortality. Consequently, one's first choice of options, based on an assumed simple linear causal relationship between input and output as in homeostasis, is probably wrong.

A number of scientists [29] have demonstrated that the stability of hierarchal biological networks is a consequence of the interactions among the elements of the network. Furthermore, there is an increase in stability resulting from the nesting of networks within networks—organelles into cells, cells into tissue, tissues into organs and so on up from the microscopic to the macroscopic. Each network level confers additional stability on the overall fractal structure. The fractal nature of the network suggests a basic variability in the way networks are coupled together. For example, the interaction between cardiac and respiratory cycles is not constant, but adapts to the physiologic challenges being experienced by the body.

The empirical evidence overwhelmingly supports the interpretation of the time series analysis that fractal stochastic processes describe complex physiologic phenomena. Furthermore, the fractal nature of these time series is not constant in time but change with the vagaries of the interaction of the network with its environment and therefore these phenomena are often weakly multi-fractal. The scaling index or fractal dimension marks a physiologic network's response and can be used as an indicator of the network's state of health. Since the fractal dimension is also a measure of the level of complexity, the change in fractal dimension with disease suggests a new definition of disease as a loss of complexity, rather than the loss of regularity [12].

References

1. Gauss CF. Theoria motus corporum coelestrium [Theory of motion of heavenly bodies moving about the sun in conic sections, eng. trans]. New York/ Hamburg: Dover; 1809.
2. Adrian R. Research concerning the probabilities of the errors which happen in making observations, etc. The Analyst; or Mathematical Museum. 1809;1:93–109.
3. Laplace PS. Analytic theory of probabilities. Paris: Imprimerie Royale; 1810.
4. Poincare H. The foundations of science. New York: The Science Press; 1913.
5. Norwich KW. Le Chatelier's principle in sensation and perception: fractal-like enfolding at different scales. Front Physiol. 2010;1:17.
6. West BJ. The wisdom of the body; a contemporary view. Front Physiol. 2010;1:1.
7. Galton F. The geometric mean in vital and social statistics. Proc R Soc Lond. 1879;29:365.
8. West BJ. The lure of modern science; fractal thinking. Singapore: World Scientific; 1995.
9. Pareto V. Cours d'economie politique. Lausanne: Rouge; 1897.
10. West BJ. Why Six Sigma Health Care is oxymoronic in hospitals. In: Davidson A, editor. Nursing, caring and complexity science. Carlisle: Springer; 2011.
11. Gupta HM, Campanha JR, Chavorette FR. Power-law distribution in high school education: effect of economical, teaching and study conditions. arXir.0301523v1 (2003).
12. West BJ. Where medicine went wrong; rediscovering the path to complexity. Singapore: World Scientific; 2006.
13. Gould SJ. Full house. New York: Harmony Books; 1996.
14. Mandelbrot BB. Fractals, form, chance and dimension. San Francisco: W.H. Freeman and Co.; 1977.
15. Bassingthwaighte JB, Liebovitch LS, West BJ. Fractal physiology. New York: Oxford University Press; 1994.
16. Suki B, Alencar AM, Frey U, Ivanov PC, Buldyrev SV, Majumdar A, Stanley HE, Dawson CA, Krenz GS, Mishima M. Fluctuations, noise and scaling in the cardiopulmonary system. Fluct Noise Lett. 2003;3: R1–25.
17. Peng CK, Mistus J, Hausdorff JM, Havlin S, Stanley HE, Goldberger AL. Long-range anti-correlations and non-Gaussian behavior of the heartbeat. Phys Rev Lett. 1993;70:1343–6.
18. Ivanov PC, Rosenblum MG, Peng C-K, Mietus J, Havlin S, Stanley HE, Goldberger AL. Scaling behavior of heartbeat intervals obtained by wavelet-based time-series analysis. Nature. 1996;383:323–7.
19. West BJ. Physiology in fractal dimension: error tolerance. Ann Biomed Eng. 1990;18:135–49.
20. West BJ, Griffin LA, Frederick HJ, Moon RE. The independently fractal nature of respiration and heart rate during exercise under normobaric and hyperbaric conditions. Respir Physiol Neurobiol. 2005;145: 219–33.
21. Mutch WAC, Harm SH, Lefevre GH, Graham MR, Girling LG, Kowalski SE. Biologically variable ventilation increases arterial oxygenation over that seen with positive end-expiration pressure alone in a porcine model of acute respiratory distress syndrome. Crit Care Med. 2000;28:2457–64.
22. Altemeier WA, McKinney S, Glenny RW. Fractal nature of regional ventilation distribution. J Appl Physiol. 2000;88:1551–7.
23. Peng C-K, Metus J, Li Y, Lee C, Hausdorff JM, Stanley HE, Goldberger AL, Lipsitz LA. Quantifying fractal dynamics of human respiration: age and gender effects. Ann Biomed Eng. 2002;30:683–92.
24. West BJ, Griffin L. Biodynamics: why the wirewalker doesn't fall. New York: Wiley; 2004.
25. Goldberger AL, Rigney DR, West BJ. Chaos, fractals and physiology. Sci Am. 1990;262:42–9.
26. Mutch WAC, Lefevre GR. Health, 'small-worlds', fractals and complex networks: an emerging field. Med Sci Monit. 2003;9:MT19–23.
27. Philippe P, Garcia MR, West BJ. Evidence of "essential uncertainty" in emergency ward length of stay. Fractals. 2004;12(2):197–209.
28. Buchman TG. Physiologic failure: multiple organ dysfunction syndrome. In: Delsboeck TS, Kresh JY, editors. Complex systems science in biomedicine. Berlin: Springer; 2006.
29. Chauvet GA. Hierarchical functional organization of formal biological systems: a dynamical approach. I. The increase of complexity by self-association increases the domain of stability of a biological system. Phil Trans R Soc Lond B Biol Sci. 1993;339:425–44.

Modelling

6

Joachim P. Sturmberg, Leonid Churilov, and Geoff McDonnell

A picture is a model of reality.

Ludwig Wittgenstein

6.1 Introduction

Tacitly we all use models all the time to help us understand and operate in the world around us. Modelling is a *formal* approach to understanding the *real world* through a *simplified external and explicit representation of a mental model* which can be manipulated and tested, before being implemented back into the real world. Mikulecky described the underlying *mental processes* as summarised in Fig. 6.1 [1].

The form and details of a model depend on its purpose.[1] Pidd [2] defines a model as *an external and explicit representation of part of reality as seen by the people who wish to use that model to understand, to change, to manage and to control that part of reality* (p. 10). They are the product of human thought and ingenuity and can consist of a simple diagram or map or a complex mathematical formulation.

Models are an important part of modern science and have helped to understand and investigate important aspects of scientific and social phenomena. Examples include the billiard ball model of a gas, the Bohr model of the atom, the MIT bag model of the nucleon, the Gaussian-chain model of a polymer, the Lorenz model of the atmosphere, the Lotka–Volterra model of predator–prey interaction, the double helix model of DNA, agent-based and evolutionary models in the social sciences, or general equilibrium models of markets [3, 4].

In this chapter we firstly demonstrate that in practice we all "model" in our daily work, modelling is a natural way of thinking and acting. We then provide an outline of the principle of modelling, before describing different modelling techniques and examples of their application, covering clustering analysis, discrete event simulation, and system dynamic modelling. These examples cover clinical issues as well as hospital and broader health policy concerns.

J.P. Sturmberg (✉)
Department of General Practice, Monash University, Melbourne, VIC Australia

The Newcastle University, Newcastle, PO Box 3010, Wamberal, NSW 2260, Australia
e-mail: jp.sturmberg@gmail.com

L. Churilov
National Stroke Research, Institute—Florey Neuroscience Institutes, Melbourne Brain Centre, Austin Health, 245 Burgundy Street, Heidelberg, VIC 3084, Australia

Department of Mathematics and Statistics, The University of Melbourne, Melbourne Brain Centre, Austin Health, 245 Burgundy Street, Heidelberg, VIC 3084, Australia

Department of Medicine, The University of Melbourne, Melbourne Brain Centre, Austin Health, 245 Burgundy Street, Heidelberg, VIC 3084, Australia
e-mail: leonidc@unimelb.edu.au

G. McDonnell
Centre for Health Informatics, University of New South Wales, PO Box 308, Waverley, NSW 2024, Australia

[1] Some modelling methods are explained in greater detail at: http://www.systemswiki.org/index.php?title=Simulation_Methods.

J.P. Sturmberg and C.M. Martin (eds.), *Handbook of Systems and Complexity in Health*, DOI 10.1007/978-1-4614-4998-0_6, © Springer Science+Business Media New York 2013

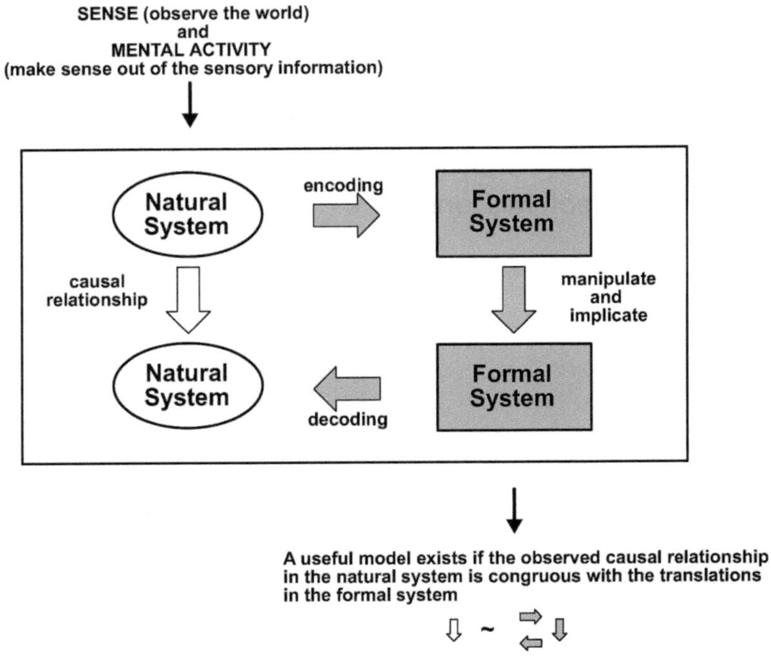

Fig. 6.1 "Real world" and "Mental world"

6.2 Concepts of Modelling

Understanding problems and finding solutions "that work" will require an appreciation of the system's agents and context. Representing problems and their solutions can involve different means, like storytelling, the use of metaphors, mathematical formulas or computational models, as illustrated in Fig. 6.2.

Donna Meadows [5] synthesised the key features of thinking in systems in the following way:

- A system is a set of elements or parts that is coherently organised and interconnected in a pattern or structure that produces a characteristic set of behaviours, often classified as its function or purpose,

and its underpinning principles include:

- A system is more than the sum of its parts
- Many of the interconnections in systems operate through the flow of information
- The least obvious part of the system, its function or purpose is often the most crucial determinant of the system
- System structure is the source of system behaviour. System behaviour reveals itself as a series of events over time. This pattern of events over time is system behaviour

We model problems before deciding on a course of action so we can avoid making big mistakes and having confidence that our actions are more likely to be effective. Guiding principles for modelling are:

- Don't solve the wrong problem
- Don't apply the wrong solution
- Don't cause worse problems
- Avoid unintended consequences
- Provide a safe place for experiments and discussion

At its most basic, modelling involves the plotting of a system diagram, influence diagram, multiple cause diagram, and a sign graph diagram, the latter identifying feedback loops within the system (Fig. 6.3). Each of the agents in a multiple cause/sign graph diagram can be given values that reflect their characteristics (stocks and flows) and behaviours (feedback loops), and running such a computational model repeatedly with different assumptions will elaborate the potentially best solution to the modelled problem (explored in detail later in this chapter).

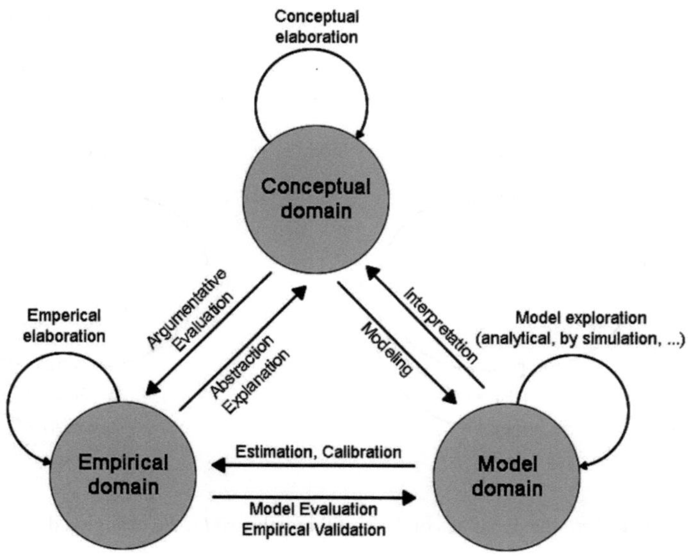

Fig. 6.2 We can **share representations of our mental models** by telling stories, drawing pictures or maps or making scale 3D models. We can represent how things change over time by using metaphors and successive snapshots or storyboards, or use mathematical or computational models to describe behaviour over time

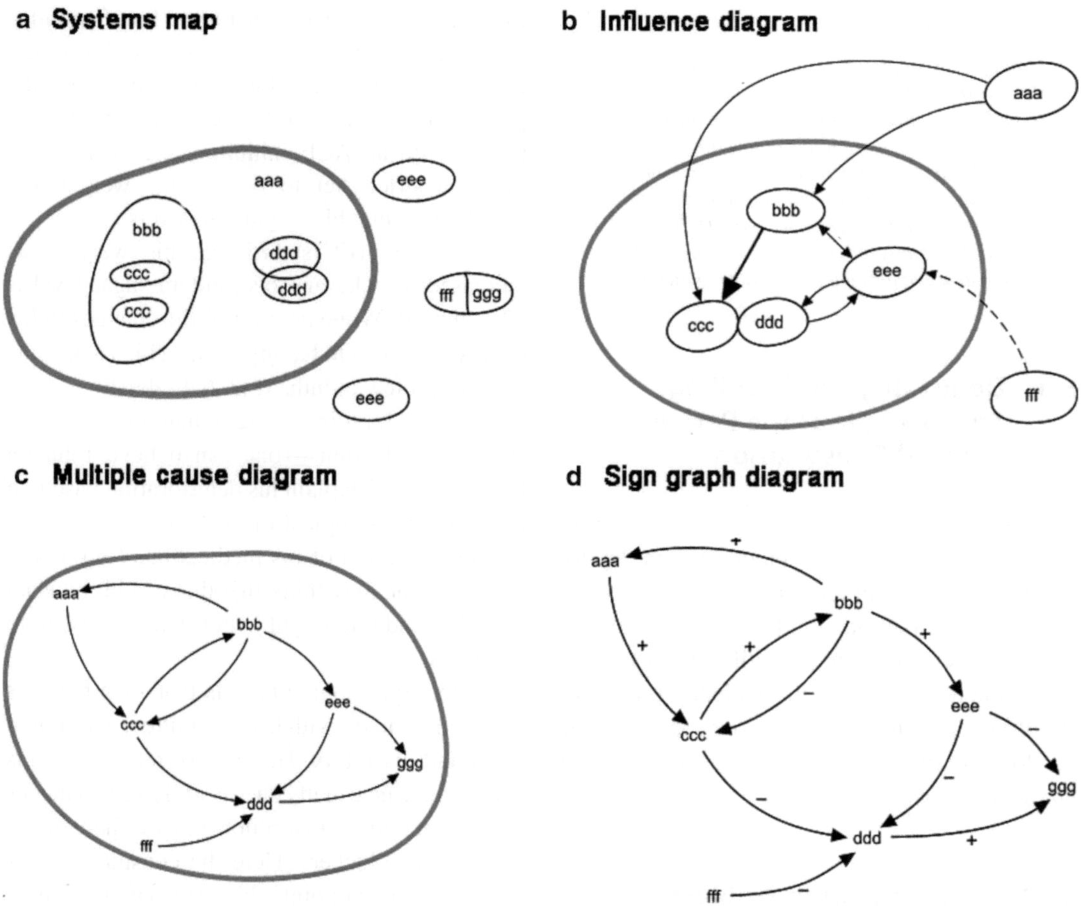

Fig. 6.3 Mapping system dynamics: a system map (**a**) provides an overview of the system and its components; the influence diagram (**b**) conceptualises the main structural features and their relationships; the multiple cause diagram (**c**) analysis main relational causes within the system; and the sign graph diagram (**d**) provides the direction of influence amongst variables, "+" indicates an influence in the same direction, "–" an influence in the opposite direction

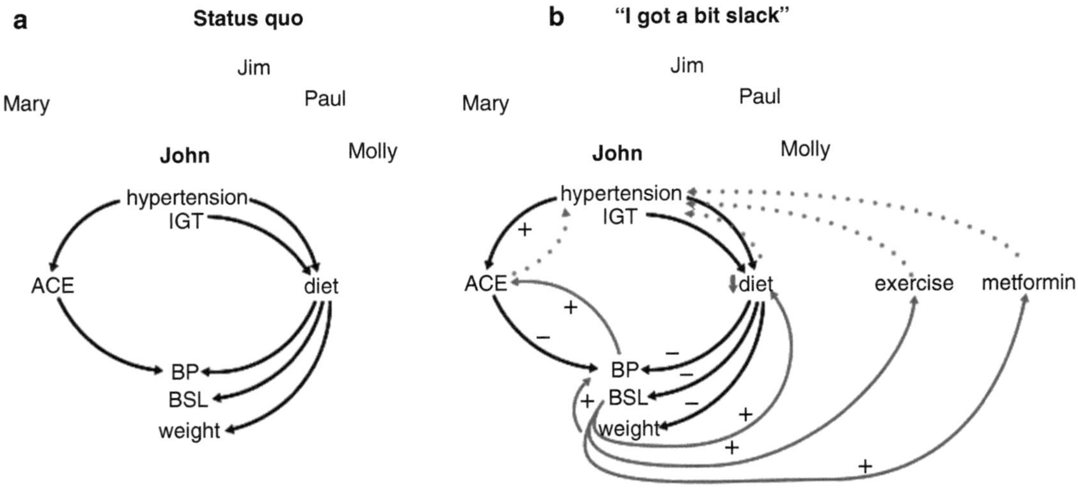

Fig. 6.4 *Left*: the dynamics of the past 9 months, *right*: the changing dynamics at the time of this consultation

In Jay Forrester's words:

> *Through an appropriate simulation model, one should know the structure causing the problem, should know how the problem is created, should have discovered a high-leverage policy that will alter behaviour, should understand the reasons why the low-leverage policies will fail, should be able to explain how strongly defended policies within the system are actually the cause of troubles, and should be able to argue for better alternative policies.*

6.3 De-mystifying Modelling: The Example of the Patient-Centred Consultation

This clinical case study illustrates how we "intuitively" model the clinical interactions in day-to-day practice. It explicitly shows how based on available information, limited time, beliefs, or intuition, we can reach quite different but legitimate conclusions about the patient's problems. Resulting outcomes depend on our sophistication in documenting "*a system* and *interpret its interconnections*".

6.3.1 The Presenting Problem

John is a 63-year-old married man who has three children—Jim, Paul and Molly, who no longer live at home. John has been retired for 18 months. He presents for his regular check-up of his hypertension and impaired glucose tolerance. For the past 9 months his blood pressure had been controlled with an ACE inhibitor, and he had followed a strict diet to control his weight and maintain normal blood glucose levels.

Today his BP is 175/105, his weight has increased by 5 kg and his random sugar level is 13.5 mmol/l. When hearing of the changes in his results he admits to having been "a bit slack" during the last few months (Fig. 6.4—left).

Drawing John in a system diagram highlights some critical points—one can believe John on face value, and explain his deterioration based on the pathophysiological mechanisms and manage him by alteration of his medications (Fig. 6.4—right), ***or*** one may think that this is a bit unusual for John, and one might better make a few more enquiries.

Further questioning reveals that John believes it must have to do with his intermittent abdominal pains and his reflux. He had not mentioned this before since he usually successfully self-manages these symptoms with a couple of over-the-counter H2-receptor blockers. Here the consultation has reached a critical point—his new symptoms may indicate a new disease which could be further investigated (Fig. 6.5—left), or it may represent just another symptom of his "true" illness.

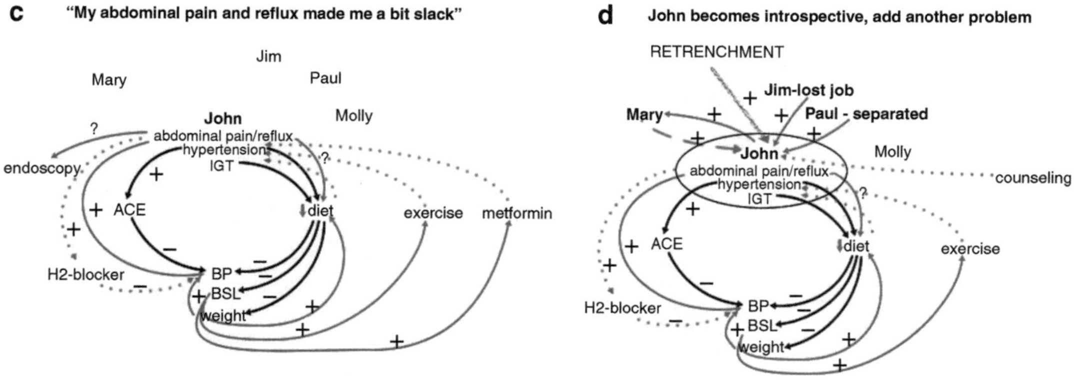

Fig. 6.5 *Left*: further symptoms complicating the clinical picture; *right*: a new insight changes the dynamics

Closer enquiries about the onset of his abdominal pains and his reflux allow John to become more reflective. He states that he is bored with his life since having been forced to retire early, he is having regular fights with his wife, and he is really worried about his two sons. John recently got retrenched from an executive position in a multi-national cooperation, and Paul has separated from his partner and children. John states that he has started smoking again and is having six standard drinks of alcohol most nights (Fig. 6.5—right).

These additional features lead to another critical point in the consultation – is the stress a separate issue, or do all of John's different complaints and his deterioration fit together?

6.3.2 Clinical Interpretation

The different perspectives of John's illness can be classified in the biomedical tradition as the biomedical mechanisms of disease, the social and mental determinants of health and illness, and the patient's construction of meaning of the health/illness experience [6]. John is a patient with multiple threats to his illness experience—peptic ulcer disease, cardiovascular disease, impaired glucose tolerance, marital problems, adjustment disorder, unhealthy lifestyle habits, and worries about his children. John's illness narrative has multiple interconnected (i.e. complex) strands.

6.3.3 System Dynamic Interpretation

As the system analysis confirms, *stress*[2] is the common focal point of all of John's problems—*retirement*, *marital problems* and *worries* relating to his son's life—even though his main complaint is abdominal pain. Increasing *stress* will increase his *marital problems*,[3] which in turn will further increase his *stress*, and vice versa decrease in *stress* will decrease his *marital problems*[4] which will decrease his *stress*. Following other relationships indicate that increased *stress* will lead to increased *alcohol consumption*, which on the one hand will increase his *carbohydrate intake* and increase his *IGT* and this in turn will increase his *stress*, and on the other it will decrease *mucosal protection*[5] which in turn will increase his *ulcer/reflux* symptoms and increase his *stress*. Following the relationship to *smoking* highlights the synergistic effects on his *ulcer/reflux* symptoms, and following the endocrine stress response shows the synergistic effects on his cardiovascular, endocrine and gastric symptoms (Fig. 6.6).

Modelling has helped to understand and communicate all of the various relationships between

---

[2] All variable names appear in *italic*.

[3] "+" sign next to the arrow indicates that the change in the variable at the tail of the arrow results in a change of the variable at the head of the arrow in the same direction.

[4] Again "+" sign as the change occurs in the same direction.

[5] Here we have a "–" sign as the change will result in a change in the opposite direction.

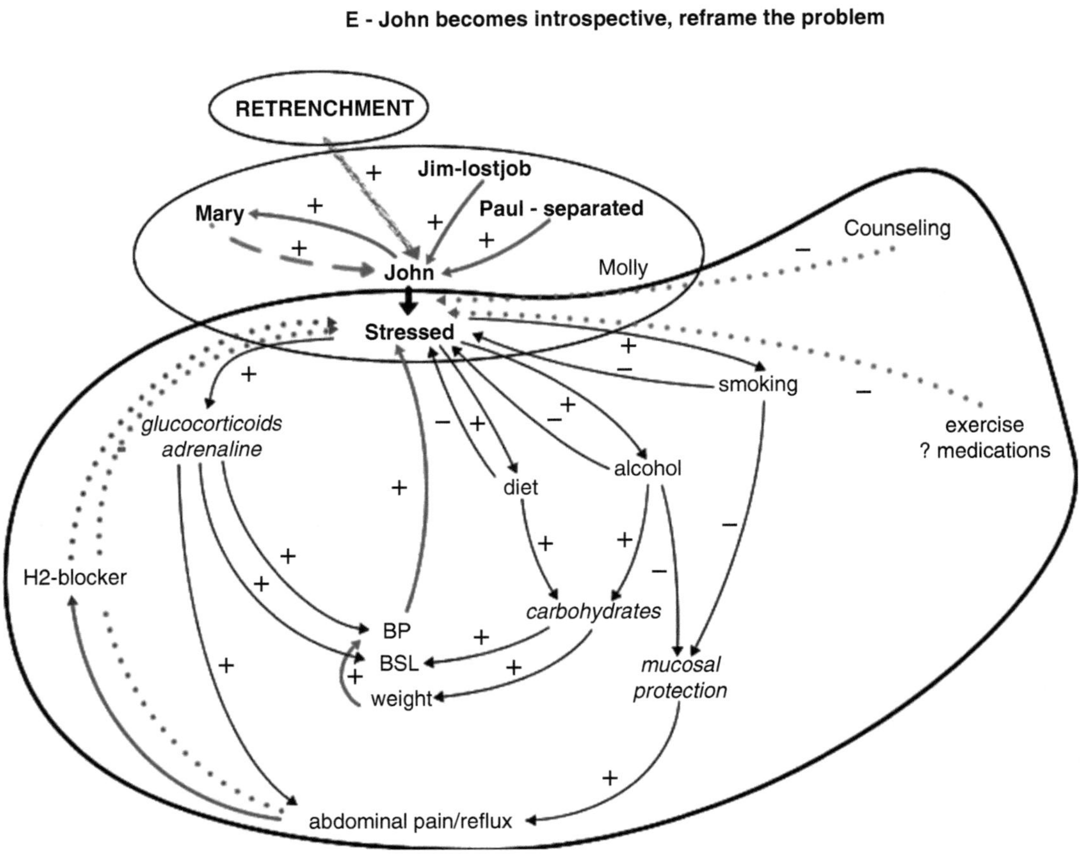

Fig. 6.6 The dynamics within the different system domains of John's illness

John's symptoms. It has allowed us to consider various possible options of acting, and in the end has allowed us to identify the correct problem and avoided unintended consequences that might have made his illness worse. All of this may seem obvious; nevertheless, it illustrates the pragmatic application of systems and complexity thinking to clinical practice.

6.4 Introduction to Dynamic Modelling

So far we have explored the structural dimensions and relationships of models; in this section we explore the dynamic interplay between the behaviours of a system's agents. *System dynamics modelling* was developed at MIT by Jay Wright Forrester in the 1950s. It arose from the insight that problem behaviours over time are produced by systemic structures.

6.4.1 Dynamic Complexity Produces Unintended Consequences

There is a persistent preoccupation with cost, quality, and access in health care, despite many practice and policy changes over the years. Sterman, in a NIH Videocast[6] has described the qualities of these persistent complex health problems which resist policy solutions as:

[6] **Videocast/Podcast** at http://videocast.nih.gov/Summary.asp?file=13712.

- Dynamic
- Tightly coupled (connected)
- Governed by feedback (with delays)
- Non-linear
- Multi-scale
- Self-organising
- Adaptive
- Evolving

This dynamic complexity refers to surprising behaviour over time—everything is connected to everything else in a meaningful way, and interactions occur on multiple timescales. Hence the hallmark of dynamic complexity is *unintended consequences*. The problems either resist all solutions (policy resistance or gridlock, where large changes have small effects), or show "tipping points" where small changes have large effects. This non-linearity can also be manifested as "sensitive dependence on initial conditions". Other writers distinguish between complex and complicated. Ravel instructed his music students, "Your playing should be complex, but never complicated." Complicated mechanisms, sometimes referred to as static or structural or operational complexity, hold few surprises, whereas dynamic or behavioural complexity is considered puzzling or *surprising*. Of course surprise depends on the understanding of the person surprised.

Another type of complexity is called analytic or evaluation complexity, where problems and causes are fuzzy and indistinct and values and views are so contested there is no way even to agree on a framework to analyse the issue. This is the territory of "*unknown unknowns*" and distinguishes uncertainty from risk. Risk is considered quantifiable, whereas uncertainty is not.

6.4.2 Mental Models Limit Learning from Shared Experience

Mental models are our way of making sense of the world, they are the beliefs inside our heads that we use to explain what we see and give us the confidence to act well. We are capable of many levels of abstraction and we act quickly using *fast but fallible* decision-making rules. Sterman again lists some of the problems with mental models as:

- Focus on "here and now"
- Stop at a single simple explanation
- Ignore feedback loops
- "Get it wrong" for
  - Chance and uncertainty
  - Time delays
  - Accumulations
  - Non-linearities

These mental models of cause and effect are learnt from our individual past experiences and from hearing and reading the stories and thoughts of others. They are therefore personal, disconnected and fail to take account of complex dynamic nonlinear feedback interactions.

Fragmented disciplines, the jargon language of management and confusion of concepts and diversity of values make these mental models difficult to describe, share and improve. To avoid embarrassment we mix with like-minded people. We value focussed analytical *tunnel vision* that ignores complexity, rather than more *imaginative synthetic* ways to reason and plan what to do in a complex world. This can lead to limiting our ways of knowing and learning by isolating parts of the world rather than exploring connections.

We tend to focus on fixing processes that fall within our narrow range of expertise and span of control rather than seeking explanations in independent interactions. As *interdependencies* increase, so does the likelihood that a given action will generate unintended consequences that may unfold over distant space and time. The more unintended consequences that are generated, the less likely it is that the intended consequences of the action will be achieved and/or sustained.

The *conflicting mental models* of health and healthcare in the heads of participants drive the health system as much as the external institutions and rules that were shaped by mental models of past leaders. From this viewpoint the health system is seen as a strife of interests, or an endless conflict among countervailing powers. Indeed it is a lot like the challenge of climate change.

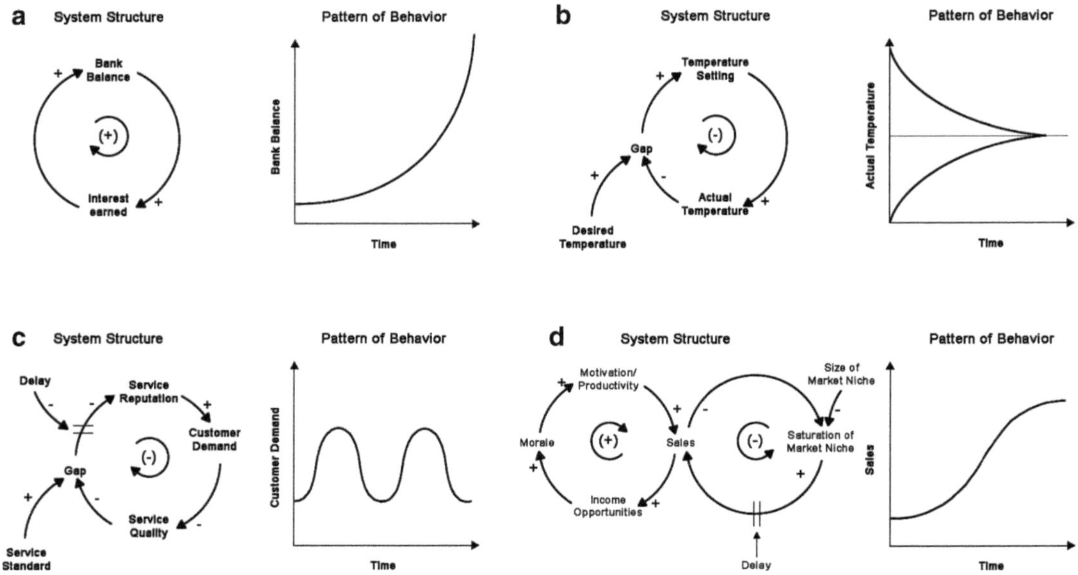

Fig. 6.7 Common system dynamic behaviours. A feedback loop is a closed chain of causal connections from a stock, through a set of decisions or rules or physical laws or actions that are dependent on the level of the stock and back again through a stock to change the stock. (**a**) Positive feedback loop—positive feedback effects are called runaway loops or reinforcing feedback, a small change over time results in large changes; (**b**) Negative feedback loop—negative feedback effects are called self-balancing, they are both a source of stability and resistance to change. Often actions taken do only show immediate results, … things take time. Delays make a system likely to oscillate between two states (**c**), and if delay is considered, changes can be anticipated and result in more controlled fashion of change (**d**)

6.4.3 Information Feedback and Circular Causation

We perceive states of the world, and we act on this information based on our beliefs or mental models, including our understanding of causes and effects. The logic people use to make decisions (converting information into action) that make sense in one part of a system may not be reasonable or desirable within a broader context or when seen as part of the wider system. So the *bounded rationality* of each actor in a system may not lead to decisions that further the welfare of the system as a whole. The system dynamics method aims to avoiding these unintended consequences of clinical policy and management interventions due to their feedback effects. Figure 6.7 illustrates some of the common system structures and their dynamic behaviour.

6.4.4 We Need Better Tools to Help Us Share Our Deep Knowledge of Cause and Effect

We need tools and methods *to shape the future and build consensus about taking effective action*, tools to help us *think clearly*, to explain, design and manage complex social and technical systems. This chapter explores the potential for using concept maps and computer models to help us agree on how to shape a challenging future. It is about computer-assisted thinking, synthesis and experimenting and *learning from virtual experience*. It introduces basic *complex systems science and engineering* methods and applies them to a range of health and health care problems using *maps* and *models* we have found useful in the past.

6.5 Application of Modelling in Healthcare

Having outlined the developments and principles of systems and modelling, we now briefly outline the application of these methods to healthcare problems. The first example describes the analysis of practice populations using cluster analysis, the second analyses the phenomenon of overcrowding of emergency departments with discrete event simulation, and the remaining two introduce system dynamics modelling in the context of chronic kidney disease and the interface of community and hospital care of the elderly.

6.5.1 Clustering: Primary Care Consultations

Little is known about the systems context of primary care consultations. A primary care practice system comprises five distinct domains—the health care system, patients and doctors as individuals, the doctor–patient interactions and consultation outcomes. In turn each domain or subsystem consists of specific variables that all interact and influence each other through feedback (Fig. 6.8) [7].

Clustering analysis was used to identify patterns of relationships between the system variables. Clustering analysis, using the Viscovery SOMine software package (Eudaptics), is based on Kohonen's Self Organising Map [8]. Self-organisation refers to a type of neural network that classifies data and discovers relationships within the dataset without any guidance during learning (unsupervised learning). The basic principle of identifying those hidden relationships is that, if input patterns are similar, they should be grouped together. Two inputs are similar if the distance between the two inputs is small. The result of this analysis is provided in Fig. 6.9 and show seven distinct patterns of distribution of the variables. Each pattern describes well-known patient characteristics and behaviours amongst different physicians [7].

Clustering provides a static, rather than dynamic, picture of the system's past behaviour and thus has limitations in terms of drawing inferences for its

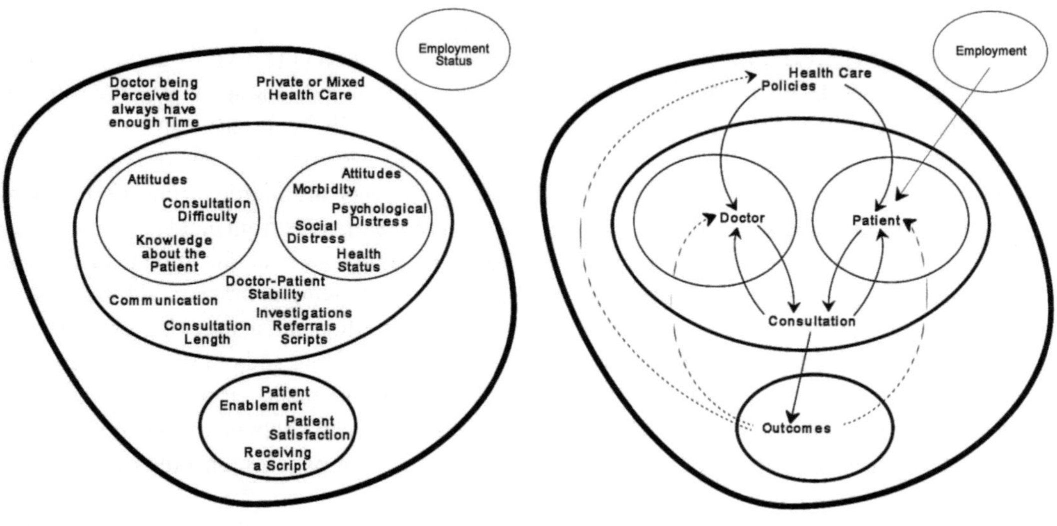

Fig. 6.8 System diagram and influence map of the Consultation System. Explanation: the system map provides a snapshot of the variables of the system at a point in time, and certain variables are grouped into subsystems. The influence diagram describes the main structural features of the system and highlights the important relationships that exist between systems variables. The employment variable belongs to a different system hence sits outside the boundaries of the "consultation system"

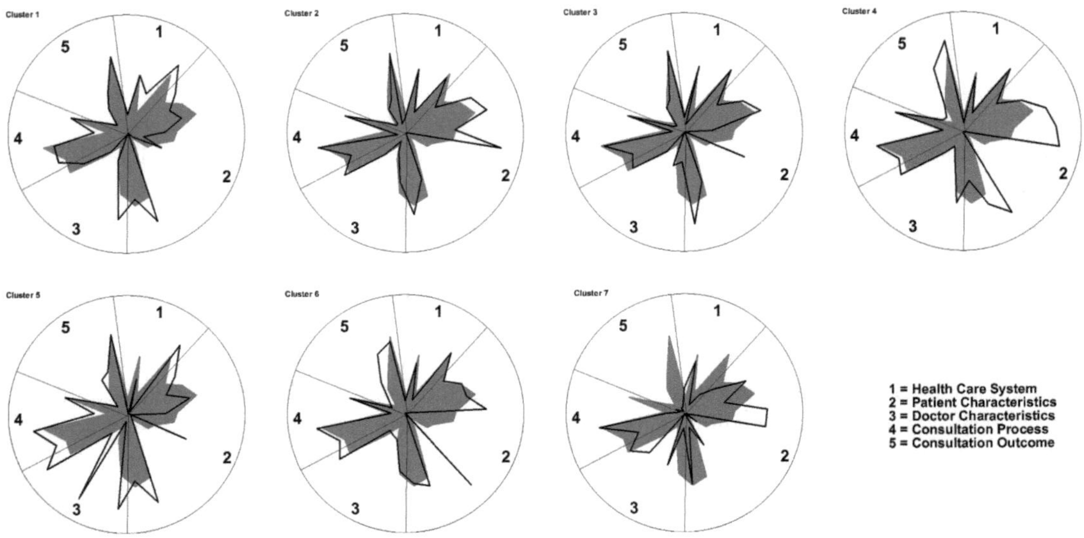

Fig. 6.9 Magnitude of cluster differences (*transparent leaves*) from the population mean (*grey leaf*). The different characteristics of the study population as a whole compared with its subgroups are easier understood in a visual fashion. The outlines of the transparent leaves show the magnitude of differences of each variable and domain compared with the whole population. The shapes of the transparent leaves highlight even more clearly how distinctively different the seven subgroups are

future behaviour. However the approach utilised in this study provides health care reformers with a basis for hypothesis formulation when considering structural and/or process changes.

6.5.2 Modelling Emergency Department Overcrowding with Discrete Event Simulation

Emergency departments (EDs) are becoming one of the dominant sources of care and an important route for admission into hospitals [9]. In recent years a large increases in presentations to emergency departments [10] has coincided with reduced healthcare budgets which has led to frequent ED blockage crises. ED blockage crises are characterised by considerably longer waiting times, ambulance diversion/bypass, and, ultimately, compromised quality of patient care. The health and political impact of these crises instigated efforts to ensure patient waiting and treatment times were minimised and ambulance "bypass" being eliminated. While such efforts have met with some success, large gaps in the understanding of ED operations remain.

The "always open" and "ready for any eventuality" nature of EDs make demand forecasting extremely complex and uncertain. While there is a well-recognised pattern to daily demand, the relative predictability of the average number of patient presentations each hour does not simplify demand estimation [11]. Even if patient numbers can be determined, the demographic mix of patients is usually wide and can vary. Patients may be of any age or either sex, have a full spectrum of ailments and injuries from life-threatening to minor and range from lucid to unresponsive [12, 13]. Ceglowski et al. [14] contextualise ED operations by looking at three main functions:

1. Availability for patients seeking care, regardless of time of day and number of patients
2. Reception and management (including treatment) of patients (both urgent and non-urgent)
3. Disposition of patients once their treatment is complete

Simulation studies have formed a large component of the drive to understand and improve emergency departments (ED) operations within the healthcare system. System dynamic simulations described earlier in this chapter, have looked at the interaction of ambulance services with the

ED, and the role of hospital policy on treatment time in ED [15]. Discrete-event simulation (DES) is particularly suitable for process systems modelling. The process systems context surrounds most of the applications of DES where effective representation of individual entities, attributes, decisions and events throughout the process of care, while explicitly modelling the randomness, are particularly important. The majority of models have used generalied distributions to describe arrival rates, lengths of stay and treatment times for ED simulation and optimisation purposes [16–18]. Jun et al. [19] surveyed the uses of Discrete Event Simulation over the past 20 years in healthcare clinics ranging from individual practices to EDs.

This case study [14] describes the experience at the Emergency Department (ED) in one of Melbourne's metropolitan teaching hospitals.[7] This emergency department is typical in setting and complexity [12]. There is a constant stream of patients into the emergency department with a range of ailments and urgencies. While the number of patients arriving each hour is reasonably well characterised, patients levels of urgency, gender, or age at any time of day and day of the year is subject to major uncertainty.

6.5.2.1 Treatment-Focused Groups of ED patients

In trying to model the uncertain nature of ED operations, one approach to simplify the situation is by grouping "similar" ED patients under the Casemix principle. Similar cases are assumed to be treated alike and to utilise a particular set of resources [20–22]. ED casemix variously suggest that cost of treating ED patients correlates to patient urgency, disposition (whether treated and discharged home or admitted to hospital) and age. However, the process-of-care grouping of patients attending emergency departments remains particularly difficult because of the broad range of demographics and clinical presentations [21].

The use of *non-parametric methods* for grouping of patients was explored by Isken and Rajagopalan [23] and Ceglowski et al. [14]. This technique, being based on data for every patient, suffers less from the depth and breadth limitations of traditional data- or knowledge-sampling approaches, and can identify non-obvious groupings of patients,

Ceglowski et al. [14] obtained 56,906 de-identified records of all ED presentations of 1 year at one Melbourne metropolitan hospital. The records contained demographic information as well as details of the visit such as "presentation problem", key time points, disposition, and of medical procedures undergone by patients during that visit.

After some preliminary data investigations, a hypothesis was formed that patients could be grouped according to the medical procedures most often performed together. A non-parametric method called self-organising maps (SOM) [24] was employed to find groups of patients with minimal intra-group diversity and maximal inter-group separation. SOM generally employs large data sets, works well with many input variables and produces arbitrarily complex models unlimited by human comprehension [25]. SOMs provide a visual understanding of patterns in data through a two-dimensional representation of all variables. Viscovery SOMine, the software tool used in this analysis, employs a variant of Kohonen's Batch-SOM [24] guided by Ward's classic Hierarchical Agglomeration algorithm [26] to determine the optimal number of clusters.

Figure 6.10 shows the distinct groups of patients who underwent particular groups of procedures. These groups of medical procedures represented the core treatment pathways. Nineteen groups of procedures accounted for treatment of all patients whose treatment involved two or more procedures. Each of the groups, or clusters, represents a pattern of treatment.

The resulting clustering model underwent extensive validation which provided the confidence that the groups of patients reflected true clinical presentations and provided a good representation of treatment activities within the ED. The obtained treatment groups were then incorporated into a Discrete Event Simulation model.

[7] Interested readers are referred to the publications by Ceglowski et al. [14].

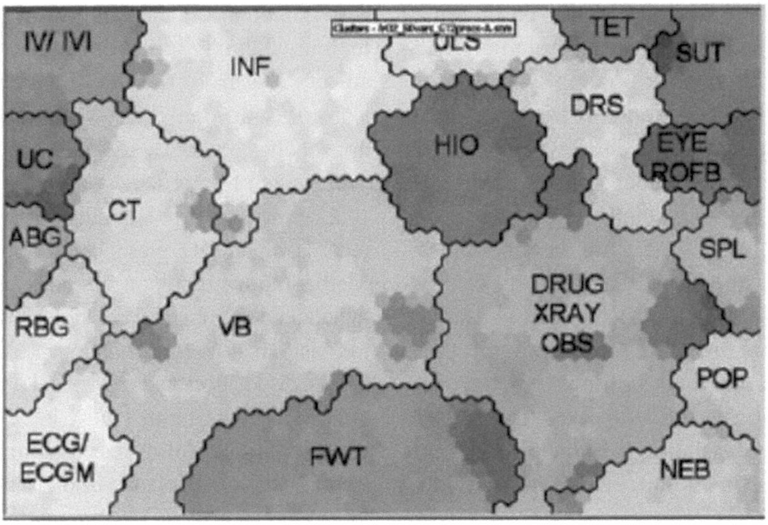

Fig. 6.10 Screenshot of the SOM treatment clusters in Viscovery SOMine. Input variables have been compressed into two dimensions and separated by boundaries. The clusters are labelled according to the procedure that is dominant (typically all patients in that cluster have that procedure and other, allied procedures).
Treatments on the right-hand side relate to accident victims, with treatments including tetanus injections (TET), dressings (DRS), sutures (SUT), eye injuries (EYE, ROFB), splints (SPL) and Plaster of Paris (POP). Those on the left relate more to illness. Examples are treatments that include tests of arterial blood gases (ABG) or random blood glucose (RBG), monitoring of echocardiograms (ECG/ECGM) and intravenous drug infusion (IVI)

6.5.2.2 Treatment-Focused Discrete Event Simulation

Discrete Event Simulation studies in EDs commonly break the ED into sub-units, assign patients to urgency categories and use these to prioritise access to resources. They generally approximate patient arrival rates and regulate patient flow by events such as completion of triage, admittance to an ED bed and review by doctors [27–34]. In trying to model the uncertain nature of ED operations, analysts have simplified the situation by grouping ED patients, developing unique process charts for each patient group (often including the duration of investigative activities such as imaging and tests, and the frequency of connections between the activities), and using generalised distributions to describe arrival rates, lengths of stay and treatment times in simulation and optimisation models [16, 17].

The model described by Ceglowski et al. [14] seeks to complement conventional scale models by providing a high-level, abstracted view of ED operations. The treatment-focused Discrete Event Simulation approach encourages a systems-wide view by concentrating on how patient and treatment differences affect queue times. The treatment grouping introduced earlier is useful in this abstraction of ED utilisation.

Since patient registration and triage are well understood and largely optimised, it is reasonable to model only the stage between patient placement in a treatment bed and their physical departure from the ED. This simplifies the system to consideration of whether treatment sites (most commonly ED beds) are physically occupied. Queues develop if all sites are occupied. The benefit of this simplified "state-based" view is that many variables become extraneous. For instance, patient bed times may vary according to the people involved in the treatment (interns or experienced doctors, for example), or admission of patients to virtual "short stay units" within the ED which may result in the ED meeting its performance obligations. Variability owing to doctors' differences is difficult to cater for in a conventional ED model. By using total bed time, it becomes unnecessary to gather these data, provided a high-level view of the ED is acceptable.

The model was designed to generate a large variety of patient types according to urgency, treatment and disposal. The use of urgency and disposal variables were occasioned by Casemix studies that had indicated the importance of these on patient grouping [35]. Patients arrived in the ED bed queue at rates dictated by the data. They were apportioned urgencies and disposal within urgency according to historic distribution. Patients of each urgency/disposal type were streamed into one of the 20 treatment pathways according to the distribution profiles noted for that urgency/disposal/treatment combination. At this point, the patient carried urgency, treatment and disposal labels that jointly defined the patient type. Patient type provided a framework for building the model and subsequent analysis. Discrete distributions were specifically developed for 161 patient types (99% of patients) and generalised distributions were used for the remaining 1% of patient types that occurred rarely (Fig. 6.11).

In the model, as in real life, patients queued for suitable beds if all beds were occupied. If access to treatment has been compromised at any time, a queue for beds develops. Waiting time may then exceed the thresholds stipulated by the national triage scale for a given triage category. Patient bed time (the total time for which they occupy a bed) was drawn from historic distributions for that urgency, treatment and disposal combination. Bed time was an input to the system and queue time was regarded as an output of the system. Generalised distributions had to be developed for bed turnover based on expert opinion.

The model was implemented in Simul8 (Version 11 from the Simul8 Corporation) through sequences of virtual workstations and queues and underwent extensive validation and verification (see footnote 7).

Data Mining led to patients being grouped by similarity of treatment. These groups were used in an abstract representation of the ED as a system that was either available or full. Ceglowski et al. [14] identified factors that impacted on access to treatment by analyzing what happened when the system was full. The queues formed often but were generally not long either in duration or in number of patients. However, in several instances long queues formed analogous to those experienced in the ED when the system became blocked (unable to accept any new patients for treatment). In studying these instances, it became apparent that system blockage depended on the combination of patient types within the system. Patient types that were characterised by long bed times were implicated in the blockage, as would be expected.

The most important finding was that the *combination* of number of patients and long bed time was significant. A simple weighting of the number of patients in each patient type with the average bed time for that patient type showed that certain patient types were occupying ED beds for a disproportionate time. It is notable that the heaviest users were all awaiting admission to a hospital ward. The data records that the decision to admit these patients was made early in their treatment, but the ED was forced to continue treating them because of the delay in moving them to a hospital ward. The treatment and symptoms of these patients give an indication of which wards were implicated in the admission delay.

6.5.3 System Dynamics Modelling

System dynamics is a formal method of computer modelling using stocks, flows and information feedback loops.[8] Using the example of dialysis we first introduce the principles of system dynamics modelling.

6.5.3.1 Stocks and Flows

Suppose someone asked you the question: Please explain how the number of people on long term renal replacement therapy will change over the next 20 years. How would you answer?

[8] For a brief introduction and references see http://www.systemswiki.org/index.php?title=System_Dynamics; http://www.systemswiki.org/index.php?title=System_Dynamics_Methodology.

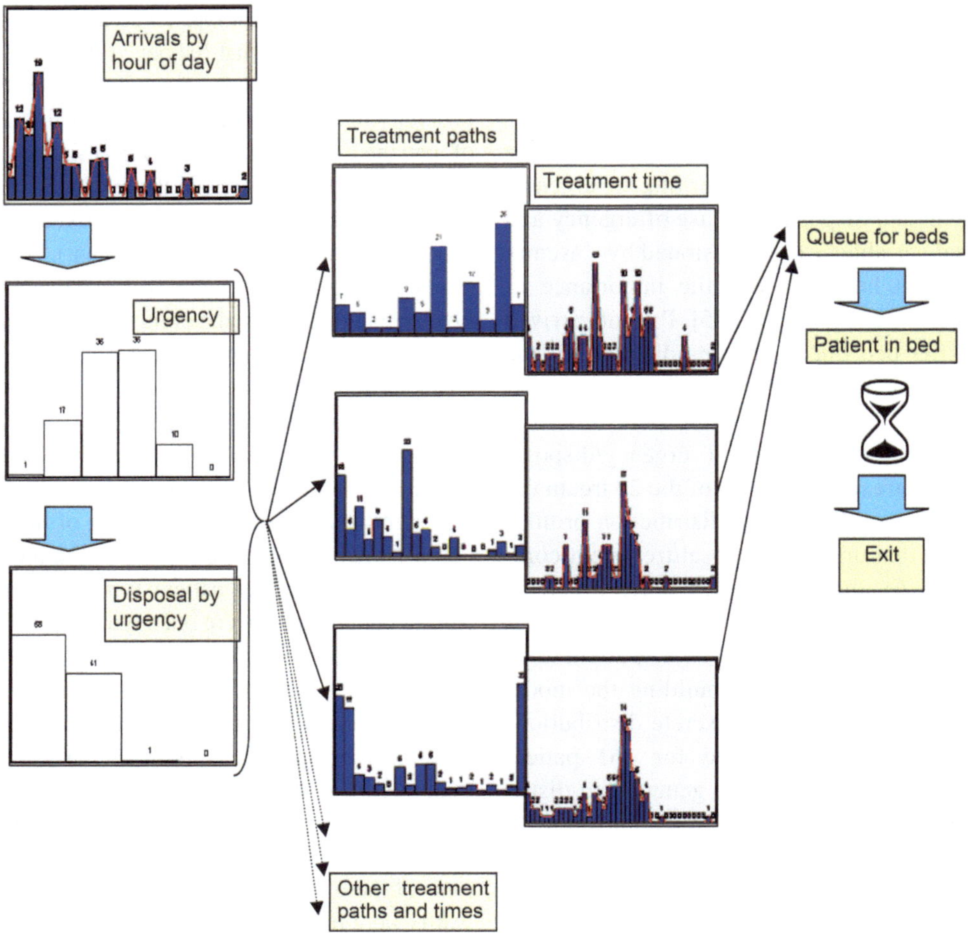

Fig. 6.11 Schematic of the simulation. Rather than following the physical movement of patients the simulation tracks the state of ED treatment sites as being "occupied" or "free". Queues result when all treatment sites are occupied (irrespective of other resource considerations). The bulk of the simulation is dedicated to allocation of appropriate urgency, disposal, treatment cluster and bed time labels to patients. While patient types are generic by urgency, disposal and treatment cluster, patient bed time is individual

This is the way a system dynamics thinker and modeller might answer.

You can consider the number of people now on dialysis as a bathtub of water with an inflow tap of new dialysis patients per year and an outflow drain of deaths per year on dialysis. Similarly, consider the current number of transplanted patients as the level of water in another bathtub. Most people who flow into the transplant bathtub are an outflow drain from the dialysis bathtub. Some people may also be transplanted without being dialysed, particularly live donor transplants. This extra inflow into transplants is represented as a tap of transplants with no dialysis flowing in each year. Now the outflow from the transplant bathtub can again be transplant deaths per year. But also people with transplants can flow back to the dialysis bathtub at the rate of the number of graft failures per year (Fig. 6.12).

This bathtub thinking, called stock-flow thinking, is a key component of system dynamics.

6.5.3.2 A SD Simulation Model of Dialysis and Transplant Patients

We will now construct a simple computer model based on real world data.

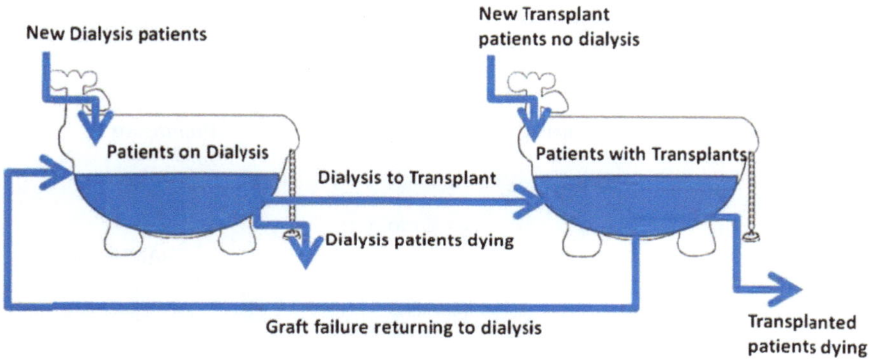

Fig. 6.12 Bathtub model of stocks and flows

Fig. 6.13 Patient flow of patients requiring renal replacement therapy—Australia 1997–1998

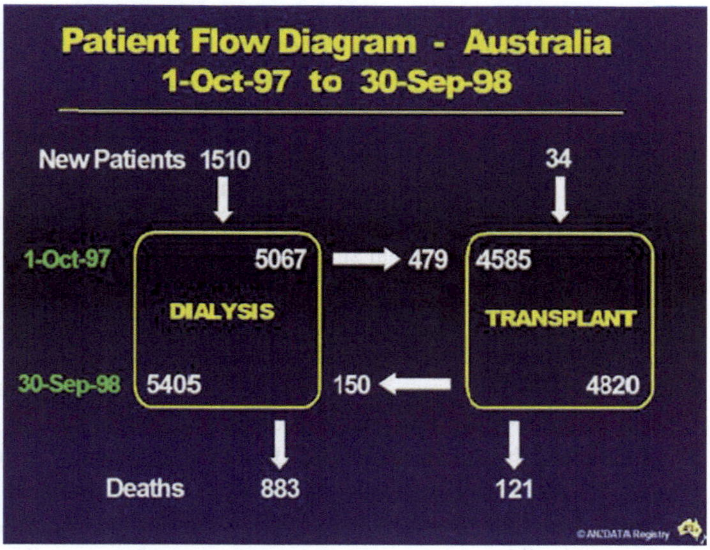

A stock flow representation has been used by the Australasian dialysis and transplant data registry, ANZDATA (http://www.anzdata.org.au), for many years in their annual reports, as a patient flow diagram (Fig. 6.13).

The calculations are as follows:

1. The stock of dialysis patients at the end of year (5405) = Number at beginning of year (5067) plus inflows of new patients during the year (1510) plus inflows from transplant to dialysis during the year (150) minus the outflows of Deaths during the year (883) minus the outflows from dialysis to transplant during the year (479)
2. $5045 = 5067 + (1510 + 150 - 883 - 479)$
3. This can be written as a differential equation:

$$DIALYSIS(t) = DIALYSIS(t - dt) + (New\_Patients + Tx\_Failures - Deaths\_Dx - Dx\_to\_Tx) \times dt$$

Several SD modelling tools are available to convert stock flow maps into model equations. Here is a stock-flow map of Renal Replacement Therapy produced using ithink/STELLA software (Fig. 6.14).

We can generalise this pattern of calculations by calculating the flows in terms of the fractional change in the relevant population. New patient and transplant rates are usually reported in rates per million of the general population per year and death and graft failure rates are usually reported as fractions of the stock of dialysis or transplant patients per year. We use connectors (between

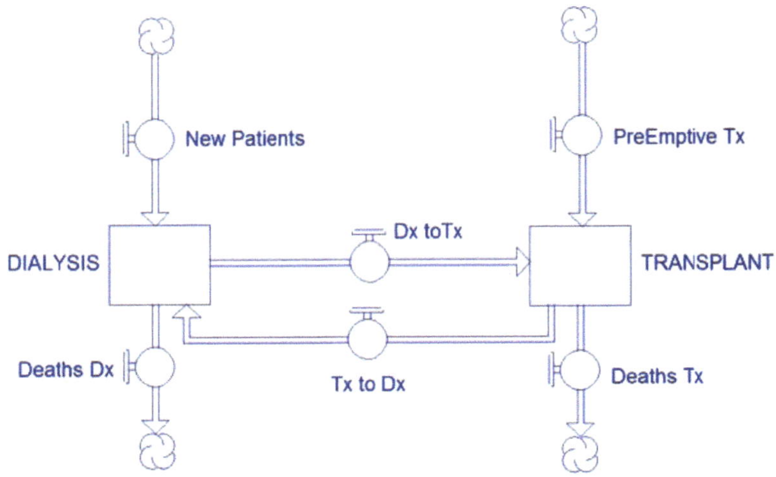

Fig. 6.14 Stock-flow map of renal replacement therapy

stocks) and converters (which alter the flow rates) to show these relationships on the stock flow map (Fig. 6.15).

We can add some additional structure to perform calculations of these key parameters. Here we have added a population stock and population increase flow to calculate the size of the future population (Fig. 6.16).

Calibrating the model initial stock and parameter values from historical data since 1994 provides us with an executable model that produces behaviour over time. We can then add a user interface with sliders to vary the parameters in the model and to conduct what-if virtual experiments. The results of two sets of virtual experiments are show below.

Modelling the Effect of Changing Acceptance Rate of New Patients Onto Dialysis

First we explore the effect of varying the percentage growth in the acceptance rate above and below the historical rate of 6% per year from 2003. There are four simulation runs shown (Fig. 6.17):

1. No change 6% growth rate (the base case) (blue)
2. Decrease to 3% growth rate (brown)
3. Decrease to 0% growth rate (mauve)
4. Increase to 10% growth rate (green)

The results from these four runs show a spread of numbers on dialysis at mid 2010 from 8,000 to 16,000 and a spread of acceptance per million population in 2010 from 90 to 210 per year.

Modelling the Effect of Varying Transplant Rates

In another set of virtual experiments we vary the transplant rate, taking a period of five years to reach the new rate. The four simulation runs shown are (Fig. 6.18):

1. No change in the rate of 26 kidney transplants per million population pa (blue)
2. Increase to 34 kidney transplants per million population pa (brown)
3. Increase to 50 kidney transplants per million population pa (mauve)
4. Decrease to 17 kidney transplants per million population pa (green)

If we take into account that the quality of life and annual cost of treatment are better for transplanted patients than patients on dialysis then the better course to follow is to increase the kidney transplant rate.

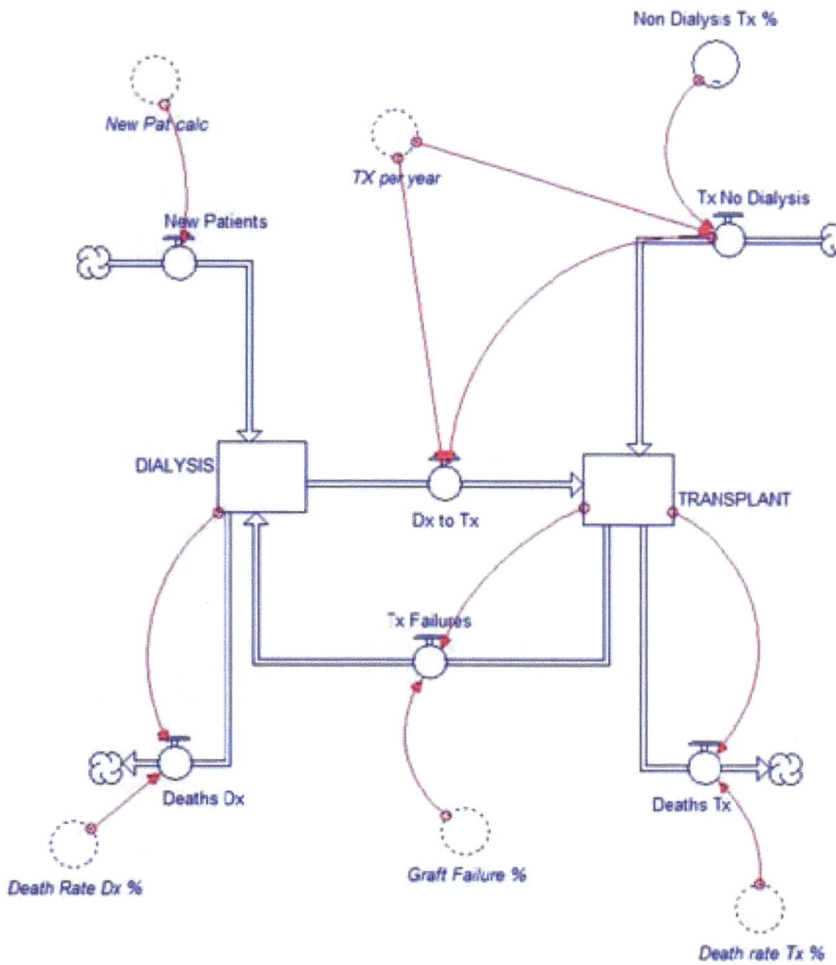

Fig. 6.15 A visual representation of the inflows and outflows (*thick arrows*) of patients on renal replacement therapies, stocks or boxes on dialysis or with a functioning transplant. The flow rates are calculated using the round circle auxiliary variables, and the *red* connectors show the variables used to calculate the flow rates

6.5.3.3 Modelling at Different Scales

Dialysis and Transplant dynamics

Here we will represent our stock flow model of renal replacement therapy as a causal loop diagram, using the online tool Insightmaker (http://insightmaker.com/insight/317) Firstly we construct the individual links inherent in the above bathtub model. Note that an inflow has a same link to its stock, and outflow has an opposite link to its stock (Fig. 6.19).

Note in the above diagram, taken from the SD model, the dialysis death rate and the acceptance rates are represented as exogenous time trends or forcings. Can you identify any other exogenous causes or variables in the diagram? (Graft failure rate, Organ donor rate and Population).

Another key feature of a good SD model is that all changes are endogenous. Another way to describe this is that the model is causally closed. For example, we can simply hypothesise that an

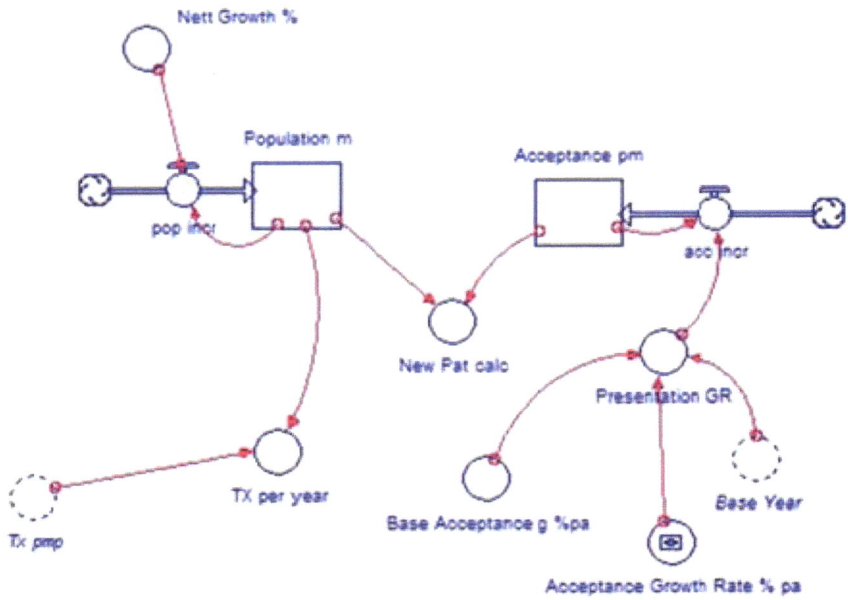

Fig. 6.16 Additional structure used to calculate growth in flow rates based on per million population (pmp). New patients accepted onto dialysis are driven by changes in the acceptance rate pmp from a base year and by the change in underlying population. The acceptance growth rate % pa can be varied using a slider to perform what-if simulation runs

increase/decrease in dialysis death rate will tend to reduce/augment the acceptance rate as dialysis becomes less/more attractive. Here is one way to close some of the loops, keeping the focus on deaths in this diagram. The diagram above suggests some additional possibilities for producing a causally closed system, by considering the influences that might change organ donor rates and the links between acceptance rate and dialysis death rate. These extra loops can be considered as dynamic hypotheses, and the exact loops we explore depend on the surprising situation we are trying to explain. For instance in Australia the growth in acceptance rate slowed and the number of new transplants with no dialysis increased more than expected. Closing the loops can provide potential explanations of these non-linear effects. This plausible explanation is described as a dynamic hypothesis (Fig. 6.20).

Here we have explicitly labelled some plausible balancing loops and reinforcing loops as possible explanations for observed non-linear trends in people on renal replacement treatment. Several other simple balancing loops are left unlabelled. Note we have explicitly shown the stocks as boxes.

Labelled Reinforcing Loops

Dialysis and transplant learning effects: The dialysis death and graft failure rates fall over time as the techniques improve with experience. This increases the number of people on renal replacement therapy over time. Another reinforcing loop is the possibility of successive grafts, which tends to increase the number of people on both dialysis and transplant. The final reinforcing loop is the delayed effect on organ donation rate of the number of people living with transplants, labelled as transplant success diffusion.

Labelled Balancing Loops

To explain what limits the number of people on dialysis we have introduced the concepts of session treatment time as the limiting resource. To manage this resource we can either ration the number of people accepted on to dialysis, labelled as rationing places, or we can reduce the time

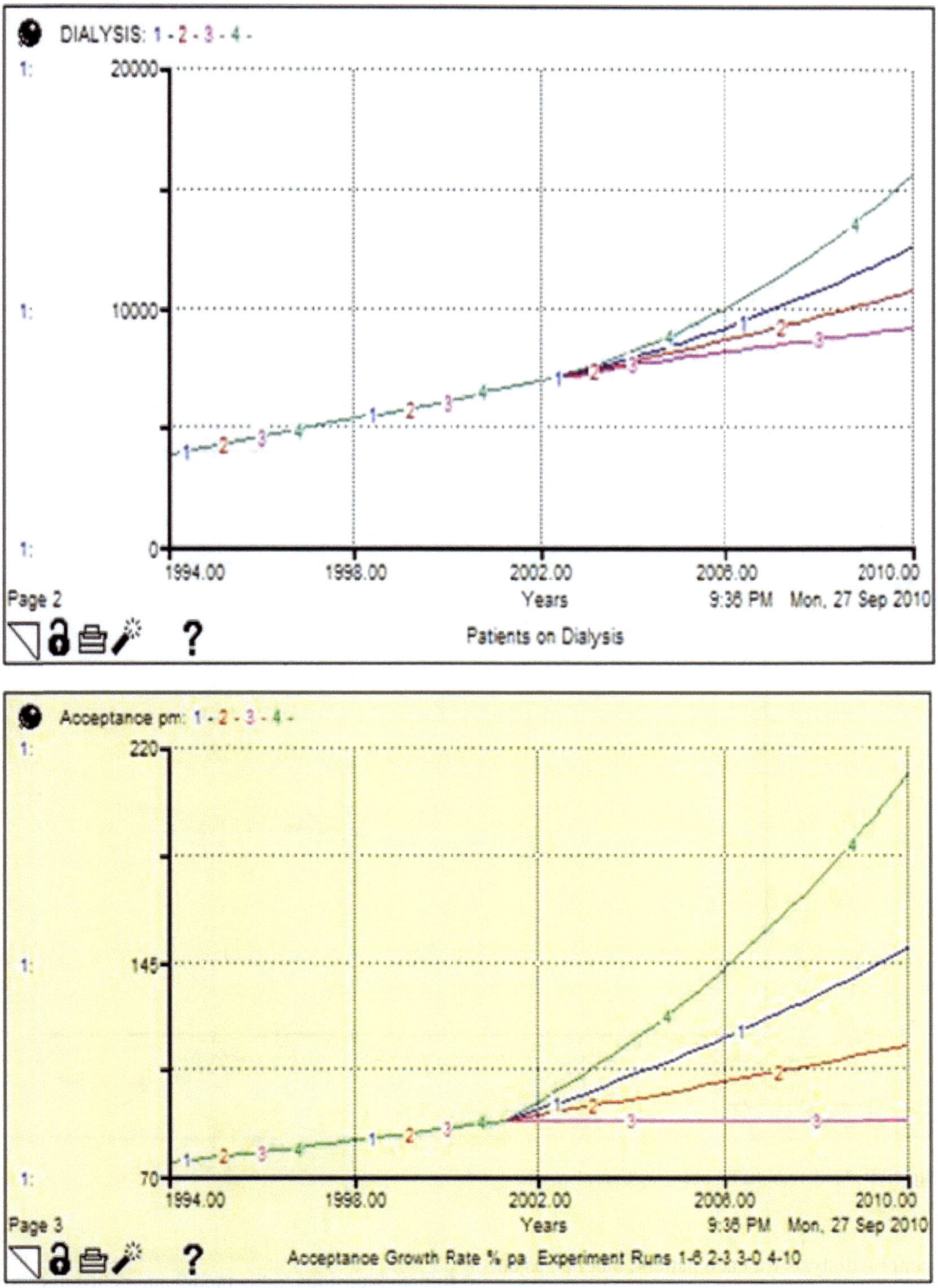

Fig. 6.17 Model results for number of patients on dialysis and acceptance rate onto dialysis per million population by calendar year. The acceptance growth rate is varied at 2001. Graph run 1 remains at 6% growth pa, Graph run 2 is at 3% growth rate, Graph 3 is at 0% growth rate and Graph 4 is at 10% growth rate. This results in a range of patients on dialysis in 2010 from 8,000 to 16,000

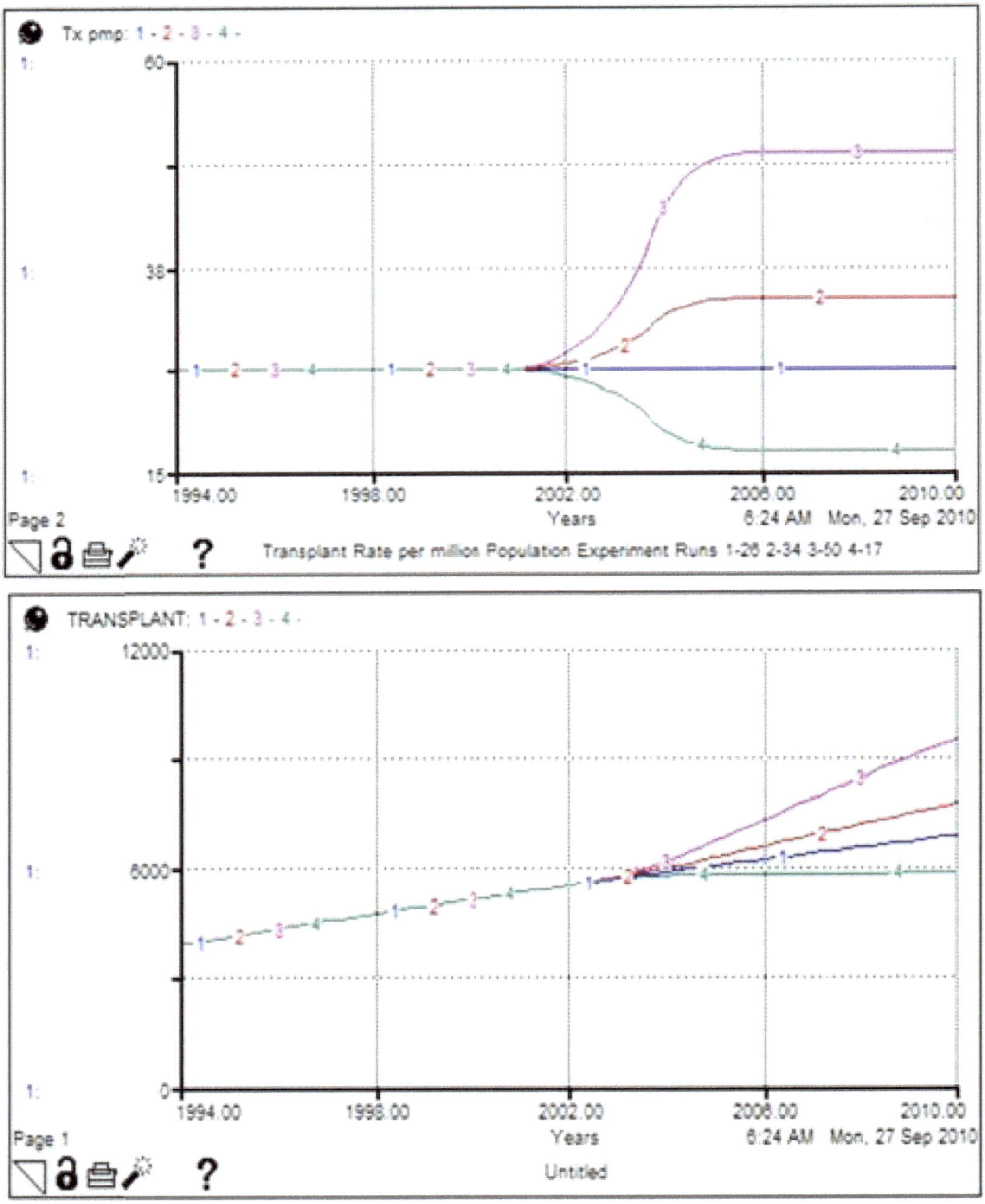

Fig. 6.18 Model results for changing transplant rates

spent on dialysis for each patient, labelled as cutting corners.

We can use these causal loop diagrams to explain the results of our models or to develop plausible dynamic hypotheses which can guide future extensions to the model and data gathering to test these models empirically.

However there are many other concepts that could be used to represent the dynamics of this complex system. We can zoom out to include

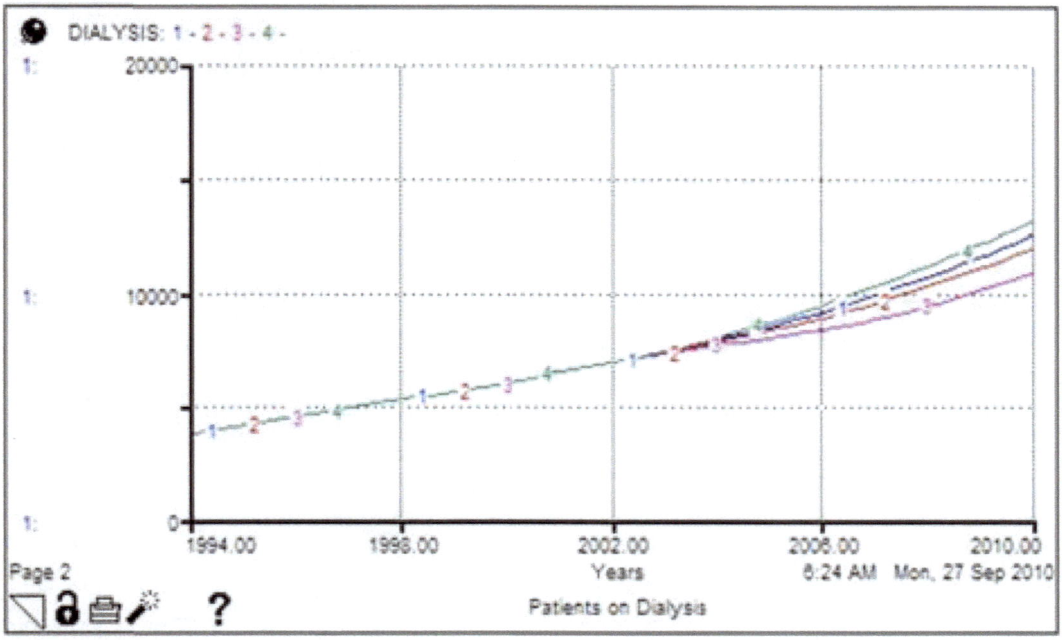

Fig. 6.18 continued

people with progressing chronic kidney disease prior to being accepted on to a dialysis programme. People involved in the technical detail of dialysis might want to explore the contribution of dialysis adequacy. Therefore the purpose of the model is an important determinant of the way we choose to represent the dynamics of a system.

Chronic Kidney Disease Dynamics

Of course the need for Dialysis is mostly driven by the number of people who have progressive forms of kidney disease, including glomerulonephritis and diabetes. The flow of people through early and late stages of chronic disease can also be represented using system dynamics models.

The key chronic kidney diseases that produce end stage renal failure in many countries are glomerulonephritis and diabetic nephropathy and their onset and progression can be delayed by screening and effective management of risk factors including hypertension, proteinuria and glycaemic control. The above generic pattern can be adapted to these specific diseases and interventions, similar to Motohashi's approach [36]. Rather than the original stock-flow diagram, the model here is represented as causal loops and explicit stocks[9] (Fig. 6.21).

You may wish to identify flows and label more loops and add the effects of other limited resources, such as funding.

Organ Donation and Transplantation Dynamics

Where possible the preferred renal replacement therapy seems to be a combination of self-managed dialysis and kidney transplantation. Of course the transplantation rate is limited by the availability of live and deceased donors. Here we will introduce a slightly different conceptualisation to explore ways to increase the transplantation rate. In this view we consider the stock of transplantable organs in the general population. These can be added to by births and in migration of transplantable organs or by advances in technology that make more organs capable of being transplanted. Perhaps transplantable organs may also be grown from stem cells in the future.

[9] More detail of the model is available online at http://insightmaker.com/insight/1003.

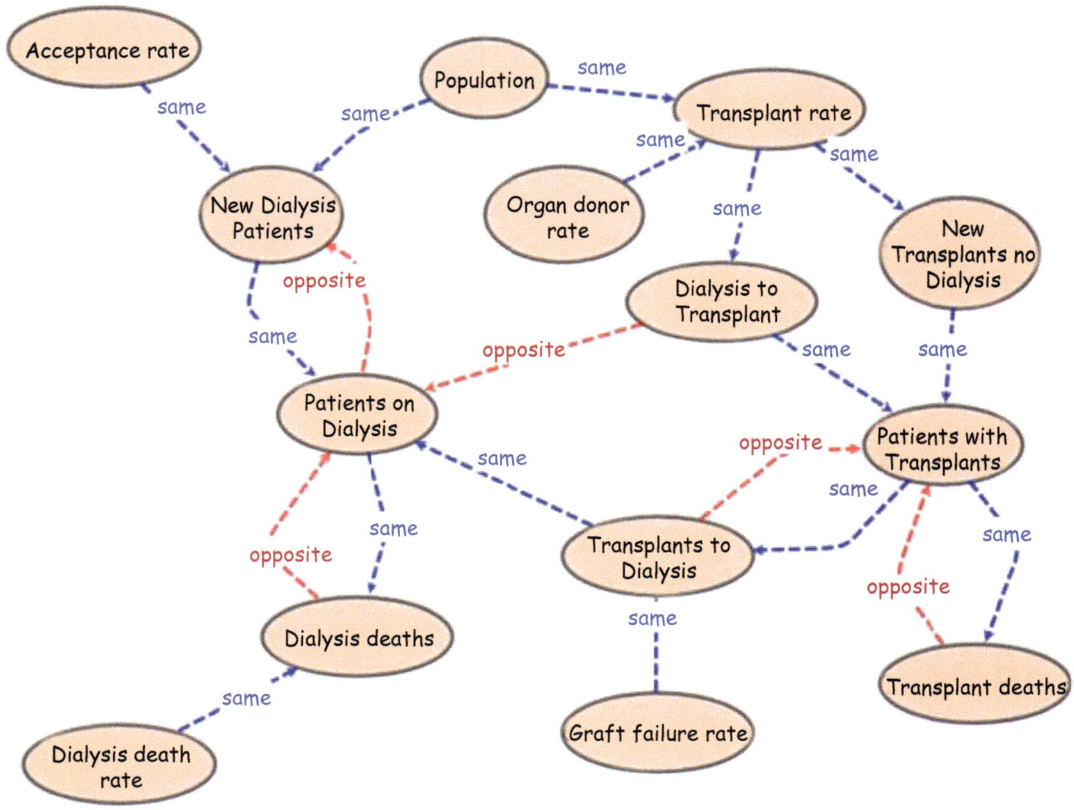

Fig. 6.19 Causal loop diagram of renal replacement therapy

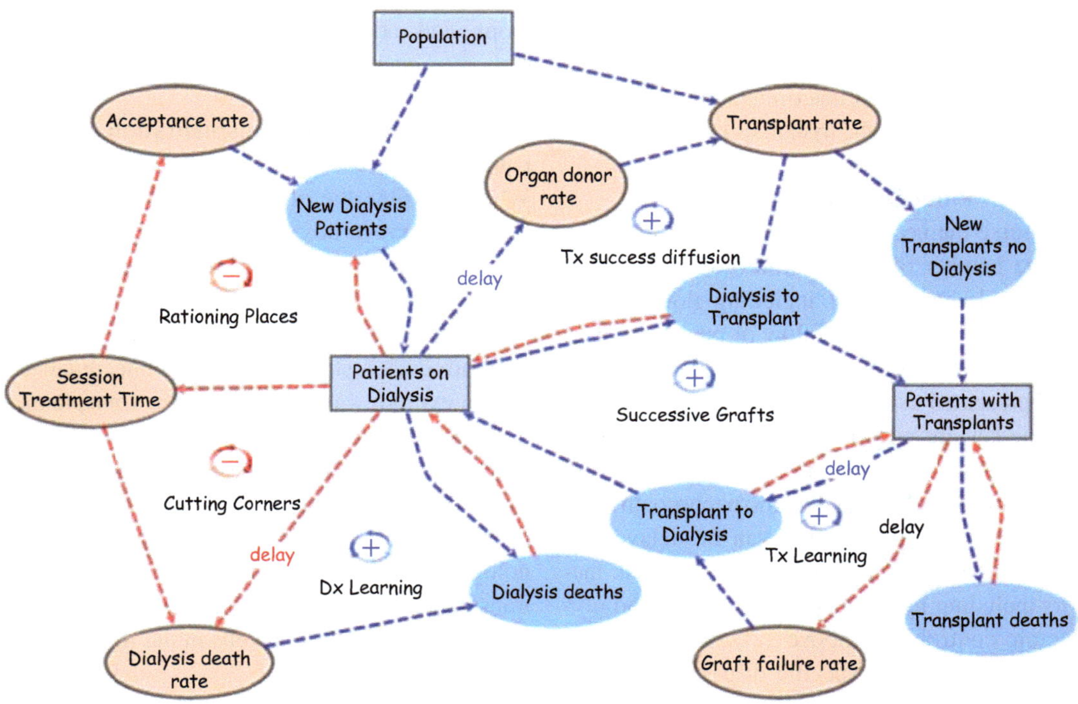

Fig. 6.20 Identifying reinforcing (positive) and balancing (negative) feedback loops

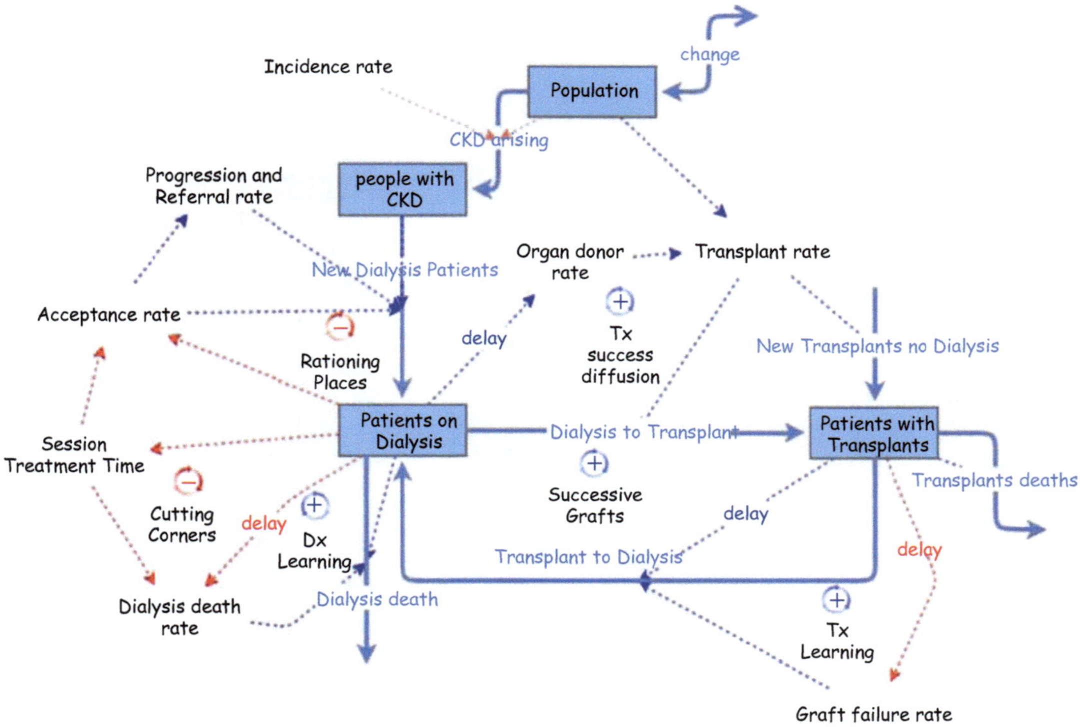

Fig. 6.21 Renal transplant dynamics

These organs mostly become no longer transplantable by the organs ageing beyond the time they are considered transplantable, or by dying without being donated. This pool of transplantable organs in people is also depleted by organs failing and people becoming potential transplant recipients. This stock of organs failed in transplantable recipients can be removed by death or by replacing organs.

By the act of donating by live and deceased donors, organs can flow outside the body, be potentially stored and then flow into the bodies of recipients. They will remain there until the death of the recipient or the failure of the graft. Graft failure takes the organ back to the organs failed in transplantable recipients, where they are again removed by dying or being retransplanted.

The diagram below shows these organ flows and the potential feedback effect of increasing donor age reducing the life of the transplanted organ[10] (Fig. 6.22).

This organ flow representation shows the ways transplantable organs are generated and are consumed by ageing, death and organ failure. Transplantable organs can flow from a donor to a recipient, with a variable time spent outside the body. This forms a basis for discussing places to intervene to promote the flow of organs to recipients and to increase the time organs spend as functioning and transplantable entities.

Zooming in on Dialysis Modality Selection

Consider a model whose purpose is to explore how to best match the supply of dialysis facilities to the demand for dialysis.[11]

The broad context, as we have previously shown, includes understanding the drivers of demand for dialysis, including both population

[10] More detail of the model is available online at http://insightmaker.com/insight/323.

[11] This section is taken from unpublished work of my NZ colleague David Rees and Ahmad Azars's papers and conference presentations (GM).

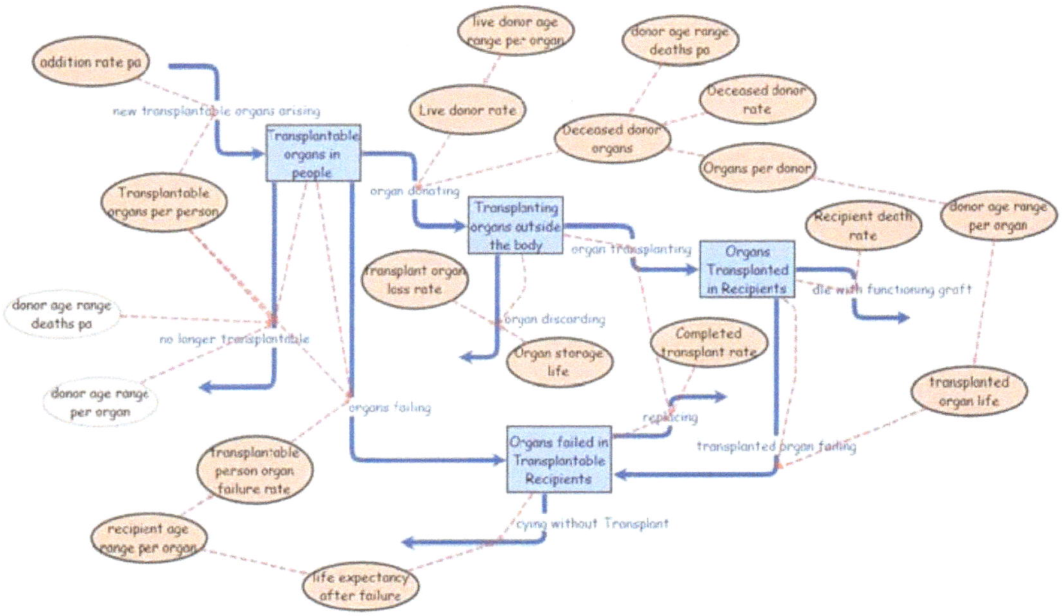

Fig. 6.22 Organ donation and transplant dynamics

dynamics and kidney disease dynamics including diabetes and other risk factors. On the supply side, the availability and configuration of resources, including technologies, specialised staff and funding determines the quality of services and therefore the patient and medical preferences for different dialysis options. Key decisions include provision of shared facilities, the availability of resources for prevention of progression of chronic kidney disease, early specialist referral and vascular access surgery which interact with the age, co-morbidity and social conditions of the population that need dialysis. Again, the availability of live and deceased kidney donors will also affect dialysis treatments and outcomes.

At a more detailed level, dialysis adequacy affects the morbidity, quality of life and mortality and attractiveness of different modality options. Intradialytic session length and filtration interact with interdialytic management of fluid balance, nutrition and anaemia. Some of these concepts are included in the following diagram (Fig. 6.23).

Another constraint which could be added to the above diagram is the interaction between costs benefits and resources. This is indicated above by the links among population, resources for prevention and dialysis resources. You may wish to modify this diagram to add loops, stocks and flows.[12]

More Detailed Models: Pros and Cons

The models already presented contain only a few stocks. Like all compartmental models we assume perfect mixing within each stock. If we are interested in the differences within stocks, we can divide or array the stock into multiple dimensions. One common way to array stocks is by age and gender, since in epidemiology and public health we often have detailed data by age and gender. This increases the accuracy of our model, but it may detract from understanding the feedback dynamics of the situation. Age-specific mortalities

[12] More detail of the model is available online at http://insightmaker.com/insight/318.

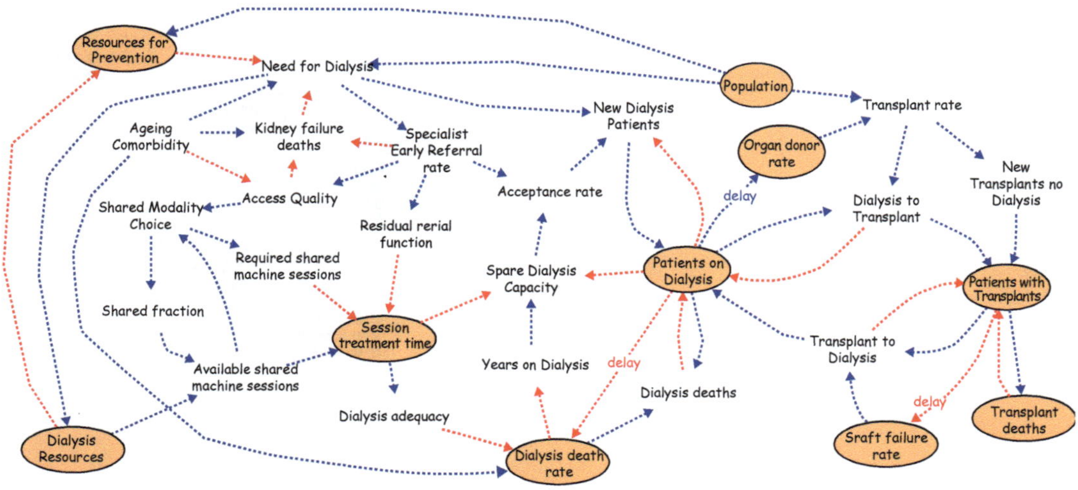

Fig. 6.23 Detail of the interactions among resources for dialysis and prevention and the drivers of the number of patients on dialysis. Key management variables are the session treatment time, the choice of shared dialysis modality (e.g. centre-based hemodialysis rather than home-based) and the referral and acceptance rates. *Blue arrows* represent same and *red arrows* represent opposite influences

are important in quantifying costs and benefits, particularly using quality adjusted life years and health adjusted life expectancy as measures of population health. Specifically modelling policies for accepting elderly people on dialysis may require this more detailed level of analysis. Other problems may require zooming out to include a much broader context, including both drivers of demand and constraints on supply. In general, the interacting components include the population, people with health conditions, patients in care, clinical services workload, workforce, facilities, technology and funding sources [37].

6.5.4 Understanding the Flows of Older Patients Between Hospital and Aged Care

6.5.4.1 A Hospital View

The simple view of improving patient flows through care locations is that more beds are needed. However when more beds were added to emergency departments, flows became worse [38]. In hospital wards the available beds are generally constrained by staff costs, together with inflexible budgets and staffing practices. Beds are perceived to be blocked by older patients waiting weeks for discharge into aged care residences. Control of aged care places generally belongs outside health care, in the aged or social care sector. One solution to hospital congestion is to give priority to admitting patients from hospitals into aged care residences. However this is resisted by aged care proprietors, since these patients are often the most unprofitable. The funding arrangements are designed to provide an acceptable level of care and constrain the growth in government expenditure. Eric Wolstenholme in the UK has described many patient flow improvements as fixes that fail (http://bit.ly/u1KwVv),[13] and these failures lead to chronically unsafe care, which he calls "coping but not coping" [39]. The key interactions are represented in the following causal loop diagram using Insightmaker (Fig. 6.24).

A Stock and Flow diagram of formal and informal coping policies is shown below, from http://bit.ly/vKuRFk. Formal adjustments to

[13] An unfolding of the arguments in the paper and link to the Insight is available on the Systemwiki website.

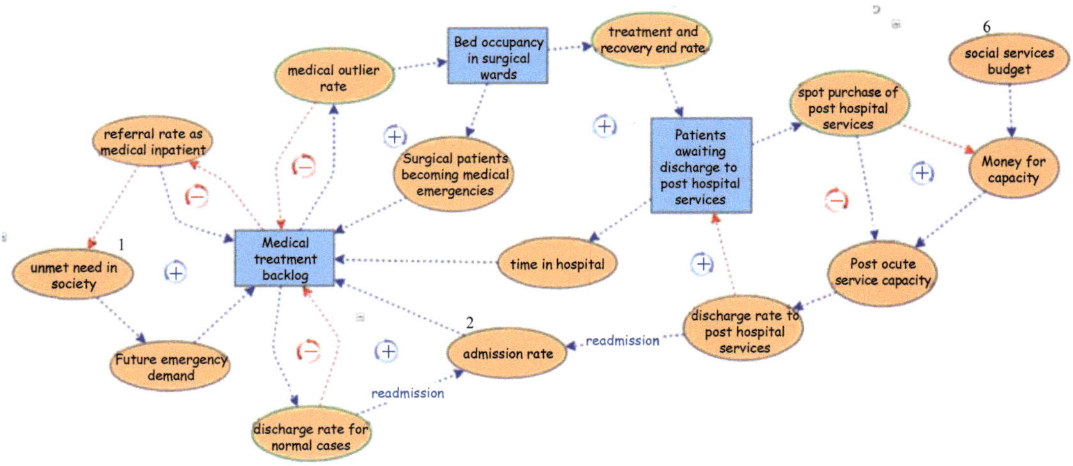

Fig. 6.24 Causal loop representation of variables (*ovals*) that influence the main stocks (*blue boxes*) involved in adjusting to changes in demand for medical inpatient treatment. The variables operate at the pre-hospital, in-hospital and post-hospital phase of care

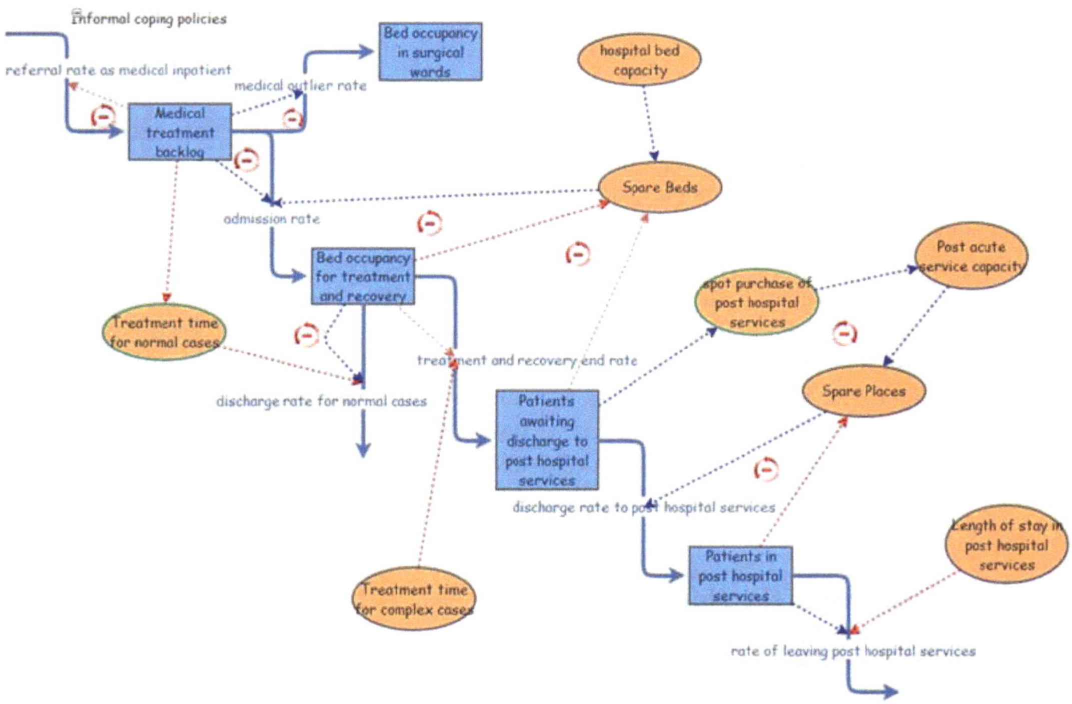

Fig. 6.25 A stock and flow diagram representation of formal and informal coping policies to managing changes in medical inpatient demand

capacity and flow rates interact with informal workarounds, including changes in referral and discharge thresholds, and placement of medical outliers in surgical wards. These workarounds then delay the use of formal long term adjustments in capacity (Fig. 6.25).

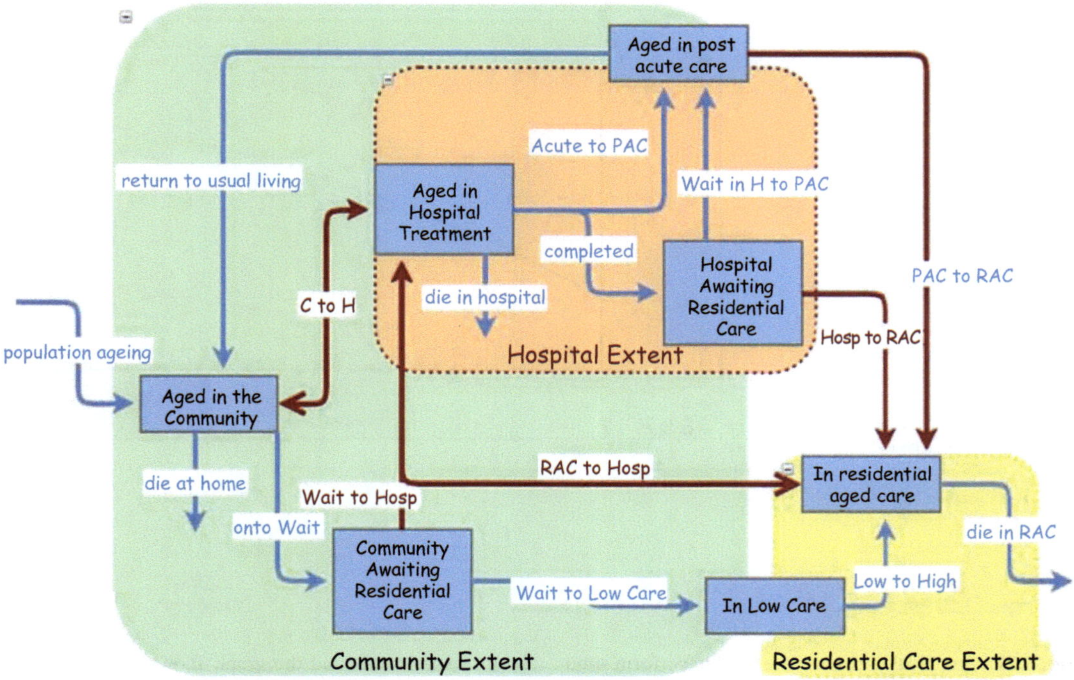

Fig. 6.26 A stock and flow diagram of flows of aged care patients between the community, acute hospital, post acute care (PAC) and residential aged care (RAC)

6.5.4.2 A Systems View of Aged Care

Based on experience with national, regional and district models of the acute aged care interface [40] we constructed a simplified model to help people understand the downstream effects associated with population ageing swamping the current systems of care across the community, hospitals and aged care sectors. A Stock and Flow diagram can be found at http://bit.ly/sLMfp8 (Fig. 6.26).

We have two groups of people waiting for aged care places, one in hospital and the other in the community (at home). The key downstream dilemma is to manage these inflows into residential aged care. This becomes increasingly difficult if quality improvements within aged care prolong life and reduce the death rate outflow from residential aged care (RAC). From the hospital point of view, the best short term fix is for RAC to admit patients from hospital as a priority. However this causes increased waits in the community and eventual increased flows of older people into hospital for treatment. Another perverse incentive is that in order to remain financially viable, aged care residences must have a flow of people through low care and so prefer to upgrade an existing resident from low to high care rather than admit a new high care patient from hospital. Virtual experiments show that intermediate post acute care (PAC) options only have lasting effects if they increase the rate of flow of return to usual living in the community. Hence the increasing demand due to baby boomers and reduced informal carers in the community requires a focus on managing expectations and services around what constitutes usual cared living in the community. The increasing complex detail is also unfolded at http://bit.ly/sLMfp8. A downloadable ithink simulation model is at http://bit.ly/u3zQn8. Detailed models are calibrated with data from many sources and include an interactive user interface which can be used to perform virtual "what-if" experiments.

Ithink/STELLA Model Output Showing Living Arrangements Output, Control Panel Options and Cost Changes over time and ability to perform what-if experiments (Figs. 6.27 and 6.28).

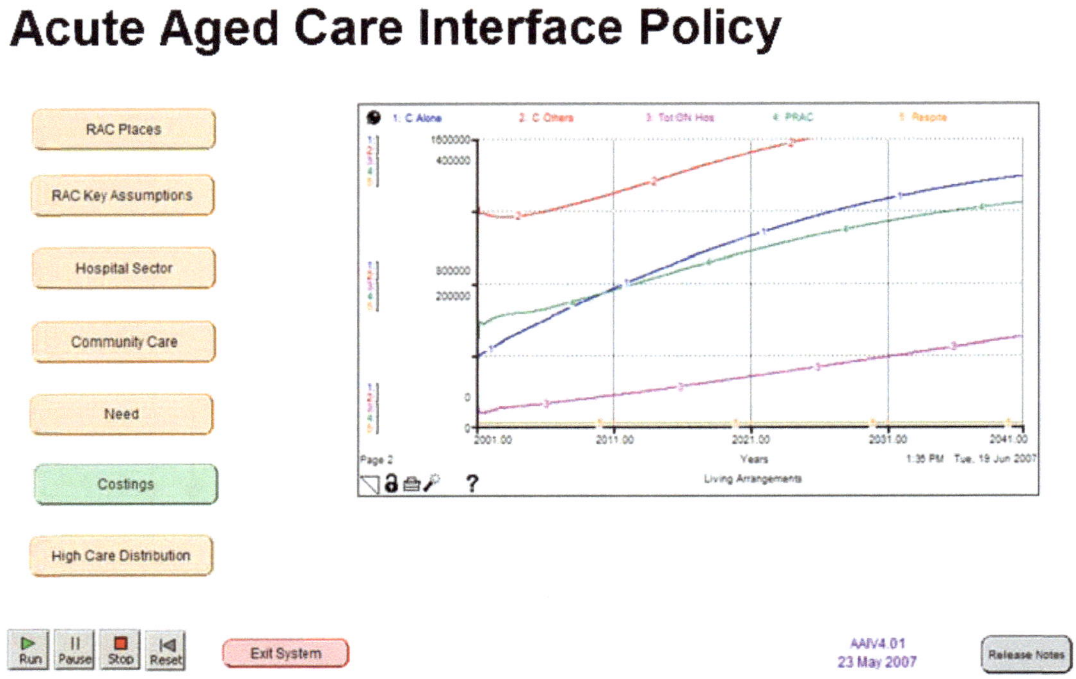

Fig. 6.27 Graphical user interface used to perform simulation experiments for exploring policies at the acute aged care interface. The left hand buttons link to detailed sector experiments and the graph shows the number of people in community, hospital, permanent and respite care by calendar year

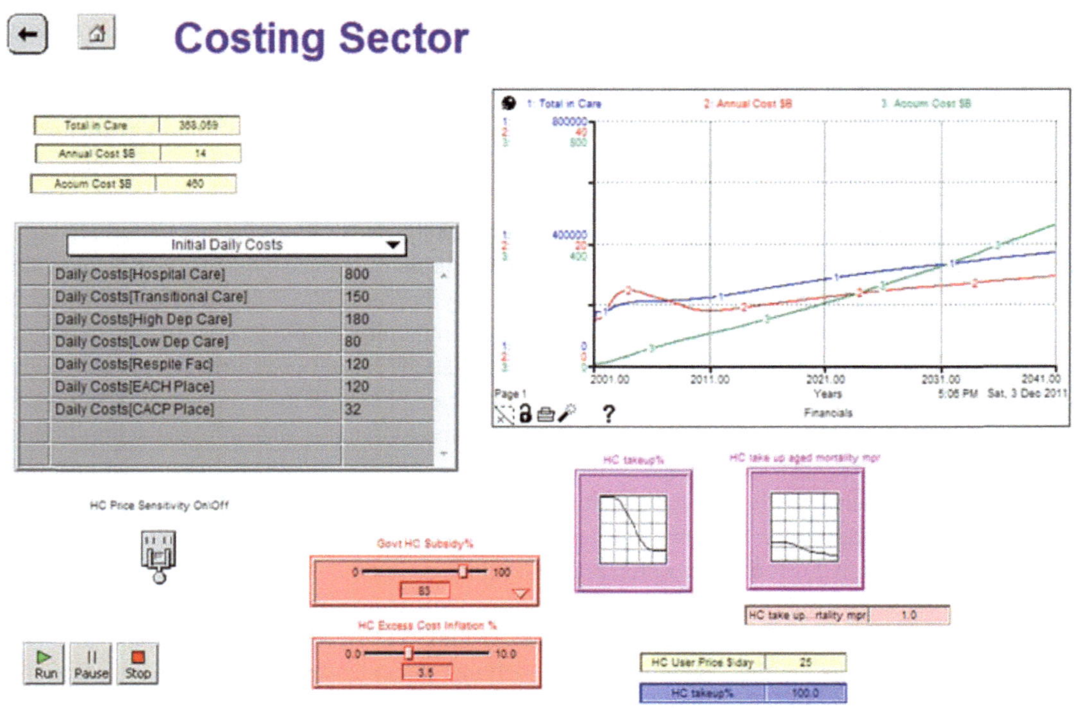

Fig. 6.28 The Costing Sector Aged Care Policy User Interface showing grey tables, pink graphical and red slider variables than can be modified to show the effect on the graphical outputs of total people in care, annual and accumulated costs by calendar years. (http://www.systemswiki.org/index.php?title=Acute_to_Aged_Care_ithink_Models)

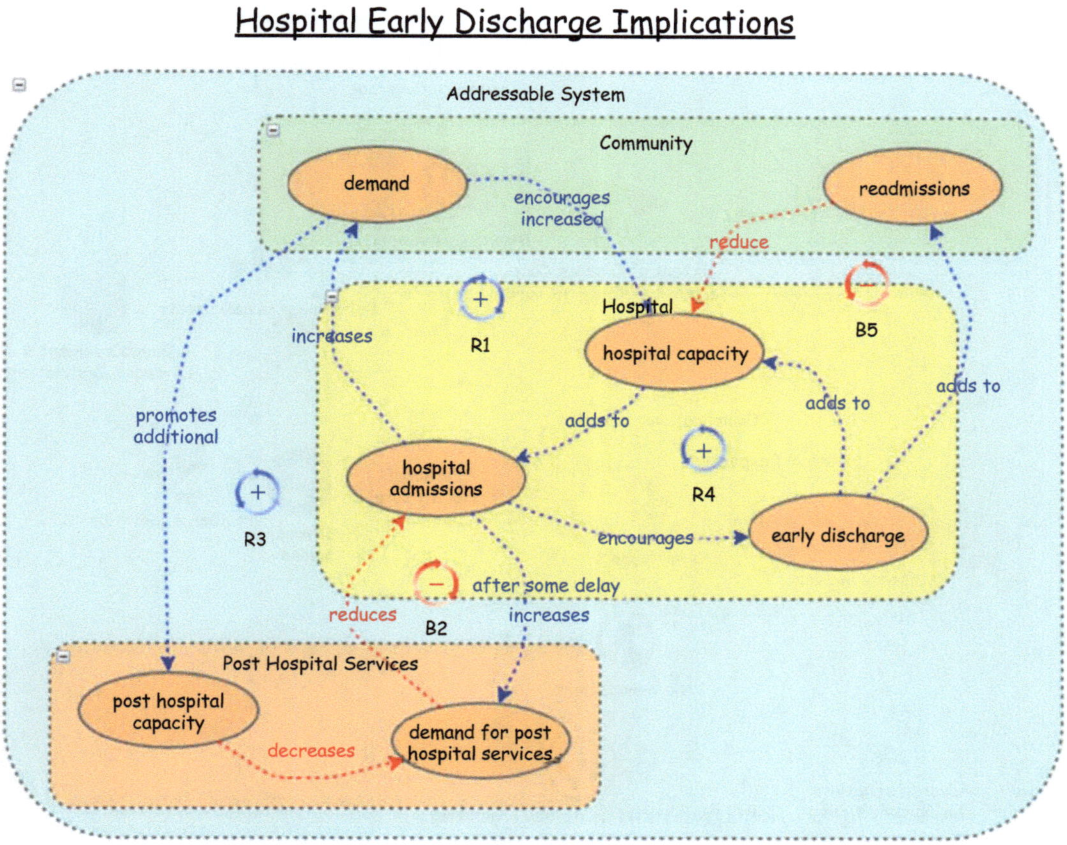

Fig. 6.29 Eric Wolstenholme's generic archetype of the implications of early hospital discharge described at http://bit.ly/svAofc

Based on these kinds of experiences in many projects, the insights gained from virtual experiments and real world experience can be expressed in insightful causal loop diagrams, such as the following generic archetype from Eric Wolstenholme [39] (Fig. 6.29).

The prevent versus treat dilemma can be extended to the whole health system, and communicated using Rich Picture diagrams, as in the following example adapted from Jack Homer, Gary Hirsch and Bobby Milstein's [41] US work on chronic illness in a complex health economy (Fig. 6.30).

6.6 Conclusions

"Models are not perfect," says Syd Levitus. "Data are not perfect. Theory isn't perfect. We shouldn't expect them to be. It's the combination of models, data, and theory that lead to improvements in our science, in our understanding of phenomena."

http://earthobservatory.nasa.gov/Features/OceanCooling/page5.php accessed Nov 12 2008

Forrester has set the standard for system dynamics models in his books on industrial, urban and world dynamics. He recently described what makes a good system dynamics model as the following:

1. The description starts with a clear statement of the system shortcoming to be improved.
2. It displays a compact model that shows how the difficulty is being caused.
3. It is based on a model that is completely endogenous with no external time series to drive it.
4. It argues for the model being generic and descriptive of other members of a class of systems to which the system at hand belongs.

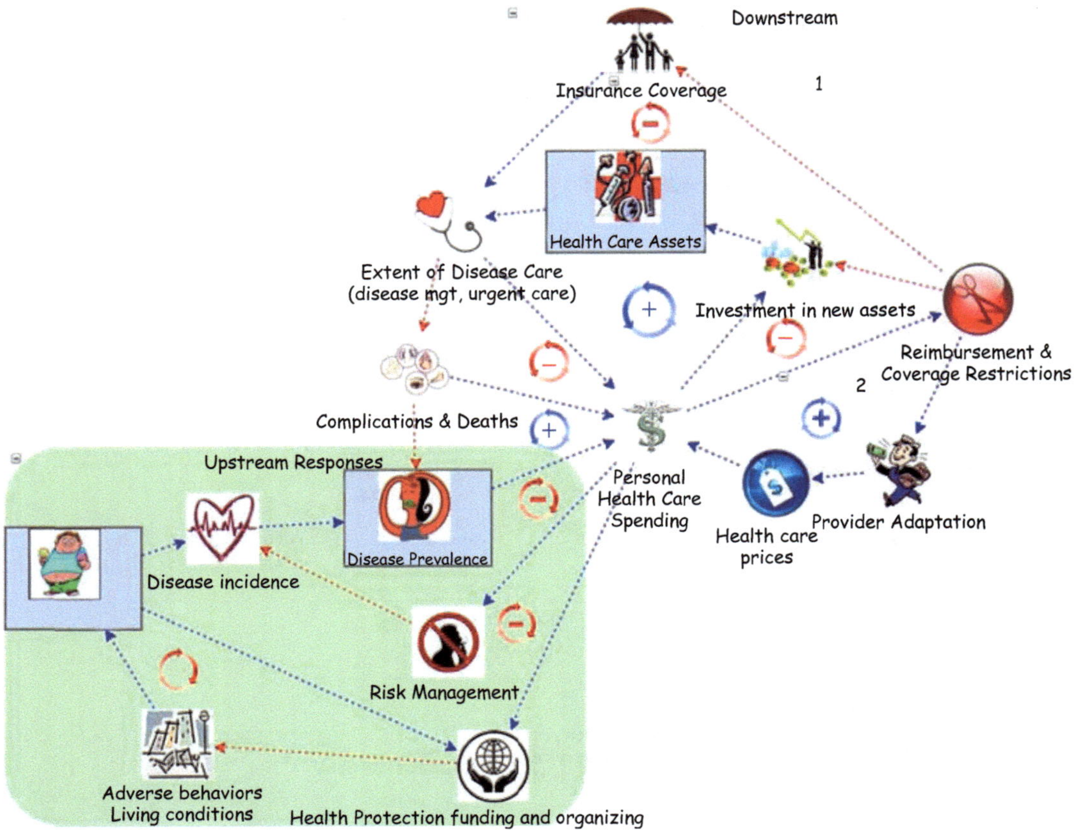

Fig. 6.30 A rich picture representation of the upstream and downstream interventions for managing chronic care in the US Health Care System. This example is unfolded progressively at http://bit.ly/wUnNxq

5. It shows how the model behaviour fits other members of the class as policies followed by those other members are tested.
6. It arrives at recommended policies that the author is willing to defend.
7. It discusses how the recommended policies differ from past practice.
8. It examines why the proposed policies will be resisted.
9. It recognises how to overcome antagonism and resistance to the proposed policies.

Forrester ISDC Plenary Session Boston 2007 and SD List 12 Feb 2008.

References

1. Sturmberg JP. The foundations of primary care. Daring to be different. Oxford, San Francisco: Radcliffe Medical Press; 2007.
2. Pidd M. Tools for thinking: modelling in management science. 3rd ed. Chichester, UK: Wiley; 2010.
3. Frigg R, Hartmann S. Models in science: Stanford Encyclopedia of Philosophy 2012, http://plato.stanford.edu/entries/models-science/#Con.
4. Rosenblueth A, Wiener N. The role of models in science. Philos Sci. 1945;12(4):316–21.
5. Meadows DH, Wright D. Thinking in systems: a primer. White River Junction, VT: Chelsea Green Publishing Company; 2009.
6. Greenhalgh T, Hurwitz B. Narrative based medicine: why study narrative? Br Med J. 1999;318:48–50.
7. Sturmberg JP, Siew E-G, Churilov L, Smith-Miles K. Identifying patterns in primary care consultations: a cluster analysis. J Eval Clin Pract. 2009;15(3): 558–64.
8. Deboeck G, Kohonen T. Explorations in finance with self organizing maps. London: Springer; 1998.
9. Bolton P, Mira M, Sprogis A. Oranges are not the only fruit: the role of emergency departments in providing care to primary care patients. Aust Health Rev. 2000;23(3):132–6.
10. Cooke M, Wilson S, Pearson S. The effect of a separate stream for minor injuries on accident and emergency department waiting times. Emerg Med J. 2002;19(1):28–30.
11. Jiang H, editor. Predicting emergency department status. Proceedings of International Conference on Optimization: Techniques and Applications Pacific

Optimization Research Activity Group. Ballarat, Australia, 2004.
12. Coleridge JT, Cameron PA, White JB, Epstein J. Profile of emergency department attendances. Emerg Med. 1993;5(1):18–27.
13. Liaw S-T, Hill T, Bryce H, Adams G. Emergency and primary care at a Melbourne hospital: reasons for attendance and satisfaction. Aust Health Rev. 2001; 24(2):120–34.
14. Ceglowski R, Churilov L, Wasserthiel J. Combining data mining and discrete event simulation for a value-added view of a hospital emergency department. J Oper Res Soc. 2007;58(2):246–54.
15. Lane DC, Monefeldt C, Rosenhead JV. Looking in the wrong place for healthcare improvements: a system dynamics study of an accident and emergency department. J Oper Res Soc. 2000;51(5):518–31.
16. Bagust A, Place M, Posnett JW. Dynamics of bed use in accommodating emergency admissions: stochastic simulation model. Br Med J. 1999;319(7203):155–8.
17. Hoffenberg S, Hill MB, Houry D. Does sharing process differences reduce patient length of stay in the emergency department? Ann Emerg Med. 2001;38(5): 533–40.
18. Harper PR. A framework for operational modelling of hospital resources. Health Care Manage Sci. 2002; 5(3):165–73.
19. Jun JB, Jacobson SH, Swisher JR. Application of discrete-event simulation in health care clinics: a survey. J Oper Res Soc. 1999;50(2):109–23.
20. Schneider KCMPH, Lichtenstein JLMD, Freeman JLP, Newbold RCMM, Fetter RBDBA, Gottlieb LMD, et al. Ambulatory visit groups: an outpatient classification system. J Ambul Care Manage. 1988; 11(3):1.
21. Jelinek G. A Casemix information system for Australian Hospital Emergency departments. Perth, Western Australia: A report to the Commissioner of Health; 1995.
22. Bond M, Erwich-Nijhout M, Phillips D, Baggoley C. Urgency, disposition and age groups: a Casemix model for emergency medicine. Emerg Med. 1998; 10(2):103–10.
23. Isken MW, Rajagopalan B. Data mining to support simulation modeling of patient flow in hospitals. J Med Syst. 2002;26(2):179–97.
24. Kohonen T. Self-organizing maps. Berlin: Springer; 1995.
25. Kennedy RL, Lee Y, Van Roy B, Reed CD, Lippman RP. Solving data mining problems through pattern recognition. Englewood Cliffs, NJ: Prentice-Hall; 1998.
26. Ward J. Hierarchical grouping to optimise an objective function. J Am Statist Assoc. 1963;58(301): 236–44.
27. Baesler F, Jahnsen H, DaCosta M, editors. The use of simulation and design of experiments for estimating maximum capacity in an emergency room. New Orleans, LA: Winter Simulation Conference INFORMS; 2003.
28. Brailsford S, Churilov L, Liew S, editors. Treating ailing emergency departments with simulation: An integrated perspective, 36th Annual Simulation Symposium—Society for modeling and simulation international. Orlando, FL: IEEE Computer Society; 2003.
29. Mahapatra S, Koelling CP, Patvivatsiri L, Fraticelli BMP, Eitel D, Grove L, editors. Pairing emergency severity index 5-level triage data with computer aided system design to improve emergency department access and throughput. New Orleans, LA: Winter Simulation Conference INFORMS; 2003.
30. Miller MJ, Ferrin DM, Szymanski JM, editors. Simulating Six Sigma improvement ideas for a hospital emergency department. New Orleans, LA: Winter Simulation Conference INFORMS; 2003.
31. Miró Ò, Sánchez M, Espinosa G, Coll-Vinent B, Bragulat E, Millá J. Analysis of patient flow in the emergency department and the effect of an extensive reorganisation. Emerg Med J. 2003;20(2):143–8.
32. Samaha S, Armel WS, Starks DW, editors. The use of simulation to reduce the length of stay in an emergency department. New Orleans, LA: Winter Simulation Conference INFORMS; 2003.
33. Wiinamaki A, Dronzek R, editors. Using simulation in the architectural concept phase of an emergency department design. New Orleans, LA: Winter Simulation Conference INFORMS; 2003.
34. Connelly LG, Bair AE. Discrete event simulation of emergency department activity: a platform for system-level operations research. Acad Emerg Med. 2004;11(11):1177–85.
35. Cleary MI, Ashby RH, Jelinek GA, Lagaida R. The future of Casemix in emergency medicine and ambulatory care. Med J Aust. 1994;161(Suppl):S30–3.
36. Motohashi Y, Nishi S. Prediction of end-stage renal disease patient population in Japan by systems dynamics model. Int J Epidemiol. 1991;20(4):1032–6.
37. Masnick K, McDonnell G. A model linking clinical workforce skill mix planning to health and health care dynamics. Hum Resour Health. 2010;8(1):11.
38. Paul SA, Reddy MC, DeFlitch CJ. A systematic review of simulation studies investigating emergency department overcrowding. Simulation. 2010;86(8-9): 559–71.
39. Wolstenholme E, Monk D, McKelvie D, Arnold S. Coping but not coping in health and social care: masking the reality of running organisations beyond safe design capacity. Syst Dyn Rev. 2007;23(4):371–89.
40. Gray L, Broe GA, Duckett SJ, Gibson DM, Travers C, McDonnell G. Developing a policy simulator at the acute-aged care interface. Aust Health Rev. 2006; 30(4):450–7.
41. Homer J, Hirsch G, Milstein B. Chronic illness in a complex health economy: the perils and promises of downstream and upstream reforms. Syst Dyn Rev. 2007;23(2–3):313–43.

General System Dynamics and Health References

- Bellinger G. Systemic perspective. 2010.
- http://www.systemswiki.org/index.php?title=Systemic_Perspective.

A Methodology for the Analysis of Medical Data

7

A. Tsanas, M.A. Little, and P.E. McSharry

7.1 Introduction and Terminology

Imagine a subject going to the clinic for a medical diagnosis, for example to assess the functionality of his cardiovascular system. The doctor requests a number of clinical tests (for example, stress test to obtain the electro-cardiogram (ECG) and Doppler ultrasound), takes into account a number of other factors (for example the demographics of the subject), and makes his final diagnosis using the current data and his prior knowledge. For his diagnosis, the doctor will usually compute some *characteristics* of the original raw signal. For example, when the raw signal is the ECG, clinicians may want to use the mean heart rate or the heart rate variability (these characteristics may also be readily provided by medical software) because experience has taught them these characteristics are useful in diagnosis.

The discipline of *statistical machine learning* (informally data analysis) offers a framework which allows researchers to decipher what the computed characteristics reveal, and how these characteristics could be used to offer a *decision support tool*. A further aim is to investigate whether additional characteristics, which may have been previously ignored, could or should be taken into account. A guide on detecting patterns is outside the scope of this study; instead we will focus on the case where a number of characteristics have been collected (as indicated earlier, these characteristics might have been extracted from the original raw signals, demographic data, values of genes, concentration of a particular component in a given area, and others). Characteristics which are *qualitative* can be assigned to an *ordinal scale*.[1] For example, in the case that medical practitioners characterize cells as having (a) low concentrations, (b) moderate concentrations, and (c) large concentration, we could define a scale that would read:

$$1 = \text{low concentration},$$
$$2 = \text{moderate concentration},$$
$$3 = \text{large concentration}.$$

A. Tsanas (✉)
Department of Engineering Science,
Systems Analysis, Modelling and Prediction (SAMP),
University of Oxford, Oxford, UK

Oxford Centre for Industrial and Applied Mathematics
(OCIAM), University of Oxford, Oxford, UK
e-mail: tsanas@maths.ox.ac.uk; tsanasthanasis@gmail.com

M.A. Little
Oxford Centre for Industrial and Applied Mathematics
(OCIAM), University of Oxford, Oxford, UK

Media Lab, Massachusetts Institute of Technology,
Cambridge, MA, USA
e-mail: maxl@mit.edu; max.a.little@gmail.com

P.E. McSharry
Smith School of Enterprise and the Environment,
University of Oxford, Oxford, UK
e-mail: patrick.mcsharry@smithschool.ox.ac.uk;
patrick@mcsharry.net

[1] The term *ordinal scale* refers to a hierarchical ordering of values to differentiate the different possible outcomes, where the difference between successive values within the scale is not equal.

J.P. Sturmberg and C.M. Martin (eds.), *Handbook of Systems and Complexity in Health*,
DOI 10.1007/978-1-4614-4998-0_7, © Springer Science+Business Media New York 2013

For reasons that will become clear later it is advisable to use progressively increasing values of the ordinal scale starting with healthy condition and characterizing pathological situations with higher values. Each characteristic is represented by a single scalar value.

We have purposefully avoided the use of mathematical terms so far, but here we need to define some terms that are commonly used in statistical settings: *explanatory variables* (or *features*), and *response variable* (or simply *response*). The term "feature" is equivalent to the computed *characteristic*, and the term "response variable" can be thought to be equivalent to the *diagnosis* or the *clinical outcome*. In most medical settings, the diagnosis or clinical outcome can take a small range of possible values. For example, the final diagnosis of a clinician may simply be a "yes" or "no" to a question (e.g. whether a subject has cancer), and might also include a third clinical outcome, e.g. "possibly". These two or three possible outcomes can be represented using an ordinal scale as indicated previously for the case of the characteristics, i.e. the response variable takes the following possible values: $1 = \text{no}'', 2 = \text{possibly}'', 3 = \text{yes}''$. This example can be generalized to a more general setting, where for example a number of explanatory variables are used to assign subjects to different pathologies. The possible values that the response variable can have are simply known as *categories* or *classes*. When the response variable can take any number of finite classes the problem of predicting the response variable is known as *classification*.[2] When the response variable can take any real value (any possible number from $-\infty$ to ∞), the problem is known as *regression*. Classification problems are met considerably more frequently in medical applications, and hence we focus exclusively on those cases.

It is important to stress that accurate statistical inference is only possible when a relatively large number of data *samples* are collected. A good rule of thumb is to use at least 15 data samples from each class in the response variable.

Data in most medical applications can be represented in the form:

$$\mathbf{X} = \begin{bmatrix} x_{11} & \cdots & x_{1M} \\ \vdots & \ddots & \vdots \\ x_{N1} & \cdots & x_{NM} \end{bmatrix}, \mathbf{y} = \begin{bmatrix} y_1 \\ \vdots \\ y_N \end{bmatrix}$$

$\mathbf{X}$ is known as the *design matrix* (or *data matrix*), where each row includes the M explanatory variables for each subject. That is, each row contains the concatenated vector of the explanatory variables which characterize the subject's condition. For example x_{11} could be the age of the first subject, x_{12} the gender, x_{13} the mean heart rate, x_{14} the heart rate variability and so on. Effectively, $\mathbf{X}$ simply summarizes the explanatory variables M for the number of observations (*samples*) N (each row usually refers to a different subject). Each column in $\mathbf{X}$ contains the values of one explanatory variable across *all* samples and is indicated as $\mathbf{f}_j = (x_{1j},\ldots,x_{Nj}),\ j\epsilon 1,\ldots,M$. The response variable $\mathbf{y}$ is believed to be associated with $\mathbf{X}$ based on prior knowledge in the given problem. It is populated with the outcomes for each sample, for example for the first sample we could have $y_1 = 0$ to denote healthy state, $y_2 = 1$ to denote pathological state for the second subject, and so on. Once the data is summarized in a format like the one presented earlier with $\mathbf{X}$ and $\mathbf{y}$, the aim is to decipher the concealed information. Questions such as the following are frequently met in medical contexts (the list is only indicative):

1. How can we associate $\mathbf{X}$ and $\mathbf{y}$? That is, what is the relationship between the explanatory variables and the response variable?
2. Is there a convenient way to estimate the response variable when presented with the explanatory variables of a subject?
3. Which of the explanatory variables are useful in actually determining the response variable?
4. What is the relationship between the explanatory variables? Is it possible that some of the

[2] When the response variable can only be one of two classes, the problem is referred to as *binary* classification; when there are more than two classes, the problem is known as *multi-class* classification. Binary classification problems are met very frequently in medical applications, for example to differentiate whether the patients live or die.

explanatory variables are redundant and need not be computed?

We will demonstrate that when analysis is confined only to reporting statistical significance values (*p-values*) these clinically important questions cannot be adequately answered.

7.2 Data Exploration and Statistical Analysis

Usually, the first step in data analysis is to explore the statistical properties of the data and to produce some plots to get an "intuitive feel". Initially the probability densities of the explanatory variables can be plotted, and the simplest approach is to use histograms.[3] Histograms provide a nice overview of the distribution of values for each explanatory variable and for the response variable. They use a number of *bins* (for example 10) which span the range of possible values of the investigated variable, and count the number of data samples that fall into the range of each bin, thus providing a general impresion of the spread of the values for this variable. In addition to density plots, we suggest using *scatter plots*: a scatter plot has on the 'x-axis' one explanatory variable, and on the 'y-axis' the response variable. Scatter plots are useful to visualize whether there is any obvious relationship between the investigated explanatory variable and the response variable. Scatter plots can be used for each of the explanatory variables to present very simply the $\{\mathbf{f}_j, \mathbf{y}\}$ N points in a figure.

Visual inspection of density plots and scatter plots is usually followed by formal statistical tests in order to determine qualitatively and quantitatively how well the explanatory variables are related to the response variable. *Correlation analysis* offers a good indication of the association between each explanatory variable and the response variable, and between explanatory variables (pairwise correlations). However, we emphasize that "correlation does not necessarily suggest *causation*" (change in the values of the explanatory variable affecting the response variable) in general [1]. Correlation coefficients are regarded as a valuable hint indicating a *potential* relationship between the explanatory variable and the response. We endorse the use of the Spearman rank correlation coefficient, which can account for general *monotonic* relationships and which in general is preferable compared to the linear (Pearson) correlation coefficient (which is more appropriate in *linear* settings). Strictly speaking, formal statistical hypothesis tests (see the following paragraph) should be used to check whether the data follow *normality* (one example of normality is data that have histograms resembling a bell shaped curve). In practice, medical data will typically deviate from normality, and hence the Spearman correlation coefficient should generally be used.

Both the Spearman rank correlation and the linear correlation coefficient lie in the numeric range $[-1\ldots1]$ and are interpreted using (a) the sign of the correlation coefficient which denotes the direction of the relationship, and (b) the magnitude (absolute value) of the correlation coefficient. Negative sign indicates that the direction of the relationship between the variables is opposite: the increase in the values of one variable leads to the decrease in the values of the other. The larger the magnitude of the correlation coefficient, the stronger the statistical relationship between the variables is. There is no general rule to determine when a relationship is *statistically strong*; it depends on the specifics of the application [5]. In medical contexts, statistical relationships are fairly weak and typically the magnitude of the correlation coefficient is lower than 0.3 (once again, we stress that this value can only be used as guidance and referring to relationships between variables as *statistical strong* when the magnitude of the correlation coefficient is above a certain threshold is considered arbitrary). To differentiate the relationships between feature and response, and between features, we introduce some additional terminology. The correlation coefficient between the feature $\mathbf{f}_j$ and the

[3] In general, histograms are considered a simple but rather crude approach. *Kernel density estimation* is typically preferable [see Hastie et al. [8] for more details], and can be thought of as a smoothened version of histograms.

response variable **y** is denoted by $I(\mathbf{f}_j;\mathbf{y})$ and is known as *relevance*; similarly, the correlation coefficient between the feature $\mathbf{f}_j$ and another feature $\mathbf{f}_s$ is denoted by $I(\mathbf{f}_j;\mathbf{f}_s)$ and is known as *redundancy*.

Statistical hypothesis tests are commonly used in data analysis applications to determine whether the observed result follows a certain hypothesis, which in statistical terminology is known as the *null hypothesis*. Often, the null hypothesis is the opposite of what we aim to demonstrate; therefore in practice the objective is often met when we can *reject* the null hypothesis in favour of the *alternative hypothesis*. Statistical hypothesis tests compute significance values, the well-established *p*-values, which can be interpreted as the probability of obtaining a similar result by chance if the null hypothesis is true. The null hypothesis is rejected when the *p-value* is lower than a prespecified *significance level*, typically 0.05 or 0.01, and the result is deemed to be *statistically significant* at the significance level chosen. Thus, for example, $p<0.05$ denotes a statistically significant result at the 5% significance level (i.e. there is less than 5% probability that the observed relationship is due to chance).

7.3 Reducing the Number of Explanatory Variables

A common problem in data analysis applications arises when using a large number of explanatory variables, and is known as the *curse of dimensionality*: potentially, using fewer explanatory variables could lead to a simpler model which may allow more accurate estimation of the response variable [8]. This initially puzzling assertion (one could imagine that collecting as much information as possible in the form of explanatory variables can only be positive to infer properties from the data) occurs because in practice we do not have an infinite number of samples. The problem is exacerbated in the cases where the number of explanatory variables is larger than the number of samples (e.g. in microarray settings where the number of genes is typically in the order of thousands and the number of samples in the order of 100-200 samples). Moreover, in practice, some explanatory variables contribute little information to predicting the response variable. In other scenarios, some explanatory variables can be considered redundant in the presence of other explanatory variables (i.e., they contribute little *additional* information towards predicting the response variable, when some other explanatory variables are already used). There are two fundamentally different approaches to reduce the number of features: *feature transformation* and *feature selection*. Feature transformation aims to transform the original features into new features, which may be more appropriate for quantifying the information in the dataset towards predicting the response variable. However, feature transformation is problematic in settings with a very large number of features [14] and is not easily interpretable because the physical meaning of the original features cannot be retrieved. On the contrary, feature selection is particularly desirable in many disciplines because the originally computed features typically quantify some characteristic which is interpretable to experts in that domain. There is a substantial body of literature addressing the topic of feature selection from many angles, and we refer to Guyon (2006) for a more extensive discussion.

For our purposes, we suggest using a simple technique which aims to select a subset of the original (large) pool of explanatory variables. Informally, we aim to select a small number of columns from the design matrix **X** and delete the remaining columns. The new design matrix will have the same number of samples *N*, but a lower number of explanatory variables: $m<M$, where m remains to be decided. It is reasonable then to select those explanatory variables which are highly correlated with the response variable. However, one problem with this approach is that potentially some explanatory variables will be highly correlated between them, which means that they will be redundant (as stated earlier, they would have little additional contribution towards predicting the response variable). Therefore, we need to find a compromise to account both for (a) including the most relevant explanatory variables towards predicting the response variable, and (b)

Table 7.1 Incremental feature selection steps in mRMR

1.	(Selecting the first feature index) include the feature index j: $\max j \in Q\left(I\left(\mathbf{f}_j;\mathbf{y}\right)\right)$ in the initially empty set S, that is $\{j\} \rightarrow S$
2.	(Selecting the next $m-1$ features, one at each step, by repeating the following) apply the criterion in Eq. (7.1) to incrementally select the next feature index j, and include it in the set: $S \cup \{j\} \rightarrow S$
3.	Obtain the feature subset by selecting the features $\{\mathbf{f}_j\}_{j=1}^{m}$, $j \in S$ from the original data matrix $\mathbf{X}$

excluding the most redundant explanatory variables. Although there are many feature selection techniques in the literature, we endorse using a conceptually simple and intuitively appealing idea proposed by Peng et al. [13], which we modify slightly here for simplicity. Specifically, mRMR relies on an intuitive heuristic criterion compromising between feature relevance and feature redundancy and can be expressed with the following equation:

$$\text{mRMR} \stackrel{\text{def}}{=} \max_{j \in Q-S}\left[\underbrace{I\left(\mathbf{f}_j;\mathbf{y}\right)}_{Relevance} - \underbrace{\frac{1}{|S|}\sum_{s \in S} I\left(\mathbf{f}_j;\mathbf{f}_s\right)}_{Redundancy}\right] \tag{7.1}$$

where $\mathbf{f}_j$ denotes the jth feature amongst the M initial features and $\mathbf{f}_s$ is a feature that has been already selected in the feature index subset S (s is an integer, Q contains the indices of all the features in the initial feature space, that is $1,\ldots,M$, S contains the indices of selected features and $Q-S$ denotes the indices of the features not in the selected subset). Peng et al. [13] have used a more complicated criterion to quantify relevance and redundancy instead of the correlation coefficient used here, but the conceptual idea remains the same. The steps used to incrementally select features are described in Table 7.1.

We remark that when there is a single dataset used to train and test the classifier, the features should be selected using a cross-validation setting (see the following section). That is, a subset of the original data samples should be selected and the feature selection process should be run on this subset. It is advisable to repeat this feature selection process a number of times, where each time a different sample subset is drawn from the original dataset (we suggest using 100 iterations where on each iteration we randomly select 90% of the data and use this subset for selecting the features). Theoretically, in all the subsets the selected features should be identical: this would be the "true" ordering. In practice, however, it is likely that a different feature ordering will result for the sample subsets drawn from the original data matrix. Then, we can either select the feature subset that occurs most often, or select individually the features which appear consecutively most often. One robust mechanism to select features in this case is described in Tsanas et al. [18].

7.4 Mapping Explanatory Variables to the Response Variable

As we have mentioned in the beginning of this chapter, in a wide range of problems we are interested in determining a functional mapping of the explanatory variables to the response variable, that is find a function f which uses $\mathbf{X}$ to predict $\mathbf{y}$: $f(\mathbf{X})=\mathbf{y}$. This can be achieved in two ways (a) we can impose a structure in the functional form of f and aim to determine the parameters of that functional form (*parametric* setting) or (b) allow the data itself determine the structure and the parameters of that structure (*non-parametric* setting). Both approaches have the right to exist, and there is considerable interest amongst statisticians regarding the merits of either approach [3]. One example of the parametric setting has the form $y_i = a_1 x_{i,1} + \cdots + a_M x_{i,M}$ ($i = 1,\ldots,N$), where the vector with the parameters $\mathbf{a}=(a_1,\ldots,a_M)$ needs to be estimated. We remind the reader that M represents the number of explanatory variables in the design matrix we use. In case we have selected a smaller subset of the explanatory variables *m* (e.g. using the feature selection technique mRMR described in the previous section), then the parametric model form would use *m* parameters **a**. Parametric settings are generally simpler than non-parametric settings and may be more easily interpretable. If the functional form (*model structure*) is known a priori, then parametric settings can be very useful. However, imposing an inaccurate model structure may lead to false interpretation of the properties of the data. Hence, in practice using a non-parametric functional form may often be more appropriate. We will now briefly introduce some widely used classifiers which our readers may already be familiar with, and one more complicated classifier, which often works very well in practice. For specific algorithmic details, we refer to Bishop [2] and Hastie et al. [8].

The first classifier is known by the name Logistic Regression (LR), which may be considered a misnomer since, by definition, it works on *classification* problems [2]. This classifier is frequently used by clinicians when constructing the functional form $f(\mathbf{X})=\mathbf{y}$ to identify the effects of features on the response [3]. Originally LR was proposed for binary classification settings, but it has been generalized for multi-class classification problems as well. Conceptually, this classifier aims to provide a model which relies *linearly* on the explanatory variables. LR models have found extensive use in medical applications where the aim is to understand how the explanatory variables affect the response variable (i.e. it can be considered as a conceptual extension of using the correlation coefficients we have previously mentioned) by looking at the coefficients associated with each feature. Nevertheless, practice has shown that LR models can lead to faulty conclusions in the presence of correlated features [2], and hence we suggest extreme caution in interpreting the values of the LR coefficients. LR models require a relatively large number of samples compared to the number of explanatory variables in order to have confidence in the computed results [2,8].

Random Forests (RF) is a powerful *non-parametric* classifier, which can provide a model where the explanatory variables combine *nonlinearly* to estimate the response variable [4]. It is constructed by combining many base experts, the *trees* (by default 500 trees), and then uses majority voting from the trees to decide on the final output. We will not go into the mathematical details for the construction of the trees, since they are readily available elsewhere [4,8]; instead we will provide an intuitive overview of this powerful machine learning approach. Interestingly, the way trees are built is not dissimilar to the mindset of clinicians, where there are successive binary splits of the data before reaching a conclusion on how to classify a new sample. Effectively, trees partition the data based on a single feature at each decision point (node), to split the population. This approach could be compared to the method that clinicians use in deciding the course of optimal treatment for a patient. For example, their first criterion could be age, where optimal treatment is different for people who are over 50 years old. Then, for those patients who are less than 50 gender may be crucial and different therapy is going to be applied. Similarly, for those who are over 50, possibly gender is not important, but there is some other parameter that the clinicians would consider before deciding on the treatment. This scenario can be schematically presented in Fig. 7.1, and this is how trees actually work (note that a particular feature can be used more than once, and it is possible that some features will not be used at all). Similarly to a clinician, the final decision of the tree is reached when we follow the nodes in the tree and we are confident there are no more useful splits. In practice, the trees are grown until we have reached a pre-specified minimum number of samples assigned to each node.

The algorithmic 'trick' to grow diverse trees is to limit the number of features that can be used in each node by each tree. The binary splits used in the tree generation process lead to *nonlinear* combinations

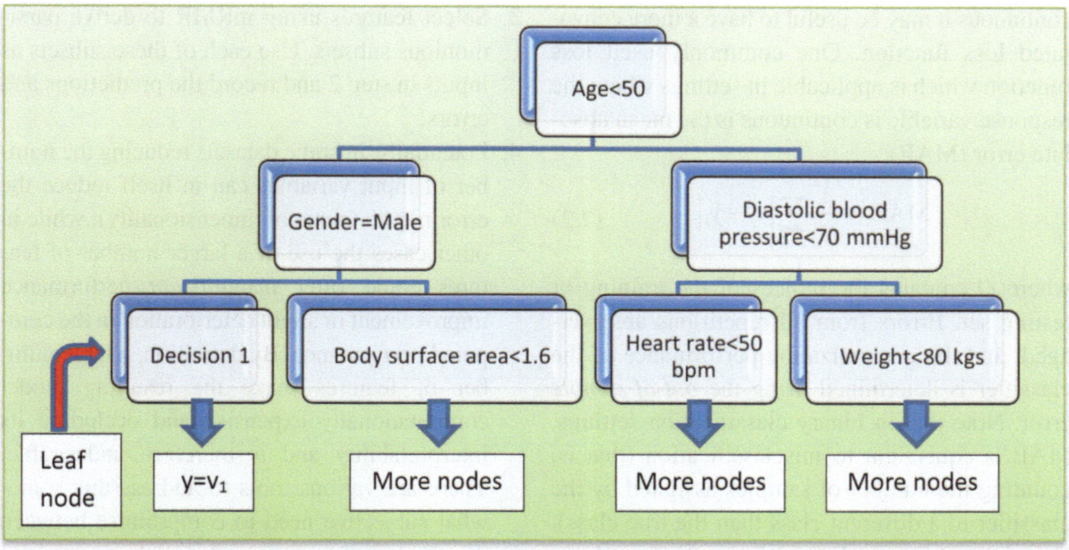

Fig. 7.1 Example of a tree. In each node, the decision is to go to the left side of the tree if the statement is true, and go to the right side of the tree if the statement is false. The point where a decision is made to assign the output y to a specific value v is known as leaf. Here, we have used variables which may be familiar to clinicians over which we split the data. Many dissimilar trees constitute the random forest

between the features, and hence RF often outperform linear classifiers such as LR (we will see specific examples in the following sections).

7.5 Model Validation and Generalization

Once the functional form f has been determined, we need to establish how accurate the mapping $f(\mathbf{X})=\mathbf{y}$ could be if a new dataset with similar properties to the dataset used to obtain f is collected. This is known as the *generalization* performance of the model which is typically estimated using (a) *cross-validation*, (b) *bootstrapping*, or (c) an *additional dataset*, which has not been used to train the model (i.e. in the determination of f). We endorse the use of cross-validation (CV), a well-known statistical re-sampling technique [19], because it is usually the simplest approach.

Specifically, in CV the original dataset is split into a *training* subset, which is used to determine f, and a *testing* subset, which is used to assess the classifier's generalization performance. The ratio of the training subset over testing subset (number of samples in each subset) is determined by the researcher and is known as K-fold cross-validation, with typical ratios being fivefold (5:1) and tenfold (10:1) [8]. The model parameters are determined using the *training* subset, and errors are computed using the *testing* subset (*out-of-sample error* or *testing error*). The process should be repeated a large number of times (e.g. 100–1,000), where the dataset is randomly permuted in each run prior to splitting in training and testing subsets, in order to obtain statistical confidence in this assessment. Depending on the requirements of the problem, different *loss functions* can be introduced. In all cases, on each repetition we record an error which has the form $L(\{y_i, \hat{y}_i\}_{i=1}^{N_t})$, where N_t represents the number of samples in the training or testing subset, y_i is the true class and $\hat{y}_i$ is the estimated class of the ith sample.

The choice of loss function is critical and depends on the demands of the application. The simplest loss function is misclassification, i.e. identifying the number of samples incorrectly assigned to a different class compared to the true class. In multi-class classification settings where the response variable classes are ordinal or continuous it may be useful to have a more convoluted loss function. One commonly used loss function which is applicable in settings where the response variable is continuous is the mean absolute error (MAE):

$$\text{MAE} = \frac{1}{N_t}\sum_{i \in Q}\left|y_i - \hat{y}_i\right|, \quad (7.2)$$

where Q contains the indices of the training or testing set. Errors from all repetitions are averaged, and the generalization performance of the classifier is determined using the *out-of-sample* error. Note that in binary classification settings, MAE is equivalent to misclassification (that is, counting the number of samples assigned by the classifier to a different class than the true class). For convenience, we have expressed all results in percentage scores, i.e.

$$\text{MAE}(\%) = 100 \times \frac{1}{N_t}\sum_{i \in Q}\left|y_i - \hat{y}_i\right|. \quad (7.3)$$

7.6 Summary of the Proposed Methodology

This chapter provided a succinct data analysis guide, pruning away all mathematically complicated details and distilling only the essential knowledge for the successful practical application of the algorithmic tools. We summarize the proposed methodology in the following steps:

1. Apply various statistical tests to the data. These could include testing for the assumption of Gaussianity, determining p-values and correlation coefficients, and understanding the underlying structure of the quantities involved in the analysis. Produce density plots and scatter plots to visualize the data. These plots may inspire the use of transforming some of the features, for example by using the log-transformation if some features are not well spread out.
2. Apply standard classification algorithms (e.g. logistic regression) and also use more complicated, nonlinear methods such as random forests. All the explanatory variables are used to predict the response variable(s).
3. Select features using mRMR to derive parsimonious subsets. Use each of these subsets as inputs in step 2 and record the predictions and errors.
4. Potentially, in some datasets reducing the number of input variables can in itself reduce the error metric (curse of dimensionality); while in other cases the use of a larger number of features could offer *insignificant* performance improvement or slight deterioration in the computed performance. By definition, a large number of features make the resulting model computationally expensive and occluding its interpretability and is therefore undesirable. There are various tools to address this somewhat subjective need to compromise between the number of features and model performance (we want the classifier to give as accurate results as possible). One approach is to use information criteria, and another is to use the one-standard-error rule. We refer to Hastie et al. [8] for more details regarding both approaches.
5. Use new data or more likely use tenfold cross-validation with at least 100 repetitions to ensure the results are robust.

The list can easily be modified and is purposefully general, so that it is applicable to a wide range of medical applications. The field of exploratory data analysis and knowledge discovery cannot be possibly covered adequately here; we refer to the survey of Kurgan and Musilek for a relatively recent authoritative overview [11].

7.7 Example Applying the Proposed Methodology in a Medical Problem

To demonstrate the proposed methodology in a practical problem we use the Hepatitis dataset, which has been widely studied in the literature. The dataset is available for download from the UCI machine learning repository at http://archive.ics.uci.edu/ml/datasets/Hepatitis. The problem is to investigate whether a set of features can be used to predict whether the patient lives or dies. The design matrix has 155×19 elements, that is,

Table 7.2 Correlations between features and the response variable for the Hepatitis dataset

Feature name	Spearman correlation coefficient	Statistical significance of the correlation (p-value)	Samples used
ALBUMIN	0.480	$p<0.001$	139
ASCITES	0.478	$p<0.001$	150
PROTIME	0.450	$p<0.001$	88
SPIDERS	0.398	$p<0.001$	150
BILIRUBIN	−0.372	$p<0.001$	149
VARICES	0.369	$p<0.001$	150
MALAISE	0.338	$p<0.001$	154
HISTOLOGY	−0.338	$p<0.001$	155
FATIGUE	0.309	$p<0.001$	154
AGE	−0.245	0.002	155
SPLEEN PALPABLE	0.239	0.003	150
ALK PHOSPHATE	−0.189	0.034	126
SEX	0.173	0.031	155
STEROID	0.135	0.095	154
ANOREXIA	0.132	0.102	154
ANTIVIRALS	−0.130	0.106	155
SGOT	−0.100	0.222	151
LIVER BIG	−0.078	0.353	145
LIVER FIRM	0.063	0.452	144

The last column in this table indicates the number of samples available for that feature. In practice, some characteristics may not be measured which explains why we do not have 155 samples for all features

it comprises 155 samples and 19 features. Each of the 19 features quantifies some characteristic which the researchers who collected the data believed that affects the response. The response in this problem can be one of two possible values which denote whether the patient lives or dies, i.e. we have a binary classification problem.

Diaconis and Efron [6] reported 17% misclassification, whilst Breiman [3] reported 13% misclas-sification using tenfold cross-validation.

To get an intuitive feel for the data, we generate density plots using histograms; in addition we generate scatter plots to visualize the relationship between each explanatory variable and the response. Next we compute the Spearman correlation coefficient between the explanatory variables and the response, which are presented in Table 7.2. The histograms and the scatter plots of the five most highly correlated features appear in Fig. 7.2. These results give a good indication that some of the features in the dataset are well correlated with the response. The fact that the spread of the features is not bell-shaped may inspire some transformation so that the distributions become more evenly spread. For example, we could use some simple transformation of the features, e.g. the log-transformation: we compute the logarithm of all samples in a feature that we want to transform (see Tsanas et al. [16]). Here, we will not experiment further with any transformation of the features.

This preliminary analysis concludes the first step in the proposed methodological guide. The subsequent steps will be integrated in the following section, where in addition to the Hepatitis dataset, additional indicative medical datasets are investigated.

7.8 Comparison of Logistic Regression and Random Forests in Various Datasets

In this section we introduce additional datasets from real-world medical problems to demonstrate how predictive modelling is influenced when using LR and RF. Moreover, we investigate the effect of using feature selection to obtain a robust parsimonious dataset and feed this subset into the

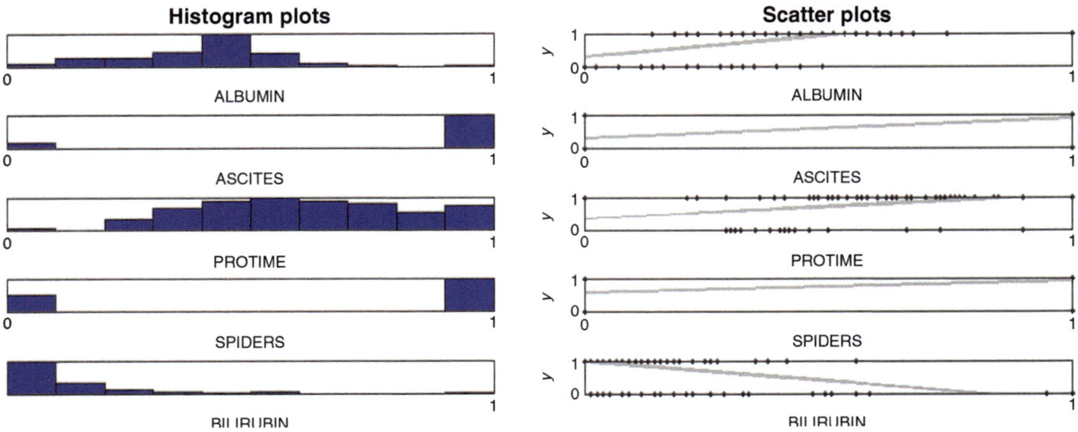

Fig. 7.2 Histogram and scatter plots of the five most correlated features with the response for the Hepatitis dataset. The horizontal axes in the scatter plots are the normalized features to facilitate direct comparison, and the vertical axes correspond to the response variable. The grey lines are the best linear fit of the data, giving a visual impression of the behaviour of the feature

Table 7.3 Summary of datasets

Dataset	Design matrix	Associated task	Feature type
Hepatitis [6]	155×19	Classification (2 classes)	C (17), D (2)
Parkinson's [12]	195×22	Classification (2 classes)	C (22)
Liver	345×6	Classification (2 classes)	D (6)
Lymphography	148×18	Classification (4 classes)	D (18)
Breast tissue [9]	106×9	Classification (6 classes)	C (9)
SRBCT [10]	63×2308	Classification (4 classes)	C (2308)
	20×2308		

The design matrix is in the form $N \times M$, where N is the number of samples and M is the number of features. Samples with missing values were amputated. 'Feature type' denotes whether the features in the design matrix are continuous (C) or discrete (D)

LR or RF classifier. All the datasets are available from the UCI machine learning repository,[4] with the exception of one dataset which was downloaded from http://stat.stanford.edu/~tibs/ElemStatLearn/data.html; due to space constraints we keep description of those datasets to a minimum and refer the reader to the original studies cited in Table 7.3 and to the UCI machine learning repository for further details.

The **Parkinson's** dataset was generated in Little et al. [12]. The aim is to characterize speech signals computing some distinctive characteristics (features) in the voices of people with Parkinson's disease versus healthy controls (binary classification problem). An extension of this concept is inferring Parkinson's disease symptom severity using speech signals [15,17]. Little et al. [12] reported approximately 10% misclassification cases when using a subset of only four features from the 22 originally computed features. The **Liver** dataset is also a binary classification dataset, where the explanatory variables refer to blood tests and the number of drinks per day. The **Lymphography** dataset focuses on a four class classification problem (the four classes are: healthy control, metastases, malign lymph, fibrosis) where the explanatory variables are various characteristics that oncologists consider relevant such as lymphatics, changes in lymphoma type, node characteristics. The

[4] The UCI machine learning repository hosts many datasets which are freely available at http://archive.ics.uci.edu/ml/.

Table 7.4 Comparison of LR and RF when using all explanatory variables

Dataset	Misclassification (%) with LR	Misclassification (%) with RF	Misclassification difference LR-RF	Relative improvement (%)	Validation scheme
Hepatitis	17.25 ± 11.48	**11.55 ± 11.03**	5.70	33.04	10-fold CV
Parkinson's	14.75 ± 8.63	**8.90 ± 6.65**	5.85	39.66	10-fold CV
Liver	32.91 ± 8.14	**25.94 ± 7.02**	6.97	21.18	10-fold CV
Lymphography	18.24	**12.83**	5.41	29.66	LOO
Breast tissue	33.02	**30.19**	2.83	8.57	LOO
SRBCT	75.00	**10.00**	65.00	86.66	Test set

The misclassification (%) results are reported in the form mean ± standard deviation. LR stands for logistic regression and RF for Random Forests. The misclassification difference between LR and RF was in all cases statistically significant ($p<0.001$) using the Mann–Whitney statistical hypothesis test. The relative improvement expresses in % terms the performance boosting when using RF over LR. It was defined as:

$$100\%\left[\text{Misclassification}(\%)\text{with LR} - \text{Misclassification}(\%)\text{with RF}\right] / \left[\text{Misclassification}(\%)\text{with LR}\right].$$

The closer the relative improvement is to 100%, the greater the improvement RF provides over LR. The validation scheme we used depended on the number of samples and the number of classes in the dataset: we used cross-validation (CV) when we had a relatively large number of samples for the number of classes in the dataset (at least 15 samples for each class), and leave-one-sample-out (LOO) when we had less than 15 samples per class. When 10-fold CV was used to validate the accuracy of the classifier, we have also used 100 repetitions for statistical confidence.

Breast tissue dataset has nine features from impedance measurements to predict the type of tissue such as carcinoma and adipose tissue (six classes). The **SRBCT** dataset contains 83 samples and 2,308 gene expression values. The response variable denotes the tumour type (4 classes). We have downloaded the dataset from http://stat.stanford.edu/~tibs/ElemStatLearn/data.html which has split the original dataset into two subsets: a *training* subset with 63 samples and a *testing* subset with 20 samples. The SRBCT dataset is particularly challenging because the number of features (the genes in this case) is considerably larger compared than the number of samples. This is known to be a scenario where LR often fails to generalize well.

Table 7.4 presents the misclassifications computed for each dataset (we report the results in the dataset kept for testing). All the features are used to obtain these results. We also experiment with mRMR to select features and feed the $1,\ldots,K$ features in the classifier (LR or RF). These results appear in Fig. 7.3. Collectively, these findings suggest that RF consistently and significantly ($p<0.001$) outperform LR, with a mean relative improvement across datasets of about 26% (without including the SRBCT dataset where LR massively underperforms). Moreover, in settings where the number of features is larger than the number of samples the improvement with RF is even more impressive. Interestingly, in some datasets a lower number of features leads to a *lower* misclassification error. This is the manifestation of the curse of dimensionality, where additional features increase the signal-to-noise ratio in the data and are detrimental for the performance of the classifier. We remark that RF is fairly robust and that LR is particularly sensitive to the number of features with respect to the number of samples, thus verifying previous reports [3,8].

Overall, the aim of this study was to encourage researchers to investigate data beyond simply reporting correlations and statistical significance values. We believe that following the simple steps outlined in this study provides a concise guide towards inferring key properties of the examined dataset.

Acknowledgments A. Tsanas gratefully acknowledges the financial support of Intel Corporation and the Engineering and Physical Sciences Research Council (EPSRC). M.A. Little acknowledges the financial support of the Wellcome Trust, grant number WT090651MF. P.E. McSharry acknowledges the financial support of the European Commission through the SafeWind project (ENK7-CT2008-213740). We also want to thank all the researchers who deposited their datasets in the UCI machine learning repository.

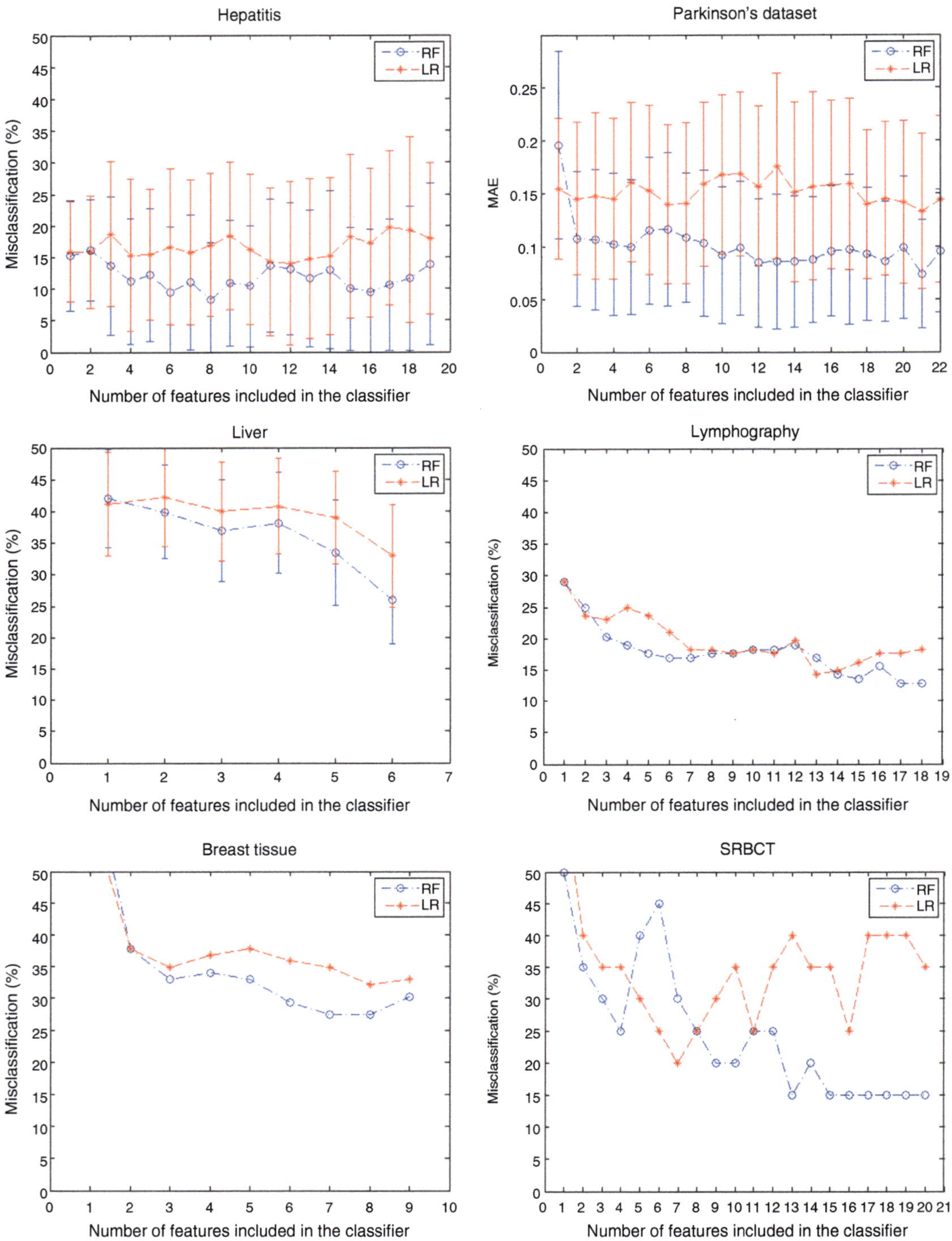

Fig. 7.3 Misclassification percentages as a function of the number of the features selected using mRMR are included in the classifier. These results suggest that Random Forests (RF) consistently outperform Logistic Regression (LR) across a wide variety of medical datasets

References

1. Aldrich J. Correlations genuine and spurious in Pearson and Yule. Stat Sci. 1995;10(4):364–76.
2. Bishop C. Pattern recognition and machine learning. Heidelberg: Springer; 2007.
3. Breiman L. Statistical modelling: the two cultures. Stat Sci. 2001;16(3):199–231 (with comments and discussion).
4. Breiman L. Random forests. Mach Learn. 2001; 45:5–32.
5. Cohen J, Cohen P, West SG, Aiken LS. Applied multiple regression/correlation analysis for the behavioral sciences. 3rd ed. New York: Routledge/Academic; 2002.
6. Diaconis P, Efron B. Computer intensive methods in statics. Sci Am. 1983;248:116–31.
7. Guyon I, Gunn S, Nikravesh M, Zadeh LA, editors. Feature extraction: foundations and applications. New York: Springer; 2006.
8. Hastie T, Tibshirani R, Friedman J. The elements of statistical learning: data mining, inference, and prediction. 2nd ed. New York: Springer; 2009.
9. Jossinet J. Variability of impedivity in normal and pathological breast tissue. Med Biol Eng Comput. 1996;34:346–50.
10. Khan J, Wei JS, Ringner M, Saal LH, Ladanyi M, Westermann F, Berthold F, Schwab M, Antonescu CR, Peterson C, Meltzer PS. Classification and diagnostic prediction of cancers using gene expression profiling and artificial neural networks. Nat Med. 2001;7:673–9.
11. Kurgan LA, Musilek P. A survey of knowledge discovery and data mining process models. Knowl Eng Rev. 2006;21(1):1–24.
12. Little MA, McSharry PE, Hunter EJ, Spielman J, Ramig LO. Suitability of dysphonia measurements for telemonitoring of Parkinson's disease. IEEE Trans Biomed Eng. 2009;56(4):1015–22.
13. Peng H, Long F, Ding C. Feature selection based on mutual information: criteria of max-dependency, max-relevance, and min-redundancy. IEEE Trans Pattern Anal Mach Intell. 2005;27(8):1226–38.
14. Torkkola K. Feature extraction by non-parametric mutual information maximization. J Mach Learn Res. 2003;3:1415–38.
15. Tsanas A, Little MA, McSharry PE, Ramig LO. Accurate telemonitoring of Parkinson's disease progression by non-invasive speech tests. IEEE Trans Biomed Eng. 2010;57:884–93.
16. Tsanas A, Little, MA, McSharry PE, Ramig LO. Enhanced classical dysphonia measures and sparse regression for telemonitoring of Parkinson's disease progression. IEEE International Conference on Acoustics, Speech and Signal Processing (ICASSP), Dallas, TX, US, pp. 594–597, 14–19 March 2010.
17. Tsanas A, Little MA, McSharry PE, Ramig LO. Nonlinear speech analysis algorithms mapped to a standard metric achieve clinically useful quantification of average Parkinson's disease symptom severity. J R Soc Interface. 2011;8:842–55.
18. Tsanas A, Little MA, McSharry PE, Spielman J, Ramig LO. Novel speech signal processing algorithms for high-accuracy classification of Parkinson's disease. IEEE Trans Biomed Eng. 2012;59(5): 1264–71.
19. Webb A. Statistical pattern recognition. New York: Wiley; 2002.

Psychological Rhythmicities

8

Eric E. Hessler, Patrick H. Finan, and Polemnia G. Amazeen

8.1 Introduction

Psychopathological and psychosomatic disorders are typically diagnosed according to a standardized set of criteria that are intended to reflect stable behavioral, cognitive, and emotional processes over time. However, decades of research have indicated that time-varying psychological, biological, and social influences interact to shape the trajectory of symptoms of psychopathological and psychosomatic disorders [1–3]. Disorders that vary as a function of temporal and environmental dynamics may have meaningful dynamical structure. Dynamical structure refers to the time-variant, sinusoidal form that individual and coupled processes take across repeated observations. For example, bipolar disorder II, which is characterized by rapid cycling between manic and depressive states, displays an oscillatory pattern in the manifestation of those symptoms over time [4].

E.E. Hessler (✉)
Department of Psychology, University of Minnesota Duluth,Duluth, MN 55812, USA
e-mail: eehessle@d.umn.edu

P.H. Finan
Department of Psychiatry and Behavioral Sciences, Johns Hopkins University School of Medicine, Baltimore, MD 21287-7413, USA
e-mail: patrick.h.finan@gmail.com

P.G. Amazeen
Department of Psychology, Arizona State University, Tempe, AZ 85287-1104, USA
e-mail: nia@asu.edu

Although dynamically structured disorders are common, they are often not treated as dynamical in theory or analysis [5]. Instead, the majority of studies have relied on means-based approaches to describe symptom variation.

Cross-sectional designs that rely on single-occasion assessment are limited not only by the inability to detect changes in disease states over time but also by error in participant recall of stressful or emotional experiences [6, 7]. Prospective designs address those problems by repeatedly assessing specific disease symptoms or other relevant physiological or psychological processes over time, thereby increasing the reliability of measurement and, in the case of microlongitudinal designs [8–10], reducing recall bias. A prospective design is needed to detect dynamical structure, but classic approaches to analyzing prospective data, including multilevel random coefficient regression [11] and growth modeling [12], do not treat variables dynamically in as much as they do not estimate periodicity in the data. That is, in prospective studies, researchers typically evaluate the average of changes in variables over time rather than the change itself.

That distinction can be appreciated in the literature on stress and depression. The experience of stressful major life events increases the risk of developing depression in the future [13]. Depressed individuals who have experienced a higher number of severe stressors exhibit more intense depression symptoms than those with fewer severe stressors [14]. Less severe stressors that are experienced chronically may, too,

J.P. Sturmberg and C.M. Martin (eds.), *Handbook of Systems and Complexity in Health*, DOI 10.1007/978-1-4614-4998-0_8, © Springer Science+Business Media New York 2013

predict the future onset of depression [15, 16]. At a finer grain of analysis, microlongitudinal studies allow researchers to examine processes associated with such risk predictions. In a study of patients with chronic pain, Zautra and Smith [17] employed multilevel random coefficient modeling to show that a higher number of negative life events and greater reactivity to stress in a given week resulted in elevated depression symptoms in the same week. Those results tell us that chronic stressors that challenge the individual's capacity to respond to environmental threats are associated with depression symptoms. They do not tell us, however, the levels and duration of chronic stress required to initiate a bout of major depression. Furthermore, they do not tell us whether the dynamical structure of observations of chronic stress, such as their rate of change, meaningfully influences the experience of depression in daily life.

8.2 Dynamical Data

Behavior must be observed for a sufficient period of time for the dynamical structure to be revealed and reliably measured [18]. The goal in examining dynamical data is that the current state of a dependent variable is linked in some way to a previous state. That is, there is a historical quality to the data in which no particular data point is isolated from another. There are a variety of different patterns that may be exhibited. For example, some dynamical patterns may be difficult to distinguish visually from a random pattern, but data points over time are, in fact, related [19]. The focus of this chapter is on a particular type of meaningful change over time: cyclical or oscillatory motion in which fluctuations are periodic (e.g., [20–22]). To understand what such processes might look like, think of seasonal fluctuations in temperature. Temperature oscillates, more or less, with the season, although the severity of such oscillations depends on the local climate.

A general category of models called oscillator models, originally developed in physics to capture meaningful dynamical structure, is used to model cyclical motion. Psychological (e.g., negative and positive affect) and physiological (e.g., heart rate) factors can also adhere to a dynamical pattern over time as opposed to a fixed mean state. For example, circadian rhythm disorders arise when individuals go to sleep or wake up at times outside of their natural rhythm [23]. As people develop a "typical" schedule for daytime activities, be it work, exercise, or other forms of activity, sleep onset latency, time asleep, number of mid-sleep awakenings, and time awake are expected to rhythmically ebb and flow in a similar pattern from day to day [24, 25]. Disruptions to one's "typical" schedule, such as working night shifts in an unpredictable pattern, have been shown to alter circadian rhythms, thereby compromising one's ability to achieve quality sleep [26]. Restoration of normal circadian rhythms has been shown to improve sleep [27].

Individual differences in characteristic dynamics may also predict outcomes. For example, manic and depressive mood states that oscillate toward an equilibrium state may predict better long-term functioning than oscillations that remain sustained over time. The different characteristic dynamics from patient to patient might also be moderated by particular individual difference variables, such as genetic factors or personality styles [28–30]. Beyond examining oscillations and moderating factors, we will explore the possibility that there can be multiple interacting processes in dynamical diseases and, more generally, in psychological phenomena.

8.3 Differential Equations and the Damped Linear Oscillator Model

There are two types of dynamical models—difference equations and differential equations—that have been used to model rhythmic processes, like predator–prey population cycles and bimanual coordination. The defining characteristic of those two model types is the assumption of either discrete or continuous time steps. For example, the logistic map is a difference equation that bases predictions about a species' population at the next time step ($t+1$) in part on the population at a previous time step (t) [31]. In contrast, ordinary differential equations base predictions about momentary change, as captured by velocity, on knowledge of the current position. One well-known example from the dynamical literature in psychology is the Haken et al. [32] model of bimanual coordination: the change in relative phasing of the limbs (ϕ) is based on both the current relative phase (ϕ) and the relevant control parameters, like speed and limb asymmetries (see summary in [33]). The models that are typically applied to oscillatory phenomena [29] are ordinary differential equation models that capture the relation among the measured variable (x) and the first ($\dot{x}$ or dx_t/dt) and second ($\ddot{x}$ or d^2x_t/dt^2) derivatives, that is, the change in the measured variable (velocity) and the rate of change of that change (acceleration).

The local linear approximation method is a reliable derivative estimation procedure that has been used in the modeling of oscillatory motion [20, 34, 35]. Relative to other derivative estimation procedures, which require more repeated measurements (e.g., [36]), local linear approximation can be applied to as few as three repeated measurements. In general, however, the collection of many more repeated measurements is recommended. For a variable x_t, velocity at x_2 is approximated as the average of the slopes between x_1 and x_2, and x_2 and x_3. The same process is repeated across all data points in a series to get a continuous measurement of velocity. To calculate acceleration, the process is repeated with velocity as the input.

A damped linear oscillator model is an ordinary differential equation that is used to model oscillatory motion. In the damped linear oscillator model, acceleration is predicted as a function of velocity and the original variable:

$$\frac{d^2 x_t}{dt^2} = \zeta \frac{dx_t}{dt} + \eta x_t. \quad (8.1)$$

The other terms in the model are used to characterize the nature of the oscillatory process. Zeta (ζ) is the linear damping coefficient, which represents linear changes in the amplitude of oscillations over time, and eta (η) is the squared frequency of oscillation. For a continuously oscillating process of constant amplitude (e.g., a sine wave), $\zeta = 0$. A nonzero ζ indicates that a process is losing ($\zeta < 0$) or gaining ($\zeta > 0$) amplitude over time.[1] When estimated, if η is significant, then the process is oscillatory. The square root of η is the frequency of oscillation (in radians). Larger magnitudes of η are indicative of more rapid oscillation.

8.3.1 Oscillation in Psychological and Physical Processes

The damped linear oscillator model can function as a model of any oscillatory system. That is, the damped linear oscillator model is a model of a system's dynamics independent of the physical substrate that generates the dynamics [37]. Psychological processes can exhibit known dynamics like oscillations that have a constant amplitude (e.g., positive and negative emotionality [22]), lose amplitude over time (e.g., emotional well-being [38]), and gain amplitude over time (e.g., adolescent drinking and smoking [20]). Oscillations in psychological processes can also exhibit faster or slower frequencies depending on the value of particular individual difference variables (e.g., psychiatric symptoms and violence [39]). Therefore, the damped linear oscillator model is well suited for modeling many psychological processes (for an early example, see [20]). The different characteristic patterns that are found in the literature can also be useful for the creation of therapy or intervention.

The damped linear oscillator model was originally developed to model purely physical systems such as pendular motion in which a mass swings back and forth on the end of a rod (see Fig. 8.1). If the mass is pulled to one side and released, then it will start swinging. Examples of different patterns of pendular motion are depicted in Figs. 8.2 and 8.3. Without friction, the pendulum

[1] Most studies have employed the damped linear oscillator model as in (8.1) with positive signs for each of the terms. Alternatively, negative signs are sometimes used for each term [21, 29]. With a negative sign for the damping coefficient (ζ), be aware that the interpretation would be opposite to the interpretation provided here.

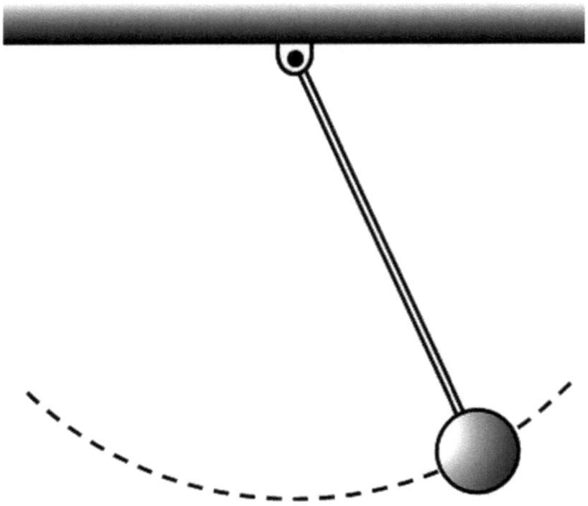

Fig. 8.1 A pendulum swings back and forth along the path depicted by the *dashed line*

swings continuously at a constant amplitude ($\zeta=0$; see Fig. 8.2a). Under more typical conditions, the pendulum slowly loses amplitude (referred to as damping; $\zeta<0$) and eventually comes to rest in a vertical position as a result of friction (see Fig. 8.2b). If there is a source constantly pumping energy into the pendulum (imagine a child starting a playground swing by pumping his/her legs [40]), then the pendulum gains amplitude over time (referred to as amplification; $\zeta>0$; see Fig. 8.2c). Pendulums tend to swing at natural frequencies that are related to their length and mass [41, 42]. Longer, heavier pendulums have slower natural frequencies (smaller η; see Fig. 8.3a) than shorter, lighter pendulums (larger η; see Fig. 8.3b). In trying to grasp the origins of meaningful patterns in psychological processes, it is helpful to consider these patterns of cyclic tendency as exemplified in known physical systems like the pendulum.

For physical systems, there is no inclination to attribute complex behavioral patterns to cognitive mechanisms [41, 43, 44]. In contrast, in traditional cognitive psychology, complex patterns in human behavior are assumed to arise from cognitive mechanisms that historically have had no physical basis [45]. A question that arises is whether that approach results in a description of the same phenomenon at a different level (from the behavioral patterns observed to cognitive mechanisms). If so, then little is added to the understanding of the behavior. The reduction from behavioral patterns to cognitive mechanisms presents another philosophical problem. When a complex cognitive mechanism is posited to explain behavior, then one is inclined to go to an even deeper level to explain the complex cognitive mechanism. Although there have been important treatment discoveries as a result of the search for "underlying" neurobiological substrates to complex behavioral disorders, such ventures have been costly and inconsistent. With those considerations in mind, we argue that a focus on cyclic psychological processes observed through oscillator models provides an important complementary research agenda to those seeking to define the pathways from molecule to behavior. In many cases, the structure of the dynamics is also a useful predictor of relevant psychological outcomes.

In one longitudinal study, college students completed self-report measures of 24 different positive and negative emotions over 52 consecutive days [22]. Chow et al. [22] identified two positive (joy and love) and four negative (sadness, fear, anger, and shame) emotion factors to examine using a damped linear oscillator model. The data were fit with structural equation modeling, and all of the emotions exhibited sustained oscillations over time (as in Fig. 8.2a). Multilevel modeling can also be used to fit the damped linear oscillator model, producing estimates that are generally the same [21]. Fluctuations in joy and love, and sadness, fear, and shame were characterized by a 7-day cycle (e.g., from high emotion to low emotion and back again to high emotion). Anger also oscillated, but at a cycle time that was slightly faster than 7 days. Positive emotionality repeatedly peaked during the weekend and dropped sharply on Mondays, whereas negative emotionality generally peaked during the middle of the week. Those results suggest that the "blue Monday" phenomenon could be attributed to the drop in positive emotions on Mondays.

A different pattern of oscillation was observed in the emotional well-being of recently bereaved widows when they sought emotional support [38]. Participants filled out biweekly diaries about emotional well-being over a 98-day study period.

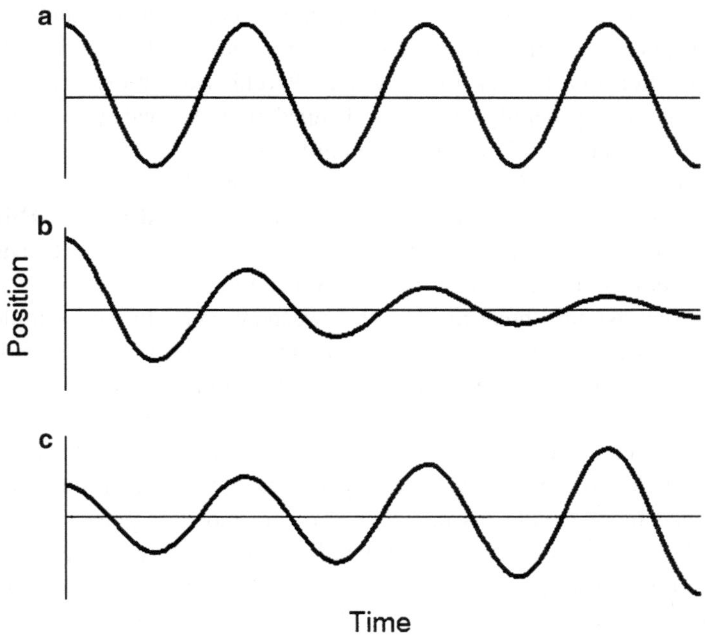

Fig. 8.2 Patterns of pendular motion that can be captured using the linear damping coefficient (ζ) in oscillator models include (a) swinging at a constant amplitude, (b) losing amplitude, and (c) gaining amplitude over time

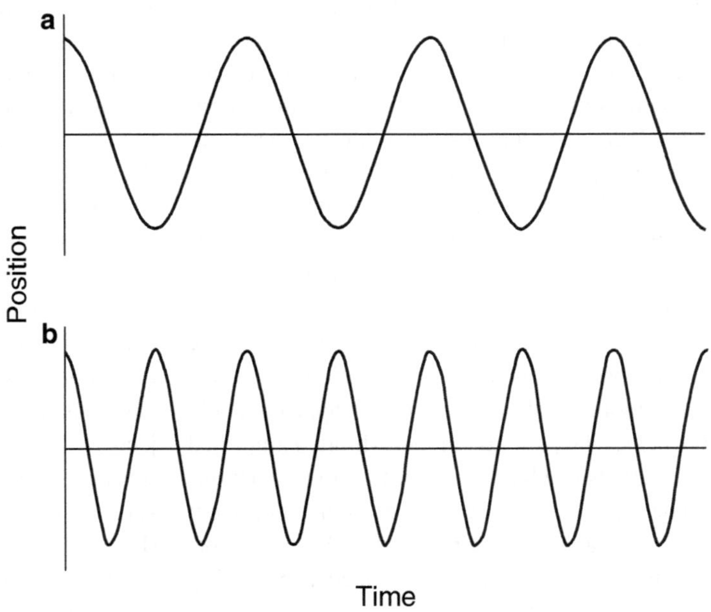

Fig. 8.3 The squared frequency coefficient (η) in oscillator models can capture (a) slower and (b) faster pendular motion over time

In comparison to the positive and negative emotions studied in college students, well-being oscillated slowly with the loss of a spouse, lasting about 47 days per cycle (e.g., from negative emotional states to positive emotional states and back again to negative emotional states). The amplitude of those fluctuations was reduced over time in the presence of a strong support structure. This characteristic pattern of damped oscillations is depicted in Fig. 8.2b.

A broader range of characteristic patterns was identified by Odgers et al. [39] as they sought to understand individual differences in violent behavior. The criminal records of patients who visited a psychiatric hospital emergency room were used to identify individuals with a documented history of violence. Participants completed a general assessment of psychiatric symptoms each week for 26 weeks. Psychiatric symptoms generally displayed damped oscillations (as in Fig. 8.2b), although amplified oscillations (as in Fig. 8.2c) were observed in a limited number of cases. Participants with an amplified symptom pattern were more likely to be involved in serious violent incidents (e.g., physical injury; sexual assault; threat made with a weapon; and aggressive act involving a weapon) than other participants. Odgers et al. noted individual differences in the frequency of oscillations as well (between about 7–10 weeks per cycle). Participants whose psychiatric symptoms oscillated more rapidly (as in Fig. 8.3b) were more likely to be involved in any violent incident than those whose psychiatric symptoms oscillated more slowly (as in Fig. 8.3a). Those examples demonstrate how aspects of dynamical structure, not just mean levels, can predict outcomes.

Amplified oscillations have been observed in adolescent substance abuse [20]. Cigarette and alcohol use swung back and forth between periods of heavy use and periods of low use. The concerning aspect of adolescent substance abuse was that small changes in either cigarette or alcohol use were followed by amplified changes, that is, extreme fluctuations between use and nonuse. In a study of patients diagnosed with bipolar disorder II [4], patients rated their mood on a scale from most depressed to most manic twice a day for 3 months. Mood cycled over approximately 6 weeks. The patients' mood exhibited amplified oscillations in which the amplitude of fluctuations tended to increase over time. Both adolescent substance abuse and rapid cycling bipolar disorder may, therefore, be characterized as becoming increasingly unstable over time.

The limited evidence accrued from studies that have employed damped linear oscillator models in psychology suggest that psychological processes may oscillate over time and that oscillation patterns may be differentiated by other psychological moderators. Importantly, the application of dynamical structure in such models permits an examination of hypotheses centered on the periodicity of data that cannot otherwise be made under traditional means-based analytic frameworks. There are several analytic considerations in the estimation from real data of dynamical systems models that we turn to next.

8.4 Multilevel Modeling: Estimating Coefficients in Oscillator Models

In the examples above, local linear approximation was used to calculate the velocity and acceleration of psychological processes [20, 34, 35]. The damped linear oscillator model [see (8.1)] can be considered as a regression equation with coefficients for damping and squared frequency [34, 35], and an error term (e_t). The typical approach is to fit the damped linear oscillator model to the data (the psychological variable and its associated derivatives) in order to obtain estimates of the damping and squared frequency coefficients. Interpretation of those estimates is generally consistent with the interpretation provided for the terms in (8.1), although caution should be observed because the interpretation can differ depending on how the model is fit [21, 29].

An appropriate method for estimating the coefficients in the damped linear oscillator model is multilevel modeling [21, 28, 29]. Multilevel modeling was developed to account for dependencies in data without violating the assumptions of common statistical techniques like ordinary least squares regression, *t*-tests, and between-subjects analyses of variance [21]. Longitudinal data sets take on a multilevel structure: repeated observations (Level 1 units) are nested within people (Level 2 units). Multilevel analyses permit the simultaneous modeling of within- and between-person parameters to answer questions about what happens when a particular event occurs and who is affected by the event. Such an analysis may pertain to discrete events, such as life stressors (e.g., [11]), or to periods during which a cognition, emotion, behavior, or physiological variable is elevated relative to an individual's mean.

As repeated measurements sampled in close proximity to one another are likely to include dependencies, it is necessary to carefully evaluate the structure of the covariance matrix. Multilevel modeling is flexible in this regard, providing the user with a variety of options for accounting for autoregressive properties of nested models. For example, individuals with chronic pain tend to report at least some pain on a daily basis. The repeated measurements of pain report are typically not independent. That is, an individual's pain report on day t typically covaries with their report on day $t+1$. Such a phenomenon is considered autoregressive and requires mathematical adjustments to the covariance matrix; these adjustments are accommodated under a multilevel modeling framework (for a review, see [46]).

Another advantage of this approach is that both fixed and random effects can be estimated. This provides an opportunity to evaluate the extent to which error not accounted for in an unconditional (null) model, and presumed to be random, can be explained through the addition of theoretically relevant variables at either level of a multilevel model. In contrast, traditional ordinary least squares regression assumes that the effects being modeled are fixed and do not vary randomly in the population. For example, individuals with chronic pain may demonstrate a relation between positive affect and pain from one day to the next. A multilevel modeling approach allows one to evaluate whether the variance in that relation can be better explained through the addition of random variables that are not currently included in the model. If it is determined that the positive affect-pain relation does, in fact, randomly vary in the population, then one may choose to model other theoretically relevant variables to account for that random variation, such as optimism.

The multilevel modeling approach has been extended to account for relations observed in a dynamical context [21, 47]. In data that exhibit oscillatory structure, there are dependencies between data points over time. Oscillatory structure exhibited at an individual level is referred to as intraindividual (within person) variation in multilevel modeling. It is likely that the oscillatory structure will differ between individuals. At Level 2 of a multilevel model, we examine interindividual (between persons) variation in oscillatory structure. Individual difference variables that may account for different oscillatory patterns can also be modeled at Level 2.

Multilevel models can be represented as systems of equations that make their multilevel nature explicit. In the multilevel model of a damped linear oscillator model, the oscillatory structure (intraindividual variation) is described at Level 1,

$$\frac{d^2 x_{tj}}{dt^2} = \zeta_j \frac{dx_{tj}}{dt} + \eta_j x_{tj} + e_{tj}, \tag{8.2}$$

in which velocity and displacement predict acceleration [the model in (8.1)] plus a term accounting for error in the prediction of acceleration (e_{tj}). The subscript j is provided in order to specify the multilevel nature of the model. A separate damped linear oscillator model is estimated for each individual j. Therefore, ζ_j is the damping coefficient for the jth individual and η_j is the squared frequency coefficient for the jth individual.

In order to account for interindividual variation in the damping and squared frequency terms, those terms are predicted at Level 2 of the model:

$$\zeta_j = \gamma_1 + u_{1j} \tag{8.3}$$

$$\eta_j = \gamma_2 + u_{2j}. \tag{8.4}$$

In the Level 2 equations, damping and squared frequency are expressed as a function of the average damping and squared frequency values across individuals (γ_1 and γ_2) and random error components representing error in the prediction of damping and squared frequency (u_{1j} and u_{2j}). The full multilevel model is fit with both the Level 1 and Level 2 terms included in the same equation (through substitution).

An important aspect of this type of multilevel modeling is the ability to account for moderation in the oscillatory structure by individual difference variables:

$$\zeta_j = \gamma_1 + \gamma_{11} W_j + u_{1j} \tag{8.5}$$

$$\eta_j = \gamma_2 + \gamma_{21} W_j + u_{2j}. \tag{8.6}$$

The Level 2 equations presented in (8.5) and (8.6) are similar to those in (8.3) and (8.4), but they include an added predictor (W_j). The utility of this approach is that a researcher can estimate the extent to which a particular psychological characteristic (W_j) influences the damping (γ_{11}) and squared frequency (γ_{21}) parameters. For example, on average, participants might exhibit damped oscillations in a behavior (see Fig. 8.2b), but depending on the value of an individual difference variable, a person in the same study could exhibit amplified oscillations (see Fig. 8.2c). Adding predictors changes the interpretation of the intercepts. They now represent average damping (γ_1) and average squared frequency (γ_2) across individuals, controlling for the average of predictor W_j. Presuming that W_j has predictive value, estimates of the Level 2 random error components (u_{1j} and u_{2j}) would be expected to decrease.

The multilevel modeling approach helps to account for the effects of particular moderating variables, such as life stressors or a period of elevated emotion. Despite this feature, researchers often employ other approaches to account for moderation effects. Chow et al. [22] used a damped linear oscillator model to examine oscillation in college students' negative and positive emotions. They observed interindividual variation in the squared frequency parameter, which suggested that individuals had different emotional cycle lengths. In their models, gender, affect intensity, extraversion, and neuroticism predicted the squared frequency of various emotions in a series of multiple regression analyses. Gender was a significant predictor of variation in the squared frequency parameter for sadness, with women displaying faster oscillations in sadness than men. Approaches such as multiple regression are popular, but they introduce new challenges to modeling. For example, the practice of pooling error into a single term in multiple regression could result in correlated errors across measurement occasions, a condition that violates the assumptions of multiple regression. In contrast, correlated error is handled appropriately in multilevel modeling [48].

8.5 Considering a Psychological Process in Context: Coupling

The damped linear oscillator model, fit using a multilevel model, is appropriate for modeling psychological rhythmicities [29]. The approach also allows researchers to account for individual difference variables that moderate the frequency and damping characteristics of those psychological cycles. Moderation in a dynamical model provides evidence of interaction between levels on one psychological variable and the oscillatory motion of a psychological process. It could be true, however, that oscillations in one psychological process are coupled with the oscillations of another psychological process, such that changes in amplitude and/or velocity in one variable are met with parallel or opposing changes in another variable. It is, therefore, important to consider a psychological process in the context of other psychological processes.

Research on coupling was pioneered in the area of behavioral physiology. The now classic example of coupling, first detailed by von Holst [49], is the coupling of rhythmically moving fins in decerebrated fish. In von Holst's [49] observations, fish fins typically moved neither independently nor in an entirely fixed relation. There was a tendency for each fin to maintain its own natural frequency and a tendency for each fin to impose its natural frequency on the other fin. Coupling was manifested in a variety of ways. The net amplitude of fins coordinated together was the sum of the individual fin amplitudes in isolation. When the fins moved opposite each other, the net amplitude was lower than either of the individual fin amplitudes in isolation. The position of one fin would change suddenly at times, typically matching the position of the other fin. From a traditional perspective, one might assume that such behavioral complexity is evidence that coupling is regulated by a cognitive mechanism. Complexity in a simple system like coupled fish fins demonstrates that that need not be the case.

In practice, it may be difficult to separate the dynamics of an individual psychological process from the influence of other interacting processes. One way to understand the stability of any process is to momentarily disrupt, or perturb, it. That perturbation has a bigger effect on an unstable system than on a stable system [50–53]. Perturbations have been used effectively to explore the stability of coupling between rhythmic movements. When participants coordinated movement of the index fingers, a brief torque was introduced to one finger and relaxation time, the amount of time taken to regain stable performance following perturbation, was used as an index of stability [51, 53]. In addition to magnifying the dynamics of coupled psychological processes, perturbations could be used to clarify the dynamics of a particular psychological process [54]. For example, bereavement may be considered a perturbation that triggers oscillations between positive and negative emotionality in the well-being of recently bereaved widows [38].

Although coupled dynamics have not been investigated in the chronic pain literature, clinical research suggests that the daily experience of pain in disorders such as fibromyalgia, rheumatoid arthritis, and osteoarthritis may be altered by perturbations from a wide range of comorbid symptoms. For example, chronic pain and insomnia are highly comorbid [55] and bidirectionally related [56], and recent evidence suggests that sleep disturbance perturbs the supraspinal regulation of pain [57]. However, sleep and pain may also be coupled processes that oscillate in tandem over time. Factors that may perturb a sleep–pain system among people with comorbid chronic pain and insomnia include stress [58], altered immune processing [57], and increased symptoms of depression [59].

8.6 Coupled Damped Linear Oscillator Model

Coupled processes have been conceptualized as two separate but interacting damped linear oscillator models (e.g., [20, 60]). For the sake of simplicity, we present the coupled damped linear oscillator model without the multilevel modeling subscripts (Note that estimates of the coefficients can be obtained through multilevel modeling.). A more complex coupled model will be discussed later [21].

$$\frac{d^2 x_t}{dt^2} = (\zeta_x \frac{dx_t}{dt} + \eta_x x_t) + \kappa_{yx}(\zeta_y \frac{dy_t}{dt} + \eta_y y_t) \quad (8.7)$$

$$\frac{d^2 y_t}{dt^2} = (\zeta_y \frac{dy_t}{dt} + \eta_y y_t) + \kappa_{xy}(\zeta_x \frac{dx_t}{dt} + \eta_x x_t) \quad (8.8)$$

In each of the two equations, the left-hand and first right-hand terms comprise a damped linear oscillator model. Equations (8.7) and (8.8) identify a damped linear oscillator model for variable x_t and variable y_t, respectively. Both of the oscillators have associated linear damping (ζ_x and ζ_y) and squared frequency (η_x and η_y) coefficients. The oscillators are coupled to one another with the addition of coupling terms of strength κ. Each κ is given a separate subscript, representing the influence of the other oscillator; for example, κ_{yx} is the influence of the y_t oscillator on the x_t oscillator. A convenient aspect of this type of modeling is that the relative influence of one process on another can be differentiated from the reverse influence because each coupling term is estimated separately [21, 28].

8.6.1 Different Types of Coupling

Coupling has been modeled in two different ways: by making it a function of both position and velocity or by just basing it on position. Imagine two pendulums swinging from side to side that are connected by two springs ([60, 61]; see Fig. 8.4). The two springs represent the coupling components in the model, the influence of each pendulum on the other. The independent movement of each pendulum is described by the damped linear oscillators in the model. However, because the two pendulums are coupled by springs, the swinging of one pendulum influences the other pendulum to a certain degree. Figure 8.5 depicts a time series for each of the two coupled pendulums. The push and pull of each pendulum on the other generates complex changes in the position of each pendulum over time. This bidirectional influence is observed in the changes that occur in each time series from one cycle to the next. The amount of push or pull depends on the stiffness of each spring. A stiff spring (e.g., a spring-based shock absorber on a car) will transfer more influence than an elastic spring (e.g., a Slinky™). In modeling, the acceleration in one pendulum may be predicted in part by the velocity and position of the other pendulum and vice versa. In the coupled damped linear oscillator model described in (8.7) and (8.8), the velocity and position of one variable influenced the prediction of acceleration in the other variable [20, 60].

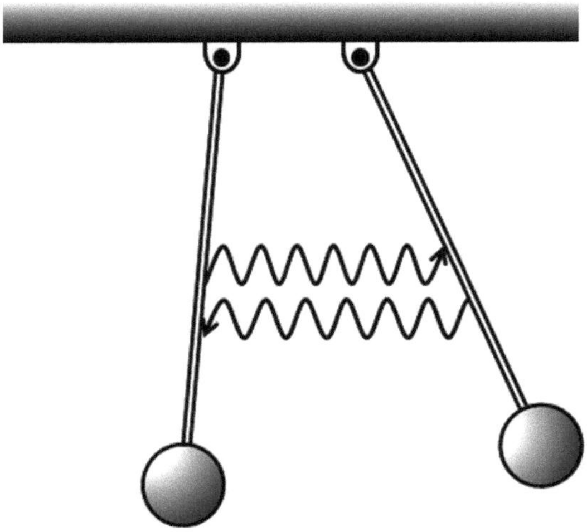

Fig. 8.4 Two pendulums swing back and forth at different points in their cycles. The pendulums are coupled together through two springs so that motion of one pendulum affects motion of the other pendulum. The pendulums appear to affect each other symmetrically in this schematic depiction, but influence is likely to be asymmetric in real systems

An alternate explanation is that coupling is dependent on the sheer distance of the pendulums from each other: $\kappa_{xy}(\eta_x x_t - \eta_y y_t)$ [21, 28]. In this conceptualization, κ changes as a function of the relative displacement of a variable from zero. The greater the displacement is from zero, the more one variable (coupled process) pulls on the other variable (coupled process). An application of this type of coupling to motor behavior will be discussed later [21]. The conceptualization of coupling a researcher chooses depends on whether coupling is expected to be a function of both velocity and position or position alone. If statistical power is a concern, then one consideration is that coupling via position alone involves fewer terms in the model.

With either conceptualization of coupling, the strength of coupling (or the stiffness of a spring linking the two pendulums) is expressed by the value of κ. When each κ is zero, the oscillators behave independently. Each pendulum swings on its own accord as if the other pendulum were not there. When a κ is nonzero, there is coupling

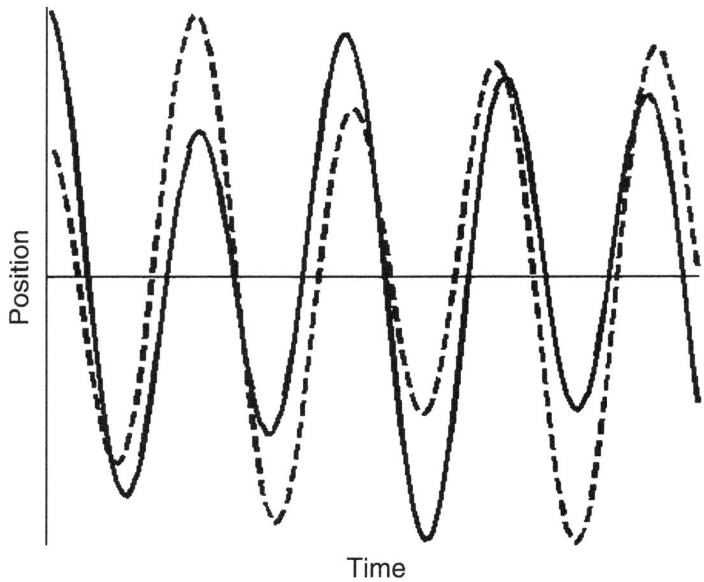

Fig. 8.5 The complex motion of two coupled pendulums (*solid* and *dashed lines*) in which the swinging of each pendulum influences and is influenced by the swinging of the other pendulum

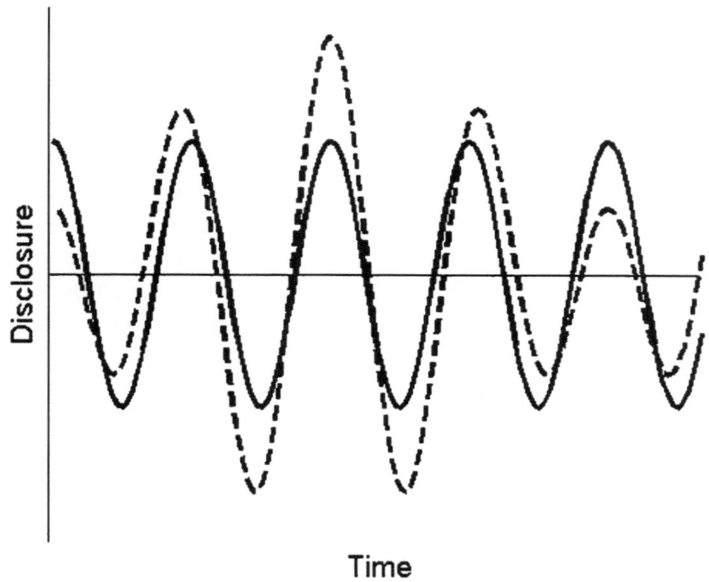

Fig. 8.6 A simulation of a unidirectional influence of husband disclosure (*solid line*) on wife disclosure (*dashed line*). Husband disclosure oscillates at a constant amplitude, uninfluenced by wife disclosure, whereas wife disclosure sometimes gains amplitude and sometimes loses amplitude over time, depending on husband disclosure

between the oscillators. The separate κ estimates associated with each direction of influence (κ_{yx} and κ_{xy}) represent the two springs connecting the pendulums. Because there are two separate κ estimates, the stiffness of each spring can vary. This variation makes different coupling relations possible [21, 60]. If the κ estimates are nonzero and equal to each other (springs of equal stiffness), then there is an equal and bidirectional influence between the pendulums. If the κ estimates are nonzero and different (springs of different stiffness), then there is an asymmetric influence between oscillators. One pendulum exerts more influence on the other than vice versa. Last, if one $\kappa=0$ and the other $\kappa\neq 0$ (effectively, one spring between the pendulums), then there is a unidirectional influence from one oscillator to the other. In practice, differentiating between asymmetric and unidirectional coupling is a matter of statistical judgment.

8.6.2 Coupled Psychological Processes and Moderating Influences

There are a limited number of examples of the use of the coupled damped linear oscillator model to investigate the dynamics between psychological processes. For example, Boker and Laurenceau [60] examined the coupled dynamics of married partners' intimacy (emotional closeness as opposed to sexual intimacy) and disclosure (sharing facts, thoughts, and feelings) by having them fill out daily diaries over 42 days. Intimacy and disclosure oscillated in a sustained fashion for both married partners with an average period of approximately 6 days. There were differences in the patterns of intimacy coupling and disclosure coupling. Intimacy coupling was bidirectional and symmetric, indicating that husband and wife intimacy influenced each other equally. Disclosure coupling, on the other hand, was unidirectional. Husband disclosure was not influenced by wife disclosure, but wife disclosure was influenced by husband disclosure.

Although exact coupling estimates were not provided in Boker and Laurenceau [60], coupling between husband and wife intimacy might follow the dynamical patterns depicted in Fig. 8.5, in which there is a bidirectional symmetric influence. Disclosure coupling, rather, would have a different graphical depiction. We have simulated that finding in Fig. 8.6, in which husband disclosure (solid line) displays oscillations of constant amplitude, uninfluenced by wife disclosure, and wife disclosure (dashed line) displays both amplified and damped oscillations, depending on the influence of husband disclosure.

In another study [62], separate models were used to examine coupling between mother's depressive symptoms, and children's internalizing and externalizing behavior. Coupling was unidirectional, with mother's depressive symptoms driving but not being influenced by children's internalizing and externalizing behavior. Helm et al. [63] examined coupling between romantic partners' physiological parameters by fitting separate coupled models for partners' respiration and heart rate. In general, across different social interaction conditions, partners' respiration was bidirectionally coupled. Coupling between heart rates was also observed, but the exact coupling relation differed widely across social interaction conditions. Coupling has also been explored between psychological processes exhibited by a single individual. In a 56-day daily diary study [30], participants reported levels of stress and negative affect. Both variables exhibited oscillations over the course of the study and were bidirectionally coupled. In sum, a variety of coupled oscillatory patterns have been observed both between and within individuals, which is an argument for modeling two separate coupling parameters.

The examples identified above illustrate that coupling between psychological processes is a common phenomenon. As with the damped linear oscillator model, researchers have used multilevel modeling to account for moderation in oscillatory structure by individual difference variables in the coupled model. Of particular interest with the coupled model is that Level 2 equations can be written to account for moderation in the coupling parameters [30, 60, 63]:

$$(\kappa_{yx})_j = \gamma_1 + \gamma_{11} W_j + u_{1j} \tag{8.9}$$

$$(\kappa_{xy})_j = \gamma_2 + \gamma_{21} W_j + u_{2j}. \tag{8.10}$$

In the Level 2 equations, W_j is the individual difference variable that is used as a predictor of the different coupling parameters (κ_{yx} and κ_{xy}). A researcher can estimate how much W_j influences the coupling parameters (γ_{11} and γ_{21}). The intercepts represent the average coupling influence from y_t to x_t (γ_1) and from x_t to y_t (γ_2), controlling for the average of predictor W_j. Error in the prediction of the different coupling parameters is also reflected in each equation (u_{1j} and u_{2j}).

8.6.3 Examples of Moderating Influences on Coupling

In three of the examples described above, specific psychological variables moderated the coupling processes. Intimacy coupling between married partners was moderated by marital satisfaction [60]. Higher marital satisfaction was associated with a stronger coupling influence from husband to wife intimacy and from wife to husband intimacy. Avoidance, anxiety, relationship satisfaction, and length of relationship were examined as moderators of coupling between physiological signals [63]. Among other significant moderating effects, when romantic partners were asked to imitate each other's physiological signals, higher anxiety levels for either partner were associated with stronger heart rate coupling. Dispositional resilience, friend support, and family support were examined as moderators of coupling between stress and negative affect [30]. Coupling from stress to negative affect was moderated by dispositional resilience and family support. For more resilient persons and those with more family support, stress had less of an influence on negative affect.

Using the conceptualization of coupling in which κ changes as a function of the relative displacement of a variable from zero, Butner et al. [28] examined coupling between the emotional processes of romantic partners. The researchers measured participants' positive affect, negative affect, and psychological predictors, including avoidance and anxiety. Between romantic partners, coupling was evident for positive affect but not negative affect. Higher avoidance was generally associated with

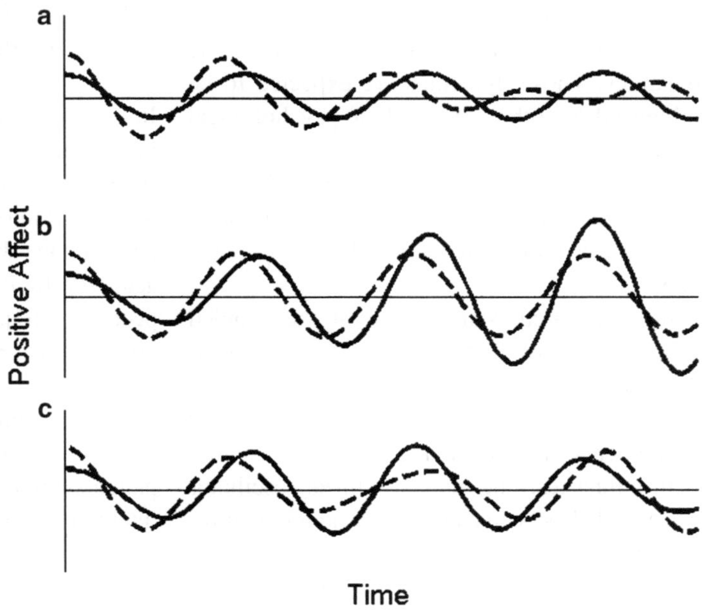

Fig. 8.7 A simulation of positive affect coupling between male (*solid lines*) and female (*dashed lines*) romantic partners. (a) Male positive affect influences female positive affect; (b) female positive affect influences male positive affect; and (c) there can be a bidirectional influence between male and female positive affect

less positive affect coupling. Butner et al. [28] suggest that this could result from avoidant individuals paying less attention to their partner's emotional cues. More detailed analyses suggested that positive affect coupling relations were rather nuanced. Asymmetric coupling was typical, with male positive affect driving female positive affect. Females only drove positive affect coupling when they were low on anxiety. For males who were low on avoidance or high on anxiety, and females who were high on avoidance, positive affect coupling tended to be bidirectional and symmetric.

Precise coupling estimates were not provided in Butner et al. [28], but the three characteristic forms of coupling were simulated in Fig. 8.7. Male-driven asymmetric coupling is depicted in Fig. 8.7a: male positive affect (solid line) is unaffected by female positive affect, and female positive affect (dashed line) displays damped oscillations, depending on male positive affect. The opposite scenario is depicted in Fig. 8.7b: female positive affect is unaffected by male positive affect and male positive affect displays amplified oscillations, depending on female positive affect. Figure 8.7c depicts bidirectional symmetric coupling: a mutually dependent influence between male and female positive affect. Different patterns of results, therefore, can be captured by different coupling relations.

8.7 Nonlinearities in Oscillatory Motion: Escapements

The previous examples demonstrate how the dynamics of a psychological process, not just mean levels, can change under different conditions. However, when the oscillations are treated as linear, as was the case with the damped linear oscillator model (8.1) and the coupled damped linear oscillator model [(8.7) and (8.8)], the result is uniform damping or amplification effects on all data points. In reality, however, and particularly with psychological phenomena, linear systems are unusual [21, 29]. Our physical model for a nonlinear system is the grandfather clock. Grandfather clocks are wound so that a mechanism called an escapement injects energy into the clock's pendulum to keep the pendulum moving [64]. The injection of energy is nonlinear in the sense that it occurs at a particular point in each swing of the pendulum, producing momentary changes in the amplitude and frequency of the swing. With respect to psychological processes, each particular psychological process will have its own characteristic dynamics. Those dynamics are expected to be influenced by other physical or psychological processes that interact with it and they may do so in a nonuniform manner over time. For two closely associated periodic processes, escapements are evidence of their influence on each other.

In some circumstances, use of the coupled damped linear oscillator model might oversimplify the coupling process. Before describing an alternate coupled model, we will identify how escapements are modeled. In order to capture such influences, the damped linear oscillator model (8.1) has been extended by adding nonlinear terms, creating the damped *nonlinear* oscillator model [65, 66]. Again, for simplicity, we present the model without the multilevel modeling subscripts:

$$\frac{d^2x_t}{dt^2} = \zeta\frac{dx_t}{dt} + \eta x_t + \rho\left[\frac{dx_t}{dt}\right]^3 + \nu\left[\frac{dx_t}{dt}\right]x_t^{\,2} + \delta\left[x_t\right]^3 + \mu\left[\frac{dx_t}{dt}\right]^2 x_t \tag{8.11}$$

The left-hand term (acceleration) and the first two right-hand terms (velocity and displacement) of the model constitute a damped linear oscillator model. The rest of the terms are nonlinear terms, higher order terms that are composed of the products of velocity and/or displacement: Rayleigh (ρ), van der Pol (ν), Duffing (δ), and π-mix odd (μ). Rayleigh and van der Pol are nonconservative terms, meaning that energy is pumped into or out of the system. The result is changes in amplitude (a wider pendulum swing) that are a function of velocity alone or both position and velocity, respectively. Duffing and π-mix odd are conservative terms, meaning that the total energy stays the same but frequency (how fast the pendulum swings) changes within a cycle. That change is a function of position alone or both position and velocity, respectively. Together, all of those terms capture the variety of shapes an oscillation can assume as the result of the presence of escapements.

8.7.1 Nonlinearities in Pain Predictions

In our previous work [29], we examined the pain prediction process for patients diagnosed with rheumatoid arthritis using the damped nonlinear oscillator model. Rheumatoid arthritis is a chronic autoimmune disease of the synovial joints characterized by disabling pain that can flare unexpectedly [67]. For patients with rheumatoid arthritis, accurate prediction of future pain can reduce the perceived averseness of a pain episode [68]. We examined whether overpredictions and underpredictions of pain (i.e., the extent to which patients predicted more or less next-day pain intensity than they actually reported experiencing on the following day) occurred systematically in an oscillating pain prediction process. Participants had an overall tendency to damp toward accurate predictions of next-day pain over the course of a month in which they recorded their data in daily diary style. The oscillation pattern was also nonlinear, suggesting that escapements influenced the pain prediction process.

We examined whether the pain prediction process varied as a function of negative affect, positive affect, and perceived control over pain. To accomplish this, we predicted the various linear and nonlinear terms as a function of those three individual difference variables. Moderation of nonlinear terms can be examined in a manner that is consistent with moderation of linear terms [see (8.5) and (8.6)] and coupling terms [see (8.9) and (8.10)]. The pain prediction process was moderated by negative affect, positive affect, and pain control. Negative affect and pain control had a nonconservative influence, with faster damping observed for participants with lower negative affect and higher pain control. In contrast, positive affect had a conservative influence, affecting the nonlinear frequency of oscillations. Participants with higher positive affect progressed more slowly, or lingered, through inaccurate pain predictions. One interpretation of these data is that positive affect is recruited during periods of pain prediction inaccuracy, possibly to serve as a buffer against the frustration that could emerge when experienced pain surpasses one's expectation.

Together, the results suggest that adaptive psychological characteristics have implications for the self-management of chronic pain. To the extent that patients may be trained to better recognize the conditions that bring about pain exacerbations and better regulate their cognitive and affective responses to pain, they may experience pain as more predictable and less stressful.

8.8 Coupled Nonlinear Oscillators

To the authors' knowledge, there has been only one previous study to date [21] in which nonlinear coupling has been examined in the psychological sciences. Both nonlinearity and coupled systems are sufficiently common so as to make nonlinear coupled modeling both ecologically relevant and likely the most accurate representation of the complexity observed in real-world systems. As seen earlier, linear models and even coupled linear models have a tendency to oversimplify real-world systems with their assumptions of persistent change. The coupled damped nonlinear oscillator model is composed of two damped nonlinear oscillator models that are coupled together to allow for each oscillator to influence the other [21]. Again, estimates of each term in the coupled damped nonlinear oscillator model can be estimated through the multilevel modeling approach.

$$\frac{d^2x_t}{dt^2} = (\zeta_x \frac{dx_t}{dt} + \eta_x x_t + \rho_x \left[\frac{dx_t}{dt}\right]^3 + \nu_x \left[\frac{dx_t}{dt}\right] x_t^2 + \delta_x [x_t]^3 + \mu_x \left[\frac{dx_t}{dt}\right]^2 x_t) + \kappa_{yx}(\eta_y y_t - \eta_x x_t) \quad (8.12)$$

$$\frac{d^2y_t}{dt^2} = (\zeta_y \frac{dy_t}{dt} + \eta_y y_t + \rho_y \left[\frac{dy_t}{dt}\right]^3 + \nu_y \left[\frac{dy_t}{dt}\right] y_t^2 + \delta_y [y_t]^3 + \mu_y \left[\frac{dy_t}{dt}\right]^2 y_t) + \kappa_{xy}(\eta_x x_t - \eta_y y_t) \quad (8.13)$$

The damped nonlinear oscillators for the variables x_t and y_t appear in the first part of each equation. They are the linear terms, damping and squared frequency, and the complete set of nonlinear terms: Rayleigh, van der Pol, Duffing, and π-mix odd. The damped nonlinear oscillators are linked together through the coupling terms, each one representing the influence of one oscillator on the other (κ_{yx} and κ_{xy}), as in (8.7) and (8.8). In the Butner et al. [21] model, coupling is represented as the difference in displacements of the two oscillators. One could also imagine using the more complex conceptualization of coupling employed by Boker and colleagues (e.g., [20, 60]).

8.8.1 Coupled Nonlinear Oscillators in Motor Coordination

Butner et al. [21] used a coupled damped nonlinear oscillator model to characterize a motor task in which individuals coordinated the movements of two handheld pendulums of different lengths. The shorter (faster natural frequency) pendulum was held in the right hand and the longer (slower natural frequency) pendulum was held in the left hand. Frequency estimates from the model were consistent with the frequencies calculated based on the physical characteristics of each pendulum: 0.78 Hz for the right-hand pendulum and 0.70 Hz for the left-hand pendulum. Estimates for nonconservative and conservative terms were significant for each hand, indicating that there were within-cycle variations in amplitude and frequency, respectively. Coupling was unidirectional, in which movement of the left-hand pendulum influenced but was not influenced by movement of the right-hand pendulum. Moderation of the squared frequency and coupling terms was examined by including handedness as a predictor at a higher level of the multilevel model. Consistent with the literature on the effects of handedness on bimanual coordination (e.g., [69, 70]), participants who were more right handed demonstrated a weaker coupling influence of the left hand on the right hand. This finding helps to validate the form of this coupled model for future use in the psychological literature.

8.8.2 Coupled Nonlinear Oscillators and the Dynamic Model of Affect

Recall that we previously observed nonlinearities in the pain prediction process for patients diagnosed with rheumatoid arthritis [29]. Those nonlinearities suggested that there were escapements or energy inputs from another process or other processes. In that research, we treated negative affect and positive affect as mean states for the sake of simplicity during that first modeling effort. However, there is evidence that variables like negative affect and positive affect display oscillatory motion. Negative affect was explored in coupled damped *linear* oscillator models by Montpetit et al. [30] and Butner et al. [28]. In Montpetit et al. [30], stress and negative affect exhibited coupled oscillatory motion over a 56-day time course. The strength of that coupling was reduced for both resilient persons and individuals with more family support. In Butner et al. [28], oscillations were apparent in negative affect and positive affect. There was between-partner coupling in positive affect but not negative affect, and avoidance and anxiety had moderating influences on positive affect coupling. We suggest that a logical next step is to explore the coupling of negative and positive affect oscillations within an individual and investigate how that coupling might be influenced by psychological stress or another aversive process.

The relative balance of positive and negative affect experienced by an individual at a given point in time may be dependent on the situational context. Positive and negative emotions are thought to exist and interact within an affective space [71]. A bipolar view of affective space holds that as an individual's negative affect increases, his/her positive affect should decrease, thereby increasing the degree of correlation between the two affects in a negative direction [72]. Another conceptualization of affective space considers positive affect and negative affect as separate, bivariate dimensions existing on a three-dimensional plane whose shape can be modified by aversive perturbations to the system [73]. The latter view, known as the Dynamic Model of Affect [74], holds that it is possible, if not common, to experience affective independence rather than affective correlation when stress or other aversive states are diminished or absent. In contrast, when stress or other aversive states are present, people have greater difficulty differentiating between the two affects, resulting in an increased negative correlation.

To visualize an affective space that allows for affective differentiation, imagine a two-dimensional Cartesian space with positive affect on one axis and negative affect on the other axis. As an individual's negative affect increases in this model, his/her positive affect may or may not change, reflecting a degree of independence between the two affects [75]. Aversive states like stress create a third dimension in the affective space and serve to contort its shape, causing the space to shrink and affect ratings to fall to opposite poles of the affective distribution. Zautra et al. [76] tested the hypothesis that stress would narrow the space between positive and negative affect in a sample of healthy workers. People were randomly alerted to provide affect and event ratings ten times per day for five consecutive days. Within-person estimates of the correlation between positive and negative affect were observed to be more negative during moments when a stressful event was reported than during non-stressful moments.

Why would stress impact the relation between positive and negative affect? Stress has been shown to increase uncertainty, which places demands on the information processing system [77]. Under such conditions, affective processing becomes limited and, consequently, positive affect and negative affect become more inversely correlated [78]. During times of acute stress, this is an adaptive process; the body must recruit energy to escape the most pertinent perceived threat, and complex affective processing consumes energy. Thus, our affective complexity diminishes in order to minimize energy expenditure, escape threat, and regain homeostatic balance. Uncertainty facilitates this process by motivating the individual to attend to the affective valence that is most closely tied to a stressor: negative affect. The individual must work considerably harder to maintain positive affect, and so it is expected to diminish during aversive states that promote uncertainty.

It may be particularly advantageous to apply the coupled nonlinear oscillator model to this type of data because of the complex dynamics of positive and negative affect and their susceptibility to systematic perturbations [73, 74, 76]. Data of the proper form for oscillator modeling are currently being collected. Conceptually, the modeling takes the following form: positive affect and negative affect, which have already been shown to oscillate [28, 30] are entered into Level 1 of the model. Coupling between those variables is conceptualized as in Butner et al. [21], as previous research identifies the importance of relative levels of positive affect and negative affect [76]. Stress has been shown to change the relation between positive and negative affect [76], and could be entered into Level 2 of a model to predict the different coupling parameters. With low stress, positive and negative affect are expected to oscillate independently. With high stress, oscillations in positive and negative affect should be coupled such that troughs in positive affect correspond to peaks in negative affect.

The coupled nonlinear oscillator model presented here is a more accurate representation of the true complexity involved in the positive and negative affect relation. An additional advantage for theory building in the Dynamic Model of Affect is that it allows us to conceptualize these processes in a fundamentally different manner. The model allows for the estimation of two different coupling terms, one reflecting the influence from positive affect to negative affect, and the other reflecting the influence from negative affect to positive affect. The dynamic coupling of affects could be unidirectional, asymmetric, or bidirectional. One might expect asymmetric coupling when stress is high, so that negative affect has a greater influence on positive affect than vice versa. Stress could also change the shape of affect oscillations through nonlinear influences. For example, high stress might result in more time being spent at both extreme high and low negative affect than in states of more desirable affect regulation. The coupling between positive and negative affect and the presence of various forms of nonlinear influence are both open research questions.

8.9 Conclusion

The thesis of this paper is that modeling and theory that were originally developed to describe physical processes (e.g., pendular motion) can be applied successfully to psychological processes. This strategy suggests that physical and psychological processes share relevant dynamical properties despite differences in material substrate. The oscillations that are exhibited by a variety of psychological processes, from the emotional well-being of recently bereaved widows [38, 54] to psychiatric symptoms of individuals with a documented history of violence [39], are remarkably similar to those of a simple pendulum. When pendulums are coupled by springs, then it becomes possible to accommodate additional psychological processes, such as married partners' intimacy and disclosure [60] and romantic partners' positive affect and negative affect [28]. The similarities between physical and psychological processes challenge traditional psychological theory by demonstrating that behavioral complexity can be displayed by simple systems characterized by the manner in which variables change and interact over time.

References

1. Bolger N, Davis A, Rafaeli E. Diary methods: capturing life as it is lived. Annu Rev Psychol. 2003;54:579–616.
2. Myin-Germeys I, Oorschot M, Collip D, Lataster J, Delespaul P, van Os J. Experience sampling research in psychopathology: opening the black box of daily life. Psychol Med. 2009;39(9):1533–47.
3. Stone AA, Shifman S. Ecological momentary assessment (EMA) in behavioral medicine. Ann Behav Med. 1994;16(3):199–202.
4. Boker SM, Leibenluft E, Deboeck PR, Virk G, Postolache TT. Mood oscillations and coupling between mood and weather in patients with rapid cycling bipolar disorder. Int J Child Health Hum Dev. 2008;1(2):181–203.
5. Butler EA. Temporal interpersonal emotion systems: the "TIES" that form relationships. Pers Soc Psychol Rev. 2011;15(4):367–93.
6. Kahneman D. Objective happiness. In: Kahneman D, Diener E, Schwarz N, editors. Well-being: the founda-

tions of Hedonic Psychology. New York: Russell Sage; 1999. p. 3–25.
7. Robinson MD, Clore GL. Belief and feeling: evidence for an accessibility model of emotional self-report. Psychol Bull. 2002;128(6):934–60.
8. Green AS, Rafaeli E, Bolger N, Shrout PE, Reis HT. Paper or plastic? Data equivalence in paper and electronic diaries. Psychol Methods. 2006;11(1):87–105.
9. Shiffman S, Stone AA, Hufford MR. Ecological momentary assessment. Annu Rev Clin Psychol. 2008;4:1–32.
10. Tennen H, Affleck G, Armeli S. Personality and daily experience revisited. J Pers. 2005;73(6):1465–83.
11. Nezlek JB. Multilevel random coefficient analyses of event- and interval-contingent data in social and personality psychology research. Pers Soc Psychol Bull. 2001;27(7):771–85.
12. Singer JD, Willett JB. Applied longitudinal data analysis: modeling change and event occurrence. New York: Oxford University Press; 2003.
13. Mazure CM. Life stressors as risk factors in depression. Clin Psychol Sci Pract. 1998;5(3):291–313.
14. Monroe SM, Hadjiyannakis K. The social environment and depression: focusing on severe life stress. In: Gotlib IH, Hammen CL, editors. Handbook of depression. New York: Guilford Press; 2002. p. 314–40.
15. Hammen C. Stress and depression. Annu Rev Clin Psychol. 2005;1:293–319.
16. Kessler RC. The effects of stressful life events on depression. Annu Rev Psychol. 1997;48:191–214.
17. Zautra AJ, Smith BW. Depression and reactivity to stress in older women with rheumatoid arthritis and osteoarthritis. Psychosom Med. 2001;63(4):687–96.
18. Thelen E, Smith LB. A dynamic systems approach to the development of cognition and action. Cambridge, MA: MIT Press; 1996.
19. Gilden DL, Thornton T, Mallon MW. 1/f noise in human cognition. Science. 1995;267(5205):1837–9.
20. Boker SM, Graham J. A dynamical systems analysis of adolescent substance abuse. Multivar Behav Res. 1998;33(4):479–507.
21. Butner J, Amazeen PG, Mulvey GM. Multilevel modeling of two cyclical processes: extending differential structural equation modeling to nonlinear coupled systems. Psychol Methods. 2005;10(2):159–77.
22. Chow SM, Ram N, Boker SM, Fujita F, Clore G. Emotion as a thermostat: representing emotion regulation using a damped oscillator model. Emotion. 2005;5(2):208–25.
23. Reid KJ, Zee PC. Circadian rhythm disorders. Semin Neurol. 2004;24(3):315–25.
24. Ancoli-Israel S, Cole R, Alessi C, Chambers M, Moorcroft W, Pollak CP. The role of actigraphy in the study of sleep and circadian rhythms. Sleep. 2003; 26(3):342–92.
25. Krauchi K, Wirz-Justice A. Circadian clues to sleep onset mechanisms. Neuropsychopharmacology. 2001; 25(5 Suppl):S92–6.
26. Quera-Salva MA, Defrance R, Claustrat B, De Lattre J, Guilleminault C. Rapid shift in sleep time and acrophase of melatonin secretion in short shift work schedule. Sleep. 1996;19(7):539–43.
27. Van Someren EJ, Lijzenga C, Mirmiran M, Swaab DF. Long-term fitness training improves the circadian rest-activity rhythm in healthy elderly males. J Biol Rhythms. 1997;12(2):146–56.
28. Butner J, Diamond LM, Hicks AM. Attachment style and two forms of affect coregulation between romantic partners. Pers Relationship 2007;14(3):431–55.
29. Finan PH, Hessler EE, Amazeen PG, Butner J, Zautra AJ, Tennen H. Oscillations in daily pain prediction accuracy. Nonlinear Dynamics Psychol Life Sci. 2010;14(1):27–46.
30. Montpetit MA, Bergeman CS, Deboeck PR, Tiberio SS, Boker SM. Resilience-as-process: negative affect, stress, and coupled dynamical systems. Psychol Aging. 2010;25(3):631–40.
31. May RM. Simple mathematical models with very complicated dynamics. Nature. 1976;261:459–67.
32. Haken H, Kelso JA, Bunz H. A theoretical model of phase transitions in human hand movements. Biol Cybern. 1985;51(5):347–56.
33. Amazeen PG, Amazeen EL, Turvey MT. Breaking the reflectional symmetry of interlimb coordination dynamics. J Mot Behav. 1998;30(3):199–216.
34. Boker SM. Differential structural equation modeling of intraindividual variability. In: Collins LM, Sayer AG, editors. New methods for the analysis of change. Washington, DC: American Psychological Association; 2001. p. 5–27.
35. Boker SM, Nesselroade JR. A method for modeling the intrinsic dynamics of intraindividual variability: recovering the parameters of simulated oscillators in multi-wave panel data. Multivar Behav Res. 2002; 37(1):127–60.
36. Gu C. Adaptive spline smoothing in non-guassian regression models. J Am Stat Assoc. 1990;85(411): 801–7.
37. Kelso JAS. Dynamic patterns: the self-organization of brain and behavior. Cambridge, MA: MIT Press; 1995.
38. Bisconti TL, Bergeman CS, Boker SM. Emotional well-being in recently bereaved widows: a dynamical systems approach. J Gerontol B Psychol Sci Soc Sci. 2004;59(4):158–67.
39. Odgers CL, Mulvey EP, Skeem JL, Gardner W, Lidz CW, Schubert C. Capturing the ebb and flow of psychiatric symptoms with dynamical systems models. Am J Psychiatry. 2009;166(5):575–82.
40. Post AA, de Groot G, Daffertshofer A, Beek PJ. Pumping a playground swing. Mot Control. 2007; 11(2):136–50.
41. Kugler PN, Turvey MT. Information, natural law, and the self-assembly of rhythmic movement. Mahwah, NJ: Lawrence Erlbaum Associates; 1987.

42. Rosenblum LD, Turvey MT. Maintenance tendency in co-ordinated rhythmic movements: relative fluctuations and phase. Neuroscience. 1988;27(1): 289–300.
43. Michaels CF, Carello C. Direct perception. Englewood Cliffs, NJ: Prentice-Hall; 1981.
44. Solé R, Goodwin B. Signs of life: how complexity pervades biology. New York: Basic Books; 2000.
45. Chomsky N. Aspects of the theory of syntax. Cambridge, MA: MIT Press; 1965.
46. Bryk AS, Raudenbush SW. Hierarchical linear models: applications and data analysis methods. Newbury Park, CA: Sage; 1992.
47. Boker SM, Ghisletta P. Random coefficients models for control parameters in dynamical systems. Multilevel Model Newslett. 2001;13(1):10–7.
48. Luke DA. Multilevel modeling. In: Lewis-Beck MS, editor. Series: Quantitative applications in the social sciences. Thousand Oaks, CA: Sage; 2004. p. 1–79.
49. Von Holst E. Relative coordination as a phenomenon and as a method of analysis of central nervous system function. In: Martin R, editor. The collected papers of Erich Von Holst: the behavioral physiology of animal and man. Coral Gables, FL: University of Miami Press; 1973. p. 33–135.
50. Kelso JA. Phase transitions and critical behavior in human bimanual coordination. Am J Physiol. 1984; 246(6):R1000–4.
51. Kelso JAS, Schöner G, Scholz JP, Haken H. Phase-locked modes, phase transitions and component oscillators in biological motion. Phys Scripta. 1987;35(1): 79–87.
52. Post AA, Peper CE, Daffertshofer A, Beek PJ. Relative phase dynamics in perturbed interlimb coordination: stability and stochasticity. Biol Cybern. 2000; 83(5): 443–59.
53. Scholz JP, Kelso JA. A quantitative approach to understanding the formation and change of coordinated movement patterns. J Mot Behav. 1989;21(2): 122–44.
54. Boker SM. Consequences of continuity: the hunt for intrinsic properties within parameters of dynamics in psychological processes. Multivar Behav Res. 2002; 37(3):405–22.
55. Kelly GA, Blake C, Power CK, O'Keeffe D, Fullen BM. The association between chronic low back pain and sleep: a systematic review. Clin J Pain. 2011;27(2): 169–81.
56. Smith MT, Haythornthwaite JA. How do sleep disturbance and chronic pain inter-relate? Insights from the longitudinal and cognitive-behavioral clinical trials literature. Sleep Med Rev. 2004;8(2):119–32.
57. Smith MT, Quartana PJ, Okonkwo RM, Nasir A. Mechanisms by which sleep disturbance contributes to osteoarthritis pain: a conceptual model. Curr Pain Headache Rep. 2009;13(6):447–54.
58. Hamilton NA, Affleck G, Tennen H, Karlson C, Luxton D, Preacher KJ, et al. Fibromyalgia: the role of sleep in affect and in negative event reactivity and recovery. Health Psychol. 2008;27(4):490–7.
59. O'Brien EM, Waxenberg LB, Atchison JW, Gremillion HA, Staud RM, McCrae CS, et al. Negative mood mediates the effect of poor sleep on pain among chronic pain patients. Clin J Pain. 2010; 26(4):310–9.
60. Boker SM, Laurenceau J. Dynamical systems modeling: an application to the regulation of intimacy and disclosure in marriage. In: Walls TA, Schafer JL, editors. Models for intensive longitudinal data. New York: Oxford University Press; 2006. p. 195–218.
61. Boker SM, Laurenceau J. Coupled dynamics and mutually adaptive context. In: Little TD, Bovaird JA, Card NA, editors. Modeling contextual effects in longitudinal studies. Mahwah, NJ: Lawrence Erlbaum Associates; 2007. p. 299–324.
62. Nicholson JS, Deboeck PR, Farris JR, Boker SM, Borkowski JG. Maternal depressive symptomatology and child behavior: transactional relationship with simultaneous bidirectional coupling. Dev Psychol. 2011;47(5):1312–23.
63. Helm JL, Sbarra D, Ferrer E. Assessing cross-partner associations in physiological responses via coupled oscillator models. Emotion 2011:1–15. doi:10.1037/a0025036.
64. Abraham RH, Shaw CD. Dynamics: the geometry of behavior. Redwood City, CA: Addison-Wesley; 1992.
65. Beek PJ, Beek WJ. Tools for constructing dynamical models of rhythmic movement. Hum Movement Sci. 1988;7(2–4):301–42.
66. Beek PJ, Schmidt RC, Morris AW, Sim MY, Turvey MT. Linear and nonlinear stiffness and friction in biological rhythmic movements. Biol Cybern. 1995;73(6): 499–507.
67. Anderson KO, Bradley LA, Young LD, McDaniel LK, Wise CM. Rheumatoid arthritis: review of psychological factors related to etiology, effects, and treatment. Psychol Bull. 1985;98(2):358–87.
68. Rachman S, Arntz A. The overprediction and underprediction of pain. Clin Psychol Rev. 1991;11(4): 339–55.
69. Treffner PJ, Turvey MT. Handedness and the asymmetric dynamics of bimanual rhythmic coordination. J Exp Psychol Hum. 1995;21(2):318–33.
70. Treffner PJ, Turvey MT. Symmetry, broken symmetry, and handedness in bimanual coordination dynamics. Exp Brain Res. 1996;107(3):463–78.
71. Cacioppo JT, Gardner WL, Berntson GG. The affect system has parallel and integrative processing components: form follows function. J Pers Soc Psychol. 1999;76(5):839–55.
72. Russell JA, Carroll JM. On the bipolarity of positive and negative affect. Psychol Bull. 1999;125(1): 3–30.
73. Davis MC, Zautra AJ, Smith BW. Chronic pain, stress, and the dynamics of affective differentiation. J Pers. 2004;72(6):1133–59.
74. Zautra A, Smith B, Affleck G, Tennen H. Examinations of chronic pain and affect relationships: applications of a dynamic model of affect. J Consult Clin Psychol. 2001;69(5):786–95.

75. Ong AD, Bergeman CS, Bisconti TL, Wallace KA. Psychological resilience, positive emotions, and successful adaptation to stress in later life. J Pers Soc Psychol. 2006;91(4):730–49.
76. Zautra AJ, Berkhof J, Nicolson NA. Changes in affect interrelations as a function of stressful events. Cogn Emot. 2002;16(2):309–18.
77. Ursin H, Olff M. Psychobiology of coping and defence strategies. Neuropsychobiology. 1993;28(1–2): 66–71.
78. Linville PW. Self-complexity and affective extremity: don't put all of your eggs in one cognitive basket. Soc Cogn. 1985;3(1):94–120.

9 Modeling Illness and Recovery with Nonlinear Dynamics

Stephen J. Guastello

9.1 Single Single-Cause Models versus Complex Systems

People like simple explanations, even if they are wrong or misleading. The single-cause model is the simplest epidemiological logic: "If condition X exists, then Y has an elevated likelihood of occurrence." By implication, "If X is removed, Y does not happen." The limitations of the reasoning become apparent when research reports pile up. By the early 1980s we had the popular quip, "Everything causes cancer," which was indicative that something was wrong with our information about cancer. By 1990 public service messages were convincing the general public that cigarette smoking caused everything, and by implication if tobacco were to disappear, so would the major causes of death. However, even though tobacco consumption in the USA in 2006 had dropped to 38% of its 1967 level [1] (Fig. 9.1) and users dropped from 26% of the adult population in 1991 to 20.5% in 2006, healthcare costs have skyrocketed during the same period, and cancer incidence rates have dropped by only a small amount [2] (Fig. 9.2), which might be attributed to medical advances instead.

Of course one could respond that the foregoing comparison is unfair because there are many variables that contribute to cancer in some way, and all the other major causes of death. There are also many variables unrelated to the diseases themselves that contribute to the rising healthcare costs. That is the point.

The single-cause reasoning in medicine did not have an unreasonable beginning. In the early twentieth century, the leading causes of death were infectious diseases [3]. Most of those diseases became curable or vaccinated out of existence. Death did not disappear, however. Instead, infectious diseases were replaced by heart failure, cancer, stroke, and accidents as the leading causes of death in the USA and the world. The reduction in heart-related deaths in the last 20 years appears to be met with an increased proportion of cancer deaths or deaths by other sources. Deaths by Alzheimer's disease increased 46% from 2000 to 2006 [4] and now ranks as the sixth leading cause of death in the USA [5].

A complex systems perspective on health outcomes would include the following: The body is a complex system with complex interrelated subsystems. Each subsystem has its limits for how long it can possibly last, all other things being equal, and for how much intrusion from external sources it can withstand. Life and death are thus a race among the body systems to the finish line. If a miracle cure takes one system out of the race, then the winner is the next system in line.

The race to the finish line is also shaped by economic, scientific, and technological events. The incidence rates of accidental occupational

S.J. Guastello (✉)
Marquette University, P. O. Box 1881,
Milwaukee, WI 53201-1881, USA
e-mail: stephen.guastello@marquette.edu

J.P. Sturmberg and C.M. Martin (eds.), *Handbook of Systems and Complexity in Health*,
DOI 10.1007/978-1-4614-4998-0_9, © Springer Science+Business Media New York 2013

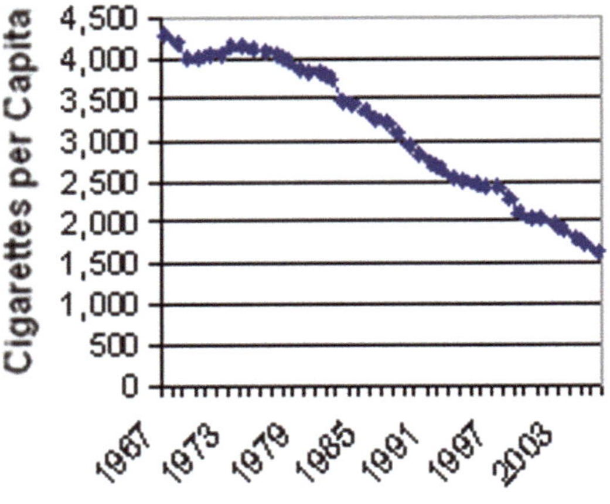

Fig. 9.1 Cigarette consumption per capita, USA, 1967–2006. From Center for Disease Control, 2010

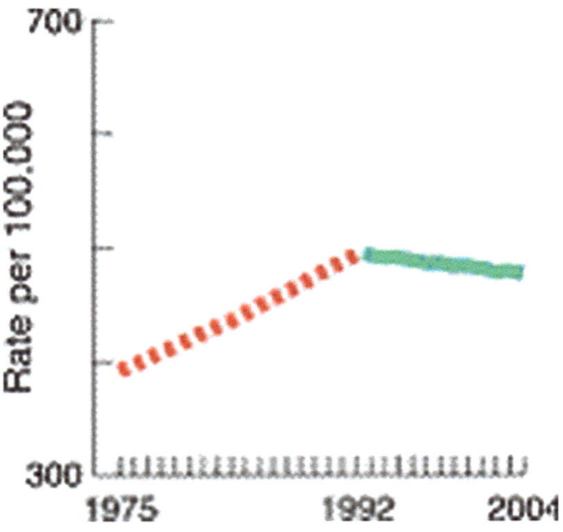

Fig. 9.2 Trends in cancer-related deaths in the USA. From National Cancer Institute, 2009

deaths in the USA has dropped over the last century due to technological advances that control on-the-job events and exporting many of the hazards jobs outside the USA. Some technologies, on the other hand, have been linked to new types of roadway accidents that never used to exist, e.g., distractions from cell phones and texting while driving. Advances in diagnostic technologies can introduce shifts in disease recognition and hence the classifications of causes of death. The age distribution of the population also affects the distribution of causes of death.

Although medical personnel might have been overly trained in the limited and reductionist single-cause reasoning [6–8], accident analysis and prevention specialists have known alternative models for quite some time [9–12]. The options include the domino, Swiss cheese, factorial, fault tree, dynamic fault tree, and NDS approaches to accident modeling and analysis. The topic now turns to one NDS solution that is more complicated than some of the alternatives yet concise and tractable.

9.2 Highlights of Catastrophe Theory

A few highlights of catastrophe theory [13] would be helpful here. According to the singularity theorem, all discontinuous changes of events can be explained by one of seven elementary topological forms. The forms differ in their levels of complexity. The cusp is the most commonly used catastrophe model overall and the model used in the applications presented here.

According to the classification theorem, given a fixed number of control parameters there is only one response surface associated with them. Similarly, if we know the number of stable states of the behavioral surface, we know how many control parameters are operating. The number of control variables and the complexity of the behavioral spectrum are unbreakable packages. The models or packages allow us to infer a global structure from a local structure. In other words, if we know how many behavioral states there are, we know the number of control parameters in the process and what they do.

The cusp model (Fig. 9.3) shows two stable states. There is a folded portion of the surface separating them; this is a repellor where very few points fall. The parts are joined together at a cusp point, which is a saddle; it is the most unstable point on the surface. A system located at the cusp point could change behavior in any direction relatively easily.

There are two control parameters in the cusp model. *Bifurcation* (b) explains large versus small differences. *Asymmetry* (a) explains proximity to the threshold of change. The equation for the response surface is:

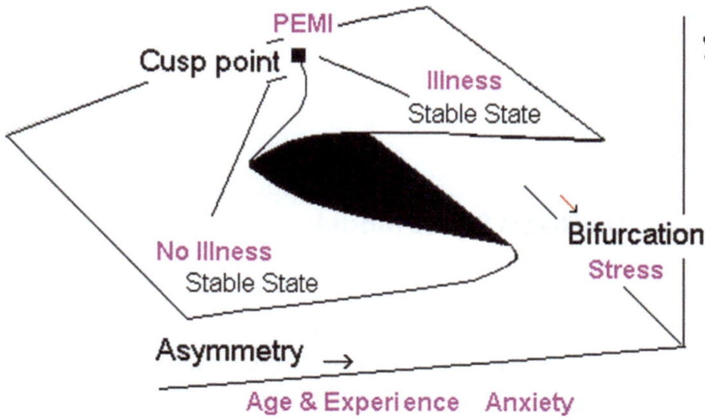

Fig. 9.3 Cusp catastrophe model for stress-related medical illness. *PEMI* pre-employment medical illness

$$df(y)/dy = y^3 - by - a. \quad (9.1)$$

The bifurcations set for the cusp generates gradients of behavior change that are shown on the cusp response surface [14]. Catastrophes also describe phase shifts, in the sense of water turning to ice or vapor, or vice versa, quite literally. Thus, they are also good explanations for the phase shifts associated with self-organizing phenomena [15, 16].

Other features of catastrophe models include hysteresis, the delay rule, and the Maxwell convention. *Hysteresis* is where behavior changes back and forth, or up and down the manifold. The presence of hysteresis strongly suggests a cusp dynamic. The *delay rule* runs as follows: Once the control parameters are in place to promote a behavior change, there could be a small time delay before the change occurs. This is analogous to the common observation of a "calm before the storm." The *Maxwell convention* addresses the case where one gradient may be stronger than the other. The system is likely to remain in the stable state that is most probable; systems can be expected to take longer to go into new states that they have not visited before.

The old school concept of "equilibrium" is more differentiated than it used to be. In NDS it might still be used to refer to stable, unstable states, attractors or saddles, all of which represent different dynamics. More commonly, the word is avoided altogether and replaced by the more specific concepts. In catastrophe models, the stable states are attractors, which may be fixed points or limit cycles. For further background on these basic forms of dynamics, see Guastello and Liebovitch [16] or Sprott [17].

Catastrophe models can be tested empirically through polynomial regression, nonlinear regression, or other optimization methods [18]. Inasmuch as the example that is explained in the next section of this chapter was done by polynomial regression, that method is expanded further here. The first step is to construct a cusp model:

$$\Delta z = \beta z_1{}^3 + \beta z_1{}^2 + \beta b z_1 + \beta a, \quad (9.2)$$

where the dependent measure y has been corrected for location (λ) and scale (σ), and the bifurcation (b) and asymmetry (a) variables have been similarly transformed:

$$z = (y - \lambda)/\sigma. \quad (9.3)$$

Location is *usually* the lowest observed value of the dependent measure, y. Scale is *usually* its standard deviation. A similar transformation is made on the control variables a and b before using them in the polynomial regression equation. The results are then compared against two linear models:

$$\Delta y = \beta b + \beta a, \quad (9.4)$$

$$y_2 = \beta y_1 + \beta b + \beta a. \quad (9.5)$$

Ideally the R^2 for the cusp model would be greater than the R^2 for both comparison models, and all the elements of the cusp model would be statistically significant.

9.3 Diathesis Stress Models

The respective roles of asymmetry and bifurcation variables correspond to what early epidemiologists called background variables and trigger variables, respectively [9, 19]. Industrial accidents do not usually occur in rubber rooms; stress could produce errors and problems, but injuries would be unlikely. On the other hand, a reasonably skilled worker in a hazardous environment might avoid harm well enough until some unexpected event occurs, or work pace and demands become too high. Here the demands induce human errors that could have serious consequences. Similarly, biological predispositions can remain dormant until they are triggered by stressors [20].

In a diathesis-stress model, an underlying physiological precondition is exacerbated by psychological or other exogenous exposures. The following example was part of a series of studies based on the experiences of 238 transit operators in a major Midwestern city [9, 21, 22]. Participants completed a survey that took an inventory of occupational safety and health risks, and included a checklist of medical disorders. The medical disorders included heart conditions, high blood pressure, kidney disorder, cancer, nervousness, insomnia, carpal tunnel, and ulcers. All eight conditions are known to be stress-related [23]. Participants indicated whether they had any of the disorders at the time of the survey, and whether they had them prior to starting employment with the bus company. They also indicated their age and how many years they had worked for the bus company. The sum of eight medical conditions, which ranged from 0 to 4 pre-employment and from 0 to 6 post-employment, was used as the dependent measure.

The hypothesized model appears in Fig. 9.3. The pre-employment medical conditions are regarded as the biological preconditions, and an inventory of job-related stress as the exacerbating factor, which was tested as a bifurcation effect. Age and job experience, which were treated as one variable, was tested as an asymmetry variable. Other variables in the study that were tested in the cusp model included safety management, transit-related hazards, and anxiety.

The results appear in Table 9.1. The overall R^2 for the cusp model was 0.70. Job-related stress was a significant contributor to the bifurcation parameter. Age and experience, anxiety, and job-related stress (again) were significant contributors to the asymmetry parameter. Although the bifurcation effect was weaker than the other parts of the model, it made a significant contribution nonetheless. The combination of the cubic and quadratic terms together accounted for more variance than the quadratic term alone. Thus, the model was classified as a cusp, and as such there must be a bifurcation variable. Therefore, the decision to retain job stress as a bifurcation variable was further supported on logical grounds.

A bootstrapping procedure was also used to assess the stability of the cusp model. For 10 samples of 200 cases randomly drawn from an expanded database of 15,590 cases, the mean r^2 was 0.62 and the median value was 0.64. Other analyses [22] showed that the same cusp relationship held up when each of the medical disorders is considered separately.

The model tells us, first, that the onset of medical disorders is a cusp catastrophe process. The discontinuous change between health and having one or more of the medical conditions is governed by two parameters. The asymmetry parameter, which is a combination of age, stress level, and anxiety, brings a person closer to the point of critical change. Stress makes the discontinuity greater than what it would ordinarily be. The specific job stressors that were cited most often by the transit operators were change in working hours, loss of sleep or change in sleeping hours, and dubious job security. The specific anxiety symptoms were extreme fatigue, migraine headaches, diarrhea or constipation, and chronic back pain. Anxiety was significantly correlated ($r=0.41$) with transit-specific hazards, which in

Table 9.1 Results for stress-related illnesses among transit operators. Reprinted from Guastello ([9], p. 248) with permission of Taylor and Francis

Predictor	r	R^2 (step)	t (weight)	F (model)
Cusp				
$z_1^{\,3}$	–0.61	0.37	4.86****	
$z_1^{\,2}$	–0.71	0.60	–6.69****	
z_1* Job-related stress	0.20	0.60	–1.77*	
Age and experience	0.16	0.63	5.23****	
Anxiety	0.24	0.69	5.34****	
Job-related stress	0.20	0.70	2.46**	88.74****
Difference control model				
Anxiety	0.35	0.12	4.74****	
Age and experience	0.25	0.20	5.00****	
Job-related stress	0.28	0.23	2.90***	23.85****
Pre–post control model				
Anxiety	0.36	0.13	4.92****	
Age and experience	0.26	0.21	5.26****	
Time-1 illnesses	0.19	0.25	3.43***	
Job-related stress	0.29	0.28	3.17***	22.29****

*$p<0.10$, **$p<0.05$, ***$p<0.01$, ****$p<0.001$

turn could be interpreted as further sources of stress: insults from the passengers, having to reprimand passengers for radio-playing, smoking, or loud talking, having to break up fights between passengers, having to assist another operator in trouble, and being attacked personally.

There are results on record to indicate improvements in stress symptoms resulting from occupational interventions. Stress management programs appear successful for reducing stress and improving job attitudes [24–26]. Murphy [25] noted, however, that stress management programs are not designed to remove the sources of stress, only to teach workers how to cope. Exercise programs have been effective for reducing stress injuries in physically demanding jobs [27, 28]. Occupational stress management programs seem to produce a reduction in actual accident rates by 15% [10]. Zwerling et al. [29] reported a medical management program that was designed to help employees make self-directed improvements on personal health issues; the program appeared to reduce accident rates by 29%. It is not possible to discern at the present time how much of an impact these programs would have on major medical outcomes, however.

9.4 Buckling and Resilience

Given that a person has incurred an illness or an accident, some bounce back more quickly than others, hence the concept of resilience has recently become a focus of study. In the context of occupational accidents, resilience of an organization involves a focus on its buffering capacity to forestall an actual accident, its degree of flexibility versus stiffness, maintenance of its margin of safety versus cutting it close, and responsiveness to the situation itself to minimize the casualties. Resilience involves both downward influence from management and its policies and upward influence from the operators "on the sharp end" [30]. Resilience effects are thus emergent processes and examples of supervenient effects resulting from the organization's nature as a complex adaptive system (CAS), although the CAS was not mentioned explicitly [31].

The theme of stiff versus flexible systems is a metaphor from material science and also emphasized as a feature of resilience [32]. A piece of material that is subjected to sufficient amounts of stress will show a certain amount of deformity, or

Fig. 9.4 Buckling of an elastic or rigid beam when weight is applied vertically

strain. Rigid materials will break, but flexible materials will rebound. The amount of deformity induced by stress is the stress–strain ratio. Figure 9.4 shows a beam of relatively stiff material that is pin-jointed at both ends. A weight is placed on the beam. If the material is rigid, and the weight is not supercritical, there will be little visible buckling. When the vertical weight becomes too large, the beam will snap. If, on the other hand, the material has a high degree of elasticity, the weight will cause the beam to waffle, more weight could cause it to waffle more, but the beam would not snap.

The connection between material strain and the cusp catastrophe was made some time ago with regard to physical materials [14] and humans performing physical labor [33]. The beam–buckling relationship is characterized as a cusp catastrophe model (Fig. 9.5). The amount of vertical weight is the asymmetry parameter. The elasticity of the material is the bifurcation factor, with low elasticity located at the high end of the bifurcation axis.

For human performance systems, the system begins on the lower sheet of the surface where the time to complete the task and the error rate are relatively low. As increased work demands are piled on, no change in work performance occurs until suddenly it does, whereupon work completion time or error rate increases dramatically. For physical labor, some physiological characteristics were identified that distinguished people who were more or less adaptive to changes in the workload. Besides making the point that a connection between human performance and material science existed, the study opened up numerous possibilities for determining the facets of elasticity that would pertain to other types of work or situations. In fact, one could consider what allows some people bounce back after a traumatic event, while other people would experience serious emotional impairment [34, 35], or why some people would make poor decisions under load stress while others would hold up just fine [36, 37].

The asymmetry parameter could be any of the usual sources of workload, demands for work speed, or environmental stressors such as extreme temperatures. Thompson [37] characterized the bifurcation variable as emotional intelligence. Stress can bring out extreme emotional reactions that interfere with clear thinking in an individual or a work team. Rigidity as a means of coping with stress only goes so far to maintain efficiency. Beyond a certain point individuals need to recognize their conditions and rely on a broader repertoire of coping strategies. One can then translate the problem into medical terms by asking what feature of the person's health or mental state contributes to elasticity or resilience.

9.5 Dynamical Diseases

Medical symptoms are not always steady. They come and go quasi-periodically or fluctuate in severity. Mackey and Glass [38] introduced the concept of dynamical diseases to describe conditions that have temporal signatures. Analysis of the temporal dynamics for NDS properties could go a long distance toward diagnosis and formulating effective treatments. If the patient could provide a history of the critical symptoms along with ambient conditions surrounding them, one might then apply a variety of nonlinear time series techniques to discern the underlying dynamics. As a possible example, Fig. 9.6 shows the history of epileptic seizures for a patient over a 2-year period in 7-day intervals [39]. This particular time series of events is confounded with medication changes and adjustments. Some means of untangling the influence of medications or other treatments that are not delivered under experimentally controlled conditions needs to be built into the time series analysis.

The idea of analyzing temporal dynamics of symptoms has carried over to psychological disorders as well [40, 41]. For instance, chaotic evolution of symptoms has been found for some, but

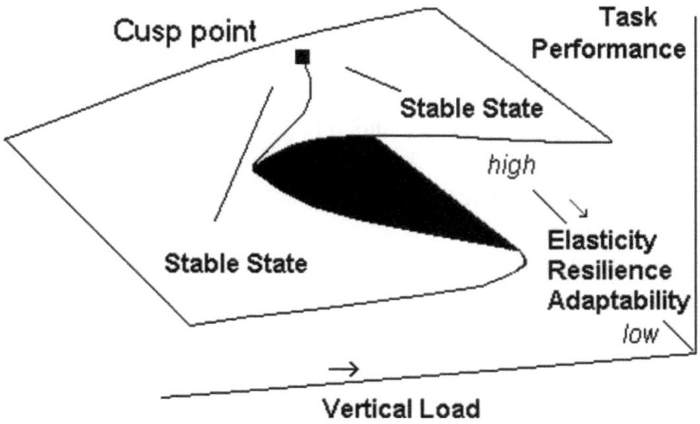

Fig. 9.5 Catastrophe model for the buckling of an elastic or rigid beam

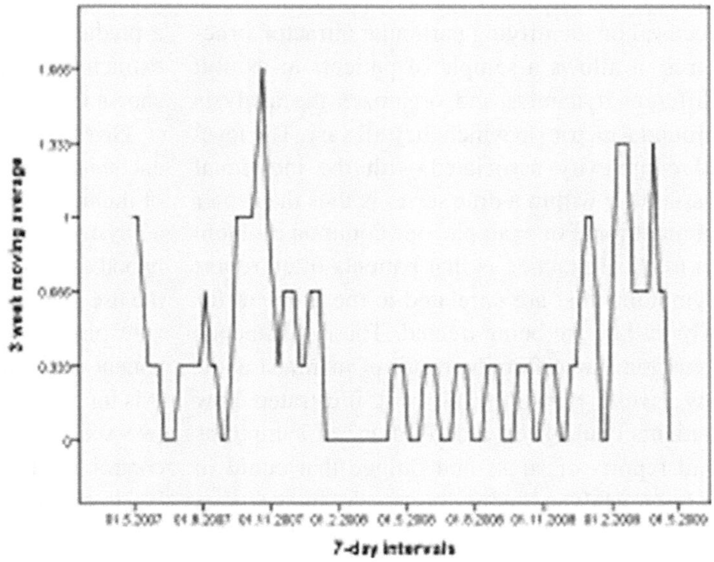

Fig. 9.6 Temporal unfolding of one patient's epileptic seizures during 2 years of treatment

not all schizophrenic patients. Chaotic patterns of mood have been observed in cases of bipolar depression [42] and unipolar depression [43]. Other research strategies have been preoccupied less with the presence of chaos per se and more so with differences among patient groups in dynamical indicators, such as those inherent in EEGs. For instance, Pezard et al. [44] found different dynamics in the episodes of severe depression among patients who never returned with another episode, and first episodes and repeated episodes for those who did have more than one, and control subjects.

A different question is whether attractors are formed at all, or whether the dynamics are mostly transients piled onto each other. Katerndahl and Wang [45] studied the temporal covariation of anxiety and depression for depressed patients, patients with panic disorders and controls. Conventional research often reported correlations between anxiety and depression but did so in cross-sectional studies where dynamics could not be ascertained. Katerndahl and Wang reported that nonclinical control subjects exhibited a narrow range of anxiety and depression values over time in a manner that suggested a single attractor. Patients with panic disorder displayed patterns that suggested multiple distinct attractors. Patients with major depressive episode, however, showed a great deal of variability in both variables, suggesting that no attractor was present, and that their levels of anxiety and depression fluctuated as if they were series of transients. Note here that chaos in the formal sense of a strange attractor was neither hypothesized nor concluded. Rather, the concern was whether a stable attractor of any type was present, and that multiple transients were a possible alternative outcome. The multiple transients associated with depressed patients showed some qualitative resemblance to the form of chaos reported by Gottschalk et al. [42] and Heiby et al. [43].

Another approach to defining a time series study of medical experiences is to frame the study in terms of entropy or variability in symptoms over time. Although the approach is less precise dynamically than approaches that focused on identifying particular attractor structures, it allows a sample of patients to exhibit different dynamics and organizes the analysis around a metric on which they all vary. The level of complexity associated with the individual variability within a time series is thus the center of attention. For example, one common problem in medical practice is that patients often report symptoms that are unrelated to the problem for which they are being treated. These extraneous symptoms are often the result of manifest anxiety levels. Burton et al. [46] illustrated how patients could keep a daily diary of symptoms and reports of stress and fatigue that could in turn be analyzed using the metric approximate entropy [47]. Burton et al. found that their patients' logbooks exhibited significant dynamical structure. For a sample of 26 patients who successfully kept their logbooks, individual differences in entropy of somatic symptoms, ratings of stress and mood, and a baseline measure of trait anxiety were all correlated; more anxious people exhibited greater variability in symptom levels over time.

In the example from Burton et al., the patients were all being treated for something different. There was no supposition that they had dynamical diseases in the sense of bipolar disorder where the temporal variation was associated with the disease itself. Rather, the concern was with the systematic distortion associated with individual differences in anxiety. The temporal dynamics of anxiety are highly variable across individuals.

9.6 Analysis of Seizures

The onset of an epileptic seizure is marked by hypersynchronous neuron firing. It is now possible to predict seizures with reasonable accuracy using nonlinear dynamics, although the prediction horizon is on a scale of hours rather than days or weeks, and is more often confined to epilepsies arising from damage to a single site. Seizures arising from damage to multiple sites are substantially more difficult to predict. A patient's self-management objectives could benefit from a system of prediction that was valid over a week time horizon. Such a prediction system would require, at minimum, extracting a rule from a time series such as the one shown in Fig. 9.6.

Given all the possible dynamics that one could encounter in a time series, including the possibility of multiple dynamics, researchers should approach a mystery time series with one or more cogent hypotheses about the dynamics. We [39] explored the use of the cusp and fold catastrophes on a seizure history of 124 weeks from one adult male patient with damage to two sites. A theoretical analysis for a cusp [48] had already been published; the two states were seizure and non-seizure, but the control variables were not yet specified. The following is a synopsis of the analytic strategy and the results. Further details about the patient, the analysis, and theory can be found in the original article.

We used linear autoregression to ascertain the best lag length. We found no prediction whatsoever for linear or cusp models, however, until we transformed the seizure record into 3-week moving averages, where a given observation was the average of itself, the week before, and the week afterwards. Autoregression identified two lag functions at 2 and 4 weeks, which we used as surrogates for control variables in the catastrophe models. We also tried the fold catastrophe model as an alternative because seizure states might not really be stable; the fold model contains one stable state, one unstable state, and one control parameter that moves the system between the two states. The analyses for the catastrophes were performed using the probability density function (pdf) method [18, 49]. The pdf method involves treating the time series of points as if they were a static distribution then fitting the distribution to a function that is uniquely associated with the cusp using nonlinear regression:

$$\text{Pdf}(z) = \xi \exp[\theta_1 z_1^{\,4} + \theta_2 z_1^{\,3} + \theta_3 B z_1^{\,2} + \theta_4 A z_1], \quad (9.6)$$

where A and B are the control variables (lag terms), and θ_i are nonlinear regression weights. The analogous model for the fold is:

$$\text{Pdf}(z) = \xi \exp[\theta_2 z_1^{\,3} + \theta_3 z_1^{\,2} + \theta_4 A z_1], \quad (9.7)$$

where both lag terms were tested as control variable A.

Strong degrees of fit were obtained for both the cusp and fold catastrophe models ($R^2=0.92$ and 0.88 respectively), which were more accurate than the counterpart linear model with the complex lag function ($R^2=0.58$). Prediction of future states was possible using a polynomial regression version of the fold catastrophe [analogous to (9.2)], and the lag 4 term (only) as the control variable, but the results were somewhat compromised ($R^2=0.47$) because of the nonstationary nature of the data and uncertainties regarding the meaning of the lag function.

9.7 Conclusions

Several points can be isolated from the foregoing discussion. First, the onset of an illness is qualitatively discreet with an underlying continuity, and lends itself to a cusp catastrophe analysis. Second, the underlying medical propensities (background variables) can be exacerbated by mental contributions or exogenous sources of shock, stress, or irritation (trigger variables). Third, stress sources play roles in bringing a person up to the critical point where an illness occurs and in making the nature of the illness better or worse. Two such processes were described here, one for diathesis stress dynamics and the other for buckling-resilience dynamics.

Fourth, recovery from an illness is thought to be a function of resilience. The full scope of biomedical resilience factors is not well defined at this time, but psychological resilience has become better understood recently. The cusp catastrophe, nonetheless, is a viable analytic model for assessing facets of resilience in different situations.

Finally, medical conditions, once they have started, follow a natural time course. There may be several possible time courses associated with a particular disorder if the underlying problem is left untreated. Medications and other forms of treatment may result in the reversal of a condition, in the sense of a cusp catastrophe, or might alter the temporal dynamics of the unfolding of the disorder. Numerous nonlinear time series techniques are potentially useful for isolating temporal dynamics and evaluating the impact of treatment attempts, although catastrophe models were viable once again in the example. Nonlinear time series analysis presents some challenges, however, that are not inherent in two observation models.

References

1. Centers for Disease Control (2010). Total and per capital yearly consumption of manufactured cigarettes and percentage change in per capita consumption – United States 1900–2006. http://www.cdc.gov/data_statistics/tables/economics/consumption/index.htm. Retrieved 11 May 2011.
2. National Cancer Institute (2009). Cancer incidence 1975–2004. http://www.cancer.gov. Retrieved 9 May 2009.
3. National Safety Council. International accident facts. 3rd ed. Lyle, IL: Author; 2002.
4. Alzheimer's Association. (2010). Alzheimer's disease facts and figures. http://www.alz.org/alzheimers_disease_facts_and_figures.asp. Retrieved 12 Jan 2011.
5. Centers for Disease Control. (2007). Number of deaths for leading causes of death. http://www.cdc.gov/nchs/fastats/lcod.html. Retrieved 12 Jan 2011.
6. Gregson RAM. Conceptual problems in cardiological prediction. Nonlinear Dynamics Psychol Life Sci. 2009;13:207–21.
7. Katerndahl D. Is your practice really that predictable? Nonlinearity principles in family medicine. J Fam Pract. 2005;54:970–7.
8. West BJ. Where medicine went wrong. Singapore: World Scientific; 2006.
9. Guastello SJ. Chaos, catastrophe, and human affairs: Applications of nonlinear dynamics to work, organizations, and social evolution. Mahwah, NJ: Lawrence Erlbaum Associates; 1995.
10. Guastello SJ. Human factors engineering and ergonomics: a systems approach. Mahwah, NJ: Lawrence Erlbaum Associates; 2006.
11. Reason J. Managing the risks of organizational accidents. Brookfield, VT: Ashgate; 1997.

12. Rowe WD. Anatomy of a risk. New York: Wiley; 1977.
13. Thom R. Structural stability and morphogenesis. New York: Benjamin-Addison-Wesley; 1975.
14. Zeeman EC. Catastrophe theory: selected papers (1972-1977). Reading, MA: Addison-Wesley; 1977.
15. Gilmore R. Catastrophe theory for scientists and engineers. New York: Wiley; 1981.
16. Guastello SJ, Liebovitch LS. Introduction to nonlinear dynamics and complexity. In: Guastello SJ, Koopmans M, Pincus D, editors. Chaos and complexity in psychology: theory of nonlinear dynamical systems. New York: Cambridge University Press; 2009. p. 1–40.
17. Sprott JC. Chaos and time series analysis. Oxford: New York; 2003.
18. Guastello SJ, Gregson RAM, editors. Nonlinear dynamical systems analysis for the behavioral sciences using real data. Boca Raton, FL: CRC Press; 2011.
19. Fine PEM. John Brownlee and the measurement of infectiousness: an historical study in epidemic theory. J R Stat Soc A. 1979;142:347–62.
20. Sun X, Zou Y, Nikiforova V, Kurths J, Walther D. The complexity of gene expression dynamics revealed by permutation entropy. Bioinformatics. 2010;11: 607–21.
21. Guastello SJ. Psychosocial variables related to transit accidents: a catastrophe model. Work Stress. 1991;5:17–28.
22. Guastello SJ. Accidents and stress-related health disorders: forecasting with catastrophe theory. In: Quick JC, Hurrell JJ, Murphy LM, editors. Work and well-being: assessments and interventions for occupational mental health. Washington, DC: American Psychological Association; 1992. p. 262–9.
23. Quick JV, Quick JD. Organizational stress and preventative management. New York: McGraw-Hill; 1984.
24. Ivancevich JA, Matteson MT, Freedman SM, Phillips JS. Worksite stress management interventions. Am Psychol. 1990;45:252–61.
25. Murphy LR. Occupational stress management: a review and appraisal. J Occup Psychol. 1984;1984(57): 1–16.
26. Murphy LR, Debois D, Hurrell JJ. Accident reduction through stress management. J Bus Psychol. 1990; 1:5–17.
27. Cady LD, Thomas PC, Karwasky RJ. Program for increasing health and physical fitness of fire fighters. J Occup Med. 1985;27:110–4.
28. Gebhardt DL, Crump CE. Employee fitness and wellness programs in the workplace. Am Psychol. 1990; 45:262–72.
29. Zwerling C, Baltroy LH, Fine IJ, Johnston JJ, Melius J, Silverstein BA. Design and conduct of occupational injury intervention studies: A review of evaluation strategies. Am J Ind Med. 1997;32:164–79.
30. Woods DD. Essential characteristics of resilience. In: Hollnagel E, Woods DD, Leveson N, editors. Resilience engineering. Burlington, VT: Ashgate; 2006. p. 21–33.
31. Woods DD, Hollnagel E. Prologue: resilience engineering concepts. In: Hollnagel E, Woods DD, Leveson N, editors. Resilience engineering. Burlington, VT: Ashgate; 2006. p. 1–6.
32. Sheridan TB. Risk, human error, and system resilience: fundamental ideas. Hum Factors. 2008;2008(50):418–26.
33. Guastello SJ. Euler buckling in a wheelbarrow obstacle course: a catastrophe with complex lag. Behav Sci. 1985;30:204–12.
34. Pincus D, Metten A. Nonlinear dynamics in biopsychosocial resilience. Nonlinear Dynamics Psychol Life Sci. 2010;14:353–80.
35. Tedeschi RG, McNally RJ. Can we facilitate posttraumatic growth in combat veterans? Am Psychol. 2011;66:19–24.
36. Guastello SJ, Boeh H, Shumaker C, Schimmels M. Catastrophe models for cognitive workload and fatigue. Theor Issues Ergonomics Sci. 2012;13:586–602.
37. Thompson H. The stress effect. San Francisco, CA: Jossey-Bass; 2010.
38. Mackey MC, Glass L. Oscillation and chaos in physiological control systems. Science. 1977;197:287–9.
39. Guastello SJ, Boeh H, Linn M. Long-range prediction of epileptic seizures with nonlinear dynamics. Nonlinear Dynamics Psychol Life Sci. 2011;15:377–88.
40. Nandrino J-L, Leroy F, Pezard L. Dynamics as a heuristic framework for psychopathology. Chaos Complex Lett. 2005;1:145–78.
41. Tschacher W, Junghan U. Psychopathology. In: Guastello SJ, Koopmans M, Pincus D, editors. Chaos and complexity in psychology: theory of nonlinear dynamical systems. New York: Cambridge University Press; 2009. p. 307–34.
42. Gottschalk A, Bauer MS, Whybrow PC. Evidence of chaotic mood variation in bipolar disorder. Arch Gen Psychiatry. 1995;52:947–59.
43. Heiby EM, Pagano IS, Blaine DD, Nelson K, Heath RA. Modeling unipolar depression as a chaotic process. Psychol Assess. 2003;15:426–34.
44. Pezard L, Nandrino J-L, Renault B, Massioui FE, Allaire JF, Varela FJ. Depression as a dynamical disease. Biol Psychiatry. 1996;1996(39):991–9.
45. Katerndahl D, Wang C-P. Dynamic covariation of symptoms of anxiety and depression among newly-diagnosed patients with major depressive episode, panic disorder, and controls. Nonlinear Dynamics Psychol Life Sci. 2007;11:349–66.
46. Burton C, Heath RA, Weller D, Sharpe M. Evidence of reduced complexity in self-report data from patients with medically unexplained symptoms. Nonlinear Dynamics Psychol Life Sci. 2010;14:15–26.
47. Pincus SM. Approximate entropy as a measure of system complexity. Proc Natl Acad Sci USA. 1991;88: 2297–301.
48. Cerf R. Catastrophe theory enables moves to be detected towards and away from self-organization: the example of epileptic seizure onset. Biol Cybern. 2006;94:459–68.
49. Cobb L. Parameter estimation for the cusp catastrophe model. Behav Sci. 1981;26:75–8.

Part II

Dynamics in Human Health

The complex dynamic behaviours in nature are obvious in the anatomy and physiology of the human body. This section provides an overview of the dynamic self-stabilising properties of physiological systems, much of which is supported by the fractal nature of its underlying anatomy.

The loss of normal fractal behaviour is characteristic of disease. Later sections describe specific conditions in these terms. Here we provide an overview of the fractal nature of many physiological systems, whose function is supported by its underlying fractal anatomy. However, the complex dynamic behaviour in physiology is not limited to the macroscopic scale. It is most obvious at the genetic level, where the interactions among genes are responsible for the phenotypical observations; understanding the genome and its interaction with the environment is more important than understanding the role of an individual gene.

10 Homeostasis: The Dynamic Self-Regulatory Process that Maintains Health and Buffers Against Disease

George E. Billman

True stability results when presumed order and presumed disorder are in balance. A truly stable system expects the unexpected, is prepared to be disrupted, waits to be transformed.

Tom Robbins (American Novelist, b. 1936) [1]

10.1 Introduction

Homeostasis, as currently defined, is a self-regulating process by which biological (or mechanical) systems maintain stability while adjusting to changing conditions. This concept explains how an organism can maintain more or less constant internal conditions that allow it to survive in the face of a changing and often hostile external environment. Our awareness of homeostasis has slowly emerged over the centuries and has become the central tenet of physiology. If one does not understand this self-regulating process, then it is not possible to comprehend fully the function of the body in health and disease. The disruption of homeostatic mechanisms is what leads to disease, and effective therapy must be directed toward re-establishing these homeostatic conditions, working with rather than against nature. The purpose of this essay is to describe the evolution of our understanding of homeostasis and the role of physiological regulation and dysregulation in health and disease.

G.E. Billman, Ph.D., F.A.H.A., F.H.R.S. (✉)
Department of Physiology and Cell Biology, The Ohio State University, 304 Hamilton Hall, 1645 Neil Avenue, Columbus, OH, 43210-1218, USA
e-mail: billman.1@osu.edu

10.2 Homeostasis: A Historical Perspective

The concept that a bodily regulation is required for health can be traced back to the ancient Greeks. The Greek physician/philosopher Alcmaeon of Croton (fl. 500 BC) proposed what can be called a "balance of opposites" to explain health and disease. He used a political analogy to define health and disease stating that: "Health is the equality of rights of the functions, wet-dry, cold-hot, bitter-sweet and the rest; but single rule of either pair is deleterious" [2]. Thus, inequality of power leads to tyranny in a political system and disease in the body. This concept was expanded by Hippocrates of Kos (ca. 460–ca. 377 BC) who proposed that health was the product of the balance and mixture of four body fluids or humors: blood, phlegm, yellow bile, and black bile. He wrote that:

> Health is primarily that state in which these constituent substances are in correct proportion to each other, both in strength and quantity and are well mixed. Pain occurs when one of these substances presents either a deficiency or excess, or is separated in the body and not mixed with the others [3].

Thus, medicine became a process "of subtraction and addition: subtraction of what is in excess, addition of what is wanting" [4]. Hippocrates further recognized the role of nature's helping hand in the healing process (*vis medicatrix naturae*), the ability of the body to heal itself [5]. It was the

J.P. Sturmberg and C.M. Martin (eds.), *Handbook of Systems and Complexity in Health*,
DOI 10.1007/978-1-4614-4998-0_10, © Springer Science+Business Media New York 2013

Fig. 10.1 Portrait of Jean Fernel (ca. 1497–1558). He is the individual who coined the term physiology. Source: National Library of Medicine (the history of medicine public domain image files)

role of the physician to clear the path so that nature could take its course. This concept became the basis for medicine in the ensuing centuries up to the dawn of the modern era.

Implicit in this concept of the "healing power of nature" is the assumption that the subunits of the body act in a cooperative manner to restore health when the normal state of the organism has been disturbed. Physiology, as a discipline dedicated to understanding how the parts of the body work together to maintain health, has its origins in the sixteenth century. The term physiology was first introduced by Jean Francois Fernel (ca. 1497–1558, Fig. 10.1) in 1542 [*De Naturali Parte Medicinae* (on the natural part of medicine)] as the study of the function of the healthy body as distinguished from pathology, the study of disease [5]. William Harvey (1578–1657) was the first individual to use carefully designed human and animal experiments to establish the function of a major bodily organ system with his description of the circulation of the blood. This application of physiology is illustrated in the following brief quotation from his seminal publication "Exercitatio Anatomica De Motu Cordis et De Circulatione Sanguinis in Animalibus" 1628 (Anatomical exercises on the motion of the heart and the circulation of blood in living creatures, first English translation 1653):

> It has been shown by reason and experiment that blood by the beat of the ventricles flows through the lungs and is pumped to the whole body … the blood in the animal body moves around in a circle continuously, and … the action or function of the heart is to accomplish this pumping. This is the only reason for the motion and beat of the heart [6].

Over the ensuing centuries, the concept of physiology has evolved and a central tenet has emerged that unites the various sub-disciplines of physiology: the quest to understand how the various components of the organism work together to maintain a healthy state. It is only by understanding normal bodily function that the disruptions that lead to disease can be determined and ultimately corrected so as to restore the healthy state.

As we have seen, a rudimentary understanding of the regulation and control of bodily function can be traced back to 6th century BC Greece. Despite sporadic progress over the centuries [7], it was not until the 19th century that systematic physiological investigation produced major advancements on this concept. Our modern understanding of physiological regulation rests firmly on the shoulders of two giants in the field: Claude Bernard (Fig. 10.2) and Walter Cannon (Fig. 10.3) who described regulations in terms of the constancy of the internal environment and homeostasis, respectively.

The French Physiologist, Claude Bernard (1813–1878) ,who is often referred to as the founder of modern experimental physiology, was perhaps the first to appreciate fully that living systems possess an internal stability that buffers and protects the organism against a constantly changing external environment [8]. He recognized that the body possesses mechanisms that operate in a coordinated fashion to maintain a relatively constant temperature and blood glucose concentration and this internal stability was vital for the health of the organism. He concluded that: "La fixité du milieu intérieur est la condition de la vie libre, independante" [9]. [The fixity (i.e., constancy or stability) of the internal environment is the condition for the free, independent life]. Although Bernard was highly honored and was the most famous French scientist during his lifetime, his hypothesis that the stability of the internal environment was independent of the external conditions, first articulated in 1854, was largely ignored for the next 50 years. Gross [10] has proposed three reasons to explain the delay between the publication of Bernard's ideas and their acceptance: (1) Pasteur's exciting discoveries in bacteriology that had immediate application in the prevention and treatment of disease came to dominate biological investigations; (2) the gap between evolutionary thought and general physiology—it took time to appreciate that natural selection provided the means by which regulatory control could evolve; and (3) the technology necessary to measure the internal environment was not yet available.

Fig. 10.2 Photograph of Claude Bernard (1497–1558). He developed the concept of "*a fixité du milieu intérieur*," that is, organisms maintain a stabile internal environment despite changing external conditions. Source: National Library of Medicine (the history of medicine public domain image files)

However, by the late nineteenth century and early twentieth century several investigators embraced Bernard's ideas, both as a central explanatory concept and as a program for research in physiology. Among those influenced by Bernard were such physiological luminaries as William M. Bayliss, Ernest H. Starling, Joseph Barcroft, J.S. Haldane and C.S. Sherrington in England, and L.J. Henderson and Walter B. Cannon in America [7, 8, 10]. Starling, in fact, coined the phrase "the wisdom of the body" to describe the maintenance of a constant internal environment [8]. Walter Cannon later popularized this phrase when he used it as the title for his book in which he introduced the concept of homeostasis. In 1900, Charles R. Richet (1850–1935), a student of Bernard who later won the Nobel Prize in Physiology and Medicine, stressed the dynamic stability of the internal environment. This is a statement, we shall see, presaged the definition supplied by Walter Cannon.

Fig. 10.3 Photograph of Walter B. Cannon (1871–1945). He built upon the work of Claude Bernard and coined the word homeostasis to describe a self-regulating process by which biological (or mechanical) systems maintain stability while adjusting to changing conditions. Source: National Library of Medicine (the history of medicine public domain image files)

> The living system is stable … it must be in order not to be destroyed, dissolved or disintegrated by colossal forces, often adverse, which surround it. By an apparent contradiction, it maintains its stability only if it is excitable and capable of modifying itself according to external stimuli and adjusting its response to the stimulation. In a sense, it is stable because it is modifiable—the slight instability is the necessary condition for the true stability of the organism [11].

This concept of a constant internal environment (*milieu intérieur*) was expanded by the American Physiologist, Walter Cannon (1871–1945) [8]. He coined the term homeostasis from the Greek words ὅμοιος (hómoios) "similar" and στάσις (stásis) "standing still" (together to mean staying similar and not staying the same) to describe the self-regulating processes by which a biological system maintains stability while adjusting to changing environmental conditions. Homeostasis is often mistakenly taken to mean unchanging or stagnant, as in hemostasis when the circulation (i.e., blood flow) stops. However, Cannon purposely selected the Greek word for similar, "hómoios," rather than the word for same, "homo," to express the idea that internal conditions could vary; that is, they are similar but not identical (stability but within range of values that allows the organism the freedom to adapt). Monotony breeds ennui while variety is the spice of life. Homeostasis, then, is the tendency of a system to maintain an internal stability as the result of the coordinated response of its parts to any situation or stimulus that disturbs normal conditions or function. Thus, the term homeostasis attempts to convey two ideas: (1) an internal stability within a range of values, and (2) the coordinated dynamic response that maintains this internal stability. As he explained in the following quote from his highly influential monograph, "The Wisdom of the Body," published in 1932:

> The coordinated physiological processes which maintain most of the steady states in the organisms are so complex and peculiar to living beings – involving, as they may, the brain and nerves, the heart, lung, kidneys and spleen, all working cooperatively – that I have suggested a special designation for these states, homeostasis. The word does not imply, something set and immobile, a stagnation. It means a condition – a condition which may vary, but is relatively constant [12].

As emphasized by Cannon, homeostasis is not static; it is, rather, a dynamic self-adjusting system that maintains viability in the face of changing environmental demands.

The final piece of the homeostasis puzzle was supplied by the application of control theory from systems engineering to explain self-regulation in biological systems. The "constancy" of internal physiochemical conditions is then largely maintained by the often complex interaction of multiple negative (and positive) feedback systems. Thus, from its inception physiological investigations have been directed toward understanding the organism (be it microbe, plant, animal or man) as a single functional entity.

10.3 Feedback Regulation: The Process that Underlies Homeostasis

> "Nam deteriores omnes sumus licentiate." (We all degenerate in the absence of control)
> Terence (Heauton Timorumenos, line 483)

As we have seen a critical feature of homeostasis is that an organism's internal environment is held within a narrow range of values via a self-adjusting system. Feedback is the mechanism by which homeostasis is obtained. A feedback system is a closed loop structure in which the results of past actions (changes in the internal environment) of the system are fed into the system (via information, feedback) to control future action; the system affects its own behavior (modified from Forrester, [13]). There are two types of feedback systems: negative feedback that seeks a goal and responds as a consequence of failure to meet this goal (maintains stable range of values) and positive feedback that produces growth processes wherein the actions build on the results that then generate still greater action (a growth cycle). These feedback systems are themselves subject to higher levels of control. Homeostasis is the result of the complex interaction and competition between multiple negative and positive feedback systems and provides the basis for physiological regulation.

Once again we can trace the origin of self-regulatory systems to the ancient Greeks.

The first documented device that employed the principle of self-regulation was a water clock (clepsydra) invented by Ktesibios (or Ctesibius, Greek Κτησίβις) of Alexandria (fl. 285–222) [14]. A water clock depends upon a steady flow of water to measure an unvarying flow of time. If the water level is not relatively constant, the water outflow will vary depending on the height of the water column supplying the clock (faster with a full container and slower as the water level in the container falls). The water clock designed by Ktesibios used a float valve (similar to that used in the modern flush toilet) to maintain a constant water level in the clock water reservoir. Thus, as water levels fall, the float also falls thereby opening a valve that allows water to flow into the clock reservoir and to replenish the water level. As the water returns to the desired level, the float rises and closes the valve. Thus, the clock water reservoir could be regulated such that there is no net gain or loss in the water level and thereby it maintains a constant water outflow rate from which an accurate estimate of time can be obtained. The accuracy of this type of water clock was not supplanted until the seventeenth century when a pendulum was employed to regulate the clock mechanism.

A number of other self-regulatory devices were invented in the ancient and medieval periods but it was not until the late eighteenth century, with the invention of the steam engine that the study of devices that incorporated "corrective feedback" for regulation became a subject for systematic investigation. A major limitation of early steam engines was that their speed was affected by both the steam pressure generated by the boiler and workload placed upon the engine. James Watt (1736–1819) vastly improved the efficiency and safety of the steam engine by the development of a centrifugal feedback valve that controlled the speed of the engine [15]. This "governor" employed a pair of metal balls spinning on each side of a rotating vertical shaft aligned in such a manner that as the engine speed increased so also did the spinning rate of metal balls (called flyweights) and, as a consequence of increased centrifugal force, the balls would spread apart. This, in turn, opened a valve to decrease the flow of steam into the engine and a slower speed was restored. Conversely, as the engine speed decreased, so also would the rotation of the flyweights, thereby decreasing the outward centrifugal force. The flyweights would drop (pulled down by gravity) closer together, closing the steam valve so more steam could enter into the engine and increase its speed. As with the water clock and its water reservoir level, a constant engine speed could be maintained despite fluctuating steam pressure and changing workload without the constant supervision of a human monitor.

Later in the nineteenth century, James Clerk Maxwell (1831–1879) published a mathematical analysis of Watt's governor that established the principles for understanding self-regulating devices and became the foundation upon which control theory is built [16]. In 1927, Harold S. Black (1898–1983) applied feedback regulation to electrical circuits to amplify transatlantic telephone signals [17]. His negative feedback amplifier (patented in 1937) can be considered to be one of the most important developments in the field of electronics. Further advances in systems control theory were achieved during World War II with the development of servo-control (negative feedback) mechanisms for anti-aircraft weapons.

In 1943, two influential papers were published that established that the mathematical principles of control theory, as first described by Maxwell, could be applied to explain behavior in living organisms. Arturo Rosenblueth, Norbert Wiener, and Julian Bigelow's paper entitled "Behavior, Purpose and Teleology" [18] and Warren McCulloch and Walter Pitts', "A Logical Calculus of the Ideas Immanent in the Nervous Activity" [19] were the first to establish a link between the self-regulating nature of physiological processes in living animals and negative-feedback systems designed by engineers. Interestingly, Rosenblueth worked closely with Cannon and undoubtedly was influenced by his ideas. A few years later, Wiener (1894–1964) introduced the term cybernetics [from kybernetes (κυβερνήτης), the Greek word for governor (as in steersman or pilot)] to describe the study of self-regulatory control and communication in the animals as well as in machines [20]. In his book Cybernetics [20], Wiener developed the first formal mathematical analysis of feedback control in biological systems: concepts that have subsequently been extensively applied in modeling physiological systems as, for example, by Arthur Guyton (1919–2003) and his many students with regards to cardiovascular regulation. Thus, feedback regulation has become accepted as the mechanism by which homeostasis is achieved and, as such, this section of the essay will close will a brief discussion and explanation of the principles of negative feedback.

The water clock and centrifugal steam governor described in the preceding paragraphs provide classic examples of negative feedback systems. As we have seen for the water clock, the opening and closing of the float/valve creates a cycle where information about the water level can be fed back into the system to affect changes to maintain the water level at some constant predetermined value. Thus, the float simultaneously affects the water levels and is affected by water level forming a circular causality or a cycle of causation. It is important to emphasize that this is an automatic self-regulatory system, meaning that it requires no external adjustment once the operating level around which the variable is regulated has been set.

A simplified general form of a closed loop feedback system is illustrated in Fig. 10.4. The illustrated cycle consist of four main components, (1) the variable (or set of variables) that are to be controlled, (2) a sensor that monitors the variable of interest, (3) a comparator or central processing unit (mathematically, the transfer function—the input/output relationship) where the information provided by the sensor (afferent or sensory pathway) is fed back into the system. The information is compared with the "desired" state (set point or operating point) to detect any error (difference between the desired state and the prevailing state), and (4) effectors (efferent or motor pathways) that are activated to correct any error. Effector activity opposes and thereby buffers against changes in the variable. A solid line is used in this diagram to indicate a direct relationship (increase leads to increase, decrease leads to decrease) between the components while a dashed line represents an inverse relationship (increase leads to a decrease and vice versa). Negative feedback regulation must contain an odd number of dashed lines in order to maintain the variable within a narrow range of the desired value.

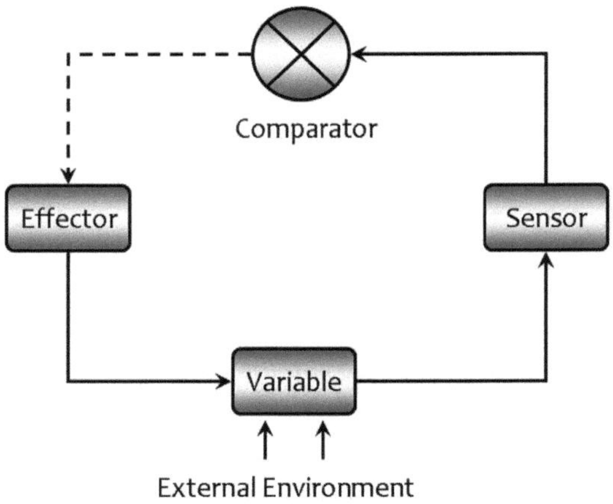

Fig. 10.4 A schematic representation of negative feedback regulation. A *solid line* indicates that the connected components are directly related (an increase in one component leads to increase the connected component, while a decrease will lead to decrease in the connected components). A *dash line* indicates the connected components are inversely related (an increase in one component leads to a decreases in the connected component while a decrease will lead to an increase in the connected component). An *odd number of dashed lines* are a necessary condition for any negative feedback cycle of causation. Negative feedback acts to maintain the controlled variable within a narrow range of values (see text for a detailed description)

A commonly used example of negative feedback is the of regulation room temperature by a thermostatically controlled heating and cooling system as displayed in Fig. 10.5. Room temperature is the variable, the sensor is a thermometer, the comparator is the thermostat—the device that compares the desired temperature (operating point) with the actual temperature (error detection) and the effector is the heating or cooling system. In this example, an increase in outside heat is detected by the sensor and the information is conveyed to the thermostat. The temperature information is compared to operating point and if there is sufficient difference between actual and desired temperature, the cooling system is acti-

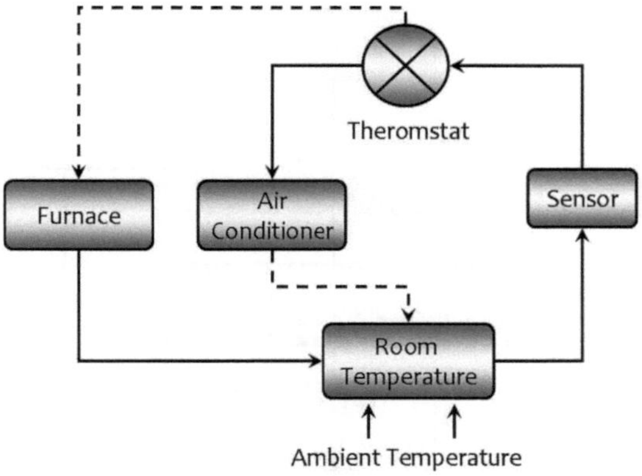

Fig. 10.5 A schematic representation of the regulation of room temperature to illustrate the concept of negative feedback regulation. A *solid line* indicates that the connected components are directly related (an increase in one component leads to increase the connected component, while a decrease will lead to decrease in the connected components). A *dash line* indicates the connected components are inversely related (an increase in one component leads to a decreases in the connected component while a decrease will lead to an increase in the connected component). Negative feedback acts to maintain the room temperature within a narrow range of values despite changes in ambient temperature (see text for a detailed description)

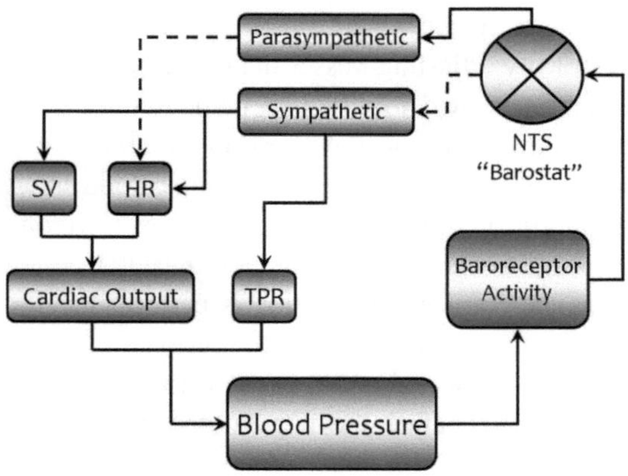

Fig. 10.6 A simplified schematic representation of the regulation of arterial blood pressure as a physiological example of negative feedback regulation. A *solid line* indicates that the connected components are directly related (an increase in one component leads to increase the connected component, while a decrease will lead to decrease in the connected components). A *dash line* indicates the connected components are inversely related (an increase in one component leads to a decreases in the connected component while a decrease will lead to an increase in the connected component). Negative feedback regulation acts to maintain the arterial blood pressure within a narrow range of values (see text for a detailed description). *NTS* nucleus tractus solitarius, the site where sensory information is processed and the efferent response is initiated. It acts as a "barostat" analogous to the "thermostat" in room temperature regulation. *SV* stroke volume (the amount of blood ejected by the heart with each ventricular contraction), *HR* heart rate, the number of beats (ventricular contractions) per minute, *TPR* total peripheral resistance, the resistance to the forward movement of blood (inversely related to the blood vessel diameter)

vated and the heating system is inactivated (reducing the error signal). The converse would happen if environmental temperature should fall, the cooling system would be turned off and the heating units activated. Thus, stable room temperatures can be maintained despite a wide range of fluctuating external conditions.

The concept of self-regulation in biological system is illustrated by the regulation of blood pressure. As early as the mid nineteenth century, it became obvious that arterial blood pressure was maintained within a narrow range of values via the activation of neurally mediated reflex adjustments [7]. However, it was not until to 1960s that the principles of negative feedback were applied to explain the homeostatic regulation of arterial blood pressure. A detailed description of intricacies of blood pressure regulation is beyond the scope of the present essay. Nonetheless, a simplified feedback cycle, analogous to the one we used for room temperature is seen in Fig. 10.6.

Before we can discuss this figure, we first must mathematically define arterial pressure using Ohm's Law expression (for a hydraulic rather than for an electrical circuit). Algebraically, blood pressure (BP, analogous to voltage, E, in an electrical circuit) is the produce of the cardiac output (CO, analogous to current, I, in an electrical circuit) and systemic vascular resistance as know as total peripheral resistance (TPR, analogous to resistance, R, in an electrical circuit). Cardiac output is itself the product of the amount of blood ejected per beat (stroke volume, SV) multiplied by the number of beats per minute (heart rate, HR).

So that, $BP = SV \times HR \times TPR$ (analogous to $E = IR$ for an electrical circuit).

It is evident that changes in arterial blood pressure can be countered by corrective changes in either the output from the heart, (SV or HR) or resistance to movement of blood through blood vessel (by adjusting vessel diameter, diameter is inversely related to TPR) or both. Returning to Fig. 10.6, the sensors are receptors (baroreceptors) located in arterial blood vessels (e.g., carotid sinus and aortic arch) that respond to changes in arterial pressure (increase in BP increase receptor activity). The comparator function is performed by a cluster of nerve cells within the medulla of brain (nucleus tractus solitarius, NTS) where the signal is processed to affect the output of the effector system. It acts as a "barostat" a function analogous to the thermostat in the regulation of room temperature shown in Fig. 10.5. The motor output consists of two sets nerves to the heart, parasympathetic nerves that decrease HR and sympathetic nerves that increase HR and SV. The sympathetic nerves also go to the blood vessels, the activation of which decreases vessel diameter and thereby increases TPR. Thus, if BP should increase, the so-called baroreceptor reflex is activated. An increase in parasympathetic activity coupled with a decrease in sympathetic activity would reduce cardiac output and decrease TPR. The opposite changes would occur if blood pressure should decrease. Thus, negative feedback regulation buffers against transitory changes and thereby helps maintain a stable blood pressure on a beat-by-beat basis throughout the day despite changing environmental or behavioral conditions.

The simple negative feedback schema described in the preceding paragraph, cannot adequately convey the complexity of the homeostatic process that allows an organism to function and adapt to changing environmental conditions [21]. For example, the operating point (or more accurately the operating range) of the negative feedback regulation can be adjusted or even overridden by higher levels of control [22]. These adjustments of the automatic (e.g., feedback) regulation allow the organism to adapt and respond appropriately to changing external conditions. This hierarchical control is multi-level multi-goal seeking system as shown in Fig. 10.7 (modified from Goodman [22]). In this schematic diagram, the first level represents the physiochemical processes, the organ and tissue functions, the component parts upon which homeostasis acts. The second level is autonomous

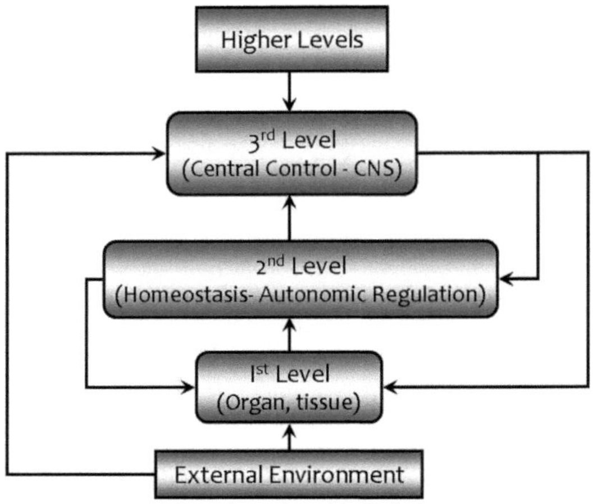

Fig. 10.7 A simplified schematic representation of the higher order control of homeostatic regulation. This hierarchical control results in a finer level of control and a greater flexibility that enables the organism to adapt to changing environmental conditions (see text for details). *CNS* central nervous system

(self) regulation, homeostasis (e.g., baroreceptor reflex). Here changes in a given variable are sensed and adjustments of the first level processes are initiated without input from higher levels of control. The third level is found in the central command and control centers (central nervous system) that process the information transmitted from the second level and integrates it with information from other sensory inputs to coordinate the physiological and behavioral response to changing environmental conditions. The higher centers can "intervene," making the adjustments as required to support the autonomic (i.e., autonomous and automatic) processes. This control can occur either at the conscious or unconscious level. An example of a conscious intervention would be the initiation of behaviors to cope with changing room temperature—adding or removing clothing, opening or closing windows seeking shade or sun, etc.—while an example of subconscious control would be the adjustments in blood pressure regulation during exercise (a shift in the operating point of the baroreceptor reflex so that HR increases despite increases BP as compared to resting conditions [23]). Thus, the third level coordinates behavioral and physiological responses to the external environment in order to maintain comfort and to ensure survival. However, it must be emphasized that higher level control is not possible if the first level components do not function properly. Finally, one could also envision even higher levels of control, factors outside of the organism.

The "autopilot" in a modern jet airliner can be used to illustrate the levels of control [22]. Once the preferred heading, attitude, and airspeed have been set, the autopilot will maintain level flight within acceptable degrees of roll, pitch, and yaw, despite changes in wind speed or minor turbulence. However, take off and landing (at least with the present technology) require the direct intervention of the human pilot. Thus, the first level consists of the components of the airliner, the jet engines and the airframe (fuselage, wings, flaps, rudder, etc.), the second level is the autopilot, and third level is the human pilot. In this example, a fourth level of control of the airplane is exerted by the air traffic controllers who provide directions to the pilot while an even higher level of control would reside in the Federal Aviation Administration, FAA, that sets the policy followed by the air traffic controllers.

The cardiorespiratory response to exercise provides a physiological example of this hierarchical control of homeostatic regulation. The first level consists of the tissues and organs that form the cardiovascular and respiratory system (heart, lung and blood vessels, but also the kidneys and endocrine glands that regulate salt and water retention and thereby blood volume), the second level of control is the baroreceptor (direct effect) and cardiorenal reflexes (indirect via regulation of blood volume), the third level of regulation takes place within the medulla (NTS) of the central nervous system where the sensory information is processed and the efferent response initiated. The medullary structures are themselves regulated by higher centers (e.g., hypothalamus and motor centers) in the brain. As previously mentioned, HR and BP are simultaneously elevated during exercise demonstrating that baroreceptor reflex regulation has been altered. These adjustments are required in order to increase oxygen delivery so that it can match the increased metabolic demand of the exercising muscles. Raven and associates [23] have demonstrated that these adjustments result from shifting the baroreceptor reflex to a higher operating point rather than from an inhibition of this reflex. Both feedback (sensory information for the exercise muscle, the so-called exercise pressor reflex) and feedforward (central command: for example, anticipation of the onset of exercise, such as visualizing the race before it is run, will increase HR, BP and skeletal muscle blood flow) contribute to these reflex adjustments. Finally, higher levels of control include the starter who determines when the race will begin, the event organizers who determine what races are run, and the sports regulatory agencies (Olympic committee, NCAA, etc.) that set the rules that govern the event.

Homeostatic control of the internal environment, therefore, involves much more than simple negative feedback regulation [20]. The hierarchical levels of command and control allow the organism to adjust its internal conditions to respond, to adapt, and to meet the challenges placed upon it by a changing and often hostile environment.

10.4 Homeostasis: Implication for Reductionism

> ...All the kings' horses and all the kings' men
> Could not put Humpty Dumpty together again
> Traditional English Nursery Rhythm (earliest published version 1803) [24]

The concept of homeostasis has important implications with regards to how best to understand physiology in intact organisms. In recent years, reductionist (attempts to explain the nature of complex phenomena by reducing them to a set of ever smaller and simpler components; the view that the whole is merely the sum of its parts), rather than holistic approaches have become dominant, not only in physiology, but in science in general. The earliest glimmerings of reductionist thought can be found in the surviving fragmentary writings of Thales and other pre-Socratic Greek philosophers who speculated that all matter was composed of various combinations of four key elements: earth, air, fire and water (the four humors of the body correspond to these elements) [5]. The pinnacle of Greek reductionism is found in the work of Leucippus and his student Democritus who proposed that all things consist of an infinitely large number of indivisibly small particles that they called atoms [5]. The modern application of reductionism in science can be traced to Francis Bacon (1561–1620) and Rene Descartes (1596–1650). Bacon incorporated reductionism as a central component, along with inductive reasoning, in his new empirical method (*Novum Organum* 1620, as opposed to Aristotle's *Organon* a treatise on logic and syllogism, i.e., deductive reasoning) [25] for the attainment of knowledge in natural philosophy, what has subsequently become known as the scientific method. Descartes likewise embraced reductionism as the pathway to knowledge, albeit with an emphasis on deduction (rationalism) rather than induction (empiricism) as advocated by Bacon. In his "Discourse on the Method of Rightly Conducting One's Reason and Seeking Truth in Science" [26], Descartes introduced two concepts that would have profound impact on biological investigations. In this, his most influential treatise, he described four precepts to arrive at knowledge. The 2nd and 3rd precepts, in particular, exemplify the reductionist's approach as follows:

> The second to divide each of the difficulties under examination into as many parts as possible and as might be necessary for its adequate solution
> The third to conduct my thoughts in such order that, beginning with those objects that are simplest and most readily understood, I ascend little by little, and as it were, step by step, to the knowledge of the more complex… [26]

His second and more far reaching conclusion was that the body was merely a machine. Thus, it was assumed that by applying Cartesian reductionism, one could deduce the complex physiology of the intact organism by understanding the presumably simpler functions of the individual organs and their constituent parts (from the molecular level to subcellular organelles to tissue to organ and finally back to the intact organism).

There can be no denying the power of this approach. In only a few decades after DNA was identified as the molecule of inheritance, its sequence of the some three billion base pairs has been mapped for man and other species, the genetic "code" for protein synthesis has been broken, and between 20,000 and 25,000 human genes that regulate a multitude of proteins have been determined. Humpty Dumpty quite literally has been smashed into a billion pieces.

However, reductionism rests upon the unstated assumption that the parts somehow entail the whole, that complexity is merely the product of incomplete understanding. The salient question is then whether this assumption is correct? Although we have sequenced the genome for many species, we have little understanding of the process by which the genome becomes an organism. We now know, in intricate detail, the basis for neuronal action potentials and synaptic transmission but do not understand how these electrical and chemical events give rise to consciousness. Complexity may not be the illusion it once naïvely was thought to be. The whole is greater than the sum of the parts!

The grand challenge faced by contemporary physiology in this post-genomic era is to integrate and to translate this deluge of information obtained in vitro into a coherent understanding of function in vivo. Although a machine may consist of many parts, the parts in isolation do not make the machine. Anyone who has tried to assemble a child's bicycle on Christmas Eve can testify that the parts do not a machine make. In an analogous fashion, while men are made of molecules, molecules are not men. The concept of one gene, one protein, one function is woefully inadequate to explain the dazzling complexity and startling beauty of the living organism—the intricate dance of homeostatic mechanisms necessary for a "free and independent life." A sequence of base pairs in the DNA molecule can no more explain the complexities of life than a series of 1 s and 0 s on a compact disc recording can explain the emotional response to music [27]. Man and other organisms are not mere vehicles for the perpetuation of genes, selfish or otherwise. The days for reductionist deconstruction are numbered; more holistic and integrated systems approaches will be required to put Humpty Dumpty back together again.

10.5 Summary

Our understanding of physiological regulation has evolved over time from the Greek idea concerning the balance between the body humors, through Claude Bernard's "milieu intérieur" to Walter Cannon's formulation of the concept of homeostasis and the application of control theory (feedback regulation) to explain how a constant internal environment is achieved. Homeostasis has become the central unifying concept of physiology and is defined as a self-regulating process by which a system (natural or synthetic) can maintain internal stability while adjusting to changing external conditions. Homeostasis is not static and unvarying; it is a dynamic process that can change internal conditions as required to survive external challenges. This is made clear by the care Cannon used when coining the word homeostasis. He deliberately selected Greek words, that when combined, meant "staying similar" rather than "staying the same" to emphasize that internal conditions could vary yet still produce stability (within a range of values rather than a single value). Thus, homeostasis does not mean "stagnation." It is also important to note that homeostatic regulation is not merely the product of a single negative feedback cycle but reflects the complex interaction of multiple feedback systems that can be modified by higher control centers. This hierarchical control and feedback redundancy produces both a finer level of control and a greater flexibility that enables the organism to adapt to changing environmental conditions. The health and vitality of the organism can be said to be the end result of homeostatic regulation of the internal environment; an understanding of normal physiology is not possible without an appreciation of this concept. Conversely, it follows that disruption of homeostatic mechanisms is what leads to disease, and effective therapy must be directed toward re-establishing these homeostatic conditions, working with rather than against nature.

Acknowledgments Portions of this essay appeared in a much condensed form and have been included with the permission of the author [28].

References

1. http://www.brainyquote.com/quotes/quotes/t/tomrobbins404093.html.
2. Freeman K. Ancilla to the pre-socratic philosophers: a complete translation of the Fragments Diels Fragmente der Vorsokratiker. Cambridge MA: Harvard University Press; 1948. Paperback edition 1983, p. 40.
3. Chadwick J, Mann WN. The medical works of Hippocrates: a new translation from the original Greek made especially for English readers. Oxford: Blackwell; 1950. p. 207.
4. Jones WHS (translator). Hippocrates Vol II: Prognostics. Loeb Classical Library No 148. Cambridge, MA: Harvard University Press. Breaths – page 219 (par. 1); 1923.
5. Hall TS. History of general physiology: 600 B.C. to A.D. 1900, vol. I from pre-socratic times to the enlightenment. Chicago: University of Chicago Press; 1975.
6. Harvey W. In: Keynes G, editor. The anatomical exercises: De Motu Cordis and De Circulatione Saguinis in english translation. New York, NY: Dover Publications, Inc.; 1995.
7. Adolph EF. Early concepts of physiological regulations. Physiol Rev. 1961;41:737–70.
8. Cooper SJ. From Claude Bernard to Walter Cannon. Emergence of the concept of homeostasis. Appetite. 2008;51:419–27.
9. Bernard C. Introduction à l'étude de la Médecine Expérimentale. Paris: J.B. Baillière et Fils; 1865 (English Translation by Greene HC, Dover Publications, Inc., New York, NY, 1957).
10. Gross CG. Three before their time: neuroscientists whose ideas were ignored by their contemporaries. Exp Brain Res. 2009;192:321–34.
11. Richet CR, editor. Dictionnaire de Physiologie, vol. 4. Paris: F. Alcan; 1900. p. 721.
12. Cannon WB. The wisdom of the body, revised and enlarged edition (first published 1939). New York, NY: W.W. Norton & Co., Inc.; 1963.
13. Forrester JW. Principles of systems. 2nd preliminary ed. Cambridge, MA: Wright-Allen Press., Inc.; 1976.
14. Landels JG. Engineering in the ancient world, revised edition (first pulbished1978, revised 2000). Berkeley, CA: University of California Press; 2000.
15. Rosen W. The most powerful idea in the world: a story of steam, industry, and invention. New York, NY: Random House, Inc.; 2010.
16. Maxwell JC. On governors. Proc Roy Soc. 1868;16:270–83.
17. Black HS. Stabilized feedback amplifiers. Electr Eng. 1934;53:114–20.

18. Rosenblueth A, Wiener N, Bigelow J. Behavior, purpose and teleology. Philos Sci. 1943;10S:18–24.
19. McCulloch W, Pitts W. A logical calculus of the ideas immanent in the nervous activity. Bull Math Biophys. 1943;5:115–33.
20. Wiener N. Cybernetics second edition: or the control and communication in the animal and the machine. Cambridge, MA: The MIT Press; 1961.
21. Carpenter RHS. Homeostasis: a plea for a unified approach. Adv Physiol Educ. 2004;28:180–7.
22. Goodman L. Regulation and control in physiological systems: 1960-1980. Ann Biomed Eng. 1980;8:281–90.
23. Raven PB, Fadel PJ, Ogoh S. Arterial baroreflex resetting during exercise: a current perspective. Exp Physiol. 2006;91:37–49.
24. Opie I, Opie P, editors. The oxford dictionary of nursery rhymes. New York, NY: Oxford University Press; 1997.
25. Bacon F (1620) In: Jardine L, Silverstone M, editors. The new organon. New York, NY: Cambridge University Press; 2000.
26. Descartes R. Discourse on the method of rightly conducting one's reason and seeking truth in science; 1637 (Original Title in French: Discours de la méthode pour bien conduire sa raison et chercher la vérité dans les sciences]. Available online as part of the Project Gutenberg at http://www.literature.org/authors/descartes-rene/reason-discourse/chapter-02.html.
27. Noble D. The music of life: biology beyond genes. New York, NY: Oxford University Press; 2006.
28. Billman GE. The grand challenge of physiology: to integrate function from molecules to man. Front Physiol. 1:5, 2010. doi:10.3389/fphys.2010.00005. published 27 April 2010.The world will not evolve past its current state of crisis

Fractals in Physiology and Medicine 11

Joachim P. Sturmberg and Bruce J. West

11.1 Fractals

Calculus is a method of reasoning by computation of symbols and in medicine this has traditionally followed the path laid out by the physics of the nineteenth and twentieth century, with its smooth continuous functions and differential equations to make predictions. In the latter part of the twentieth century physical scientists began to look in earnest at complex phenomena and discovered to their surprise that the analytic functions they had touted for so long were not adequate for characterising the variations in any but the simplest of processes. This particular failing was discussed from a statistics perspective in an earlier chapter. It is now time to squarely face the general limitations of the traditional modelling techniques in medicine and address a calculus of medicine that is able to incorporate nonlinearity into its descriptions.

J.P. Sturmberg (✉)
Department of General Practice, Monash University, Melbourne, VIC, Australia

The Newcastle University, Newcastle, PO Box 3010, Wamberal, NSW 2260, Australia
e-mail: jp.sturmberg@gmail.com

B.J. West
Army Research Office, 4300 S. Miami Blvd., Research Triangle Park, NC, 27709, USA
e-mail: bruce.j.west.civ@mail.mil

Edwald Weibel suggested that fractals are a design principle in biological organisms. He showed that a fractal design is structurally and functionally efficient as it requires little energy consumption to sustain itself [1]. It is hence not surprising that fractal structures are found everywhere in nature. Examples of naturally occurring fractals include clouds, river networks, fault lines, mountain ranges, snowflakes, crystals, lightning, cauliflower, broccoli, and ocean waves. Physiological systems such as blood vessel trees, pulmonary airways and vessels, DNA, heartbeat, and gait patterns all show fractal characteristics as well.

A fractal is "a rough or fragmented geometric shape that can be split into parts, each of which is (at least approximately) a reduced-size copy of the whole" [2]. This property is called *self-similarity*. Figure 11.1 (left) illustrates the principle where one repeatedly replaces the terminal branches of a "Y" with another "Y." This repeated branching pattern results in the well-known pattern of a tree branch. This process is geometric in nature with each new generation being geometrically identical to the previous generation.

The right side of Fig. 11.1 illustrates the dynamics of a self-similar pattern in a fluctuating time series. This latter kind of fractality is not geometrical but is statistical in nature. If the fluctuations are fractal the statistics of the entire time series is repeated in every interval of the time series. Consider the red section of the top curve in Fig. 11.1 and its magnified version in the centre curve. The two curves are essentially statistically identical. The same can be said for the

J.P. Sturmberg and C.M. Martin (eds.), *Handbook of Systems and Complexity in Health*,
DOI 10.1007/978-1-4614-4998-0_11, © Springer Science+Business Media New York 2013

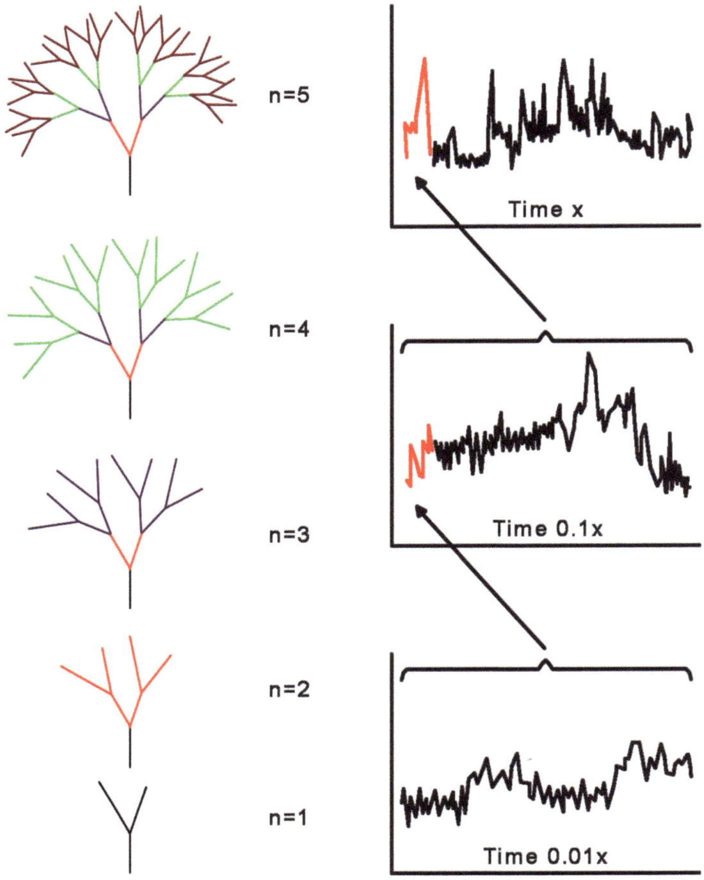

Fig. 11.1 *On the left* is a sketch of generating a self-similar geometrical fractal structure by adding segments from the *bottom* to the *top*. *On the right* is a sketch of a statistical fractal time series. The statistical self-similarity is indicated by magnifying sections of the top time series as indicated by the factors of 10 and 100

centre curve and the one in the lower panel. When the magnified regions of the time series do in fact reproduce the statistics of the original time series the statistics are fractal. The fractal nature of the distribution is manifest as a scaling property in the probability density as we shall discuss.

The term fractal was coined by the mathematician Benoît Mandelbrot in 1975 after a decade of studying the statistical properties of phenomenon that eluded description using the Normal distribution [2]. In an effort to be precise in his definition of the properties of fractals he created a mathematical fractal based on an iterated discrete equation,[1] iteration being a form of recursive feedback. The solution to the iterative equation in the footnote graphs as the *Mandelbrot Set*. These beautiful pictures in Fig. 11.2 illustrate very clearly the property of self-similarity; each smaller scale is a close copy of the whole.

One of the authors mentioned in the 2010 inaugural article in *Frontiers in Fractal Physiology* [3] that for a number of years the study of fractals and its application to physiology was restricted to the determination of the fractal dimension of structure, in particular, the static structure of objects and the scaling of time series. He goes on to emphasise that it is now time to explore the dynamics of fractal processes and apply the nonlinear techniques for analysing the dynamics of fluctuating time series to both regular and stochastic physiologic processes.

There are three types of fractals that appear in the life sciences: geometrical fractals, which determine the spatial properties of the tree-like

[1] $z_{n+1}=z_n^2+c$, where z_n and c are complex numbers, $z_0=0$, and c is a point on the plain. The formula is iterated until $|z_n|$ (the magnitude of z) is greater than or equal to the bailout value 2.

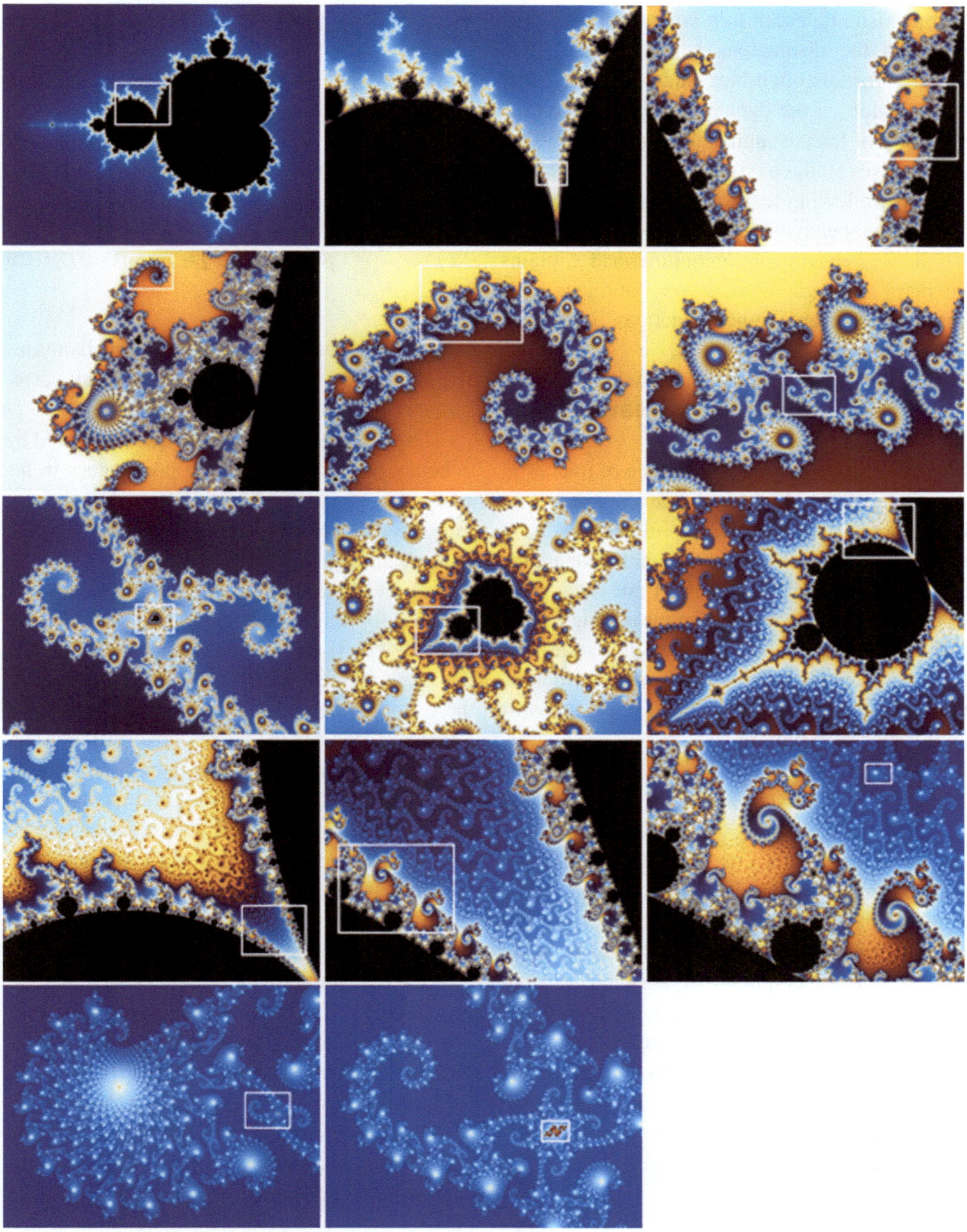

Fig. 11.2 A Mandelbrot set at increasingly smaller scale. (Created by Wolfgang Beyer, reproduced under Creative Commons Attribution-Share Alike 3.0 Unported license from Wikimedia Commons)

structures of the mammalian lung, arterial and venous systems, and other ramified structures; statistical fractals, which determine the properties of the distribution of intervals in the beating of the mammalian heart, breathing, walking, and in the firing of certain neurons; and finally dynamical fractals, which determine the nonlinear dynamical properties of networks that have a large number of characteristic timescales. In complex physiologic networks the distinctions between these three kinds of fractals often blur, and herein we focus our attention on the dynamics rather than on the geometry of fractals; although in general physiology involves all three types.

In the following sections we present examples of nonlinear behaviour in physiology and disease. Nonlinear behaviour is ubiquitous and explains the many "unpredicted outcomes" of linear interventions in current medical practice.

11.2 Measures of Nonlinearity

We have emphasised the importance of fractals in describing physiological dynamics. Here we hypothesise that the fractal dimension is a significantly better indicator of health than are the traditional averages such as heart rate. In support of this hypothesis we report on the scaling behaviour of a great many physiological time series.

In complex physiological networks we denote the zero-centred time series by $X(t)$ and the variance scales with time as

$$\left\langle X(t)^2 \right\rangle \propto t^{2H}. \tag{11.1}$$

The exponent H is a real constant, often called the Hurst exponent, and is related to the fractal dimension by $D=2-H$ [2]. In a complex physiologic network $X(t)$ is expected to depart from the entirely random condition of a simple random walk model, because real fluctuations are expected to have memory and correlation quantified by H to be different from 1/2 or D different from 3/2. Anomalous diffusion ($H \neq 0.5$) is associated with phenomena with long-term memory such that the two-time autocorrelation function is [4, 5]:

$$C(t_1,t_2) = \left\langle X(t_1) X(t_2) \right\rangle \propto |t_1 - t_2|^{\mu}. \tag{11.2}$$

Here the power-law index is given by $\mu = 2H-2$ and the two-point autocorrelation function is assumed to depend only on the time difference; the underlying process is stationary. The autocorrelation function is an inverse power law in time because $0 \leq H \leq 1$ implying that the correlation between data points decreases in time with increasing time separation. Note that inverse power law loss of memory is much slower than the exponential decay that is often assumed. This scaling is also manifest in the spectrum, the Fourier transform of the autocorrelation function, which is an inverse power law in frequency f:

$$S(f) \propto \frac{1}{f^{\beta}}; \quad \beta = 2\mu + 1 = 2H - 1 = 3 - 2D. \tag{11.3}$$

Note that for $H \leq 1/2$ the spectrum increases with frequency and for $H = 1$ the spectrum corresponds to $1/f$ noise.

The phenomenon of $1/f$ noise was discovered by Schottky [6] at the turn of the last century in his study of electrical conductivity. Over the intervening years this spectral form has been rediscovered in biological, economic, linguistic, medical, neurological, and social phenomena as well as in physics [7]. The spectra of such complex phenomena are given by (11.3) and the spectral index falls within the range $0.5<\beta<1.5$. Complex phenomena span the dynamic range from the macroscopic behavioural level down to the microscopic level. It is evident that $1/f$ variability appears in body movements such as walking, postural sway, and movement in synchrony with external stimulation such as a metronome; also such variability resides in physiologic networks as manifest in heart rate variability (HRV) [8], human vision [9], the dynamics of the human brain [10, 11], and in human cognition [12, 13]; also $1/f$ noise is measured at the level of single-ion channels [14, 15] and in single neuron adaptation to various stimuli [16]. Each of these psychophysical phenomena manifests $1/f$ variability [7]. The original assertion that $\beta=1$ was shown in these subsequent studies to extend the spectral index to the broader range indicated.

Now let us distinguish between the statistics of the underlying process and the correlations within the process. The probability that a dynamic variable $X(t)$ is in the phase space interval $(x, x+dx)$ at time t is given by $P(x, t)dx$. If the probability density scales then one can in general write

$$P(x,t) = \frac{1}{t^{\delta}} F\left(\frac{x}{t^{\delta}}\right), \tag{11.4}$$

where $F(.)$ is an unknown function of its argument. For classical diffusion this solution is well known to be

$$P(x,t)=\frac{1}{\sqrt{4\pi Dt}}\exp\left[-\frac{x^2}{4Dt}\right] \quad (11.5)$$

so that $\delta=1/2$ and the unknown function is the Normal distribution

$$F\left(\frac{x}{\sqrt{t}}\right)=\frac{1}{\sqrt{4\pi D}}\exp\left[-\frac{1}{4D}\left(\frac{x}{\sqrt{t}}\right)^2\right]. \quad (11.6)$$

Another example is fractal Brownian motion where $\delta=H$ and the unknown function is again a Normal distribution:

$$F\left(\frac{x}{t^H}\right)=\frac{1}{\sqrt{4\pi D}}\exp\left[-\frac{1}{4D}\left(\frac{x}{t^H}\right)^2\right]. \quad (11.7)$$

In general the statistics need not be Normal and the scaling parameter δ can be quite general.

Such scaling of the probability density can be empirically determined using the Shannon/Wiener Entropy

$$S(t)=-\int_\Omega P(x,t)\log P(x,t)dx, \quad (11.8)$$

where Ω is the domain of the variate. Suppose the time series consists of a sequence of discrete values $\{Z(t_1), Z(t_2), \ldots\}$ a random walk or diffusion processes is constructed by taking the sum $X(t)=\sum_{n=1}^{N} Z(t_n)$ with $t=t_N$. If the histogram for the probability density of the diffusion time series is inserted into (11.8) and the histogram scales as indicated by (11.4) the diffusion entropy becomes

$$S(t)=S_0+\delta\log t, \quad (11.9)$$

where the additive constant is defined in terms of the unknown function

$$S_0\equiv-\int_\Omega F(y)\log F(y)dy. \quad (11.10)$$

Note that a graph of the diffusion entropy versus the logarithm of time yields a straight line with positive slope whose value is given by the scaling index if the underlying process scales.

Patterns in time series contain information about the process being examined, and by ignoring such patterns in physiologic time series, information is discarded that may be of value to the physician. In fact it is quite possible that the information contained in the variability pattern may be more important than the information given by the average values. For example the scaling revealed by (11.9) can be due either to correlations as we have discussed, or it can be due to a more general statistical process with uncorrelated data point in time as we discuss subsequently.

A pattern in HRV time series, a time series constructed from the RR intervals, can be extracted in a simple manner by calculating the information contained within the time series [8] using (11.8). All the information is determined by the scaling exponent, which is the slope of the information (entropy) versus logarithm of time curve. An uncorrelated random process would have a slope of 0.50, a slope greater than this indicates a long-time memory in the time series. The information contained in an HRV time series is depicted in Fig. 11.3. The slope of the original HRV time series in Fig. 11.3 is 0.68 > 0.5 and consequently has a positive correlation. This positive correlation in the HRV data means that a long-time interval is probably followed by an even longer time interval and a short-time interval is followed by an even shorter time interval, this is the phenomena of persistence in HRV. The persistence is characteristic of a normal healthy individual.

To determine whether this scaling is in fact due to correlations within the data, in this case the persistence, the position of each data element within the sequence is randomly changed. This shuffling of the data removes any scaling due to correlations. The shuffled data are also depicted in Fig. 11.3 and are seen to have a slope very nearly 0.50 from which it is safe to say that any structure in the original HRV data is due to the time ordering of the data points. However it must be emphasised that this is not always the case since rearranging the data points does not change its statistics. Therefore if the slope of the information curve did not change after shuffling it could be concluded that the scaling was due to statistics and not correlations. This does not

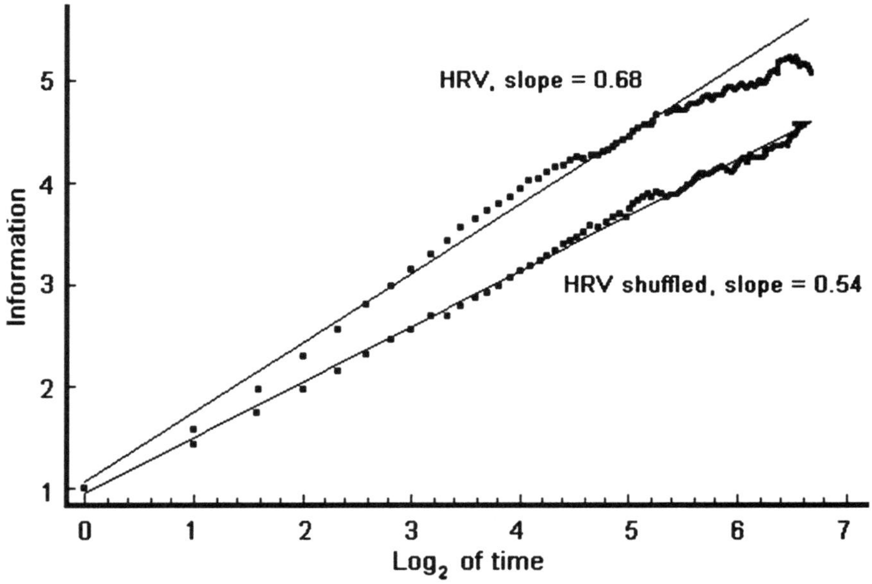

Fig. 11.3 The information as measured by the diffusion entropy is generated as a function of time by the heart beat interval time series or HRV for a typical healthy individual has a slope of 0.68, consistent with the scaling index obtained by other scientists using a variety of techniques. This information curve is compared with that generated by the same data shuffled to suppress the long-time correlation

happen often in physiologic data sets but it is common in complex physical processes [7].

The HRV fractal dimension obtained for the HRV time series is a measure of the complexity of the cardiovascular network and has a well-defined range for normal healthy individuals $1.1 \leq D \leq 1.3$ [17]. The fractal dimension changes as we age, becoming closer to three-half the older we get, indicating the loss of coherence in the generation of sequential time intervals. On the other hand the fractal index approaches one as the variability in heartbeat is lost as depicted in Fig. 11.4. Three individuals can have the same average heart rate; the first at the top of the figure is normal with a fractal dimension in the normal range; the second individual has a fractal dimension of nearly one and is dying by being overly organised in congestive heart failure, lower left of figure; the third individual has a fractal dimension near to three-half and is dying by disorganisation due to atrial fibrillation, lower right of figure. In short, the heart rate does not distinguish between the three cases but the scaling index does. This is the nonlinear world in which the complexity of a healthy physiologic network is captured by the scaling index (the fractal dimension).

The argument used for the HRV time series has also been successfully applied to the respiratory network using breathing rate variability (BRV) time series for the time intervals between breaths; the motor control network using stride rate variability (SRV) time series for inter-stride intervals; the regulatory network for body temperature using body temperature variability (BTV) time series; and finally gastrointestinal rate variability (GRV) using gut rumbling. Each of these complex networks presents a scaling index (fractal dimension) as a better indicator of the state of health of the underlying network than the associated average value, see for example West [17] for a complete discussion and references to the appropriate literature.

11.3 Nonlinearity in Physiology

Exact self-similar fractals are a mathematical abstraction and are not found in the real world. What nature generates are scaling phenomena that have a largest scale and a smallest scale and whether the fractal concept is useful in their analysis is how many decades exist between these

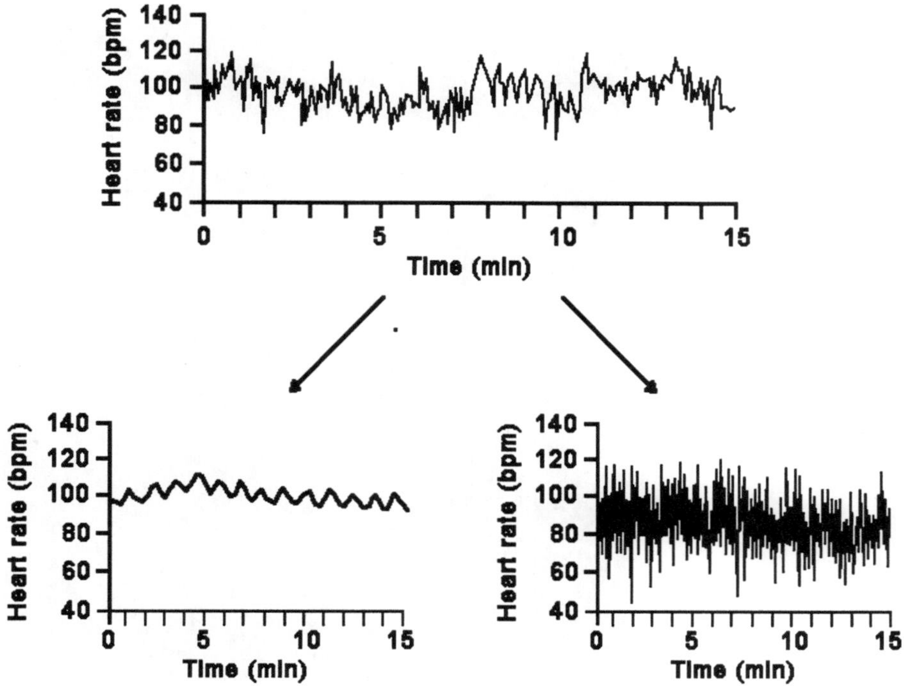

Fig. 11.4 Loss of fractal characteristics of HRV. *Top*—Normal healthy individual with a fractal dimension in the healthy interval; *Bottom left*—Sinusoidal pattern in severe heart failure; *Bottom right*—Atrial fibrillation (reproduced from [18] with permission)

extremes. Consequently natural fractals only exist within an extended, but finite, range of scales.

11.3.1 The Lung

The lung shows a typical branching structure for its airways as well as its vascular tree as depicted in Fig. 11.5. Consecutive branches vary in diameter and length, and after a given number of generations terminal branches end in alveoli. On average the adult airway has 23 generations, with the last 8–10 being alveoli. The pulmonary artery shows the same pattern; however, the latter networks continue to branch in the alveolar space to enhance gas exchange. On average the pulmonary artery has 28 generations.

Measurements have shown that branching follows consistent rules. The average length to diameter ratio is 3.25, and at each branch point the diameter of the branch reduces by a factor of ~0.86 and its length by a factor of ~0.62. This equates to a reduction by a factor of ~1.26, which has been shown to reduce the flow resistance in branching tubes to a minimum. This design results in a laminar flow at the alveolar level, reducing the cost of transport and mixing between air and blood [1].

The solid red curve in Fig. 11.6 depicts classical exponential scaling where the average diameter of a bronchial airway decreases by the same amount between generation z and generation $z+1$. Classical scaling seems to hold for approximately 10 generations after which a second scaling mechanism must be invoked. On the other hand, the entire range of bronchial diameters is fit by the solid black curve that is the result of the fractal model of the lung introduced by West et al. [20] using renormalisation group theory.

The fractal model [20] of the mammalian lung was the first to formally recognise that it is the average diameter that scales and not the diameter, which varies substantially within a given generation. In fact the diameter of a bronchial airway is a random quantity at each generation. Consequently they introduced a distribution

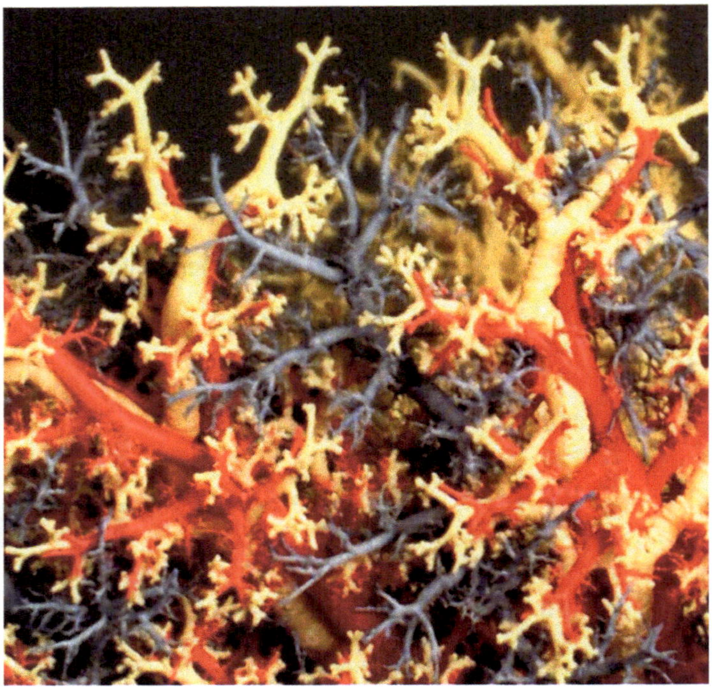

Fig. 11.5 Cast of human airways (*yellow*) and pulmonary vessels. Pulmonary arteries (*red*) accompany airways, whereas veins (*blue*) take an intermedial course. Note the irregular tree structure. (Courtesy Institute of Anatomy, University of Bern, Prof. Ewald R. Weibel)

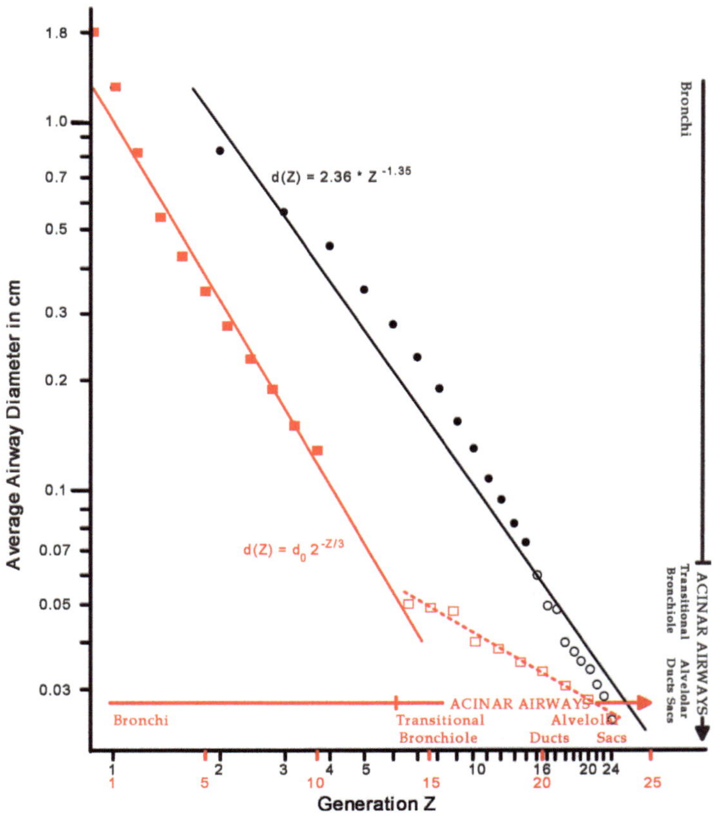

Fig. 11.6 Superimposed plots of airways diameter (logarithmic scale) against branching generations—*black scale logarithmic*, *red scale linear*. The log/log representation illustrates the power-law relationships between airway diameters and branching levels (adapted from [1, 19])

function for the fluctuations in the diameter within a generation and calculated that the average satisfies the scaling relation:

$$d(ax) = bd(z), \tag{11.11}$$

where a and b are constants. The solution to this renormalisation group scaling relation yields the average diameter as an inverse power law of the generation number z:

$$d(z) = \frac{A(z)}{z^{\nu}}; \quad \nu = \frac{\log(1/b)}{\log a} > 0. \tag{11.12}$$

The exponent in (11.12) has the value $\nu = 1.35$ in Fig. 11.6. The function $A(z)$ is a periodic function in the logarithm of the parameter b, which accounts for the slight bowing of the data in Fig. 11.6 [20]. However the curvature in the data was not fit by the solid black curve in Fig. 11.6 so that coefficient function was assumed to be constant $A(z) = 2.36$ in the figure.

This fractal model of bronchial airways provides an excellent fit to the average bronchial tube diameter in four distinct species: dogs, rats, hamsters, and humans [21]. The form of (11.12) strongly suggests that the renormalisation group relation (11.11) captures a fundamental property of the structure of the lung and that is its fractal nature.

11.3.2 The Heart

Important structures of the heart—coronary system, His-Purkinje conduction system and the chordae tendineae—all show fractal characteristics as sketched in Fig. 11.7.

The anatomical arrangements sketched in Fig. 11.7 are matched by fractal physiological function. For example, Goldberger et al. [22] showed that the nonlinear structure of the His-Purkinje system is linked to the nonlinear distribution of the electrical current resulting in the ventricular depolarisation of the myocardium. The electrical current entering the main bundle activates a new current at the branching point of the main right and left bundle, each bundle creates another two currents at their next bifurcation point and so forth. The arrival times of an impulse at the myocardium are not uniform as the length of each segment is different and varies in accordance with the length at each scale. The outcome is represented as the QRS complex in the ECG. Decomposing the frequency spectrum of the QRS complex shows a power-law distribution in Fig. 11.8 consistent with the fractal nature of a normally functioning conduction system.

The coronary tree structure, similar to the bronchial tree, allows for scale-free distribution of blood flow to the myocardium, i.e., the anatomical design of the vascular tree allows for minimal friction loss and metabolic power dissipation proportional to the transported blood volume as shown in Fig. 11.9. Murray [23] first described this relationship for the bifurcation of a vessel which states that the cube of the radius of a parent vessel equals the sum of the cubes of the radii of the daughter vessels. The relation generally associated with Murray was first expressed in general by Leonardo da Vinci some 400 years earlier and had to do with plants and rivers and not physiology, but the principle was the same. A more general rule would relate the diameter or cross-sectional area of a mother vessel to the total length or volume of its distal tree.

In Fig. 11.10 four measures of blood distribution in the coronary tree using the symmetric tree model indeed show a power-law relationship between the cross-sectional area of the main vessel and the cumulative arterial length of the tree perfusing a myocardial region [24, 25]. It should be emphasised that the scaling behaviour in this figure span as little as two and one-half decades to as much as eight decades. This latter size is like comparing the rate of growth of bamboo to the speed of a bullet or making a comparison in the same units, the width of fat human hair to the length of a football field.

11.4 Loss of Nonlinearity

As discussed previously nonlinearity is a necessary part of normal physiological function. The corollary, the loss of nonlinearity, either in the form of increasing orderliness (loss of complexity) or increasing disorderliness (increase in

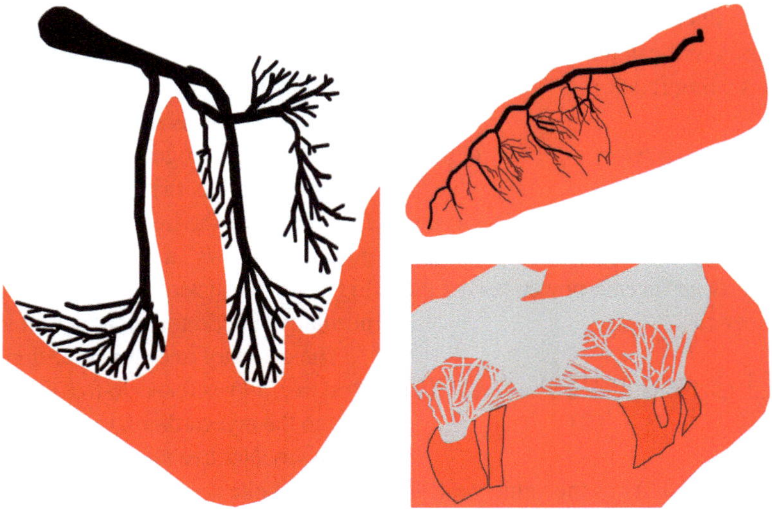

Fig. 11.7 Schematic drawings of the fractal structure of the His-Purkinje conduction system on the *left*, coronary vasculature on *top right*, and the chordae tendineae of the mitral valve on the *bottom right*

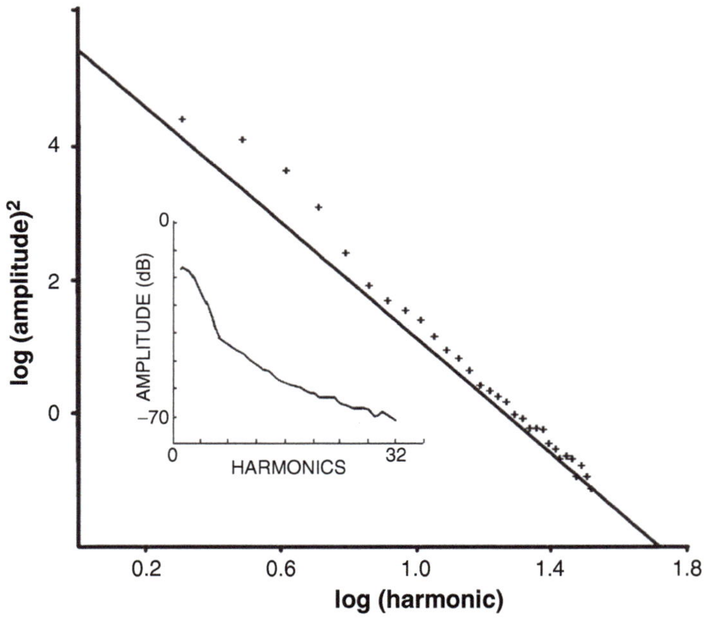

Fig. 11.8 The power spectrum (inset) of normal ventricular depolarisation (QRS) waveform shows a broadband distribution with a long, high-frequency tail. Power-law scaling is demonstrated by replotting the same data on a log–log graph. The straight line is the linear regression to a power-law power spectrum (reproduced from [22] with permission)

complexity), indicates ageing and impending or established disease.

Vaillancourt [26] postulates that the directional deviation from a nonlinear pattern in physiological systems covaries with the dimension of the intrinsic dynamic that organises the system's output as sketched in Fig. 11.11. The key feature of physiological dynamic networks is to achieve an outcome demanded by the network's environment, either to maintain a property within a narrow range or to achieve an oscillating outcome. These outcomes describe two types of intrinsic dynamics.

The first type of intrinsic dynamics is a fixed-point attractor behaviour and has zero dimensions; it is a point in phase space. Such a system maintains equilibrium through negative feedback from a receptor; the efferent pathways regulate the system's output back around to its fixed point in the presence of physiological fractal noise. The system requires more complexity in the out-

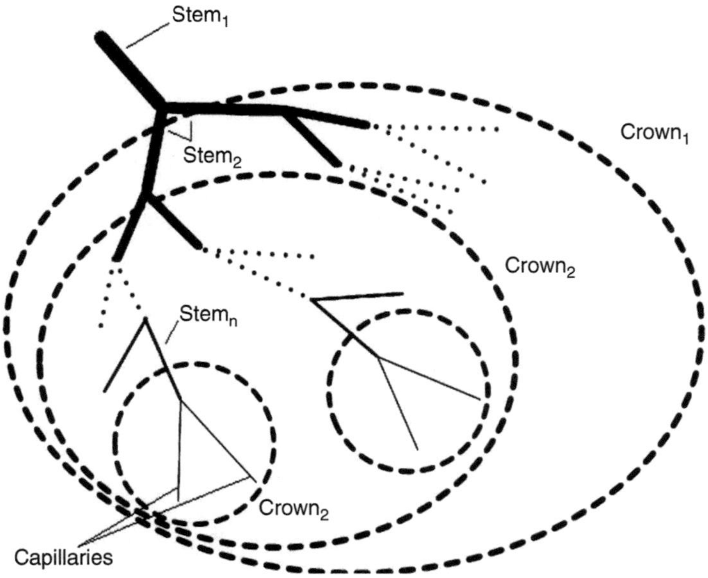

Fig. 11.9 Schematic illustration of the branching patterns of the vascular tree. Three stem-crown units are illustrated, with the smallest unit corresponding to an arteriole-capillary or venule-capillary unit (based on data from Murray [23])

put to perform optimally. With ageing and disease there is a reduction in the performance of the system, which is the system's ability to maintain the long-range order as measured by the fractal dimension of the fluctuation. This loss of fractal control results in decreased complexity.

The second type of intrinsic dynamics is a limit cycle attractor generating oscillatory behaviour and is one dimensional. Here the system requires less complexity in its output as it follows the internal dynamics in the presence of physiological fractal noise. With ageing and disease the fluctuations increase around the intrinsic dynamics, which results in an apparent increase of the system's complexity, see Fig. 11.11. On the other hand an increase in fluctuations does not necessarily increase complexity. If the fluctuations are uncorrelated random noise the fractal dimension approaches 1.5 and complexity is lost.

11.5 Clinical Examples of Loss of Complexity

Complexity is an elusive quantity that has been associated with an increasing influence of nonlinearity as well as with the entropy associated with increasing fluctuations. However we find that dynamic complexity is not a monotonic function of the fractal dimension. Neither a completely regular process with unvarying predictability nor an uncorrelated random process with no structure whatsoever satisfies the intuitive notion of complexity. A regular one-variable linear process has a fractal dimension $D = 1$ and is simple. As nonlinearity and more variables are included the process becomes more complex. On the other side of the ledger an infinite dimensional uncorrelated random process has a fractal dimension $D = 1.5$ and is also simple. As the number of variables decreases and correlations are introduced into the time series the process also becomes more complex. A fractal dimension midway between the regular and the random, where the measured time series is actually a blend of the two, is what appears over and over again in physiological time series. For example, this is what was observed in Fig. 11.4.

On the other hand static or spatial complexity is a monotonic function of the fractal dimension. As the fractal dimension increases between integer dimensions the complexity also increases. For example a two-dimensional surface is simple as is a three-dimensional volume but an object such as the mammalian lung has a fractal dimension midway between a surface and a volume and is more complex than either. Here again we might

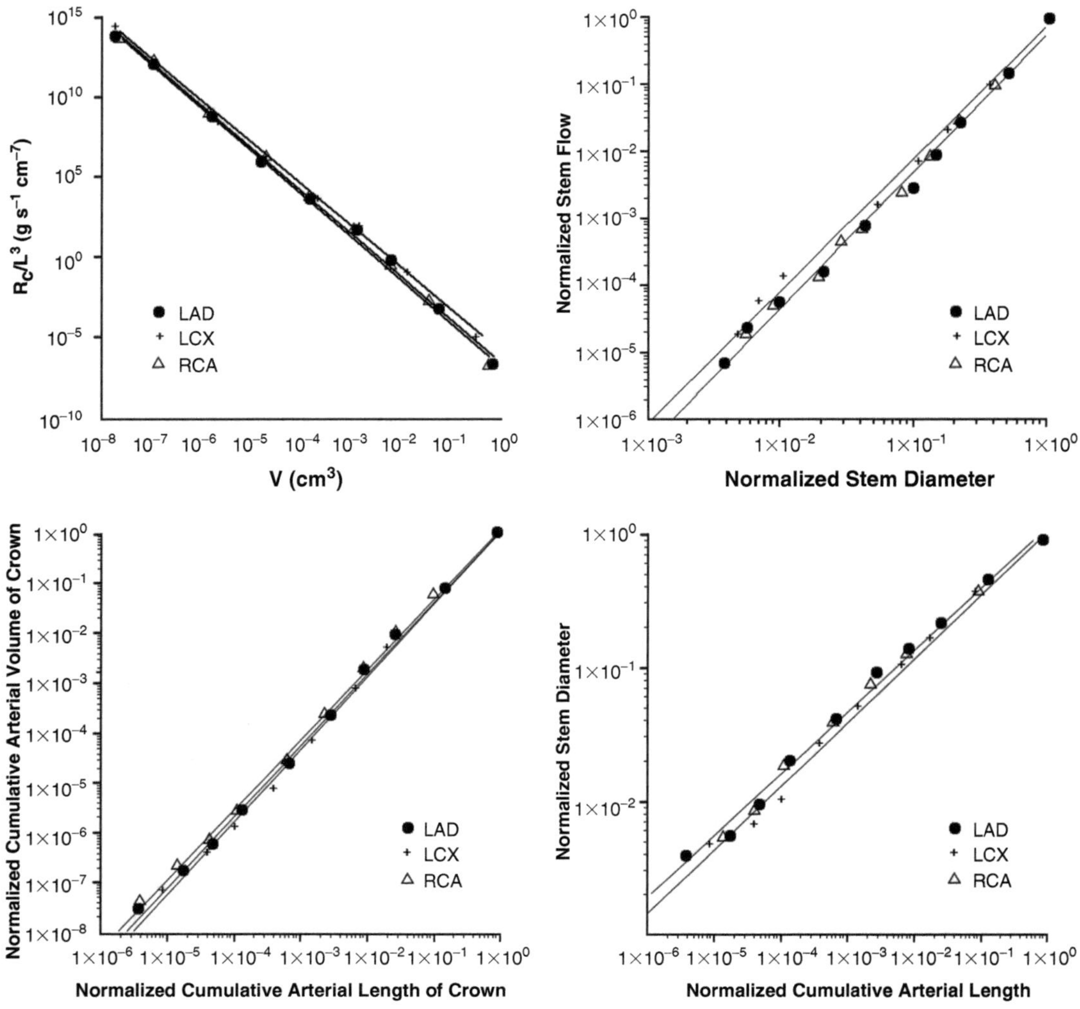

Fig. 11.10 Power-law relationships of four measures of blood distribution in the coronary tree based on the symmetric tree model. *Top left*: Relationship between the crown's equivalent resistance (R_c) divided by the length of the crown (L) cubed and the volume of the crown (V); *top right*: Relationship between normalised stem flow and stem diameter (flow and diameter are normalised with respect to their values at the most proximal segment); *bottom left*: Relationship between normalised arterial volume and arterial length of crown (volume and length are normalised with respect to the total volume and length of the respective arterial tree); *bottom right*: Relationship between normalised stem diameter and arterial length of crown (stem diameter is normalised with respect to the diameter at the most proximal segment while crown length is normalised with respect to the total length of the arterial tree. All data are curve fitted with power-law expressions using least squares (adapted from [24])

say the maximum complexity lies midway between the two extremes of simplicity.

11.5.1 Ageing

Ageing is associated with a variety of nonlinear changes in anatomical and physiological complexity [26, 27]. These changes affect neurones [28], bone trabeculae [29], HRV [27, 30, 31], blood pressure variability [31], gait [32], pulsatile thyrotropin release [33], LH and testosterone secretion [34], Cushing's disease [35], Conn's syndrome [36], acromegaly [37], EEG-evoked potentials [38], as well as hearing loss [39]. An example of anatomical loss of fractal dimension is seen in the

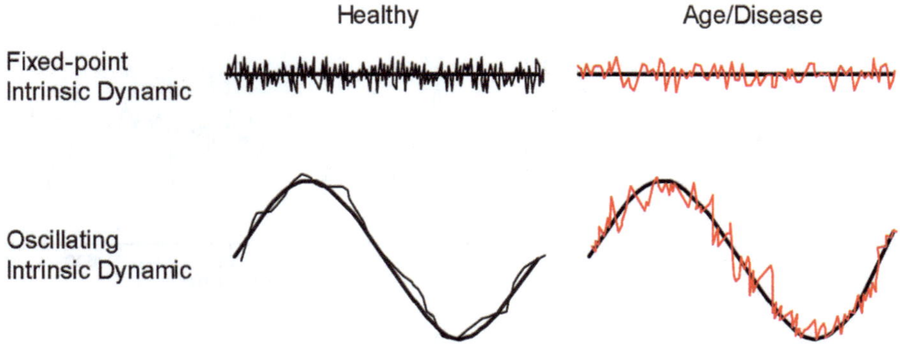

Fig. 11.11 Schematic illustration of the fixed-point and rhythmical intrinsic dynamics postulate. *Top*: The fixed-point intrinsic dynamic is consistent with homeostatic models in physiology and behaviour where the system attempts to regulate output around a constant mean state in the presence of noise. *Bottom*: The rhythmical intrinsic dynamic is consistent with a limit cycle or higher dimensional attractor where the mean output changes over time in the presence of physiological noise (reproduced from [26] with permission)

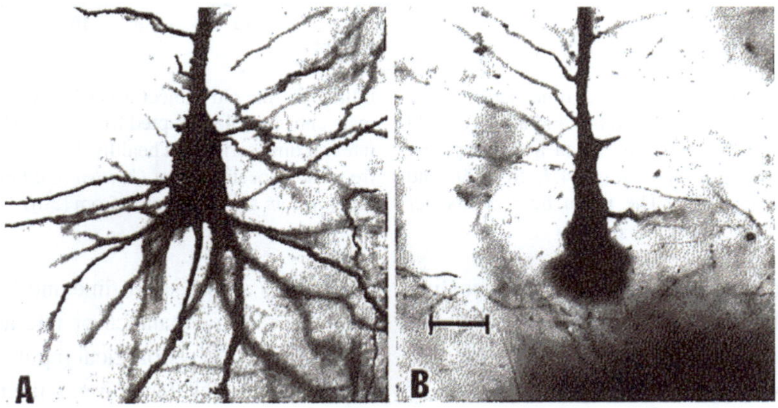

Fig. 11.12 Characteristic age-related changes in the structure of a giant pyramidal cell of Betz in the motor cortex. (**a**) Normal Betz cell in young adult male, (**b**) a Betz cell in a 65-year-old male. Most dendritic spines and virtually all of the basilar dendrites have been lost (reproduced from [28])

ageing brain. The histological image in Fig. 11.12 clearly demonstrates the loss of dendritic structure [26]; most of the dendritic spines and virtually all of the basilar dendrites have been lost.

Vaillancourt and Newell [26] suggested that these ageing changes can be associated with a decrease as well as an increase in complexity. Figure 11.13 illustrates this duality in the context of HRV where short-term beat-to-beat variability (left panels) clearly decreases complexity [27], however long-range variability shows an increase in complexity (right panels). The observed characteristics indicate a change in the fractal scaling of beat-to-beat variability in "healthy ageing." Short-range variations are caused by autonomic and respiratory influences, whereas the long-range variations are considered to reflect the influences of volume shifts, effects of the endocrine system, and metabolic and other processes [30].

Despite the nearly identical mean values and standard deviations (SDs) of heart rate for the two time series shown in Fig. 11.13 the "complexity" of the signal from the older subject is markedly reduced from that of the younger. Note the crossover behaviour resulting in two scaling regions in the frequency spectra and fluctuation plots. The SD $F(n)$ of the beat-to-beat fluctuation

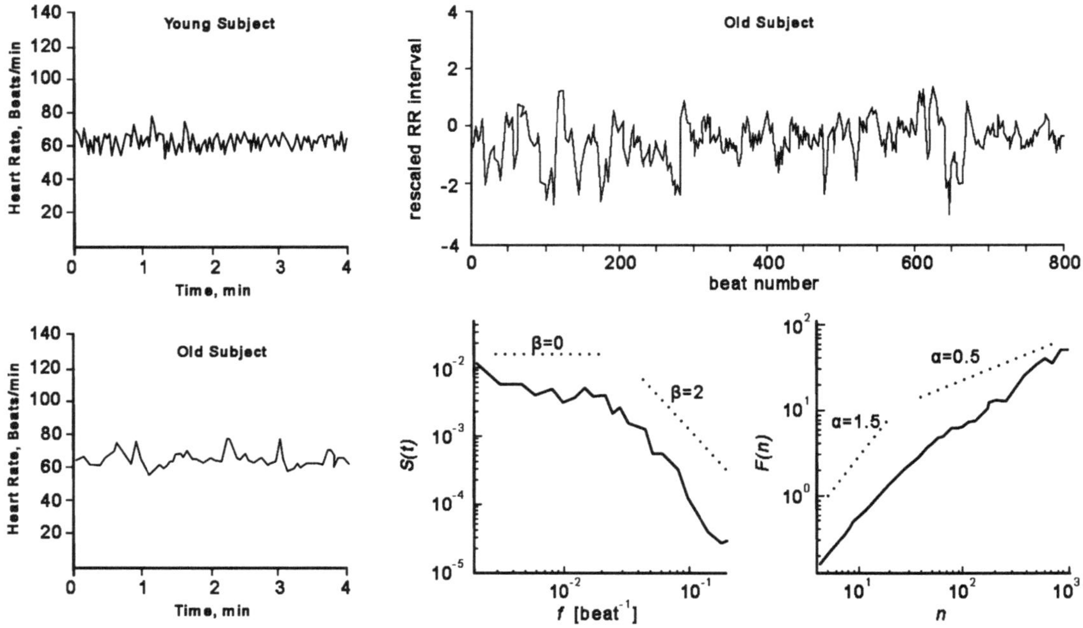

Fig. 11.13 *Left panels*—Heart rate time series for a 22-year-old female subject (*top panel*) and a 73-year-old male subject (*bottom panel*). The mean heart beats per minute for the young subject was 64.7, SD 3.9, and approximate entropy 1.09. The mean heart beats per minute for the old subject was 64.5, SD 3.8, and approximate entropy 0.48 (adapted from [27]); *Right panels*—RR interval time series of a healthy 77-year-old woman (*top*), frequency spectrum plot (*bottom middle*) and fluctuation plot (*bottom right*) (adapted from [30])

scales with the time n as $F(n) \propto n^{\alpha}$ with scaling exponent α. The frequency spectrum corresponding to this scaling is inverse power law $S(f) \propto 1/f^{\beta}$ with power-law exponent $\beta = 2\alpha - 1$. In Fig. 11.13 the exponent $\alpha = 1.5$ (dotted line) for short-range scaling resulting in $\beta = 2$ (dotted line) for the high frequency (short times) and indicates Brownian motion or a random walk. The exponent $\alpha = 0.5$ (dotted line) for long-term scaling indicating white noise with uncorrelated random dynamics with $\beta = 0$ (dotted line) that indicates equal power at each of the low frequencies (long times). Note that walks are determined by summing over processes with fractal statistics and their index increases by 1. Consequently an uncorrelated random process with $H = 0.5$ can be summed to generate a random walk with $\alpha = 1.5$ as observed for early times in Fig. 11.13, for longer times the individual fluctuations and not their sum determines the process and the underlying uncorrelated nature of the process appears [40].

In contrast to the stride interval time series for the elderly, young adults show no difference in complexity across short-time and long-time intervals, indicating a balance of interaction between many different physiological inputs [30]. The random walk model that mimics their behaviour is one having persistence in the steps with a scaling index in the interval $0.5 < \alpha < 1$ or equivalently a fractal dimension in the interval $1 < D < 1.5$. This is discussed more completely by West [17].

Gait similarly has fractal dynamics, and stride interval fluctuations show a decrease in the scaling component for elderly in Fig. 11.14 [32]. Strides become more random and less predictable with ageing, the fractal dimension approaches 1.5 and the time series is less complex [26] than those of the young.

It is also worth pointing out that most physiological time series change their properties over time and consequently it is not possible to strictly assign a single scaling index or fractal dimension to them. We saw such change in the two scaling domains of the HRV time series for the elderly. More generally the fractal dimension of a physiological time series changes continuously in time

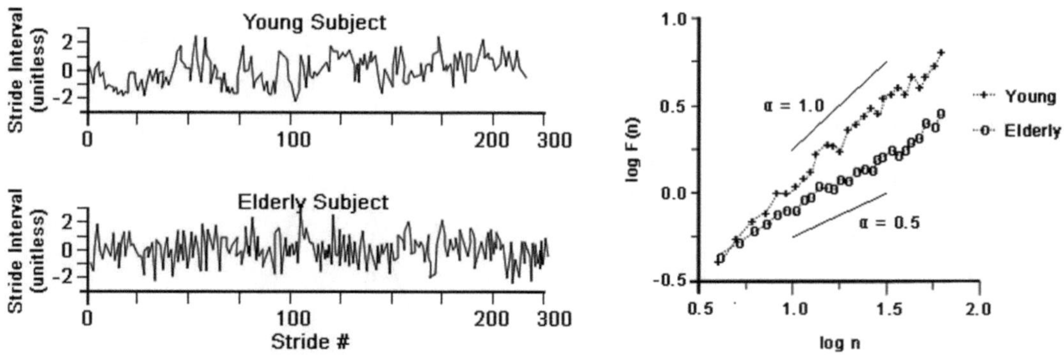

Fig. 11.14 Stride interval time series are shown in *top* and fluctuation analysis is shown in *bottom panel* for a 71-year-old elderly subject and a 23-year-old young subject. For this elderly subject, fluctuation analysis shows that stride-interval fluctuations [$F(n)$] increase more slowly with time scale n. This indicates a more random and less correlated time series. Indeed, scaling index (α) is 0.56 for this elderly subject and 1.04 for this young subject (adapted from [32])

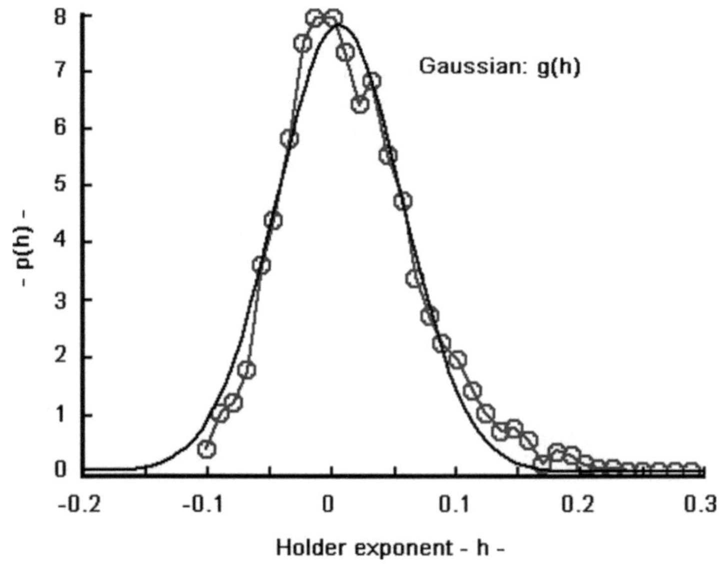

Fig. 11.15 Histogram and probability density estimation of the Holder exponents $h = H - 1$ and H is the Hurst exponent. The fitting curve is a Gaussian centred in $h_0 = 0.007 \pm 0.001$ and with a width $\sigma = 0.051 \pm 0.001$. The mean Hurst exponent in this case is therefore $H = 1$ and correspond to the maximum of the distribution (from [40] with permission)

and the resulting time series are called multifractal. In Fig. 11.15 the fractal dimension for gait is depicted as a spectrum of fractal dimensions and for technical reasons the authors [40] have used the Hölder exponent h that is related to the Hurst exponent by $h = H - 1$ and the distribution is centred on $H = 1$.

To show that this distribution of fractal dimensions is robust the time series for a cohort group was analysed in the same way [40]. Each member of the group was instructed to walk at a normal pace for 15 min, then on a separate occasion to walk faster than normal for 15 min, and finally on a third occasion to walk slower than normal for the same length of time. In Fig. 11.16 the three distributions are indicated and it is evident that by changing the gate mode from slow to normal the mean Hölder exponent h_0 decreases but from normal to fast it increases. Thus, the fractal dimension of the walk is dependent on feedback with

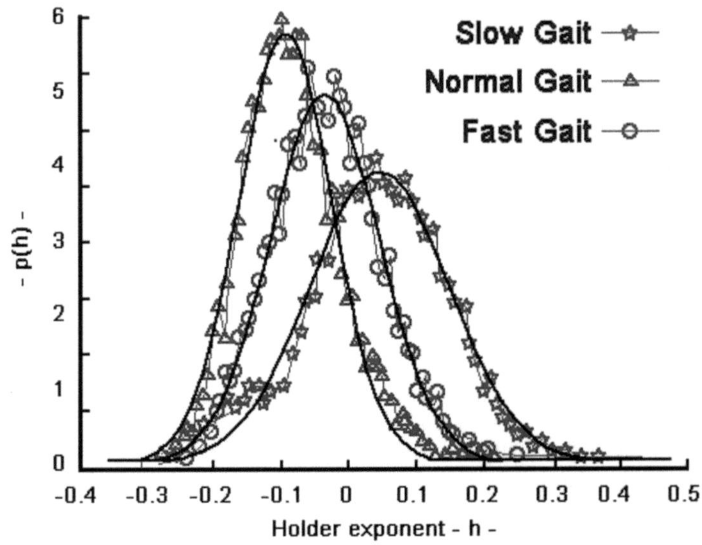

Fig. 11.16 Histogram and probability density estimation of the Hölder exponents ($h=H-1$) for three walking groups are shown: slow – *star*, normal – *triangle*, and fast – *circle gait*. By changing the gate mode from slow to normal the mean Hölder exponent h_0 decreases but from normal to fast it increases. There is also an increasing of the width of the distribution σ by moving from the normal to the slow or fast gait mode. The fitting curves are Gaussian functions, and the mean value h_0 and the SD σ are recorded in [40] (from [40] with permission)

the nervous system. A mathematical model developed by West and Scafetta [41] adapted the van der Pol oscillator, historically the first self-excitatory nonlinear oscillator, and was able to capture the distribution of scaling indices, its width and the ordering of the peaks due to internal feedback.

11.5.2 Disease

Changes in normal variability are seen in many diseases. Common cardiac conditions [18, 42], like severe heart failure or atrial fibrillation, show decrease in complexity as shown in Fig. 11.4. Gait disorders that occur in such neurodegenerative diseases as Huntington's disease (HD) show an increase in stride-interval fluctuations. Indeed, the scaling index is 0.40 for a person with HD compared with 0.92 for the young control subject (top right panel) in Fig. 11.17. This relationship between HD and randomness of gait is thought to be determined by changes in basal ganglia function. Functional disability correlates well with the scaling index with the most severe HD being associated with the least correlated random gait time series as shown in the lower bottom panel in Fig. 11.17 [32].

A number of scientists [32, 43] showed that neurodegenerative disease patients present altered fractal dynamics of gait characterised by reduced stride-interval correlations, that is, the walking of these individual becomes more random. The randomness increases with the severity of the neurodegenerative impairment, as Fig. 11.18 shows for several patients with HD. The peak of the multifractal distribution that gives the mean Hurst exponent is graphed in Fig. 11.18 as a function of the severity index of HD where it is clear that the greater the severity of HD the more random the gait time series.

The multiple distributions for scaling indices resulting from internal control depicted in Fig. 11.16 was extended to include external control using a metronome set to the average pace for normal, slower, and faster walking. In Fig. 11.19 the influence of a metronome on a group of walkers is shown. The nonintuitive shifts in the locations of the peaks of the three distributions are shown and this shifting has also been explained using the West–Scafetta model [41]. Another comparison shown in this figure is with the distributions for an elderly person and another with Parkinson's disease, both of which have peaks shifted towards increased randomness and lower complexity [44] as we have come to expect.

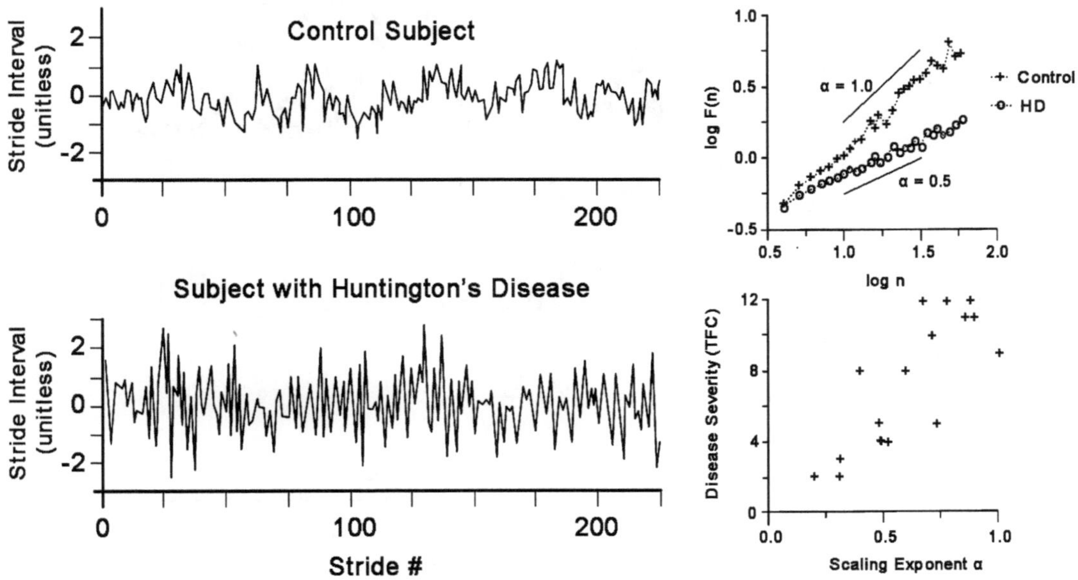

Fig. 11.17 Stride interval dynamics for a patient with Huntington's disease (HD). Stride interval fluctuations [$F(n)$] increase more slowly with timescale n, indicating a more random and less correlated time series. Randomness increases in line with severity, as measured on the Unified Huntington's Disease Rating Scale. TFC = total functional capacity, 0 = most impairment, 14 = no impairment (adapted from [32])

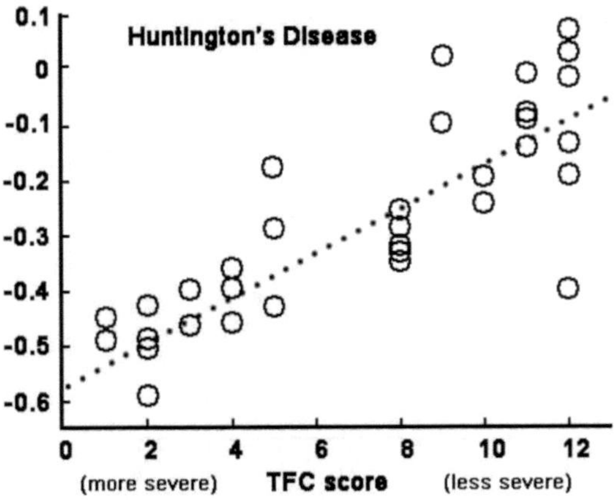

Fig. 11.18 Empirical relationship between the mean Hölder exponent and the total functional capacity score of Unified Huntington's Disease Rating Scale (0 = most impairment; 13 = no impairment). The mean Hölder exponent decreases (i.e., the sequences become more random) as the disease severity increases. The two measures are highly correlated $r^2 = 0.64$, $P \leq 0.005$ (from [18] with permission)

In Cushing's disease, not only is the secretion of ACTH and cortisol markedly increased, it also is more irregular as shown in Fig. 11.20. The asynchrony of ACTH–cortisol secretion is also significantly greater. The secretion of ACTH occurs without the coordinate control of the hypothalamic hormones CRH and vasopressin, and the negative feedback is diminished due to loss of cortisol sensitivity. Bihormonal asynchrony is also observed in growth hormone secreting pituitary [37] and aldosterone-secreting adrenal tumours [36] and may represent a general class of tumours [35].

The development of diabetes is also characterised by a progressive loss of complexity in the glycaemic time series, showing firstly metabolic syndrome before becoming frankly diabetic. Complexity as measured by a method for determining the scaling parameter called detrended fluctuation analysis (DFA) [45] was introduced in the analysis of HRV. Using this procedure the scaling parameter increased significantly from 1.17 in healthy individuals to 1.40 in metabolic syndrome and 1.53 in diabetes. There is evidence emerging that with increasing severity the loss of complexity increases further [46].

Mental illness also shows evidence of fractal behaviour. Katerndahl [47] showed that patients

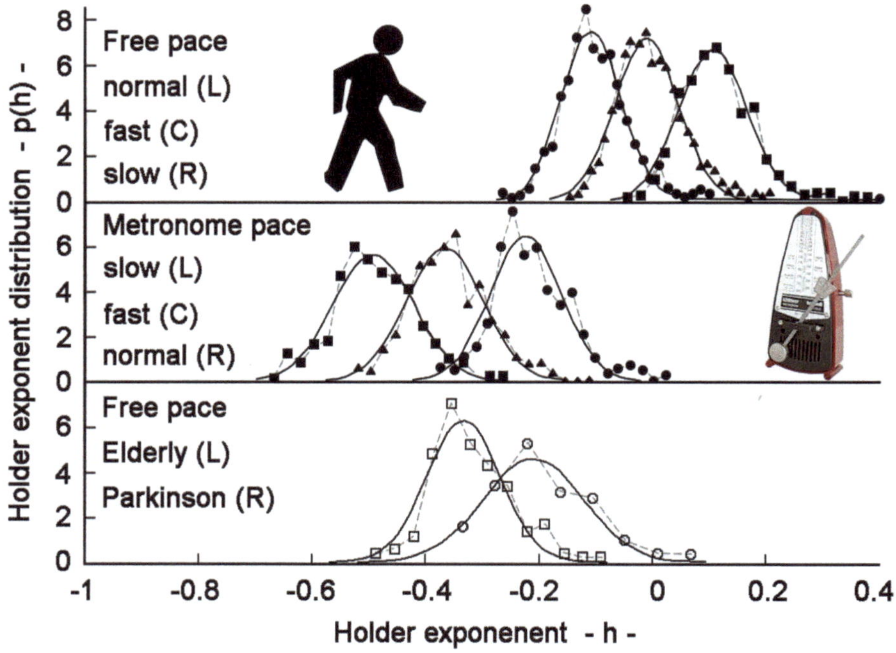

Fig. 11.19 Typical Hölder exponent histograms for the stride interval sequences during free walking and metronome-paced conditions for normal, slow, and fast paces and for a normal elderly person and a subject with Parkinson's disease. The histograms are fitted with Gaussian functions (*L* Left, *C* Centre, *R* Right). Data from http://www.physionet.org (from [44] with permission)

Fig. 11.20 Time series serum concentration of ACTH (*continuous line*) and cortisol (*dotted line*) in a patient control subject (*top*) and a patient with Cushing's disease (*bottom*) (adapted from [35])

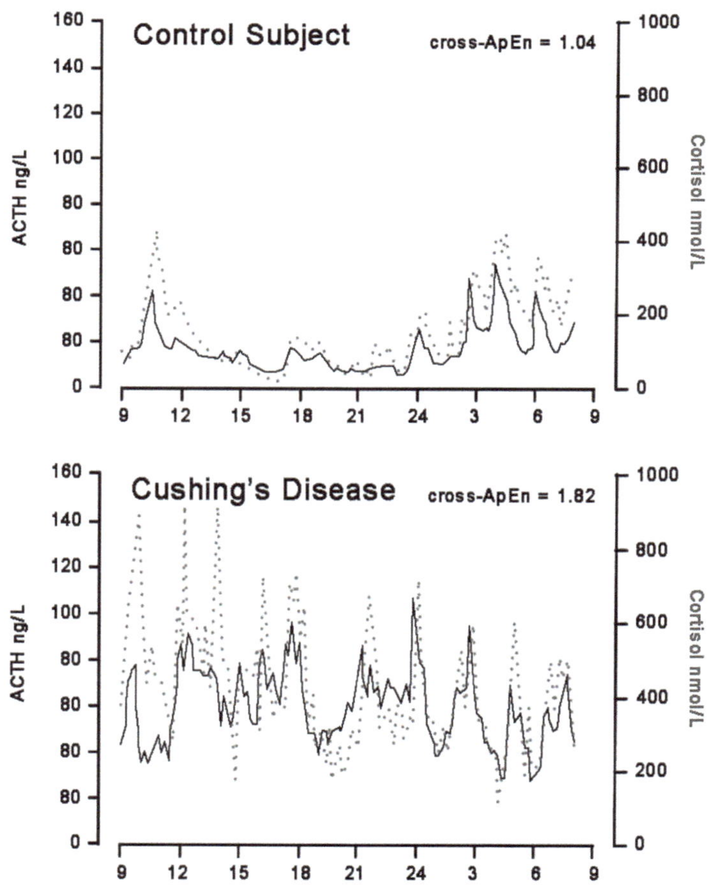

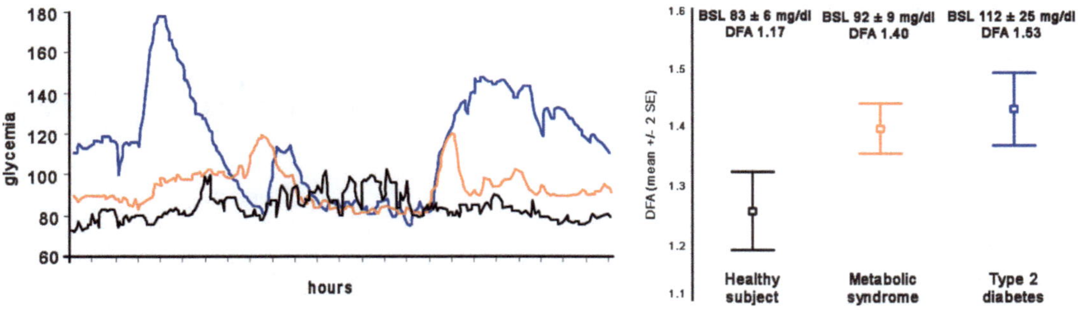

Fig. 11.21 Glucose profiles of a healthy person (*black line*), a patient with metabolic syndrome (*orange line*), and diabetes (*blue line*). The *right panel* shows the loss of complexity measured by DFA. *BSL* blood sugar levels (adapted from [46])

with major depression and panic disorder exhibit power-law relationships between their anxiety and depression symptoms. A more detailed discussion about the dynamics in panic disorder can be found in Chap. 23. Bipolar disorder is another condition that is a manifestation of complex dynamics. In general mood follows a nonlinear pattern, with patients affected by bipolar disorder having significantly more organised mood pattern as is indicated in the middle and lower sections of Fig. 11.22. The spectral slope of the patient is greater than that of the control person implying a reduction in variability in the time series. This loss of variability is also evident using the attractor reconstruction technique (ART) [17] where $X(n)$ and $X(n+\tau)$ are considered different variables with n the discrete time and τ an integer. An uncorrelated stochastic process would randomly fill the region of state space like that of a random walker even with $\tau = 1$. This is what we see for the control, but the restriction to a finite region of state space suggests chaos rather than randomness. The difference between the two is often subtle and will not be taken up here.

On the other hand, there is structure buried in the patient data indicating some underlying correlations in the variations of the bipolar time series. The information contained in the dynamic attractors in the bottom of Fig. 11.22 [48] may well provide guidance on the proper medications required for a patient.

11.6 Summary

Nonlinearity, fluctuations, and fractals are ubiquitous in nature. This chapter provides a glimpse into the structural and functional fractal dynamics of normal physiology, and the breakdown of these structures and function with normal ageing and specific disorders. It is beyond the scope of this book to go into detailed descriptions of each of the examples and interested readers are referred to the references and may seek specific publications in their area of interest. At the risk of appearing self-serving there is a new journal *Frontiers in Fractal Physiology* that was launched for the purpose of such ideas to have a specified home.

One cannot but help to make the observation that much of current medical practice and research has ignored these "new" insights to its peril. In particular these findings challenge much of the explanation of disease causation and therapeutic interventions. Other chapters in this book also describe that the nonlinear behaviour of nature extends into the organisational and interactional dimensions of the health professions, another insight has been widely missed.

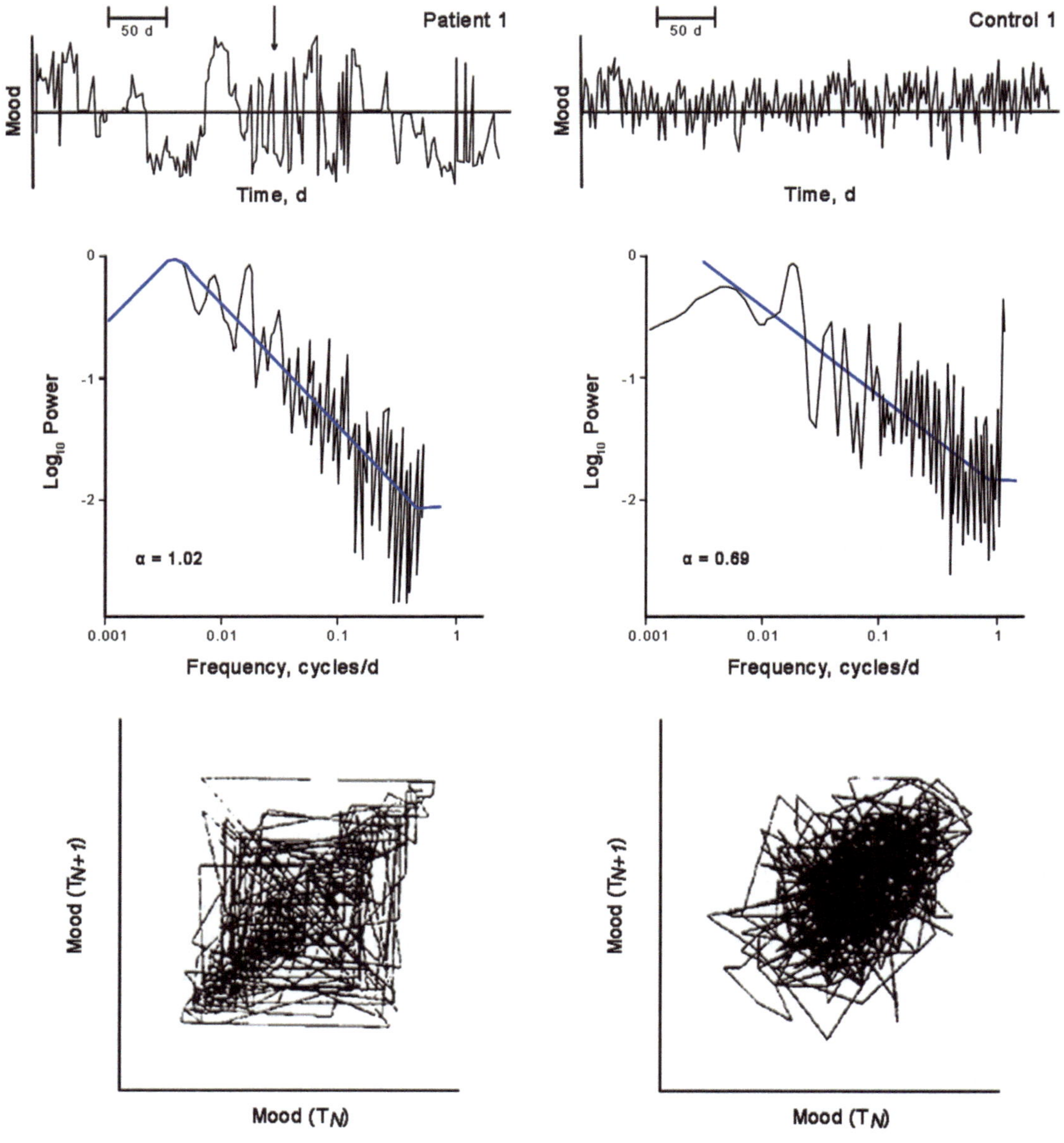

Fig. 11.22 Mood profiles of a patient with bipolar disorder. *Top*—2-year time period—series of mood in a patient and control—the bracket indicates a period of "pseudocycles" which if observed separately would suggest a periodic component to the mood in bipolar disorder; *Middle*—Power spectral densities show no well-defined spectral peak which would indicate a predominantly periodic process, the *blue line* shows the log–log plot with a slope of $\alpha = 1.02$ for the patient and $\alpha = 0.69$ for controls; *Bottom*—Phase space trajectories (adapted from [48])

References

1. Weibel ER. Fractal geometry: a design principle for living organisms. Am J Physiol. 1991;261(6):L361–9.
2. Mandelbrot BB. The fractal geometry of nature. New York: W.H. Freeman and Company; 1982.
3. West BJ. Fractal physiology and the fractional calculus: a perspective. Front Physiol. 2010;1:12.
4. Bassingthwaighte JB, Liebovitch LS, West BJ. Fractal physiology. New York: Oxford University Press; 1994.
5. Beran J. Statistics for long-memory processes. New York: Chapman & Hall; 1994.

6. Schottky W. Uber spontane Stromschwaukungen in verschiedenen Elektrizitattslei-tern. Ann Phys. 1918;362:541–67.
7. West BJ, Bologna M, Grigolini P. Maximizing information exchange between complex networks. Phys Rep. 2008;468:1–99.
8. Task force of the European Society of Cardiology and the North American Society of Pacing and Electrophysiology. Heat rate variability: standards of measurement, physiological interpretation, and clinical use. Eur Heart J. 1996;17:354–81. and references cited therein.
9. Alvarez-Ramirez J, Ibarra-Valdez C, Rodriguez E, Dagdug E. 1/f Noise structures in Pollocks's drip paintings. Phys A. 2008;387:281–95.
10. Grigolini P, Aquino G, Bologna M, Lukovic M, West BJ. A theory of 1/f noise in human cognition. Phys A. 2009;388:4192–204.
11. Gilden DL. Cognitive emissions of 1/f noise. Psych Rev. 2001;108:33–56.
12. Van Orden GC, Holden JG, Turvey MT. Human cognition and 1/f scaling. J Exp Psychol Gen. 2005;134:117–23.
13. Kello CT, Beltz C, Holden JG, Van Orden GC. The emergent coordination of cognitive function. J Exp Psychol Gen. 2007;136:551–68.
14. Liebovitch LS, Krekora P. The physical basis of ion channel kinetics: the importance of dynamics. In: Layton HE, Weinstein AM, editors. Institute for Mathematics and its Applications Volumes in Mathematics and its Applications, Membrane Transport and Renal Physiology, vol. 129. New York: Springer; 2002. p. 27–52.
15. Roy S, Mitra I, Llinas R. Non-Markovian noise mediated through anomalous diffusion with ion channels. Phys Rev E. 2008;78:041920.
16. Das M, Gebber GL, Bauman SM, Lewis CD. Fractal properties of sympathetic nerve discharges. J Neurophysiol. 2003;89:833–40.
17. West BJ. Where medicine went wrong; rediscovering the path to complexity. Singapore: World Scientific; 2006.
18. Goldberger A. Non-linear dynamics for clinicians: chaos theory, fractals, and complexity at the bedside. Lancet. 1996;347:1312–4.
19. Haefeli-Bleuer B, Weibel ER. Morphometry of the human pulmonary acinus. Anat Rec. 1988;220(4):401–14.
20. West BJ, Bargava V, Goldberger AL. Beyond the principle of similitude: renormalization in the bronchial tree. J Appl Physiol. 1986;60:1089–98.
21. Nelson TR, West BJ, Goldberger AL. The fractal lung: universal and species-related scaling patterns. Experientia. 1990;46:251–4.
22. Goldberger AL, Bhargava V, West BJ, Mandell AJ. On a mechanism of cardiac electrical stability. The fractal hypothesis. Biophys J. 1985;48(3):525–8.
23. Murray CD. The physiological principle of minimum work. I. The vascular system and the cost of blood. Proc Natl Acad Sci USA. 1926;12:207–14.
24. Kassab GS. Scaling laws of vascular trees: of form and function. Am J Physiol Heart Circ Physiol. 2006;290(2):H894–903.
25. Zhou Y, Kassab GS, Molloi S. On the design of the coronary arterial tree: a generalization of Murray's law. J Phys Med Biol. 1999;44(12):2929–45.
26. Vaillancourt DE, Newell KM. Changing complexity in human behavior and physiology through aging and disease. Neurobiol Aging. 2002;23(1):1–11.
27. Lipsitz L, Goldberger A. Loss of 'complexity' and aging. potential applications of fractals and chaos theory to senescence. JAMA. 1992;267(13):1806–9.
28. Scheibel AB. Falls, motor dysfunction, and correlative neurohistologic changes in the elderly. Clin Geriatr Med. 1985;1(3):671–6.
29. Mosekilde L. Age-related changes in vertebral trabecular bone architecture - assessed by a new method. Bone. 1988;9(4):247–50.
30. Iyengar N, Peng CK, Morin R, Goldberger AL, Lipsitz LA. Age-related alterations in the fractal scaling of cardiac interbeat interval dynamics. Am J Physiol. 1996;271(4):R1078–84.
31. Kaplan DT, Furman MI, Pincus SM, Ryan SM, Lipsitz LA, Goldberger AL. Aging and the complexity of cardiovascular dynamics. Biophys J. 1991;59(4):915–49.
32. Hausdorff JM, Mitchell SL, Firtion R, Peng CK, Cudkowicz ME, Wei JY, et al. Altered fractal dynamics of gait: reduced stride-interval correlations with aging and Huntington's disease. J Appl Physiol. 1997;82(1):262–9.
33. Greenspan SL, Klibanski A, Rowe JW, Elahi D. Age-related alterations in pulsatile secretion of TSH: role of dopaminergic regulation. Am J Physiol. 1991;260(3):E486–91.
34. Pincus SM, Mulligan T, Iranmanesh A, Gheorghiu S, Godschalk M, Veldhuis JD. Older males secrete luteinizing hormone and testosterone more irregularly, and jointly more asynchronously, than younger males. Proc Natl Acad Sci. 1996;93(24):14100–5.
35. Roelfsema F, Pincus SM, Veldhuis JD. Patients with Cushing's disease secrete adrenocorticotropin and cortisol jointly more asynchronously than healthy subjects. J Clin Endocrinol Metab. 1998;83(2):688–92.
36. Siragy HM, Vieweg WV, Pincus S, Veldhuis JD. Increased disorderliness and amplified basal and pulsatile aldosterone secretion in patients with primary aldosteronism. J Clin Endocrinol Metab. 1995;80(1): 28–33.
37. Hartman ML, Pincus SM, Johnson ML, Matthews DH, Faunt LM, Vance ML, et al. Enhanced basal and disorderly growth hormone secretion distinguish acromegalic from normal pulsatile growth hormone release. J Clin Invest. 1994;94(3):1277–88.
38. Frolkis VV, Bezrukov VV. Aging of the central nervous system. Interdiscip Top Gerontol. 1979;16: 87–9.
39. Mader S. Hearing impairment in elderly persons. J Am Geriatr Soc. 1984;2(7):548–53.
40. Scafetta N, Griffin L, West BJ. Holder exponent spectra for human gait. Phys A. 2003;328:561–83.

41. West BJ, Scafetta N. Nonlinear dynamical model of human gait. Phys Rev E. 2003;67:051917-1.
42. Goldberger AL. Fractal mechanisms in the electrophysiology of the heart. IEEE Eng Med Biol Mag. 1992;11(2):47–52.
43. Jian-Jun Z, Xin-Bao N, Xiao-Dong Y, Fang-Zhen H, Cheng-Yu H. Decrease in Hurst exponent of human gait with aging and neurodegenerative diseases. Chin Phys B. 2008;17:852.
44. Scafetta N, Moon RE, West BJ. Fractal response of physiological signals to stress conditions, environmental changes and neurodegenerative diseases. Complexity. 2005;12:13–7.
45. Peng C-K, Havlin S, Stanley HE, Goldberger AL. Quantification of scaling exponents and crossover phenomena in nonstationary heartbeat time series. Chaos. 1995;5:82.
46. Churruca J, Vigil L, Luna E, Ruiz-Galiana J, Varela M. The route to diabetes: Loss of complexity in the glycemic profile from health through the metabolic syndrome to type 2 diabetes. Diabetes Metab Syndr Obes. 2008;1:3–11.
47. Katerndahl DA. Power laws in covariability of anxiety and depression among newly diagnosed patients with major depressive episode, panic disorder and controls. J Eval Clin Pract. 2009;15(3):565–70.
48. Gottschalk A, Bauer MS, Whybrow PC. Evidence of chaotic mood variation in bipolar disorder. Arch Gen Psychiatry. 1995;52(11):947–59.

Bio-Complexity: Challenging Reductionism

12

Henry H.Q. Heng

12.1 Introduction

The success of current molecular biology coupled with impressive technologies has greatly influenced biomedicine. The common viewpoint is that in the future, medicine will become more molecular oriented and that most common diseases will be dissected by molecular mechanisms which will then serve as the specific target for treatment. In an era of gene-based medicine, the identification of specific genetic defects in most diseases and the development of individual genetic profiles have promised to provide a foundation for molecular diagnosis and intervention. And it is true that impressive progress has been made in genomic research, including technologies of gene profiling of various diseases, specific molecular targeting of cancer, and the ultimate Human Genome Project that promised to change medicine forever. Unfortunately, over the past 10 years since the completion of the Human Genome Sequence Project (which started in the 1980s and officially declared started in 1990 in the USA), none of the primary clinical goals have been fulfilled [1]. In contrast, increasingly unexpected findings have paradoxically challenged not only these initially promising approaches, but the whole rationale of treating diseases as arrays of molecular targets (isolated parts) rather than as a whole system [2]. The following examples illustrate this point:

Example 1: Type 2 diabetes has been linked to abnormal levels of various molecules such as high blood glucose and elevated "bad" cholesterol (LDL). The treatment options derived from this molecular reasoning seem straightforward. If the disease condition is caused by high blood sugar, then drastically reducing the blood sugar level should return the patient's system to normal. However, in a recent landmark clinical trial of over 10,000 participants from 77 medical centers in the USA and Canada, it was found that rigorously lowering blood sugar to near-normal levels increased the risk of death for participants relative to those in the standard blood sugar control group. Also surprising, the combinational therapy to lower "bad" cholesterol (LDL) and increase "good" cholesterol (HDL) did not lower the risk of heart attack, stroke or death compared to non-combined therapy. Furthermore, studies found that while lowering blood pressure to normal levels using a variety of medications may reduce the risk of stroke, it does not significantly reduce the risk of cardiovascular events overall [3–5]. Together, these unexpected findings challenge the notion that medical treatment can restore the system by focusing on defective or altered components.

H.H.Q. Heng (✉)
Center for Molecular Medicine and Genetics, Wayne State University School of Medicine, 540 E. Canfield, 3226 Scott Hall, Detroit, MI 48201, USA

Karmanos Cancer Institute, Wayne State University School of Medicine, Detroit, MI 48201, USA

Department of Pathology, Wayne State University School of Medicine, Detroit, MI 48201, USA
e-mail: hheng@med.wayne.edu

J.P. Sturmberg and C.M. Martin (eds.), *Handbook of Systems and Complexity in Health*,
DOI 10.1007/978-1-4614-4998-0_12, © Springer Science+Business Media New York 2013

Example 2: Molecular medicine states that gene defects represent the molecular mechanism underlying diseases. The comparison of allelic sequences of normal individuals and patients identifies abnormal genetic patterns that serve as a target for medical diagnosis and intervention. Despite many powerful technologies such as whole genome scanning and deep sequencing, it has been very difficult to identify common patterns that are responsible for many common diseases. For example, there are over 10,000 different genetic variants that have been linked with Schizophrenia, each of them is relatively rare and confers only a tiny increase in disease risk to the carrier [6]. When these hard-to-identify markers are used for clinical screening and prediction, the results are equally disappointing as they show little clinical significance. Recent disappointing results have been observed in many trials including heart disease and cancer research. When 101 well characterized genetic markers were used in a clinical setting, none were of significant medical value [7]. Clearly, there is a disconnection between disease phenotype and the genetic basis of a disease at the DNA level [8–10].

Example 3: In many current cancer chemotherapies and other target-specific treatments, tumors often shrink in size in response to the initial treatment. However, success is short lived as most tumors quickly grow back. More importantly, after painful and expensive treatment, the survival time is only moderately improved [11], with the exception of a few types of cancer, such as CML (imatinib treatment prior to the crisis stage) and testicular cancer. In fact, "extensive clinical experience has shown that while chemotherapy of advanced epithelial cancers frequently leads to dramatic tumor response, these do not translate into commensurate improvement in patient survival" [12] and no trial has yet to show that a change in treatment, which increases the response rate of the primary tumor will also improve long-term survival. Our studies have illustrated that initial drug treatment can eliminate dominant clonal populations but at the same time induce genome chaos—a powerful way to achieve rapid evolution under the stress of drug treatment that ultimately brings more harm to the system ([13]; Heng et al., unpublished observation). It has been suggested that chemotherapy might favor cancer evolution [14, 15]. Standard chemotherapy works best in tumors with low diversity at the genome level, including some pediatric cancers with balanced translocations/gene fusions and limited cytogenetic diversity. Unfortunately, however, this doesn't apply to most cancers, as these less heterogeneous types represent the exception to most cancers. Thus, there seems to be a contradiction between the rationale (elimination of cancer) and the consequences of current chemotherapy regimens (promotion of cancer evolution), as well as a conflict between treating tumors and the overall benefit to the individual.

Example 4: the Cancer Genome Sequence Project (Cancer Genome Atlas Project) was launched to systematically categorize gene mutations important to cancer and particularly commonly shared gene mutations [16]. According to current cancer theory, cancer is a genetic disease caused by the sequential accumulation of mutations of oncogenes and tumor suppressor genes [17]. If large numbers of cancer samples could be sequenced, the pattern of gene mutations should emerge since a limited number of driver mutations should exist for most cancers of the same type, and the troublesome heterogeneity among samples should be washed out. The initial sequencing data came pouring in. But contrary to what people had expected, the more samples that were sequenced, the more diversity was detected, which was predicted by our analysis based on system complexity and the stochastic evolution of cancer [18]. The data are very convincing: no hidden universally shared patterns were identified that could be used in the clinical setting. Each sample displayed high numbers of genetic changes, and the vast majority of them did not overlap among patients. Furthermore, current sequencing technologies use the averaging of sequences in a population for analysis. Unless one uses individual cell analysis where most of the cancer cells are shown to be of the non-clonal type, the high diversity at both the DNA and genome level is under-detected. The conclusion is the emergent evolutionary properties of cancer can be achieved by many different factors and each of them has limited predictability value [9, 10, 15].

All these surprising examples reflect the challenge that exists when approaching complex medical problems and diseases (systems) with reductionist's methods [2, 19]. Specifically, there are conflicts between considering the parts or the whole, between the cause and effect, and between the action and reaction of complex systems. Traditionally, we believed that all complex processes could be dissected and understood, and any complicated issue could be reduced to fundamental components that could be better understood one part at a time. We also have many good examples of successful attempts in the past, but why is this approach no longer working for those examples?

I would argue that, many previously successful stories may have represented simpler cases or so called low-hanging fruit scenarios where medical interventions either worked due to the more or less linear relationship or did not trigger system chaos. In these cases, the reductionist approach seemed workable. Such approaches fail in more complex cases since key features of complex systems conflict with reductionist's efforts of understanding and/or intervention. Despite the difficulty of predicting which cases might work and which will not, it is essential to appreciate that there is a real difference between simple and complex cases. If one compares simple and complex matters to be different levels of the system, it is then easy to understand why there are different principles involved, and a new holistic approach is needed [20]. To illustrate this important concept, it is necessary to briefly outline some key features of the complexity of biological systems, and why the traditional ways of reductionism are ultimately flawed with regard to fundamental features of complex systems and thus have significant limitations.

12.2 Key Features of Complex Systems

Understanding complex systems is important for future medicine as most medical problems (especially common diseases) are associated with complex systems. With the transition from classical determinist thinking to dynamic system thinking, increasing attention must focus on the behavior of complex systems. To date, however, most of the models of complex systems are based on nonliving systems due to their relative simplicity. Traditionally, people have thought that there is no crucial difference between living and nonliving forms of complexity despite the different degree of complexity. This issue has been debated between physicists and biologists. I would like to emphasize here that there are fundamental differences between them, where different rules are involved. One key difference is called inheritance [20]. Inheritance can serve as a new selective platform for emergent properties that are unique to a given biological system. In contrast, nonliving systems can also have novel properties emerge, but they cannot reproduce like bio-systems and therefore will have less heterogeneity. We have recently suggested that genomic topology ensures the boundary of system dynamics as genome-mediated inheritance defines the platform on which self-organization occurs [9, 10, 18, 20]. A unique aspect of inheritance is that the system can evolve (reflected by gene mutations and epigenetic alterations) but remains stable or static at the genome-defined system level. This process will result in the accumulation of complexity through evolution over long periods of time. The following key features of biological complex systems are listed.

12.2.1 Parts Versus the Whole

For many complex systems—particularly biological systems—the whole is greater than the sum of the parts. For example, the genome cannot be considered just a collection of genes or all sequences [10]. According to the recently introduced genome theory, an individual gene is not an isolated informational unit because its meaning is defined by the genome context where both the gene set and the genome topology (defined by the 3D interactive genomic arrangement map) are important. When the genome topology is different (such as different species whose similar genes

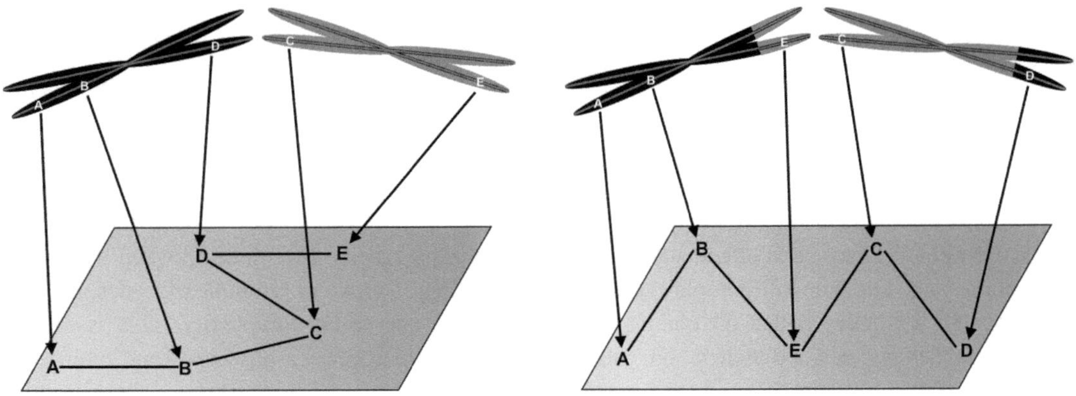

Fig. 12.1 A diagram showing the relationship among genes, the genome and genetic networks and how genome level changes determine the structure of networks (diagram modified from [20]). To simplify the presentation, only two chromosomes were drawn to illustrate the genome. When a chromosome translocation occurs, the genome topology is altered among chromatin domains, which changes the genetic network structure. The protein network changes according to the new genome topology. Genes determine the specific part (protein); however, individual genes/proteins cannot determine the network. It is the higher system, such as the genome topology that determines and controls the network structure by providing the foundation for self-organization. The network structure for each system is then preserved by maintaining the genome context

exhibit different karyotypes), there are diverse emergent properties. For example, identical genes can have different functions in different species and that when the genome is altered experimentally, that gene no longer displays the initial function [21]. There is also ample evidence that a gene function is defined by chromatin domains and chromosomal positions [22, 23]. Without genetic topology, the list of genes only represents the list of potential parts. Similarly, without interactive relationships within cells, the current proteomics approach amounts to the study of a protein parts list.

Currently, most molecular efforts are focused solely on the parts and there is no conceptual framework or technical platform to address the issue of the whole system. The Human Genome Project illustrates this well where all efforts have been to characterize and decode the information stored within DNA, with little regard to understanding genetic coding above the DNA which is responsible for biological functions. If information on genome operations is not stored within the DNA as the genome theory suggests, then the current Human Genome Sequencing Project will never deliver its medically related promises. Clearly, the DNA "map" is just a parts list rather than the blueprint of life, as there is no direct information regarding the genetic network and its functions which are achieved through 3D interactions and coded at the genome level [10, 20] (Fig. 12.1).

12.2.2 Multiple Levels of Information and Control Systems

Most biological systems are composed of multiple levels of subsystems [20], from single cell organisms to a human being. It is easy to accept that the study of humans can be partitioned by physiological systems (neurological systems, cardiovascular systems, etc.) and individual organs (brain, heart) and also through specific tissues and cells. It is however important to realize that there are different genetic information systems and control systems even within a cell. According to the genome theory, genetic information is coded at both the DNA and genome level, with DNA coding the parts/tools and genome coding for network assembly and dynamics [10, 20]. As this issue is under-appreciated and it is directly related to the rationale of the genome project, it is worth discussing further.

For decades, the genetic information was thought to be coded within DNA. As genetic information is responsible for bio-functions, it

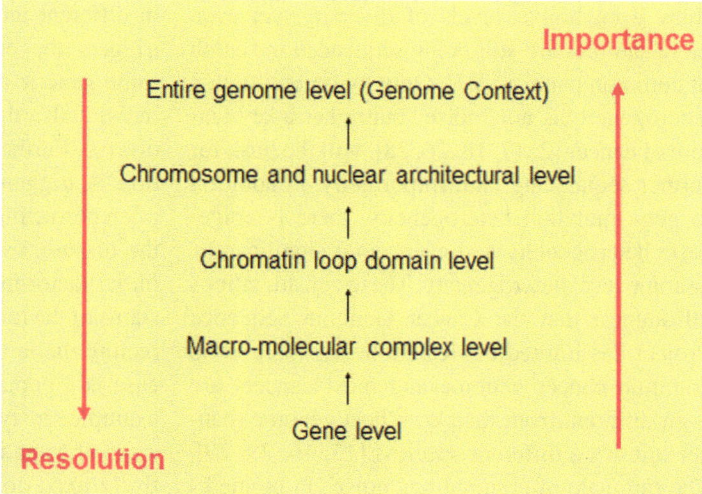

Fig. 12.2 A diagram illustrating the multiple levels of genetic organization. With an increase in the resolution of study (reductionism's approach), the system can be dissected further. However, this decreases the biological importance, which is particularly relevant to the control of the system. This analysis questions the reductionist's approach of focusing primarily on the gene level or even lower levels (higher resolution)

was thought that by sequencing DNA one should be able to understand the mystery of life. That was the rationale of various sequencing projects. Now, following the sequence of a large number of genomes of various species, it seems that the more we sequence DNA, the less we understand life, as overwhelming sequence information no longer makes sense to us. But what are we missing? It turns out that we have missed appreciating the fundamental differences among levels of genetic organization by assuming that there is a simple relationship between genes (part) and the genome (the whole functional system).

Genome theory maintains there are three main heritable coding systems within cells: DNA coding, epigenetic coding and genome coding. DNA codes the parts or tools for potential biological functions, epigenetic codes the short-term modifications of network dynamics prior to fixation or replacement by DNA level alterations, and the genome codes the pattern of overall interactions of the network by providing the matrix upon which the self-organization principle acts. Different coding systems have different control functions. Genes control their potential products if they successfully produce proteins, genomes organize the interactions among protein networks and define their reactive boundary. Understanding only the potential of individual genes certainly will not help with the understanding of a specific bio-system, particularly when the gene content is similar among related species and between normal and abnormal cells (Fig. 12.2). As illustrated, increasing resolution is coupled with decreasing importance, which is clearly contrary to reductionist thinking that places a high emphasis on increasing resolution in genetic studies.

12.2.3 Heterogeneity Is Not "Noise" But a Key Feature of Biological Systems

To make things more complex, there is a high level of heterogeneity across all levels of a biological system [24–26]. Heterogeneity is a phenomenon disliked by many biologists. In the course of finding patterns, heterogeneity has often been considered insignificant noise, and there are many statistical means to eliminate this "noise" in order to emphasize the important patterns. The elimination of "noise" has become a rationale to analyze large numbers of clinical samples. For example, when the initial cancer genome sequencing data became available, the high level of heterogeneity compellingly gave proof that most mutations are not shared among patients. Many researchers were not convinced and argued that samples size must be increased. The ever increasing numbers of samples that have been sequenced show even higher levels of diversity, yet even more samples are still being sequenced in search of common patterns [27]. Only by accepting that heterogeneity is not "noise" but rather a key feature of cancer [2, 17, 18, 26, 28], will the push for further sequencing end. Importantly, in addition to gene mutation heterogeneity, there is epigenetic heterogeneity and even more significantly, genome level heterogeneity. These considerations all suggest that the Cancer Genome Sequence Project has limited clinical value, as there is no common cancer genome and most cancers are even different from their own host genome (cancer mimics a different species) [15, 16, 28, 29]. The rationale of eliminating "noise" in biomedical research is also favored by the tradition of only being interested in the importance of bio-specificity, not knowing the ultimate importance of lower specificity in biology. High specificity is advantageous in certain conditions but could also be disadvantageous or even harmful in other conditions. The interaction between high specificity and low specificity mediated flexibility serves as a key factor for bio-robustness. Genome theory considers heterogeneity to be highly important as evolutionary selection is based on the genome as a complete package rather than an isolated specific part.

12.2.4 The Evolutionary Process Occurs in Many Disease Conditions

One additional reason that biological systems, particularly diseases, display high levels of heterogeneity is the involvement of the evolutionary process. Transitions from germ lines to somatic cells, somatic cell maintenance, the transition from physiological to pathological conditions all involve somatic cell evolution which occurs at all stages [8, 10]. During the aging process, for example, system stability decreases. Coupled with an accumulation of genetic damages that can further promote somatic cell evolution, this is one of the key reasons why increased cancer rates are linked to the aging process. Somatic cell evolution is a stochastic process and the initial condition that triggers the process in a disease can vary in different individuals. Even if the same factor triggers the evolutionary process, such as the same genetic defect in the germline of different individuals, the disease outcome can be very diverse. Furthermore, even if somatic cell evolution is triggered by a similar dominant factor, often correcting the causative factor does not cure the disease, as evolution is a process where the initial factor may or may not be important to the status of the tumor in later stages. The idea of correcting the initial contributing factor to cure disease is a popular idea but an incorrect one. For example, in typical cancer treatment, the early stage of somatic cell evolution can involve a relatively fixed dominant pathway. Since the system has not reached the later diverse genome stage, targeting this dominant pathway will likely slow down the progression of the micro evolutionary phase (this likely is the case for imatinib used in CML during the chronic phase). However, if cancer progression passes the point of no return where high levels of genome alterations predominate, targeting a specific pathway will be too little too late, as somatic cell evolution has now reached the genome level of selection. This is why it has been challenging to treat solid tumors using the same concept of specific pathway targeting. When solid tumors reach the later stages, any dominant pathway can be easily altered by applying stress. During later stages, the ultimate fate is the same even in blood cancers. This explains why magic bullet drugs like imatinib have failed to effectively treat late stage CML. The imatinib story also suggests that, in the micro-evolution phase, this type of dominant pathway targeting sometimes works, but at the macro-evolutionary phase, targeting a pathway is no longer effective [29, 30].

12.2.5 The Unavoidable Stochasticity

All factors discussed so far contribute to the stochastic nature of disease conditions including differences in progression, diagnosis and treatment response, as there is no truly identical initial condition for diseases among individuals. Even in the same individual, cells from different tissues and even in the same tissue due to different aging, display genetic and epigenetic heterogeneity resulting from genome dynamics. The evolutionary process further adds another layer of stochasticity. Therefore, all diversity or heterogeneity leads to unavoidable stochasticity. Interestingly, as long as the genome system and the germline of a given species is preserved (by passing the same set of chromosomes on to progeny), a stochastic alteration in a somatic cell can be tolerated and can even be favored by evolutionary selection, as somatic cell dynamics are useful for adaptation provided the framework of the germline genome is maintained such that the system/species is perpetuated [31–34].

Another stochastic feature occurs due to the relationship between the population and individuals. A given species is defined by the framework of a genome, while different individuals have many unique small genetic alterations such as specific gene mutations. For a species to survive, the key is to preserve the genome while the distribution of many rare mutations among individuals may stochastically occur. Interestingly, mutations may come and go which provide short-term adaptation advantages but will be unlikely to change the system [8, 35].

High levels of stochasticity pose the ultimate challenge to medical intervention. Most medical issues are not caused by a linear based system, and the initial conditions are drastically different even in patients who display the same symptoms. Their response to the same drug can be very different as well. Due to this unpredictable response, any treatment might be beneficial for some but detrimental to others. It is also possible that the same treatment response may vary in the same individual at different times and under different conditions.

12.2.6 Is There a Causative Relationship at the Molecular Level?

Due to the complex nature of biomedical systems, the rationale of searching for the causative relationship becomes more challenging and less certain. In some linear progression systems, the causative relationship can be established. However, in many complex systems due to a non-linear relationship, the chance of establishing a true causative relationship is extremely low. Despite this fact, biomedical researchers have been focused on the identification of a causative relationship when studying disease conditions.

One of the most frequently asked questions at biological seminars is: "is this a correlation or causation?", implying that only a causative relationship matters. I would argue that in complex systems, only correlations are possible as a universal causative relationship rarely exists in complex systems. We recently investigated a number of causative relationships (different factors that can be linked to cancer and different pathways that can be linked to mitotic cell death). We demonstrated that the main causative relationship is system stress and the stochastic response of the system, rather than the specific molecular causative factor that many have claimed. For example, there have been many reports that conclude that a specific molecular mechanism is the primary mechanism of mitotic cell death, including gene defects, viral infections and various drug treatments. By linking these factors to system stress, our study suggests that these previously established molecular mechanisms only represent one of many contributing factors, and might not be the universal causative factor [29, 30, 36]. Similarly, we have established a general evolutionary mechanism of cancer which consists of three general elements (stress, population diversity and genome mediated macro-evolution) and can be described as connecting all individual molecular mechanisms [9, 10, 15]. Each individual factor can trigger system stress and contribute to genome evolution leading to cancer, but is not the causative mechanism for cancer in general. In addition, even if one could establish a causative relationship in a given case (such as identifying the key gene mutations involved, or genomic instability), it is difficult to predict what is going to happen under medical intervention. Similar to the study of history, it is easy to pinpoint fundamental events and their causes, but it is hard to foresee and predict key events that will happen in the future and how such events will be triggered.

One important difference exists between experimentally defined conditions and real life or natural situations. In some artificially defined conditions, a weak causation factor is selected and isolated for the purpose of study, and many other variables are eliminated, reducing a complex system to more or less a linear system. In a sense, these model systems isolate and accentuate a causative relationship which does not exist in nature where too many variables demolish the causative relationship that people believe exists. This analysis explains why in nature there are less dominant factors detected and it is harder to find the causative relationship. Unfortunately, in molecular biology, due to the methodology of profile averaging, there is a reduction of system dynamics plus a tendency to cherry pick patterns within a defined condition, leading to apparently successful examples of causative relationships in experimental systems. In our view, however, knowledge that is based on an artificial system will have very limited clinical value [29].

12.3 Traditional Approaches: Successes and Limitations

Following the identification of these basic features of complex biomedical systems, we now have a greater understanding of both the many cases of medical success as well as the fundamental limitations. Many of the successful medical interventions that seem to work have common features and are due to the fact that these diseases are significantly different systems (such as due to exogenous or foreign agents such as bacteria) or diseases that modify certain functions without triggering system chaos. To illustrate this point, the following comparisons will be briefly discussed.

12.3.1 Distinguishing Between Treating Infectious Diseases and Cancer Drug Therapy

Using antibiotics to control infectious diseases represents the most successful medical intervention in modern history. Despite a range of adverse effects, the overwhelming benefits of antibiotics are obvious. A variety of antibiotics have been developed to effectively kill infectious agents by interfering with the formation of cellular components (such as a bacterium's cell wall), or stopping or slowing down proliferation by interfering with DNA replication, protein production, or other aspects of their cellular metabolism. The causative relationship seems clear and relatively simple. Infectious diseases are caused by bacteria, and their elimination is a logical solution. However these successful strategies have been challenging when trying to apply them to cancer therapy. In fact, the concept of using antibiotics in battling infectious diseases has been used as a treatment model for cancer where a "magic bullet" has been sought to target and kill cancer cells. However, this transfer of thought has misled cancer treatment for decades [13].

Cancer cells are drastically different from infectious bacteria. First, despite a dissimilar genome context, the same cellular machineries are shared by normal and cancer cells, and targeting methods can affect both cancer and normal cells (in contrast, many antibiotics act specifically on bacterial systems). Previously, differential cell division has been used for the differential killing of cancer cells. However, it is now clear that most drugs also act on nondividing or slower dividing normal cells. Second, when subjected to the stress of chemotherapy, the cancer genome often changes drastically through genome chaos, which can generate high levels of new genomes suitable for genome-mediated macro-evolution (in contrast, drug resistance in bacteria usually occurs through preexisting mutations) [10, 15, 30, 38]. Third, various treatments can function as a type of evolutionary pressure that favors cancer cell evolution (while normal cells can only evolve into abnormal cells). Obviously, cancer treatment is a much different disease process than infectious disease (Fig. 12.3).

12.3.2 Distinguishing Single Gene Diseases from Complex Common Diseases

Certain single gene diseases have been identified that have shown a high correlation between the genotype and phenotype and have also displayed

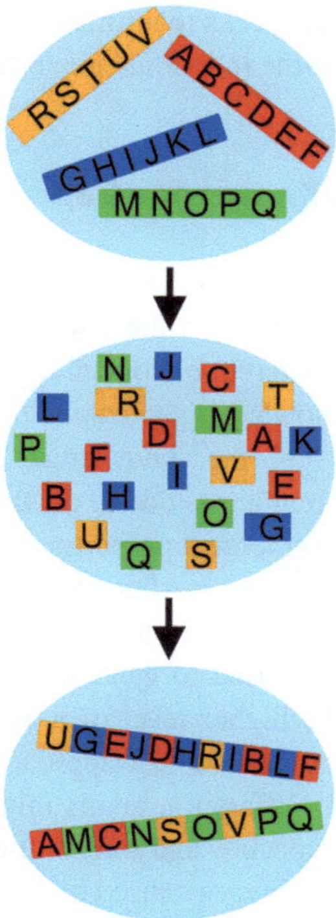

Fig. 12.3 A model of how genome chaos can drastically re-organize the genome to form new genome systems that are essential to the progression of cancer evolution during crisis. (*top panel*) shows relatively normal chromosomes prior to crisis. The four chromosomes are shown with the original order of genes along chromosomes; (*middle panel*) shows the chromosome fragmentation stage induced by crises such as drug treatment or the immortalization process. The chromosomes are cut into pieces; (*bottom panel*) shows chromosomal fragments forming two chaotic chromosomes by the seemingly random rejoining of fragments to form new chromosomes. Massive numbers of different new chromosomes with new gene orders are generated. This process represents a powerful way to create new systems by re-shuffling the existing genome

high penetration within a patient population. In these situations, the causative relationship between disease genes and disease seems obvious particularly under experimentally controlled conditions. Genomic research has identified some very good examples of this including Cystic Fibrosis gene mutations. However, most common diseases are influenced by a large number of different genetic and environmental factors affecting patient populations and because of this, it is challenging to classify patients into defined subgroups as the correlation between genotype and phenotype is hard to establish. Increasing the number of samples analyzed or creating more powerful technologies and statistics may never reveal a single universal causative relationship for these diseases due to their complexity. Rather, consideration of the degree of complexity might require alternate concepts and strategies. The disappointing results of current efforts using whole genome scanning to identify common and clinically useful genetic loci have forcefully illustrated this point. Following the expenditure of a great deal of effort, the genomic community is now increasingly asking, "Where is the missing heritability in common diseases? [8, 35]"

The new reality illustrated by above examples seems to be in direct contrast with our past experience where technological advancements have continuously changed the world we live in. I would like to point out that most of our previous successes and failures fall into two categories. In the first, the overall approach works if the intervention targets a different system or modifies a part of the system but does not alter the entire system in a harmful way. Of this category, technical developments often generate impressive results. Examples include the use of antibiotics, vaccines, organ transplantation, pace maker implantation, hip replacement, and plastic surgery. In contrast, the second category, if the intervention causes overall stress leading to system chaos, or there is a strong conflict between the parts of the system and the whole, or the function of the whole relies on close interactions among the parts (the condition of non-dissected components), we will not be successful and a new approach is needed. For example, initial organ transplant rejection stimulates system chaos, but by sero-typing patients to determine the most appropriate donor–recipient match and through the use of immunosuppressant drugs, organ transplantation has become a medical reality (despite the side effect of lowering a patient's immune-capability). The emerging field of regenerative medicine has further promised to create organs that can be regrown using a patient's own cells. The challenge here is again to prevent overall system changes (such as genome alterations during the in vitro manipulation and in vivo somatic cell evolution) which could result in cancer evolution. Increasing reports have illustrated that altered genomes have been detected from induced stem cells [39]. Even standard stem cell therapy can lead to cancer [40]. By appreciating the conflict that exists between different levels, it becomes understandable that even a seemingly simple case of using antibiotics to treat an infection can become a very complicated issue. Antibiotics save many individuals; however repetitive antibiotic use has accelerated the emergence of super bugs which might 1 day potentially endanger the very existence of the human race if this issue is not eventually addressed.

In a nutshell, medical therapies of complex diseases should not continue to focus only on singular universal "magic bullets". All treatments are a double edged sword with benefits and hazards inherent to complex systems and it is therefore not surprising when new promising therapies result in unexpected detrimental effects when this complexity is not taken into consideration.

12.4 New Medical Challenges

Undoubtedly, medical science will eventually move forward in spite of the newly identified complexity of many diseases. However, the correct framework is essential to ensure that this process will yield productive results based on realistic goals and thereby avoid decades of the types of setbacks we have seen in the cancer field. In addition to further understanding the biological basis as well as the technological advancements accompanying holistic research, some important issues are worth discussing. In particular, following decades of favoring reductionist's approaches, and the accumulation of massive amounts of molecular information, a new conceptual synthesis is needed. Despite the fact that there is no clear vision for what a future holistic medical approach would look like, the overall trend should not continue to push solely molecular approaches. There should be a more balanced strategy that places greater weight on higher level information, and seeks to ultimately cure or manage diseases at the individual phenotypic level.

12.4.1 The Knowledge Gap

Due to the high complexity of many medical issues, we must explore the knowledge gap that exists among different biological levels of organization. The knowledge gap can be defined as the discontinuity of the knowledge relationship between different levels due to the unpredictability of different emergent properties that cannot be linearly accumulated or dissected. For example, in many cases, information collected from lower levels cannot be used to generate crucial understanding at the higher level of a system [9]. In such cases, the reductionist approach will be of limited value. Different systems might require a variety of approaches and interventions, such as the disease phenotypes caused by bacterial infection versus cancer. In addition, each level of a system often has its own concepts or laws, and using laws that apply to one level in an attempt to understand other related levels can result in confusion [20]. It is critical to determine whether a knowledge gap exists. If a knowledge gap does not exist then continuing to push for greater technological resolution and collecting more information would be reasonable. However, if a knowledge gap does exist for a specific medical problem, research is needed to directly identify the determining influential level and associated dynamics, rather than continue to apply routinely used methodologies (such as sequencing) directed at easily accessible levels since knowledge obtained at these levels will not transfer to other levels and ultimately have limited clinical value [10].

One example of this is the recent global efforts to sequence the cancer genome. If cancer is caused by genes, and most patients share the same gene mutations or a handful of common gene mutations as previously believed, then the sequencing of large numbers of samples is the correct approach to identify the key targets [18]. If, however, as predicted by the genome theory and illustrated by initial cancer genome sequencing data, large numbers of mutations can be detected from each patient, and it is genome alterations rather than gene mutations that are the key to driving cancer progression, then just increasing the number of sequencing samples will not help us to understand the mechanism of cancer [41]. Under these circumstances, the sequencing information will be less useful, as the information encoded within DNA is fundamentally different from the system assembly information encoded by the genome context [10]. Clearly, there is a knowledge gap that exists between the gene and genome, as they represent different levels of the system. Future effort should now focus on the issue of genome evolution and its implications to cancer.

Another example is the current effort to identify the genetic basis of many common diseases using whole genome scanning. This approach is based on the success of identifying genetic causes for Mendelian diseases. Based on this "simpler" genetic disease system, defective genes are the main cause of diseases and should be readily identifiable. In contrast, most common diseases involve many genetic loci and also include environmental factors [8]. As such, these different features will affect changes to the entire system compared to Mendelian diseases. Methods reflecting this fundamental difference must be adopted rather than merely modifying strategies that deal only with the quantitative differences. When large numbers of factors are involved, the presence of complexity might invalidate traditional methodologies. In other words, when complexity is in, linear studies should be out. The identification of unlimited numbers of small factors becomes less meaningful in the clinical setting despite the continuing academic interest in this data [10]. It is the main reason for switching our focus from individual genes to genome alterations when monitoring common complex diseases. In some purely Mendelian diseases (which are rare), linear reasoning might work. But in the majority of common complex diseases, a linear causative relationship is not detectable in nature, even though under controlled artificial experimental systems that attempt to factor out stochastic effects, causal relationships can be demonstrated [37].

One critical issue needing clarification is the process of medical validation [22]. In infectious diseases, the identification of a specific infectious agent and the effect of a specific drug can be validated in patient populations. However, this approach does not apply to medical validation of most cancer types, as many cancers are not caused by common and highly penetrating genetic factors. If the majority of patients of the same type of cancer can be linked to many different factors, how can we validate multiple identified causes and determine effective medical intervention against them all? The current thinking is to study large numbers of samples. However, due to the high heterogeneity of most common diseases, the more samples we analyze, the more diversity will emerge, creating greater difficulty when seeking medical therapeutic approaches. It is for these reasons that a new system is urgently needed for cancer validation. The comparison between validation of infectious diseases and cancer illustrates the importance of determining the level of the system involved before applying therapy.

The knowledge gap can also apply to different systems with different degrees of complexity. One interesting example is the fundamental difference between different cancer types and their response to treatments. In some blood cancers, particularly when there is little diversity at the genome level, the treatment is often effective and can be lifesaving. In contrast, by adding a layer of genome complexity and heterogeneity, the knowledge we have accumulated from blood cancers no longer effectively applies to many solid tumors with diverse genomes [30].

12.4.2 Balanced Medical Intervention

In the future, medicine must take a more balanced approach with respect to the following issues:

12.4.2.1 A Balanced Approach Is Needed Between Prevention and Treatment

Traditional medicine always seeks to cure diseases and focuses on treatment rather than prevention. However, in many common complex diseases such as cancer and obesity, no magic cure has been identified as the causes are very diverse and the phenotype cannot be easily changed by targeting a specific well-defined cause. The key in these cases then becomes prevention. Highly publicized medical advancements that focus only on a small percentage of successful therapies have created feel-good stories and as a result, the general public has come to expect that medicine can fix anything. This message has overshadowed the importance of healthy lifestyles and prevention. The attitude now is that doctors can fix anything so why bother changing bad unhealthy habits? One effective way to educate the public is to discuss the medical reality that there probably *is* no magic cure for many diseases and there is a greater need for each individual to apply preventive practices. Only when medicine stops perpetuating the myth that a "cure" for complex diseases is imminent will people start to change their lifestyles. This realization will also result in the added benefit of more research resources being allocated to medical prevention.

12.4.2.2 A Balance Is Needed Between Medical Intervention Directed at the Parts Versus the Whole System, Between Short-Term Gains and Long-Term Overall Benefits, As Well As Between Radical Aggressive Treatments and Measured Restrained Treatments

Current molecular medicine seems to favor treating the parts of a system more than the whole system, as many molecular mechanisms deal only with dissected elements, and it is too complicated dealing with individual levels with so many variations. Many chemotherapy drugs, for example, work well when initially reducing the size of a tumor. However, in general, these agents have contributed little to prolonging patients' lives, or significantly improving the quality of life as the treatment usually effectively targets a specific clonal population. Yet increased drug-mediated stress leads to accelerated cancer cell evolution. This type of drug induction causes cancer cells to be quickly selected through a process called genome chaos, forming large numbers of new genomes which results in treatment failure. When these newly selected cancer cells become dominant, the cancer often becomes highly resistant to therapy. New generations of target specific drugs have shown some promise, particularly those that boost the immunological system to fight cancer. The success of these new approaches that actually prolong patient's lives questions the rationale of continuously using chemotherapy in some cancers which do not prolong a patient's life despite positive tumor response. A debate among the medical community is needed to reevaluate the overall benefit of chemotherapy in many cancers. In particular, the long held notion that it is best to hit cancer hard, attempting to eradicate as many cancer cells as possible, while debilitating patients immunologically really needs reevaluation as well.

The issue of short- and long-term treatment/benefits also requires discussion with regard to treating all common diseases. A gradual and gentler treatment approach might be more beneficial in many diseases as opposed to drastic treatments, particularly in individuals with a weaker system status, as treatment also represents a stress that can further destabilize the whole system. In addition, in patients where a disease condition has already gradually altered the system status over time, the strategy of drastically restoring the "normal" status without considering the entire system can often result in harmful stress that will trigger system chaos. The correct way to treat many disease conditions in cases that have resulted in gradual adaptation is also gradual medical improvement rather than a quick and radical intervention.

12.4.2.3 Converting complexity into simplicity

Interestingly, by changing the focus to different levels of complexity, predictability can be improved and this seems to transition the complex to become simpler. This particularly applies to switching the focus from a specific individual mechanism to the overall system behavior. For example, there are large numbers of research papers that study different molecular mechanisms of cancer. By comparing the specific molecular mechanisms, the picture becomes extremely complex due to the different mechanisms. It is challenging to apply these highly diverse mechanisms to medical prediction. However, if we switch from monitoring genes to focusing on the genome level, all different molecular mechanisms can be unified under the more simplified concept of degrees of stress induced genome heterogeneity. Such heterogeneity can be directly measured and used to predict the evolutionary potential of cancer [9–10, 15, 24, 42]. The molecular level of complexity has now been transferred into genome level simplicity. Similarly, when tracing molecular contributions to obesity, many factors can be linked to the disease and it appears to be a challenge to understand it. However, by simply implementing a healthy lifestyle, diverse patients with different gene profiles will all benefit. Future medicine needs to think more along the lines of such strategies as opposed to treating a single disease molecular pathway.

Such strategies are in line with the concept of evolutionary medicine. In the future, it will be crucial to apply evolutionary theory and methodologies to understand and solve medical issues, as the progression of many diseases represents an evolutionary process [43].

12.4.2.4 New Considerations Regarding Treatment/Over-Treatment, Early Diagnosis/Over Diagnosis

The advance of molecular medicine has pushed reductionism to a new high. With various highly sensitive molecular tools available for detection, many isolated/dissected factors at the molecular level can be studied and considered for potential medical targets. Accordingly, there are so many molecular parts that can now be monitored and used as therapy targets such as the activity of certain enzymes, the status of mitochondria, and "good" or "bad" cholesterol molecules. Recently, there was a report highlighting the detection of cancer based on single abnormal cells among millions of normal ones. The trend of over diagnosis and treatment is clearly being pushed by reductionist thinking. However, in complex disease systems, the relationship of an individual part and the symptom is again not a linear one, which questions the practice of tracing "the earliest" events if predictability based on these events is very low. In addition, there is a high level of variation among molecules and there are dynamics at the molecular level, and high level systems can tolerate lower level variation. Also, most diseases result from an evolutionary process where there is no one-to-one causative relationship to trace. The initial rationale of trying to achieve the earliest detection possible in cancer diagnosis is based on the faulty assumption that cancer is caused by common molecular defects such as key cancer gene mutations. The logic was that detecting and eliminating gene mutations prior to their progression into full blown cancer, will halt cancer "development". Unfortunately, however, cancer progression is a stochastic process where many different mutations are involved and most are of limited prediction and diagnostic value. In fact, non-clonal chromosome aberrations can be detected even in normal individuals. Under such situations, the identification of single abnormal cells will not offer any useful information but will bring unnecessary stress to patients. Furthermore, earlier intervention could be harmful in some cases, as the treatment itself might function as a stress to accelerate cancer evolution. Even the emotional stress of knowing the presence of a possible disease can bring harm to many.

The issue of over diagnosis and treatment has recently attracted the public's attention [33, 44]. We strongly suggest a "wait and see" approach with regard to many common complex diseases, while introducing lifestyle changes with gentle medical interventions. It is necessary to educate patients that in regard to therapy for many common diseases, less is sometimes more. Patients should not be influenced by drug companies to demand the latest or newest drugs from physicians. There should be a tendency toward a minimalist approach when considering therapy in complex diseases.

12.4.3 Personalized Medicine

Personalized medicine now has become the latest medical buzzword. The hope now is to treat patients according to their genetic profiles. The decreased cost of sequencing individual patients seems to support this idea. However, based on the complex relationship between genetic makeup and disease phenotype, we would argue that for most common diseases, such a genetic profile will have limited value as the diverse genetic factors among patients do not have a dominant influence, and represent only potential factors of a complex evolutionary process where environmental factors and historic contingency play more dominant roles.

The importance of the personalized concept in medicine is applicable in a much broader sense. Instead of focusing on difficult-to-identify molecular patterns, from a conceptual level we need to realize the conflict between personalized medicine and evidenced-based medicine when validating specific drugs or protocols, and the conflict between individual and population benefits. Dealing with patients who display high levels of heterogeneity contrasts with efforts to standardize medicine. Balancing these two considerations needs to be further evaluated. For example, what is the tolerance of acceptable benefits versus risks for a new drug? What percentage morbidity/mortality is acceptable due to side effects? To what extent do we treat terminally ill patients? Should we continuously push the development of new antibiotics despite the risk of "creating" superbugs?

There are no easy answers to these questions which reflect the complexity of medicine. Timely debate is necessary. In this book, many chapters will deal with such important issues. It is our hope that by accepting the reality of medical complexity, new generations of physicians will finally adopt a holistic approach to face the challenges of disease complexity. New medical strategies will certainly emerge following a change in our way of thinking [2, 8, 10, 35, 37, 43, 44].

Box 1

Be aware of the difference between parts and the whole. Do not assume that a good strategy at the molecular level will certainly translate to the individual patient level. When treatment benefits conflict, carefully weigh the benefits with the hazards and treat patients not molecules.

Be aware of the long- and short-term benefits and the harm caused by medical intervention. There are benefits of not treating or less aggressively treating, as well as the harm of over diagnosing.

Be aware of the problems with current medical validation. There is a conflict between standardized care and individualized needs. A drug can be good for many but harmful for some or vice versa.

Be aware of not over promising and encourage patients to change their lifestyle rather than rely on a magic bullet strategy to disease that is non-existent.

Be aware of the importance of evolutionary medicine.

Acknowledgments This chapter is part of a series of studies entitled, "The mechanisms of somatic cell and organismal evolution". I would like to thank Barbara Spyropoulos and Steve Bremer for their help in editing. Thanks also go to Drs. Gloria Heppner, Markku Kurkinen and Christine J Ye for discussions, and Dr. Joshua Stevens for help with the diagrams. This work was partially supported by grants from SeeDNA Biotech Inc., the Susan G. Komen Breast Cancer Foundation, the DOD (GW093028), from National CFIDS foundation, and The Nancy Taylor Foundation for Chronic Diseases.

References

1. Evans JP, Meslin EM, Marteau TM, Caulfield T. Deflating the genomic bubble. Science. 2011;331:861–2.

2. Heng HH. The conflict between complex systems and reductionism. JAMA. 2008;300:1580–1.
3. Action to Control Cardiovascular Risk in Diabetes Study Group, Gerstein HC, Miller ME, Byington RP, Goff Jr DC, Bigger JT, Buse JB, Cushman WC, Genuth S, Ismail-Beigi F, Grimm Jr RH, Probstfield JL, Simons-Morton DG, Friedewald WT. Effects of intensive glucose lowering in type 2 diabetes. N Engl J Med. 2008;358:2545–59.
4. ACCORD Study Group, Ginsberg HN, Elam MB, Lovato LC, Crouse III JR, Leiter LA, Linz P, Friedewald WT, Buse JB, Gerstein HC, Probstfield J, Grimm RH, Ismail-Beigi F, Bigger JT, Goff Jr DC, Cushman WC, Simons-Morton DG, Byington RP. Effects of combination lipid therapy in type 2 diabetes mellitus. N Engl J Med. 2010;362:1563–74.
5. ACCORD Study Group, Cushman WC, Evans GW, Byington RP, Goff Jr DC, Grimm Jr RH, Cutler JA, Simons-Morton DG, Basile JN, Corson MA, Probstfield JL, Katz L, Peterson KA, Friedewald WT, Buse JB, Bigger JT, Gerstein HC, Ismail-Beigi F. Effects of intensive blood-pressure control in type 2 diabetes mellitus. N Engl J Med. 2010;362:1575–85.
6. The International Schizophrenia Consortium. Common polygenic variation contributes to risk of schizophrenia and bipolar disorder. Nature. 2009;460:748–52.
7. Paynter NP, Chasman DI, Paré G, Buring JE, Cook NR, Miletich JP, Ridker PM. Association between a literature-based genetic risk score and cardiovascular events in women. JAMA. 2010;303:631–7.
8. Heng HH. Missing heritability and stochastic genome alterations. Nat Rev Genet. 2010;11:813.
9. Heng HH, Stevens JB, Bremer SW, Ye KJ, Liu G, Ye CJ. The evolutionary mechanism of cancer. J Cell Biochem. 2010;220:538–47.
10. Heng HH, Liu G, Stevens JB, Bremer SW, Ye KJ, Abdallah BY, Horne SD, Ye CJ. Decoding the genome (rather than sequencing linear DNA): the next phase of genomic research. Genomics. 2011;98:242–52.
11. Fojo T, Grady C. How much is life worth: cetuximab, non-small cell lung cancer, and a $440 billion question. J Natl Cancer Inst. 2009;101:1044–8.
12. Mittra I. The disconnection between tumor response and survival. Nat Clin Pract Oncol. 2007;4:203.
13. Heng HH, Liu G, Stevens JB, Bremer SW, Ye KJ, Ye CJ. Genetic and epigenetic heterogeneity in cancer: the ultimate challenge for drug therapy. Curr Drug Targets. 2010;11:1304–16.
14. Maley CC, Reid BJ, Forrest S. Cancer prevention strategies that address the evolutionary dynamics of neoplastic cells: simulating benign cell boosters and selection for chemosensitivity. Cancer Epidemiol Biomarkers Prev. 2004;13:1375–84.
15. Heng HH, Bremer SW, Stevens JB, Liu G, Ye CJ. Evolutionary mechanisms and diversity in cancer. Adv Cancer Res. 2011;112:217–53.
16. Collins FS, Barker AD. Mapping the cancer genome. Pinpointing the genes involved in cancer will help chart a new course across the complex landscape of human malignancies. Sci Am. 2007;296:50–7.
17. Vogelstein B, Kinzler KW. Cancer genes and the pathways they control. Nat Med. 2004;10:789–99.
18. Heng HH. Cancer genome sequencing: The challenges ahead. BioEssays. 2007;29:783–94.
19. Sturmberg JP. Systems and complexity thinking in general practice: part 1 - clinical application. Aust Fam Physician. 2007;36:170–3.
20. Heng HH. The genome-centric concept: resynthesis of evolutionary theory. Bioessays. 2009;31:512–25.
21. Rancati G, Pavelka N, Fleharty B, Noll A, Trimble R, Walton K, Perera A, Staehling-Hampton K, Seidel CW, Li R. Aneuploidy underlies rapid adaptive evolution of yeast cells deprived of a conserved cytokinesis motor. Cell. 2008;135:789–91.
22. Heng HH, Chamberlain JW, Shi XM, Spyropoulos B, Tsui LC, Moens PB. Regulation of meiotic chromatin loop size by chromosomal position. Proc Natl Acad Sci U S A. 1996;93:2795–800.
23. Heng HH, Goetze S, Ye CJ, Liu G, Stevens JB, Bremer SW, Wykes SM, Bode J, Krawetz SA. Chromatin loops are selectively anchored using scaffold/matrix-attachment regions. J Cell Sci. 2004;117:999–1008.
24. Heng HH, Bremer SW, Stevens JB, Ye KJ, Liu G, Ye CJ. Genetic and epigenetic heterogeneity in cancer: a genome-centric perspective. J Cell Physiol. 2009;220:538–47.
25. Davidoff F. Heterogeneity is not always noise lessons from improvement. JAMA. 2009;302:2580–6.
26. Heppner GH. Tumor heterogeneity. Cancer Res. 1984; 44:2259–65.
27. Lander ES. Initial impact of the sequencing of the human genome. Nature. 2011; 470:187–97.
28. Ye CJ, Liu G, Bremer SW, Heng HH. The dynamics of cancer chromosomes and genomes. Cytogenet Genome Res. 2007; 118:237–46.
29. Heng HH. 4D-Genomics: genome dynamics and constraint in evolution. Springer (in press).
30. Horne SD, Stevens JB, Abdallah BY, Liu G, Bremer SW, Ye CJ, Heng HH. Why imatinib remains an exception of cancer research. J Cell Physiol 2012; doi:10.1002/jcp.24233.
31. Heng HH. Elimination of altered karyotypes by sexual reproduction preserves species identity. Genome. 2007; 50: 517–524.
32. Wilkins and Holliday. The evolution of meiosis from mitosis. Genetics. 2009; 181: 2–12.
33. Gorelick R and Carpione. J. Origin and maintenance of sex: the evolutionary joys of self sex. Biol. J. Soc. 2009; 98: 707–728.
34. Gorelick R and Heng HH. Sex reduces genetic variation: a multidisciplinary review. 2011; Evolution. 64: 1088–1098.
35. Heng HH. Preface: Back to the future. Cytogenet Genome Res 2012; (in press).
36. Stevens JB, abdallah BY, Liu G, Ye CJ, Horne SD, Wang G, Savasan S, Shekhar M, Krawetz SA, Huttemann M, Taiinsky MA, Wu GS, Xie Y, Zhang K, and Heng HH. Diverse system stresses: common mechanisms of chromosome fragmentation. Cell Death and Disease. 2011; 2:e178. Doi:10.1038/cddis.2011.60

37. Heng HH, Stevens JB, Liu G, Bremer SW, Ye KJ, Reddy PV, Wu GS, Wang YA, Tainsky MA, Ye CJ. Stochastic cancer progression driven by non-clonal chromosome aberrations. J Cell Physiol. 2006; 208: 461–472.
38. Stevens JB, Abdallah BY, Liu G, Horne SD, Bremer SB, Ye KJ, Kurkinen M, Ye CJ, Heng HH. Heterogeneity of cell death. Cytogenet Genome Res 2012; (in press).
39. Lister R, Pelizzola M, Kida YS, Hawkins RD, Nery JR, Hon G, Antosiewicz-Bourget J, O'Malley R, Castanon R, Klugman S, Downes M, Yu R, Stewart R, Ren B, Thomson JA, Evans RM, Ecker JR. Hotspots of aberrant epigenomic reprogramming in human induced pluripotent stem cells. Nature. 2011 Mar 3;471(7336):68–73.
40. Amariglio N, Hirshberg A, Scheithauer BW, Cohen Y, Loewenthal R, et al. 2009 Donor-Derived Brain Tumor Following Neural Stem Cell Transplantation in an Ataxia Telangiectasia Patient. PLoS Med 2009; 6: e1000029.
41. Heng HH, Stevens J, Lawrenson L, Liu G, Ye KJ, Bremer, S and Ye, CJ. Patterns of genome dynamics and cancer evolution. Cellular Oncology 2008; 30:513–4.
42. Ye CJ, Stevens JB, Liu G, Bremer SW, Jaiswal AS, Ye KJ, Lin MF, Lawrenson L, Lancaster WD, Kurkinen M, Liao JD, Gairola CG, Shekhar MP, Narayan S, Miller FR, Heng HH. Genome based cell population heterogeneity promotes tumorigenicity: the evolutionary mechanism of cancer. J Cell Physiol. 2009; 219: 288–300.
43. Nesse RM. Ten questions for evolutionary studies of diseases vulnerability. 2011; Evolutionary Applications 4:264–277, 2011
44. H. Gilbert Welch, Lisa Schwartz, and Steve Woloshin. 2011. Overdiagnosed: making people sick in the pursuit of health. Beacon Press.

Part III

Concepts of Health

Central to the whole endeavour of health care must be the nature of health. What is health that healthcare systems aim to deliver? The current and prevailing operationalization of health through its inverse—the absence of disease—is no longer sustainable within an environment of escalating healthcare costs that recognizes the complex and dynamic systems. This section examines health through three different system's lenses and concludes that all contribute to a coherent systems based characterization.

13 Trying to Make Sense of Health

Stephen Lewis

Having a concept of 'health' is important in the identification, treatment and non-treatment of a wide range of physical and mental conditions. Knowing when to treat, when not to treat and when to cease treating is important for the provision of proper ethical patient care. Appropriate non-intervention can be as important as appropriate intervention. Furthermore, wastefulness—not to mention patient distress—ensues when responses are made needlessly to subtle deviations from the norm. A better understanding of health may help limit unnecessary interventions and limit some of the unrealistic expectations increasingly being placed upon healthcare providers. Furthermore, the goals and limits of individual therapies and of medicine as a whole may become clearer.

However, defining health appears to be a simple, even unnecessary, matter that does not spark the interest of most healthcare workers. Everybody knows what health is—that is, until they are asked to provide a clear and concise definition of the word. Not only is it hard to provide such a definition but, if past experience is anything to go by, those devised are unlikely to stand up to critical scrutiny well enough to gain general approval. However, that is no reason to stop trying. Perhaps the approaches taken in the past have been misdirected.

S. Lewis, B.Sc., M.A., Ph.D. (✉)
Department of Biological Sciences, University of Chester, Chester, UK
e-mail: stevelewis55@gmail.com

13.1 Concerted Attempts

Under the editorship of Richard Smith, the British Medical Journal, in press and online, has, over a number of years, continued to raise the question of how health might be described or defined. As Smith notes in a BMJ blog:

> Last week about 30 of us spent 36 hours in The Hague discussing whether we could produce a new definition of health—and eventually deciding that we couldn't. [1]

A brief account of these deliberations eventually appeared in the British Medical Journal [2] where, taking a lead from environmental science, the authors suggested that health might be '*the ability to adapt and to self manage*'. Or, at least, '*[t]his could be a starting point for a … fresh twenty-first century way of conceptualising human health*'.

As Smith also noted, when it comes to asking what health is '*(f)or most doctors that's an uninteresting question*'. As he goes on to state '*[d]octors are interested in disease, not health. Medical textbooks are a massive catalogue of diseases.*' So, perhaps thinkers have been coming at this problem from the wrong angle. 'Health' and 'disease' are the two most fundamental terms in medicine. Intuitively, they form the polar opposites between which medicine is practised. Combating one and ensuring or restoring the other is the mainstay of clinical practice. So perhaps the question of health might be assailable via disease.

J.P. Sturmberg and C.M. Martin (eds.), *Handbook of Systems and Complexity in Health*,
DOI 10.1007/978-1-4614-4998-0_13, © Springer Science+Business Media New York 2013

However, defining what disease might be seems equally difficult as studies also published in the BMJ have discovered [3, 4]. Offered a list of named conditions with which clinicians frequently deal, different groups of people—including medical academics and general practitioners—were asked which they considered to be diseases and which they thought were not. The specific results are, perhaps, less significant than the overall findings that, noticeably, complete agreement was lacking. There were differences of opinion both within and between the groups surveyed. Clearly, deciding whether something merits being called a 'disease' is not a simple proposition.

13.2 Each End of a Line?

Often, when health and disease are represented graphically, they are shown at opposite ends of a line. A horizontal line typically has disease to the left and health to the right. A vertical line has disease at the bottom and health at the top. In adopting this approach, there is clearly a tacit inclusion of value judgement: health is being equated with the conventionally more positive ends of these lines; disease with the less positive, or negative, ends.

In these representations, there is also an underlying assumption that disease is continuous with health. For this to be the case, both disease and health must be commensurable. That is, they must be capable of being measured by the same common standard. That this is the case has never been established. What can be said to be commensurable and so part of a continuity is how one feels: one's experience of ease or unease [5]. One may say that one feels 'well' or 'unwell' or perhaps somewhere in between.

Here it is important to make a distinction between experience and physical state. This seems to be easily and frequently overlooked. Both are integral to being an individual. One may feel well but have potentially life-threatening physical lesions. One may feel quite unwell but without any discernible physical reason [6]. In the former scenario, one is able to go about one's daily life unhampered. In the latter, one will not be able to go about one's daily life quite so well. How life is led is determined to a large extent by how one feels rather than by how one might be physically. Any definition or conceptualisation of health and disease must, therefore, take this into account.

13.3 A Wider Debate

Among philosophers of medicine there has been considerable debate about the definitions of both 'health' and 'disease'. Philosophical interest began, arguably, within the medical profession in the years which followed the founding of the WHO and was perhaps inspired by its founding definition that

> health is a state of complete physical, mental and social well-being and not merely the absence of disease or infirmity. [7]

However, a more intense debate about the definitions of 'health' and 'disease' has continued since the mid-1970s when Christopher Boorse published a brief series of articles [8–10]. Contrary to the WHO definition, Boorse saw no problem in defining health as the absence of disease. In so doing, he also proposed the Bio-Statistical Theory of disease (BST), which posits disease as a statistical deviation from an average functional state.

However, despite mounting a vigorous defence [11], most of those who have responded to his ideas have responded negatively to the point that it has been proposed that the BST should be considered to have been refuted [12]. For all the negative comment and claims of refutation, the BST somehow lingers on. The reason for this may lie in the fact that it reflects something of prevailing medical thought. It was Boorse's intention to extract a definition of disease from what clinicians were thinking. His was primarily a philosophical task. If the definition Boorse gives in the BST is flawed, it is perhaps because the thinking he has crystallised is also flawed. This is not to imply that medical thinking is entirely wrong but that it might be improved upon.

13.4 Schools of Thought

There are many identifiable nuances of opinion regarding the concepts of 'health' and 'disease' which may be teased out from writings on the subject [13]. However, Nordenfelt [14] has suggested that there are two broadly identifiable schools of thought. There are those who hold that objective (value-free) definitions of 'health' and 'disease' can be found via a scientific approach to the workings of the body. It holds that 'health' and 'disease' can be understood in largely physical terms. This school of thought, often described as 'naturalism', is represented by Boorse. Alternatively, others hold that such definitions are inherently non-scientific, value judgements. This school of thought, often described as 'normativism', has come to be associated with the work of Lennart Nordenfelt himself [12].

Nordenfelt has approached the definition of 'health' from the perspective of action theory. Of central importance here is the individual's ability to achieve various 'vital goals' associated with daily living. By 'vital goals', Nordenfelt is referring to those of an individual's needs that have the highest priority. Thus, his definition of health becomes that state '*[w]hen the individual A is in a bodily and mental state which is such that A has the second-order ability to realize all his or her vital goals given a set of standard or otherwise reasonable conditions*' [15]. Although philosophically more precise, this definition is less memorable than the much criticised WHO definition.

While there is a clear contrast between Boorse's and Nordenfelt's approaches, importantly, the bipartite division between their two schools of thought is not rigid. It does not follow that one is exclusively a 'naturalist' or a 'normativist' for all medical conditions. It has been suggested that one can be a naturalist regarding physical conditions and a normativist regarding psychiatric conditions [16]. Boorse focuses upon the function of parts of the body whereas Nordenfelt focuses upon the ability of an individual to live out their life. Considered separately, there is in each something with which clinicians can find agreement. However, when considered together, one perhaps begins to notice how distinct each approach is. Thus, it appears that both the schools of thought have something to offer. Yet, a unification of the two approaches has not been possible, so far.

13.5 To Define or to Conceive

Much emphasis has been placed upon trying to find a verbal definition of 'health' and 'disease'. Here 'health' and 'disease' tend to be treated objectively as if they were, in philosophical terms, 'natural kinds'. A 'natural kind' is something that exists in nature independent of human categorisation. One of the core questions that needs to be asked is whether that debate has adopted the most appropriate approach. Instead of verbal definitions there might yet be an alternative approach which may prove to be more successful in revealing what the words 'health' and 'disease' represent. An approach that is able to provide something akin to a definition without being burdened with the difficulties that worded formulations experience.

If clinicians are unable to offer neat and precise verbal definitions of 'health' and 'disease', it may not necessarily be because they do not know intuitively what they are dealing with. Instead, it may perhaps be because providing definitions of this kind is the wrong approach. Not everything is done in accordance with verbal codifications. Much relies upon mental images—pictures and conceptualisations. Instead of trying to describe 'health' or 'disease' in words, perhaps one should be trying to paint a picture of what they are. Here the image of humanity prevailing at any given moment in time can have a direct bearing upon how health and disease might be understood.

13.6 The View from History

Although there is a continuity with the healing traditions in the ancient world, the scientific bases upon which modern Western medicine are built are of only fairly recent origin. Furthermore, the roots of medicine go much deeper than is usually realised. While the Greco-Roman world gave written form to much of what was to influence European medicine for over 1,500 years, medical practice, in its broadest form, has existed since human pre-history. Dating from around 6500BC, archaeological evidence exists for the practice of trepanning. This early form of quite hazardous surgery was performed skilfully with only rudimentary instruments. The openings produced in the skull were carefully sculpted and were clearly not fatal. Subjects certainly survived for some time following surgery as bone edges have been found which show evidence of new bone formation. Furthermore, it was a practice for which evidence can be found in both the old and new worlds suggesting perhaps a very ancient common ancestry. Exactly why such a procedure was performed is not entirely clear—skull trauma does not seem to have been the reason. The practice points instead, perhaps, to the existence of medical cosmologies lost in pre-history. Significantly, it points to a system of care within early human groups and to the existence of individuals who had acquired considerable skills. It may also point to the beginnings of the enduring notion that disease is a distinct entity which needs to be encountered—even extracted or released from the body. The trepanned skulls may have been opened to allow something to escape. By this one does not necessarily mean a subdural haematoma or an embedded bone fragment. Instead it may have been some supposed entity such as a controlling spirit which required exit.

Throughout the whole of human history, others have been called upon to provide assistance with what might be described as 'quality of life' issues and with staying alive in general. That continues to be the case throughout the world today. Individuals identified as healers are known in all societies whether they be shamans, witch doctors, medicine men or members of the medical and allied professions. These are those from whom help is sought for a whole host of issues associated with the problems of life.

13.7 Ascent to Science

Auguste Comte (1798–1857) suggested that there were three phases to the development of any society or field of knowledge. These were the theological, the metaphysical (or abstract) and finally the positive (or scientific) stage. Using Comte's phases as a rough guide, it is possible to envisage the history of medicine along similar lines. However, instead of considering medicine to have begun in the ancient world of Greece and Rome—which, instead, more closely equates to the start of Comte's second, metaphysical, phase—the pre-historical period with its shamanism and other appeals to the supernatural belong more properly to the theological phase. It was then that appeals to supernatural realms and explanations in terms of oppressive spirits or demons were not uncommon. Although echoes of this persisted into Ancient Greek medical thought—and even into the modern day—much of medicine's Comtean theological phase has been lost to pre-history.

Medicine's metaphysical phase was largely spent in the shadow of Galen (129–199/217). It was not a period entirely in the doldrums but neither was it a period of significant technical or conceptual advancement. Medicine's metaphysical phase was one also coloured by astrology and alchemy. At its end, change was to come about largely as a result of the work of the physician-academics of the sixteenth and seventeenth centuries. In particular, the observation-based anatomy of Andreas Vesalius (1514–1564) and the nascent physiology of William Harvey (1578–1657) set the scene for what was soon to follow. Observation and experimentation began to take preference over the customary deference to the authority of the ancients.

Two important conceptual innovations of particular note in medicine occurred during the seventeenth and eighteenth centuries. In the mid-seventeenth century, Thomas Sydenham (1624–1689)—the so-called English Hippocrates for his mastery of medicine—formulated a concept of disease that still circulates in one form or another today. His view—for it could not be based upon rigorous scientific evidence at that time—was that diseases were distinct entities [17]. His conceptual system of disease—his nosology—held that diseases were natural kinds; they could be classified into a taxonomic system in much the same way that Linnaeus [Carl von Linné (1707–1778)] classified plants and animals. Indeed, Linnaeus was also to propose such a system in his 'Genera Morborum' (1763). This attitude, which was probably held in various forms from pre-historic times, now found formal and respectable expression.

13.8 The Dawn of Modern Medicine

Sydenham was particularly innovative in seeking to give a full description or picture of the objective characteristics of each disease as presented by the patient. Thus, he advocated what might be described as a 'natural history' approach to disease, detailing how each progressed in different patients. Significantly, it is to the work of Giovanni Battista Morgagni (1682–1771) that modern pathology as a scientific discipline can be traced. Indeed, Morgagni has been called the father of modern pathology. In 1761, Morgagni published 'De sedibus et causis morborum per anatomem indagatis'. The title translates as 'On the seats and causes of disease investigated by anatomy'. In it, Morgagni also took a 'natural (or medical) history' approach reporting case studies of patients upon whom autopsies were performed. His intention was to relate post mortem findings with the symptoms reported during life. In so doing, Morgagni provided the physical foundation that Sydenham's nosology had lacked. It was reckoned that within the body of the patient, there was something which could be physically located and which could be held responsible for their ailments. From here, it was a simple step to assuming that physical lesions were tantamount to diseases per se. Subsequently, in the nineteenth century, the focus shifted from the visible lesions that Morgagni had located to alterations in tissues, as described by Marie François Xavier Bichat (1771–1802) using a magnifying glass and then to cells, as observed by Rudolf Ludwig Karl Virchow (1821–1902) using microscopy.

During the twentieth century, this increasingly reductionist approach reached gene level. Where individuals, organs, tissues and cells could once be described as diseased—and the word 'healthy' is applied in a similar way—the same could not be applied to genes. It is clearly inappropriate to refer to 'diseased genes'—which are, after all only chemical strings—in the same way that one refers to a 'diseased liver', for example. Disease, at least, is a term that can be applied only to something that is alive. The situation is less clear for health. Foods and certain forms of behaviour have increasingly come to be described as 'healthy' or 'unhealthy', perhaps complicating the problem of definition.

13.9 What's in a Word?

Etymologically, the word 'disease' literally means a loss or lack of ease—that is, 'un-ease'. It previously described what is now more often implied by the word 'illness'. This contrasts interestingly with what the WHO definition counsels against: that health not be seen as merely an absence of disease. Disease was originally an absence: an absence of ease. In English, the earliest use of the word 'disease' dates from the early fourteenth century and simply meant 'discomfort', having been derived from the old French 'desaise' which meant much the same. By the late fourteenth century, however, it had already come to be used in the sense of being unwell or ailing but its literal sense of general discomfort seems to have continued until the early seventeenth century [18].

Recognising the difference between disease as a 'loss-of-ease' and disease as something with some sort of physical basis is an important metaphysical issue for medicine for it relates directly to what may be understood as health. If disease is seen as an entity which merely resides at a specific location within an otherwise healthy individual, then removal of the offending lesion is all that is necessary for the restoration of the patient to full health. Here, health is seen as an individual's default or normal state. However, if somebody is suffering because of what emanates or results from a lesion, then this is something with broader effect. In this case, removal of the lesion may only be part of a wider approach to treatment and cure requiring a more general systemic realignment of a range of interconnected bodily processes.

13.10 Conclusion

When trying to define health, what appears at first sight to be quite simple, soon becomes quite complex and seemingly intractable. In the past, to feel well was taken to be much the same as being healthy. Now, with the introduction of more analytical scientific thinking which seeks material explanations within the substance of the body, attention has shifted. While this has been going on, no accompanying conceptual change has occurred. Instead, word usage appears to have become more complicated rather than simplified by greater understanding.

In trying to understand what health might be, the first task must be to recognise that this Gordian knot is made up of different threads. Health has an experiential dimension in how one feels, it has a physical dimension in how one is constituted and it has a behavioural dimension in how one is able to go about the activities of daily life. To some extent, the three-part core of the WHO definition recognises this. In another sense, it does not go deeply enough and much remains to be done.

Acknowledgements I would like to thank Annette Lewis for her help in the preparation of the manuscript of this chapter.

References

1. Smith R (2009) Trying to redefine health from BMJ Group Blogs 2009. http://blogs.bmj.com/bmj/2009/12/14/richard-smith-trying-to-redefine-health/. Accessed 11 Nov 2011.
2. Huber M, Knottnerus JA, Green L, van der Horst H, Jadad AR, Kromhout D, Leonard B, Lorig K, Loureiro MI, van der Meer JWM, Schnabel P, Smith R, van Weel C, Smid H. How should we define health? BMJ. 2011;343:d4163. doi:10.1136/bmj.d4163.
3. Campbell E, Scadding J, Roberts R. The concept of disease. BMJ. 1979;2:757–62.
4. Smith R. In search of "non-disease". BMJ. 2002; 324:883–5.
5. Antonovsky A. Health, stress, and coping. San Francisco, CA: Jossey-Bass Publishers; 1980.
6. Marinker M. Why make people patients? J Med Ethics. 1975;1:81–4.
7. WHO. Constitution of the World Health Organization. 2006. http://www.who.int/governance/eb/who_constitution_en.pdf.
8. Boorse C. On the distinction between disease and illness. Philos Public Aff. 1975;5:49–68.
9. Boorse C. What a theory of mental health should be. J Theory Soc Behav. 1976;6:61–84.
10. Boorse C. Health as a theoretical concept. Philos Sci. 1977;44:542–73.
11. Boorse C. A rebuttal on health. In: Humber J, Almeder K, editors. What is disease? Totowa, NJ: Humana Press; 1997. p. 3–134.
12. Khushf G. An agenda for future debate on concepts of health and disease. Med Health Care Philos. 2007;10(1):19–27.
13. Hofmann B. Complexity of the concept of disease as shown through rival theoretical frameworks. Theor Med Bioeth. 2001;22(3):211–36.
14. Nordenfelt L. Health and disease: two philosophical perspectives. J Epidemiol Community Health. 1986;40(4):281–4.
15. Nordenfelt L. Animal and human health and welfare: a comparative philosophical analysis. Wallingford, CT: CABI; 2006.
16. Murphy D. Concepts of disease and health. In: Zalta EN, editors. The Stanford encyclopedia of philosophy (Summer 2009 Edition). http://plato.stanford.edu/archives/sum2009/entries/health-disease/. Accessed 11 Nov 2011.
17. Bynum W. Nosology. In: Bynum W, Porter R, editors. Companion encyclopedia of the history of medicine, vol. 1. London: Routledge; 1993. p. 335–56.
18. Lewis S. A pathological misunderstanding. Wellcome Hist. 2011;46:6–7.

Picture Health

14

Stefan Topolski

14.1 Introducing the Problem

What is health? How do we understand it? Where does it come from? Why does it go away? Health is not like sunlight, for instance, which we can see, nor is health like the sun's warmth which we can touch. While philosophical dualism such as Descartes' can put doubt to our senses' feelings of hot and cold, of light and dark, health itself remains inscrutable. We cannot put our finger on it, touch it, point towards it nor measure it directly. Discussing subjective sensations of health reminds us of other human experiences of happiness and pain, sensations which we have built scales to measure while there always remains some ambiguity and doubt.

Doubt is inseparable from knowledge unless we are omnipotent. Science thus depends on objective measurement, validation, the opportunity to test knowledge in the external physical world to disprove it. Medicine claims to be such a scientific process … and yet medicine has no testable understanding of health. How can I ever be sure that you see and feel the health that I feel? How do physicians treat something they cannot agree on? This is our problem. While traditional scientific reductionism would pretend to eliminate uncertainty by shooting a healthy buffalo herd to dissect its parts ad infinitum, what have we then? A dead herd and a big mess. So we cannot eliminate the natural uncertainty in our world with traditional scientific reductionism. Instead, with a deeper and wiser understanding of complexity, we may corral our jostling herd of philosophical uncertainties slightly more, hem them in, reduce their spread and perhaps feel closer to a valid scientific understanding of health.

S. Topolski (✉)
University of Massachusetts Medical School,
1105 Mohawk Trail, Shelburne Falls, MA, USA
e-mail: public@cottagemed.org

14.2 Health's Epistemology

Imagining health springs from experiencing illness. First we feel pain, "Ow!" and then we make meaning around it—"why does that hurt? … why do *I* hurt?" We can imagine our first human beings' surprise when they no longer felt how they had felt before. They might have wondered why, and here now we see the myth-making, meaning-making, semiotic drive of our uniquely human species. Death comes from illness. Illness often precedes death. Health exists where illness does not. Health often is simply described as a lack of illness. While observations of death and concepts of afterlife could be experienced by hunter-gatherer family groups, the gift of speech then allowed a sharing of the subtler feelings of illness itself. The question "does anyone else feel like me?" could begin to be asked. Today we call this process theorizing about disease, but in the end from the beginning our spirits have walked heavy in life accompanied by illness as we all slowly approach the specter of death.

J.P. Sturmberg and C.M. Martin (eds.), *Handbook of Systems and Complexity in Health*,
DOI 10.1007/978-1-4614-4998-0_14, © Springer Science+Business Media New York 2013

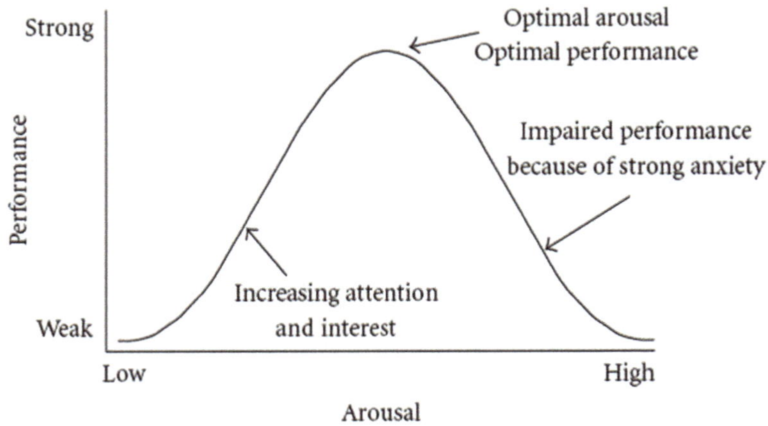

Fig. 14.1 Shannon information theory's inverted U curve reflected in the Hebbian version of a Yerkes-Dodson stress vs. performance curve (reproduced under Wikimedia Commons)

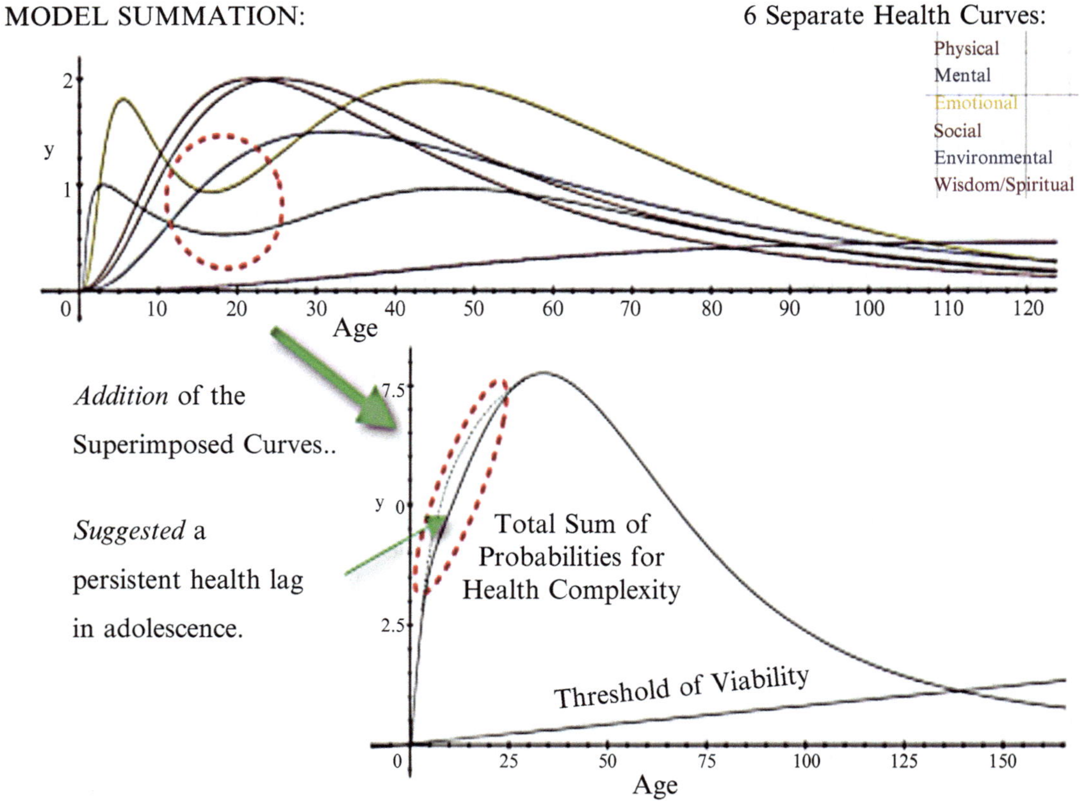

Fig. 14.2 Complex health trajectory model 2D: Log-normal approximations of three pairs of internal–external factors in human health in aggregate produce a picture of how potential health complexity rises rapidly from birth; health is predicted to sag during adolescence (a relative decline) and peak in early adulthood; health then declines towards older age while rising entropy with age produces a rising threshold of viability; although clinicians observe patients living briefly after falling below a threshold of viability, this state is not compatible with human life

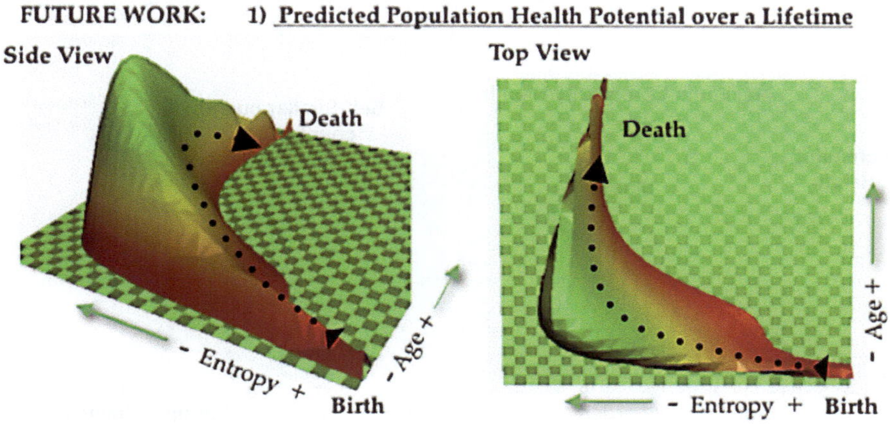

Fig. 14.3 Complex health trajectory model 3D: Human beings are born with high entropy and low functional health complexity; as we grow older we gain healthy function by building body structure and reducing entropy as much as possible; health (and fecundity) peak in early adulthood; with further aging our health's functional complexity and survival potential decline until we fall below the entropy threshold of survivability where we die

Historically human beings have approached understanding by asking "Why" before "What" and then often answering it with "Who." *What* spirit hurt me, *What* god punished me, *Who* was the actor responsible for *Why* there is an observation of illness or health in a person. Presuming that health changing to illness is an action requiring an actor for explanation is a most ancient of human heuristics. Experimental science has shown how these older semiotic methods shortchanged our ability to understand the world we live in. Animism, spiritism, religion can be overused as explanatory mechanisms … a belief preserved today in the proverb "*when you only have a hammer everything looks like a nail*."

The scientific revolution rectified this to the exclusion of all non-material interpretations of the natural world. In a world that seems to know no balance, our pendulum has now swung to the other extreme, the extreme of science, an extreme where we can no longer ask the deeper "Why" questions of religion and philosophy. "Why" questions lie outside the realm of our scientific method. Our materialism ignores such deep semiotic (meaning making) questions while settling for the simpler material questions answerable through physical experiment and shared observation. We and Descartes accept the presumption that sensory information represents an external reality [1]. Within this logical perspective we seek to break the tautology between health and illness. Physicians cannot define health as the absence of illness, as the absence of non-health [2]. For a modern scientific practice of medicine we hope to more precisely and reproducibly describe health on its own terms.

What remain of our myths become "models" of scientific understanding. Newton had delivered on Francis Bacon's scientific promise by reducing the mystifying movements of heavenly bodies to simple mathematical equations. The scientific revolution raced ahead from there revolutionizing human understanding. Here we begin discussing the relatively undeveloped field of theoretical medicine—theories of health, illness, and disease, their causes, character, and effects. In horribly broad strokes one may paint medical theorizing as a wavering progress from early spirit theories to those of bad humors and more recent, more acceptable, still inadequate germ theories of health.

William Harvey's mechanistic modeling of heart circulatory function [3] in the early 1600s laid the medical groundwork which grew with scientific and industrial revolution into our severely reductionistic twentieth century medical models of human body as machine, parts to be replaced, healing as fixing what was broken [4]. Medicine promised that the more we could dissect human beings into tinier and tinier parts then the more we would understand how they work and the more powerful our technological fixes could become.

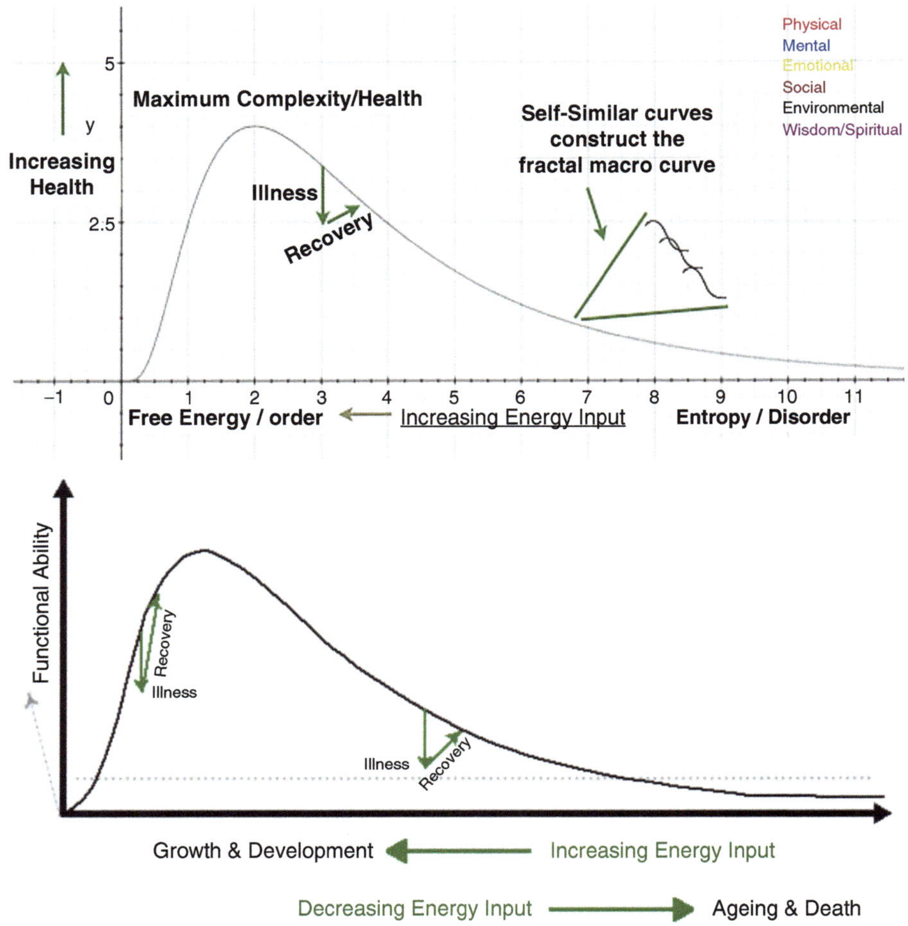

Fig. 14.4 2D curve describing illness with *arrows*: Human beings are born and later die with higher entropy to the *right end of the graph*; with time and effort (energy expended) we grow in functional health towards the *left end of the graph*; while most people live on the *right side* below their peak health potential some do overexert in sport or life and lie partly to the *left* of their peak health potential; overexertion illnesses on the *left hand side* of peak health do heal and recover at a higher level of function with a lack of exertion (rest of the injured overused body part); most human beings underachieve their peak health, stand to the *right* of peak health, and when ill they rest and recover at a lower level of function than their baseline before illness

We still live in the midst of this biomedical paradigm. We have seen the great flower of its promise in our technology-laden hospitals which bring some patients back from almost certain death. What we never see, however, are those critical junctures where our days have become numbered, our paradigms cease to deliver, our world views are failing; never do we see these truths until after an economy crashes, a dictator is overthrown, a criticality is reached where the trusted assumptions of what does, has, and will always work no longer do. Material reductionist science, and the industrial Taylorian assembly line controlitis built upon its foundation, are paradigms whose end is near. Nonlinear dynamics, the quantum physics of the world we can see around us, bears the revolutionary standard of the new age in science.

14.3 Nonlinear Science

Nonlinear processes abound in the physical and biological worlds. Productive quantitative study of nonlinear dynamics began in the 1960s with

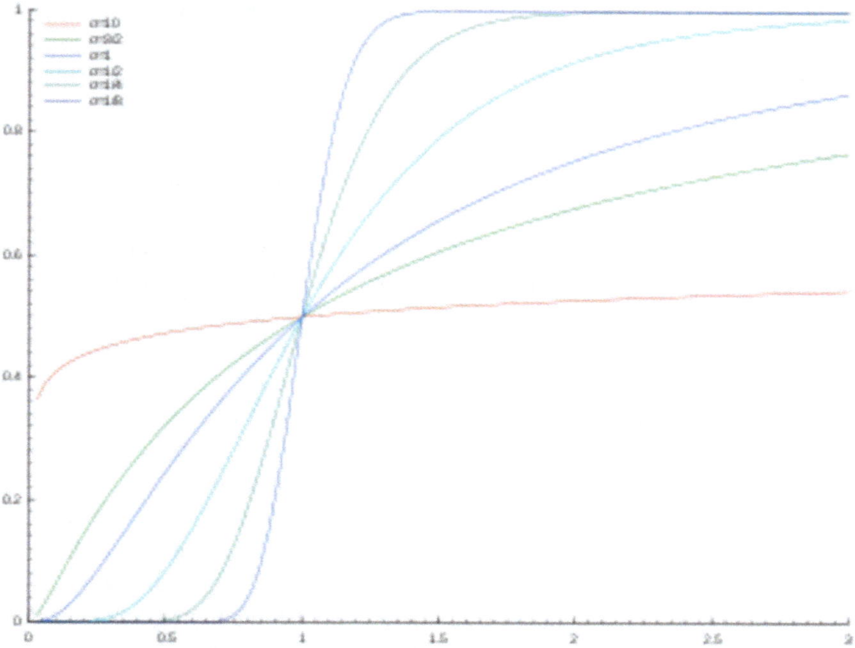

Fig. 14.5 Life history accumulation of complexity/meaning/semiotic wisdom as derived from a general cumulative log-normal distribution. Log-normal summation curves from Wiki—The two in the middle, $\sigma=½$ (steeper) and $s\sigma=1$ (less steep)—borrowed from Wikipedia—show the skewness of the log-normal curves used in our models

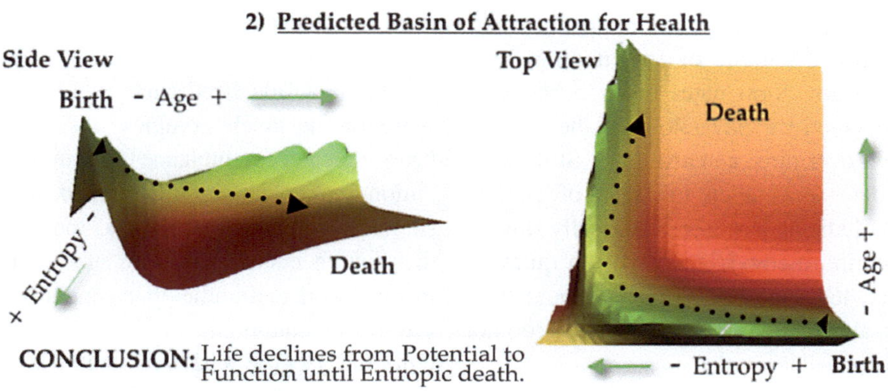

Fig. 14.6 Predicted health basins of attraction: higher and greener are healthier, lower and redder are less healthy with a higher probability of death; we are born with high health potential; energy is easily expended early in life to move towards less entropy and greater function in adolescence; with older age it becomes harder to export entropy from a human system to climb to greater healthy function; the result is increasing entropy as we decline and approach death

the advent of powerful microcomputer technology. Physicians, however, work in a vastly qualitative realm of people, culture, and society. This "art" of medicine, however, is not divorced from science. Art and life's complexity are complimentary [5]. Our medicine can be advanced with explorations of nonlinear methods.

The last 50 years of scientific research show an exponential increase in the publication of medically relevant nonlinear science since the late 1980s [6].

Early research showed logarithmic scales of self-similarity in anatomic structures known as fractal patterns. Since then log-normal distributions in biologic data sets and the related presence of power-law relationships have become commonplace [7]. Analytical time series studying EKG, EEG, gait, and other patterns have followed but remained more limited due to difficulties in qualitative data collection. The high prevalence of nonlinear features in biological systems has actually led to a glut of repetitive studies more valuable for their reliable ability to garner grant funding than for their novel creative contributions to the nonlinear field.

The field of nonlinear dynamics has since diversified into fractal structure analysis, chaos and catastrophe theories, network dynamics, and complex systems research. They all borrow insights from information theory, evolutionary development theory, game theory, and more. After five decades nonlinear dynamics has made useful contributions to every field of quantitative and qualitative scientific enquiry as well as literature and the arts [8]. The core tool driving this paradigm shift is the blazing speed of an electronic algorithmic machine—the microcomputer.

Computer models are familiar to the traditional medical discipline of epidemiology, and modeling in any form dates back to Harvey's original work on the physiology of the circulatory system over three centuries ago. Still, computer models seem alien to much of clinical medicine. Medicine has been historically slow to adopt scientific methods and then too quick to ape the overdue sciences without sufficient critical appraisal. "White coat and Petri dish" do not a scientific physician make.

I suggest that medicine's insufficiently critical approach is inadequately scientific. It has landed U.S.-type health care into a technological quagmire of high-energy quick-fix technologies which we cannot afford—technologies which are delivered to damage as much as to help patient health [9]. Now that a new science of complexity has appeared on the horizon, physicians do run the same risk of again aping overdue scientific complexity methods just as we uncritically accepted past scientific theories. However, nonlinear dynamics does seem at least as useful and appropriate as the respected philosophical speculations of such widely esteemed physician authors as Pellegrino [10] and Fine [11]. Complexity deserves its place at the table.

Complex systems studies offer many computer modeling methods adding insight and posing new questions. Computer modeling is criticized as divorced from reality. However, the goal of modeling is not to capture reality en toto. This is logically impossible and neither necessary nor desirable. On the contrary, a model that simplifies reality while maintaining some of the subject reality's pattern and behavior can reveal causes and relationships not perceived in the overwhelming whole [12].

Some of the current complex systems concepts in modeling health and illness and disease (these terms all having distinctly different meanings) may be mentioned here. Shannon information theory gives us an inverted U curve which is a core concept in describing the relationship between chaos, complexity, energy, and entropy. This curve already existed in the sports medicine literature [13] where information complexity and competitive performance in sport are allegorically equivalent, but physicians have not appeared to realize its larger scientific theoretical context.

This is an idealized curve whereas its representation in much complex systems research bears a greater resemblance to asymmetric distribution curves [14]. Often called long tailed, skewed, or log-normal curves, these frequency distribution curves are more ubiquitous in our medical work and studies than generally accepted. Physicians commonly fit skewed data into bell ("normal") curve statistical methods such that their inconvenient long tails with the richest qualitative data are lost.

Further nonlinear modeling methods abound with application to medical theory and practice. Network dynamics aid the tracking and pattern of infection in epidemiologic populations [15]. Time series have made large inroads where reams of real time physiologic data are available, namely EKG, EEG, and human gait measurements over time periods of minutes to hours [16]. Two such tracings may appear the same to a traditional clinician's eye yet prove significantly different when measuring their subtle nonlinear self-similarity. Complex systems methods can separate identical heart tracings and even identify mental illness in physical body measurements.

What proves very fascinating is the broad predictive value of such measurements. Presumably all organ systems are closely connected and communicate continually in a living human being. Declining complexity in heart rhythm can predict increased risk of geriatric stroke as well as neonatal intracranial hemorrhage [17]. Complexity in heart rhythms changes with temporal lobe epilepsy [18]. Such predictive patterns have been found crossing between other organ systems. Strange attractors and behavior attraction basins can be found in the body's methods of regulating a healthy heart beat and other body rhythms [19]. Even solitons, circular regenerating patterns that persist, can be found in modeled human data [20]. The human body is a truly complex event.

One more abstract modeling method to discuss would be cellular automata (CA). CA are familiar from epidemiologic studies and epitomized in the longstanding computer game of "Life" with its "blinkers" and "gliders" and other homeodynamic behavior patterns. Complex systems study demonstrates the core observation that complex patterns can emerge from few ***very*** *simple rules reiterated many times over.*

This is the crux of CA modeling and the reason why nonlinear dynamics has taken off so wildly after the advent of the microcomputer. Never before could *simple rules* be run over and over again *to observe large phase spaces of behavior*. This brief survey of complex systems methods leaves much of complexity untouched. Worthy of further study, we can carefully apply them appropriately to our highly qualitative fields of health and healing. We already are.

14.4 A Complex Model of Health

Complex systems models help portray to understand human health and improve medical practice. A complex systems bio-psycho-social model of health has been advanced with theoretical rigor, but its application to human health requires clarification. Practicing clinicians' professional training and continuing education remain largely tradition-oriented. They lack rigorous scientific method. Lacking familiarity with current complexity concepts leaves physician readers unable to apply current theoretical projects to clinical medicine. With care, however, a model of approximations describing health as one aspect of a richly nuanced and deeply symbolic human life-health as a potential health trajectory over time-may still describe to physicians useful aspects of and new predictions for the care, meaning and measure of human health.

Health had traditionally been defined as homeostasis with an absence of disease [21]. That twentieth century biomedical model leaned heavily on Descartian reductionism to supercede earlier theories of humors, meridians, germ theory, and more. Recently George Engel returned to Osler's earlier counsel to emphasize the connection between person and environment which has become the standard bio-psycho-social model of health [22]. We have extended Engel's bio-psycho-social health model based on complex systems research demonstrating that biological life is of essence nonlinear [23]. This complex systems model of health is quite descriptive. It packs a tremendously nuanced practical punch in its seemingly abstract picture.

The three proposed core factors of health—physical, emotional, and mental—are consistent with views as ancient as the platonic ideal of the human being as spirited (acting), appetitive (feeling), and rational (thinking) [24]. A fourth Hellenic virtue of moderation can be represented by the balance between internal and external aspects of these factors—body and environment *physical* health, spirit and society *emotional* health, and internal cognition with external experience *mental* health expressed as a semiotic wisdom.

As discussed earlier, illness and death appear related through human history. Health accompanies life. We appear to grow healthier as we grow up, then health declines as we grow older. There is a minimum amount of health needed for survival. Clinicians observe patients—premature infants, victims of accident and war, the elderly after hip fractures—who are alive but are not healthy enough to survive. There appears to be a threshold of viability below which health is incompatible with life. Such patients may remain alive briefly after death has become inevitable. This health analysis recapitulates an evolutionary development perspective of health as successful viability producing genetic offspring. Health appears directly related to viability, now what is the threshold of viability related to?

Thermodynamics drove scientific advance in the nineteenth century. While even Einsteinian relativity or quantum chromodynamics may be overturned by future science, the laws of thermodynamics appear immutable. Entropy always increases. Every premedical student studies entropy, and to most students entropy appears as irrelevant as most Flexnerian premedical preparation. The inevitable gradual increase in disorder with time is entropy. The human body grows from embryonic potential to healthy structure with a fine fractal detail. Time then takes its toll and, as we age, our bodies begin to break down. Health declines as entropy rises. Tissue and cellular patterns degrade, become more random and less predictable, less functional. When tissues cannot complete their functions we die. Our health has descended below a threshold of viability described by rising entropy as we age. Increasing entropy drives increasing mortality. Entropy can define the threshold of viability.

Describing the potential for health to change over a human lifetime is one of the most novel features of a complex systems model of health. While it resembles the light-cone of possibility within space and time in the physical sciences, this health model is asymmetric to better model biological life through commonly observed symmetry breaking which is essential to biological life [12].

14.5 Understanding Health Complexity

This complex definition of health seems very abstract, but it is the first health model to describe many aspects of healthy human life *and* illness.

We begin at the beginning. Individuating biological life begins at fertilization, arousing a quiescent cell to logarithmic change from a senescent tangent to the horizon at the model's origin. That this distribution curve never again declines to its original depth reflects a human being's vast growth from pluripotent single cell fertilization to an orders of magnitude larger size and function through the remainder of life. The model's preponderance of physical (and not mental or emotional) development in utero derives directly from the observed earlier peaking of physical development among young adults later in life.

Complexity describes many health phenomena. Health complexity curves are lowest in the very young and very old. The period immediately before and after birth is marked by great fragility when challenged by illness or neglect. We equate neglect with illness because an infant's health cannot be viewed in isolation from their mother, family, and social context. This fragility of the young resembles that of the very old. Neither infant nor elder thrive isolated from community. What is different is the new infant's rapid rise in robustness within a few months standing in stark contrast to the many years that elderly individuals suffer declining robustness when often ill and neglected. The very young and very old also share less caloric intake to maintain viability. They experience less physical variation in function with less adaptation to exercise compared with much healthier adults. These observations are all represented in the height, width, and slope of the model.

Infants develop into children. Childhood shows increased variability among individuals. Exercise effects more developmental change compared to infancy. Overuse injuries begin to appear. There is generally higher subjective health reported by children compared to elderly adults no matter where a person stands across the population distribution. Children also show greater flexibility, adaptation, and rapid recovery from environmental injury compared to older individuals. Since trajectories of probability multiply into the future, this model emphasizes the importance of early choices and childhood stimuli to maximize a person's health probability and performance potential in later adulthood.

Adolescence varies most across world cultures. While this health model can change to represent different socioeconomic cultures, it cannot represent all diversities at once. In order to respect diversity and minimize cultural bias, the current model only reflects peculiarities of adolescent development in more open and complex western industrial society. There would be less sag in adolescent emotional and environmental health curves in a more rigid or agricultural society. By defining health as a fixed potential distributed among different factors, this model allows for different human experiences while it makes clear that not all factors of health can be simultaneously maximized. For example, a young adult could be a better student or a better athlete, but one cannot be the best student or athlete you can possibly be at the same time. Adolescence thus is a crucial time where many health futures and life potentials become possible or inaccessible due to volition, values, experience, and opportunities lost and found near adulthood's beginning.

14.6 Predictions of Health Complexity

This model suggests many interesting health qualities to be observed. Acute disease at either the beginning or end of life occurs on a health curve lower and closer to the threshold of viability reflecting the higher mortality of the very young and very old. Healthy adults by comparison contract disease nearer their peak of complexity and health far from the viability threshold; they have lower mortality. Most people also lie to the right of peak health in the "couch potato" excess entropy portion of the curve rather than to the left of peak health in the "pathologic jogger's" excess energy portion.

The model identifies external and internal loci of health to emphasize where public health resources may be better spent—external health in good nutrition, sufficient exercise, safer environment, and nurturing social structures. Such external moneys are less likely to effect change in harder to reach internal loci of body, emotion and mind. These are the areas of health left most to the individual patient. The model calls for healthcare resources to be poured into the early years of life for maximum multiplied effect over a long productive human lifetime of development and experience.

This model may controversially suggest that fewer resources should be available to our oldest individuals where the potential return on health is least. An ethically acceptable compromise may be to temper the gross disparity between the excessive costs near life's end and paltry sums spent in infancy and childhood. A society with more wisdom through meaning-making (semiotics) might better navigate this challenging responsibility to share resources from young for old and old for young.

The emphasis on health as function and viability draws attention to something we have neglected in medicine. The process of sacrificing flexible potential for structural function redefines human development. Childhood development is not simply growth, growth as more and more, more height, more weight, more words, and more coordination. Other potentials are lost—more languages, other talents, and different body shapes. In the end, however, early rapid development produces maximum health change early in life and arrives at the widest range of human health potential with lowest relative entropy in early adulthood.

The curve models the plateaus observed in health over time by being a fractal summation of smaller self-similar curves. If one zoomed in for a much closer view of any portion the model would have a scale-free property common to biological data sets. To the far left it shows an accelerating decline in health when athletes overtrain. To the right the model curves have gentler slope and more surface area above the threshold of viability—plenty of room for the less healthy who are the rest of us. With more time and older age one sees a slow decline in health experienced by mature adults no matter how healthy and active their lifestyles.[1] The model demonstrates the opposite phenomenon in younger individuals

[1] Also known as "loss of functional reserve."

where health increases until early adulthood even without active exercise programs.

The model surprisingly suggests that infants are unexpectedly unhealthy, health in early adulthood is far higher than in childhood and old age, and that overall health and range of health potential decline sooner in adulthood than currently believed. It may be that the health of a vibrant 70-year-old and a bed-bound 70-year-old are not as different as supposed. The limited healing response and higher mortality of either when suffering similar life-threatening illness and injury belies the mirage of subjective good health. While optimism and idealism do prolong life [2,25], this model states their limits. Such surprises suggest a helpful role for this new model of health.

These curves are not static—they can grow and shrink with patient effort or inactivity. Potential health generally increases with effort and declines with rest. When acutely ill a health insult pushes a person from their original healthy state. This requires effort/energy to return to their prior health curve's height. The farther from peak health a person falls when struck ill, the more time and energy may be expended in the return. In the case of chronic overuse injury driving an individual's physiology towards chaos the model predicts the opposite, and this explains their paradox of rest—inactivity—producing health and recovery.

Complexity modeling describes a further paradox near peak health function when time to recover is longer, resembling the highly entropic sick individual. Finely tuned athletes may report that a week's pause in training requires several weeks of retraining to regain their peak performance. This paradox only occurs near the flattened apogee of peak health, whether physical, mental, or emotional. This is analogous to other aspects of human health previously described and consistent with the clinical observation that recovery lags behind sickness. Recovery from illness is slower and lasts much longer than the initial rapid decline into illness.[2]

Time to recover is generally (with exceptions) quicker in viral and slower in bacterial infection. How does one explain this difference? The model explains that viral changes to cellular metabolism may return to equilibrium of health faster than structural organ damage and increased entropy caused by destructive inflammation of bacterial disease. Such a model prediction may be tested by measuring declines in physiologic and anatomic fractal structure during illness and recovery through various laboratory techniques. This objective change in health is easier to measure and test than subjective changes in health; we can measure the patient's body we see easier than we can measure their feelings which we cannot.

Subjective health and objective health are not the same. This is a more subtle feature of the health-as-complexity model. As mentioned earlier, individuals may report the paradox of feeling healthy despite being ill. Borrowing from Von Neumann game theory, we propose that the rate of change in health is generally more important to a person's self-reported feeling of health than the absolute quantity of that change. Thus the slope (differential) of the potential health curve becomes a measure of subjective health. It is a quantitative representation of humanity's endless optimism.

This brings our discussion to the paradox of silent disease in the younger patient—chronic diseases like tuberculosis, diabetes, or cancer. Rather than quickly and unarguably dragging a person from their prior state of health they instead appear to progress upon an individual slowly and quietly before declaring their presence. Chronic illness slowly damages the anatomy and physiology of body and mind. Chronic illness does not allow a rebound to full health. The slope and height of the curve models shrink with time.

The health potential's range of possible states also narrows after middle age. With age's chronic illnesses we observe steeper sides to the entropy–complexity curve. This represents a loss of flexibility and higher risks of sudden large failures under stressors with advancing age. Steepened curve slopes describe the inevitable often stepwise loss of function from injury in our elderly. Even while it describes our converging health trajectories towards death, however, it still

[2] For example, a healthy person's lung consolidation with pneumonia will appear within days, clear within a week of treatment, yet his energy levels and functional abilities often take another month to recover.

allows the information content of individual life stories to grow and continue diverging with time.

This proposed picture of health is worth many a 1,000 words of clinical phenomena.

14.7 Potential Criticism

Modeling natural phenomena is an inherently incomplete process rife with difficulty. A model by definition abstracts and simplifies the natural world. One ponders where to bound a studied system, what scale to observe it at, and how to relate it to the real world. Qualitative phenomena such as human health exponentially increase the potential data load. Computational resources are limited and detail must be sacrificed. While detail must be sacrificed to model nature, this simplifying can reveal hidden patterns. This novel complex systems health model has revealed paradoxes which may improve our understanding of human health.

Not every computer model in this world is worth its weight in phosphor. Measurement and validation are crucial to the utility of any model. Current healthcare industry models of "health" measures—blood pressure, blood sugar, height, and weight—do not begin to capture this model's deeply qualitative measures of health. Simple measures of embryonic and infant growth speed up and slow down and shift from organ to organ over time all the way through puberty to maturity. This makes it challenging to use documented rates of cell growth in the first days, weeks, or months of fetal development—highly exponential in terribly short windows of time—to aid the calibration of a model thousands of times larger.

It can be argued that the model over-simplifies when it diverges from this published human growth data. On the other hand, perhaps its applied information theory calls for us to recognize significant aspects of human health which are not represented by traditional concepts of health. Physical health may also be flexibility, robustness, coordination, strength, speed, and global function. A physical health curve can represent more than merely physical growth.

While pediatric growth curves include pauses in rate of growth, complexity studies suggest that quality of growth and connection continue in the pauses in between where the quality of anatomy and physiology becomes more important than the quantity. Pauses in quantitative growth measures do not equate with pauses in healthy physical development. These growth pauses allow integration, tuning, and remodeling of organ systems. This brings us back to our argument that organisms build structure and lose flexibility. The balance between both becomes important to health. Ultimately physical health is a balance between flexibility lost and structure gained while exporting entropy from a living system. Health at your doctor's visit is no longer simply a measure of amassed physical shape and size alone.

Cultures also vary in their structure and flexibility. Some argue that human culture and health are too varied to be represented by one model. However the model's six different health curves do allow for flexible adaptation to represent varied historic periods, cultures, and degrees of economic–industrial development. Still others may argue that this relativism reduces validity, yet with good anthropologic ethnography a contextually accurate model can result. We attempt to minimize bias by applying the same normalized lognormal distribution function to each factor only changing the maximum value and time of the maximum value. Available epidemiologic data on mortality suggests a more peaked log-normal distribution of health complexity. The model can easily incorporate these changes. Our research has shown that while individual curves can vary, the final aggregate health potential curve is fairly robust. When health is measured as the opposite of mortality, the distribution of health across ages and cultures changes little.

Traditional societies value wisdom and faith more than western industrialized society may. This leaves room to wrestle with the definition of the semiotic wisdom curve. Some with a semiotic world view suggest that cognition through worldly experience develops virtue, wisdom, or spirituality long into old age even as creativity and imagination may decline. How do we differentiate between the human traits of spirituality, religion, and cult behaviors which attempt to make meaning of the world, and to what degree might they be adaptive and healthy or unhealthy and unwise? Some posit it to be the only curve to develop into old age without decline while others do not believe it exists at all. The model is obliged to reduce the wisdom curve's power and peak to reduce potential bias while this argument goes on.

Everyone also argues about entropy. Von Neumann joked that nobody really knew what entropy was. Information, probability, and entropy are at the core of this health model, but any measure of entropy is indirect and barely approximate. By recognizing that effort (energy) must be invested to maintain and improve health over a lifetime, we expect that only a small minority will achieve maximum health on our *x–y* axis. The probability of anyone achieving maximum health is scant, and the probability of achieving maximum health in all aspects of life is impossible. To describe instead the health experience of most people we derive a *basin of attraction* for health. The inverse of one's potential health complexity becomes the relative probability of achieving that maximum health. Summing this inverse of our original *x–y* function with our remaining *x–z* function relating health complexity to entropy produces a basin of attraction for human health which resembles the more likely human experience: health grows easily as we grow early in life and export entropy while later in life peak health is harder and harder to achieve and we slide back into increased entropy and eventual death.

These health potential curves are intended to be a first stab at conservative approximations. They are likely not so smoothly idealized, and further research will undoubtedly produce different proportions. For example the threshold of viability representing increasing entropy over time need not be linear. Probability theory instead suggests that entropy and the threshold of viability curve exponentially cove upwards. This change would improve the agreement between calculated and observed maximum human life span. Experiments on the simplest physical systems also suggest that complexity and entropy do not exist in a one-to-one relationship. Instead one level of entropy may support different stepped levels of complexity with gaps in between. This strongly suggests that the proposed health potential surfaces (manifolds) are likely fractal volumes beyond our ability to directly calculate.

14.8 Conclusion

Human health is robust and complex—we adapt very well to change. The human body holds an inherent ability to "spring back" from insult and injury like a spring under tension. We term this crucial aspect of health measurement a *dialectic tensor*. A destructive chronic disease, however, may disrupt *fractal organ structures* and *feedback systems* in human physiology weakening the body's spring. Disease may also reduce organ function efficiency increasing the friction in human physiology. As a diseased spring bounces back more slowly, a diseased human being also heals more slowly.

If the weakened spring of the human body can no longer respond to environmental stressors the chronic disease has played its role and fulfilled our worst expectations. These observations are analogous to increasing the model's viability threshold slope where more energy is required to export entropy and maintain the negentropy necessary for life. In the 2D model this reduces the distance between health potential curve and threshold of viability thus raising mortality and shortening life span on the graph. Complex systems modeling is aiding deeper understandings of human health.

This model also offers many novel theoretical approaches to measuring health. Health may be defined as maximum complexity, as the difference between curve height and viability threshold, as the fractal measure of the curved surface itself, and as the change in slope of the curve (subjective health) versus the quantitative maximum of the curve (objective health). The potential advances and advantages of this model are as follows:

1. Applies observable quantitative properties of a physical world to build an understanding of the qualitative properties of human life.
2. Adds key complexity principle of asymmetry to the light-cone model of physical state potentiality.
3. Offers a flexible yet robust approach.
4. Allows cross-cultural diversity with equality.
5. Changes a nebulous concept to something potentially measurable and verifiable.
6. Identifies that current "health measures" do not measure health.
7. Suggests means of better prioritizing limited health care resources.
8. Defines curves consistent with biological theories of survival and procreation.
9. Predicts that a human's true growth peak is later than the end of puberty.
10. Clearly defines different forms or classes of illness within one unified model.
11. Suggests several novel aspects of defining and measuring health.
12. Suggests basins of attraction for probability of health derived directly from a definition of maximum health.
13. Helps frame and encourage discussion of the core subject of medicine.
14. Allows the first testing and validation of a theoretical model of health.
15. Connects person to environment and structure to function indicating that quality is more important than quantity.
16. Begins to bridge a gap between physical science and human life, between the incredibly simple reductionist microscopic material scientific view and the massively complex qualitatively rich human experience.
17. Offers a conceptual explanation of this paradox—how terminal disease may be silent illness, that is, how people with very dangerous disease do not feel ill while those with very little disease can suffer great dis-ease.

Inclusion and robustness are hallmarks of this health model. It does not discard the old, yet it suggests many new qualities and measurements of health. It reflects distributions of health within a population and over time. It describes three distinct classes of disease—under-energetic, over-energetic, and diminished spring from permanent damage. Objective health resides in the cumulative heights of health potential curves. Subjective illness may be found in the slope and rate of change in the curves over time. The field of complexity provides a concise bundling of many observations and some paradoxes in clinical medicine.

Still we cannot over promise nor fully deliver on our claims for these methods. Ideas have always come and will forever go. As long as humanity survives, however, the doctor and patient will endure. Several criticisms have been raised in the spirit of scientific discourse to discovery. An improved model better representing available data has resulted. The shortfalls of this health model have also become clearer.

To model health is to attempt the impossible. Any and all modeling efforts must simplify, approximate and ignore details in order to illuminate a certain aspect, angle, or understanding of a subject. The theoretical lens used brings into view those aspects of health congruent with that lens. To model health with complex systems methodology produces a very different, imperfect but useful model of health.

Epidemiological validation of a model purporting to describe the vast possibilities of human health seems a task beyond Herculean and yet, within the limitations of any simplified model, we have found quantitative epidemiologic data that begins to test and validate this model for robust reproducible behavior [27]. Further testing of model derivations such as a basin of attraction for health will likely reveal further validations and deviations.

Aristotle defined the purpose of life as Eudaemonia—JOY—a full flowering of the Hellenic virtue of the golden mean as a balance and interplay between health factors, a dialectic view of health as the product of conscious action and continual change between person and environment. In a modest manner health defined as a fractal dimensioned curve of many parts allows us in some small way to discover, observe and measure this old anew.

Acknowledgments Joachim Sturmberg and Kathleen Burgess provided valuable advice, support, and encouragement in the completion of this manuscript.

References

1. Baker GP, Morris KJ. Descartes' dualism. New York, NY: Psychology Press; 1996.
2. Preamble to the Constitution of the World Health Organization as adopted by the International Health Conference, New York, 19-22 June, 1946; signed on 22 July 1946 by the representatives of 61 States (Official Records of the World Health Organization, no. 2, p. 100) and entered into force on 7 April 1948.
3. Willius F. A history of the heart and the circulation. Philadelphia, PA: Saunders; 1948.
4. Ottensen J. Mathematical modelling in medicine. Philadelphia, PA: IOS Press; 2000. p. 244.
5. Bar-yam Y. Dynamics of complex systems. Boulder, CO: Westview Press; 2003. p. 864.
6. Topolski S. Clinical application of chaos and complexity theory in family medicine. Eighth International Conference on Complex Systems, Quincy, MA: NECSI; 2010.
7. Walleczek J. Self-organized biological dynamics and nonlinear control: toward understanding complexity, chaos and emergent function in living systems. Cambridge: Cambridge University Press; 2000. p. 442.
8. Redies C. A universal model of esthetic perception based on the sensory coding of natural stimuli. Spat Vis. 2007; 21(1–2):97–117. doi:10.1163/156856807782753886.
9. Illich I. Medical nemesis. New York, NY: Pantheon; 1982.
10. Pellegrino E, Thomasma D. A philosophical basis of medical practice. New York, NY: Oxford University Press; 1981.
11. Fine M, Peters JW, Lawrence RS. (2007). The nature of health: how america lost, and can regain, a basic human value. Radcliffe Publishing.
12. Bar-Yam Y. Making things work: solving complex problems in a complex world. 1st ed. Cambridge, MA: Knowledge Press; 2005.
13. Williams JM, Landers DM, Boutcher SH. Arousal-performance relationships. In: Williams JM, editor. Applied sport psychology: personal growth to peak performance. Palo Alto, CA: Mayfield Publishing; 1993. p. 170–84.
14. Zurek WH. Complexity, entropy, and the physics of information. Boulder, CO: Westview Press; 1990. p. 227.
15. Potterat JJ, Rothenberg RB, Muth SQ. Network structural dynamics and infectious disease propagation. Int J STD AIDS. 1999; 10(3):182–5. doi:10.1258/0956462991913853.
16. Costa M, Goldberger AL, Peng C-K. Multiscale entropy analysis of complex physiologic time series. Phys Rev Lett. 2002;89(6):068102.
17. Tuzcu V, Nas S, Ulusar U, Ugur A, Kaiser JR. Altered heart rhythm dynamics in very low birth weight infants with impending intraventricular hemorrhage. Pediatrics. 2009;123(3):810–5. doi:10.1542/peds.2008-0253.
18. Ansakorpi H, Korpelainen JT, Huikuri HV, Tolonen U, Myllylä VV, Isojärvi JIT. Heart rate dynamics in refractory and well controlled temporal lobe epilepsy. J Neurol Neurosurg Psychiatry. 2002;72(1):26–30. doi:10.1136/jnnp. 72.1.26.
19. Kanters JK, Højgaard MV, Agner E, Holstein-Rathlou NH. Short- and long-term variations in non-linear dynamics of heart rate variability. Cardiovasc Res. 1996;31(3):400–9.
20. Aslanidi OV, Mornev OA, Skyggebjerg O, Arkhammar P, Thastrup O, Sørensen MP, Christiansen PL, et al. Excitation wave propagation as a possible mechanism for signal transmission in pancreatic islets of Langerhans. Biophys J. 2001;80(3):1195.
21. Guyton AC, Hall JE. Textbook of medical physiology. 10th ed. Philadelphia, PA: W.B. Saunders Company; 2000. p. 1064.
22. Engel GL. The need for a new medical model: a challenge for biomedicine. Science. 1977;196:129–36.
23. Topolski S. Understanding health from a complex systems perspective. J Eval Clin Pract. 2009;15(4):749–54.
24. Rees DA. Bipartition of the soul in the early academy. J Hell Stud. 1957;77:112–8. doi:10.2307/628642.
25. Idler EL, Benyamini Y. Self-rated health and mortality: a review of twenty-seven community studies. J Health Soc Behav. 1997;38(1):21–37.
26. Jylhä M. What is self-rated health and why does it predict mortality? Towards a unified conceptual model. Soc Sci Med. 2009;69(3):307–16.
27. Topolski S, Sturmberg J. Validation of a complex systems model of health. 8th International Conference on Complex Systems. Quincy, MA: NECSI; 2011.

15 Health: A Personal Complex-Adaptive State

Joachim P. Sturmberg

'Good health–it depends on who you are, where you live, the society around you–and in particular on the person you are. There is no common denominator. What is good for one, cannot help 2000'.

(Arctic Fisherman) [1]

Defining health has been a long-term endeavour, each attempt taking a particular perspective that emphasises one aspect of the experience of health over others. It is notable that only the WHO definition mentions disease—the "enemy" that needs to be wiped out—as part of the health definition; all others emphasise personal aspects that result in the experience of *personal health* (Table 15.1). The experience of health is essentially personal and has been equated to well-being and happiness. The dynamics between personal internal and external factors determine the *experience of health*, be it good health or poor health, or be it in the presence or the absence of discrete diseases. People can report poor health in terms of illness or disease rather than their specific conditions (diseases).

15.1 The Evaluative Nature of Health

The term "health" arose from the old English word "hal" meaning "whole"; health is wholeness, and a person being healed is one who has "become whole" again.

The notion of "wholeness" in the conceptualisation of health is not new. Plato already allude to it describing health as being impacted upon by the interrelationship of physical, psychological, social and spiritual dimensions and functions of an individual and society [2, 3]. Health, illness and disease are interdependent concepts that can only be fully understood if the personal dimensions are included [4]—health is a state of the "whole person".

Kant commented on the subjective nature of health stating: *Every man has his particular way of being in good health.* This view is corroborated in the words of one of my patients, a 23-year-old totally bedridden dying woman, who a week before passing away remarked: "*Doctor, I know I'm going to die, but can you explain to me why I'm feeling so healthy?*"

The holistic nature of health is also reflected in the WHO's 1986 Ottawa Charter for Health Promotion: *Health is ... seen as a resource for everyday life, not the objective of living. Health is a positive concept emphasizing social and personal resources, as well as physical capacities* [5]. Fundamental interrelated conditions to achieving health include peace, shelter, education, food, income, a stable eco-system, sustainable resources, social justice, and equity.

15.1.1 Health and the Medical Perspective

As physicians, we encounter people without diagnosable disease complaining about poor health, while others with severe disease state that

J.P. Sturmberg (✉)
Department of General Practice, Monash University, Melbourne, VIC, Australia

The Newcastle University, Newcastle, PO Box 3010, Wamberal, NSW-2260, Australia
e-mail: jp.sturmberg@gmail.com

J.P. Sturmberg and C.M. Martin (eds.), *Handbook of Systems and Complexity in Health*, DOI 10.1007/978-1-4614-4998-0_15, © Springer Science+Business Media New York 2013

Table 15.1 Conceptualisations of health

Year	Concept/model	Description
	A multidimensional state of the whole person	Health as an application to human nature in all its parts, operations, levels, and dimensions—the physical, psychological, social, and spiritual. (Plato [3])
1911	A holistic ability to function well in the lifeworld	Health is a holistic ability to relate properly to and function well in the whole lifeworld in all its aspects, and disease a disturbance of this ability, on any of a variety of levels or in any of a variety of dimensions. (Husserl [6])
1946	Health as an ideal state	Health is a state of complete, physical, mental and social well-being and not merely the absence of disease and infirmity. (WHO [7])
1951	Health involves effective compliance with expected or normal roles	Health in this sense becomes a prerequisite and a resource for maintenance of the social system. (Parson [8])
1960	Adaptation model	Health and happiness are the expressions of the manner in which the individual responds and adapts to the changes that he meets in everyday life. (Dubos [9])
1968	Existentialist model	Ultimate health is obtained through self-realisation. Man must search for meaning on his own grounds and live in accordance with his own values, skills and free dispositions. (Maslow [10])
1972	A self-evaluative state	… a summary statement about the way in which numerous aspects of health, both subjective and objective, are combined within the perceptual framework of the individual respondent. (Tissue [11])
1975	Role analysis of health and disease	Health is the capacity for human development and self discovery and the transcendence of alienating social circumstances. (Kehlman [12])
1976	A positive state across the life span	Health is a positive state that dynamically spans across the stages of life—"The ability to adapt to changing environments, to growing up and to ageing, to healing when damaged, to suffering and to the peaceful expectation of death". (Illich [13])
1979	Sense of coherence	The sense of coherence is a global orientation that expresses the extent to which one has a pervasive, enduring though dynamic feeling of confidence that one's internal and external environments are predictable. (Antonovsky [14])
1984	Sense of personal integrity/indigenous concept	Health is the outcome of a complex interplay between the individual, his territory of conception and his integrity: his body, his land and his spirit. (Reid [15])
1986	Social medical explanation model	Health is equivalent to the state of the set of conditions which fulfill or enable a person to work to fulfill his or her realistic chosen and biological potentials. (Seedhouse [16])
1986	A resource for everyday life	Health is … seen as a resource for everyday life, not the objective of living. Health is a positive concept emphasizing social and personal resources, as well as physical capacities. (WHO Ottawa Charter for Health [5])
2006	Dignity, sharing in community, culture and place of belonging	Health depends on many interconnected aspects of life: belonging to one's local environment/land, the sense of freedom, cultural and spiritual belonging, and the sense of dignity and security. (Ingstad [17])
2007	Balance between body, mind, environment and sense-making	Health is a dynamic balance within a complex adaptive somato-psycho-socio-semiotic framework. (Sturmberg [18])

their health is good or at least good except "this bit". These types of subjective statements have been largely dismissed as irrelevant in the reductionist objectified world of biomedical thinking. However, the importance of health perception, or self-rated health—the subjective

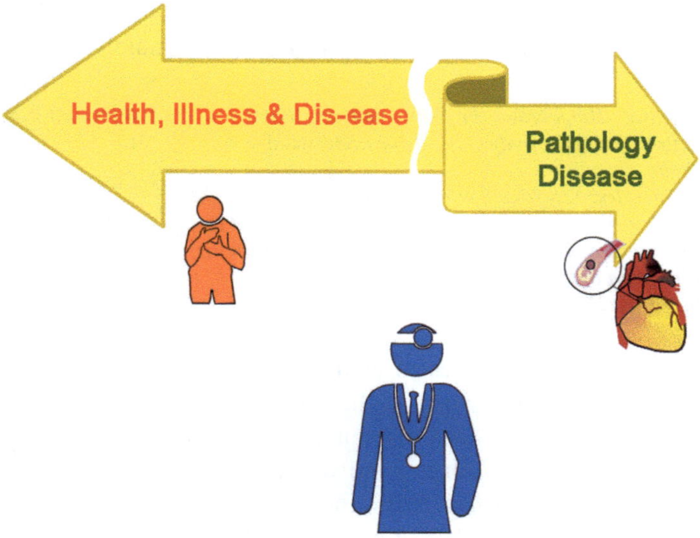

Fig. 15.1 Health, illness, and dis-ease versus pathologies. The ***clinical encounter*** is the meeting place of the subjective experience of the patient and the objective world of the pathologist. The doctor emerges as the intermediary between these two worlds

interpretation of one's health and disease experience [19], has been highlighted by Idler and Benyamini, who showed that this *subjective* measure predicts mortality and morbidity better than objective clinical assessments [20].

These observations about the experiential nature of health and disease should not detract from the undoubted benefits of our enhanced understanding of and ability to diagnose and treat specific diseases; however, the utility of our medical advances has to be balanced with our patients' perceptions, understandings and needs. In Baron's words: [These medical advances] *derive their significance from what they mean for human beings and what effect they have on suffering and individual capability* (p. 608) [21]. Fugelli reminds physicians about the subjectiveness of the object "disease" this way: ... *disease does not exist, only the experience of disease* [does] (p. 185) [22]. These insights should encourage health professionals to approach their profession as much from a *narrative* as *biomedical* perspective.

15.1.2 Health–Illness–Dis-ease

Historically and culturally people always have embraced a "balance notion" in their explanatory models of health and disease [23]. All incorporate habitual, environmental and spiritual components and acknowledge the complex interactions between the different aspects of human existence. The objective inter-relationships of health and disease co-exist with the subjective and evaluative personal nature of health and dis-ease and are deeply connected to the conceptions of medicine as a discipline. Pellegrino and Thomasma concluded:

> ... the principal conception of medicine, health, and disease are necessarily related to, and acquire their meaning from, the epistemological features of clinical interaction. Both health and disease are essential conceptions of medicine as a discipline. To the objection that health and disease are definientia only of organ systems, one must counter with the large body of evidence that both concepts are evaluative; that is, they include in their meaning the values of patients, societies, and cultures (p. 63) [23].

Hence health, illness, and dis-ease reflect different points on the same *subjective* scale of a patient's health experience, and need to be distinguished from the *objective* findings of the organ, cell, or sub-cellular changes seen by the pathologist (Fig. 15.1) [24]. The doctor's role is that of a translator, helping the patient to make sense of the subjective experiences and their potential underlying objective bodily changes. As a result, the consultation provides legitimacy to the person's experience; having been validated, society accepts sickness and provides sick people with certain privileges [18].

15.2 A Systems Model of Health

Phaedrus: Hippocrates the Ascelepiad says that the nature even of the body can only be understood as a whole.

Socrates: Yes, friend, and he is right – still we ought not to be content with the name of Hippocrates, but to examine and see whether his argument agrees with his conception of nature.
Phaedrus, 270

Despite all the successes of the biomedical reductionist model when dealing with well-defined discrete dis-ease, its limitations when dealing with health, illness, and dis-ease were noted as far back as the 1970s. The "new thinking" reconnected with the age-old views of health and dis-ease being complex, being a state in time resulting from the interactions between the myriad of personal, social, and environmental circumstances a person finds himself in.

In the 1970s and 1980s, three notable system models emerged, each integrating emerging knowledge and insights (Fig. 15.2). Engel [25] proposed a—still rather linear—biopsychosocial model of health, which was readily embraced especially by the primary care profession. Riedl (1985), coming from a biological systems perspective, described a hierarchical layered model based on their degree of complexity [26]. Riedl's model proposed that each layer is materially influenced by the layer below, and structurally and functionally influenced by the layer above [27].

Developments in physiology showed that the interactions between body parts occur by way of transmitters like hormones or electrical potentials, and are controlled by interconnected feedback loops. Uexküll and Pauli [28] suggested that the material flow of hormones through the bloodstream, or of electrical potentials along a nerve or muscle, solely provides signals for response. They distinguished the transmission of information from that of its content (or meaning). Meaning arises from the overall functioning of the entire human organism, just as the phonemes of an utterance in a foreign language get their meaning only from the context of the speaker's language and way of life. The understanding of this distinction led them to propose the need for *biosemiotic* thinking, implying one has to separate the interpretation of signals from the meaning assigned to them. Understanding the meaning of a signal in turn allows a person to respond, e.g. changing his behaviour in light of an adverse health event.

Adding a semiotic dimension to the health and illness construct extends the earlier models and has important pragmatic implications to understanding health and health care (see below). Expanding the network of explanatory variables overcomes, many of the limitations of the prevailing views. Pauli et al. expressed their concerns about the status quo as: *Contemporary science, by reducing all life phenomena to their physical or biochemical mechanisms, and their roles in communication to that of carriers of communication, profoundly limits the understanding of disease and even more so of health* (p. 167) and argued one has to ... *conceptualize health and disease as due in part not just to our material circumstances (e.g. genes and germs) but also to our life situations and the meanings we assign to these and other non-material circumstances* (p. 169) [29].

15.3 Health: A Dynamic State

Historically, despite all the progress in disease-specific management, the notion of health has maintained two stable elements: health being a balance between various aspects of a person's life experience, and the need to *make sense* of, or find *meaning*, in one's illness experience.

People's health experience has objective as well as subjective characteristics. Health (and dis-ease) therefore is neither solely an individual construction nor solely a reflection of societal attributes.[1] Societal understanding of health is

[1] A detailed discussion on the framework of self-perception of health has been provided by Marja Jylhä. She proposes a contextual framework of evaluation of self-rated health encompassing culturally and historically varying conceptions of "health"; resorting to reference groups, comparison with earlier health experiences, health expectations, positive or negative dispositions, and depression; and cultural conventions in expressing positive and negative opinions and in the use of a rating scale [19].

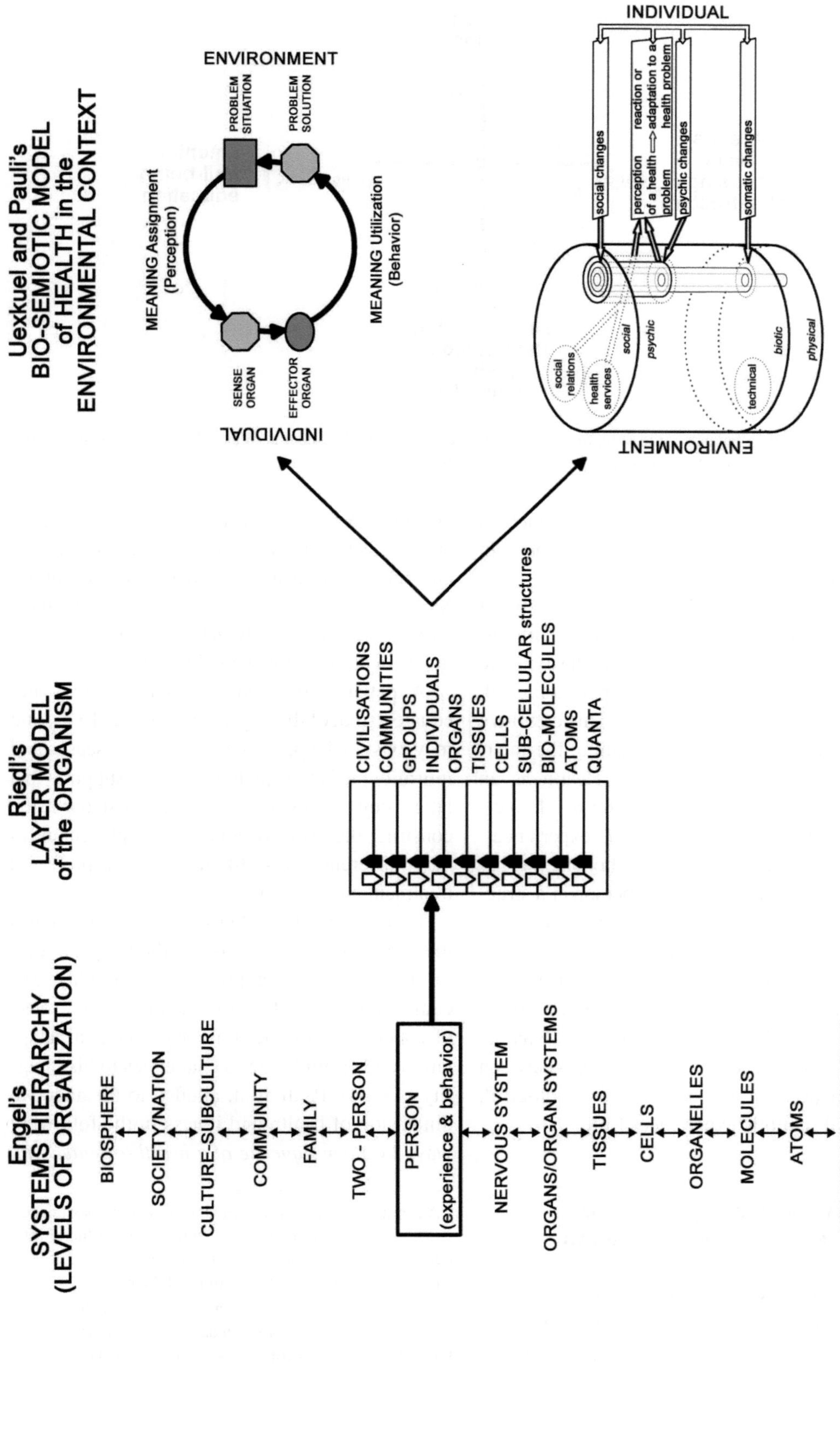

Fig. 15.2 The emergence of system based health models. The increasingly complex models from Engel, Riedl and Uexküll & Pauli

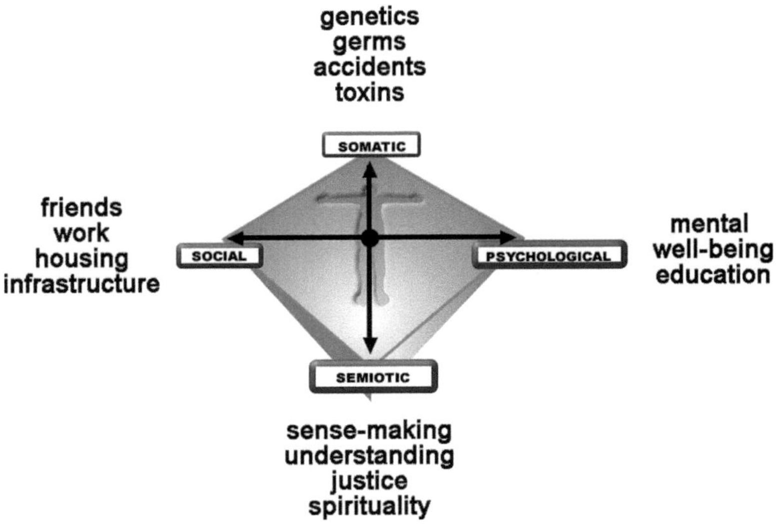

Fig. 15.3 Health as a balance—the somato-psycho-socio-semiotic model of health. When the influences of the four domains of health are in relative balance, a person experiences *health*

consciously enacted by individuals, as much as social structures subconsciously influence an individual's health perception.

These interdependencies are characteristic of complex adaptive systems, suggesting health, illness, and dis-ease being states within the dynamic interactions between somatic, psychological, social, and cognitive experiences. This cognitive process is called "semiotic", meaning the interpretation of signs and symbols. Health as an experience arises from *making sense* of one's physical, emotional, and social experiences. Health, illness, and dis-ease is a holistic experience that affects all parts of the person at a structural and functional level. Such a dynamic understanding better reflects the levelled reality of the human experience and leads to a somato-psycho-socio-semiotic model of health as illustrated in Fig. 15.3. Health then is a dynamic balanced state—***a basin of attraction***—between the interactions of the somatic, psychological, social, and semiotic experiences [30].

15.3.1 A Somato-Psycho-Socio-Semiotic Model of Health

In this somato-psycho-socio-semiotic model of health, each corner represents one of the four domains of health and some of its characteristic elements. A strict delineation of each component is not possible (and not necessary) as the domains are representational and symbolic in nature. Metaphorically, health is achieved when the individual perceives the interactions between the elements in all domains being in balance.

A patient's perception of balance may not equate to that of the physician (see Fig. 15.4), and may exist in the presence of specific diseases and infirmity [7, 31]. A qualitative study supports this latter point, patients with knee osteoarthritis construct the need for total knee replacement in very different ways[2] to justify their preferred treatment choice [32].

In addition, the balance model incorporates the concepts of homeostasis[3]—the body's ability to self-regulate to a position of stability and equilibrium in the face of challenge—and sense of coherence—the person's psychological adaptation and assimilation to challenges to his integrity [33, 34]. Pauli et al. alluded to the adaptive dimension of health and illness in the following way: *As a consequence of a health-oriented and*

[2] Adoption of the medical model, a person's social network, pain, functional loss, feelings of vulnerability, dependency, low mood and fatigue, ideas related to disease progression, and expectations of TKR.

[3] Homeostasis refers to Cannon's concept of maintaining a stable internal state within boundaries which are maintained through regulatory feedback loops (also see Chap. 10).

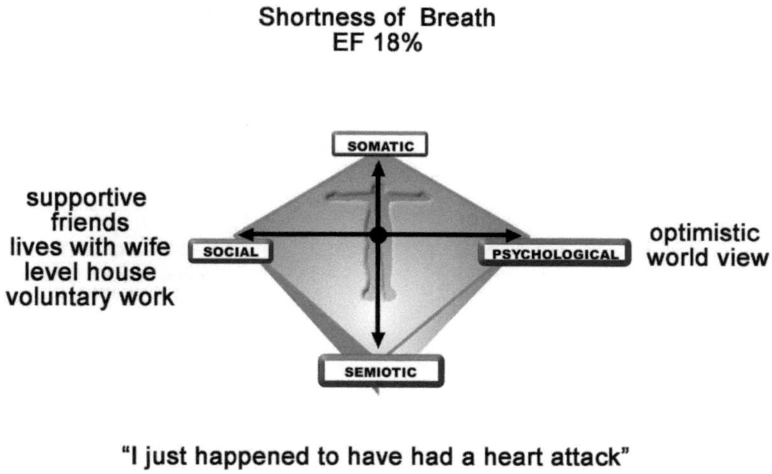

Fig. 15.4 A patient experiences *health* following a myocardial infarction. The patient has regained a balance between the four domains of health, despite the limitations caused by the myocardial damage

expanded medical model, illness and disease can be seen, in part, as a person's or an organism's failure to find or to generate meaning from her or his internal and external environmental experiences and opportunities (p. 171) [29].

15.3.2 Health: A Dynamically Balanced State

Having shown that the experience of health, illness, and dis-ease represent a dynamic and adaptive state, it becomes clear that they can be experienced as much in the *absence* as in the *presence* of identifiable pathologies. Potential mechanisms have been described in Jylhä's paper [19]. In fact, clinical experience would suggest that the length of a patient's problem list is often inversely related to his health experience.

Recapping Idler and Benyamini's findings: the personal health experience encompasses more than the "physical" health status and includes psychological well-being, health behaviours, social support, self-confidence, and so forth, and predicts future mortality and morbidity better than objective clinical assessments [20]. Figure 15.4 illustrates the somato-psycho-socio-semiotic representation of adaptation of a patient who had suffered an "acute coronary event" and is now experiencing "*good health*".

15.3.3 Illness and Dis-ease—Loss of Dynamic Balance

Figure 15.5 shows four patients with an illness and disease experience, illness secondary to a now resolved physical disease, mental illness, social illness and semiotic illness. Looking at these figures highlights that independent of the predominant source of the patient's complaint illness always has an important semiotic component, which typically is described by the patient in metaphorical terms.

Illness and disease result from a disturbance of the balance within the somato-psycho-socio-semiotic frame. Notably all illness alters perception and understanding, and the experienced physical changes are typically of a secondary nature. The physical changes are triggered by physiological changes in the neuro-endocrine system and these mechanisms are more closely described in Chap. 19.

15.3.4 Health: The Attractor that Maintains Survival

Health is a necessary attractor for survival in the complex dynamic challenges to a person's life (for a more detailed exploration on this notion see Lewis). Though health varies in time, the health attractor maintains a dynamic equilibrium

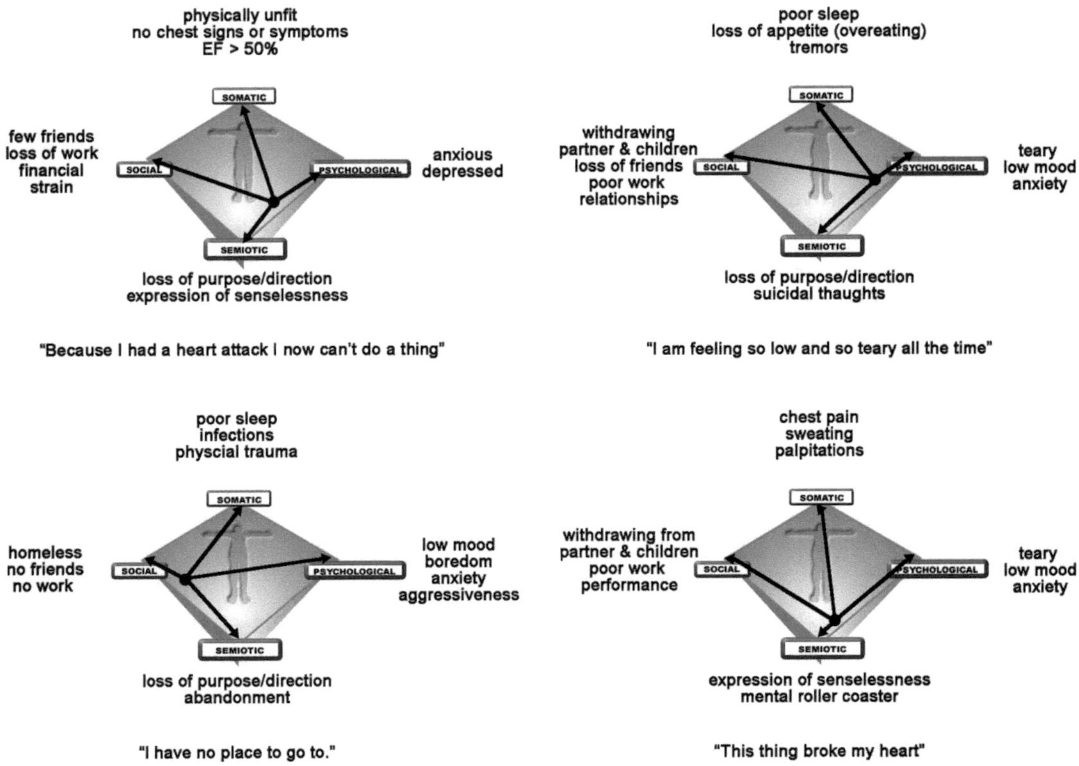

Fig. 15.5 The experience of illness due to somatic, psychological, social, and semiotic causes. Note that a shift towards any one corner is associated with changes in the other three

(Fig. 15.6). This equilibrium is constraint by the external environment on the one hand, and the internal circumstances (genetics, limitation due to previous disease) on the other.

A system dynamic understanding necessarily results in understanding health as an attractor state. It follows that health cannot be "one" particular state, but rather a relative point in a phase space, i.e. a chaotic attractor. Three attractor patterns emerge; a health attractor around the centre of the somato-psycho-socio-semiotic phase space, with illness, dis-ease, and [acute self-limiting] disease occur on "more distant orbits" of the attractor; a chronic disease attractor centred towards the somatic corner of the phase space; and a "psychosomatic attractor" whose dynamics swap between two phase space areas, the physical and the psycho-semiotic (Fig. 15.7).

Engelbert and Frith arrived at the same conclusions, and are quoted here to illustrate the point in a different way. Engelbert [35] wrote:

From a systems-theoretical aspect, the following comprehensive view of health is proposed. The dynamic equilibrium of health previously described may be considered to be of overall importance with respect to the processes of synthesis within the whole of the several system levels, representing as it were the driving force in life. Disturbance of the dynamics of this health equilibrium will leave the system as a whole vulnerable. On a personal level, the loss of synthesis can be manifested in a deficient feed-forward and feed-back into the 'outer' world, the reality of the environment. Next, feed-back towards the 'inner' world (the body) will be affected, the personal level just existing on the basis of feed-back from the body level. Hence forward disintegration of the integrity of the system will proceed until the living person has been reduced to an existence on a mere organismal level, left at the mercy of the vulnerable autonomy of his sub-systems. The person becomes a patient, no longer able to restore the integrity of his system by himself. A system so disintegrated cannot cope with confrontation with harmful conditions, either painful-external-events or 'painful' feed-back from a deranged 'inner' world system. In the course of the aforementioned, different status of system-level direction associated with the unfolding of the

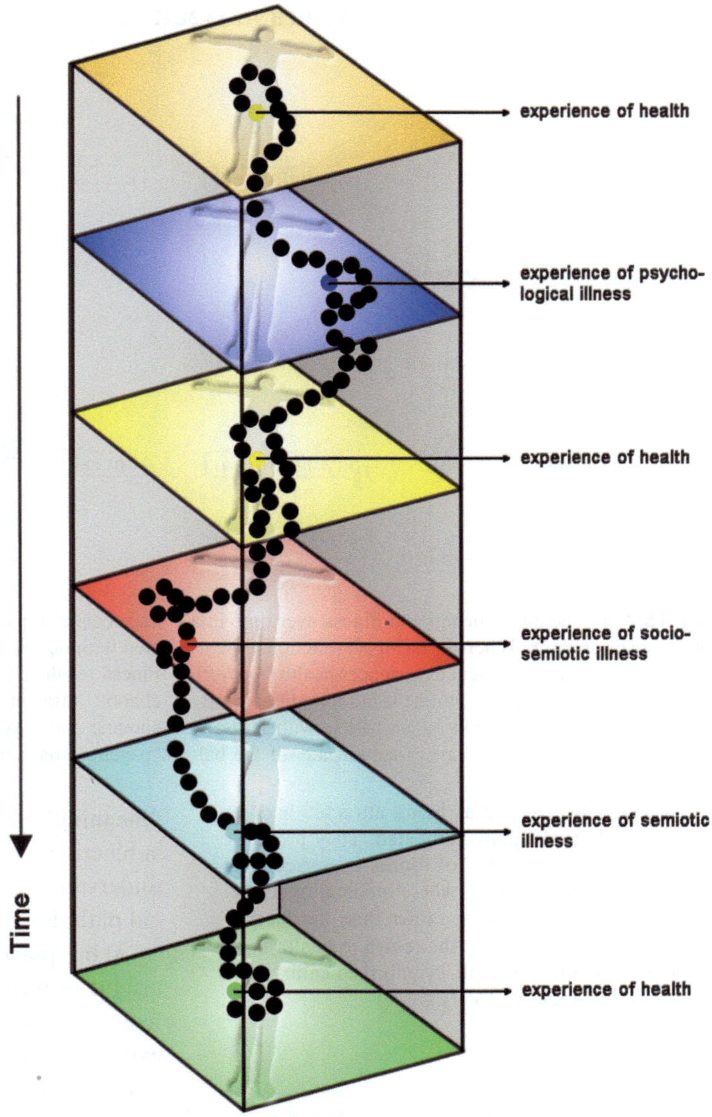

Fig. 15.6 The dynamic changes of health and illness over time. The basin of attraction at the centre ensures that people over time maintain a *good health experience*

pain function, a gradual return towards the 'outer' world experience will not be possible, prevented by the pre-existing deficient feed-forward towards that reality, eventually leading to fixation and further disintegration of the 'living' system. In other words, if the world is shared on the basis of a stable, dynamic equilibrium of health, the pain function initiated by some painful event will follow a physiological course. The onset and unfolding of the pain function then represent a healthy reaction of the body inhabited by a healthy individual. In the unhealthy individual, however, the existing unstable equilibrium will lead to a pathological pain reaction in which the course of the pain function becomes stagnant; 'chronic' pain results as a manifestation of the unhealthy state and is essentially unrelated to the initiating painful event (pp. 1390–1391).

Frith [36] eluded to health being an attractor this way:

> Let us suppose that "health" is an attractor (probably chaotic) which must coexist with "death" (a fixed point attractor) in some multidimensional phase space which we do not need to even try to imagine, but which could be given a schematic representation as in figure 2. "Illness" would then be a state in which the system had got out of the health attractor because of infection, injury, or whatever. What is the prognosis? Most ill states lead to recovery (even without medical attention). That is a consequence of health being an attractor state, which is in turn an inevitable consequence of evolution: those "fittest" species which prosper have robust healthy states. Thus, we could picture ill states

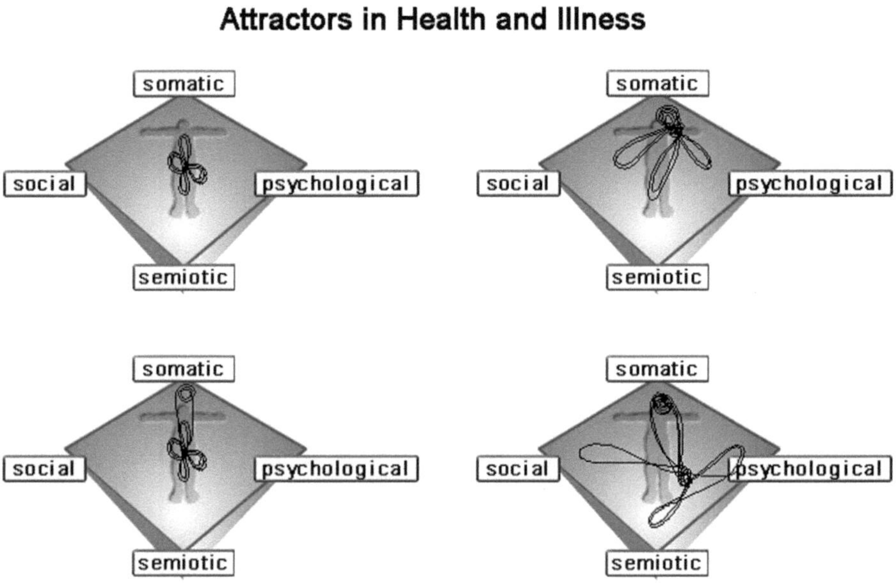

Fig. 15.7 Projection of health and illness over time in a phase space diagram. When the basin of attraction remains around the centre, the patient experiences health (*top left*), an acute biomedical illness shifts the attractor temporarily to the somatic corner (*bottom left*), a patient with a usually stable chronic illness will have a stable state off the balance centre, typically towards the somatic corner, associated with occasional shifts when the consequences of the illness result in distressing limitations (*top right*), and a chaotic attractor with two predominant states—in the somatic and psycho-semiotic corners—in a patient with "psychosomatic illness" (*bottom right*)

> A and B in figure 2 as being attracted back to health along the paths indicated. They probably correspond to the kind of illness I referred to at the start of this article, where medical treatment affects mainly the rate, rather than the fact, of recovery. Furthermore, the return to health is by a fairly predictable path, even if the healthy state is itself chaotic (p. 1567).

15.4 Implications for Health Care

15.4.1 Implications for Health Professionals

The somato-psycho-socio-semiotic model of health shifts the emphasis of health care back on the *person*. The model emphasises the interrelationships between the somatic, psychological, social, and sense-making dimensions of the human experience. As such, it dispels the fallacy of objectiveness "inherent in modern medical thinking and acting". This point is no more obvious than in the transformation of "meaning" of the term "disease"—its original subjective origin being "in dis-ease" is now understood as the objective state of "unequivocal pathology".

At this point, one should reflect on the historical origins of medicine. Health care first emerged in hunter and gatherer societies in response to the experience that caring for the sick and injured improved the group's survival. As caring for the sick became more sophisticated, the role of the shaman/medicine man evolved to include three functions—curing the sick, directing communal sacrifice, and escorting the dead to another world, functions that "fragmented" into three professions: the doctor, the priest, and the undertaker [37].

Although these functions are separated in modern society, our patient's still *intuitively* regard them as inherent in their healing relationship with *their personal* doctor. Medicines' foundations, and thus its abilities to predictably alter the trajectory of a patient's life journey, have not changed. The doctor's and every other health professionals' role remains unchanged—

to cure sometimes, to relieve often, to comfort always.[4] Despite our "ability to cure" a small number of diseases,[5] our main tool remains *personalised care*, care for the existential experience of a patient's illness and dis-ease experience [18, 38–40].

15.4.2 Implications for Health Care Services

People's subjective health, illness, and disease experience and individualistic responses to these experiences are the antithesis to the objective reductionist project of "conquering disease" which has arisen since Descartes "successfully" split the body from the mind. It should by now be clear that the quest for objectification and certainty are the great illusion of the twentieth century [41]. The rationalist project of biomedicine has led to a "medical industry" paradigm. Suffering without the objective identification of a disease has lost its legitimacy, affecting the patient and the doctor alike. Objectivity has been coupled to accountability, resulting in reimbursement for medical services in many countries being linked to defined discrete disease entities [42]. The current state is clearly not only unsustainable, but harmful and demeaning to our patients.

Objectification, i.e. the overwhelming focus on diseases, distorts the purpose and the function of *health care* delivery. The consequences of the objectified disease focus have been summarised by Barbara Starfield [43]: *diseases* (1) *are professional constructs,* (2) *can be and are artificially created to suit special interests* [with the peculiar outcome that] *the sum of deaths attributed to diseases exceeds the number of deaths,* (3) *do not exist in isolation from other diseases and are, therefore, not an independent representation of illness, and* (4) *are but one manifestation of ill health.*

[4] Originally attributed to Hippocrates.

[5] Many infectious diseases (tuberculosis and malaria however remain a major cause of mortality), vaccination preventable diseases, cataracts, joint replacement, and most atrial and ventricular septal defects.

All involved in healthcare need to refocus on patient need resulting from his/her health and illness experience. Deteriorating illness is more often than not an environmental problem, be it the physical environment, the lack of social support or unresolved concern and grief (ref to Martin, Chap. 27).

Chapter 47 will take up these themes in a discussion about patient-centred health system reform, the transformative step required to honour our patients trust in the healing professions.

References

1. Fugelli P. The healing dance. Why society and medicine need the general practitioner. 2008. http://www.racgp.org.au/Content/NavigationMenu/News/Conferencesandevents/asc20071/Abstracts07/Specialpresentations/ASC07Fugelli.pdf.
2. Moes M. Plato's dialogue form and the care of the soul. New York: Peter Lang Press; 2000.
3. Moes M. Plato's conception of the relationship between moral philosophy and medicine. Perspect Biol Med. 2001;44(3):353–67.
4. Rosenberg CE. The tyranny of diagnosis: specific entities and individual experience. Milbank Q. 2002;80(2):237–60.
5. WHO. Ottawa charter for health promotion. First International Conference on Health Promotion. Ottawa, 21 Nov 1986. WHO/HPR/HEP/95.1. http://www.who.int/hpr/NPH/docs/ottawa_charter_hp.pdf.
6. Husserl E. The Basic Problems of Phenomenology: From the Lectures, Winter Semester, 1910–1911 Dordrecht, The Netherlands: Springer; 2006.
7. WHO. Declaration of Alma-Ata. International Conference on Primary Health Care, Alma-Ata, USSR, 6-12 September 1978. Geneva: World Health Organisation 1978.
8. Parsons T. The Social System. Glencoe: Free Press; 1951.
9. Dubos R. The Mirage of Health. London: Allen Unwin; 1960.
10. Maslow HA. Toward a Psychology of Being. New York: Van Nostrand Reinhold Company; 1968.
11. Tissue T. Another look at self-rated health among the elderly. J Gerontol. 1972;27(1):91–4.
12. Kehlman S. The Social Nature of the Denition Problem in Health. Int J Health Serv. 1975 5:625–42.
13. Illich I. Limits to Medicine. Medical Nemesis: The Expropriation of Health. London: Marion Boyars Book; 1976.
14. Antonovsky A. Health, Stress and Coping. San Francisco: Jossey-Bass; 1979.

15. Reid J. Body, land and spirit. St Lucia: Queensland University Press; 1984.
16. Seedhouse D. Health: The Foundations for Achievement. New York: John Wiley Sons; 1986.
17. Ingstad B, Fugelli P. "Our Health Was Better in the Time of Queen Elizabeth": The Importance of Land to the Health Perception of the Botswana San. In: Hitchcock RK, Ikeya K, Lee RB, Biesele M, editors. Updating the San: Image and Reality of an African People in the 21st Century (Senri Ethnological Studies No 70). Osaka: Senri; National Museum of Ethnology; 2006.
18. Sturmberg JP. The Foundations of Primary Care. Daring to be Different. Oxford San Francisco: Radcliffe Medical Press; 2007.
19. Jylhä M. What is self-rated health and why does it predict mortality? Towards a unified conceptual model. Soc Sci Med. 2009;69(3):307–16.
20. Idler EL, Benyamini Y. Self-rated health and mortality: a review of twenty-seven community studies. J Health Soc Behav. 1997;38(1):21–37.
21. Baron R. An introduction to medical phenomenology: i can't hear you while i'm listening. Ann Intern Med. 1985;103(4):606–11.
22. Fugelli P. Clinical practice: between Aristotle and Cochrane. Schweiz Med Wochenschr. 1998;128:184–8.
23. Pellegrino E, Thomasma D. A philosophical basis of medical practice. Towards a philosophy and ethic of the healing professions. New York: Oxford University Press; 1981.
24. Lewis S. Exploring the biological meaning of disease and health. 2003. http://sites.google.com/site/sjlewis55/presentations/vienna2003. Last accessed 14 May 11.
25. Engel GL. The need for a new medical model: a challenge for biomedicine. Science. 1977;196:129–36.
26. Riedl R: Die Spaltung des Weltbildes. Berlin-Hamburg, Parey, 1985.
27. Feigl W, Bonet EM. Systemtheorie in der Medizin und Biologie. Wien Med Wochenschr. 1989;139(5):87–91.
28. Uexküll T, Paul HG. The mind-body problem in medicine. Advances J Inst Adv Health. 1986;3(4):158–74.
29. Pauli HG, White KL, McWhinney IR. Medical education, research, and scientific thinking in the 21st century (part two of three). Educ Health (Abingdon). 2000;13(2):165–72.
30. Sturmberg JP, Martin CM, Moes M. Health at the centre of health systems reform - how philosophy can inform policy. Perspect Biol Med. 2010;53(3):341–56.
31. Bentzen N. WONCA international glossary for general practice/family medicine. Fam Pract. 1995;12(3):341–69.
32. Toye F, Barlow J, Wright C, Lamb S. Personal meanings in the construction of need for total knee replacement. Soc Sci Med. 2006;63(1):43–53.
33. Cannon WB. Stresses and strains of homeostasis. Am J Med Sci. 1935;189:1–14.
34. Antonovsky A. The structure and properties of the sense of coherence scale. Soc Sci Med. 1993;36(6):725–33.
35. Engelbart HJ, Vrancken MAE. Chronic pain form the perspective of health: a view based on systems theory. Soc Sci Med. 1984;19(13):1383–92.
36. Frith WJ. Chaos - predicting the unpredictable. BMJ. 1991;303(6817):1565–8.
37. May W. The physician's covenant: images of the healer in medical ethics. Philadelphia, PA: The Westminster Press; 1983.
38. McWhinney IR. A textbook of family medicine. New York: Oxford University Press; 1989.
39. Scott JG, Cohen D, DiCicco-Bloom B, Miller WL, Stange KC, Crabtree BF. Understanding healing relationships in primary care. Ann Fam Med. 2008;6(4):315–22.
40. Suchman AL. A new theoretical foundation for relationship-centered care. Complex responsive processes of relating. J Gen Intern Med. 2006;21 Suppl 1:S40–4.
41. Sturmberg JP. The illusion of certainty – a deluded perception? J Eval Clin Pract. 2011;17(3):507–10.
42. Rosenberg CE. Disease in history: frames and framers. Milbank Q. 1989;67 Suppl 1:1–15.
43. Starfield B. Diseases, comorbidity, and multimorbidity in primary care. 2011. http://ocw.jhsph.edu/courses/starfieldcourse/PDFs/Starfield08_Diseases,%20Co-% & % Multimorbidity % FINALSlides. pdf. Last accessed 2 May 2011.

16 Understanding Health By Building Better Bio-Medical Models

Stephen Lewis

Current medical thinking is the product of many years of intellectual evolution with influences coming from a number of different sources. As noted earlier (Chap. 13), medicine has passed through stages comparable with the theological and metaphysical (or abstract) stages proposed by Comte. Now, medicine very much exists within a scientific (or positive) framework. In passing through different phases in its development, medicine has been subject to different modes of thinking. These have coloured the explanations and descriptions of what was being experienced by suffering individuals. Keating and Cambrosio [1] have proposed that such ways of thinking—that is, medical models—play a key part in all clinical thinking. They go so far as to suggest that '*... the object of medicine is not the body* per se *but, rather, models of the body*' [1]. The human body is not necessarily seen as it really is but rather as it appears in terms of the conceptual models that have emerged over time. As a consequence, these models form intellectual frameworks within which one's professional duties are conducted. It is not necessarily the case that the models by which many work are being deliberately or explicitly followed. Rather, a mental image or impression builds up tacitly over time as a result of a number of different influences. It is important to continually strive for the development of the best models possible and not to rely upon some traditional way of thinking merely because it appears to work.

One's model of the human organism can act as a significant guide to how diagnosis, treatment, etc., are approached. Furthermore, such models influence how the notions of 'health' and 'disease' are understood. It may even be argued, in keeping with Keating and Cambrosio [1], that it is models of the body that get treated rather than suffering persons. For example, until the nineteenth century, the prevailing model of the body in Western medicine was based upon the ancient notion of humouralism. How well or unwell one's patient felt was thought to be the product of the way in which four supposed bodily humours—black bile, yellow bile, phlegm and blood—were in proportion to each other. Therapies and treatments of questionable merit were delivered in accordance with how various observations were interpreted in terms of humoural theory. If a patient's ailment was deemed to be related to an excess of the humour blood, for example, this excess was alleviated by the process of blood-letting. The effects of any anaemia that may have resulted from this process seem to have gone largely unnoticed.

16.1 Thinking More Fully

So far, although largely focused upon what is known of the body's anatomy and physiology, no medical model has drawn fully on biological

S. Lewis, B.Sc., M.A., Ph.D. (✉)
Department of Biological Sciences, University of Chester, Chester, UK
e-mail: stevelewis55@gmail.com

J.P. Sturmberg and C.M. Martin (eds.), *Handbook of Systems and Complexity in Health*,
DOI 10.1007/978-1-4614-4998-0_16, © Springer Science+Business Media New York 2013

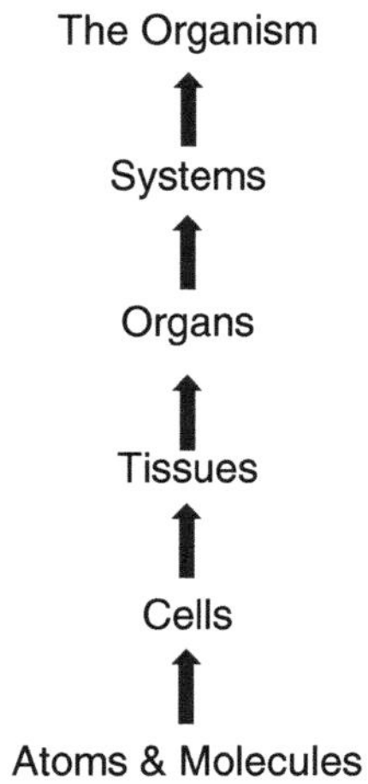

Fig. 16.1 The **'ladder of levels'** model

theory in its broadest sense. There is, for example, no reference to the theory of evolution. This is a significant omission. The theory of evolution is the foundation upon which all biology is now based. It is to biology what the illusive 'Grand Unifying Theory' is to physics. Famously, Theodosius Dobzhansky [2] stated that '*nothing in biology makes sense, except in the light of evolution*'. For any medically useful model to be said to have a rigorously developed biological basis, it too must include something of the evolutionary foundations of biology. Echoing Dobzhansky, Nesse and Williams [3, 4] have suggested that '*nothing in medicine makes sense except in the light of evolution*'. Nesse and colleagues have been at pains to emphasise that an understanding of evolutionary theory should form part of medical thinking about the conditions with which it has to deal. However, despite concerted encouragement [5–8], medical schools have been reluctant to adopt the teaching of evolutionary theory within their already crowded curricula.

This may not be entirely surprising. The emphasis in modern evolutionary biology is to think largely in terms of groups and populations. Much of the focus has been on how the genetic characteristics of populations change over time with particular attention given to how gene frequencies change. This, perhaps understandably, has diverted attention away from the individual. It is with the process of evolving that evolutionary biology is primarily concerned. Individual organisms are evolved, not evolving, objects; they are steps along the way between the past and the future. This does not fit well with medicine which is focused primarily on the individual. There is clearly a gap that needs to be bridged. There may yet be an alternative approach. Rather than a detailed technical understanding of evolutionary theory, an appreciation of the fundamental characteristics of biological survival as it relates to the individual as an organism within an evolutionary context may be more appropriate.

16.2 The 'Ladder of Levels' Model

To understand the notions of health and disease, it is not unreasonable to first try to understand the organism to which these terms are applied—typically using models. One frequently used in the introductory chapters of modern textbooks of anatomy and physiology used by students of clinical and non-clinical subjects alike may be described as the 'ladder of levels' model. It typically takes the form of a list of the structural levels which make up the human body. These range from the atomic and molecular level, via cellular, tissue and organ levels to that of the whole organism (Fig. 16.1).

In textbooks of anatomy and physiology, this model is used to introduce and demarcate what will be discussed in subsequent chapters. Set out by physiological system, each chapter typically focuses on the organs that compose these systems, their structure and what they do. The organism level is usually as far up in the hierarchy that such books are willing to go.

While a 'ladder of levels' is an interesting and, in some cases, useful model, it should be subjected to a more thorough scrutiny than it typically attracts. This is important given its prominence in standard textbooks and the influence it is, therefore, likely to have on understanding not only the organism as a whole but also what constitutes 'health' and 'disease'.

There is also something rigidly linear and one-dimensional about the ladder. Each rung is a supposedly different level of organisation. However, in terms of the size of the things considered, the rungs are not equidistant. Scale is something that is rarely emphasised and a sense of the number of components at each level is largely missing. Also, the rungs of the ladder are not all of the same kind. Most significantly, between organ level and that of the organism, is often a level entitled 'System'—or sometimes 'Organ System' divided in accordance with the way in which the rest of the chapters are divided. The notion of a 'system' is not primarily a structural one. Instead, it is something quite abstract: one based upon notions of commonality of purpose and of cooperation between constituent organs. One is not denying that there are organs arranged in particular ways which allow them to work together. What is at issue, however, is that while the other levels of the ladder represent structural hierarchies, a system is not of the same kind. In the body, the separation of organs into distinct systems is not necessarily obvious. Their association into a system is mainly via their mode of operation. Furthermore, different textbooks are often at variance over how to make these divisions into systems and even over the names the different systems should be given. Often organs do a variety of things that can be said to contribute to more than one system. The kidneys, for example, are best known as organs producing urine but this is really a combination of waste excretion and water balance. In addition, they have key roles to play in acid–base balance and blood pressure regulation amongst other things. Simply belonging to the urinary—that is, 'urine-producing'—system underplays and may even detract from their wider importance. More generally, the nerves and blood vessels which textbooks typically consider within their respective chapters actually form an integral part of the structure and operation of all organs. Overall, as a result of the typical way of describing the body, a sense of bodily integration is easily lost. Indeed, the approach is one of deliberate de-construction.

16.3 Levels and Qualities

It may be necessary for practical purposes to divide the body in this way but it does not necessarily give the best impression of what it is to be a living entity experiencing, amongst other things, 'health' and 'disease'. To say that an organism is more than the sum of its parts is not wrong. However, at the same time, this is perhaps a somewhat hackneyed adage. It may be better to suggest that the whole is not just more than the sum of its parts but that the whole is qualitatively different to the sum (or any other combination) of its parts.

Instead of trying to understand a phenomenon at the lowest level, it may be more accurate to suggest that certain phenomena belong only to given levels—that is, only to specific rungs on the ladder. This includes that of the organism as a whole. Instead of taking a vertical approach and applying the 'ladder of levels' to the whole organism and separately to each system, one might also look horizontally at each level across all systems. Particular phenomena may only be evident at specific levels. Furthermore, it may be that some things are better understood at just one level. Just as 'the time' is not a property of the structure of a clock but is what is shown on the clock face, so too 'health' and 'disease' may be qualities which belong only to a particular level. What is going on at other levels is, of course, important but it is not necessarily of the same kind as what presents via, say, the organism as a whole.

16.4 The 'Humpty Dumpty Problem'

While the 'ladder of levels' model can be useful, at the same time it suffers from what might be called the 'Humpty Dumpty Problem'. As presented in textbooks of anatomy and physiology, the 'ladder of levels' model is used as a prelude to the ensuing de-construction of the human body into its component parts. At the end of this process, after having described the various systems and their component organs and processes, there is no form of or attempt at re-construction. Humpty isn't put back together again. The level of the whole organism—the level of the person—is not described. There is perhaps a tacit assumption that somehow the reader knows what an organism is before they even start to read because they are one, living in a world full of others of the same kind. However, this omission of some form of reconstruction may point to a serious shortcoming. The intention of de-constructing the organism is to understand it better. If that understanding cannot be used to make a more meaningful reconstruction, it is questionable whether the aims of that de-construction were fully realised.

16.5 No Theory of the Organism

What is very much missing from the 'ladder of levels' model is a theory of the organism as a whole entity [9]. Indeed, currently there is nothing in either biology or medicine that might be described as a 'theory of the organism' as such. Yet, it is at this level that each human being primarily exists. It is at this level that human beings interact with each other and with the world in general. Furthermore, it is at this level that 'health' and 'disease' have a bearing upon how life is lived out [10].

Not only is Humpty not put back together again but it is quite likely that, given the present approach and the lack of a theory of the organism, there is no way of knowing how to even go about doing this. Where all the bits belong is well known as are many of the associated physiological and biochemical processes. However, this is only part of what is needed to understand human beings as whole integrated entities. Without a theory of the whole, a knowledge of the components and merely what they do is insufficient.

16.6 Information Flows

Importantly, human beings are integrated entities, systems within which there is a constant information flow. The integration of the whole is reliant upon internal communication. Neural and chemical messages have a bearing on the organism beyond their measurable physical properties. Fast acting nerve impulses and slower acting hormonal and chemical signals act to integrate the operation of disparate organs. However, measuring the physical properties displayed by these signals is not the same as understanding the information content they contain or convey. What they bring about in terms of the effect on the organism as a whole is qualitatively different from their physical properties. For example, the cogs inside a clock may move in an understandable way. However, knowing this is not the same as understanding the information content that they convey. A single turn of one cog may signify the passing of a minute while a single turn of another seemingly identical cog may signify the passing of an hour. Indeed, as noted earlier, what happens inside a clock is quite different to the meaning that is conveyed at the clock face.

Within an organism, much of the information flow occurs below the level of consciousness. While the body is known to be self-regulating, this is understood largely in terms of changes in measurable physical parameters. However, there is another sense in which the organism might be described as being 'self- (or 'auto-)referential'. By this, one is not merely referring to conscious self-awareness. There is a range of internal feedback mechanisms and other responses which result from changes within the body about which one is never conscious. Those that do reach the level of consciousness do so as a variety of different experiences. Importantly, for the organism to remain viable, these frequently provoke behavioural responses which make good any internal physical deficiencies. For example, one becomes aware of the internal physiological state of dehydration via an experience of thirst and is prompted to drink. At an experiential level, this may slake the thirst and may also be associated with the pleasurable experience of taste. At a physiological level, the water balance of the body is restored. It is via the experience of thirst that the body ensures that it has an adequate fluid intake. As has been said in relation to food, '*[f]ruit tastes sweet, not nutritious*' [11]. What is necessary at a physical level is ensured via experiential surrogates. Feeling ill may be such a surrogate.

16.7 The Subjective and the Objective

In 1943, the medically trained philosopher, Georges Canguilhem (1904–1995) posed an important medical and philosophical puzzle [12]. He describes a man who was going home one day but as a result of, for example, a car accident or murder, died on the way. The precise details are not significant; what is important is what ensued from his sudden death. An autopsy was performed where it was found that the man had a tumour on one of his kidneys. In life, that man knew nothing of this condition; he had experienced no symptoms or ill effects associated with that lesion. Canguilhem asked whether that man should be considered as having had a disease or not. Put another way, should the man have been considered healthy even though there was a symptom-free lesion within?

In one sense, this is a story about how the man feels. He feels fine. If asked how he was, he would have said that he was well. He was able to go about his everyday life unimpeded until it was tragically cut short. In another sense, this story is both a corollary to the work of Morgagni (see Chap. 13) and a presaging of the modern problem of defining 'health' and 'disease'. Although the man feels well, he has within him a serious lesion. How should he be classified in a modern scientifically medical sense?

With modern Western medicine, there are many precautionary procedures which are now performed routinely. Various screenings are performed, not because one is in obvious need of medical assistance but because it is deemed wise to assess the physical state of certain groups of people for some reason. In the past, how one felt was the main and sometimes only reason for seeking assistance from others. If one felt a need within oneself which could not be met by retiring to one's bed for a period of time, then one called upon someone else for remedial help. The man in Canguilhem's puzzle may well have had his lesion spotted had he gone for one of today's routine screenings or check-ups. However, these were not available when Canguilhem posed the question. Many decades old, his question remains pertinent. Importantly, it demonstrates that there are two quite distinct aspects of the individual which need to be considered together. One is the lived-experience of the individual. The other is the physical nature of that individual. Where once disease was literally 'dis-ease' associated simply with personal experience [10], medical advances have realigned the disease focus (post-Morgagni) primarily onto the physical constitution of the patient and onto the search for a lesion, in particular.

16.8 Feeling Alive

How one feels—one's experience of oneself—is so 'second nature' that its importance, indeed its centrality to staying alive, is easily overlooked. In order to stay alive and to ensure a comfortable quality of life, a number of things must be managed simultaneously. One must not be too hot or too cold. One must not be too thirsty or too hungry. One's bowels and bladder should not become uncomfortably full. One must not be too tired or try to do what one's body—via the aches and pains experienced—is, in effect, telling one not to do. Each of these is an experience for which there is an underlying physical basis. The body's biochemical reactions have an optimal temperature range. This coincides with a range which is correspondingly experienced as 'comfortable'. The body also has an optimal state of hydration. Dehydration coincides with a state which is correspondingly experienced as thirst. A flow of substances derived from a sufficient provision of food is necessary for the maintenance of bodily structure and the provision of energy. Experiences of hunger and intestinal emptiness help ensure that a flow of these substances is maintained. One must not go beyond the daily limits that the body sets for itself. Thus, in order for the body to continue to operate within safe limits, a range of different phenomena are experienced when those limits begin to be exceeded. These, in turn, elicit different behavioural responses from the individual such as eating, drinking, etc. Here, resting may also be viewed as a behaviour.

16.9 Feeling Pain

The phenomena mentioned earlier are typical everyday experiences. To these may be added pain. One naturally avoids being in too much pain or discomfort. Yet, even though pain and discomfort are not uncommon, they are not typically included with the other ordinary experiences mentioned earlier. Nevertheless, pain provides important signals; surely as important as the likes of hunger and thirst. Indeed, to be unable to feel pain is to risk one's whole ability to survive [13].

Without medical help, the life expectancy of those born with congenital absence of pain is much shorter than that enjoyed by those who can experience pain. Even with medical help, the effects can be quite marked. It is easy to bite the inside of one's mouth during eating. Lips are frequently bitten and the tip of the tongue not uncommonly bitten off. Small cuts and grazes go unnoticed and particles damaging to the eye go without response. Thus, pain and discomfort should be seen as having important biological roles and not just as something unpleasant [13].

If the benefits of something as marked as pain can be easily overlooked, the benefits of being able to experience seemingly trivial discomforts and irritations are missed almost completely. For example, one frequently shifts one's body position in response to feelings that are barely noticeable. Indeed, this even occurs during sleep. Not to do so can prove highly damaging and has the potential to prove fatal. To remain constantly in the same position results in joint and muscle stiffness. If severe, this could progressively impede one's ability to move. Failure to move often enough can also lead to decubitus ulcers. In extreme cases, these can become infected leading to septicaemia and death.

The experience of those with leprosy further demonstrates the value of being self-aware (or auto-aware) in this way. Nerve damage in leprosy leads to loss of peripheral sensation. What are, to the unaffected person, minor cuts and grazes typically go unnoticed. This can prove highly damaging to the extent that the loss of fingers and toes typical of leprosy is not as a direct product of the infection but of the inadvertent damage resulting from the loss of peripheral sensation and the secondary infections which set in leading to auto-amputation.

Pain and discomfort are not valued for their potential benefits. Instead, perhaps unsurprisingly, they are simply dis-valued for their unpleasantness and avoided accordingly. Indeed, when in pain or discomfort, what is known intellectually about these experiences may be of little or no consolation, given the distress being felt.

The ability to experience pain and discomfort is perhaps a key physiological feature of being an organism. This is exemplified by the fact that, like any other physiological mechanism, it can sometimes operate in a way that is counterproductive for the individual concerned. Pain can become chronic and can, as a result, become the object of an individual's suffering, in and of itself. It can even continue to be experienced without apparent cause.

16.10 Feeling Ill

Other important, dis-valued experiences are those feelings associated with 'illness'. Illness is easier to define than either 'health' or 'disease' [14]. The word 'illness' now means what the word 'disease', when it literally meant 'dis-ease', used to mean. Thus, illness is a particular sense of unease. It is possible to describe illness as a subjective, personal experience; a feeling of being unwell. As such, illness is a characteristic of human life. Everybody has felt ill at some point or another. When somebody says that they are feeling ill, all can equate with that experience indirectly. However, being able to describe with precision what a given bout of illness—or of being 'unwell'—is like can prove very difficult. Associated with illness is a variety of feelings such as malaise, nausea, tiredness, headache, etc. Illness is not one specific thing or a particular set of named phenomena. Instead, it is a variable constellation of negatively valued inner experiences grouped under a single term.

The experience of illness is important in that it provokes remedial responses. Just as phenomena like hunger and thirst elicit important behavioural responses which ensure the survival of the individual, so too does the experience of illness. For what one considers as minor illness, all that is usually required is rest and warmth. Brought on by minor infections, for example, this allows the immune system greater call on the body's otherwise shared resources in combating that infection.

Body temperature may be raised, quickening biochemical processes and providing a less hospitable environment for the infective agent. Simply going to bed may be enough until the feeling of illness passes. On other occasions, one may instinctively feel that one's illness is more serious than one can manage on one's own or that it will not pass off. It is then that the help of others may be sought.

16.11 The Silence of the Body

Many of the experiences one has of one's body are subtle. Furthermore, one typically responds to them before they become too severe. For example, one does not wait until one's hunger or thirst become overtly distressing before deciding to eat or drink. Such responses ensure that the operation of one's body continues virtually unnoticed. Indeed, for much of the time that one is going about one's daily business, one is largely unaware of one's own body and what is going on internally. In effect, the body only draws attention to itself when a need must be met. The French surgeon René Leriche (1879–1955) even went so far as to consider health as 'life lived in the silence of the organs' [12]. This was, in effect, how life appeared to the man in Canguilhem's problem. There were no 'noises' requiring his attention. As a result, he went about his daily life without distraction.

Strictly speaking, one might argue that Leriche was defining health as an absence of an experience of illness and that he was only focusing upon experiential aspects. Here there is little or no direct reference to the inner workings of the body. From this, it would reasonably follow that feeling 'well' is not so much having specific positively valued feelings but having a silent body: having no negatively valued feelings of illness, pain or discomfort.

Thus, one may discern two distinct dimensions: the 'lived' and the modern (scientifically) medical. The 'lived' dimension is that of the individual who responds to the presence or otherwise of subjective experiences. The modern medical viewpoint gives particular attention to the physical condition of the patient seeking objectively discernible phenomena. It is important to make this distinction. It enables quite separate aspects of being alive to be recognised: that there are distinct physical and experiential dimensions to life. For a fuller understanding of 'health' and 'disease', both dimensions should be taken into account.

16.12 The Struggle for Life

Darwin recognised something that is now easily overlooked by those living in the relative comfort enjoyed in the developed world—that life is a struggle. The latter part of the subtitle of the book which launched the modern understanding of evolution—'On the Origin of Species by Means of Natural Selection, or the Preservation of Favoured Races in the Struggle for Life'—spells this out [15]. Darwin here, and throughout the book, draws attention to the fact that only some are 'preserved'; many perish. Where there are those who perish, there is also suffering. Nowhere is this more vivid than in the phrase 'survival of the fittest'. Darwin did not coin this phrase. Neither is it to be found in the earliest editions of 'On The Origin Of Species'. It began to appear from the fifth edition (1869) after Herbert Spencer (1820–1903) first used the expression in his 'Principles of Biology' (1864) [16]. Now the phrase has become inextricably linked with the notion of evolution.

The survival Spencer had in mind was that of the individual. This is important to remember. Writing in 1864, his was not the emphasis on populations or gene frequencies that has come to dominate modern evolutionary biology. Extending this idea, it is noticeable that even among those who do survive, there are some who fare better than others. Even among those who are surviving, there are those who will, from time to time, seek help from others—not least the medical profession. In this, they are, in effect, seeking help with various aspects of ensuring their survival and enhancing their biological fitness. It is this focus upon the individual from an evolutionary perspective that may well help forge a greater integration of biological and medical thinking [9]. Here, the theoretical framework that allows biology to make sense [2] and medicine's focus on the individual meet [10]. For an improved and truly biomedical model, the notion of individual survival—the personal 'struggle for life'—is central.

16.13 Conclusion

Currently, the survival perspective is a direct but poorly explored bridge between biology and medicine. As described earlier, it is how the individual feels that is of particular importance in how life is lived out. One's experience of one's own body—whether the organs are silent or 'noisy'—determines how one goes about one's daily life. In a lived context, therefore, one's experience of oneself is perhaps more important than biological science tends to realise. Traditionally, biological science has not given much emphasis to the inner feelings of the individual. These are, after all, inaccessible to direct investigation. Yet, experiences such as illness and the behavioural changes that ensue from this have a direct bearing upon individual survival chances. At the same time, feelings of illness, pain and discomfort form the starting point for medical involvement. The help offered by the medical and allied professions is essentially help with individual survival. In so doing, the focus is directed largely on the physical state of the body. However, when seeking to build better biomedical models, it is not only the physical state of individuals that matters. It may well be that the notion of survival, with its physical and experiential dimensions, is what bridges biology and medicine better and how the notion of health may come to be better understood.

Acknowledgments I would like to thank Annette Lewis for her help in the preparation of the manuscript of this chapter.

References

1. Keating P, Cambrosio A. Biomedical platforms - realigning the normal and the pathological in late-twentieth-century medicine. Cambridge, MA: MIT; 2003.
2. Dobzhansky T. Nothing in biology makes sense except in the light of evolution. Am Biol Teach. 1973;35:125–9.
3. Nesse R, Williams G. Why we get sick - the new science of Darwinian medicine. New York: Times Books; 1994.
4. Nesse R, Williams G. Evolution and healing - the new science of Darwinian medicine. London: Weidenfeld and Nicolson; 1995.
5. Nesse R, Williams G. Evolutionary biology in the medical curriculum - what every physician should know. Bioscience. 1997;47:664–6.
6. Nesse R. Medicine's missing basic science. The New Physician (December); 2001. p. 8–10.
7. Nesse R, Schiffman J. Evolutionary biology in the medical school curriculum. Bioscience. 2003;53(6):585–7.
8. Nesse R, Stearns S, Omenn G. Medicine needs evolution. Science. 2006;311:1071–3.
9. Lewis S. Seeking a new biomedical model. How evolutionary biology may contribute. J Eval Clin Pract. 2009;15(4):745–8.
10. Lewis S. Conceptual models of the human organism: towards a new biomedical understanding of the individual. In: Komorowska M, Olsztynska-Janus S, editors. Biomedical engineering, trends, research and technologies. Vienna: Intech; 2011.
11. Cronin H. The ant and the peacock. Cambridge: Cambridge University Press; 1991.
12. Canguilhem G. On the normal and the pathological (C Fawcett, Trans.). Dordrecht: D Reidel Publishing Company; 1966 (trans 1978).
13. Melzack R, Wall P. The challenge of pain. 2nd ed. Harmondsworth: Penguin; 1988.
14. Marinker M. Why make people patients? J Med Ethics. 1975;1:81–4.
15. Darwin, CR. On the origin of species by means of natural selection, or the preservation of favoured races in the struggle for life. London: John Murray; 1859. http://darwin-online.org.uk/contents.html#origin. Accessed 11 Nov 2011.
16. Spencer H. Principles of biology. London: Williams and Norgate; 1864.

17 Health: A Systems- and Complexity-Based Definition

Joachim P. Sturmberg, Stefan Topolski, and Stephen Lewis

The previous three chapters have explored the dimensions of health through different lenses—the multiplicative agents of one's health potential, the subjective experience arising from sense-making, and a functional state that allows for survival and reproduction—the struggle for life. Together, they provide the basis for a congruent systems-based definition of *health.*

> Human health is a *balanced state* between *physical, emotional, social* and *cognitive/sense-making* domains. Within any *local environmental context*, a health state exists within a *multidimensional phase space* of *physical integrity, functional performance and subjective experience* producing an *entropic state* most consistent with *viability*.

J.P. Sturmberg (✉)
Department of General Practice, Monash University, Melbourne, VIC, Australia

The Newcastle University, Newcastle, PO Box 3010, Wamberal,, NSW, 2260, Australia
e-mail: jp.sturmberg@gmail.com

S. Topolski
University of Massachusetts Medical School, 1105 Mohawk Trail, Shelburne Falls, MA, USA
e-mail: public@cottagemed.org

S. Lewis
Department of Biological Sciences, University of Chester, Chester, UK
e-mail: stevelewis55@gmail.com

17.1 Definitions May Not Necessarily Be Complete But They Can Always Be Useful

Definitions are an attempt to provide a *MODEL* within which to frame an issue. Being a *FRAME*, a definition provides a *BROAD BOUNDARY*, and a boundary is almost always more important than the details entailed within.

17.2 What Makes This Definition Useful?

This definition describes health as bounded by anatomy (or physical integrity), function (or physiology) and experience (interpersonal, social and environmental). The interactions of this interplay can be expressed as the entropy of a person's system.

Health is not dependent on one dimension alone; rather health is present in many different "shapes" given by the near infinite possible combinations of these three dimensions, and in many different "amounts". Overall health, measured in "the least" entropy, is bounded within different shapes and different amounts. When the total entropy exceeds a critical level, a person enters an illness state. When a person reaches his personal maximal entropy, i.e. has used up all of their health potential, death occurs (Fig. 17.1).

J.P. Sturmberg and C.M. Martin (eds.), *Handbook of Systems and Complexity in Health*, DOI 10.1007/978-1-4614-4998-0_17, © Springer Science+Business Media New York 2013

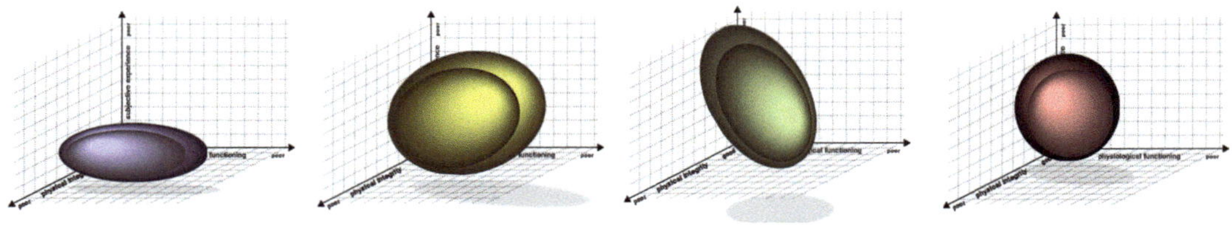

Fig. 17.1 The different "shapes" and "amounts" of health. The figures illustrate how health can be made up of "infinite" combinations of the three dimensions of health—anatomy (or physical integrity), function (or physiology) and experience (interpersonal, social and environmental). The total amount of health, measured as "least entropy", is contained within the smaller shape, the *larger shape* indicates the amount of entropy that is associated with illness but still compatible with life. All states outside the larger shape are not compatible with life. Note that the entropy can increase to that level with a significant deterioration in any one direction, i.e. loss of anatomical structure, loss of physiological function and/or loss of experience

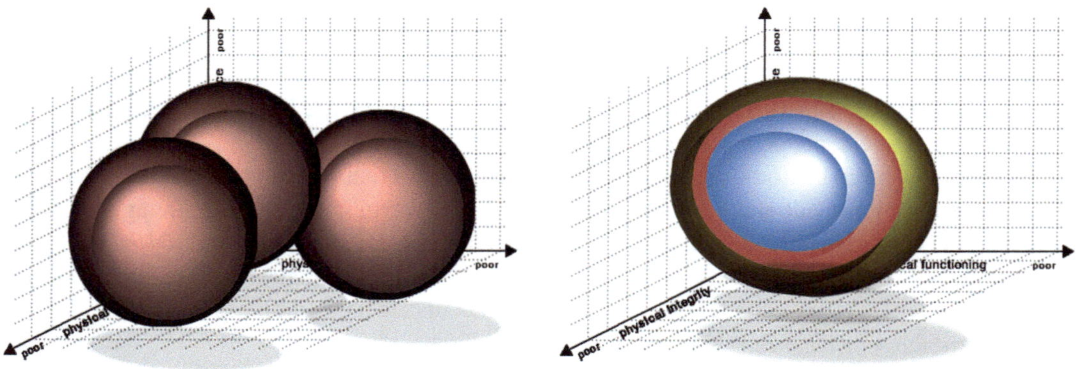

Fig. 17.2 The "same" shape and amount of health can exist as a result of different combinations of the three dimensions of health, and thus is represented in different spaces of a phase spaces (*left*). On the other hand, the "same" shape of health can be associated with different "amounts" of health

Health can have the same shape and amount but nevertheless be constituted of different combinations of dimensions, and health can have the same shape but have different amounts (Fig. 17.2).

These boundaries make *health* a far greater concern by taking account of the many factors that patients talk about when describing their health and complaining about their illness experiences.

It also recognises that we all have a limited amount of health (sufficient for survival, before becoming sick and die), and that—even small—disturbances in any of the three dimensions can substantially increase our entropy, resulting in premature morbidity and mortality, accounting for our many different individual health and illness trajectories.

17.3 How Is This Definition Justifiable?

Quantitative and qualitative findings support the notion that health, like illness and disease, is a subjective state. People's experiences and interpretations integrate physical, social, emotional and cognitive events in time and across the life span. Health is a *complex adaptive state* in so far as the inner workings at a psycho-neuro-immunological level maintain a certain homeostasis for self-sustaining physiological function. This in turn allows for the development and maintenance of a person's health potential, providing the robustness to succeed in the physical and social environment and to successfully reproduce and raise offspring. The experience of health is maintained over the life span, even though ageing is a slow process of *adaptive decline*. Health is maintained by adapting certain failures (technically referred to as diseases) into one's self in order to maintain one's role in society until such point where one's decline is no longer compatible with life. This state can be reached without having sustained any discrete diseases, can be accelerated by disease processes or can be abrupt in the case of an accident.

From a biological perspective, the *utility* of health is *the struggle for life* itself, the ability to constantly *adapt* to changing internal and external challenges, and being able to sustain oneself within the *bounds* of one's physical and cultural environment with dignity and social connectedness.

17.4 Consequences of This Definition

Health being *bounded* by personal *biological*, *social* and *environmental constraints* allows for a wider appreciation of people's health behaviours and life choices. The notion of health being *bounded* embraces the importance of some issues that in different cultures are viewed as not important or existentially important "for being healthy". The sense of belonging/loss of belonging to their land for most indigenous and agrarian people is a notable example.

The *complex adaptive state* of health is supported by psycho-neuro-immunological feedback loops that contribute to the maintenance of physiological homeostasis and overall resilience along the emerging life trajectory. Adaptation slows the rise of entropy to levels associated with morbidity and mortality.

This is the first scientific definition that predicts objective outcomes. Such a systemic health definition has fundamental consequences for healthcare planning, financing and delivery. In turn observed/measured health system/care outcomes must be appreciated as the result of the overall functioning of people's *bounded constraints*.

Thus, healthcare planning most importantly requires this global view, i.e. to take account of all the local and global constraints experienced by individuals and communities. Without a health maintaining physical and social environment, health care can ever only be a band aid intervention. In this context, people's views about their health and disease provide important *feedback* to understanding local health constraints, which in turn provide necessary information on appropriate resource allocation.

Healthcare delivery, at the grassroots level, then has a wider purview. *Health care* is a targeted perturbation of the "patient system" that facilitates the system "to do the necessary adaptive work" to achieve (self-)healing. This stands in stark contrast to the prevailing "healthcare" approach of targeted "curative interventions" done to the patient by others (known as healthcare providers). *Health care* becomes a participatory encounter between two experts who work together to achieve a common understanding of the patient's disease in the quest to succeed in *the struggle for life*, and[1] at times, this may entail the "passive" endurance of interventions aimed to maintain one's life.

[1] In general terms, systems do not have either/or behaviours, they usually show "and" behaviours of a mutually agreeable nature.

Part IV

Clinical Applications

Complexity sciences are increasingly applied to many different clinical problems ranging from the individual consultation and specific diseases through to the management of the frail elderly.

The chapters in this section have one common theme; they all highlight the highly interconnected nature of the patient and his/her environment, be it internally within the body, externally in their local environment, or in their relationship with their healthcare provider in the consultation. These chapters allude to the surprising impacts that the appreciation of interconnectedness can have on the "clinical picture", the "clinical approach", and the "clinical outcome".

Complexities of the Consultation

18

John G. Scott

18.1 Introduction

For the purposes of this discussion, it is important to define the clinical consultation. What is meant by this term is the interaction between a clinician and a patient in the privacy of the consultation room or examining room. That interaction may include family members or friends as well if they are present.

The purpose of this chapter is to further develop a theoretical model of the clinical consultation using complexity theory, informed particularly by the work of Ralph Stacey on complex responsive processes of relating [1]. This model will also incorporate a conceptual model of healing relationships developed by my colleagues and I [2] and the work of Ronald Epstein on mindfulness in medicine [3, 4]. Some familiarity with complexity theory on the part of the reader will be assumed, since other chapters in this volume will deal more specifically with complexity theory itself.

First, the literature will be reviewed, focusing on studies and ideas that have led to understanding the clinical consultation as a complex system. Considerable space will then be devoted to explicating Stacey's work on complex responsive processes of relating. A thorough understanding of his conceptual framework is essential to appreciating how it underlies all aspects of the consultation. A description of our healing relationship model and Epstein's principles of mindful practice will follow. Finally, all of these threads will be pulled together to form an integrated theory of the consultation as a set of patterns of complex responsive processes of relating that are analogous to, but not identical to, computer models of complex adaptive systems. Clinical examples will be used to illustrate the potential usefulness of the theory to practitioners in the real world. The chapter will conclude with some reflections on how this theory of the consultation could improve both patient experience and health outcomes.

18.1.1 History

The clinical consultation and the clinician–patient relationship have been the subject of much attention from many disciplines over the years. A complete review of this voluminous literature is beyond the scope of this chapter. The understanding of the consultation as a complex system, however, has been built upon work by a number of pioneer researchers and thinkers. A review of some of this seminal work follows below.

The traditional view of the clinical consultation, a view that is still taught in most medical

J.G. Scott, M.D., Ph.D. (✉)
Robert Wood Johnson Medical School, University of Medicine and Dentistry of New Jersey,
1 Robert Wood Johnson Place, New Brunswick, NJ, USA

Northeastern Vermont Regional Hospital,
195 Industrial Parkway, Lyndonville, VT 05851, USA
e-mail: john.glenn.scott@gmail.com

J.P. Sturmberg and C.M. Martin (eds.), *Handbook of Systems and Complexity in Health*,
DOI 10.1007/978-1-4614-4998-0_18, © Springer Science+Business Media New York 2013

schools [5], is based on a form of Cartesianism, the idea that the body is a machine and that medical professionals are technicians whose job it is to repair that machine [6]. In this model, the purpose of the clinical consultation is to diagnose (name the disease) and to provide evidence-based treatment, the most reliable evidence being that from randomized controlled clinical trials. Interviewing and relational skills are used to elicit information from the patient that will be useful in diagnosis and treatment. Although this model has been discredited and challenged from many quarters, it is still the dominant paradigm in most of medicine.

Perhaps, the first person to challenge this physician-centered linear model was Michael Balint in the 1950s [7]. Balint recognized that the personality of the clinician was part of the patient's treatment. Although still physician-centered, Balint's model called attention to the importance of the quality of the clinician–patient interaction in the treatment of the patient.

Another milestone in understanding the clinical consultation came from the work of George Engel in the late 1970s [8]. Engel challenged the prevailing biomedical model of illness and proposed that human illness had biological, psychological, and social dimensions, and that all three dimensions were important in treating human illness. His biopsychosocial model was a step forward in appreciating the complexity of human illness, but was still physician-centered and failed to address the complexity of the clinician–patient relationship.

In the 1990s, Moira Stewart and colleagues developed the idea of patient-centered care, and demonstrated that clinicians who practiced patient-centered care had better outcomes [9]. Patient-centered care was defined as care that "(a) explores the patients' main reason for the visit, concerns, and need for information; (b) seeks an integrated understanding of the patients' world—that is, their whole person, emotional needs, and life issues; (c) finds common ground on what the problem is and mutually agrees on management; (d) enhances prevention and health promotion; and (e) enhances the continuing relationship between the patient and the doctor" [10].

This focus on the whole rather than the parts and the emphasis on the clinician–patient relationship further laid the groundwork for understanding the clinical consultation as a complex system.

A final development, which grew out of the patient-centered care model, was relationship-centered care, which was initially articulated in a report by the Pew–Fetzer task force [11]. The authors maintained that "Practitioners' relationships with their patients, their patients' communities, and other health care practitioners are central to health care and are the vehicle for putting into action a paradigm of health that integrates caring, healing, and community" ([11], page 24).

Relationship-centered care focuses on neither clinician nor patient, but on the interaction itself and sets the stage for introducing complexity theory as a useful lens through which to view the clinical consultation. The connection between complexity theory and relationship-centered care was made explicit by Anthony Suchman, who used Ralph Stacey's ideas about complex responsive processes of relating as a theoretical basis for understanding relationship-centered care [12]. There will be much more to say later about the importance of Stacey's ideas in understanding the clinical consultation as analogous to a complex adaptive system.

There have been a number of other authors who have used complexity theory as a way of understanding the clinical consultation [13–21].

Innes et al. [15] and Hassey [13] exemplify one approach. The clinician and patient are seen as agents who interact with other external agents such as the lawyer (threat of lawsuit), the media (health beliefs), the statistician (evidence-based medicine), and a computer (electronic medical record). These externalities have agency through the individuals that are physically present. Through the interaction of agents, a complex system has the capacity to show emergent self-organization; that is, patterns or structures that cannot be predicted from the behavior of the individual agents. If the consultation can be considered a complex system, then it also has the potential for self-organization. This implies that outcomes of the consultation may be under the control of neither the clinician nor the patient, and that those outcomes have the potential to be novel and/or unexpected. Another property of the consultation as a complex system is non-linearity; that is, small changes may have profound effects, while large changes may fail to change existing patterns. The consultation as a complex system also shows coevolution; that is, all the agents affect each other in an iterative fashion such that behavior of the agents and thus the system changes over time. Innes et al. and Hassey use a Stacey matrix [22] to characterize the type of consultation that may exhibit the characteristics of complex systems at the edge of chaos. Not all consultations, they maintain, have these characteristics, especially those where there is certainty about the correct course of action and high agreement of the agents involved. Innes et al. briefly discuss Stacey's complex responsive processes of relating, but consider these and the themes generated by these processes as other independent agents that participate in the complex adaptive system of the consultation.

Aita et al. [14] take a slightly different approach. In a qualitative analysis of ethnographic observation of clinician–patient encounters, they found that the ability of clinicians to practice patient-centered medicine was affected by multiple factors, including physician characteristics, patient characteristics, practice culture, and community culture. Using a complexity science model, they viewed these factors as attractors in a complex system. These attractors acted as constraints (or facilitators) on the practice of patient-centered medicine by the clinicians. Implied in this complexity model is the idea that the attractors identified were qualitatively different from each other.

Katerndahl and Parchman [23] used a quantitative approach to try to measure the complexity of clinician–patient encounters and to correlate the complexity of the consultation with change in medication for diabetic patients. They used a mathematical technique called orbital decomposition analysis to measure the non-linearity of clinician–patient interactions. There was a slight trend toward non-linearity in consultations in which there was a change of medications, but this did not reach statistical significance. This approach assumes that quantitative study of human interactions can be undertaken using the same mathematical constructs that are used in computer models of complex systems.

Sturmberg has written extensively about complexity theory and the clinical consultation [17–20, 24]. Two papers take a systems theory approach to understanding the consultation as a complex system [18, 24]. Systems theory takes a bounded system and attempts to identify the variables that make up that system and how they interact and affect each other. Larger systems may contain subsystems. Variables and subsystems affect each other through multiple feedback loops. Sturmberg uses an example of a patient consultation to illustrate these ideas. He creates a multiple cause systems diagram that details physiological, psychological, and social subsystems that interact in multiple ways through feedback loops to cause the patient to lose his previously good control of hypertension, weight, and diabetes [18]. This approach does a good job of illustrating the non-linearity of the consultation and the inappropriateness of linear solutions to a non-linear problem. Systems theory, like the constructs of Innes et al., posits that different variables and subsystems act as independent, but coupled agents, and that the interaction of those agents self-organize to form coherent patterns.

Another particularly interesting Sturmberg paper discusses the nature of knowing and knowledge in medicine [20]. He introduces the idea that knowledge is dynamic, "a thing and a flow, constructed as well as in constant flux" ([20], page 767). In the context of the consultation, this means that knowledge is not transferred from the clinician's head into the patient's head, but that knowledge is constantly being constructed in the interaction of the two. This is very similar to Stacey's ideas about knowledge, although Stacey's conception is even more at odds with conventional ideas of knowledge than Sturmberg's. There will be more to say about Stacey's concept of knowledge in the following section.

18.2 Complex Responsive Processes of Relating

In this section, it will be argued, based on Stacey's work, that there is only one class of agents present in the consultation (or any other form of human interaction), a class that Stacey calls complex responsive processes of relating. All the other agents or attractors described by Innes et al., Aita et al., and Sturmberg are emergent patterns or themes generated by complex processes of relating and have no agency in and of themselves, and no independent existence outside of continuously being produced by the interaction of human bodies. It will also be argued that *all* clinical consultations exhibit properties analogous to complex adaptive systems at the edge of chaos; that is, the paradoxical simultaneous presence of stability and potential for radical change. This section will consider Stacey's ideas about complex responsive processes of relating in some detail.

The essence of Stacey's argument draws on the work of sociologist George Herbert Mead [25]. Mead held that the individual mind and the social structures in which individual minds exist co-create each other through the interaction of human bodies using symbols. Symbols are defined as bodily "gestures" and responses to those gestures. A gesture is any action by a human body that evokes a response in another human body. There are three kinds of symbols: *protosymbols* ([1], page 102), *significant symbols* ([1], page 106), and *reified symbols* ([1], page 108). Interactions between human bodies always involve at least the first two.

18.2.1 Protosymbols

Protosymbols are feeling states. Based on the work of Stern [26] and Damasio [27], Stacey posits that body rhythms constitute the feeling states that create awareness of the self. "The millisecond time scale of neuronal firing in the brain; the second by second time scale of the heart; the slower rhythms of the metabolic and endocrine systems; the hourly dynamics of the digestive system; and the daily, weekly, monthly and yearly rhythms of body energy … mesh into a symphony of rhythms having particular time contours marked by beat, duration and variations in intensity" ([1], page 102–103). Each person has a unique pattern of these bodily time contours. This pattern is directly perceived by others and "seems to be what enables humans to yoke together sight, sound, touch, smell and taste into the experience of some whole person" ([1], page 103). Because others can perceive and respond to these body rhythms, human bodies are constantly conducting a sort of protoconversation at an unconscious level and this conversation both constitutes meaning and co-creates a sense of identity in each individual body. Protoconversation consists of "mirroring, echoing and resonating with each other's temporal feeling dynamic and thereby empathically attuning ourselves to each other" ([1], page 104). Stacey is careful to point out that this form of communication is not rudimentary or primitive, but is simultaneously present in all other forms of conversation. This will be an especially important concept in the later discussion of the application of Stacey's work to the clinician–patient consultation.

18.2.2 Significant Symbols

Significant symbols have an additional feature not found in protosymbols. A significant symbol (gesture) is one that creates a similar response in both the maker of the gesture and the other. This allows for the possibility of the maker of the gesture to be aware of the meaning of the gesture (symbol). Although significant symbols can include facial or body gestures, language creates the opportunity for complex communication in significant symbols, because the speaker can hear vocal symbols in the same way as the listener. "The action of speaking throws together a vocal gesture by one person with a a similar response in that person as in the other. The meaning is in the dual nature of this response" ([1], page 106). Ordinary conversation is communication in significant symbols.

18.2.3 Reified Symbols

Reified symbols are used to communicate using abstractions. In contrast to protosymbols and significant symbols, which take place in the context of the living present and the history of lived experience, reified symbols refer to abstract, systematic frameworks that have their own historical development. Medical science is such a framework within which people talk about the functioning of the human body. Reified symbols, either spoken or written, point to, or stand for, a particular abstract-systematic framework of explanation.

It is important to understand that even when we are communicating using reified symbols, we are also using significant symbols and protosymbols, and therefore these simultaneous conversations are likely to be contradictory. Let us use a hypothetical example of a clinical consultation. I, the clinician, say to you, the patient, "You have type II diabetes mellitus." This vocal bodily gesture I make is a pattern of reified symbols that only make sense in the abstract framework of medical science. I say these words, however, in a particular tone of voice, with a particular body posture at a specific place and time. You, the patient, however, will respond not simply in the abstract terms of medical science, but also in terms of my tone and posture and the context we are in, a context that includes the history of previous interactions with others and with me. You may respond to my words using abstract terms, "Thank you, doctor, for making this diagnosis," while at the same time feeling attacked, upset, and angry. I may find some similar responses called forth in me (significant symbol) but I may be surprised at some of your responses (protosymbol). These multiple aspects of interaction make responses to any gesture (symbol) unpredictable.

18.2.4 Analogies with Computer Simulations

Stacey calls these circular patterns of gesture and response among people who are different, *complex responsive processes of relating*. He uses the analogy of complex adaptive systems, and maintains that complex responsive processes are analogous to digital symbols arranged as algorithmic rules in computer simulations of complex adaptive systems. At critical ranges of parameters for the digital "agents" in such models, "a dynamic between stability and randomness arises and this takes the form of attractors (self organizing coherent patterns) that are paradoxically stable and unstable at the same time" ([1], page72). This is the so-called "edge of chaos" with attractors that are "similar to the strange and fractal attractors of chaos theory" ([1], page 72). If random mutation and/or crossover replication (diversity) are introduced into such models, the "agents" will evolve in a radically unpredictable manner. At the edge of chaos, the stability of attractors is fostered by redundancy, loose coupling, and the power law ([1], page 142), while at the same time, instability is fostered by amplification of small differences.

Human interaction has analogous properties. In everyday conversation, people often say the same thing in several different ways, thus demonstrating redundancy. Furthermore, it is not necessary to understand every word or concept another person says in order to have a reasonable understanding of what they mean. This could be considered a form of loose coupling. Finally, in ordinary conversation small misunderstandings are frequent, but large misunderstandings are rare, analogous to systems that follow the power law. These properties foster the stability of meaning, but are always in tension with the possibility that small misunderstandings may escalate and produce new and unexpected conversational patterns.

Stacey is careful to point out that there are limits to the analogy. There are no simple rules for complex responsive processes of relating, and there is no outside designer. There is no "outside" perspective at all, since we are all constantly and continuously participating in complex responsive processes. Complex responsive processes of relating do, however, have other characteristics of complex adaptive systems at the edge of chaos.

One such characteristic is the fractal nature of complex responsive processes of relating. The essential property of fractal phenomena is scale invariance. A fractal pattern always shows similar degrees of irregularity, no matter the degree of detail of examination of the pattern, and no degree of detail is more fundamental than another. Complex responsive processes of relating have this property of scale invariance. At the level of the self, each of us constantly participates in a silent role play in which we carry on a conversation with ourselves. These private conversations include all the types of symbols (protosymbols, significant symbols, and reified symbols) that characterize our public interactions with others. This private role play forms and is formed by our interactions with others. The complex responsive processes, like fractal patterns, look the same whether viewed at the level of the individual, the dyad, the group, the society, or the culture.

Stacey maintains that when the perspective expands from the dyad to multiple interactions between many people, each with a unique history of interacting with themselves and with others, then the analogy with the computer models of complex adaptive systems suggests that these interactions will produce attractors (coherent patterns) that have the paradoxical features of continuity and novelty at the edge of chaos. These attractors, however, are not like the mathematical attractors of formal chaos theory. They are conversational themes that "are perpetually constructing the future" ([1], page 142).

An important consequence of this way of characterizing human interaction is that all social structures are simply coherent patterns formed by human bodies interacting in complex responsive processes. This means that "shared, repetitive and enduring values, beliefs, traditions, habits, routines and procedures" ([1], page 95) are simply repetitive and predictable patterns of interaction. They do not exist in any meaningful way apart from their continued reproduction through the interaction of people.

Even these repetitive patterns, however, are rarely communicated in exactly the same way. This leaves open the possibility of variance and change. Because of another property of complex systems at the edge of chaos, small variations in repetitive patterns can be amplified so that novel patterns are always possible. The best recent social example of this is the radical political change in Egypt. An established pattern of relating (the autocratic Mubarak regime) was suddenly transformed because the act of a single individual in Tunisia, the self-immolation of the street vendor Mohamed Bouazizi, became amplified in ways that were impossible to predict beforehand.

At a different level of detail, the same process operates in the clinical consultation. For example, I once had a young woman patient whose life had essentially come to a standstill for almost a year because of her grief and guilt over the death of her father (destructive repetitive pattern of interaction with herself and others). She had recurrent dreams about her father, and a particularly vivid image in the dreams was of her father's bare feet. I had just been learning about Gestalt therapy (abstract systematic framework—reified symbols), so I asked her to pretend to be her father's toes in the dream and speak from the perspective of the toes (introduction of a novel pattern of communication). As she described the tension and stiffness of the toes, she suddenly made the connection with her own patterns of relating with herself and her family. This small change in her repetitive pattern of relating was amplified to such an extent that her life was transformed. She immediately put her grief behind her and resumed her former productive and rewarding life with her family and friends.

In work that will be discussed later, master clinicians seem to have the ability to reflect upon ongoing interactions with patients, and recognize when the patterns they are co-creating with patients are unhealthy, repetitive and unproductive, or dynamic, creative, and growth promoting. These clinicians often are willing to risk stepping into the unknown to change the conversation (and that conversation includes using the tools of medical technology) in the hope of generating new, dynamic, and growth-promoting patterns. The risk, of course, is that emergent patterns are inherently unpredictable, and may be destructive as well as positive.

There are several other aspects of Stacey's work that are important to understanding how complexity shapes the clinical consultation. These are the importance of power dynamics, the nature of knowledge, and the role of history and narrative in interactions.

18.2.5 Power

Interaction with other humans is necessary for an individual (or social group, or society) to survive. Consequently, we must depend on others for something we desperately need, indeed for our very identity.

Stacey argues that the nature of human gesture and response involves turn-taking and turn-making, and that themes emerge in the process of turn-taking/turn-making. These themes have a history in the individual, the group of which the individual is a part, and the wider communities and societies of which the individual is also a part. Some of these themes are ideological, in that they establish membership categories that determine who may take a turn and when and how they may do so. The process of turn-taking/turn-making therefore of necessity establishes power differences in which some people are included and others are excluded. These power dynamics are an inevitable property of human interaction, because "when one person takes a turn, others are at that moment excluded from doing so" ([1], page 149).

Because our identity depends on continued interaction with others, any threat of exclusion arouses a deep existential anxiety, and that anxiety creates emergent themes in individuals and groups to deal with that anxiety. Thus, any ideological theme that maintains power differentials creates opposing themes (shadow themes) that are opposed to the existing power differentials. These power dynamics can not only be disruptive or even destructive, to collaborative activity, but they also create the possibility of emergent novelty.

It is worth emphasizing again that power dynamics and power differentials emerge from the nature of human interaction and cannot be eliminated no matter how noble one's social intentions may be. "Communicative interaction is a process in which people account for their actions and negotiate their next actions. This is a political process, the exercise of power" ([1], page 151). Because of the fractal nature of complex responsive processes of relating, power dynamics are a part of an individual's private role play as well as his/her interactions with others.

In the clinical consultation then, power dynamics are an inseparable part of the interaction between clinician and patient. An existing ideological theme, clinician as the medical authority, may threaten the patient and induce anxiety, triggering a theme of opposition. If neither clinician nor patient recognize these dynamics, then the clinician may label the patient as "non-compliant," or the current politically correct euphemism for the same thing, "non-adherent." The patient may justify his/her resistance as inability to comply with the treatment plan. On the other hand, if the clinician and/or the patient are conscious of power dynamics, then there exists the possibility that those power dynamics can be used in ways that are constructive for the patient and for the clinician–patient relationship.

18.2.6 Knowledge

Stacey maintains that from the perspective of complex processes of relating, knowledge is the process of making meaning in the act of relating. This means that knowledge cannot be stored, but is perpetually created in acts of relating.

In the clinical consultation, then, clinician and patient create knowledge, that is, meaning, in the ongoing interaction. The patient shares his/her story of illness. The clinician responds with his/her story of how the abstract framework of medical science may be of help to the patient. At the same time, both patient and clinician are communicating emotionally, through protosymbols, and reflectively, through significant symbols. These interactions form and are formed by the silent role play of both clinician and patient. This meaning-making activity cannot be stored in an electronic medical record. It exists and is maintained only in continued communicative interaction.

The technology we use to store "knowledge," such as books, databases, or electronic medical records, actually only store reified symbols as artifacts, "that is 'abstracted' themes describing past interactions and the qualities that emerged in those interactions" ([1], page 164). These artifacts only contribute to knowledge as tools we use in communicative interaction in the living present. Unfortunately, most current use of electronic medical records in the consultation interferes with, rather than facilitates, the process of meaning making, and therefore knowledge creation [28]. A complexity science-informed perspective, therefore, should be useful in designing electronic medical records and ways to use them in the consultation that would enhance the essential process of meaning making between clinician and patient.

18.2.7 History

All interactions between human bodies are constrained by the history of previous interactions of each individual and the social group. This history is what gives the stability to social structures. Furthermore, the historical nature of human interaction means that human experience has a narrative-like structure. These experiences can be recounted in the form of stories, and stories are important tools for negotiating communicative interaction in the living present.

A patient in a clinical consultation, therefore, reflects on his/her narrative-like experience of illness, and creates a story of illness that is formed by and forms the interaction with the clinician in the living present. The story is therefore in some ways co-created by patient and clinician together. As will be discussed later, listening carefully and responsively to that story is important in both diagnosis and healing.

18.2.8 Summary

Based on Stacey's work on complex responsive processes of relating as described above, it is proposed that the clinical consultation forms and is formed by the complex responsive processes of relating taking place between the clinician and patient and in the silent role plays of each individual. Coherent patterns emerge in these interactions that are constrained by the history of interaction of each of the individuals involved, and by the ideological themes of power dynamics. While many of these patterns are repetitive and predictable, the potential for radical change is always present, and may be manifested by small changes that become amplified in ways that produce entirely new and unpredictable patterns. The proper use of medical technologies is as tools in the interactive communicative processes between clinician and patient.

18.3 Healing Relationships and the Clinical Consultation

In this section, work on healing relationships by my colleagues and I will be described. Using Stacey's framework discussed above, the processes, outcomes, and competencies that make up our model can be seen as a set of conversational themes and patterns that lead to the emergence of the experience of healing.

The clinical consultation has a specific purpose, and that purpose is to create an experience of healing. There are several complementary definitions of healing in recent literature [2, 29, 30]. For the purpose of this chapter, the definitions developed by Hsu et al. and by Egnew will be combined.

> Healing is a multidimensional process with physical, emotional, and spiritual dimensions. The key themes are as follows: (1) healing is multidimensional and holistic; (2) healing is a process, a journey; (3) the goal of healing is recovery or restoration; (4) healing requires the person to reach a place of personal balance and acceptance; and (5) relationships are essential to healing. ([30], page 306)
>
> Healing may be operationally defined as the personal experience of the transcendence of suffering. ([29], page 255)

The addition of Egnew's definition is important, because all too frequently recovery and restoration are not possible and yet patients with chronic or incurable illnesses are still capable of experiencing healing.

It will be argued that healing, as defined above, is an emergent pattern of complex responsive processes of relating, and that the clinician–patient relationship is an important, though not exclusive, part of that pattern.

If healing is the goal of the clinical consultation, and if it is also an emergent pattern of complex responsive processes of relating (and therefore unpredictable), how then do a clinician and patient go about promoting healing?

My colleagues and I attempted to answer this question in a qualitative interview study of master clinicians and their patients [2]. We found that there were three processes or patterns of relating that seemed to make healing more likely to emerge. These processes led to three relational outcomes that were associated with the experience of healing. We also found that there were certain clinician competencies (historical patterns of relating) that were necessary for clinicians to participate in the three processes. These processes, outcomes, and competencies will be briefly outlined and the case will be made that this healing relationship model (Fig. 18.1) fits well within the larger framework of complex responsive processes of relating.

18.3.1 Processes

18.3.1.1 Valuing

Master clinicians valued their patients by relating in three ways. First, they approached every patient in a non-judgmental manner, treating every patient as a person of worth. Second, they found some way to connect to each patient as a person, finding some personal resonance that helped them to make an emotional connection with the patient. Using Stacey's framework, much of this connection was likely made unconsciously, in conversation using protosymbols. Third, master clinicians were fully present for patients in the consultation. They devoted their full attention to the ongoing interaction with the patient in the living present. They participated with patients in co-creating a narrative of the patients' illnesses.

18.3.1.2 Appreciating Power

As Stacey points out, power dynamics are inescapable in human interactions, and the consultation is no exception. Ideological themes of authority, counter themes of resistance to authority, past experience of racism, and cultural stereotypes, among others, may all be operative in the consultation. Challenging existing power dynamics in the consultation is intensely anxiety provoking, and therefore all too often avoided by both clinicians and patients. Master clinicians in our study, however, were acutely aware of these power dynamics, and made use of them when possible for the patients' benefit. Usually, this involved empowering patients by engaging them as partners in diagnosis and treatment plans, and by serving as guides to the world of medicine, thus giving patients tools to take control of their own treatment. Sometimes, however, clinicians used their authority to push patients (usually gently) to do things they needed to do for their health, but were somewhat resistant to doing.

18.3.1.3 Abiding

We tend to think of the clinical consultation as an isolated event, but much of the time, particularly in primary care, there is a historical dimension as well. Clinician–patient relationships develop over time [19], and our study found that master clinicians practiced what we call abiding. First, they were accessible. This did not mean being available 24 h a day, but being there for major health events over time. These clinicians also made it clear to patients that whether treatment succeeded or failed, they were committed to continue the relationship and would not abandon patients in their suffering. Finally, abiding included the accumulation of caring actions; small acts of kindness that demonstrated to patients that the clinician cared what happened to them. These accumulated acts were in part responsible for the emergence of trust, one of the outcomes of the healing relationship processes.

In Stacey's framework, these aspects of abiding would be seen as historical themes that pattern the ongoing interaction between clinician and patient.

18.3.2 Outcomes

We identified three outcomes of the processes of healing relationships: hope, trust, and a sense of being known as a person. Stacey's framework would suggest that these outcomes are emergent patterns of relating, both with the clinician and in the patient's own private role play.

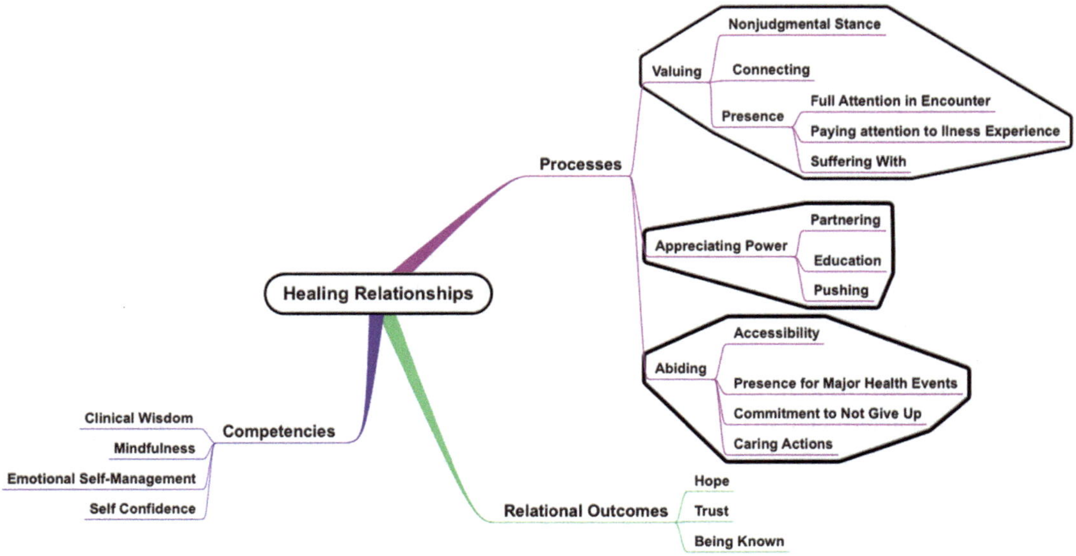

Fig. 18.1 Healing relationship model

18.3.3 Competencies

There were four clinician competencies that seemed to be necessary for clinicians to participate in healing relationship processes: mindfulness, emotional self-management, confidence, and clinical wisdom. Using Stacey's framework, these competencies developed as historical patterns of relating, and exist only in the continued interaction between clinicians and their patients, and within the minds of clinicians as continued private role play. These competencies will be briefly described below.

18.3.3.1 Mindfulness

Mindfulness is the ability to be aware of the emotional effect on one's self of the ongoing interaction, while at the same time being aware of the effect of the interaction on the patient. From the standpoint of complex responsive processes of relating, mindfulness is the ability to be conscious of all three simultaneous modes of symbolic communication: protosymbols, significant symbols, and reified symbols. It is mindfulness that allowed the master clinicians to be fully present for their patients during the consultation. Mindfulness is also an important competency in order to avoid errors in diagnosis and treatment [3]. There will be more to say about mindfulness in a discussion of Epstein's conception of mindful practice in a later section in the chapter.

18.3.3.2 Emotional Self-Management

Emotional self-management is the ability to calibrate one's emotional response to the patient's story of illness in such a way that the patient benefits. It requires mindfulness, because in order to calibrate one's emotional response, one has to be aware of it in the first place. Emotional self-management requires the clinician to maintain enough emotional distance to analyze the patient's illness narrative from the perspective of medical science, while at the same time maintaining an emotional connection with the patient as a person. From the perspective of complex responsive processes of relating, the master clinician is able to monitor and regulate consciously the ongoing protosymbolic interaction with the patient.

18.3.3.3 Confidence

Confidence is the ability to project to the patient confidence that the clinician will be able to help the patient heal. This confidence cannot be faked, but arises from the clinician's experience with other patients and knowledge of scientific medicine. It is communicated not only with words (significant symbols), but also with body posture, tone of voice, and facial expressions (protosymbols). The projection of confidence can be seen as an attractor that leads to the emergence of hope.

18.3.3.4 Clinical Wisdom

Clinical wisdom requires the clinician to be fully versed in the abstract framework of scientific medicine and to be able to apply that framework in a way that benefits a particular patient in the specific context that the patient brings to the consultation. In short, it is the ability to use scientific medicine in the service of healing this particular patient at this particular time in this particular place. Scientific medicine, then, becomes a tool to be used to facilitate the interaction characterized by the healing relationship.

18.3.4 Summary

Healing is an emergent property of a particular coherent pattern of relating between clinician and patient characterized by valuing, appreciating power dynamics, and abiding. These processes can be seen as analogous to attractors in complex systems at the edge of chaos. In order to participate effectively in this pattern, clinicians must have developed competencies in the history of other interactions and in their own private role play of mindfulness, emotional self-management, projection of confidence, and clinical wisdom. These competencies have no independent existence, but are constantly enacted through interactions with others and in private role play. Trust, hope, and a sense of being known as a person are also emergent properties of the healing pattern of relating.

18.4 Mindful Practice

This section will be devoted to Ronald Epstein's work on mindful practice [3, 4, 31]. Describing the consultation as analogous to a complex adaptive system at the edge of chaos is of little use unless that perspective helps clinicians and patients in improving patient care, relieving patient suffering, and promoting healing. Epstein's ideas about mindful practice may enable clinicians and patients to participate in the processes of healing relationships and make the best use of the potential for change embodied in the idea of a complex system at the edge of chaos.

Epstein defines mindfulness as "a purposeful, non-anxious, reflective presence that can be applied to any aspect of practice" [4]. It involves purposely paying attention to one's own internal state during everyday tasks. Using Stacey's framework, this means paying attention to one's own internal dialog at all three levels of symbolic interaction: protosymbols (what are my feelings), significant symbols (what am I thinking about this interaction), and reified symbols (how am I categorizing this story in the abstract framework of scientific medicine). In addition to being aware of one's own internal conversations, mindfulness also involves being aware of the multiple levels of conversation with the patient and the problem he/she presents. The purpose of mindful practice is to bring clarity and insight into one's everyday clinical work. Epstein identifies four habits of mindful practice.

18.4.1 Attentive Observation

This includes observation of oneself, the patient, and the problem. There is a non-judgmental quality to this kind of observation and therefore it means observing for the unexpected as much as for the familiar. The clinician practicing attentive observation becomes aware of how he is filtering the information presented by the patient, and the biases that she may bring to this interaction.

18.4.2 Critical Curiosity

Critical curiosity means attempting to see the world (and oneself) as it is, and not as one would like it to be. Being critically curious invites and accepts doubt and uncertainty. This is an especially useful habit for participating in the complex system of the consultation. Emergent patterns are always unpredictable and uncertain, and being comfortable with uncertainty avoids attempting to force a linear solution in a non-linear system.

18.4.3 Beginner's Mind

Beginner's mind is the ability to set aside preconceived categories and approach each patient and each interaction with a fresh perspective. It requires the ability to be reflective about what one's preconceived categories are and to consciously suspend the practice of categorization. Eventually, some categorization will be necessary, but the habit of beginner's mind helps to keep that categorization from taking place prematurely.

18.4.4 Presence

In our healing relationship model, we include presence as part of valuing, because presence requires a connection between the clinician and the patient. When a clinician is present, she is devoting her undistracted attention to the ongoing interaction with the patient. Her total being is in the here and now, the living present. She is not thinking about the next patient, worrying about her ill mother, or looking forward to her vacation next week. A clinician who is present can pay attention to small nuances in the ongoing co-creation of the patient's story of illness, nuances that may be important leverage points in creating new and healthier attractors for the patient.

18.5 Putting It All Together: A Complexity Science-Based Theory of the Consultation

The integrated theoretical model of the clinical consultation is illustrated in Fig. 18.2. Complex responsive processes of relating form and are formed by communication between human bodies using protosymbols, significant symbols, and reified symbols. Healing relationships are emergent patterns of the complex responsive processes of relating. The tools of medical technology alter the physiologic context and therefore the nature of the protosymbolic communication. The use of tools is governed by healing relationships, but the tools of medical technology also affect the formation of healing relationships. The discipline of mindful practice is a competency necessary to form healing relationships, and to use medical technology appropriately to facilitate the experience of healing and healing relationships. In the sections that follow, the components of the model and the analogies with complex systems will be explained in more detail.

18.5.1 Agency

To quote Stacey, "human agency means human bodies doing something … and what they do to survive is interact with each other" [1]. Complex responsive processes of relating are patterning processes forming and being formed by a continuous iterative circular process of relating between human bodies. Thus, the only "agents" in the clinical consultation are the complex processes of relating going on between clinician and patient and the private role play of each individual mind. The patterning process is constrained by historical conversational themes and ideologies, which are themselves patterns of other complex responsive processes of relating. These themes and ideologies have no independent existence, but are maintained by ongoing interaction between human bodies. Therefore, social themes, ideologies, cultural beliefs, health beliefs, organizational structures, etc. are not agents in the complex system of the clinical consultation, but are simply repetitive and relatively stable patterns of complex responsive processes of relating. Because each human body has a unique pattern of physiological biorhythms, and each mind has formed and been formed by a unique history of interacting with other human bodies, complex responsive processes of relating have the necessary diversity to self-organize into coherent patterns (attractors).

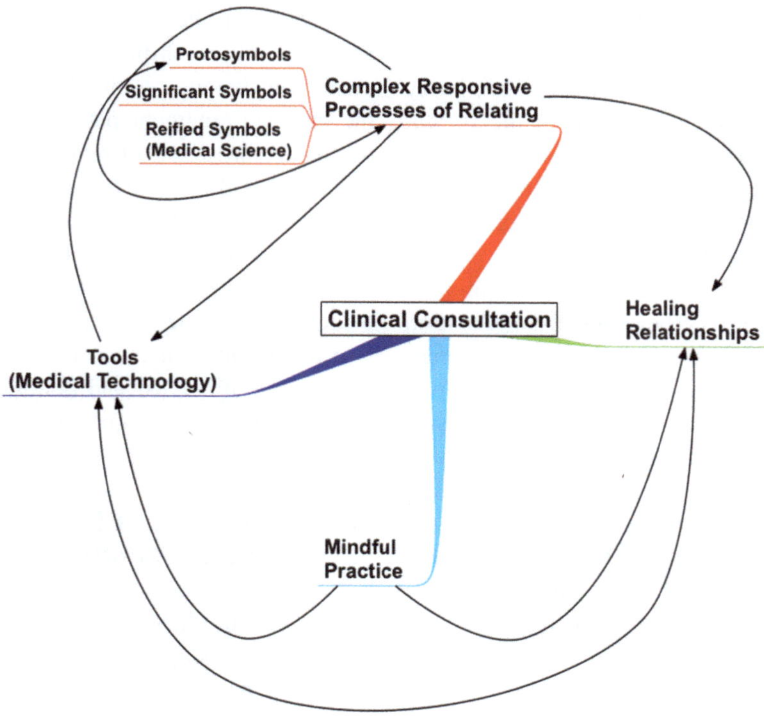

Fig. 18.2 Conceptual model of clinical consultation

18.5.2 The Edge of Chaos

Complex responsive processes of relating in the consultation have analogous properties to computer models of complex adaptive systems at the edge of chaos. That is, the attractors (self-organizing patterns) are paradoxically stable and unstable at the same time. According to Stacey, the themes constraining or facilitating the interaction are stable because they have analogous properties to complex adaptive systems, namely, redundancy, loose coupling, and follow the power law as described in Sect. 18.2.4. Redundancy means that there are many historical pathways to get to the same patterns. In loosely coupled systems, one interaction does not depend in any direct way on other interactions, and so parts of the system can fail without much effect on other parts. The power law leads to stability because systems that follow the power law have few large system-changing events and many small ones. Stable themes or attractors in the consultation can include patterns of reimbursement, power dynamics in the health system, and clinical guidelines, among others. From the patient standpoint, such themes could be unhealthy lifestyle habits such as overeating, sedentary lifestyle, and smoking. This stability is in tension with instability, however. Human interactions always contain the potential for misunderstanding, such that small misunderstandings may escalate and trigger new and unexpected themes. An offhand comment by a clinician about the monetary cost of smoking, for example, might trigger a change in the patient's silent conversation with him/herself that escalates into a radical change in a behavior pattern, i.e., giving up tobacco.

18.5.3 Coevolution

Patients come to the consultation with a narrative-like experience of illness, a sequence of feeling states that occur over time. These feeling states constitute an internal dialog conducted in protosymbols. By asking an open-ended question, the clinician invites the patient to convert this narrative-like experience into a story of illness. This story will be communicated in protosymbols, significant symbols, and reified symbols, and will be constrained by the history of conversational themes in which the patient participates and has participated, such as cultural patterns, health beliefs, and power dynamics. The story will also be formed by the clinician's history as well as his interaction with the patient, which also takes place in all three kinds of symbolic communication. The patient's story, in turn, also influences the clinician's interaction pattern as well as the clinician's internal dialog. New patterns emerge as the complex responsive processes of relating change each other during the consultation. The goal of this coevolutionary process in the consultation is that the story of illness evolves into a story of healing. Whether coevolution occurs in the course of one consultation or over a series of consultations, it requires time. As Sturmberg points out, this argues for "a certain slowness" in the consultation [19].

18.5.4 Tools

In the context of the clinical consultation, the term tool signifies any medical technology that may be used in the clinical consultation for diagnosis and/or treatment, including pharmaceuticals. Tools are, in a sense, extensions of human bodies, and are used "in the transformation of the human and non-human contexts within which people live" [1]. The proper use of medical technology, then, is to transform the physiologic context in patients' bodies, and thereby contribute to the transformation of a story of illness into a story of healing. Since the human body itself is a complex system [32], it is useful to think of a medical technological intervention as analogous to a significant symbol in Mead's framework [25]. The clinician has some idea what the effect of the intervention will be on the patient's physiology, but the intervention also has the potential to create unpredictable emergent physiologic patterns, and therefore unexpected (and perhaps unwelcome) feeling states. For example, the use of antibiotics for pneumonia in a particular patient may cure the pneumonia, but may also lead to diarrhea from alteration of gut flora, or even a life-threatening c-difficile colitis. The effects of medical technology tools, then, become part of the ongoing emergent patterns of the complex responsive processes of relating in the consultation, and are always context-dependent. Tools are not an end in themselves, but only a means to help create the experience of healing, as defined in Sect. 18.3. The use of clinical guidelines and evidence-based medicine in a prescriptive, one size fits all linear manner, therefore ignores the importance of context and is often not likely to contribute to the experience of healing.

18.5.5 Healing Relationships as Emergent Patterns

The goal of the clinical consultation is to create a healing relationship between clinician and patient so that the patient experiences healing. The processes of healing relationships are emergent themes formed by and forming the complex responsive processes of relating between clinician and patient and in the private conversation of each individual with himself. The tools of medical technology are part of these complex responsive processes of relating. Because no consultation (and no body) is exactly like any other, there can be no prescriptive way to insure that a healing relationship will emerge in the consultation. It is possible, however, to describe in general terms what a healing relationship looks and feels like, even though there will be a unique path to that emergent pattern for each clinician and patient. Having a clear goal, however, is important. To quote a sage of our time, Yogi Berra, "If you don't know where you are going, you might wind up someplace else" [33].

As discussed in Sect. 18.3, the emergent pattern of a healing relationship is characterized first by the process of valuing. Clinicians can make this process more likely to occur, first by becoming aware of their own prejudices and judgments about people, and then consciously suspending those judgments as they enter the consulting room. By doing so, clinicians create a space for a new internal conversation, and therefore for a new pattern to emerge. Note that considerable self-reflection is required on the part of the clinician in order to bring unconscious patterns of ideology to consciousness. The second part of valuing is for the clinician to find some way to connect emotionally to the patient, so that what happens to this individual patient matters. Most of this connection must of necessity happen at the protosymbol (feeling) level of communication, again requiring clinician self-reflection as well as bringing to consciousness the ongoing protosymbolic conversation between clinician and patient. Only when the clinician can approach every patient as a person of worth and form an emotional resonance with each particular patient is it possible for the clinician to be completely present for the patient in the consultation.

Healing relationships also require of the clinician an appreciation of the inevitable themes of power dynamics that operate in the consultation. A clinician who is aware of an ongoing theme of clinician authority and a shadow theme of patient resistance can move the interaction closer to the edge of chaos, thereby increasing the likelihood of the emergence of a new and more productive theme of cooperation. The clinician may do this by treating the patient as an expert (on his body and feelings), by valuing information the patient brings to the consultation (from the internet, for example) and by providing the patient with information that will help her to independently manage her illness. Another way of changing the power dynamics in the consultation is to use the techniques of motivational interviewing [34].

Coevolution of a story of illness toward a story of healing requires time, but just as important is how that time is spent. We have called this "abiding," by which we mean being present in an ongoing way and demonstrating caring over time. Primary-care clinicians have the advantage of multiple visits over time, but much can also happen in a single consultation. It is particularly important for specialists who may see patients only once or twice to be fully present for patients in the consultation. Our research has shown the clinicians who give their full attention and listen carefully in the consultation are perceived by patients as spending more time, even though there is no difference in clock time [2].

It also appears that the emergent pattern of healing relationships has beneficial effects for the clinician as well as the patient. Despite being in practice in the same location for many years, none of the clinicians in our study experienced the phenomenon of burnout, and in fact derived a great deal of pleasure from their long-term healing relationships with patients.

18.5.6 Mindful Practice and the Management of Uncertainty

If, as has been argued, the clinical consultation is a complex system exhibiting many of the characteristics of computer models of complex systems at the edge of chaos, then uncertainty exists in even the most routine consultations. Established and routine patterns of complex responsive processes of relating may change suddenly and without warning into entirely new patterns. Giving trimethoprim–sulfamethoxazole for uncomplicated cystitis, for example, almost always results in the disappearance of the patient's symptoms, but for each patient there exists the possibility that she may develop life-threatening Stevens–Johnson syndrome. Many clinicians, used to thinking linearly, believe that uncertainty can be decreased or even eliminated by gathering more data, thus the endless stream of ever more sophisticated blood tests and ever more detailed imaging modalities. A major contribution of the complexity science approach is the understanding that non-linear processes such as the clinical consultation are inherently unpredictable and that therefore outcomes of complex responsive processes of relating using the tools of medical technology have irreducible uncertainty.

Mindful practice, as described in Sect. 18.4, is a way for clinicians to make use of the inherent uncertainty in the clinical consultation in order to increase the probability of the experience of healing as an emergent pattern.

18.5.6.1 Attentive Observation of Themes in the Consultation

Clinicians practicing attentive observation pay attention to the themes and patterns of interaction in the living present. Recognizing the themes, both in private role play and in the interaction between clinician and patient, is a prerequisite to trying to change unhealthy or unproductive themes. This also, of course, includes attentive observation of the patient's physiology as manifested by the patient's narrative-like experience of illness as well as the results of diagnostic tools of medical technology.

18.5.6.2 Critical Curiosity and Acceptance of Uncertainty

A clinician who practices critical curiosity is open to the unexpected, whether it be a laboratory value that does not fit the expected pattern, a story of illness that does not fit into a preconceived category of disease, or unexpected emotional responses from either patient or clinician. The critically curious clinician investigates these phenomena and is willing to let go preconceived patterns that do not fit what is being observed.

18.5.6.3 Beginner's Mind Before the Consultation

Beginner's mind is a practice that should take place before the clinician opens the door of the examination room or consulting room. It involves consciously setting aside expectations about what will happen in the interaction to come, and preparing to listen to the patient as if one were seeing him for the first time. This may allow the clinician to see patterns and themes that were not appreciated in previous consultations or by other clinicians. Beginner's mind is also an important practice prior to reviewing information from diagnostic testing. Once again, patterns may be appreciated that were missed because of premature diagnostic closure. Beginner's mind at the start of the consultation is a prerequisite for attentive observation and critical curiosity.

18.5.6.4 Presence and Emergence

The clinical consultation is in some ways like a dance, and good dancing requires total attention from both partners. Being fully present in the consultation means being attuned to all three levels of communication between and within clinician and patient. In order to be present, the clinician must also be attentively observant, have critical curiosity, and enter the consultation with beginner's mind. Being present allows the clinician to be aware of the ongoing themes of complex responsive processes of relating and therefore to be able to risk introducing novelty to bring about new emergent patterns. Although these self-organizing patterns are unpredictable, presence allows the clinician to constantly adjust to the changing themes in an iterative fashion. Because of the irreducible uncertainty inherent in complex responsive processes of relating, not every patient will experience healing, but a clinician who practices presence makes the emergent pattern of healing more likely.

18.6 Clinical Examples

The following clinical examples are based on my interviews with patients and clinicians describing consultations. Pseudonyms are used for all patients and clinicians.

18.6.1 Maria

In the interview excerpt below, Maria describes a consultation with a pediatric cardiologist:

> I had a daughter with a heart condition. And she [the doctor] assumed that here I am sitting here with all these little brown kids running around that I was totally overwhelmed and had no clue what she was going to talk about. So she sat in front of me and she went, "I'd like your attention." And then proceeded to talk to me like I was deaf too, because she talked very loudly and said one word at a time and drew a picture of a heart that looked like a valentine heart and said, "There is a hole here and there's a hole here," and I went home and I called Susan [her family doctor] at home. I said, "Alright, this is what she said, what does it mean?" Because this doctor had assumed that I was so undereducated that she didn't tell me anything. She said the name of it, but she didn't tell me about it, because she just figured I wouldn't know. So I mean, I'm panic-stricken, here's this baby with this

> problem … yeah, she's the 5th one, but that doesn't matter. And they just treated it like I was an idiot.

It appears that there are several ideological themes playing out in this consultation. The clinician's private internal conversation (probably mostly not conscious) was likely something like the following.

This is a young white woman with lots of biracial children. She is not educated. She could not be a very good parent. She lacks the intellectual capacity to understand the implications of her child's congenital heart defect. I need to treat her like a child to have any hope of communicating.

The power dynamics of this interaction are also clear. I, the clinician, am the authority. I will talk and you, the patient, will listen. Maria's narrative about the consultation suggests that there was also a counter theme (shadow theme) of resistance, but despite this, Maria's anxiety made her unable or unwilling to challenge the balance of power in her direction.

At the protosymbolic level, Maria must have been feeling (and therefore communicating) anger, dismay, fear for her child, a feeling of being attacked, disgust, etc. The clinician was receiving these protosymbols, but likely did not bring them to consciousness. She was not mindful, and did not practice presence.

Maria obviously did not experience healing in this consultation. The established and repetitive themes of authority and prejudice played out with little opportunity for novel patterns to emerge.

18.6.2 George

In the interview excerpts below, George tells the story of his first consultation with his primary-care doctor, Tom. George was a Christian Scientist, and had been suffering from severe progressive congestive heart failure for 6 months without consulting a physician. He had so much edema that Tom, in his interview with me, described George as looking like "the Michelin tire man." The consultation that George describes below was a home visit.

> Tom made it so easy for me … the point is, Tom the M.D. was talking to George the guy that was sick who didn't know from third base about medicine or doctors. He was explaining to me, not only in laymen's terms, but in the … the ABC's … he made me feel so at ease. That's what was important. That's the point that I want to make. He made me feel so at ease. He said, "Now, look, I'd like to do a couple of things. I'd like to listen to your heart and I'd like to listen to your lungs and I'd just like to take a look at you all over. How are you with that?" I said, "Doctor, you do whatever you want to do." And right up front he said to me, he said, "Let's get this relationship started on an informal, comfortable basis. I'd like you to call me Tom, if you're comfortable with that." I said, "Well, I think you've earned the right to be called doctor." He said, "Maybe I have, maybe I haven't, but do you feel good calling me Tom?" I said, "Tom, I'll call you Tom." I said, "If you'll call me George." He said, "O.K. George … George it is." So from that day on we've been Tom and George.… Tom said, "O.K. Here's what my recommendation is." He said, "I understand you have a Christian Science background." And he said, "I respect that." And I was really surprised to hear him say that and I said, "Well, I appreciate that. That's very gracious of you to say. Seeing me in this condition and knowing that I've brought it on myself because I hadn't talked to a doctor about it and here I'm bringing you in at the 11th hour." And he said, "Don't worry about it. We can take care of you and you're going to be fine. I want to assure you that you're going to be fine." Well, my whole attitude changed… really…I went from the darkest despair to bright hope. And that was because of him and what he said to me and how he said it and just the caringness of him.
>
> I was swollen up like a sow ready for slaughter. My arms were humongous. And he said, "Now, I'd like to draw some blood." And he said, "We're going to take it out of your arm here." The confidence that Tom instilled in me, "I can do this." He could tell I was afraid. He said, "You're going to feel a little prick, but it's not going to hurt. O.K.?" He said, "You're just going … make a fist and relax your arm. Just relax it and let it go. Well, the confidence I just felt. Finally, someone is here to take care of me. Somebody is here … I have suffered through 6 months and here is someone who wants to take care of me".

The first thing to note about this rich description is the context. This consultation took place in the patient's home. This in itself has the potential to alter the power dynamics that were observed in Maria's consultation. Furthermore, the clinician consciously does several things to alter the balance of power. He insists that the patient call him by his first name, and he asks the patient's permission to examine him. Also, in contrast to Maria's consultation, George reports that Tom translated medical jargon into language that Tom could understand, but did so with full respect for George's intellectual capacity. All of this served to undermine the doctor-as-authority theme that so dominated the consultation with Maria. A new theme, that of partnership begins to emerge.

On the other hand, the clinician uses his genuine authority, a projection of confidence in his ability as a physician, to inspire hope and confidence in this very ill and suffering patient. He states, "We can take care of you and you're going to be fine." What George says next demonstrates the emergence of a new pattern, a transformation from his story of illness and despair to the beginning of a story of restoration and healing. "I went from the darkest despair to bright hope."

It was, of course, more than Tom's words that led to this transformation. There is a clue about this in George's next sentence. "And that was because of him and what he said to me and how he said it and just the caringness of him." Tom was communicating in protosymbols as well as significant and reified symbols, and the protosymbolic message, a genuine feeling of care and concern for this suffering man, was received at the protosymbolic level by George.

The story of the drawing of blood can also be understood at two levels. Tom was demonstrating his competence in the tools of medical technology by the technically difficult task of drawing blood from this tremendously edematous patient. He was also introducing a novel interaction, because physicians do not usually draw blood themselves. This act reinforced the protosymbolic message of caring and concern. George's response shows how he received the messages at both levels. "Somebody is here. I have suffered through 6 months and here is someone who wants to take care of me."

Through the use of beginner's mind and attentive observation, Tom was able to approach this patient who held a belief system very different from his own, in a non-judgmental and respectful manner. It is clear that he did this consciously, because he brought it to the significant symbol level by explicitly saying, "'I understand you have a Christian Science background.' And he said, 'I respect that.'"

By practicing mindfulness and by understanding the processes of healing relationships, Tom and George together helped something new to emerge, the experience of healing. Note that this outcome was not guaranteed, but was made more likely as each party in the consultation responded to the complex responsive processes of relating and allowed coevolution to happen.

18.6.3 George and the Surgeon

One more clinical example is provided below to illustrate that the emergence of healing is not limited to consultations with primary-care clinicians. The interview excerpts below come from another part of my interview with George, the same patient described in Sect. 6.2. George's congestive heart failure was the result of severe mitral insufficiency due to almost complete incompetence of the mitral valve. He had to have a mitral valve replacement, and the following interview excerpts are from his description of his two consultations with his heart surgeon.

Consultation 1

And anyway I sit there in the waiting room waiting about a half an hour, and no complaints because he got us in. And that's the first thing I said. I said, "Doctor, I sure appreciate your seeing us today." And he said, "Well, I sure don't want you to have to make that round trip…a second round trip. That's entirely unnecessary." He said, "We doctors can sometimes be a little self-important and think that we have to…we have to space our time out and can't possibly see 2 patients in one hour," or whatever he said. But he said, "I'm human, but I can do that." And when we shook hands he said, "I'm Mark Spencer." I said, "Doctor, it's good to meet you." And he repeated, "I'm Mark Spencer." I said, "O.K. Thank you. I'm George Johnson." So we spoke. He told me the advantages and disadvantages of a porcine valve, or actually the advantages of a porcine valve over the mechanical one. He said, "Which you will hear clicking every second for the rest of your life." And I said, "Mmmm… sounds like a no brainer to me." He says, "It is to me too, but a lot of people don't like the idea of having a part of a pig implanted in their bodies."

> Consultation 2
> And I got in and got admitted to a room, and Dr. Spencer came in to see me with his girl Friday, his assistant, and the…he said, "Do you mind if I sit on the end of your bed?" And I said, "Please do." And…
> Interviewer: How did that make you feel?
> Oh, it made me feel good. It made me feel like…close…closeness. And Dr. Smith, the cardiologist…I was there a couple of weeks in the hospital…he never did that. He'd come in and he'd sit in the chair. (Chuckle) We had to keep a distance…a distance between him and me, not as far as I was concerned, but as far as he was concerned. So Dr. Mark sat at the end of the bed and said to his assistant, Susan, "This is Susan." "Hi, Susan. I'm George." "Hi, George." I said, "Why don't you sit on the other side of the bed?" And she said, "Oh, fine." So, I said, "I'd feel better if you'd hold my hand." So the whole atmosphere was just all of a sudden was relaxed. And Dr. Spencer asked me a lot of questions.
> Except for my experience in Jefferson County [hospital] for 2 days, that was the only time I'd been in a hospital in my life. But here in a major medical center to have my heart operated on. I was…oh, I was scared. I was truly scared and I told Dr. Spencer, "I am scared." He said, "Well, let me put you at rest." He said…I don't know the number, but it was like, "I've done 2200 of these." I said, "That's really not my concern, how many you've done. Honestly, how many have not survived?" And he looked at Susan and he said, "I don't think we've lost anybody, have we?" And she said, "Not that I recall." And he looked at me and he said, "She knows everything. She knows all the answers." He said, "This is a routine procedure." I said, "Easy for you to say. It's not your heart." He said, "I do it all the time. I do it 4 times a day." I said, "You do?" He said, "Oh, yeah." So…
> Interviewer: Did that help you come down off the ceiling a little bit?
> Oh, yeah. His whole thing was calm the patient down, make them feel comfortable, put them at ease.

In consultation 1, Mark, like Tom, consciously changes the power dynamic by introducing himself by his first name, not preceded by "Dr." Interestingly, George seems to have more difficulty referring to his surgeon by his first name than his family doctor. He refers to him variously as Dr. Spencer or Dr. Mark. Mark respects this preference, and does not insist on being addressed by his first name.

Also in consultation 1, Mark uses significant symbols (ordinary conversation) to demonstrate caring and concern by expressing his willingness to work George in for a same day visit in order to help George avoid another long drive to the medical center. George's account suggests that the protosymbolic (feeling) communication was congruent with the significant symbol conversation. Mark, like Tom, translates the reified symbols of medical science into language that George can understand and gives him the information he needs to be a partner in the decision about which kind of valve to use.

In consultation 2, Mark again changes the authority power dynamic by asking permission to sit on George's bed. He addresses his assistant, Susan, as an equal, and in fact suggests that she has superior knowledge about Mark's surgical outcomes. On the other hand, he uses the authority power dynamic, by pointing out that this operation is routine for him, as a way of easing George's fear and making him more confident about the operation.

George expresses his fear of the surgery in significant symbols, but he also must have been communicating his fear and anxiety in protosymbols as well. By responding in the way he did, Mark was clearly picking up on this protosymbolic communication, and responded to it in a flexible way, defusing some of the anxiety with humor, but also providing reassurance about his technical competence. The result of Mark's mindful practice was the transformation of a pattern of fear and anxiety to one of comfort and ease, thereby increasing the probability that George would successfully weather the stress of surgery and recovery.

George also participated actively in this process by asking Susan to sit on the bed as well and to hold his hand, increasing his sense of comfort.

18.7 Conclusion

The clinical consultation demonstrates many of the properties of a complex adaptive system. Complex responsive processes of relating are the interacting agents in the consultation, and these processes are sufficiently diverse to generate self-organization. The patterns, or themes that thus emerge in the consultation have the paradoxical quality of stability and instability at the same time, similar to computer models of complex systems at the edge of chaos. The complex responsive processes of relating interact with each other in a circular and iterative fashion, so that coevolution takes place in the emergent patterns generated in the process of the consultation (see Fig. 18.2).

Having the patient experience healing is the goal of the consultation, but healing is an emergent pattern of complex responsive processes of relating and, as such, is inherently unpredictable. Clinicians can make the emergence of healing more likely by being able to recognize and participate in the processes of healing relationships: valuing, appreciating power, and abiding. Clinicians who are able to do this demonstrate certain competencies, namely, clinical wisdom, mindfulness, emotional self-management, and projection of confidence in their use of the tools of medical technology.

Mindful practice, characterized by attentive observation, critical curiosity, beginner's mind, and presence, is necessary for clinicians to be aware of the nuances of emergent patterns in the consultation, and to recognize when it is necessary to risk introducing novelty in the complex responsive processes of relating that may (or may not) transform repetitive unproductive and unhealthy patterns into ones that promote healing.

As the clinical examples demonstrate, the understanding of clinical consultations as analogous to complex systems at the edge of chaos has the potential to improve the quality of medical care clinicians provide and patients receive in the course of their interaction together.

Acknowledgments The author would like to thank Curt Lindberg, who introduced the author to Stacey's ideas of complex responsive processes of relating and read several early drafts of the manuscript, providing much helpful input on the work of Ralph Stacey.

References

1. Stacey RD. Complex responsive processes in organizations: learning and knowledge creation. London: Routledge; 2001.
2. Scott JG, Cohen D, Dicicco-Bloom B, Miller WL, Stange KC, Crabtree BF. Understanding healing relationships in primary care. Ann Fam Med. 2008;6(4):315–22.
3. Epstein RM. Mindful practice. JAMA. 1999; 282(9):833–9.
4. Epstein RM. Mindful practice in action (I): technical competence, evidence-based medicine, and relationship-centered care. Fam Syst Health. 2003;21(1):1–9.
5. Boudreau J, Cassell EJ, Fuks A. A healing curriculum. Med Educ. 2007;41(12):1193–201.
6. Scott JG, Scott RG, Miller WL, Stange KC, Crabtree BF. Healing relationships and the existential philosophy of Martin Buber. Philos Ethics Humanit Med. 2009;4:11.
7. Balint M. The doctor, his patient, and the illness. New York, NY: International Universities Press; 1957.
8. Engel GL. The need for a new medical model: a challenge for biomedicine. Science(New York). 1977;196(4286):129–36.
9. Stewart M. Patient-centered medicine: transforming the clinical method. Thousand Oaks, CA: Sage Publications Inc.; 1995.
10. Stewart M. Towards a global definition of patient centred care. BMJ. 2001;322(7284):444–5.
11. Tresolini C, The Pew-Fetzer Task Force. Health professions education and relationship centered care. San Francisco, CA: Pew Health Professions Commision; 1994.
12. Suchman AL. A new theoretical foundation for relationship-centered care. Complex responsive processes of relating. J Gen Intern Med. 2006;21 Suppl 1:S40–4.
13. Hassey A. Complexity and the clinical encounter. In: Sweeney K, Griffiths FE, editors. Complexity and health care: an introduction. Abingdon: Radcliffe Medical Press Ltd; 2002. p. 59–74.
14. Aita V, McIlvain H, Backer E, McVea K, Crabtree B. Patient-centered care and communication in primary care practice: what is involved? Patient Educ Couns. 2005;58(3):296–304.
15. Innes AD, Campion PD, Griffiths FE. Complex consultations and the 'edge of chaos'. Br J Gen Pract. 2005;55(510):47–52.
16. Katerndahl DA. Is your practice really that predictable? Nonlinearity principles in family medicine. J Fam Pract. 2005;54(11):970–7.
17. Martin C, Sturmberg J. Complex adaptive chronic care. J Eval Clin Pract. 2009;15(3):571–7.
18. Sturmberg JP. Systems and complexity thinking in general practice: part 1 - clinical application. Aust Fam Physician. 2007;36(3):170–3.
19. Sturmberg JP, Cilliers P. Time and the consultation–an argument for a 'certain slowness'. J Eval Clin Pract. 2009;15(5):881–5.
20. Sturmberg JP, Martin CM. Knowing–in medicine. J Eval Clin Pract. 2008;14(5):767–70.
21. Thygeson M, Morrissey L, Ulstad V. Adaptive leadership and the practice of medicine: a complexity-based approach to reframing the doctor-patient relationship. J Eval Clin Pract. 2010;16(5):1009–15.
22. Stacey RD. Strategic management and organisational dynamics: the challenge of complexity. 3rd ed. Harlow: Financial Times; 2000.

23. Katerndahl D, Parchman ML. Dynamical difference in patients encounters involving uncontrolled diabetes: an orbital decomposition analysis. J Eval Clin Pract. 2010;16(1):211–9.
24. Sturmberg JP, Siew E-g, Churilov L, Smith-Miles K. Identifying patterns in primary care consultations: a cluster analysis. J Eval Clin Pract. 2009;15(3): 558–64.
25. Mead GH, Morris CW. Mind, self & society from the standpoint of a social behaviorist. Chicago, IL: The University of Chicago press; 1934.
26. Stern DN. The motherhood constellation: a unified view of parent-infant psychotherapy. New York, NY: Basic Books; 1995.
27. Damasio AR. The feeling of what happens: body and emotion in the making of consciousness. 1st ed. New York, NY: Harcourt Brace; 1999.
28. Frankel R, Altschuler A, George S, Kinsman J, Jimison H, Robertson NR, et al. Effects of exam-room computing on clinician-patient communication: a longitudinal qualitative study. J Gen Intern Med. 2005;20(8):677–82.
29. Egnew TR. The meaning of healing: transcending suffering. Ann Fam Med. 2005;3(3):255–62.
30. Hsu C, Phillips WR, Sherman KJ, Hawkes R, Cherkin DC. Healing in primary care: a vision shared by patients, physicians, nurses, and clinical staff. Ann Fam Med. 2008;6(4):307–14.
31. Epstein RM. Mindful practice in action (II): cultivating habits of mind. Fam Syst Health. 2003;21(1):11–7.
32. Goldberger AL, Amaral LA, Hausdorff JM, Ivanov P, Peng CK, Stanley HE. Fractal dynamics in physiology: alterations with disease and aging. Proc Natl Acad Sci USA. 2002;99 Suppl 1:2466–72.
33. Berra Y. Yogi Berra Qotes. Brainy Qote (2011) Available from: http://www.brainyquote.com/quotes/authors/y/yogi_berra_2.html. Cited 20 Mar 2011
34. Rollnick S, Miller WR, Butler C. Motivational interviewing in health care: helping patients change behavior. New York, NY: Guilford Press; 2008.

Inflammation Through a Psychoneuroimmunological Lens

19

Jeanette M. Bennett, Brandon L. Gillie, Monica E. Lindgren, Christopher P. Fagundes, and Janice K. Kiecolt-Glaser

19.1 Psychoneuroimmunology

Psychoneuroimmunology (PNI) is a field in which researchers investigate the intersections among behavior, the nervous system, and the immune system. Its development over the past 35 years followed the realization that the immune system does not function in isolation [1, 2]. The initial focus on biological mechanisms encouraged the use of animal models to uncover pathways through which the brain and behavior affect immune activity. Once biological pathways were established, researchers started replicating similar results in humans with broad applications in areas such as infectious diseases, cardiovascular disease, autoimmunity, and cancer. Today, the transdisciplinary field of PNI continues to unravel the complex connections among behavior, immune function, and health.

In this chapter, we use a PNI lens to understand and describe the complex influences of biology and psychology on inflammation. Inflammation is an underlying etiological factor in many chronic diseases. A brief description of brain–immune communication is first introduced as background, followed by a summary of inflammation's effect on health. The biological, psychological, and psychosocial influences on inflammation are then discussed, followed by a review of inflammation and cellular aging.

J.M. Bennett (✉)
Division of Oral Biology, College of Dentistry, The Ohio State University, Columbus, OH, USA

Institute for Behavioral Medicine Research, College of Medicine, The Ohio State University, Columbus, OH, USA

Department of Psychology, University of North Carolina at Charlotte, 9201 University City Blvd, Colvard 4049, Charlotte, NC 28223, USA
e-mail: jeanette.m.bennett@gmail.com

B.L. Gillie
Department of Psychology, The Ohio State University, 1835 Neil Avenue, Columbus, OH 43210, USA
e-mail: gillie.6@osu.edu

M.E. Lindgren
Department of Psychology, The Ohio State University, 460 Medical Center Drive, Columbus, OH 43210-1228, USA

Institute for Behavioral Medicine Research, College of Medicine, The Ohio State University, 460 Medical Center Drive, Columbus, OH 43210-1228, USA

C.P. Fagundes
Institute for Behavioral Medicine Research, College of Medicine, The Ohio State University, 460 Medical Center Drive, Columbus, OH 43210-1228, USA
e-mail: fagundes.1@osu.edu

J.K. Kiecolt-Glaser
Institute for Behavioral Medicine Research, Comprehensive Cancer Center, Department of Psychiatry College of Medicine, The Ohio State University,

Department of Psychology, The Ohio State University, 460 Medical Center Drive, Columbus, OH 43210-1228, USA
e-mail: Janice.Kiecolt-Glaser@osumc.edu

J.P. Sturmberg and C.M. Martin (eds.), *Handbook of Systems and Complexity in Health*,
DOI 10.1007/978-1-4614-4998-0_19, © Springer Science+Business Media New York 2013

19.2 Neuroendocrine–Immune Communication

The two major stress systems include the sympathetic–adrenal–medullary (SAM) axis and the hypothalamic–pituitary–adrenal (HPA) axis. Both systems influence inflammation and affect immune cells through adrenergic and glucocorticoid receptors; the end products of SAM and HPA axes can modulate immune functioning [3]. Of note, neuroimmune communication is not limited to these two pathways; however, an in-depth review of the bidirectional communications between the nervous and immune systems is beyond the scope of this chapter. There are several thorough reviews that address the ways in which neuroimmune communications occur and the observed effects [4–6].

The SAM axis connects the brain directly to the adrenal medulla via sympathetic innervations. Upon stimulation, the adrenal medulla releases catecholamines, epinephrine and norepinephrine. Although catecholamines have short half-lives and are metabolized quickly, they can regulate many facets of the immune system [4]. Therefore, chronic sympathetic activation can lead to immune dysregulation.

Epinephrine increases interleukin (IL)-6 and tumor necrosis factor-alpha (TNF-α) production during stress [6]. In addition, norepinephrine promotes nuclear factor-kappa B (NF-κB) activation [7]. NF-κB is a transcription factor that regulates the gene expression of several proinflammatory mediators, such as IL-6 and IL-8 [8, 9]. NF-κB activation increases the gene expression of inflammatory mediators, which in turn enhances inflammation [7]. Therefore, epinephrine and norepinephrine can induce proinflammatory cytokine production.

Although inherently slower than the SAM axis, the HPA axis provides a more sustained response following activation. It begins with the release of corticotropin-releasing hormone (CRH) from the hypothalamus. CRH triggers the release of adrenocorticotropic hormone (ACTH) from the anterior pituitary into the blood stream. In turn, ACTH stimulates the adrenal cortex, its target organ, to produce cortisol, a glucocorticoid [10]. A negative feedback loop regulates HPA axis activation. Cortisol binds to glucocorticoid receptors in the hippocampus which inhibit the production of CRH and ACTH from the hypothalamus and anterior pituitary, respectively [11]. Other neuroendocrine hormones influence the HPA axis including androgens, estrogens, and posterior pituitary hormones, vasopressin and oxytocin [12–14].

Cortisol can inhibit immune cell activity by binding to glucocorticoid receptors; this process inhibits activation and release of proinflammatory cytokines [15, 16]. However, chronic stress can lead to hippocampal damage and HPA axis dysregulation resulting in increased cortisol production [17]. Chronically elevated cortisol can induce glucocorticoid insensitivity where immune cells downregulate the expression of glucocorticoid receptors [18, 19]. As a result, inflammation is increased due to unregulated immune cells producing proinflammatory cytokines [20].

Neuroendocrine–immune communication is not unidirectional. The immune system communicates with the brain via cytokines. For example, IL-1 receptors are located throughout the brain, especially in the hypothalamus. In turn, IL-1 can stimulate CRH secretion from the hypothalamus, leading to increased HPA axis activity [21]. Peripheral cytokines induce sickness behavior, behavioral changes that are associated with fever, decreased energy, decreased appetite, and changes in sleep [22]. Proinflammatory cytokines can access the brain through a variety of pathways including the leaky regions in the blood–brain barrier (e.g., circumventricular organs) and cytokine-specific transport molecules expressed on brain endothelium [5]. In addition, the vagus nerve detects cytokine levels in the periphery and relays this information to the brain via afferent fibers [23, 24]. This bidirectional communication not only allows an integrated response to occur, but also increases the opportunity for dysregulation when one system is disrupted (Fig. 19.1).

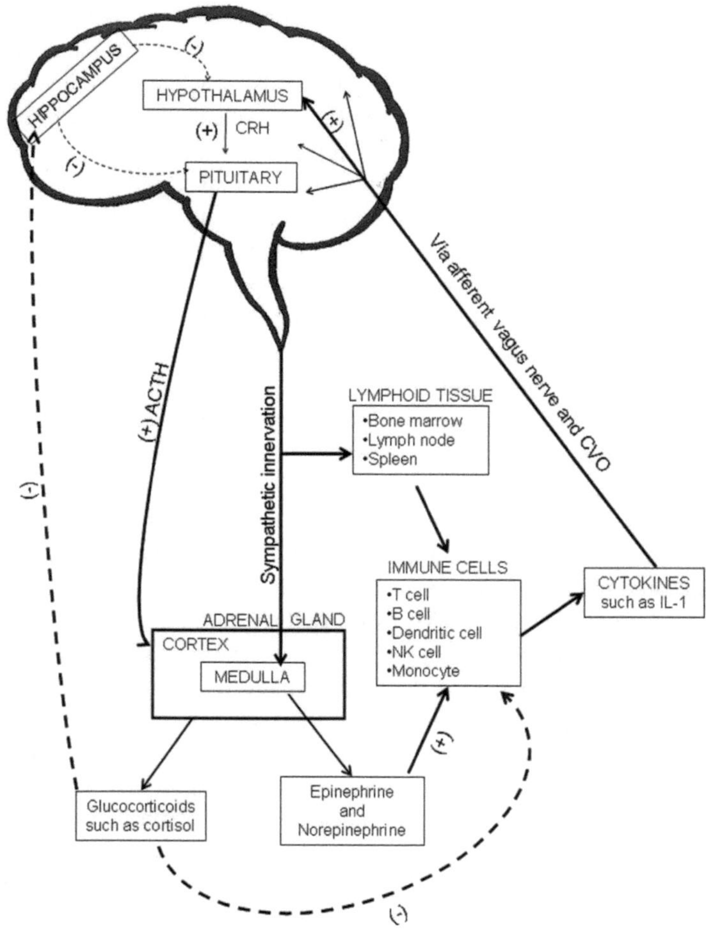

Fig. 19.1 Neuroendocrine–immune bidirectional communication. Hypothalamic-pituitary-adrenal (HPA) axis and the sympathetic nervous system influence the immune cells through glucocorticoid and adrenergic receptors. Immune cells can communicate with the brain via peripheral cytokine levels surveyed by the circumventricular organs and the afferent vagus nerve. *CRH* corticotropin-releasing hormone, *ACTH* adrenocorticotropic hormone, *NK* natural killer, *IL* interleukin, *CVO* circumventricular organs

19.3 Health Consequences of Inflammation

Inflammation is an immune response to infection or injury that aids in the removal of foreign pathogens and promotes wound healing. Acute inflammation is beneficial; however, chronic low-grade inflammation is harmful. Chronically high levels of inflammation are found in a number of age-related diseases including cardiovascular disease and cancer [25–28].

Inflammation can be measured by assessing serum or plasma levels of acute-phase proteins and proinflammatory cytokines. C-reactive protein (CRP) is the most commonly studied acute-phase protein; acute infections and tissue damage increase IL-6 levels that in turn induce liver production of CRP [29]. CRP can bind to foreign or damaged cells and lead to cell destruction. Many cells throughout the body including immune cells, adipocytes (fat cells), and damaged cells, produce proinflammatory cytokines such as IL-6 and TNF-α that then recruit and stimulate additional immune cells to clear and repair tissue. In addition, IL-6, TNF-α, and IL-1 levels follow a diurnal rhythm such that peak levels occur during the early night and reach a nadir in the morning [30, 31]. CRP, however, does not appear to vary across the day [32].

Outside of acute infection and tissue injury, CRP is considered clinically relevant as a nonspecific biomarker of inflammation; minor elevations have been linked to cardiovascular disease risk [33]. For example, individuals with CRP greater than 3 mg/L are at higher risk for developing cardiovascular disease. Unlike CRP, there are no clinically relevant standards for proinflammatory cytokines. Therefore, a typical research strategy is to compare individuals with higher proinflammatory cytokines to those individuals with lower levels or unhealthy patient populations with healthy controls. In addition, researchers may also investigate within-person changes in proinflammatory cytokine levels following a study manipulation such as an intervention or laboratory stressor.

Individuals who have higher levels of inflammation are at greater risk for many diseases including cancer, cardiovascular disease, type 2 diabetes, Alzheimer's disease, osteoporosis, rheumatoid arthritis, and periodontal disease. Elevated inflammation is associated with greater all-cause mortality risk [34]. We briefly review how inflammation contributes to cardiovascular disease, cancer, and type 2 diabetes, three diseases that account for the majority of deaths in developed countries [35].

In the case of cardiovascular disease, proinflammatory cytokines facilitate early atherogenesis and clinical vascular events [36]. Inflammation contributes to atherosclerosis by reducing vascular endothelial cells' capacity to resist leukocyte (white blood cell) adhesion. When leukocytes adhere to vascular endothelial cells, they proliferate, and enhance cytokine production. Elevated inflammation has been implicated in the onset of clinical vascular events because they weaken fibrous caps. Weak fibrous caps are more likely to rupture leading to a heart attack or stroke [37].

Inflammation is also linked to cancer incidence and progression [38]. Chronic inflammation is a contributing factor in at least 15% of all cancers and also influences tumor survival, proliferation, invasion, angiogenesis, and metastases [38–40]. When proinflammatory cytokines enter tumor cells, they promote uncontrolled growth and subsequent metastasis. Furthermore, when macrophages are activated during the inflammatory response, they release many different cancer-promoting messengers including growth and angiogenic factors, proteases, and reactive oxygen species [40].

Individuals with type 2 diabetes are insulin-resistant, which means they either cannot produce enough insulin or the body cannot use the insulin adequately. Inflammatory cytokines can mediate insulin resistance. For example, elevated inflammation impairs blood glucose control by suppressing insulin signal transduction [41, 42]. Furthermore, TNF-α is the major proinflammatory cytokine implicated in this process [43].

In sum, elevated inflammation has been linked to disease progression. Yet, it is unknown whether higher cytokine levels cause the disease, or if the disease results in greater proinflammatory cytokine production. However, we do know several factors that influence inflammation. The following sections describe how biology and behavior affect proinflammatory mediators.

19.4 Biological Influences on Inflammation

19.4.1 Age

Proinflammatory cytokine levels rise with age and have known ties to a number of age-related illnesses [27, 44]. Circulating IL-6, soluble IL-6 receptor (sIL-6r), TNF-α, soluble TNF receptor II (sTNFR-II), and IL-1 receptor antagonist (IL-1ra) increase with age [44–47]. A recent review describes the relationship between age and inflammation as linear, but evidence has not established the age when the relationship can first be detected [48]. For example, in studies with middle-age and older adults (≥40 years old), inflammation increases with age [45, 46, 49, 50]. However, among young adults (≤30 years old), the linear relationship between inflammation and age does not appear consistently [46, 49], suggesting that young adults' health behaviors may have more salient influences on inflammation than age.

Epidemiological studies in healthy older adults indicate a twofold higher risk of all-cause mortality in those who had IL-6 levels in the highest quartile compared to those in the lowest IL-6 quartile, independent of known health risks [34]. When compared to those in the lowest tertile, elderly individuals whose IL-6 levels were within the highest tertile range were nearly 2 times more likely to develop mobility-related disability, and 1.5 times more likely to develop additional disability related to activities of daily living [51].

Interleukin-6 promotes CRP production by the liver [52]. In a group of healthy participants, older adults [75.4 ± 6.8 years (±SD)] had higher CRP than young adults [31.6 ± 7.7 years (±SD)] [53]. In several large population-based studies, CRP increased as men and women aged, even after controlling for possible pre-existing conditions and sub-acute illnesses such as cardiovascular risk factors and disease [46, 49, 54]. High CRP levels are clinically significant; particularly when predicting cardiovascular disease risk [29, 55, 56]. In a recent meta-analysis, individuals with CRP levels >3.0 mg/L were 1.54 times more likely to experience a cardiovascular event than those with <1.0 mg/L CRP [57].

19.4.2 Obesity

Obesity is characterized by elevated circulating proinflammatory cytokines; hence, obese individuals experience a state of chronic inflammation. In epidemiological studies, obese individuals had higher CRP compared to those not obese, even after controlling for negative health behaviors and disease status [58]. Similarly, obese individuals had higher CRP per unit increase in weight, body mass index (BMI), and waist circumference compared to normal weight individuals over a 10-year span [59]. Circulating IL-6, as well as IL-6 produced from abdominal adipose tissue, increases with adiposity [60]. In addition, IL-6 released from abdominal adipose tissue accounts for an estimated 30% of systemic IL-6 in healthy, overweight subjects [60]. Among premenopausal women, obese women had higher IL-6 levels before and after public speaking stress compared to non-obese women [61].

Obesity-induced inflammation has been linked to the development of insulin resistance. Increased obesity was associated with greater CRP, IL-6, and TNF-α. Higher CRP was also related to insulin resistance, suggesting that elevated inflammation may underlie the progression of metabolic syndromes including type 2 diabetes [62]. In participants with obesity-related insulin resistance, abdominal adipose tissue expression of TNF-α and plasma IL-6 were elevated compared to insulin-sensitive participants [63]. Interestingly, the two groups were matched for BMI, suggesting that being insulin-resistant elevates inflammation beyond that observed in obese individuals.

Diseases with an inflammatory component can be exacerbated by insulin resistance. For example, hepatitis-C-infected patients with comorbid type 2 diabetes had higher TNF-α levels than patients without type 2 diabetes [64]. In addition, TNF-α inhibitors significantly improved insulin sensitivity in patients with rheumatoid arthritis [65]. The infusion of TNF-α lowered insulin-mediated glucose uptake and induced IL-18 gene expression in human muscle tissue [66], which demonstrates the relationship between these two inflammatory mediators and their effects on insulin resistance.

Weight loss lowers inflammation. For example, a diet-induced weight loss intervention reduced circulating levels of CRP, IL-6, and sTNFR-1 in a sample of older adults, regardless of physical activity, suggesting that weight reduction is independently associated with reduced proinflammatory cytokines [67]. Serum TNF-α levels in obese individuals fell ~25% after an average weight loss of 12 kg [68]. Two years after a diet and exercise intervention in obese women, the treatment group had lower IL-6, IL-18, and CRP levels related to weight loss than obese women in the control group [69]. In another study, weight loss reduced plasma IL-18 and increased insulin sensitivity [70].

Measures of relative fat mass composition may partially account for the relationship between physical activity and inflammation. For instance, more physical activity resulted in lower IL-6, CRP, and sTNFR than less physical activity; however, when adjustments were made for BMI and leptin levels, physical activity no longer was related to decreased inflammation [71]. During a 3-year follow-up period, increased low-grade inflammation was associated with greater adiposity, but not physical fitness [72]. Therefore, although physical activity is associated with lower inflammation, this relationship may result from less obesity in physically active people.

19.4.3 Sex

Sexual dimorphic immune responses can be readily observed in human populations. For example, women are more like to suffer from an autoimmune disease; however, men are disproportionately affected by Parkinson's disease and early-onset cardiovascular disease [73, 74]. Gonadal hormones (e.g., estrogen, progesterone, and testosterone) may partially account for the differences observed between males and females. Androgen and estrogen receptors are present on immature immune cells in the thymus and bone marrow [75–77]. However, sex differences in gonadal hormones do not fully account for disparities in circulating inflammatory markers between males and females.

Levels of most inflammatory markers do not differ consistently between sexes, although CRP levels are one exception. In large population-based studies, females have higher CRP levels than males [78–80]. During the follicular phase of the menstrual cycle, women had lower levels of CRP compared to those in the luteal phase [81]. Post-menopausal women have higher CRP than premenopausal women [82]. In addition, women using oral contraceptives or hormone replacement therapy (HRT) have increased CRP levels compared to age-matched women not taking hormones [49, 83–87].

Unlike the reliable CRP difference, proinflammatory cytokines such as IL-6 and TNF-α are not always different between the two sexes [88, 89]. It remains unclear whether menstrual cycle phase and menopausal status impact proinflammatory cytokines. The follicular phase may be associated with higher IL-6 levels compared to the luteal phase [90]. However, several studies suggest that inflammation is greater during the luteal phase compared to the follicular phase [91–93]. Neither menstrual cycle phase nor oral contraceptive use affects proinflammatory cytokine levels [87, 94–96]. The use of HRT inconsistently affects proinflammatory cytokines, with studies showing decreases, increases, and no change [83, 84, 97, 98]. These discrepant findings may be due to relatively small sample size; the majority of the proinflammatory cytokine studies include 68 women or less.

19.5 Psychological Influences on Inflammation

19.5.1 Depression

Patients with inflammatory-related diseases including cardiovascular disease and cancer have higher rates of depression compared to healthy individuals [99, 100]. Both syndromal depression and depressive symptoms are associated with heightened levels of proinflammatory mediators including IL-1, IL-6, and CRP [101–105]. Additionally, depression severity and inflammation appear to have a dose–response relationship; as depressive symptoms worsen, inflammatory markers increase [104, 106]. While these findings demonstrate an association between depression and circulating levels of proinflammatory cytokines, it is important to consider factors that influence inflammation and covary with depression including antidepressant use, sex, BMI, and comorbid symptoms of anxiety [48, 107].

Not only do depressed people have higher inflammatory levels, they also have a greater inflammatory response to stress. For example, compared to nondepressed males, those with major depression show greater IL-6 and NF-κB activity in response to acute psychosocial stress [108]. Clinically depressed individuals also display decreased sensitivity to the anti-inflammatory properties of glucocorticoids, resulting in greater production of IL-6 and TNF-α compared to their nondepressed counterparts [109, 110]. Thus, excessive NF-κB activity and decreased responsiveness to glucocorticoids may enhance and sustain the production of proinflammatory cytokines in individuals with depression.

Growing evidence suggests that the relationship between depression and inflammation is bidirectional. Administration of interferon-alpha and other cytokine inducers produces depression-like symptoms including low mood, fatigue, and psychomotor slowing in otherwise healthy volunteers [111, 112]. Cytokines appear to influence the production and metabolism of mood-relevant neurotransmitters such as serotonin, dopamine, and norepinephrine [113]. Moreover, clinically depressed individuals who receive anti-inflammatory medication in addition to antidepressants show greater symptomatic reduction than those who receive a combination of antidepressant and placebo [114, 115]. Elevated inflammation affects not only physical health, but also emotional well-being, including anxiety.

19.5.2 Anxiety

Laboratory-based and cross-sectional studies in healthy and patient populations have been used to investigate the relationship between anxiety and inflammation. In the laboratory setting, stress-induced increases in anxiety and anger enhanced IL-6 production following stress [116]. These associations varied by sex; for women, anxiety was more strongly associated with IL-6 responses, while anger in men was related to IL-6 production [116]. Administration of endotoxin, a substance used to mimic an actual infection, increased anxiety as well as circulating levels of TNF-α, IL-6, and IL-1ra [111].

Cross-sectional studies indicate that anxiety can influence inflammation outside the laboratory. During an examination, anxious medical students produced more proinflammatory interferon-gamma (IFN-γ) and less anti-inflammatory IL-10 and IL-4 compared to non-anxious medical students [117]. More anxious adults had higher CRP, IL-6, and TNF-α than less anxious ones [118].

Anxiety may exacerbate inflammatory responses in people with allergies. In patients with allergic rhinitis (AR), anxiety enhanced the effects of stress on late-phase responses assessed 24-h after a skin prick test (SPT), and was associated with higher IL-6 production [119]. Therefore, continued inflammation that occurs during late-phase allergic responses may "prime" hyperresponsiveness to irritant triggers and other allergens, especially in anxious AR patients. In addition, anxious AR patients' lymphocytes had greater Concanavalin A (ConA)-stimulated IL-6 production compared to those who were not anxious [119].

Chronically ill individuals may be especially susceptible to anxiety's effect on inflammation. For instance, leukocytes from anxious hemodialysis (HD) patients produced significantly higher in vitro levels of IL-6 compared to less anxious HD patients [120]. This anxiety-related increase within the HD patient group was over and above the already observed higher inflammation in the HD patients compared to healthy controls [120], suggesting that anxiety may have an additive effect on inflammation in patient populations.

19.6 Psychosocial Influences on Inflammation

19.6.1 Socioeconomic Status

Epidemiological data demonstrate consistent and striking effects of socioeconomic status (SES) on health outcomes [121, 122]. Measures of SES often include income, education, and occupational prestige as the three main components. Lower SES individuals have higher rates of all-cause mortality and a lower life expectancy [123–125]. In particular, one estimate indicates that those with a lower SES have a lifespan 4.5 years shorter than their higher SES counterparts [126]. Furthermore, health disparities increase with each step down the SES ladder [122].

An individual's SES can shape their life course and lead to a number of lifestyle choices, many of which may contribute to the observed association between SES and health. For instance, individuals with low SES are more likely to engage in behaviors such as smoking, excessive alcohol use, reduced physical activity, and they are more likely to experience stress and depression, all of which can negatively impact health [127]. Despite these associations, the relationship between low SES and mortality persists even when these factors are statistically controlled [128].

Heightened inflammation may provide one link between low SES and poor health outcomes. In fact, a number of acute and chronic medical conditions are associated with both elevated levels of inflammatory markers and low SES. Compared to higher SES individuals, lower SES individuals have higher IL-6, TNF-α, and CRP [55, 129–132]. The individual components of most composite SES measures, such as income and education, show similar negative associations with proinflammatory cytokines [133, 134]. While informative, these associations do not explain the mechanisms through which low SES promotes inflammation and, by proxy, poor health outcomes.

Different inflammatory responses to psychological stress may partly account for health disparities between SES groups. Compared to high SES individuals, lower SES individuals show greater increases in IL-6 and CRP that persist longer in response to acute mental stress [135, 136]. Thus, lower SES individuals tend to have maladaptive responses to stress, an attribute which may play a role in maintaining higher levels of inflammation. While the pathways through which low SES individuals develop negative health outcomes remains unclear, increased inflammation represents an attractive possibility.

19.6.2 Social Support

Close relationships have clear ties to better health and reduced inflammation may account for these associations. Social support refers to the degree that one believes that support would be available if and when it is needed [137]. In one study, older women who had more satisfying interpersonal relationships had lower IL-6 compared to those who had less satisfying relationships [138]. In another study, women with ovarian cancer who reported greater social support had lower circulating IL-6 levels compared to women who reported less social support [139]. Furthermore, gynecologic cancer survivors who sought more support at diagnosis had lower circulating IL-6 one year later compared to those who sought less support [140].

19.6.3 Marriage

Married individuals' mortality rates are lower than those of their unmarried counterparts [141]. Inflammation may be one possible mechanism for these findings. In a population-based study of community-dwelling older adults, being married was associated with reduced CRP for both sexes; these effects were particularly pronounced in men [142]. The absolute magnitude of the risk reduction for married men was equivalent to being a nonsmoker, having normal blood pressure, and having a healthy BMI [142].

While marriage typically has positive health benefits, marital quality has important health implications [143]. Marital interaction studies demonstrate the relationship between marital quality and immune function. Hostile marital interactions have particularly important negative physiological consequences. Both younger and older couples who were more hostile to their spouse during marital problem discussions produced more epinephrine, norepinephrine, and ACTH than their less hostile counterparts [144]. In another study in which couples engaged in a supportive discussion and a marital problem discussion across two separate sessions, those couples who were more hostile produced more IL-6 after the conflict discussion than the supportive discussion (113 vs. 45%). In contrast, less hostile couples' IL-6 production was similar after both discussions (70 vs. 65%) [88].

Cognitive engagement (the use of cognitive processing words) during a marital disagreement is associated with a dampened inflammatory response. More cognitively engaged individuals produced less IL-6 and TNF-α in the following 24 h after a disagreement compared to less cognitively engaged individuals [145]. In addition, those who were more cognitively engaged had lower

Table 19.1 Summary of key characteristics and health behaviors that influence inflammation

Individual characteristics/health behaviors	Effects on inflammation
Aging	↑ IL-6, TNF-α, CRP [44–47, 49, 50, 53, 54]
Obesity/higher BMI	↑ CRP, IL-6, TNF-α [49–63, 71]
Weight loss	↓ CRP, IL-6, IL-18, TNF-α [64–67, 69]
Sex	CRP: females>males [78–80]
Depression	↑ IL-1, IL-6, CRP [101–105, 108–110]
Anxiety	↑ TNF-α, IL-6, CRP [116–120]
Low social economic status	↑ TNF-α, IL-6, CRP [55, 129–132]
Low social support/poor martial quality	↑ IL-6, CRP [88, 138–140, 142]
Smoking	↑ TNF-α, IL-6, IL-8, CRP [79, 151–167]
Exercise	Immediate: ↑ IL-6, IL-8, IL-15 [168–172] Long term: ↓ CRP, IL-1, IL-6, IFN-γ [174–178, 180, 181] ↑ IL-10 [176, 177]
Poor diet	↑ CRP, IL-1, TNF-α, IL-6 [190–197]
Poor sleep	↑ IL-6, TNF-α, and CRP [211–215] ↓ IL-10 [215]

absolute levels of IL-6 and TNF-α than those who were less cognitively engaged at baseline [145].

Marital stress may be particularly detrimental if combined with other known health risk factors, such as BMI or sagittal abdominal diameter. Women with larger waists showed a stronger positive association between marital stress and CRP than women with smaller waists [146]. Given that having higher levels of CRP raises cardiovascular disease risk [147], the combination of marital stress and having a large waist may be particularly prognostic for heart problems. See Table 19.1 for a summary of characteristics and health behaviors that affect inflammation.

19.7 Health Behaviors and Inflammation

19.7.1 Smoking

Smoking tobacco has been linked to the development of many chronic diseases, such as heart disease, stroke, diabetes, cancer, and chronic airway inflammation such as chronic obstructive pulmonary disease and continues to be the most preventable cause of illness and death in the United States [148]. On average, adults who smoke cigarettes die 14 years earlier than nonsmokers [149]. Smokers' greater inflammatory state may underlie the increased risk of developing chronic diseases and premature death [150].

Smoking appears to elevate CRP [151–153]. In large-scale, population-based studies across several countries, male and female smokers had higher CRP than nonsmokers [79, 154–157]. CRP levels increase with smoking exposure in a dose-dependent manner [158, 159]. Furthermore, CRP remained higher in former smokers even 10–20 years following smoking cessation compared to those who have never smoked [154, 160, 161]. Lifetime smoking exposure elevates CRP levels in both smokers and former smokers [152, 162]; specifically, greater smoking exposure is associated with higher CRP levels in smokers and slower CRP decline after smoking cessation.

Smoking also enhances IL-6 and TNF-α production. Male and female smokers had substantially higher IL-6 compared to former smokers and nonsmokers [155, 163–165]. Similar to the relationship between CRP and smoking exposure, the greater number of cigarettes smoked per day, the higher circulating IL-6 in current smokers [164]. In former smokers, IL-6 remained elevated compared to nonsmokers and decreased significantly as abstinence increased [164]. Male smokers had higher TNF-α than nonsmokers; among smokers, greater tobacco exposure (i.e., pack years) was associated with more TNF-α [166]. An additional study suggested that smokers may also have higher IL-8 and monocyte chemotactic protein (MCP)-1 than nonsmokers [167].

19.7.2 Exercise

Exercise increases proinflammatory cytokine production [168]. Acute IL-6, IL-8, and IL-15 increases during and following exercise have been consistently demonstrated [169–172]. In the laboratory, endotoxin was administered to young healthy males during rest, following exhaustive exercise, or after an injection of IL-6 [173]. In response to the endotoxin, the exercise and IL-6 groups' plasma TNF-α rise was attenuated compared to the rest group [173]. These results suggest that exercise-induced elevations of IL-6 may have anti-inflammatory effects.

Many studies have shown that increased physical activity lowers inflammation. In population-based studies, more physically active adults had lower serum CRP levels, even when controlling for possible demographic confounds and health behaviors [174, 175]. Among older men, higher fitness levels were associated with lower IL-6 and higher IL-10 [176].

Longitudinal studies also demonstrate the anti-inflammatory benefits of exercise. In a 12-week study, coronary heart disease patients who underwent an intense aerobic training program had lower IL-6, IL-1, and IFN-γ levels and higher levels of the anti-inflammatory cytokine IL-10 compared to their baseline levels [177]. Furthermore, at the end of the study, CRP levels had improved significantly in all participants; among those at the highest risk for developing type 2 diabetes, CRP was 46% lower [177].

CRP levels dropped following a 2-month exercise training program in women [178]. However, women in the moderate weight-reduction quartile showed the most significant CRP decreases, even over those in the largest weight-reduction quartile. These data suggest that women who had the greatest weight loss may have been the result of overtraining, which can lead to increased inflammation [178].

Patients undergoing an exercise and pharmacological (i.e., pravastatin) intervention trial had similar reductions in MCP-1, regardless of exercise assignment [179]. However, the combination group's IL-8 levels decreased significantly more than the drug use only group [179], suggesting that exercise provided additional anti-inflammatory benefits beyond the pharmacological intervention.

Yoga practice also may reduce inflammation. For example, yoga reduced IL-6 and CRP levels in patients with chronic heart failure compared to pre-yoga baseline levels [180]. In a study of healthy participants, expert yoga practitioners had 41% lower serum IL-6 levels compared to novice yoga practitioners [181]. In addition, the novice group was 4.75 times as likely to have detectable CRP levels compared to the expert group. Following an acute stressor, stimulated IL-6 production in the expert group was lower compared to the novice group, suggesting that extended yoga practice may buffer stress-induced proinflammatory cytokine elevations [181].

19.7.3 Nutrition

Large-scale epidemiological studies demonstrate relationships among diet, health, and inflammation. Diets that are high in refined grains, processed meat, sugar, saturated and *trans*-fatty acids, and low in fruits, vegetables, and whole grains promote inflammation and increase the risk for cardiovascular disease and type 2 diabetes [182–185]. Diets are becoming increasingly less healthy, therefore it is important to understand the ways dietary components can elevate inflammation.

The intake of certain macronutrients may produce oxidative stress and lead to inflammation. Oxidative stress results from the metabolism of food and can promote inflammation through activation of the NF-κB pathway [186]. In particular, ingestion of glucose is associated with greater oxidant production and increased NF-κB activity [187, 188]; intravenous administration of glucose raises circulating levels of IL-6 and TNF-α [189, 190]. Moreover, metabolism of high-fat meals begets increased levels of glucose and triglycerides that can enhance oxidative stress and promote increases in IL-6 and CRP [191]. In contrast, higher fruit and vegetable intake is associated with lower oxidative stress and inflammation, which may counteract the proinflammatory responses to high saturated fatty meals [190, 192].

Some dietary components are the molecular precursors of proinflammatory cytokines. For instance, the omega-6 (n–6) polyunsaturated fatty acid (PUFA), arachidonic acid (AA), found in refined vegetable oils, such as corn, sunflower, and safflower, is a major substrate in the synthesis of eicosanoids, molecules that help regulate the intensity and duration of the inflammatory response [193]. Overconsumption of n–6 PUFAs increases the production of IL-1, TNF-α, and IL-6 [194, 195]. In contrast, the omega-3 (n–3) PUFAs found in fish, fish oil, and flax seed decrease the production of inflammatory eicosanoids and cytokines [193, 195]. Two key n–3 PUFAs, eicosapentaenoic acid (EPA) and docosahexanoic acid (DHA), can decrease NF-κB activity and TNF-α transcription in response to endotoxin exposure [196, 197].

19.7.4 Sleep

Sleep is essential for good health. Short sleep duration (<7 h/night), poor sleep quality, and extended sleep latency are associated with higher risk for all-cause mortality [198–200]. Sleep disruptions also play a role in inflammatory-related diseases and conditions. For example, disrupted sleep is thought to advance the onset of type 2 diabetes [201] and is a prominent feature of major depressive disorder [202].

The relationship between sleep and proinflammatory cytokines is complex and bidirectional. Circulating levels of IL-6, TNF-α, and IL-1 exhibit a diurnal rhythmicity such that peak levels occur during the early night and reach a nadir in the morning [30, 31]. Cortisol and growth hormone also exhibit a circadian rhythm, suggesting that the effect of sleep on the immune system may be mediated in part through changes in hormones [203]. Thus, cytokine levels are linked to the onset of sleep. Although it is unclear why these variations in cytokine levels occur with the onset of sleep, nonrapid eye movement (NREM) sleep may serve to reallocate energy resources from wakefulness activities to immune responses, which combat latent infections [204].

Consistent with this idea, healthy volunteers injected with endotoxin show increases in the amount and intensity of NREM sleep [205, 206]. Additionally, cytokines themselves produce alterations in normal sleep functions. For example, the administration of IL-1, TNF-α, and IL-6 produces increases in NREM sleep and decreases in rapid-eye movement (REM) sleep for both animals and humans [207, 208]. Taken together, these findings suggest that cytokines are not only influenced by sleep but also actively regulate sleep activities.

Disruptions of sleep and sleep disorders may affect health through elevations of proinflammatory cytokines. As few as 4 h of sleep loss results in greater NF-κB activation and higher morning levels of IL-6 and TNF-α compared to a night of uninterrupted sleep [209, 210]. Similarly, extensive total sleep deprivation (e.g., staying awake for 88 or more consecutive hours) elevates IL-6 and CRP [211, 212].

The immunomodulatory effects of chronic sleep loss are observed in patients with obstructive sleep apnea (OSA). Patients with OSA exhibit higher nighttime levels of plasma TNF-α and IL-6, which increase after each nighttime episode of OSA, and lower levels of the anti-inflammatory marker, IL-10, compared to control patients [213]. This activation of the inflammatory response during sleep may partially account for the elevated levels of CRP and increased risk for cardiovascular disease observed in patients with OSA [214].

19.7.5 Inflammation and Cellular Aging

Burgeoning data suggest that psychological stress accelerates the cellular aging of the immune system [215–219]. A telomere is a group of nucleoprotein complexes that cap chromosomes to protect and stabilize their integrity across the lifespan [220]. Telomere length is a proxy measure of the biological age of a cell. Shorter telomeres limit the amount of cellular replication, which indicates how close the cell is to death [220, 221]. Young women who were more stressed had shorter telomeres compared to those who were less stressed [215, 219]. Dementia caregiving, a chronic stressor, in older adults was also associated with shorter telomeres and higher TNF-α production [218]. Thus, it appears that stress contributes to accelerated cellular aging.

The evidence-linking stress and telomere shortening suggests that inflammation could be a common biological pathway [222]. In fact, higher levels of inflammation can activate T-cell proliferation, a process that in turn leads to shorter telomeres [223]. Chronic stress also increases oxidative stress [215], which promotes telomere shortening during replication [223]. Because inflammation plays a role in cellular aging, it seems plausible that biological, psychological, and psychosocial factors, as well as health behaviors, may affect telomere length.

Psychological factors like mood disorders and psychosocial factors such as negative childhood experiences are related to shorter telomeres. In a study comparing individuals with mood disorders to healthy, age-matched controls, those with a diagnosed mood disorder had shorter telomeres [216]. Another study showed that patients with major depression had shorter telomeres compared to those without major depression [224]. Childhood maltreatment in young adults has been associated with shorter telomeres [225]. Older adults who experienced an adverse event during childhood had shorter telomeres and higher IL-6 levels than those who did not [226]. These findings suggest that mood disorders as well as negative events from childhood can have lasting effects on cellular aging.

Biological factors and some health behaviors can also modify telomere length. Aging has been associated with shorter telomeres [219, 227] Age-related telomere shortening has been linked to age-related diseases and mortality [228]. Obese women and smokers also have shorter telomeres [219, 227]. Less physically active participants had shorter telomeres than more physically active ones [229]. Therefore, factors that are known to increase inflammation also shorten telomere length. These data support the conclusion that inflammation may be a biological pathway linking stress and cellular aging.

19.8 Conclusion

Understanding inflammation requires knowledge of multiple biological, psychological, and psychosocial factors, as well as health behaviors. Higher proinflammatory cytokines are associated with aging, obesity, depression, anxiety, poor quality relationships, smoking, poor diet, exercise, and sleep habits. These factors independently impact inflammation but they can also coincide to have additive effects on inflammation.

Socioeconomic status provides an excellent example of how several factors converge to affect inflammation. Individuals with low SES are more likely to smoke, abuse alcohol, be sedentary, have poorer diets, sleep less, and experience more stress and depression [127]. The relationship between low SES and poor health still exists despite statistically controlling for these negative factors, suggesting that the sum effect is greater than its parts.

Health behaviors may buffer these negative psychosocial factors and ameliorate harmful effects on inflammation. Increasing physical activity lowers proinflammatory mediators [177–179]. Restorative yoga participation lessened inflammation in chronic heart failure patients [180]. Weight loss without increasing physical activity also lowered inflammation [67, 68, 71]. Smoking cessation appears to reduce elevated cytokines relatively quickly; however, CRP levels in former smokers may take 10–20 years to drop to those of nonsmokers' [154, 160, 161]. In addition, diets high in fiber and low in saturated fats are also associated with lower inflammation [190, 192].

Because inflammation contributes to many chronic diseases such as cardiovascular disease and diabetes, controlling or reducing inflammation is important. Clearly individual characteristics including age and sex cannot be modified; however, helping individuals change their poor behavioral habits and reducing inflammation may both extend longevity and increase the quality of life.

Taken together, inflammation is a transactional process; many factors can overlap and have additive effects. The observation that proinflammatory cytokines accelerate immune cell replication and cellular aging is a recent finding, suggesting a new and exciting direction for further PNI-focused research. Thus health research using a PNI lens will continue to pioneer novel and integrative investigations into the factors that influence the relationship between inflammation and disease.

Important information researchers should consider when studying inflammation:
• Inflammation is the result of interactions among many biological pathways including the autonomic nervous system, hypothalamic–pituitary–adrenal axis, and the innate immune system
• Psychological stress induces proinflammatory cytokine release
• Psychosocial factors and health behaviors impact chronic inflammation through direct and indirect pathways
• Age is positively associated with inflammation
• Inflammation is important for physical and psychological health; the pathway between the brain and the immune system is bidirectional

Important information clinicians should consider:
• Inflammation is the result of interactions among many biological pathways including the autonomic nervous system, hypothalamic–pituitary–adrenal axis, and the innate immune system
• Controlling/reducing chronic inflammation is important for good physical and psychological health
• Although clinicians cannot change a patient's sex, age, or SES, creating an action plan based on the patient's needs could include
– Weight reduction
– Increasing exercise
– Making better nutritional choices
– Increasing hours slept at night
– Providing mental health referrals

Acknowledgments Work on this chapter was supported in part by the following grants: National Institute on Aging (AG029562), National Institute of Dental & Craniofacial Research (DE014320), National Cancer Institute (CA126857 and CA131029) and American Cancer Society Postdoctoral Fellowship Grant PF-11-007-01-CPPB awarded to CPF. The content is solely the responsibility of the authors and does not necessarily represent the official views of the American Cancer Society or National Institute of Health or any division providing financial support.

References

1. Ader R, Cohen N. Behaviorally conditioned immunosuppression. Psychosom Med. 1975;37:333–40.
2. Solomon GF. Stress and antibody response in rats. Int Arc Allergy. 1969;35:97–104.
3. Glaser R, Kiecolt-Glaser JK. Stress-induced immune dysfunction: implications for health. Nat Rev Immunol. 2005;5:243–51.
4. Wrona D. Neural-immune interactions: an integrative view of the bidirectional relationship between the brain and immune systems. J Neuroimmunol. 2006;172(1–2):38–58.
5. Maier SF, Watkins LR. Cytokines for psychologists: implications of bidirectional immune-to-brain communication for understanding behavior, mood, and cognition. Psychol Rev. 1998;105:83–107.
6. Black PH. Stress and the inflammatory response: a review of neurogenic inflammation. Brain Behav Immun. 2002;16:622–53.
7. Bierhaus A, Wolf J, Andrassy M, Rohleder N, Humpert PM, Petrov D, et al. A mechanism converting psychosocial stress into mononuclear cell activation. Proc Natl Acad Sci USA. 2003;100(4):1920–5.
8. Barnes PJ, Karin M. Nuclear factor-kappaB: a pivotal transcription factor in chronic inflammatory diseases. New Engl J Med. 1997;336(15):1066–71.
9. Baldwin AS. The NF-kappa B and I kappa B proteins: new discoveries and insights. Ann Rev Immunol. 1996;14:649–83.
10. Chrousos GP, Gold PW. The concepts of stress and stress system disorders. JAMA. 1992;267:1244–52.
11. Sapolsky R, Krey L, McEwen B. Glucocorticoid-sensitive hippocampal neurons are involved in terminating the adrenocortical stress response. Proc Natl Acad Sci USA. 1984;81(19):6174.
12. Da Silva JAP. Sex hormones and glucocorticoids: interactions with the immune system. In: Cutolo M, Masi AT, Bijlsma JWJ, Chikanza IC, Bradlow HL, Castagnetta L, editors. Neuroendocrine immune basis of the rheumatic diseases. New York, NY: New York Acad Sciences; 1999. p. 102–18.
13. Kajantie E, Phillips D. The effects of sex and hormonal status on the physiological response to acute psychosocial stress. Psychoneuroendocrinology. 2006;31(2):151–78.
14. Taylor SE, Gonzaga GC, Klein LC, Hu P, Greendale GA, Seeman TE. Relation of oxytocin to psychological stress responses and hypothalamic-pituitary-adrenocortical axis activity in older women. Psychosom Med. 2006;68(2):238–45.

15. Barnes PJ. Anti-inflammatory actions of glucocorticoids: molecular mechanisms. Clin Sci. 1998;94(6): 557–72.
16. Brattsand R, Linden M. Cytokine modulation by glucocorticoids: mechanisms and actions in cellular studies. Aliment Pharmacol Ther. 1996;10:81–90.
17. Sapolsky R, Krey L, McEwen B. Prolonged glucocorticoid exposure reduces hippocampal neuron number: implications for aging. J Neurosci. 1985;5(5):1222.
18. Webster JC, Cidlowski JA. Downregulation of the glucocorticoid receptor. a mechanism for physiological adaptation to hormones. Ann N Y Acad Sci. 1994;746:216–20.
19. Webster JI, Tonelli L, Sternberg EM. Neuroendocrine regulation of immunity. Ann Rev Immunol. 2002;20:125–63.
20. Miller G, Cohen S, Ritchey A. Chronic psychological stress and the regulation of pro-inflammatory cytokines: a glucocorticoid-resistance model. Health Psychol. 2002;21(6):531–41.
21. Sapolsky R, Rivier C, Yamamoto G. Interleukin-1 stimulates the secretion of hypothalmic corticotropin-releasing factor. Science. 1987;238:522–6.
22. Kelley KW, Bluthe RM, Dantzer R, Zhou JH, Shen WH, Johnson RW, et al. Cytokine-induced sickness behavior. Brain Behav Immun. 2003;17 Suppl 1:S112–8.
23. Maier SF, Goehler LE, Fleshner M, Watkins LR. The role of the vagus nerve in cytokine-to-brain communication. Ann N Y Acad Sci. 1998;840(1):289–300.
24. Dantzer R, O'Connor JC, Freund GG, Johnson RW, Kelley KW. From inflammation to sickness and depression: when the immune system subjugates the brain. Nat Rev Neurosci. 2008;9(1):46–56.
25. Kiecolt-Glaser JK, Gouin J-P, Hantsoo L. Close relationships, inflammation, and health. Neurosci Biobehav Rev. 2010;35(1):33–8.
26. Mills PJ, Parker B, Dimsdale JE, Sadler GR, Ancoli-Israel S. The relationship between fatigue and quality of life and inflammation during anthracycline-based chemotherapy in breast cancer. Biol Psychol. 2005;69(1):85–96.
27. Ershler WB, Keller ET. Age-associated increased interleukin-6 gene expression, late-life diseases, and frailty. Annu Rev Med. 2000;51:245–70.
28. Bazzano L, He J, Muntner P, Vupputuri S, Whelton P. Relationship between cigarette smoking and novel risk factors for cardiovascular disease in the United States. Ann intrn med. 2003;138(11):891.
29. Pepys MB, Hirschfield GM. C-reactive protein: a critical update. J Clin Invest. 2003;111(12):1805–12.
30. Bauer J, Hohagen F, Ebert T, Timmer J, Ganter U, Krieger S, et al. Interleukin-6 serum levels in healthy persons correspond to the sleep-wake cycle. Clin Investig. 1994;72(4):315.
31. Moldofsky H, Lue FA, Eisen J, Keystone E, Gorczynski RM. The relationship of interleukin-1 and immune functions to sleep in humans. Psychosom Med. 1986;48(5):309–18.
32. Meier-Ewert HK, Ridker PM, Rifai N, Price N, Dinges DF, Mullington JM. Absence of diurnal variation of C-reactive protein concentrations in healthy human subjects. Clin Chem. 2001;47(3):426–30.
33. Black S, Kushner I, Samols D. C-reactive protein. J Biol Chem [Review]. 2004;279(47):48487–90.
34. Harris T, Ferrucci L, Tracy R, Corti M, Wacholder S, Ettinger WJ, et al. Associations of elevated interleukin-6 and C-reactive protein levels with mortality in the elderly. Am J Med. 1999;106:506–12.
35. Eyre H, Kahn R, Robertson RM. Preventing cancer, cardiovascular disease, and diabetes: a common agenda for the American Cancer Society, the American Diabetes Association, and the American Heart Association. Stroke. 2004;35(8):1999.
36. Hansson G. Inflammation, atherosclerosis, and coronary artery disease. New Engl J Med. 2005;352(16):1685.
37. Libby P. Inflammation in atherosclerosis. Nature. 2002;420(6917):868–74.
38. Coussens LM, Werb Z. Inflammation and cancer. Nature. 2002;420(6917):860–7.
39. Aggarwal BB, Shishodia S, Sandur SK, Pandey MK, Sethi G. Inflammation and cancer: how hot is the link? Biochem Pharmacol. 2006;72(11):1605–21.
40. Marx J. Inflammation and cancer: the link grows stronger. Science. 2004;306:966–8.
41. Duncan B, Schmidt M, Pankow J, Ballantyne C, Couper D, Vigo A, et al. Low-grade systemic inflammation and the development of type 2 diabetes. Diabetes. 2003;52(7):1799.
42. Hildreth C. Inflammation and diabetes. JAMA. 2008;300(21):2476.
43. Hotamisligil G, Peraldi P, Budavari A, Ellis R, White M, Spiegelman B. IRS-1-mediated inhibition of insulin receptor tyrosine kinase activity in TNF-alpha-and obesity-induced insulin resistance. Science. 1996;271(5249):665.
44. Bruunsgaard H, Pedersen M, Pedersen BK. Aging and proinflammatory cytokines. Curr Opin Hematol. 2001;8(3):131–6.
45. Bruunsgaard H, Andersen-Ranberg K, Jeune B, Pedersen AN, Skinhoj P, Pedersen BK. A high plasma concentration of TNF-alpha is associated with dementia in centenarians. J Gerontol A Biol Sci Med Sci. 1999;54(7):M357–64.
46. Ferrucci L, Corsi A, Lauretani F, Bandinelli S, Bartali B, Taub DD, et al. The origins of age-related proinflammatory state. Blood. 2005;105(6): 2294–9.
47. Shurin GV, Yurkovetsky ZR, Chatta GS, Tourkova IL, Shurin MR, Lokshin AE. Dynamic alteration of soluble serum biomarkers in healthy aging. Cytokine. 2007;39(2):123–9.
48. O'Connor MF, Bower JE, Cho HJ, Creswell JD, Dimitrov S, Hamby ME, et al. To assess, to control, to exclude: effects of biobehavioral factors on circulating inflammatory markers. Brain Behav Immun. 2009;23(7):887–97.

49. Hung J, Knuiman MW, Divitini ML, Davis T, Beilby JP. Prevalence and risk factor correlates of elevated C-reactive protein in an adult Australian population. Am J Cardiol. 2008;101(2):193–8.
50. Baggio G, Donazzan S, Monti D, Mari D, Martini S, Gabelli C, et al. Lipoprotein(a) and lipoprotein profile in healthy centenarians: a reappraisal of vascular risk factors. FASEB J. 1998;12(6):433–7.
51. Ferrucci L, Harris TB, Guralnik JM, Tracy RP, Corti MC, Cohen HJ, et al. Serum IL-6 level and the development of disability in older persons. J Am Geriatr Soc. 1999;47(6):639–46.
52. Lau DCW, Dhillon B, Yan H, Szmitko PE, Verma S. Adipokines: molecular links between obesity and atherosclerosis. Am J Physiol Heart Circ Physiol. 2005;288(5):H2031–41.
53. Ballou SP, Lozanski GB, Hodder S, Rzewnicki DL, Mion LC, Sipe JD, et al. Quantitative and qualitative alterations of acute-phase proteins in healthy elderly persons. Age Ageing. 1996;25(3):224–30.
54. Worns MA, Victor A, Galle PR, Hohler T. Genetic and environmental contributions to plasma C-reactive protein and interleukin-6 levels - a study in twins. Genes Immun. 2006;7(7):600–5.
55. Danesh J, Wheeler JG, Hirschfield GM, Eda S, Eiriksdottir G, Rumley A, et al. C-reactive protein and other circulating markers of inflammation in the prediction of coronary heart disease. N Engl J Med. 2004;350(14):1387–97.
56. Pearson TA, Mensah GA, Alexander RW, Anderson JL, Cannon III RO, Criqui M, et al. Markers of inflammation and cardiovascular disease: application to clinical and public health practice: A statement for healthcare professionals from the Centers for Disease Control and Prevention and the American Heart Association. Circulation. 2003;107(3):499–511.
57. Buckley DI, Fu R, Freeman M, Rogers K, Helfand M. C-reactive protein as a risk factor for coronary heart disease: a systematic review and meta-analyses for the U.S. preventive services task force. Ann Intern Med. 2009;151(7):483–95.
58. Visser M, Bouter LM, McQuillan GM, Wener MH, Harris TB. Elevated C-reactive protein levels in overweight and obese adults. JAMA. 1999;282(22):2131–5.
59. Fransson EI, Batty GD, Tabak AG, Brunner EJ, Kumari M, Shipley MJ, et al. Association between change in body composition and change in inflammatory markers: an 11-year follow-up in the Whitehall II Study. J Clin Endocrinol Metab. 2010;95(12):5370–4.
60. Mohamed-Ali V, Goodrick S, Rawesh A, Katz DR, Miles JM, Yudkin JS, et al. Subcutaneous adipose tissue releases interleukin-6, but not tumor necrosis factor-alpha, in vivo. J Clin Endocrinol Metab. 1997;82(12):4196–200.
61. Benson S, Janssen OE, Hahn S, Tan S, Dietz T, Mann K, et al. Obesity, depression, and chronic low-grade inflammation in women with polycystic ovary syndrome. Brain Behav Immun. 2008;22(2):177–84.
62. Yudkin JS, Stehouwer CDA, Emeis JJ, Coppack SW. C-reactive protein in wealthy subjects: associations with obesity, insulin resistance, and endothelial dysfunction - a potential role for cytokines originating from adipose tissue? Arterioscler Thromb Vasc Biol. 1999;19(4):972–8.
63. Kern PA, Ranganathan S, Li C, Wood L, Ranganathan G. Adipose tissue tumor necrosis factor and interleukin-6 expression in human obesity and insulin resistance. Am J Physiol Endocrinol Metab. 2001;280(5):E745–51.
64. Knobler H, Zhornicky T, Sandler A, Haran N, Ashur Y, Schattner A. Tumor necrosis factor-alpha-induced insulin resistance may mediate the hepatitis C virus-diabetes association. Am J Gastroenterol. 2003;98(12):2751–6.
65. Gonzalez-Gay MA, De Matias JM, Gonzalez-Juanatey C, Garcia-Porrua C, Sanchez-Andrade A, Martin J, et al. Anti-tumor necrosis factor-alpha blockade improves insulin resistance in patients with rheumatoid arthritis. Clin Exp Rheumatol. 2006;24(1):83–6.
66. Krogh-Madsen R, Plomgaard P, Moller K, Mittendorfer B, Pedersen BK. Influence of TNF-alpha and IL-6 infusions on insulin sensitivity and expression of IL-18 in humans. Am J Physiol Endocrinol Metab. 2006;291(1):E108–14.
67. Nicklas BJ, Ambrosius W, Messier SP, Miller GD, Penninx BW, Loeser RF, et al. Diet-induced weight loss, exercise, and chronic inflammation in older, obese adults: a randomized controlled clinical trial. Am J Clin Nutr. 2004;79(4):544–51.
68. Dandona P, Weinstock R, Thusu K, Abdel-Rahman E, Aljada A, Wadden T. Tumor necrosis factor-alpha in sera of obese patients: fall with weight loss. J Clin Endocrinol Metab. 1998;83(8):2907–10.
69. Esposito K, Pontillo A, Di Palo C, Giugliano G, Masella M, Marfella R, et al. Effect of weight loss and lifestyle changes on vascular inflammatory markers in obese women: a randomized trial. JAMA. 2003;289(14):1799–804.
70. Bruun JM, Stallknecht B, Helge JW, Richelsen B. Interleukin-18 in plasma and adipose tissue: effects of obesity, insulin resistance, and weight loss. Eur J Endocrinol. 2007;157(4):465–71.
71. Pischon T, Hankinson SE, Hotamisligil GS, Rifai N, Rimm EB. Leisure-time physical activity and reduced plasma levels of obesity-related inflammatory markers. Obes Res. 2003;11(9):1055–64.
72. Hamer M, Steptoe A. Prospective study of physical fitness, adiposity, and inflammatory markers in healthy middle-aged men and women. Am J Clin Nutr. 2009;89(1):85–9.
73. Jacobson DL, Gange SJ, Rose NR, Graham NMH. Epidemiology and estimated population burden of selected autoimmune diseases in the United States. Clin Immunol Immunopathol. 1997;84(3):223–43.

74. Czlonkowska A, Ciesielska A, Gromadzka G, Kurkowska-Jastrzebska I. Estrogen and cytokines production - the possible cause of gender differences in neurological diseases. Curr Pharm Des. 2005;11(8):1017–30.
75. Rosen JL, Tran HT, Lackey A, Viselli SM. Sex-related immune changes in young mice. Immunol Invest. 1999;28(4):247–56.
76. Grimaldi CM, Cleary J, Dagtas AS, Moussai D, Diamond B. Estrogen alters thresholds for B cell apoptosis and activation. J Clin Invest. 2002;109(12):1625–33.
77. Tanriverdi F, Silveira LFG, MacColl GS, Bouloux PMG. The hypothalamic-pituitary-gonadal axis: immune function and autoimmunity. J Endocrinol [Review]. 2003;176(3):293–304.
78. Khera A, McGuire DK, Murphy SA, Stanek HG, Das SR, Vongpatanasin W, et al. Race and gender differences in C-reactive protein levels. J Am Coll Cardiol. 2005;46(3):464–9.
79. Nazmi A, Oliveira IO, Victora CG. Correlates of C-reactive protein levels in young adults: a population-based cohort study of 3827 subjects in Brazil. Braz J Med Biol Res. 2008;41(5):357–67.
80. Lakoski SG, Cushman M, Criqui M, Rundek T, Blumenthal RS, D'Agostino RB, et al. Gender and C-reactive protein: data from the Multiethnic Study of Atherosclerosis (MESA) cohort. Am Heart J. 2006;152(3):593–8.
81. Jilma B, Dirnberger E, Loscher I, Rumplmayr A, Hildebrandt J, Eichler HG, et al. Menstrual cycle-associated changes in blood levels of interleukin-6, alpha 1 acid glycoprotein, and C-reactive protein. J Lab Clin Med. 1997;130(1):69–75.
82. Muzzio ML, Berg G, Zago V, Basilio F, Sanguinetti S, Lopez G, et al. Circulating small dense LDL, endothelial injuring factors and fibronectin in healthy postmenopausal women. Clin Chim Acta. 2007;381(2):157–63.
83. Eilertsen AL, Hoibraaten E, Os I, Andersen TO, Sandvik L, Sandset PM. The effects of oral and transdermal hormone replacement therapy on C-reactive protein levels and other inflammatory markers in women with high risk of thrombosis. Maturitas. 2005;52(2):111–8.
84. Lacut K, Oger E, Le Gal G, Blouch MT, Abgrall JF, Kerlan V, et al. Differential effects of oral and transdermal postmenopausal estrogen replacement therapies on C-reactive protein. Thromb Haemost. 2003;90(1):124–31.
85. Zegura B, Guzic-Salobir B, Sebestjen M, Keber I. The effect of various menopausal hormone therapies on markers of inflammation, coagulation, fibrinolysis, lipids, and lipoproteins in healthy postmenopausal women. Menopause. 2006;13(4):643–50.
86. Vitale C, Cornoldi A, Gebara O, Silvestri A, Wajngarten M, Cerquetani E, et al. Interleukin-6 and flow-mediated dilatation as markers of increased vascular inflammation in women receiving hormone therapy. Menopause. 2005;12(5):552–8.
87. van Rooijen M, Hansson LO, Frostegard J, Silveira A, Hamsten A, Bremme K. Treatment with combined oral contraceptives induces a rise in serum C-reactive protein in the absence of a general inflammatory response. J Thromb Haemost. 2006;4(1):77–82.
88. Kiecolt-Glaser JK, Loving TJ, Stowell JR, Malarkey WB, Lemeshow S, Dickinson SL, et al. Hostile marital interactions, proinflammatory cytokine production, and wound healing. Arch Gen Psychiatry. 2005;62:1377–84.
89. Sadeghi M, Daniel V, Naujokat C, Weimer R, Opelz G. Strikingly higher interleukin (IL)-1 alpha, IL-1 beta and soluble interleukin-1 receptor antagonist (sIL-1RA) but similar IL-2, sIL-2R, IL-3, IL-4, IL-6, sIL-6R, IL-10, tumour necrosis factor (TNF)-alpha, transforming growth factor (TGF)-beta(2) and interferon IFN-gamma urine levels in healthy females compared to healthy males: protection against urinary tract injury? Clin Exp Immunol. 2005;142(2):312–7.
90. Souter I, Janzen C, Martinez-Maza O, Breen EC, Stanczyk F, Chaudhuri G, et al. Serum levels of soluble vascular cell adhesion molecule-1 are decreased in women receiving oral contraceptives compared with normally menstruating women: implications in atherosclerosis. Fertil Steril. 2005;83(5):1480–8.
91. Willis C, Morris JM, Danis V, Gallery EDM. Cytokine production by peripheral blood monocytes during the normal human ovulatory menstrual cycle. Hum Reprod. 2003;18(6):1173–8.
92. Abrahamsen B, Stilgren LS, Rettmer E, Bonnevie-Nielsen V, Beck-Nielsen H. Effects of the natural and artificial menstrual cycle on the production of osteoprotegerin and the bone resorptive cytokines IL-1 beta and IL-6. Calcif Tissue Int. 2003;72(1):18–23.
93. O'Brien SM, Fitzgerald P, Scully P, Landers A, Scott LV, Dinan TG. Impact of gender and menstrual cycle phase on plasma cytokine concentrations. Neuroimmunomodulation. 2007;14(2):84–90.
94. Yasui T, Uemura H, Yamada M, Matsuzaki T, Tsuchiya N, Noguchi M, et al. Associations of interleukin-6 with interleukin-1 beta interleukin-8 and macrophage inflammatory protein-1 beta in midlife women. Cytokine. 2008;41(3):302–6.
95. Kluft C, Leuven JA, Helmerhorst FM, Krans HM. Pro-inflammatory effects of oestrogens during use of oral contraceptives and hormone replacement treatment. Vascul Pharmacol. 2002;39:149–54.
96. Giraldo E, Hinchado MD, Garcia JJ, Ortega E. Influence of gender and oral contraceptives intake on innate and inflammatory response. Role of neuroendocrine factors. Mol Cell Biochem. 2008;313(1-2):147–53.
97. Straub RH, Hense HW, Andus T, Scholmerich J, Riegger GAJ, Schunkert H. Hormone replacement therapy and interrelation between serum interleukin-6 and body mass index in postmenopausal women: a population-based study. J Clin Endocrinol Metab. 2000;85(3):1340–4.
98. Abrahamsen B, Bonnevie-Nielsen V, Ebbesen EN, Gram J, Beck-Nielsen H. Cytokines and bone loss in a 5-year longitudinal study - hormone replacement

therapy suppresses serum soluble interleukin-6 receptor and increases interleukin-1-receptor antagonist: the Danish Osteoporosis Prevention Study. J Bone Miner Res. 2000;15(8):1545–54.
99. Musselman DL, Evans DL, Nemeroff CB. The relationship of depression to cardiovascular disease: epidemiology, biology, and treatment. Arch Gen Psychiatry. 1998;55(7):580–92.
100. Massie MJ. Prevalence of depression in patients with cancer. J Natl Cancer Inst Monogr. 2004;32:57–71.
101. Maes M, Meltzer HY, Bosmans E, Bergmans R, Vandoolaeghe E, Ranjan R, et al. Increased plasma-concentrations of interleukin-6, soluble interleukin-6, soluble interleukin-2 and transferrin receptor in major depression. J Affect Disord. 1995;34(4):301–9.
102. Bremmer MA, Beekman AT, Deeg DJ, Penninx BW, Dik MG, Hack CE, et al. Inflammatory markers in late-life depression: results from a population-based study. J Affect Disord. 2008;106(3):249–55.
103. Lesperance F, Frasure-Smith N, Theroux P, Irwin M. The association between major depression and levels of soluble intercellular adhesion molecule 1, interleukin-6, and C-reactive protein in patients with recent acute coronary syndromes. Am J Psychiat. 2004;161(2):271–7.
104. Suarez EC, Krishnan RR, Lewis JG. The relation of severity of depressive symptoms to monocyte-associated proinflammatory cytokines and chemokines in apparently healthy men. Psychosom Med. 2003;65(3):362–8.
105. Ford DE, Erlinger TP. Depression and C-reactive protein in US adults: data from the Third National Health and Nutrition Examination Survey. Arch Intern Med. 2004;164(9):1010–4.
106. Glaser R, Robles TF, Sheridan J, Malarkey WB, Kiecolt-Glaser JK. Mild depressive symptoms are associated with amplified and prolonged inflammatory responses after influenza virus vaccination in older adults. Arch Gen Psychiatry. 2003;60(10):1009–14.
107. Kenis G, Maes M. Effects of antidepressants on the production of cytokines. Int J Neuropsychoph. 2002;5(4):401–12.
108. Pace TWW, Mletzko TC, Alagbe O, Musselman DL, Nemeroff CB, Miller AH, et al. Increased stress-induced inflammatory responses in male patients with major depression and increased early life stress. Am J Psychiatry. 2006;163(9):1630–2.
109. Miller GE, Rohleder N, Stetler C, Kirschbaum C. Clinical depression and regulation of the inflammatory response during acute stress. Psychosom Med. 2005;67(5):679–87.
110. Miller GE, Freedland KE, Carney RM. Depressive symptoms and the regulation of proinflammatory cytokine expression in patients with coronary heart disease's. J Psychosom Res. 2005;59(4):231–6.
111. Reichenberg A, Yirmiya R, Schuld A, Kraus T, Haack M, Morag A, et al. Cytokine-associated emotional and cognitive disturbances in humans. Arch Gen Psychiatry. 2001;58(5):445–52.
112. Bryclon L, Harrison NA, Walker C, Steptoe A, Critchley HD. Peripheral inflammation is associated with altered substantia nigra activity and psychomotor slowing in humans. Biol Psychiatry. 2008;63(11):1022–9.
113. Anisman H, Merali Z, Hayley S. Neurotransmitter, peptide and cytokine processes in relation to depressive disorder: comorbidity between depression and neurodegenerative disorders. Prog Neurobiol. 2008;85(1):1–74.
114. Mendlewicz J, Kriwin P, Oswald P, Souery D, Alboni S, Brunello N. Shortened onset of action of antidepressants in major depression using acetylsalicylic acid augmentation: a pilot open-label study. Int Clin Psychopharmacol. 2006;21(4):227–31.
115. Muller N, Schwarz MJ, Dehning S, Douhe A, Cerovecki A, Goldstein-Muller B, et al. The cyclooxygenase-2 inhibitor celecoxib has therapeutic effects in major depression: results of a double-blind, randomized, placebo controlled, add-on pilot study to reboxetine. Mol Psychiatry. 2006;11(7):680–4.
116. Carroll JE, Low CA, Prather AA, Cohen S, Fury JM, Ross DC, et al. Negative affective responses to a speech task predict changes in interleukin (IL)-6. Brain Behav Immun. 2011;25(2):232–8.
117. Maes M, Song C, Lin A, De Jongh R, Van Gastel A, Kenis G, et al. The effects of psychological stress on humans: increased production of pro-inflammatory cytokines and a Th1-like response in stress-induced anxiety. Cytokine. 1998;10(4):313–8.
118. Pitsavos C, Panagiotakos DB, Papageorgiou C, Tsetsekou E, Soldatos C, Stefanadis C. Anxiety in relation to inflammation and coagulation markers, among healthy adults: the ATTICA study. Atherosclerosis. 2006;185(2):320–6.
119. Kiecolt-Glaser JK, Heffner KL, Glaser R, Malarkey WB, Porter K, Atkinson C, et al. How stress and anxiety can alter immediate and late phase skin test responses in allergic rhinitis. Psychoneuroendocrinology. 2009;34(5):670–80.
120. Montinaro V, Iaffaldano GP, Granata S, Porcelli P, Todarello O, Schena FP, et al. Emotional symptoms, quality of life and cytokine profile in hemodialysis patients. Clin Nephrol. 2010;73(1):36–43.
121. Adler N, Boyce T, Chesney M, Cohen S, Folkman S, Kahn R, et al. Socioeconomic status and health. Am Psychol. 1994;49:15–24.
122. Adler N, Rehkopf D. US disparities in health: descriptions, causes, and mechanisms. Annu Rev Public Health. 2008;29(1):235.
123. Haan M, Kaplan GA, Camacho T. Poverty and health - prospective evidence from the Alameda County Study. Am J Epidemiol. 1987;125(6):989–98.
124. Bucher HC, Ragland DR. Socioeconomic indicators and mortality from coronary heart-disease and cancer - a 22-year follow-up of middle-aged men. Am J Public Health. 1995;85(9):1231–6.
125. Lantz PM, House JS, Lepkowski JM, Williams DR, Mero RP, Chen JM. Socioeconomic factors, health behaviors, and mortality - results from a nationally

representative prospective study of US adults. JAMA. 1998;279(21):1703–8.
126. Singh GK, Siahpush M. Widening socioeconomic inequalities in US life expectancy, 1980–2000. Int J Epidemiol. 2006;35(4):969–79.
127. Marmot MG, Smith GD, Stansfeld S, Patel C, North F, Head J, et al. Health Inequalities among British Civil-Servants - the Whitehall-Ii Study. Lancet. 1991;337(8754):1387–93.
128. Marmot MG, Shipley MJ, Rose G. Inequalities in death - specific explanations of a general pattern. Lancet. 1984;1(8384):1003–6.
129. Koster A, Bosma H, Penninx BWJH, Newman AB, Harris TB, van Eijk JTM, et al. Association of inflammatory markers with socioeconomic status. J Gerontol A Biol Sci Med Sci. 2006;61(3):284–90.
130. Jousilahti P, Salomaa V, Rasi V, Vahtera E, Palosuo T. Association of markers of systemic inflammation, C reactive protein, serum amyloid A, and fibrinogen, with socioeconomic status. J Epidemiol Community Health. 2003;57(9):730–3.
131. Gruenewald TL, Cohen S, Matthews KA, Tracy R, Seeman TE. Association of socioeconomic status with inflammation markers in black and white men and women in the Coronary Artery Risk Development in Young Adults (CARDIA) study. Soc Sci Med. 2009;69(3):451–9.
132. Steptoe A, Owen N, Kunz-Ebrecht S, Mohamed-Ali V. Inflammatory cytokines, socioeconomic status, and acute stress responsivity. Brain Behav Immun. 2002;16(6):774–84.
133. Ranjit N, Diez-Roux AV, Shea S, Cushman M, Ni H, Seeman T. Socioeconomic position, race/ethnicity, and inflammation in the multi-ethnic study of atherosclerosis. Circulation. 2007;116(21):2383–90.
134. Friedman EM, Herd P. Income, education, and inflammation: differential associations in a national probability sample (The MIDUS Study). Psychosom Med. 2010;72(3):290–300.
135. Brydon L, Edwards S, Mohamed-Ali V, Steptoe A. Socioeconomic status and stress-induced increases in interleukin-6. Brain Behav Immun. 2004;18(3):281–90.
136. Owen N, Poulton T, Hay FC, Mohamed-Ali V, Steptoe A. Socioeconomic status, C-reactive protein, immune factors, and responses to acute mental stress. Brain Behav Immun. 2003;17(4):286–95.
137. Uchino BN. Understanding the links between social support and physical health: a life-span perspective with emphasis on the separability of perceived and received support. Perspect Psychol Sci. 2009;4(3):236–55.
138. Friedman EM, Hayney MS, Love GD, Urry HL, Rosenkranz MA, Davidson RJ, et al. Social relationships, sleep quality, and interleukin-6 in aging women. Proc Natl Acad Sci USA. 2005;102(51):18757–62.
139. Costanzo ES, Lutgendorf SK, Sood AK, Anderson B, Sorosky J, Lubaroff DM. Psychosocial factors and interleukin-6 among women with advanced ovarian cancer. Cancer. 2005;104(2):305–13.
140. Lutgendorf SK, Anderson B, Sorosky JI, Buller RE, Lubaroff DM. Interleukin-6 and use of social support in gynecologic cancer patients. Int J Behav Med. 2000;7(2):127–42.
141. Lillard L, Waite L. 'Til death do us part: Marital disruption and mortality. Am J Sociol. 1995;100(5):1131–56.
142. Sbarra DA. Marriage protects men from clinically meaningful elevations in C-reactive protein: results from the national social life, health, and aging project (NSHAP). Psychosom Med. 2009;71(8):828–35.
143. Kiecolt-Glaser JK, Newton TL. Marriage and health: his and hers. Psychol Bull. 2001;127(4):472–503.
144. Kiecolt-Glaser JK, Glaser R, Cacioppo JT, Malarkey WB. Marital stress: immunologic, neuroendocrine, and autonomic correlates. Ann N Y Acad Sci. 1998;840:656–63.
145. Graham JE, Glaser R, Loving TJ, Malarkey WB, Stowell JR, Kiecolt-Glaser JK. Cognitive word use during marital conflict and increases in proinflammatory cytokines. Health Psychol. 2009;28(5):621–30.
146. Shen BJ, Farrell KA, Penedo FJ, Schneiderman N, Orth-Gomer K. Waist circumference moderates the association between marital stress and C-reactive protein in middle-aged healthy women. Ann Behav Med. 2010;40(3):258–64.
147. Kardys I, Knetsch A, Bleumink G, Deckers J, Hofman A, Stricker B, et al. C-reactive protein and risk of heart failure. The Rotterdam Study. Am Heart J. 2006;152(3):514–20.
148. US Department of Health and Human Services. The Health Consequences of Smoking: A Report of the Surgeon General. 2004; Atlanta, GA: Centers for Disease Control and Prevention.
149. Doll R, Peto R, Boreham J, Sutherland I. Mortality in relation to smoking: 50 years' observations on male British doctors. Br Med J. 2004;328(7455):1519.
150. Cross CE, Traber M, Eiserich J, van der Vliet A. Micronutrient antioxidants and smoking. Br Med Bull. 1999;55(3):691–704.
151. Ridker PM, Hennekens CH, Buring JE, Rifai N. C-Reactive protein and other markers of inflammation in the prediction of cardiovascular disease in women. New Engl J Med. 2000;342(12):836–43.
152. Tracy RP, Psaty BM, Macy E, Bovill EG, Cushman M, Cornell ES, et al. Lifetime smoking exposure affects the association of C-reactive protein with cardiovascular disease risk factors and subclinical disease in healthy elderly subjects. Arterioscler Thromb Vasc Biol. 1997;17(10):2167–76.
153. Das I. Raised C-reactive protein levels in serum from smokers. Clin Chim Acta. 1985;153(1):9–13.
154. Wannamethee SG, Lowe GDO, Shaper AG, Rumley A, Lennon L, Whincup PH. Associations between cigarette smoking, pipe/cigar smoking, and smoking

cessation, and haemostatic and inflammatory markers for cardiovascular disease. Eur Heart J. 2005;26(17):1765–73.
155. Bermudez EA, Rifai N, Buring JE, Manson JE, Ridker PM. Relation between markers of systemic vascular inflammation and smoking in women. Am J Cardiol. 2002;89(9):1117–9.
156. Bo S, Gentile L, Ciccone G, Baldi C, Benini L, Dusio F, et al. The metabolic syndrome and high C-reactive protein: prevalence and differences by sex in a southern-European population-based cohort. Diabetes Metab Res Rev. 2005;21(6):515–24.
157. Haddy N, Sass C, Maumus S, Marie B, Droesch S, Siest G, et al. Biological variations, genetic polymorphisms and familial resemblance of TNF-alpha and IL-6 concentrations: STANISLAS cohort. Eur J Hum Genet. 2005;13(1):109–17.
158. Bazzano LA, He J, Muntner P, Vupputuri S, Whelton PK. Relationship between cigarette smoking and novel risk factors for cardiovascular disease in the United States. Ann Intern Med. 2003;138(11):891–7.
159. Mendall MA, Strachan DP, Butland BK, Ballam L, Morris J, Sweetnam PM, et al. C-reactive protein: relation to total mortality, cardiovascular mortality and cardiovascular risk factors in men. Eur Heart J. 2000;21(19):1584–90.
160. Frohlich M, Sund M, Lowel H, Imhof A, Hoffmeister A, Koenig W. Independent association of various smoking characteristics with markers of systemic inflammation in men - results from a representative sample of the general population (MONICA Augsburg Survey 1994/95). Eur Heart J. 2003;24(14):1365–72.
161. Lowe GDO, Editorial B. Why do smokers have higher plasma fibrinogen levels than non-smokers? Clin Sci. 2001;101(2):209–10.
162. Hastie CE, Haw S, Pell JP. Impact of smoking cessation and lifetime exposure on C-reactive protein. Nicotine Tob Res. 2008;10(4):637–42.
163. Woodward M, Rumley A, Tunstall-Pedoe H, Lowe GDO. Associations of blood rheology and interleukin-6 with cardiovascular risk factors and prevalent cardiovascular disease. Br J Haematol. 1999;104(2): 246–57.
164. Wannamethee SG, Whincup PH, Rumley A, Lowe GDO. Inter-relationships of interleukin-6, cardiovascular risk factors and the metabolic syndrome among older men. J Thromb Haemost. 2007;5(8):1637–43.
165. Helmersson J, Larsson A, Vessby B, Basu S. Active smoking and a history of smoking are associated with enhanced prostaglandin F-2 alpha, interleukin-6 and F-2-isoprostane formation in elderly men. Atherosclerosis. 2005;181(1):201–7.
166. Petrescu F, Voican SC, Siolosi I. Tumor necrosis factor-alpha serum levels in healthy smokers and nonsmokers. Int J Chron Obstruct Pulmon Dis. 2010;5:217–22.
167. Berrahmoune H, Lamont JV, Herbeth B, FitzGerald P, Visvikis-Siest S. Biological determinants of and reference values for plasma interleukin-8, monocyte chemoattractant protein-1, epidermal growth factor, and vascular endothelial growth factor: results from the STANISLAS cohort. Clinl Chem. 2006;52(3): 504–10.
168. Pedersen BK, Febbraio MA. Muscle as an endocrine organ: focus on muscle-derived interleukin-6. Physiol Rev. 2008;88(4):1379–406.
169. Febbraio MA, Pedersen BK. Muscle-derived interleukin-6: mechanisms for activation and possible biological roles. FASEB J. 2002;16(11):1335–47. Official Publication of the Federation of American Societies for Experimental Biology.
170. Fischer CP. Interleukin-6 in acute exercise and training: what is the biological relevance? Exerc Immunol Rev. 2006;12:6–33.
171. Nielsen AR, Pedersen BK. The biological roles of exercise-induced cytokines: IL-6, IL-8, and IL-15. Appl Physiol Nutr Metab. 2007;32(5):833–9.
172. Pedersen BK, Steensberg A, Fischer C, Keller C, Keller P, Plomgaard P, et al. Searching for the exercise factor: is IL-6 a candidate? J Muscle Res Cell Motil. 2003;24(2):113–9.
173. Starkie R, Ostrowski SR, Jauffred S, Febbraio M, Pedersen BK. Exercise and IL-6 infusion inhibit endotoxin-induced TNF-alpha production in humans. FASEB J. 2003;17(3):884–6.
174. Borodulin K, Laatikainen T, Salomaa V, Jousilahti P. Associations of leisure time physical activity, self-rated physical fitness, and estimated aerobic fitness with serum C-reactive protein among 3,803 adults. Atherosclerosis. 2006;185(2):381–7.
175. Ford ES. Does exercise reduce inflammation? Physical activity and c-reactive protein among U.S. adults. Epidemiology. 2002;13:561–8.
176. Jankord R, Jemiolo B. Influence of physical activity on serum IL-6 and IL-10 levels in healthy older men. Med Sci Sports Exerc. 2004;36(6):960–4.
177. Goldhammer E, Tanchilevitch A, Maor I, Beniamini Y, Rosenschein U, Sagiv M. Exercise training modulates cytokines activity in coronary heart disease patients. Int J Cardiol. 2005;100(1):93–9.
178. Okita K, Nishijima H, Murakami T, Nagai T, Morita N, Yonezawa K, et al. Can exercise training with weight loss lower serum C-reactive protein levels? Arterioscler Thromb Vasc Biol. 2004;24(10):1868–73.
179. Troseid M, Lappegard KT, Claudi T, Damas JK, Morkrid L, Brendberg R, et al. Exercise reduces plasma levels of the chemokines MCP-1 and IL-8 in subjects with the metabolic syndrome. Eur Heart J. 2004;25(4):349–55.
180. Pullen PR, Nagamia SH, Mehta PK, Thompson WR, Benardot D, Hammoud R, et al. Effects of yoga on inflammation and exercise capacity in patients with chronic heart failure. J Card Fail. 2008;14(5):407–13.
181. Kiecolt-Glaser JK, Christian L, Preston H, Houts CR, Malarkey WB, Emery CF, et al. Stress, inflammation, and yoga practice. Psychosom Med. 2010;72(2):113–21.
182. Knoops KT, de Groot LC, Kromhout D, Perrin AE, Moreiras-Varela O, Menotti A, et al. Mediterranean diet, lifestyle factors, and 10-year mortality in elderly

European men and women: the HALE project. JAMA. 2004;292(12):1433–9.
183. Schulze MB, Hoffmann K, Manson JE, Willett WC, Meigs JB, Weikert C, et al. Dietary pattern, inflammation, and incidence of type 2 diabetes in women. Am J Clin Nutr. 2005;82(3):675–84. quiz 714-5.
184. Nettleton JA, Steffen LM, Mayer-Davis EJ, Jenny NS, Jiang R, Herrington DM, et al. Dietary patterns are associated with biochemical markers of inflammation and endothelial activation in the Multi-Ethnic Study of Atherosclerosis (MESA). Am J Clin Nutr. 2006;83(6):1369–79.
185. Lopez-Garcia E, Schulze MB, Fung TT, Meigs JB, Rifai N, Manson JE, et al. Major dietary patterns are related to plasma concentrations of markers of inflammation and endothelial dysfunction. Am J Clin Nutr. 2004;80(4):1029–35.
186. Calder PC, Albers R, Antoine JM, Blum S, Bourdet-Sicard R, Ferns GA, et al. Inflammatory disease processes and interactions with nutrition. Br J Nutr. 2009;101 Suppl 1:S1–45.
187. Ceriello A, Taboga C, Tonutti L, Quagliaro L, Piconi L, Bais B, et al. Evidence for an independent and cumulative effect of postprandial hypertriglyceridemia and hyperglycemia on endothelial dysfunction and oxidative stress generation: effects of short- and long-term simvastatin treatment. Circulation. 2002;106(10):1211–8.
188. Bowen PE, Borthakur G. Postprandial lipid oxidation and cardiovascular disease risk. Curr Atheroscler Rep. 2004;6(6):477–84.
189. Nappo F, Esposito K, Cioffi M, Giugliano G, Molinari AM, Paolisso G, et al. Postprandial endothelial activation in healthy subjects and in type 2 diabetic patients: role of fat and carbohydrate meals. J Am Coll Cardiol. 2002;39(7):1145–50.
190. Esposito K, Nappo F, Marfella R, Giugliano G, Giugliano F, Ciotola M, et al. Inflammatory cytokine concentrations are acutely increased by hyperglycemia in humans: role of oxidative stress. Circulation. 2002;106(16):2067–72.
191. O'Keefe JH, Gheewala NM, O'Keefe JO. Dietary strategies for improving post-prandial glucose, lipids, inflammation, and cardiovascular health. J Am Coll Cardiol. 2008;51(3):249–55.
192. Giugliano D, Ceriello A, Esposito K. The effects of diet on inflammation: emphasis on the metabolic syndrome. J Am Coll Cardiol. 2006;48(4):677–85.
193. Calder PC. n-3 polyunsaturated fatty acids, inflammation, and inflammatory diseases. Am J Clin Nutr. 2006;83(6 Suppl):1505S–19.
194. Mozaffarian D, Pischon T, Hankinson SE, Rifai N, Joshipura K, Willett WC, et al. Dietary intake of trans fatty acids and systemic inflammation in women. Am J Clin Nutr. 2004;79(4):606–12.
195. Pischon T, Hankinson SE, Hotamisligil GS, Rifai N, Willett WC, Rimm EB. Habitual dietary intake of n-3 and n-6 fatty acids in relation to inflammatory markers among US men and women. Circulation. 2003;108(2):155–60.
196. Novak TE, Babcock TA, Jho DH, Helton WS, Espat NJ. NF-kappa B inhibition by omega -3 fatty acids modulates LPS-stimulated macrophage TNF-alpha transcription. Am J Physiol Lung Cell Mol Physiol. 2003;284(1):L84–9.
197. Zhao Y, Joshi-Barve S, Barve S, Chen LH. Eicosapentaenoic acid prevents LPS-induced TNF-alpha expression by preventing NF-kappaB activation. J Am Coll Nutr. 2004;23(1):71–8.
198. Dew MA, Hoch CC, Buysse DJ, Monk TH, Begley AE, Houck PR, et al. Healthy older adults' sleep predicts all-cause mortality at 4 to 19 years of follow-up. Psychosom Med. 2003;65(1):63–73.
199. Tamakoshi A, Ohno Y. Self-reported sleep duration as a predictor of all-cause mortality: results from the JACC study, Japan. Sleep. 2004;27(1):51–4.
200. Ferrie JE, Shipley MJ, Cappuccio FP, Brunner E, Miller MA, Kumari M, et al. A prospective study of change in sleep duration: associations with mortality in the Whitehall II cohort. Sleep. 2007;30(12):1659–66.
201. Spiegel K, Knutson K, Leproult R, Tasali E, Van Cauter E. Sleep loss: a novel risk factor for insulin resistance and Type 2 diabetes. J Appl Physiol. 2005;99(5):2008–19.
202. Benca RM, Obermeyer WH, Thisted RA, Gillin JC. Sleep and psychiatric-disorders - a metaanalysis. Arch Gen Psychiatry. 1992;49(8):651–68.
203. Redwine L, Hauger RL, Gillin JC, Irwin M. Effects of sleep and sleep deprivation on interleukin-6, growth hormone, cortisol, and melatonin levels in humans. J Clin Endocrinol Metab. 2000;85:3597–603.
204. Motivala SJ, Irwin MR. Sleep and immunity - cytokine pathways linking sleep and health outcomes. Curr Dir Psychol Sci. 2007;16(1):21–5.
205. Mullington J, Korth C, Hermann DM, Orth A, Galanos C, Holsboer F, et al. Dose-dependent effects of endotoxin on human sleep. Am J Physiol Regul Integr Comp Physiol. 2000;278(4):R947–55.
206. Haack M, Schuld A, Kraus T, Pollmacher T. Effects of sleep on endotoxin-induced host responses in healthy men. Psychosom Med. 2001;63(4):568–78.
207. Spath-Schwalbe E, Hansen K, Schmidt F, Schrezenmeier H, Marshall L, Burger K, et al. Acute effects of recombinant human interleukin-6 on endocrine and central nervous sleep functions in healthy men. J Clin Endocrinol Metab. 1998;83(5):1573–9.
208. Shoham S, Davenne D, Cady AB, Dinarello CA, Krueger JM. Recombinant tumor necrosis factor and interleukin 1 enhance slow-wave sleep. Am J Physiol. 1987;253(1 Pt 2):R142–9.
209. Irwin MR, Wang M, Ribeiro D, Cho HJ, Olmstead R, Breen EC, et al. Sleep loss activates cellular inflammatory signaling. Biol Psychiatry. 2008;64(6):538–40.
210. Irwin MR, Wang M, Campomayor CO, Collado-Hidalgo A, Cole S. Sleep deprivation and activation

of morning levels of cellular and genomic markers of inflammation. Arch Intern Med. 2006;166(16): 1756–62.
211. Shearer WT, Reuben JM, Mullington JM, Price NJ, Lee BN, Smith EO, et al. Soluble TNF-alpha receptor 1 and IL-6 plasma levels in humans subjected to the sleep deprivation model of spaceflight. J Allergy Clin Immunol. 2001;107(1):165–70.
212. Meier-Ewert HK, Ridker PM, Rifai N, Regan MM, Price NJ, Dinges DF, et al. Effect of sleep loss on C-reactive protein, an inflammatory marker of cardiovascular risk. J Am Coll Cardiol. 2004;43(4):678–83.
213. Alberti A, Sarchielli P, Gallinella E, Floridi A, Mazzotta G, Gallai V. Plasma cytokine levels in patients with obstructive sleep apnea syndrome: a preliminary study. J Sleep Res. 2003;12(4):305–11.
214. Arter JL, Chi DS, Girish M, Fitzgerald SM, Guha B, Krishnaswamy G. Obstructive sleep apnea, inflammation, and cardiopulmonary disease. Front Biosci. 2004;9:2892–900.
215. Epel ES, Blackburn EH, Lin J, Dhabhar FS, Adler NE, Morrow JD, et al. Accelerated telomere shortening in response to life stress. Proc Natl Acad Sci USA. 2004;101(49):17312–5.
216. Simon NM, Smoller JW, McNamara KL, Maser RS, Zalta AK, Pollack MH, et al. Telomere shortening and mood disorders: preliminary support for a chronic stress model of accelerated aging. Biol Psychiatry. 2006;60(5):432–5.
217. Epel ES, Lin J, Dhabhar FS, Wolkowitz OM, Puterman E, Karan L, et al. Dynamics of telomerase activity in response to acute psychological stress. Brain Behav Immun. 2010;24(4):531–9.
218. Damjanovic AK, Yang Y, Glaser R, Kiecolt-Glaser JK, Nguyen H, Laskowski B, et al. Accelerated telomere erosion is associated with a declining immune function of caregivers of Alzheimer's disease patients. J Immunol. 2007;179(6):4249–54.
219. Parks C, Miller D, McCanlies E, Cawthon R, Andrew M, DeRoo L, et al. Telomere length, current perceived stress, and urinary stress hormones in women. Cancer Epidemiol Biomarkers Prev. 2009;18(2):551.
220. Blackburn EH. Switching and signaling at the telomere. Cell. 2001;106(6):661–73.
221. Blackburn EH. Telomere states and cell fates. Nature. 2000;408:53–6.
222. Kiecolt-Glaser JK, Glaser R. Psychological stress, telomeres, and telomerase. Brain Behav Immun. 2010;24(4):529–30.
223. Aviv A. Telomeres and human aging: facts and fibs. Sci Aging Knowledge Environ. 2004;51:pe43.
224. Lung FW, Chen NC, Shu BC. Genetic pathway of major depressive disorder in shortening telomeric length. Psychiatr Genet. 2007;17(3):195–9.
225. Tyrka AR, Price LH, Kao HT, Porton B, Marsella SA, Carpenter LL. Childhood maltreatment and telomere shortening: preliminary support for an effect of early stress on cellular aging. Biol Psychiatry. 2010;67(6):531–4.
226. Kiecolt-Glaser JK, Gouin JP, Weng N, Malarkey WB, Beversdorf DQ, Glaser R. Childhood adversity heightens the impact of later-life caregiving stress on telomere length and inflammation. Psychosom Med. 2011;73(1):16.
227. Valdes AM, Andrew T, Gardner JP, Kimura M, Oelsner E, Cherkas LF, et al. Obesity, cigarette smoking, and telomere length in women. Lancet. 2005;366(9486):662–4.
228. Epel ES. Telomeres in a life-span perspective: a new "Psychobiomarker"? Curr Dir Psychol Sci. 2009;18(1):6–10.
229. Cherkas LF, Hunkin JL, Kato BS, Richards JB, Gardner JP, Surdulescu GL, et al. The association between physical activity in leisure time and leukocyte telomere length. Arch Intern Med. 2008;168(2): 154–8.

Diabetes Control: Insights from Complexity Science

20

Tim A. Holt

20.1 Introduction

The latter half of the twentieth century witnessed a global expansion in the prevalence of diabetes. This expansion resulted from generally increasing life expectancy, and from 'diabetogenic' lifestyles and environments associated with reduced physical activity and increased calorie intake. All stages of the development of type 2 diabetes are associated with raised risk of cardiovascular disease, a fact that threatens to reverse the improving trend in cardiovascular mortality of recent decades in most world populations. At the same time, the incidence of the less common type 1 diabetes (which is usually triggered by autoimmune mechanisms) also increased.

This phenomenon occurred in parallel with the emergence of complexity science. This chapter will explore the possible ways that this science can inform the study of diabetes and the care of those affected, and review some of the evidence supporting this area. I will first of all describe the basic principles of diabetes care, both from clinician-centred and patient-centred perspectives. I will then describe how interactive determinants of clinical outcomes may impact on predictability and on dynamical patterns for people affected by these conditions.

T.A. Holt, PhD, MRCP, FRCGP (✉)
Department of Primary Care Health Sciences, University of Oxford, Radcliffe Observatory Quarter, Woodstock Road, Oxford, OX2 6GG, UK
e-mail: tim.holt@phc.ox.ac.uk

20.2 Basics of Diabetes Care

Diabetes is a group of conditions associated with *chronic* (long term) *hyperglycaemia* (raised blood glucose levels). Hyperglycaemia is caused by absolute or relative insulin deficiency. Insulin is the most important hormone regulating blood glucose levels, and is produced by the pancreas, a gland in the abdomen. Reduced insulin production causes blood glucose levels to rise and may lead to a diagnosis of diabetes in a number of situations. This diagnosis depends both on the detection of blood glucose levels above specified thresholds and on an assumption that the tendency for hyperglycaemia is long term [1].

One situation typically involves children and young adults, who simply fail to produce sufficient insulin, and become unwell over a period of weeks, because their insulin production falls to very low levels (*absolute* deficiency). Blood glucose levels rise substantially, enough to make the person clearly unwell. This form of diabetes is termed 'type 1' and the problem is usually due to an attack on the insulin producing 'beta cells' in the pancreas launched by the individual's own immune system (auto-immunity).

Type 2 individuals are usually older, with insulin production declining over a longer timescale, but have the added problem of insulin *resistance*. Insulin acts largely through increasing the transport of glucose from the blood to where it is needed (e.g. as a fuel for muscle tissue) or to where it can be stored for future use (e.g. the

J.P. Sturmberg and C.M. Martin (eds.), *Handbook of Systems and Complexity in Health*, DOI 10.1007/978-1-4614-4998-0_20, © Springer Science+Business Media New York 2013

liver). Resistance to insulin means that despite adequate production these tissues do not respond and the glucose tends to remain in the blood. Insulin resistance is usually associated with central obesity and inactivity, and may develop years or more before blood glucose levels become substantially raised. In an attempt to maintain normal blood glucose levels, the insulin resistant individual produces more insulin than normal (although in some cases not enough to achieve truly normal blood glucose levels, i.e. *relative* deficiency). Finally, the pancreas fails to keep up with the excess demand, insulin production fails more obviously, and overt hyperglycaemia develops, often gradually over a period of years. One of the results of this is the development of macro-vascular disease (i.e. disease of the larger blood vessels). The mechanisms triggering macro-vascular complications are complex, but part of the process involves the raised insulin levels of the 'pre-diabetes' condition. Those diagnosed with type 2 diabetes are often affected by established macro-vascular disease at the outset, even though blood glucose levels themselves may have become raised relatively recently.

In the last 20 years, the distinction between type 1 and type 2 diabetes has become less clear than it was in prior decades [2]. This is an interesting phenomenon, and relates to increased incidence both of obesity and inactivity in young people, and of auto-immune and other factors in older people. Type 2 diabetes is sometimes diagnosed in teenagers, a finding almost unheard of during the early and mid-twentieth century (although more often in certain ethnic minority groups), whilst older people are sometimes seen running a 'type 1' pattern of absolute deficiency over a period of months. In some people, both problems may be relevant. In any case, diagnostic definitions assume that an affected individual has at least a long-term (and probably life-long) tendency towards raised blood glucose levels and are based either on current glycaemia (blood glucose levels) or on estimated mean levels over the previous 2–3 months.

Despite (and perhaps because of) the modern blurring of type 1 and type 2 stereotypes, the need (and the opportunity) to distinguish the patterns of blood glucose variability that characterise them lends itself to a dynamical approach based on time series analysis of glucose profiles. This issue and its implications will be explored in this chapter. But the first question to ask is: what are the patterns of variability that might distinguish type 1 and type 2 diabetes, and why should we attempt to identify and define them?

20.3 Convergence and Divergence

Let us return to the stereotypes for a moment. The 'classical' type 1 individual has little or no endogenous insulin production, and is utterly dependent upon injected insulin. Classical type 2 diabetes involves a certain level of endogenous insulin (albeit insufficient, and declining over time) and insulin resistance.

The presence of residual insulin production in type 2 diabetes allows the system to self-correct to some extent. Whilst the blood glucose level may always be higher than it should be, there is a tendency for it to converge back to the same baseline value. This value tends to occur reliably in the fasting state—i.e. the 'before breakfast' situation where the individual has had no carbohydrate intake for over 8 h. This state tends to produce reasonably consistent measurements of blood glucose, consistent enough that fasting blood glucose can be used as a means of diagnosing type 2 diabetes.

In the type 1 situation, where endogenous insulin production is very low or absent, the physiological tendency to convergence is also absent and the pattern depends almost entirely on conscious behavioural mechanisms. The two patterns might be represented by imagining a bowl with a marble sitting within it (Fig. 20.1). In the classical early stage type 2 condition, a stable equilibrium is achieved and the marble returns to the middle of the concave bowl, whichever direction it is pushed. Displacement results in movement back towards the prior position. In type 1 diabetes, the shape of the bowl could be anything from concave, to flat, to frankly convex (inverted). In the latter situation, even though in theory there is an equilibrium point at the top,

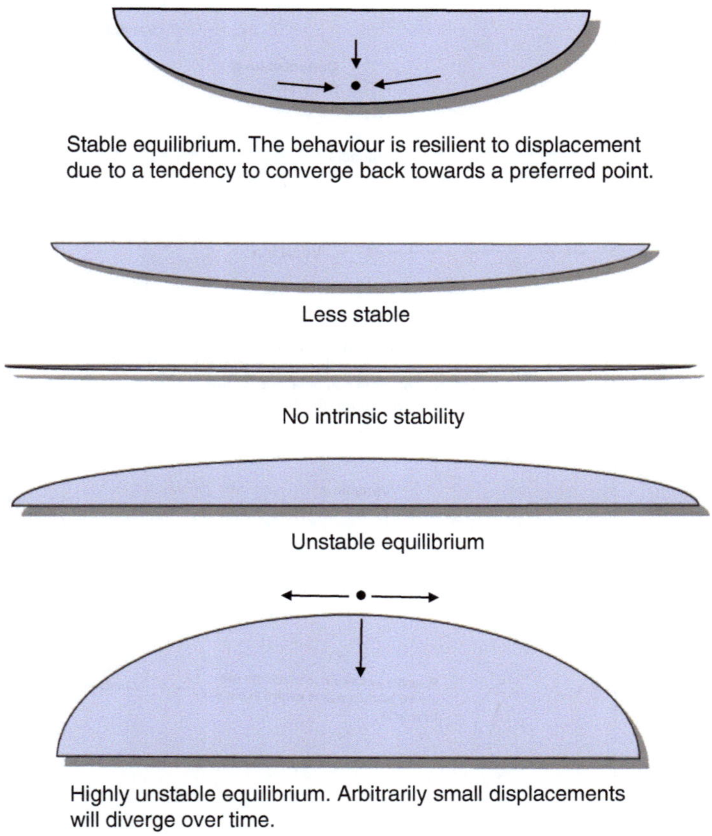

Fig. 20.1 In a simple bowl and marble model, the stability of the system depends on the concavity/convexity of the bowl. This determines whether the behaviour is convergent or divergent

the equilibrium is unstable with a tendency for divergence, and stasis is an unrealistic goal. Arbitrarily small differences in starting position, or nudges in any direction will result in outcomes that become more (rather than less) different over time. So a question to be addressed is: can we measure the concavity/convexity of the 'bowl' in individuals' cases? What implications does this have for management of diabetes? How can we help individuals to achieve their target dynamical pattern?

20.4 A Phase Space of Glycaemic Determinants

Imagine if we were to plot the individual's blood glucose, insulin, and physical activity levels as separate axes, to produce a 'phase space' plot (Fig. 20.2). The fasting state in type 2 diabetes would represent a point attractor—a position to which the system tends to return on an at least 24-h basis. Phase space plots have been used to model many different types of system including physiological [3] and ecological [4] examples. The existence of this point attractor suggests a convergent tendency in the dynamics. Displacement away from it (through carbohydrate ingestion, associated insulin production, and physical activity during the day) will produce movement of the system in this space, with a tendency to move back to the attracting point each morning. The point may be different from that of a person without diabetes (as the glucose level is higher), but the system nevertheless returns to it reliably. Individuals with highly regular patterns of activity (mealtimes, exercise schedules, etc.) will produce periodic patterns in this space, with sequences of value combinations repeating over time including a daily return to the fasting state. Displacement (e.g. through eating more carbohydrate than usual) will nevertheless result in convergence back towards the attracting point. The approach to management (at least in the

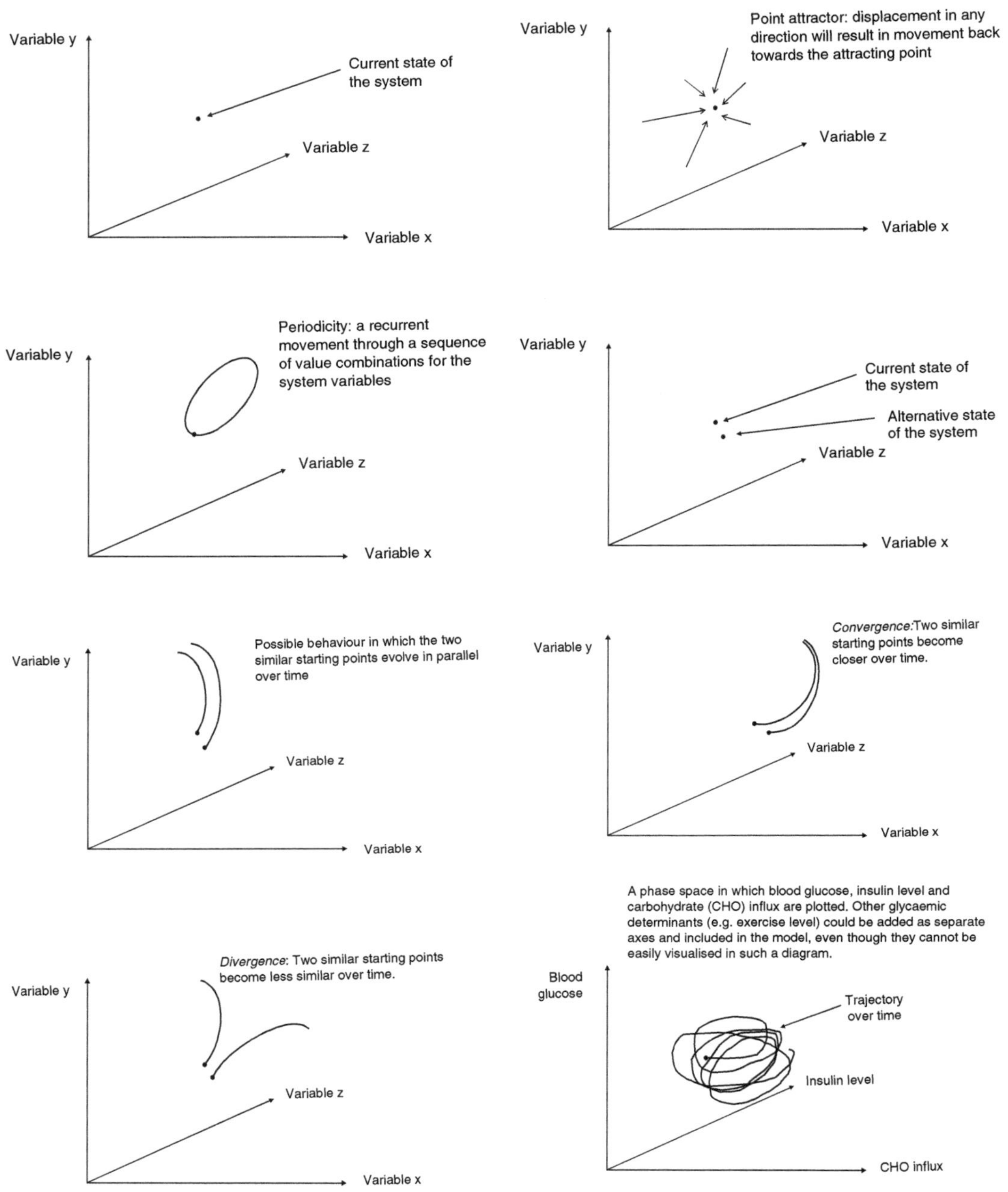

Fig. 20.2 Plotting a system's major variables in phase space enables us to identify the dynamical characteristics of its behaviour. As the value of each variable changes over time, the position of the system in this phase space moves as a trajectory. The position or pattern to which the system settles over time, or returns to if displaced, is known as its 'attractor'. The dynamical properties of the system are related to the geometry of the attractor, which can be explored through time series analysis. Static systems with stable equilibrium points can be represented by point attractors. Periodic behaviour involves a repeated movement through a sequence of variable value combinations. Chaotic behaviour is inherently unpredictable, but is more orderly than random variation as trajectories move in similar directions from similar starting positions in phase space before diverging, reflecting the underlying determinism. Short-term divergence implies sensitivity to the initial conditions but is followed by folding back as the values of the system variables are bounded. The system therefore has longer term stability despite short to medium-term unpredictability. For more on sensitivity to initial conditions visit: http://www.exploratorium.edu/complexity/java/lorenz.html. (Copyright 1996, James P. Crutchfield. All rights reserved)

early stages) is largely that of resetting this point attractor. This is achieved through lifestyle changes (which reduce weight and increase sensitivity to the endogenous insulin) and medication, which reduces fasting glucose levels and the magnitude of displacements.

One immediate implication is that type 2 individuals with convergent dynamics do not need to interfere with the system once their treatment schedule has been optimised (although type 2 diabetes is a progressive condition that almost inevitably requires escalation of treatment as insulin production falls). This supports the current policy in the United Kingdom (and elsewhere) away from self-monitoring of blood glucose for most early stage type 2 individuals, who have been shown not to benefit on the whole [5], as the system has a natural tendency to self-correct without conscious action on the individual's part.

For type 1 diabetes the situation is very different. The traditional approach to management is that of insulin replacement, but as well as lacking insulin, the individual has lost something almost as important—the subconscious regulatory mechanism that controls both blood glucose and insulin levels in the physiological state. A person with no history of diabetes does not have to think about regulating blood glucose, except perhaps to eat when hungry to avoid unpleasant but generally harmless mild hypoglycaemia. Changes in blood glucose are detected by the pancreas and corrected largely through responses in insulin and other hormone levels in the blood stream. For some homeostatic mechanisms, a combination of subconscious and conscious mechanisms is important, such as body temperature regulation. Most of the time we manage to maintain a narrow range of core temperature values without thinking about it, but there is very definite danger if we find ourselves in situations where behavioural mechanisms are unavailable (such as access to shade or water in a dessert). In the case of type 1 diabetes, conscious behavioural mechanisms predominate for maintaining blood glucose within reasonable limits, and behaviour becomes a key to successful control. The person has a still partially intact means of raising blood glucose when it is falling (though production of hormones that oppose the action of insulin) and will usually feel unwell, indicating the need to consume rapid acting carbohydrate. But when levels are rising, the only way of correcting the situation is to inject more insulin, eat less carbohydrate, or carry out more exercise, all conscious behavioural actions largely unassisted by subconscious physiology.

Corrective behaviour involves negative feedback to detect displacement of blood glucose (in whichever direction) and take appropriate action. Where symptoms of hypoglycaemia are present, appropriate correction is possible without actual blood glucose measurement. The target blood glucose levels (around 4–8 mmol/L) are close to the hypoglycaemic range of <4 mmol/L where (for most individuals) symptoms will trigger the need to act. However, the blood glucose level has to rise well above the target range (perhaps to 14 mmol/L or more) before most individuals become symptomatically aware of it. Hyperglycaemic symptoms are typically vaguer than those of hypoglycaemia, and it is unsafe to take corrective insulin on the basis of symptoms alone. Correcting moderate rises therefore requires a further component—self-monitoring of blood glucose, usually undertaken by finger prick capillary blood glucose measurements.

Successful control in type 1 diabetes therefore requires three components: replacement of insulin; awareness of displacement through symptoms or self-monitoring; and appropriate corrective behaviours. Let's look at these from a dynamical systems perspective.

1. *Replacement of insulin.* The 'cure' for type 1 diabetes was the discovery of insulin in the early 1920s, following experiments on pancreatectomised dogs, leading to extraction and purification of the responsible endocrine hormone [6]. The initial task was simply to replace the missing insulin. Children dying in a state of coma were revived by this miraculous new therapy. Then came the more subtle challenge of replacing insulin over different times of the day, when requirement varies. The most obvious distinction is that between

the ‘post-prandial’ state (i.e. after a meal, when digested carbohydrate is entering the blood stream from the gut as glucose) and the ‘post-absorptive’ state, where such absorption has finished, and insulin replacement simply needs to meet the background requirement. The development since its original discovery of insulin formulations acting over different timescales greatly improved such replacement approaches, particularly towards the end of the twentieth century [7], when insulin analogues were developed that were either rapid acting (mimicking the abrupt post-prandial rise in natural insulin levels that follows a meal); or long acting, producing a constant background level that mimics the baseline (post-absorptive) component.

2. *Awareness of displacement through symptoms or self-monitoring*. The dogs that were also ‘cured’ of their (experimental) diabetes through injection of the extracted pancreatic hormone in the early 1920s, like their human type 1 counterparts, had a deficiency of insulin and an on-going dependence on injections for survival. Measuring a profile of blood glucose levels in such a dog over several days might *retrospectively* identify fairly consistent patterns of variation in insulin requirement either in response to food, or perhaps through variation in the baseline component. This could be used by the human researcher to tailor the replacement accordingly, with more given with meals for instance. The dog would be unable of course to report symptoms related to displacements, although these might have been observable as clinical signs at the extremes. For human patients, it was only through the later technology of self-monitoring (becoming available during the 1970s and 1980s) that individuals were able to detect and respond to the more minor displacements, moving the control of diabetes away from the clinician and towards the patient.
3. *Appropriate corrective behaviours*. Awareness of displacement is only any use if it is coupled with restorative mechanisms. In the physiological state, subconscious homeostatic processes provide these mechanisms and require very little or no conscious effort. The type 1 individual requires experience and training in responding to these displacements. The wide variation in the ‘glycaemic stability’ of such people is partly the result of variation in this skill. ‘Brittleness’ is a long established concept in diabetes and much debate has centred on whether it is a physiological phenomenon [8]. More likely, it results from an interaction between physiology (variation in blood glucose levels and the symptoms this produces) and the individual’s disordered behavioural response to this variation, a response influenced by psychological factors. Ensuring that restorative behaviours are adaptive and appropriate is important because there is huge potential for the untrained individual to do more harm than good in correcting glycaemic excursions (Fig. 20.3).

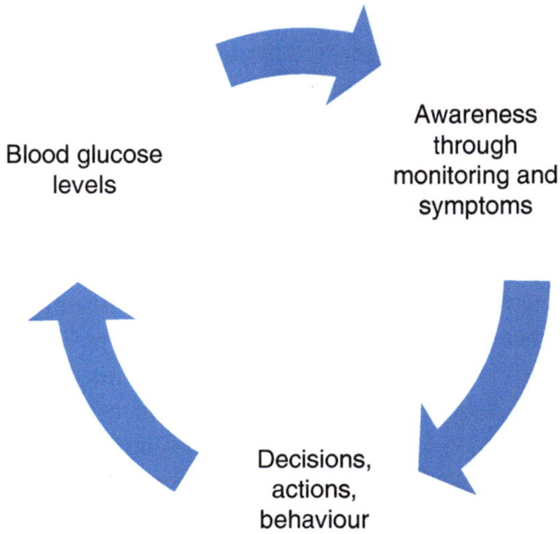

Fig. 20.3 In type 1 diabetes, feedback between awareness of blood glucose and actions influencing future blood glucose determines the intrinsic stability or instability of the system. Depending on the skill and training of the individual, this feedback may promote either tight control through adaptive decision making, or instability through ‘tampering’

During the late twentieth century patient-centred models were developed that would promote adaptive behaviours and take us beyond the ‘retrospective’ approach to the use of self-monitored data. Clinicians examining historical blood glucose profiles may not recognise the influence emerging data have had on the patient's behaviour, an influence that does not apply in the 'veterinary' setting, nor in the cases of human patients who record their blood glucose values purely for their clinician's later benefit. For many if not most, awareness of the data influences immediate behaviour, and the modern patient expects to use this information in real-time to control the system.

Prior to current self-monitoring technology, the retrospective approach was of course the most effective (and indeed the only available) strategy, and is still appropriate in some cases. The early recipients of insulin in the 1920s felt fortunate enough to have their lives saved through replacement of this hormone. Fine tuning the system to account for hour-to-hour variation in requirement was a detail. But more recently, the benefits of tight control on microvascular complications have become evident, and quality of life has increased substantially with the move towards individual self-management. The most successful educational model is the DAFNE (Dose Adjustment for Normal Eating) programme [9], which has proven beneficial in training type 1 people to respond to frequently self-monitored blood glucose results by tailoring insulin doses to unrestricted carbohydrate intake through a personalised algorithm derived over the course of a week [10]. The principles underlying it are that appropriate mealtime (short acting) insulin dose requirements are directly proportional to carbohydrate intake (although the ratio of one to the other needs adjustment, for instance depending on time of day) and that there is a stable 'baseline' insulin requirement provided by a different type of insulin (or by the basal rate of an insulin pump) that once optimised, should not need altering very often. Accurate carbohydrate counting and frequent monitoring are the basis for this technique, allowing dietary freedom.

This programme is an apparent victory for linear approximation and for an equilibrium-based model. It has been shown to improve glycaemic control without increasing frequency of hypoglycaemia, improving autonomy, and increasing quality of life. The mean improvement in HbA1c is less than generally desired in poorly controlled individuals, and a follow-up study has shown that whilst quality of life is maintained at 4 years, glycaemic control is slipping back towards baseline values [11]. Nevertheless, it is the best educational intervention available for promoting glycaemic control in type 1 diabetes, and has been widely adopted by the UK National Health Service.

20.5 Insights from Complexity Science

Can complexity science add to this discussion? As discussed above, we know that external observers (experimenting on dogs or helping to manage type 1 patients) are limited in their ability to close the feedback loop in real time (i.e. respond to displacements with appropriate correction), and rely on the *retrospective* examination of historical glucose profiles over at least several days in order to detect patterns consistent enough to inform adjustment of insulin doses. Whilst this approach is very much better than nothing, it may feel to the patient like 'driving a car by looking through the rear view mirrors'.[1] This was the standard approach until quite recently, but it is increasingly recognised that in fact individuals also measure blood glucose levels to support 'real-time' *prospective* actions and that (to continue the metaphor) any good driver will use both the rear view mirrors and the front windscreen views simultaneously to control the vehicle [12].

Complexity science has been used to challenge the traditional assumption that all variation is undesirable [13], because some variation is required for the natural flexibility inherent in complex adaptive systems. In diabetes, we must accept that glycaemic variation (as well as hyperglycaemia itself) may indeed be detrimental as it may promote the pathophysiological mechanisms

[1] An 'off the record' comment from a qualitative research participant with type 1 diabetes, University of Warwick, 2004.

of micro-vascular disease [14]. However, a major goal is the achievement of satisfactory HbA1c levels with infrequent hypoglycaemia and good quality of life, and the avoidance of catastrophic events requiring hospitalisation. Meeting this challenge might in principle require an ability to harness variation rather than extinguish it, particularly where it cannot in practice be extinguished completely. A sail-board enthusiast cannot remain upright indefinitely without actively harnessing the board's natural movement, and repeatedly changes sides to maintain control. Tennis players often sway naturally as the opponent serves the ball, as this makes it more likely they can respond to the unpredictable direction of the serve. How might this principle apply to the individual aiming to control glycaemia?

20.6 Sensitivity to Initial Conditions

Henri Poincare was the mathematician most associated with a phenomenon now identified as one of the roots of nonlinear dynamics, known as the three body problem [15]. Poincare recognised that whilst two gravitational bodies orbiting each other describe a periodic orbit, which is in principle predictable into the indefinite future, the inclusion of a third such body changes the dynamics in a fundamental way. Not only is the motion more complicated, in the sense that it is no longer periodic, but Poincare found that the equations predicting future positions of the bodies were actually *unsolvable*. This was surprising, because the laws of motion and gravitational forces acting between the bodies are completely understood. The reason for this unpredictability is that even small uncertainties become amplified over time. Arbitrarily small errors in measurement of starting conditions rapidly result in unknowable future trajectories. This future uncertainty is not a result of inaccurate measurements—even the best available can only be taken to a finite number of decimal points, and are therefore never known 'perfectly'. Neither is it a result of a lack of understanding over how the system works mechanistically, which was established at the much earlier time of Newton. It is a fundamental property of the dynamical system, and the unpredictability arises intrinsically within it. *Sensitivity to initial conditions* is the term given to this property (see Fig. 20.2), and is one of the reasons why meteorologists are unable to predict with confidence what will happen beyond a few days into the future. Weather forecasters may study initial starting conditions as accurately as they can, but the *predictive time horizon* is still limited—a perhaps insurmountable problem—because the dynamics of weather systems are *divergent* at least over the days or weeks of interest.

Later on, in the twentieth century the American meteorologist Edward Lorenz demonstrated this property by describing the way the trajectories of a simplified weather system (described by just three model variables and three equations) became divergent over time as a result of arbitrarily small errors in measurement (Fig. 20.4). This brought home the fact that even when the future positions are completely determined by the starting conditions and the equations (i.e. there is no external 'noise' entering the system from outside) the future is unknowable. In other words, even deterministic systems may be unpredictable. Contrasting this dynamical property (which has come to be known as *deterministic chaos*) with that of periodic systems and with static systems, he titled his paper 'Deterministic non-periodic flow' [16]. Prior to this discovery, it was assumed that if a natural system were dynamic (i.e. changed over time through a space of possible values for its variables), it would tend to return to a preferred position (a 'point attractor'), or settle into a periodic trajectory. In fact, the existence of chaotic behaviour became increasingly recognised in robust naturally occurring systems, and chaotic dynamics became the basis for understanding the flexibility and adaptability of such systems. Whilst most are 'open', that is set in a wider context of external influences (unlike Poincare's idealised gravitational bodies, or Lorenz's low dimensional weather model), the conclusion is the same: an element of the unpredictability arises as an *intrinsic* property of the dynamics, which are divergent in the short or

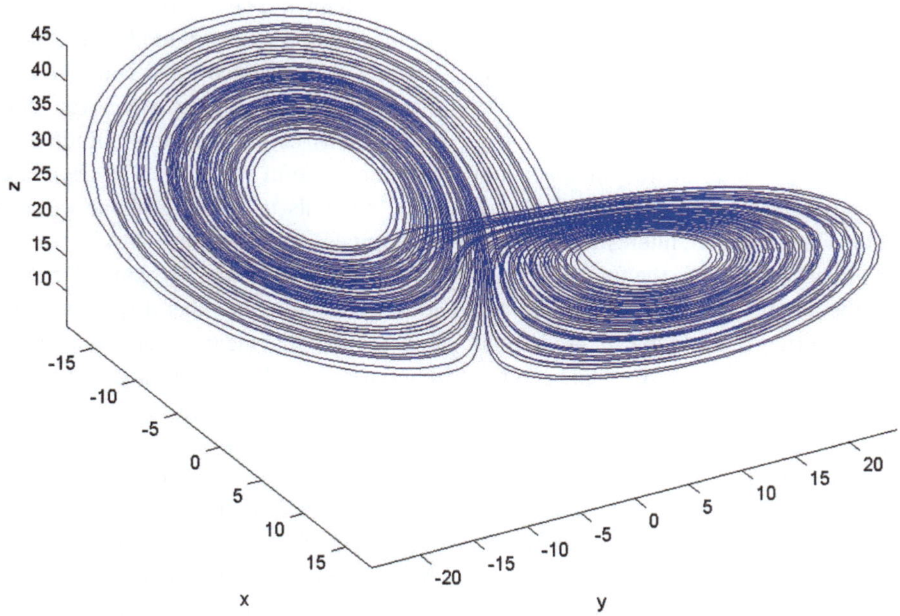

Fig. 20.4 The Lorenz butterfly, a classical example of a chaotic ('strange') attractor. This is a model weather system described by just three differential equations and three system parameters (σ, ρ, and β). The future trajectory (i.e. sequence of values of x, y, and z) is purely determined by the initial starting conditions, but is unpredictable and will never repeat itself exactly. There are no external sources of 'noise' and the unpredictability arises as an intrinsic property. Real life weather systems are set in a wider context of other influences but are also inherently unpredictable in the medium term due to the sensitivity to initial conditions. The stability of the Lorenz system can be modified by tuning the parameters. Whilst it is not possible to control the weather, other nonlinear systems may be stabilised through similar tuning as well as by protection from external disruption (Reproduced with kind permission from Ricardo Carretero, San Diego State University http://www-rohan.sdsu.edu/~rcarrete/teaching/M-637/lectures/lectures.html)

medium term. For people with diabetes trying to understand and control blood glucose levels, this may set a limit on our ability to predict future behaviour based on current starting conditions. For those at times perplexed by inexplicable blood glucose measurements, this recognition in itself may be quite liberating. But how can it be used to improve control and hopefully quality of life?

In addition to the realisation that even simple, deterministic systems may produce unpredictable behaviour, a further insight was achieved through these developments that is relevant to diabetes. This involves the importance of *interaction* between the determinants of blood glucose outcomes. During the 1970s, the 'weighing scales' model developed as a means of representing the 'balance' required for control of blood glucose. 'Balance' was the name given for the patient's journal of the British Diabetic Association (later renamed Diabetes UK [17]). The weighing scales were typically displayed as the traditional two tray device and the implication was that two commodities at a time (carbohydrate vs. insulin dose, or carbohydrate vs. exercise), needed to be balanced to avoid blood glucose displacement. But as people living with diabetes know, these three factors are often operating simultaneously, so that a three tray scales is a more appropriate model (still failing to recognise other more subtle factors, but nevertheless useful). Despite being a simple system, this is considerably more difficult to balance in practice. Worse than that, the three respective factors are not, in the case of blood glucose control, as separate and independent as a three tray model suggests, but in fact interactive. Whilst carbohydrate intake can be 'balanced' against insulin requirement (a fundamental basis for the DAFNE approach), and in theory against exercise (a more difficult commodity to quantify), exercise in fact not only increases carbohydrate requirement, it also increases *sensitivity* to insulin. This changes things very significantly, because rather like an interest free bank account recording deposits and withdrawals that has suddenly developed compound interest, predicting the balance of the account into the future depends not only on the amount of money deposited and withdrawn, but very much on their *timing* in relation to each other. Dynamics once again becomes the focus.

I have described how blood glucose control in type 1 diabetes, to go beyond the 'retrospective' approach, requires a coupling of appropriate detection of displacement with adaptive restorative behavioural mechanisms. These mechanisms involve the insulin injecting, dose adjusting, carbohydrate taking (or restricting), and exercising behaviours of the individual. This sounds complicated, but in fact involves just a few determinants of blood glucose, most within the person's control. Others are not as easily controlled, such as psychological stress, which is important as a determinant for some people quite frequently and for most occasionally. However, in type 1 diabetes we appear to have a dynamical system whose most important determinants are within the control of the individual. We have the technology to detect displacement through self-monitoring (when awareness through symptoms fails). What is needed is an adaptive behavioural mechanism to ensure that displacements lead to restoration of normal levels and patterns rather than worsening of control. We are aiming to produce *convergent* rather than *divergent* behaviour. How can this be achieved, and can we measure it? Let's first of all look at what is meant by 'stability' in this context.

20.7 Towards a Dynamical Definition of Glycaemic Stability

Adequacy of control in diabetes is usually measured using the glycosylated haemoglobin level, otherwise known as HbA1c, a term used earlier in this chapter. This measures the proportion of haemoglobin (present in red blood cells) that has become 'glycosylated', and is directly related to the average blood glucose level during the 2–3 months of the red cells' lifetimes. It is an extremely useful measure of glycaemia over this time, but tells us nothing about variation. Two individuals with similar HbA1c values could have grossly different dynamical patterns, one somewhat static, the other displaying wide variation around the mean value. Indeed, Fig. 20.5 demonstrates this limitation using a continuous glucose monitor applied to two volunteers with similar HbA1c levels.

Continuous glucose monitoring devices measure subcutaneous glucose values that are strongly correlated with blood glucose values, although there is a lag between fluctuation in one and the other [18]. They enable us to measure such variation, and produce as their outputs a mean value, the range, and standard deviation. The standard deviation is important because high values around a satisfactory mean glucose level suggests that at times the value is unacceptably low or high. Such devices are increasingly used in clinical practice (often to detect nocturnal hypoglycaemia) [19].

One might assume that these measures (HbA1c, mean, range, and standard deviation) would be all that we need to describe the blood glucose dynamics. HbA1c is established as a predictor of microvascular complications, so provided this and the mean value are satisfactory and the standard deviation is not high, what more do we need?

For people with type 2 diabetes, this might indeed be all that is required, although post-prandial blood glucose excursions are also known to predict cardiovascular risk independently of HbA1c [20]. The convergent dynamics typical of this condition (at least in its early stages) means that self-monitoring is less useful, as discussed above. But a type 1 individual must take more conscious control of blood glucose levels, responding appropriately to displacements, steering the system from one state to the next. This skill is important in achieving control, but is not measured through the parameters described so far. An ability to measure this skill might be a first step towards enabling us to teach it as part of an educational intervention.

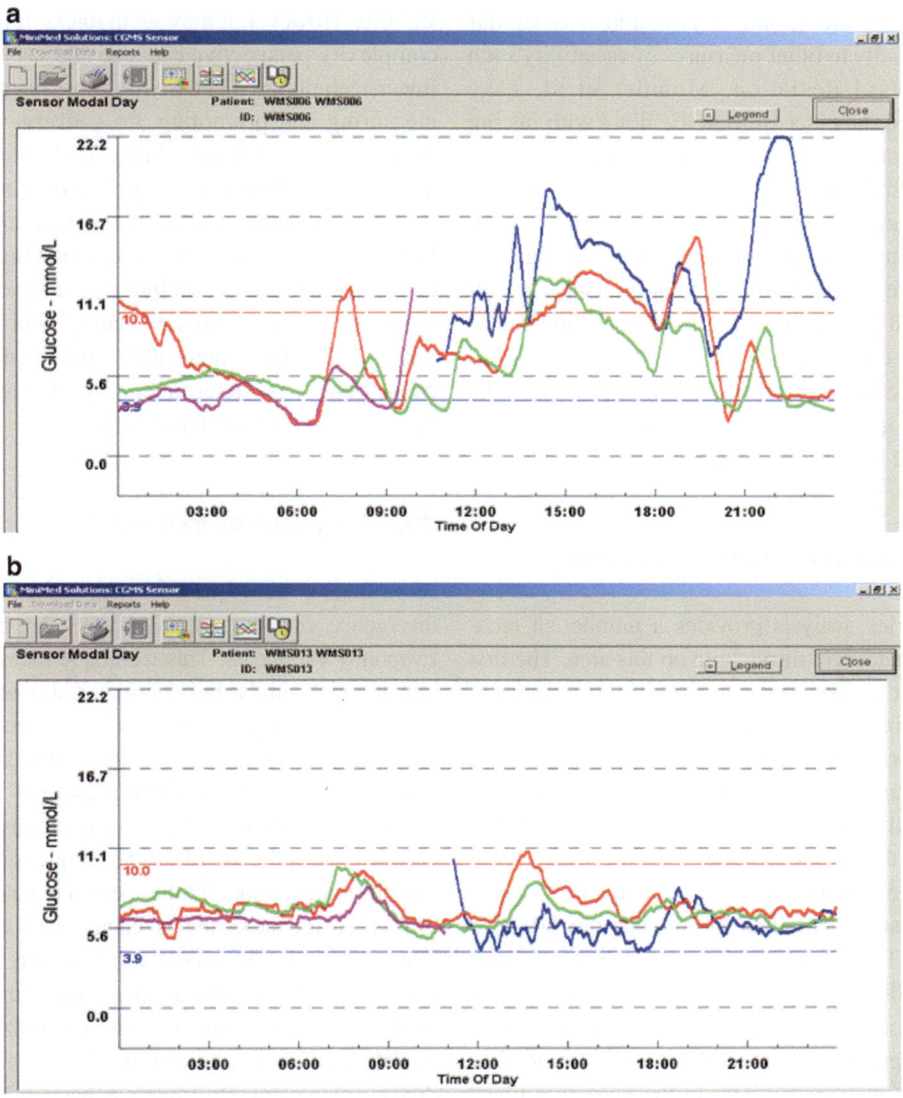

Fig. 20.5 CGMS profiles (**a** and **b**) from two people with similar HbA1c levels but clearly different dynamical patterns

A commonly used continuous glucose monitoring system (CGMS) measures subcutaneous values every 5 min for 72 h (3 days), via a probe inserted under the skin. This yields 864 consecutive glucose values. The mean, range, and standard deviation are automatically calculated by the device. But in determining the dynamical stability of the system we need to understand something that is not captured by these values. We need to understand the way one state moves to the next, the smoothness of variation, and the degree to which each value is influenced by preceding values. Imagine taking a dataset from the CGMS device, and randomly shuffling the 864 glucose values. In the shuffled dataset, the mean, range, and standard deviation are the same as before shuffling. What is lost in this process is the relationship between neighbouring values, and this is a missing component in traditional models of glycaemic stability.

The mean amplitude of glucose excursions (MAGE) has been suggested [21], and takes us significantly further towards understanding the structure of the variation, although it is similar conceptually to other measures of variability such as standard deviation. Monnier et al. have discussed the place of MAGE, along with fasting blood glucose and post-prandial glucose level in the determination and minimisation of glycaemia. They discuss its likely relevance to the development of micro-vascular complications, and make the point that fluctuation itself, as well as overall glycaemia, is likely to be detrimental in terms of complication risk as it promotes oxidative stress [22]. However, this measure still falls short of quantifying the way the system moves from one state to the next over time.

20.7.1 Autocorrelation Function

Time series analysis provides a number of techniques that might throw light on this area. The first is a linear method well established in many areas of engineering: the autocorrelation function [23]. Autocorrelation refers to the linear correlation between neighbouring values in a time series, and the way this correlation decays as the time interval between the values increases. Glucose values separated by 5 min are highly correlated, a relationship that is less evident after 30 min, and much less so after 6 h. This information is completely lost through the shuffling process described above. How does this rate of decay differ between individuals, and to what extent might this reflect their ability to skilfully move from one state to the next over time? Might educational interventions such as the DAFNE programme impact on these indices? Highly correlated profiles might be more 'stable' but less flexible, so the target values for such measures is unclear and requires more research. This approach might allow us to measure an individual's skill and then tailor such programmes accordingly. The autocorrelation function has been applied to blood glucose time series but an obstacle to this approach is the non-stationary nature of such data, that is the tendency of statistical properties such as mean values and standard deviation to vary significantly between different intervals within the profile [24].

Autocorrelation takes us significantly closer to a *dynamical* understanding and definition of stability. However, it may seem unexciting from a complexity perspective because it is very much a linear technique. What we are most interested in measuring and promoting are *convergent* rather than divergent tendencies in blood glucose behaviour. As described above, type 2 individuals tend to have naturally 'convergent' dynamics (due largely to the presence of some remaining endogenous insulin production that can vary according to need, albeit less than adequately). For the type 1 individual, this is unavailable and the degree of convergence is determined by behaviour in response to self-monitored values.

20.7.2 Lyapunov Exponents

A promising technique for identifying consistent divergence in phase space is the derivation of Lyapunov exponents. This technique measures the rate at which similar but non-identical points in the space diverge. A positive Lyapunov exponent indicates exponential divergence and is a hallmark of chaotic behaviour. It is a direct measure of 'sensitivity to initial conditions'. However, in the case of glycaemic control the use of this technique is probably inappropriate (although it has been attempted [25]). This is because the system is 'open', that is not determined purely by the model variables, but set in an effectively infinite (or certainly high) dimensional context of potential influences. In the Lorenz weather model, and other artificial models of nonlinear dynamical system, the future behaviour is determined solely by the model variables.

20.7.3 Detrended Fluctuation Analysis

Detrended fluctuation analysis (DFA) is a technique that can assess the 'smoothness' of variation in a time series, a property potentially relevant to the issue of glycaemic stability. The technical details are beyond the scope of this chapter, but are described elsewhere [26, 27]. In outline, the time series is first of all integrated, then divided up into different window boxes of length *n*. A least squares line is drawn through each box to determine the local trend. The fluctuations from the least squares

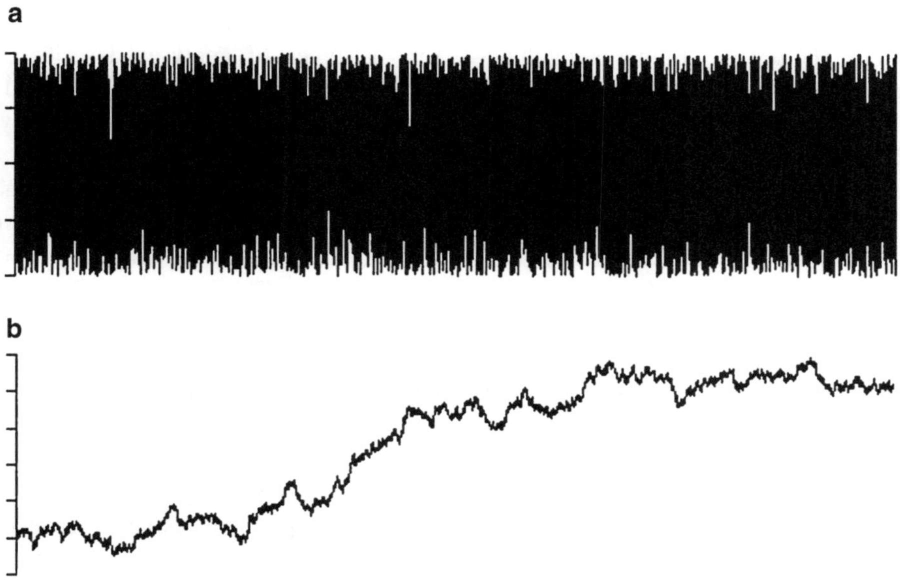

Fig. 20.6 White noise (**a**) and Brownian noise (**b**). White noise is completely uncorrelated, that is each value in the series is completely independent of all other values. In blood glucose time series, there is a component of the variation due to measurement error that has this property, but it is small for high quality monitoring devices. Brownian motion involves a 'random walk' in which each increment occurs in a random direction, but starts off at the position of the previous value, and is therefore highly correlated with it. White noise signals have $\alpha=0.5$, while Brownian noise has $\alpha=1.5$. Some natural processes exhibit fractal scaling with α values around 1.0 (Reproduced with kind permission from Paul Bourke, University of Western Australia http://paulbourke.net/fractals/noise/)

line are estimated over the whole time series to derive $F(n)$. Generally speaking $F(n)$ is positively correlated with n. For processes producing power law scaling the log–log plot of n versus the $F(n)$ will produce a straight line, and the gradient of this line is derived as the DFA scaling exponent α. Power law scaling occurs in a range of naturally occurring processes. DFA identifies the presence or absence of power law scaling and α is related broadly speaking to the 'smoothness' of variation.

For random 'white noise' variation the value of α is 0.5 indicating a very 'rough' pattern—each value is entirely independent of all other values in the series. Higher values of α indicate positive correlation between neighbouring values and therefore a 'memory' in the data. A value of $\alpha=1.5$ suggests 'Brownian noise' typical of a 'drunken walk', where each new step occurs in a random direction but is highly determined by the previous step or steps, producing relatively 'smooth' variation (Fig. 20.6). Between these extremes, a value of $\alpha=1.0$ is typical of some naturally occurring processes (e.g. healthy heart rate variability) in which there is a compromise between complete independence of neighbouring values and the opposite situation of highly correlated structure.

A 2006 paper has described the use of DFA applied to CGMS data from individuals with and without diabetes [28]. According to the authors, negatively correlated long range (>2 h) behaviour is seen in healthy controls due to physiological responses to glucose flux. Higher than average glucose levels are likely to be followed by lower than average values after this time. These responses are part of the physiological control mechanism maintaining homeostasis. They appear to break down in diabetes, where positive correlation tends to persist. This group used DFA as well as MAGE, and reported lower values of α for controls than for volunteers with diabetes. Comparing these values with the 'Brownian motion' value of 1.5, the control values underwent a cross-over at around 2 h, before which values were higher, and after which lower than this threshold. The α values from those with diabetes were >1.5 in both the short and longer range, with a less obvious reduction in value at around 3 h. These interesting findings require further exploration to determine the clinical implications for glycaemic control. The interpretation of DFA in this situation is difficult and further work on it is underway [29].

Glycaemic control in diabetes requires a stable but flexible state, adaptable to changing needs whilst observing reasonable limits of variation. Stability is not only a statistical concept defining acceptable limits to this variation, but also a dynamical concept reflecting both smoothness and predictability, and equally importantly, a *resilience to external disruption*. A stable system is not only one that is 'well behaved' and reasonably predictable. It is also one that can withstand disruption and resume its previous behaviour if displaced. This property is important in diabetes and is related to the issue of convergence/divergence described above. The identification of attractor patterns in dynamical systems including the point attractor of the equilibrium state, the periodic attractors of many natural processes governed by diurnal sleep/wake cycles, and the strange attractors of chaotic systems lends itself well to the individualised approach to modern diabetes care. Each individual has his or her own 'attractor' pattern, whose properties determine their glycaemic stability and reflect their personal need for flexibility. The attractor is the pattern to which the system returns if displaced. Attempts to describe such attractors in geometrical terms using dynamical indices are still in their infancy, but this approach has great potential to support personalised models of control.

20.8 What Is the Target Dynamic for Optimum Glycaemic Control?

Research into the issues explored in this chapter is still in an early phase. Whilst CGMS datasets have opened a new window into the dynamical world of the individual with diabetes, it is still uncertain how adequate such data are (in quality, volume, and profile duration) for the types of analyses described, or the extent to which they can reflect blood (rather than merely subcutaneous) glucose dynamics for the purposes of such analyses.

However, one implication naturally suggests itself through this discussion and from the discoveries of nonlinear dynamical systems over recent decades. This is the realisation that whilst chaotic behaviour is less stable than periodic or static behaviour, it is significantly *more orderly* than the random variation that it superficially resembles. For individuals attempting to control blood glucose variation, stasis may be unachievable as the system may have an inevitable tendency to 'keep moving'. Such a tendency might be due to physiological factors, such as variation in insulin absorption, or to behavioural factors. Whilst we accept that variation should be minimised, the equilibrium state that is the basis for traditional models of control (including DAFNE) may be difficult to achieve in reality. The creation of this state, in which movement is the exception rather than the rule may in practice be unrealistic. Accurate carbohydrate counting and appropriate dose adjustment are of course a good thing and reduce the magnitude of displacements. The use of long acting insulin analogues that produce fairly flat background insulin levels also helps enormously, as the dose adjustments required to counteract carbohydrate intake are much more effective on the basis of a stable baseline. Where background insulin requirement varies significantly over 24 h, the use of insulin pumps programmed to deliver this variable requirement have also proven very useful. But for some individuals, the static equilibrium state may simply be out of reach. Some degree of movement is inevitable and awareness of where the system is going as well as where it is now becomes vital. This observation was in fact demonstrated in 1990, in a study of intravenous insulin dose titration during labour [30]. An approach in which trends in glucose level as well as glucose levels themselves were taken into account improved control by avoiding over-shooting. To those experienced in using insulin, this is an everyday reality: corrective doses of short acting insulin designed to restore target levels need to take account not simply of the difference between current and desired values, but of the current movement in the system, that is the way the various glycaemic determinants (not just glucose) are already moving. This is particularly relevant in the setting of frequent monitoring and corrective actions in the DAFNE approach. It is also the basis for a new generation of insulin pumps which require the input of information into the device not only on carbohydrate intake and blood glucose measurements but very much on their *timing* in relation to each other. In other words, the dynamical characteristics of the system are recognised and included in the model, although exercise and other determinants that are more difficult to quantify are not. For those unequipped with such devices, or in whom the un-included factors are particularly important, an intuitive approach that accounts for several determinants simultaneously and adjusts decision making accordingly is the only option [31]. This approach may require 'nudging' of the system on to trajectories that will converge towards the desired point rather than forcing the system directly towards the desired point itself, to avoid the repeated over-correction known as 'chasing the tail'. Poorly designed and over-simplified control algorithms can do more harm than good [32]. This repeated nudging technique has been shown to control the chaos of isolated cardiac conduction tissue in vitro [33], and is assumed to work for the same basic reason, that over-shooting is avoided.

20.9 Summary

The past 30 years have seen the focus of diabetes control move away from the clinician and towards the patient. This has required a fundamental change in approach, from the external control techniques (using retrospective analysis of blood glucose data) to the patient-centred approach that also requires healthy 'real-time' responses to displacement and prospective actions. A lesson from complexity science is that adaptability and flexibility are as important as regularity for the promotion of stability, and this applies at least as much to glycaemic control as to any other area of health care. Chaotic behaviour (in its strictly defined sense) is widespread in healthy, robust, and adaptable complex systems. Whilst this finding may appear to contradict the common-sense notion of stability, it is possible that a *pattern* rather than a fixed point represents the attractor most likely in practice to minimise glycaemic variation whilst maintaining flexibility, particularly for type 1 patients whose dynamics are so heavily dependent on behaviour. Successful control becomes a case of adopting this pattern through adaptive responses to self-monitoring. This chapter has been necessarily exploratory as much of the basic theory remains untested in practice. The implications of these insights for diabetes care set an exciting research agenda for the coming decade.

References

1. World Health Organization. Definition, diagnosis and classification of diabetes mellitus. Report of a WHO consultation. Geneva: WHO; 1999.
2. Farmer A, Fox R. Diagnosis, classification, and treatment of diabetes. BMJ. 2011;342:d3319. doi:10.1136/bmj.d3319.
3. Glass L, Kaplan D. Time series analysis of complex dynamics in physiology and medicine. Med Prog Technol. 1993;19(3):115–28.
4. Schaffer WM. Order and chaos in ecological systems. Ecology. 1986;66(1):93–106.
5. Farmer A, Wade A, Goyder E, et al. Impact of self monitoring of blood glucose in the management of patients with non-insulin treated diabetes: open parallel group randomised trial. BMJ. 2007;335:132.
6. Tattersall R. Diabetes: the biography. Oxford: Oxford University Press; 2009.
7. Bolli GB. Physiological insulin replacement in type 1 diabetes mellitus. Exp Clin Endocrinol Diabetes. 2001;109 suppl 2:S317–32.
8. Shima K, Tanaka R, Morishita S, Tarui S, Kumahara Y, Nishikawa M. Studies on the etiology of 'brittle diabetes'. Relationship between diabetic instability and insulinogenic reserv. Diabetes. 1977;28:717–25.
9. http://www.dafne.uk.com/.
10. DAFNE Study Group. Training in flexible, intensive insulin management to enable dietary freedom in people with type 1 diabetes: dose adjustment for normal eating (DAFNE) randomised controlled trial. BMJ. 2002;325:746.
11. Speight J, Amiel SA, Bradley C, Heller S, Oliver L, Roberts S, Rogers H, Taylor C, Thompson G. Long-

term biomedical and psychosocial outcomes following DAFNE (Dose Adjustment For Normal Eating) structured education to promote intensive insulin therapy in adults with sub-optimally controlled type 1 diabetes. Diabetes Res Clin Pract. 2010;89:22–9.
12. Holt T, Kumar S, editors. ABC of diabetes. 6th ed. Oxford: Wiley Blackwell; 2010. p. 44–5.
13. Plsek PE, Wilson T. Complexity science: complexity, leadership, and management in healthcare organisations. BMJ. 2001;323:746.1.
14. Monnier L, Colette C. Glycemic variability: should we and can we prevent it? Diabetes Care. 2008;31 Suppl 2:S150–4.
15. Flake GW. The computational beauty of nature: Computer explorations of fractals, chaos, complex systems, and adaptation. Cambridge: MIT Press; 1996. p. 226–7.
16. Lorenz E. Deterministic non-periodic flow. J Atmos Sci. 1963;20:130–41.
17. http://www.diabetes.org.uk/.
18. Sparacino G, Facchinetti A, Cobelli C. Smart continuous glucose monitoring sensors: on-line signal processing issues. Sensors. 2010;10:6751–72.
19. Pickup JC, Freeman SC, Sutton AJ. Glycaemic control in type 1 diabetes during real time continuous glucose monitoring compared with self monitoring of blood glucose: meta-analysis of randomised controlled trials using individual patient data. BMJ. 2011;343:d3805.
20. Ceriello A, Hanefeld M, Leiter L, Monnier L, Moses A, Owens D, Tajima N, Tuomilehto J. Postprandial glucose regulation and diabetic complications. Arch Intern Med. 2004;164:2090–5.
21. Monnier L, Colette C, Owens DR. Glycemic variability: the third component of the dysglycemia in diabetes. Is it important? How to measure it? J Diabetes Sci Technol. 2008;2:1094–100.
22. Monnier L, Mas E, Ginet C, Michel F, Villon L, Cristol JP, Colette C. Activation of oxidative stress by acute glucose fluctuations compared with sustained chronic hyperglycemia in patients with type 2 diabetes. JAMA. 2006;295(14):1681–7.
23. Kaplan D, Glass L. Understanding non-linear dynamics. New York, NY: Springer; 1995. p. 295–7.
24. Bremer T, Gough DA. Is blood glucose predictable from previous values? A solicitation for data. Diabetes. 1999;48:445–51.
25. Kroll MH. Biological variation of glucose and insulin includes a deterministic chaotic component. Biosystems. 1999;50:189–201.
26. http://www.physionet.org/physiotools/dfa/.
27. Peng CK, Havlin S, Stanley HE, Goldberger AL. Quantification of scaling exponents and crossover phenomena in nonstationary heartbeat time series. Chaos. 1995;5:82–7.
28. Ogata H, Tokuyama K, Nagasaka S, Ando A, Kusaka I, Sato N, Goto A, Ishibashi S, Kiyono K, Struzik ZR, Yamamoto Y. Long-range negative correlation of glucose dynamics in humans and its breakdown in diabetes mellitus. Am J Physiol Regul Integr Comp Physiol. 2006;291:R1638–43.
29. Khovanova NA, Khovanov IA, Sbano L, Griffiths F, Holt TA. Characterization of linear predictability and nonstationarity of subcutaneous glucose profiles. Computer Methods and Programs in Biomedicine, (in press).
30. Lean MEJ, Pearson DWM, Sutherland HW. Insulin management during labour and delivery in mothers with diabetes. Diabet Med. 1990;7:162–4.
31. Wilson T, Holt T, Greenhalgh T. Complexity science: complexity and clinical care. BMJ. 2001; 323:685.1.
32. Queale WS, Seidler AJ, Brancati FL. Glycemic control and sliding scale insulin use in medical inpatients with diabetes mellitus. Arch Intern Med. 1997;157: 545–52.
33. Garfinkel A, Spano ML, Ditto WL, et al. Controlling cardiac chaos. Science. 1992;257:1230–5.

21 Complexity and the Onset of Psychosis

Rajkumar Kamatchi and Matthew R. Broome

21.1 Psychosis

Psychotic illness refers to a group of psychiatric disorders that can be severe and enduring. The core symptoms of delusions and hallucinations affect the person's ability and judgement on how to behave or how to respond to the world around them. The psychoses are usually grouped into three major categories: schizophrenia and schizophrenia-like illness, bipolar affective disorders (BPAD) and other affective psychoses and organic psychotic disorders. However, the psychotic symptoms can be seen in many other psychiatric and physical disorders as well as in normal healthy individuals.

21.2 Onset of Psychosis

No one model clearly explains the onset of psychosis yet. Despite extensive research and huge investment of money and energy, researchers still struggle to establish a model which can solely explain the disease process of psychosis and how it manifests. It has become clear now that psychosis is a phenomenon of severe complexity and it needs a systems approach to understand. Complexity, in psychiatry, means there is no simple relationship between the causes and the syndrome we see clinically.

Psychosis, like most mental disorders, is not a straightforward problem to deal with scientifically. It is difficult to understand psychosis by simple linear or conventional reductionist approaches. The complexity arises right from the causes of psychosis, the emergence of symptoms, and how it evolves as a full-blown condition in a person and the interactions between the various factors involved. In this chapter, we focus on the complexity around the onset of psychosis. We describe the various models of causes for psychosis and the emergence of symptoms from the at-risk stage to full-blown psychosis.

21.2.1 The Neurodevelopmental Model

It has been apparent for almost two decades that there is a developmental component to schizophrenia. In its simple form, this model postulates that genes involved in neurodevelopment [1] and/or environmental insults in early life lead to aberrant brain development, which in turn predisposes to the later onset of psychosis [2–4]. However, more recent formulations incorporate the role of social factors [5] such as urban upbringing, social isolation and

R. Kamatchi (✉)
Coventry and Warwickshire Partnership NHS Trust, The Caludon Centre, Coventry, UK

Warwick Medical School, University of Warwick, Coventry, CV4 7AL, UK
e-mail: rajkumaranjali@yahoo.com

M.R. Broome
Warwick Medical School, University of Warwick, Coventry, CV4 7AL, UK

Coventry and Warwickshire Partnership NHS Trust, Coventry, UK
e-mail: M.R.Broome@warwick.ac.uk

J.P. Sturmberg and C.M. Martin (eds.), *Handbook of Systems and Complexity in Health*, DOI 10.1007/978-1-4614-4998-0_21, © Springer Science+Business Media New York 2013

migration, and point to an interaction between the biological and psychological factors in a cascade of increasingly deviant development [6].

Prospective studies show that children who later develop schizophrenia are more likely than peers to show subtle developmental delays and cognitive impairments; they also tend to be solitary and socially anxious [7, 8]. Some evidence suggests that individuals destined to develop schizophrenia fail to learn new cognitive skills as they enter adolescence, thus appearing to show a relative decline compared with their peer group [8, 9]. The combination of neurocognitive and emotional deviance increases the likelihood of developing minor quasi-psychotic symptoms; indeed, a prospective study in Dunedin showed preschizophreniform individuals to be more likely to manifest such symptoms as early as age 11 years [7, 10]. It is postulated that in those destined to develop psychosis the strength, frequency, and associated distress of the odd ideas and experiences increase, and at some ill-defined point, the individual crosses a threshold into the pre-psychotic or prodromal phase. Recent genetic studies have postulated that psychoses should be seen as a member of a group of neurodevelopmental syndromes where genetic and environmental factors interact in causing the symptoms. However, a pure neurodevelopmental approach may not explain the timing of the onset of psychosis.

21.2.2 The Neurochemical Model: Dopamine as the Wind of Psychotic Fire

While the neurodevelopmental hypothesis can explain neurocognitive and emotional deficits in preschizophrenic children, and some at least of the neuropsychological and neuroanatomical abnormalities found in those with established schizophrenia, neither it nor epidemiological or high-risk studies explain the biological processes that accompany the onset of frank psychosis. In considering this issue, it is worth examining what we know concerning the neurochemical basis of the positive symptoms experienced in acute psychosis.

Hemsley [11] described how in the acutely psychotic individual "*Meaningful connections are created between temporary coincident external impressions ... or perceptions with thoughts that happen to be present, or events and recollections happening to occur in consciousness at the same time*". He and his colleagues [12] pointed out that, normally, input from the hippocampus controls the mesolimbic dopamine system, and speculated that damage to it causes a loss of this control, with the resultant heightened dopamine transmission facilitating the formation of "meaningful connections" between coincident events. Relatively little attention was paid to this paper at the time for the dopamine hypothesis of schizophrenia was in the doldrums. Then, Laruelle and colleagues [13], who demonstrated increased striatal dopamine release following amphetamine challenge in acute schizophrenic patients, provided evidence demonstrating the long suspected link between dopamine dysregulation and psychosis. Furthermore, the degree of dopamine release correlated positively with the extent of acute positive symptoms and with subsequent response to dopamine blockers [14, 15].

In 2003, Kapur [16] elaborated on the link between dopamine and the positive symptoms of psychosis and drew on theories of the role of mesolimbic dopamine in the healthy brain derived from studies using experimental animals. It has been argued that, in the normal individual, mesolimbic dopamine acts to provide significance or salience, transforming an affectively neutral mental representation of a stimulus into an attractive or aversive one [17]. Thus, mesolimbic dopamine activity may determine whether an intrusion into awareness from either external perceptual or internal mental sources receives a positive or negative "hedonic vector", and thus "grabs the attention" of the individual. If psychosis were associated with increased, often stimulus-independent, release of dopamine, salience would be granted to what would otherwise be relatively innocuous events and stimuli. In this way, it is argued, dopamine provided "*the wind of psychotic fire*" [18]. A recent study [19] confirms that the dopamine overactivity in striatal areas predates the onset of schizophrenia and seen in individuals with prodromal symptoms. It has been correlated with the severity of symptoms and the neurocognitive dysfunction.

In normal health, interplay between the hippocampus and the amygdala helps to maintain the individual in emotional balance. As Gray et al. [20] outlined, the hippocampus maintains focus on a task, and sets current environmental stimuli in the context of previous experience, allowing only response patterns that are appropriate to a given context to impact on mesolimbic dopamine; however, the amygdala can provide an emotional or affective override to this information. Grace [21] points out that normally the prefrontal cortex provides a supervisory input to the hippocampus and amygdala that tempers their reactions to stimuli and ensures that the responses are appropriate to the particular circumstances. However, in schizophrenia, this cortical–limbic circuitry malfunctions; many patients show deficits in prefrontal or executive functions, and much evidence demonstrates that the hippocampus and amygdala are decreased in volume in schizophrenia [22]. Grace suggests that either excessive input from these limbic structures or a loss of the normal prefrontal "brake" on the limbic system causes increased mesolimbic dopamine that results in the increased salience and overreaction to emotional stimuli, which leads ultimately to paranoia and psychosis. Illicit drugs have been found to disrupt the modulation of ventrostriatal function and interact with the dopamine system in causing psychotic experiences. Also, the stress associated with prodromal and early psychotic phases increases cortisol levels, which in turn induces structural brain changes and dopamine dysregulation, thus causing/maintaining the psychotic symptoms. As such, dopamine can be seen as the wind that sets the final stage in the onset of psychotic fire.

21.2.3 Genetics

Since a high proportion of the variance in liability to schizophrenia is genetic, we might suppose that at least some of the susceptibility genes influence the dopamine system. Recently, there has been some success in identifying genes that increase the risk of schizophrenia, particularly neuregulin (NRG1), dysbindin (DTNBP1) and regulator of G-protein signalling 4 (RGS4) [23]. These have been noted to have effects on the glutamate system [23–30], which is well known to regulate dopamine, while another candidate gene, catechol-*O*-methyl transferase gene (COMT), is involved in the breakdown of prefrontal dopamine and the latter's effect on cognition [31–33]. But a recent meta-analysis doubts the association between COMT and the schizophrenia [34]. Two other genes, disrupted in schizophrenia 1 (DISC1) [35, 36] and Neurexin 1 (NRxN1) [37], have also been linked to increased risk of schizophrenia in some families. Genetic factors as such could not substantiate the cause of psychosis as not all of them who have these genetic changes experience psychosis. Gene–environment interaction is crucial in the onset of psychosis.

21.2.4 Early Environmental Insults

Environmental factors also have a role to play. Individuals exposed to a range of obstetric hazards [38] are at increased risk of later schizophrenia, and schizophrenic subjects who have been exposed to obstetric complications are particularly likely to show decreased volume of the hippocampus [39, 40]. Animal studies have modelled such insults. For example, Lipska and colleagues [41–44] showed that lesioning the ventral subiculum in the neonatal period produced adult rats with a mesolimbic dopaminergic system prone to overreaction to an amphetamine challenge. Neonatal rats and guinea pigs, which have been subject to caesarean section and global anoxia, show dopaminergic abnormalities, and such early insults effect how dopamine is regulated in response to stress in adulthood [45]. Flagstad and colleagues [46] suggest that late gestational disruption of neurogenesis in rats leads not only to behavioural changes that mimic positive and negative psychotic symptoms, but also to a dysregulation of subcortical dopamine transmission.

21.2.5 The Role of Illicit Drugs

Many studies have shown that animals repeatedly exposed to amphetamines become sensitised rather than tolerant to the drug, and with successive exposures release increasing amounts of dopamine. There is some evidence that a similar process of dopamine sensitisation occurs in humans exposed to intermittent amphetamine in experimental conditions [47]. Of course, the common situation of human exposure to amphetamine is in the context of abuse of the drug, which is well known to induce schizophrenia-like psychosis [47–49]. Such evidence has given rise to the notion that not only amphetamine-induced psychosis [50] but also psychosis in general is the consequence of sensitisation in mesolimbic–cortical-striatal circuits mediated by dopamine [50–52].

There is increasing evidence that heavy use of cannabis in adolescence can increase the risk of later schizophrenia [53] and that effects on dopamine also mediate this. A study [54, 55] has shown that liability to psychosis induced by cannabis is strongly influenced by a polymorphism in the COMT gene that determines the rate of catabolism of frontal dopamine. Researchers have noted that people who smoke higher potency cannabis like skunk for longer periods and at greater frequency are more prone to psychosis and thus added to the evidence that Δ9-tetrahydrocannabinol (THC) is the active ingredient in cannabis increasing the risk of psychosis [56].

Recent studies have also noted that Δ9-THC disrupts the modulation of ventrostriatal function [57] and the communication between the two frontal lobes of the brain [58], and these are associated with psychotic symptoms. We can see that illicit drugs interact with dopamine system and with brain functions leading to psychosis. Although the association between illicit drug use and psychosis is very strong, it becomes complicated whether it causes or precipitates psychosis in someone susceptible to have psychosis secondary to genetics or other reasons. Also we know that people self-medicate with illicit drugs to alleviate the distress in the prodromal phase. This complex interaction between illicit drugs, genetics and dopamine system in the onset of psychosis is not easily understandable.

21.2.6 The Nuture Model: Role of Social Factors

Any plausible aetiological model of schizophrenia needs to incorporate the growing evidence that social factors can also modulate risk. Urban birth and upbringing [5] are risk factors, and several cohort studies have shown the quality of maternal–child relationship to be a predictor of risk of later schizophrenia [7, 8]. A recent meta-analysis [59] has confirmed that migration increases the risk of schizophrenia, the most striking findings being the evidence that Africans and African–Caribbeans living in the UK show an incidence of schizophrenia at least 6 times that of the native white population [60, 61]. Sharpley, Boydell and colleagues [62, 63] demonstrated that the risk was especially high where the migrant group was a small minority, suggesting that social isolation and lack of social support may play a role. A recent study (AESOP—aetiology and ethnicity in schizophrenia and other psychoses) looking at this issue has found that the perception of social disadvantage and strong ethnic identification may be contributing factors to high rates of psychosis in the African–Caribbean population in UK [64, 65]. They also found that unemployment, social isolation and achievement–expectation mismatch are important environmental factors associated with increased risk of psychosis [66].

Boydell and colleagues [5] also note that isolation rearing of rats leads to a sensitised dopamine system [67] and argue by analogy that social factors may have a similar impact on humans. Another proposal suggests that factors such as urbanicity, ethnic minority status and low IQ operate by subjecting the individuals repeatedly to the experience of social defeat [68, 69]; in animal studies, repeated social defeat leads to an enhanced behavioural response to dopamine agonists. Selten and Cantor-Graae applied this to humans and suggested that chronic experience of social defeat may lead to sensitisation and/or increased baseline activity of mesolimbic dopamine system, thereby increasing the risk of schizophrenia [70].

In an interesting animal analogy, a primate's standing in the social hierarchy can influence occupancy at D2 receptors. Using positron emission tomography (PET), Morgan et al. [71] found that, upon being transferred from individual to social housing, socially dominant macaque monkeys show an increase and more subordinate monkeys no change, in availability of dopamine D2 receptors. The authors suggest that this is because individually housed and socially subordinate monkeys have high levels of synaptic dopamine, whereas those who are able to attain dominance in social housing are able to return to "normal" dopamine levels. Hence, living alone or being in a lower position in the social hierarchy may be, at least for macaque monkeys, associated with a hyperdopaminergic state. Recently, Morgan proposed an integrated model—a sociodevelopmental pathway to psychosis emphasising the importance of environmental risk factors in the origin of psychosis [72].

21.2.7 Structural Brain Changes

There has been renewed interest in the possibility of progressive brain changes in schizophrenia [73]. These have been most clearly demonstrated in childhood onset cases, but several groups have claimed that progressive processes occur at, or immediately before, the onset of the first episode of psychosis [74–78]. While there is a degree of agreement that some brain changes can be seen on MRI scans, there is much dispute as to what they represent [79, 80]. The findings could reflect developmental changes occurring in late adolescence and early adult life, be degenerative, or even in some cases be consequent upon antipsychotic medication.

One possibility is that the changes may be consequent upon the stress that accompanies the onset of psychosis. Cotter and Pariante [81] point out that stress is well known to induce elevated cortisol and this may induce secondary brain changes such as hippocampal volume reduction. Further observations have suggested that the HPA axis is elevated in psychotic patients who are acutely unwell but normal in patients who are clinically remitted and receiving medication [82, 83]. Increased volume of the pituitary on MRI imaging is a marker of HPA activity, and Pariante and colleagues [82] demonstrated that those in a first episode of psychosis had larger pituitary volumes, and those with a chronic illness, smaller volumes than controls. The temporally closer a prodromal subject was to the onset of psychosis, the larger was the pituitary. In the acute phase of the psychosis, the increased pituitary volume could represent a *consequence* of the distress and arousal associated with the psychotic experience; alternatively, it could represent an increased activation of the stress response *preceding* the development of psychosis, for an increased susceptibility to daily life stress, an increased level of independent stressors, or both [84, 85]. This cross-sectional study does not allow a clarification of this point; however, but recent data suggest that the enlarged pituitary volume *precedes* the onset of psychosis, in a group of subjects at ultra-high risk of developing psychosis, and hence seems to support the latter model [86].

Mondelli, Pariante and their colleagues [87] noticed smaller left hippocampal volume in first episode psychosis patients and associated this with higher cortisol levels secondary to stressors in the prodromal stage. As explained previously when discussing the neurochemical model, the hippocampus maintains the emotional balance in humans and thus hippocampal damage upsets the dopamine regulation that leads to the onset of psychotic symptoms.

Recently, the researchers have found structural brain abnormalities in individuals with an at-risk mental state and associated few of these brain changes to increased vulnerability to psychosis but most other changes are associated with the development of psychosis [88]. Morgan and his colleagues [89] have noticed structural abnormalities—enlarged ventricular volumes, both in first episode schizophrenia and first episode affective psychosis. Also they found grey matter abnormalities in different regions of the brain in schizophrenia compared to affective psychosis. These new findings add to the evidence that brain changes occur just before the onset of psychosis and also brain changes continue throughout the course of the illness.

21.2.8 Psychological Mechanisms

Most psychological effort has been expended in demonstrating that people with schizophrenia show deficits on neuropsychological tasks. However, in recent years, an increasing number of cognitive models of positive psychotic symptoms have been proposed. These can be roughly divided into those that attempt to understand the cognitive basis of anomalous perceptual experiences and those that attempt to explain abnormal belief systems; the former tend to be more biologically grounded than the latter.

Anomalous perception. Garety et al [90] propose that the most common route to psychosis is via a "*basic cognitive dysfunction*" or disturbance of automatic processing. Possible mechanisms include difficulties in (a) integrating information into its temporal and spatial context or (b) in the self-monitoring of intentions and actions that lead to the individual's own actions being experienced as alien. These disturbances are assumed to lead to anomalous conscious experiences, such as heightened perception, actions experienced as unintended, thoughts appearing to be broadcast, thoughts experienced as voices and events that are unconnected appearing to be causally linked.

In the light of our earlier discussion of the role of dopamine, it may be that Garety's "*basic cognitive dysfunction*" is a cognitive parallel of heightened mesolimbic dopamine transmission [16, 20]. Hippocampal dysfunction may not only contribute to this, as described earlier, but it may also contribute directly to the misinterpretation of incoming stimuli since it plays a crucial role in the comparison of the past and present environment. As Gray and colleagues [12, 20] point out, damage to the hippocampus will therefore result in a "*weakening of the influence of stored memories of regularities of previous input on current perception*", and this in turn will lead to ambiguous and unstructured sensory input [11].

A recent model suggests that traumatic experience may also result in a transient increase in the processing of information via the amygdala relative to the hippocampus [91]. Information processed in this way is less integrated into a temporal and spatial context, and consequently, the information intrudes into awareness unexpectedly.

The role of attention is central to this and many cognitive models. In a classic description of the subjective experience of acute psychosis, McGhie and Chapman [92] quote a patient as stating,

> My thoughts get all jumbled up…. Things are coming in too fast. I lose my grip and get lost. I am attending to everything at once and as a result I do not attend to anything.

Normally, individuals become aware of only those aspects of the internal and external environment to which attentional resources are deployed. When a stimulus intrudes into awareness, this implies that attentional resources have been devoted to the content of that intrusion. Intrusions are normal phenomena experienced by all people, and include external stimuli, body state information or cognitive state information (intrusive thoughts and images) [92–94]. Morrison's [95] model of psychosis emphasises that such intrusions into awareness become problematic when they are appraised as a threat, as this may result in emotional, cognitive and behavioural responses that in turn increase the frequency of further intrusions. Since the relative salience of information that might potentially enter awareness is influenced by dopamine, increased dopaminergic activity will result in attention being deployed inappropriately, with the consequence that inappropriately salient intrusions intrude into awareness. Inappropriate deployment of attention might also be explained by poor contextual integration because this would result in a decrease in the influence of temporal context on attentional control. Thus, theories implicating impaired contextual integration and abnormal appraisal on the one hand and dopamine dysregulation on the other may be attempts at explaining the same processes at the different levels of information-processing and neurochemistry respectively.

Abnormal Beliefs. Several groups have gone on to attempt to explain how anomalous perceptual experiences are transformed into psychotic symptoms. Maher [96] pointed out that the experiences are often puzzling and associated with intense emotion, and may seem extremely personally relevant to the individual; not surprisingly, they trigger a search for explanation as to their cause. Kapur [16] makes a similar point that the experience of being bombarded by apparently salient stimuli causes great anxiety and arousal in the acutely psychotic individual, and this in turn might lead to a continued disruption of contextual integration and consequent vulnerability to intrusions into awareness. One way to resolve this anxiety and arousal is to develop a delusional explanation of the experiences.

> This is all happening to me because they are conspiring against me.

However, resolving anxiety and arousal by developing a delusional explanation is counterproductive. In an epidemiological sample of non-psychotic individuals, Krabbendam et al. [97] found that the presence of delusional ideation in combination with hallucinatory experience significantly increased the risk of developing a psychotic disorder and needing psychiatric care relative to hallucinatory experience alone.

Biased conscious appraisal processes may contribute to a judgement that the anomalous experiences are externally caused. Garety et al. [90] describe several cognitive biases that may adversely influence how the perceptual experience is appraised. First, there is a "**jumping-to-conclusions**" data-gathering bias [98]. If this bias operates, individuals prematurely terminate their search for an explanation of their experience. Such a bias seems to be present in subjects with an ARMS (at-risk mental state), prior to the onset of frank delusions [99]. A second bias is that of "**externalising attributions**", which causes individuals to attribute negative events to an external cause. Bentall et al. [100] found that people with persecutory delusions are more likely than normal controls to attribute negative events that they experience to other people than circumstances or fate. Frith's [101] ideas of theory of mind represent a third possible bias. Thus, a failure to construct accurate representations of the contents of other people's minds may cause the individual to misrepresent the intentions of others in a paranoid manner.

Morrison et al. [95] claim that any factor that increases the probability of making a culturally unacceptable appraisal of an intrusion will increase the risk of psychosis. The way in which individuals appraise intrusions will be influenced by their experience, and certain types of experience, for example bullying, victimisation, racism and alienation from mainstream culture, increase the probability of culturally unacceptable appraisals and thus the risk of psychosis. Several groups have pointed out that social isolation reduces access to alternative and normalising explanations for anomalous experiences [102, 103] and that the failure to be part of a normalising social network distinguishes those who develop psychosis from those who have anxiety and mood disorders. As we noted earlier, in animal experiments, isolation and subordination increases dopamine transmission.

More recently, Strumwasser [104] proposed a new model that psychosis can be considered as the impossibility of conflict resolution among the special neuronal assemblies that separately mediates feelings of attachment versus abandonment, security versus anxiety, calmness versus anger, fulfilment versus helplessness and satisfaction of sexual needs. This idea has to be explored further.

21.2.9 Recent New Model: The Neurophysics of Psychiatric Diagnosis—Clinical Brain Profiling

Peled [105] has recently proposed the new exciting model of Neurophysics. It argues that a comprehensive theoretical framework for the re-conceptualisation of mental disorders as real brain disorders, called "Clinical brain profiling", can be generated to make testable predictions about the aetiopathology of psychiatric disorders. According to this hypothesis, all mental

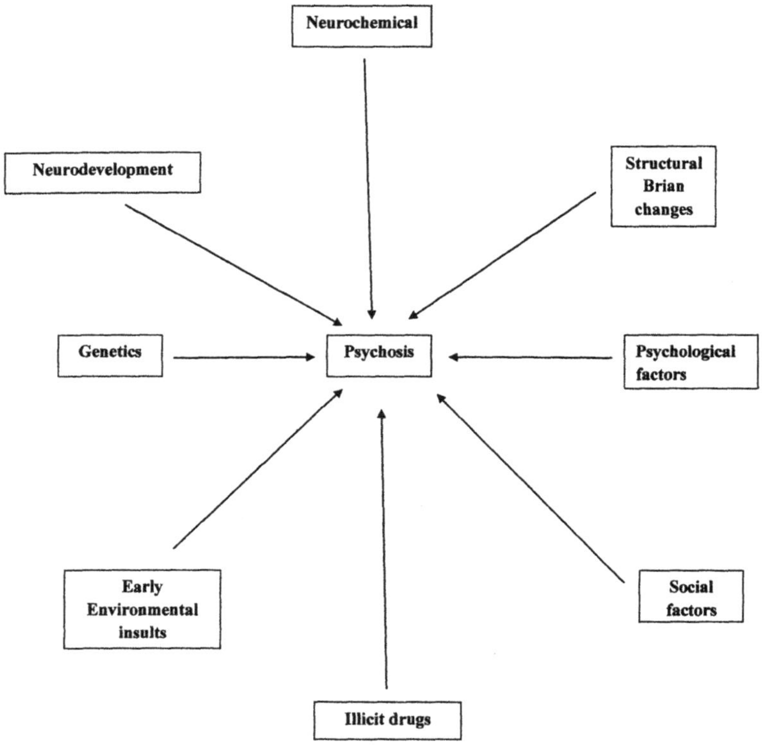

Fig. 21.1 The causes of psychosis

disturbances can be defined in a three-dimensional space of brain disturbances: (1) neural complexity organisation, (2) to neural resilience optimisation dynamics and (3) to connectivity constructs for context and internal representations. Accordingly, someone suffering psychosis will have disturbance in the nigra–striatum–cortex circuitry that can be picked up by appropriate signal-processing imaging. This model gives a totally different perspective of psychosis and needs to be tested scientifically further.

Thus, there are many theories and models that explain the probable causes of onset of psychosis (Fig. 21.1). These factors are complex in nature, as are the interactions between them in the process of onset of psychosis. The genetics, the early and late neurodevelopmental factors, the role of illicit drugs and social factors are all inter-related and interact with each other at different stages in the onset of psychosis. Further, structural brain changes and psychological mechanisms play their part in the cause of psychosis. All these models reach the end point that the dopamine dysregulation causes the onset of psychosis and how it occurs varies from person to person. Thus, no one model can explain solely the onset of psychosis and no one factor can be regarded as the cause of psychosis.

A further complication in the subject is that psychosis is not seen just as a discrete problem but the psychotic experiences continue into normal people. In the following section, we discuss the continuum of psychosis and how it progresses from at-risk state to full-blown problem.

21.3 Continuum of Psychosis

Currently, there is an increasing move away from studying solely discrete, categorical psychotic illnesses and more into a continuum model of psychosis. As well as casting the nosological status of established diagnostic categories such as Schizophrenia in doubt, it also impacts upon the status of the prodromal phase or at-risk mental state, both described below.

21.3.1 The Continuum and Prodromal Phase of Psychosis

It is now clear that not only many children but also a proportion of the general adult population experience brief or isolated psychotic phenomena without coming into contact with psychiatric services [106]. For example, in the Dunedin cohort, 25% of the entire population reported having experienced isolated or transient delusions or hallucinations at the age of 26 years, though only 3.7% met criteria for a schizophreniform illness. Initially, there was considerable scepticism that such minor symptoms bore any relation to the frank and persistent hallucinations and delusions experienced by schizophrenic patients. However, van Os and his colleagues reported that the very same risk factors that are associated with clinical schizophrenia (single state, unemployment, urban living, etc.) are also associated with the occurrence of minor psychotic symptoms in the Dutch general population. Johns et al. [6] confirmed these findings in a large sample ($n=8{,}580$) of the British general population. Factors independently associated with psychotic symptoms were lower IQ, poorer educational qualifications, cannabis dependence, alcohol dependence, victimisation, stressful life events and neurotic symptoms. Such findings have suggested that psychosis is best considered as a dimension extending well into the general population [107].

21.3.1.1 The Prodrome

Several research groups have identified characteristics of young people thought to be at "**ultra high risk**" of developing frank psychosis. These individuals are described as experiencing cognitive dysfunction as the earliest detectable anomaly, followed by attenuated "negative" symptoms such as decreased motivation and socialisation [108]. Later, positive psychotic symptoms develop but are not sufficient in intensity or duration to meet formal criteria for frank psychotic illness. This constellation of symptoms has been combined with having: (a) a first degree relative with psychosis or (b) a diagnosis of schizotypal personality disorder plus a decline in function, to create what is termed an "**at-risk mental state**" or ARMS [109–111]. If one accepts psychosis as a dimension, then the "at-risk mental state" is likely to cover a segment of it, and the point of transition to frank psychosis is somewhat arbitrary.

Nevertheless, Yung and colleagues reported that the presence of the "at-risk" mental state in their clients predicted a 40% transition rate to frank psychosis within 12 months. Morrison and colleagues [95] noted that 22% of those they identified became psychotic, while Cornblatt and colleagues [108] studied a group of 15 individuals with "schizophrenia-like psychoses" of whom one-third developed schizophrenia within 6 months. Hambrecht et al. [112] reported that, at 15-month follow-up, five (9.8%) of 51 individuals identified as ultra high risk met ICD-10 criteria for a psychotic disorder. Part of this variation may be due to different criteria for determining those at clinical high risk, as well as differences in pathways to care/sampling. Specifically, the declining transition rate in Melbourne has been linked to detecting at-risk clients sooner, with a shorter duration of attenuated symptoms [113]. Given that not all those who are at risk go on to develop psychosis, Yung and colleagues [109] advocate that the term "prodrome" be reserved for retrospective use only.

21.3.1.2 The Prodrome in Context

Van Os and Delespaul [114] point out that the highest transition rates occur in those "at-risk" populations which have been most highly selected, and that the process of screening and referral into these populations makes a major contribution to the success of researchers in identifying individuals at such high risk of transition. Thus, the application of ARMS criteria to the general population may have much less predictive power than in a clinic to which individuals suspected of being in a pre-psychotic phase have been specially referred. One reason for this loss of power is the evidence discussed above that psychosis exists as a continuous phenotype in the general population. A second reason is that the commonest outcome of subclinical psychotic symptoms in a general population is remission [115].

The strategy of "sample enrichment" is utilised by most prodromal services to increase their ability to predict transition in an at-risk population. However, such enhanced predictive power may be falsely attributed to the psychopathological measure employed [114]; thus, the high predictive value may be consequent upon how the patients who make up a prodromal service are selected, rather than their mental state. This is not necessarily because those so detected are any less at risk, but rather that the greater proportion of those in the general population who would meet "at-risk" criteria for developing psychosis will never make it through the various selection procedures that account for the sample enrichment. However, such epidemiological criticisms may be more valid for some services than others that have less "filters" between the person experiencing early psychotic symptoms and the clinician [116].

This is an obvious example of the more general problem that findings from cohort, epidemiological and at-risk studies have not yet been fully integrated. Thus, neurodevelopmental theorists have struggled to explain what converts a developmentally impaired or socially isolated adolescent with odd ideas and experiences into a psychotic individual. Similarly, it is not yet clear what differentiates an individual in the community who experiences hallucinations and holds delusional beliefs but never sees a psychiatrist from the individual who reaches a specialised clinic for those at risk of psychosis where he/she is considered as prodromal. The work of Hanssen et al. [115] suggests that the intensity of the experiences is important but so too, they suggest, is depression. Escher et al. [117] also believe that the co-existence of affective disturbance is a major factor in determining whether young people who experience minor psychotic symptoms will progress to psychotic disorder that requires care.

Of course, cohort studies have consistently shown that preschizophrenic children have an excess of depression and social anxiety. Retrospective studies show that the first noticeable psychological disturbances in individuals who later become psychotic include depression and anxiety; indeed most patients with a first episode of schizophrenia will have had a depressed mood, and at least one frank episode of depression, in the year prior to hospitalisation [118]. From studies of those with the ARMS, it is clear that prior to the onset of frank psychosis, there are prominent mood symptoms, many of which reach DSMIV diagnostic criteria. Such affective symptomatology may play a role in the genesis of positive symptoms such as hallucinations and delusions. Freeman and Garety [119] suggest that anxiety also facilitates the development of aberrant cognitive schema and beliefs, and influences the generation of anomalous experiences and then the maintenance of delusions once formed.

21.3.2 Normativity and Delusions

Another argument for conceiving psychosis as a continuum as opposed to a gradation of discrete stages comes from reflecting about the nature of psychopathology itself. Not only may the criteria of disorders themselves be not wholly dependent on neuroscientific or other biological variables, but the psychopathological symptoms used in the definitions of the disorders, or the items in the assessment tools we use to characterise the prodrome, are also contingent upon wider, normative, issues that are unlikely to be able to be reduced or correlated with brain function. A clear example, and one crucial in the prodrome, is that of delusions and disorders of thought content. In the case of delusions, a merely neurobiological investigation becomes problematic early on. Unlike some other symptoms of psychosis, delusions are not discrete either temporally or in terms of their demarcation from other mental states—it would seem inconceivable to instruct a subject to button-press when deluded and when non-deluded, for example, in a neuroimaging experiment [120].

Delusions are usually thought of as false beliefs that are maintained in the face of strong counterevidence, but as any clinician can attest, there is a lot more to a delusion than merely being wrong. Most contemporary accounts of delusion view them as non-discrete mental states, a symptom observed when a number of differing dimensional attitudes to a belief are adopted. Characteristics of delusions include having an implausible content, being reported with conviction, being unfounded, being distressing, causing preoccupation and not being shared by others [121]. Delusions may lead to the subject's whole experience of themselves and the world to be altered. What was once banal, and beneath conscious attention, becomes salient and self-referential. The normative, socially conditioned, rules for linking reasons, causes and explanations can be disrupted, and we are left with the hallmark of delusion: namely, that the reasons the person gives for holding his/her delusional beliefs either do not look like reasons or are not regarded as intersubjectively good reasons [122–124]. The effect of an inappropriate dopamine-driven generation of salience to otherwise neutral representation may lead to the private creation of affect-laden meaning and new reason-relations that cannot be shared or recognised by others as valid.

Delusions held without doubt are extremely resistant to counterevidence or counterargument, contributing to the isolation of the person reporting and believing them. This shows how delusions manifest behaviourally and interpersonally: it is by observing how the person behaves with respect to their beliefs, and by witnessing such behaviour in the process of the giving and asking of reasons that one suspects delusions, not in viewing a brain scan or a genetic sequence. In other words, the diagnosis of delusions is based on the observation of behaviour that violates accepted norms (e.g. of rationality for belief reports). Listing these features of delusions helps us realise that the idea of what is pathological in delusion cannot be fully captured without referring to psychological notions and an interpersonal dimension. This does not mean that it is a priori impossible to reduce the concept of delusion to its biological underpinnings or to arrive at its physical aetiology, but that focusing on local brain dysfunctions would not begin to give us a sense of why the delusion is a disorder, and why both the clinically trained and lay interlocutor can spot that something is awry when conversing with a person with delusions—delusions stretch our folk-psychological categories and practices.

Odd beliefs and delusions are a remarkably heterogeneous phenomena and it has been notoriously difficult to offer an all-inclusive definition or to operationalise criteria to enable their reliable detection. This is not so much due to conceptual woolliness or confusion but rather due to the fact that the term "delusion" does not pick out or index a discrete psychological state or natural kind: such difficulty carries over into the "disorders of thought content" delineated by tools such as the CAARMS (Comprehensive Assessment for At-Risk Mental State), where attenuated symptoms of abnormal thought content are distinguished from full-blown delusions using problematic criteria such as intensity, frequency and distress.

Over the years, there have been numerous ways to subclassify delusions: primary or autochthonous ("delusions") or secondary ("delusion-like ideas") [1], based upon whether the interviewer's empathic skills (*Verstehen*) can be used to understand the narrative genesis of the belief or not, Schneiderian, by theme or content, by congruity with mood, or by degree of fixity and conviction. Even if one focuses on a subtype of delusion defined by theme, for example persecutory delusions, these difficulties do not go away. Despite an explicit definition being offered, persecutory delusions are still a psychologically complex phenomenon. Nevertheless, clinicians can usually agree whether delusions are present or not, and despite being difficult to define, they serve as important criteria in the diagnosis of mental illness in the psychiatric classifications. This may be because delusions show themselves when a belief tends to have characteristics that lie at the extreme end of several related, but not necessarily dependent, dimensions. These dimensions include such things as plausibility, foundation of the belief, conviction, and level of distress incurred, preoccupation evoked and degree of being shared by others [121].

The problem in defining delusions becomes clear: how are the dimensions weighted? Are they normally distributed? Do they correlate with one another? How "low" can one score on one dimension if score "highly" on the others and still be deluded? What are the cut-offs for each dimension for "delusionality" and are they altered by the scores on other dimensions? Which elements are necessary to be deluded? Part of the answers to these questions, particularly those of cut-off, threshold and interaction, will lie outside of natural science: what does society view as being deluded? This cannot be determined by biological or psychological research in itself as the dimensions integral to delusions are themselves defined using epistemic criteria, some of which may be more socially constructed (for example, plausibility) whereas others link into more basic and profound issues in rationality and judgment (foundedness). Further, certain beliefs have special rules of justification which apply to them (such as religious belief).

Cognitive neuropsychology will have a crucial role in understanding the various processes, including perceptual and affective, as well as information processing, which may underpin shifts up and down the dimensions linked to "delusionality". These issues of cut-off, threshold and interaction not only demarcate delusion threshold but also challenge our ideas around the attenuated symptoms used to characterise clinical high-risk states for psychosis. Given this framework, one can appreciate the role of functional neuroimaging and cognitive neuroscience. Although on this analysis delusions are not amenable to a smooth reduction from symptom to circuits, as hoped for by the movement of biological realism in psychiatry, they can be deconstructed into several discrete phenomena. Imaging, and cognitive neuroscience more broadly, will have a crucial role in determining the neural correlates of the processes that may impact upon shifts in these dimensions, and with PET, may also allow progress into the biological physiology of such dimensions (for example, dopaminergic salience in relation to preoccupation, foundedness and distress). A further methodological advance is linking longitudinal research with imaging: fMRI when used in high-risk studies (whether clinical or genetic studies) is a powerful method of linking biological variables (neural activations) with cognitive variables (such as reasoning biases) and the evolution of psychopathology, such as delusions, and the transition to psychosis.

21.4 The Complexity of Psychosis: Conclusions

It is now widely accepted that Schizophrenia has a multifactorial aetiology in which genes and early environmental brain insults interact to cause neurodevelopmental impairment and set pre-schizophrenic children on a trajectory of increasing deviance. However, the neurodevelopmental hypothesis has struggled to explain the timing of the onset of psychosis. The initial view was that an early brain abnormality interacts with normal events during adolescence which includes hormonal changes, axonal myelination and the social factors we discussed above—drug misuse and social stress. Thus, the genes, early and late neurodevelopmental problems interacting with psychological and social factors lead to the onset of psychosis.

As we discussed above, there is an increasing move away from studying solely discrete, categorical psychotic illnesses and more into a continuum model of psychosis. The clinical staging model, which has been widely used in other fields of medicine but virtually ignored in psychiatry, may provide a coherent clinicopathological framework which will help us to understand the transition from at-risk state to prodromal phase to full-blown psychosis. Raballo and Laroi [125] described at least four stages as a premorbid phase, then early prodromal and late prodromal phases which lead to full-blown psychosis.

Further, recent research indicates genetic overlap between schizophrenia and childhood neurodevelopmental disorders such as autism spectrum disorders, intellectual disability and attention-deficit hyperactivity disorder. Owen et al [126] suggest that psychoses should be viewed as members of a group related and overlapping syndromes that occur as an effect of combination of genetic and environmental factors on brain development.

In conclusion, psychosis is such a complex phenomenon not only in the various hypotheses about its aetiology but also the complexity in its onset and the emergence of symptoms, how it gets organised in a person into a full-blown episode and also how various mechanisms interact with each other in the process. And in this brief survey, we have not attempted to discuss prognosis and treatment, topics that share similar complexity. Thus, psychosis needs the systems approach to understand better and to provide a holistic care for the sufferer.

Acknowledgments The authors would like to thank Taylor and Francis for permission to use some material from Broome M, Dale J, Marriott C, Merino C, Bortolotti L. Neuroscience, continua and the prodromal phase of psychosis. In: Fusar-Poli P, Borgwardt S, McGuire PK, editors. Vulnerability to psychosis: from psychopathology to neurosciences: Routledge; 2012. p. 1–22.

References

1. Jones P, Murray RM. The genetics of Schizophrenia is the genetics of neurodevelopment. Br J Psychiatry. 1991;158:615–23.
2. Bullmore ET, Woodruff PW, Wright IC, Rabe-Hesketh S, Howard RJ, Shuriquie N, Murray RM. Does dysplasia cause anatomical disconnectivity in Schizophrenia? Schizophr Res. 1998;30:127–35.
3. Mc Donald C, Fearon P, Murray R. Neurodevelopmental hypothesis of Schizophrenia 12 years on: data and doubts. In: Rapoport J, editor. Childhood onset of "adult" psychopathology. Washington: American Psychiatric Press; 1999.
4. Murray RM, Lewis SW. Is Schizophrenia a neurodevelopmental disorder? Br Med J (Clin Res Ed). 1987;295:681–2.
5. Boydell J, Van Os J, Mckenzie K, Murray RM. The association of inequality with the incidence of schizophrenia—an ecological study. Soc Psychiatry Psychiatr Epidemiol. 2004;39:597–9.
6. Johns LC, Cannon M, Singleton N, Murray RM, Farrell M, Brugha T, Bebbington P, Jenkins R, Meltzer H. Prevalence and correlates of self- reported psychotic symptoms in the British population. Br J Psychiatry. 2004;185:298–305.
7. Cannon M, Caspi A, Moffit TE, Harrington H, Taylor A, Murray RM, Poulton R. Evidence for early—childhood, pan-developmental impairment specific to schizophreniform disorder: results from a longitudinal birth cohort. Arch Gen Psychiatry. 2002;59:449–56.
8. Jones P, Rodgers B, Murray R, Marmot M. Child development risk factors for adult schizophrenia in the British 1946 birth cohort. Lancet. 1994;344:1398–402.
9. Fuller R, Nopoulos P, Arndt S, O'Leary D, Ho BC, Andreasen NC. Longitudinal assessment of premorbid cognitive functioning in patients with schizophrenia through examination of standardised scholastic test performance. Am J Psychiatry. 2002;159:1183–9.
10. Poulton R, Caspi A, Moffit TE, Cannon M, Murray R, Harrington H. Children's self-reported psychotic symptoms and adult schizophreniform disorder: a 15-year longitudinal study. Arch Gen Psychiatry. 2001;57:1053–8.
11. Hemsley DR. A simple (or simplistic?) cognitive model for schizophrenia. Behav Res Ther. 1993;31: 633–45.
12. Gray J, Feldon J, Rawlins J, Hemsley D, et al. The neuropsychology of schizophrenia. Behav Brain Sci. 1991;14:1–84.
13. Laruelle M, Abi-Dargham A, Van Dyck CH, Gil R, D'Souza CD, Erdos J, McCance E, Rosenblatt W, Fingado C, Zoghbi SS, Baldwin RM, Seibyl JP, Krystal JH, Charney DS, Innis RB. Single photon emission computerized tomography imaging of amphetamine-induced dopamine release in drug free schizophrenic subjects. Proc Natl Acad Sci USA. 1996;93:9235–40.
14. Abi-Dargham A, Rodenhiser J, Printz D, Zea-Ponce Y, Gil R, Kegeles LS, Weiss R, Cooper TB, Mann JJ, Van Heertum RL, Gorman JM, Laruelle M. Increased baseline occupancy of D2 receptors by dopamine in schizophrenia. Proc Natl Acad Sci USA. 2000;97:8104–9.
15. Laruelle M, Abi-Dargham A, Gil R, Kegeles L, Innis R. Increased dopamine transmission in schizophrenia: relationship to illness phases. Biol Psychiatry. 1999;46:56–72.
16. Kapur S. Psychosis as a state of aberrant salience: a framework linking biology, phenomenology, and pharmacology in schizophrenia. Am J Psychiatry. 2003;160:13–23.
17. Berridge KC, Robinson TE. What is the role of dopamine in reward: hedonic impact, reward learning, or incentive salience? Brain Res Rev. 1998;28:309–69.
18. Laruelle M, Abi-Dargham A. Dopamine as the wind of the psychotic fire: new evidence from brain imaging studies. J Psychopharmacol. 1999;13:358–71.
19. Howes OD, Montgomery AJ, Asselin M, Murray RM, Valli I, Tabraham P, et al. Elevated striatal dopamine function linked to prodromal signs of schizophrenia. Arch Gen Psychiatry. 2009;66:13–20.
20. Gray JA, Joseph MH, Hemsley DR, Young AM, Warburton EC, Boulenguez P, Grigoryan GA, Peters SL, Rawlins JN, Taib CT, et al. The role of mesolimbic dopaminergic and retrohippocampal afferents to the nucleus accumbens in latent inhibition: implications for schizophrenia. Behav Brain Res. 1995;71:19–31.
21. Grace A. Developmental dysregualtion of the dopamine system and the pathophysiology of schizophrenia. In: Keshavan MS, Kennedy J, Murray R, editors. Neurodevelopment and schizophrenia. Cambridge: Cambridge Press; 2004.
22. Wright IC, Rabe-Hesketh S, Woodruff PW, David AS, Murray RM, Bullmore ET, et al. Meta-analysis of regional brain volumes in schizophrenia. Am J Psychiatry. 2000;157:16–25.

23. Harrison PJ, Owen MJ. Genes for schizophrenia? recent findings and their pathophysiological implications. Lancet. 2003;361:417–9.
24. Bray NJ, Preece A, Williams NM, Moskvina V, Buckland PR, Owen MJ, O'Donovan MC. Haplotypes at the dystrobrevin binding protein 1 (DTNBP1) gene locus mediate risk for schizophrenia through reduced DTNBP1 expression. Hum Mol Genet. 2005;14:1947–54.
25. Carter CJ. Schizophrenia susceptibility genes converge on interlinked pathways related to glutamatergic transmission and long-term potentiation, oxidative stress and oligodendrocyte viability. Schizophr Res. 2006;86:1–14.
26. Gu Z, Jiang Q, Fu AKY, Ip NY, Yan Z. Regulation of NMDA receptors by neuregulin signalling in prefrontal cortex. J Neurosci. 2005;25:4974–84.
27. Harrison PJ, Law AJ. Neuregulin 1 and schizophrenia: genetics, gene expression and neurobiology. Biol Psychiatry. 2006;60:132–40.
28. Harrison PJ, Weinberger DR. Schizophrenia genes, gene expression and neuropathology: on the matter of their convergence. Mol Psychiatry. 2005;10:40–68. image 5.
29. Munafo MR, Thiselton DL, Clark TG, Flint J. Association of the NRG1 gene and schizophrenia: a meta-analysis. Mol Psychiatry. 2006;11:539–46.
30. Williams NM, O'Donovan MC, Owen MJ. Is the dysbindin gene (DTNBP1) a susceptibility gene for schizophrenia? Schizophr Bull. 2005;31:800–5.
31. Egan MF, Te G, Kolachana BS, Callicott JH, Mazzanti CM, Straub RE, Goldman D, Weinberger DR. Effect of COMT Val108/158 Met genotype on frontal lobe function and risk for schizophrenia. Proc Natl Acad Sci USA. 2001;98:6917–22.
32. Malhotra AK, Kestler ND, Mazzanti C, Bates JA, Goldberg T, Goldman D. A functional polymorphism in the COMT gene and performance on a test of prefrontal cognition. Am J Psychiatry. 2002;159:652–4.
33. Rosa A, Peralta V, Cuesta MJ, Zarzuela A, Serrano F, Martinez-Larrea A, Fananas L. New evidence of association between COMT gene and prefrontal neurocognitive function in healthy individuals from sibling pairs discordant for psychosis. Am J Psychiatry. 2004;161:1110–2.
34. Okochi T, Ikeele M, Kishi T, Kawashima K, Kiroshita Y, Kitajima T, et al. Meta-analysis of association between genetic variants in COMT and Schizophrenia: an update. Schizophr Res. 2009;110:140–8.
35. Ross CA, Margolis RL, Reading SA, Pletnikov M, Coyle JT. Neurobiology of schizophrenia. Neuron. 2006;52:139–53.
36. Williams HJ, Owen MJ, O'Donovan MC. Schizophrenia genetics: new insight from new approaches. Br Med Bull. 2009;91:61–74.
37. Rujescu D, Ingasin A, Cichon S, Pietilainen OPH, Barnes MR, Toulopoulou T, et al. Disruption of the Neurexin 1 gene is associated with schizophrenia. Hum Mol Genet. 2009;18:988–96.
38. Cannon M, Jones PB, Murray RM. Obstetric complications and schizophrenia: historical and meta-analytic review. Am J Psychiatry. 2002;159:1080–92.
39. Verdoux H, Geddes JR, Takei N, Lawrie SM, Bovet P, Eagles JM, Heun R, McCreadie RG, McNeil TF, O'Callaghan E, Stober G, Willinger MU, Wright P, Murray RM. Obstetric complications and age at onset in schizophrenia: an international collaborative meta-analysis of individual patient data. Am J Psychiatry. 1997;154:1220–7.
40. Schulze K, McDonald C, Frangou S, Sham P, Grech A, Toulopoulou T, Walshe M, Sharma T, Sigmundsson T, Taylor M, Murray RM. Hippocampal volume in familial and nonfamilial schizophrenic probands and their unaffected relatives. Biol Psychiatry. 2003;53:562–70.
41. Stefanis N, Frangou S, Yakeley J, Sharma T, O'connell P, Morgan K, Sigmundsson T, Taylor M, Murray R. Hippocampal volume reduction in schizophrenia: effects of genetic risk and pregnancy and birth complications. Biol Psychiatry. 1999;46:697–702.
42. Lipska BK, Jaskiw GE, Weinberger DR. Postpubertal emergence of hyper responsiveness to stress and to amphetamine after neonatal excitotoxic hippocampal damage: a potential animal model of schizophrenia. Neuropsychopharmacology. 1993;9:67–75.
43. Lipska BK, Luu S, Halim ND, Weinberger DR. Behavioural effects of neonatal and adult excitotoxic lesions of the mediodorsal thalamus in the adult rat. Behav Brain Res. 2003;141:105–11.
44. Weinberger DR, Lipska BK. Cortical maldevelopment, anti-psychotic drugs and schizophrenia: a search for common ground. Schizophr Res. 1995;16:87–110.
45. Boksa P, El-Khodor BF. Birth insult interacts with stress at adulthood to alter dopaminergic function in animal models: possible implications for schizophrenia and other disorders. Neurosci Biobehav Rev. 2003;27:91–101.
46. Flagstad P, Mork A, Glenthoj BY, Van Beek J, Micheal-Titus AT, Didriksen M. Disruption of neurogenesis on gestational day 17 in the rat causes behavioural changes revelant to positive and negative schizophrenia symptoms and alters amphetamine- induced dopamine release in nucleus accumbens. Neuropsychopharmacology. 2004;29: 2052–64.
47. Curran C, Byrappa N, McBride A. Stimulant psychosis: systematic review. Br J Psychiatry. 2002;185: 196–204.
48. Chen CK, Lin SK, Sham PC, Ball D, Loh EW, Hsiao CC, Chiang YL, Ree SC, Lee CH, Murray RM. Premorbid characteristics and co-morbidity of methamphetamine users with and without psychosis. Psychol Med. 2003;33:1407–14.
49. Soulsby B, Dazzan P, Chitnis X, Wood S, Suckling J, McGuire P, McGorry P, Murray R, Velakoulis D, Phillips L, Yung A, Pantelis C. Grey matter abnormalities before the onset of psychosis: an automated MR image analysis. In: The 12th Scientific Meeting and

Exhibition of the International Society for Magnetic Resonance in Medicine; 2004, Kyoto, Japan
50. Tsapakis E-M, Guillin O, Murray RM. Does dopamine sensitization underlie the association between schizophrenia and drug abuse? Curr Opin Psychiatry. 2003;16 Suppl 2:S45–52.
51. Lewis DA, Lieberman JA. Catching up on Schizophrenia: natural history and neurobiology. Neuron. 2000;28:325–34.
52. Lieberman JA, Perkins D, Belger A, Chakos M, Jarskog F, Boteva K, Gilmore J. The early stages of schizophrenia: speculations on pathogenesis, pathophysiology, and therapeutic approaches. Biol Psychiatry. 2001;50:884–97.
53. Arseneault L, Cannon M, Witton J, Murray RM. Causal association between cannabis and psychosis: examination of the evidence. Br J Psychiatry. 2004;184:110–7.
54. Caspi A, Moffit TE. Gene-environment interactions in psychiatry: joining forces with neuroscience. Nat Rev Neurosci. 2006;7:583–90.
55. Caspi A, Moffit TE, Cannon M, McClay J, Murray R, Harrington H, Taylor A, Arseneault L, Williams B, Braithwaite A, Poulton R, Craig IW. Moderation of the effect of adolescent-onset cannabis use on adult psychosis by a functional polymorphism in the catechol-O-methyltransferase gene: longitudinal evidence of a gene X environment interaction. Biol Psychiatry. 2005;57:1117–27.
56. Di Forti M, Morgan C, Dazzan P, Pariante C, Mondelli V, Marques TR, Handley R, et al. High Potency cannabis and the risk of psychosis. Br J Psychiatry. 2009;195:488–91.
57. Bhattacharyya S, Fusar-Poli P, Borgwardt S, Martin-Santos R, Nosarti C, O'Carroll C, et al. Modulation of mediotemporal and ventrostriatal function in humans by Δ9-THC. A neural basis for the effects of Cannabis Sativa on learning and psychosis. Arch Gen Psychiatry. 2009;66:442–51.
58. Morrison PD, Nottage J, Stone JM, Bhattacharyya S, Tunstall N, Brenneisen R, et al. Disruption of frontal lobe coherence by Δ9-THC is associated with positivepsychoticsymptoms.Neuropsychopharmacology. 2011;36:827–36.
59. Cantor-Graae E, Selten J. Schizophrenia and migration: a meta-analysis. Schizophr Res. 2004;67:63.
60. Fearon P, Kirkbride JB, Morgan C, Dazzan P, Morgan K, Lloyd T, Hutchinson G, Tarrant J, Fung WLA, Holloway J, Mallett R, Harrison G, Leff J, Jones PB, Murray RM, Group AS. Incidence of schizophrenia and other psychoses in ethnic minority groups: results from the MRC AESOP study. Psychol Med. 2006;36:1541–50.
61. Fearon P, Morgan C. Environmental factors in schizophrenia: the role of migrant studies. Schizophr Bull. 2006;32:405–8.
62. Boydell J, Van Os J, McKenzie K, Allardyce J, Goel R, McCreadie RG, Murray RM. Incidence of schizophrenia in ethnic minorities in London: ecological study into interactions with environment. BMJ. 2001;323:1336–8.
63. Sharpley M, Hutchinson G, McKenzie K, Murray RM. Understanding the excess of psychosis among the African-Caribbean population in England. Review of current hypotheses. Br J Psychiatry Suppl. 2001;40:S60–8.
64. Reininghaus UA, Craig TKJ, Fisher HL, Hutchinson G, Fearon P, Morgan K, Dazzan P, Doody GA, Jones PB, Murray R, Morgan C. Ethnic identity, perceptions of disadvantage and psychosis: findings from the AESOP study. Schizophr Res. 2010;124:43–8.
65. Cooper C, Morgan C, Byrne M, Dazzan P, Morgan K, Hutchinson G, Doody GA, Harrison G, et al. Perceptions of disadvantage, ethnicity and psychosis. Br J Psychiatry. 2008;192:185–90.
66. Reininghaus UA, Morgan C, Simpson J, Dazzan P, Morgan K, Doody GA, Bhugra D, Leff J, Jones P, Murray R, et al. Unemployment, social isolation, achievement-expectation mismatch and psychosis: findings from the AESOP study. Soc Psychiatry Psychiatr Epidemiol. 2008;43:743–51.
67. Heidbreder CA, Foxton R, Cilia J, Hughes ZA, Shah AJ, Atkins A, Hunter AJ, Hagan JJ, Jones DN. Increased responsiveness of dopamine to atypical, but not to typical antipsychotics in the medical prefrontal cortex of rats reared in isolation. Psychopharmacology (Berl). 2001;156:338–51.
68. Selten J-P, Cantor-Graae E, Kahn RS. Migration and schizophrenia. Curr opin Psychiatr. 2007;20:111–5.
69. Selten J-P, Cantor-Graae E. Social defeat: risk factor for schizophrenia? Br J Psychiatry. 2005;187:101–2.
70. Selten J-P, Cantor-Graae E. Hypothesis: social defeat is a risk factor for schizophrenia. Br J Psychiatry. 2007;181:s9–12.
71. Morgan D, Grant KA, Gage HD, Mach RH, Kaplan JR, Prioleau O, Nader SH, Buchheimer N, Ehrenkaufer RL, Nader MA. Social dominance in monkeys: dopamine D2 receptors and cocaine self-administration. Nat Neurosci. 2002;5:169–74.
72. Morgan C, Charalambides M, Hutchinson G, Murray RM. Migration, ethnicity and psychosis: toward a sociodevelopmental model. Schizophr Bull. 2010;36:655–64.
73. Woods BT. Is schizophrenia a progressive neurodevelopmental disorder? Toward a unitary pathogenic mechanism. Am J Psychiatry. 1998;155:1661–70.
74. Lawrie SM, Whalley HC, Abukmeil SS, Kestelman JN, Miller P, Best JJ, Owens DG, Johnstone EC. Temporal lobe volume changes in people at high risk of schizophrenia with psychotic symptoms. Br J Psychiatry. 2002;181:138–43.
75. Pantelis C, Velakoulis D, McGorry PD, Wood SJ, Suckling J, Phillips LJ, Yung AR, et al. Neuroanatomical abnormalities before and after onset of psychosis: a cross-sectional and longitudinal MRI comparison. Lancet. 2003;361:281–8.
76. Phillips LJ, Velakoulis D, Pantelis C, Wood S, Yuen HP, Yung AR, Desmond P, Brewer W, McGorry PD.

Non-reduction in hippocampal volume is associated with higher risk of psychosis. Schizophr Res. 2002;58:145–58.
77. Whalley HC, Simonotto E, Flett S, Marshall I, Ebmeier KP, Owens DG, Goddard NH, Johnstone EC, Lawrie SM. fMRI correlates of state and trait effects in subjects at genetically enhanced risk of schizophrenia. Brain. 2004;127:478–90.
78. Cahn W, Pol HE, Lems EB, Van Haren NE, Schnack HG, Van Der Linden JA, Schothorst PF, Van Engeland H, Kahn RS. Brain volume changes in first episode schizophrenia: a 1 year follow-up study. Arch Gen Psychiatry. 2002;59:1002–10.
79. Weinberger D, Marenco S. Schizophrenia as a neurodevelopmental disorder. In: Hirsch s, Weinberger D, editors. Schizophrenia (second edition). Oxford: Blackwell press; 2003.
80. Weinberger DR, Mcclure RK. Neurotoxicity, neuroplasticity, and magnetic resonance imaging morphometry: what is happening in the schizophrenic brain? Arch Gen Psychiatry. 2002;59:553–8.
81. Cotter D, Pariante CM. Stress and the progression of the developmental hypothesis of schizophrenia. Br J Psychiatry. 2002;181:363–5.
82. Pariante CM, Vassilopoulou K, Velakoulis D, Phillips L, Soulby B, Wood SJ, Brewer W, et al. Pituitary volume in psychosis. Br J Psychiatry. 2004;185:5–10.
83. Tandon R, Mazzara C, Dequardo J, Craig KA, Meadorwoodruff JH, Goldman R, Greden JF. Dexamethasone suppression test in schizophrenia: relationship to symptomatology, ventricular enlargement, and outcome. Biol Psychiatry. 1991;29:953–64.
84. Bebbington P, Wilkins S, Jones P, Foerster A, Murray R, Toone B, Lewis S. Life events and psychosis. Initial results from the Camberwell Collaborative Psychosis Study. Br J Psychiatry. 1993;162:72–9.
85. Myin-Germeys I, Van Os J, Schwartz JE, Stone AA, Delespaul PA. Emotional reactivity to daily life stress in psychosis. Arch Gen Psychiatry. 2001;58: 1137–44.
86. Garner B, Pariante CM, Wood SJ, Velakoulis D, Phillips L, Soulby B, Brewer WJ, Smith DJ, et al. Pituitary volume predicts future transition to psychosis in individuals at ultra high risk of developing psychosis. Biol Psychiatry. 2005;58:417–23.
87. Mondelli V, Pariante C, Navari S, Aas M, D'Albenzio A, Di Forti M, Handley R, Hepqul N, et al. Higher cortisol levels are associated with smaller left hippocampal volume in first episode psychosis. Schizophr Res. 2010;119:75–8.
88. Borgwardt SJ, McGuire PK, Aston J, Berger G, Dazzan P, Gschwandtner UTE, Pfluger M, D'Souza M, et al. Structural brain abnormalities in individuals with an at-risk mental state who later develop psychosis. Br J Psychiatry. 2007;191:569–75.
89. Morgan KD, Dazzan P, Orr KG, Hutchinson G, Chitnis X, Suckling J, Lythgoe D, et al. Grey matter abnormalities in first episode schizophrenia and affective psychosis. Br J Psychiatry. 2007;191: s111–6.
90. Garety PA, Kuipers E, Fowler D, Freeman D, Bebbington PE. A cognitive model of the positive symptoms of psychosis. Psychol Med. 2001;31: 189–95.
91. Holmes E, Steel C. Schizotypy: a vulnerability factor for traumatic intrusions. J Nerv Ment Dis. 2004;192:28–34.
92. McGhie A, Chapman J. Disorders of attention and perception in early schizophrenia. Br J Med Psychol. 1961;34:103–16.
93. Rachman S, De Silva P. Abnormal and normal obsessions. Behav Res Ther. 1978;16:233–48.
94. Pm S, Harrison J. Abnormal and normal obsessions-a replication. Behav Res Ther. 1984;22:549–52.
95. Morrison AP, Bentall RP, French P, Walford L, Kilcommons A, Knight A, et al. Randomised controlled trials of early detection and cognitive therapy for preventing transition to psychosis in high-risk individuals. Study design and interim analysis of transition rate and psychological risk factors. Br J Psychiatry Suppl. 2002;43:S78–84.
96. Maher B. Anomalous experience and delusional thinking: The logic of explanations. In: Oltmanns TF, Maher BA, editors. Delusional beliefs. Chichester: Wiley; 1988. p. 15–33.
97. Krabbendam L, Myin-Germeys I, Hanssen M, Bijl RV, De Graaf R, Vollebergh W, Bak M, Van Os J. Hallucinatory experiences and onset of psychotic disorder: evidence that the risk is mediated by delusion formation. Acta Psychiatr Scand. 2004;110: 264–72.
98. Garety P, Hemsley D. Delusions: investigations into the psychology of delusional reasoning, Maudsley monograph. Hove: Psychology press; 1994.
99. Broome MR, Johns LC, Valli I, Wolley JB, Tabraham P, Brett C, Valmaggia L, Peters E, McGuire PK. Delusion formation and reasoning biases in those at clinical high risk of psychosis. Br J Psychiatry. 2007;191:s38–42.
100. Bentall RP, Kinderman P, Kaney S. The self, attributional processes and abnormal beliefs: Towards a model of persecutory delusions. Behav Res Ther. 1994;32:331–41.
101. Frith CD. The Cognitive Neuropsychology of Schizophrenia. LEA: Hove, UK. 1992.
102. Hodges A, Byrne M, Grant E, Johnstone E. People at risk of schizophrenia. Sample characteristics of the first 100 cases in the Edinburgh high-risk study. Br J Psychiatry. 1999;174:547–53.
103. Van Os J, Hanssen M, Bijl RV, Ravelli A. Strauss (1969) revisited: a psychosis continuum in the general population? Schizophr Res. 2000;45:11–20.
104. Strumwasser F. The human mind: building bridges between neuroscience and psychiatry. Psychiatry. 2003;66:22–31.
105. Peled A. The neurophysics of psychiatric diagnosis: clinical brain profiling. Med hypothesis. 2011;76(1): 34–49.
106. Johns LC, Van Os J. The continuity of psychotic experiences in the general population. Clin Psychol Rev. 2001;21:1125–41.

107. Verdoux H, Van Os J. Psychotic symptoms in non-clinical populations and the continuum of psychosis. Schizophr Res. 2002;54:59–65.
108. Cornblatt B, Lencz T, Obuchowski M. The Schizophrenia prodrome: treatment and high-risk perspectives. Schizophr Res. 2004;67:63.
109. Yung AR, McGorry PD. The prodromal phase of first-episode psychosis: past and current conceptualisations. Schizophr Bull. 1996;22:353–70.
110. Yung AR, Phillips LJ, Pd MG, McFarlane CA, Francey S, Harrigan S, Patton GC, Jackson HJ. Prediction of psychosis-a step towards indicated prevention of schizophrenia. Br J Psychiatry. 1998;172:14–20.
111. Yung AR, Phillips LJ, Yuen HP, Francey SM, McFarlane CA, Hallgren M, McGorry PD. Psychosis prediction: 12-month follow up of a high-risk ("prodromal") group. Schizophr Res. 2003;60:21–32.
112. Hambrecht M, Lammertink M, Klosterkotter J, Matuschek E, Pukrop R. Subjective and objective neuropsychological abnormalities in a psychosis prodrome clinic. Br J Psychiatry Suppl. 2002;43: s30–7.
113. Yung AR, Yuen HP, Berger GE, Francey S, Hung T-C, McGorry P. Declining transition rate in ultra high risk (prodromal) services: dilution or reduction of risk? Schizophr Bull. 2007;33:673–81.
114. Van Os J, Delespaul P. Toward a world consensus on prevention of schizophrenia. Dialogues Clin Neurosci. 2005;7:53–67.
115. Hanssen M, Bak M, Bijl R, Vollebergh W, Van Os J. The incidence and outcome of subclinical psychotic experiences in the general population. Br J Psychiatry. 2005;44:181–91.
116. Broome MR, Woolley JB, Johns LC, Valmaggia LR, Tabraham P, Gafoor R, Bramon E, McGuire PK. Outreach and support in South London (OASIS): implementation of a clinical service for prodromal psychosis and the at risk mental state. Eur Psychiatry. 2005;20:372–8.
117. Escher S, Romme M, Buiks A, Delespaul P, Van Os J. Independent course of childhood auditory hallucinations: a 3-year follow-up study. Br J Psychiatry Suppl. 2002;43:s10–8.
118. An Der Heiden W, Hafner H. The epidemiology of onset and course of schizophrenia. Eur Arch Psychiatry Clin Neurosci. 2000;250:292–303.
119. Freeman D, Garety P. Comments on the content of persecutory delusions: does the definition need clarifying? Br J Clin Psychol. 2000;39:407–14.
120. Broome MR, McGuire PK. Imaging and delusions. In: Freeman D, Bentall R, Garety PA, editors. Persecutory delusions: assessment, theory and treatment. Oxford: Oxford university press; 2008.
121. Freeman D. Suspicious minds: the psychology of persecutory delusions. Clin Psychol Rev. 2007;27: 425–57.
122. Bortolotti L, Broome MR. Delusional beliefs and reasoning giving. Philos Psychol. 2008;21:821–41.
123. Bortolotti L, Broome MR. The future of scientific psychiatry. In: Broome M, Bortolotti L, editors. Psychiatry as cognitive neuroscience: philosophical perspectives. Oxford: Oxford University Press; 2009.
124. Bortolotti L, Broome MR. A role of ownership and authorship in the analysis of thought insertion. Phenomenol Cogn Sci. 2009;8:205–24.
125. Raballo A, Laroi F. Clinical Staging: a new scenario for the treatment of psychosis. Lancet. 2009;374: 365–7.
126. Owen MJ, O'Donovan MC, Thapar A, Craddock N. Neurodevelopmental hypothesis of Schizophrenia: reappraisal. Br J Psychiatry. 2011;198:173–5.

22 Understanding Intimate Partner Violence Through Its Dynamics

David A. Katerndahl

22.1 Introduction

Intimate partner violence (IPV) has far-reaching effects. Since the landmark National Crime Victimization Survey of 1992, American society has known the frequency and severity of the abuse experienced by woman at the hands of their intimate partners. In fact, the survey found that more than 2,000 people were murdered by intimate others [1]. One half of all serious injuries seen in women who are treated in emergency departments are the result of battering. Often such injuries involve multiple sites, especially the face, chest, breasts, and abdomen [2].

But the psychological injuries may outweigh the physical ones. Not only are battered women at risk for posttraumatic stress disorder (PTSD), but women in abusive relationships have problems with coping and trust, often experiencing depression, anger, confusion, fearfulness, paranoia, and low self-esteem [3–5]. In addition, these women often turn to psychoactive substances [5] or suicide [6, 7] for escape. In fact, the fewer their resources and social connections, the greater is the psychological risk [8, 9].

D.A. Katerndahl, M.D., M.A. (✉)
Department of Family and Community Medicine, University of Texas Health Science Center at San Antonio, 7703 Floyd Curl Drive, San Antonio, TX 78229-3900, USA
e-mail: katerndahl@uthscsa.edu

Consequently, women in abusive relationships often develop chronic health problems in addition to the acute injuries, resulting in higher health care utilization and costs than seen in non-abused women [10]. Along with vague, unremitting symptoms [11], battered women commonly complain about insomnia, fatigue, chronic pain, anemia, and gastrointestinal or premenstrual symptoms [11–13]. Finally, battered women experience more negative pregnancy outcomes (miscarriages, stillbirths, and low birthweight newborns) [14] and negative health behavior outcomes (eating disorders, sexually transmitted diseases, and HIV) than non-abused women [11].

But the woman is not the only victim in the abusive relationship; children in such households are more likely to experience physical abuse themselves [15]. Such children tend to exhibit poorer adjustment, health, and sleep than children from nonviolent families. As adolescents, these children are often aggressive and anxious, while, as adults, they often have violent relationships themselves [16]. In fact, the most consistent predictor of becoming a batterer or battered wife is witnessing parental violence as a child [15].

If we are ever to break this cycle and avoid such consequences, we must understand the development and evolution of such violent relationships. Although decades of research have yielded many theories and predictors, our understanding of the dynamics of these relationships is still rudimentary. This chapter will summarize a study of 16 women in violent relationships who were studied daily for 8 weeks, suggesting the

J.P. Sturmberg and C.M. Martin (eds.), *Handbook of Systems and Complexity in Health*,
DOI 10.1007/978-1-4614-4998-0_22, © Springer Science+Business Media New York 2013

need to view IPV as a dynamical problem within a complex systems framework.

22.2 Intimate Partner Violence as a Complex Systems Problem

Linear study of IPV has yielded results. Yet, it is this multiplicity of varied, multilevel correlates along with the nonlinear implications of Family Systems Theory (FST) that imply a complex systems approach is needed if we are to understand the phenomenon of IPV.

22.2.1 Intimate Partner Violence Is a "Many-Splendored" Phenomenon

IPV is a multidimensional problem [17], which depends upon cultural, oppression, historical, and situational constructs [18]. This multidimensional, multi-construct phenomenon has produced a plethora of multilevel correlates of IPV. Based upon Hotaling and Sugarman's [15] extensive literature review, a variety of consistent and inconsistent markers of IPV have been found.

At the individual level, witnessing violence as a child was the only consistent marker in abused wives while eight different markers from demographics to violence towards children were consistently found in abusive husbands. In addition, studies have inconsistently found an additional 12 markers in women and six markers in men. At the couple level, Hotaling and Sugarman reported five consistent markers of IPV, including argument frequency and religious incompatibility as well as three inconsistent markers.

Furthermore, recent studies have found correlates of IPV at the level of the woman's social network. Not only is social isolation a correlate of IPV [19, 20], but women experienced mixed reactions within their social network [21], resulting in a nonlinear pattern of help seeking [22]. In fact, women in abusive relationships cope through a process of reappraisal and refocus for themselves and their partners, often reaching out to family, friends, health professionals, and community organizations [23].

Thus, IPV is a multifactorial, multilevel problem involving markers and coping strategies at all levels.

22.2.2 Nonlinearity Inherent in Family Systems Theory

FST is an application of General Systems Theory in an attempt to understand family process. General Systems Theory includes three basic assumptions, which clearly apply to families. First, systems theory serves as a unifying framework for understanding the components and interactions of the system, often relying upon concepts from information theory (such as entropy) to conceptualize family interaction. Second, the system must be understood as a whole, not merely the sum of its components. In fact, without this system-wide perspective, emergent properties of the system would not be recognized. Finally, systems theory assumes that human systems are self-reflexive, seeking to evaluate themselves and their family functioning to explain behavior.

FST has been used to understand a diversity of family processes from family communication and conflict to cohesion and adaptability. The FST perspective has led to an understanding that, due to the wealth of family interdependencies, simple cause-and-effect mechanisms rarely apply. In addition, family systems are hierarchical with individual family members nested within couples nested within households nested within communities. Thus, interaction within marriages is often a product of more than the issues inherent to the couple, but includes spouse-specific factors as well as household and community influences. Thus, FST includes many of the concepts that define complexity science [24].

Based upon a research base that found many multilevel correlates of IPV and a theoretical framework (FST) that views families as indivisible systems of interdependencies, circular causation, and hierarchical interactions, IPV could be viewed as a dynamical problem within a complex adaptive system.

22.3 Dynamical Conditions in Mental Health

If we accept that the presence of varied, multilevel correlates of abuse along with the theoretical framework as described in FST, then we should anticipate that relationships involving intimate partners represent complex adaptive systems and that IPV may display nonlinear dynamics.

22.3.1 Dynamical Psychosocial Conditions

When systems appear to have a multiplicity of endogenous and exogenous causes, it raises the possibility that, in fact, no definitive causes really exist and that the dynamics of the phenomenon is actually a product of the system's inherent attempts to self-regulate. Disorders that exhibit dynamic variability without an obvious, simple cause are termed dynamical disorders.

For example, the nonlinear patterns observed in mood variability may result from the minute-to-minute attempts the brain makes to adjust to varying internal and external factors. This complex interaction between stressors and adaptations results in a nonlinear pattern in mood variability. Healthy dynamics may, in fact, be a combination of regular deterministic diurnal variation combined with the instantaneous adaptations to stressors; when such nonlinear dynamics is lost due to insufficient adaptability, psychopathology and linear dynamics may result.

Because many mental disorders can be controlled but not cured, these dysfunctions may represent dynamical disorders. Thus, bipolar disorder, chronic depression, schizophrenia, and panic disorder may be considered dynamical disorders with symptomatology due in part to changes in their underlying mood variability [25].

Because such dynamical disorders are more about problems with variation than with specific stressors, we may learn more about these disorders by understanding their dynamics than by understanding their stressors. Thus, to fully understand IPV and its complex day-to-day reality, we must understand the dynamics of the violence as well as the milieu of its internal and external environment.

22.3.2 Measuring Dynamics in Psychosocial Conditions

There are three predominant dynamic patterns seen in complex systems. Periodic dynamics, in which the system cycles its behavior, results when actions and outcomes are tightly coupled, and when current behavior is dependent on previous behavior. Thus, periodic systems have strong attractors limiting their possible behaviors and are insensitive to small changes in their state. Periodic systems are predictable and respond predictably to interventions. Chaotic dynamics, in which the overall pattern of behavior recurs but the specific path is unpredictable, results when actions and outcomes are separated in time, and when feedback within the system varies in strength and direction. Thus, chaotic systems also have attractors limiting their behavior but are sensitive to small changes (sensitive to initial conditions) in terms of the specific path they follow. Chaotic systems are unpredictable and do not respond predictably to interventions. Finally, a type of random dynamics (criticality) is also common in complex systems. Criticality results from constant stress on a system composed of interdependent components with varying predilections to respond, yielding a random pattern of responses of varying intensity. Systems characterized by criticality have no attractors limiting their behavior, and may or may not be sensitive to initial conditions [26]. Thus, these systems are unpredictable in behavior and in response to intervention. The dynamic pattern of violence may thus be a marker for the process at work in the abusive relationship.

Both chaotic and random dynamics are said to be "nonlinear" because the output of such systems is not proportional to the input and, hence, unpredictable. Three types of complexity measurements are available and we used one example of each type in our study [27]. First, LZ complexity, an information-based measure, measures of algorithmic complexity (the amount of information needed to describe the data) [28]. Second, approximate entropy, a regularity-based measure, assesses the regularity (or the lack of it) in time series [29]. Finally, a chaos-based measure of sensitivity to initial conditions (speed with which two adjacent points diverge over time) was assessed with the largest Lyapunov's exponent [30]. Thus, when studying nonlinear systems, not only can the specific dynamic pattern be determined, but measures of nonlinearity can be calculated.

22.4 Intimate Partner Violence as a Dynamical Condition

If we accept that the violent relationship represents a complex system, often consisting of more than two interacting members (i.e., children, extended family) and vulnerable to multiple influences at a variety of levels (scales), then we should expect that IPV is a dynamical disorder. IPV would then be defined by its dynamics and its dynamics can reveal much about the nature of the condition.

As a dynamical condition generated by a complex system, IPV should exhibit nonlinear dynamics. However, that is not to say that nonlinear, complex systems cannot generate linear dynamics. In fact, even simple systems can display a variety of dynamical patterns depending upon their constraints, resources, and interconnections [26]. In the logistic map seen in Fig. 22.1, a simple, nonlinear system can exhibit varying dynamics with differing numbers of possible states based upon the current state of the system. If dynamics is viewed on a continuum from linear to nonlinear, then the dynamics displayed depends upon the constraints, resources, interconnections, and feedback exerting influence on the state of the system (see Fig. 22.2). Hence, a healthy, nonlinear system may show linear dynamics if it is heavily constrained or lacks sufficient resources or connections.

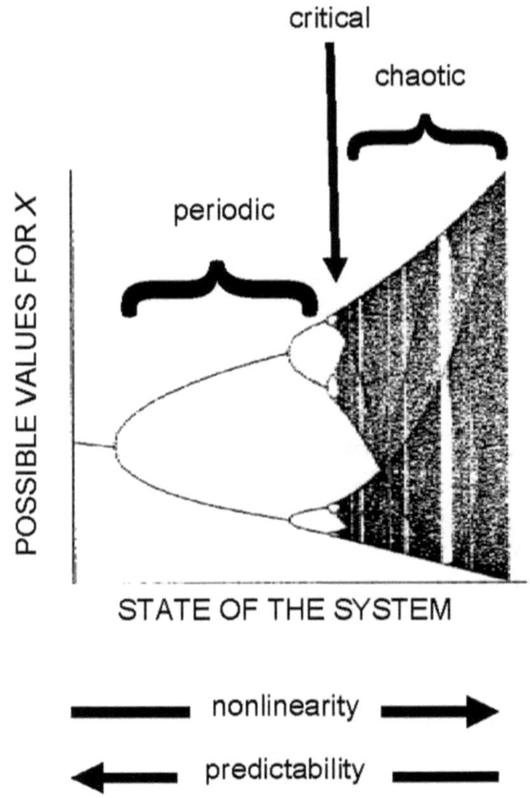

Fig. 22.1 Varying dynamics of simple systems

22.4.1 Prior Dynamical Assessments in Intimate Partner Violence

What little research has been conducted on the dynamics of IPV supports the complex nature of such dynamics. Both quantitative and qualitative attempts have been made to understand the evolution of these violent relationships.

22.4.1.1 Dynamics in the Relationship

Umberson et al. [31] compared the daily diaries documenting relationship dynamics, stress, and emotional state kept by 22 men who had a history of domestic violence versus 23 men without a violent history. Over the 14-day period of study, nonviolent men were more emotionally reactive to relationship dynamics and stress than their violent counterparts.

In another daily diary study of men seen at a drug treatment center, this time over a 15-month period, 149 victims of male-to-female violence

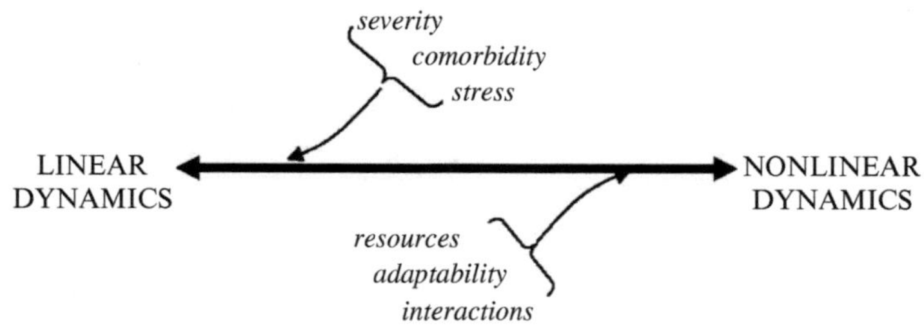

Fig. 22.2 Effects of resources and constraints upon dynamics

recorded their partners' substance use and any physical aggression that occurred. After controlling for the man's antisocial personality and the couple's global assessment of relationship distress, violence was more likely to occur when the man used alcohol or cocaine [32].

Ristock [33] interviewed 80 women who were in violent lesbian relationships and found a theme of shifting power dynamics. Yet, six focus group discussions of 45 service providers found that assessment of such power dynamics was particularly difficult for these providers.

These studies suggest that patterns of violence may have both predictable and unpredictable components: predictable in their relationship to substance use by the perpetrator and the perpetrator's lack of emotional reaction to relationship dynamics and stress, but unpredictable in the shifting power dynamics within the relationship.

22.4.1.2 Dynamics of Help-Seeking and Interventions

Chang et al. [34] conducted qualitative interviews of 20 women with a history of IPV, mapping the actions the women took to deal with the violence within the stages-of-change framework. This study found that women moved through stages at various rates in a nonlinear fashion, influenced by a variety of internal and external factors, until a "turning point" was reached, which enabled them to progress.

Hovmand and Ford [35] used prior data to model the implementation of three potential community interventions: (1) mandatory perpetrator arrest, (2) victim advocacy, and (3) changes in level of cooperation. While the rate of initial arrests of perpetrators was independent of when the advocacy and cooperation interventions were implemented, the rate of victim arrests displayed nonlinear relationships to implementation of the cooperation and advocacy interventions. The advocacy intervention was sensitive to the sequence of implementation, while the Wolf-Smith and LaRossa [36] interviewed 50 women residing in a battered women's shelter concerning three incidents of abuse. Although descriptions of the perpetrators' accounts and actions showed fewer apologies and justifications given by perpetrators over time, the victims were decreasingly likely to honor perpetrators' accounts over time.

These studies suggest that help-seeking is a nonlinear process in victimized women, perhaps necessitating the crossing of a "turning point," reached through the progressive decline in acceptance of the perpetrator's account. Community interventions to assist abused women must be carefully sequenced and timed due to the nonlinear nature of the phenomenon.

22.4.2 Dynamical Theories of Intimate Partner Violence

As the limited research cited suggests, the dynamics of IPV appears to have both linear and nonlinear components. In fact, although there are many theories concerning causes of IPV, only three models speak to the dynamics of IPV, each suggesting a different dynamical pattern.

22.4.2.1 The Cycle Theory of Violence

Based upon interviews with self-identified battered women, this theory states that battered women are not constantly abused, nor is their abuse inflicted at totally random times. Instead, *battering appears to recur in cycles* (see Fig. 22.3a). The battering cycle has three

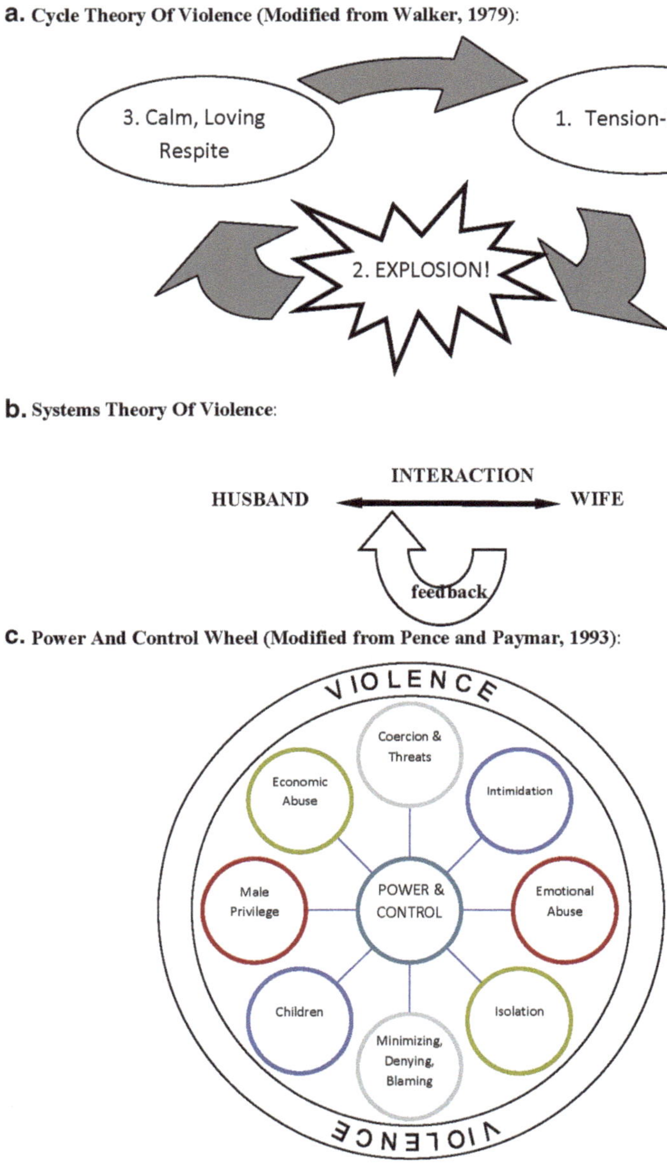

Fig. 22.3 Three theories of violence dynamics, modified from [37]

distinct phases, which vary in length and pattern across couples. In the Tension-Building Phase (Phase 1), minor battering events may occur, but the woman alters her behavior to keep the peace. Many couples remain in this phase for long periods of time, but eventually, tension builds, leading to the explosion. Phase 2 is the Acute Battering Incident (Explosion), characterized by high severity and brutality, batterers' lack of control, and brevity (usually a few hours). These episodes are usually triggered by an external event, and the timing may be unpredictable to the victim. Following the explosion is the Calm, Loving Respite Phase (Phase 3), where the batterer knows he has gone too far and tries to compensate for his violent behavior with loving kindness. This behavior is usually successful at pulling the victim back into the relationship, where she remains vulnerable to future victimization. Eventually, tension builds, and the couple moves into Phase 1 again [38, 39]. Just as Pincus et al. [40] found that conflict among members of a group decreases the nonlinearity of their interaction, so too does the Cycle Theory of Violence suggest that violent relationships should be more linear than in healthy relationships. Copel [41] supported this model, suggesting that low self-esteem and negative thinking keep women in their violent relationships. Further, Wolf-Smith and LaRossa [36] found that all women who do leave went through the same phases, but that the content of the third phase was modified over time.

Under the Cycle Theory, the three phases should yield a cyclic or periodic pattern. Although the period would vary among couples due to variable phase length, within each couple the constellation of constraints and history would keep the periodicity fairly constant. Periodic systems are predictable and respond predictably to interventions. Because periodic dynamics is so predictable, it is said to be "deterministic" (predictable over the long term).

22.4.2.2 Systems Theory

Although little research on FST has been conducted as it applies to IPV, previous work has supported its role in the divorce process [42] as well as in the households of abused and troubled adolescents [43, 44]. In this application of FST to IPV, Systems Theory focuses on wife *battering as an ongoing interaction pattern* resistant to change (see Fig. 22.3b). The first event of violence in a relationship is generally not severe, and the hitter is usually contrite, so the event does not drive the victim away. When hitting begins, a boundary breaks—the unspoken rule against using violence. The hitter has "stretched" the boundary, and the relationship held [45]. In laboratory experiments, aggressive acts increase the likelihood of the person being aggressive again [46]. At the first act of violence, perpetrators are distressed and contrite about their own behavior. However, with repetition, they become desensitized, and the shock and self-reproof extinguish over time. Being aggressive toward another seems to create a need to justify the violence by degrading the victim. This denigration may then support further violence as victims now seem to deserve the treatment they are receiving. At the same time, victims may acclimate to the punishment and react less dramatically, which leads the aggressor to work harder and harder to achieve the same effect. Escalation of aggression is not necessarily tied to the performance of the victim, but seems to be more contingent on the abuser's having begun aggressive acts in the first place. Over time, the physical aggression will suppress resistance, while the denigration will damage the victim's self-concept and self-efficacy. The victim becomes trapped in this interaction [45]. Ristock's [33] findings of shifting power dynamics in violent lesbian relationships support the Systems Theory through its emphasis upon the relational interaction and may explain the mutual violence that is sometimes observed.

Under the Systems Theory, the violence depends on the variable feedback loops between victim and batterer, theoretically yielding a chaotic pattern. Chaotic systems are unpredictable over the long term and do not respond predictably to interventions; yet, chaotic systems can be deterministic over the immediate short term.

22.4.2.3 The Power and Control Wheel Theory

Supported by interviews and quantitative cohort studies [47, 48], this theory posits that violence is used to control people's behavior. In contrast to the theory that battering occurs in cycles, authors posit that *abuse is a constant force* in battered women's lives (see Fig. 22.3c). Using information from group interviews with more than 200 women, Pence and Paymar developed the Power and Control Wheel which depicts eight key nonphysical abusive behaviors exhibited by men who batter: coercion and threats, intimidation, emotional abuse, isolation, denying, blaming or minimizing the violence, using the children, evoking male privilege, and economic control. It illustrates that violence is part of a pattern of controlling behaviors, not simply isolated incidents of physical violence or cyclical explosions of pent-up anger. A batterer's use of physical assaults may be infrequent, but the assaults reinforce the power of other controlling tactics. These tactics eventually undermine the partner's ability to act autonomously [49]. Other studies have also found that IPV involves "mindful manipulation" [50], experienced as "being controlled" and "a loss of self" [51]. Brewster's [52] study of 187 women being stalked by former intimate partners found that most stalking relationships involved the use of several power and control strategies (from financial and social to physical and psychological) both before and after the stalking had begun. Most stalkers included threats of violence against the woman or her contacts as well as against the children or himself; almost half of the stalkers involved the children in their schemes to exert control.

Under the Power and Control Wheel, the constant threat of abuse should produce constant stress within the relationship, occasionally erupting in violence. The use of multiple strategies for control resembles the interdependent components with varying predilection to respond, typical of critical systems under constant stress. These conditions would lead to occasional random violent catastrophes of varying intensity and the dynamics of criticality. This form of random dynamics is common in complex systems. Thus, these systems are unpredictable in behavior and in response to intervention.

22.5 The Face of Intimate Partner Violence Dynamics

Despite the varied theories and complex mathematical underpinnings inherent in this approach, ultimately our understanding about the dynamics of IPV rests upon the unique constellation of factors that influence the couple locked in violence. In many ways, dynamics of IPV is specific to the individual relationship due to its pattern of correlates and behaviors that shape meaning for the victim [53]. To quantitatively study the dynamics of IPV, we identified 16 women who experienced IPV during the previous month and asked them to complete a daily assessment of the previous day's violence and household environment via telephone for 8 weeks. Details of the methodology can be found in Katerndahl et al. [37]. To demonstrate the nonlinear facets in a violent relationship, consider the following case (adapted from Katerndahl et al. [54]):

A.F. was a middle-aged, white housewife with some college education and an annual household income of over $60,000. Although initially depressed, her mood improved once doctors diagnosed her with fibromyalgia, explaining the vague symptoms plaguing her. As a child, A.F. suffered physical and sexual abuse at the hands of her father, and frequently observed IPV in her home.

She was married in a church wedding more than 20 years ago. Although she and her husband wanted children, they were unable to conceive. Two years ago, her husband began his abuse, often throwing objects but never striking her. During the 53 days on which she reported, violence occurred more than half of the time (28 days); while all of these violent days included throwing objects, her husband insulted or threatened her 10 times and pushed or slapped her once.

Figure 22.4 demonstrates the dynamics of IPV in this case along with the fluctuations in the household environment. As the figure shows, her husband drank heavily every day. The levels of stress and hassles varied considerably. Although half of the time (26 days) days were argument-free, she reported 10 days with frequent arguments.

The IPV in this relationship showed a chaotic pattern, including a 12-day period of daily abuse. This extended interval of violence began with spikes in hassles and arguments; although the arguments persisted, they did steadily decline in frequency during this period.

Prior-week violence and arguments were associated with violent events, often preceded by a day of frequent arguments but a sense of closeness as well as 1–2 days of increased violence. Table 22.1 shows the complex pattern of associations over 3-day periods among violence and other factors. This table demonstrates the interdependencies among these variables; perceived stress has a particularly complex

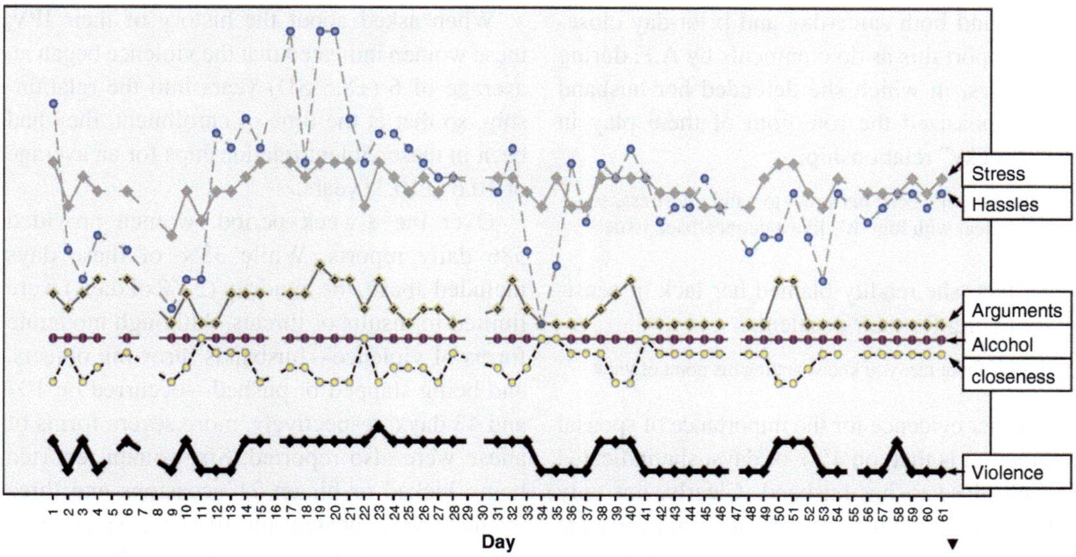

Fig. 22.4 Daily variability of violence and environment for subject A.F.

Table 22.1 Interdependencies among factors for subject A.F. (b-coefficient from vector autoregression)

Outcomes					
Predictor lag	Violence	Hassles	Arguments	Stress	Closeness
Violence					
1	0.46#	−0.39	0.33	0.56***	−0.66****
2	0.46#	0.94*	−0.39	0.09	−0.07
3	−0.07	0.13	0.00	0.61***	0.04
Hassles					
1	−0.03	0.45****	0.03	−0.00	−0.06
2	0.07	−0.04	0.06	0.09*	0.02
3	0.00	0.21*	−0.00	−0.08#	0.00
Arguments					
1	0.90**	1.79****	0.62*	0.93****	−0.19
2	−0.13	−0.58	0.02	−0.31#	0.14
3	−0.18	0.48	0.08	−0.65****	0.29
Stress					
1	−0.04	−1.21***	−0.23	−0.25	−0.12
2	0.06	−0.72#	−0.23	0.11	−0.17
3	−0.04	−1.05**	−0.15	0.31*	−0.05
Closeness					
1	0.76*	0.69	0.15	0.95****	−0.49#
2	0.22	0.23	0.02	0.21	−0.20
3	−0.03	−0.00	−0.18	−0.19	0.25

#$p \leq 10$, *$p \leq 05$, **$p \leq 01$, ***$p \leq 005$, ****$p \leq 001$

relationship with violence, hassles, and arguments. Same-day triggers included increased hassles, stress, and arguments, presumably followed by a loss of closeness.

The link between a chaotic pattern of IPV and a systems theory explanation is supported by the emphasis placed by A.F. upon the marital interaction. First, the relationship between violent events, and both same-day and prior-day closeness support this as do comments by A.F. during interviews, in which she defended her husband and emphasized the role both of them play in their "rocky" relationship:

> with him, it's been his stress too and my stress, so I can deal with him. It's like a bounce-back issue.

In fact, she readily blamed her lack of sensitivity to him for their problems:

> it's all about me, you know, getting his point of view

Further evidence for the importance of spousal interaction is that, on 45% of days, she indicated feeling close to her husband. Clearly, not only was her sense of closeness an important correlate of violence, but she accepted a role for herself in their dysfunction.

In addition, several factors promoted nonlinearity. The fact that she was educated with financial resources and recognized the support she received from her husband reinforces the potential for nonlinearity of the relationship. In fact, during the study, A.F. also sought support through spiritual guidance as well and was ultimately baptized.

Thus, in this example, a rich qualitative description of a violent relationship which emphasized the resources available and the husband–wife interaction was supplemented by quantitative measures showing same-day triggers as well as prior-day and prior-week correlates. Although each woman's story is unique in many ways, displaying its own patterns of dynamics and correlates, the phenomenon of IPV and its dynamics may also hold certain commonalities, inherent dynamical patterns observed across settings and victims.

22.6 Dynamics of Intimate Partner Violence

Drawn from an inner city primary care clinic, the low-income women in the study, many (56%) with less than a high school education, were predominantly Hispanic (75%) with a mean age of 37 (±12 SD) years. As a group, they had been married for 14 (±10 SD) years and had an average of 1.44 (±1.37) children.

When asked about the history of their IPV, these women indicated that the violence began an average of 6 (±8.5 SD) years into the relationship, so that at the time of enrollment, they had been in these violent relationships for an average of 10.6 (±10.3) years.

Over the 8-week period, women provided 586 daily reports. While 33% of these days included abuse, the majority (27% of days) were limited to insults or threats. Although moderate forms of violence—husbands throwing objects, and being slapped or pushed—occurred on 104 and 43 days, respectively, more severe forms of abuse were also reported. Six women reported being kicked or hit on 21 occasions and three women were beaten on five occasions; one woman reported being threatened with a weapon on 31 days.

22.6.1 Violence as a Group Experience

The dynamics of IPV is understood as one of intra-individual constancy with interindividual differences; dynamics are specific to the individual [53]. However, even though IPV is an individual experience and each relationship has a unique set of stressors and characteristics, there may be observations about the phenomenon that can be made, defining the community of IPV.

22.6.1.1 The Week Leading to Violence

Looking over the week prior to a violent experience, (see Table 22.2), several daily factors seem to predispose to the violence. First, not only does violence during the previous week correlates with another violent event, but also do the levels of arguments, hassles, and sense of closeness in the relationship. While the association with preceding arguments is understandable, the role of hassles and closeness/distance as predisposing factors is less clear. It may suggest that violent men, being less emotionally reactive to relationship dynamics and stress [31], cannot deal with the emotions of fluctuating closeness and irritating hassles, eventually leading to a violent outburst. Yet, the only prior-day correlate of violence was the level of arguments.

Table 22.2 Correlates of violence, modified from [37]

	Dynamic patterns			
Variable	Total	Periodic	Chaotic	Random
Same-day correlations (r_s)				
Hassles	0.22**	0.21#	0.31***	0.14
Arguments	0.54****	0.36****	0.48**	0.65****
Alcohol	–0.22	–0.15	0.19	–0.30
Stress	0.36****	0.23****	0.40	0.41**
Closeness	–0.37***	–0.39	–0.47***	–0.22
Prior day predictors (b)				
Violence	0.058	0.081	0.168	–0.024
Hassles	0.018	0.038	0.009	–0.010
Arguments	0.179****	0.099	0.718****	0.194*
Alcohol	–0.009	–0.023	0.605****	0.001
Stress	0.001	0.021	0.062	0.008
Closeness	–0.040	–0.078	0.006	0.017
Causality (Z)				
Violence	5.94****	2.22*	4.56****	2.59**
Hassles	4.14****	2.94***	1.81#	1.57
Arguments	4.60****	1.27	2.82***	2.81**
Alcohol	0.48	1.44	–0.87	0.36
Stress	1.55	2.23*	0.10	–0.78
Closeness	3.18****	0.48	2.08*	1.48

Correlates of abuse:
Same-day [r_s]
Prior-day [(b) = b-coefficient from vector autoregression]
Prior-week [Causality (Z) = Granger causality (Z-statistic)]
#$p \leq 10$, *$p \leq 05$, **$p \leq 01$, ***$p \leq 005$, ****$p \leq 001$

On the day of violence, however, several factors were associated with the event. While the positive correlations with hassles, stress, and arguments suggest that these factors may trigger the violence, the inverse association with closeness may reflect the victim's emotional state subsequent to the outburst [37].

22.6.1.2 Patterns of Patterns

Considering environmental patterns across days and women, the dynamics appear to be nonlinear, best described in triplets of three consecutive days. Most commonly, triplets involving physical or verbal abuse consisted of 3 days of abuse, which were either preceded or followed by another day of abuse; however, triplets involving physical violence were *not* preceded by days of verbal abuse only. In addition to verbal abuse triplets sometimes not including 3 days of abuse, verbal abuse triplets differed from physical abuse triplets in the consistency of their links to other factors. Although arguments and high stress were commonly associated with both verbal and physical abuse, verbal abuse was associated with heavy alcohol intake by the perpetrator while physical violence was linked to feelings of distance and the presence of many hassles [55].

Thus, not only does violence lead to violence, but verbal abuse alone differs from physical violence in its correlates and persistence of abuse. In addition, the observation that violence occurs in 3-day triplets suggests that the unit of study for violence (especially, physical violence) should perhaps be the 3-day interval. This may reflect a predisposition to violence during the weekend (or its equivalent for men who work on weekends).

22.6.2 Individual Violent Experiences

Within this generalized framework characterizing the IPV phenomenon of 3-day chunks of violence associated with arguments, hassles, stress, and emotional distance, often building over the prior week, individual patterns of IPV are seen, each unique in its dynamics and correlates. For example, while some individual experiences were associated with many prior-week factors, others reported no such relationship. While the violent event occurred in association with a variety of same-day factors for some women, others reported violence apparently without triggers.

In addition to considerable variability in the frequency and severity of violence experienced, the dynamics of that violence also varied from woman-to-woman. Figure 22.5 shows the ranges of measures of IPV nonlinearity found among these women. While the chaos-based measure (Lyapunov exponent) and the regularity-based measure (approximate entropy) both found that the dynamics clustered around periodic and chaotic norms, the algorithmic complexity measure (LZ complexity) suggested that the dynamics were more consistent with randomness. Thus, from a regularity perspective, IPV appears to vary as chaotic systems might. However, from an informational-content perspective, no simple algorithms are likely to adequately describe the violence for most violent relationships.

These measures of nonlinearity correlated with some relational descriptors, suggesting that dynamical patterns may impact violence-associated parameters. While algorithmic complexity was inversely related to the correlation between prior-day arguments and violence, it was also inversely related to the duration of the marriage; the longer a couple had been married, the less information it took to describe their violent dynamics. The other two nonlinearity measures correlated with the frequency of violence, especially that of insults and threats; just as these two measures suggested chaotic dynamics, so too they correlated with violence frequency. However, these two measures differed in their relationships with predictors of violent events; while approximate entropy in violence correlated with an association of violence with hassles, Lyapunov exponent correlated with an association with prior-day arguments.

One conclusion of this analysis is that, although there was some overlap among these measures of nonlinearity in terms of the dynamics suggested and their associations with violence, each measure also displayed some unique associations and could potentially play a unique role in our understanding. Second, LZ complexity and Lyapunov exponent were both significantly correlated with the relationship between prior-day arguments and a violent event. Perhaps this suggests that, when violence varies in a nonlinear fashion, prior-day arguments predispose to subsequent violence through its perturbation effects. However, when a violent relationship loses the informational variability (becomes routinized), it also predisposes to an association between prior-day arguments and subsequent violence [56].

22.6.3 Dynamic Patterns of Intimate Partner Violence

Using the Lyapunov exponent and whether the correlation dimension saturates as the embedding dimension increases, indicating the presence of an attractor, time series data can be classified into dynamic patterns: periodic, chaotic, random. As mentioned above, from a theoretical standpoint, each of these patterns may correspond to a different theory of violence dynamics. Although we anticipated identifying one predominant pattern, which would in turn suggest that one theory was more likely to be correct over the other two theories, this was not what we found. Instead, all three dynamic patterns were found in similar frequencies, randomness identified most often and periodicity found least often. Corroborating the dynamic patterns as classified, ARIMA 1,0,0 models and the maximal significant lagged predictors were seen in the women with periodic dynamics while ARIMA 0,0,0 models were seen predominantly with random dynamics. Also, the mean approximate entropy was maximal in

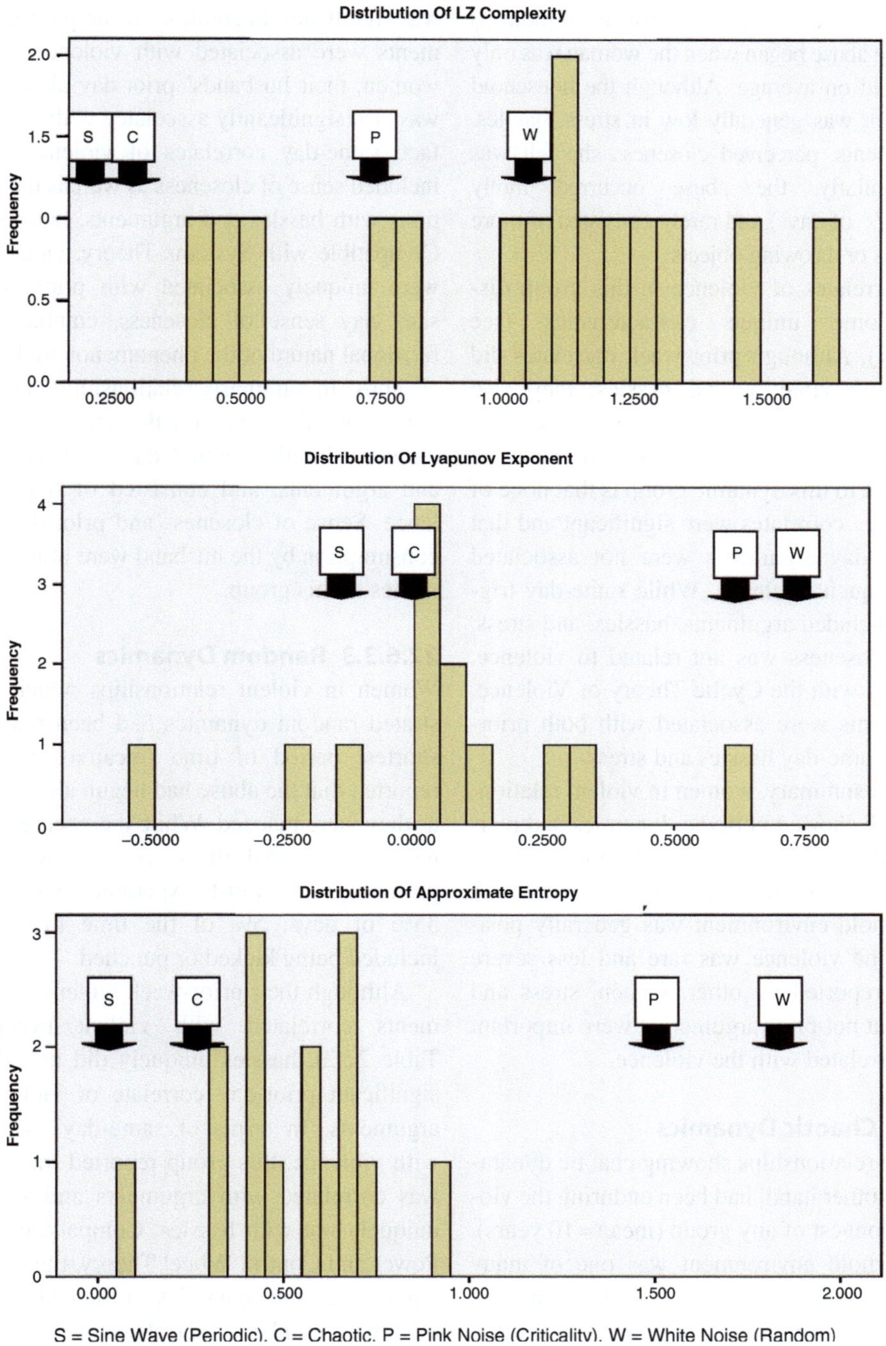

Fig. 22.5 Distributions of nonlinearity measures with standardized comparisons

random dynamics (0.671), minimal in periodic dynamics (0.437), and intermediate in chaotic dynamics (0.538). Thus, there is evidence to support the classification of these dynamic patterns [56].

22.6.3.1 Periodic Dynamics

From a clinical standpoint, the violent relationships so classified differed in potentially important ways. Women in violent relationships which demonstrated periodic dynamics reported that they had married soon after the relationship began (mean = 4 months) and that the abuse began when the woman was only 25 years old on average. Although the household environment was generally low in stress, hassles, and arguments, perceived closeness she felt was high. Similarly, the abuse occurred rarely (mean = 12% of days) and rarely consisted of more than threats or throwing objects.

The correlates of violence in this group displayed some unique characteristics (see Table 22.2). Although prior-week correlates did include both violence and hassles, they also included stress rather than arguments (both features unique to those with periodic dynamics). Also unique to this dynamic group is that none of the prior-day correlates were significant and that even prior-day arguments were not associated with subsequent violence. While same-day triggers still included arguments, hassles, and stress, sense of closeness was not related to violence. Compatible with the Cyclic Theory of Violence, violent events were associated with both prior-week and same-day hassles and stress.

Thus, in summary, women in violent relationships which showed periodic dynamics had married quickly after starting the relationship with the violence beginning at a young age. Although the household environment was generally positive, and the violence was rare and less severe than that reported by other women, stress and hassles (but not prior arguments) were important factors correlated with the violence.

22.6.3.2 Chaotic Dynamics

Women in relationships showing chaotic dynamics, on the other hand, had been enduring the violence the longest of any group (mean = 10 years). The household environment was one of many hassles and arguments as well as frequent episodes of violence (mean = 66% of days), often including pushing or slapping behaviors (mean = 14% of days).

Again, the pattern of correlates of violence was unique (see Table 22.2). In addition to prior-week correlates of violence that included violence, hassles, and arguments, women in relationships with chaotic dynamics in the violence were the only women for which prior-week sense of closeness was significant. In addition, while prior-day arguments were associated with violence, for these women, their husbands' prior-day alcohol intake was also significantly associated with violence. In fact, same-day correlates of violence uniquely included sense of closeness as well as the associations with hassles and arguments, but not stress. Compatible with Systems Theory, violent events were uniquely associated with prior-week and same-day sense of closeness, emphasizing the relational nature of the phenomenon in this group.

Thus, in summary, relationships demonstrating chaotic dynamics for the abuse tended to be longer in duration, included high levels of hassles and arguments, and consisted of frequent violence. Sense of closeness and prior-day alcohol consumption by the husband were important correlates in this group.

22.6.3.3 Random Dynamics

Women in violent relationships which demonstrated random dynamics had been married the shortest period of time (mean = 8 years) and reported that the abuse had begun almost as soon as they were married. While their sense of closeness was minimal, the severity of violence was, on average, the most; experiencing violence on 35% of days, 5% of the time that violence included being kicked or punched.

Although their prior-week violence and arguments correlated with violent events (see Table 22.2), hassles uniquely did not; the only significant prior-day correlate of violence was arguments. In terms of same-day associations with violence, this group reported that violence was correlated with arguments and stress, but uniquely not with hassles. Compatible with the Power and Control Wheel Theory, the group displaying random dynamics to their violence had the fewest number of correlates overall.

Thus, in summary, women in violent relationships with random dynamics had been married the shortest time but reported that the severe levels of violence had begun almost immediately after marrying. Feeling little closeness with their husbands, these women reported few correlates of violence with that violence uniquely unrelated to daily hassles [37].

22.6.4 Unanticipated Effects of Involvement in Research

In a final proof that IPV is a complex systems problem, we identified several unexpected effects of participation in the IPV investigation. Qualitative interviews with the participants showed that, not only did these women feel safe during the study, but many reported other positive effects of being involved.

Of the six themes identified during qualitative analysis, six women reported that they felt that the study served as an outlet for them, allowing them to share unpleasant feelings and stress. Second, nine women felt that the study served as a reality check, allowing them to gain insight into their situation. Five women actually felt that, through participation, they changed their outlook on life and ultimately their behavior. Fourth, three women felt that their daily reporting served as a safety net because someone outside the home knew what was happening. Finally, while eight women felt that participation gave them a sense of support and provided resources, viewing it as a form of counseling, five women felt empowered, gaining confidence, strength, and self-esteem through participation.

In addition, although women were selected for enrollment *because* they indicated that they had no immediate plans to leave the relationship, 35% of the women in this small study did indeed leave their partners during or after participation. Such action may have reflected the empowerment, support, and outlook change they experienced via participation. Comparing women who left their partners with those who stayed in the relationship, those who left had significantly more education but lower income. In addition, they had generally not been married in a church or civil ceremony. From a violence standpoint, women who left reported a lack of a relationship between prior-day arguments or same-day hassles, and violence; their Lyapunov exponents were negative, suggesting periodic dynamics.

Thus, the women who left were generally more educated women in lower income relationships *not* characterized by a marriage ceremony; such educated women may have been more likely to question their decision to remain when the income was lower and there was no legal or religious bond. From a violence perspective, the women who left were in relationships with more predictable violence but less reason to blame external stressors (hassles) or escalating arguments for the violent events; they were experiencing regular violence with fewer ways of rationalizing it. Although these are possible explanations for why some women left, it may have been the insight provided through daily reporting and the empowerment experienced through regular contact with study personnel that enabled these women to gain that insight and the strength to act upon it [57].

22.7 Implications of a Dynamics Perspective

If we assume that the findings presented here are confirmed in the large-scale studies under way, what are the implications for our understanding of IPV and for intervention? Clearly, no one theory was exclusively supported by these findings, but, just as clearly, IPV indeed displayed nonlinear dynamics with multiple prior-day and prior-week correlates as well as interventional surprise.

22.7.1 Understanding Intimate Partner Violence

What does the observation that longitudinal study of these relationships found support for all three dynamical patterns congruent with the three theories along with significant prior correlates of violence mean for our understanding of the IPV phenomenon? At least three potential interpretations apply.

22.7.1.1 IPV as a Dynamical Disorder

One possibility is that none of the theories reflect the "cause(s)" of IPV because IPV has no cause in the classical sense; it is a dynamical disorder. Our identification of patterns and multiple prior correlates, in this case, is spurious and misleading, a product of the multiplicity of the data collected and our inherent desire for pattern recognition.

Thus, IPV may simply represent a dynamical disorder brought on, not by 1–2 basic cause(s), but by the relationship's inability to adapt in a timely fashion to the micro-internal and external stressors of daily life. IPV here is its own cause as "violence begets violence." In this situation, IPV was in the marital toolkit from the inception of the relationship as a learned behavior from childhood and its dynamics reflects the day-to-day fluctuations in the household environment and the propensity for violence inherent in the relationship itself.

22.7.1.2 IPV as a Flexible, Nonlinear Disorder

Another possibility is that all three theories are correct but play out at different times in the relationship. In this scenario, just as dynamics can flow from chaotic-to-random (critical)-to-periodic based upon changes in support, resources, and/or constraints, so too the motivations driving the violence may change over time. Brewster [52] indeed found that the power and control dynamics evolved, but could be triggered by stressors. In addition, Flinck and Paavilainen [58] found that the meanings men attributed to their violence varied from denial and panic to self-defense and need for control to admission and self-blame. Such studies suggest that no one theory may be applicable throughout the course of the relationship.

One possible story of evolution could begin with the initiation of violence as described by Systems Theory. During a stressful encounter, violence unexpectedly erupts, taking both parties by surprise. Although shocked and repentant, the husband finds that the marriage has held despite the violence (and, in fact, the "winning" of the argument has rewarded it), lowering the threshold for subsequent violence. Each time that violence occurs without the breaking of the bond, the acceptability of further violence is reinforced and becomes more likely.

Over time, the husband unconsciously learns that IPV is a viable response, typically resulting in "getting his way." At some point, the motivation for his behavior becomes less about the quality of their relationship and more about control. At this point, he learns that violence is only one strategy that can enhance his control of the relationship. As he employs a greater diversity of these strategies, violence frequency may diminish and dynamics may transition from the chaos of Systems Theory to the criticality (randomness) of the Power and Control Wheel.

Finally, over the years, the woman's defenses and resilience become depleted, and she is unable to protest any more. The need for violence to maintain control diminishes because the wife can no longer resist and has become physically and emotionally isolated. In this phase, IPV assumes a periodic pattern, erupting only occasionally as a "reminder," no longer needed for dominance.

Although this continuous scenario could account for the unification of theories, transitions between dynamical patterns, and is compatible with the decreased frequency of IPV over time [59], it is certainly not the only way of combining them.

22.7.1.3 IPV as a Conglomerate Disorder

A third possibility is that IPV is a heterogeneous phenomenon and that there is no *one* correct dynamic theory; different couples in different environments may yield different dynamics. Not only may particular sets of childhood and life experiences create different predilections within husbands, but sets of interdependent environmental stressors may alter the threshold for violence on a daily basis.

Hence, there may be three (or more) distinct groups of couples with different backgrounds, environments, and dynamics. In fact, Johnson [60] proposed that two forms of IPV existed based upon the role the woman plays in the violence. In such circumstances, no one theory can explain the phenomenon because there are in fact multiple interdependent causes for IPV.

If we are to make progress in our understanding of IPV, we must consider all of these possible explanations and use detailed, nonlinear approaches to glean the interdependencies from the noise, using rich, multimethod, longitudinal datasets.

22.7.2 Dynamics-Related Intervention

Understanding IPV is not enough; we must be able to translate such understanding into intervention. Based upon the modest success of current interventions, we need to consider whether dynamics-related approaches could yield improved results whether defined as relationship extrication, diminished violence frequency or severity, or enhanced victim resilience.

22.7.2.1 Dynamics-Specific Approaches

Based upon theoretical considerations, we should expect that different dynamic patterns will be amenable to different interventional approaches. IPV exhibiting periodic dynamics should be the most predictable in its treatment response. With its deterministic underpinning and predictability, periodic IPV should respond in predictable ways. If we assume that the Cyclic Theory of Violence is the most likely explanation for this pattern, then interventions focused on the stressful triggers and the "knee-jerk" mechanism of response may be most successful.

IPV displaying chaotic dynamics, on the other hand, will be deterministic only over the short term. With attractors resisting change and sensitivity to initial conditions magnifying change, chaotic IPV may be particularly challenging when intervening. However, recognizing and using these "drivers of change" may produce results, but only through in-depth knowledge of the violent system. First, the attractors inherent in the system must be identified. "Joining" is a strategy that seeks to enhance desirable existing attractors to change the system. But the system can also be changed by transforming attractors. "Hammering" seeks to nudge an attractor through coercion and "wedging" pushes the attractor (and the system) to the edge of chaos where major system changes can spontaneously occur [61]. Second, if potential bifurcation points can be recognized, then small, well-timed perturbations in the system could potentially put the relationship on a nonviolent trajectory. The possibility that chaotic IPV may depend heavily upon the marital interaction for its generation suggests that marital counseling may facilitate recognition of attractors and bifurcation points as well as implementation of such chaos-specific interventions.

Finally, IPV displaying random dynamics is perhaps the most challenging. Such random dynamics will not respond in predictable ways, especially if truly driven by the husband's need for control and independent of other dependencies and emotional bonds. However, two approaches may yield change. First, the implementation of multiple, independent interventions may provide a change in the violence milieu. Such a multifaceted approach may be particularly effective if the interventions are multilevel in nature, perturbing not only the husband and wife but the household, extended family, and community as well. Although challenging to implement, this milieu approach provides an environment for change, enabling the system to respond to a variety of approaches. Second, if such randomness indeed reflects criticality, then the potential for self-organization and emergence of new patterns is even greater than in periodic or chaotic dynamics. Interventions to promote self-organization might include exposing the perpetrator and/or victim to new agents which possess positive qualities (i.e., role models), facilitating reorganization of the violent relationship around new, positive attractors.

22.7.2.2 Nonspecific Approaches

Regardless of the specific dynamic pattern observed, there may be nonspecific but dynamic-related approaches. First, because different dynamic patterns may differ in the frequency and/or severity of IPV in characteristic ways, changing the dynamics of the IPV may ultimately result in less frequent or less severe violence. Attempting to change th dynamic pattern itself is a novel approach and could be orchestrated via two potential mechanisms. As demonstrated in Fig. 22.2, the specific dynamic pattern that a nonlinear system displays may depend, at least in part, upon the support, resources, and constraints at work. Hence, by imposing or providing constraints or resources, we may be able to alter the dynamics of IPV in a relationship and thus alter the frequency or severity of violence occurring. Alternatively, previous studies of "stable" nonlinear systems have shown that timed control or anti-control perturbations can move the system along the periodic-chaotic continuum [62–64]. Hence, an intervention consisting of periodic perturbations (i.e., contacts with a therapist or trained family member) may succeed in nudging a violent system into a dynamic pattern with less frequent violent events.

A second nonspecific approach may rely upon the mindfulness paradigm [65]. Homework exercises in which the victim regularly "journals" the violence and household environment produces a longitudinal record of the violence and its potential triggers, making the unconscious conscious. Just as the women in our study reported insight-enhancement through the daily reporting activities of the study, so too may victims become self-aware of the reality of their experience through such exercises. When coupled with periodic discussion with a counselor, such journaling can have a powerful effect upon some women, convincing them of the negativity of their situation and the need for action.

Studying the dynamics of IPV may yield important insights into the mechanism(s) at work within the violent relationship as well as promoting our understanding of the phenomenon. In addition, this knowledge may lead to both novel dynamics-specific and nonspecific interventions for IPV.

22.8 Conclusion

While the importance of IPV to society and the family is evident, our linear-minded investigations have yielded a limited understanding of the phenomenon. The evidence for multiple, multilevel correlates of IPV, for the importance of dynamics for understanding psychosocial problems, and for the inherent nonlinearity of FST suggest that we need to study IPV as a complex adaptive system. The interdependency of factors, existence of turning points, and unanticipated surprises reinforce this notion. Not only can understanding IPV through its dynamics suggest the mechanism(s) at work within the violent relationship, but such understanding can lead to novel dynamics-related interventions

References

1. United States Department of Justice. Violence between intimates. Washington DC: Bureau of Justice Statistics; 1994 (NCJ-149259).
2. Stark E, Flitcraft A, Frazier W. Medicine and patriarchal violence: the social construction of a "private" event. Int J Health Serv. 1979;9(3):461–93.
3. Carmen EH, Rieker PP, Mills T. Victims of violence and psychiatric illness. Am J Psychiatry. 1984;141(3):378–83.
4. Gellen MI, Hoffman RA, Jones M, et al. Abused and nonabused women: MMPI profile differences. Pers Guid J. 1984;62(10):601–4.
5. Rosewater LB. Battered or schizophrenic? Psychological tests can't tell. In: Yllo K, Bograd M, editors. Feminist perspectives on wife abuse. Newbury Park CA: Sage; 1988.
6. Bergman B, Brismar B. A 5-year followup study of 117 battered women. Am J Public Health. 1991;81:1486–9.
7. Kurz D, Stark E. Not-so-benign neglect: the medical response to battering. In: Yllo K, Bograd M, editors. Feminist perspectives on wife abuse. Newbury Park CA: Sage; 1988.
8. Astin MC, Lawrence KJ, Foy DW. Posttraumatic stress disorder among battered women: risk and resiliency factors. Violence Vict. 1993;8(1):17–28.
9. Mitchell RE, Hodson CA. Coping with domestic violence: social support and psychological health among battered women. Am J Community Psychol. 1983;11(6):629–54.
10. Koss MP, Woodruff WJ, Koss PG. Relation of criminal victimization to health perceptions among women medical patients. J Consult Clin Psychol. 1990;58(2):147–52.
11. Koss MP, Heslet L. Somatic consequences of violence against women. Arch Fam Med. 1992;1:53–9.
12. Haber JD, Roos C. Effects of spouse abuse and/or sexual abuse in the development and maintenance of chronic pain in women. Adv Pain Res Ther. 1985;251:3259–64.
13. Kerouac S, Taggart ME, Lescop J. Dimensions of health in violent families. Health Care Women Int. 1986;7:413–26.
14. Helton AS, McFarlane J, Anderson ET. Battered and pregnant: a prevalence study. Am J Public Health. 1987;77(10):1337–9.
15. Hotaling GR, Sugarman DB. An analysis of risk markers in husband to wife violence: the current state of knowledge. Violence Vict. 1986;1(2):101–24.
16. Jaffe PG, Wolfe DA, Wilson SK. Children of battered women. Newbury Park CA: Sage; 1990.
17. Weaver TL, Resnick H. Toward developing complex multivariate models for examining the intimate partner violence-physical health relationship. J Interpers Violence. 2004;19:1342–9.
18. Lindhorst T, Tajima E. Reconceptualizing and operationalizing context in survey research on intimate partner violence. J Interpers Violence. 2008;23:362–88.

19. Farris CA, Fenaughty AM. Social isolation and domestic violence among female drug users. Am J Drug Alcohol Abuse. 2002;28:339–51.
20. James SE, Johnson J, Raghavan C. I couldn't go anywhere. Viol Women. 2004;10:991–1014.
21. Trotter JL, Allen NE. Good, the bad, and the ugly. Am J Community Psychol. 2009;43(3–4):221–31.
22. Trotter JL. A dynamic investigation of domestic violence survivors' experiences with their informal social networks. Diss Abstr Int B Sci Eng. 2009;69(11-B):7154.
23. Zink T, Jacobson Jr CJ, Pabst S, et al. A lifetime of intimate partner violence. J Interpers Violence. 2006;21:634–51.
24. Whitchurch GG, Constantine LL. Systems theory. In: Boss PG, Doherty WJ, LaRossa R, Schumm WR, Steinmetz SK, editors. Sourcebook of family theories and methods. New York, NY: Plenum; 1993.
25. Heath RA. Complexity and mental health. In: Holt TA, editor. Complexity for clinicians. Abingdon: Radcliffe Publishing; 2004.
26. Morrison F. Art of modeling dynamic systems. New York, NY: Wiley; 1991.
27. Land B, Elias D. Measuring the "complexity" of a time series. 2005. http://www.nbb.cornell.edu/neurobio/land/PROJECTS/Complexity/index.html
28. Ziv J, Lempel A. Compression of individual sequences via variable-rate coding. IEEE Trans Inf Theory. 1978;IT-24:75.
29. Pincus SM. Approximate entropy as a measure of irregularity for psychiatric serial metrics. Bipolar Disord. 2006;8:430–40.
30. Wolf A, Swift JB, Swinney HL, et al. Determining Lyapunov exponents from a time series. Physica. 1985;16D:285–317.
31. Umberson D, Anderson KL, Williams K, et al. Relationship dynamics, emotion state, and domestic violence. J Marriage Fam. 2003;65:233–47.
32. Fals-Stewart W, Golden J, Schumacher JA. Intimate partner violence and substance abuse. Addict Behav. 2003;28:1555–74.
33. Ristock JL. Exploring dynamics of abusive lesbian relationships. Am J Community Psychol. 2003;31:329–41.
34. Chang JC, Dado D, Ashton S, et al. Understanding behavior change for women experiencing intimate partner violence. Patient Educ Counsel. 2006;62:330–9.
35. Hovmand PS, Ford DN. Sequence and timing of three community interventions to domestic violence. Am J Community Psychol. 2009;44:261–72.
36. Wolf-Smith JH, LaRossa R. After he hits her. Fam Relat. 1992;41:324–9.
37. Katerndahl D, Burge S, Ferrer R, et al. Complex dynamics in intimate partner violence. Prim Care Companion J Clin Psychiatry. 2010;12(4):e1–12.
38. Walker LE. The battered woman. New York, NY: Harper and Row; 1979.
39. Walker LE. Victimology and the psychological perspectives of battered women. Victimology. 1983;8:82–104.
40. Pincus D, Fox KM, Perez KA, et al. Nonlinear dynamics of individual and interpersonal conflict in an experimental group. Small Group Res. 2008;39(2):150–78.
41. Copel LC. Partner abuse in physically disabled women. Perspect Psychiatr Care. 2006;42:114–29.
42. Madden-Derdich DA, Leonard SA, Christopher FS. Boundary ambiguity and coparental conflict after divorce. J Marriage Fam. 1999;61:588–98.
43. Franck KL, Buehler C. A family process model of marital hostility, parental depressive affect, and early adolescent problem behavior. J Fam Psychol. 2007;21:614–25.
44. Slesnick N, Bartle-Haring S, Gangamma R. Predictors of substance use and family therapy outcome among physically and sexually abused runaway adolescents. J Marital Fam Ther. 2006;32:261–81.
45. Giles-Sims J. Wife battering: a systems theory approach. New York, NY: Guilford Press; 1983.
46. Browne A. When battered women kill. New York, NY: Free Press; 1987.
47. Ehrensaft MK, Langhinrichsen-Rohling J, Heyman RE, et al. Feeling controlled in marriage. J Fam Psychol. 1999;13:20–32.
48. Sonis J, Langer M. Risk and protective factors for recurrent intimate partner violence in a cohort of low-income and inner-city women. J Fam Violence. 2008;23:529–38.
49. Pence E, Paymar M. Education groups for men who batter. New York, NY: Springer; 1993.
50. Queen J, Brackley MH, Williams GB. Being emotionally abused. Issues Ment Health Nurs. 2009;30:237–45.
51. McCosker H, Barnard A, Gerber R. Phenomenologic study of women's experiences of domestic violence during the childbearing years. Online J Issues Nurs. 2004;9(1):10913734.
52. Brewster MP. Power and control dynamics in prestalking and stalking situations. J Fam Violence. 2003;18:207–17.
53. Nurius PS, Macy RJ. Heterogeneity among violence-exposed women. J Interpers Violence. 2008;23:389–415.
54. Katerndahl DA, Burge SK, Ferrer RL, Becho J, Wood R. Understanding intimate partner violence dynamics using mixed methods. Fam Syst Health. 2012;30(2):141–53.
55. Katerndahl DA, Burge SK, Ferrer RL, et al. Recurrent patterns of daily intimate partner violence and environment. Nonlinear Dynamics Psychol Life Sci. 2010;14(4):511–24.
56. Katerndahl D, Burge S, Ferrer R, et al. Dynamic patterns of husband-to-wife abuse (final report). Washington, DC: National Science Foundation; 2007.

57. Tinitigan M, Becho J, Burge SK, et al. "Serendipitous intervention" among women experiencing intimate partner violence. Fam Med. 2010; 42(online supp2). http://www.stfm.org/fmsup/napcrg/fmconferencesupplement.cfm?confid=133.
58. Flinck A, Paavilainen E. Violent behavior of men in their intimate relationships, as they experience it. Am J Mens Health. 2008;2:244–53.
59. Kim J, Gray KA. Leave or stay? J Interpers Violence. 2008;23:1465–82.
60. Johnson MP. Patriarchal terrorism and common couple violence: two forms of violence against women. J Marriage Fam. 1995;57(2):283–94.
61. Sharp LF, Priesmeyer HR. Tutorial: chaos theory. Qual Manag Health Care. 1995;3(4):71–86.
62. Christini D, Collins J, Linsay P. Experimental control of high dimensional chaos. Phys Rev E Stat Nonlin Soft Matter Phys. 1996;54:4824–7.
63. Christini DJ, Stein KM, Markowitz SM, et al. Nonlinear-dynamical arrhythmia control in humans. Proc Natl Acad Sci USA. 2001;98:5827–32.
64. Weiss JN, Garfinkel A, Spano ML, et al. Chaos and chaos control in biology. Chaos Biol. 1994;93:1355–60.
65. Baer RA, editor. Mindfulness-based treatment approaches: clinician's guide to evidence base and applications. San Diego, CA: Elsevier Academic; 2006.

Multilevel Interdependencies and Constraints in Panic Disorder: Many Triggers, Few Responses

23

David A. Katerndahl

Panic disorder is unique among mental disorders because it is defined by an intermittent, dramatic, time-limited event—the panic attack. The apparent pervasive calm is periodically punctuated by moments of spontaneous terror, wreaking brief, autonomic and psychological havoc upon the individual. The "sting" of panic disorder afflicts almost 5% of the population, 45% of which rate the phenomenon as "serious" [1]. Not only does panic disorder lead to impaired sense of well-being and quality of life [2], but it often compromises work productivity and attendance [3]. In fact, 75% of the economic cost of panic disorder is due to its effect on productivity. A recent estimate of the adjusted per capita cost for panic disorder found that panic disorder cost over $11,000 annually per person, more than other anxiety or mood disorders, alcoholism, or chronic somatization [4].

In this chapter, we will take the wealth of research data available about the panic phenomenon and examine it through the lens of complexity science, documenting the interdependencies, constraints, and nonlinearities at work. Perhaps via this perspective we can understand why the diversity of nondescript "triggers" and reactions crystallize as a "panic attack" with its limited help-seeking patterns. Although the sting of the panic attack is uniform and disabling, its potential sparks are protean. Just as the classic movie "The Sting" describes the evolution of a complex system of interdependent factors which build to a climax, so too is panic disorder such a complex phenomenon.

D.A. Katerndahl, M.D., M.A. (✉)
Department of Family & Community Medicine, University of Texas Health Science Center at San Antonio, 7703 Floyd Curl Drive, San Antonio, TX 78229-3900, USA
e-mail: katerndahl@uthscsa.edu

23.1 The Overview

Panic disorder is defined by the spontaneous panic attack, a discrete episode of anxiety associated with a particular constellation of somatic and psychological symptoms, which begins without conscious warning and peaks in intensity within 10 min of onset. Such attacks are so intense that the person either takes action or withdraws.

Although the onset of the disorder typically commences during young adulthood, it can begin anywhere from childhood to middle-age or beyond [5]. Based upon daily diaries, patients may experience daily panic attacks, some situational, some unexpected [6]. However, panic symptoms are only part of the problem; the fear and expectancy of panic (all intercorrelated) contribute to its disability [7].

Uncomplicated panic disorder often resolves within 3–5 years [3, 8]. However, rarely is panic disorder uncomplicated. Classically, panic disorder is associated with agoraphobia as fear of panic leads to withdrawal from external activities [7]. In fact, the presence of agoraphobia is associated with decreased recovery from panic disorder [8, 9],

J.P. Sturmberg and C.M. Martin (eds.), *Handbook of Systems and Complexity in Health*,
DOI 10.1007/978-1-4614-4998-0_23, © Springer Science+Business Media New York 2013

adding an additional \$3,000 (per person) to the annual economic burden for those with panic disorder [4].

In addition, panic disorder is often associated with major depressive disorder (MDD). The presence of MDD is linked to increased disability, suicidality, substance use, and chronicity [10].

In fact, Noyes et al. [11] proposed that the co-occurrence of either agoraphobia or MDD with panic disorder represented more severe forms of the disorder, with worse outcomes and chronicity. In addition, MDD and agoraphobia may be interrelated in panic disorder; not only does the presence of agoraphobia increase the odds of MDD [10], but the presence of MDD in panic disorder increases avoidant behavior [11].

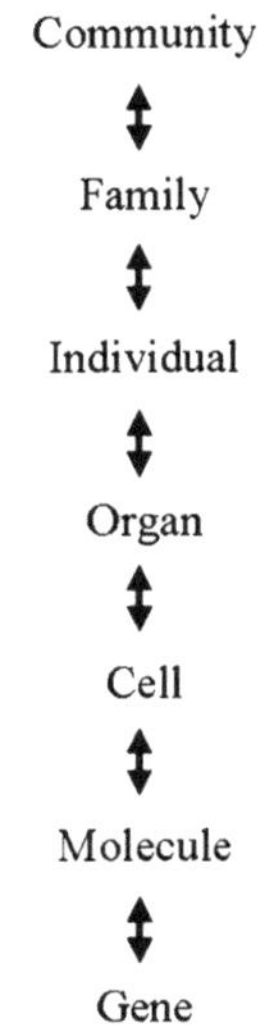

Fig. 23.1 Multi-level reality

23.2 The Setup

Much of the vast literature on panic disorder has focused on its predispositions. These correlates occur across the spectrum from genetics to the community.

23.2.1 Multilevel Dynamics

In many ways, panic disorder is descriptively a multilevel, multisystem disorder (see Fig. 23.1). At its most basic level, panic disorder may be a genetic disorder. Not only is it familial, but panic disorder has been linked to a variety of chromosomes. However, it is unlikely that panic disorder is due to a simple autosomal dominant or recessive trait [12, 13]. Similarly, biochemical abnormalities noted during spontaneous panic attacks, such as elevated prolactin, cortisol, growth hormone, and norepinephrine levels, raise the possibility that panic disorder represents a biochemical disorder; in fact, the intensity of the panic attack correlates with the extent of the elevation in serum prolactin [14].

At the organ level, changes in the electrophysiology of the heart have long been studied in panic disorder. Most studies have found that, compared with healthy controls, adult patients with panic disorder display increased variability in the QT interval [15–17], but decreased variability in heart rate in both adults [18–23] and children [24]. In fact, medications known to be effective in treating panic disorder have been shown to increase heart rate variability (HRV) in those suffering from the disorder [23, 25, 26]. However, the fact that recovered panic disorder patients continue to display diminished HRV compared with controls suggests that such decreased variability may indeed predispose to the disorder [27].

Above that, at the individual level, panic disorder is associated with a number of comorbid medical disorders. Not only are childhood respiratory disturbances associated with the development of panic disorder during adolescence or young adulthood [28], but panic disorder has been linked to mitral valve prolapse [29], hyperthyroidism [30], temporal lobe epilepsy [31], sleep disorders [32], and drug effects [5]. However, these associations cannot be explained as simple extensions of the biochemical or electrophysiological changes noted earlier.

But individuals do not exist in a vacuum; they are members of families and communities. Not only does a family history of mental disease, in general [33, 34], and panic disorder, in particular [35], predispose the individual to the development of panic disorder, but patients with panic attacks report other family issues. While patients with infrequent panic attacks indicate elevated levels of both current and childhood family stress as well as more violent family events in the prior year than controls [36], adults with panic disorder also report increased family stress and perceived family dysfunction with diminished family support [37].

But even the community may impact panic disorder. Not only does community income deprivation correlate with levels of mental illness, in general, mitigated by social cohesion [38], but panic disorder, in particular, may be sensitive to community characteristics. Although census tract accounts for little of the variance in quality of life and psychiatric symptoms among adults with panic disorder, it does explain 13% of the variance in mental health utilization for panic symptoms [39]. In fact, while census tract economic descriptors correlate with psychiatric comorbidity, sense of panic control, and hospitalization for panic symptoms, housing descriptors correlate with levels of substance abuse, disability, and hospitalization; community educational descriptors are associated with both ambulatory and hospital utilization among those with panic disorder [40].

Thus, panic disorder is a multilevel phenomenon. From genes and molecules to organs and the individual, panic disorder has multiple, potentially interrelated factors. Even external factors at the levels of the family and community potentially impact panic disorder, its manifestation, and its outcomes. How do these multi-level factors impact the psyche of the patient with panic disorder?

23.2.2 Anxiety Sensitivity

Is there a psychological foundation that must be in place before the individual will manifest panic attacks? Not only do those with panic disorder often have coexisting mental disorders in addition to MDD and agoraphobia [41, 42], especially prior anxiety disorders, but certain personality disorders (avoidant, dependent, obsessive–compulsive, histrionic) are particularly prevalent [35, 43].

Childhood seems to be the training ground for the development of panic disorder. Childhood physical and sexual abuse has been repeatedly linked to the development of adulthood panic disorder [44]. Not only is personality established early in childhood, but particular psychological traits appear in the childhoods of those destined to develop panic. Emotional reactivity seems to predispose to later panic disorder [28]. In particular, children who display separation anxiety are at risk for the development of a variety of adult mental disorders, including panic disorder and agoraphobia [45]. One consequence of separation anxiety is the development of anxiety sensitivity (the belief that harm will result if they experience anxious feelings).

Anxiety sensitivity predicts panic severity in longitudinal studies [46]. In fact, even controlling for distress or discomfort tolerance, anxiety sensitivity uniquely predicts physical and cognitive panic symptoms [47]. Although the degree of anxiety sensitivity is variable among those with panic disorder, such variation is predictive of clinical status over the next 6 months and is related to differences in comorbidity, medications, and symptom variability [48]. Further supporting the role of anxiety sensitivity in the development of panic disorder is evidence linking anxiety sensitivity to correlates of panic disorder. Hence, the findings that anxiety sensitivity is associated with a family history of mental disease as well as childhood trauma (separation from parents and divorce, sexual abuse, household violence) reinforce the anxiety sensitivity-panic disorder linkage [49]. In addition, as in panic disorder, Melzig et al. [22] found that patients with anxiety sensitivity displayed reduced HRV compared with controls.

Thus, between the prevalence of personality disorders and anxiety sensitivity providing a fertile psychic field, and the multilevel, internal, and external factors observed in panic disorder, we might expect that the vulnerable individual will, when exposed to the right trigger, develop panic attacks which exhibit highly individualized, nonlinear patterns over time. Is there evidence that panic disorder represents a complex phenomenon?

23.2.3 Panic Disorder as a Complex Phenomenon

The idea that mental disorders represent dynamical disorders and should be studied via nonlinear means is not new. Tschacher [50] advocated the

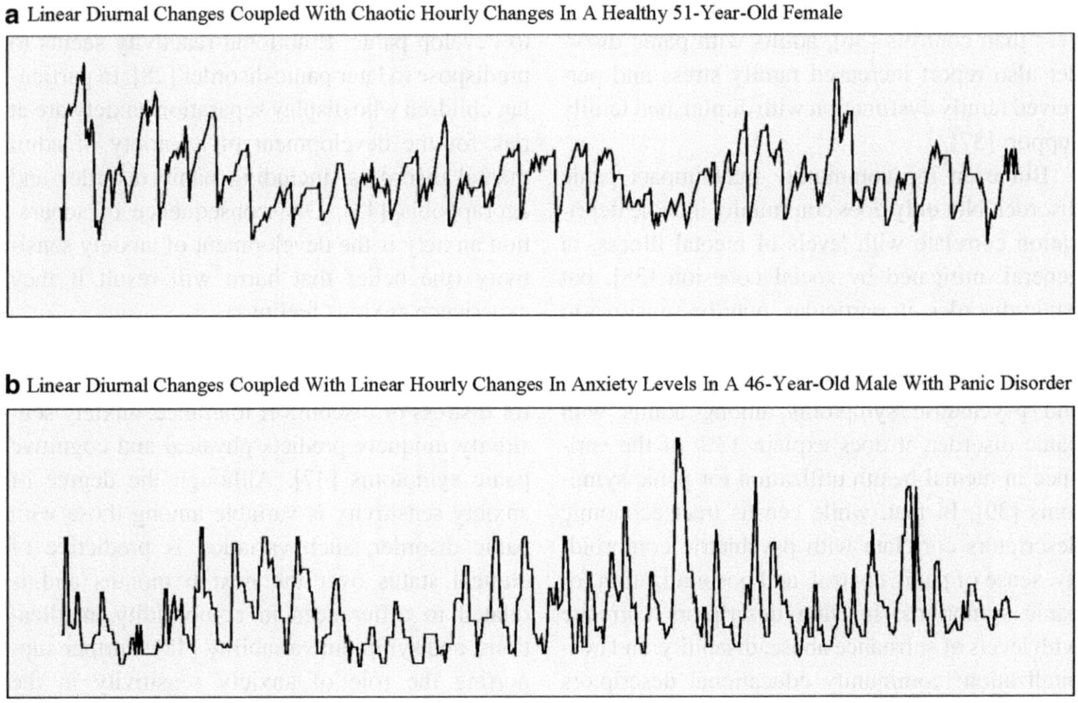

Fig. 23.2 Hourly changes in self-rated anxiety levels

use of a dynamic systems approach in psychiatry and psychology based upon his study of psychiatric inpatients. Ovid'ko [51] suggested that, unlike brainless animals, human thought and feeling should generate chaotic dynamics in action. In general, it is the loss of complexity and nonlinearity that suggests disease as unhealthy systems lose their ability to adapt and respond to microchanges internally and externally; in fact, the loss of variability may facilitate diagnostic categorization [52]. The finding that patients with panic disorder exhibit increased white matter connectivity in cingulate regions, which mediate fear and learning, and that such connectivity correlates with panic severity [53] reinforces the expectation that panic disorder should display features of highly interconnected and interdependent complex systems.

First, panic disorder is a dynamical disease. Such disorders show dynamical variability without obvious cause, can be controlled but not cured, and involve complex interactions of endogenous and environmental processes [54]. Panic disorder indeed demonstrates impaired dynamics. Not only is there decrease HRV in panic disorder, but measures of nonlinearity are lower as well [24, 55]. However, nonlinearity of QT intervals [15] and EEG activity is actually increased in panic disorder; EEG nonlinearity correlates with anxiety levels [56]. The dynamics of mood in panic disorder is even more complex. Healthy controls demonstrate a circadian rhythm for changes in anxiety level with an overlying chaotic pattern (see Fig. 23.2a) [57]. Such nonlinearity agrees with Combs et al. [58] who found chaotic patterns in 30-min ratings of emotions in healthy controls. Panic subjects, however, continue to display circadian variation but often lose the overlying chaos (see Fig. 23.2b) [57]. Such findings raise the possibility that the nonlinear system of mood variation is capable of exhibiting a range of dynamic patterns (see Fig. 23.3) from periodic to chaotic based upon the state of the system, and its constraints and resources (see Fig. 23.4). For example, in the 46-year-old male panic subject who provided the anxiety diary shown in Fig. 23.2b, he retained his linear diurnal pattern but lost the embedded hourly chaotic pattern, instead displaying a more linear-on-linear dynamic. His low level of education (less than high school diploma), lack of employment, and low income further reduced his resources and adaptability, encouraging linear dynamics. Gustafson and Meyer [59] documented the course of panic disorder in one subject, demonstrating nonlinearity coupled with sudden changes in affect. When anxious and depressive mood changes are considered simultaneously in panic disorder, there is less correlation between these moods than in controls and the attractors they define in state space (see Fig. 23.5) differ from the small, strong attractors seen in controls; attractors in panic disorder tend to be smaller and less influential [60]. Hence, panic disorder appears to qualify as a dynamical disorder exhibiting changes in dynamics which correspond to clinical state.

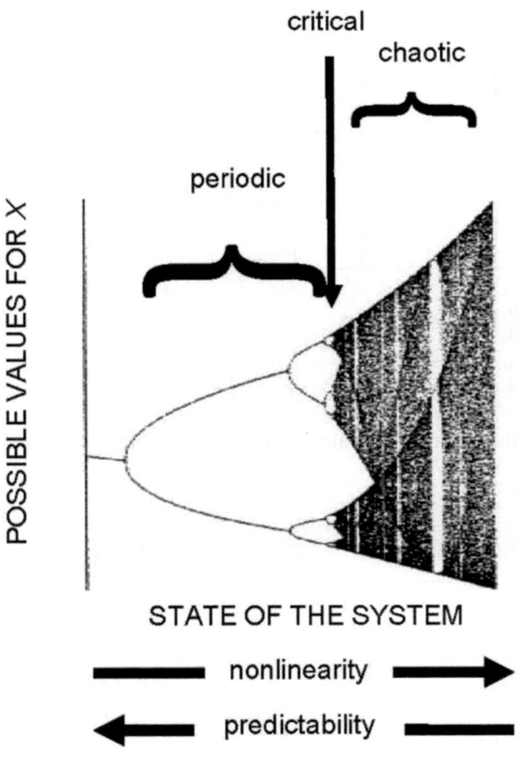

Fig. 23.3 Varying dynamics of simple systems

Second, interdependencies among factors and patients are seen. Many of the correlates mentioned earlier are themselves interrelated. Perhaps the strongest evidence of interdependence is the observance of power law dynamics at work. When phenomena follow power law distributions, log–log plots reveal linear trends. For example, lab mice demonstrating escape panic follow power law dynamics [61]. In fact, when anxiety and depression levels are considered simultaneously, the hour-to-hour variations show power law dynamics as do controls (see Fig. 23.6). However, while the log–log plots of controls show slopes consistent with "persistence" in which there is long-term memory and feedforward processes, the slopes for panic subjects showed "anti-persistence" in which deviation from a group "norm" is quickly brought back into line [62]. The presence of a power law suggests that the mood changes within the person function as a "system".

In fact, interdependencies in complex systems often make "causation" meaningless. There is indeed evidence for such circular causation in panic disorder. While anxiety sensitivity appears to set the stage for experiencing a panic attack, Schmidt et al. [63] found that spontaneous panic attacks increase anxiety sensitivity, especially if their underlying anxiety and depression levels are low. In addition, the familial association may involve factors other than genetics or household environment. Acs et al. [64] found a relationship between *untreated* maternal panic disorder and congenital abnormalities in their infants, sug-

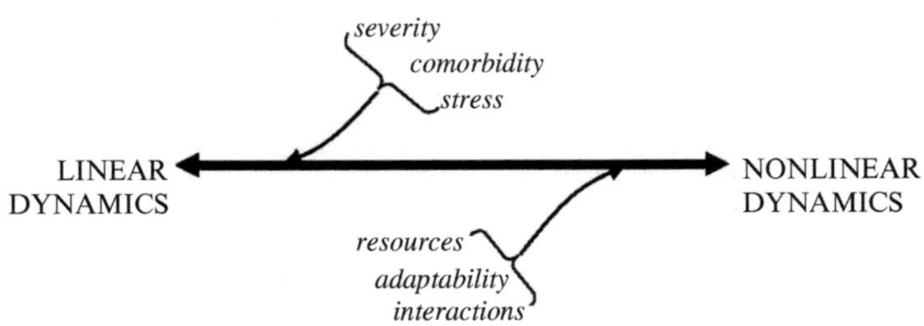

Fig. 23.4 Effects of resources and constraints upon dynamics

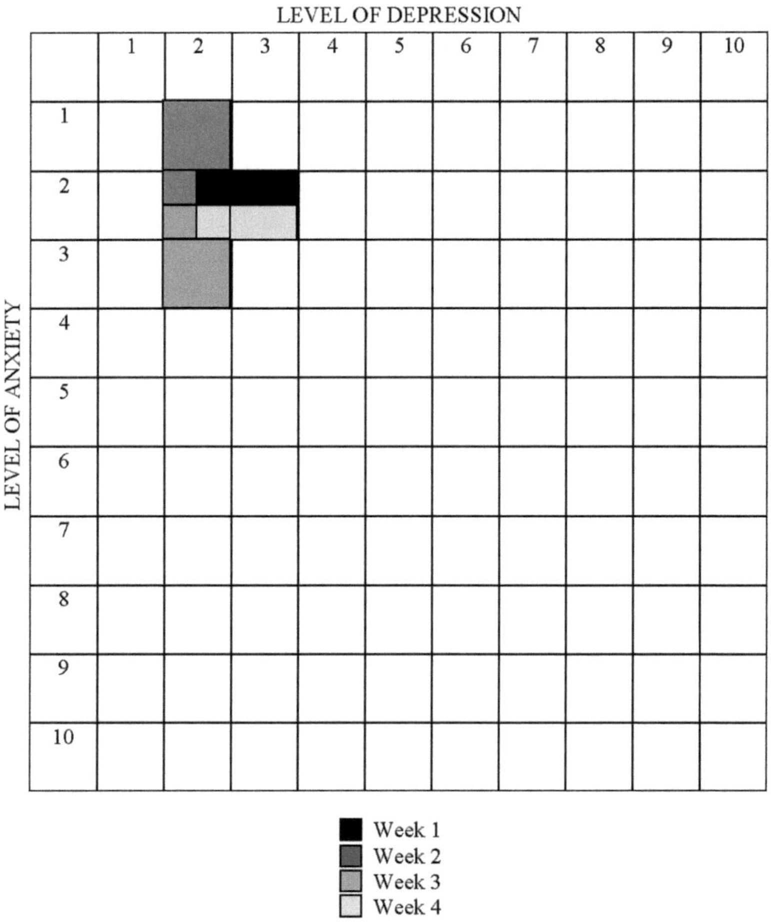

Fig. 23.5 Primary attractors observed over 4-week period in the patient with panic disorder shown in Fig. 23.2b

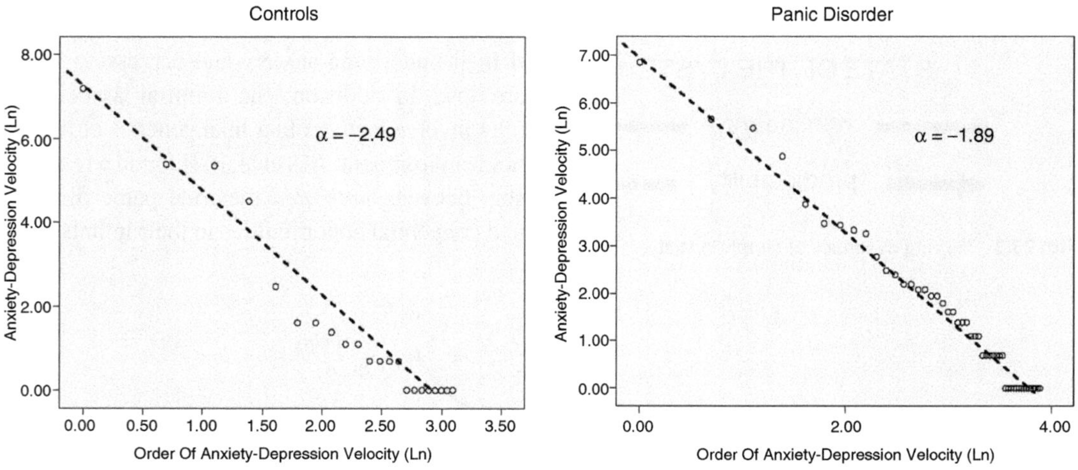

Fig. 23.6 Log–log plots of anxiety–depression changes among patients with panic attacks and controls, originally published in [62]: from Fig. 2. Reprinted with permission from Journal of Evaluation in Clinical Practice

gesting a biological effect or lifestyle interaction. However, Bandelow et al. [49] found that children with childhood illnesses or handicaps were also more likely to develop adult panic disorder. These studies suggest that causation in panic disorder is a circular, multifactorial phenomenon.

Thus, even before a panic attack is triggered, the individual is exhibiting complex dynamics at multiple levels, responding to familial and community factors as they interact with endogenous characteristics, making the person sensitive to anxiety and its anticipated harmful effects. In such a milieu, a variety of triggers can set off a panic attack.

23.3 The Trigger(s)

Anxiety sensitivity represents the primary stage in coping with panic as patients start by appraising the consequences of whatever potential threat appears. Their secondary appraisal focuses on how to handle the threat and only then do they call upon whatever coping strategies they possess [65]. But those with panic disorder are more than just sensitive to potential effects of anxiety; they are generally sensitive to unanticipated events [66]. Not only are those with panic disorder more embarrassed by symptoms [67] and misread emotions in others [68], but their sensitivity extends to anxiety-provoking concepts as well. Patients with panic disorder learn panic-related words faster [69] and more readily make subliminal inferences about physical threats than others do [70].

23.3.1 Triggers

In such a heightened, vigilant state, it is little wonder that a diversity of stimuli can trigger a panic attack. In addition to infusion studies which implicate lactate [71], beta agonists [72], and serotonin [73] as potential triggers, induction of panic can occur via hyperventilation [74] (implicating hypocalcemia and alkalosis) and carbon dioxide inhalation [75]. Similarly, panic attacks are associated with chemical intolerance, raising the possibility that chemical or odor exposure triggers panic attacks in some individuals [76]. In fact, panic attacks may be triggered by the use of stimulatory drugs or withdrawal from depressant drugs [77]. In addition, panic disorder may be associated with several genetic polymorphisms, suggesting that such individuals could react abnormally to an exposure to which most of us would not react [78–87]. Even sleep deprivation and fluorescent lighting could trigger attacks [88]. The fact that panic disorder patients display changes in HRV and that such patients are particularly accurate in heart rate perception [89] raises another potential mechanism for triggering panic; sudden drops in HRV may be perceived by individuals and elicit a panic attack. In fact, such atypical reactions to physiological events may cause those with panic attacks to cue on such symptoms, reacting to any trigger that produces a sensation reminiscent of the panic-eliciting symptom; these are termed "interoceptive cues" [90].

In addition to internal factors, panic attacks may be triggered by stressful events. This is particularly true of those in whom panic disorder did not begin until adulthood [35]. In fact, anger outbursts are associated with panic [91], possibly due to misinterpretation of anger in others [68]. Batelaan et al. [9] also found that panic attack frequency increased when positive life events were lacking.

Finally, as a dynamical disorder, panic disorder may elicit its own symptoms in response to prior changes in mood. For example, a 49-year-old male who was unemployed due to a disability experienced panic attacks when his levels of both anxiety and depression were maximal, and his level of depression had increased in the previous hour. Such sensitivity to prior mood and mood change may be supported by the findings of Mavissakalian and Guo [92], in which 2-month relapse in panic was associated with linear changes in generalized anxiety, fearfulness, and disability.

Thus, although patients often cannot identify the triggers of their panic attacks, a variety of agents may cause their occurrence from internal molecules and inhaled chemicals to perceived heart rate and mood alterations. However, whatever the specific trigger may be, they all may act through interoceptive mechanisms.

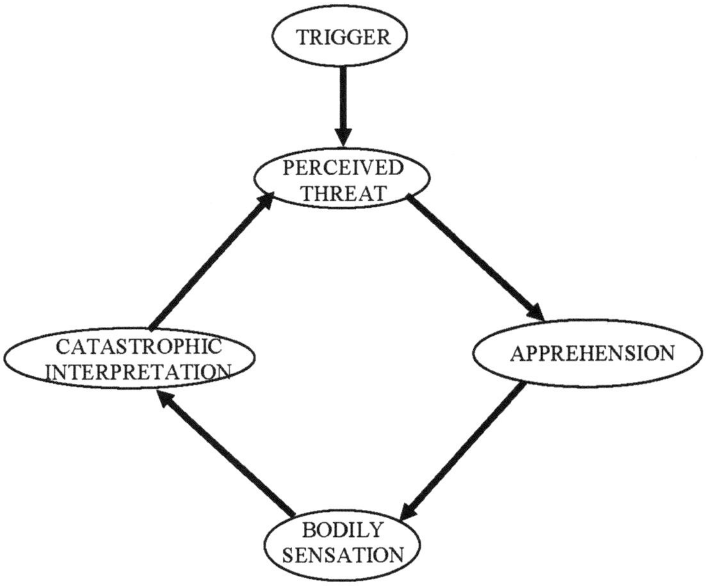

Fig. 23.7 Cognitive cycle in panic disorder, modified from [94]

23.3.2 Catastrophe

Many believe that panic disorder is a learned phenomenon [93], crystallized through a process of catastrophic misperception of threat (see Fig. 23.7). Under this model, once the various triggers elicit a perceived threat, apprehension develops. In addition to the panic expectancy mentioned earlier, panic apprehension is related to dysfunctional beliefs and poor self-efficacy [95]. Such panic expectancy correlates with both the fear of panic attacks as well as their perceived aversiveness [7]. It is the combination of the perceived threat and apprehension that leads to the bodily sensations and their ability to serve as interoceptive cues.

Unfortunately, the cuing of such sensations leads to catastrophic cognitions in those with panic attacks. First, patients with panic disorder tend to overestimate the association between panic stimuli and negative consequences [96] and have more negative beliefs concerning danger [97]. Second, they have a cognitive bias towards negative information, predisposing them to bodily misinterpretation and catastrophic thoughts [98]. Third, while patients with panic are slower to correctly identify general anxiety-provoking stimuli, they are faster at identifying panic-related stimuli [99], suggesting a bias in their sensitivity to stimuli. Finally, compared with patients with other anxiety disorders, those with panic disorder report more feelings of impending doom [100]. In fact, successful treatment of panic disorder is associated with fewer catastrophic cognitions [65], especially if the treatment focused on safety-seeking behaviors [101].

Thus, despite a variety of potential triggers, their result is often one of catastrophic misinterpretation and sense of impending doom. Unfortunately, coping strategies employed by those with panic disorder tend to be dysfunctional, probably reflecting their disease conviction and phobia, fear of death, and bodily preoccupation [102] as well as an underlying personality disorder. Such misinterpretations and dysfunctional coping are a setup for a panic attack.

23.4 The Sting

Whatever the trigger or stimulus, a discrete, scripted panic attack results. What constrains the plethora of triggers and their resulting sensations into the reproducible, categorical, fear-laden phenomenon called "a panic attack"? The variety of triggers and multilevel factors should foster nonlinearity (as in Fig. 23.4), so what constraints prevent this and coalesce the symptomatic result into a predictable, linear panic attack?

23.4.1 Constraining Multilevel Triggers

A number of factors serve to constrain the symptom constellation that those with panic attacks experience. First, because panic attacks are a recurrent phenomenon in those with panic disorder, memory of prior attacks and their correlates can potentially affect response to current triggers. Unfortunately, there is evidence for dysfunctional memory in those with panic. Not only do those with panic disorder uniquely have difficulty recalling fearful memories when anxious [103], but they appear to retrieve sensory and emotional memories in fragments [93]. In fact, although remembered imagery is important to maintenance of agoraphobic behavior, the sounds and smells associated with such memories are often impaired, dependent upon the age of the individual at onset and its situational context [104]. Consequently, patients with panic attacks exhibit increased decision-making search strategies to compensate for memory impairment [105], yet the expected increase in EEG nonlinearity when faced with such stress is less than in controls [56], suggesting their more restricted options.

In addition, personality and life experiences may constrain how individuals perceive the somatic interoceptive cues. The presence of histrionic, obsessive–compulsive, and avoidant personality disorders would suggest that, when confronted by aversive stimuli, those with panic disorder would be inclined to react dramatically, yet routinized, triggering avoidant behavior. Those who were victims of childhood abuse would tend to perceive noxious triggers as reminiscent of the violence they experienced as children [106]. In addition, if victims of current domestic abuse, their learned reactions may be limited to "perpetrator-acceptable" frameworks, such as "heart attacks" and "strokes".

Thus, once exposed to symptom-producing triggers which cause catastrophic interpretations, memory dysfunction may constrain the possible constellations to a limited few. Personality may enhance and distort these, while exposure to current and past violence may favor the linkage to patterns experienced during such trauma or "permitted" by others.

23.4.2 The Constrained Attack

The constrained result is the "panic attack" (see Table 23.1). In addition to the presence of these symptoms, particular symptoms occur earlier than others. Hence, dyspnea, palpitations, chest discomfort, and chills/hot flushes tend to occur early in attacks while fear of dying, going crazy, and losing control as well as paresthesias occur late in the attack [108]. In addition, catastrophic cognitions typically are present during the attacks [65] and correlate with their intensity [109].

Although the symptoms in Table 23.1 characterize the panic attack, subsets of attacks have been suggested. Moynihan and Gevirtz [110] suggest that patients with predominantly respiratory versus cognitive symptoms should to treated differently. Similarly, the presence of particular symptoms during attacks may differ depending upon whether phobic anxiety is present [111]. In addition, panic attacks that typically occur at night (nocturnal panic) may differ from daytime attacks in that they may exhibit more panic-related symptoms [112], higher levels of fear [113], and greater severity [114].

However, despite such subsets, the panic attacks seem to be homogeneous within patient in terms of symptom pattern and sequence as well as abatement pattern and precipitating event [115]. In fact, there is significant homogeneity in symptom sequence across patients as well [116], suggesting that, not only is the panic attack a distinct phenomenon within each patient but across patients as well.

Yet, the abatement of panic attacks may reflect a diminishing sensitivity to interoceptive input of their biochemical mediators as habituation develops [117] or the dysfunctional memory characteristic of panic disorder patients as they lose the sensory associations linked to recurrent, maintaining memories [104]. Unfortunately, it may be the residual expectancy of panic that not only lowers the threshold for the occurrence of a panic attack but increases its sense of danger, catastrophic thoughts, and avoidance, maintaining the disorder [118].

Thus, despite the varied triggers, dysfunctional memory, personality, and the history of

Table 23.1 Definition of a panic attack (modified from [107])

A discrete period of intense fear or discomfort, in which at least four of the following symptoms developed abruptly and peak within 10 min of onset:
1. Palpitations, pounding heart, or accelerated heart rate
2. Sweating
3. Trembling or shaking
4. Dyspnea or smothering
5. Feeling of choking
6. Chest pain or discomfort
7. Nausea or abdominal distress
8. Feeling dizzy, unsteady, lightheaded, or faint
9. Derealization (feelings of unreality) or depersonalization (being detached from oneself)
10. Fear of losing control or going crazy
11. Fear of dying
12. Paresthesias (numbness or tingling sensations)
13. Chills or hot flushes

violence may constrain the symptomatic experience into the discrete panic attack characteristic of the disorder, consistent within and among patients. Although habituation and dysfunctional memory may limit panic attacks, the expectancy of another attack serves to maintain the disorder.

23.5 The Reaction(s)

Although panic attacks are clearly defined phenomena, consistent across time and patients, those who experience them have a variety of reactions. While fear of dying, going crazy, and losing control are considered part of the panic attack experience, occurring late in the attack [108], patients interpret the experience differently and the resultant action taken (if any) depends upon a wealth of factors.

23.5.1 Meaning in Panic

How a patient interprets their panic attack is important, yet difficult to assess. While most such patients define their self-described "worst" panic attack as that which has the most severe symptoms (96%), a particular troubling symptom (95%), and that of longest duration (86%) [119], it may be how they explain the phenomenon that is most important. Although the most common explanation for their worst attack is "anxiety or stress" (31%), 22% believe that they are having a heart attack or stroke while 9% believe that they are dying; 10% have no explanation for their symptoms. In fact, having no explanation is associated with a rapid onset of agoraphobia [120]. In addition, the meaning assigned to the panic attack is important to the patient's help-seeking behavior [121].

23.5.2 Factors in Help Seeking

Beyond the meaning of the panic attack, there are a variety of factors important in help seeking for panic. Rogler and Cortes [122] believe that help seeking begins with a sense of distress. Then psychosocial and cultural factors interact with the psychiatric diagnosis and its severity to determine the direction and duration of help seeking. "Distress" is reflected in the reasons for seeking or not seeking care provided by those who experience panic. While 44% sought care due to their cognitions about the attack with 25% and 21% care seeking due to the somatic and psychological symptoms they experienced, respectively, 33% of those who never sought care did so because of how they interpreted the experience [121]. Such reasons for not seeking care during the worst attack coalesce into two factors—lack of perceived seriousness and belief that it was not a medical issue.

Once distressed, a wealth of factors is associated with care seeking. In addition to panic characteristics, comorbidity and quality of life issues, symptom perceptions, illness attitudes and behaviors, coping style, and access to care are linked to presenting for care, 37 different associations were observed in one study [121]. In addition, different sites of care were associated with different predictive factors [123]. In fact, if we look at the contemplation of care-seeking for their worst attack, we find that numerous factors are relevant, particularly if considering different sites of care. Katerndahl [119] found that panic symptoms, their perception, and interpersonal sensitivity were important overall when contemplating care seeking for the worst attack. However, when deciding between different medical sites, attack explanation, symptom presence and perception, ethnicity and comorbidity, attack appraisal, coping via support, and family utilization were all significant factors. In addition, when including use of family and friends, self-treatment, and alternative sites, other factors correlate with contemplation. Here, in addition to attack explanation and symptom perception, new factors are relevant, including education level, access to care, family life events and adaptability, and illness attitudes. Such studies suggest that many factors enter into the patient's consideration of how to react to the panic attack.

We found that those who did not seek care for their worst attack apparently endorsed the Rogler and Cortes model; "seriousness" was associated with psychosocial (symptoms interfering with activities, coping through blame), cultural (Hispanic origin) and psychiatric (number of prior attacks) issues, while "a non-medical label" was associated with psychosocial (fear perception, religious coping), family (family size and cohesion) and psychiatric (chest pain severity, comorbid mental disorders) issues. Similarly, Rogler and Cortes [122] found that social networks and race/ethnicity were important to mental health care utilization and treatment.

Finally, in addition to associations with multilevel factors, there is evidence of interdependencies. Although panic attack appraisal was rarely associated with either contemplating or seeking care [121], such appraisals showed power law relationships with hospitalization, ambulatory utilization, and family physician use (see Fig. 23.8). While all three sites had a power law distribution with panic attack novelty, only ambulatory utilization showed power law distributions with appraisals of transience, self-causation, and predictability. Power laws were also noted between attack importance and hospitalizations as well as between attack controllability and both hospitalizations and family physician usage. Such power law relationships were not observed with symptom perceptions. These findings suggest nonlinear, interdependent relationships between care-seeking and attack appraisal [124].

Although these studies suggest that numerous factors may be related to how patients react to their attacks, especially in terms of care-seeking, when applying the Anderson Model to care-seeking in panic disorder, Goodwin and Anderson [125] found only a few factors predicted utilization (generally comorbidity associated with need for care). The fact that Katerndahl and Parchman [126] found that the panic population is vulnerable to poor health care outcomes and that adding such vulnerability domains to the Anderson Model explained more of the variance suggests that vulnerability may constrain how patients respond to panic.

23.6 The Response

Despite the many factors that could evoke care-seeking in those with panic attacks and evidence for nonlinearity at least in the relationships with attack appraisal, health care utilization for panic is modest and needs are unmet. This implies that the forces that should drive utilization are constrained.

23.6.1 Health Care Utilization

Only 57% of those with panic attacks ever seek care for them [121]. If they do seek care, then tend to do it soon after onset, more so than in other mental disorders [127]. Although 51% who

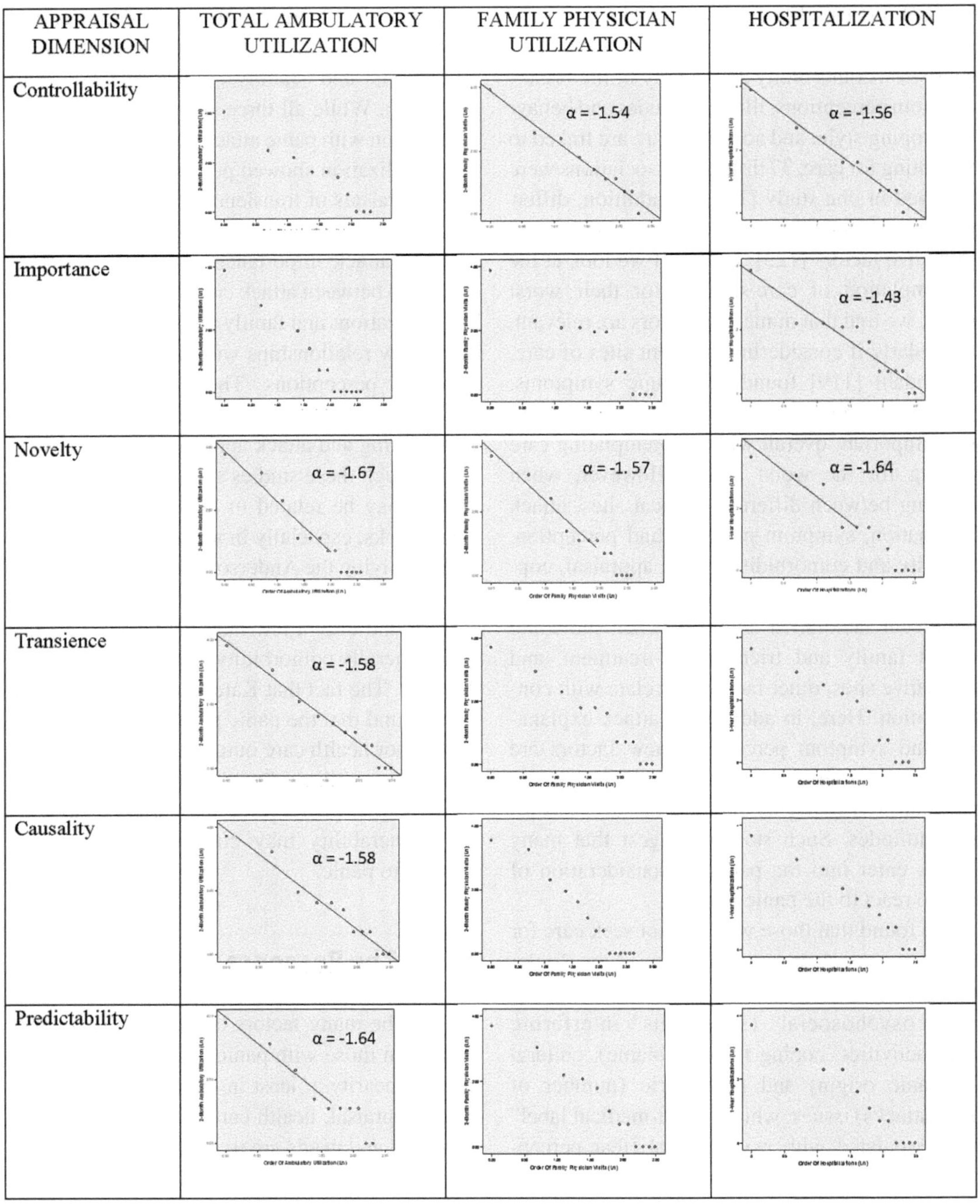

Fig. 23.8 Log–log plots of utilization and panic appraisals, originally published in [124]: Fig. 1. Reprinted with permission from Journal of Evaluation in Clinical Practice

seek care do so for their first panic attack, more commonly they seek care for their worst attack (78%) and/or when the attacks increase in frequency (67%) [128]. Yet, the number of possible sites from which they seek that care is limited, especially initially. Although eventually, 49% of people with panic attacks will seek care from a general medical setting (most commonly either the family physician's office [35%] or the emergency department [32%]), when first seeking

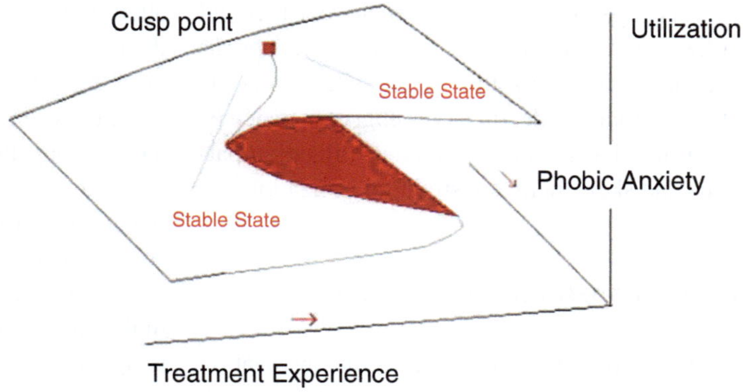

Fig. 23.9 Cusp relationship of health care utilization in panic disorder

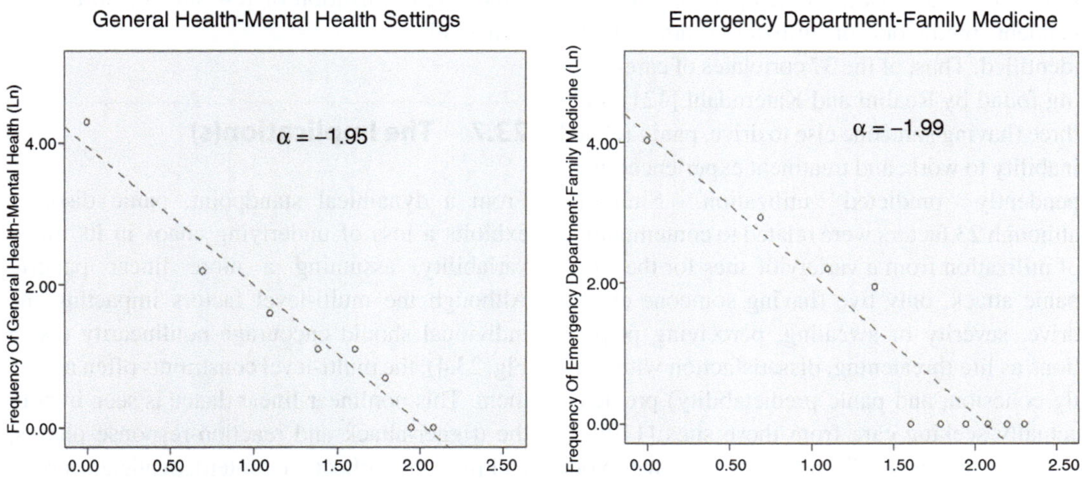

Fig. 23.10 Log-log plots of 2-month utilization of complementary sites among panic subjects

care, it is generally from a medical setting (85%) with the emergency department used more often than the family physician's office (43% versus 35%). Although the psychiatrist is typically the source used within the mental health sector, only 26% of people with panic ever seek care from a mental health facility [123].

One explanation for the constrained use of sites is that utilization may assume a cusp catastrophe pattern in which, despite predisposing factors, utilization tends to remain stable at the extremes of either high or low utilization, depending upon a bifurcation factor (see Fig. 23.9).

This, in fact, may be the case, at least when it comes to utilization of mental health settings and self treatment (and potentially emergency department usage), in which phobic anxiety serves as the bifurcation factor, splitting people into high and low utilizers if they suffer from agoraphobia; their severe disability causes some to repeatedly seek care while the extreme agoraphobic fear causes others to avoid care-seeking despite the disability [129]. Such high utilizers use the system specifically for their panic symptoms despite having poorer health insurance and just as many barriers to access as healthy controls [130].

Another factor that may constrain the diversity of utilization is co-dependency among sites. In fact, although distributions of utilization for particular sites do not show power law distributions, power laws have been found for co-utilization of general health and mental health settings as well as emergency department and family physician office settings (see Fig. 23.10).

This suggests that independencies exist around utilization of such pairs of sites, perhaps as patients weigh utilization of one site versus the other, perhaps reflecting ongoing unmet needs. Such interdependencies can alter utilization patterns and skew them, constraining utilization.

23.6.2 Constraining Utilization

Further evidence for constraint is the observation that, although many factors are associated with utilization or at least its contemplation, few independent predictors of utilization are actually identified. Thus, of the 37 correlates of care-seeking found by Realini and Katerndahl [121], only three (having someone else to drive, panic-related inability to work, and treatment experience) independently predicted utilization. Similarly, although 23 factors were related to contemplation of utilization from a variety of sites for the worst panic attack, only five (having someone else to drive, severity of sweating, perceiving palpitations as life threatening, dissatisfaction with family cohesion, and panic predictability) predicted actually seeking care from those sites [119]. In fact, feeling free to discuss symptoms with your physician accounted for much of the variance in care-seeking [131].

The fact is that many of the factors that constrained the personal phenomenon experienced by patients into a panic attack also constrain their care-seeking responses. Their personality disorders limit how they interpret their attack while their treatment experience with the health care system may limit their care-seeking options. Similarly, the lack of community and socioeconomic resources as well as husband-imposed restrictions on mobility may further limit those care-seeking behaviors. Finally, the comorbid disorders that may define severe panic disorder—MDD and agoraphobia—may further constrain their responses to panic. The lack of energy and interest may retard their reaction to the panic attack or inhibit their ability to respond; in fact, thoughts of death inherent in MDD may lead the person with self-perceived, life-threatening panic to intentionally avoid care in a passive suicidal attempt. Conversely, agoraphobic fear may limit their ability to seek care from the outside world even though they may recognize their need for it.

Thus, even though those experiencing panic attacks may have a range of interpretations of their experience and a plethora of factors that would encourage care-seeking from a variety of sources, patients respond in a limited number of ways with few predicting variables. Although multi-level factors may lead the patient to contemplate utilization of many sites, constraints and interdependencies produce restrictive patterns of either high utilization of few sites or no utilization at all.

23.7 The Implication(s)

From a dynamical standpoint, panic disorder exhibits a loss of underlying chaos in its mood variability, assuming a more linear pattern. Although the multi-level factors impacting the individual should encourage nonlinearity (as in Fig. 23.4), the multi-level constraints often negate them. This nonlinear-linear dance is seen in both the trigger-attack and reaction-response phases, and probably reflects the interdependencies relevant to these phases, confirmed by the power law distributions. Such patterns suggest certain management strategies in panic disorder.

23.7.1 Multiple Treatment Modalities but Unmet Needs

If indeed the system(s) responsible for the development and maintenance of panic disorder are complex as proposed, we would expect that the context in which interventions occur would be important to response. We would anticipate that many, even unrelated, treatments could be effective, but that the timing of such treatments would matter and that, despite effective treatment, unmet needs would still be expressed.

In fact, there is evidence of all of these in the management of panic disorder. First, the context of the intervention seems to affect the response; how the therapist interacts with the patient having panic attacks changes the form of the attacks [132]. Second, a variety of treatments have been shown to be effective in panic disorder. Not only are certain GABA-mediated drugs (such as high-potency benzodiazepines) effective, but medications that affect serotonergic and norepinephrinergic neurotransmitters (such as SSRI's, NSRI's and TCA's) treat panic disorder. In addition, medications that seem to work via other mechanisms (such as MAOI's and valproic acid) are also effective. In addition, several non-drug interventions have been shown to treat panic disorder. Thus, in addition to therapist-mediated cognitive behavioral therapy (CBT) and exposure therapy, everything from hypnosis [133] to inositol ingestion [134], from education to self-help programs [135, 136] works. In fact, not only does aerobic exercise prevent induction of panic attacks [137], but a program of diffuse lifestyle changes targeting caffeine, diet, exercise, nicotine and alcohol has also been shown to be effective [138]. In addition, not only does computer-based treatment work [139], but it is just as effective as face-to-face therapy [140] and is equally effective for general physicians and psychologists [141]. In fact, panic disorder has a particularly high placebo response rate in drug trials among anxiety disorders [142]. Third, time-sensitive intervention has been advocated once the patient has been appropriately prepared for therapy [143]; such time-sensitivity may explain the effectiveness that even a brief, 2-day CBT intervention can have [144]. Finally, despite the variety of effective management options available, those with panic disorder report more unmet needs (26%) than those with other mental disorder. When identifying such needs, providers report that similar numbers of patients (around 40%) need additional behavioral and/or drug treatments, while 11% need more counseling or self-help experience and 36% have other social needs [145].

Thus, treatment studies in panic disorder further support its complex nature. How does such realization lead to a different treatment approach, one based upon the dynamics rather than a static, linear framework?

23.7.2 Multi-level Intervention

Intervening in complex systems is a challenge and differs from intervening in complicated, linear systems. In addition, as described above, the dynamics and challenges differ depending upon the point at which intervention is to occur. If we assume that the underlying dynamics of chaos superimposed upon periodicity is optimal for mental health and that the hourly changes in anxiety–depression levels display power law distributions, then particular dynamics-focused interventions are suggested. Similarly, if we assume that the nonlinear-linear tension of the trigger-attack and reaction-response phases, often displaying power law distributions, is problematic due to the constraints limiting nonlinearity, then other dynamics-focused interventions may be appropriate. Finally, the recognition of the presence of a cusp catastrophe in the process of self-treatment and the use of mental health settings may again necessitate different approaches.

23.7.2.1 Treating the Underlying Dynamics

How might we intervene to re-establish the underlying chaotic hour-to-hour variability in mood that is characteristic of healthy individuals? First, monitoring the dynamics of mood variability may have therapeutic benefits. Not only may a change in dynamics signal a pending relapse, but particular dynamic patterns may suggest which targeted treatment is optimal for an individual patient. The use of technology such as personal digital assistants (PDAs), mood diaries or internet reporting may enable rapid identification of dynamic patterns and their sudden change [57]. In addition, counseling strategies may permit recognition of dynamic patterns via narrative analysis [146]. This may explain why behavioral methods such as CBT appear to be more effective long-term in panic disorder than drug therapy [98].

Second, short-course, well-timed interventions may induce change in dynamics, taking advantage to the sensitivity to conditions inherent in complex systems. Such chronotherapeutic approaches would necessitate knowledge of real-time dynamics via monitoring. Previous studies have shown that pulsed intervention can change dynamic patterns [147–149]. Thus, if monitoring revealed a patient's mood dynamics were slipping from its healthy underlying chaos into periodicity, a well-timed pulse of a medication or therapist-delivered cognitive "boost" could forestall such a change and maintain emotional health.

Finally, if part of the vulnerability in panic disorder is due to a weakening of its anxiety–depression attractors as suggested by Katerndahl and Wang [60], then strategies that reinforce positive attractors may also forestall a detrimental change in dynamics [150]. Again, this would necessitate real-time knowledge of mood dynamics among patients. Hence, if we could monitor underlying mood dynamics, it could enable us to intervene in novel ways to alter detrimental changes or reinforce positive mood states, preventing unhealthy vulnerability among panic disorder patients.

23.7.2.2 Treating the Nonlinear-Linear Tension

If we accept that nonlinear dynamics reflective of the sensitivity suggested by the multiplicity of triggers and reactions in the trigger-attack and reaction-response pairs discussed above is desirable, then Fig. 23.4 suggests that the maintenance on nonlinearity could be reinforced through the provision of resources and interactions while minimizing constraints. Resources, in this case, typically reflect support and management options. This suggests that case managers may be a useful adjunct in panic disorder. In addition, social interactions should be encouraged. In fact, although group therapy may be viewed as threatening to those with panic disorder, statistically its outcomes were no different than those of individual CBT [151]. Such resources and interactions could reinforce the nonlinear diversity of potential reactions to panic attacks.

However, to truly minimize the tendency towards the linear dynamics of the panic attack itself or its help-seeking, we need to attack the constraints producing them. The fact that CBT is so effective may be its power to decrease catastrophic cognitions [65] or increase an internal sense of control [152]. Just as important to minimizing constraints in those with panic disorder is addressing the "demons" of the past (personality and violence) as well as the psychopathology of the present (depression and agoraphobia). Again, a case manager could mobility community and health system access if these are constraints.

In addition, power law distributions were observed repeatedly along the vulnerability-to-utilization pathway discussed above. Such power laws typically imply self-organization and critical dynamics. Although a variety of mechanisms could produce such distributions [153], such self-organized systems act as a single system with its agents behaving interdependently, often through the shared maintenance and release of tension within the system. The slopes observed in Figs. 23.6 and 23.8 support such criticality. While such power law dynamics may buffer the system against large changes, representing the most efficient state of the system [154], it may also thwart attempts at intervention, partly because all of the components are acting as a single system. Thus, the subjects used in the study represented in Fig. 23.8 may have made "joint" decisions concerning utilization; while joint decision-making within families is understandable, within communities, such decision-making may reflect indirect communication via community events, media presentations, or word-of-mouth [124]. Such critical systems display a form of random dynamics. This is especially true in the analysis of utilization pairings as seen in Fig. 23.10. While both pairings show power law distributions, their slopes are more compatible with random walks than with critical dynamics. However, both represent forms random dynamics.

Intervening in systems displaying randomness typically necessitates multi-factorial approaches. First, the system should be viewed as a meshwork of interdependent elements contributing to the balance of inputs and outputs, costs and benefits. Intervening via internal and external relationships may be productive. In addition, we must recognize the potential for an intervention to ripple through such an interdependent system with untoward effects [155]. That said, interventions designed to address patients displaying random dynamics should be individualized and multidimensional (from molecular and individual to family and community) as well as time- and space-sensitive. Seek to tailor interventions to specific moments (i.e. tipping points) and settings (i.e. reactive places) which may yield the maximal response [156].

Finally, it is also possible that the power law distributions observed do not reflect self-organized criticality, but rather a highly organized tolerance (HOT). Such HOT systems are engineered (sometimes by biological systems) to provide robust performance despite uncertain environments. The resulting power laws seen are due to tradeoffs among costs, tolerances, and outcomes. However, HOT systems differ from critical systems in that they involve more interconnectedness, are more resilient when exposed to designed-for uncertainties, and are more sensitive to unanticipated flaws and perturbations [157]. If indeed the power laws observed in panic disorder reflect a HOT system, then intervention may be more difficult. However, it may explain some of the catastrophic developments observed in panic disorder as a highly organized tolerant system of mood regulation is disrupted by extreme (unanticipated) perturbations such as childhood violence, leaving the system flawed and vulnerable to minor triggers.

23.7.2.3 Treating the Cusp Catastrophe of Utilization

The observation that "feeling free to discuss symptoms with your physician" was a primary predictor of utilization emphasizes the potential impact of the doctor–patient relationship on utilization for panic disorder. Yet, the existence of a cusp acting on utilization does not necessarily require intervention; if a patient is persistently seeking mental health care for panic attacks, that is a positive activity. On the other hand, if a patient is persistently *not* utilizing mental health services or is persistently using non-productive self-treatments (such as alcohol, illicit drugs, or PRN medications), then such persistent behavior needs to change. In such case, intervention could assume two possible approaches. First, we could focus on moving the individual centered on the dysfunctional persistent behavior (i.e. avoidance of mental health services) to its functional persistent counterpart (i.e. persistent use of mental health services). The delay convention states that individuals delay such change in a cusp catastrophe situation until it is no longer reasonable not to change. In this case, we would need to approach the patient locked in the dysfunctional behavior by convincing them that such behavior is not and never will produce the results they want, in light of their ongoing unmet needs [158]. Once converted over to the more positive extreme, the cusp would tend to maintain such behavior.

An alternative approach would be to target the bifurcation variable (phobic anxiety); if the bifurcation variable is reduced, then utilization becomes linearly related to predisposition (treatment experience) and predictable. Although improvement in phobic anxiety typically requires improvement in panic attacks, systematic desensitization could potentially lessen the phobic anxiety enough such that the cusp no longer dominates behavior. Once the phobic anxiety is reduced such that the relationship between treatment experience and utilization is linear, then intervention targeting treatment experience could be used to increase or decrease utilization. Although Katerndahl [129] did not clearly demonstrate that cusp catastrophe modeling could account for utilization of emergency departments, it is likely that one exists but dependent upon a bifurcation variable other than phobic anxiety. If true, then reducing excessive emergency department utilization by some panic disorder patients will require an intervention targeting this cusp phenomenon.

23.7.3 Possibility of Emergence

If indeed many of the phenomena within panic disorder display self-organized criticality as suggested by their observed power laws, then the system has the potential for emergent behavior. Emergence means that novel and unexpected structure, pattern or process can arises spontaneously in self-organizing systems. This implies that synergistic activity is possible when the system with its interdependencies is in reactive, energized situations or combinations. In fact, not only does panic disorder occasionally resolve spontaneously or in response to mere education or self-help methods, but there is evidence that combined interventions may be more effective than either treatment alone. This appears to be true of combination therapy involving psychotherapy and antidepressant medication, at least during acute phases [159], as well as the addition of placebo tablets to CBT [160]. In addition, despite the importance of deeply-ingrained, maladaptive cognitive schemas to the maintenance of panic disorder, sudden, dramatic results have been observed. For example, short-course cognitive therapy can reduce panic symptomatology without altering maladaptive schemas [161] and sudden gains have been reported in group CBT programs [162]. These findings suggest the emergence of adaptive mechanisms in reactive combinations of treatments or individuals, despite the presence of persistent, maladaptive inclinations.

Thus, a final approach to the management of panic disorder may be to foster the opportunity for emergence through the encouragement of reactive situations, such as combination therapy, group therapy, or the utilization of "reactive" moments during the course of illness to promote epiphany development. Such situations would encourage self-organization and emergence within individuals. Although such self-organization cannot be directed by the therapist and could potentially lead to worsening disability, such an approach could be potentially useful in recalcitrant or extreme cases in which standard therapies have not worked.

23.8 The Conclusion

Panic disorder with its characteristic "sting" shows clear signs of complexity in its multi-level interdependencies, underlying dynamics, and its balance between diversity and constraint. Such recognition carries a number of implications. First, it explains some of the unexpected, nonlinear findings in the literature. Second, it provides potential interventions for the underlying dynamics to possibly decrease vulnerability. Third, it suggests potential interventions to diminish constraints and enhance nonlinearity to possibly promote adaptability. Fourth, it provides guidance for those wishing to maintain or reduce the impact of the cusp present in the manifold describing utilization. Finally, it suggests that the system is ripe for emergence, which could be harnessed to treat recalcitrant cases of panic disorder.

References

1. Kessler RC, Wang PS. Descriptive epidemiology of commonly occurring mental disorders in the United States. Annu Rev Public Health. 2008;29:115–29.
2. Cramer V, Torgerson S, Kringlen E. Quality of life and anxiety disorders. J Nerv Ment Dis. 2005;193: 196–202.
3. Andersch S, Hetta J. 15-year follow-up study of patients with panic disorder. Eur Psychiatry. 2003;18:401–8.
4. Batelaan N, Smit F, de Graaf R, et al. Economic costs of full-blown and subthreshold panic disorder. J Affect Disord. 2007;104:127–36.
5. Katerndahl DA. Intrapatient agreement on phenomenology of panic attacks. Psychol Rep. 1996;79(1):219–24.
6. Gurguis GNM, Cameron OG, Ericson WA, et al. Daily distribution of panic attacks. Compr Psychiatry. 1988;29(1):1–3.
7. deBeurs E, Chambliss DL, Goldstein AJ. Measurement of panic disorder by a modified panic diary. Depress Anxiety. 1997;6:133–9.
8. Francis JL, Weisberg RB, Dyck IR, et al. Characteristics and course of panic disorder and panic disorder with agoraphobia in primary care patients. Prim Care Companion J Clin Psychiatry. 2007;9:173–9.
9. Batelaan NM, de Graaf R, Spijker J, et al. Course of panic attacks in individuals with panic disorder and subthreshold panic disorder. J Affect Disord. 2010;121:30–8.
10. Simon NM, Fischmann D. Implications of medical and psychiatric comorbidity with panic disorder. J Clin Psychiatry. 2005;66 Suppl 4:8–15.
11. Noyes Jr R, Reich J, Christiansen J, et al. Outcome of panic disorder. Arch Gen Psychiatry. 1990;47: 809–18.
12. Crowe RR, Goedken R, Samuelson S, et al. Genomewide survey of panic disorder. Am J Med Genet. 2001;105:105–9.
13. Knowles JA, Fyer AJ, Vieland VJ, et al. Results of a genome-wide genetic screen for panic disorder. Am J Med Genet. 1998;81:139–47.
14. Cameron OG, Lee MA, Curtis GC, et al. Endocrine and physiological changes during "spontaneous"

panic attacks. Psychoneuroendocrinology. 1987; 12(5):321–31.
15. Yeragani VK, Rao KAR. Nonlinear measures of QT interval series. Psychiatry Res. 2003;117:177–90.
16. Yeragani VK, Pohl R, Balon R, et al. Twenty-four-hour QT interval variability. Neuropsychobiology. 2002;46:1–6.
17. Yeragani VK, Pohl R, Jampala VC, et al. Increased QT variability in patients with panic disorder and depression. Psychiatry Res. 2000;93:225–35.
18. McCraty R, Atkinson M, Tomasino D, et al. Analysis of twenty-four hour heart rate variability in patients with panic disorder. Biol Psychiatry. 2001;56:131–50.
19. Yeragani VK, Pohl R, Balon R, et al. Heart rate in panic disorder (letter). Acta Psychiatr Scand. 1991;83:79.
20. Aikens DE, Craske MG. Sleep-based heart period variability in panic disorder with and without nocturnal panic attacks. J Anxiety Disord. 2008;22:453–63.
21. Yeragani VK, Balon R, Pohl R, et al. Decreased R-R variance in panic disorder patients. Acta Psychiatr Scand. 1990;81:554–9.
22. Melzig CA, Weike AI, Hamm AO, et al. Individual differences in fear-potentiated startle as a function of resting heart rate variability. Intl J Psychophysiol. 2009;71:109–17.
23. Garakani A, Martinez JM, Aaronson CJ, et al. Effect of medication and psychotherapy on heart rate variability in panic disorder. Depress Anxiety. 2009;26:251–8.
24. Srinivasan K, Ashok MV, Vaz M, et al. Decreased chaos of heart rate time series in children of patients with panic disorder. Depress Anxiety. 2002;15:159–67.
25. Slaap BR, Boshuisen ML, van Roon AM, et al. Heart rate variability as predictor of nonresponse to mirtazapine in panic disorder. Intl Clin Psychopharmocol. 2002;17:69–74.
26. Baker B, Khaykin Y, Devins G, et al. Correlates of therapeutic response in panic disorder presenting with palpitations. Can J Psychiatry. 2003;48:381–7.
27. Middleton HC. Cardiovascular dystonia in recovered panic patients. J Affect Dis. 1990;19:229–36.
28. Craske MG, Lang AJ, Tsao JCI, et al. Reactivity to interoceptive cues in nocturnal panic. J Behav Ther Exp Psychiatry. 2001;32:173–90.
29. Katerndahl DA. Panic and prolapse. Meta-analysis. J Nerv Ment Dis. 1993;181(9):539–44.
30. Katerndahl DA, Vande Creek L. Hyperthyroidism and panic attacks. Psychosomatics. 1983;24(5):491–6.
31. Weilburg JB, Schachter S, Worth J, et al. EEG abnormalities in patients with atypical panic attacks. J Clin Psychiatry. 1995;56:358–62.
32. Edlund MJ, McNamara E, Millman RP. Sleep apne and panic attacks. Compr Psychiatry. 1991;32(2):130–2.
33. Hayward C, Wilson KA, Lagle K, et al. Parent-reported predictors of adolescent panic attacks. J Am Acad Child Adolesc Psychiatry. 2004;43(5):613–20.
34. Albert U, Maina G, Bergesio C, et al. Nocturnal panic and recent life events. Depress Anxiety. 2005;22:52–8.
35. Venturello S, Barzega G, Maina G, et al. Premorbid conditions and precipitating events in early-onset panic disorder. Compr Psychiatry. 2002;43(1):28–36.
36. Dumas CA, Katerndahl DA, Burge SK. Familial patterns in patients with infrequent panic attacks. Arch Fam Med. 1995;4(10):863–7.
37. Katerndahl DA, Realini JP. Family characteristics of subjects with panic attacks. Fam Med. 1997;29(8):563–7.
38. Fone D, Dunstan F, Lloyd K, et al. Does social cohesion modify the association between area income deprivation and mental health? Int J Epidemiol. 2007;36:338–45.
39. Katerndahl DA. Community effects on mental health outcomes in subjects with and without panic attacks: results from a population-based study in San Antonio. Texas Prim Care Companion J Clin Psychiatry. 2007;9:429–32.
40. Katerndahl DA. Community correlates of outcomes in subjects with panic attacks. Depress Anxiety. 2001;13(4):194–7.
41. Goodwin RD, Hamilton SP. Early-onset fearful panic attack as a predictor of severe psychopathology. Psychiatry Res. 2002;109:71–9.
42. Katerndahl DA, Larme AC, Palmer RF, et al. Reflections on DSM classification and its utility in primary care: case studies in "Mental Disorders". Prim Care Companion J Clin Psychiatry. 2005;7(3):91–9.
43. Ampollini P, Machesi C, Signifredi R, et al. Temperament and personality features in patients with major depression, panic disorder and mixed conditions. J Affect Disord. 1999;52:203–7.
44. Goodwin RD, Fergusson DM, Horwood LJ. Childhood abuse and familial violence and the risk of panic attacks and panic disorder in young adulthood. Psychol Med. 2005;35:881–90.
45. Biederman J, Petty CR, Hirshfeld-Becker DR, et al. Developmental trajectories of anxiety disorders in offspring at high risk for panic disorder and major depression. Psychiatry Res. 2007;153:245–52.
46. Plehn K, Peterson RA. Anxiety sensitivity as a predictor of the development of panic symptoms, panic attacks, and panic disorder. Anxiety Disord. 2002;16:455–74.
47. Kutz A, Marshall E, Bernstein A, et al. Evaluating emotional sensitivity and tolerance factors in the prediction of panic-relevant responding to a biological challenge. J Anxiety Disord. 2010;24(1):16–22.
48. Schmidt NB, Bates MJ. Evaluation of a pathoplastic relationship between anxiety sensitivity and panic disorder. Anxiety Stress Coping. 2003;16(1):17–30.
49. Bandelow B, Spath C, Tichauer GA, et al. Early traumatic life events, parental attitudes, family history, and birth risk factors in patients with panic disorder. Compr Psychiatry. 2002;43(4):269–78.
50. Tschacher W. Dynamics of psychosocial crises. J Nerv Ment Dis. 1996;184(3):172–9.

51. Ovid'ko IA. Order and chaos in thoughts, feelings, and actions. Psychol Rep. 1999;84(1):87–106.
52. Goldberger AL. Fractal variability versus pathologic periodicity. Perspect Biol Med. 1997;40:543.
53. Han DH, Renshaw PF, Dager SR, et al. Altered cingulated white matter connectivity in panic disorder patients. J Psychiatr Res. 2008;42:399–407.
54. Heath RA. Complexity and mental health. In: Holt TA, editor. Complexity for clinicians. Abingdon: Radcliffe; 2004.
55. Rao RKA, Yeragani VK. Decreased chaos and increased nonlinearity of heart rate time series in patients with panic disorder. Auton Neurosci. 2001;88:99–108.
56. Bob P, Kukleta M, Riecansky I, et al. Chaotic EEG patterns during recall of stressful memory related to panic attack. Physiol Res. 2006;55 suppl 1:S113–9.
57. Katerndahl DA, Ferrer RL, Best RG, et al. Dynamic patterns in mood among newly-diagnosed patients with major depressive episode, panic disorder, and normal controls. Prim Care Companion J Clin Psychiatry. 2007;9(3):183–7.
58. Combs A, Winkler M, Daley C. Chaotic systems analysis of rhythms in feeling states. Psychol Rec. 1994;44:359–68.
59. Gustafson JP, Meyer M. Non-linear model of dynamics. Psychodyn Pract. 2004;10(4):doi: 10.1080/14753630412331313703.
60. Katerndahl DA, Wang C. Dynamic covariation of symptoms of anxiety and depression among newly-diagnosed patients with major depressive episode, panic disorder, and controls. Nonlinear Dynamics Psychol Life Sci. 2007;11(3):349–65.
61. Saloma C, Perez GJ, Tapang G, et al. Self-organized queuing and scale-free behavior in real escape panic. Proc Natl Acad Sci USA. 2003;100(21):11947–52.
62. Katerndahl DA. Power laws in covariability of anxiety and depression among newly diagnosed patients with major depressive episode, panic disorder and controls. J Eval Clin Pract. 2009;15:565–70.
63. Schmidt NB, Lerew DR, Joiner Jr TE. Prospective evaluation of the etiology of anxiety sensitivity. Behav Res Ther. 2000;38:1083–95.
64. Acs N, Banhidy F, Horvath-Puho E, et al. Maternal panic disorder and congenital abnormalities: a population-based case-control study. Birth Defects Res A Clin Mol Teratol. 2006;76:253–61.
65. Bakker A, Spinhoven P, Van Balkom AJLM, et al. Relevance of assessment of cognitions during panic attacks in the treatment of panic disorder. Psychother Psychosom. 2002;271:158–61.
66. Grillon C, Lissek S, Rabin S, et al. Increased anxiety during anticipation of unpredictable but not predictable aversive stimuli as a psychophysiologic marker of panic disorder. Am J Psychiatry. 2008;165:898–904.
67. Katerndahl DA. Symptom severity and perceptions in subjects with panic attacks. Arch Fam Med. 2000;9(10):1028–35.
68. Kessler H, Roth J, Von Wietersheim J, et al. Emotion recognition patterns in patients with panic disorder. Depress Anxiety. 2007;24:223–6.
69. Neidhardt E, Florin I. Memory bias for panic-related material in patients with panic disorder. Psychopathology. 1999;32:260–6.
70. Lim SL, Kim JH. Cognitive processing of emotional information in depression, panic, and somatoform disorder. J Abnorm Psychol. 2005;114(1):50–61.
71. Sheehan D, Carr DB, Fishman SM, et al. Lactate infusion in anxiety research. J Clin Psychiatry. 1985;46:158–65.
72. Frohlich ED, Tarazi RC, Duston HP. Hyperdynamic beta adrenergic circulatory state. Arch Intern Med. 1969;123:1–7.
73. Charney DS, Woods SW, Goodman WK, et al. Serotonin function in anxiety. Psychopharmacologia. 1987;92:14–24.
74. Gorman JM, Askanazi J, Liebowitz MR, et al. Response to hyperventilation in a group of patients with panic disorder. Am J Psychiatry. 1984;141:857–61.
75. Perna G, Barbini B, Cocchi S, et al. 35% CO_2 challenge in panic and mood disorders. J Affect Disord. 1995;33:189–94.
76. Tinitigan R, Ali F, Tinitigan M, et al. Prevalence, comorbidity, and outcomes of chemical Intolerance in primary Care Settings. Fam Med. 2011;43(Suppl 1). http://www.stfm.org/fmsup/napcrg/fmconference-supplement.cfm?confid=135.
77. Katerndahl DA, Realini JP. Relationship between substance abuse and panic attacks. Addict Behav. 1999;24(5):731–6.
78. Maron E, Hettema JM, Shlik J. Advances in molecular genetics of panic disorder. Mol Psychiatry. 2010;15(7):681–701.
79. Erhardt A, Lucae S, Kern N, et al. Association of polymorphisms in the angiotensin-converting enzyme gene with syndromal panic attacks. Mol Psychiatry. 2008;13(3):242–3.
80. Unschuld PG, Ising M, Erhardt A, et al. Polymorphisms in the galanin gene are associated with symptom-severity in female patients suffering from panic disorder. J Affect Disord. 2008;105:177–84.
81. Okamura N, Hashimoto K, Iyo M, et al. Gender-specific association of a functional coding polymorphism in the Neuropeptide S receptor gene with panic disorder but not with schizophrenia or attention-deficit/hyperactivity disorder. Biol Psychiatry. 2007;31(7):1444–8.
82. Ishii T, Akiyoshi J, Hanada H, et al. Association between the obestatin and BDNF gene polymorphism and panic disorder. Psychiatr Genet. 2009;19(3):159.
83. Erhardt A, Lucae S, Unschuld PG, et al. Association of polymorphisms in P2RX7 and CaMKKb with anxiety disorders. J Affect Disord. 2007;101:159–68.
84. Maron E, Shlik J. Serotonin function in panic disorder. Neuropsychopharmacology. 2006;31(1):1–11.
85. Maron E, Nikopensius T, Koks S, et al. Association study of 90 candidate gene polymorphisms in panic disorder. Psychiatr Genet. 2005;15(1):17–24.

86. Hamilton SP, Slager SL, deLeon AB, et al. Evidence for genetic linkage between a polymorphism in the Adenosine 2A receptor and panic disorder. Neuropsychopharmacology. 2004;29(3):558–65.
87. Ho HP, Westberg L, Annerbrink K, et al. Association between a functional polymorphism in the progesterone receptor gene and panic disorder in women. Psychoneuroendocrinology. 2004;29(9):1138–41.
88. Roy-Byrne PP, Uhde TW. Exogenous factors in panic disorder. J Clin Psychiatry. 1988;49:56–61.
89. Van der Does AJW, Antony MM, Ehlers A, et al. Heartbeat perception in panic disorder. Behav Res Ther. 2000;38:47–62.
90. Craske MG, Poulton R, Tsao JCI, et al. Paths to panic disorder/agoraphobia. J Am Acad Child Adolesc Psychiatry. 2001;40(5):556–63.
91. Fava M, Anderson K, Rosenbaum JF. "Anger attacks": possible variants of panic and major depressive disorders. Am J Psychiatry. 1990;147:867–70.
92. Mavissakalian MR, Guo S. Early detection of relapse in panic disorder. Acta Psychiatr Scand. 2004;110:393–9.
93. Jacobs WJ, Nadel L. First panic attack. Can J Exp Psychol. 1999;53(1):92–107.
94. Casey LM, Oei TPS, Newcombe PA. Integrated cognitive model of panic disorder. Clin Psychol Rev. 2004;24:529–55.
95. Bouchard S, Gauthier J, Nouwen A, et al. Temporal relationship between dysfunctional beliefs, self-efficacy and panic apprehension in the treatment of panic disorder and agoraphobia. J Behav Ther Exp Psychiatry. 2007;38:275–92.
96. Wiedemann G, Pauli P, Dengler W. *A priori* expectancy bias in patients with panic disorder. Anxiety Disord. 2001;15:401–12.
97. Morrison AP, Wells A. Comparison of metacognitions in patients with hallucinations, delusions, panic disorder, and non-patient controls. Behav Res Ther. 2003;41:251–6.
98. Goldberg C. Cognitive processes in panic disorder. Psychol Rep. 2001;88:139–59.
99. Poloni C, Riquier F, Zimmermann G, et al. Perceptual defense in anxiety disorders. Percept Mot Skills. 2003;97:971–8.
100. Holt PE, Andrews G. Provocation of panic. Behav Res Ther. 1989;27(3):253–61.
101. Salkovskis PM, Clark DM, Hackmann A, et al. Experimental investigation of the role of safety-seeking behaviours in the maintenance of panic disorder with agoraphobia. Behav Res Ther. 1999;37:559–74.
102. Katerndahl DA. Illness attitudes and coping process in subjects with panic attacks. J Nerv Ment Dis. 1999;187(9):561–5.
103. Berksun OE. Panic disorder and memory. Eur Psychiatry. 1999;14(1):54–6.
104. Day SJ, Holmes EA, Hackmann A. Occurrence of imagery and its link with early memories in agoraphobia. Memory. 2004;12(4):416–27.
105. Ludewig S, Paulus MP, Ludewig K, et al. Decision-making strategies by panic disorder subjects are more sensitive to errors. J Affect Disord. 2003;76:183–9.
106. Tinitigan R, Katerndahl DA, Miller CS. Relationship of chemical intolerance and childhood abuse to the onset of panic attacks. Fam Med. 2010;42(Suppl 2). http://www.stfm.org/fmsup/napcrg/fmconference-supplement.cfm?confid=133.
107. American Psychiatric Association. Diagnostic and statistical manual of mental disorders (DSM-IV-TR). 4th ed. Washington, DC: American Psychiatric Association; 2000.
108. Katerndahl DA. The sequence of panic symptoms. J Fam Pract. 1988;26(1):49–52.
109. Hedley LM, Hoffart A, Dammen T, et al. Relationship between cognitions and panic attack intensity. Acta Psychiatr Scand. 2000;102:300–2.
110. Moynihan JE, Gevirtz RN. Respiratory and cognitive subtypes of panic. Behav Modif. 2001;25(4):555–83.
111. Katerndahl DA, Gabel LL, Monk JS. Comparative symptomatology of phobic and nonphobic panic attacks. Fam Pract Res J. 1986;6(2):106–13.
112. Norton GR, Norton PJ, Walker JR, et al. Comparison of people with and without nocturnal panic attacks. J Behav Ther Exp Psychiatry. 1999;30:37–44.
113. Craske MG, Lang AJ, Mystkowski JL, et al. Does nocturnal panic represent a more severe form of panic disorder? J Nerv Ment Dis. 2002;190:611–8.
114. Merritt-Davis O, Balon R. Nocturnal panic. Depress Anxiety. 2003;18:221–7.
115. Katerndahl DA. Panic attacks and panic disorder. J Fam Pract. 1996;43(3):275–82.
116. Katerndahl DA. Factors associated with persons with panic attacks seeking medical care. Fam Med. 1990;22(6):462–6.
117. Van den Hout MA, de Jong P, Zandbergen J, et al. Waning of panic sensations during prolonged hyperventilation. Behav Res Ther. 1990;28(5):445–8.
118. Kenardy J, Taylor CB. Expected versus unexpected panic attacks. J Anxiety Disord. 1999;13(4):435–45.
119. Katerndahl DA. Factors influencing care seeking for a self-defined worst panic attack. Psychiatr Serv. 2002;53(4):464–70.
120. Katerndahl DA. Predictors of the development of phobic avoidance. J Clin Psychiatry. 2000;61(8): 618–23.
121. Realini JP, Katerndahl DA. Factors affecting the threshold for seeking care: the Panic Attack Care-Seeking Threshold (PACT) Study. J Am Board Fam Pract. 1993;6(3):215–23.
122. Rogler LH, Cortes DE. Help-seeking pathways. Am J Psychiatry. 1993;150:554–61.
123. Katerndahl DA, Realini JP. Where do panic attack sufferers seek care? J Fam Pract. 1995;40(3):237–43.
124. Katerndahl DA. Power law relationships between health care utilization and symptom assessment among people with panic attacks. J Eval Clin Pract. 2010;16:421–6.

125. Goodwin R, Anderson RM. Use of the Behavioral Model of Health Care use to identify correlates of use of treatment for panic attacks in the community. Soc Psychiatry Psychiatr Epidemiol. 2002;37:212–9.
126. Katerndahl DA, Parchman ML. Understanding ambulatory care use by people with panic attacks: testing the behavioral model for vulnerable populations. J Nerv Ment Dis. 2002;190(8):554–7.
127. Kessler RC, Olfson M, Berglund PA. Patterns and predictors of treatment contact after first onset of psychiatric disorders. Am J Psychiatry. 1998;155:62–9.
128. Katerndahl D. Initial care seeking for panic attacks. Psychiatr Serv. 2003;54(8):1168.
129. Katerndahl DA. Explaining health care utilization for panic attacks using cusp catastrophe modeling. Nonlinear Dynamics Psychol Life Sci. 2008;12(4):409–24.
130. Katerndahl DA, Realini JP. Use of health care services by persons with panic symptoms. Psychiatr Serv. 1997;48(8):1027–32.
131. Katerndahl DA. Panic is panic: homogeneity of the panic attack experience. Fam Pract Res J. 1990;9(2):147–55.
132. Zane MD. Contextual approach to panic. In: Baker R, editor. Panic disorder: theory, research and therapy. Oxford: Wiley; 1989.
133. Iglesias A, Iglesias A. Awake-alert hypnosis in the treatment of panic disorder. Am J Clin Hypn. 2005;47:249–57.
134. Benjamin J, Levine J, Fux M, et al. Double-blind, placebo-controlled, crossover trial of inositol treatment for panic disorder. Am J Psychiatry. 1995;152:1084–6.
135. Baillie AJ, Rapee RM. Predicting who benefits from psychoeducation and self help for panic attacks. Behav Res Ther. 2004;42:513–27.
136. Barlow JH, Ellard DR, Hainsworth JM, et al. Review of self-management interventions for panic disorders, phobias and obsessive-compulsive disorders. Acta Psychiatr Scand. 2005;111:272–85.
137. Strohle A, Feller C, Onken M, et al. Acute antipanic activity of aerobic exercise. Am J Psychiatry. 2005;162:1376–8.
138. Lambert RA, Harvey I, Poland F. Pragmatic, unblended randomized controlled trial comparing an occupational therapy-led lifestyle approach and routine GP care for panic disorder treatment in primary care. J Affect Disord. 2007;99:63–71.
139. Kiropoulos LA, Klein B, Austin DW, et al. Is internet-based CBT for panic disorder and agoraphobia as effective as face-to-face CBT? J Anxiety Disord. 2008;22:1273–84.
140. Marks IM, Kenwright M, McDonough M, et al. Saving clinicians' time by delegating routine aspects of therapy to a computer. Psychol Med. 2004;34:9–18.
141. Shandley K, Austin DW, Klein B, et al. Therapist-assisted, internet-based treatment for panic disorder. J Med Internet Res. 2008;10(2):e14.
142. Huppert JD, Schultz LT, Foa EB, et al. Differential response to placebo among patients with social phobia, panic disorder, and obsessive-compulsive disorder. Am J Psychiatry. 2004;161:1485–7.
143. Budman SH, Gurman AS. Time-sensitive model of brief therapy. In: Budman SH, Hoyt MF, Friedman S, editors. The first session in brief therapy. New York: Guilford; 1992.
144. Deacon B. Two-day, intensive cognitive-behavioral therapy for panic disorder. Behav Modif. 2007;31:595–615.
145. Wittchen HU. Met and unmet need for interventions in community cases with anxiety disorders. In: Andrews G, Henderson S, editors. Unmet need in psychiatry. Cambridge: Cambridge University Press; 2000.
146. Greenhalgh T, Hurwitz B. Why study narrative? Br Med J. 1999;318:48–50.
147. Garfinkel A, Spano ML, Ditto WL, et al. Controlling cardiac chaos. Science. 1992;257:1230–5.
148. Christini D, Collins J, Linsay P. Experimental control of high dimensional chaos. Phys Rev E. 1996;54:4824–7.
149. Regalado A. Gentle scheme for unleashing chaos. Science. 1995;268:1848.
150. Miller WL, Crabtree BF, McDaniel R, et al. Understanding change in primary care practice using complexity theory. J Fam Pract. 1998;46:369–76.
151. Sharp DM, Power KG, Swanson V. Comparison of the efficacy and acceptability of group versus individual cognitive behavior therapy in the treatment of panic disorder and agoraphobia in primary care. Clin Psychol Psychother. 2004;11(2):73–82.
152. Bakker A, Spinhoven P, van der Does AJW, et al. Locus of control orientation in panic disorder and the differential effects of treatment. Psychother Psychosom. 2002;71:85–9.
153. Morel B, Ramanujam R. Through the looking glass. Organ Sci. 1999;10(3):278–93.
154. Smethurst DP, Williams HC. Are hospital waiting lists self-regulating? Nature. 2001;410:652–3.
155. Viney LL, Clarke AM, Benjamin YN. General systems approach to the patient, hospital staff, family, and community. Behav Sci. 1986;31:239–53.
156. Ahn AC, Tewari M, Poon CS, et al. Clinical applications of a systems approach. PLoS Med. 2006;3(7):e209.
157. Carlson JM, Doyle J. Highly optimized tolerance. Phys Rev E. 1999;60:1412–27.
158. Zeeman EC. Mathematical model for conflicting judgments caused by stress, applied to possible misestimations of speed caused by alcohol. Br J Math Stat Psychol. 1976;29:19–31.

159. Furukawa TA, Watanabe N, Churchill R. Psychotherapy plus antidepressant for panic disorder with or without agoraphobia. Br J Psychiatry. 2006;188(4):305–12.
160. Furukawa TA, Watanabe N, Omori IM, et al. Can pill placebo augment cognitive-behavior therapy for panic disorder? BMC Psychiatry. 2007;7:73.
161. Rusinek S, Graziani P, Servant D, et al. Cognitive therapy and cognitive schemas. Eur Rev Appl Psychol. 2004;54(3):173–7.
162. Clerkin EM, Teachman BA, Smith-Janik SB. Sudden gains in group cognitive-behavioral therapy for panic disorder. Behav Res Ther. 2008;46: 1244–50.

Nonlinear Relations of Cardiovascular Risk Factors to Neuropsychological Function and Dementia

24

Carrington R. Wendell, Leslie I. Katzel, and Shari R. Waldstein

Cardiovascular diseases are the leading cause of morbidity and mortality in the United States and most westernized nations [1]. Cardiovascular risk factors and diseases confer substantial increase in risk for ischemic and hemorrhagic stroke [2]. Yet, outside the context of clinical stroke, the brain is an under-recognized target organ of a spectrum of cardiovascular diseases. Although it has long been known that cardiovascular risk factors and diseases contribute to the development of vascular (previously known as multi-infarct) dementia, we now know that similar risk is conferred for Alzheimer's disease (AD) [3]. Importantly, long before clinical manifestations of stroke or dementia are apparent, cardiovascular risk factors are also known to negatively impact the brain and neurocognitive function. Evidence suggests that cardiovascular risk factors elevate risk of concurrent cognitive dysfunction, as well as accelerated cognitive aging.

Traditionally, cardiovascular risk factor associations with neuropsychological outcomes have been considered linearly. That is, as the severity or level of a certain risk factor rises, so does risk of cognitive impairment. For example, as average daily systolic blood pressure rises, risk of damage to end organs (e.g., brain) and corresponding diseases (e.g., dementia) also increases. Though these associations are valid, emerging evidence suggests that such characterizations of risk factor relations may only tell part of the story. More specifically, risk factor values that lie outside the normal range, whether too high *or* too low, may confer elevated risk for poor cognitive function and brain disease. Nonlinearity, in the form of various curvilinear relations (e.g., U-shaped, J-shaped, inverse U- or J-shaped), may therefore help explain current inconsistencies and areas of confusion in the literature. For instance, a significant nonlinear relation between a particular cardiovascular risk factor and neurocognitive outcome could easily be overlooked among nonsignificant analyses containing only linear terms.

In this chapter, we first briefly review classification of cardiovascular risk factors and diseases, followed by a corresponding section on

C.R. Wendell, Ph.D.
Department of Psychiatry, Johns Hopkins University School of Medicine, 600 N Wolfe St, Baltimore, MD 21287, USA
e-mail: cwendell@jhmi.edu

L.I. Katzel, M.D., Ph.D.
Division of Gerontology & Geriatric Medicine, Department of Medicine, University of Maryland School of Medicine, Baltimore, MD, USA

Geriatric Research Education & Clinical Center, Baltimore VA Medical Center, Room 4B-189, Baltimore, MD 20201, USA
e-mail: lkatzel@grecc.umaryland.edu

S.R. Waldstein, Ph.D. (✉)
Department of Psychology, University of Maryland, Baltimore County, 1000 Hilltop Circle, Baltimore, MD, 21250, USA

Division of Gerontology & Geriatric Medicine, Department of Medicine, University of Maryland School, of Medicine, Baltimore, MD, USA

Geriatric Research Education & Clinical Center, Baltimore VA Medical Center, Room 4B-189, Baltimore, MD 20201, USA
e-mail: waldstei@umbc.edu

J.P. Sturmberg and C.M. Martin (eds.), *Handbook of Systems and Complexity in Health*,
DOI 10.1007/978-1-4614-4998-0_24,

neuropsychological function and diseases. Next, we discuss the complex associations between cardiovascular risk factors and neuropsychological function. We have selected three risk factors—blood pressure, lipid levels, and body composition—to illustrate how nonlinearity and other complexities exist in their associations with neuropsychological outcomes. For each risk factor, we provide general background information regarding definitions and epidemiology, present and summarize evidence for nonlinear associations, and discuss relevant mechanisms. Lastly, we provide suggestions for practitioners and researchers regarding consideration of nonlinear cardiovascular risk in their future work.

24.1 Cardiovascular Disease Classification

To help ensure standardization of disease reporting, epidemiologists and clinicians in the United States and elsewhere typically classify cardiovascular and other diseases based on the International Classification of Diseases (ICD) codes published by the World Health Organization [4]. Cardiovascular disease is part of the broadly defined diseases of the circulatory system (ICD-10 I00–I99, Q20–Q28). Many studies use the definitions initially adopted by the Framingham Heart Study which define cardiovascular disease as comprised of: (1) coronary heart disease (CHD) (coronary death, myocardial infarction (MI), coronary insufficiency, and angina); (2) cerebrovascular disease (ischemic stroke, hemorrhagic stroke, and transient ischemic attack); (3) peripheral artery disease; and (4) heart failure [5, 6]. For the purposes of this chapter, a cardiovascular risk factor represents any biomedical or behavioral characteristic that enhances risk for any or all of these cardiovascular diseases. Myriad cardiovascular risk factors have been identified, including, but not limited to, hypertension, dyslipidemia, obesity, insulin resistance, glucose intolerance, oxidative stress, renal function, homocysteine, smoking, excessive alcohol consumption, poor diet, and physical inactivity. These risk factors operate through both overlapping and disparate mechanisms, including genetically and non-genetically driven physiological pathways, as well as behavioral and lifestyle patterns and choices.

24.2 Neuropsychological Function and Disease Classification

Diagnostically, cognitive disorders are usually classified according to the standards of the current edition of the *Diagnostic and Statistical Manual of Mental Disorders*, published by the American Psychiatric Association (DSM-IV-TR) [7]. Additional guidelines are also widely accepted for fine-tuned diagnoses [8–11]. The most common cognitive disorders include mild cognitive impairment (MCI) and dementia, with Alzheimer's disease accounting for the bulk of dementia prevalence. MCI is characterized as the presence of a significant deficit in one or more cognitive domains, with minimal to no interference with activities of daily living. Vascular cognitive impairment (VCI) is comparable to MCI, except that a primary vascular etiology is suspected, and it is associated with a different pattern of affected cognitive domains. More severe cognitive impairments, typically causative of pronounced functional limitations (e.g., caregiver dependence), are often indicative of a dementia diagnosis.

To assess cognitive status, neuropsychological assessments involve characterization of an individual's relative pattern of cognitive strengths and weaknesses. A cognitive profile is comprised of a variety of major domains of performance, including attention, learning and memory, executive functions, visuospatial and visuoconstructional skills, psychomotor abilities, perceptual skills, and language [12]. These domains are assessed most thoroughly by a battery of neuropsychological tests, although under select circumstances such as initial dementia assessment, gross estimates can be derived from screening tests such as mental status exams. Taken together, results from a neuropsychological assessment help distinguish between patterns of normal and abnormal cognitive function and provide important information about complex brain–behavior associations. Herein, we describe available evidence relevant to multiple neuropsychological outcomes, including concurrent cognitive function, longitudinal trajectories of cognitive function, and dementia.

24.3 Nonlinear Relations of Cardiovascular Risk Factors to Neuropsychological Function

Of the numerous cardiovascular risk factors described earlier, we selected blood pressure, cholesterol, and body composition for several reasons. First, these risk factors are particularly predictive of neuropsychological outcomes, as will become readily apparent in the ensuing sections. In addition, these three risk factors are highly common and relatively representative of the larger group of traditional cardiovascular risk factors. Lastly, because of the novelty of studying nonlinearity in this literature, there is comparatively less relevant research to report regarding other risk factors at this time.

24.3.1 Blood Pressure

24.3.1.1 Definitions and Epidemiology

Hypertension, an elevation in blood pressure that places individuals at increased risk for end-organ damage in a number of vascular beds, including the heart, brain, kidneys, retina, and large conduit arteries, is the most common risk factor for cardiovascular disease [13]. Diagnostically, hypertension is defined as a systolic blood pressure ≥140 mmHg, diastolic blood pressure ≥90 mmHg, taking antihypertensive medicine, or having been told at least twice by a physician or other health professional that one has hypertension [5]. Applying this definition, about one third of adults have hypertension. More than 90 % of those affected have primary or idiopathic hypertension. About 10 % have secondary hypertension where there are underlying diseases (such as renovascular disease) that cause hypertension. There is also growing awareness of the health importance of "prehypertension" defined as untreated systolic blood pressure of 120–139 mmHg or untreated diastolic blood pressure of 80–89 mmHg (and not having been told on two occasions by a health professional that one has hypertension) [5]. It is estimated that 37.4 % of the US population >20 years of age has prehypertension [1]. Prehypertension markedly increases the risk for the development of overt hypertension and cardiovascular disease.

Hypertension has a major impact on morbidity and mortality. It is estimated that hypertension is associated with 5 years reduced overall life expectancy [14]. Yet, awareness of hypertension and adequate treatment and blood pressure control in known hypertensives remains inadequate. Data suggest that perhaps 40 % of all hypertensives do not meet their blood pressure goals with resistant or difficult-to-control systolic hypertension being more common in older patients [15]. This is of major clinical importance as patients with poorly controlled hypertension are more likely to develop end-organ damage (e.g., heart failure, stroke, MI, and renal failure) and have a substantially higher long-term cardiovascular disease risk than patients with well-controlled blood pressure.

Hypotension, or unusually low blood pressure, is generally defined as a systolic blood pressure <90 mmHg or diastolic blood pressure <60 mmHg. Hypotension is most frequently an acute consequence of bodily trauma or blood loss, but other conditions such as clinical cardiac disease (e.g., heart valve problems, heart failure), endocrine disease, and sustained dehydration can cause chronic hypotensive states. Below, we focus primarily on continuous low blood pressure levels instead of diagnosable hypotension, because chronic hypotension is quite unusual among healthy adults and older adults. Consequently, the majority of research examining low blood pressure levels and cognitive function has focused on low normotensives levels, rather than true hypotension.

24.3.1.2 High Blood Pressure: Linear Relations

Hypertension and continuous blood pressure levels have been studied quite extensively in relation to pre-stroke and pre-dementia cognitive performance. A multitude of case–control studies contrasting the performance of persons with diagnosed hypertension to that of normotensives (i.e., those with normal levels of blood pressure) reveal that hypertensives perform more poorly than normotensives on tests of executive function, learning and memory, attention, perceptuo-motor (or psychomotor) speed, motor function, and visuospatial abilities [16, 17]. These differences have been documented across the life span in age cohorts ranging from children [18] to the elderly [19, 20], although relatively fewer studies have focused on older adults per se. Cross-sectional studies of blood pressure levels in older adults similarly report that higher pressures are related to diminished cognitive performance [21–24].

It is unclear whether the magnitude or patterning of association between hypertension or blood pressure and neurocognition differs across the life span. Complicating the integration of findings from case–control and cross-sectional studies that have directly examined the interaction of age and hypertension (or blood pressure) is the contrast of different age cohorts defined variously. Although "young" is often defined as less than 40 years, "middle-aged" as 40–60 years, and "older" as 60+ years, investigators have used alternative age cut scores. Nonetheless, several earlier studies indicated that younger cohorts of persons with diagnosed hypertension showed greater neurocognitive difficulties than middle-aged hypertensives [25]. However, studies utilizing continuous blood pressure measures revealed somewhat different findings. In that regard, results of one recent study suggested an association of higher levels of blood pressure with reduced executive functions and speed of information processing in older adults but not young or middle-aged adults [26]. In contrast, another investigation found an association of higher blood pressure with lowered neurocognitive performance in middle-aged rather than older persons [27]. Still others suggested that the "young old" might be more affected by higher blood pressure than middle-aged or "old old" [21]. Some studies have found no age interactions [28]. In a recent review, Anson and Paran [29] suggested that mildly elevated blood pressure may be related to optimal cognitive function in older adults. Importantly, case–control and cross-sectional studies are prone to healthy survivor effects, and persons with substantial risk factor or disease comorbidity are often excluded. More generally, these studies may be influenced by various unmeasured age cohort effects.

Longitudinal investigations have revealed that higher levels of blood pressure are associated with persistently lowered levels of cognitive function, and/or cognitive decline among initially middle-aged to older adults [20, 30, 31], although findings are mixed. The chronicity of exposure to high levels of blood pressure may be a particularly important determinant of poor prospective cognitive outcomes [30]. However, onset and duration of hypertension are notoriously difficult to capture, particularly given the issue of limitations in awareness of hypertension. Further, higher blood pressure at midlife predicts poorer cognitive performance during older age [32].

With respect to longitudinal age interactions, data from the Baltimore Longitudinal Study of Aging indicated cognitive decline among those with higher blood pressure in the 80-year-old cohort, but not 60- or 70-year-old cohorts [33]. However, Elias and colleagues reported blood pressure-related cognitive decline in both their younger and older age cohorts in the Maine-Syracuse study [34]. No age interactions were noted in the Framingham data [34].

Regarding dementia, it is well known that hypertension contributes significantly to the pathogenesis of stroke and vascular dementia [35], and a growing literature links hypertension with AD [36, 37]. Cross-sectional studies of late-life blood pressure and AD are confounded by the characteristic decline in blood pressure levels in premorbid AD, and accordingly findings are mixed as to relations of high versus low blood pressure to AD [38]. Evidence is strongest for a relation between midlife blood pressure and future development of AD, presumably due to the cumulative impact of longstanding hypertension [39]. Perhaps surprisingly given the traditional focus on vascular dementia, several studies have suggested that midlife hypertension confers a similar degree of increased risk (approximately 3–4 times) for both vascular dementia and AD [35, 40, 41].

24.3.1.3 Low Blood Pressure: Linear Relations

Though the bulk of the literature has examined relations of hypertension and high blood pressure to cognitive outcomes, mounting evidence suggests examinations of low blood pressure are also of importance when studying blood pressure–brain associations, particularly among older adults. For example, among a sample of elderly Swedish men, Axelsson and colleagues [42] found significant associations between low blood pressure and performance decrements on a battery of neuropsychological tests, including measures of visuoconstruction, verbal reasoning, verbal memory, visual memory, psychomotor speed, and the Mini Mental State Examination (MMSE), a cognitive screening test. Reduced cognitive function was characterized as poor performance on at least one test in the battery, and poor performance was defined as a score at or below one standard deviation of the sample mean. Specifically, participants with average systolic blood pressures <130 mmHg (averaged over 24 h, daytime, or nighttime) were at higher risk of reduced cognitive function [ORs = 2.6 (1.1–6.5), 2.6 (1.1–6.6), 3.6 (1.1–11.3), respectively] than individuals with average systolic pressures at or above 130 mmHg. In another larger study of Norwegian elderly, low systolic and diastolic blood pressures were also significantly associated with poorer MMSE performance [43]. Importantly, both of these studies were cross-sectional, assessed elderly samples (>80 years old), and utilized very gross measures of cognitive function. As mentioned earlier, cross-sectional studies of late-life blood pressure and cognitive function can be confounded by prototypical declines in blood pressure in premorbid dementia. In fact, older adults with systolic blood pressure ≤140 mmHg or diastolic blood pressure ≤75 mmHg are more likely to have a concurrent dementia diagnosis than those with higher pressure values [44]. However, continued debate exists regarding whether low blood pressure (a) is independently causative of cognitive decline and dementia, (b) exacerbates the neurodegenerative process, (c) is a byproduct of premorbid dementia, or (d) any combination of these roles. Maule and colleagues [45] have proposed that low blood pressure is more likely to play a causal role in neurodegenerative diseases among the medically frail, whereas low blood pressure may actually be neurologically protective among the healthy elderly.

Longitudinal research focusing on low blood pressure and cognitive outcomes is quite limited. Interestingly, Swan and colleagues found that, among older adults, those who maintained higher blood pressure levels over approximately 30 years prior displayed poorer performance on tests of verbal learning and memory, whereas those who showed a pattern of declining blood pressure had slower psychomotor speed [46]. Longitudinal trajectories of increasing versus decreasing blood pressure over adulthood may therefore confer differential risk for performance changes across different cognitive domains of function.

24.3.1.4 Direct Evidence of Nonlinearity

Separate studies reporting evidence for relations between either low or high blood pressure and neuropsychological outcomes only provide indirect support of nonlinear associations. Fortunately, recent work has directly examined nonlinearity of blood pressure in relation to cognitive function, particularly among the elderly. Molander and colleagues [47] identified a significant cross-sectional inverse U-shaped association between systolic blood pressure and cognitive function among 575 85+ year olds, such that both high and low systolic pressures were associated with poorer MMSE performance. Approximately 40 % of the sample had dementia, and stratified analyses were not performed, so it remains unclear to what extent dementia pathology drove the nonlinearity, particularly on the low blood pressure end of the curve. Further, the MMSE provides limited information regarding cognitive function, as it was designed only as a screening measure.

Additional cross-sectional evidence supports nonlinearity of blood pressure–cognition associations for neuropsychological tests that target specific domains of function. Among 495 Israeli 75–85-year-olds, significant inverse J-curve relations were identified between blood pressure and performance on measures of verbal and visual memory [48]. Memory performance improved with increasing blood pressure levels only until a certain point, beyond which performance plateaued or declined. Comparatively, performance on the MMSE and a concentration measure were related to blood pressure only in a linear fashion in this study. In another larger study (>5,000 participants) of 65+ year olds, modest inverse U-shaped associations were found for both systolic and diastolic blood pressure and performance on the MMSE and measures of immediate memory, delayed recall, and perceptual speed/executive function [49]. Individuals with a systolic pressure of 100 mmHg demonstrated cognitive scores 2–5 percentiles lower than those at the sample mean systolic pressure of 140 mmHg, across all tests. Systolic pressure of 180 mmHg was associated with a lower cognitive score as well, but only at <1 percentile difference from the mean. A similar pattern emerged for diastolic blood pressure, such that a diastolic pressure of 60 mmHg and 100 mmHg were associated with lower cognitive scores by 2–3 percentiles and 1–2 percentiles, respectively (compared to the sample mean of 77 mmHg). Though these performance differences are relatively small, the magnitudes are quite meaningful on a population level. Further, a performance decrement of 2–5 percentiles can be clinically significant and noticeable on an individual level, particularly at extreme ends of the curve.

Longitudinal studies have also examined nonlinear blood pressure directly. At least two studies have identified U-shaped associations between blood pressure and number of errors on the Short Portable Mental Status Questionnaire (SPMSQ), another cognitive screening test [50, 51]. Bohannon et al. [50] showed this nonlinear relation between systolic blood pressure and 3-year change in error scores among 4,000+ community-dwelling individuals aged 65–105 at baseline. That is, individuals with low and high systolic blood pressure committed significantly more errors on the SPMSQ from Time 1 to Time 2 than individuals with systolic pressure levels near the sample mean. Glynn and colleagues' [51] findings are quite similar, except extended over a 9-year follow-up period. In their analyses, baseline systolic blood pressures below 130 mmHg and above 160 mmHg were associated with 9 % and 7 % higher error rates at 9-year follow-up, respectively, when compared with the 130–160 mmHg referent group.

Our group has extended these longitudinal findings to include performance on a comprehensive neuropsychological battery over 11 years of follow-up [33]. Among 847 community-dwelling participants aged 39–96 enrolled in the Baltimore Longitudinal Study of Aging and free of cerebrovascular and neurologic diseases, significant linear effects of blood pressure on cognitive function were nearly always qualified by significant nonlinear effects. Cross-sectional analyses, in which cognitive scores were collapsed across all testing sessions, showed both high and low diastolic blood pressure to be associated with poorer performance on tests of (a) executive function and language (i.e., confrontation naming) among less-educated individuals, (b) perceptuo-motor speed and executive function among unmedicated (antihypertensive) participants, and (c) executive function among older participants. Longitudinal models demonstrated similar U- and J-shaped associations between systolic blood pressure and worsening performance over time on tests of visual memory and confrontation naming.

24.3.1.5 Mechanisms

We have proposed previously that the mechanisms underlying hypertension (or BP) and neurocognition associations may differ across the life span [16, 52]. Particularly in middle-aged to older adults, these include neurophysiological factors such as reduced regional or global cerebral blood flow or metabolism, disruption of the blood–brain barrier, endothelial dysfunction, or other aspects of cellular dysfunction, all of which have been associated with hypertension. Neuroanatomical findings in hypertension include increased cerebral white matter disease, silent brain infarction, and brain atrophy, in addition to macrovascular disease. We and others have suggested that early alterations in cerebral perfusion may set the stage for changes in brain morphology with increasing age. Low blood pressure may also operate via perfusional mechanisms. In that regard, particularly among older adults, lower levels of blood pressure or decreasing blood pressure may in part reflect myocardial dysfunction. de la Torre [53] has argued that cerebral hypoperfusion may also play a critically important role in triggering the development of Alzheimer-type brain pathology.

24.3.1.6 Summary

Evidence from both linear and nonlinear examinations of blood pressure and neuropsychological function clearly demonstrates support for pervasive nonlinearity among these associations. Both high and low blood pressure levels are related to decrements in concurrent and prospective cognitive performance on both screening tests and measures of specific neuropsychological domains. Currently the literature is too limited to draw strong conclusions regarding particularly affected domains. However, decrements across multiple domains including memory, executive function, perceptuo-motor speed, and language have been identified, lending support to the hypothesis of a neuroanatomically nonspecific etiology such as diffuse cerebral hypoperfusion. Though significant nonlinear blood pressure–dementia associations also exist, ongoing debate regarding directionality of effects (i.e., pathological blood pressure declines causing, exacerbating, or resulting from dementia) precludes conclusions regarding low blood pressure as a risk-enhancing physiological state vs. a secondary marker of the neurodegenerative process.

24.3.2 Lipid Levels

24.3.2.1 Definitions and Epidemiology

Dyslipidemia encompasses a range of disorders of lipoprotein lipid metabolism that include both abnormally high and low lipoprotein concentrations, as well as abnormalities in the composition of these lipoprotein particles. Dyslipidemias are clinically important because of their role in the pathogenesis of cardiovascular disease. In clinical practice, a lipid or cholesterol panel commonly measures total cholesterol, low-density lipoprotein (LDL) cholesterol, and high-density lipoprotein (HDL) cholesterol. Higher levels of LDL cholesterol promote atherosclerosis, whereas higher levels of HDL cholesterol are in part protective against atherosclerosis.

The consensus treatment guidelines [54] for the management of dyslipidemia are continually being reevaluated, and the consensus guidelines have made the target lipoprotein concentrations more stringent for individuals with cardiovascular disease. Epidemiological and clinical trial data suggest that the optimal concentration for total cholesterol is <200 mg/dl (<5.17 mmol/l), though the relative contributions of LDL and HDL cholesterol are critical to consider. LDL cholesterol should ideally be <100 mg/dl (<2.58 mmol/l), or even as low as 70 mg/dl (<1.81 mmol/l) for some high-risk patients with known CHD. It is recommended that HDL cholesterol be ≥40 mg/dl (≥1.00 mmol/l), with optimal concentrations as high as >60 mg/dl (>1.55 mmol/l).

In the National Health and Nutrition Examination Survey III (NHANES III), an epidemiologic investigation of US adults conducted in the 1990s, 59 % of the population demonstrated undesirable cholesterol levels in some form [55]. Regarding total cholesterol, 29 % of participants had borderline high levels (200–239 mg/dl, 5.17–6.18 mmol/l)) and 18 % had demonstrably high levels (≥240 mg/dl, ≥6.20 mmol/l). More recent NHANES data (1999–2004) indicate that prevalence of high LDL cholesterol among US adults is 25 % [56]. Treatment of dyslipidemia most often involves lifestyle modification approaches (e.g., increased physical activity, dietary changes), with statin[1] medication use for refractory lipid levels.

[1] The safety of statin medications is currently under review. The FDA (http://www.fda.gov/Drugs/DrugSafety/ucm256581.htm#Simvastatin_;06-08-2011) has warned against the use of high dose simvastatin as the risk of muscle injury increases exponentially with dose increase, though the cardiovascular outcomes do not differ for low or high dose use (Lancet. 2010; 376(9753):1658–69). As statins involve the CYP3A4-pathway the risk issues potentially relate to all statins.

24.3.2.2 High Total and LDL Cholesterol: Linear Relations

Numerous investigations have studied linear relations of total cholesterol and LDL cholesterol to cognitive function. Of these, several have identified cross-sectional associations between high total and LDL cholesterol and poor performance on cognitive screening measures among the elderly [57, 58]. In their population-based cohort of 1,711 older adults, Carlsson and colleagues [57] found the highest quartile of LDL to be associated with increased odds of MMSE-defined cognitive impairment (odds ratio: 2.06, CI: 1.07–3.98). Further, in a sample of 30–80 year olds, serum levels of cholesterol precursor molecules (lathosterol and lanosterol) were negatively correlated with concurrent performance on verbal learning and memory measures [59].

Analogous to the blood pressure literature, longitudinal evidence for midlife lipid levels as risk factors for late-life cognitive function is particularly compelling. In a systematic review and meta-analysis of total cholesterol and cognitive decline, high midlife total cholesterol was consistently associated with late-life cognitive impairment across 18 studies [60]. In one of the most comprehensive studies to date examining cholesterol and longitudinal neuropsychological performance among non-demented individuals, Solomon and colleagues [61] reported an association between high midlife total cholesterol and poorer late-life performance on measures of memory and psychomotor speed. These analyses were conducted among 1,382 participants over a mean follow-up period of 21 years. Reynolds and colleagues [62] have also pointed out important moderating effects of age and sex in these relations. Specifically, they showed that lipid levels *prior* to the age of 65 (vs. levels after age 65) were more predictive of cognitive performance trajectories on a comprehensive neuropsychological battery over 16 years of follow-up. Further, women appeared to be more susceptible than men to the detrimental effects of high total or LDL cholesterol on longitudinal cognitive performance.

Relatively limited research has examined the relation between lipid levels and dementia [38]. This paucity is surprising, given the established associations of APOE genotype with both lipid metabolism and AD [63]. Studies have demonstrated both significant and nonsignificant relations of high levels of total and LDL cholesterol with increased risk of dementia [64–67]. Similar to hypertension, the evidence is strongest for an effect of midlife dyslipidemia on prospective dementia incidence [41, 60, 68]. Moreover, and also similar to the blood pressure literature, conflicting literature regarding concurrent lipid levels and dementia prevalence is confounded by characteristic declines in cholesterol levels in premorbid dementia [69].

24.3.2.3 Low Total and LDL Cholesterol: Linear Relations

Cross-sectional associations between low total or LDL cholesterol and cognitive function have been observed across a range of ages. Low total cholesterol and slowed psychomotor speed have been linked in university students [70]. Similarly, Muldoon and colleagues [71] found low total cholesterol to be associated with poorer performance on a test of visuoconstruction and psychomotor speed among healthy adults. Regarding LDL cholesterol, Henderson and colleagues [72] found parallel relations between low LDL and poorer verbal memory performance among women aged 52–63 years. Both low total and LDL cholesterol were associated with slowed visuomotor speed in male NHANES III participants aged 20–59 [73]. Cross-sectional findings in the non-demented elderly are rather limited, but at least one study has noted associations between low total and LDL cholesterol and poorer memory function among community-dwelling 85+ year olds [74]. These studies are difficult to carry out due to the prevalence of comorbid frailty and sarcopenia (reduced muscle mass) in the elderly, which are independently associated with both low cholesterol levels and cognitive impairment due to reduced nutritional parameters and other factors. Among persons with dementia, serum lipid profiles have been found to be both elevated [75] and lowered [76].

Longitudinal associations between low total or LDL cholesterol and poor cognitive function have also been identified. For instance, recent data from the Framingham Heart Study cohort (55–88-year-olds) showed relations of lower levels of total cholesterol to poorer performance on measures of abstract reasoning, attention/concentration, executive function, and word fluency over 16–18 years of follow-up [77]. Specifically, participants with total cholesterol levels <200 mg/dl (<5.17 mmol/l) (the recommended cutpoint for optimal total cholesterol) performed worse than individuals with total cholesterol levels exceeding 200 mg/dl (5.17 mmol/l). In the National Heart, Lung, and Blood Institute Twin Study, among middle-aged monozygotic twin pairs discordant for longitudinal decline on the Digit Symbol Substitution Test (a test of visual scanning and psychomotor speed), "decliner" twins had significantly lower total cholesterol values than their "non-decliner" co-twins [78].

Studies tracking trajectories of total cholesterol change over time also suggest an interesting pattern of findings. Individuals with decreasing total cholesterol levels over adulthood demonstrate greater cognitive difficulty later in life. In a cohort of >1,300 individuals followed over 21 years, accelerated trajectories of decreasing total cholesterol from midlife to late-life were associated with greater likelihood of impaired late-life MMSE performance, as well as poorer late-life episodic memory and category fluency [61, 79]. High midlife total cholesterol predicted likelihood of late-life cognitive impairment in the same cohort. A highly similar pattern of findings was identified in the Honolulu-Asia Aging Study, such that men who eventually developed dementia demonstrated more steeply declining slopes of total cholesterol starting at least 15 years prediagnosis [69].

24.3.2.4 HDL Cholesterol: Linear Relations

Increased attention has recently been paid to LDL cholesterol's major counterpart—HDL cholesterol—for its protective role in the development of cardiovascular disease. As its protective functions have become better understood, HDL cholesterol has also been studied in relation to cognitive function and dementia, though research in this area remains limited. In a cross-sectional study of centenarians, higher HDL levels were associated with higher MMSE scores [80]. However, higher late-life HDL levels have also been associated with more dementia neuropathology (i.e., cortical neuritic plaques and neurofibrillary tangles) at autopsy in a population-based sample (35 % with dementia) [81]. In the absence of additional research, the meaning of cross-sectional HDL cholesterol levels remains unclear, though these findings suggest that HDL likely signifies different risk profiles in "normal" versus exceptional agers (i.e., centenarians) due to a healthy survivor effect or other factors.

Longitudinal findings regarding HDL cholesterol are currently more consistent than the cross-sectional findings described above. In a small population-based sample of 101 60–70-year-old women, low baseline HDL cholesterol levels were significantly associated with poor memory at 12-year follow-up [82]. More specifically, with every one standard deviation decrease in HDL cholesterol, risk of poor memory increased by nearly 50 %. Singh-Manoux and colleagues [83] extended these findings in the Whitehall II cohort of >3,000 participants, aged 55–61 and followed for 5 years. Results showed that low HDL cholesterol at baseline, as well declining trajectories of HDL over time, were significantly associated with worse memory performance and declining memory performance over time, respectively. Similarly, a prospective study of baseline cognitively normal individuals demonstrated that high HDL cholesterol conferred decreased risk of incident Alzheimer's disease over 4,469 person-years of follow-up [84].

24.3.2.5 Direct Evidence of Nonlinearity

Very little, if any, research has directly examined nonlinearity in relations of total cholesterol, LDL cholesterol, or HDL cholesterol to cognitive function or dementia. In the previously discussed Elias et al. [77] study that demonstrated longitudinal associations between low total cholesterol and poor cognitive function, quadratic cholesterol terms failed to reach statistical significance. However, null linear findings abound [85], and the extent of conflicting findings described in previous sections suggests that nonlinear effects require additional investigation. Indeed, our group has recently identified nonlinear cholesterol effects across two samples. Among 190 healthy older adults, significant age-modified quadratic effects of total and LDL cholesterol were identified across multiple neuropsychological tests of attention, memory, and executive function [86]. In particular, 70+ year-old performed better at high and low levels of total cholesterol than at midrange total cholesterol (U-shaped), whereas the <70 group performed worse at high and low levels of total cholesterol than at midrange total cholesterol (inverted U-shaped). A comparable pattern of results arose for LDL cholesterol, though no significant HDL cholesterol effects were identified. In a separate sample of up to 1,500 participants from the Baltimore Longitudinal Study of Aging (aged 19–93), we demonstrated similar results over up to 19 years of follow-up [87]. In general, higher total cholesterol was associated with poorer midlife cognitive performance, but better late-life cognitive performance on measures of global mental status, attention, verbal learning, executive function, and language.

24.3.2.6 Mechanisms

Biological mechanisms linking high versus low cholesterol levels to cognitive function may differ. As reviewed by Muldoon and colleagues [88], cholesterol is an important constituent of neuronal and glial membranes and of myelin sheaths. It provides structural integrity, modulates membrane fluidity, and is important for synaptic function, neurotransmission, and the transport of nutrients to the brain. Brain lipids are indeed vulnerable to serum lipid levels. Cholesterol is also a precursor of steroid hormones (e.g., estrogen) involved in brain function. Therefore, it is possible that lower levels of cholesterol may negatively impact the brain's microstructure and function. Further, cholesterol may act as an antioxidant. Yet, higher levels of cholesterol play a major role in the development of atherosclerosis, which may lead to macrovascular disease and associated structural and functional changes in the brain prior to stroke. In addition, in vitro studies have suggested that increased cholesterol levels may lead to increased formation of beta-amyloid from amyloid-precursor proteins [89]. The relation of the APOE polymorphism to cholesterol is beyond the scope of this chapter, but this association could be pertinent to cognitive decline and dementia [90]. Additional mechanisms linking lipid levels with dementia include atherosclerosis, modulation of beta amyloid protein, and oxidative stress [38, 91]. Declines in cholesterol levels immediately prior to dementia diagnosis create grounds for debate regarding lipids as risk-enhancing, risk-indicating, or both.

24.3.2.7 Summary

Overall, there is currently evidence to support relations of both high and low total, LDL, and HDL cholesterol with both poor and improved cognitive function. The strongest evidence exists for (a) concurrent associations between low total and LDL cholesterol and decrements in cognitive function and (b) longitudinal associations between midlife cholesterol levels and risk of late-life cognitive dysfunction, impairment, or dementia. Different mechanisms likely explain these findings (i.e., reduced nutrient delivery versus atherosclerosis, respectively). However, the quantity of contradictory findings elsewhere in the literature implies the necessity of increased attention to nonlinear effects. Our group's recent investigations suggest the detectability of nonlinear J- and U-shaped effects in both cross-sectional and longitudinal samples. Future research is warranted regarding replication of these effects and their potential extension to dementia. At a minimum, the relations between total cholesterol and its components to cognitive outcomes are decidedly more complex than originally hypothesized.

24.3.3 Body Composition

24.3.3.1 Definitions and Epidemiology

Overweight is defined as a body mass index (BMI) of 25–29.9 kg/m^2 and obesity as a BMI of ≥30 kg/m^2. The degree of obesity is often further broken down into subcategories; a BMI of

30–34.9 is classified as class 1 obesity; 35–39.9 kg/m^2 is classified as class 2 obesity, and >40 as morbid obesity. A BMI less than 18.5 is underweight, and a BMI of 18.5–24.9 is considered normal. BMI is an easily measurable index of overweight and obesity. Other adiposity measures such as waist circumference, waist to hip ratio assessment of total body fat by DXA scan, and measurement of intra-abdominal and subcutaneous fat by computed tomography (CT) scan may correlate more strongly with metabolic abnormalities that mediate the association between obesity and CHD. These findings lead many to advocate the inclusion of measures of body fat distribution such as waist circumference in conjunction with measurement of BMI. An increased waist circumference (>102 cm in men and >88 cm in women) is used as a measure of central obesity and is included in the definition of the metabolic syndrome [92]. Some also advocate use of the waist–hip ratio of >0.95 and >0.88 for men and women respectively as index of abdominal obesity [92, 93].

Approximately 97 million US adults are overweight or obese [94]. Obesity is associated with significantly increased risk of morbidity and mortality, particularly in younger populations. Further, recent Framingham data demonstrate that greater BMI is predictive of first CHD event (angina, MI, or cardiac death), and first cerebrovascular event (stroke, transient ischemic attack, and stroke-related death) [95]. There is strong evidence that weight loss in overweight and obese individuals reduces risk factors for diabetes and cardiovascular disease.

24.3.3.2 Overweight and Obesity: Linear Relations

Results of a rapidly growing number of case–control and cross-sectional investigations have shown relations of obesity (and sometimes overweight) to lower levels of cognitive performance in non-demented, stroke-free cohorts ranging from children to older adults, following adjustment for correlated risk factors such as hypertension and diabetes [96, 97]. Affected measures typically include executive function and memory. Age interactions have been explored but not noted [98]. Examining participants from the Framingham Study, Elias and colleagues [99] reported associations of obesity to executive function and memory in men only. These investigators also reported a significant cumulative effect of obesity and hypertension on several memory measures. Our group has reported significant interactions of BMI (or waist circumference) with BP level [100]. Those with higher BMI and BP showed diminished performance on tests of motor speed and manual dexterity, and executive function (i.e., response inhibition). In contrast, Kuo [101] recently found that overweight persons performed better than normal weight persons on tests of reasoning and visuospatial speed of processing. Obese persons were also better than normal weight individuals on the latter measure.

Recent prospective data indicated that midlife central obesity, in conjunction with hypertension, was associated with decreased executive function and visual memory 12 years later [102]. Yet, the relation between central obesity and cognitive function is diminished after adjustment for physical activity [103]. Further, a recent investigation [104] found evidence for associations between obesity and both better and worse cognitive performance in the same sample, depending on body composition measure utilized (i.e., BMI, waist circumference, or waist–hip ratio), cognitive domain assessed, and specified model terms (i.e., cross-sectional versus longitudinal). The evidence was strongest for obesity as a predictor of accelerated cognitive decline on measures of global mental status, memory, and executive function. Nevertheless, the complexity of the overall pattern of findings was extensive.

The obesity–dementia literature is similarly complex at first glance [105], though ostensibly conflicting findings are likely explained by the decreased validity of adiposity measurement in the elderly, as well as the increased rate of adiposity decline immediately before dementia onset [106–108]. Taking these methodological issues into consideration, the bulk of the research supports a relation between obesity and increased risk of incident dementia (AD, vascular dementia, and all-cause) [109]. Central obesity at midlife may be an especially potent risk factor for dementia [110].

24.3.3.3 Underweight: Linear Relations

Research has also uncovered linear relations of underweight to cognitive function in multiple samples, though findings are limited. Underweight has been found to be associated with poor memory performance in young (ages 6–19) females [111]. In elderly samples, leanness has been related to lower concurrent and prospective change in MMSE scores [112, 113]. More extensive research has examined underweight and dementia. There is relative consensus that declining trajectories of BMI are associated with increased risk of incident Alzheimer's disease and all-cause dementia [114, 115]. Though weight loss precedes formal dementia onset [116], controversy exists regarding whether declining body weight is an early consequence of the premorbid neurodegenerative process versus a vulnerability factor for future diagnosis.

24.3.3.4 Direct Evidence of Nonlinearity

Similar to the lipid-cognition literature, minimal research has examined nonlinear relations of body composition indices to cognitive outcomes. However, emerging evidence suggests that nonlinear associations exist. Quadratic cross-sectional associations between BMI and a composite measure of cognitive function were recently reported in a large biracial community-dwelling population [117]. Individuals on both the upper and lower ends of BMI performed most poorly on cognitive measures, in comparison to normal weight participants. Similarly, Sabia and colleagues [118] reported quadratic cross-sectional associations between BMI and cognitive function, such that both underweight and obesity were associated with poorer late midlife (mean age = 61) cognitive function. This pattern of findings arose for performance on the MMSE, as well as tests of executive function. Further, when the authors examined cumulative impact of BMI on cognitive function across three time points (early adulthood, early midlife, and late life), individuals who were either underweight or obese at ≥2 time points demonstrated reduced late midlife executive function. Long-term obesity additionally predicted decrements in MMSE and memory performance. Results from both of these studies demonstrate the possibility of nonlinear body composition effects in other studies and populations.

24.3.3.5 Mechanisms

Jagust [119] has reviewed potential mechanisms linking obesity to the brain. These include metabolic, inflammatory, vascular, degenerative, and lifestyle (e.g., exercise) factors. Increased BMI or waist–hip ratio have been associated with temporal lobe or hippocampal atrophy [119], greater overall brain atrophy [120], and greater white matter disease [119]. There is some suggestion that the frontal lobes may be particularly affected [121].

Central obesity may also negatively affect the brain via neuroendocrine disturbances such as hypercortisolemia and low levels of sex steroid and growth hormones [122]. Both central and total obesity have been associated with other hormonal abnormalities such as hyperleptinemia (i.e., high serum levels of leptin—a hormone that plays a major role in fat metabolism), which has known central effects [123]. These hormonal abnormalities have been related to enhanced sympathetic nervous system activity [122, 123] that may promote silent cerebrovascular disease [107, 110]. Both central and total obesity have also been associated with enhanced proinflammatory factors [109, 110]. Sweat [124] recently found that C-reactive protein was associated with decreased frontal lobe function among overweight or obese women (but not men). Obesity may also operate, in part, via correlated cardiovascular risk factors such as the metabolic syndrome.

Underweight likely impacts cognitive function and the brain through different mechanisms than overweight and obesity. Among nondemented elderly, underweight may be the result of presymptomatic dementia. It has been posited that deposition of dementia-related neuropathology (e.g., plaques and tangles) in the areas of the brain responsible for weight regulation may lead to lower BMI [117]. In the context of frank dementia, underweight may also result from changes in eating habits due to impaired olfaction, forgetting to eat, abulia, or other factors. However, there appear to be additional operating mechanisms, given that weight loss often precedes dementia onset and underweight–cognition associations exist at younger ages. In that regard, leanness can be a sign of poor health, malnutrition, medical comorbidity, or physical frailty, all of which independently impact cognitive function. Relevant mechanisms in the absence of one of these factors are currently poorly understood.

24.3.3.6 Summary

Both underweight and obesity are associated with cognitive outcomes in a nonlinear fashion. However, limited direct evidence of nonlinearity exists within samples. Current linear evidence supports both underweight and obesity as prospective risk-indicators of cognitive decline and dementia, though the mechanisms are likely disparate. Underweight is most often linked with poorer cognitive function cross-sectionally, whereas cross-sectional associations between obesity and cognitive function appear to vary greatly across samples and neurocognitive domains. Characteristic declines in body mass prior to dementia onset, as well as confounding of underweight by poor health, significantly complicate the study of nonlinear BMI–cognition associations.

24.4 Summary, Conclusions, and Implications

Taken together, findings across all three discussed risk factors—blood pressure, lipid levels, and body composition—demonstrate the presence of nonlinearity in their associations with neuropsychological outcomes, including both cognitive function and dementia diagnosis. Paradoxically, the bulk of evidence for nonlinear associations derives primarily from linear analyses, though emerging research directly documenting nonlinearity provides further support for these hypotheses. Of the three risk factors, curvilinear blood pressure associations with cognitive outcomes have garnered the most extensive attention and support, though the lipid and body composition literatures appear to be following suit.

Several common themes can be gleaned from the findings discussed in this chapter. First, associations of blood pressure, lipid levels, and body composition to neuropsychological outcomes are best characterized as too complex for solely linear examination. Linear data analytic strategies may neglect to observe significant nonlinear relations, and nonlinearity could therefore help explain ostensibly conflicting findings in the literature. Researchers are thus urged to include nonlinear terms in their statistical models, and clinicians should be similarly aware of nonlinear risk relations when following patients over time. Past literature also requires careful interpretation in light of variable consideration of nonlinearity. Second, prototypical pre-dementia declines in blood pressure, lipid levels, and body composition similarly confound investigations of their associations with cognitive function and decline. Both clinicians and researchers should be cognizant of this issue and take measures to identify and account for this confounding among their patients and participants. Third, effect modifiers further complicate these associations. Interactive age effects are particularly important to consider, but a host of other vulnerability/resilience factors, including sex, race/ethnicity, socioeconomic status, genetics, and behavioral characteristics, require additional investigation. Fourth, the mechanisms that explain associations of high and low values of a given risk factor and cognitive outcomes are likely disparate. Continued research is necessary to better understand these mechanistic relations and develop corresponding methods of intervention.

Lastly, interactive effects within and across cardiovascular risk factors are likely to exist. Aggregation of multiple risk factors, as would be the case in a person with high blood pressure, high total cholesterol, and high BMI, can result in additive or synergistic effects on overall level of cardiovascular risk. Consideration of nonlinear effects in the context of such risk potentiation will be a challenging but necessary task for future researchers. Further, for cardiovascular risk factors with multiple indicators, interaction among these measures is also probable. For example, waist–hip ratio and BMI interact to predict incident cognitive impairment [125], and simultaneous examination of multiple lipid levels (e.g., both LDL and HDL cholesterol) could yield interactive effects as well. Furthermore, these risk factors often exist within the context of subclinical (and unidentified) vascular disease and/or clinical manifestations of CVD, all of which have neurocognitive consequences [17]. Although clinical CVD is often (although not always) adjusted in the studies cited herein, the likelihood of residual confounding remains. In addition, relations of CVD risk factors to neurocognitive endpoints may differ, in part, as a function of the presence of occult or overt CVD.

In conclusion, nonlinearity in the associations between blood pressure, lipid levels, and body composition and neuropsychological outcomes is present and deserves increased consideration by researchers and practitioners alike. Though this chapter focused on only three risk factors, patterns of nonlinearity across other cardiovascular risk factors remain plausible. Importantly, improvements in our understanding of these complex associations will only bolster our efforts to delay or prevent neurocognitive dysfunction and decline with aging.

Acknowledgements Adapted in part from: (a) Waldstein SR, Wendell CR, Hosey MM, Seliger SL, Katzel LI. Cardiovascular disease and neurocognitive function. In: Armstrong C, Morrow LA, editors. Handbook of medical neuropsychology: applications of cognitive neuroscience. New York: Springer; 2010, (b) Waldstein SR, Wendell CR, Katzel LI. Hypertension and neurocognitive function: blood pressure and beyond. In: Whitfield K, editor. Annual review of gerontology and geriatrics. New York: Springer; 2010. p. 113–132, and (c) Waldstein SR, Wendell CR, Hosey MM. Applications of neurocognitive assessment in behavioral medicine. In: Steptoe A, editor. Handbook of behavioral medicine. New York: Springer; 2010. p. 125–36.

References

1. Rosamond W, Flegal K, Furie K, Go A, Greenlund K, Haase N, et al. Heart disease and stroke statistics – 2008 update: a report from the American Heart Association Statistics Committee and Stroke Statistics Subcommittee. Circulation. 2008;117(4):e25–146.
2. Neaton JD, Wentworth DN, Cutler J, Stamler J, Kuller L. Risk factors for death from different types of stroke. Multiple Risk Factor Intervention Trial Research Group. Ann Epidemiol. 1993;3(5):493–9.
3. Meyer JS, Rauch GM, Rauch RA, Haque A, Crawford K. Cardiovascular and other risk factors for Alzheimer's disease and vascular dementia. Ann N Y Acad Sci. 2000;903:411–23.
4. International statistical classification of diseases and related health problems, 10th revision. World Health Organization. 2007. http://www.who.int/classifications/apps/icd/icd10online/.
5. Chobanian AV, Bakris GL, Black HR, Cushman WC, Green LA, Izzo Jr JL, et al. Seventh report of the Joint National Committee on Prevention, Detection, Evaluation, and Treatment of High Blood Pressure. Hypertension. 2003;42(6):1206–52.
6. Cupples LA, D'Agostino RB. Section 34: some risk factors related to the annual incidence of cardiovascular disease and death in pooled repeated biennial measurements. In: Kannel WB, Wolf PA, Garrison RJ, editors. Framingham Heart Study: 30 Year Follow-up. Bethesda, MD: US Department of Health and Human Services; 1987.
7. American Psychiatric Association. Diagnostic and statistical manual of mental disorders. Washington, DC: American Psychiatric Association; 2000.
8. Albert MS, DeKosky ST, Dickson D, Dubois B, Feldman HH, Fox NC, et al. The diagnosis of mild cognitive impairment due to Alzheimer's disease: recommendations from the National Institute on Aging-Alzheimer's Association workgroups on diagnostic guidelines for Alzheimer's disease. Alzheimers Dement. 2011;7(3):270–9.
9. Hachinski V. Vascular dementia: a radical redefinition. Dementia. 1994;5(3–4):130–2.
10. McKhann GM, Knopman DS, Chertkow H, Hyman BT, Jack Jr CR, Kawas CH, et al. The diagnosis of dementia due to Alzheimer's disease: recommendations from the National Institute on Aging-Alzheimer's Association workgroups on diagnostic guidelines for Alzheimer's disease. Alzheimers Dement. 2011;7(3):263–9.
11. Sperling RA, Aisen PS, Beckett LA, Bennett DA, Craft S, Fagan AM, et al. Toward defining the preclinical stages of Alzheimer's disease: recommendations from the National Institute on Aging-Alzheimer's Association workgroups on diagnostic guidelines for Alzheimer's disease. Alzheimers Dement. 2011;7(3):280–92.
12. Lezak MD, Howieson DB, Loring DW, editors. Neuropsychological assessment. 4th ed. New York: Oxford University Press; 2004.
13. Lloyd-Jones D, Adams RJ, Brown TM, Carnethon M, Dai S, De Simone G, et al. Executive summary: heart disease and stroke statistics – 2010 update: a report from the American Heart Association. Circulation. 2010;121(7):948–54.
14. Franco OH, Peeters A, Bonneux L, de Laet C. Blood pressure in adulthood and life expectancy with

cardiovascular disease in men and women: life course analysis. Hypertension. 2005;46(2):280–6.
15. Moser M, Setaro JF. Clinical practice. Resistant or difficult-to-control hypertension. N Engl J Med. 2006;355(4):385–92.
16. Waldstein SR, Katzel LI. Hypertension and cognitive function. In: Waldstein SR, Elias MF, editors. Neuropsychology of cardiovascular disease. Mahwah, NJ: Erlbaum; 2001. p. 15–36.
17. Waldstein SR, Wendell CR, Hosey MM, Seliger SL, Katzel LI. Cardiovascular disease and neurocognitive function. In: Armstrong C, Morrow LA, editors. Handbook of medical neuropsychology: applications of cognitive neuroscience. New York: Springer; 2010.
18. Lande MB, Kaczorowski JM, Auinger P, Schwartz GJ, Weitzman M. Elevated blood pressure and decreased cognitive function among school-age children and adolescents in the United States. J Pediatr. 2003;143(6):720–4.
19. Harrington F, Saxby BK, McKeith IG, Wesnes K, Ford GA. Cognitive performance in hypertensive and normotensive older subjects. Hypertension. 2000;36(6):1079–82.
20. Elias MF, Wolf PA, D'Agostino RB, Cobb J, White LR. Untreated blood pressure level is inversely related to cognitive functioning: the Framingham Study. Am J Epidemiol. 1993;138(6):353–64.
21. Euser SM, van Bemmel T, Schram MT, Gussekloo J, Hofman A, Westendorp RG, et al. The effect of age on the association between blood pressure and cognitive function later in life. J Am Geriatr Soc. 2009;57(7):1232–7.
22. Kilander L, Nyman H, Boberg M, Hansson L, Lithell H. Hypertension is related to cognitive impairment: a 20-year follow-up of 999 men. Hypertension. 1998;31(3):780–6.
23. Kuo HK, Sorond F, Iloputaife I, Gagnon M, Milberg W, Lipsitz LA. Effect of blood pressure on cognitive functions in elderly persons. J Gerontol A Biol Sci Med Sci. 2004;59(11):1191–4.
24. Whitfield KE, Allaire JC, Gamaldo A, Aiken-Morgan AT, Sims R, Edwards C. Blood pressure and memory in older African Americans. Ethn Dis. 2008;18(2):181–6.
25. Waldstein SR. Hypertension and neuropsychological function: a lifespan perspective. Exp Aging Res. 1995;21(4):321–52.
26. Bucur B, Madden DJ. Effects of adult age and blood pressure on executive function and speed of processing. Exp Aging Res. 2010;36(2):153–68.
27. Knecht S, Wersching H, Lohmann H, Bruchmann M, Duning T, Dziewas R, et al. High-normal blood pressure is associated with poor cognitive performance. Hypertension. 2008;51(3):663–8.
28. Elias PK, D'Agostino RB, Elias MF, Wolf PA. Blood pressure, hypertension, and age as risk factors for poor cognitive performance. Exp Aging Res. 1995;21(4):393–417.
29. Anson O, Paran E. Hypertension and cognitive functioning among the elderly: an overview. Am J Ther. 2005;12(4):359–65.
30. Elias MF, Robbins MA, Elias PK, Streeten DH. A longitudinal study of blood pressure in relation to performance on the Wechsler Adult Intelligence Scale. Health Psychol. 1998;17(6):486–93.
31. Starr JM, Deary IJ, Inch S, Cross S, MacLennan WJ. Blood pressure and cognitive decline in healthy old people. J Hum Hypertens. 1997;11(12):777–81.
32. Launer LJ, Masaki K, Petrovitch H, Foley D, Havlik RJ. The association between midlife blood pressure levels and late-life cognitive function. The Honolulu-Asia Aging Study. J Am Med Assoc. 1995;274(23):1846–51.
33. Waldstein SR, Giggey PP, Thayer JF, Zonderman AB. Nonlinear relations of blood pressure to cognitive function: the Baltimore Longitudinal Study of Aging. Hypertension. 2005;45(3):374–9.
34. Elias PK, Elias MF, Robbins MA, Budge MM. Blood pressure-related cognitive decline: does age make a difference? Hypertension. 2004;44(5):631–6.
35. Roman GC. Vascular dementia prevention: a risk factor analysis. Cerebrovasc Dis. 2005;20 Suppl 2:91–100.
36. Staessen JA, Richart T, Birkenhager WH. Less atherosclerosis and lower blood pressure for a meaningful life perspective with more brain. Hypertension. 2007;49(3):389–400.
37. Skoog I, Gustafson D. Update on hypertension and Alzheimer's disease. Neurol Res. 2006;28(6):605–11.
38. Stampfer MJ. Cardiovascular disease and Alzheimer's disease: common links. J Intern Med. 2006;260(3):211–23.
39. Qiu C, Winblad B, Fratiglioni L. The age-dependent relation of blood pressure to cognitive function and dementia. Lancet Neurol. 2005;4(8):487–99.
40. Launer LJ, Ross GW, Petrovitch H, Masaki K, Foley D, White LR, et al. Midlife blood pressure and dementia: the Honolulu-Asia aging study. Neurobiol Aging. 2000;21:49–55.
41. Whitmer RA, Sidney S, Selby J, Johnston SC, Yaffe K. Midlife cardiovascular risk factors and risk of dementia in late life. Neurology. 2005;64(2):277–81.
42. Axelsson J, Reinprecht F, Siennicki-Lantz A, Elmstahl S. Low ambulatory blood pressure is associated with lower cognitive function in healthy elderly men. Blood Press Monit. 2008;13(5):269–75.
43. Hestad K, Kveberg B, Engedal K. Low blood pressure is a better predictor of cognitive deficits than the apolipoprotein e4 allele in the oldest old. Acta Neurol Scand. 2005;111(5):323–8.
44. Guo Z, Viitanen M, Fratiglioni L, Winblad B. Low blood pressure and dementia in elderly people: the Kungsholmen project. BMJ. 1996;312(7034):805–8.
45. Maule S, Caserta M, Bertello C, Verhovez A, Naso D, Bisbocci D, et al. Cognitive decline and low blood pressure: the other side of the coin. Clin Exp Hypertens. 2008;30(8):711–9.
46. Swan GE, Carmelli D, Larue A. Systolic blood pressure tracking over 25 to 30 years and cognitive performance in older adults. Stroke. 1998;29(11):2334–40.

47. Molander L, Gustafson Y, Lovheim H. Low blood pressure is associated with cognitive impairment in very old people. Dement Geriatr Cogn Disord. 2010;29(4):335–41.
48. Paran E, Anson O, Reuveni H. Blood pressure and cognitive functioning among independent elderly. Am J Hypertens. 2003;16(10):818–26.
49. Morris MC, Scherr PA, Hebert LE, Bennett DA, Wilson RS, Glynn RJ, et al. Association between blood pressure and cognitive function in a biracial community population of older persons. Neuroepidemiology. 2002;21(3):123–30.
50. Bohannon AD, Fillenbaum GG, Pieper CF, Hanlon JT, Blazer DG. Relationship of race/ethnicity and blood pressure to change in cognitive function. J Am Geriatr Soc. 2002;50(3):424–9.
51. Glynn RJ, Beckett LA, Hebert LE, Morris MC, Scherr PA, Evans DA. Current and remote blood pressure and cognitive decline. JAMA. 1999;281(5): 438–45.
52. Iadecola C, Davisson RL. Hypertension and cerebrovascular dysfunction. Cell Metab. 2008;7(6):476–84.
53. de la Torre JC. Is Alzheimer's disease a neurodegenerative or a vascular disorder? Data, dogma, and dialectics. Lancet Neurol. 2004;3(3):184–90.
54. Third Report of the National Cholesterol Education Program (NCEP) Expert Panel on Detection, Evaluation, and Treatment of High Blood Cholesterol in Adults (Adult Treatment Panel III) final report. Circulation. 2002;106(25):3143–421.
55. Johnson CL, Rifkind BM, Sempos CT, Carroll MD, Bachorik PS, Briefel RR, et al. Declining serum total cholesterol levels among US adults. The National Health and Nutrition Examination Surveys. JAMA. 1993;269(23):3002–8.
56. Hyre AD, Muntner P, Menke A, Raggi P, He J. Trends in ATP-III-defined high blood cholesterol prevalence, awareness, treatment and control among U.S. adults. Ann Epidemiol. 2007;17(7):548–55.
57. Carlsson CM, Nondahl DM, Klein BE, McBride PE, Sager MA, Schubert CR, et al. Increased atherogenic lipoproteins are associated with cognitive impairment: effects of statins and subclinical atherosclerosis. Alzheimer Dis Assoc Disord. 2009;23(1):11–7.
58. Yaffe K, Barrett-Connor E, Lin F, Grady D. Serum lipoprotein levels, statin use, and cognitive function in older women. Arch Neurol. 2002;59(3):378–84.
59. Teunissen CE, De Vente J, von Bergmann K, Bosma H, van Boxtel MP, De Bruijn C, et al. Serum cholesterol, precursors and metabolites and cognitive performance in an aging population. Neurobiol Aging. 2003;24(1):147–55.
60. Anstey KJ, Lipnicki DM, Low LF. Cholesterol as a risk factor for dementia and cognitive decline: a systematic review of prospective studies with meta-analysis. Am J Geriatr Psychiatry. 2008;16(5):343–54.
61. Solomon A, Kareholt I, Ngandu T, Wolozin B, Macdonald SW, Winblad B, et al. Serum total cholesterol, statins and cognition in non-demented elderly. Neurobiol Aging. 2009;30(6):1006–9.
62. Reynolds CA, Gatz M, Prince JA, Berg S, Pedersen NL. Serum lipid levels and cognitive change in late life. J Am Geriatr Soc. 2010;58(3):501–9.
63. Grodstein F. Cardiovascular risk factors and cognitive function. Alzheimers Dement. 2007;3:S16–22.
64. Reitz C, Tang MX, Luchsinger J, Mayeux R. Relation of plasma lipids to Alzheimer disease and vascular dementia. Arch Neurol. 2004;61(5):705–14.
65. Dufouil C, Richard F, Fievet N, Dartigues JF, Ritchie K, Tzourio C, et al. APOE genotype, cholesterol level, lipid-lowering treatment, and dementia: the Three-City Study. Neurology. 2005;64(9):1531–8.
66. Tan ZS, Seshadri S, Beiser A, Wilson PW, Kiel DP, Tocco M, et al. Plasma total cholesterol level as a risk factor for Alzheimer disease: the Framingham Study. Arch Intern Med. 2003;163(9):1053–7.
67. Kalmijn S, Foley D, White L, Burchfiel CM, Curb JD, Petrovitch H, et al. Metabolic cardiovascular syndrome and risk of dementia in Japanese-American elderly men. The Honolulu-Asia aging study. Arterioscler Thromb Vasc Biol. 2000;20(10): 2255–60.
68. Kivipelto M, Ngandu T, Fratiglioni L, Viitanen M, Kareholt I, Winblad B, et al. Obesity and vascular risk factors at midlife and the risk of dementia and Alzheimer disease. Arch Neurol. 2005;62(10):1556–60.
69. Stewart R, White LR, Xue QL, Launer LJ. Twenty-six-year change in total cholesterol levels and incident dementia: the Honolulu-Asia Aging Study. Arch Neurol. 2007;64(1):103–7.
70. Benton D. Do low cholesterol levels slow mental processing? Psychosom Med. 1995;57(1):50–3.
71. Muldoon MF, Ryan CM, Matthews KA, Manuck SB. Serum cholesterol and intellectual performance. Psychosom Med. 1997;59(4):382–7.
72. Henderson VW, Guthrie JR, Dennerstein L. Serum lipids and memory in a population based cohort of middle age women. J Neurol Neurosurg Psychiatry. 2003;74(11):1530–5.
73. Zhang J, Muldoon MF, McKeown RE. Serum cholesterol concentrations are associated with visuomotor speed in men: findings from the third National Health and Nutrition Examination Survey, 1988-1994. Am J Clin Nutr. 2004;80(2):291–8.
74. West R, Beeri MS, Schmeidler J, Hannigan CM, Angelo G, Grossman HT, et al. Better memory functioning associated with higher total and low-density lipoprotein cholesterol levels in very elderly subjects without the apolipoprotein e4 allele. Am J Geriatr Psychiatry. 2008;16(9):781–5.
75. Sabbagh M, Zahiri HR, Ceimo J, Cooper K, Gaul W, Connor D, et al. Is there a characteristic lipid profile in Alzheimer's disease? J Alzheimers Dis. 2004;6(6):585–9. Discussion 673–81.
76. Lepara O, Valjevac A, Alajbegovic A, Zaciragic A, Nakas-Icindic E. Decreased serum lipids in patients with probable Alzheimer's disease. Bosn J Basic Med Sci. 2009;9(3):215–20.
77. Elias PK, Elias MF, D'Agostino RB, Sullivan LM, Wolf PA. Serum cholesterol and cognitive performance

in the Framingham Heart Study. Psychosom Med. 2005;67(1):24–30.
78. Swan GE, LaRue A, Carmelli D, Reed TE, Fabsitz RR. Decline in cognitive performance in aging twins. Heritability and biobehavioral predictors from the National Heart, Lung, and Blood Institute Twin Study. Arch Neurol. 1992;49(5):476–81.
79. Solomon A, Kareholt I, Ngandu T, Winblad B, Nissinen A, Tuomilehto J, et al. Serum cholesterol changes after midlife and late-life cognition: twenty-one-year follow-up study. Neurology. 2007;68(10): 751–6.
80. Atzmon G, Gabriely I, Greiner W, Davidson D, Schechter C, Barzilai N. Plasma HDL levels highly correlate with cognitive function in exceptional longevity. J Gerontol A Biol Sci Med Sci. 2002;57(11):M712–5.
81. Launer LJ, White LR, Petrovitch H, Ross GW, Curb JD. Cholesterol and neuropathologic markers of AD: a population-based autopsy study. Neurology. 2001;57(8):1447–52.
82. Komulainen P, Lakka TA, Kivipelto M, Hassinen M, Helkala EL, Haapala I, et al. Metabolic syndrome and cognitive function: a population-based follow-up study in elderly women. Dement Geriatr Cogn Disord. 2007;23(1):29–34.
83. Singh-Manoux A, Gimeno D, Kivimaki M, Brunner E, Marmot MG. Low HDL cholesterol is a risk factor for deficit and decline in memory in midlife. The Whitehall II Study. Arterioscler Thromb Vasc Biol. 2008;28(8):1556–62.
84. Reitz C, Tang MX, Schupf N, Manly JJ, Mayeux R, Luchsinger JA. Association of higher levels of high-density lipoprotein cholesterol in elderly individuals and lower risk of late-onset Alzheimer disease. Arch Neurol. 2010;67(12):1491–7.
85. Arntzen KA, Schirmer H, Wilsgaard T, Mathiesen EB. Impact of cardiovascular risk factors on cognitive function: the Tromso study. Eur J Neurol. 2011;18(5):737–43.
86. Wendell CR, Katzel LI, Rosenberger WF, Plamadeala V, Hosey MM, Waldstein SR. Plasma lipid levels and neuropsychological function: nonlinear relations and effect modification by age. J Int Neuropsychol Soc. 2011;17(Suppl S1):121.
87. Wendell CR, Waldstein SR, Ferrucci L, Zonderman A. Nonlinear associations between trajectories of total cholesterol and cognitive change over time: what's good for the heart may not always be good for the brain. Psychosom Med. 2010;72(Abstract Suppl):A43.
88. Muldoon MF, Flory JD, Ryan CM. Serum cholesterol, the brain, and cognitive functioning. In: Waldstein SR, Elias MF, editors. Neuropsychology of cardiovascular disease. Mahwah, NJ: Erlbaum; 2001. p. 37–59.
89. Kojro E, Gimpl G, Lammich S, Marz W, Fahrenholz F. Low cholesterol stimulates the nonamyloidogenic pathway by its effect on the alpha-secretase ADAM 10. Proc Natl Acad Sci USA. 2001;98(10): 5815–20.
90. Hauser PS, Narayanaswami V, Ryan RO. Apolipoprotein E: from lipid transport to neurobiology. Prog Lipid Res. 2011;50(1):62–74.
91. Rosendorff C, Beeri MS, Silverman JM. Cardiovascular risk factors for Alzheimer's disease. Am J Geriatr Cardiol. 2007;16(3):143–9.
92. Grundy SM, Cleeman JI, Daniels SR, Donato KA, Eckel RH, Franklin BA, et al. Diagnosis and management of the metabolic syndrome: an American Heart Association/National Heart, Lung, and Blood Institute Scientific Statement. Circulation. 2005;112(17):2735–52.
93. Despres JP, Lemieux I, Prud'homme D. Treatment of obesity: need to focus on high risk abdominally obese patients. BMJ. 2001;322(7288):716–20.
94. Executive summary of the clinical guidelines on the identification, evaluation, and treatment of overweight and obesity in adults. Arch Intern Med. 1998;158(17):1855–67.
95. Wilson PW, Bozeman SR, Burton TM, Hoaglin DC, Ben-Joseph R, Pashos CL. Prediction of first events of coronary heart disease and stroke with consideration of adiposity. Circulation. 2008;118(2):124–30.
96. Cserjesi R, Molnar D, Luminet O, Lenard L. Is there any relationship between obesity and mental flexibility in children? Appetite. 2007;49(3):675–8.
97. Gunstad J, Paul RH, Cohen RA, Tate DF, Gordon E. Obesity is associated with memory deficits in young and middle-aged adults. Eat Weight Disord. 2006;11(1):e15–9.
98. Gunstad J, Paul RH, Cohen RA, Tate DF, Spitznagel MB, Gordon E. Elevated body mass index is associated with executive dysfunction in otherwise healthy adults. Compr Psychiatry. 2007;48(1):57–61.
99. Elias MF, Elias PK, Sullivan LM, Wolf PA, D'Agostino RB. Lower cognitive function in the presence of obesity and hypertension: the Framingham heart study. Int J Obes Relat Metab Disord. 2003;27(2):260–8.
100. Waldstein SR, Katzel LI. Interactive relations of central versus total obesity and blood pressure to cognitive function. Int J Obes (Lond). 2006;30(1):201–7.
101. Kuo HK, Jones RN, Milberg WP, Tennstedt S, Talbot L, Morris JN, et al. Cognitive function in normal-weight, overweight, and obese older adults: an analysis of the Advanced Cognitive Training for Independent and Vital Elderly cohort. J Am Geriatr Soc. 2006;54(1):97–103.
102. Wolf PA, Beiser A, Elias MF, Au R, Vasan RS, Seshadri S. Relation of obesity to cognitive function: importance of central obesity and synergistic influence of concomitant hypertension. The Framingham Heart Study. Curr Alzheimer Res. 2007;4(2):111–6.
103. Dore GA, Elias MF, Robbins MA, Budge MM, Elias PK. Relation between central adiposity and cognitive function in the Maine-Syracuse Study: attenuation by physical activity. Ann Behav Med. 2008;35(3):341–50.
104. Gunstad J, Lhotsky A, Wendell CR, Ferrucci L, Zonderman AB. Longitudinal examination of obe-

sity and cognitive function: results from the Baltimore longitudinal study of aging. Neuroepidemiology. 2010;34(4):222–9.
105. Rosano C, Newman AB. Cardiovascular disease and risk of Alzheimer's disease. Neurol Res. 2006;28(6): 612–20.
106. Gustafson D. Adiposity indices and dementia. Lancet Neurol. 2006;5(8):713–20.
107. Whitmer RA. The epidemiology of adiposity and dementia. Curr Alzheimer Res. 2007;4(2):117–22.
108. Luchsinger JA, Mayeux R. Adiposity and Alzheimer's disease. Curr Alzheimer Res. 2007;4(2):127–34.
109. Barrett-Connor E. An introduction to obesity and dementia. Curr Alzheimer Res. 2007;4(2):97–101.
110. Whitmer RA, Gustafson DR, Barrett-Connor E, Haan MN, Gunderson EP, Yaffe K. Central obesity and increased risk of dementia more than three decades later. Neurology. 2009;72(11):1030–1.
111. Gunstad J, Spitznagel MB, Paul RH, Cohen RA, Kohn M, Luyster FS, et al. Body mass index and neuropsychological function in healthy children and adolescents. Appetite. 2008;50(2–3):246–51.
112. Auyeung TW, Lee JS, Kwok T, Woo J. Physical frailty predicts future cognitive decline – a four-year prospective study in 2737 cognitively normal older adults. J Nutr Health Aging. 2011;15(8):690–4.
113. Sakakura K, Hoshide S, Ishikawa J, Momomura S, Kawakami M, Shimada K, et al. Association of body mass index with cognitive function in elderly hypertensive Japanese. Am J Hypertens. 2008;21(6): 627–32.
114. Buchman AS, Wilson RS, Bienias JL, Shah RC, Evans DA, Bennett DA. Change in body mass index and risk of incident Alzheimer disease. Neurology. 2005;65(6):892–7.
115. Stewart R, Masaki K, Xue QL, Peila R, Petrovitch H, White LR, et al. A 32-year prospective study of change in body weight and incident dementia: the Honolulu-Asia Aging Study. Arch Neurol. 2005;62(1):55–60.
116. Barrett-Connor E, Edelstein S, Corey-Bloom J, Wiederholt W. Weight loss precedes dementia in community-dwelling older adults. J Nutr Health Aging. 1998;2(2):113–4.
117. Sturman MT, de Leon CF, Bienias JL, Morris MC, Wilson RS, Evans DA. Body mass index and cognitive decline in a biracial community population. Neurology. 2008;70(5):360–7.
118. Sabia S, Kivimaki M, Shipley MJ, Marmot MG, Singh-Manoux A. Body mass index over the adult life course and cognition in late midlife: the Whitehall II Cohort Study. Am J Clin Nutr. 2009;89(2):601–7.
119. Jagust W. What can imaging reveal about obesity and the brain? Curr Alzheimer Res. 2007;4(2): 135–9.
120. Ward MA, Carlsson CM, Trivedi MA, Sager MA, Johnson SC. The effect of body mass index on global brain volume in middle-aged adults: a cross sectional study. BMC Neurol. 2005;5:23.
121. Gazdzinski S, Kornak J, Weiner MW, Meyerhoff DJ. Body mass index and magnetic resonance markers of brain integrity in adults. Ann Neurol. 2008;63(5):652–7.
122. Bjorntorp P, Rosmond R. Neuroendocrine abnormalities in visceral obesity. Int J Obes Relat Metab Disord. 2000;24 Suppl 2:S80–5.
123. Ren J. Leptin and hyperleptinemia – from friend to foe for cardiovascular function. J Endocrinol. 2004;181(1):1–10.
124. Sweat V, Starr V, Bruehl H, Arentoft A, Tirsi A, Javier E, et al. C-reactive protein is linked to lower cognitive performance in overweight and obese women. Inflammation. 2008;31(3):198–207.
125. Kerwin DR, Gaussoin SA, Chlebowski RT, Kuller LH, Vitolins M, Coker LH, et al. Interaction between body mass index and central adiposity and risk of incident cognitive impairment and dementia: results from the Women's Health Initiative Memory Study. J Am Geriatr Soc. 2011;59(1):107–12.

25 Pain and Complex Adaptive System Theory

Cary A. Brown

25.1 Introduction

Chronic pain is an expensive and growing problem within industrialised countries [1–4]. The ever-expanding volume of research in this area ranges across physiological and bio-psychosocial models and runs parallel to changes in social attitudes and beliefs towards health [5]. In a bio-psychosocial model, pain is conceptualised as a perception based on the interaction between physiology (specific pain receptors) and emotional, motivational and cognitive modifying factors [6]. There is an apparent reluctance of many medical professionals to abandon this body/mind dualism so the search for a cause and cure continues, as does the confusion and frustration of people seeking help for chronic pain. Chronic pain research is often unidisciplinary, lacking the interdisciplinary focus to generate integrated theory and interventions. To describe pain research, Sullivan and Lewin use the analogy of a group of people in a dark room who describe an elephant based on what they can touch [7]; many disciplines touch a piece of the elephant that is called pain, but most cannot tell whether they have the trunk or the tail.

This chapter will first apply a complex adaptive system (CAS) theory to explore the consequences of ongoing resistance to reframing chronic pain as a system as opposed to a singularly biological event. The second half of the chapter will focus on applying the principles of complex adaptive system to identify strategies and simple rules to modify the current enduring system attractor of chronicity.

25.1.1 CAS, Healthcare and Chronic Pain

The experience of pain and pain management occur within the wider context of the healthcare system. To understand pain as a CAS, first we need to explore how CAS theory is relevant for framing healthcare. In the beginning of the twenty-first century, the application of CAS thinking to healthcare was gaining momentum in a few industrialised nations. For example, David Fillingham, previous director of the National Health System (NHS) Modernisation Agency in the United Kingdom (UK), stated: "*The NHS is the epitome of a complex adaptive system. Such systems do not always respond well to mechanistic formulae*" [8]. A growing number of theorists and researchers echoed this sentiment in relation to not just healthcare systems but also to patients. Clinicians and theorists who understood the potential power of applying adaptive systems thinking to healthcare increasingly pointed out that people are complex biological systems that

C.A. Brown, F.H.E.A., Ph.D. (✉)
Faculty of Rehabilitation Medicine,
University of Alberta, 2-64 Corbett Hall,
Edmonton, AB T6G 2G4, Canada
e-mail: cary.brown@ualberta.ca

J.P. Sturmberg and C.M. Martin (eds.), *Handbook of Systems and Complexity in Health*,
DOI 10.1007/978-1-4614-4998-0_25, © Springer Science+Business Media New York 2013

do not behave in a linear fashion [9–12] and that effective healthcare for the growing number of chronic disease and lifestyle issues such as chronic pain must be grounded in a non-reductionist paradigm focused on an understanding of relationships and a flexible approach to problem solving [13–16]. Chapman et al. [17] have predicted that for pain research, "*...the principle challenge will not be technological advancement but rather the generation of a theoretical framework that can guide complex scientific inquiry*" ([17], p. 138).

25.2 Defining Complex Adaptive Systems

> A Complex Adaptive System (CAS) is a collection of individual agents with freedom to act in ways that are not totally predictable, and whose actions are interconnected so that one agent's actions change the context for other agents. [18]

Cilliers [19] suggests that attempting to reduce and define a multifaceted concept such as a CAS is contradictory. Instead Cilliers leads the reader to explore the concept as a dynamic composite of multiple elements. Zimmerman et al. [20] suggest that the concept of a CAS is best understood through examining the significance of each term: "complex" implies diverse relationships between many elements, "adaptive" reinforces the dynamic capability for change and "system" highlights the interconnectedness of the individual elements.

Complexity theory places events within a matrix recognising that while simple relationships between two elements can occur, more complex interrelationships between multiple dynamic influences can also occur. Within this matrix very chaotic events can also emerge, often with undetected and far reaching influence. Simple events have a linear relationship in which the cause of an event is directly proportional to the effect; the greater the magnitude of the causative agent, the larger will be the effect. The body's response to external temperature is an example of this effect. The body's thermal regulatory system increases in activity as the external temperature deviates from the body's normal temperature. In linear relationships, there tends to be a high agreement among stakeholders about the correct course of action (e.g. a child chokes on a sweet, you remove the sweet).

On the other hand, complex events are multidimensional and have the features of high uncertainty and disagreement. An example of this took place in 1998 when a groundswell of panic arose in the UK when Wakefield and his research team published research findings concluding a link between the combined vaccination for measles, mumps and rubella (MMR) and autism. This single article based on a small and biased sample of nine children with pre-existing health problems resulted in 100,000s of children not receiving any immunisations for measles, mumps or rubella. Twelve years later the findings were demonstrated to be fraudulent and Wakefield was struck from the UK medical registry [21].

However, public opinion seems difficult to reset and many parents continue to express more concerns over vaccinations than over the significant consequences of childhood measles, mumps and rubella [22]. As these events are embedded in many contexts (social, political, cultural, economic, temporal, affective, etc.) they require non-linear strategies to understand and guide decision making [12]. In a CAS a small input can have an unanticipated large effect. For example, charitable donations to fund spinal cord research after one person's injury such as for the actor Christopher Reeves are often cited as an example of this complex phenomenon.

Complexity theorists propose that the system becomes immobile and destabalised when linear thinking and searches for the one best solution are applied to complex problems [11, 14, 15]. The one best solution imposes too many controls to allow for the flexible responses necessary to incorporate new information as it becomes available and to provide feedback to the existing problem. Complex problems require a range of strategies that facilitate timely implementation of good enough solutions that can be modified as additional feedback and information become available. Searching for a global best solution is seen as a futile, counterproductive exercise; this process can take so long to formulate a solution

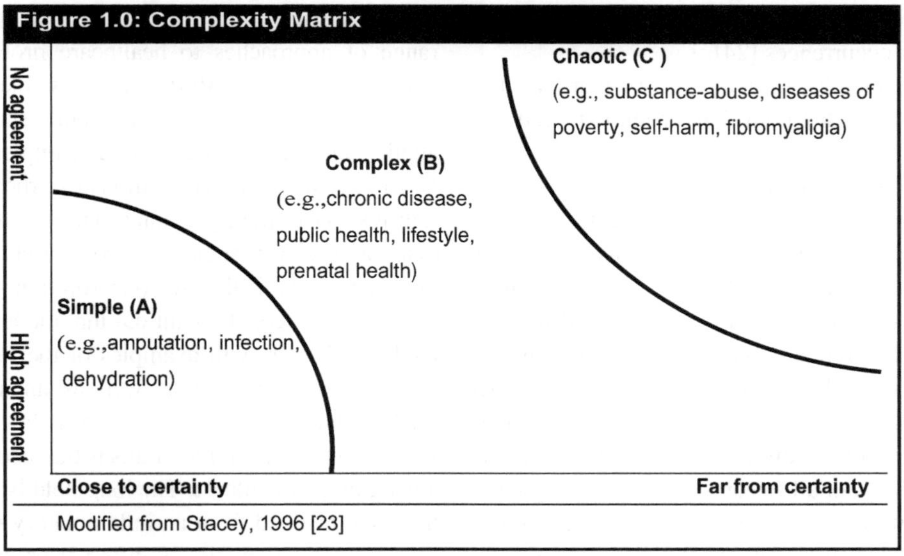

Fig. 25.1 Complexity matrix

that the dimensions of the original problem have long since evolved and have often compounded into a new issue.

The agreement/certainty matrix model of complexity (Fig. 25.1), generally attributed to Stacey [23], has been widely used to illustrate the range of events that can take place when relationships shift from predictable and agreed upon towards high uncertainty and a range of idiosyncratic opinions. Events with high agreement and certainty about the outcome respond to the conventional scientific model with a linear notion of causality (Point A in Fig. 25.1). As events become less predictable and the number of contextual influences to be considered increases the event becomes complex (Point B in Fig. 25.1). Complex events are still governed by basic rules (e.g. people with diabetes have problems regulating insulin production) but are also influenced by a myriad of conditions, features and forces. For example, a day in EuroDisney, eating unhealthy food to avoid being singled out by peers as different, combined with fatigue consequent to travel, time change and the high energy levels exerted for a prolonged period in a theme park present a very different situation for a child with diabetes than the usual predictable school day. Environment, social pressures and parental awareness all become important influences that modify and, in turn, are modified by one child's diabetes.

Complex events are understood by focusing on the relationships between the multitude of influences as opposed to seeking reductionist rules. Complex systems are constantly being influenced by and adapting to their environment. They respond to and at the same time act upon the extensive range of elements that comprise their unique context [12]. A system can also become chaotic (Point C), where no discernible pattern will emerge and the system appears to randomly fluctuate and no longer adapts to the environmental context.

25.3 History of Complexity and Healthcare

> If things were that simple, word would have gotten round.
> Derrida (1988) in Cilliers [19]

In the eighth century B.C., the Greek poet Hesiod wrote of how the earth and order emerged from chaos and order comes from disorder [24]. However, the concept of chaos and its relationship to order was not widely explored until there were the mid-twentieth century advances in the ability of computers to manipulate previously large and complicated calculations and examine irregular occurrences [24].

The principles of chaos theory that began to emerge from this point have been explored and refined in a diverse range of disciplines. An example of the application of chaos theory is in Robert Mays' application in the field of biology and population dynamics [24]. This application of the principles is credited with demonstrating that the size of a population in the wild will not necessarily remain predictable dependent on predators, food and the environment, but rather will oscillate periodically between two states. This concept of bifurcation demonstrates that systems are not all linear in nature. Putting more food into the system (input) will not necessarily result in more fish (output). Rather, some systems will increase to a certain point and then split or cycle [24, 25].

Examples of non-linear systems emerged from a range of different scientific disciplines and their application to real life problems. These examples have become collectively referred to as complex adaptive systems theory [24]. Contemporary theorists see chaos and complexity theory as a revolutionary advancement in science, allowing for the interdisciplinary integration of non-reductionist creativity and ideas required to deal with the problems of the twenty-first century. "*Chaos breaks across the lines that separate scientific disciplines. Because it is a science of the global nature of systems, it has brought together thinkers from fields that have been widely separated*" [26].

The growing acceptance of post-modernist thinking in western industrialised culture has fostered the application of CAS thinking to a range of contemporary social issues such as healthcare. Since the mid-1990s, theorists have applied the principles of complexity and chaos theory [12, 15, 20] to healthcare. Theorists propose that while the scientific model of illness (consequent to the historical foundations of philosophers and scientists like Newton and Descartes) is effective for understanding certain disease states, it is an inappropriate framework to apply to many contemporary lifestyle and chronic health problems [12, 27, 28].

Complexity science maintains that a flexible range of approaches to healthcare problems is essential. For straightforward issues (like a ruptured appendix) a scientific model should be applied and conventional healthcare implemented. However, other health conditions like rheumatoid arthritis have multiple interrelating influences (such as lifestyle, social context and chronicity) and there is typically not one solution to these conditions. Theorists point out that the history of medicine is rich with examples of the failure of scientific reductionism as well as an accepted medical fact being proven to be a flawed construct. A recent example of this is the reconceptualising of gastric ulcers. The long-held belief was that they resulted from an inflammatory process. However, the dramatically different opinion currently held is that ulcers are a bacterial infection [29] and clinical practice has shifted accordingly.

Post-modernist thinking assumes that there are a multitude of truths dependent on the viewer's context, influences [30] and construction of their personal reality. Taken to the extreme, post-modernist thinking is seen as a rejection of orthodox medicine and can ironically contribute to both the over-zealous pursuit of a positivist-generated evidence-base and the exclusion of all other forms of evidence [31–33]. On the opposite side of the same coin, post-modernist thinking can lead to resistance to implementing evidence-based change in practice [34]. Complex adaptive systems theory fosters the view that healthcare decision making occurs within a matrix of influences and certain approaches are more effective at certain times in certain conditions.

Martin [30] proposes that orthodox medicine's response to challengers will result in new ideas either changing practice or being labelled heresy and marginalised. There is a growing weight of evidence that traditional medical establishments are open to change and new ways of thinking about chronic health conditions. For example, the UK Medical Research Council (MRC) has stated: "*The greater the difficulty in defining precisely what, exactly, are the 'active ingredients' of an intervention and how they relate to each other, the greater the likelihood that you are dealing with a complex intervention*" ([35], p. 1).

To guide medical researchers in addressing these issues, the UK MRC produced the document ***A Framework for Development and Evaluation of RCTs for Complex Interventions to Improve Health*** [35]. The document falls short of a clear call for abandoning the traditional hierarchical positioning of randomised control studies (RCTs) as the gold standard for many healthcare issues. However, as the MRC is the largest public sponsor of RCTs in the UK, the guidelines have a strong potential to legitimise researchers' efforts to address complex problems from a wider perspective. Other widely respected and influential voices of the medical establishment have expressed support for reconceptualising health and illness within a complexity framework. The 2001 British Medical Journal (BMJ) series on complexity science [18, 27, 36, 37] and the World Health Organisation's call for a paradigm shift for understanding healthcare needs of the twenty-first century in the publication: ***Innovative Care for Chronic Conditions: Building Blocks for Action*** [38] demonstrate that many in the medical establishment are ready to look at a new framework for contemporary healthcare needs. The following section outlines the characteristics of complex adaptive systems in healthcare and provides examples illustrating how chronic pain is consistent with the characteristics of a complex adaptive system.

25.4 Properties of Complex Adaptive Systems

> Our meddling instinct mis-shapes the beauteous form of things we murder to dissect.
> Wordsworth ([39], lines 26–28)

Complex adaptive systems are more than the sum of their parts. Both poets and theorists warn of the danger of taking elements in isolation as their magnification will result in an obscured perspective. For example, the researcher who focuses only on serotonin uptake in nerves will not develop an understanding of living with pain. The multidimensionality of complex systems dictates that it is essential to have a way to organise and guide examination of complex and adaptive systems. The range of research and theoretical paradigms on chronic pain are continuously growing and if examined from a CAS perspective, the paradigms can reveal patterns to explain the persistent resistance of chronic pain to remediation.

Griffiths [40] proposes that complex adaptive systems theory should be seen as a meta-theory that provides a way of organising and relating a range of other theories. In this way, complexity theory reconciles potentially conflicting analytical models by demonstrating how each theory is of value in certain contexts, at certain times and for certain people. Griffiths compares complexity theory to the development of the periodic table in the mid-1800s. Before that time, there were many techniques for examining minerals, but the theory of essential elements (earth, air, fire, water) employed by alchemists was flawed. With the development of atomic mass theory, existing techniques were employed to quickly gain new insights into chemistry. Griffiths proposes that a similar process will occur when existing healthcare research data is viewed from within a complexity framework, thus allowing new and revised insights to emerge from existing theories.

Plsek [15] organises CAS into eight key properties:

1. Relationships are central to understanding the system
2. Complex systems are described by their structure, processes and patterns
3. Actions are based on internalised simple rules and mental models
4. Underlying attractor patterns explain complex behaviour
5. Complex systems are in constant adaptation
6. Experimentation and pruning must occur for the system to adapt
7. Complex systems are inherently non-linear
8. Complex systems are embedded within other systems and co-evolve

The following sections will examine research theory and findings in the area of chronic pain as they relate to these eight properties. Evidence for the proposal that chronic pain is not so much an experience or syndrome but rather a CAS will be developed in more depth in the following section.

25.5 Chronic Pain and the Characteristics of CAS

25.5.1 Relationships Are Central to Understanding the System

Behaviour is generated in a CAS by interaction between the stakeholders. Stakeholders are guided by their own personal meaning constructs for the event and as such their behaviour is not always predictable to others [19]. Additionally, miscommunication is highly possible in a CAS. The reasons for the miscommunication may be a result of the following situations: stakeholders do not share the same beliefs, culture or opinions; stakeholders come from different linguistic groups [41]; they do not feel the need/comfort/ability to overtly share these opinions; stakeholders may not be allowed to speak for themselves [42], and they may be suffering iatrogenic effects of past miscommunications [43].

An ant colony is a frequent example of competitive and cooperative interrelationships within systems. Complex organisations in an ant colony are maintained by the dynamic interplay between competition for reproduction and cooperation for food. Understanding how these forces are balanced to ensure maximum system integrity offers insights that can be applied to other complex systems [44]. In regard to change, relationships can be negatively affected by the idiosyncratic beliefs and coping style of the agents in the system. Many of the interactions concerning chronic pain focus on changes in behaviour and belief. Research demonstrates that the approach taken by the change agent needs to be consistent with what has been termed the recipient's stage of readiness for change [45]. Efforts to affect change without considering barriers and fears about change can be met with resistance and potentially serve to hinder the establishment of effective relationships [46, 47]. For example, this can be seen when a service provider tries to assist a service user adopt the belief that pain is not best dealt with by bed rest.

The chronic pain experience is highly interactive. As well, stakeholder interactions are high because of chronic pain's impact on productivity, social roles and public resources. Communication issues are well documented [48–50], as is the significance of personal meaning construction within the chronic pain experience [51–55].

25.5.2 Structures, Processes and Patterns

Structure, processes and patterns are the three components of a CAS and must be considered highly interactive. Changing one component will result in a change in the others. Assuming that consistency in two components can predict consistency in the third is flawed thinking. In complexity literature, an example of this finding is the Cuban missile crisis. Although the structure (government's decision-making mechanisms) and processes (protocols for progressing between stages of military preparedness and pre-emptive strike) during the Bay of Pigs crisis were stable and predictable, the third element of the system (pattern) was not. The Russian leaders' patterns of behaviour were perceived by President Kennedy to be inconsistent with that to which the structures and processes were pointing [44]. Consequently his response was not consistent with what would have been the logical behaviour in a linear system, where the assumption exists that if "A" and "B" are true, then "C" must also be true.

We see the same variability in the three components of the system when looking at chronic pain. For example in the UK the Clinical Standards Advisory Group's (CSAG) [56] review of services for people with pain concluded that how services are delivered (structures) for these individuals are quite varied. They also identified that treatment options range from single modality and condition-specific clinics to inpatient intensive multidisciplinary programmes and also that the availability of human and material resources within these options are widely divergent. Additionally, the processes for accessing services differ from region to region, the routes for referral are unclear and public versus private sector options are not clearly apparent. Structures and processes guiding this communication have many tiers and include government policy, healthcare directives, healthcare structures and process, and the ethno-social realities of multicultural service users and providers [57]. The result is that this confusion is further compounded by an often lack of enforcement of existing regulations and guidelines intended to ensure access to healthcare information for a range of stakeholders [57].

There are barriers to changing the patterns of interaction that emerge from the power dynamics and socio-political forces. Many chronic pain interactions continue to occur within an exclusively biomedical paradigm despite guidelines from the British Pain Society [58], the International Society for the Study of Pain [59] and growing evidence-base strongly supporting a bio-psychosocial approach to chronic pain. Service providers continue to be inculcated with the professional values of objectivity and pursuit of reductionism or the correct answer to health-related problems [60–62].

Research has also demonstrated that some healthcare professionals are slow to change clinical practice to incorporate new scientific findings. Reasons for this include the following factors: difficulty accessing research literature [63]; lack of time or knowledge required to search, read, interpret, and evaluate relevant reports [64, 65]; the belief that experience and expert opinion are more important to competency than knowing the results of research [66]; failure to recognise the existence of guidelines based on evidence; and institutional barriers to changing current practice [67]. As well, for certain clinical conditions such as chronic pain the literature is diverse, extensive and at times, contradictory. When this happens the problem of accessing and evaluating the evidence is compounded and presents a further barrier to modifying traditional professional/patient pattern of interaction.

The traditional patient/professional relationship places a high premium on objectivity and is strongly evident in interactions between service users and providers. However, there is increasing discussion about the importance for healthcare providers to employ reflective practice for their patients and for the interests of their own health [68–70]. People with chronic, non-malignant pain can bring and generate a range of emotions to the clinical encounter [71, 72] and these are also related to the wide variety of processes and structures where each service is offered and received. When working with service users who have complex problems requiring frequent attendance, healthcare providers are at particular risk from the heavy emotional content of these interactions [60]. Anecdotal evidence suggests that any number of people may attend a clinic over the course of a day with the same age, gender and diagnosis, but each encounter will be unique for both the service provider and service user.

25.5.3 Internalised Simple Rules and Mental Models

Simple, clear purpose and principles give rise to complex, intelligent organisational behaviour. Complex rules and regulations give rise to simple, stupid organisational behaviour.
Dee Hock in Pierce [73]

The game tic-tac-toe illustrates the concept that simple rules can result in very complex outcomes and is a commonly used as an example by complexity theorists. In the case of this 3-by-3 game, there are in excess of 50,000 possible legal configurations [25]. If this complexity is possible from the simple rules of a children's game, the implication for interactions between people and organisations is vast.

In human interactions, internalised rules are not static and linear but are reconstructed with experience and reflective cognitions. The significance of this personal meaning construction in chronic pain has been demonstrated in recent research. Literature supports that the biomedical model is still a widely held assumption for people with chronic pain, and service user and provider patterns of behaviour are strongly influenced by the search for cure metaphor [74]. However, as this search becomes more extensive, so do the search results become more and more complicated and fragmented. Simple internalised rules such as the doctor will give me a pill to take the pain away no longer operate or offer solutions. The search for a cure leads people into an ever-expanding field of practices and processes, policies and procedures. New and complicated information about waiting lists, treatment regimes, locations of clinics and claims forms presents additional demands on the resources of service users, family members and service providers. Navigating the healthcare system when one has chronic pain requires concentration and rapid new learning skills at a time when, ironically people with pain are least able to complete cognitive tasks [75].

There is also a confusing variety of rationales offered for each treatment. It has been pointed out that even highly trained and specialised pain management programme teams may not share a uniform understanding of the primary goal of intervention [76]. Some team members believe that the primary goal of intervention is pain relief, while others state the goal is improved physical function, while others may operate with a focus on self-expression and psychological insight. Service users themselves have internalised expectations and goals about treatment and when these varied and often uncommunicated beliefs collide, the treatment process becomes complicated and frustrating. This conflict has been shown to contribute to iatrogenic problems for people with chronic pain [77, 78].

25.5.4 Attractor Patterns

The concept of attractor patterns comes from the field of mathematics and refers to underlying forces that attract behaviours towards a consistent outcome regardless of what type of input is exerted [24]. "*An attractor is the area that a system moves towards and where it will tend to stay*" [79]. A simple analogy is that of pouring pebbles into a bowl. The pebbles, regardless of a multitude of factors, will move towards the lowest point in the bowl which is the bottom. The common cold viral infection is a biological attractor. Most common colds in western industrialised society follow the same trajectory; regardless of the input exerted, the cold symptoms eventually resolve.

Attractors are not always straightforward and the term hidden attractor is applied to systems where the likely point of outcome is not necessarily evident [79]. For example, a social service agency may believe that its attractor is equity of service delivery. However, the reality is that the limited financial resources and provision of services needs to be rationalised based on age, disability or employment potential and the formal attractor of equity of service is not correct for the agency. As a result, the system will always fluctuate in terms of equity of service delivery across the community. The hidden attractor is actually fiscal restraint and an attempt to modify the wrong attractor results in wasted input. As Burton points out, the hidden attractor may have been there all along and "*it just wasn't obvious until we looked in the right way*" [79].

In the history of illness there are numerous examples where resources were misdirected and did not actually target or influence the true attractor in the problem. To determine the attractor in an event, complexity science focuses on examining patterns and relationships. However, relationships and how we look at them are socio-culturally embedded and change with the thinking of the time. Often social-political forces actually work against uncovering hidden attractors and it is not until thinking changes that new ways of looking at events are possible. With those new ways of looking what was once obscure becomes clearly apparent.

A good example of how attractors in a system change depending on the acceptable social thought of an era can be traced by the view of medicine about cholera. When religion was the tool through which society understood events and constructed meaning, cholera was seen as a divine punishment. However, over time people came to see that cholera was indiscriminate in afflicting both sinners and saints. Therefore, people concluded, cholera must be the output of some other system, not grounded in spirituality. Also, by the sixteenth century, secular leaders were emerging who wanted to be seen as scientific and modern as opposed to guided by religious leaders they perceived as superstitious and too powerful. Cholera then came to be viewed as a disease found only among the poor and labourers.

Thinking shifted again when it crossed social classes and affected the affluent. This coincided with the vast rural out-migrations and the subsequent overcrowding in urban areas. Social and political leaders speculated that cholera was a consequence of the poor being in closer proximity to the affluent. This proximity resulted in the forced sharing of the common resource of air (miasmatic theory). In 1854 an Italian doctor, Filippo Pacini, identified the cholera bacteria but his findings were rejected as fanatical by the medical community who still ascribed to the class conscious miasmatic theory. In the same year, the English doctor, John Snow, identified the now famous Broad Street pump in London as a source of cholera and, although it was the same year and the same disease, Pacini and Snow's contexts were different and the outcomes were not the same. In response to Snow's findings the scientific gaze slowly shifted once more [80]. Understanding of cholera changed as technology, public sentiment, political power and scientific orthodoxy refocused. Efforts to prevent cholera aimed at improving the quality of air were probably somewhat effective because people moved from miasmic areas to the countryside where the water supply would also change. Efforts aimed at the wrong problem inadvertently still resulted in positive consequences. The attractor for cholera remained hidden until such time as a complex array of social, political and cultural forces interacted to facilitate a new way of looking.

A parallel situation could be proposed for current approaches to pain management. Chronic pain appears to continue to have hidden attractors and the pattern of the system remains elusive. Regardless of the input to the system of an extensive array of pharmaceutical, surgical, rehabilitative and other types of interventions, chronic pain remains an expanding, unresolved health and social issue. The efforts may be less than effective because the hidden attractors are obscured in current ways of thinking about and understanding of chronic pain. Current thinking about chronic pain is firmly entrenched in the biomedical model. However, over the past 30 years the focus has shifted to the importance of functional outcomes as a treatment goal [59, 81]. As the pervasive influence of biomedical thinking re-adjusts and professionals and the public construct new models of health and illness, hidden attractors may be uncovered.

25.5.5 Constant Adaptation

A complex adaptive system is functional when it is adaptable. When strict regulation and linear problem solving are imposed the system becomes less adaptive. Theorists propose that central control can stifle adaptive responses and throw a functional system into stagnation, inertia and ultimately chaos.

Complexity theorists use the example of the Red Queen's comments in ***Alice through the Looking Glass***, where she explains to Alice that they must keep running just to stay in one place. To stop running is to fall behind [97]. The argument here is that too much central control in the form of rules and regulations can prevent participants from not only adapting to new demands on the system but from even maintaining their current position.

This situation occurs in treatment for chronic pain where efforts remain focused on arriving at the best solution across the board and on the national standardisation of intervention programmes [56, 58, 81–84]. While these initiatives are important, these types of policy statements often make little or no recognition of the need for problem solving within a community context. Because health issues, like chronic pain, are deeply embedded within multiple contextual dimensions, high degrees of central control may be an undesirable barrier to creative, pragmatic problem solving. In addition to requiring flexibility, problem solving within a CAS is iterative, requiring access to high amounts of information and the resources for feedback and reflection. These features are challenging to attain on a national, decontextualised level.

25.5.6 Experimentation and Pruning

> Ships can't steer if they are not moving, and living systems—such as organisations—cannot survive without change, challenge, variety and surprise.
> Flower and Guillame [85]

CAS theory is rich with examples of how experimentation and pruning have resulted in the emergence of exciting and effective innovations. Experimentation with putting telephone lines and binary code together set the stage for the Internet. Pruning of less robust contenders in the same product system (like the Betamax recorder and the 5 1/4″ computer disk) resulted in superior product design and efficiency [23, 24]. The pruning component of innovation is considered a critical and yet often overlooked issue in healthcare.

Zimmerman [86] proposes that a current problem for healthcare systems is letting go of the outmoded as new approaches are forced into pre-existing structures and processes. A system needs to be able to adjust to maintain its balance and conserve energy. Systems that have the freedom to take on new methods and abandon others allow for this dynamic interplay and can self-regulate. Loss of self-regulation, because rigid external control mechanisms have been applied, prevents the system from being adaptive. Petros argues that this type of external control, imposed consequent to a linear approach to healthcare, creates dysfunction and iatrogenic disease [87]. The balance in health has been lost through exclusive application of the scientific reductionist model, and other complex influences have been devalued. He suggests the evidence that this scientific biomedical approach has failed is due to the increasing numbers of patients who are "*…voting with their feet, to iridologists, reflexologists and any other 'ologist' who can supply their needs*" [87].

The reluctance to abandon routine practice and to implement alternative ways of thinking is evident within chronic pain. For example, there is often a resistance of stakeholders to uptake new information about the use of opioids and self-administered medication systems for non-malignant pain because of persistent beliefs (e.g. fear of addiction in the face of evidence to the contrary) [88, 89].

This finding is consistent with other studies where lack of uptake among healthcare professionals for evidence-based interventions has been repeatedly highlighted [34, 90, 91], and practices that lack an evidence-base continue to be viewed as desirable treatment components. For example, a review of the back pain literature revealed that relaxation, acupuncture, homeopathy and biofeedback all remain commonly employed forms of intervention despite systematic reviews which concluded that there was no evidence-base for the treatments' effectiveness [92–95].

25.5.7 Inherent Non-linearity

In linear systems effort and return have a direct relationship as compared to a CAS where small events can effect major and not necessarily predictable change. Conversely, large highly regulated and centrally controlled efforts will bear little result [12, 27, 96] and can even have negative, iatrogenic effects [87]. Kauffman illustrates this point with the example of automobile innovation. A wide diversity of sizes, shapes, wheels and drive linkages existed in the early twentieth century and creative efforts resulted in innovative improvements. Over time, the actual improvements required more and more effort to be input related to the system itself. The result of this need for increased effort was that the effort/output relationship was no longer linear in nature but rather required dramatic shifts in thinking to achieve any true change beyond aesthetics [97].

Perhaps the best example of non-linear treatment of chronic pain is the outcome of multidisciplinary programmes (MDPs). Although routinely recommended as the gold standard for chronic pain treatment, MDPs are resource and labour intensive for both the service provider and the person with pain. Also the evidence-base for selecting a multidisciplinary approach over other less intensive forms of service delivery is not strong [98] and the degree of effectiveness of different features of these programmes remains unclear [99]. A systematic review of 14 studies concluded that the outcome methodologies employed were for the most part flawed and no claim of economic effectiveness could be made in relation to the multidisciplinary team approach for chronic pain intervention [2].

Thunberg [100] cautions that poorly functioning teams, consequent to poor communication and a lack of shared values, can be less effective than sole practitioners. In Thunberg's review the healthcare providers in one chronic pain setting had concluded that effective treatment requires an organisation that is "*… loosely coupled,* *decentralized and organic in design…the character of a problem should determine the knowledge that is needed—not who has most power*" [100]. These service providers' conclusions and the features of other studies highlighted by Thunberg et al. are all characteristics of a complex adaptive system.

Treatment interventions do not follow a linear model for people with chronic pain. For example, a consultant who interrupts a patient to take a phone call at the wrong time is a very small action but can nudge the event towards large-scale positive or negative outcome. There are examples in public health of large-scale, centrally coordinated initiatives resulting in very little return. An example widely cited in the health promotion literature is that of the growing number of homosexual males who currently engage in unsafe sex. A range of explanations have been proposed and include the following mediators: increased scepticism in healthcare, resistance to authority, an overall lack of personal responsibility in society, message over kill and trait theory [101]. Crossley proposes that resistance to health promotion cannot be reduced to a single explanatory model but rather is embedded in the context of particular situations and interactions. For example, information about safe sex needs to be individualised to the group and venue. A pamphlet about condom use may be received differently by a group of 14 year-olds in a school health class than by sexually active homosexual men in a nightclub.

25.5.8 Systems Are Embedded Within Other Systems and Co-evolve

Sardar and Abrams [24] use examples from economics to explain the concept of embedded, co-evolving systems. They discuss how computers and electronic transactions have created a society where currency has become virtual; the printed monetary note is no longer the tangible representation of pension dividends, stock premiums, salaries and savings. This state was made possible through advances in computer technology. This computer technology meant that criminals no longer focused on physically stealing notes but instead applied this advancement to divert/steal other people's virtual money.

Law enforcement agencies were forced to evolve along with the new technology to prevent computer theft. Financial institutions are among the major funders of educational institutions training students and researchers to develop anti-theft software. In turn, the software can be applied for either legal or illegal gain. As one aspect of the system evolves, so do the other aspects. It is the interaction between the aspects that allows for creative new behaviours to emerge and creates a system greater than the sum of its parts.

The proposition that systems are embedded and co-evolve is strongly evident in the arm of pain research that focuses on the bio-psychosocial framework to conceptualise the chronic pain experience. Emotions [102], race and ethnicity [103], psychological well-being [104], the influences of partners [105] and employers [106] have all been shown to impact the experience of pain. Roberts' [41] study of Welsh speakers with pain and their interactions with English-speaking care providers offer some interesting insights into the erroneous assumptions about the language of the clinical encounter and chronic pain. Kalvemark et al. [107] suggest that advanced medical technology, organisational reform, the evolving business ethos of healthcare and a more educated and consumer-focused population have contributed a new and growing element of an ethical dilemma regarding the present complex health services delivery system. The list of elements embedded within systems pertaining to healthcare and chronic pain is extensive and ever expanding as new research is reported.

Time, place, cultural expectations, financial situation, social roles and responsibilities, learned behaviours from childhood and access to healthcare services have all been identified as features acting on, and in turn, being influenced by the person with chronic pain. The recent World Health Organisation publication ***Innovative Care for Chronic Conditions: Building Blocks for Action*** [38] highlights the growing imperative to reframe healthcare. The challenge of the twenty-first century is to manage chronic, lifestyle-related diseases.

25.6 Effecting Change

The preceding section has described the growing awareness within the community of health researchers and policy makers to refocus medical care within a CAS framework. The section also highlighted ways in which CAS characteristics apply to the chronic pain experience. Complexity theory may offer a route to reconcile and legitimise the diverse range of theoretical perspectives evident in the field of chronic pain management. Applying a complexity theory framework can also help stakeholders avoid the counterproductive sparring inherent in linear thinking that is based on either/or choices. Complex adaptive systems theory stresses that systems, made up of a myriad of interacting and idiosyncratic elements, require a range of explanatory models depending on the circumstances and context. The focus shifts to an orientation of this/as well as that and on an understanding of the relationships that move these systems in certain directions. A theory however is only as useful as its application and CAS theory offers many examples of how innovative and pragmatic management strategies can be applied to produce a positive outcome.

Emerging management strategies for influencing outcomes in complex systems focus on flexible simple rules as opposed to highly structured and detailed engineering of solutions. The concepts of having a good enough vision that balances control and flexibility, adjusts to the needs of both safety and risk, values diversity and free flow of information, accepts paradox and dissent as opportunities for innovating new ideas and implements small actions as opposed to applying one big solution are all seen as tools for effecting change in complex adaptive systems. Additional tools include accepting the power of, and working with informal organisational systems (e.g. conversations around the coffee machine) and seeking solutions through local level initiatives as opposed to central control [11, 20, 28, 36, 108–110]. The literature emphasises developing simple rules that can be creatively applied in ways that accommodate local context and circumstances.

25.7 Effecting Change in the Complex Adaptive System of Chronic Pain

> Pattern and creativity are the two poles of action. … It is precisely this ability to discern and manipulate patterns unknown to the ordinary person that makes the followers of Tao so formidable. When unpredictable things happen, those who follow Tao are also skilled at improvisation. … To avoid confusion they still discern the patterns of the situation and create new ones, much like a chess player at the board.
>
> Deng Ming-Dao [111]
>
> Miss Marple is able to solve difficult crimes not only because of her shrewd intelligence, but because St. Mary Meade, over her lifetime, has put on a pageant of human depravity rivalled only by that of Sodom and Gomorrah. No crime can arise without reminding Miss Marple of some parallel incident in the history of her time.
>
> The Free Dictionary [112]

At first glance, there seems little commonality between one of Agatha Christie's popular fictional characters, a contemporary Taoist philosopher and this chapter's aim of applying CAS theory to chronic pain intervention. But in actuality Miss Marple, Taoist philosophy and complexity theory all share a fundamental characteristic that the key to solving new problems is to look for patterns. Identifying patterns allows one to employ an element of past learning and experience. Anxiety and fear are reduced when recognisable patterns emerge and this, in turn, frees up energy for creative problem solving, as opposed to taking a protective stance towards the perceived threat of the unknown. Complexity theory reiterates that many contemporary problems are a consequence of highly interactive contexts and agents and cannot be reduced to a single cause-and-effect analysis. Complex problems can only be successfully addressed through strategies focused on understanding patterns and interrelationships.

Positivist, linear thinking for chronic pain management has led to repeatedly asking the question: what is the best approach? I would argue that this question actually contributes to the problem of chronicity and keeps the system fixed in a perpetual state of search for cure. From a CAS perspective, we should be asking the question: what is keeping the system in its current state (i.e. chronic unresolved pain)? Complexity scientists propose that healthcare services remain resistant to change mainly because the wrong tools are being applied to a flawed construct of the problem [14, 113–115]. Management styles that were effective in addressing previous, more linear problems in healthcare are now perceived to be a major barrier to deal with the emergent health needs of the twenty-first century [37]. Zimmerman et al. [20] propose that there are nine leadership principles to influence complex systems to facilitate adaptive and positive change:

1. Apply a complexity framework.
2. Provide minimum specifications, rather than exhaustive criteria and protocols.
3. Balance planning and acting, risk and safety.
4. Foster the right degree of information flow, volume and exchange.
5. Expect and employ surprise as opposed to suppressing paradox and tension.
6. Recognise that action is necessary for a recursive system.
7. Recognise the importance of "informal" systems.
8. Allow complex systems to emerge from the inter-relationship of simple, independent systems.
9. Mix cooperation with competition [20].

This section will examine each principle within the context of service delivery for people with chronic pain to identify factors that facilitate or impede the system's movement away from the current attractor of chronicity.

Principle 1: Apply a complexity framework

Current healthcare services are a product of a historically positivist paradigm. Healthcare is also described in predominantly militaristic and mechanical metaphors such as the war on cancer, fighting to his last breath, the invasion of bacteria, the heart is a pump in the circulatory system and food to fuel the body. This perspective is deeply entrenched and many stakeholders such as managers [116] and high-status service providers [117] may have a clear interest in maintaining the status quo. Some theorists have proposed that the current drive for evidence-based medicine runs a significant risk of being subverted into a tool to maintain the existing power structure and devalue alternative perspectives [118]. It has also been suggested that the existing healthcare system has a depersonalised culture that does not allow for reflexivity and experimenting with new manners of relating and behaving [119, 120]. Rather, it deals with the inherent conflict generated by its increasing inability to manage through the imposition of yet more structure and partialisation in the guise of professionalism [108].

A recent hospital-based study in the United States found that over 60% of the managers in the study supported the concept that healthcare is a complex system. However, at the same time, half of the participants felt that successful leadership needs to "*control the parts of the machine*", and one-third thought that successful leadership required "*strong direction and control*" [28]. The researchers concluded that, although many participants recognised that healthcare had the features of a complex system, "*the same leader who is frustrated when trying to control the chaotic, complex healthcare environment is also uncomfortable giving up control and allowing the complex adaptive system to adapt and evolve toward an unpredictable outcome*" [28].

Although in healthcare, there exists a well-established force to apply a reductionist lens to healthcare, a contrary movement is growing. Authoritative voices, which cannot be marginalised or discounted as heretical, have emerged from within respected establishments of the medical profession. Examples exist that complexity science is growing in acceptance and value within the healthcare system. For example, the British Medical Journal (BMJ) ran a series of four articles focusing on aspects of complexity science: what it is, why it is relevant to healthcare, educating practitioners and how to manage within complex systems [18, 27, 36, 37]. As well, a book about the complex world of the general practitioner (***Friends in Low Places*** [1]) was written with the support of a sabbatical approved by the NHS Executive ([121], p. 2), by the provost of the Wessex Faculty of the Royal College of General Practitioners. Highly regarded universities and scholars have established complexity organisations and networks,[2] and have taken the initiative to disseminate information about healthcare and complexity through a range of publications (e.g. ***Complexity and Healthcare: an Introduction*** [12]), newsletters and conferences. These legitimising forces lay a firm foundation for applying a CAS framework to chronic pain.

Principle 2: Provide minimum specifications

> Even in traditional 3-by-3 tic-tac-toe, the number of distinct legal configurations exceeds 50,000 and the ways of winning are not immediately obvious. ([25], p. 23)

Simple rules serve to favour adaptation and survival of a complex system. A common example in the complexity literature is that of a flock of migrating geese. Geese on migration have three simple rules: fly at the same speed, stay as close to the centre as possible and avoid collisions. Complex behaviours emerge when these rules are followed. It is easy to see why a more linear style of migration where there is a CEO goose, a head catering goose and a security goose charged with watching out for hunters would not work, particularly if any of the geese with specialist roles were unsuccessful in avoiding the hunters in their flight path.

Healthcare has also been criticised for having become overspecialised, lacking in the ability to understand the relationship between the parts and to respond flexibly to emerging and unanticipated events. There is a difference between systems requiring clock-work organisation (e.g. a car assembly line) and complex adaptive systems that are most successful when swarm-work principles are in place. For complex systems, central control and micro-management impede responsiveness and adaptation to change.

Theorists propose that three types of simple rules exist within CAS: those that set direction (pointing), those that set boundaries (prohibition) and those having to do with required resources (permission) [15, 122]. Kernick [122] offers examples of simple rules that can be applied to developing more responsive healthcare services:

1. Accept that death, sickness and pain are part of life.
2. Medicine has limited power, particularly to solve social problems, and is risky.
3. Doctors do not know everything—they need decision-making and psychological support.
4. We are all in this together.
5. Patients cannot leave problems to doctors.
6. Doctors should be open about their limitations.
7. Politicians should refrain from extravagant promises and concentrate on reality.

These is some evidence that these types of rules are gaining credibility and emphasis within the current healthcare system. The WHO has clearly stated that medicine is limited in its power to address social and lifestyle problems which are fast becoming the major health concerns of the twenty-first century [38] and they are actively structuring programmes around this philosophy. The rise of public health initiatives (like anti-smoking campaigns) echoes many of the sentiments in Kerrnick's rules, as does the NHS's efforts to develop a culture of shared responsibility between service users and providers such as with the Expert Patient [123] and Concordance programmes [124]. Additionally, service providers

1 "Focusing on the inadequacy of models to reflect or predict the infinite subtlety of human behaviour, and the false promises of the Evidence-Based Medicine movement, Willis encourages us to have faith in our own intuitions as doctors, teachers, managers, or in whatever roles we play in relating to other human beings". Review by Douglas Jeffries, GP, on the Amazon.UK website (http://www.amazon.co.uk/exec/obidos/ASIN/1857754042/qid=1093000452/sr=1-6/ref=sr_1_8_6/026-6750948-5058031).

2 Centre for Complexity Research—http://www.liv.ac.uk/ccr/, The Complexity Society—http://www.complexity-society.com/, Plexus Institute—http://www.plexusinstitute.org/services/Fractal_Networks.cfm, Complexity in Primary Care—http://www.complexityprimarycare.org, and Health Complexity Group—http://www.healthcomplexity.net.

are receiving much more extensive training in communication skills and reflective practice, both of which can contribute to the acceptability of and willingness to abandon the all knowing-all powerful professional image. Of particular importance to the area of chronic pain, the International Association for the Study of Pain (IASP) [125] has written communication skills into its recommended curriculum for service providers [125] and is currently involved in updating and expanding these recommendations.

The IASP and the British Pain Society [58] have also published guidelines for good practice in the delivery of services for people with chronic pain. The Pain Society good practice guidelines outline nine objectives for chronic pain services and include the following:

1. Alleviation of pain. This is not always possible because any pain that is described as chronic has already proved resistant to treatment.
2. Alleviation of psychological and behavioural dysfunction and distress.
3. Reduction of disability and restoration of function.
4. Rationalisation of medication.
5. Reduction of utilisation of healthcare services including consultations in primary and secondary care, surgical operations and treatments such as physiotherapy.
6. Attention to social, family and occupational issues.
7. Education for nursing, medical staff and other allied healthcare professionals.
8. Continuing audit and evaluation of the service and the needs of patients. Outcome measures for patients with chronic pain should include assessment of physical functioning, psychological status, medication consumption, utilisation of healthcare resources and work record in addition to measurement of pain intensity.
9. Research into the epidemiology, causes and management of chronic pain.

These nine objectives may appear complicated and they are predominately focused on process. As well, the direction pointing and setting of boundaries/prohibitions that characterise simple rules is not easily apparent. However, as Plsek points out healthcare is already a complex system and all complex adaptive systems follow simple rules. The challenge is to identify what those rules are [126]. Just because the system's rules are not overt does not mean they are not present, but rather simply that they are unknown.

It appears from this review that the existing IASP and British Pain Society objectives for chronic pain currently reflect the prevalent biomedical framework whose simple rules promote as opposed to preclude chronicity. The IASP and British Pain Society guidelines do however make some important departures from the biomedical paradigm. For example, item number 6 highlights that there is a social and occupational context to the chronic pain experience. A re-examination from a complex adaptive system perspective could serve as a catalyst to change the system's rules and attractors from chronicity towards resolution.

Principles 3, 4 and 6: Planning and acting, information and the iterative process

Three CAS leadership principles are inherently interdependent: principle 3—balance planning and acting, risk and safety; principle 4—foster the right degree of information flow, volume, and exchange; and principle 6—recognise that action is necessary for a recursive system. Consequently, this section will deal with these three principles together.

Successful modification of a CAS depends on an iterative cycle of action, feedback, modification and new action. As Kernick points out, this requires a shift in values from the prevalent model of highly prescriptive healthcare planning that attempts to control for every contingency towards a good enough vision. Good enough to address basic safety and risk concerns by setting basic parameters while at the same time allowing timely action and flexibility to modify actions as new information becomes available [122]. Successfully balancing planning and doing within a good enough vision can only be possible when feedback information is available. Without action the system lacks feedback. Without feedback the system cannot modify action. In a 2004 study [127], participants with chronic pain reported that they believed new strategies were hard to implement in the healthcare system and that old practices were hard to abandon. These participants also reported experiencing differential access to the type and amount of information that was made available to them about chronic pain.

There are a number of developments within healthcare service delivery that can facilitate a shift in vision, allowing for more rapid cycles of planning, acting and modifying. For example, the idea that there are different types of evidence required for different types of questions has become an overt and legitimate debate taking place within mainstream research publications [117, 128–130]. This idea differs from the positivist positioning of the randomised control trial at the top of the hierarchy of evidence. Some researchers have proposed that standardisation of trials, a key feature of RCTs, to prevent any contextually influenced fluctuations in process can actually result in the treatment erroneously appearing to be ineffective. Because the study design is uniform across sites, this serves as a deterrent to the usual and necessary, local level adaptation within complex systems and therefore could actually be one of the reasons for the failure of an intervention [129]. The British Medical Research Council (MRC) has responded to this concern by developing guidelines for designing RCTs for complex health interventions [35]. These types of initiatives not only provide guidance for those wishing to engage in research but also serve to alert funding bodies about the need for more inclusive definitions of high-quality research.

The editor of the British Pain Society's newsletter demonstrated how the challenge of applying national resources at a local level in response to a local need can be taken up. He told readers that: "*At a meeting last week, our pain management services came under threat for the first time …. together with varicose vein surgery, breast reduction surgery and a 'dizziness clinic' run by the ENT surgeons, we have been asked to establish our worth*" [131]. Going on to highlight that there are excellent national reports clearly identifying the high cost of chronic pain to both the individual and the public [56, 82] he points out however, "… *we can publish all the documents we like about how important our service is and how wonderful we are but if the commissioners don't even understand what we do on a day-to-day basis then we are sunk*" (p. 2). Instead of calling for a study of why the commissioners do not understand and asking the membership of the British Pain Society to strike a committee to plan a survey of who does not understand what, and how best to tell them as well as a committee to fundraise for this survey and planning, Ward suggests several simple, local activities that each member can participate in with the least amount of time and financial burden:

1. Send copies of important national reports to the local Primary Care Trusts (PCT).
2. Ask service users to write to the PCTs about the benefits of the service.
3. Invite the PCT service development officer to attend a clinic.
4. Request that Member of Parliaments write to the Secretary of State asking why action has not been taken on these reports.

Because chronic pain and healthcare are CASs, there is no one correct action within this list and different activities will have different effects given the local context and relationships between the agents. Different members of the British Pain Society can select from a range of simple options based on what they know about their own local context. They can balance the need for action against the degree of risk and identify local agents of change, while knowing who are the people to target within both formal and informal local systems. CAS theory maintains that predictions about the outcome of these interventions can be made with some accuracy based on an understanding of the patterns within the inter-related systems. For example in this case, one can predict that issues brought forward by constituents to their MPs will receive attention because retaining political position depends on public votes. Likewise, the PCT chairperson will be influenced by what local service users are asking for. The author's use of his position as editor of a national, multidisciplinary newsletter to reach a wide constituency is itself an example of applying a CASs approach to change. The multidimensional nature of the membership precludes a linear approach to the problem of poor understanding of pain services. Rather than the British Pain Society issuing a directive to its members specifying who to contact dependent on your profession and rank in the organisation, what form the communication should take dependent on whether you have access to e-mail or a day-time phone number, and when the action should occur dependent on your shift rotation, sittings of the House of Parliament or annual leave, Ward has provided a simple goal: "*... we need to do some thing about our profile and make them understand*". The action is left to the local level to strategise, who, where, how and when.

Principles 5 and 9: Paradox and tension

> Needless to say, surprise and complexity are the norm and not the exception.
> Crabtree [132]

Two principles, principle 5—expect and employ surprise as opposed to suppressing paradox and tension and principle 9—mix cooperation with competition, are presented by many complexity theorists as key tools to effect change. Kernick states that when paradox and disagreement are suppressed people will become self-protective and not engage in the types of risk taking and information sharing necessary for creative problem solving and recursive organisational learning [122]. He goes on to propose that increased attention to the process and interaction of health services delivery can facilitate the organisational trust and learning needed to adapt to complex events. Other theorists stress the importance of making dissent and conflict overt [30, 116]. This attitude to conflict does not come easily to many current health service providers and will interfere with their ability to take on new policies and behaviours [133]. For example, a study of interprofessionals working across practices in the same health authority concluded that an interprofessional culture was lacking and that "*it seems probable that it will take a new generation of health professionals to bring about an interprofessional culture in the NHS*" [134].

The area of pain management presents interesting paradoxes. One paradox is when situations exist that can be used either to attempt reconciliation of service providers' beliefs and values or to facilitate the creative process that can emerge from a non-prejudiced sharing of ideas. For example, the British Pain Society has an interdisciplinary membership and has stated that pain management programmes need to be multidisciplinary in nature. Depending on the level of trust and respect within these groups, ideas can either be exchanged or normed towards a party line. However, as service providers in the area of chronic pain are members of the larger healthcare community it is not surprising to find that they share much of the communication problems, culture and ethos of that predominantly reductionist system.

A second paradox is that despite numerous educational strategies used to implement the evidence-base in pain management, authors have concluded that there is little evidence that clinical practice has actually been much enhanced [135]. Gordon and Dahl [135] speculate that this failure to see change proportionate to the effort exerted is because the wrong questions are being asked. They propose that it is not the technical aspects of pain management that require attention but rather the system itself. "*Quality pain management depends on a host of complex relationships and processes... little is known about the relationships among these processes and how they impact patient outcomes*" [135]. They propose that continuing to examine systems' problems with clinical tools is akin to trying to break the sound barrier by tinkering with a Model T Ford. The important role of issues that arise within the delivery system, as opposed to clinical pathology, has also been identified in the chronic pain literature as "*black and blue flags*" [136]. The flag system (red is for biomedical, yellow is for psychological, blue is for socio-economic and black if for occupational) models the complex nature of chronic pain and can help service users and providers move away from a linear understanding of the chronic pain experience.

In summary, paradox and disagreement exist within chronic pain. Traditionally, efforts have mirrored the values of the wider healthcare systems and focused on reducing disagreement and suppressing the expression of open dissent. Dealing constructively with disagreement can generate positive and creative changes but these inter-relational skills are not typically part of a service provider's preparation. Employing the warning flag model would help facilitate a more open appreciation of the complexity of chronic pain.

Principle 7: The importance of informal systems

Healthcare and chronic pain are complex systems. Complexity theorists stress that the human and organisational agents that inter-relate within these systems are CASs in themselves. In other words, complex systems are embedded and have overlapping boundaries with other systems. Complexity scientists stress that this highly interactive nature must be properly understood and managed to effect positive change [20, 137]. Shaw proposes that one of the key opportunities for managing change exists in the informal, shadow system. The shadow system is created through unofficial communication networks (e.g. lunch break conversations and waiting in line for the photocopier) and expresses the felt beliefs of agents in the system. This is opposite to the authorised, official position that is communicated through more legitimised channels such as staff meetings and policy statements. The shadow system and the formal organisation may actually hold divergent beliefs, sending conflicting and destructive messages. These messages in turn are subject to the idiosyncratic interpretation of individual agents within the context of their own social, political and psycho-dynamic processes [138]. To ignore the shadow system is potentially destructive to all change efforts. Traditionally healthcare has attempted to deal with the informal system through increased command and control efforts, taking a position of attempting to control behaviours and drown out dissenting voices with the volume of scientific evidence [130].

However, the shadow system can also be a creative force, allowing novel ideas to emerge from the dynamic interplay of multiple interpretations and perspectives made possible by the system's inherent lack of formalised rules and structure. Zimmerman et al. [20] state that the coexistence of both systems create diversity, tension and paradox, which in turn generates ideal circumstances for change within a CAS. To achieve positive adaptation, both systems' energy needs to be focused on listening to and working with as opposed to battling against each other.

Principle 8: Allow complex systems to emerge

Complexity theorists use the Internet as an example of a successful CAS. The Internet is an emergent system, evolved from many components that combine and interact in an iterative manner. What works is integrated into the system, what does not work is abandoned and the system moves on. The Internet like other complex systems was not designed; it evolved as linkages were made between the smaller components. The smaller components in themselves (e.g. the telephone line and the video display screen) were designed for entirely separate purposes but, when combined in new ways, resulted in unanticipated results. The successful or good enough combination of components are retained in the system, and the obsolete are abandoned (e.g. the 5 1/4 floppy disk). It is proposed that the most successful approach to building and refining complex systems requires the bottom-up application of a series of small chunks (simple, independent systems) with careful attention to the patterns of interaction that emerge.

A number of these chunks are evident when looking at chronic pain as a CAS and can include the following: the pharmaceutical industry, National Health Service, British Pain Society, local PCTs, treatment team, person with pain, public transport system that takes people to and from appointments or a malfunctioning automatic coffee machine that leaves service users in the overcrowded waiting room without refreshment. Zimmerman et al. [20] propose that adopting this perspective helps the system's stakeholders think in new and creative ways and move from an immobilising and futile search for the problem and its solution towards taking small actions that provide relatively rapid feedback and opportunity for re-evaluation and refinement.

Freyer (http://www.trojanmice.com) uses the imagery of the Trojan horse to illustrate the difference between a linear, large-scale, command

and control approach to complex problems and Trojan mice, small, well-focused and easily maneuverable initiatives focused on building creative relationships between existing chunks. Trojan mice reflect the CAS's principle that change is not incremental; small inputs can have unexpectedly large return.

Many of the initiatives such as the NHS Modernisation Agency, multidisciplinary nature of the British Pain Society and IASP, the legitimisation of complexity thinking in highly regarded journals like the BMJ and peoples' ever-growing access to information available through the electronic media can be used to facilitate an examination of the relationship between the chunks and to empower people to take action, regardless of how small. "*You cannot reflect on anything until you do something. So start small, but do start*" ([20], p. 41).

This paradigm shift is essential because the existing perspective no longer works for many dominant issues. A CAS paradigm presents a radical challenge to the prevalent biomedical model, and also affords a non-threatening approach through conceptualising a meta-framework grounded in the assumption that a range of theoretical models are required. However, the challenges are high to broaden the vision of health service delivery, so that both linear and complex approaches can be employed as the situation indicates.

Examples of potentially overwhelming barriers exist in areas such as political agendas, healthcare organisation and with healthcare workers. The political agenda that pervades healthcare planning creates unrealistic expectations of the pace at which change can occur [11]. The traditionally hierarchical organisation of service delivery is resistant to perceived challenges to power and control, and pragmatically the cognitive dissonance, consequent to equally important but contradictory demands, experienced by many healthcare workers are several examples of potentially overwhelming barriers.

Also the tools for understanding the patterns of systems and how to change them are found not only in Cochrane reviews and the labyrinth-like government web pages but also with reassuring accessibility in the stories about Miss Marple, Taoism, metaphors and imagery, children's games and the migration of flocks of geese. Because nature is a CAS in itself, the opportunities to understand patterns and interrelationships are commonplace. Kernick suggests that a starting point for this paradigm shift is to change the metaphor we apply to health services. Instead of the modernist view of a machine that can be engineered, we should view healthcare as an ecosystem. "*Each agent cannot be understood in isolation. All parts adapt by learning to survive in a topography that is provided by coexisting and changing parts. Ecosystems cannot be engineered, … . However, they can be nurtured*" [122]. The final section of this chapter will present recommendations on how this nurturing process of change can be facilitated in particular reference to the area of chronic pain.

25.8 Recommendations

The following recommendations identify strategies for affecting change, in relation to these three conclusions, based within CAS theory.

25.8.1 Identify Existing Patterns

> …all clinicians, whatever their degree of subspecialisation, have to manage an irreducible uncertainty. It is not the consequence of ignorance of fact or failure of logic. It is rather the inevitable consequence of all those givens addressed by 'complexity'—the multiplicity, the multilevel, diverse, interconnectedness and dynamism, of events.
> Marinker [139]

To manage uncertainty we need to seek patterns and learn lessons from what is already known about how complex agents interact within embedded systems. There is a growing body of healthcare literature that people concerned with chronic pain can refer to, for example of how CAS principles have been successfully applied. Patterns and examples of good practice in creative strategies for intervention are not condition specific and can be drawn from across a range of other chronic illness. Recent publications contain examples from clinical areas as diverse as community regeneration, public health, acute-care hospital bed control [140], diabetes, cardiac care and mental health service delivery [141]. Exploring these examples will help chronic pain service providers identify patterns within their own area of expertise. Likewise, lessons from this research can be applied to develop a deeper understanding of the issues across a range of chronic healthcare conditions sharing similar features with chronic pain.

Primary care research offers another route to identify patterns that serve to maintain chronicity as an attractor in the pain experience. Research that utilises the spectrum of methods needs to be supported and legitimised. As evidenced by the range of scientific approaches presented at the British Pain Society Annual Scientific Conferences and the content of the IASP publication Pain this shift has already begun. But attention is still needed to withstand the counter forces that tie research funding and professional credentials to more linear types of inquiry. The pharmaceutical industry is a major source of funding for pain research and allocates funding based on a commercial agenda. In academia, research expertise is recognised through audit-driven initiatives such as the British system of Research Assessment Evaluation (RAE). The RAE is funded by the government's Higher Education Funding Council for England (HEFCE) and uses a researcher's publication in high-rated scientific journals as an indicator of quality. Journals with the high citation ratings desired by the HEFCE predominately reflect a positivist research agenda. This systemic valuation of positivist research over other approaches makes it difficult for questions of a more complex nature to be addressed.

Identifying existing patterns allows the simple rules maintaining chronicity to become overt. The research concerning differences of opinion between and within groups of service users and providers also demonstrates a pattern that efforts to shift beliefs towards those held by the most powerful in the hierarchy are often disproportionately low in relation to the effort exerted. Research also shows that people in a range of healthcare situations are selective in what they disclose and often do not deal with conflicting opinions in an overt way [50, 142, 143]. A linear perspective applied to these three examples would pose the question: How can we get people to agree more? Using CAS theory the question is how do we make it possible for people to express their beliefs so we can communicate and collaboratively develop a way forward? The simple rule that emerges when linear problem solving is applied to complex issues is that there is a right way that everyone must follow.

In the CAS perspective the rule is that open communication facilitates problem solving. An illustration of this point would be a chronic pain programme, operating with rules from a linear problem-solving lens where service users are informed that attendance is mandatory for all sessions regardless of the service user's belief in the value of each individual component. Research has shown that operant behavioural approaches may be effective in the short term but the behaviour is extinguished once outside of the treatment environment and its reinforcement system. For example, if a person does not value relaxation therapy, forced participation is more likely to have negative consequences (hostility, frustration, disruptive behaviour) and provide fuel for counter-productive positioning in the shadow system. The simple rule that everyone must agree generated by applying linear thinking to complex problems serves to reinforce chronicity.

25.8.2 Encourage the Expression of Dissent

To understand the true patterns and simple rules at play in a system people must freely express their beliefs. Much has been written about the need for more collaborative communication within health services delivery and a number of initiatives exist to teach people the necessary skills. Dissent and expression of alternative views to the status quo generate energy that can either be used destructively or to facilitate creative change. The healthcare system is often resistant to change and historically we see that shifting course requires skilled leadership [11, 14, 126, 137]. It has been proposed that the biggest barriers to achieving new ways of relating in the current healthcare system are the incumbent leaders who achieved their current positions of power and status through success within a system based on command and control-based hierarchy [37]. That being the case, initiatives like the NHS Modernisation Agency's ***Improvement Leaders' Guidebook*** series [144] were important tools because they offered both practical information and perhaps most importantly, they legitimised efforts to increase information flow and flatten decision-making structures. Similarly strategies developed by the ***Expert Patient Programme and Involve*** [145, 146], can be employed to facilitate open exchange of ideas and opinions by involving service users in the actual design and evaluation aspects of programming.

25.8.3 Create New Simple Rules

New rules are needed to move chronic pain away from chronicity. Identifying existing patterns and encouraging new forms of communication and information flow are important parts of that process. Equally the process of creating new simple rules is key. The NHS Modernisation Agency can serve as an example of this. They set out five simple rules to guide service delivery in healthcare [147]:

1. See things through the patient's eyes.
2. Find a better way of doing things.
3. Look at the whole picture.
4. Give frontline staff the time and tools to tackle the problems.
5. Take small steps as well as big leaps.

Providing simple rules is not sufficient. A mechanism by which to make simple rules meaningful at a local level and within the context of chronic pain is also required. Without this additional component rules easily become rhetorical, lacking sufficient specificity to fulfill their purpose of setting direction, prohibitions and providing permission [126]. Plsek proposes that there are key questions to ask when designing simple rules and that taking a questioning approach will facilitate the adoption of good-enough strategies and the openness required to evaluate, modify and nurture the system's evolution.

There are a number of mechanisms in place for service users and providers to ask and discuss questions specific to chronic pain within a complexity framework. The various national and international pain society meetings provide an excellent venue for multidisciplinary sharing of ideas and the opportunity to identify common patterns of and participate in facilitated application of CASs principles to emerging issues. Complexity science theory can be incorporated into healthcare research and teaching so as to widen awareness of issues across disciplines and increase the acceptability of the framework. Involving other disciplines holds the potential to stimulate new ways of thinking and would also validate the relevance to healthcare of the principles of CASs theory. Local chronic pain programmes that incorporate patient and stakeholder feedback sessions and public forums are another potential source of ideas exchange.

It is beyond the scope of this research to suggest what the simple rules should be for people with pain. However, questions that can guide that process should include attention to all three of the features identified by Plsek: direction, prohibition and permission.

For example, possible questions guiding the process include:

1. What are the rules currently preventing people from accepting increased function as a goal?
2. What are the rules keeping people focused on searching for a cure?
3. What can be done to make it possible for people to express their opinions openly?
4. What are the linear components of the system that need to be managed in a traditional way (e.g. determining the dosage of analgesics) and what are the more complex components that would benefit from less control and more flexibility to respond (e.g. which components of a chronic pain management program can the service user opt to participate in)?
5. What do service users and providers stand to lose by taking on new ways of communicating

and interacting (e.g. time, status, self-protection, control)?

6. How can service users and providers avoid a culture of blame and promote willingness to try new ideas and abandon outdated ideas?
7. What are some achievable, good enough steps to start the change happening?

The preceding list is by no means inclusive and, given the iterative nature of working in complex systems, it will likely change with each reading. Similarly this chapter is not the final word on complexity and pain. Rather, its usefulness lies in being a Trojan mouse prompting us to think about chronic pain from a different perspective that recognises paradox, conflict and uncertainty as givens and most importantly, as tools to facilitate positive change in complex and ever-changing human systems.

References

1. Maniadakis N, Gray A. The economic burden of back pain in the UK. Pain. 2000;84(1):95–103.
2. Thomsen AB, et al. Economic evaluation of multidisciplinary pain management in chronic pain patients. a qualitative systematic review. J Pain Symptom Manage. 2001;22(2):688–98.
3. Waddell G. Low back pain: a twentieth century health care enigma. Spine. 1996;21(24):2820–5.
4. Loeser JD. Economic implications of pain management. Acta Anaesthesiol Scand. 1999;43(9):957–9.
5. Rey R. The history of pain. London: Harvard University Press; 1993.
6. Melzack R, Taenzer P. Concepts of pain perception and therapy. Geriatrics. 1977;32(11):44–8.
7. Sullivan S, Lewin ME. The economics and ethics of long-term care and disability. AEI studies. 467. Washington, DC: American Enterprise Institute for Public Policy Research; 1988. p. xii, 145.
8. Fillingham D. Open space. Take five. Health Serv J. 2002;112(5791):27.
9. Dershin H. Nonlinear systems theory in medical care management. Physician Exec. 1999;25(3):8–13.
10. Griffiths F, Byrne D. General practice and the new science emerging from the theories of 'chaos' and complexity. Br J Gen Pract. 1998;48(435):1697–9.
11. Kernick D. The demise of linearity in managing health services: a call for post normal health care. J Health Serv Res Policy. 2002;7(2):121–4.
12. Sweeney K, Griffiths F. Complexity and healthcare: an introduction. Oxon: Radcliffe; 2002.
13. Dent EB. Complexity science: a Worldview shift. Emergence. 1999;1(4):5–19.
14. Miller WL, et al. Understanding change in primary care practice using complexity theory. J Fam Pract. 1998;46(5):369–76.
15. Plsek P. Complexity and the adoption of innovation in health care. In: Accelerating quality improvement in health care. Washington, DC; 2003.
16. Zimmerman B. Complexity science: a route through hard times and uncertainty. Health Forum J. 1999;42(2):42–6, 69.
17. Chapman RC, Nakamura Y. Consciousness, complexity, and causality: concerns and conundrums. Pain Forum. 1999;8(3):136–8.
18. Plsek PE, Greenhalgh T. Complexity science: the challenge of complexity in health care. BMJ. 2001;323(7313):625–8.
19. Cilliers P. Complexity and postmodernism. London: Routledge; 1998.
20. Zimmerman B, Lindberg C, Plsek P. Edgeware: insights from complexity science for health care leaders. Irving, TX: VHS; 2001.
21. Godlee F, Smith J, Marcovitch H. Wakefield's article linking MMR vaccine and autism was fraudulent. BMJ. 2011;342:c7452.
22. Wikipedia. Andrew Wakefield. 2001. http://en.wikipedia.org/wiki/Andrew_Wakefield#Epidemics_and_effects. Cited 16 Apr 2011.
23. Stacey RD. Complexity and creativity in organizations. 2nd ed. London: Pitman; 1996.
24. Sardar Z, Abrams I. Introducing chaos. Cambridge: Icon; 1999.
25. Holland JH. Emergence: from chaos to order. Oxford: Oxford University Press; 1998.
26. Gleick J. Chaos. London: Vintage; 1998.
27. Wilson T, Holt T, Greenhalgh T. Complexity science: complexity and clinical care. BMJ. 2001; 323(7314): 685–8.
28. Burns JP. Complexity science and leadership in healthcare. J Nurs Adm. 2001;31(10):474–82.
29. Gillett G. Clinical medicine and the quest for certainty. Soc Sci Med. 2004;58(4):727–38.
30. Martin B. Dissent and heresy in medicine: models, methods, and strategies. Soc Sci Med. 2004; 58(4): 713–25.
31. Frost DP. Medicine, postmodernism, and the end of certainty. Complex systems result in a new kind of fundamental uncertainty. BMJ. 1997;314(7086): 1045.
32. Mays N, Pope C. Qualitative research in health care. Assessing quality in qualitative research. BMJ. 2000; 320(7226):50–2.
33. Tickle-Degnen L, Bedell G. Heterarchy and hierarchy: a critical appraisal of the "levels of evidence" as a tool for clinical decision making. Am J Occup Ther. 2003;57(2):234–7.
34. Freeman AC, Sweeney K. Why general practitioners do not implement evidence: qualitative study. BMJ. 2001;323(7321):1100–2.
35. Medical Research Council. A framework for development and evaluation of RCTs for complex

interventions to improve health. London: Medical Research Council; 2000. p. 1–18.
36. Fraser SW, Greenhalgh T. Coping with complexity: educating for capability. BMJ. 2001;323(7316): 799–803.
37. Plsek PE, Wilson T. Complexity, leadership, and management in healthcare organisations. BMJ. 2001; 323(7315):746–9.
38. WHO. Innovative care for chronic conditions: building blocks for action. Geneva: World Health Organisation; 2002.
39. Wordsworth W. The table turned. In: Matthew Arnold, editor. The poems of Wordsworth. London: MacMillan; 1947.
40. Griffiths F. Conclusions. In: Sweeney K, Griffiths F, editors. Complexity and healthcare: an introduction. Oxon: Radcliffe; 2002. p. 167–76.
41. Roberts G, et al. Describing chronic pain: towards bilingual practice. Int J Nurs Stud. 2003;40(8): 889–902.
42. Carter B. Chronic pain in childhood and the medical encounter: professional ventriloquism and hidden voices. Qual Health Res. 2002;12(1):28–41.
43. Hafner C. The information we give may be detrimental. In: Gifford L, editor. Topical issues in pain. Falmouth: CNS; 2002. p. 101–18.
44. Kurtz C, Snowden D. The new dynamics of strategy: sense-making in a complex and complicated world. IBM Syst J. 2003;42(3):462–83.
45. Prochaska JO, Velicer WF. The transtheoretical model of health behavior change. Am J Health Promot. 1997;12(1):38–48.
46. Dijkstra A, et al. Readiness to adopt the self-management approach to cope with chronic pain in fibromyalgic patients. Pain. 2001;90(1–2):37–45.
47. Lindbladh E, Lyttkens CH. Habit versus choice: the process of decision-making in health-related behaviour. Soc Sci Med. 2002;55(3):451–65.
48. Stevenson FA, et al. GPs' recognition of, and response to, influences on patients' medicine taking: the implications for communication. Fam Pract. 2000;17(2):119–23.
49. Salmon P, Peters S, Stanley I. Patients' perceptions of medical explanations for somatisation disorders: qualitative analysis. BMJ. 1999;318(7180):372–6.
50. Ong LM, et al. Doctor-patient communication: a review of the literature. Soc Sci Med. 1995;40(7): 903–18.
51. Hilbert R. The acultural dimensions of chronic pain: flawed reality construction and the problem of meaning. Soc Probl. 1984;31(4):365–78.
52. Johansson EE, et al. The meanings of pain: an exploration of women's descriptions of symptoms. Soc Sci Med. 1999;48(12):1791–802.
53. Kleinman A, Kleinman J. Suffering and its professional transformation: toward an ethnography of interpersonal experience. Cult Med Psychiatry. 1991;15(3):275–301.
54. Morris D. Sociocultural and religious meanings of pain. In: Gatchel RJ, Turk DC, editors. Psychosocial factors in pain: critical perspectives. New York: Guilford; 1999. p. 118–31.
55. Schussler G. Coping strategies and individual meanings of illness. Soc Sci Med. 1992;34(4):427–32.
56. Clinical Standards Advisory Group. A summary of the CSAG report on services for NHS patients with acute and chronic pain. CSAG Support Team, Unit of Health-Care Epidemiology, Institute of Health Sciences, 2000.
57. Caldron J, Beltran R. Pitfalls in health communication: healthcare policy, institution, structure and process. MedGenMed. 2004;6(1):9.
58. Society P. Pain management services: good practice. The Royal Collage of Anaesthetists and The Pain Society (British Chapter of ISASP): London; 2003.
59. IASP. Desirable characteristics of multidisciplinary pain centers. Seattle: International Association for the Study of Pain; 1990.
60. Bellon JA, Fernandez-Asensio ME. Emotional profile of physicians who interview frequent attenders. Patient Educ Couns. 2002;48(1):33–41.
61. Zinn WM. Doctors have feelings too. JAMA. 1988;259(22):3296–8.
62. Dempsey OP, Bekker HL. 'Heads you win, tails I lose': a critical incident study of GPs' decisions about emergency admission referrals. Fam Pract. 2002;19(6):611–6.
63. Dubouloz C, et al. Occupational therapists' perception of evidence-based practice. Can J Occup Ther. 1999;65:136–43.
64. Pollock A. Barriers to achieving evidence-based stroke rehabilitation. Clin Rehabil. 2000;14:611–7.
65. Gervais I, et al. Attempting to use a Cochrane Review: experience of three occupational therapists. Am J Occup Ther. 2002;56(1):110–3.
66. Belanger A. There are 'continuing education alternatives' to 'alternative continuing education' in physiotherapy. Physiother Can. 1997;49:249–52.
67. Funk SG, et al. Administrators' views on barriers to research utilization. Appl Nurs Res. 1995;8(1): 44–9.
68. Adler HM. The sociophysiology of caring in the doctor-patient relationship. J Gen Intern Med. 2002;17(11):874–81.
69. Novack DH, Epstein RM, Paulsen RH. Toward creating physician-healers: fostering medical students' self-awareness, personal growth, and well-being. Acad Med. 1999;74(5):516–20.
70. Freemantle N. Are decisions taken by health care professionals rational? A non systematic review of experimantal and quasi experimental literature. Health Policy. 1996;38:71–81.
71. Fernandez E, Turk DC. The scope and significance of anger in the experience of chronic pain. Pain. 1995;61(2):165–75.
72. Sanders SH. Integrating practice guidelines for chronic pain: from the Tower of Babel to the Rosetta Stone [e-journal] 2000;10(6). http://www.ampainsoc.org/pub/bulletin/nov00/clin1/htm. Cited 21 Nov 2002.

73. Peirce JC. The paradox of physicians and administrators in health care organizations. Health Care Manage Rev. 2000;25(1):7–28.
74. Jensen MP, Karoly P. Pain-specific beliefs, perceived symptom severity, and adjustment to chronic pain. Clin J Pain. 1992;8(2):123–30.
75. Dick BD, Rashiq S. Disruption of attention and working memory traces in individuals with chronic pain. Anesth Analg. 2007;104(5):1223–9. Tables of contents.
76. Flor H, Fydrich T, Turk DC. Efficacy of multidisciplinary pain treatment centers: a meta-analytic review. Pain. 1992;49(2):221–30.
77. Freeman M, Miller C, Ross N. The impact of individual philosophies of teamwork on multiprofessional practice and the implications for education. J Interprof Care. 2000;14(3):237–47.
78. Main CJ, Richards HL, Fortune DG. Why put new wine in old bottles: the need for a biopsychosocial approach to the assessment, treatment, and understanding of unexplained and explained symptoms in medicine. J Psychosom Res. 2000;48(6):511–4.
79. Burton C. Introduction to complexity. In: Sweeney K, Griffiths F, editors. Complexity and healthcare: an introduction. Oxon: Radcliffe; 2002. p. 1–18.
80. Halliday S. Death and miasma in Victorian London: an obstinate belief. BMJ. 2001;323(7327):1469–71.
81. Pain Society. Desirable criteria for pain management programmes. London: Pain Society, The British & Irish Chapter of the International Association for the Study of Pain; 1997.
82. Dr Foster and the Pain Society. Adult chronic pain management services in the UK. London: Dr Foster; 2003.
83. Health and Welfare Canada. Chronic pain programs: report of the Subcommittee on Institutional Program Guidelines. Ottawa: Health and Welfare Canada; 1990.
84. Pain Society. Recommendations for nursing practice in pain management. London: The Pain Society; 2002.
85. Flower J, Guillaume P. Surfing the edge of chaos. Health Forum J. 2002;March:17–20.
86. Zimmerman B, Dooley K. Mergers versus emergers: structural change in health care systems. Emergence. 2001;3(4):65–82.
87. Petros P. Non-linearity in clinical practice. J Eval Clin Pract. 2003;9(2):171–8.
88. Gardner JR, Sandhu G. The stigma and enigma of chronic non-malignant back pain (CNMBP) treated with long-term opioids (LTO). Contemp Nurse. 1997;6(2):61–6.
89. Haythornthwaite JA, et al. Factors associated with willingness to try different pain treatments for pain after a spinal cord injury. Clin J Pain. 2003;19(1):31–8.
90. Barnard S, Wiles R. Evidence-based physiotherapy: physiotherapists attitudes and experiences in the Wessex area. Physiotherapy. 2001;87(3):115–24.
91. Upton DJ. How can we achieve evidence-based practice if we have a theory-practice gap in nursing today? J Adv Nurs. 1999;29(3):549–55.
92. Ernst E. Homeopathic prophylaxis of headaches and migraine? A systematic review. J Pain Symptom Manage. 1999;18(5):353–7.
93. Ezzo J, et al. Is acupuncture effective for the treatment of chronic pain? A systematic review. Pain. 2000;86(3):217–25.
94. McQuay HJ, et al. Systematic review of outpatient services for chronic pain control. Health Technol Assess. 1997;1(6):1–135.
95. van Tulder M, Koes B. Low back pain and sciatica: chronic. Clin Evid. 2002;7:1032–48.
96. Redfern S, Christian S. Achieving change in health care practice. J Eval Clin Pract. 2003;9(2):225–38.
97. Kauffman S. Escaping the Red Queen effect. McKinsey Q. 1995;1:118–29.
98. Karjalainen KA, et al. Multidisciplinary biopsychosocial rehabilitation for neck and shoulder pain among working age adults. Cochrane Database Syst Rev. 2010;(3):CD002194.
99. Uhlig T, Finset A, Kvien TK. Effectiveness and cost-effectiveness of comprehensive rehabilitation programs. Curr Opin Rheumatol. 2003;15(2):134–40.
100. Thunberg KA, Hallberg LR. The need for organizational development in pain clinics: a case study. Disabil Rehabil. 2002;24(14):755–62.
101. Crossley M. Resistance to health promotion: a preliminary comparative investigation of British and Australian students. Health Educ. 2002;102(6):289–99.
102. Bruehl S, Chung OY, Burns JW. Differential effects of expressive anger regulation on chronic pain intensity in CRPS and non-CRPS limb pain patients. Pain. 2003;104(3):647–54.
103. Green CR, et al. The unequal burden of pain: confronting racial and ethnic disparities in pain. Pain Med. 2003;4(3):277–94.
104. Ong KS, Keng SB. The biological, social, and psychological relationship between depression and chronic pain. Cranio. 2003;21(4):286–94.
105. Miller J, Timson D. Exploring the experiences of partners who live with a chronic low back pain sufferer. Health Soc Care Community. 2004;12(1):34–42.
106. DeGood DE, Kiernan B. Perception of fault in patients with chronic pain. Pain. 1996;64(1):153–9.
107. Kalvemark S, et al. Living with conflicts-ethical dilemmas and moral distress in the health care system. Soc Sci Med. 2004;58:1075–84.
108. Anderson RA, McDaniel RR. Managing health care organizations: where professionalism meets complexity science. Health Care Manage Rev. 2000;25(1):83–92.
109. Lefebvre E, Letiche H. Managing complexity from chaos: uncertainty, knowledge and skills. Emergence. 1999;1(3):7–15.
110. Murphy M, et al. Consensus development methods, and their use in clinical guideline development. Health Technol Assess. 1998;2(3):i–iv, 1–88.
111. Ming-Dao D. 365 Tao: daily meditations. New York: HarperCollins; 1992.
112. TheFreeDictionary.com. Definition: Miss Marple. 2004. http://encyclopedia.thefreedictionary.com/Miss%20Marple. Cited 20 Sept 2004.

113. Keene A. Complexity theory: the changing role of leadership. Ind Commerc Train. 2000;32(1):15–8.
114. McKenna SD. Leveraging complexity: the middle manager's dilemma. Learn Organ. 1994;1(2):6–14.
115. Smith PA. Implications of complexity and chaos theories for organizations that learn. Learn Organ. 2003;10(6):321–4.
116. Nixon B. Creating the future s we desire-getting the whole system into the room: part 1. Inst Commerc Train. 1998;30(1):4–11.
117. Dean K. The role of methods in maintaining orthodox beliefs in health research. Soc Sci Med. 2004;58:675–85.
118. Byrne D. Complexity theory and social sciences: an introduction. London: Routledge; 1998.
119. Barr H. Collaboration and conflict. J Interprof Care. 2001;15(1):5–6.
120. Brodbeck P. Complexity theory and organisational procedure design. Bus Process Manage J. 2002;8(4):377.
121. Willis J. Friends in low places. London: Radcliffe; 2001.
122. Kernick D. Complexity and healthcare organisation. In: Sweeney K, Griffiths C, editors. Complexity and Healthcare: an introduction. Oxon: Radcliffe; 2002. p. 93–122.
123. Government of UK. The expert patient: a New approach to chronic disease management for the 21st century. 2002. http://www.ohn.gov.uk/ohn/people/expert.htm. Cited 17 Apr 2002.
124. Concordance. Welcome to the concordance website. 1999. http://www.concordance.org. Cited 20 Dec 1999.
125. Fields HL. Core curriculum for professional education in pain: a report of the Task Force on Professional Education of the International Association for the Study of Pain. Seattle: IASP Press; 1995. p. i–123.
126. Plsek P. Appendix B: Redesigning healthcare with insights from the science of complex adaptive systems. In: National Academy of Science, editor. Crossing the quality chasm: a new health system for the 21st century. Washington, DC: National Academy Press; 2002.
127. Brown CA. The beliefs of people with chronic pain in relation to 'important' treatment components. Europ J Pain. 2004;8:325–33.
128. Hyde P. Fool's gold: examining the use of gold standards in the production of research evidence. Br J Occup Ther. 2004;67(2):89–94.
129. Hawe P, Shiell A, Riley T. Complex interventions: how "out of control" can a randomised controlled trial be? BMJ. 2004;328(7455):1561–3.
130. Jones R. Schism and heresy in the development of orthodox medicine: the threat to medical hegemony. Soc Sci Med. 2004;58:703–12.
131. Ward S. Editorial. The British Pain Society Newsletter 2004;Summer.
132. Crabtree B. Primary care practices are full of surprises! Health Care Manage Rev. 2003;28(3):279–83.
133. Gillespie R, Florin D, Gillam S. How is patient-centred care understood by the clinical, managerial and lay stakeholders responsible for promoting this agenda? Health Expect. 2004;7(2):142–8.
134. Elston S, Holloway I. The impact of recent primary care reforms in the UK on interprofessional working in primary care centres. J Interprof Care. 2001;15(1):19–27.
135. Gordon DB, Dahl JL. Quality improvement challenges in pain management. Pain. 2004;107(1–2):1–4.
136. Main CJ, Williams AC. Musculoskeletal pain. BMJ. 2002;325(7363):534–7.
137. Stacey RD. Strategic management and organisational dynamics: the challenge of complexity. 4th ed. Harlow: Pearson; 2003.
138. Shaw P. Intervening in the shadow systems of organizations: consulting from a complexity perspective. J Organ Change Manage. 1997;10(3):235–50.
139. Marinker M. Forward. In: Holt T, editor. Complexity for clinicians. Oxford: Radcliffe; 2004. p. v–vi.
140. Kernick D, editor. Complexity and healthcare organisation: a view from the street. Oxford: Radcliffe; 2004.
141. Holt T, editor. Complexity for clinicians. Oxford: Radcliffe; 2004.
142. Elwyn G, Edwards A, Kinnersley P. Shared decision-making in primary care: the neglected second half of the consultation. Br J Gen Pract. 1999;49(443): 477–82.
143. Bell RA, et al. Unmet expectations for care and the patient-physician relationship. J Gen Intern Med. 2002;17(11):817–24.
144. NHS. Managing the human dimension of change: working with individuals. In: Modernisation Agency, editor. Improvement leaders' guide, Vol. Series 3. London: Crown Printer; 2004.
145. The Expert Patients Programme - accessed Sept 2012 at http://www.expertpatients.co.uk/.
146. Involve accessed Sept 2012 at http://www.involve.org.uk/.
147. NHS. Achieving real improvement for the benefit of patients: NHS Modernisation Agency Annual Review 2002/2003. London: Department of Health; 2003.

Complexity in Movement Disorders: A Systems Approach to Intervention

26

S. Lee Hong

Too often overlooked, normal motor function is one of the most critical components of the human existence. The ability to move rests at the core of quality of life, due to the freedom that independent mobility offers. Despite its central role in everyday life, motor function is sometimes viewed as independent from and subsidiary to cognitive function (see [1]). As a result, there has been relatively less attention paid to the deficits in motor function that arise due to disease. However, a growing body of evidence points to movement disorders as being a central issue in a variety of neurological diseases and disorders, even ones that were considered as exclusively "mental" disorders in the past. In this chapter, we will explore a systems approach to motor dysfunction. The chapter is laid out in the following way. First, the chapter will briefly review the ubiquity of similar patterns of behaviour in physics and biology as an overarching framework. The ubiquity of findings across a wide range of complex systems forms the central theme of this chapter. I will also highlight similarities across findings in a broad range of areas of study that are often considered to be disparate fields of science.

Based on a complex systems approach, similar to that presented in West [2], the discussion of disease and disorder will be centred on the loss of fractal complexity, and different underlying mechanisms for these changes in behavioural and physiological dynamics will be explored. This section will highlight evidence that many similarities across diseases exist, if examined through the lens of complex dynamics. In human movement, however, the task and environment have been shown to directly impact the dynamics of the motor behaviour. In the study of movement disorders and motor dysfunction, the task and environment are not often entered into the equation during the development of clinical interventions. Thus, I will present the "*uncertainty conservation hypothesis*" as a potential framework that encompasses task, organism, and environment, as the potential basis for future development of therapeutic interventions in movement disorders. Continuing the central theme of similarity and ubiquity, I hope to provide ideas that are still in their infancy that could extend towards other areas such as cognition and physiology.

26.1 Uniqueness and Ubiquity: Scaling Laws in Physics, Biology, and Beyond

Over the past few decades, there has been a growth in the use of dynamical systems theory in the study of motor control. Fundamentally, dynamical systems theory is based on general systems theoretic approaches to science, where

S.L. Hong (✉)
Ohio University, Irvine Hall 246, 1 Ohio University, Athens, OH 45701, USA
e-mail: hongs@ohio.edu

J.P. Sturmberg and C.M. Martin (eds.), *Handbook of Systems and Complexity in Health*,
DOI 10.1007/978-1-4614-4998-0_26, © Springer Science+Business Media New York 2013

systems can be classified based on inferences made from their dynamics (see [3]). As a result, a sizeable portion of the literature on complex dynamical systems focuses on similar patterns of behaviour in a variety of naturally occurring systems. What Bak [4] sought to demonstrate was that a broad range of natural systems exhibit scaled behaviour (refer also to Chap. 1.10, [5]). Without going into too much detail, the concept of a critical system, one resting on a knife's edge, was unstable, yet in balance at the same time. Bak [4] demonstrated this phenomenon in a wide range of data sets, from the occurrence of earthquakes through cotton prices, showing that a single scaling law could be used to describe their patterns of occurrence and variability. This scaling law related the occurrence with magnitude of change. In the now seminal demonstration of the earthquakes, the number of occurrences scaled as power law relationship with the magnitude of the earthquake. Here, the occurrence of earthquakes can be scaled as a function of the Richter scale, which is scaled on base of 10 (i.e. a Richter value of 2 is 10 times greater magnitude than a 1; 3 is a 100 times greater than 1, and so on). Essentially, the occurrence is a direct function of its magnitude.

26.1.1 Scaling Laws and Criticality in Motor Patterns

The motor patterns of healthy young individuals, much like the systems reviewed by Bak [4], have also been found to exhibit similar scaling properties. Often termed *1/f* dynamics, these patterns reflect a scaling of signal amplitude to frequency. This relationship can be characterized by the power law equation:

$$A = f^{\beta} \tag{26.1}$$

where A is the amplitude and f is the frequency with the scaling exponent, β. The unique pattern is one where the exponent is −1, or *1/f*. Thus, the amplitude of the fluctuations is scaled as a direct function of the frequency on which it occurs. As the frequency is increased, there is a relative decrease in amplitude, but, on a single scale. This dynamic pattern has been observed across various domains of human motor function. *1/f* patterns have been found in human cognition [6, 7] and in a variety of human motor behaviours. Table 26.1 provides examples of common findings of *1/f* dynamics and scaling laws in the human movement literature.

Table 26.1 Presented here is a list of various studies that have demonstrated the common pattern of *1/f* dynamics in different areas of human movement

Study	Motor behaviour
Hausdorff et al. [8]	Walking
Jordan et al. [9]	Running
Hong et al. [10]	Sitting posture
Duarte and Zatsiorsky [11]	Standing posture
Blesic et al. [12]	Tremor
Aks et al. [13]	Eye movements
Nakamura et al. [14]	Physical activity

The source and biological mechanisms that give rise to the *1/f* patterns that have been detected will likely remain an issue of scientific debate. That being said, it does not diminish or negate the findings of similar dynamics across multiple systems. In fact, as we will see in the following section, the patterns of change to physiological and behavioural dynamics as the result of disease and disorder also share many commonalities.

26.2 Ubiquity of Motor Dysfunction in Disease and Disorder

Interestingly, much like the ubiquity of scaling laws in motor output, movement disorders are just as prevalent, occurring even in disorders that are primarily considered to be cognitive in nature. Motor dysfunction has been observed in a variety of developmental disorders, such as 22q11 Deletion Syndrome (e.g. [15]), AD/HD (e.g. [16]), Down syndrome (e.g. [17]), and autism (e.g. [18]). Degenerative disorders such as Huntington's (e.g. [19]), Parkinson's (e.g. [20]), and Alzheimer's (e.g. [21]) also present a variety of symptoms of motor dysfunction. Movement disorders have also now been detected in psychopathologies such as schizophrenia [22, 23] and bipolar disorder [24, 25]. Whether motor dysfunction is a core symptom or due to side effects of medications, the fact that movement disorders consistently presents itself in so many diseases and disorders warrants further attention. Interestingly, common across many of these movement disorders is a change in the dynamics of the motor output. In the following sections, I will review the common direction of change and the proposed underlying mechanisms.

26.2.1 The Loss of Complexity Hypothesis

Conventionally, homeostasis is viewed as the healthy physiological state. Hence, the closer a physiological process is to the mean, the more stable the system. A similar view has long been held in the study of motor behaviour, as lower magnitude or amount of variability had always been considered to be indicative of a healthier system with less "neural noise." At a conceptual level, a less variable system is often equated with lower levels of random brain activity, i.e. a system with less neural noise [26, 27]. This approach places its emphasis on magnitude of variability, and based on assumptions of a normal distribution and independent, random samples does not account for sequences that are present in the data.

A new approach that has developed over the past two decades was the concept of a "*loss of complexity*" in disease and disorder. Lipsitz and Goldberger [28] introduced this new approach to the study of cardiovascular physiology. Originally, fluctuations in physiological data had often been considered as noise that needed to be eliminated in order to allow the "true" signal to be obtained, through filtering or averaging across trials and subjects. The key finding that provided initial support for the loss of complexity hypothesis was a demonstration that the heartbeat patterns of young and old could be distinguished through dynamic analyses. Effectively, Lipsitz and Goldberger [28] were able to demonstrate that although two individuals (one young and one old) could possess similar heartbeat means and standard deviations, their dynamics were characteristically different. By using Approximate Entropy [29], they were able to demonstrate that the young person's heartbeat fluctuated much more irregularly and unpredictably than that of the old.

Over the past two decades, empirical studies have continued to uncover evidence of a loss of complexity in many different diseases and disorders. Much like the ubiquity of the scaling laws, a variety of studies have observed declines in complexity in a variety of different physiological outputs and motor dysfunction. As a result, beyond the seminal example of the loss of complexity in heart rate [28], there are now demonstrations of decreased complexity in behavioural, physiological, and psychological dynamics across a range of disorders.

26.2.1.1 Loss of Complexity in Movement Disorders

As with the widely documented $1/f$ dynamics in human motor behaviour, declines in complexity have also been widely recorded in the movement domain as a result of disease and disorder. Since the loss of complexity hypothesis was first supported in aging, a majority of the literature in movement declines has focused on seniors and the elderly. Across a wide range of different motor functions, a decrease in the complexity of the motor output of the elderly has been observed. To name a few, aging has been shown to result in a loss of complexity in motor behaviour, such as force production tasks [30], postural sway [31], and finger tremor [32]. Interestingly, there are other conditions where this effect is also apparent. One example is Down syndrome, where the postural sway dynamics during sitting are much less complex than that of the age- and body size-matched controls. In fact, the sway dynamics of the Down syndrome subjects were, in fact, more similar to the rhythmic body rocking dynamics of the controls [33]. Lower complexity in motor output has also been observed in people with bipolar disorder [24], although, whether this is a core feature of the disorder or is a side effect of the medications (or both) will require further examination. A detailed description of loss of motor output complexity has been provided by

Newell et al. [34] and Vaillancourt and Newell [35]. In addition, the loss of complexity has been widely observed across a variety of different domains that extend from movement patterns to physiological output.[1]

26.2.2 Proposed Sources and Mechanisms Underlying the Loss of Complexity

There are, however, a variety of different viewpoints regarding the mechanisms that cause a decline in the complexity of physiological output. When the loss of complexity hypothesis was initially proposed, a view that healthy physiological systems could be characterized using fractals was presented. As such, the branching of anatomical structures such as alveoli, nerve fibres, and blood vessels, for example, could be represented as a self-similar repeating pattern that is invariant across measurement scales. With aging, it is often the anatomical structures at the smallest scales that are lost to time. Lipsitz and Goldberger [28] highlighted this phenomenon through images of nervous tissue, that the smallest nerve endings are lost while larger nerves are preserved. The analogue of degeneration at the smallest spatial scales, when translated to time, is that fluctuations at higher frequencies will be the first to be eliminated during the aging process. As a result, one would predict that higher frequency or shorter timescale fluctuations will be absent or diminished as a result of a given disease. While the loss of complexity hypothesis has been widely supported (see [38] for a review), the source of these age-related declines remains in question. The following subsections present some of the theorized sources of the loss of complexity.

26.2.2.1 System Isolation and Decoupling

One of the early conceptualizations of the underlying source of the loss of complexity was the idea that with aging or disease, the many physiological systems become isolated from one another [39]. In such a situation, the connections between subsystems or inputs that generate the complex physiological output are diminished, resulting in a signal that evolves on fewer timescales [40]. What results is a system that is less adaptive to change in the environment and more susceptible to perturbations. Based on the isolation approach, a loss of system components or diminished coupling will result in a physiological output that is less complex. However, as Assisi et al. [41] demonstrated, a reduction in coupling strength between system components is not the only reason for a loss of complexity. When the components within a system are too tightly coupled, the complexity of its output will also be diminished.

This phenomenon/mechanism is exemplified in the motor behaviour of children that is generally less complex than that of adults [42, 43]. From this perspective, one would view that the infant reflects a system with tightly coupled components, and that the developmental processes result in a gradual differentiation of the components within the overall system. Assuming that this process continues throughout the lifespan, complexity would peak during early adulthood, where there is an optimal level of coupling [42]. An example of these changing motor dynamics as a function of development can be seen in Fig. 26.1, where postural sway traces of children and an adult are provided. At a certain point, this optimum is exceeded, with observable declines in complexity as the components of the system steadily isolate as a result of the aging process.

26.2.2.2 Loss of Adaptability

It is generally common to equate disease with decline or a lack or loss due to a "malfunctioning" component of the entire system, that is, the organism as a whole. Yet, there is evidence that a loss of fractal complexity in motor and other physiological output does not necessarily reflect functional decline(s) that arise as a result of the diseased states [35]. In particular, there are situations where a more complex output is observed in

[1] For a broader overview, see Chap. 11.

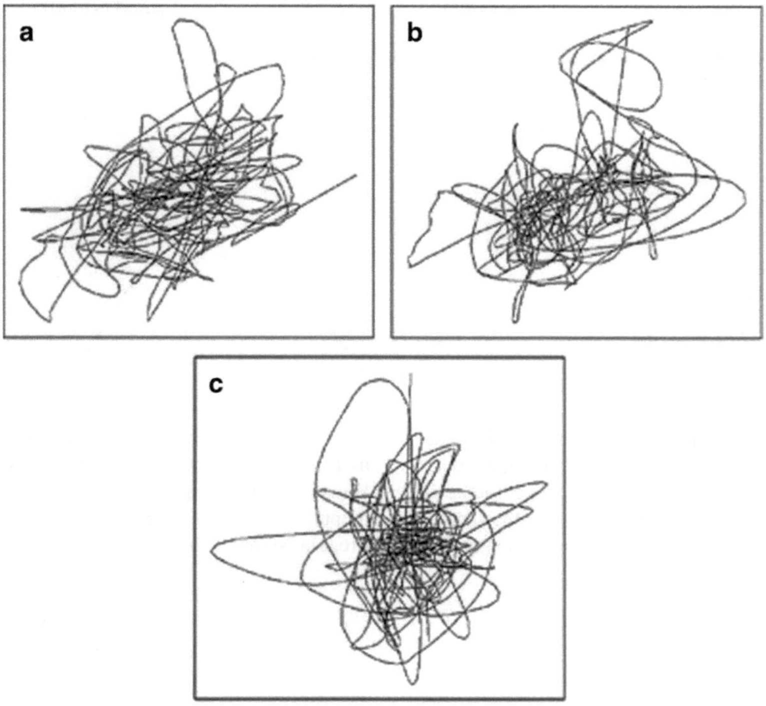

Fig. 26.1 Postural sway profiles from children aged 6 (panel **a**) and 10 (panel **b**) years and an adult over the age of 18 years (panel **c**). Of note is how the sway pattern in the children seems to move along a diagonal, showing a strong coupling between side-to-side and fore-aft motion. The adult exhibits clearly distinct forward and rearward movements that are quite independent from the sideways motion. From [10]

diseased subjects in comparison to the controls. In their review, Vaillancourt and Newell [35] presented a variety of examples of increased complexity in the motor behaviour of patients with Huntington's disease and schizophrenia. More importantly, they proposed that the demands of the motor task also had an important role to play in the dynamics of the observed motor output. When the task demands were to maintain a constant force level with a finger, Parkinson's disease (PD) patients generate a predictable more rhythmic pattern [44]. This "less complex" pattern results in increased error (i.e. distance from the target line) and is distinct from the irregularly pattern of force production of the controls that allows them to remain close to the target line. On the other hand, when required to produce a sine wave, the output of the PD patient instead becomes more complex than that of a healthy control. Here, the PD patient is now unable to generate the smooth force output that allows them to accurately trace the sine-wave target (see Fig. 26.2). Effectively, the output of the PD patients was "more complex" and irregular than that of the controls. The interesting but seemingly paradoxical finding is that the "simple" task of maintaining a constant force level requires the PD patient to generate a complex motor output. On the other hand, the higher dimensional task of the sine wave requires that the subject be able to damp out the fluctuations in muscle force that unfolds along the shortest timescales. Even though the complexity of the force output for PD patients, as measured by Approximate Entropy, moves in opposite directions than that of the controls, both situations are, however, consistent with a loss of fractal complexity. In the case of the constant force, the PD patient has increased (though unwanted) contributions of slow timescale patterns, while during the sine-wave condition, the PD patient exhibits undesired short timescale fluctuations in muscle force. In both

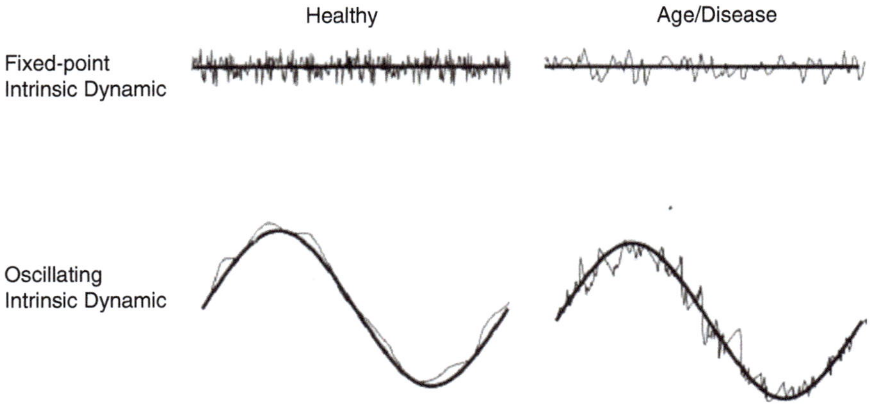

Fig. 26.2 Schematic illustration of the role of the "intrinsic" task dynamics on the complexity of motor output. When asked to maintain a steady-state force, the healthy subjects are more able to remain close to the target by generating a complex signal that evolves on many timescales. The old/diseased individual generates a slower, more predictable pattern to approximate the steady-state. However, when required to produce a sine-wave force output, the old/diseased individual generates a jagged, irregular signal. From [35]

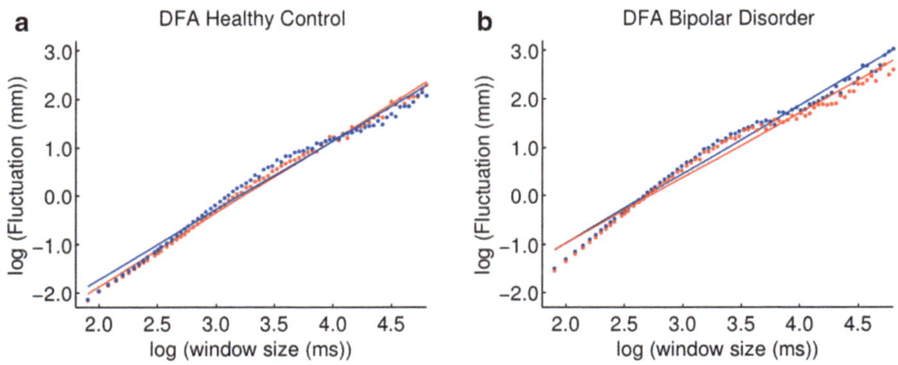

Fig. 26.3 This plot shows the group differences between bipolar subjects (*diamonds*) and controls (*circles*). Each data point represents a single subject. The *left panel* shows the differences in sway area, that is, the space of the ground covered by the sway movements. Group differences are clearly observable in both conditions. On the *right panel*, the differences in Detrended Fluctuation Analysis (DFA) values can be observed. During the eyes open condition, there is no significant difference between groups. However, this difference becomes significant when the eyes are closed. From [24]

cases, performance on the task is poor and the desired force pattern cannot be produced.

Effectively, the healthy system is not just able to generate a more complex output, but, it is able to greatly restrict its output when required by the task demands. There is further evidence to support the idea that different levels of adaptability are apparent when patients with disease or disorder are tested across a range of tasks rather than a single condition. A reduced ability to adapt to different tasks has been demonstrated in the aging literature [30, 32, 45] and in people with Down syndrome [33].

Interestingly, this phenomenon also arises in movement disorders associated with psychopathology [24]. In a recent study, we examined the postural sway of individuals with bipolar disorder and age-, height-, and weight-matched controls under four different task conditions: (1) eyes open-feet apart; (2) eyes closed-feet apart; (3) eyes open-feet together; and (4) eyes closed-feet together. We found that patients with bipolar disorder had poorer postural control in general as the patients swayed over a larger area than the controls (Fig. 26.3). The bipolar patients also

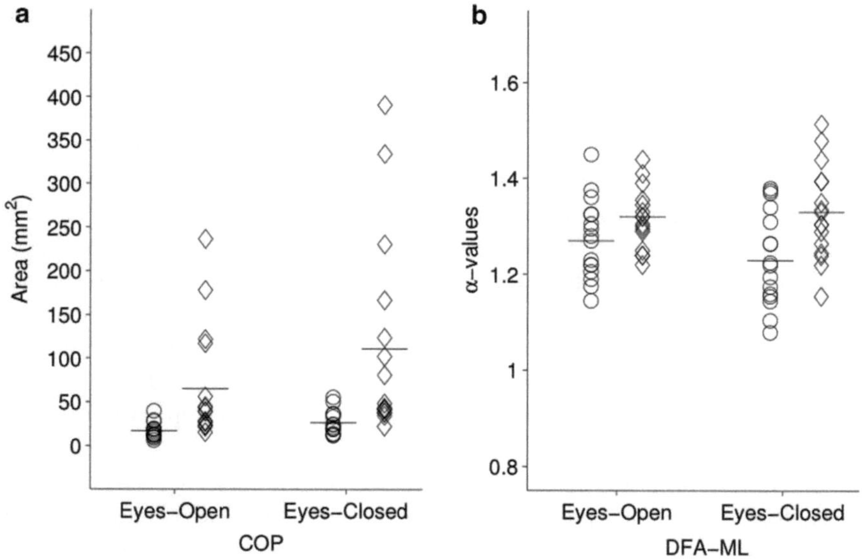

Fig. 26.4 Exemplar plot of the DFA of the mediolateral (side-to-side) postural sway from a single bipolar and control subject, respectively. The lines reflect a log–log relationship between fluctuation magnitude and the timescale (window size) on which it occurred. The eyes open condition is represented with a *blue line*, while the eyes closed conditions is represented with a *red line*. From [24]

increased their sway when asked to close their eyes. When detrended fluctuation analysis (DFA, [46]) was used to measure the complexity of the sway dynamics, we found that the main difference between the groups was during the conditions with their eyes closed (Figs. 26.3 and 26.4) on the mediolateral direction of motion (i.e. side-to-side). This is where the loss of complexity became apparent in their sway dynamics, indicative of a reduced ability to adapt to the deprivation of visual information. Yet, the effects of a change in stance did not yield any effects, suggesting that the loss of adaptability might not necessarily be a universal phenomenon in disease and disorder, but specific to certain aspects of the system.

Overall, the loss of adaptability approach hypotheses that the disordered individual is less capable of altering his or her motor behaviour to adapt to different task demands. Here, the loss of complexity is not necessarily a representation of a decline in physical function, but, rather, the inability to alter the dynamics of the behaviour to respond to task demands is the better marker. A synthesis of the complex systems approach to movement disorders converges on the idea that more or less complexity in motor output alone cannot be used as an indicator of disease or disorder. But, the underlying mechanism that gives rise to the observed loss of fractal complexity due to aging and disease remain a matter of scientific debate. Potentially, both a loss of adaptability and the decoupling of subsystems may account for the empirical phenomenon of changes in complexity in motor output. These aberrant patterns of behaviour are reflected in a shift away from fractal dynamics, where the magnitude of fluctuations in the motor system is no longer in direct proportion to the timescale on which they occur. Yet, these putative mechanisms do not provide clear paths towards novel clinical approaches to intervening or remitting changes in complexity in motor output, physiology, or psychology that arise due to disease or disorder.

26.2.3 Uncertainty Conservation in Human Motor Control

Simply stating that complexity can increase, decrease, or stay the same depending on the task demands is an unsatisfying answer to a complex problem. In order to address this issue, an explanation of how motor patterns adapt to different task demands and under different levels of available information from the environment is needed. Originally, the concept of task–organism–environment relationships stemmed from Newell [47], who proposed that human movement is not the product of the person alone, but is instead a reflection of the "confluence of constraints" at the level of the task, organism, and environment. In this sense, any goal-directed action is a product of a larger system that encompasses task demands and the information that the environment provides.

The goal of Newell [47] was to begin to address the "degrees of freedom" problem presented by Bernstein [48]. Simply, the number of controllable components, i.e. degrees of freedom, of the neuromotor system is extremely high, with the added complexity of having myriad ways to organize those degrees of freedom to generate a single movement. Not only does the motor system have an inordinate number of independently controllable components, it also has the added problem of redundancy. One example of redundancy is that a variety of different configurations of muscle and body movements allow the same goal to be achieved. Take for instance, the scenario of reaching out to grasp an object. Both underhand (palm up) and overhand (palm down) grasps would suffice, and thus, multiple arm configurations have the same capacity to achieve the same goal. At the same time, more than a single goal can be achieved by the same movement, like touching my nose or scratching it.

As with the dynamical systems concepts of human motor control (cf. [49, 50]) it became apparent that there were inherent constraints at the level of the organism. Beyond the tendons, ligaments, and muscles themselves, human motor control was also substantially restricted by coordination patterns that served to reduce the number of independent degrees of freedom that needed to be controlled. In this sense, the motor system has a tendency to naturally favour certain patterns of movement over others and even has difficulty learning some movements. For one, the natural tendency of movements is to "tune into resonance," that is, moving at speeds that reflect the natural mechanical properties of the limbs, prescribed by its mass and length (cf. [49]). Effectively, the motor system "self-selects" the movement patterns based on the mechanics of a pendulum in order to obtain maximal return in elastic energy from the musculature. Such a natural tendency to optimize behaviour for the purpose of achieving mechanical resonance has been proposed to be an explanation for the self-selection of comfortable walking speeds based on limb length [51]. It has been demonstrated that walking at the resonance speed results in the lowest energy consumption costs, while increasing or decreasing the walking pace away from this preferred speed increases energy consumption [51].

The natural tendency to self-determine movement speed is not the only inherent constraint on motor behaviour. Our motor system also has the tendency to favour some movement patterns and synchronized rhythms above others. For example, the seminal work of Haken et al. [52] demonstrated that human movement has two stable coordination modes of in-phase and anti-phase patterns. The former is a situation where both limbs extending and flexing at the same time, while the latter is akin to gait, with one limb extended while the other is flexed. Haken et al. [52] found that the "default" coordination mode of the human motor system is an in-phase pattern, as movements initiated in anti-phase transitioned to an in-phase pattern when the movement was sped up sufficiently. Transitions in the opposite direction, however, did not occur. As a result, learning other coordination patterns in between in-phase and anti-phase has been shown to be difficult (cf. [53, 54]). In sum, these studies provide evidence of constraints upon coordination and movement within the organism itself.

However, the key issue that Newell [47] highlighted is that not all of the constraints on motor behaviour lie at the level of the organism, and not all redundancies are created equally. Depending on the context in which the behaviour is being performed affects the number of ways in which it can be achieved. If we return to the task of reaching and grasping an object in front of us, it becomes immediately apparent that the ensuing action is constrained in various ways. At the level of the task, the distance required for the reach plays an integral role. If the object is close, I would have a greater number of options to grasp the object. This is due to the fact that varied combinations of movements at my shoulder, elbow, and/or wrist would suffice to allow me to make contact with the object. However, if the object is far away, the number of movement options at my disposal would be reduced. To cover the distance and reach the object I would have to lean forward and use my shoulder and elbow to the fullest to make contact with my target object.

Similar constraints would arise depending on the shape of the object or the environment in which the reaching and grasping action was being performed. If the object were either small or fragile, I would be able to only engage a few digits, perhaps restricted to the pinch grip of my thumb and index finger. A large or heavy object would require that I employ a whole hand grasp, and perhaps even both hands. Similarly, if the object were placed on a cluttered surface or if I had to perform this movement in a dark room, I would be prevented from using a variety of different movements to achieve the goal. Such restrictions would be necessary to avoid obstacles or inadvertently knocking over another object. Once again, the movement patterns available to me are reduced by the constraints outside that of those of the body itself.

What had not yet been provided in 1986 was a hypothesis regarding the directionality of the effects of a constraint at one level on another. Originally, the process of movement production was approached from the perspective of open systems or dissipative structures (cf. [55]). While an appropriate description of the organism alone, the open systems approach to human motor control resulted in an open-ended view of the role of task and environmental constraints upon motor production. As a result, the prevailing view was that somehow, the confluence of task, organismic, and environmental constraints interacted to generate a given movement [47]. But how they interacted remained in question.

This is because the constraint-based approach did not provide clear hypotheses as to what were (1) the units of constraint and (2) the direction of change as the constraints at one level or another were increased or decreased. The next step was to develop a conceptual framework that allowed the task, environment, and organism to be captured as a single system. To achieve this, we had to return to the most fundamental issue in the control of human movement, which is the "degrees of freedom problem," first presented by Bernstein [48].

With this central problem in mind, we envisioned the task–organism–environment system as sharing a single pool of degrees of freedom. Conserved, as in a zero-sum game, the task and environment shape the ultimate behaviour by "taking away" or "borrowing" degrees of freedom from the organism. As with the common concept of constraints, the task and environment reduce the number of available degrees of freedom at the level of the organism. What this framework provided was a testable hypothesis regarding how the uncertainty in the motor output should change as a function of the task and environment. Here, as the uncertainty in either task and/or environment increase, the uncertainty in the behaviour of the organism should decrease, where the total amount of uncertainty within the task–organism–environment system is held at constant (see Fig. 26.5).

Next, a common unit that provides a measure of the degrees of freedom within the system and the constraints placed upon them was necessary [56, 57]. To achieve this, I returned to a concept central to the loss of complexity hypothesis, *entropy*. While entropy is often viewed as a measure of uncertainty or information or unpredictability, it also has other descriptions in thermodynamics. Boltzmann's original demonstration was a mathematical proof that the number of spatial configurations within a molecular structure can be represented as entropy where:

$$S = k \log W = -k \log p \qquad (26.2)$$

Here, the entropy, S is equal to the number of ways, W, that the atoms can be configured. Assuming a flat or equal distribution (i.e. maximum entropy) one can then relate this to the

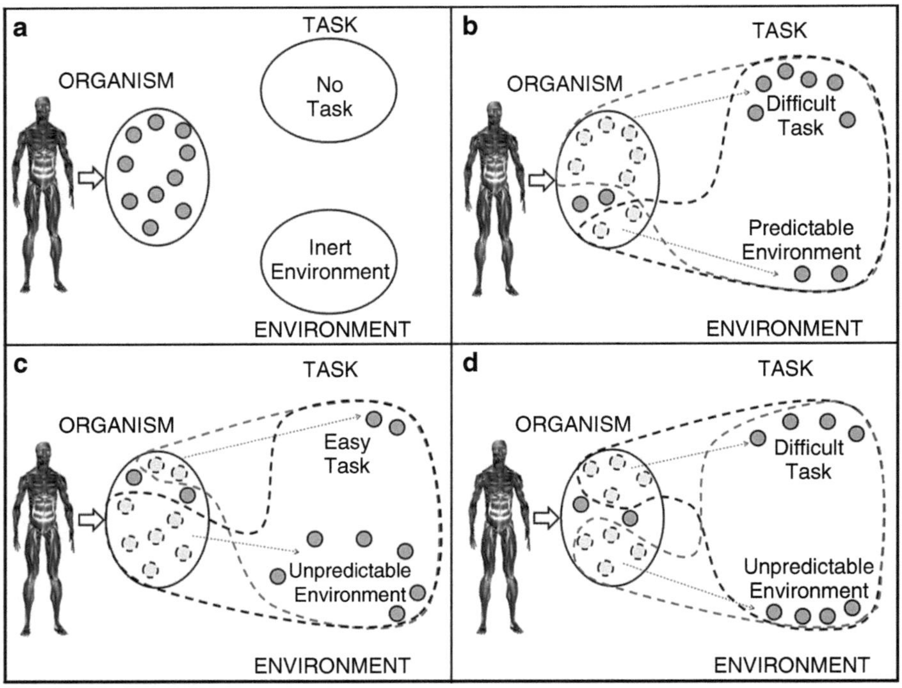

Fig. 26.5 A schematic illustration of the process of entropy conservation and compensation across task, organism, and environment as the (re-)distribution of degrees of freedom. The following conditions are represented: (**a**) a "resting" state where no goal-directed movement is being performed; (**b**) high task entropy with low environmental entropy; (**c**) low task entropy with high environmental entropy; and (**d**) high task and environmental entropy. From [57]

likelihood of finding an atom in a given place, as measured by the probability p. The Boltzmann constant, is represented as k. In a similar vein, constraints link the individual degrees of freedom and reduce the number of spatial configurations, thus resulting in lower entropy (see [58] for a more in depth explanation). Now, based on this, both the directions of change can be predicted using a measurable unit that accounts for the changes in variability in the behaviour.

26.2.3.1 Empirical Evidence

To test the uncertainty conservation hypothesis, we employed a series of simple experiments involving isometric force control [59, 60]. The particular benefit of this experimental protocol is that isometric force is virtually impossible to maintain without visual feedback, as the force level produced degrades within less than half a second [61]. As a result, the subject is almost completely dependent on information from the environment. In this experiment the force level to be generated at a given time unit was presented as a dot on a computer screen. The task was simple: maintain the force produced by the finger as close to a target line as possible. We manipulated the difficulty of the task by narrowing or enlarging the error tolerance bounds around the target. The amount of information provided by the environment to the subject was manipulated by changing the amount of time that passed between each presentation of the dot on the screen. This experimental protocol was conducted under two conditions: (1) single finger (Fig. 26.6); and (2) two fingers coordinating to produce a total force (Fig. 26.7). In the second condition, only the total force of the two fingers was provided to the subject through visual feedback.

Across both experiments, we observed that the entropy in the force output declined as (1) the amount of information in the environment was reduced; and (2) the task difficulty increased by

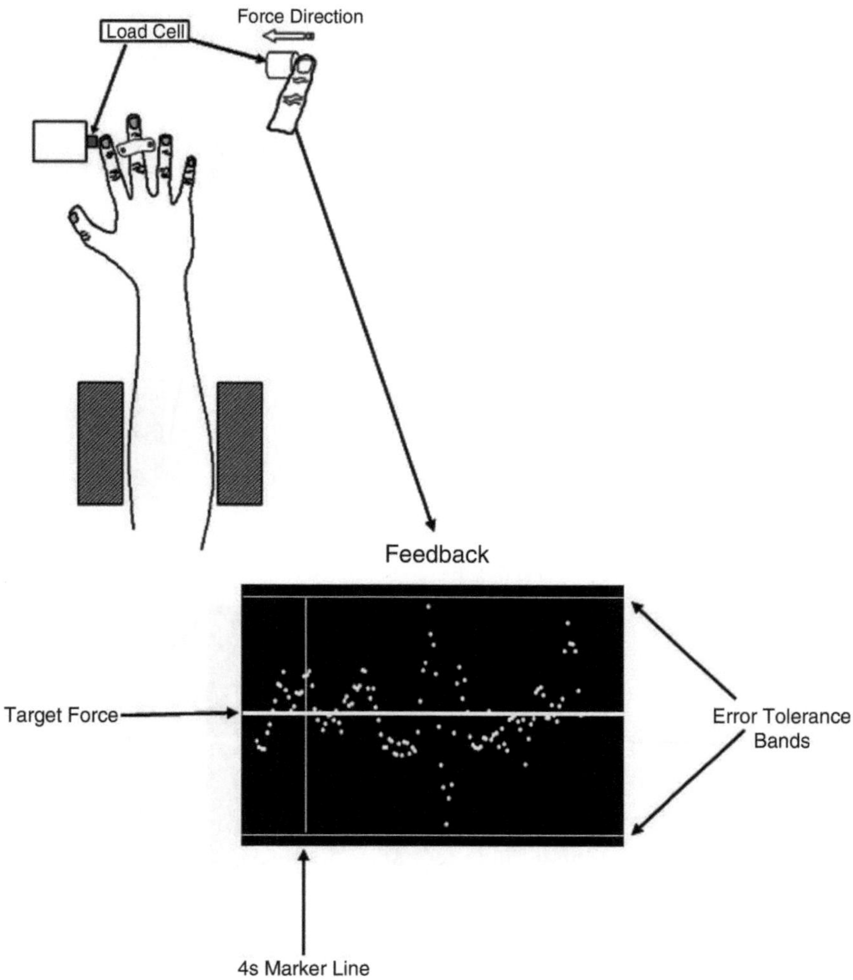

Fig. 26.6 Illustration of the experimental setup for the single finger isometric force production experiment. From [60]. Note, the *4s marker line* denotes the "grace period" that allowed the subject to approximate the target force. These data were not included in the analysis

reducing the relative size of the error bands. Interestingly, the effects of changing the task and environment conditions were found to be compensatory, where the effects of a difficult task in conjunction with an information-rich environment were the same as an easy task being performed in a low-information environment. The relationship between task, organism, and environment could be captured with a quadratic surface (see Figs. 26.8 and 26.9). Across both studies the findings were consistent when the target variable was (1) the Approximate Entropy of the force generated by the single finger (Fig. 26.8) or (2) information entropy of the relative phase of the forces produced by the two fingers simultaneously (Fig. 26.9). These findings were consistent as long as the entropies were made conditional upon successful completion of the task (i.e. staying within the boundaries of the error bands). What this achieved was the "idealization" of the results, restricting this to situations where the behaviour successfully satisfied the goal.

From the perspective of the behaviour, when faced with the constraints of a challenging task and environment, the force output is adjusted by reducing the entropy of the motor output. In the task that required the force production of a single finger, the force output time series exhibits

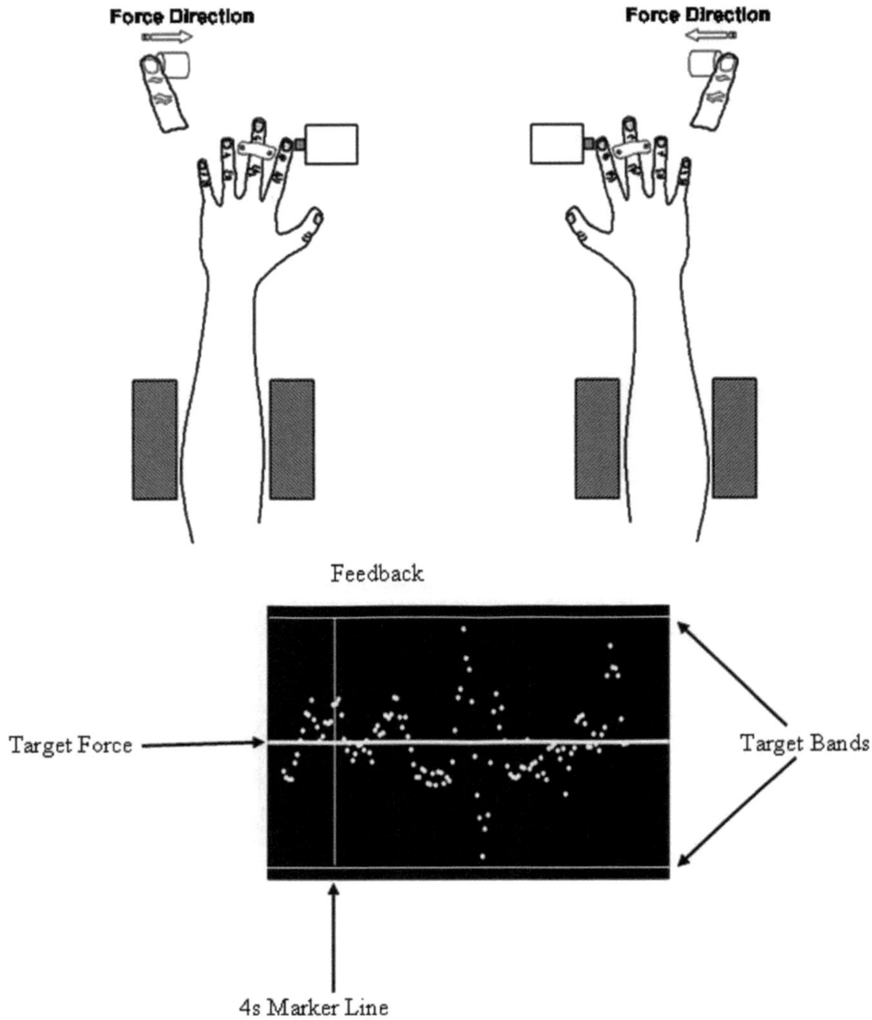

Fig. 26.7 Illustration of the experimental setup for the coordinated, two finger isometric force production experiment. From [59]

a gradual shift from what is a broad spectrum signal to what is more similar to a sine wave that has an extremely predictable pattern. In the case of the two-finger coordinated force output, we observed that the distribution of the number of coordination (relative phase) patterns is reduced. When the demands of the task are high and the environment provides little information, the force output is restricted to fewer coordination patterns, and the distribution of relative phase values becomes peaked. This was a marked shift from the broad distribution of relative phase values during the easy task and high environmental information conditions.

Such a conservation rule also holds in the study of cognitive response and eye fixation patterns. In a recent study [62], we measured the effects of temporal and spatial uncertainty on response times and eye fixation durations when reacting to a single stimulus (Fig. 26.10). In this task, subjects were seated facing a computer monitor with their head braced to allow for accurate eye tracking to take place. They held a video game controller in hand and were asked to push either of the buttons on the controller as soon as they see a red square appear on the computer screen. While the number of stimuli remained constant at one, we altered the

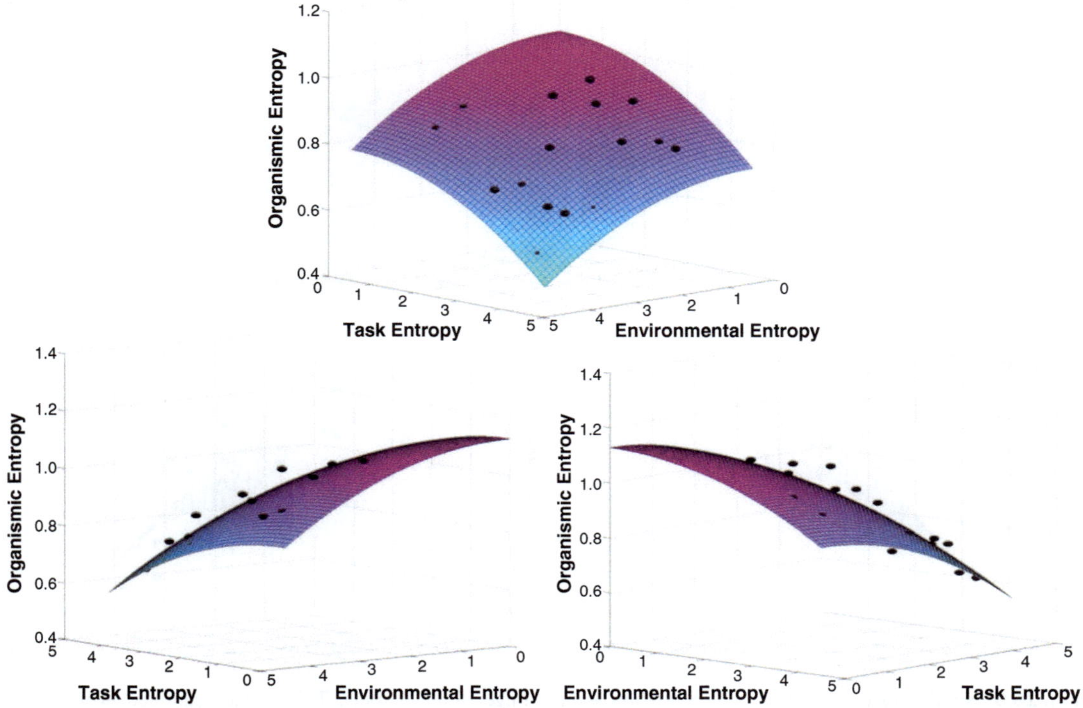

Fig. 26.8 Quadratic surface representation of the task–organism–environment entropy conservation relationship. The surface is generated using the equation: $H_{\mathrm{ORGANISM}} = k - aH_{\mathrm{TASK}}^2 - bH_{\mathrm{ENVIRONMENT}}^2$, where k is the intercept, with a and b as free obtained through a least-squares fit. The entropy of the organism is represented by approximate entropy [39] values of the force dynamics, made conditional on the probability that the force level remained within the target bands. This surface captured 92% of the total variance. From [60]

uncertainty of when (temporal uncertainty) and where (spatial uncertainty) the stimulus would appear in the various conditions. This was achieved by simply increasing the number of stimulus locations and time between response and subsequent stimulus from 1 to 2 to 4. The average interval, however, was maintained at 1,250 ms. Interestingly, we found that while the response times became more unpredictable as a result of increased stimulus uncertainty (Fig. 26.11), the eye fixation durations were more predictable (Fig. 26.12).

Both patterns of change in entropy of the response times and fixation durations could be captured using a single quadratic surface, requiring only a reversal of sign in the equation (Fig. 26.13). The interesting aspect to this finding is that the pattern and not the amount of time of the response and fixation durations changed as a function of stimulus uncertainty. Furthermore, there seems to be a compensatory relationship between the "input" eye fixations and the "output" of the response times. When the stimulus was more unpredictable, the eye fixations became restricted to more similar durations, and in turn, the response times became more evenly distributed. Much like finger force production [63], we found that spatial and temporal uncertainty had compensatory effects on the response time and eye fixation patterns. Essentially, the effects of not knowing when and where the stimulus would appear were similar for both response times and eye fixation durations.

With empirical evidence for uncertainty conservation in human motor and cognitive behaviour,

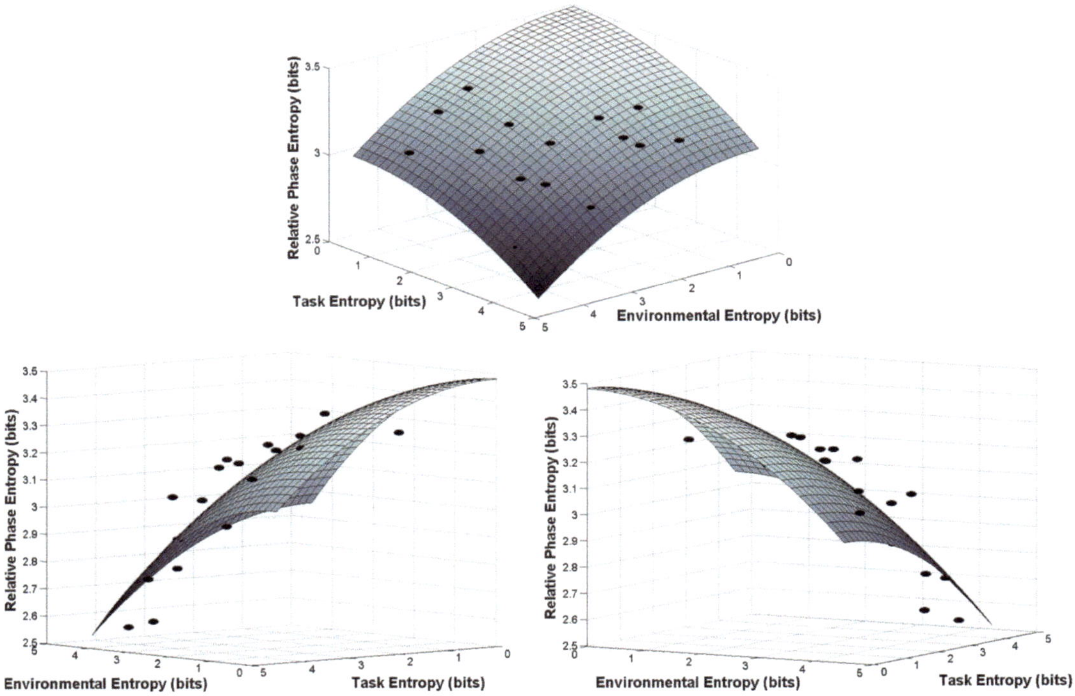

Fig. 26.9 Similar quadratic surface representation as in Fig. 26.8, with the entropy of the organism represented by the Shannon entropy of the relative phase between the two finger force outputs. This surface captured 80% of the total variance. From [59]

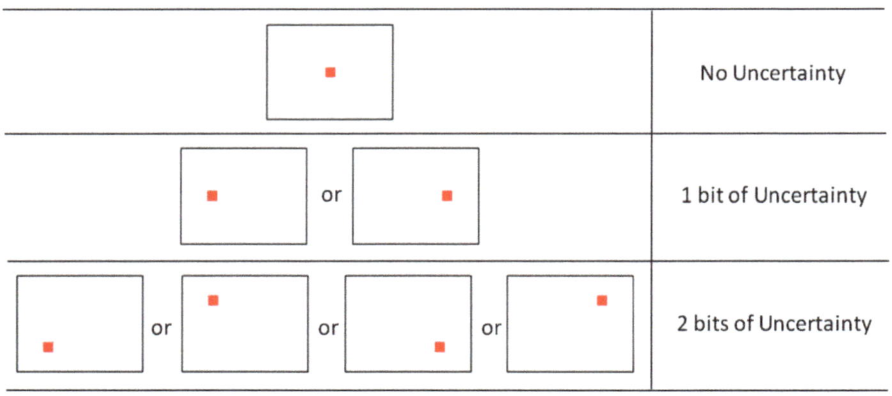

Fig. 26.10 Illustration of the stimulus presentation and resulting levels of uncertainty in bits. Subjects were required to push a button on a video game control pad as soon as the target was visible. From [62]

it provides a hypothesis that views the task, organism, and environment as parts of a larger system. It is possible that this hypothesis can be used to predict the direction of change in motor function based on manipulations of the task and environment. The following sections provide some initial ideas on how the uncertainty conservation framework can be applied to clinical settings.

26.3 Uncertainty Conservation as Framework for Care and Clinical Interventions

Despite the ubiquity of motor dysfunction across a broad range of diseases and disorders, systems approaches are rarely used in this domain. Dominated by the use of contraries in contemporary

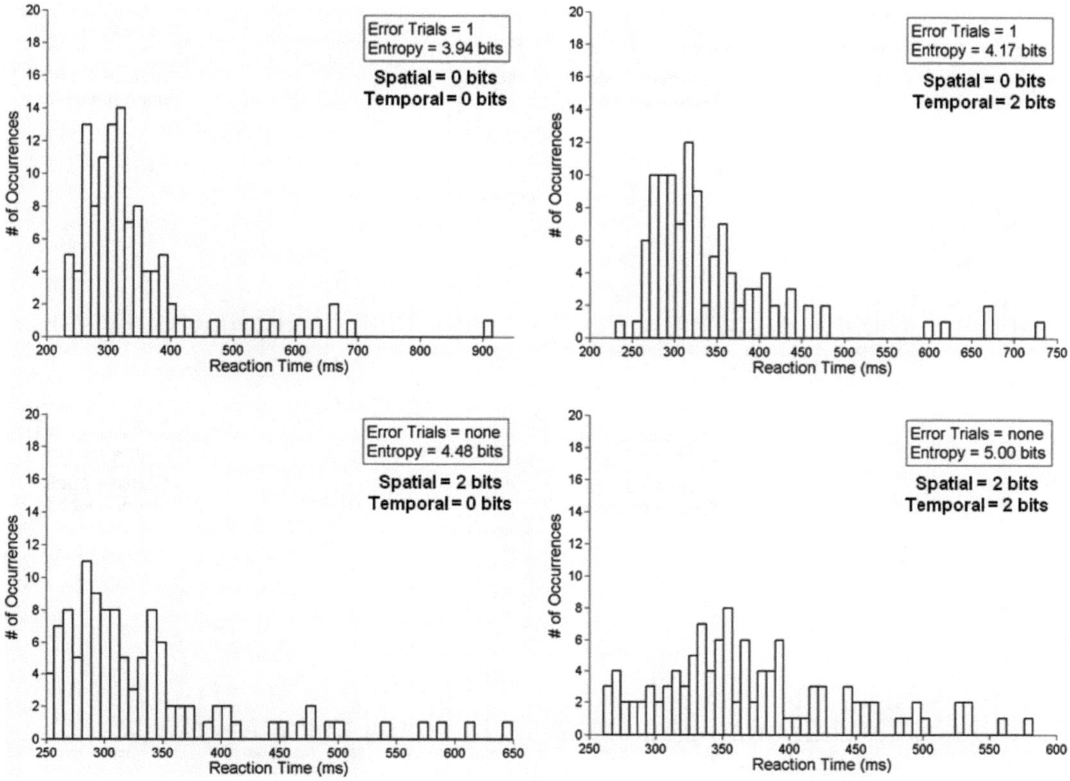

Fig. 26.11 Frequency histogram of the distribution of response times as the result of the four extreme spatial and temporal uncertainty conditions. These data are obtained from a single subject. Note the flattening of the distribution as the uncertainty in the conditions is increased. From [62]

science (see [64]), medicine and healthcare also generally target problems using unidimensional interventions [2]. For example, addressing motor issues through movement rehabilitation or exercise has been exclusively the domain of physical or occupational therapists, while addressing motor dysfunction from a neurological perspective is most often achieved through pharmacology or surgery. The lack of overlap in the aforementioned approaches arises due to the constant separation of central from peripheral, cognitive from motor, structure from function, and physical from mental. Obscuring the complex connections between physiological subsystems through categorization, it has resulted in treatments that are more symptom-based, rather than targeting the underlying problems. Sometimes, this can also prevent biological mechanisms of a given disorder from being uncovered, because, simply, movement disorders are rarely "motor system" problems alone.

A good example is Parkinson's disease (PD), a disorder that is viewed almost exclusively as a movement disorder. It primarily presents motor symptoms of tremor and freezing, and slowness of movements, especially in gait. Anatomically, PD presents itself as a loss of dopaminergic neurons in the substantia nigra pars compacta, a region of the basal ganglia. As a result, the targeted pharmacological therapy has been L-DOPA, as a means of supplementing declines in the dopamine levels in the brain. While effective at reducing complications in the early stages of PD, L-DOPA also results in a variety of often debilitating side effects as the disease progresses (see [65]). PD, however, is now acknowledged as being far from only a "movement disorder." Not only are there motor issues in PD, but also cognitive declines [66, 67], with additional symptoms of depression [68] and even psychosis [69].

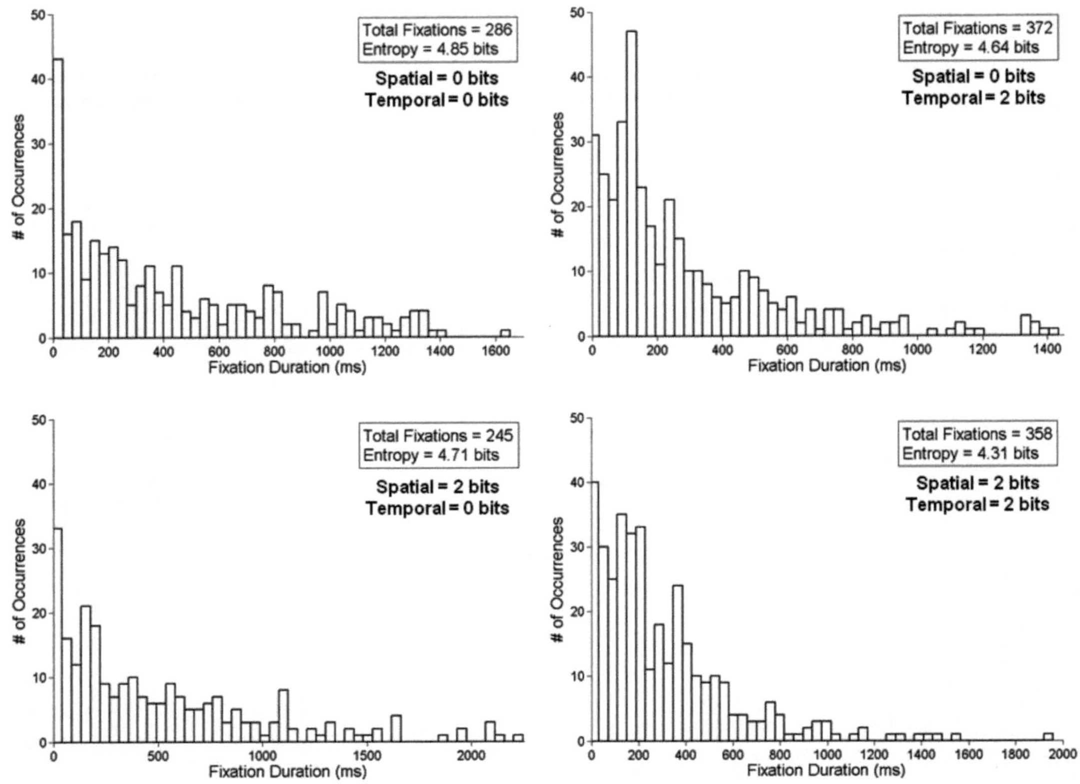

Fig. 26.12 Obtained from the same subject in Fig. 26.11, this frequency histogram illustrates the distribution of eye fixation durations across the same four conditions. Here, the increasing uncertainty in the experimental conditions resulted in a narrowing of the distribution. From [62]

This multidimensional nature of movement and cognitive/psychological disorders raises the need for a multifaceted approach that is based on complementarity, rather than contraries [64]. Separating symptoms of the disorder into its cognitive and motor components and addressing them independently may not be the most viable approach. However, what is needed is a complementary viewpoint that views cognition and action as components of a single system. The uncertainty conservation framework provides an approach to change the complexity of the motor output in a given individual. By altering the amount of information available in the environment or changing the demands of the task, the desired change in behaviour can be elicited. This framework provides predictions regarding the behaviour of the system's output in response to task demands and environmental information. At the very least, if no adaptive change in motor behaviour is obtained it would allow for further experimentation to gain insight as to why the system no longer adapts to the task and environmental constraints placed upon the behaviour.

One example of a successful intervention that would fall within the uncertainty conservation framework has been the application of the "stochastic resonance" concept to motor output. Simply, stochastic resonance occurs through the addition of further random interference or "noise" to a highly variable nonlinear system and enhances the information content through an improved signal-to-noise ratio. This phenomenon is evident in both man-made and naturally occurring signals (see [70]). External assistive devices have been applied successfully, improving posture in older individuals. Using vibratory insoles worn on the feet [71] and a mini electrical stimulator placed at the knee [72] have shown improvements in balance and postural control, as well as

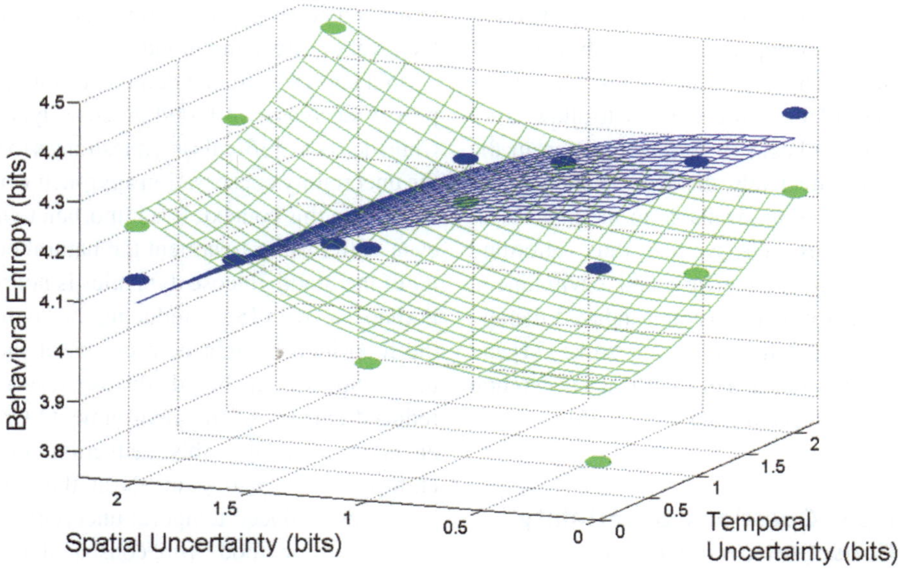

Fig. 26.13 Quadratic surface representation of entropy conservation across space, time, and behaviour. The entropy of the eye fixation durations, represented by the *blue surface* is captured using a similar function to Figs. 26.8 and 26.9: $H_{\text{FIXATIONDURATION}} = k - aH^2_{\text{SPACE}} - bH^2_{\text{TIME}}$, with two free parameters and the intercept. To obtain the surface for the response times, the signs are simply reversed to: $H_{\text{RESPONSETIME}} = k + aH^2_{\text{SPACE}} + bH^2_{\text{TIME}}$. Both surfaces captured approximately 80% of the total variance. From [62]

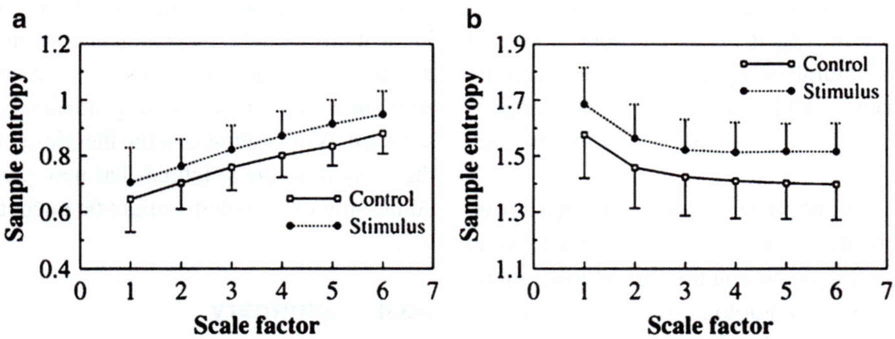

Fig. 26.14 Increased complexity in the postural sway resulting from the noisy/random stimulus delivered using a mini electrical device. Panels (**a**) and (**b**) represent the change in Sample Entropy along a variety of different timescales along the anteroposterior (fore-aft) and mediolateral (side-to-side) axes, respectively. The higher Sample Entropy values across all of the scales indicate an increase in complexity of the postural sway due to the stochastic resonance phenomenon. From [73]

an increase in complexity of the sway pattern [73], and is illustrated by the increased Sample Entropy values along the various timescales in Fig. 26.14. These studies demonstrate that the complexity of the postural sway signal can be increased by improving the information from the environment through enhanced cutaneous sensation from the soles of the feet. Here, reducing the uncertainty at the level of the environment afforded an increase in the complexity in the motor output.

In other realms of physical and occupational therapy, similar approaches have been employed and are currently still in use. Known as constraint-induced therapy, it is a clinical intervention originally developed for the rehabilitation of stroke patients. This therapeutic approach restricts the movement of the unaffected limb to allow structured practice to begin to restore function in the affected limb. Effectively, this reduces the number of degrees of freedom available to the motor system, reducing its entropy, and thus, affording increased entropy in the task and environment. There is evidence that constraint-induced therapy improves motor function following stroke [74] and there is growing evidence of its efficacy in cerebral palsy and hemiplegia in children [75, 76].

26.3.1 Open Questions: Uncertainty Conservation in Clinical Interventions

While these clinical interventions and approaches were developed before and independent of the uncertainty conservation framework, it does suggest its potential utility as a broader theoretical basis for clinical interventions of motor dysfunction. If one were to take a general systems theoretic viewpoint [3], it opens the possibility that using the natural compensation of uncertainty and entropy could lead to novel therapeutic approaches to a variety of diseases and disorders.

As an example, instead of focusing on addressing movement disorders alone, movement-based rehabilitation has the potential to improve mood states and psychopathology. Known as interpersonal and social rhythm therapy or ISRT [77–79], this approach uses social interactions, such as feeding, sleep, and exercise to improve clinical outcomes in people with bipolar disorder. From a systems perspective and the uncertainty conservation hypothesis, this approach reduces the uncertainty in the behavioural routines in order to increase the complexity of the shifting mood states. With evidence of reduced complexity of mood fluctuations [37], a compensatory relationship between cognition, action, and emotion might explain the reported therapeutic benefits of ISRT. Perhaps, increasing the rhythmicity of one's interaction with the environment, which reduces its entropy, leads to increased entropy at the level of affect and mood.

In a similar vein, the demonstration that complexity decreases in blood glucose dynamics by Churruca et al. [36] raises a question as to whether the uncertainty conservation approach would also be valid in this context. It is important to note that even with insulin treatment the natural complexity of the blood glucose dynamics is not restored. Perhaps, as with ISRT, reducing the "entropy" of the food intake might be a means of increasing the complexity in blood glucose levels via the reduced entropy of the environment? One could envision achieving this reduction in dietary entropy by (1) eating meals at the same time every day (reduced temporal uncertainty) and/or (2) consuming food with consistent nutritional composition.

Overall, there is evidence, although tangential, that the uncertainty conversation approach has the potential to restore the behavioural dynamics and complexity through task demands and environmental influences. At this stage, however, these ideas remain speculative. However, the consistency and ubiquitous nature of patterns of change in so many complex dynamical systems should compel the clinical sciences to at least attempt new interventions. By taking advantage of the similarities in systems and the theories underlying them, there is the potential that new methods of addressing old problems might be uncovered.

26.4 Summary

The overall goal of this chapter was to review the complexity and complex systems-based approaches to motor function and dysfunction. Based on this framework and that of complementarity, an approach that bridges different modalities of cognition and action is needed, even though cognition and action are often viewed as contraries. Through the uncertainty conservation hypothesis, one can conceive a new approach that connects these sometimes disparate components of a larger whole in a manner that could lead to the development of novel approaches to the treatment of movement disorders. Hopefully, this will result in a move towards the consideration of holistic, multidomain therapeutic interventions that work alongside conventional, unimodal symptom-based clinical targets. With a hypothesis-based approach through uncertainty conservation, it is possible that the more speculative elements of this chapter are tested and evaluated in the future.

Acknowledgements This research was supported in part by grants from the National Institutes on Aging (R21AG035158 and 1R21AG039818).

References

1. Rosenbaum DA. The Cinderella of psychology: the neglect of motor control in the science of mental life and behavior. Am Psychol. 2005;60(4):308–17.
2. West BJ. Where medicine went wrong: rediscovering the path to complexity. Hackensack: Wold Scientific; 2006.
3. Guastello SJ. Managing emergent phenomena: nonlinear dynamics in work organizations. Mahwah, NJ: Lawrence Erlbaum; 2002.
4. Bak P. How nature works: the science of self-organized criticality. New York: Copernicus; 1996.
5. Bar-Yam Y. Dynamics of complex systems. New York: Westview; 1997.
6. Gilden D, Thornton T, Mallon M. 1/*f* noise in human cognition. Science. 1995;267(5205):1837–9.
7. Van Orden GC, Holden JG, Turvey MT. Human cognition and 1/f scaling. J Exp Psychol Gen. 2005;134(1):117–23.
8. Hausdorff JM, Purdon PL, Peng CK, Ladin Z, Wei JY, Goldberger AL. Fractal dynamics of human gait: stability of long-range correlations in stride interval fluctuations. J Appl Physiol. 1996;80(5):1448–57.
9. Jordan K, Challis JH, Newell KM. Long range correlations in the stride interval of running. Gait Posture. 2006;24(1):120–5.
10. Hong SL, James EG, Newell KM. Age-related complexity and coupling of children's sitting posture. Dev Psychobiol. 2008;50(5):502–10.
11. Duarte M, Zatsiorsky VM. On the fractal properties of natural human standing. Neurosci Lett. 2000; 283(3):173–6.
12. Blesic S, Stratimirovic D, Milosevic S, Maric J, Kostic V, Ljubisavljevic M. Scaling analysis of the effects of load on hand tremor movements in essential tremor. Phys A Stat Mech Appl. 2011;390(10):1741–6.
13. Aks DJ, Zelinsky GJ, Sprott JC. Memory across eye-movements: 1/*f* dynamic in visual search. Nonlinear Dynamics Psychol Life Sci. 2002;6(1):1–25.
14. Nakamura T, Kiyono K, Yoshiuchi K, Nakahara R, Struzik ZR, Yamamoto Y. Universal scaling law in human behavioral organization. Phys Rev Lett. 2007;99(13):138103.
15. Roizen NJ, Higgins AM, Antshel KM, Fremont W, Shprintzen R, Kates W. 22q11.2 deletion syndrome: are motor deficits more than expected for IQ level? J Pediatr. 2010;157(4):658–61.
16. Piek JP, Pitcher TM, Hay DA. Motor coordination and kinaesthesis in boys with attention deficit–hyperactivity disorder. Dev Med Child Neurol. 1999;41(3):159–65.
17. Henderson SE, Morris J, Frith U. The motor deficit in Down's syndrome children: a problem of timing? J Child Psychol Psychiatry. 1981;22(3):233–45.
18. Ming X, Brimacombe M, Wagner GC. Prevalence of motor impairment in autism spectrum disorders. Brain Dev. 2007;29(9):565–70.
19. Smith MA, Brandt J, Shadmehr R. Motor disorder in Huntington's disease begins as a dysfunction in error feedback control. Nature. 2000;403(6769):544–9.
20. Berardelli A, Rothwell JC, Day BL, Marsden CD. Movements not involved in posture are abnormal in Parkinson's disease. Neurosci Lett. 1984;47(1):47–50.
21. Manckoundia P, Pfitzenmeyer P, d'Athis P, Mourey F. Impact of cognitive task on the posture of elderly subjects with Alzheimer's disease compared to healthy elderly subjects. Mov Disord. 2006;21(2):236–41.
22. Carroll CA, O'Donnell BF, Shekhar A, Hetrick WP. Timing dysfunctions in schizophrenia as measured by a repetitive finger tapping task. Brain Cogn. 2009; 71(3):345–53.
23. Marvel CL, Schwartz BL, Rosse RB. A quantitative measure of postural sway deficits in schizophrenia. Schizophr Res. 2004;68(2–3):363–72.
24. Bolbecker AR, Hong SL, Kent JS, Klaunig MJ, O'Donnell BF, Hetrick WP. Postural control in bipolar disorder: increased sway area and decreased dynamical complexity. PLoS One. 2011;6(5):e19824.
25. Bolbecker AR, Hong SL, Kent JS, Forsyth JK, Klaunig MJ, Lazar EK, et al. Paced finger-tapping abnormalities in bipolar disorder indicate timing dysfunction. Bipolar Disord. 2011;13(1):99–110.
26. Allen PA, Namazi KH, Patterson MB, Crozier LC, Groth KE. Impact of adult age and Alzheimer's disease on levels of neural noise for letter matching. J Gerontol. 1992;47(5):P344–9.
27. Li S-C, Lindenberger U, Sikström S. Aging cognition: from neuromodulation to representation. Trends Cogn Sci. 2001;5(11):479–86.
28. Lipsitz LA, Goldberger AL. Loss of 'complexity' and aging. Potential applications of fractals and chaos theory to senescence. JAMA. 1992;267(13):1806–9.
29. Pincus SM. Approximate entropy as a measure of system-complexity. Proc Natl Acad Sci USA. 1991;88:2297–301.
30. Sosnoff JJ, Newell KM. Age-related loss of adaptability to fast time scales in motor variability. J Gerontol B Psychol Sci Soc Sci. 2008;63(6):P344–52.
31. Duarte M, Sternad D. Complexity of human postural control in young and older adults during prolonged standing. Exp Brain Res. 2008;191(3):265–76.
32. Hong SL, James EG, Newell KM. Coupling and irregularity in the aging motor system: tremor and movement. Neurosci Lett. 2008;433(2):119–24.

33. Hong SL, Bodfish JW, Newell KM. Power-law scaling for macroscopic entropy and microscopic complexity: evidence from human movement and posture. Chaos. 2006;16:013135.
34. Newell KM, Vaillancourt DE, Sosnoff JJ. Aging, complexity and motor performance: health and disease states. In: Birren JE, Schaie KW, editors. Handbook of the psychology of aging. Amsterdam: Elsevier; 2006. p. 163–82.
35. Vaillancourt DE, Newell KM. Changing complexity in human behavior and physiology through aging and disease. Neurobiol Aging. 2002;23(1):1–11.
36. Churruca J, Vigil L, Luna E, Ruiz-Galiana J, Varela M. The route to diabetes: loss of complexity in the glycemic profile from health through the metabolic syndrome to type 2 diabetes. Diabetes Metab Syndr Obes. 2008;1:3–11.
37. Gottschalk A, Bauer MS, Whybrow PC. Evidence of chaotic mood variation in bipolar disorder. Arch Gen Psychiatry. 1995;52:947–59.
38. Lipsitz LA. Dynamics of stability: the physiologic basis of functional health and frailty. J Gerontol A Biol Sci Med Sci. 2002;57:B115–25.
39. Pincus SM. Greater signal regularity may indicate increased system isolation. Math Biosci. 1994;122(2):161–81.
40. Lipsitz LA. Physiological complexity, aging, and the path to frailty. Sci Aging Knowledge Environ. 2004;16:pe16.
41. Assisi CG, Jirsa VK, Kelso JAS. Synchrony and clustering in heterogeneous networks with global coupling and parameter dispersion. Phys Rev Lett. 2005;94:4.
42. Hong SL. The dynamics of structural and functional complexity across the lifespan. Nonlinear Dynamics Psychol Life Sci. 2007;11(2):219–34.
43. Newell KM, Liu Y-T, Mayer-Kress G. A dynamical systems interpretation of epigenetic landscapes for infant motor development. Infant Behav Dev. 2003;26(4):449–72.
44. Vaillancourt DE, Slifkin AB, Newell KM. Regularity of force tremor in Parkinson's disease. Clin Neurophysiol. 2001;112(9):1594–603.
45. Morrison S, Sosnoff JJ. Age-related changes in the adaptability of neuromuscular output. J Mot Behav. 2009;41(3):274–83.
46. Peng CK, Havlin S, Stanley HE, Goldberger AL. Quantification of scaling exponents and crossover phenomena in nonstationary heartbeat time-series. Chaos. 1995;5(1):82–7.
47. Newell KM. Constraints on the development of coordination. In: Wade MG, Whiting HTA, editors. Motor development in children. Amsterdam: Nijhoff; 1986. p. 341–61.
48. Bernstein NA. The co-ordination and regulation of movements. Oxford: Pergamon; 1967.
49. Kugler PN, Turvey MT. Information, natural laws, and self-assembly of rhythmic movement. Hillsdale, NJ: Erlbaum; 1987.
50. Kelso JAS. Dynamic patterns: the self-organization of brain and behavior. Cambridge, MA: MIT Press; 1995.
51. Holt KG, Hamill J, Andres RO. The force-driven harmonic oscillator as a model for human locomotion. Hum Mov Sci. 1990;9(1):55–68.
52. Haken H, Kelso JAS, Bunz H. A theoretical model of phase-transitions in human hand movements. Biol Cybern. 1985;51(5):347–56.
53. Zanone PG, Kelso JA. Evolution of behavioral attractors with learning: nonequilibrium phase transitions. J Exp Psychol Hum Percept Perform. 1992;18(2):403–21.
54. Zanone PG, Kelso JAS. Coordination dynamics of learning and transfer: collective and component levels. J Exp Psychol Hum Percept Perform. 1997;23(5):1454–80.
55. Kugler PN, Kelso JAS, Turvey MT. On the concept of coordinative structures as dissipative structures I. Theoretical lines of convergence. In: Stelmach GE, Requin J, editors. Tutorials in motor behavior. Amsterdam: Elsevier; 1980.
56. Hong SL, Newell KM. Entropy conservation in the control of human action. Nonlinear Dynamics Psychol Life Sci. 2008;12(2):163–90.
57. Hong SL. The entropy conservation principle: applications in ergonomics and human factors. Nonlinear Dynamics Psychol Life Sci. 2010;14(3):291–315.
58. Müller I. A history of thermodynamics: the doctrine of energy and entropy. Berlin: Springer; 2007.
59. Hong SL, Newell KM. Entropy compensation in human motor adaptation. Chaos. 2008;18:013108.
60. Hong SL, Newell KM. Motor entropy in response to task demands and environmental information. Chaos. 2008;18:033131.
61. Vaillancourt D, Russell D. Temporal capacity of short-term visuomotor memory in continuous force production. Exp Brain Res. 2002;145(3):275–85.
62. Hong SL, Beck MR. Uncertainty compensation in human attention: evidence from response times and fixation durations. PLoS One. 2010;5(7):e11461.
63. Hong S, Brown A, Newell K. Compensatory properties of visual information in the control of isometric force. Atten Percept Psychophys. 2008;70(2):306–13.
64. Kelso JAS, Engstrøm DA. The complementary nature. Cambridge, MA: MIT Press; 2006.
65. Jenner P. Molecular mechanisms of L-DOPA-induced dyskinesia. Nat Rev Neurosci. 2008;9(9):665–77.
66. Gotham AM, Brown RG, Marsden CD. 'Frontal' cognitive function in patients with Parkinson's disease 'on' and 'off' levodopa. Brain. 1988;111(2):299–321.
67. Lees AJ, Smith E. Cognitive deficits in the early stages of Parkinson's disease. Brain. 1983;106(2):257–70.
68. Starkstein SE, Preziosi TJ, Bolduc PL, Robinson RG. Depression in Parkinson's disease. J Nerv Ment Dis. 1990;178(1):27–31.
69. Factor SA, Feustel PJ, Friedman JH, Comella CL, Goetz CG, Kurlan R, et al. Longitudinal outcome of Parkinson's disease patients with psychosis. Neurology. 2003;60(11):1756–61.
70. Moss F, Ward LM, Sannita WG. Stochastic resonance and sensory information processing: a tutorial and review of application. Clin Neurophysiol. 2004;115(2):267–81.

71. Priplata A, Niemi J, Salen M, Harry J, Lipsitz LA, Collins JJ. Noise-enhanced human balance control. Phys Rev Lett. 2002;89(23):238101.
72. Gravelle DC, Laughton CA, Dhruv NT, Katdare KD, Niemi JB, Lipsitz LA, et al. Noise-enhanced balance control in older adults. Neuroreport. 2002;13(15): 1853–6.
73. Costa M, Priplata AA, Lipsitz LA, Wu Z, Huang NE, Goldberger AL, et al. Noise and poise: enhancement of postural complexity in the elderly with a stochastic-resonance-based therapy. Europhys Lett. 2007;77(6):68008.
74. Wolf SL, Blanton S, Baer H, Breshears J, Butler AJ. Repetitive task practice: a critical review of constraint-induced movement therapy in stroke. Neurologist. 2002;8(6):325–38.
75. Charles JR, Wolf SL, Schneider JA, Gordon AM. Efficacy of a child-friendly form of constraint-induced movement therapy in hemiplegic cerebral palsy: a randomized control trial. Dev Med Child Neurol. 2006;48(08):635–42.
76. Gordon AM, Charles J, Wolf SL. Methods of constraint-induced movement therapy for children with hemiplegic cerebral palsy: development of a child-friendly intervention for improving upper-extremity function. Arch Phys Med Rehabil. 2005;86(4): 837–44.
77. Frank E, Swartz HA, Kupfer DJ. Interpersonal and social rhythm therapy: managing the chaos of bipolar disorder. Biol Psychiatry. 2000;48(6):593–604.
78. Frank E, Kupfer DJ, Thase ME, Mallinger AG, Swartz HA, Fagiolini AM, et al. Two-year outcomes for Interpersonal and Social Rhythm Therapy in individuals with bipolar I disorder. Arch Gen Psychiatry. 2005;62(9):996–1004.
79. Frank E, Soreca I, Swartz HA, Fagiolini AM, Mallinger AG, Thase ME, et al. The role of Interpersonal and Social Rhythm Therapy in improving occupational functioning in patients with bipolar I disorder. Am J Psychiatry. 2008;165(12):1559–65.

Avoidable Hospitalizations in Older Adults

27

Applying Complexity Science Principles and Machine Learning Approaches

Carmel M. Martin, Carl Vogel, Lucy Hederman, Kevin Smith, Atieh Zarabzadeh, Deirdre Grady, and Jing Su

One unintended outcome of the modern transformation of the medical care system is that it does just about everything to divert the practitioner's attention away from the experience of illness.

Arthur Kleinman [1]

27.1 Introduction

Avoidable hospitalizations are the subject of considerable interest to decision makers, because such admissions are deemed to be expensive, unhelpful to patients, and reflect underperformance of health systems' organization [2–7]. Hospitalizations that might have been averted by health service interventions for older people are of particular concern. This chapter examines the nature of the problem and limitations of current approaches. It identifies the need for a comprehensive conceptual framework that addresses the complex human systems of aging, being ill, and dying in contemporary society. Avoidable hospitalizations conceptualized as reflecting biological, psychosocial, organizational, and social phenomena opens up a complex adaptive systems approach to the problematical issue of hospitalizations for older people. It focuses on the current and potential contributions of informatics and computational science to this field. A real-time informatics system based on patients' narratives of their wellness and illness: the Patient Journey Record system provides an example of an adaptive system that addresses avoidable hospitalizations using a complex systems framework.

C.M. Martin (✉)
Department of Public Health and Primary Care, Trinity College, Dublin, College Green, Dublin D2, Ireland
e-mail: martinc7@tcd.ie

C. Vogel • L. Hederman • J. Su
School of Computer Science and Statistics, Trinity College, Dublin, College Green, Dublin D2, Ireland
e-mail: Carl.Vogel@scss.tcd.ie; Lucy.Hederman@scss.tcd.ie

K. Smith
Department of Medicine, University College Cork, Cork, Ireland
e-mail: kevinallansmith@gmail.com

A. Zarabzadeh
National University of Ireland, Galway, Ireland

D. Grady
Patient Journey Record Campus Company, Trinity College Dublin, College Green, Dublin D2, Ireland
e-mail: Deirdre.OGrady@ucd.ie

27.2 What Is an Avoidable Hospitalization?

Considerable variations in terminology are used to describe models and components of models designed to reduce avoidable hospitalizations, their context of care and settings. Even definitions of the terms "acute", "hospital", and "admissions" and "readmissions" vary, internationally. The US Agency for Healthcare Research and

J.P. Sturmberg and C.M. Martin (eds.), *Handbook of Systems and Complexity in Health*, DOI 10.1007/978-1-4614-4998-0_27,

Quality defines potentially avoidable hospitalizations for quality assessment as "*potentially avoidable hospitalizations are admissions for inpatient care of chronic illnesses that could be averted if the patients had good quality outpatient care. Without such care, the risk of complications requiring hospitalization is greater*" [8]. For example, it focuses on admissions of heart failure for people who have been diagnosed with heart failure and produced league tables as state performance indicators. This approach is followed by other countries such as the UK, Canada, and Australia.

However, from a clinical perspective, most people with heart failure invariably have comorbidities, and focusing on one condition may detract from whole person care and other avoidable problems. Thus a broader definition is appropriate. Avoidable hospitalizations, in the broader sense, represent a range of conditions for which hospitalizations could be avoided because the disease or condition has been prevented from occurring and/or deteriorating, and because individuals have had access to timely and effective care [9]. A recent definition of avoidable readmissions summarizes their multifactorial nature—"*a preventable readmission as an unintended and undesired subsequent postdischarge hospitalization, where the probability is subject to the influence of multiple factors*" [10–12].

27.3 Avoidable Hospitalizations: A Complicated Picture in the Health Services Literature

There are two main themes in the health services research literature—avoiding readmissions within 30 or 60 days postdischarge from hospital and avoiding preventable hospitalizations from ambulatory patients in the community [9, 13, 14]. In addition, distal or upstream predictors of avoidable admissions identified in the literature are socioeconomic factors (low income, decreasing social networks, health illiteracy, and poor neighborhoods) and poor management of preventable predisposing conditions and health risks. Immediate or proximal predictors of avoidable hospitalizations are deteriorations in chronic disease (cardiac and respiratory in particular) and poor social support [12].

High-risk groups in particular are older and sicker patients, with multimorbidity mostly comprising chronic disease and frailty or senescent syndromes or major psychosocial issues. Subacute conditions that could have been addressed earlier are implicated at times [15].

Hospital-based risk prediction models are mainly focused at 30-day readmissions. Readmissions soon after hospital discharge appear highly dependent upon hospital management particularly within the first 24–72 h [16, 17]. Ambulatory or Primary Care sensitive conditions (ACSCs) are those conditions for which hospital admission could be prevented by interventions in community settings [18]. The problem of ACSCs is increasing in size, as the population ages; this underscores the importance of developing, adopting, and implementing innovative programs [19]. ACSCs have both demand and supply side access components and require multiple strategies. Community-based programs address self-assessed health, support, psychosocial, and environmental issues and disease factors [20] while hospital programs focus on diagnostic groups, disease severity, length of stay, and physiological variables [20, 21].

Overlap among hospital and community programs [10] is demonstrated in prolonged postdischarge programs by the work of Courtney [22] and others [23, 24]. Case management and transitional care programs may successfully address disease and personal issues in hospital, the emergency department, and community settings [25, 26]. The "intermediate" sector—case management, disease management, telehealth, home healthcare—are all implicated as being solutions to avoidable hospitalizations [27–30]. Postdischarge readmissions and transitions may be the easiest targets to achieve [3], yet there is increasing evidence that primary care programs can achieve admission avoidance [31].

27.4 Building a Conceptual Framework Using Complexity Theory

Avoidable hospitalizations are demonstrably a problem for health systems, and also for the society at large. There is a lack of a coherent conceptual framework in which to view the problem. Currently the problem is seen as one of hospital or community care inadequacies without examining the nature of the human systems implicated. In health care systems, numerous factors are implicated—resource inputs, structure, process, and agency. These include: hospital and primary care financial incentives; integration, timing, and coordination of personnel, services, and medication across hospital and ambulatory care [9, 12, 32].

In addition there is a lack of agreement as to how best to manage very complex patients efficiently and how to deal with both aging and chronic disease. Moreover health services and prevention take place in complex, dynamic, and often unpredictable systems [33, 34]. These involve multiple layers and interdependencies. Influences range from evolutionary legacies in human biological systems, political economy and population health determinants, health and social systems, and ultimately individual personal journeys. Philosophical, cultural, and ethical considerations are important. All influence the contemporary phenomenon that we label "*avoidable hospitalizations*", from upstream and distal to downstream and proximal levels to the individual at the center of care. Figure 27.1 describes a high level overview of a conceptual framework for systems that will be explored in subsequent paragraphs.

27.4.1 Evolutionary and Systems Knowledge

Evolutionary medicine and systems medicine provide perspectives through which to see human health as "*an evolved, complex system that strives to maintain homeostasis and promote individual survival and genetic reproduction*" [35, 36]. Upstream knowledge including the fields of phylogeny, life history theory, genetics, biochemistry, and evolutionary medicine help us understand the nature of human aging, illness, and dying [37, 38]. Intra-system integrity of the person, their genome, their physical, social, and environmental stressors and supports, and their epigenetic adaptation is central to a conceptual framework. Candidate genes, epigenetics, and transcription are likely to interact with social experiences over many decades of life and perhaps previous generations to predict health [39, 40].

Aging, implicated in both frailty or aging syndromes and chronic disease progression, can also be understood as a fractal process. Frailty has been understood to be a uniform process of aging where the whole body gradually loses adaptive capacity, while chronic disease and dementia reflects particular organ or system failing ahead of the rest of the body. Depending on the pattern of frailty, dementia, and chronic disease different patterns emerge [41, 42].

Both types of conditions result in reduced homeostasis, resilience and adaptability, vulnerability to disease and organ failure, and reduced social and environmental resources [43]. Telomere research, for example, provides potential strategic direction for more direct interventions to slow aging and chronic disease [44, 45]. The roles of infection, inflammation, and nutrition in human development, disease, and dying are prominent. While aging, disease, and death are inevitable, health systems need to be continually adapting to and learning about the nature of aging and chronic life-limiting disease and their complex interdependencies, particularly in the current context of avoidable hospitalizations.

27.4.2 Cultural Perspectives

Avoidable hospitalizations predominantly occur among the oldest and most disadvantaged people in society [46]. There are many uncertainties about boundaries between states of health and illness and aging, and how health care and society deal with these [47].

Most people die in hospital. This happens in spite of end of life care strategies to support the

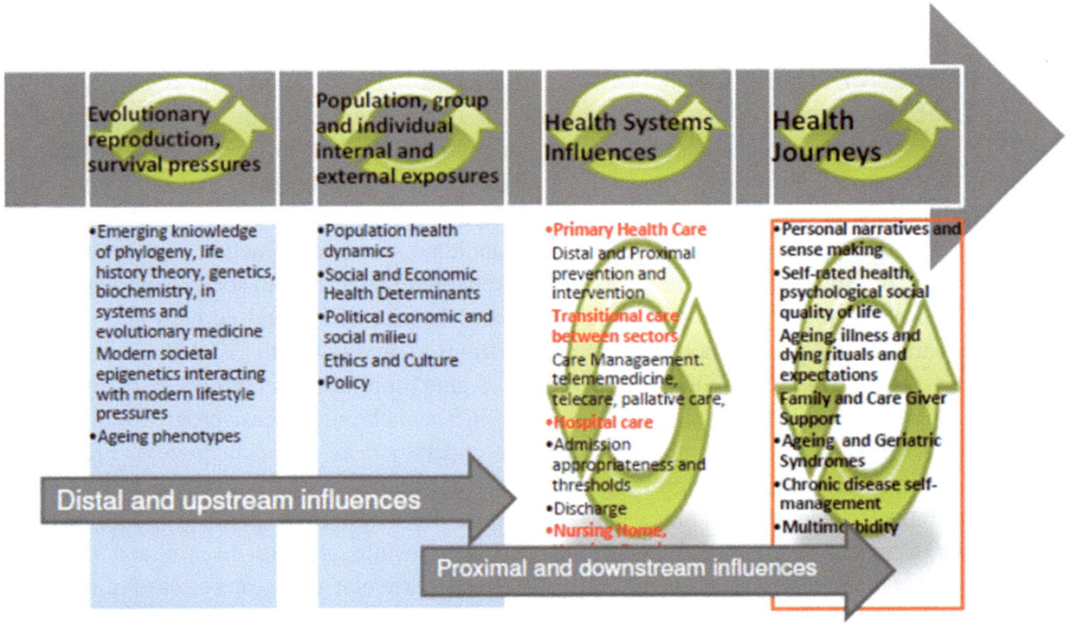

Fig. 27.1 Conceptual framework for avoidable hospitalizations—multiple layers of distal upstream and downstream proximal influences subject to nonlinear dynamics in complex systems. Evolutionary, reproductive, and survival pressures interact with modern lifestyles to set up aging, disease, and illness biological trajectories in populations. Population health dynamics and societal systems shape biopsychosocial health and quality of life trajectories in different groups. Health systems through its sectors of primary and community care; and institutions of hospital, hospice and nursing homes shape health care trajectories of groups. All these shape individual personal health journeys. At critical times instabilities in personal journeys lead to a "crisis" which results in hospitalization. Such a hospitalization is avoidable if at any layer interventions are possible to prevent or ameliorate the instability in physical or psychosocial domains or to find alternative social or health pathways to deal with them

majority of people who wish to die at home [48, 49]. Avoidable admissions for individuals mainly occur in the last 2–3 years of life, and the highest costs occur in hospital in the last 6 months of life [3, 50]. Hospital care may not be appropriate for many [51]. Transitions from active management to supportive end of life care are gray areas that entail balancing probabilities of survival with quality of life [47, 52, 53]. Risk and dignity are often finely balanced. End of life care planning and decision making are important cultural changes which are slowly becoming main stream. However one size does not fit all and cultural and social expectations of death and dying of each individual are highly personal within their social milieu [54].

Avoidable hospitalizations for older people, at a societal level, represent complex cultural attitudes to aging, illness, and dying. Human rights and ethics maintain that there is a duty of care for the older, frail, and ill [55]. Yet collectively, how do we best ameliorate the adverse effects of disease and senescence, particularly if they are "ill-defined" [42] and how do we support individuals in their journeys through disease, dying, and death and the phases of life that go before [56]? It is clear that the many problems facing health services reside in different knowledge domains and thus will require different strategies to synthesize practical knowledge for decision making [54, 57]. However, most strategies appear to focus on specific diseases. Disease-based care is seen to be legitimately within the scope of medical care.

27.4.3 Population Health Dynamics

Avoidable hospitalizations are attributable to chronic diseases in most cases [9]. Obesity, alcohol use, smoking, poverty, health insurance status, ethnicity, and location have been causally associated in population health dynamics [8, 58]. These are linked to diabetes complications and circulatory and respiratory conditions as the most common diagnoses [8]. Reducing avoidable hospitalizations, thus, may involve paying attention to social determinants of health in policy and practice reforms directed at health and welfare sectors.

Yet, in one region, an in-depth study found that "ill-defined" or "non-disease" symptoms as reasons for admissions in older patients via the emergency department reached 58 % [59], although most were coded with cardiac or respiratory diagnoses. Classifying the reason for hospital admission or attendance at emergency departments by this "ill-defined" type of destabilization as cardiac or respiratory conditions underplays the complexity of these patients' needs and may result in inappropriate deployment of services [58, 60]. Other studies also confirm that multimorbidity [61] and aging syndromes rather than a single disease is associated with need for care and greater service use [62, 63]. End of life trajectories in older adults tend to have common patterns of disability, apart from those with severe dementia [41, 42]. Any destabilization related to unplanned service use, including avoidable admissions, is multifactorial and related to social and environmental disruptions and nonspecific aging syndromes, as much or more than disease-related conditions predominantly cardiac or respiratory [64].

27.4.4 Health Systems

Health promotion acts on both upstream and downstream influences, chronic disease, aging syndromes and multimorbidity, family and caregiver support, aging and dying rituals, and expectations [65]. Primary Health Care addresses both distal and proximal prevention and intervention [66]. Hospital care, with its responsibility for admission appropriateness and thresholds and discharge processes, is central to avoidable hospital readmission [67].

Interventions with significant potential for improvement in transitional care between sectors includes case management [22, 68], intermediate care [28], telemedicine and telecare [69, 70], and respite [71] and home care [72]. Accessible pathways to quality palliative care, hospice care, post acute care and nursing home care are important means that influence avoidable hospitalizations for individual patients [73].

27.4.5 Individual Health Journeys

At the center of health care are the individuals experiences and quality of life and that of their care giver/s or family. Personal narratives focus directly upon individuals' experiences within the contexts of their daily lives. Numerous studies demonstrate that personal experiences of illness (sometimes termed lay experiences as opposed to professional judgments), including those operationalized as self-rated health metrics, are sensitive to a broad range of internal bodily processes [74], micro- and macro-contextual influences, and is highly predictive of health care costs and outcomes [75, 76].

The illness experience literature is highly developed but to date has not been implicated as significant in trajectories to avoidable hospitalizations [77]. There is the need to recognize clinical and dynamic processes in the end-of-life trajectories which form the basis for the majority of avoidable hospital admissions. Chronic illness, aging, and frailty lead to poor self-rated health and together with reduced social support and conducive environments, leads to "brittleness" [78] with reduced resilience. Lack of ability to be resilient in the physical, psychological, social, and environmental domains underpin the personal health crisis that leads to an emergency admission [79].

27.5 Complex Adaptive Systems Theory

Avoidable hospitalizations are particularly challenging because they result from a system containing a diverse set of actors, at many different levels of scale, with differing individual motivations and priorities. The expanding problem and variation in outcomes cannot be reduced to a single mechanism. Avoidable hospitalizations thus would seem to represent a *complex adaptive system* (CAS). A CAS is one composed of many heterogeneous pieces, interacting with each other in subtle or nonlinear ways [80]. A CAS is contrasted with ordered and chaotic systems by the relationship that exists between the system and the agents which act within it. In a highly controlled ordered system the level of constraint ensures that all agent behavior is limited to the rules of the system. In a chaotic system the agents are largely unconstrained and seemingly random and unmanageable, although there are underlying patterns. In a CAS, arguably an ideal for health systems, the system and the agents coevolve; the system lightly constrains agent behavior, but the agents modify the system by their interaction with it [81]. Features of the CAS system applicable to avoidable hospitalizations include:

- *Individuality*: population aging and disease dynamics and health service processes are multilevel but individual journeys are subject to and driven by decentralized, local interactions of constituent parts and personal sense making.
- *Agents that learn*: learning is understood in terms of the adaptive behaviors of phenomena (in this case why and how we define and frame and *adapt* to the phenomena that lead to "avoidable hospitalizations") that arise in the interactions of multiple agents [33]. Clinical systems and computational machines can also be viewed as agents that learn—agents will learn according to how they are constrained internally or externally.
- *Heterogeneity*: substantial diversity among actors and processes at each level—in goals, rules, adaptive capacity, and constraints—shape the dynamics of aging, illness, and dying, and society's responses in important ways.
- *Interdependence and emergence*: a CAS usually contains many interdependent interacting pieces, connected across different levels. System dynamics are often characterized by feedback and substantial nonlinearity. Unintended consequences commonly emerge (well-intentioned acute medical care may result in fruitless investigations, loneliness, and hospital-acquired infections).
- *Emergence*: personal narratives and sense-making emerges from situated, active integration within an external environment and context, rather than primarily within the internal confines of externally imposed disease criteria.
- *"Tipping points"*: nonlinearity means that the impacts caused by small changes can seem hugely out of proportion. The individual may spend long periods in a state of relative stability, yet be easily "tipped" to an avoidable hospitalization by a disturbance in one of many domains—biopsychosocial environmental—that pushes illness and social support across an individual threshold [80].

Many different synergistic, overlapping, or competitive interdependencies exist. Diversity, which has been long recognized as a key to successful evolution and adaptation in complex systems, is something to be fostered and valued. Moreover, competition and cost containment are key drivers of innovation, adaptation, and development. Looking at health care as heterogeneous systems to deal with complex human journeys through different lenses can lead to the identification of gaps and distortions in our knowledge base. Human organizations exist in silos and constraints based on different knowledge approaches and core values [82]. Hospitals and primary care have different world views.

Hospitals see individual people in terms of their disease profiles while primary care sees people who have particular diseases in terms of their well-being and quality of life and information systems and records reinforce these world views [83]. Funders and managers of health services see health episodes in terms of classifications and costs and information systems provide the basis for these judgements. However, information, ideas, or evidence that inform particular knowledge systems are never "innocent", without tacit and explicit assumptions which may or may not best serve the individuality of personal health journeys [57, 84].

Nevertheless health systems must proceed with imperfect and developing frameworks and the use of complex system principles, characterized by a "*Probe-Sense-Respond*" approach [85].

The complex environment surrounding health service responses to aging and chronic disease flux, and unpredictability precludes a "*one*" right answer approach.[1] In such systems, "*emergent*" answers reflect the development of inferential instructive patterns about unknown unknowns. On the other hand, aspects of avoidable hospitalizations related to transitions from hospital to home that appear to fit a simple or complicated approach, include creating new services comprising nurse, pharmacist or establishing a team of various health professionals following up after hospital discharge [3]. By recognizing complexity features in what may appear straight forward interventions in avoidable hospitalizations—individuality, heterogeneity, tipping points, emergence, and interdependence—it becomes apparent that there are likely to be many diverse mechanisms to achieve these improvements cost-effectively [3].

The validity and predictive capacity of expression of where one is in one's health journey through narrative has been well documented in the health care literature [86]. Thinking outside the box, e.g. applying principles of artificial intelligence and Machine Learning, would enable us to view narrative and illness experience as showing potential for greater understanding. Embodiment holds that the nature of the human thinking is largely determined by the state of the human body and the human body is shaped by mental processes. All aspects of cognition are shaped by aspects of the body. The aspects of cognition include high-level mental constructs and human performance such as reasoning and sense-making about the body. The aspects of the body include the physical motor system, the body's interactions with the environment, and the assumptions about the world that are built into the body and the brain [87].

Embodied cognition principles reflect complex adaptive systems theories, and reconnect our conception of humanity to an evolutionary continuum. Cognition and judgement are not rule-based manipulation of symbols, but rather are an emergent phenomenon. Language representing a particular state of embodied cognition such as narratives about their illness and judgement of their self-rated health can be appreciated by cutting-edge artificial intelligence [87], as well as showing immense potential in enhancing human intelligence through computation [88].

The field of avoidable hospitalizations must be informed by broadly based knowledge beyond the literature based on discrete health service interventions. Adopting approaches that acknowledge the complex systems underpinning avoidable hospitalizations recognizes the many competing ideas and fosters diverse approaches to be tried and developed.

27.6 Who Is *at Risk* in Contemporary Health Services?

How do we identify who is at risk of avoidable hospitalizations and how do we predict when they are at risk in what *after all is the inevitable* journey towards the end-of-life in order to optimize personalized care? What patterns should we be observing—health care utilization, disease, and biometrics—that identify pathophysiological status pathways? Or the individual's narratives reflecting the embodiment of their health state expressed in their own language and the narrative terms of self-rated health, anxiety, pain, and daily functioning?

Clinical knowledge can predict current risks and short-term risks, but additional resources are needed to improve risk prediction over time [89]. Tools to help identify people at high risk of future avoidable emergency admission are based on the analysis of databases using a mixture of personal and service use information [25, 90–92]. Contemporary approaches are reviewed, and innovative approaches are discussed.

27.6.1 Threshold Criteria

The objective is to optimize the participant selection process for care/case management (CM) programs to reduce avoidable admissions. The

[1] The "*one*" right answer principle reflects the classical Newtonian cause-and-effect model.

threshold criteria approach uses rule-based case finding and identifies those at high risk of future admissions who meet a set of criteria at a point in time. Case finding has used threshold metrics from hospital data, such as repeated emergency admissions, chronic disease registers, and prolonged length of stay, as a marker to identify those at high risk of future admissions [58]. However, they have a short-term perspective as admission rates and bed use among high-risk patients fall to the mean rate for older people over time as the individuals re-stabilize at a lower risk or die after a crisis [93]. Unpredictable instabilities may arise with many admissions from individuals initially screened out because they were below a threshold which is set, in order to make case management practical and affordable [94]. Alternatives such as identifying patients at high risk through a questionnaire administered by a primary care practice [91] also do not take account of changing trajectories in individual journeys unless repeated regularly [23, 24, 95], and may result in resources directed to those whose journeys have stabilized.

27.6.2 Predictive Modeling to Identify High-Risk Patients

Predictive modeling uses retrospective datasets to identify disease cut off points and then predict patients who fall outside those criteria as being high risk and requiring intense care management [96]. For example, in order to augment the ability to prospectively identify high-risk patients, organizations such as American Healthways use predictive modeling technology—data from multiple patient clinical and administrative systems to predict the likelihood for short-term health problems. They identify high-risk individuals in order to intervene to avert deteriorations and improve outcomes [97].

Techniques used are statistical and computational. In fact complexity theory does not assume that it is possible to accurately predict outcomes in complex systems, although methods such as Bayesian probability theory, cusp catastrophe theory, and other mathematical approaches can provide better estimates than simple linear models.

Ongoing access to hospital and community-based service data to update risk profiles is needed to address changing population and service profiles. However, one needs to appreciate that individual health trajectories following hospitalization regress to the mean, and that other unpredictable factors not recorded in community or hospital services may become more important in the long run [89, 93, 98]. More sophisticated data analytics of combined databases form the basis of predictive scoring to help identify people at high risk of future avoidable emergency admission [25, 90–92]. An example of such predictive models on admissions related to heart failure in the US, learned from large-scale hospital data sets, have been used to identify patients who are at the highest risk of being hospitalized within a short time after they are discharged.

27.6.3 Limits to Current Approaches

This then raises the problem with current risk assessment tools, whether they are threshold or predictive techniques—they are not adaptive. They are cross-sectional with static or short-term predictions while people's at-risk status is complex and nonlinear and over symptom-free time regresses to the mean [93, 95]. Also, while there are promising and innovative interventions to reduce hospital admissions, they are not necessarily cost-effective [29, 99]. Arguably this is due to the complex system dynamics of the patient journey beyond short-term observation such as within 30 days of discharge.

How to distribute intense resources to people only when they are needed, rather than employ intense interventions en masse on a population risk basis at a particular time or a hospitalization event? After all, individuals navigate their unique personal health and illness journeys through periods of both illness and wellness even when dying.

There is no *right* solution. Health services research and service innovations have a tendency to look for facts, best practice, and simple solutions rather than adopt a philosophy that allows new and diverse locally adapted service patterns to emerge [100]. There is the desire for the accelerated resolution of problems or exploitation of opportunities in the environment of financial pressures in health systems.

This is a dynamic field, for example, knowledge is emerging that self-rated health, quality of life, symptoms, and anxiety appear to reflect underlying disturbances in homeostasis in individual journeys, as much as biometric measures or health service patterns [101–103].

27.7 Predicting Avoidable Admissions: Illness and Nonlinear Dynamics Systems Theories

Complexity science and chaos theory open up new ways to conceptualize ambulatory care sensitive and avoidable hospitalizations within a wider conceptual framework. Information strategies need to identify embodied cognition and sense-making in order to address the influences that take place at multiple levels. These may include genes, epigenetics, neurobiology, psychology, family structure and influences, social context and social norms, environment, markets, and public policy, as well as health services. While these levels entail very different pathways and diverse methodologies for study in a broad range of domains, they can be understood as interdependent and interconnected systems (Fig. 27.2).

Even where mechanisms of effect are clear, the *linkages and feedback between these mechanisms* are not well studied or well understood. Furthermore, no single mechanism appears to be able to account for all that we know about aging, disease, and dying. Neurobiological and genetic mechanisms help explain resilience at individual and group levels but have difficulty identifying the mechanisms to improve the end of life for individuals. Such complex adaptive systems are increasingly serviced by advances in modern computational sciences.

Sophisticated informatics is a major component of evolutionary and systems biology, health systems, and population health information processing. Bioinformatics with its interaction between computational science and biological sciences has Machine Learning as a basic method for automated analysis of DNA, protein arrays, genomics, and epigenetics applying complex systems principles. Population health and social determinants such as in obesity research use complex adaptive systems and multilevel informatics systems [104]. Health informatics is being increasingly used to support clinical services and health service organization [105]. Horvitz of Microsoft Corporation sees potential for using predictive modeling to address the challenges of chronic disease management, and for combining prediction with decision analysis in the allocation of resources [105]. A recent analysis applied Machine Learning to a large multiyear data set of patient hospitalizations in the Greater Washington, DC, Metropolitan area. The resulting predictive models infer the likelihood of hospitalization within 30 days of discharge. Predictions of risk can be coupled with automated decision analyses to weigh the costs and benefits of intervening with special postdischarge support programs [106].

At the individual patient level, sensors and biometric data input systems have been used to detect falls, ECG anomalies and symptoms of heart failure, diabetes deterioration, and respiratory dysfunction using a real-time monitoring technology and Machine Learning techniques. This is a rapidly developing field with continual developments and refinements, mostly linking with disease-specific approaches to avoidable hospitalization prevention and is the major direction of most research and implementation.

27.8 Predicting Deteriorations in Personal Journeys

Predicting deteriorations in personal journeys can be made sense of through the complex systems conceptual framework. While multiple factors are independent predictors of emergency department attendances and emergency medical admissions, personal experience is a gestalt of these phenomena and others not measured. This is a feature of the human capacity to make sense of their personal health state [107]. Sensors and objective biometrics can describe physical and bodily changes yet they do not have the capacity to tap into this higher level of human functioning and self-awareness. Self-awareness is an emergent feature in evolution that humans uniquely express through language [108]. Thus it is difficult to monitor "ill-defined" states of illness. Yet informatics in this domain is very important to ensure that people are supported

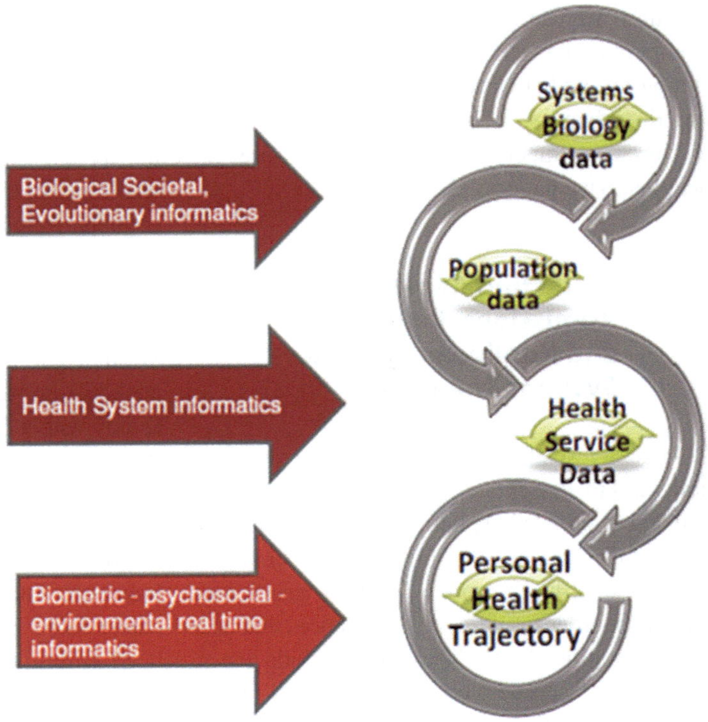

Fig. 27.2 Nonlinear dynamic systems and health informatics conceptual framework

in their individual illness and end of life journeys through the multiple layers of health and social systems. Embodied cognition is a field that can link concepts of narratives, self-awareness, illness experience such that they are amenable to information systems in health [108].

27.9 Applying Artificial Intelligence and Machine Learning to Linguistic and Outlook Analysis of Real-Time Narratives: Interdisciplinary Approaches

Personal Health Trajectories are central to real time analytics of patients outcomes and predicted future needs.

27.9.1 Language, Emotion, Metaphor, and Outlook

On the personal level, people often prefer to express themselves using language (this is not a truism, since gesture is at times also preferred, for example). People are capable of stating directly in natural language the true generalizations about themselves that Machine Learning or expert inference applied to sensor data is meant to achieve. People also make utterances that are available for indirect interpretation, just as for sensor data. This gives rise to the need for reliable interpretation processes, and falls within the remit of Natural Language Processing (NLP). Artificial Intelligence and its related disciplines provide an important tool to analyze patient journeys in real time.

Analysis of the expression of personal health states through narratives is rooted in the ancient fields of hermeneutics and philosophy. Semiotics, for example, is the general study of signs and symbols both within language and without, and the centrality of semiotics of the biopsychosocial model of health has been recognized [109]. Computing that analyses semantics is an interdisciplinary field that builds on hermeneutics, philosophy, political science, and sociology, and is now a key area of research in computer science. Computational linguistics is an interdisciplinary field dealing with the modeling of natural language from a computational perspective. Applied computational linguistics focuses on the practical outcome of modeling human language use, such as including the analysis of outlook from semantic analysis. Outlook analysis is a form of predictive modeling based on linguistic features text. "Looking forward to my sister coming" for example represents a positive outlook and predicts that the health journey is not towards an unplanned emergency event.

27.9.2 Artificial Intelligence, Natural Language Processing, and Machine Learning

Artificial Intelligence is a discipline that studies systems which exhibit behaviors associated with human intelligence, whether embodied solely in computer software, or as embedded software in a physical structure like a robot. Artificial Intelligence has developed concepts, methods, and techniques relevant to Natural (human) Language Processing (NLP)[2] and Computational Linguistics that are increasingly being recognized as predictive tools in medicine and health care. In particular, Machine Learning has been developed as a means to induce concepts and regularities (the grammar rules of natural language are just one sort of regularity approach) from large data sets, structured or unstructured. The essential elements of Machine Learning are an automated approach to learning patterns from empirical data using training examples, usually large data bases, with the development of a Machine Learning algorithm which when applied to new examples creates accurate prediction.

Machine Learning, as a scientific discipline applied to the problem of preventing avoidable hospitalizations, is concerned with identification of linguistic, meta-linguistic, and nonlinguistic patterns in a wide range of data that can assist in quantifying and predicting risk of hospitalization [110, 111]. Such data include quantitative databases or textual corpora such as in medical records, and personal reporting or personal health records. The challenge that Machine Learning addresses is to identify patterns of interest in data sets that encode extremely large numbers of factors and with vast numbers of instances from which generalizations are to be formed. Conversely, there is the difficulty in the need to generalize from relatively small numbers of known instances of recognized patterns to new possibilities, without over-fitting the data.

It is common to distinguish supervised and unsupervised methods, even though the label "unsupervised" is slightly misleading in that for any method to work, a machine-readable data set must be encoded and this means some level of individuation of features and their values for individual cases (a case is a complete description of values for all features measured for any instance; this is a total description of a data point). However, where methods are explicitly supervised, this tends to mean that there is a designated target feature and values within that feature such that training can be tuned to identify with greater levels of supervision. Error analysis from initial training is fed back into the training process. Figure 27.3 provides a schematic picture of the two processes, and Table 27.1 characterizes common Machine Learning methods in these terms. The default is for a method to be supervised, so an indication is provided if the method has "unsupervised" applicability. In the case of Machine Learning for predicting unnecessary hospitalization events, target features in the training data might be hospitalization, other urgent unplanned health service use, or features closely correlated with such events.

27.10 Natural Language Processing

NLP can be understood as an engineering approach to language, complementing the scientific perspective of computational linguistics. The difference, to the extent that there is one, may be seen in how models are evaluated: whether they successfully model human competence and performance, or whether they achieve a task at hand, without regard to their approximation of cognitive reality. Many contemporary approaches in computational linguistics induce rules and regularities (corresponding to linguistic competence of humans) from bodies of texts, also known as corpora, which may have been annotated with attributes and values to be learned. Because

[2] Natural language processing, as a discipline, is also known as natural language engineering and human language technology. NLP in this sense is not to be confused with neuro-linguistic programming.

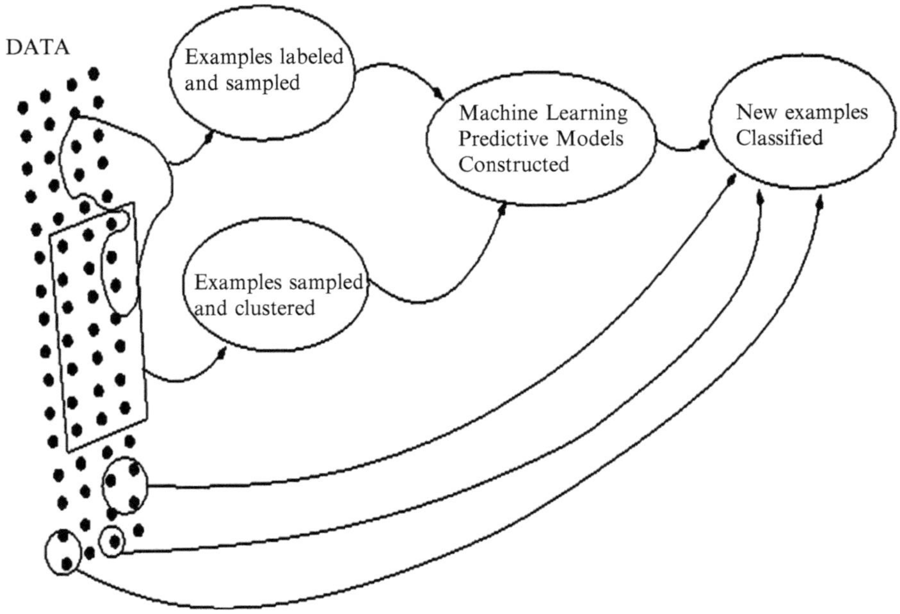

Fig. 27.3 This figure depicts two sorts of learning paradigms: supervised learning, where sampling is based on distributions of features of interest among examples labeled with their values for those features; and unsupervised learning, where features of interest are not identified in advance. In either case, various learning algorithms are available for separating the data according to relevant features, towards achieving models that can be used to predict values for unseen examples

corpora record "live" data, they instantiate a kind of performance data (in Chomsky's sense), potentially of multiple writers or speakers, or even a single speaker in different attitudes, states of well-being, or personal history, and the rules induced tend to have exceptions.

27.10.1 Examples of NLP

An example of an artifact of corpus-driven language processing that many are familiar with is the notion of predictive text completion. This may or may not, in any implementation, be driven by statistical likelihood of valid completions. If the method weighs likely completions, it is on the basis of training of a language model—assessments of likely (as opposed to possible) completions on the basis of text typed so far within a finite window of typing memory. The more complex the predicting task (for example, predicting sentence completions rather than word completions) the more vast the training data set must be, and correspondingly, the amount of time necessary to train the language model. Models are evaluated with respect to the training data itself and data sets that are kept aside, not visible to training, specifically for evaluation purposes (if the test data is taken into account for training, this may result in over-fitting of the model—the test data is isolated as a proxy approximation of arbitrary selections of unseen data).

Decision trees are one sort of Machine Learning tool for inducing generalizations about data sets in which each individual case is described using the same finite set of possible features. One feature is the target and one hopes to generalize from training data what combinations of values for the other features lead to particular values for the feature of interest as a target feature. This typically involves rating individual features in terms of how informative they are in separating the data set with respect to values of the target feature: possible values for the most informative feature create branching contexts for consideration of the next most informative features, and so on, until one may disregard some of the less informative features. The result is a tree of choice points, a

Table 27.1 Machine Learning methods—an overview of statistical and computational methods used to analyze text and make predictions. Sheskin [112], Vapnik [113], Quinlan [110], Manning et al. [114], Mitchell [115], Jain et al. [116], Dunteman [117], Cryer and Chan [118], Witten et al. [119]

Objective						
Method	Reference	Data mining	Strength of association	Classification	Hypothesis testing	Prediction
Regression (linear, multivariate, logistic, etc.)	Sheskin (2004) [112]	Supervised		Supervised	Supervised	Supervised
Support vector machines	Vapnik (1998) [113]	Supervised		Supervised		Supervised
Decision trees	Quinlan (1993) [110]	Supervised		Supervised		Supervised
Naïve Bayes classifiers	Manning et al. (2008) [114]	Supervised		Supervised		Supervised
Neural networks	Mitchell (1997) [115]			Supervised		Supervised
Cluster Analysis	Jain et al. (1999) [116]			(Un)supervised		(Un)supervised
Principal components analysis	Dunteman (1989) [117]	Supervised		Supervised		
Time series analysis	Cryer and Chan (2008) [118]	Supervised	Supervised	Supervised		Supervised
Association rules	Witten et al. (2011) [119]		(Un)supervised			(Un)supervised

decision tree that may be used to classify previously unseen cases in terms of their values for the informative features. In this way, the distributions within training data are used to construct a predictive model (see Fig. 27.3).

In the case of linguistically oriented Machine Learning for avoiding unnecessary hospitalization, the potential linguistic features are the words and phrases used by patients and their carers in describing the patient's individual state of health or general concerns. Meta-linguistic features are summary statistics based on this kind of language use, abstracting over the entire set of utterances: qualities like the average word length for any conversation or the variety of word usage in a conversation relative to the total number of words uttered, etc. Non-linguistic features may be objective (e.g. physical location, date, weather conditions, sensor data readings, etc.) or subjective (e.g. self-rated health, statements or compliance with medication dosages, etc.).

27.11 Patient Journey Narratives

Humans have an innate capacity to make sense of their own health status through narratives, which reflect the sense-making of their complex biopsychosocial personal health system [107, 120]. This capacity reflects an innate ability to promote survival that has emerged through evolution [108].

The patient journey concept recognizes that hospital admissions take place in journeys through stages of health and illness, which are strongly influenced by the social and nonsocial determinants of health [121]. The individual patient journey is shaped by their biological state and disease process, and their health care, social and environmental milieu. The need for hospitalization is strongly linked to feeling ill and whether one has supportive care at a personal level [122]. Self-care and the work of managing the illness increasingly require informational and practical support as illness become unstable. People journey through phases of health and illness, stages of care including health promotion and prevention, risk management, diagnosis, treatment, and self-management. Enablement through support, coaching, and feedback is a key concept [103]. Key elements include: addressing health anxiety and barriers to seeking help, and enabling people to self-manage and seek help in a timely and as needed basis [123]. Such an approach creates more directed support and recommendations through real-time monitoring and intervening where necessary.

27.12 The Patient Journey Record Approach

Individuals have individual journeys or trajectories through illness, aging, and dying. Risk of hospitalization remains low while the individuals' physical, psychological, social state, and their overall sense making are functional and in balance. Using text summarizing the person's journey reports in key domains identified in phone calls that a person has with a formal care provider creates a linguistic pattern of that individual's risk profile. As important controls (self-rated health and social support, for example) gradually change, the individual and their network adapt. Once control variables reach a critical point there is a sudden change to high risk and a potentially avoidable hospitalization/ACSC adverse event occurs. Subcritical risk levels are early indicators of potential transitions to high-risk states and tipping points. Because of the unpredictability of the timing of triggers and enablers of decline and the individual trajectory towards a catastrophe (an avoidable hospitalization/ACSC adverse event), real-time intervention around early detection of triggers and enablers of decline is modeled. Our case study demonstrates the triggers and instabilities in real-time patient journeys demonstrated in Fig. 27.4.

Such an approach continually samples the personal narratives summarized with brief phone calls in order to have sufficient timely information for Machine Learning to deal with early detection of root causes for deteriorations, false positives, and rapid regression or early resolution.

The real-time monitoring of the personal narratives provides the linguistic signals to trigger alerts to the need for services to address fluctuations in health and health concerns when

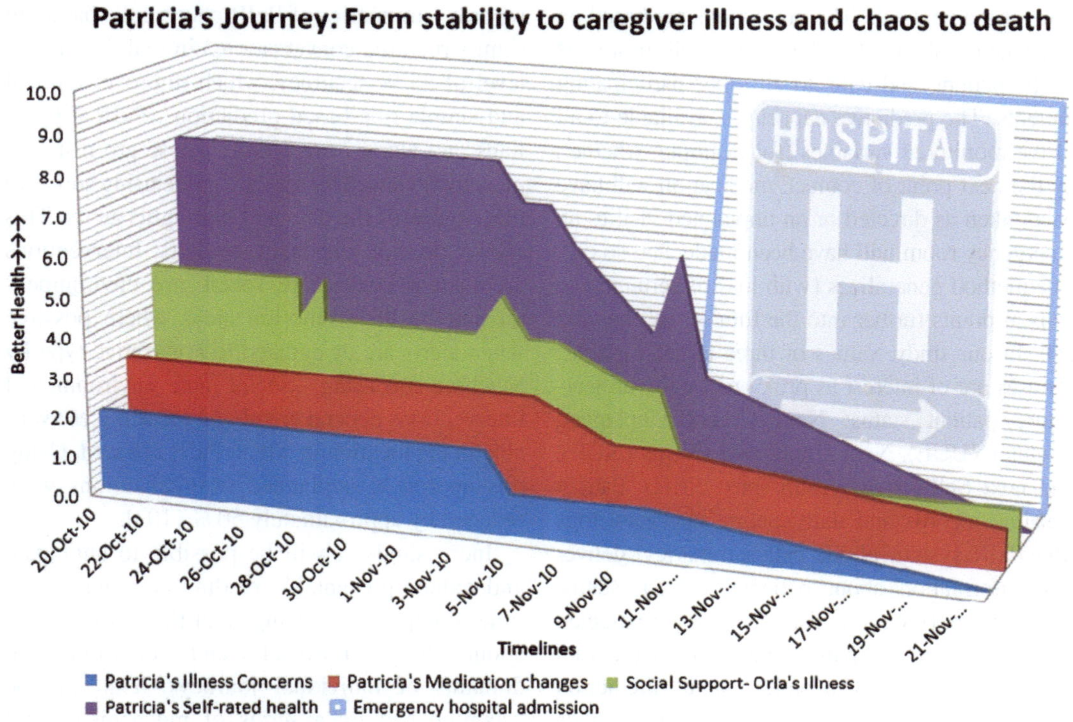

Fig. 27.4 This figure demonstrates the trajectories of concerns, illness, self-rated health, social support, medication, and health services reported through frequent reports of narratives of P5 (pseudonym Patricia) and her daughter (pseudonym Orla) culminating in an unplanned hospitalization and Patricia's death. Patricia, who had multiple chronic conditions hypertension, noninsulin-dependent diabetes, osteoarthritis, and early dementia, was stable and well with no concerns. However, Orla who was her live-in caregiver was gradually becoming stressed and overwhelmed by the burden of care, particularly her mother's inability to sleep at night. The trigger to Patricia's sudden decompensation was Orla's high blood pressure and a chest infection on the anniversary of Patricia's husband's death. Orla who was managing her own business had not succeeded in leveraging sufficient practical, instrumental, and social support from the services. Orla felt acutely unable to cope and considered sending her mother into a respite care situation. Patricia became increasingly agitated and breathless and confrontational with her family. She was sent to respite care unwillingly and was extremely agitated. Her self-rated health deteriorated and she was transferred to hospital in an emergency and died in a very distressed state. Orla blamed herself and became clinically depressed for some months following the death of her mother, around the anniversary of Orla's husband's death

they are needed. Daily concerns, self-rated health, and other narratives are elicited to identify these deteriorations early and create positive feedback through health promotion and other interventions to avert deterioration.

27.12.1 Machine Learning and Predictive Modeling

The patient journey Machine Learning component predicts deteriorating patient status based on patient conversations with callers that include structured questions and unstructured spontaneous expressions of personal narratives. Other methods of eliciting narratives are possible with technologies in the future, but the personal conversation is highly acceptable and provides unmeasured support [103].

The PaJR system applies Machine Learning methods to predict patient status in the near future. Given that points of contact with patients are frequent, we use individual points of contact with patients to model the flow of time: in predicting the status of a variable at some future point, an estimate is made of the value of that variable at the time of the next call. The predictions are based on information obtained at the current call, past calls to the patient, and calls to all other participating patients. The predictive capacity is available to all of the monitored variables (for example, whether, by the next point of contact, medication will have been taken as directed or an unplanned visit to an emergency room will have been made, and so on). The method generalizes (with loss of certainty) to calls at points further into the future than the next call. In our study, values of three target variables statuses are of interest as proxy measures of near-future patient status: Next Urgent Unplanned Events (NUUE), Next Unplanned Events (NUE), and next Self-Rated Health (next SRH). Patient baseline records and daily online interview logs (the PaJR system phone call summaries) deliver rich information about patients' current status. Linguistic and meta-linguistic features are extracted together with current patient status, in order to train prediction models. The patient status prediction system is constructed in two phases, and it responds to requests in nearly real time. The two phases are: offline training module and online prediction module. The offline training module utilizes the newest patient interviews and re-trains decision models within a few hours, while the online prediction module runs over the latest successful model, and it takes only moments to deliver prediction results. To predict the binary value of NUUE a decision tree is used based on a highly refined set of features organized hierarchically as rules.

Error analysis shows that the system is effective at learning, spotting 85 % of true positives, with only 29 % false positives. We take false negatives to be an order of magnitude more expensive than false positives. A false positive prediction might trigger a phone call or visit, but its cost is much less than a false negative prediction, where a true danger is overlooked.

27.13 PaJR Findings, Impacts, and Opportunities

The PaJR system can predict patients' future health and risk of unplanned events, with Machine Learning using semantic analysis of conversation records (summaries of PaJR phone calls based on semi-structured conversations) in real time to the level of 85 % accuracy, which improves clinical and simple rule-based predictions. PaJR triggers flags and alerts in real time. These are instantaneously reviewed by the care guide using software that organizes the data and highlights alerts. This allows prompt responses to alerts. Interventions are made in health care, social care, environmental and health promotion areas where possible when alerts are designated to a narrative type by Machine Learning. Social and environmental interventions depend greatly on the services available in the location in which PaJR is based. Using this approach, unplanned hospitalizations were reduced by approximately 50 % [103].

Increasingly it will be possible to anticipate and reduce preventable readmissions with real-time computation using real-time knowledge about biopsychosocial and environmental domains of individual journeys. This can be accomplished by a series of mechanisms that effectively define and monitor patient risk groups and individual trajectories, using dynamical data of current status in key health-related domains and historical data. Such data can assist to proactively manage at-risk patients at discharge, through their admission and in the community by assessing patterns in key indicators to identify and address root causes of readmissions. Using qualitative sense-making techniques based on explicit and tacit knowledge, the clinician will have real-time information to make informed decisions.

The unprecedented pace of progress in informatics, automatic voice and speech technologies, and Machine Learning seen at present means that any current assessment of the opportunities and challenges of risk prediction is likely to require re-evaluation within a short time frame. Moreover, in the phases of acute illness and end of life, sensitivity to personal capacities and respect for dignity should prevail to ensure quality of life in any technology deployed.

Which admissions are avoidable and *which* admissions serve a valuable health and social purpose for *which* individuals reflect complex system dynamics. Clinical knowledge, threshold criteria, and predictive modeling at a particular point in time can only make static and short-term predictions which may be suitable for limited interventions in transitions. Yet personal journey interactions with health and socio-cultural systems cannot be anticipated and resources shifted promptly based on static predictions of changing trajectories. While close clinical monitoring might address this problem through case management, it is a resource intensive exercise that would be made more adaptive, efficient, and effective if supported by real-time informatics.

27.14 Conclusion

Avoidable hospitalizations are the subject of considerable interest to health service decision makers, because of cost pressures, concerns for patients, and a desire for continuing to improve health systems organization.

Age and social disadvantage puts those with chronic conditions and aging syndromes in the last 2–3 years of life at high risk of hospitalization. Upstream influences range from evolutionary and systems biology to population health and social and cultural approaches to illness and dying. Downstream or immediate influences are individual health states and support in the context of health service quality. Many unplanned admissions in older patients might be avoided. There is a wide range of solutions, for which the evidence is variable. Chronic illness, aging, and frailty lead to poor self-rated health and together with reduced social and environmental support, reduced resilience and capacity for homeostasis. Lack of ability to be resilient in any or all of the physical, psychological, social, and environmental domains of later life is at the root cause of the personal health crisis that leads to many emergency admissions. However these complicated underlying reasons for the admission are obscured by an admission diagnosis of cardiac or respiratory disease decompensation.

Case finding of those at high risk of avoidable hospitalizations can be conducted clinically, by use of thresholds based on screening instruments or by using computational predictive modeling.

Machine Learning predictive modeling is a highly promising tool for healthcare and has been applied to develop knowledge in the related fields of bioinformatics, population and organizational informatics, and health systems data that influence avoidable hospitalizations at upstream or downstream or proximal levels. This is a rapidly growing field.

Ongoing developments in Machine Learning can provide a way forward for patient-centered user-driven care, particularly for monitoring early deterioration patterns in real-time personal journeys to assist timely interventions. A case is made for real-time monitoring of health and its embodiment in language and narrative with computational linguistics informing ongoing Machine Learning predictive modeling. By anticipating critical points in patient journeys and providing timely and individual patient-centered interventions in the multiple domains that can destabilize, health systems can ameliorate avoidable unplanned events such as hospitalizations and visits to the emergency department.

References

1. Kleinman A. The illness narratives: suffering, healing, and the human condition. New York: Basic; 1988. p. xviii. 284 p.
2. Fleming ST. Primary care, avoidable hospitalization, and outcomes of care: a literature review and methodological approach. Med Care Res Rev. 1995;52(1):88–108.
3. Reschovsky JD, et al. Following the money: factors associated with the cost of treating high-cost Medicare beneficiaries. Health Serv Res. 2011;46(4):997–1021.
4. CARDI: Centre for Ageing Research and Development in Ireland. Stocktake of Ageing Public Policy Initiatives in Ireland, North and South. 2008. Cited 12/9/11.
5. Moloney ED, Bennett K, Silke B. Factors influencing the costs of emergency medical admissions to an Irish teaching hospital. Eur J Health Econ. 2006;7(2):123–8.
6. Crimmins EM, Hayward MD, Saito Y. Changing mortality and morbidity rates and the health status and life expectancy of the older population. Demography. 1994;31(1):159–76.
7. Royal College of Physicians of Ireland, et al. Community medical services for the older person. In: Report of the National Acute Medicine Programme; 2011.
8. U.S. Agency for Healthcare Research and Quality. AHRQ quality indicators – guide to prevention quality indicators: hospital admissions for ambulatory care sensitive conditions. 2007.

9. Vest J, et al. Determinants of preventable readmissions in the United States: a systematic review. Implement Sci. 2010;5(1):88.
10. Agency for Healthcare Research and Quality. Chronic care and disease management improves health, reduces costs for patients with multiple chronic conditions in an integrated health system. Department of Health and Human Services, Editor 2009: United States. http://innovations.ahrq.gov/content.aspx?id=1696.
11. Jimenez-Puente A, et al. Readmission rate as an indicator of hospital performance: the case of Spain. Int J Technol Assess Health Care. 2004;20(3):385–91.
12. van Walraven C, et al. Proportion of hospital readmissions deemed avoidable: a systematic review. Can Med Assoc J. 2011;183(7):E391–402.
13. Naylor MD. Transitional care for older adults: a cost-effective model. LDI Issue Brief. 2004;9(6):1–4.
14. Condelius A, Hallberg I, Jakobsson U. Hospital and outpatient clinic utilization among older people in the 3-5 years following the initiation of continuing care: a longitudinal cohort study. BMC Health Serv Res. 2011;11(1):136.
15. Martin M, Hin PY, O'Neill D. Acute medical take or subacute-on-chronic medical take? Ir Med J. 2004; 97(7):212–4.
16. Baer RB, Pasternack JS, Zwemer Jr FL. Recently discharged inpatients as a source of emergency department overcrowding. Acad Emerg Med. 2001;8(11):1091–4.
17. Pines JM, et al. Postdischarge adverse events for 1-day hospital admissions in older adults admitted from the emergency department. Ann Emerg Med. 2010;56(3):253–7.
18. Brunjes GB. The search for meaning in the face of death. J Health Care Chaplain. 1987;1(1):63–9.
19. Barnard A, Hollingum C, Hartfiel B. Going on a journey: understanding palliative care nursing. Int J Palliat Nurs. 2006;12(1):6–12.
20. Allaudeen N, et al. Redefining readmission risk factors for general medicine patients. J Hosp Med. 2011; 6(2):54–60.
21. Wallmann R, et al. Prediction of 30-day cardiac-related-emergency-readmissions using simple administrative hospital data. Int J Cardiol. 20 July 2011 (10.1016/j.ijcard.2011.06.119).
22. Courtney MD, et al. Improved functional ability and independence in activities of daily living for older adults at high risk of hospital readmission: a randomized controlled trial. J Eval Clin Pract. 2012;18(1):128–34.
23. Parry C, et al. Further application of the care transitions intervention: results of a randomized controlled trial conducted in a fee-for-service setting. Home Health Care Serv Q. 2009;28(2–3):84–99.
24. Bittencourt RJ, Hortale VA. Interventions to solve overcrowding in hospital emergency services: a systematic review. Cad Saude Publica. 2009;25(7):1439–54.
25. Purdy S. Avoiding hospital admissions.What does the research evidence say? London: The King's Fund; 2010.
26. Crane S, et al. Use of an electronic administrative database to identify older community dwelling adults at high-risk for hospitalization or emergency department visits: the elders risk assessment index. BMC Health Serv Res. 2010;10(1):338.
27. Spencer J, et al. Functional outcomes and daily life activities of African-American elders after hospitalization. Am J Occup Ther. 2002;56(2):149–59.
28. Rea H, et al. Integrated systems to improve care for very high intensity users of hospital emergency department and for long-term conditions in the community. N Z Med J. 2010;123(1320):76–85.
29. Armstrong CD, et al. Home-based intermediate care program vs hospitalization: cost comparison study. Can Fam Physician. 2008;54(1):66–73.
30. Basu A, Brinson D. The effectiveness of interventions for reducing ambulatory sensitive hospitalisations: a systematic review. In: Health Services Advisory Committee Technical Report, Canterbury, New Zealand; 2008. p. 245.
31. Kronman AC, et al. Can primary care visits reduce hospital utilization among Medicare beneficiaries at the end of life? J Gen Intern Med. 2008;23(9): 1330–5.
32. Linertová R, et al. Interventions to reduce hospital readmissions in the elderly: in-hospital or home care. A systematic review. J Eval Clin Pract. 2011;17(6): 1167–75.
33. Jordon M, et al. Implications of complex adaptive systems theory for interpreting research about health care organizations. Forum Syst Complex Health Care J Eval Clin Pract. 2010;16(1):228–31.
34. van Kerkhoff L, Szlezak NA. Knowledge Systems for Sustainable Development Special Feature Sackler Colloquium: The role of innovative global institutions in linking knowledge and action. Proc Natl Acad Sci USA. 2010. Epub 2010/03/03.
35. Glass DJ, Arnold SE. Some evolutionary perspectives on Alzheimer's disease pathogenesis and pathology. Alzheimers Dement. 2012;8(4):343–51.
36. Gluckman PD, Hanson MA. Living with the past: evolution, development, and patterns of disease. Science. 2004;305(5691):1733–6.
37. Weber BH. Extending and expanding the Darwinian synthesis: the role of complex systems dynamics. Stud Hist Philos Sci C Stud Hist Philos Biol Biomed Sci. 2011;42(1):75–81.
38. Shanahan MJ, Hofer SM. Molecular genetics, aging, and well-being: sensitive period, accumulation, and pathway models. In: Binstock RH, George LK, editors. Handbook of aging and the social sciences. 7th ed. San Diego, CA: Academic; 2011. p. 135–47.
39. Khushf G. Health as intra-systemic integrity: rethinking the foundations of systems biology and nanomedicine. Perspect Biol Med. 2008;51(3):432–49.
40. Van Wave TW, Scutchfield FD, Honore PA. Recent advances in public health systems research in the United States. Annu Rev Public Health. 2010;31: 283–95.
41. Lunney J, et al. Patterns of functional decline at the end of life. J Am Med Assoc. 2003;289:2387–92.
42. Gill TM, et al. Trajectories of disability in the last year of life. N Engl J Med. 2010;362(13):1173–80.

43. Goldberger AL, et al. Fractal dynamics in physiology: alterations with disease and aging. Proc Natl Acad Sci USA. 2002;99 Suppl 1:2466–72.
44. Eisenberg DTA. An evolutionary review of human telomere biology: the thrifty telomere hypothesis and notes on potential adaptive paternal effects. Am J Hum Biol. 2011;23(2):149–67.
45. Zhu H, Belcher M, van der Harst P. Healthy aging and disease: role for telomere biology? Clin Sci. 2011;120(10):427–40.
46. Glover J, et al. Atlas of avoidable hospitalisations in Australia: ambulatory care-sensitive conditions. Canberra: AIHW; 2007.
47. Glouberman S. The grey zones of birth and death. J Eval Clin Pract. 2011;17(2):394–9.
48. Cole A. Proportion of people dying in hospital is falling. BMJ 2010;341:c4722.
49. Gruneir A, et al. Where people die: a multilevel approach to understanding influences on site of death in America. Med Care Res Rev. 2007;64(4):351–78.
50. Kardamanidis K, et al. Hospital costs of older people in New South Wales in the last year of life. Med J Aust. 2007;187(7):383–6.
51. Ahearn DJ, et al. Improving end of life care for nursing home residents: an analysis of hospital mortality and readmission rates. Postgrad Med J. 2010;86(1013):131–5.
52. Murtagh FEM, et al. End of life trajectories across conditions. BMJ. 2011;342:397–8.
53. Zaninotto P, Falaschetti E, Sacker A. Age trajectories of quality of life among older adults: results from the English Longitudinal Study of Ageing. Qual Life Res. 2009;18(10):1301–9.
54. Visser G, et al. The end of life: informal care for dying older people and its relationship to place of death. Palliat Med. 2004;18(5):468–77.
55. Niederman MS, Berger JT. The delivery of futile care is harmful to other patients. Crit Care Med. 2010;38(10 Suppl):S518–22.
56. Wilson DM, et al. Age-based differences in care setting transitions over the last year of life. Curr Gerontol Geriatr Res. 2011;2011:101276.
57. Martin CM. Distortions, belief and sense making in complex adaptive systems for health. J Eval Clin Pract. 2011;17(2):387–8.
58. Yam CH, et al. Avoidable readmission in Hong Kong – system, clinician, patient or social factor? BMC Health Serv Res. 2010;10:311.
59. Walsh B, Roberts H, Nicholls P. Features and outcomes of unplanned hospital admissions of older people due to ill-defined (R-coded) conditions: retrospective analysis of hospital admissions data in England. BMC Geriatr. 2011;11(1):62.
60. Condelius A, et al. Hospital admissions among people 65+ related to multimorbidity, municipal and outpatient care. Arch Gerontol Geriatr. 2008;46(1):41–55.
61. Asarnow JR, Jaycox LH, Anderson M. Depression among youth in primary care models for delivering mental health services. Child Adolesc Psychiatr Clin N Am. 2002;11(3):477–97, viii.
62. Laniece I, et al. Incidence and main factors associated with early unplanned hospital readmission among French medical inpatients aged 75 and over admitted through emergency units. Age Ageing. 2008;37(4): 416–22.
63. Bayliss EA, Ellis JL, Steiner JF. Seniors' self-reported multimorbidity captured biopsychosocial factors not incorporated into two other data-based morbidity measures. J Clin Epidemiol. 2009;62(5):550–7.e1.
64. Martin CM, Vogel C, Grady D, Zarabzadeh A, Hederman L, Kellett J, et al. Implementation of complex adaptive chronic care: the Patient Journey Record system (PaJR). J Eval Clin Pract. 2012. Epub 2012/07/24.
65. Norman CD. Health promotion as a systems science and practice. J Eval Clin Pract. 2009;15(5):868–72.
66. Watt G, O'Donnell C, Sridharan S. Building on Julian Tudor Hart's example of anticipatory care. Prim Health Care Res Dev. 2011;12(1):3–10.
67. Kellett J. Hospital medicine (Part 2): what would improve acute hospital care? Eur J Intern Med. 2009;20(5):465–9.
68. Naylor MD, et al. Translating research into practice: transitional care for older adults. J Eval Clin Pract. 2009;15(6):1164–70.
69. Clarke M, Shah A, Sharma U. Systematic review of studies on telemonitoring of patients with congestive heart failure: a meta-analysis. J Telemed Telecare. 2011;17(1):7–14.
70. McLean S, et al. Telehealthcare for chronic obstructive pulmonary disease. Cochrane Database Syst Rev. 2011;(7):CD007718.
71. Tan HM, et al. GP and nurses' perceptions of how after hours care for people receiving palliative care at home could be improved: a mixed methods study. BMC Palliat Care. 2009;8:13.
72. McEvoy P, Escott D, Bee P. Case management for high-intensity service users: towards a relational approach to care co-ordination. Health Soc Care Community. 2011;19(1):60–9.
73. Lunney JR, et al. Patterns of functional decline at the end of life. JAMA. 2003;289(18):2387–92.
74. Jylha M, Volpato S, Guralnik J. Self-rated health showed a graded association with frequently used biomarkers in a large population sample. J Clin Epidemiol. 2006;59:465–71.
75. Nielsen AB, et al. The impact of changes in self-rated general health on 28-year mortality among middle-aged Danes. Scand J Prim Health Care. 2009;27(3):160–6. 1 p following 166.
76. DeSalvo KB, et al. Health care expenditure prediction with a single item, self-rated health measure. Med Care. 2009;47(4):440–7.
77. Lawton J. Lay experiences of health and illness: past research and future agendas. Sociol Health Illn. 2003;25(3):23–40.
78. Rockwood K. What would make a definition of frailty successful? Age Ageing. 2005;34(5):432–4.
79. Nesse RE, et al. Framing change for high-value healthcare systems. J Healthc Qual. 2010;32(1):23–8.

80. Hammond RA. Complex systems modeling for obesity research. Prev Chronic Dis. 2009;6(3):A97.
81. Beautement P, Broenner C. Complexity demystified: a guide for practitioners. Axminster, Devon: Triarchy; 2011. p. 268 p.
82. Harquail CV, Wilcox King A. Construing organizational identity: the role of embodied cognition. Organ Stud. 2010;31(12):1619–48.
83. Stange KC, Ferrer RL. The paradox of primary care. Ann Fam Med. 2009;7(4):293–9.
84. Zoloth-Dorfman L, Rubin S. The patient as commodity: managed care and the question of ethics. J Clin Ethics. 1995;6(4):339–57.
85. Kurtz C, Snowden D. The new dynamics of strategy: sense-making in a complex and complicated world. IBM Syst J. 2003;42(3):462–83.
86. Silva SA, Charon R, Wyer PC. The marriage of evidence and narrative: scientific nurturance within clinical practice. J Eval Clin Pract. 2011;17(4): 585–93.
87. Lakoff G, Johnson M. Philosophy in the flesh: the embodied mind and its challenge to Western thought. New York: Basic; 1999. p. xiv. 624 p.
88. Nelson B. A field guide to embodied cognition. 2008. http://corporature.wordpress.com/2008/12/28/a-field-guide-to-embodied-cognition/. Cited 12 Jan 2012.
89. King's Fund. Predictive Risk Project: Literature review. NHS Modernisation Agency, New York University, Health Dialog. 2005. http://www.kingsfund.org.uk/current_projects/predicting_and_reducing_readmission_to_hospital/#context. Cited 2 Sept 2011.
90. Boult C, Pacala JT, Boult LB. Targeting elders for geriatric evaluation and management: reliability, validity, and practicality of a questionnaire. Aging (Milano). 1995;7(3):159–64.
91. Lyon D, et al. Predicting the likelihood of emergency admission to hospital of older people: development and validation of the Emergency Admission Risk Likelihood Index (EARLI). Fam Pract. 2007;24(2):158–67.
92. Lindquist LA, Baker DW. Understanding preventable hospital readmissions: masqueraders, markers, and true causal factors. J Hosp Med. 2011;6(2):51–3.
93. Roland M, et al. Follow up of people aged 65 and over with a history of emergency admissions: analysis of routine admission data. BMJ. 2005;330:289–92.
94. Sidorov J, Shull R. "My patients are sicker:" using the Pra risk survey for case finding and examining primary care site utilization patterns in a medicare-risk MCO. Am J Manag Care. 2002;8(6):569–75.
95. Martin C. Complex adaptive chronic care - typologies of patient journey: a case study. J Eval Clin Pract. 2011;17(3):1–5.
96. Murphy SME, Castro HK, Sylvia M. Predictive modeling in practice: improving the participant identification process for care management programs using condition-specific cut points. Popul Health Manag. 2011;14(4):205–10.
97. Obenshain MK. Application of data mining techniques to healthcare data. Infect Control Hosp Epidemiol. 2004;25(8):690–5.
98. Diehr P, et al. Patterns of self-rated health in older adults before and after sentinel health events. J Am Geriatr Soc. 2001;49(1):36–44.
99. Gravelle H, et al. Impact of case management (Evercare) on frail elderly patients: controlled before and after analysis of quantitative outcome data. BMJ. 2007;334:31–4.
100. Snowden DJ, Boone ME. A leader's framework for decision making. A leader's framework for decision making. Harv Bus Rev. 2007;85(11):68–76, 149.
101. Biddiss E, Brownsell S, Hawley MS. Predicting need for intervention in individuals with congestive heart failure using a home-based telecare system. J Telemed Telecare. 2009;15(5):226–31.
102. Gruneir A, Silver MJ, Rochon PA. Emergency department use by older adults: a literature review on trends, appropriateness, and consequences of unmet health care needs. Med Care Res Rev. 2011;68(2):131–55.
103. Martin CM, Grady D, Deaconking S, McMahon C, Zarabzadeh A, O'Shea B. Complex adaptive chronic care - typologies of patient journey: a case study. J Eval Clin Pract. 2011;17(3):520–4.
104. Huang T, et al. A systems-oriented multilevel framework for addressing obesity in the 21st century. Prev Chronic Dis. 2009;6(3):A82.
105. Horvitz E. From data to predictions and decisions: enabling evidence-based healthcare. Series on data analytics 1. Computing Community Consortium; 2010.
106. Bayati M, et al. Predictive models and policies for minimizing rehospitalizations for congestive heart failure. 2010. in press.
107. Jylhä M. What is self-rated health and why does it predict mortality? Towards a unified conceptual model. Soc Sci Med. 2009;69(3):307–16.
108. Lewis S. Seeking a new biomedical model. How evolutionary biology may contribute. J Eval Clin Pract. 2009;15(4):745–8.
109. Sturmberg J, Martin C, Moes M. Health at the centre of health systems reform – how philosophy can inform policy. Perspect Biol Med. 2010;53(3):341–56.
110. Quinlan JR. C4.5: Programs for machine learning. Burlington, MA: Morgan Kaufmann, Elsevier; 1993.
111. Kerr WT, Lau EP et al. The future of medical diagnostics: large digitized databases. Yale J Biol Med. 2012;85(3):363–377.
112. Sheskin D. Handbook of Parametric and Nonparametric Statistical Procedures. 3 ed. Baton Rouge, Florida: CRC Press; 2004.
113. Vapnik VN. Statistical Learning Theory. New York: John Wiley and Sons, Inc; 1998.
114. Manning C, Raghavan P, Schütze H. Introduction to Information Retrieval. Cambridge: Cambridge University Press; 2008.
115. Mitchell T. Machine Learning. Liu C, editor. USA: McGraw Hill; 1997.
116. Jain A, Murty M, Flynn P. Data Clustering. A Review. http://eprints.iisc.ernet.in/273/1/p264-jain.pdf1997. (Accessed 7/10/12).

117. Duntemann G. Principal Components Analysis. Newbury Park, California: Sage Publications Inc; 1989.
118. Cryer J, Chan K. Time Series Analysis with Applications in R New York: Springer 2008.
119. Witten I, Frank E, Hall M. Data mining: practical machine learning tools and techniques, The Morgan Kaufmann series in data management systems. 3rd ed. Burlington, MA: Morgan Kaufmann, Elsevier; 2011.
120. Cheshire J, Ziebland S. Narrative as a resource in accounts of the experience of illness. In: Thornborrow J, Coates J, editors. The sociolinguistics of narrative. Amsterdam: John Benjamins; 2005.
121. Martin CM, et al. Patient Journey Record Systems (PaJR): the development of a conceptual framework for a patient journey system. In: Biswas R, Martin CM, editors. User-driven healthcare and narrative medicine: utilizing collaborative social networks and technologies. Hershey, PA: IGI; 2010. p. 75–92.
122. Boyd CM, et al. Guided care for multimorbid older adults. Gerontologist. 2007;47(5):697–704.
123. Murray SA, et al. Archetypal trajectories of social, psychological, and spiritual wellbeing and distress in family care givers of patients with lung cancer: secondary analysis of serial qualitative interviews. BMJ. 2010;340:c2581.

Continuous Multiorgan Variability Monitoring in Critically Ill Patients: Complexity Science at the Bedside

28

Christophe L. Herry, Geoffrey C. Green, Andrea Bravi, and Andrew J.E. Seely

28.1 Introduction

Complexity science has been an active subject of investigation for over five decades [1], although its roots date at least a century, when Poincaré uncovered the irreducible uncertainty associated with the three body problem. Numerous individuals have built pillars of this large yet still incomplete theoretical framework, including Shannon [2], von Bertalanffy [3], Kauffman [4], Mandelbrot [5], Bak [6], Capra [7], Glass [8], Prigogine [9], Goldberger [10], and many more. Despite a rich theoretical framework, the clinical impact of complexity science currently in use to directly improve patient care is inconsequential in comparison to the enormous impact of analytical science (basic science) and population science (epidemiology). This chapter will present a practical framework and highlight exciting clinical research studies aiming to deliver complexity science at the bedside for the direct benefit of the individual patient.

A fundamental problem facing clinicians caring for patients at risk for or with existing critical illness is the inherent uncertainties and inefficiencies associated with their care. For clinicians caring for patients admitted to an emergency room (ER), hospital ward, or intensive care unit (ICU), there is a great deal of uncertainty on whether a patient will subsequently deteriorate requiring life support, in particular during the early stages following significant injury, operation, or infection. Often, the recognition of patient deterioration is made late, well after illness and organ dysfunction have progressed. For example, there often exists uncertainty regarding the timing of removal of mechanical ventilation life support (i.e., extubation). Clinicians have to weigh the pressure to optimize bed occupancy with the necessity to avoid harm to their patients if extubation fails, requiring emergency re-institution of ventilation (i.e., re-intubation). This uncertainty leads to significant inefficiency, with patients admitted unnecessarily to the ICU, kept on life support or in the ICU longer than necessary. Complexity science teaches us that this uncertainty has an irreducible component; that is one

C.L. Herry
Ottawa Hospital Research Institute (OHRI), Ottawa, ON, Canada
e-mail: cherry@sce.carleton.ca

G.C. Green
Therapeutic Monitoring Systems, Ottawa, ON, Canada

A. Bravi
University of Ottawa, Ottawa, ON, Canada
Ottawa Hospital Research Institute (OHRI), Ottawa, ON, Canada

A.J.E. Seely (✉)
Ottawa Hospital Research Institute (OHRI), Ottawa, ON, Canada
University of Ottawa, Ottawa, ON, Canada
Division of Thoracic and Critical Care Medicine, Ottawa Hospital, General Campus, 501 Smyth Road, Box 708, Ottawa, ON, Canada, K1H 8L6
e-mail: aseely@ohri.ca

J.P. Sturmberg and C.M. Martin (eds.), *Handbook of Systems and Complexity in Health*, DOI 10.1007/978-1-4614-4998-0_28, © Springer Science+Business Media New York 2013

cannot predict the exact behavior of a complex system for any given time in the future, even with infinite knowledge of the present. Thus, embracing the reality of uncertainty as well as that of emergence, it is logical to attempt to track the complex system as a whole, and do so continuously over time, in order to track the trajectory of the system. Seemingly paradoxical, accepting uncertainty leads us to tracking the system continuously, which then may help reduce uncertainty, at least in the short term.

The technology we have pursued to track the patient as a complex system is multiorgan variability monitoring. Variability analysis can be defined as the assessment of the degree and character of patterns of variation over time intervals [11]. It provides a measure of the patterns of fluctuation occurring over an interval-in-time, in contrast to a point-in-time assessment. These patterns define properties of the system such as its information content, degree of scale invariance, high and low frequency variation, irregularity, and more. Both cardiac variation [12] and respiratory variation [13–15] were discovered not to be random, but rather correlated and containing information. Since then, numerous investigators have developed a broad and increasing array of analysis techniques that are grouped empirically into "domains" of analysis [11, 16]. Research has focused on the development of algorithms that optimally characterize healthy (physiologic) variability, and the study of how illness and aging are associated with deterioration into unhealthy (pathologic) variability. In clinical settings, variability analysis has the potential for early detection of deterioration and improved real-time prognostication [16–18]. Variability analysis has been successfully applied to clinical problems such as the prediction of mortality after acute myocardial infarction [19, 20], the detection of sleep apnoea [21, 22], the assessment of the autonomic nervous system activity [23, 24], and evaluation of the circadian rhythms regulating the body [25, 26].

Hypothesizing that multiorgan variability reflects system-level integrity, then monitoring multiorgan variability offers a means to track the emergent properties of a complex system.

The aim of this chapter is to introduce a processing framework that enables standardized comprehensive multiorgan variability analysis from continuously monitored waveforms (e.g., electrocardiogram (ECG), end-tidal capnogram ($EtCO_2$), respiration from chest impedance, photoplethysmogram, blood pressure, etc.). This processing framework has been integrated into a Continuous Individualized Multiorgan Variability Analysis (CIMVA™) software [27], about which a high-level description is provided hereafter, as an example of the type of variability analysis that can be performed in a clinical setting.

28.2 Overview of a Variability Monitoring Framework

Variability monitoring can be structured as a library of software routines that:

1. Takes as input a variety of physiological waveforms or event detections (e.g., beats or breaths) and evaluates their data quality.
2. Performs the continuous measurement of the degree and character of patterns of variation over time in selected features extracted from these physiological waveforms (i.e., windowed analysis).
3. Generates reports and output files that clinicians and researchers can use to investigate the clinical utility of these measurements in their own patient populations.

A simplified block diagram of the overall variability monitoring system is shown in Fig. 28.1, followed by a brief description of each component.

28.2.1 Input Physiological Data

The inputs to the variability monitoring system are physiological waveforms that have been acquired from a patient. In general, these fall into three categories:

1. Regularly sampled waveforms, for e.g., ECG (500 Hz typical) or $EtCO_2$ (125 Hz typical).
2. Regularly sampled numerics, e.g., heart rate or respiratory rate (1 Hz typical).

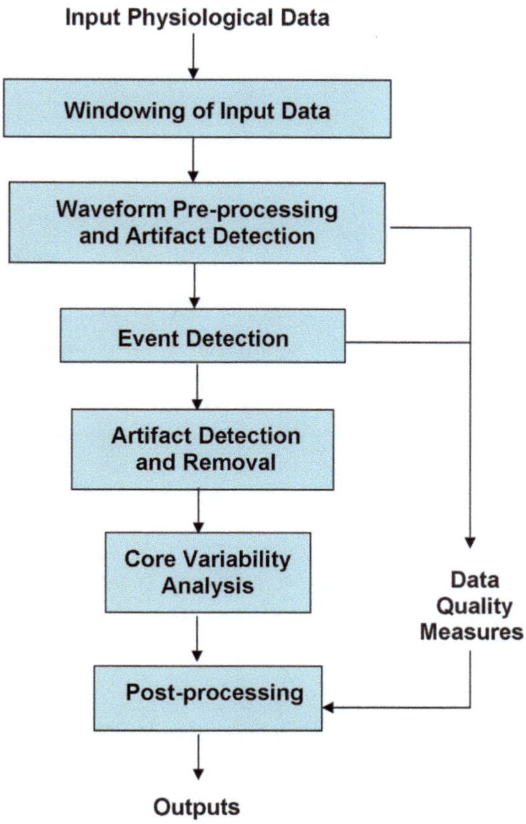

Fig. 28.1 Overall data flow of the variability monitoring system (for a regularly sampled waveform input)

3. Irregularly sampled time series, e.g., tachogram (one sample per heartbeat) or tidal volume (one sample per breath), possibly annotated with additional information (e.g., sinus rhythm).

Physiological waveforms can be harvested from the major intensive care unit (ICU) monitoring vendors as well as selected ambulatory physiological monitors.

28.2.2 Windowing of Data

The variability monitoring framework of Fig. 28.1 is designed to produce measures of physiological signal complexity and variability as a continuous time output. This is done with a moving window analysis approach as illustrated in Fig. 28.2.

The analysis window size, W_s, represents the portion of the input signal used in the current window, and the step size, Δt, represents how much the window is shifted from one iteration to the next. The output measures are calculated using the input signal values at each step, a process that results in a new set of outputs at an interval determined by the step size [27]. Typical values for W_s and Δt might be $W_s = 5$ min and $\Delta t = 2.5$ min, resulting in 5 min analysis windows overlapping by 50%.

28.2.3 Waveform Preprocessing and Waveform Artifact Detection (If Required)

Preprocessing is applied to input waveforms to condition them for subsequent analyses. Preprocessing includes noise reduction, filtering, and re-sampling.

In order to generate trustworthy results, only the highest quality input waveform data should be considered. Since the location of high data quality is not generally known a priori, several software modules should be used to identify areas where artifacts occur in the input (or preprocessed) waveform data, such as disconnection and saturation artifacts or high-amplitude artifacts.

Each of these modules provides a measure of the degree to which the current window is contaminated by these artifacts. These measures are used in the subsequent Window Quality Assessment stage (Sect. 28.2.6).

28.2.4 Event Detection

From the input waveforms, a time series of events is extracted for subsequent analysis. For example, heart rate variability (HRV) studies typically operate on the R–R interval time series (or tachogram), which requires accurate determination of R-peak locations. Other examples of event detection include extraction of inter-breath intervals from respiratory waveforms (for the calculation of respiratory rate variability, RRV), or T-wave widths from ECGs.

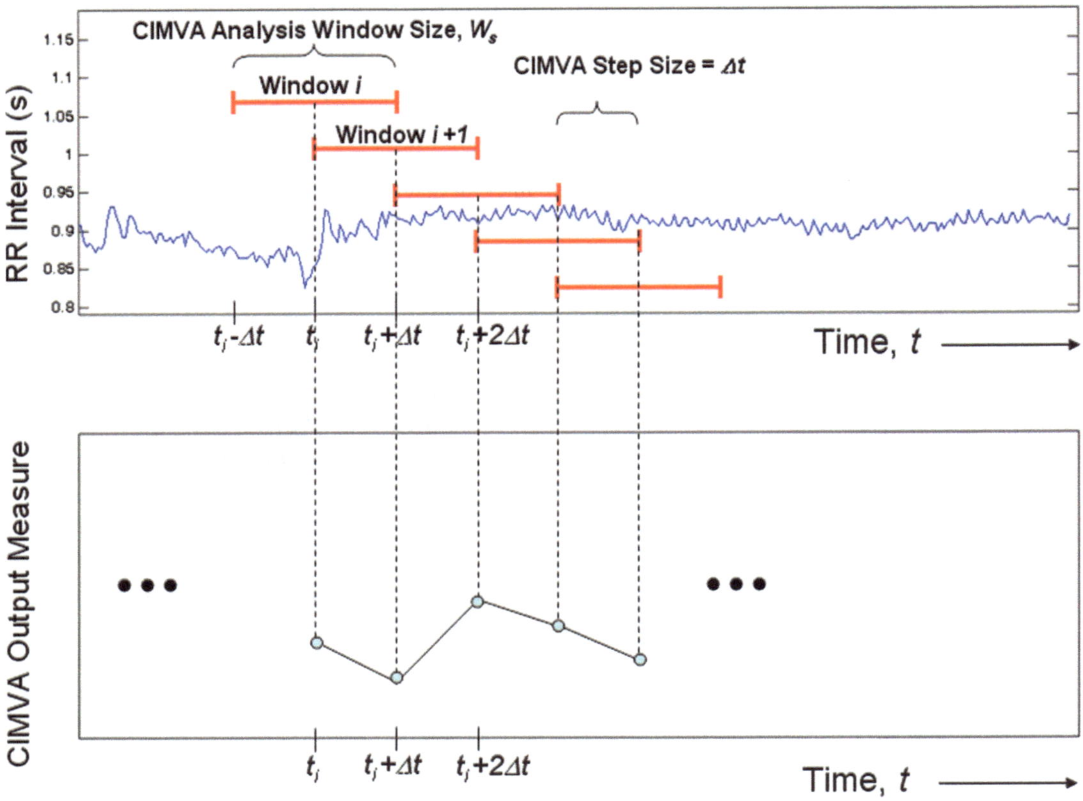

Fig. 28.2 Illustration of a window analysis (using R–R interval (RRI) event data). For window i, RRI samples between $(t_i - W_s/2) < t < (t_i + W_s/2)$ are included in the analysis window. The output measure for that window is recorded at $t = t_i$. The process is repeated by shifting the window by step size Δt, so for window $i+1$, RRI samples between $t_i < t < (t_i + W_s)$ are included. The resulting output measures have a sampling interval equal to the step size, Δt. This example shows the case of 50% overlap between adjacent windows

Figure 28.3 shows some examples of events detected.

28.2.5 Event-Based Artifact Detection and Removal

Physiological signals (and time series derived from these) are frequently contaminated by artifact, which can be due to pathological conditions (e.g., atrial fibrillation) or issues with instrumentation (spurious noise, disconnections, movement artifacts).

It is clear from the growing literature that the result of applying variability analysis to a time series of events (e.g., R–R intervals) is highly affected by the presence of physiological events (e.g., ectopic beats) and artifacts (e.g., noise spikes) in the input data [28]. Specific algorithms are used to identify and measure the prevalence of these events (and possibly exclude the window from subsequent analysis during the Window Quality Assessment stage). This is similar to the waveform-based artifact detection stage mentioned in Sect. 28.2.3, but the artifacts being detected are based on an event time series, not a waveform. The following steps are performed:

1. Detection of abnormal physiology

   In certain cases, it is possible to identify intervals of potentially abnormal patient physiology based on the input event data. A measure representing the extent to which the input data exhibits qualities associated with abnormal physiology is computed and used during Window Quality Assessment. As an example, the detection of atrial fibrillation (a-fib) can be performed at this

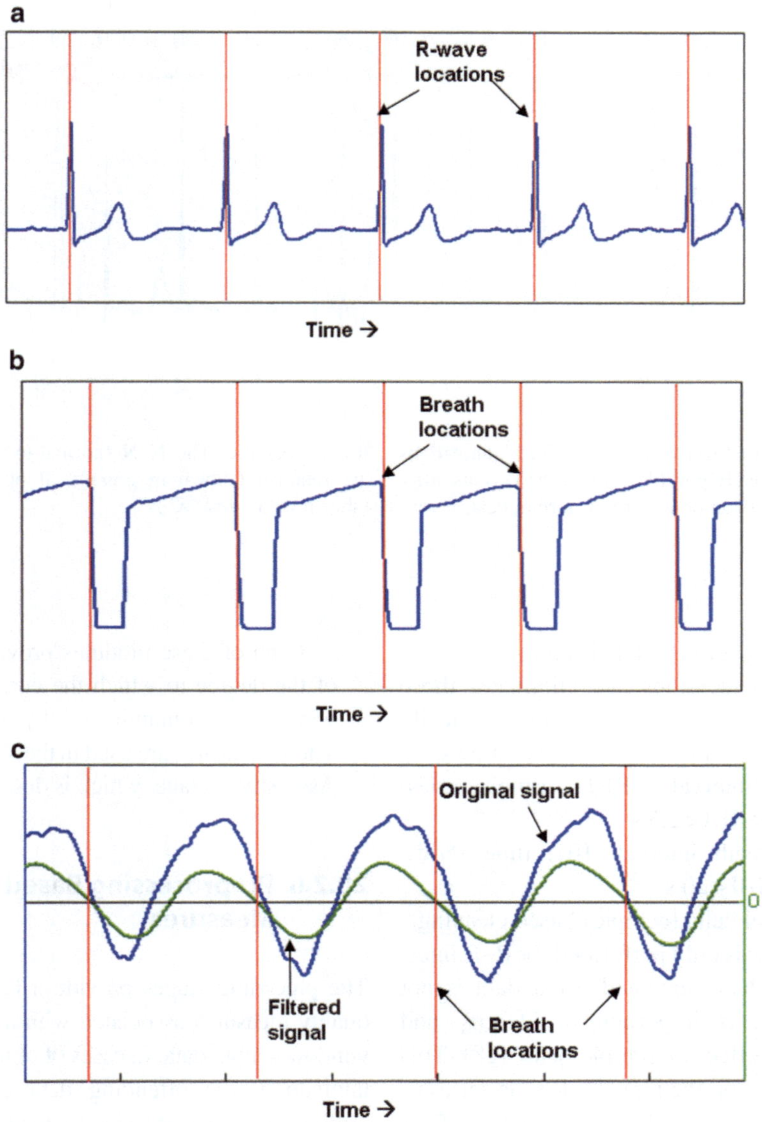

Fig. 28.3 (**a**) ECG showing R-peak locations, (**b**) CO_2 waveform showing inspiration and expiration locations, and (**c**) chest impedance signal showing zero crossings in the filtered output

step, yielding a number of (nonoverlapping) n second intervals (e.g., $n = 10$ s) that are deemed to exhibit "a-fib" characteristics, based on the algorithm described in [29].

2. Rejection of events based on annotations

If the input time series consists of annotated event data, this supplementary information can be used to filter out events which might compromise output quality. As an example, for annotated R–R intervals, each R-peak location (associated with a heartbeat) contains a label describing the classification of that beat (based on ECG morphology, local R–R interval character, etc.) Examples of beat labels include: normal, ventricular ectopic, supra-ventricular premature.

Analysis is usually restricted to heartbeats that occur during episodes of normal sinus rhythm [30]. When annotated R–R intervals are available, this is performed by simply removing all intervals that are not N–N intervals (normal-to-normal)—see Fig. 28.4.

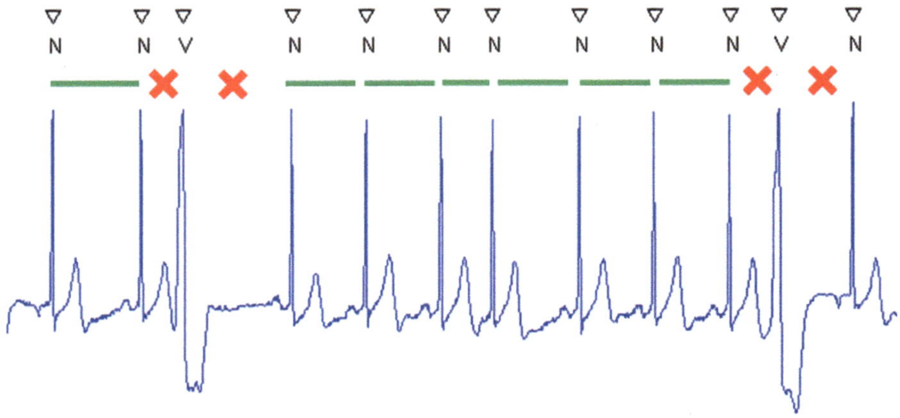

Fig. 28.4 Example ECG trace showing beat annotations (from a Philips Data Export file) and rejected events (after the event-based artifact detection). *N* normal beat, *V* ventricular ectopic. The N–N (normal-to-normal) intervals are retained (shown in *green*); all others are rejected (shown with a *red* "X")

3. Physiological limit-based cleaning

   This step identifies and eliminates those events which are deemed to be physiologically impossible. The limits used are as follows:
   (a) R–R interval (RRI) time series: $0.25\ \text{s} < \text{RRI} < 2.5\ \text{s}$
   (b) Interbreath interval (IBI) time series: $0.1\text{s} < \text{IBI} < 20\ \text{s}$.
4. Physiological delta (ectopic)-based cleaning:

   This stage is only performed for R–R interval data. When annotated event data is not available, it is important to identify and remove the effect of ectopic beats [28]. This is done based on the input R–R time series as follows: if the *i*th R–R interval (RR_i) differs from the previous one by more than 15%, then both RR_i and RR_{i+1} are removed from the analysis [31].
5. Statistical cleaning (optional)

   In order to further remove the effect of statistical outliers on the subsequent analysis, a cleaning stage is performed based on the distribution of the events in the analysis window. For instance, if the distribution of events is normal, those events which fall greater than 3 standard deviations from the mean (in a given analysis window) would be eliminated, as shown in Fig. 28.5.

Each of these modules provides a measure of the degree to which the current processing window is contaminated by these artifacts. These measures are used in the Window Quality Assessment stage, which is described next.

28.2.6 Preprocessing Based on Quality Measures

The preceding stages provide a large number of quality measures associated with a given analysis window's input data. In cases of minor artifact contamination, the offending data can simply be removed before processing by the Core Variability Analysis block from Fig. 28.1. In other instances, however, the data may be so corrupted that the outputs of the variability analysis would be untrustworthy. In such cases, it is desirable to let the user decide whether a window should be rejected or not, based on a set of quantities with adjustable threshold values. These quantities estimate for each analysis window the duration of sensor quality issues (e.g., disconnection, saturation) and the number of suspect events (ectopic, atrial fibrillation, outside physiological limits, statistical outliers). This selection process ensures that only the highest quality windows are retained for the variability analysis.

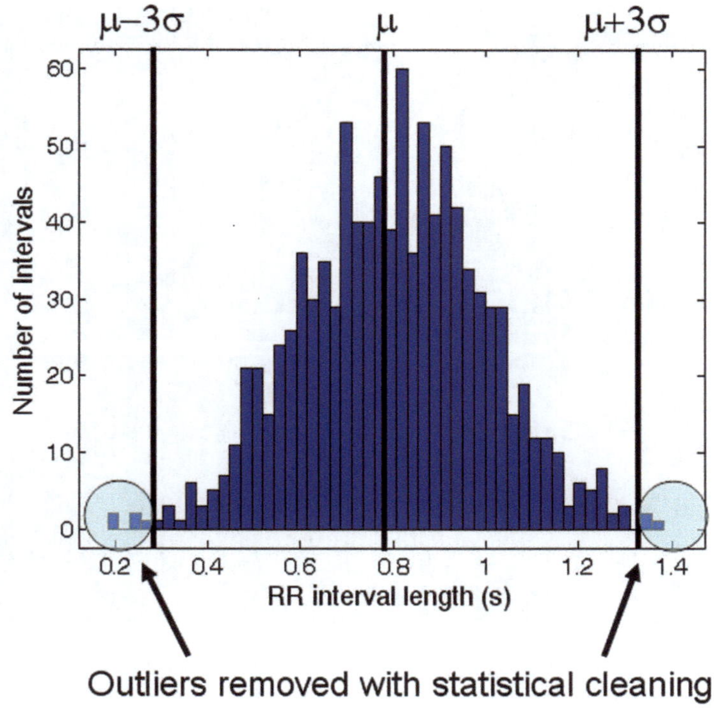

Fig. 28.5 Cleaning of event data based on statistical outliers. Events which are more than three standard deviations from the mean are removed from subsequent analysis

Table 28.1 Domains of variability and examples of associated features (modified from [11])

Domains of variability	Examples of features
Statistical [30, 32]	Standard statistical features, form factor, some symbolic dynamics features, turns count
Geometric [16]	Grid counting, heart rate turbulence, Poincaré plots features, recurrence plot features, spatial filling index
Informational [33, 34]	Approximate entropy, conditional entropy, compression entropy, fuzzy entropy, Kullback–Leibler permutation entropy, multiscale entropy, predictive-based features, sample entropy, Shannon entropy, similarity indexes, Rényi entropy
Energetic [17, 30, 35]	Frequency features, energy operators, multiscale time irreversibility, time–frequency features
Invariant [36–38]	Allan scaling exponent, correlation dimension, detrended fluctuation analysis, diffusion entropy, embedding scaling exponent, Fano scaling exponent, finite growth rates, Higuchi's algorithm, index of variability, Kolmogorov–Sinai entropy, largest Lyapunov exponent, multifractal exponents, power spectrum scaling exponent, probability distribution scaling exponent, rescaled detrended range analysis, scaled windowed variance

28.2.7 Core Variability Analysis

At this stage, a window is deemed to be of sufficient quality for the variability analysis core block, which calculates a diverse panel of variability measurements, based on the event data in the window. Many variability measures can be computed, spanning five different domains of variability and summarizing different properties of the input data. These domains are listed in Table 28.1 (see [11] for more details), along with examples of the associated measures. As an example, CIMVA™ software currently outputs 102 different variability measures [27].

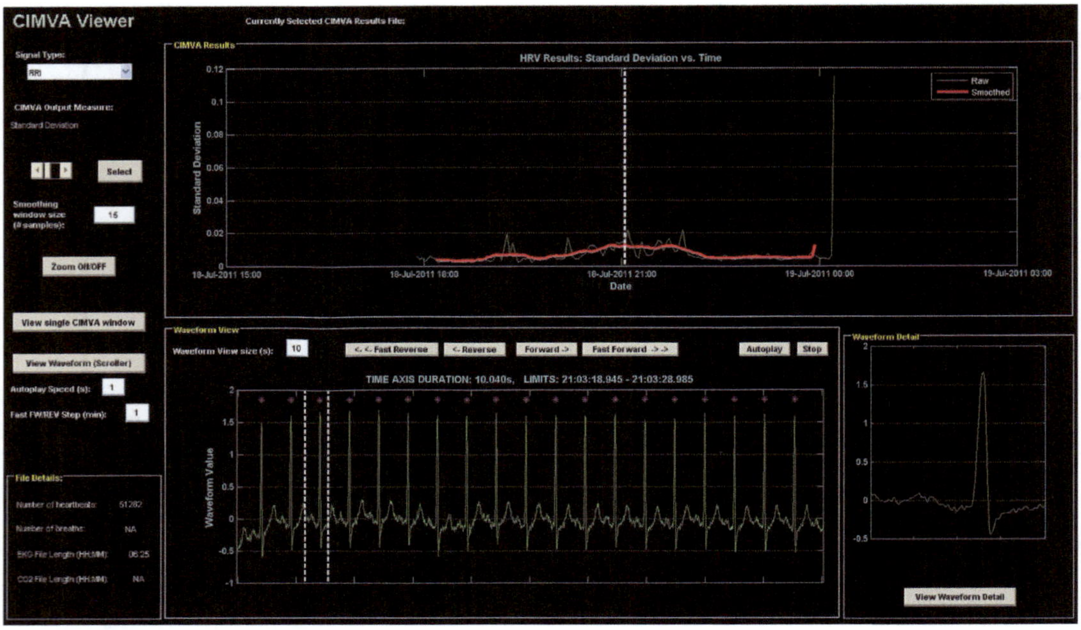

Fig. 28.6 Example of a graphical summary of variability analysis for an ECG (one measure shown)

28.2.8 Variability Analysis Outputs

The outputs from the variability analysis framework presented in Fig. 28.1 summarize the properties of each analysis window, give a description of the waveform or event data quality, highlight the data discarded by the cleaning process, and finally provide the variability measures for each window. These outputs can be exported to a variety of different formats for convenient graphical representation and statistical postprocessing. The evolution of variability measures over time for all input signals (e.g., ECG, CO_2, or RESP) can be conveniently plotted over the course of relevant clinical events (e.g., a 30 min Spontaneous Breathing Trial), alongside the quality measures (Fig. 28.6).

28.3 Clinical Applications

In this section we briefly introduce and discuss some of the early investigations achieved by applying the variability monitoring framework presented in this chapter to various clinical scenarios. Other potential clinical applications of variability monitoring are also presented.

28.3.1 Early Detection and Prognostication of Infection

Early diagnosis of sepsis leading to aggressive resuscitation involving antibiotic administration is vital to recovery and survival. Current clinical approaches for diagnosing sepsis are based on an increased absolute value of one or more vital signs (e.g., fever) in addition to laboratory tests such as blood cultures to look for evidence of a pathogen. The current standard of care for patients with infection involves a response to abnormal physiologic parameters (i.e., blood pressure) targeting improvement of that parameter as the end goal.

The identification of decreased heart rate variability has been shown to herald the development of septic shock, multiorgan failure, and mortality [39–42].

Several past and ongoing studies have tackled this issue.

28.3.1.1 AVERT

A multicenter prospective study is currently under way to assess the variability in early resuscitation and treatment for sepsis (AVERT) using a composite variability index based on multiple HRV and RRV measures. The overall objective of this study is to utilize HRV and RRV in 135 adult patients with sepsis (currently from three centers) to predict subsequent clinical deterioration and death. Recent literature continues to support the use of a predefined goal when resuscitating septic shock [43]; however, the specific goals to target are still debatable. Understanding the effect that specific interventions and resuscitation decisions have on HRV and RRV could better define effective treatment strategies and provide the support to advance this measure of prognostication into an endpoint of resuscitation.

28.3.1.2 Ambulatory High-Risk Patients

Ambulatory patients undergoing bone marrow transplantation (BMT) for the treatment of acute leukemia become neutropenic (abnormally low white blood cell count) as a side effect of the BMT. These neutropenic patients comprise a group at a high risk of sepsis (approximately 80%) and mortality (approximately 5%).

Ahmad et al. performed HRV analysis on continuously recorded heart rate waveforms of 21 ambulatory outpatients as they underwent BMT [17]. Of the 21 patients enrolled, 4 patients withdrew, leaving 17 patients who completed the study (12 ± 4 days of continuous Holter monitoring). Fourteen patients developed sepsis requiring antibiotic therapy, whereas three did not. On average, in 12 out of 14 infected patients (86%), an important (25%) reduction in several variability output measures was observed prior to the clinical diagnosis and treatment of sepsis. In infected patients, Wavelet AUC demonstrated on average a 25% drop from baseline 35 h prior to sepsis. In noninfected patients, all measures except two showed no significant reduction. For further details refer to [17].

28.3.1.3 Neonatal Sepsis

Late-onset neonatal sepsis is an important cause of morbidity and mortality in very low birth weight infants in a neonatal intensive care unit (NICU) [44]. Typically, no clear clinical presentation occurs until late in the course of the sepsis, when severe illness is already present [45]. However, altered HRV may be present in the hours to days before diagnosis of late-onset proven or clinical neonatal sepsis [46].

Moorman et al. conducted a large two-group, parallel, individually randomized controlled clinical trial of 3,003 very low birth weight (VLBW) infants in 9 neonatal intensive care units (NICU) to assess whether HRV monitoring could improve neonatal outcomes [18].

They computed a composite index score (HRC index) representing the fold-increase in risk of sepsis in the next 24 h based on variability measures (e.g., reduced baseline variability and short-lived heart rate decreases). The HRC index score was displayed continuously at the bedside for one group, whereas for the other group it was masked.

Results showed that the mortality rate was reduced significantly in VLBW infants whose HRC monitoring was displayed, and that there was a trend toward increased days alive and ventilator-free. Moorman et al. also highlights that HRC index score monitoring could provide patient-specific testing and intervention and overall improved outcomes in high-risk premature infants.

28.3.2 Sedation and Variability

Finding the appropriate level of sedation in the intensive care unit (ICU) remains a challenge for ICU clinicians [47]. Proper management of depth of sedation is important in preventing complications related to changes in the autonomic nervous system caused by sedation [48, 49] since patient outcomes are significantly influenced by inappropriate level of sedation and the choice of agent among others [50–52].

It has been shown that frequency domain measures of heart rate variability (HRV) provide a noninvasive means of quantifying ANS modulation [12, 53]. In healthy populations, these HRV measures have been used to evaluate the effects of anesthesia and sedation on autonomic nervous function [54–57]. Although there is general agreement that sedation is associated with a reduction in variability, there is still little known about the effect of sedation on ANS activity and HRV in critically ill patients. In critical illness, conditions such as multiple organ dysfunction syndrome and sepsis can alter normal ANS activity [58], thus disturbing normal physiological processes and reducing the ability to adapt to adverse changes [59].

In studies of ICU patients, there have been a few investigations on the relationship between HRV and depth of sedation, correlating HRV with both Ramsay sedation score [49] and objective measures of depth of sedation based on electroencephalography [59]. Kasaoka et al. demonstrate a real-time approach to monitoring autonomic nervous function in ICU patients and observed area under the power spectra in the low-frequency (LF) band, high-frequency (HF) band, as well as the LF/HF ratio to be significantly higher when the patients were no longer sedated or ventilated [60]. Few studies have explored the effect of sedation in ICU patients with other than frequency domain variability measures and considering both HRV and respiratory rate variability (RRV).

A recent pilot study was undertaken by our group to explore the effects of sedation on continuous HRV and RRV monitoring of 33 critically ill adult patients with respiratory and/or cardiac failure, using ECG and end tidal capnography waveforms. HRV and RRV monitoring were performed (5 min windows) prior (4 h) and during sedation interruption (SI) for an average of 11 days from admission to extubation. HRV and RRV measures were then compared to multiple organ dysfunction syndrome (MODS) scores and sedative types. Results indicated an increase in both HRV and RRV during SI with a lower increase or even a decrease for patients with high MODS. Thus, variability monitoring suggests a variable response and benefit from SI, depending on the level of organ failure. A multicenter prospective observational study currently under review will attempt to confirm these preliminary results.

28.3.3 Predicting Extubation Failure

Expeditious yet safe extubation is critically important in the care of mechanically ventilated ICU patients. Failed extubation (defined as re-intubation within 48 h) is associated with increased morbidity, mortality, and costs. Spontaneous breathing trials (SBTs), whereupon patients are subjected to brief periods of reduced ventilatory support (taking on a greater workload of breathing as a simulation of breathing after extubation), are the current gold standard of care to predict failed extubation. Nonetheless, several studies have determined that 15% patients who are extubated subsequently fail [61].

Several investigators have evaluated the presence of altered respiratory system variability or cardiovascular variability as a marker of whether a patient will tolerate extubation and unassisted breathing [62–67]. These single-center studies independently demonstrated an association between altered variability and traditional clinical assessment of readiness for extubation, as well as support its use as a predictor of extubation failure.

Our own pilot study, in which we performed continuous HRV and RRV monitoring during 125 SBTs in 60 patients, demonstrated similar results. Twelve of these patients were excluded for missing data and protocol violations, leaving a total of 48 patients—41 of these passed extubation and 7 of these failed extubation. Restricting analysis to the last SBT performed prior to extubation, patients who subsequently failed extubation had a greater absolute loss of RRV compared to patients who passed extubation as measured by several variability measures ($p<0.05$) calculated by CIMVA software [27]. Furthermore, there was a nonsignificant reduction in the change in HRV associated with failed extubation. No correlation was found between measures of variability and standard measures of readiness for extubation.

A multicenter observational study, supported by the Canadian Critical Care Trials Group, is underway to evaluate the added value of multiorgan variability monitoring (compared to standard of care) and determine thresholds of HRV and RRV that predict extubation failure.

28.3.4 Monitoring Severity of Illness in the ICU

Severity of illness is determined by clinical judgment and is never certain in critically ill patients. Although illness and injury severity scores and organ dysfunction scores exist, they represent population-based measures and are not well suited to evaluate the prognosis or severity of illness in individual patients over time.

Continuous data collection was performed for 35 patients by harvesting the ECG and $EtCO_2$ waveforms (already monitored as per standard practice). These data were collected continuously for 24 h a day (the enrolment in the study lasted on average 11 days per patient). In this pilot study we demonstrated feasibility of continuous HRV and RRV analysis in critically ill patients [68]. In addition, we observed correlation between increasing organ failure and reduced variability. Patients with low levels of organ failure had preservation of HRV and RRV, while patients with increasing degree of organ failure had progressively lower levels of variability. Decreasing HRV was observed in patients ($n=6$) as they progressed towards the onset of shock (as defined by initiation of vasopressor and/or inotrope therapy). Increasing RRV was observed in patients as they progressed towards resolution of respiratory failure and extubation ($n=15$) [69].

28.3.5 Evaluation of Cardiopulmonary Fitness

Since the 1970s, staged exercise testing has been increasingly employed in the outpatient or ambulatory care setting, providing an assessment of cardiac and respiratory functioning, determining an individual's pre-, peri-, and postoperative risk assessment, and measuring fitness level. Although cardiopulmonary exercise testing is a well-justified assessment tool, its major limitations include physical incapability, a lack of motivation and desire, reluctance to repeatedly perform exercise testing to exhaustion, and invasive and expensive test equipment. Therefore, alternative measures that are noninvasive, inexpensive, and require minimal exertion should be explored.

Thirty-nine healthy and physically active (Group 1: ≤25 years of age, $n=12$; Group 2: ≥40 years of age, $n=12$) or unhealthy and less physically active (Group 3: ≤25 years of age, $n=3$ or Group 4: ≥40 years of age, $n=12$) subjects had R–R interval data recorded from a portable Holter monitor which participants wore for 24 h prior to maximal aerobic capacity (VO_2max) test using a metabolic cart.

The average Day 1 awake HRV was found strongly correlated with cardiopulmonary fitness, such that both wavelet AUC ($r=0.83$, $p<0.001$) and detrended fluctuation analysis (DFA) ($r=0.83$, $p<0.001$) accounted for approximately 80% of the variation in cardiopulmonary fitness. Furthermore, a paired t-test demonstrated significant differences between young healthy and older unhealthy ($p<0.05$) and older healthy and older unhealthy ($p<0.05$) for measures of standard deviation and DFA [70].

28.4 Variability at the Bedside

The previous section introduced some promising clinical applications of variability analysis. However, it is essential to provide meaningful, practical variability results to clinicians, in order to impact patient care. Typically, clinicians in intensive care settings are presented with a deluge of clinical measurements [71]. Summarizing that information and sorting through the relevant key features for important clinical decisions is not a trivial task. Additional streams of measures from multiorgan variability monitoring will likely be overwhelming and hard to interpret in raw form for clinicians. However, we believe that continuous multiorgan variability data from higher level analytics or any other result should be presented to clinicians at the bedside in a clinically intelligible fashion to optimize utility.

For instance, measuring an individual patient's change in multiorgan variability with reference to an established baseline or with respect to clinical events (e.g., SBT for extubation) could be more relevant for diagnostic decisions. In addition, assessing a patient's variability with respect to normative ranges derived from population studies would help shed some light on the variability trends. These ranges of normal values for variability measures will likely require significant clinical research studies as they will most likely vary by age groups, comorbidities, and more [72–74].

Another approach is to summarize a large number of variability measures into a few key composite measures that can provide clinicians with a quick assessment of variability trends and allow them to anticipate adverse events. Composite measures are currently being developed for specific clinical outcomes.

An important aspect of bringing multiorgan variability monitoring at the bedside is translating variability results into relevant clinical information in the form of likelihood ratios of specific clinical events occurring over time. These likelihood ratios and real-time specific clinical outcome predictions based on variability results are starting to appear on the market. For example, the HeRO system (Medical Predictive Sciences Corporation) provides an odds ratio of a neonate developing sepsis using measurements of transient decelerations in heart rate variability as a score component [18, 75, 76]. However, much work remains to be done and a significant research effort is still required, building on and validating recent and on-going research studies. In particular, randomized clinical trials are needed to demonstrate what improvement in care, if any, are offered by providing clinicians with clinical decision aids rooted in variability analysis.

28.5 Conclusion

Bedside application of the science of complex systems has grown remarkably in recent decades. Rooted in this paradigm, researchers have increasingly assessed the value of variability monitoring to solve problems in medicine. As discussed, recent examples include the assessment of ICU patients' mortality through HRV [77], the prediction of neonatal sepsis through HRV [78], and the characterization of RRV during spontaneous breath trials [79]. These results highlight two principal challenges that await the evolution of this research.

The first challenge is to characterize multiorgan variability in the most efficient manner possible, focusing on independent clinically useful information within coupled physiological systems. Examples include the evaluation of interactions between the cardiac and respiratory systems through spectral analysis [80], or the evaluation of the differences between heart rate variability and blood pressure [81]. Despite the relevance of these analyses, the research in this area still needs to take advantage of novel multivariate methods and multiple signal comparison. Indeed, when assessing variability, few studies compare two signals at a time, and rarely more than two signals are simultaneously considered. Thus, multiorgan variability integration remains an ongoing challenge.

The second challenge corresponds to the exploration of the nature and meaning behind variability and its domains. While several theories exist, definitive physiological explanations for the numerous patterns of variation seen in health and their alteration in disease remain an area of active exploration. These physiological underpinnings are necessary for bedside application. The future represents an exciting voyage of discovery.

Competing Interests
Andrew J.E. Seely is founder and Chief Science Officer and Geoff Green is Product Manager of Therapeutic Monitoring Systems (TMS). TMS was founded in order to commercialize patented Continuous Individualized Multiorgan Variability Analysis (CIMVA) technology, with the objective of delivering variability-directed clinical decision support to improve quality and efficiency of care. The other authors have no conflicts of interest to disclose.

List of Abbreviations

AUC	Area under the curve
BMT	Bone marrow transplantation
CIMVA	Continuous Individualized Multiorgan Variability Analysis

DFA	Detrended fluctuation analysis
ECG	Electrocardiogram
ER	Emergency room
$EtCO_2$	End-tidal capnogram
HF	High frequency
HRC	Heart rate characteristics
HRV	Heart rate variability
IBI	Inter-breath interval
ICU	Intensive care unit
LF	Low frequency
NICU	Neonatal intensive care unit
RRI	R–R interval
RRV	Respiratory rate variability
SBT	Spontaneous breathing trial
VLBW	Very low birth weight

Acknowledgments We gratefully acknowledge funding support from the Ottawa Technology Transfer Network (OTTN) and the Canadian Institutes of Health Research (CIHR) for support for this research program.

References

1. http://upload.wikimedia.org/wikipedia/en/5/5a/Complexity-map-overview.png.
2. Shannon CE, Weaver W. The mathematical theory of communication. Urbana, IL: The University of Illinois Press; 1949.
3. Von Bertalanffy L. An outline of general system theory. Br J Philos Sci. 1950;1:134–65.
4. Kauffman SA. At home in the universe: the search for laws of self-organization and complexity. New York: Oxford University Press; 1995.
5. Mandelbrot BB. The fractal geometry of nature. New York: W.H. Freeman; 1983.
6. Bak P. How nature works: the science of self-organized criticality. New York: Copernicus; 1996.
7. Capra F. The web of life: a new scientific understanding of living systems. New York: Random House; 1997.
8. Glass L, Mackey MC. From clocks to chaos: the rhythms of life. Princeton, NJ: Princeton University Press; 1988.
9. Prigogine I. From being to becoming: time and complexity in the physical sciences. San Francisco, CA: W. H. Freeman; 1980.
10. Goldberger AL. Non-linear dynamics for clinicians: chaos theory, fractals, and complexity at the bedside. Lancet. 1996;347(9011):1312–4.
11. Bravi A, Longtin A, Seely AJ. Review and classification of variability analysis techniques with clinical applications. Biomed Eng Online. 2011;10(1):90.
12. Akselrod S, Gordon D, Ubel FA, Shannon DC, Berger AC, Cohen RJ. Power spectrum analysis of heart rate fluctuation: a quantitative probe of beat-to-beat cardiovascular control. Science. 1981;213(4504):220–2.
13. Priban IP. An analysis of some short-term patterns of breathing in man at rest. J Physiol (Lond). 1963;166:425–34.
14. Benchetrit G, Bertrand F. A short-term memory in the respiratory centres: statistical analysis. Respir Physiol. 1975;23(2):147–58.
15. Tobin MJ, Mador MJ, Guenther SM, Lodato RF, Sackner MA. Variability of resting respiratory drive and timing in healthy subjects. J Appl Physiol. 1988;65(1):309–17.
16. Seely AJE, Macklem PT. Complex systems and the technology of variability analysis. Crit Care. 2004;8(6):R367–84.
17. Ahmad S, Ramsay T, Huebsch L, et al. Continuous multi-parameter heart rate variability analysis heralds onset of sepsis in adults. PLoS One. 2009;4(8):e6642.
18. Moorman JR, Carlo WA, Kattwinkel J, et al. Mortality reduction by heart rate characteristic monitoring in very low birth weight neonates: a randomized trial. J Pediatr. 2011;159(6):900–6.e1.
19. Voss A, Kurths J, Kleiner HJ, Witt A, Wessel N, Saparin P, Osterziel KJ, Schurath R, Dietz R. The application of methods of non-linear dynamics for the improved and predictive recognition of patients threatened by sudden cardiac death. Cardiovasc Res. 1996;31(3):419–33.
20. Schmidt G, Malik M, Barthel P, Schneider R, Ulm K, Rolnitzky L, Camm AJ, Bigger Jr JT, Schömig A. Heart-rate turbulence after ventricular premature beats as a predictor of mortality after acute myocardial infarction. Lancet. 1999;353(9162):1390–6.
21. Gil E, Bailón R, Vergara JM, Laguna P. PTT variability for discrimination of sleep apnea related decreases in the amplitude fluctuations of PPG signal in children. IEEE Trans Biomed Eng. 2010;57(5):1079–88.
22. de Chazal P, Heneghan C, McNicholas WT. Multimodal detection of sleep apnoea using electrocardiogram and oximetry signals. Philos Transact A Math Phys Eng Sci. 2009;367(1887):369–89.
23. Lin C-W, Wang J-S, Chang P-C. Mining physiological conditions from heart rate variability analysis. Comp Intell Mag. 2010;5(1):50–8.
24. Eckberg DL. Sympathovagal balance: a critical appraisal. Circulation. 1997;96(9):3224–32.
25. Singh RB, Cornélissen G, Weydahl A, et al. Circadian heart rate and blood pressure variability considered for research and patient care. Int J Cardiol. 2003;87(1):9–28. Discussion 29–30.
26. Bilan A, Witczak A, Palusiński R, Myśliński W, Hanzlik J. Circadian rhythm of spectral indices of heart rate variability in healthy subjects. J Electrocardiol. 2005;38(3):239–43.
27. http://www.therapeuticmonitoring.com/product-development/cimva-overview/.
28. Clifford GD. ECG statistics, noise, artifacts and missing data. In: Clifford GD, Azuaje F, McSharry PE, editors. Advanced methods and tools for ECG data analysis. London: Artech House; 2006.

29. Logan B, Healey J. Robust detection of atrial fibrillation for a long term telemonitoring system. In: Computers in cardiology, Lyon; 2005. p. 619–22.
30. Malik M, Camm AJ, Bigger Jr JT, et al. Heart rate variability. Standards of measurement, physiological interpretation, and clinical use. Eur Heart J. 1996; 17(3):354–81.
31. Clifford GD, McSharry PE, Tarassenko L. Characterizing artefact in the normal human 24-hour RR time series to aid identification and artificial replication of circadian variations in human beat to beat heart rate using a simple threshold. In: Computers in cardiology; 2002. p. 129–132.
32. Kovatchev BP, Farhy LS, Cao H, Griffin MP, Lake DE, Moorman JR. Sample asymmetry analysis of heart rate characteristics with application to neonatal sepsis and systemic inflammatory response syndrome. Pediatr Res. 2003;54(6):892–8.
33. Richman JS, Moorman JR. Physiological time-series analysis using approximate entropy and sample entropy. Am J Physiol Heart Circ Physiol. 2000; 278(6):H2039–49.
34. Pincus SM. Approximate entropy as a measure of system complexity. Proc Natl Acad Sci USA. 1991; 88(6):2297–301.
35. Agarwal R, Gotman J. Adaptive segmentation of electroencephalographic data using a nonlinear energy operator. ISCAS. 1999;4:199–202.
36. Tibby SM, Frndova H, Durward A, Cox PN. Novel method to quantify loss of heart rate variability in pediatric multiple organ failure. Crit Care Med. 2003;31(7):2059–67.
37. Peng C-K, Buldyrev SV, Havlin S, Simons M, Stanley HE, Goldberger AL. Mosaic organization of DNA nucleotides. Phys Rev E. 1994;49(2):1685–9.
38. Kleiger RE, Stein PK, Bigger Jr JT. Heart rate variability: measurement and clinical utility. Ann Noninvasive Electrocardiol. 2005;10(1):88–101.
39. Barnaby D, Ferrick K, Kaplan DT, Shah S, Bijur P, Gallagher EJ. Heart rate variability in emergency department patients with sepsis. Acad Emerg Med. 2002;9(7):661–70.
40. Chen W-L, Kuo C-D. Characteristics of heart rate variability can predict impending septic shock in emergency department patients with sepsis. Acad Emerg Med. 2007;14(5):392–7.
41. Pontet J, Contreras P, Curbelo A, Medina J, Noveri S, Bentancourt S, Migliaro ER. Heart rate variability as early marker of multiple organ dysfunction syndrome in septic patients. J Crit Care. 2003;18(3):156–63.
42. Seely AJ, Christou NV. Multiple organ dysfunction syndrome: exploring the paradigm of complex nonlinear systems. Crit Care Med. 2000;28(7):2193–200.
43. Jones AE, Brown MD, Trzeciak S, Shapiro NI, Garrett JS, Heffner AC, Kline JA. The effect of a quantitative resuscitation strategy on mortality in patients with sepsis: a meta-analysis. Crit Care Med. 2008;36(10):2734–9.
44. Stoll BJ, Hansen N, Fanaroff AA, et al. Late-onset sepsis in very low birth weight neonates: the experience of the NICHD Neonatal Research Network. Pediatrics. 2002;110(2 Pt 1):285–91.
45. Fanaroff AA, Korones SB, Wright LL, et al. Incidence, presenting features, risk factors and significance of late onset septicemia in very low birth weight infants. The National Institute of Child Health and Human Development Neonatal Research Network. Pediatr Infect Dis J. 1998;17(7):593–8.
46. Griffin MP, Moorman JR. Toward the early diagnosis of neonatal sepsis and sepsis-like illness using novel heart rate analysis. Pediatrics. 2001;107(1):97–104.
47. Mehta S, McCullagh I, Burry L. Current sedation practices: lessons learned from international surveys. Crit Care Clin. 2009;25(3):471–88, vii–viii.
48. Schmidt H, Müller-Werdan U, Hoffmann T, et al. Autonomic dysfunction predicts mortality in patients with multiple organ dysfunction syndrome of different age groups. Crit Care Med. 2005;33(9): 1994–2002.
49. Haberthür C, Lehmann F, Ritz R. Assessment of depth of midazolam sedation using objective parameters. Intensive Care Med. 1996;22(12):1385–90.
50. Arroliga A, Frutos-Vivar F, Hall J, Esteban A, Apezteguía C, Soto L, Anzueto A. Use of sedatives and neuromuscular blockers in a cohort of patients receiving mechanical ventilation. Chest. 2005;128(2): 496–506.
51. Brook AD, Ahrens TS, Schaiff R, Prentice D, Sherman G, Shannon W, Kollef MH. Effect of a nursing-implemented sedation protocol on the duration of mechanical ventilation. Crit Care Med. 1999;27(12): 2609–15.
52. Kress JP, Pohlman AS, O'Connor MF, Hall JB. Daily interruption of sedative infusions in critically ill patients undergoing mechanical ventilation. N Engl J Med. 2000;342(20):1471–7.
53. Pomeranz B, Macaulay RJ, Caudill MA, Kutz I, Adam D, Gordon D, Kilborn KM, Barger AC, Shannon DC, Cohen RJ. Assessment of autonomic function in humans by heart rate spectral analysis. Am J Physiol. 1985;248(1 Pt 2):H151–3.
54. Kanaya N, Hirata N, Kurosawa S, Nakayama M, Namiki A. Differential effects of propofol and sevoflurane on heart rate variability. Anesthesiology. 2003;98(1):34–40.
55. Win NN, Fukayama H, Kohase H, Umino M. The different effects of intravenous propofol and midazolam sedation on hemodynamic and heart rate variability. Anesth Analg. 2005;101(1):97–102. Table of contents.
56. Ebert TJ. Sympathetic and hemodynamic effects of moderate and deep sedation with propofol in humans. Anesthesiology. 2005;103(1):20–4.
57. Galletly DC, Williams TB, Robinson BJ. Periodic cardiovascular and ventilatory activity during midazolam sedation. Br J Anaesth. 1996;76(4):503–7.
58. Schmidt HB, Werdan K, Müller-Werdan U. Autonomic dysfunction in the ICU patient. Curr Opin Crit Care. 2001;7(5):314–22.
59. Unoki T, Grap MJ, Sessler CN, Best AM, Wetzel P, Hamilton A, Mellott KG, Munro CL. Autonomic nervous system function and depth of sedation in adults receiving mechanical ventilation. Am J Crit Care. 2009;18(1):42–50.

60. Kasaoka S, Nakahara T, Kawamura Y, Tsuruta R, Maekawa T. Real-time monitoring of heart rate variability in critically ill patients. J Crit Care. 2010;25(2):313–6.
61. Esteban A, Alía I, Gordo F, et al. Extubation outcome after spontaneous breathing trials with T-tube or pressure support ventilation. The Spanish Lung Failure Collaborative Group. Am J Respir Crit Care Med. 1997;156(2 Pt 1):459–65.
62. Shen H-N, Lin L-Y, Chen K-Y, Kuo P-H, Yu C-J, Wu H-D, Yang P-C. Changes of heart rate variability during ventilator weaning. Chest. 2003;123(4):1222–8.
63. Bien M-Y, Shui Lin Y, Shih C-H, Yang Y-L, Lin H-W, Bai K-J, Wang J-H, Ru Kou Y. Comparisons of predictive performance of breathing pattern variability measured during T-piece, automatic tube compensation, and pressure support ventilation for weaning intensive care unit patients from mechanical ventilation. Crit Care Med. 2011;39(10):2253–62.
64. Wysocki M, Cracco C, Teixeira A, Mercat A, Diehl J-L, Lefort Y, Derenne J-P, Similowski T. Reduced breathing variability as a predictor of unsuccessful patient separation from mechanical ventilation. Crit Care Med. 2006;34(8):2076–83.
65. Orini M, Giraldo BF, Bailón R, Vallverdu M, Mainardi L, Benito S, Díaz I, Caminal P. Time-frequency analysis of cardiac and respiratory parameters for the prediction of ventilator weaning. Conf Proc IEEE Eng Med Biol Soc. 2008;2008:2793–6.
66. El-Khatib M, Jamaleddine G, Soubra R, Muallem M. Pattern of spontaneous breathing: potential marker for weaning outcome. Spontaneous breathing pattern and weaning from mechanical ventilation. Intensive Care Med. 2001;27(1):52–8.
67. Papaioannou V, Dragoumanis C, Pneumatikos I. Biosignal analysis techniques for weaning outcome assessment. J Crit Care. 2010;25(1):39–46.
68. Bradley B, Green GC, Batkin I, Seely AJE. Feasibility of continuous multiorgan variability analysis in the intensive care unit. J Crit Care. 2012;27(2):218.e9–20.
69. Bradley BD, Green GC, Ramsay T, Seely AJE. Impact of sedation and organ failure on continuous heart and respiratory rate variability monitoring in critically ill patients: a pilot study. Critical care Med. 2013;41(2).
70. Armstrong RG, Kenny GP, Green G, Seely AJE. Diurnal Variation in Heart Rate Variability before and after Maximal Exercise Testing. Chronobiology International 2011;28:344–51.
71. McIntosh N. Intensive care monitoring: past, present and future. Clin Med. 2002;2(4):349–55.
72. Masaoka S, Lev-Ran A, Robert Hill L, Vakil G, Hon EHG. Heart rate variability in diabetes: relationship to age and duration of the disease. Diabetes Care. 1985;8(1):64–8.
73. O'Brien IA, O'Hare P, Corrall RJ. Heart rate variability in healthy subjects: effect of age and the derivation of normal ranges for tests of autonomic function. Br Heart J. 1986;55(4):348–54.
74. Umetani K, Singer DH, McCraty R, Atkinson M. Twenty-four hour time domain heart rate variability and heart rate: relations to age and gender over nine decades. J Am Coll Cardiol. 1998;31(3):593–601.
75. Moorman JR, Delos JB, Flower AA, Cao H, Kovatchev BP, Richman JS, Lake DE. Cardiovascular oscillations at the bedside: early diagnosis of neonatal sepsis using heart rate characteristics monitoring. Physiol Meas. 2011;32(11):1821–32.
76. Medical Predictive Science Corporation HeRO (R) – Neonatal Heart Rate Observation System. http://www.heroscore.com/index.htm. Accessed 30 Mar 2012.
77. Norris PR, Ozdas A, Cao H, Williams AE, Harrell FE, Jenkins JM, Morris Jr JA. Cardiac uncoupling and heart rate variability stratify ICU patients by mortality: a study of 2088 trauma patients. Ann Surg. 2006;243(6):804–12. Discussion 812–4.
78. Moorman JR, Lake DE, Griffin MP. Heart rate characteristics monitoring for neonatal sepsis. IEEE Trans Biomed Eng. 2006;53(1):126–32.
79. White CE, Batchinsky AI, Necsoiu C, Nguyen R, Walker 3rd KP, Chung KK, Wolf SE, Cancio LC. Lower interbreath interval complexity is associated with extubation failure in mechanically ventilated patients during spontaneous breathing trials. J Trauma. 2010;68(6):1310–6.
80. Indic P, Salisbury E, Paydarfar D, Brown E, Barbieri R. Interaction between heart rate variability and respiration in preterm infants. Comput Cardiol. 2008;35: 57–60.
81. Nozawa M, Yana K, Kaeriyama K, Mizuta H, Ono T. Spontaneous variability analysis for characterizing cardiovascular responses to water ingestion. Conf Proc IEEE Eng Med Biol Soc. 2009;2009:1816–9.

Part V

Health Services

A systems and complexity perspective has important implications for understanding, designing and improving health service functions and outcomes. The chapters in this section provide some insight how system dynamics impact on the service delivery in primary, preventive and community care. There is a considerable overlap with the following section on health systems, only to highlight the close interconnectedness between various structures and functions in health care. Health systems refers to the broader social, biological and environmental systems that influence human health, while the clinical industrial complex that delivers services specifically to prevent and treat illness and disease is designated as health services.

An Overview of Complexity Theory: Understanding Primary Care as a Complex Adaptive System

29

Beverley Ellis

29.1 An Overview of Complex Adaptive Systems Theory: Elements and Features

An overview of Complex Adaptive Systems (CAS) theories and the key features that allow a system to be described as a CAS is introduced in this section. The origins of many of CAS theories are attributed in the literature to 'physicists, meteorologists, chemists, biologists, economists, psychologists and computer scientists (who) worked across their disciplines to develop new theories of systems' during the twentieth century [1]. This includes the work of Maturana and Varela [2] on cybernetics; Prigogine and Stengers [3] and Prigogine [4] on dissipative structures. To illustrate the relevance of integrating different complexity perspectives to fulfil the purpose of this chapter, the author attempts to simplify the works of key authoritative theorists. Key elements that characterise a CAS are introduced next [5–7] They form useful models of the types of social interactions between professionals looking to implement change [8–10] (Box 1)

Human agents in a CAS are observed operating according to their own internal rules in response to their environment. CAS theories provide a useful tool for interpreting responses and behaviours. It is suggested that this approach enables insight into how, by applying CAS principles, those operating within networks communicate with each other to determine actions that govern their most relevant concerns. Insight, provided by thinking of primary care as a CAS, can prove significant in terms of interpreting what is "going on" in response to change instigated by policy. For instance, a CAS framework will help you answer the following questions:

What impact has a recent policy change had on you and your team/organisation?

What would you like to happen?

How can you make it happen?

The core argument is that assumptions embedded in the rational and mechanistic approach to policy may no longer be valid and that in a networked, complex and unpredictable environment alternative approaches may need to be considered. Policy initiatives typically combine as part of a larger programme for improving healthcare systems, the context is complex, multifaceted and multidisciplinary in nature. A CAS conceptual framework will be developed further, highlighting some of the issues that may take time to resolve.

29.2 What Is Primary Care?

In the preceding section, the reader was introduced to the theories of CAS upon which this chapter is based. The purpose of the current section is to provide an indication of the main characteristic fea-

B. Ellis, M.A., Ph.D., F.H.E.A., M.B.C.S. C.I.T.P. (✉)
School of Health, University of Central Lancashire (UCLan), Brook Building 242, Preston, PR1 2HE, UK
e-mail: BSEllis@uclan.ac.uk

J.P. Sturmberg and C.M. Martin (eds.), *Handbook of Systems and Complexity in Health*,
DOI 10.1007/978-1-4614-4998-0_29, © Springer Science+Business Media New York 2013

Box 1. Key Elements That Characterise a CAS

CAS Element: Multiple components, which interact with the environment, some may represent different world-views. Such systems are open systems that can be understood by observing, as a participant, and appreciating interrelated relationships, rich interaction, feedback and behaviour among components.

CAS Element: Self-organising networks; influence is exercised both by the system on the components and by the components on the system, termed mutual causation. The pervasive nature of non-linear, interlinked interactions can be observed in the patterning of behaviours that emerge in response to change, which cannot be predicted by studying the elements of the system. There may be no central direction. Small inputs may have large effects and vice versa.

CAS Element: Co evolution and system adaptation; complex systems have a history which co-creates the present. Non-linear interactions generate new properties, known as emergent behaviours of the system. Associated principles acknowledge that there is a need to respect ecologies by avoiding disturbance of natural systems with major change, and allowing time for properties to emerge.

tures of primary care. The approach taken calls for a deeper understanding of the complex organisational and policy context of an increasingly important primary care sector. Quality improvement policy is discussed within this context. The section concludes with a discussion of the issues that influence management approaches to quality improvement programmes.

29.2.1 Definitions of Primary Care

Definitions of primary care have been described as either descriptive or normative depending on the purpose they serve [11]. A normative approach incorporates a commitment towards equitable access to care, the protection and promotion of health and on broad intersectoral collaboration supported by appropriate technology when dealing with community problems. Starfield [12] attempts to associate a descriptive purpose more with the content and the range of primary care services including its integrative function within healthcare systems, in both organising and rationing resources. These two definitions can be conceptualised through the concepts of primary health care and primary care. Keleher [13] provides a distinction between these two terms:

> Primary healthcare incorporates personal care with health promotion, the prevention of illness and community development. The philosophy of primary care includes the interconnecting principles of equity, access, empowerment, self-determination and intersectoral collaboration. It encompasses an understanding of the social, economic, cultural and political determinants of health.
>
> Primary care is more clinically focused, and can be considered a sub-component of the broader primary care system. Primary care is considered healthcare provided by a medical professional, which is a patient's first point of entry into the health system. Primary care is practiced widely by nursing and allied health, but predominately in general practice [13].

Primary care and general practice can be considered as organisational features of primary care service delivery. A characteristic of primary care is that patients present with the problems that are important to them. In general terms, primary care can be described as supporting an integrated service delivery approach that may include social care. Each primary care organisational representation represents complex perspectives that stem from unique local factors. For instance, in socio-economic terms primary care service provision is generally centrally located to a population, which often encourages the close proximity of pharmacists, opticians and dentists. Primary care is a function that benefits people and is essential to healthcare systems, emphasising the importance of information to underpin quality.

Towards the end of the twentieth century, there was a strategic shift in political priorities affecting health system organisation and delivery. A need

Box 2. Summary Characteristics of Primary Care

Principle	Meaning
Primary healthcare is essential healthcare that forms an integral part of the overall social and economic development of a community	It encompasses the full range of health problems as presented by the population; implies comprehensive "cradle to grave" approach;
Primary care is typically the main gatekeeper to accessing a range of care provision	Usually first contact, and if appropriate, referral to other sectors of the healthcare system, which may include social care
Primary care respects contractual relationships	Professionally-based, networked service delivery structures
Primary care acknowledges the potential of technology	IT infrastructure developed to support organisational features of service delivery; information sharing, for example on population-health targets

was claimed for creativity in the delivery of choice, diversity and new models of care within patient-led healthcare services. The strategic emphasis changed from the larger hospital to an enhanced role for primary or community-based delivery of services, with a balance of cost-effective preventive and curative programs. General Practitioners have come to work less in single handed practices, more frequently in partnership with other GPs and increasingly in cooperation with other healthcare professionals and practices (Box 2).

The use of technology, particularly electronic health records with an emphasis on informatics, has become essential in supporting the strategic drive for quality and efficiency within primary care. Information flows, feedback and co-evolution forms the essence of processes in primary care systems, their iterative patterning co-creates coherent behaviours and outcomes. For example, in terms of communicating details about registered populations, and providing evidence of achieving population-based health targets [14]. Assuming that change can occur in all of the interacting components of the healthcare system has implications for health service research. Thinking from a CAS perspective suggests the benefit of considering both the linear and non-linear features of the system.

29.2.2 Definition of Primary Care Informatics

There are fundamental questions to answer in response to the theme of this chapter. First, what is health informatics? There are many interpretations, each representing a different perspective. Attempts to define health informatics include 'The knowledge, skills and tools which enable information to be collected, managed, used and shared to support the delivery of healthcare and promote health.' [15] Health informatics is of relevance to all those who generate, retrieve and use information and technology to support health care [16]. The term, primary care informatics, used throughout this chapter follows the definition proposed by de Lusignan [17]:

> The scientific study of data, information and knowledge, and how they can be modelled, processed or harnessed to promote health and develop patient-centred primary medical care. Its methods reflect the bio psychosocial model of primary healthcare and the longitudinal relationships between patients and professionals. Its context is one in which patients present with unstructured problems to specially trained primary care professionals who adopt a heuristic approach to decision making within the consultation. [17]

29.3 Understanding Primary Care as a Complex Adaptive System

CAS offers new insights into many aspects of primary care. For instance, primary care organisations have internal processes that allow them to survive changes to the environment, but they remain recognisable as the system of primary care. A CAS perspective acknowledges primary care structures and move towards semi-autonomous networked organisations. The notion of structure is used in the literature in many different ways, in the context of this chapter it refers to the patterns of interaction in the system [8]. 'Complex systems are systems in process, systems that constantly evolve and unfold over time' [18]. This has implications to primary care strategists in terms of the empowerment and responsibilities of local communities, which can be thought of as ecosystems that place an emphasis on a holistic 'whole system' management perspective. This section will argue that there are limits to a reductionist focus, which merely reduce primary care to a limited number of measurable objectives, which can be captured using technology.

Many public administration systems, which include health service organisations, are attributed to a Weberian model and the logic of materialism [19, 20]. For Weber there exist prescriptive rules, regulations and structures that either predetermine, or assist in discovering optimum methods using a rationalist approach. A reductionist focus emphasises performance indicators, task effectiveness and goal attainment criteria. This approach is claimed to provide potential to reduce primary care to less than the sum of its parts [21]. It is important that Friedlander and Pickle (1968) show that organisations that perform well on a criterion preferred by one constituency tend to do poorly on a criterion favoured by another. There is no guarantee that improvements in the overall governance of quality health care will be ensured only that limited evidence is available to stimulate decision-making [21, 22].

Primary care organisations consist of a democratic collaboration of stakeholders that may include multidisciplinary primary health care teams (PHCTs), other healthcare workers, support staff, social services and lay representatives; and operate within various internal structures. Multidisciplinary representation within primary care does not commonly compete on equal terms, either in terms of professional values, or upon contractual arrangements; this suggests a need to consider the dynamics of a collaborative approach to implementing policy changes such as quality improvement programmes. Primary care systems that support quality improvement provide:

- A focus on the extended multidisciplinary PHCT as the prime means of delivery
- Systematic use of electronic health records to improve communication between professionals and support cohesive care
- Greater recognition of users' views concerning quality health care

These strands reflect principles outlined in the WHO Declaration on primary care (WHO, 1978)—a multisectoral approach, with stakeholder involvement, supported by appropriate technology [23]. These principles underpin ways of working across primary care that ensures the integration of quality improvement, working in partnership to meet agreed standards while remaining a local responsibility enabled by developments in primary care informatics.

In response to being perturbed by the introduction of new policy there may be some fusion of ideas aimed at implementing arrangements locally. Each individual may shed some of their existing ideology in conjunction with others to establish a joint attempt at responding to flux and change that may include:

- Improving efficiency and effectiveness by reducing bureaucracy, disseminating good practice, ideas and innovations.
- Shifting focus so that quality becomes the central driving force behind decision-making. Developing leadership skills necessary to plan and implement policy on quality.
- Implementation of clinical risk reduction programmes for detecting and openly investigating adverse events, and to learn from them.
- Renewing, from a patient's standpoint, a service accountable to patients, open to the public and shaped by their perspectives.

The dynamics of policy changes that support quality improvement in primary care are typically attributable to financial incentives, self-assessment and focused decision-making processes. Moreover, a participative, multi-stakeholder, evidence-based approach to service delivery, supported by appropriate technology, claim to drive these dynamics. Such an interactive and relational approach reflects primary care inputs that influence socioeconomic, technical and political activities which link to produce specific outputs that include products and services [22]. Linking networked care services via data transmission using the Internet, increases the possibilities for informing effective policy decision-making [24]. This thinking can be attributed to efforts that seek to improve the symmetry between the information available to the public, healthcare organisations and government. For instance, primary care professionals may have increased powers of self-regulation through self-assessment, whilst healthcare organisations may claim a greater role in assessing and monitoring the performance of local primary care services. At the same time, governments appear to be seeking to standardise practice across the primary care sector. Calnan [25] suggests that taking the views and opinions of various stakeholders into account through methods such as increased participation in the policy-making process, may serve as a means of democratising health services, thus making those working in health services and the state more accountable to the general populace. The new requirement for primary care clinicians to be answerable to colleagues in their practice and local populations is central to the notion of quality improvement within primary care.

Thinking from a CAS perspective offers insight into the subjective and socially constructed nature of primary care. It enables a focus on the interaction of key actors with other components in the system. For instance, a CAS conceptual framework offers the potential to understand and explain the meanings attributable to the actions of key actors and interaction with organisational relations. CAS theorists claim that such interactions are capable of producing unpredictable capabilities that self-regulate and adjust. Thinking about primary care as a CAS at this level emphasises the importance of combining positivist and ontologically objective approaches to primary care with approaches that emphasise alternative perspectives. 'It is not necessary to favour any paradigm over another because CAS models and ordinary causal models are complements, not rivals' [26]. Thus, it is proposed that primary care can be thought of as a negotiated balance between imperfectly aligned and sometimes conflicting goals within a CAS. The emergent properties that arise from micro- and macro-adaptations reflect the primary care community's interaction with the wider environment.

29.4 Quality Improvement

Quality improvement happens at the interface between policies, the rich interaction of those implementing clinical governance and the development of primary care informatics [27]. Primary care informatics enables measurement of the quality of health care. Thinking from a CAS perspective allows healthcare workers to adopt radically new ways of working that embrace, in the first instance, facets of the governance of quality improvement that include:

- *Continuous improvement*: Requiring those working at the grass roots of patient care to use quantitative data to support the critical examination of the clinical processes in which they work as part of the continuous quality development cycle.
- *Performance management*: Creating an environment where healthcare workers acknowledge and support the right of patients and taxpayers to have access to meaningful performance data, both to justify continuing investment, and to inform individual patient choice [22]. This element assumes that organisational culture will foster new perspectives and insights into problem-solving solutions and innovations.

Health informatic programmes used in primary care are able to support change to take advantage of new ways of working—clinical teams can integrate the use of information and develop information systems within a framework of a broad process of continuous quality improvement [22]. Practical informatic tools and quality methods underpin intelligent decisions; these include decisions that improve system-wide performance (organisational learning). The approach integrates technology throughout care and business processes. This enables information flows throughout primary care practice for maximum and constant learning.

Table 29.1 Conceptualisation of various governance models, themes and influence [27]

Theme	State model	Corporate model	Market model	Network model	Influence on quality
Accountability	Rationality; Rules and procedures; Inputs > outputs; standards; Professional expertise	Leadership direction Long-term planning Budget control	Maintain financial balance through market forces	Non hierarchical, based on relations and interdependencies; Governance not government Self assessment	Earned autonomy Based on relationships that evolve
Collectively binding decision-making	Public interest; Materialism; Economic performance Social inclusion pressures	Vision Harmonisation Staff focus	Fails to consider social benefits, or levy the full cost of market activity	Taxation funded but jointly provided by mix of separate private and public providers	Mutual recognition of common or complementary strategic agenda
Develop communication systems and feedback	Procedural factors; Maximise system's purpose;	Movement of parts within the system; Need to develop infrastructure	Need for high standard of information to support exchange of transactions;	Based on reciprocal information requirements - systems evolve	Established through coalitions, reducing asymmetry of information
Cultural change	Not specifically addressed	Shared national and local ownership	Based on private sector ideals;	Relationship building	Open trusting relationships
Build capacity and capability	Not specifically addressed	Ensure skills available in the workforce to meet new challenges	Based on assumptions to equip systems to support quality improvement	Emphasis on education and training, knowledge transfer	Flexible, based on knowledge transfer
Mechanistic tools	Not specifically addressed	Structured observation,	Break NHS products and services down into units of analysis	Not specifically addressed	Clinical audit; Quality assurance
Competitive strategies	Rationality	Incentives	Market forces	Self organising	Competition for resources

Reproduced with kind permission of the British Computer Society Journal of Informatics in Primary Care and Radcliffe Medical Press

Table 29.2 Typology of public sector management models [28]—an adaptation based on Ferlie [29]

New Public Management 1	The dominant form in the 1980s characterised as emphasising value for money, finance, strong management, audit, increased responsiveness to consumers, shift in power from professionals, empowering of the less bureaucratic and more entrepreneurial management. This combined with new forms of corporate governance
New Public Management 2	Characterised as developing more elaborate markets, move to management by contract, split between strategic core and operational activity, down-sizing, move from "command and control" to management by influence
New Public Management 3	Associated with the "excellence" movement that emphasises the cultural approach to managing the "learning organisation"
New Public Management 4	Represents a fusion between private and public sector management—focused on service quality, users and empowerment; joined-up initiatives and collaboration

29.5 Governance of Quality Improvement

Models of governance provide the framework within which quality improvement programmes operate [27]. Table 29.1 illustrates a conceptualisation of governance models, themes and influence on quality.

The network model accommodates the characteristics of earlier formulations, and both positive and negative influences. The network model supports the trend towards integrated organisations, evolution of primary care from single hander, to group and locality-based practice with federated models of practice.

29.6 Complex Adaptive Systems Theory: Application to Management Theory

In order to improve understanding as to what happens when policy is introduced in a human system, attention is drawn to the role and approach of management in a CAS. Towards the end of the 1980s and early 1990s, progress on many governments quality improvement in healthcare priorities was linked to incentives that emphasised primary care. In keeping with the WHO's conceptualisation [23], and in general terms, primary care can be thought of as reflecting and evolving from the economic conditions, sociocultural and political characteristics of local communities. If it is possible to accept this suggestion, the relationship between primary care and governments gains significance. Such thinking is based on addressing the main health issues of a population, organisational features of service delivery and experience. International influence on political reform can present a challenge to administrative and professional cultures that dominate many health services and moves to stimulate accountability. A typology that describes development models of managerialism is presented in Table 29.2

The implications of developments in New Public Management (NPM) shown in Table 29.2, offer potential to help understand what happens in a CAS environment. For example, each stage can be thought of as providing a separate, but potentially overlapping model.

A current policy innovation within governments is to construct networks of organisations to manage healthcare services and to think of health care as a system of interrelated systems that interact in a non-linear way. Advocates of complexity-based network management theories, exemplified by Blockson and Van Buren [30] p.64 claim that 'Societal issues can be best addressed through multi-sectoral collaboration'. A CAS management approach facilitates a move away from linear analysis to an appreciation of the importance of relationships, providing tools for interpreting responses, behaviours and iterative patterns of order among system components. Predictability is limited, i.e. the future cannot be predicted with any degree of certainty as it is being created by the interactions among components of the system and of the system on the components. Networked information flows that are continually updated support a move toward 'real-time' monitoring and learning as a strategy for change. Three types of learning have been identified [31]:

- Single-loop learning leads to incremental improvements to existing practice through the use of a feedback loop, for example clinical audit where existing practice is compared to an agreed standard.
- Double-loop learning occurs when organisations reflect upon behaviours and their simple rules. Plesk [32] identifies three types of simple rules for human systems: general direction pointing; system prohibition, i.e., setting boundaries and resource or permission providing.
- Meta-learning reflects an organisation's attempts to learn and improve its capacity to learn.

In summary, the strategic management approaches described allow a conceptualisation that for quality improvement programmes to be effective, management approaches need to take into account the context of implementation. Each component cannot be understood in isolation, each adapt by learning to survive and can exhibit linear and non-linear interactions. The overall aim is the creation of a healthcare system that can learn for itself, continuously, guided by democratically agreed standards and priorities.

29.7 Systems Based Thinking and Its Relation to Learning Theories

Open systems theories take into account the dynamic whole of the organism, its interaction with its environment and permeable boundaries. Second-order systems based thinking moves away from simple objective observation to understand humans as participants in systems that allows for the flow of energy (motivation, information and innovation) and networked interactions [16]. Learning theories identify the role of feedback in sustaining and improving human performance at work that involves single loop and double loop learning, associated with proactively challenging and influencing a range of different or conflicting perspectives [33, 34]. Similarly, learning theories suggest exploratory models of problem appreciation and problem solving by a plurality of stakeholders that reflect practical day-to-day concerns, relevant to participants' daily working lives and activities. Learners go through a cycle in which they acquire knowledge, assimilate, experiment and then normalise the learning into their daily work informed by feedback loops. Contemporary learning models manifest an understanding of the need for skills and knowledge to be embedded in experience, and allow reflection on that experience to create new meaning and enduring changes in behaviour.

It is recommended that effective health informatics learning should be set in a context of multi- and inter-professional working. This approach provides a wider perspective for learners, but also highlights the importance of inter-professional cooperation in the development, clinical management and monitoring of all health information.

29.8 CAS Thinking and Its Relation to Research

Due to the complex nature of primary care systems, new approaches to research are necessary to reflect the context in which services operate. From a CAS perspective, there are limits to rational approaches and broader alternatives may need to be considered in the design of research strategies. For instance, the heterogeneous factors of primary care suggest the importance of context. Approaches that seek to emphasise generalisability are unlikely to provide sufficient description of specific local organisation and context. CAS provides a framework, which considers varied research methods and supports a move towards greater integration of research, service development and quality health care.

29.9 Conclusion and Recommendations

Thinking from a CAS perspective has an important role to play in improving the public's confidence in the delivery and quality of primary healthcare. Informatics complements this role based on the belief that the best decisions are based on the best information that flows throughout networked systems of healthcare. Educational programmes need to be designed to support the delivery of information technology enabled change and meet the future information requirements of patients, public and healthcare service. For instance, the trust that patients and carers place in healthcare professionals can be supported by knowledge that treatment is evidence-based and effective. The challenge is to develop research strategies and systems in such a way as to embrace the multifaceted and multidisciplinary nature of primary care.

- Current health service organisational explanatory models accommodate the insights provided by the principles of complexity and CAS: allowing us to think of primary care delivery as emergent properties of iterative, recursive interactions.
- This helps us to comprehend how evidence, experience and behaviours, found in primary care systems that interact in a non-linear way, may be synthesised and shared across networks, underpinned by developments in primary care informatics, to establish mutual understanding.
- Networked information flows that are continuously updated support a move toward 'real-time' monitoring, reflective practice and learning as a strategy for change.

Quotes

1. 'physicists, meteorologists, chemists, biologists, economists, psychologists and computer scientists (who) worked across their disciplines to develop new theories of systems'. [1] p .216
2. 'Primary healthcare incorporates personal care with health promotion, the prevention of illness and community development. The philosophy of primary care includes the interconnecting principles of equity, access, empowerment, self-determination and intersectoral collaboration. It encompasses an understanding of the social, economic, cultural and political determinants of health.

   Primary care is more clinically focused, and can be considered a sub-component of the broader primary care system. Primary care is considered healthcare provided by a medical professional, which is a patient's first point of entry into the health system. Primary care is practiced widely by nursing and allied health, but predominately in general practice'. [13] p. 57
3. 'Complex systems are systems in process, systems that constantly evolve and unfold over time'. [18] p. 107
4. 'The knowledge, skills and tools which enable information to be collected, managed, used and shared to support the delivery of healthcare and promote health'. [15] p. 63
5. 'The scientific study of data, information and knowledge, and how they can be modelled, processed or harnessed to promote health and develop patient-cantered primary medical care. Its methods reflect the bio psychosocial model of primary healthcare and the longitudinal relationships between patients and professionals. Its context is one in which patients present with unstructured problems to specially trained primary care professionals who adopt a heuristic approach to decision making within the consultation'. [17] p. 304
6. 'Societal issues can be best addressed through multi-sectoral collaboration'. [30] p. 64

Acknowledgements I would like to thank Editors—Joachim P. Sturmberg and Carmel M. Martin for kindly inviting me to contribute a book chapter to the Handbook on Complexity in Health, part of Springer Science and Business Media.

Also grateful thanks to the British Computer Society Journal of Informatics in Primary Care and Radcliffe Medical Press for kindly agreeing for me to reproduce Table 29.1 Conceptualisation of various governance models, themes and influence [27].

Finally, I would like to thank my family, David, Christian and Kerry, for their patience, encouragement and support during the writing of this chapter.

References

1. Stacey RD. Strategic management and organisational dynamics the challenge of Complexity, Fourth edn. Harlow: Prentice Hall; 2003.
2. Maturana HR, Varela FJ. Autopiesis and cognition: The realization of the living. Kluvwer: Kluvwer; 1979.

3. Prigogine I, Stengers I. Order out of chaos: man's new dialogue with nature. New York: Bantam Books; 1984.
4. Prigogine I. The end of certainty: time, chaos and the new laws of nature. New York: The Free Press; 1997.
5. Reynolds CW. Flocks, herds and schools: a distributed behaviour model. Comput Graph. 1987;21(4):25.
6. Kauffman SA. Origins of order: self organisation and selection in evolution. Oxford: Oxford University Press; 1993.
7. Gell-Mann M. The Quark and the Jaguar. New York: Freeman; 1994.
8. Cilliers P. Complexity and post-modernism. London: Routledge; 1998.
9. Anderson P. Complexity theory and organization science. Organizat Sci. 1999;10:216.
10. Ellis B. Complexity in practice: understanding primary care as a complex adaptive system. Inform Primary Care. 2011;18(2):135–40.
11. Boerma WGW. Coordination and integration in European primary care. Milton Keynes, Bucks: Open University Press, McGraw-Hill Education; 2006.
12. Starfield B. Primary care: balancing health needs, services, and technology. Oxford: Oxford University Press; 1998.
13. Keleher H. Why primary care offers a more comprehensive approach to tackling health inequalities than primary care. Aust J Primary Care. 2001;7:57.
14. Fogarty L (1997) Primary care informatics development—One view through the miasma. J Inform Primary Care, Primary Health Care Specialist Group of the British Computer Society.
15. NHS Information Authority. Learning to Manage Health Information: Moving Ahead. London: HMSO; 2002. p. 63.
16. Ellis B, Howard J. Governance, education and learning to manage health information. Clinical Governance An International Journal, 2011:16(4)
17. de Lusignan S. What is primary care informatics? J Am Med Inform Assoc. 2003;10:304.
18. Arthur WB. Complexity and the economy. Science. 1999;284:107.
19. Weber M. The theory of social and economic organisation. New York: Free Press; 1947.
20. Turchin V. Cybernetics and philosophy. Salinas, California: Intersystems; 1990. p. 61.
21. Williams PH, de Lusignan S. Does a higher 'quality points' score mean better care in stroke? An audit of general practice medical records. Inform Primary Care. 2006;14:29.
22. Cowley C, Daws L, Ellis B. Bridging the gap. Inform Primary Care. 2003;4(3):207.
23. World Health Organisation (1978) Alma-Ata—Primary Health Care, Health For All, Series. No. 1
24. Urry J. Global complexity. Cambridge: Polity Press; 2003.
25. Calnan M. Why take into account patient views about healthcare? Br J Health Care Manage. 1996;2(6):328.
26. Anderson P. Complexity theory and organization science. Organization Sci. 1999;10(3):216.
27. Ellis B. Complexity in practice: understanding primary care as a complex adaptive system. J Inform Primary Care. 2011;18(2):135–40.
28. Ellis B (2008) Managing governance programmes in primary care: lessons from case studies of the implementation of clinical governance in two primary care trusts, University of Central Lancashire (UCLan).
29. Ferlie E (1994) Characterising the 'New Public Management', Paper presented at the Annual Conference of the British Academy of Management, September, 12–14.
30. Blockson L. C. a. v. B. H. J. (1999) Strategic Alliances among Different Institutions: An argument for Multisector Collaboration in Addressing Societal Issues. Independent Sector Working papers, Washington DC
31. Agyris C, Schön D. Organisational learning: a theory of action perspective. New York: McGraw Hill; 1978.
32. Plsek P. Creativity, innovation and quality. Milwaukee: ASQ Quality Press; 1997.
33. Pendleton D. Professional development in general practice: problems, puzzles and paradigms. Br J General Pract. 1995;45:377.
34. Senge PM. The fifth discipline: the art and practice of the learning organisation. New York: Doubleday; 1990.

30 Real-Time Analytics and Quality of Care

Liam Peyton, Alain Mouttham, Khaled A. Ali, Aladdin Baarah, and Hussein T. Mouftah

30.1 Introduction

Patient care, in a medical setting, is complex both in terms of assessing the naturally occurring phenomena that are presented for care and in terms of coordinating resources, information, and expertise across diverse organizations to deliver care. A large amount of information must be accessed, processed and communicated in a timely manner in order to be effective and efficient in the provision and management of care. Information technology, especially wireless, online information technology, has transformed other information-intensive industries through the use of real-time analytics to provide an order of magnitude improvement in efficiency and effectiveness. To date, however, the adoption of such technology in health care has been slow and the results have been mixed. In this chapter, we provide insights into some of the information processing complexities that make patient care challenging while introducing emerging technology and techniques behind real-time analytics that are being leveraged to overcome such challenges. To illustrate the approach, we present a scenario constructed from a series of case studies and prototype systems that have been developed in collaboration with cardiac care facilities at a number of hospitals in Canada.

L. Peyton (✉) • A. Mouttham • K.A. Ali • A. Baarah • H.T. Mouftah
School of Electrical Engineering and Computer Science, University of Ottawa, 800 King Edward Avenue, Ottawa, ON, Canada, K1N 6N5
e-mail: lpeyton@uottawa.ca; amouttham@discover.uottawa.ca; kali@eecs.uottawa.ca; aladdin_baarah@yahoo.com; mouftah@eecs.uottawa.ca

30.1.1 Complexity in Patient Monitoring

Patient monitoring is a continuous process of assessing naturally occurring phenomena that are presented for care. This assessment is usually communicated in the form of *measurements*. In health applications, monitoring patient's health requires monitoring several physiological parameters such as electrocardiography (ECG), electroencephalography (EEG), blood pressure level, etc. Many of these parameters are currently measured in an invasive way. The recent emergence of wireless sensor networks in the medical domain allows for non-invasive measurement of such physiological variables and their communication via wireless transmission. These wireless sensor networks on a patient's body are known as a Wireless Body Area Sensor Network (WBASN). Much of the sensed medical information will have different urgency levels and timing constraints that require a sensor to:

- Sense a physiological phenomenon (e.g. ECG)
- Process the sensed value
- Communicate the processed value to a central node, in a bounded and guaranteed time

J.P. Sturmberg and C.M. Martin (eds.), *Handbook of Systems and Complexity in Health*,
DOI 10.1007/978-1-4614-4998-0_30, © Springer Science+Business Media New York 2013

The importance of differentiating traffic in wireless sensor networks is growing, and guaranteeing different Quality of Service (QoS) levels is considered a key challenge for research on wireless sensor networks [1, 2]. Many new protocols are currently being proposed and tested for wireless sensor networks, but most of them focus on energy efficiency, and not on throughput and rejection rate constraints [3]. Therefore, protocols that maximize network throughput, minimize packet rejection rate and differentiate network traffic are desirable.

During the last few years, significant progress has been made in development of wearable or on-body wireless sensors [4–9]. Similarly, smartphones are frequently used as a Personal Heath Monitoring System (PHMS) and as base stations in a wireless sensor network, as described in [10, 11]. Finally, advanced machine learning techniques have been developed for data stream mining in order to process the continuous, massive, temporally ordered, and fast changing flow of data coming from wireless sensors [12–15]. There is a wide variation in patient size, shape and physiology, so the *context* must be *personalized* in order to classify what are normal measurements for a particular patient. Further, the new wireless sensors allow the patient to be mobile, so the *context* must be further *adapted* in order to classify what are normal measures for a particular patient while lying down, standing, walking or running.

30.1.2 Complexity in Coordination of Care

The purpose of any healthcare system is to coordinate resources, information, and expertise across diverse organizations to deliver care in as effective and efficient manner as possible. The 2001 Institute of Medicine (IOM) study 'Crossing the Quality Chasm: A New Healthcare System for the 21st Century' analyses the challenges and opportunities to provide consistent, high quality care to all who need it [16]. Key challenges include communication, quality improvement and the need for integration of multiple processes. There is an opportunity to leverage information technology to facilitate and improve delivery of care through integrated services [17]. Currently, a common source of miscommunication and medical errors are the handoffs that occur between different services. These errors are due to multiple factors including poorly structured information and ineffective communication exchanges [18, 19].

The need for healthcare systems to provide efficient, effective care has put an emphasis on performance management, which measures and communicates outcomes in the form of *key performance indicators* [20]. It has been estimated that a 500-bed hospital loses approximately four million dollars a year because of communication and other integration inefficiencies [16]. However, while performance management can measure outcomes and suggest policy and protocol for achieving efficiency it does not drive the actual processes [21]. In fact a key issue is that the development of healthcare analytics and the development of IT to support clinical processes have occurred in a divergent manner. The result of that divergence is that analytics may tell us that something is wrong (i.e. a bad outcome) but it does not tell us where the issues lie and more importantly how to fix the issues to improve performance. Studies have shown that gaps exist between healthcare system analytics and the front line clinical processes where care delivery takes place [22].

Complex event processing (CEP) [23] is an emerging technology and an efficient approach for processing high volumes of continuous streams of events in real time. The main goal of CEP is to process events from distributed sources in order to *identify meaningful complex events* across these sources by analysing and correlating basic events [24]. Another goal is to correlate these events to detect and respond in real time to any business critical situation [25]. Furthermore, CEP is responsible for event pattern matching, that is, the CEP engine recognizes in a cloud of events those patterns that are significant [26].

30.1.3 Complexity in Decision Support for Quality of Care

Improvements in patient monitoring and coordination of care help ensure that the information needed for patient care is available in a timely fashion to those who can act upon it, but it leaves us with a final complexity. How can one determine which course of action is the best one to take given *competing priorities* and *uncertainty* or *incomplete information* about the *impact of one's actions*? In a complex healthcare system with thousands of patients being treated simultaneously, how and when should one decide to allocate resources?

In a hospital, there are two kinds of processes:

- Clinical processes which describe what level of care should be provided to the patient day after day. An example would be the clinical pathway for handling acute coronary syndrome.
- Operational processes, which support clinical processes. Examples would be admissions/discharges/transfers, scheduling, bed management, order entry, and patient flow management.

Decision support can be provided for automating and facilitating the operational processes in support of specific clinical processes. For instance, real-time analytics and decision support for the flow management of patients with acute coronary syndrome.

Using real-time analytics, information about how well competing priorities are handled can be measured and communicated using key performance indicators that measure events as they occur. A real-time dashboard could, for example display the current status of key performance indicators and colour-code them green if they are within an acceptable range, red if they are not, and yellow if they are trending towards red without corrective action. Based on medical guidelines and in-depth processing and analysis of historical data, possible actions can be provided in context when handling particular situations. Real-time simulations can provide feedback on expected outcomes of each action, by combining data about the current situation with predictions of how the situation will evolve by simulating event trends into the future based on historical data.

Real-time analytics have the potential to improve quality of care by collecting, processing and communicating event data from diverse sources in real time to provide context, communicate situations, and support decisions with guidelines and predicted outcomes. In order to support that, information technology needs to provide:

- Analytics that convert natural phenomena into measurements and events that are personalized and adapted
- Communication protocols that ensure QoS by filtering, prioritizing and ensuring minimum response times for communication of measurements and events
- Real-time processing and integration of large volumes of disparate event data from different sources into meaningful events that can be aggregated into key performance indicators
- Real-time analytics that can provide dashboards for monitoring patients and overall key performance indicators
- Predictive analytics and decision support guidelines based on combining current data with historical data and event trend simulations

30.2 Background

In this section, we give a brief overview of background material related to patient monitoring, wireless networks, complex event processing, performance measurement and business process management.

30.2.1 Patient Monitoring

Health monitoring is a generic function in applications like chronic disease management, emergency response, homecare for the elderly and palliative care. This is a critical foundation for telemedicine. Current health monitoring systems are based on portable recorders (e.g. a Holter monitor, to continuously record the heart rhythms of cardiac patients). They record health data locally to the device and upload it to an application server after several days. The server then does the off-line data analysis. These systems are

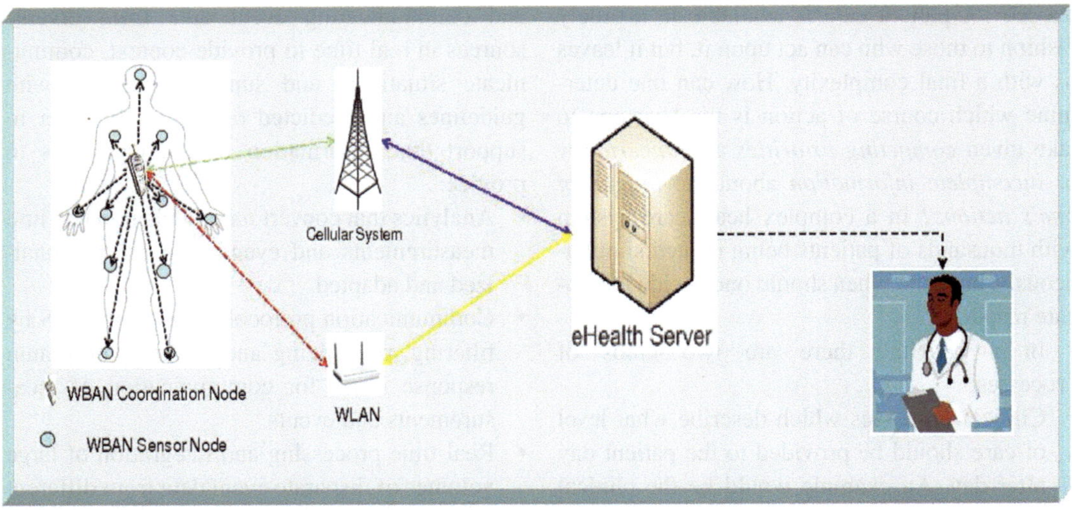

Fig. 30.1 Heterogeneous wireless medical networks

now being replaced by a new generation of continuous health monitoring systems based on wireless sensors located on the patient's body or clothing.

During the last few years, significant progress has been made in wearable or on-body wireless sensors [4–9]. Similarly, smartphones are frequently used as a base station in a wireless sensor network as described in [10, 11]. Finally, advanced machine learning techniques have been developed for data stream mining in order to process the continuous, massive, temporally ordered, and fast changing flow of data coming from wireless sensors [12–15, 27].

30.2.2 Wireless Networks

Currently, patient monitoring is limited to hospitals and elderly homecare locations which induce an extra cost for healthcare services. Continuous advancement in the field of wireless networks introduces an economically feasible solution for continuous patient monitoring both in and outside of such locations.

Wireless medical networks are heterogeneous systems that are composed of different networking technologies such as wireless body area networks (WBANs), cellular networks, Wireless Local Area Networks (WLANs), and the Internet (see Fig. 30.1). QoS is an issue in such heterogeneous networks to ensure that information is communicated consistently, effectively and in a reliable manner to ensure that messages are delivered in a timely fashion. The level of QoS provided to medical traffic can be measured at different levels:

- Access level where the several measured physiological parameters are sampled and communicated to a central control unit of the PHMS
- Mobile phone, transporting level where central control unit of the WBAN communicates with one or more of the transporting networks such as cellular systems, WLANs
- The Internet

Well-designed QoS mechanisms are required for medical traffic to remote healthcare locations such as an eHealth server.

The Wireless Personal Area Networks (WPANs) standard has been developed by IEEE [1]. This standard defines the mechanism that can be used by multiple sensor nodes allocated around the human body to communicate together. These sensor nodes can be composed of ECG sensor, temperature sensor, movement sensor, blood pressure sensor, etc. In this standard, sensor nodes are given equal probability accessing the transmission medium to communicate their traffic to the central node of the PHMS. This protocol has been extended in

Ultra Wide Band (UWB) standard to enhance nodal access mechanism to the transmission medium [2]. WPAN and UWB have been adapted by many research communities as the underlying technology of WBAN. The WPANs and UWB standards support two network topologies: star and peer-to-peer. In star topology, sensor nodes communicate with a central coordination node which can function as a gateway to other networking technologies such as Cellular networks, WLAN, Internet, etc. On the other hand, peer-to-peer network topology has no central coordination node. However, sensor nodes cooperate together to achieve a specific task in an ad hoc communication mode. Star network topology is the most suitable networking topology for WBAN.

Two communication modes are defined for the star topology: beacon and non-beacon. In beacon mode, network communications are controlled by the network coordinator which periodically transmits beacon frames for device synchronization and network association control. In non-beacon mode, network nodes randomly contend for the transmission medium access to send data to the coordinator node. The advantage of non-beacon mode is that the node's receiver does not have to regularly power up as in beacon mode to receive the beacon. The disadvantage is that the network coordinator cannot communicate with the node unless it has been invited by the node to communicate.

The WPANs' node channel access mechanism called Medium Access Control (MAC) protocol proposed in [1] has been mathematically analysed in [3]. The analysis considers a star-based network topology for WPAN. The study shows that beacon mode operation is suitable under tight data rate restrictions. It has been proven that the non-beacon operation mode provides the best performance since nodes' receivers are not required to power up for receiving the beacon frame. In [28], a QoS MAC protocol that modifies the original MAC frame structure is proposed. A Distributed Queuing Body Area Network (DQBAN) MAC protocol for healthcare applications is proposed in [29]. DQBAN requires the management of different queues as well as the fuzzy-logic system implementation in every sensor node.

In the above performance studies of the proposed WPAN and UWB systems, there are no clear sensor traffic differentiations. However, in the future wireless medical networks, sensor traffic needs to be classified into different categories to be able to prioritize critical traffic over non-critical traffic. For example, in Sect. 5, we introduce an efficient QoS-based MAC protocol for supporting different classes of traffic of different urgency levels.

30.2.3 Complex Event Processing

Complex event processing (CEP) [23] is an emerging technology and an efficient approach for processing high volumes of continuous streams of events in real time. The main goal of CEP is to process events from distributed sources in order to identify meaningful complex events across these sources by analysing and correlating basic events [24]. Another goal is to correlate these events to detect and respond in real time to any business critical situation [25]. Furthermore, CEP is responsible for event pattern matching, that is, the CEP engine recognizes in a cloud of events those patterns that are significant for the backend systems [26].

Significant research has been undertaken regarding event processing that is focused on processing and detecting complex events based on a continuous stream of radio frequency identification (RFID) events such as in [30]. To infer complex events from the streams of basic events received from distributed sources, the following steps are taken [25]:

- Event filtering of basic events to remove duplicate events
- Correlation of basic events from multiple sources in order to build a view of a higher-level (complex) event
- Complex event creation based on pre-defined rules

Event-driven architecture (EDA) provides the supporting architecture needed to feed basic events into the CEP and distribute the complex events it generates. The CEP is an event consumer that receives basic events from all the sources it is interested in and it is an event producer that propagates all its complex events to interested consumers. Upon receiving a complex event, a consumer may take appropriate actions. Often, events are delivered as messages through the use of message-oriented middleware, such as a message broker [31].

30.2.4 Performance Management and Business Process Management

The desire for healthcare delivery to be more efficient and effective in its delivery has put increased value on performance analytics. Performance analytics can be summarized in the statement "you need to measure what you want to manage" [22]. The premise of analytics is they define how organizational objectives are quantified and measured to assess one's ability to achieve a certain outcome. However, a challenge in implementing analytics is to understand the relationship between operational business processes and the outcomes the analytics measure [32].

Business Process Management (BPM) is defined as "supporting business processes using methods, techniques, and software to design, enact, control, and analyse operational processes involving humans, organizations, applications, documents and other sources of information" [33]. It facilitates the online management and tracking of business processes and their data. Often, BPM is supported by service-oriented architecture (SOA). The event-driven architecture described in this chapter is complementary to SOA. The life cycle of BPM [34] is composed of several repeated steps to enhance the business process, which are:

- Discovery and modelling of business processes
- Simulation and validation of business processes
- Deployment and execution of online business processes
- Process monitoring and performance management
- Improvement of business process by using the results of performance measurements

Business Activity Monitoring (BAM) is often integrated with BPM in order to supply real-time information about the results of business processes [35]. This real-time information is often used to track key performance indicators (KPI) that are defined to measure the performance of business processes in terms of achieving business goals which are critical to health care. Often KPIs are displayed on a dashboard, and integrated with an alerting mechanism in case of a severe difference between the actual value and the expected value for a KPI [20].

30.3 Cardiac Care Scenario for Real-Time Analytics

We have been developing an integrated case study of cardiac care processes, real-time analytics and quality of care through our collaboration with a number of hospitals in the province of Ontario in Canada. We use the following fictitious example scenario of cardiovascular care to articulate the complexities involved in care delivery with respect to patient monitoring, coordination of care, and decision support. The scenario covers two phases: the coordination of care as a patient requiring cardiac care during their stay in the hospital; and the remote monitoring of the patient when they are released from the hospital where there is the possibility of readmission.

Figure 30.2 illustrates the scenario. The stars in the figure indicate the points where performance measures are used to drive real-time analytics. A patient presents with symptoms of a heart attack, and calls 911.[1] An ambulance brings him/her to the Emergency Department of the nearest hospital. The patient has an electrocardiogram (ECG) and a blood test for cardiac markers to differentiate an Acute Myocardial Infarction (AMI) from other medical conditions. The patient can have one of two types of AMI:

- ST-Elevation Myocardial Infarction (STEMI), for which an urgent cardiac re-vascularization is critical
- Non-ST-Elevation Myocardial Infarction

[1] North America emergency call number, 000 in Australia, 112 in many parts of Europe

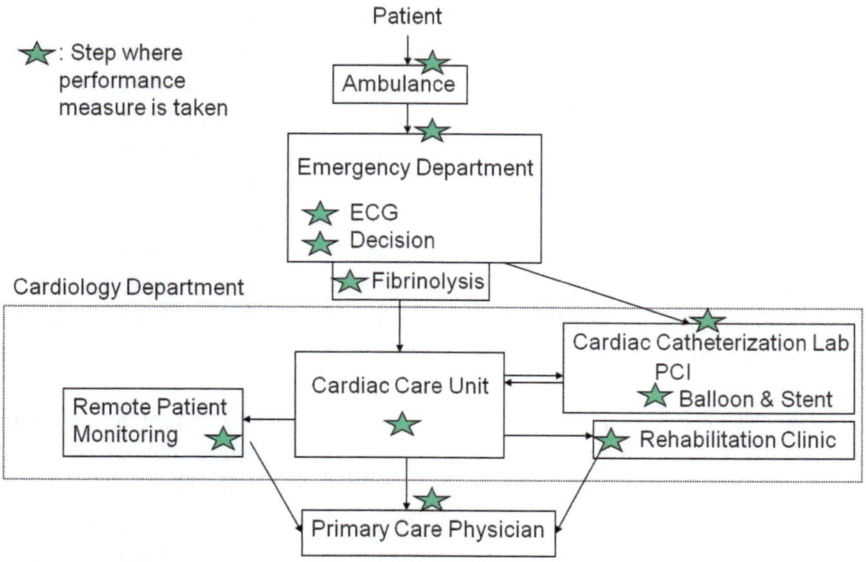

Fig. 30.2 Linking of fine-grained metrics with care processes

If the patient is diagnosed with STEMI there are three possible treatment options:

1. Fibrinolytic therapy: medication to bust the plaque obstructing coronary vessel(s). Should be done within 30 min (Door-to-Delivery time) according to medical guidelines.
2. Percutaneous Coronary Intervention (PCI): done in the Cardiac Catheterization Lab, by introducing a catheter equipped with Balloon and Stent in the coronary vessel(s), in order to enlarge them. Should be done within 90 min (Door-to-Balloon time).
3. Pharmaco-Invasive therapy: a combination of both procedures, with Fibrinolytic therapy within 30 min, followed by a PCI within 24 h.

The patient will require hospitalization in the Cardiac Care Unit. A care plan for follow-up will be developed and the patient is then discharged to the outpatient clinic where they start cardiac rehabilitation. The patient is followed by his/her Primary Care Physician (PCP), the family doctor. The patient may also be remotely monitored at home, following discharge. The patient regularly consults the PCP to evaluate progress, discuss problems and modify medication if required (according to the care plan done before discharge). After several weeks, the patient achieves a stable state, does not experience any more symptoms, and the medication dosage is satisfactory. They are encouraged to keep a regimen for healthy living.

There are several performance measures defined by the Canadian Institute for Health Information and the American Agency for Healthcare Research and Quality that are relevant to our scenario:

- Door-to-Balloon time
- Wait Times (for admission to ED, for admission to acute care, for admission to Clinics)
- Alternate Level of Care days (number of extra days)
- Percent Discharge Summaries transmitted electronically within one day of discharge
- Readmission rate within 30 days

These recommended performance measures tell us what needs to be achieved, but there is poor underlying linkage at the clinical level. They do not provide any details on the level of integration of care processes that take place as the patient moves through the cardiovascular care system. Fine grained metrics of how the various processes integrate are needed to identify bottlenecks in the system.

For example, corresponding to the top part of Fig. 30.2, when a patient with chest pain arrives at ED, an ECG must be taken, and if STEMI is diagnosed and a primary PCI is decided, they must immediately be sent to the Cardiac Catheterization Lab. The end-to-end metric for this pathway is the Door-to-Balloon time (which should be less than 90 min). We can further decompose the Door-to-Balloon time metric as

- Ambulance door-to-arrival in ED (<30 min). This metric requires integration between the ambulance system and the hospital system: not all hospitals can handle a STEMI; therefore a protocol has to be put in place and understood by all parties.
- Arrival in ED-to-ECG (<10 min). The 12-lead ECG could also be done in the ambulance and sent to the ED, therefore reducing even further the time.
- ECG-to-Decision.
- Decision-to-Communication with Cardiac Cath Lab (<10 min).
- Communication-to-arrival in Cardiac Cath Lab (<10 min).
- Cardiac Cath Lab arrival-to-Balloon (<30 min).

In the bottom part of Fig. 30.2, the patient stays in the cardiology department for several days for observation. They are usually attached to a Holter monitor in order to monitor for irregular heartbeats (arrhythmia). There would be significant improvement in quality of life and significant cost savings if patients could be released earlier with a portable monitoring device, but one would need to ensure that irregular heartbeats would be accurately recognized by the portable monitor and that hospital staff would be able to respond appropriately in a timely fashion. Some recommended high-level metrics are mortality rate and number of hospital re-admissions within 1 day, 1 week and 1 month. Some fine-grained metrics for integration of care are: the number of episodes of arrhythmia per day (which drives the decision to readmit or not); and the time in minutes for a nurse to follow up with a phone call after an arrhythmia is detected.

We will start with the second half of the scenario in our analysis. In Sect. 4, we will introduce a portable monitor that uses wireless body sensors to measure: ECG, blood pressure, temperature, heart rate, and has accelerometers that can detect vertical or horizontal body position, and speed. The low-level events from the accelerometers can provide low-level events from which one can infer: lying down (horizontal), sitting or standing (vertical), walking (movement below 5 km/h), running (movement 5–15 km h), and driving (>15 km h, but at rest). The body sensors communicate with a smartphone (attached to the waist of the person or beside the bed). The smartphone includes GPS and has a wireless internet (GPRS) connection to server at the cardiac lab in the hospital.

The complexity of remote patient monitoring will be analysed in Sect. 4, where we look at personalized and adapted analytics for measuring the heart and raising arrhythmia events, and in Sect. 5, where we look at protocols to ensure QoS in the wireless body sensor network that feeds the analytics. Then, we will come back to the first half of the scenario analysis to look at the complexity of coordination of care in a hospital. Section 6 will look at how to link complex event processing with key performance indicators in order to support real-time analytics which are discussed in Sect. 7. Finally, in Sect. 8, the complexity of decision support is analysed as we look at the potential of predictive analytics.

30.4 Personalized and Adaptive Analytics

Figure 30.3 gives an overview of a framework to support remote patient monitoring through the use of a personal health monitoring system within a network of Health Service Providers (HSP). The building blocks of our framework are:

- The sensors used to collect health-related data (e.g. ECG, pulse, blood pressure) and also context data (e.g. accelerometers for movement, thermometers for temperature).
- The smartphone which, together with the sensors, constitutes a Personal Health Monitoring System (PHMS). It also fulfils the role of Base Station for the sensors. Its mobility makes it key for context awareness.
- Different HSP systems which can receive data and event notifications from the smartphone

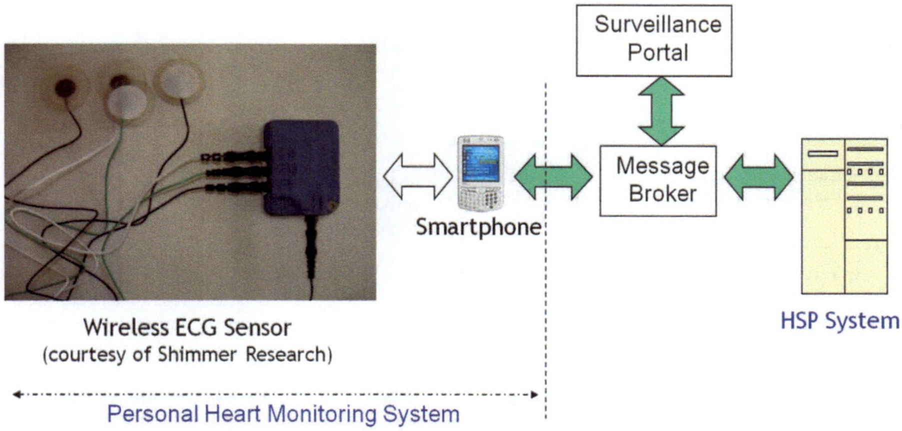

Fig. 30.3 Personal health monitoring framework

so each HSP can respond quickly and effectively as appropriate to the context.

- A Message Broker and Surveillance Portal [36], which are part of the enabling infrastructure for the interoperability between the PHMS and the HSP systems.

The smartphone offers two interfaces:

- A SOA interface for specific service requests that can be made directly by the Health Service Provider for configuration, data collection, sensor management, and communications with the Patient.
- An Event interface based on the Publish–Subscribe model that allows events to be flexibly published and subscribed between the PHMS and HSP systems through the mediation of the Message Broker. Events convey both data and context information. A Surveillance Portal can also receive all events and raise alerts

The PHMS is customized and belongs to one patient. It is the conduit of information between the HSPs and the patient. The PHMS architecture minimizes sensor–smartphone and smartphone–server communication with local intelligence in the sensor (for normal conditions) and smartphone (e.g. for alarm conditions). Each sensor has a memory card to hold up to 1 week of data if the smartphone is unavailable.

The PHMS provides personalized monitoring because it learns about the "normal" vitals of the patient. For instance, a normal heart rate and heart beat can vary from one patient to another based on gender, age, weight, physical fitness, etc. The PHMS can also provide adaptive monitoring based on the patient context (activities). If the patient is walking, the PHMS will adjust the range of monitored parameters, for instance the heart rate, and will monitor some new parameters which are meaningful only if the patient is exercising, for instance a cardiac stress test.

The HSP hosts a Health Monitoring Server that also supports the two interfaces: a SOA interface and a Publish-Subscribe Event interface. It may also include a Complex Event Processing (CEP) engine (not shown in the diagram, but described in Sect. 6) for event filtering, correlation and aggregation. There is one such server per HSP (hospital, community health centre, family doctor, ambulance, caregiver). Note that a minimal HSP is also possible, to support, for example, a caregiver with a cell phone. In that case, there would not be a Health Monitoring Server or CEP. The cell phone would simply have the capability to subscribe to events relayed through the Message Broker (e.g. by receiving SMS alerts). In Fig. 30.4, the interaction between the body sensors, smartphone and HSP is shown.

There are a number of existing projects that are piloting next-generation health system monitoring in clinical trials [4, 13, 31]. However, the typical architecture for these projects uses an

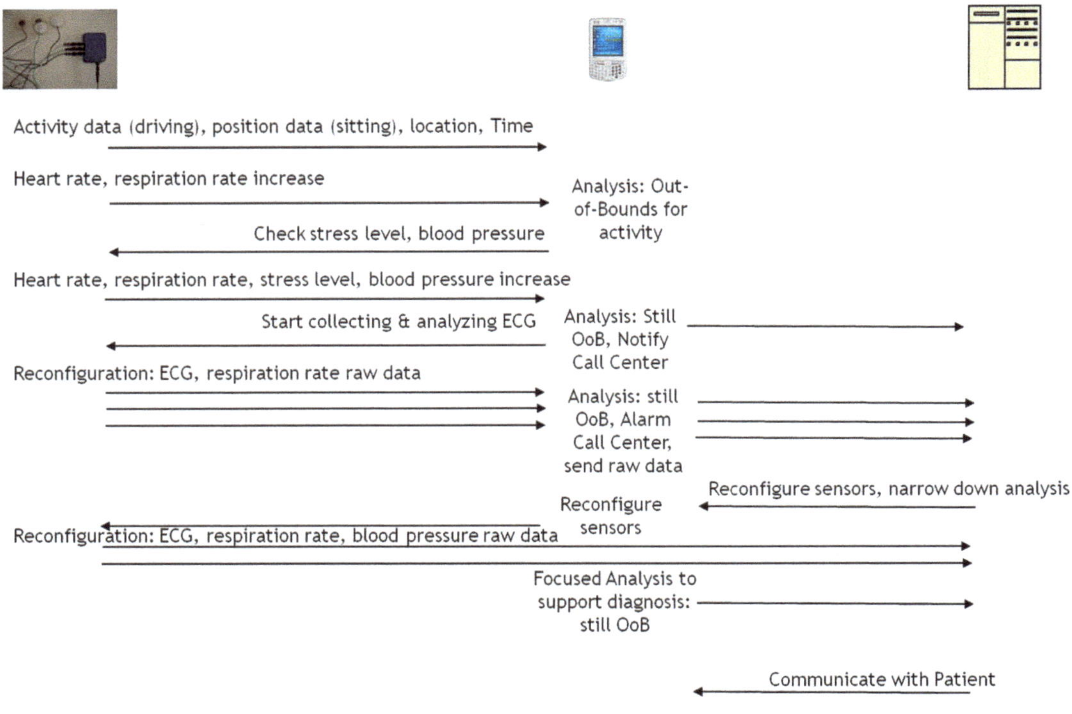

Fig. 30.4 Body sensor communication scenario

application specific architecture in which particular sensors are configured to deliver a continuous data stream to a dedicated application server. Integration with existing e-health infrastructure is on an ad hoc basis, in which only select HSPs have accessibility to the dedicated application server, and the integration is manual or batch based.

One significant aspect of the scenario is the analytics that are needed to determine whether the measures returned by the sensor are out of bound. Normal ECG and heart rate vary from person to person, as seen above. In addition, the ECG electrodes measure electric potential at the body surface and from there infer electric potential at the heart surface. As a result, different size torsos and in particular different densities of fat in different patients may result in different interpretations of arrhythmias, for instance. Finally, there may be a number of sensors attached to the patient with a large number of messages being emitted from each. In the next section, we discuss how to deal with the complexity of processing wireless messages.

30.5 Quality of Service in Messaging

Supporting QoS over any communication network; wired or wireless, requires the classification of network traffic into different classes of quantifiable parameters such as the required end-to-end delay; bandwidth, packet loss rate, etc. For network traffic to be valuable, the communication network needs to maintain its QoS parameters, which are provisioned by the network management system, within limits. Medical traffic is critical traffic since it concerns human lives and needs to be given higher priority over other network traffic types. For example, traffic related to a patient with a heart attack needs to be given the highest priority.

Generally, medical traffic can be classified into periodic and non-periodic traffic. Examples of periodic traffic are monitoring temperature, glucose level, ECG, EEG and blood pressure level at normal health conditions. Periodic traffic is more suited to centralized communication protocols in which a central network node efficiently allocates network resources to different sensor nodes. An example of a well-designed centralized communication protocol is the Time Division Multiple Access (TDMA) protocol. Non-periodic traffic such as emergency traffic when the patients' health conditions are life threatening is best suited for randomized communication protocols. Since this type of traffic is generated at irregular intervals, sensor nodes need not to wait for its allocated time slot to communicate its critical traffic as in the case of centralized communication protocols.

Contention-based Medium Access Control (MAC) protocols are distributed communication protocols where sensor nodes randomly contend for the transmission medium to communicate its sensed traffic. However, simultaneously implementing centralized and distributed communication protocols in a single communication network is not efficient and a complex task. Therefore, a distributed protocol is needed that efficiently supports periodic and non-periodic medical traffic simultaneously. The design of such protocol in a WBAN needs to be energy efficient and capable of differentiating medical traffic based on the traffic urgency levels. QoS support and energy saving features in WBAN can be exhibited by a well-designed MAC protocol which takes into consideration: collision avoidance, traffic latency, network throughput, efficient bandwidth utilisation, and control packet overhead.

30.5.1 QoS-Based MAC Protocol for WBASN

To provide QoS support in WBASN, we proposed a priority access mechanism to be incorporated with the IEEE 802.15.4 MAC protocol [37, 38]. The enhanced protocol is called QoS-based MAC Protocol. In such a protocol, a sensor node of urgent medical information such as ECG traffic of a person with a heart problem is allowed to contend for the transmission medium more than a sensor node of non-urgent medical information. In the proposed QoS-based MAC protocol, slotted Aloha channel access mechanism is assumed in which sensor nodes contend for the wireless transmission medium only at the beginning of each time slot of the MAC frame (see Fig. 30.5).

In the proposed QoS-based MAC protocol, network nodes are grouped into sensor nodes of critical health information and sensor nodes of non-critical health information. These nodes report their sensed information to the network coordination node which conveys the received information to a remote health server database at the hospital over a different network (i.e. cellular, WLAN or Internet).

Figure 30.5 illustrates the QoS-based MAC protocol when there are two classes of medical traffic, C_1 and C_2. The timeline of the wireless transmission channel is divided into timeslots. Each sensor node is allowed to transmit at the beginning of the timeslot. If there is more than one transmission at a timeslot, collision occurs

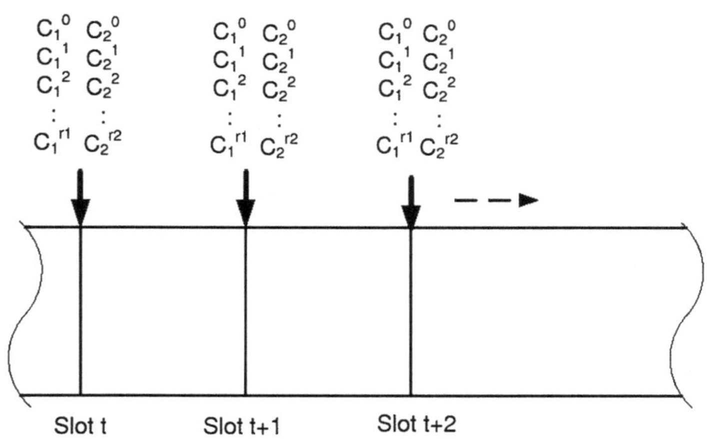

Fig. 30.5 QoS-based MAC protocol illustration

and every node has to terminate its transmission and retransmit in a later timeslot. The maximum number of allowed traffic retransmissions of C_1 and C_2 traffic is r_1 and r_2, respectively. Sensor nodes might be dedicated to sense a single traffic type or have the capability to sense multiple physiological health parameters. Based on a predefined threshold, the sensed traffic might be critical or non-critical. At the beginning of each time slot, network traffic is classified into critical and noncritical. Moreover, each traffic type is composed of traffic transmitted for the first time, C^0_x, where $|C^0_x|$ is the number of packets in such traffic set and x is the traffic class index, and traffic retransmitted for the nth time where $n = 1, 2, \ldots, r_x$. Therefore, the total number of packets G ready for transmission at the beginning of a time slot is composed of packets from sensor nodes of different classes which are transmitted for the first time and other packets from other sensor nodes of different classes that are involved in n collisions at previous timeslots. For a successful packet transmission, G must be equal to 1. To prioritize critical traffic over non-critical traffic for QoS support, the number of allowed retransmissions of the critical traffic must be configured to be more than the number of allowed retransmissions of the non-critical traffic. Hence, critical traffic contends more for the transmission medium access which maximizes its success rate over the non-critical traffic who contends less.

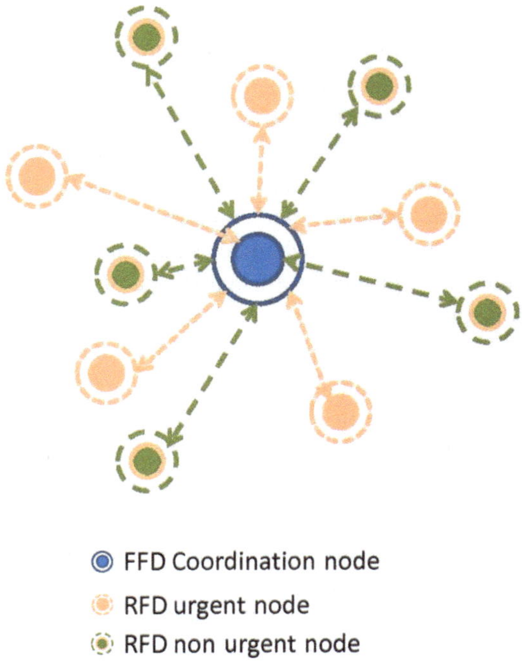

Fig. 30.6 WBAN network for QoS-based MAC protocol evaluation

30.5.2 Performance Evaluation and Results Analysis

To evaluate the performance of the proposed QoS-based MAC protocol, a star network topology composed of a single network coordinator and a number of sensor nodes is assumed as shown in Fig. 30.6. The sensor nodes are composed of Reduced-Function Devices (RFDs) while the coordinator node is only a Full-Function Device (FFD). The FFD can either be a standalone device or integrated within other devices such as a cell phone or a personal data assistant (PDA). Sensor nodes are battery powered devices in which power constraints apply while the coordinator node is assumed to have an external power supply. Wireless sensor nodes are classified into urgent and non-urgent nodes according to their assigned physiological variable to measure such as ECG, EEG, blood pressure, etc. The sensor nodes periodically report their measured values to the coordinator node. The coordinator node may process the received results before reporting them to an eHealth server over a different network.

30.5.2.1 WBAN Network Configuration and Parameters Values

The WBAN in Fig. 30.6 will be used to evaluate the performance of the proposed QoS-based MAC protocol. It is composed of N_c critical nodes and N_{nc} non-critical nodes which all communicate with a single coordination node. Different scenarios are used to evaluate the proposed QoS-based MAC protocol mathematically through system level simulation. In each scenario, the critical and non-critical nodes' packet arrival rates are set and maintained at a fixed and equal value; $\lambda_c = \lambda_{nc}$ and increased by τ for each scenario. Also, the maximum allowed non-critical packet retransmissions, r_{nc}, is set and maintained at a fixed value. Then, for each scenario, the number of critical packet retransmissions, r_c, is varied from r_c^{min} to r_c^{max}. The actual system parameter values used for the results in the following figures are: number of critical nodes $N_c = 10$ and number of non-critical nodes $N_{nc} = 10$, arrival rates $\lambda_c = \lambda_{nc}$ packets/node/timeslot; increased by $\tau = 0.005$ packets/node/timeslot for each scenario, the maximum allowed non critical packet retransmissions $r_{nc} = 1$. Then, for each scenario, the number of critical packet retransmissions r_c is varied between $r_c^{min} = 1$ and $r_c^{max} = 11$, which increased by one for each subsequent scenario. First, the system performance is evaluated analytically for the above given system parameters values. For each parameter setting, the critical and non-critical nodes throughput and their packet rejection rates are computed using the mathematical model detailed in [37, 38]. The results are shown in Figs. 30.7, 30.8, 30.9, 30.10, 30.11 and 30.12.

In addition to the analytical analysis, a discrete event simulator is used to evaluate the performance of the proposed QoS-based MAC protocol. In the simulator, a WBAN composed of N_c and N_{nc} sensor nodes communicates with a single coordinator node. The same system parameters used in the analysis are used in each simulation run which last for 10,000 timeslots each of 1 s length. Two counters are defined; ***criticalSuccessCounter*** and ***nonCriticalSuccessCounter*** for critical and non-critical nodes respectively. Whenever a packet is successfully transmitted, its corresponding counter is incremented by one. Also, packet rejection counters ***criticalRejectionCounter*** and ***nonCriticalRejectionCounter*** for critical and non-critical nodes are defined respectively which are incremented by one whenever a packet is rejected by the system. These counters are used to compute the system throughput and packet rejection rate.

Each time a packet collides, critical or non-critical, a random backoff value is generated between 0 and 15 timeslots. This value is maintained in a backoff counter which is decremented by one in each subsequent timeslot. Whenever the backoff counter reaches 0, its corresponding collided packet is retransmitted again given that the maximum allowed number of such packet retransmissions has not been reached. Otherwise, if the maximum number of the allowed retransmissions is reached, such a packet is dropped and its corresponding packet rejection counter is increased by one. The obtained simulation results are also plotted in Figs. 30.7, 30.8, 30.9, 30.10, 30.11 and 30.12.

30.5.2.2 Aggregate Throughput Analysis

Figures 30.7, 30.8, and 30.9 show the results of the aggregate system throughput, critical throughput and non-critical throughput of the QoS-based MAC protocol. As the number of critical nodes retransmission is increased while the non-critical nodes retransmission is maintained fixed, the probability for critical nodes successful packet transmission is increased. Therefore, as r_c is increased, as shown in Fig. 30.7, the aggregate throughput of critical traffic is increasing and the aggregate throughput of non-critical traffic is decreasing up to $r_c = 5$. Beyond $r_c = 5$, the throughput for critical and non-critical nodes reaches a saturation point where increasing r_c has no effect on the system throughput. This can be explained by the fact that the successful critical packet transmission is archived before reaching the maximum number of provisioned retransmissions. Similar system behaviour is shown in Fig. 30.8; however, the aggregate throughput is increased due to the increase in packet arrival rates.

As the packet arrival rates, λ_c and λ_{nc}, are increased to 0.03 packets/node/timeslot more packets are generated and the aggregate throughput varies from the above results (see Fig. 30.9). As the provisioned r_c value is increased from 1 to 4, the critical nodes throughput increases. As r_c increases beyond 4, the critical nodes throughput starts decaying due to the increased probability of more than one packet arrival in a time slot; new and previously collided. Therefore, the probability of reaching the maximum number of provisioned packet retransmission for a node; critical or non critical, is increased. Therefore, the critical traffic rejection rate is also increased which negatively affects the aggregate system throughput.

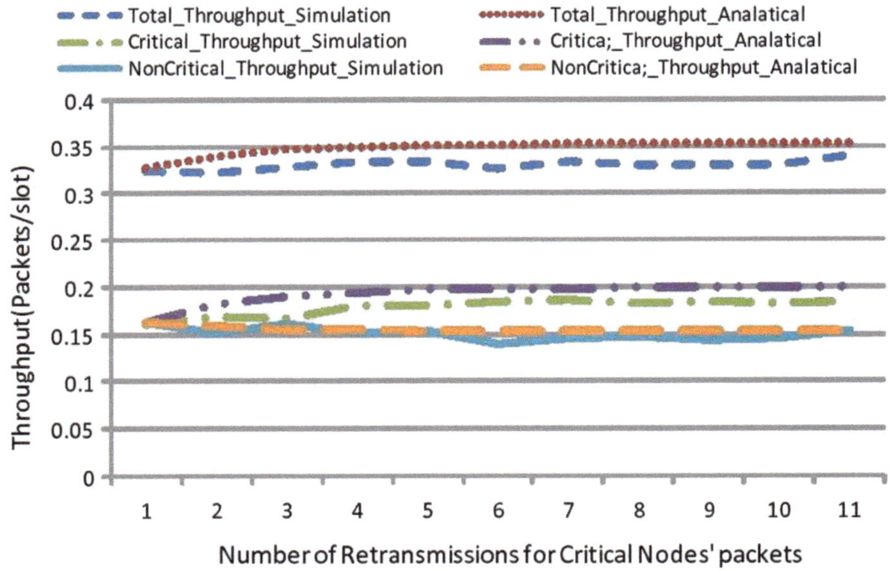

Fig. 30.7 Aggregate throughput, $\lambda_c = \lambda_{nc} = 0{:}02$

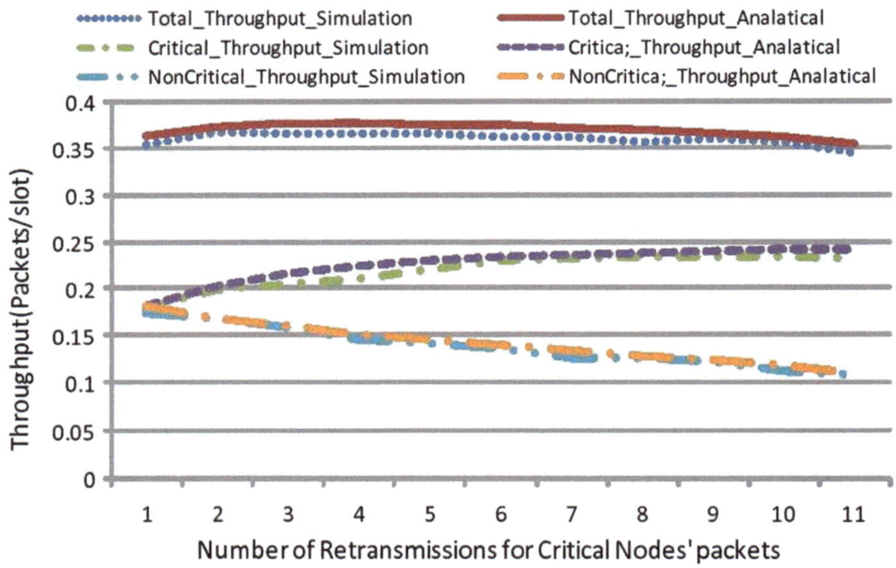

Fig. 30.8 Aggregate throughput, $\lambda_c = \lambda_{nc} = 0{:}025$

30.5.2.3 Aggregate Packet Rejection Rate Analysis

The results of the aggregate critical and non-critical packet rejection rates are shown in Figs. 30.10, 30.11, and 30.12. As r_c increases, the system throughput has similar results for 0.02 and 0.025 traffic arrival rates as shown in Figs. 30.10 and 30.11. In such results, since the critical traffic contend more for the transmission medium as the r_c is increased the critical nodes rejection rate decreases while the non-critical nodes rejection rate increases since it contend less for the transmission medium. Similar to the throughput results, the rejection rates reach stability region as r_c increased beyond 5.

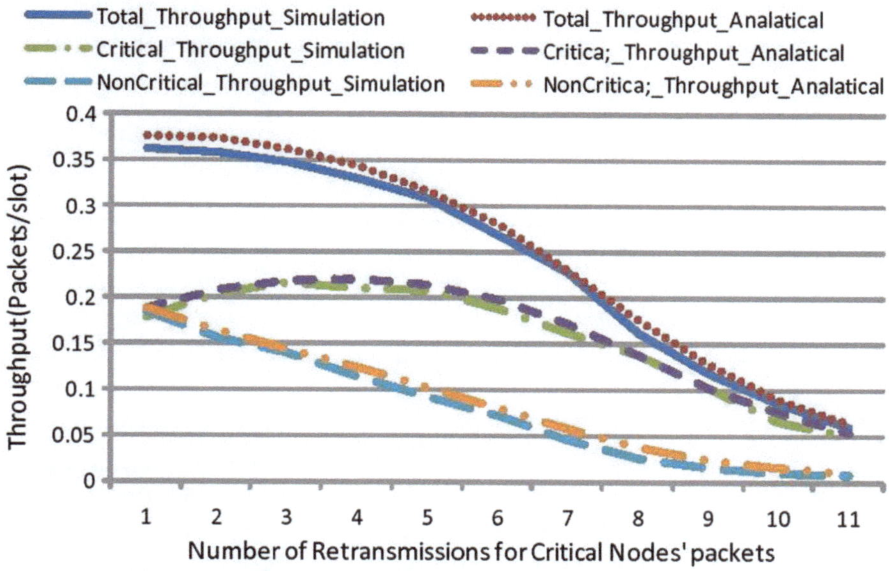

Fig. 30.9 Aggregate throughput, $\lambda_c = \lambda_{nc} = 0{:}03$

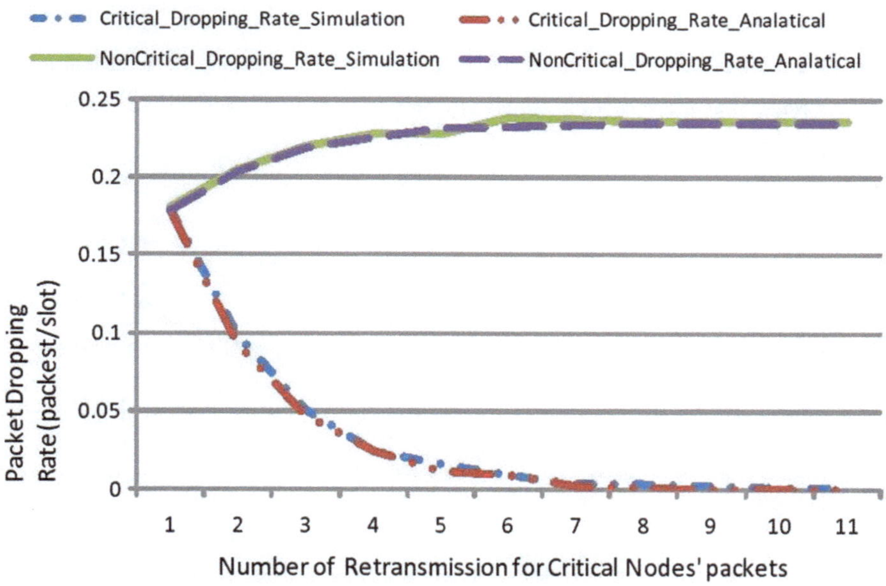

Fig. 30.10 Packet rejection rate, $\lambda_c = \lambda_{nc} = 0{:}02$

For the 0.03 packet arrival rate, the critical traffic rejection rate shows different results from one results of 0.02 and 0.025 packet arrival rates (see Fig. 30.12). As can be seen from the figure, as r_c increases beyond 5, the critical nodes rejection rate starts to increase sharply. This is due to the increased probability of more than one packet arrival per timeslot. Hence, this leads to increasing the probability of reaching the maximum provisioned retransmission values for critical and non-critical node traffic which finally results in higher packet rejection. The behaviour of the non-critical traffic rejection rate is sharply increasing from the beginning due to the same reason as mentioned for the critical traffic.

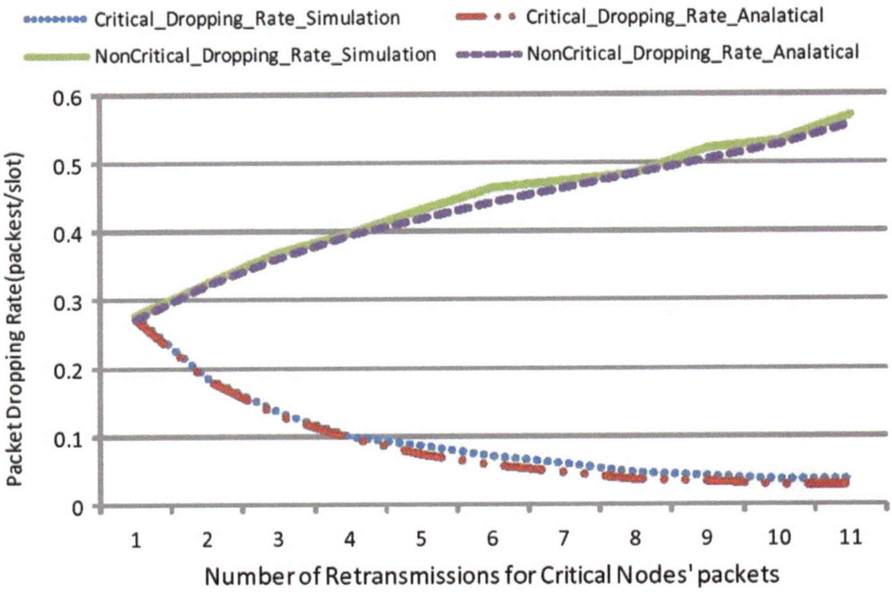

Fig. 30.11 Packet rejection rate, $\lambda_c = \lambda_{nc} = 0{:}025$

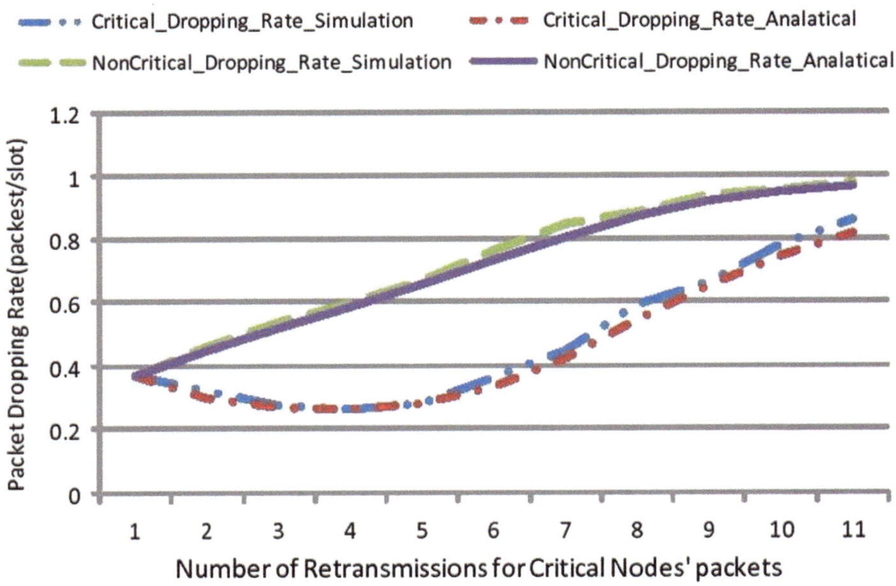

Fig. 30.12 Packet rejection rate, $\lambda_c = \lambda_{nc} = 0{:}03$

30.5.2.4 Summary

The results reveal the capability of the proposed protocol in providing differentiated services to medical wireless sensor networks. The results also show the unnecessary increase in the number of packet retransmissions for critical traffic beyond a certain value. At such a point, the critical traffic throughput either reaches a saturation point when the traffic arrival rate is moderate or sharply decreases for a high traffic arrival rate. Similar behaviour is observed for the traffic rejection rate.

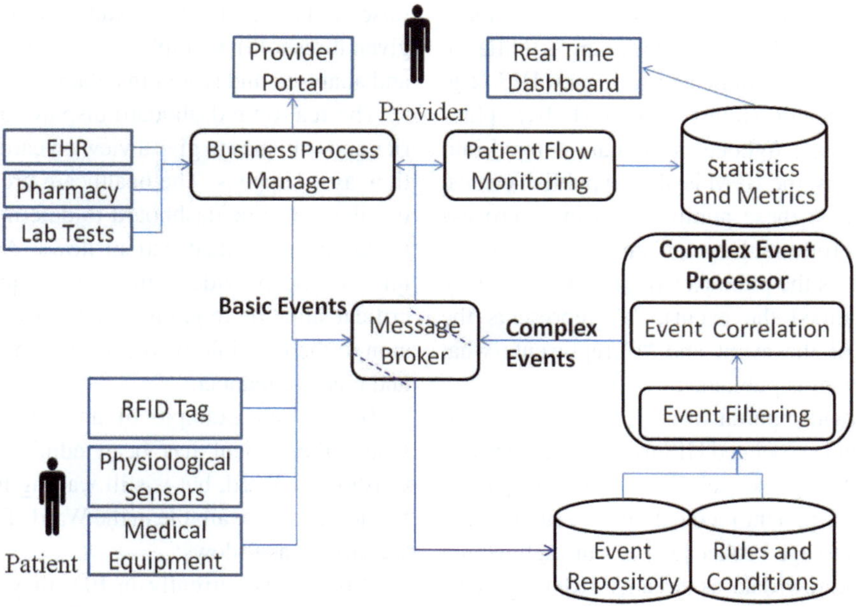

Fig. 30.13 Proposed architecture

30.6 Complex Event Processing and Key Performance Indicators

Let us now consider further the scenario from Sect. 3, by stepping beyond the requirements of a single patient, in order to see how to improve cardiac patient flow by integrating events from a wide range of sources within a hospital using complex event processing (CEP) to support fine grained monitoring of patient flows and detailed analysis of bottlenecks. Potential sources of events within the hospital include medical equipment, physiological sensors, RFID tag readers, health information systems (HIS) and BPM systems.

Event delivery architectures can collect streams of events from sensors for processing to infer critical medical events in real time [25]. However, to address cardiac patient flow, it is critical to also incorporate business process events. For example, a cardiac patient could be equipped with a bracelet that contains an RFID tag to track physically where in the hospital the patient is located. However, the data emitted by RFID tag readers throughout the hospital are low-level information which is not sufficient on its own to understand where in the cardiac patient process flow the patient is waiting. For instance, is the cardiac patient in the emergency department (ED) waiting for an ECG or are they waiting to see an ED physician to review test results? Further we would like to aggregate events (e.g. a patient waited 40 min for an ED physician to review test results) and display them on a real-time dashboard to communicate current average waiting times at key points in the cardiac process flow to help decision making.

In Fig. 30.13, there are four types of sources that can emit streams of raw data that can be delivered as basic events to the Complex Event Processor. The first source is medical equipment that run tests on patients (e.g. ECG) and emits this raw data in real time with attributes such as equipmentID, timestamp and value (measured by the equipment). The second source is physiological sensors that measure vital signs (e.g. blood pressure, temperature and heart rate) and emit this raw data with attributes such as sensorID, timestamp and value (of the vital sign measured by the sensor). The third source is RFID tag readers that read the RFID tags embedded in patient bracelets at specific locations in the hospital (e.g. ED) and emit this raw data in real time with attributes such as RFIDtagID, timestamp, and location (often as GPS coordinates).

The last source is the business process manager (BPM) which orchestrates many activities in the hospital. It is connected to legacy HIS (e.g. electronic health record system (EHR), pharmacy, lab tests). Whenever data are entered into these systems, the BPM is able to publish events. The format of these published business process events is as follows: Event_Name (ID, a, t), where ID represents the publisher ID (i.e. which legacy system sourced the event), "a" represents the attributes of the event and "t" represents what time the event is published.

The provider portal offers a web interface to the BPM and associated HIS for the clinicians and administrative personnel. They can input patient data, retrieve patient medical history, run reports, etc. The message broker [31] in our architecture receives the raw data as streams of basic events from our distributed sources and routes them to the Event Repository. It also receives complex events generated by the CEP and routes them to the Patient Flow Monitoring application or to the Business Process Management application.

All the events stored in the Event Repository are processed by the Complex Event Processor. One of the functions of CEP is to filter events, particularly, duplicate RFID events [39]. For instance, when the patient is in the waiting room and the RFID event IN_ED is emitted every minute, only the first published event of the sequence (until the patient leaves the ED) is useful for event processing, and the subsequent events have to be discarded. Filtering events is done according to the Rules and Conditions defined for the CEP (stored in the rules knowledge base).

Another task for the CEP is to correlate events from the different distributed sources in order to create complex events. It does this by correlating both basic events and complex events [30] using either logical constructors (e.g. AND, OR, Sequence) or temporal constructors (e.g. WITHIN). Rules and Conditions are defined for the event repository to create complex events by matching pre-defined rules with event patterns that use event constructors [40].

The Patient Flow Monitoring service collects all the complex events that are used either to build metrics for measuring performance or are used to raise alerts (e.g. if patient wait times rise above a given threshold). Complex events are aggregated and analysed and stored in a database.

The real-time dashboard displays alerts, metrics and statistics to give a view on cardiac patient flow as it happens. The healthcare providers can use the real-time dashboard to determine bottlenecks in the cardiac patient flows. As a consequence, the provider can respond quickly to a critical situation (e.g. high wait in ED for admission to the Cardiology Ward) in both a proactive and reactive manner.

Here is an example of a scenario analysis where the patient has been admitted into the Cardiology Ward, but is still waiting in ED for a bed to become available in the Ward. The detailed scenario is as follows:

- Patient-1 is currently in ED; they have been admitted to the Cardiology Ward, but since no bed is available, the patient must wait
- In the meantime, Patient-2 is in the Cardiology Ward, being discharged, and therefore should leave their bed (Bed_207) soon
- As soon as Patient-2 departs from the Cardiology Department, Housekeeping is called to prepare Bed-207 which is then assigned to Patient-1
- Patient-1 then moves to Bed-207 in the Cardiology Ward
- In order to track this situation in our real-time dashboard, we need to aggregate complex events which will allow us to compute the metric: Patient Wait Time in ED for admission in Cardiology Ward. This is the time interval between two complex events: Admit_CardWard and Wait_ForBed
- Admit_CardWard is the time when Patient-1 moves into Bed-207 in the Cardiology Ward
- Wait_ForBed is the time when Patient-1 starts waiting in ED for a bed to become available in the Cardiology Ward

Figure 30.14 shows the basic and complex events related to the detailed scenario above; this list is not exhaustive and does not show all attributes for sake of brevity. Here, we highlight Admit_CardWard and Wait_ForBed.

The Wait_ForBed event is the correlation of two basic events:

Complex Event	Correlation	Basic Event	Source	Attributes
		IN_ED	RFID Reader	Patient-1, ED_Location, timestamp
		Request_AdmCardWard	EHR and BPM	Patient-1, Physician_ID, timestamp
Wait_ForBed	IN_ED **AND** Request_AdmCardWard			Patient-1, ED, CardWard, timestamp
		IN_CardWard	RFID Reader	Patient -2, Bed_207, timestamp
		Discharge	EHR and BPM	Patient-2, Physician_ID, timestamp
OUT_CardWard	Discharge **AND** (**NOT** IN_CardWard)			Patient-2, timestamp
Clean_Bed	OUT_CardWard		BPM	Bed_207, timestamp
		Bed_Available	Bed Manager and BPM	Bed_207, timestamp
		Bed_Assigned	Bed Manager and BPM	Bed_207, Patient-1, timestamp
		IN_CardWard	RFID Reader	Patient -1, Bed_207, timestamp
Admit_CardWard	(Wait_ForBed **AND** Bed_Assigned **AND** IN_CardWard			Patient-1, timestamp

Fig. 30.14 Event analysis table

- IN_ED which is an RFID event saying that Patient-1 is located in an ED Examination Room.
- Request_AdmCardWard which is the Admission Order that the ED Physician has entered into an HIS requesting a bed in the Cardiology Ward. This request is mapped into an event by the BPM system.

The Admit_CardWard complex event is the correlation of three events: The Wait_ForBed event for Patient1 that was just described above plus the Bed_Assigned and IN_CardWard events.

The detailed steps for correlating those events are as follows:

- Bed-207 is currently occupied by Patient-2 who is being discharged. Their Physician has entered a Discharge Order into an HIS; this request is mapped into a Discharge event by the BPM system.
- Patient-2 then leaves Bed-207 to go home
- An OUT_CardWard complex event is created based on the Discharge event and the fact that Patient-2 is no longer in the Cardiology Ward (the condition NOT IN_CardWard is true since

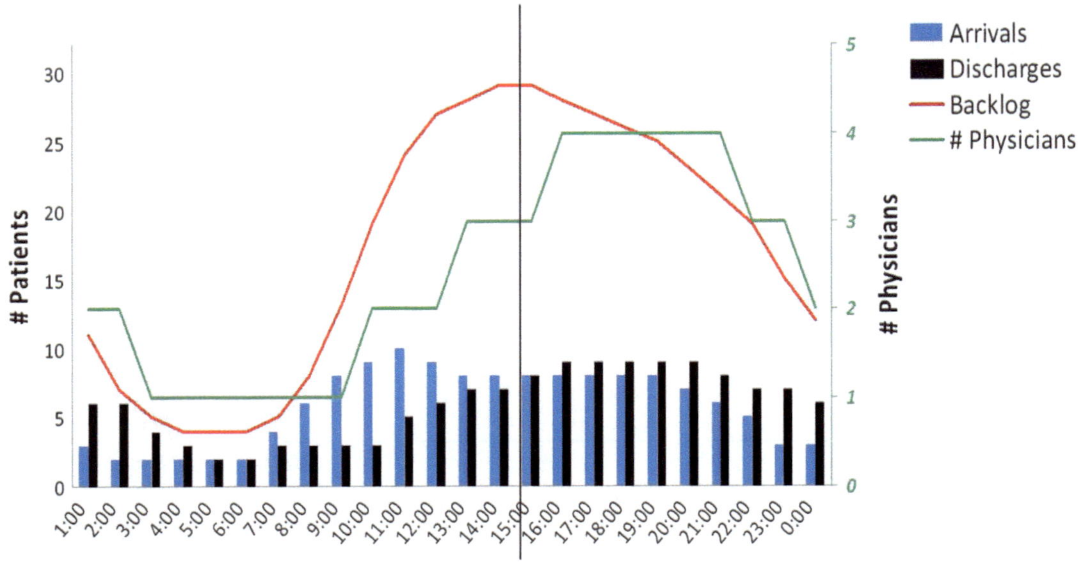

Fig. 30.15 Real-time analytics: situation in ED at 3 pm

there is an RFID event showing that Patient-2 is in the Discharge Lounge waiting for his family to pick him up)

- The BPM system automatically generates a request to Housekeeping (via the Bed Manager system) to clean Bed_207 and a Clean_Bed event is created
- When Housekeeping has finished cleaning Bed_207, the Bed Manager system is updated and a Bed_Available event is created by the BPM system
- The Admission clerk assigns Bed_207 to Patient-1 and enters the information in the Bed Manager system which creates a Bed_Assigned event
- The ED staff are informed that Patient-1 is assigned to Bed_207 in the Cardiology Department, and they transport the patient there. The RFID reader in the Cardiac Ward raises the basic event IN_CardWard for Patient-1
- The Admit_CardWard complex event is created by correlating the Wait_ForBed, Bed_Assigned and IN_CardWard events

This analysis shows how real-time events can be correlated with business process activity events in order to provide complex events to measure wait times. The measurement of different wait times can then help identify bottlenecks in patient flows.

30.7 Real-Time Analytics

The collection and aggregation of events and data as seen in Sect. 6 enable real-time analytics where the user can visualize several performance metrics on a real-time dashboard, and gains more insight into what decision to make. Figure 30.15 shows an example of a real-time dashboard in ED [41] where patient arrivals rate and discharges rate are displayed on an hourly basis. The cumulative difference between these two rates gives the backlog of patients waiting in ED (red graph). This enables the user, for instance, to assess when to bring in more ED Physicians (green graph). What is interesting here is that a 3rd ED Physician is brought in only when the backlog is peaking, at 13:00. A better decision might have been to call physicians in when the patient arrivals are peaking, at 11:00, knowing that the rate of discharges is still low. This would have avoided the patient backlog, and therefore could have reduced patient wait times.

This simple example shows how real-time analytics can bring more insight to support decision-making. Other examples of real-time metrics include:

- Patient wait times, for a bed, for a procedure, for a test. An example of how the patient wait time for a bed is computed is given in Fig. 30.14 above
- Service times
- Bed management escalation triggers
- Transport times
- Housekeeping times (to get a bed cleaned up and ready for the next patient)

30.8 Predictive Analytics and Decision Support

The power of predictive analytics and decision support can best be understood through an example in patient flow management. Some examples of decisions that Patient Flow Managers in hospitals may make in order to address bottlenecks include:

- Review current and next day Same-Day Admit (SDA) to identify possibilities of cancellation
- Off-servicing
- Transfers-out
- Placement of hallway patients
- Opening Flex Beds
- Restricting repatriations and regional referrals
- Transitioning private rooms into semi-private rooms
- Creating mixed gender rooms with patient permission
- Cohorting isolated patients
- Using unconventional hospital spaces
- Redeploying staff; temporary changing staff mix; adding staff

These decisions typically fall into one of three groups

- Shape the patient demand: transfer patients to another hospital, cancel elective procedures, for instance
- Increase capacity: open flexible beds, add more clinical staff
- Mitigate the impact of the bottleneck: do off-servicing by sending cardiac patient to a General Medicine ward instead of a Cardiology ward, for instance

Today, the decision-making process is manual, based more on gut feeling and experience than on hard data. Furthermore, the decision made is typically optimal for the unit but not necessarily for the hospital as a whole, because of the lack of a real-time view of patient flows across the continuum of care. However, there are a number of approaches that can be used to provide decision support in a more systematic manner. Decision models, based on decision tables, rules, and decision trees map the real-time performance metrics into the above decision options. Another technology used for decision support is fast-forward simulation. By giving the staff the ability to look at the situation 6 or 12 h ahead, they are able to conduct what-if scenarios, evaluate the impact of each decision option before it's executed, and decide accordingly. This predictive decision support technology is based on discrete event simulation and has been used in other industries for shop floor control [42].

One example of a decision optimal for a hospital unit but sub-optimal for the hospital as a whole, is the following:

- There is a peak in patient arrivals in ED and wait times are increasing by the hour
- The Patient Flow Manager in charge of ED decides to increase capacity and opens more beds and calls in more staff in order to care for this additional load
- The call fails to reach colleagues in the Cardiology Ward, while at the same time not realizing that the root cause for the wait time increase in ED is actually the lack of beds in the Cardiology Ward
- By increasing capacity in ED, the call has increased pressure on the Cardiology Ward when those additional ED patients are going to be admitted to that Ward, next
- A better decision, supported by real-time performance metrics, would have been to either increase capacity in the Cardiology Ward or to transfer these additional ED patients to another hospital

Table 30.1 Evaluation

Criteria	Integrated CEP architecture	Traditional data architecture
Technology	Real-time dashboard, EDA, BPM, RFID	Manual Reports, EHR, Paper Forms
Timeliness	Real time	Delayed
Process duration	Seconds	Hours or days
Metrics	Fine grained	High level
Decision support	Care delivery	Administrative

30.9 Evaluation

Metrics and other performance measurement analytics which can be displayed on a real-time dashboard are important tools in improving integration of care [22]. However, most hospitals today, have a traditional data architecture that does not include wireless patient monitoring, BPM and CEP. In traditional data architecture, the collection of data is a tedious process of extracting data from disconnected data sources (sensor logs, departmental databases) and integrating them into a data warehouse view over a sufficient period of time to give a view of what happened in the past [43]. In addition, the disjointed nature of the data often means that fine-grained monitoring of patient flows and detailed analysis of bottlenecks is not possible. Only coarse-grained metrics to support high-level administration decisions are available that are insufficient for monitoring patient flows.

We evaluate our proposed integrated CEP architecture for cardiac patient flow by comparing it to the traditional data architecture that is typically used in hospitals today. Most hospitals have a patient EHR that is at the centre of their data architecture. Typically, raw data is collected in paper documents by clinicians and clerks. Feeding the data from these document into the EHR is often done after the fact (anywhere from several hours to several days), and often there is missing or incomplete data. Complete details of vital signs and equipment readings may stay as documents which are never entered into the EHR (although they may be available in electronic data storage associated with the equipment).

So, for example, to create a report to show the average waiting time for patients to be admitted to the Cardiology Ward requires collecting information from EHR, paper files and equipment logs and then performing manual analysis which might take several hours or even days. On the other hand, the CEP architecture enables the collection and processing of event data from distributed sources in real time, and the aggregation of complex events to create metrics that support decision making in real time. Each metric is pre-defined and the complex events that measure these metrics are defined. The metric is displayed in real time in a dashboard interface. This helps healthcare providers to monitor fine-grained aspects of patient flow and react quickly to critical situations.

Table 30.1 compares the integrated CEP architecture with the traditional data architecture. The integrated CEP architecture leverages EDA, BPM, RFID and real-time dashboards. It has the potential to process distributed events in seconds and provide fine-grained monitoring of patient flow that can highlight bottlenecks as they occur and enable healthcare providers to dynamically improve care delivery. Because of its reliance on paper-based forms and manual reporting against a centralized EHR, the traditional data architecture is generally oriented toward supporting delayed, after-the fact processing that can take hours or days to process incomplete information. As a result, it is most suitable for reporting high-level metrics for administrative decisions related to provisioning of long term resources for the hospital such as overall staffing levels and number of beds supported.

It should be emphasized though that the Integrated CEP Architecture requires significant investment in new IT infrastructure for wireless, CEP, EDA, BPM, RFID readers and real-time dashboards. This also implies significant investment in IT staff with up-to-date training and skills. It also implies a change in the way clinicians carry out their work (to embrace online BPM rather than paper forms) with a focus on improvement of care delivery based on performance measurement metrics. Further research is needed to establish the return on investment for such investments as well as guidelines for change management in order to implement them effectively.

30.10 Conclusions

Emerging technology can collect, correlate and process disparate data from multiple different sources in context (location, process) and in real time. The use of analytics is critical in health care in order to process complex information into real-time responses. To do so provides opportunity for large-scale improvements in the ability to deliver enhanced quality of care. This includes large scale improvements in quality of life for those who can be mobile outside of healthcare institutions through the use of personal health monitoring systems. It also includes large-scale improvements in our ability to manage patient flow and troubleshoot bottlenecks as patients move through the healthcare system. This has been illustrated using results from a real, ongoing study in cardiac patient care done in collaboration with a large community hospital in Ontario, Canada.

Putting It into Practice

- Complexity in healthcare is too high for clinicians to manage manually, with paper forms as is done currently. A re-engineering of operational and clinical processes is necessary, along with a new generation of information systems.
- The technologies described in this chapter are available and have been used in other industries for some time already. Real-time analytics and decision support can improve patient outcomes, improve quality of care, and improve efficiency in delivery of care.
- Patients and clinical staff can be equipped with smartphones, thus giving them access to these information systems wherever they are, in real time.
- Wireless communication is gradually becoming an integral component of real-time analytics and decision support in the medical field. This facilitates tracking patients while at hospitals, homecare, or roaming in different locations.
- Differentiating medical traffic according to its urgency levels is required for efficient telehealth systems. The QoS-based MAC protocols described in this chapter can be utilized in real health systems to satisfy these system requirements.

Acknowledgements Some of the material here has been previously published [22, 31, 37, 38, 44]. We thank Peter MacKinnon for all his helpful comments and invaluable support in writing this chapter.

References

1. IEEE 802.15.4: Wireless medium access control (MAC) and physical layer (PHY) specifications for low-rate wireless personal area networks (WPA), IEEE, October 2003
2. IEEE 802.15.4a: Wireless medium access control (MAC) and physical layer (PHY) specifications for low-rate wireless personal area networks (WPA), IEEE, October 2007
3. Timmons N, Scanlon W. Analysis of the performance of IEEE 802.15.4 for medical sensor body area networking. In: IEEE Conference on Sensor and Ad Hoc Communications and Networks, p 16–24, October 2004.
4. Bannach D, Lukowicz P, Amft O. Rapid prototyping of activity recognition applications. IEEE Pervasive Computing. 2008;7(2):22–31.
5. Webster JG (2009) Medical instrumentation: application and design, 4th Edn. Wiley. ISBN: 978-0-471-67600-3
6. Axisa F, Schmitt PM, Gehin C, Delhomme G, McAdams E, Dittmar A. Flexible technologies and smartclothing for citizen medicine, home healthcare, and disease prevention. IEEE Trans Inf Technol Biomed. 2005;9(3):325–36.
7. Cleland JG, Louis AA, Rigby AS, Janssens U, Balk AH. Non-invasive home telemonitoring for patients with heart failure at high risk of recurrent admission and death. J Am Coll Cardiol. 2005;45(10):1654–64.
8. Fulford-Jones T, Wei G, Welsh M. A Portable, Low-Power, Wireless Two-Lead EKG System. In Proceedings of the 26th IEEE EMBS Annual International Conference,San Francisco, September 2004.

9. Shaltis PA. A wearable blood pressure sensor using oscillometric photo-plethysmography and microaccelerometers, PhD Thesis, MIT, June 2007
10. Chiarugi F, Zacharioudakis G, Tsiknakis M, Thestrup J, Hansen KM, Antolin P, Melgosa JC, Rosengren P, Meadows J. Ambient Intelligence support for tomorrow's Health Care: Scenario-based requirements and architectural specifications of the EU-Domain platform. In: Proceedings of the International Special Topic Conference on Informational Technology in BioMedicine.
11. Otto C, Milenkovic A, Sanders C, Jovanov E. System architecture of a wireless body area sensor network for ubiquitous health monitoring. J Mobile Multimedia. 2006;1(4):307–26.
12. Jankowski S, Szymanski Z, Piatkowska-Janko E, Oreziak A. Improved recognition of sustained ventricular tachycardia from SAECG by Support Vector Machine. Anatolian J Cardiol. 2007;7 Suppl 1:112–5.
13. Bescos C, Harris M, Bover R, Schmidt R, Perez-Villacastin J. Decision support system for the practical implementation of the chronic heart failure guidelines: the MyHeart approach. Comput Cardiol. 2007;34:57–60.
14. Han J, Kamber M. Data mining: concepts and techniques (2nd Edition), Morgan Kaufmann 2006. ISBN1-55860-901-6
15. Gama J, Gaber M. Learning from data streams. Heidelberg: Springer; 2007. ISBN 3540736786.
16. Institute of Medicine. Crossing the quality chasm: a new health system for the twenty-first century. Washington, DC: National Academies Press; 2001.
17. Stetson PD, McKnight LK, Bakken S, Curran C, Kubose TT, Cimino JJ. Development of an ontology to model medical errors, information needs, and the clinical communication space. Proc AMIA Symp 2001. pp 672–676.
18. Bates DW. Getting in step: electronic health records and their role in care coordination. J Gen Intern Med. 2010;25(3):174–6.
19. Reddy MC, Shabot MM, Bradner E. Evaluating collaborative features of critical care systems: a methodological study of information technology in surgical intensive care units. J Biomed Inform. 2008;41: 479–87.
20. Wetzstein B, Ma Z, Leymann F, Towards measuring key performance indicators of semantic business processes, lecture notes in business information processing, Springer, Vol 7, pp. 227-238, 2008
21. Krishnipillai A., Understanding Key Performance Indicators through Driver Measures, Masters Thesis, University of Ottawa, September 2009
22. Mouttham A, Peyton L, Kuziemsky C. Leveraging performance analytics to improve integration of care. Hawaii, US: Software Engineering in Healthcare; 2011.
23. Luckham D, The power of events: an introduction to complex event processing in distributed enterprise systems, Springer, 2002.
24. Wang G, Jin G, Research and design of RFID data processing model based on complex event processing. In: Proceedings of 2008 Intern'l Conference on Computer Science and Software Engineering, vol.5, pp.1396–1399, IEEE Society, Dec. 2008
25. Yao W, Chu C, Li Z. Leveraging complex event processing for smart hospitals using RFID. J Network Comput Appl. 2011;34(3):799–810.
26. Dunkel J. On complex event processing for sensor networks. In: Intern'l Symposium on Autonomous Decentralized Systems (ISADS '09), pp.1–6, 23–25 March 2009.
27. Lombriser C, Bharatula NB, Roggen D, Tröster G. On body activity recognition in a dynamic sensor network. In: Proceedings of the ICST 2nd international Conference on Body Area Networks, Florence, Italy, 2007.
28. Bucaille I, Tonnerre A, Ouvry L, and Denis B., MAC layer design for UWB LDR systems: PULSERS proposal,4th Worksshop on Positioning, Navigation and Communication 2007 (WPNC '07), pages 277–2837, March 2007.
29. Otal B, Alonso L, Verikoukis C. Highly reliable energy-saving MAC for wireless body sensor networks in healthcare systems. IEEE J Select Areas Commun. 2009;27(4):553–65.
30. Wang F, Liu S, Liu P, Bai Y. Bridging physical and virtual worlds: complex event processing for RFID data streams, 10th International conference on extending database technology (EDBT), Munich. Germany. 2006;3896:588–607.
31. Mouttham A, Peyton L, Eze B, El Saddik A. Event-driven data integration for personal health monitoring. J Emerg Technol Web Intelligence. 2009;1(2):110–8. doi:10.4304/jetwi.1.2.110-118.
32. Wilding H. Integrating care: from horizontal to vertical integration. J Integr Care. 2010;18(3):15–20.
33. Van der Aalst W, Hofstede A, Weske M. Business process management: a survey. Lecture Notes in Computer Science. 2003;2678:1–12.
34. Pourshahid A, Amyot D, Peyton L, Ghanavati S, Chen P, Weiss M, Forster AJ. Business process management with the user requirements notation. Electron Commerce Res. 2009;9(4):269–316.
35. Kang J, Han K, A Business Activity Monitoring System Supporting Real-Time Business Performance Management, Third International Conference on Convergence and Hybrid Information Technology (ICCIT '08), vol.1, pp.473–478, 11–13 Nov. 2008.
36. Middleton G, Peyton L, Kuziemsky C, Eze B, A framework for continuous compliance monitoring of eHealth processes, World Congress on Privacy, Security, Trust and the Management of Ebusiness, August, 2009
37. Ali KA, Saraker JH, Mouftah HT. QoS-based MAC Protocol for Medical Wireless Body Area Sensor Networks, 2010 IEEE Symposium on Computers and Communications (ISCC). pp 216–221, June 22-25, 2010, Riccione, Italy.

38. Ali KA, Saraker JH, Mouftah HT. Urgency-based MAC Protocol for Wireless Body Area Sensor Networks, 2010 IEEE Intern'l Conference on Communications Workshops (ICC). pp 1–6, May 23–27, 2010, Cape Town, South Africa
39. Bai Y, Wang F, and Liu P. Efficiently Filtering RFID Data Streams. In: First International VLDB workshop on Clean Databases, Seoul, Korea, pp. 50–57, 2006
40. Yao W, Chu C, Li L, and Mullen T, Leveraging Complex Event Processing for RFID Applications: A Case Study in Hospitals. In: 39th National Conference of Decision Sciences Institute, November 2008, Baltimore, MD., pp. 4341–4346
41. Kimball R, Ross M. The data warehouse toolkit. USA: Wiley; 2002.
42. Ontario Ministry of Health and Long-Term care (MOHLTC) web site on ED Process Improvement Program: http://www.patientflowtoolkit.ca/AboutPIP.htm (last accessed Oct 26th, 2011)
43. Mirdamadi S, Fontanili F, Dupont L. Discrete event simulation-based real-time shop floor control. In: Proceedings 21st European Conference on Modeling and Simulation, ECMS 2007
44. Baarah A, Mouttham A, Peyton L. Improving cardiac patient flow based on complex event processing, 2011 IEEE Jordan Conference on Applied Electrical Engineering and Computing Technologies, Amman, Jordan, December 2011

31 Medical Sociology and Case-Based Complexity Science: A User's Guide

Brian Castellani, Corey Schimpf, and Frederic Hafferty

31.1 Introduction

In this chapter, we provide readers with the first formal overview of case-based complexity science and its related methodology, case-based modeling. **Case-based modeling**, championed largely by [1], constitutes a fifth major method for modeling complex systems, offering itself as an alternative to (and also integration of) agent (rule-based) modeling, dynamical (equation-based) modeling, qualitative (idiographic) modeling, and statistical (aggregate-based) modeling. For us, as medical sociologists, case-based modeling makes sense because, fundamentally, medicine is about the case. Case-based modeling also resonates with our particular practice of a **case-based complexity science**, which can be defined as a generalist approach, grounded in the epistemological perspectives of Byrne's complex realism—which we explain later.

B. Castellani, Ph.D. (✉)
Kent State University, 3300 Lake Road West, Ashtabula, OH, USA
e-mail: bcastel3@kent.edu

C. Schimpf, M.A.
Purdue University, School of Engineering Education, Neil Armstrong Hall of Engineering, Room 1300,701 W. Stadium Avenue, West Lafayette, IN, USA
e-mail: cschimpf@purdue.edu

F. Hafferty, Ph.D.
Program in Professionalism and Ethics, Mayo Clinic, Minneapolis, MN, USA
e-mail: fredhafferty@mac.com

In terms of case-based modeling, we employ the **SACS Toolkit**. The SACS Toolkit is a new, computationally grounded, case-based method we created for modeling complex social systems as a set of cases [2, 3]. In terms of health, we use the SACS Toolkit to study communities, school systems and stress and coping issues as different types of complex systems [4, 5]; and, in terms of health care, we use it to study the complexities of medical professionalism and medical education [6, 7]. By the end of our review, interested readers should have enough knowledge of case-based modeling and the SACS Toolkit to determine its viability for their own research.

Our chapter is organized as follows. We begin with an overview of case-based method, providing a quick history of how case-based complexity science and, more specifically, case-based modeling emerged and how we position our approach relative to other ways of modeling complex systems. From here we turn to the SACS Toolkit, providing a quick overview of how it models complex systems. Detailed reviews of the SACS Toolkit currently exist—one qualitative in focus [2] and the other mathematical [3]. However, a few advances are made in the current chapter, as it is (1) our first attempt to clarify the SACS Toolkit's explicit links to case-based complexity science and (2) our first effort to integrate our two previous versions into a new, updated version. To help readers grasp a basic understanding of the SACS Toolkit, our review will draw on some examples from our research in medical sociology.

J.P. Sturmberg and C.M. Martin (eds.), *Handbook of Systems and Complexity in Health*,
DOI 10.1007/978-1-4614-4998-0_31, © Springer Science+Business Media New York 2013

31.2 Case-Based Complexity Science: An Overview

Over the last several years, Professor David Byrne of *Durham University, UK* has emerged as a leading international figure in what most scholars see as two highly promising but distinct fields of study (1) case-based method and (2) the sociological study of complex systems. An example of the former is Byrne's *Sage Handbook of Case-Based Methods* [1]—which he co-edited with Charles Ragin, the most prominent figure in case-based method. An example of the latter is his widely read *Complexity Theory and the Social Sciences* [8]. What scholars (including the current authors) are only beginning to grasp, however, is the provocative premise upon which Byrne's work in these two fields is based. His premise, while simple enough, is ground-breaking:

> Cases are the methodological equivalent of complex systems; or, alternatively, complex systems are cases and therefore should be studied as such.

With this premise, Byrne adds to the complexity science literature an entirely new approach to modeling complex systems, alongside the current repertoire of agent (rule-based) modeling, dynamical (equation-based) modeling, statistical (variable-based) modeling, network (relational) modeling, and qualitative (meaning-based) method.

Working independently of and yet in tandem with Byrne, we have used his premise to develop a case-based, computationally grounded, mixed-methods technique called the SACS Toolkit [2, 3]. However, because it is designed for studying both small-database and large-database complex systems, the SACS Toolkit makes a slight variation on Byrne's premise: it models a complex system as a *set* of cases, ranging from, at minimum, 1 case to any large number of cases. In the language of matrix algebra, these cases are k dimensional vectors (See 3 for complete mathematical overview). The goal of the SACS Toolkit is to compare and contrast and then condense and cluster databases comprised of a large number of cases to create a low-dimensional model of a complex system's structure and dynamics across time/space. To create these models, the SACS Toolkit employs a variety of computational techniques—including cluster analysis, network analysis, agent-based modeling, and artificial neural nets—as well as statistics, historiography and qualitative method.

Before we can overview the SACS Toolkit, however, it is necessary to situate it within the larger fields of case-based method and complexity science. We begin with case-based method.

31.2.1 Case-Based Method

Case-based method is an umbrella term for a somewhat varied set of techniques that have a long history in the social sciences and others fields such as biology, history, archaeology, and medicine [1].

Case-based methods, whatever the type, can be explanatory or descriptive. They can be static or longitudinal, retrospective or prospective. Despite differences, the goal of these methods is to study a case or set of cases more holistically, systematically, and ideographically. The simplest example of a case-based method is the **case study**, which is an in-depth investigation of a single case. Most approaches, however, tend to study a set of cases, engaging in what is called **case-comparative method**.

The most popular version of case-comparative method is Ragin's **qualitative comparative analysis** (QCA) [9]. Over the last decade, QCA has developed into a set of comparative techniques that allow case comparative methods to move beyond the limitations of traditional qualitative method. While case comparison is implicitly the purpose of such statistical techniques such as cluster analysis and discriminatory analysis, it is rarely couched in such terms. As such, most case-comparative methods are grounded in a qualitative tradition, focusing on a small number of cases. QCA pushes case-comparative method into a mixed-methods frame, allowing researchers to capitalize on the strengths of both qualitative and quantitative analysis, insomuch as it uses

Boolean algebra and its matrices to search for patterns and to make generalizations with larger datasets [1]. As a side note, Boolean algebra is a variant of algebra that works only with 1 s and 0 s, as truth values: cases either do or do not belong to a dominant profile identified—for example, sick or not sick patients. Ragin has also developed a fuzzy-set version of QCA which overcomes the limitations of Boolean algebra and its crisp sets to allow cases to have "degrees of membership" in the main profiles identified by a study—think, for example, of a study that allows people to be, in varying degrees, both healthy and sick [9, 10].

Regardless of the case-based method used, a **case** can be a person, event, place, concept, etc. Whatever studied, the case is the focus of the study, not the individual variables or attributes of which it is comprised. Case-based researchers would not, for example, study the impact gender has on the professional behavior of physicians. Instead, they would study how the different profiles of physicians explain their dissimilar professional behaviors, with gender being one of the key attributes examined. Case-based methods also treat the cases they study as **composites**, viewing them as comprised of an interdependent, interconnected set of variables, factors, or attributes that form some type of emergent configuration, such that the whole is more than the sum of its part. Each variable, therefore, is not an isolated factor impacting the case of study; instead, it is part of a larger, context-specific set of factors which collectively define the case of study, usually in rather nonlinear ways. For example, in Ragin's QCA, variables are treated as sets [10]. A case profile, therefore, tells us which sets (a.k.a variables) a case belongs to and in what manner or degree.

Case-based methods do, however, differ from one another in the degree to which they seek to **generalize** their findings. Byrne [11, 12] for example—the leading figure in case-based complexity science—advocates contextualized and limited forms of case-based generalization. Despite these differences, all case-based methods treat cases as particular instances, examples, occurrences, or types of some larger population.

It is its **configuration approach** to variables, however, that ultimately makes case-based method a radical departure from normative, variable-based inquiry, as defined by the majority of statistical methods used in the social sciences and, more specifically medical sociology—think here of **conventional method**. Variable-based statistics has no interest in cases or any in-depth understanding of how a set of variables collectively define or impact these cases. Instead, variable-based inquiry seeks to understand the relationship variables have with each other, and usually in the most parsimonious, reductionist, nomothetic, linear, unidirectional manner possible. To illustrate, let us go back to our example of physician professionalism. A variable-based study might examine which factor (amongst some set of supposedly independent variables) best explains the different professional behaviors of physicians. For example, which variable is more important for later misconduct? Is it the networks students hang out in or the number of times they were cited for unethical behavior? In contrast, a case-based approach would examine how the different variable-based configurations of some set of physicians (cases) account for differences in professional misconduct. For example, one may find that male students, specializing in surgery, who attended schools that failed to really punish their misconduct, and who socialized in student networks that approved of their "bad" behavior went on to practice in similar ways later in life: that is, they engaged in misconduct while working at hospitals that did little about their behavior, and they worked in physician networks that approved of their behavior.

31.2.2 Case-Based Complexity Science

As Byrne recognized in his research [1, 8, 11, 12], not only is case-based method a radical departure from variable-based inquiry but also it has strong affinity with complexity science. Going even further, it also, in some very useful ways, advances the study of complex systems.

Based on Byrne, we wish to introduce in this chapter two new terms: case-based complexity science and case-based modeling. **Case-based**

complexity science is defined as scholarly activity that seeks to actively integrate case-based method with complexity science for the purpose of modeling complex systems as cases. **Case-based modeling** is defined as the set of techniques scholars use to conduct case-based complexity science.

In addition to Byrne, scholars involved in the development of case-based complexity science and case-based modeling include 2,9,10,13, and 14. The argument is simple enough. Cases are the methodological equivalent of complex systems. If one thinks about it, complexity scientists and case-based researchers make a similar argument (1) variable-based inquiry is insufficient for modeling complex systems; (2) needed instead are methods that employ an idiographic approach to modeling, one grounded in the techniques of constant comparison; (3) the whole of a case or system is more than the sum of its part; (4) and yet, the study of parts and their complex interactions, from the ground-up, including the interactions these parts have with the case or system as a whole, is the basis to modeling. We can go on. Bottom line: cases are complex systems; complex systems are cases.

The above argument, however, is as far as the similarities go. Fact is, Byrne (as well as ourselves) set case-based complexity science as its own particular approach, distinct from the approach *en vogue* within complexity science today. To clarify this distinction, several comments are in order.

31.2.2.1 Situating Case-Based Complexity Science

In the last thirty years, Academia has witnessed the emergence of what many scholars—including Stephen Hawking—call a "new kind of science." The name of this new, massively interdisciplinary science is **complexity**. While young, complexity science (like many new scientific innovations of late) has captured part of the academic and public imagination—in this case with discussions of six-degrees of separation, swarm behavior, computational intelligence, and simulated societies. This popularity, however, has come with a price: confusion over the field's core terminology and the disciplinary divisions within it. As Mitchell explains in her popular work, *Complexity: A Guided Tour* [15], while it is popular to refer to complexity science in the singular, "*neither a single science of complexity nor a single complexity theory exists yet, in spite of the many articles and books that have used these terms*" (2009, p. 14).

If one follows Castellani and Hafferty [2], however, complexity science's confusion over terminology has less to do with its age, and more to do with its interdisciplinary and therefore interstitial (between things) character. Interstitial areas of thinking, no matter how novel, replicate the dominant intellectual divisions of academia, such as science versus theory or qualitative method versus statistics. Complexity science, given that it situates itself within the full range of academic inquiry—from the humanities and the social sciences to mathematics and the natural sciences—is replete with such divisions. As such, while oriented toward the study of complex systems in general, scholars in complexity science find themselves struggling with significant divisions regarding the complexity theories they use, the methods they employ, the epistemologies upon which they rely, and the definitions of a complex system they embrace. Given these divisions, a few clarifications are in order—all of which help us to understand better the goal of case-based complexity science.

1. The first clarification concerns the goals of science. As mentioned by Mitchell [15], complexity science is really the complexity sciences. To date, complexity science can be organized into several competing types, based on different combinations of the dominant distinctions in academia [16].

For Byrne (and for us), one of the most important distinctions is between what Morin [16] calls restricted versus general complexity science. **Restricted complexity science** is popular in economics and the natural sciences. It is defined as the empirical study of complex systems via the methods of rule-based, computational modeling. Its goal is quasi-reductionist, as it seeks to identify and explore the set of rules out of which complex systems emerge, so it can generate quasi-general laws about complex systems. In contrast is **general complexity science**, which is defined as the empirical study of complex systems via the broader methods of the humanities and the sciences. Its goal is more qualitative and holistic, seeking to model complex systems to create context-specific, grounded theoretical understandings of complex systems. **Case-based complexity science** situates itself in the latter approach.

As Klüver and Klüver make clear in their book *Social Understanding: On Hermeneutics, Geometrical Models and Artificial Intelligence* [17], most sociological phenomena are simply too complex to be reduced to the emergent consequence of rule-following. A more general approach, as Byrne explains [12], is one that that acknowledges this point: context and messiness and the mutual influence of macroscopic and microscopic structures and dynamics are crucial to understanding social systems.

2. The second clarification concerns computational modeling. A defining feature of the complexity sciences (restricted and general) is their reliance upon the latest developments in computational modeling. As Mitchell [15] explains, while the complexity sciences offer scholars a handful of new concepts (autopoiesis, self-organized criticality), their major advancement is **method**. Case in point: one can go back to the 1800s to Weber, Marx, Pareto, or Spencer to find reasonably articulate theories of society as a complex system; or, one can go back to the 1950s to systems science and cybernetics (or, more recently, social network analysis in sociology) to find many of the concepts complexity scientists use today. Despite their theoretical utility—which, albeit critically received, is widespread—all the aforementioned theories ultimately stalled in terms of the study of complex systems because (amongst other reasons) they lacked a successful methodological foundation.

Computational modeling is the usage of computer-based algorithms to construct reasonably simplified models of complex systems. There are three main types of computational models used in complexity science: agent (rule-based) modeling, network (relational) modeling, and dynamical (equation-based) modeling. Different methods yield different results. Situating itself within the latest advances in computational modeling, case-based complexity science seeks to use these tools. Byrne [12] and Uprichard [14], for example, use cluster analysis; and our own work employs agent-based modeling, cluster analysis, neural nets, and network analysis [3]. But, the focus is on comparing cases and searching for common case-based profiles, as concerns a particular health outcome. The consequence of this focus is the causal model built—not the techniques used. Focusing on cases is a search for profiles: context dependent assemblages of factors (k dimensional vectors) that seem to explain well for example different types of health outcomes. For example, one could use computational modeling to examine a set of health factors (e.g., income level, education, gender, age, and residential location) to see which case-based assemblage of these factors relate to differences in mortality rates.

3. The third clarification concerns the distinction between complexity science and complexity theory. Like complexity science, there are multiple complexity theories, which form a loosely organized set of arguments, concepts, theories, and schools of thought from across the humanities and the social sciences that various scholars use in a variety of ways to address different topics.

In terms of intellectual lineage, these theories are strongly grounded in two intersecting epistemological and theoretical traditions: the one stems from systems theory, Gestalt psychology, biological systems theory, second-order cybernetics, and ecological systems theory; while the other stems from semiotics, post-structuralism, feminism, postmodernism, constructivism, constructionism, and critical realism [2].

Complexity theories and their related epistemologies are also tied up in the substantive systems theories of sociology, anthropology, political science, economics, psychology, and managerial studies. As such, complexity theories can differ dramatically from one another. For example, Niklas Luhmann uses complexity theory to articulate a new, *metaphorical* theory of global society (a grand theory with no agents, only a communicating society); while John Holland uses complexity theory to build a bottom-up, agent-based *computational theory* of complex emergent systems.

Perhaps the sharpest distinction between complexity theory and complexity science, however, is that neither necessarily has affinity for the other. In fact, complexity theories need not be data driven, empirically grounded, computational, or scientific. They can even be anti-data, anti-empirical, anti-computation, and anti-scientific. For example, Francois Lyotard uses early *empirical* research in complexity science (mainly chaos theory) to end grand narrative and place a limit on the conditions of science, which he called post-modernity. Meanwhile, most scholars in the managerial sciences use complexity theory in a *prescriptive* manner, with almost no empirical backing whatsoever [16]. In contrast, the complexity sciences, while reliant upon key concepts from complexity theory, such as self-organization or emergence, tend to ignore theory (Mitchell 2009). For example, most rule-based complexity science is theoretically vacuous.

Given the above distinctions, the generalist approach of case-based complexity science is grounded in a post-positivistic epistemology, albeit one that has learned from the errors and shortcomings of much of postmodernism and post-structuralism. This seasoned viewpoint is best described as complex realism, which combines Bhaskar's critical realism with Cillier's understanding that knowledge and the world are complex interdependent processes. Together, these two ideas form what Byrne calls *complex realism*. Here is an all-too-short overview of its main point. For an in-depth review, see Byrne [12]. Complex realism seeks to overcome two key problems.

The first is epistemological. Why is reality so hard to comprehend? Is it because our minds cannot know reality? No, it is not because we are immured within a solipsist (simulated) mind-constructed view of the world. Complex realism explains that much of the contingency in knowing (causal modeling) is not because reality cannot be apprehended. Reality escapes us because it is fundamentally complex, both in terms of the real and the actual.

Second, in relation to this complexity, we have a methodological problem. **Quantitative modeling** (statistics) fails us because it does not know how to model complexity and is lost in a reductionist world of variables and parsimony. In turn, **qualitative modeling** limits itself because it cannot deal with generalization and often falls prey to problematic post-positivist ideas, such as post-modernism and radical post-structuralism. **Restrictive complexity** limits itself because it fails to actually address complexity, primarily in the form of context and contingency—that is, the manner in which things practiced are done so uniquely and done so in contextual frames, larger complex systems, etc. Complex systems are more than just rules. **Conventional case-comparative method** has all the methodological tools and the epistemological basis, but it does not have yet an explicit theory of complexity and complex systems. Finally, **equation-based modeling** cannot get beyond the dynamics of simple systems. So, what is the solution? Complex realism coupled with a generalist complexity theory coupled with case-comparative method—that is the solution. The link pin to this 'trifecta coupling' is the idea that cases are the methodological equivalent of complex systems. If reality and our knowledge of it is complex, then complexity is the issue to address. If complex systems are cases, then complex systems cannot be reduced to some set of rules or variables, and context has to be explicitly modeled. If cases are complex systems, then case-based researchers need a wider explicit vocabulary grounded in a wider set of methods, including computational modeling.

So, how do these clarifications help us contextualize the SACS Toolkit? The SACS Toolkit is part of the case-based complexity science agenda. It was designed to be the first explicit case-based modeling method designed for modeling complex social systems. Epistemologically speaking, it embraces a generalist complexity science and complex realism perspective, tempering this approach with an equal embrace of Michel Foucault's post-structuralism (which it uses to develop its theoretical framework) and Richard Rorty's neo-pragmatist understanding of the tool value of modeling; that is, scientific models of complex systems are true insomuch as they work, not because they gain a direct understanding of a complex system in its entirety. With these clarifications established, we turn now to a review of how the SACS Toolkit works.

31.3 The SACS Toolkit

The SACS Toolkit is a case-based, mixed-method, system-clustering, data-compressing, theoretically-driven toolkit for modeling complex social systems. It is comprised of three main components:

1. First, it is comprised of a theoretical blueprint for studying complex systems called **social complexity theory**. Social complexity theory is not a substantive theory; instead, it is a theoretical framework comprised of a series of key concepts necessary for modeling complex systems. These concepts include field of relations, network of attracting clusters, environmental forces, negotiated ordering, social practices, and so forth. Together, these concepts provide the vocabulary necessary for modeling a complex system.
2. Second, it is comprised of a set of case-based instructions for modeling complex systems from the ground up called **assemblage**. Regardless of the methods or techniques used, assemblage guides researchers through a seven-step process of model building—which we review below—starting with how to frame one's topic in complex systems terms, moving on to building the initial model, then on to assembling the working model and its various maps to finally ending with the completed model.
3. Third, it is comprised of a recommend list of case-friendly modeling techniques called the **case-based toolset**. The case-based toolset capitalizes on the strengths of a wide list of techniques, using them in service of modeling complex systems as a set of cases. Our own repertoire of techniques include k-means cluster analysis, the self-organizing map neural net, Ragin's QCA, network analysis, agent-based modeling, hierarchical regression, factor analysis, grounded theory method, and historical analysis.

As stated earlier, the SACS Toolkit is a variation on Byrne's [1, 11, 12] general premise regarding the link between cases and complex systems. For the SACS Toolkit, case-based modeling is the study of a complex system as a set of n-dimensional vectors (cases), which researchers compare and contrast, and then condense and cluster to create a low-dimensional model (map) of a complex system's structure and dynamics over time/space.

Because the SACS Toolkit is, in part, a data-compression technique that preserves the most important aspects of a complex system's structure and dynamics over time, it works very well with databases comprised of a large number of complex, multi-dimensional, multi-level (and ultimately, longitudinal) variables. Compression can be done using a variety of techniques, from qualitative to computational.

It is important to note, however, before proceeding, that the act of **data compression** is different from reduction or simplification. Data compression maintains complexity, creating low-dimensional maps that can be "dimensionally inflated" as needed; reduction or simplification, in contrast, is a nomothetic technique, seeking the simplest explanation possible. This distinction is crucial. At no point during the model building process is the full complexity of a system lost. Searching for the most common case-based configurations and patterns amongst the data is a way of generating a causal model, upon which the full complexity of a topic can be arranged, managed, and further data-mined. For example, while cluster analysis identifies the most common profiles in a database, one still knows which cases belong to which profiles and the degree to which they belong. Consider Fig. 31.1 as a demonstration, which comes from a study we conducted on community-level health disparities in a county

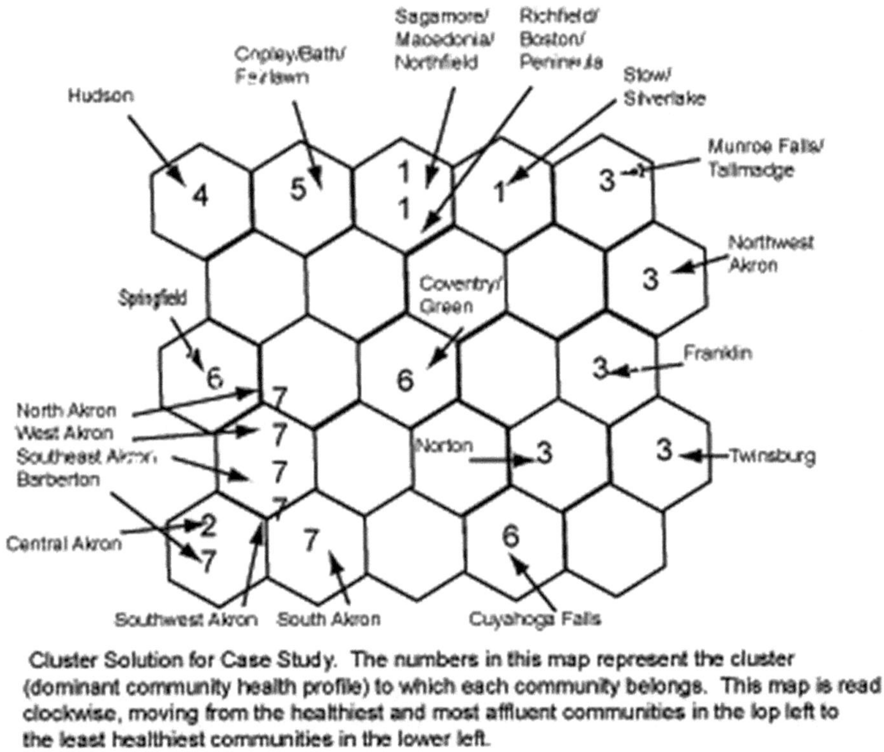

Fig. 31.1 Cluster solution for 20 communities in summit county

of 20 communities [4, 5]. In this database, we found seven clusters—each cluster represents one of the main profiles in the complex system (County) of study. Each community in Fig. 31.1 is identified by the profile (cluster) to which it belongs. One can see, however, that compression still allows us to examine every case in our database; in fact, we could (and do) go on to further cluster and differentiate any one cluster into further profile gradations, such as different types of poor communities. It all depends upon the level of granularity sought.

The SACS Toolkit is also versatile and consolidating. The strength, utility, and flexibility of the SACS Toolkit come from the manner in which it is, mathematically speaking, put together. The SACS Toolkit emerges out of the assemblage of a set of existing theoretical, mathematical, and methodological techniques, and fields of inquiry. The "assembled" quality of the SACS Toolkit, however, is its strength. While it is grounded in a highly organized and well defined mathematical framework, with key theoretical concepts and their relations, it is simultaneously open-ended and therefore adaptable and amenable, allowing researchers to integrate into it many of their own computational, mathematical, and statistical methods. Researchers can even develop and modify the SACS Toolkit for their own purposes.

31.3.1 The SACS Toolkit Updated

To date, we have written two pieces that address the SACS toolkit as a method. First, there is our book, *Sociology and Complexity Science: A New Field of Inquiry* [2], which provides a theoretical and qualitative overview, including a historically grounded case study. The second is our article *Case Based Modeling and the SACS Toolkit: A Mathematical Outline* [3], which provides a mathematical overview, including a quantitatively grounded case study. Moving from the book to the article, in addition to providing a mathematical foundation to our method, we made several major advances, most of which had to do with assembling what is called the network of attracting clusters—which we explain below. In the current chapter, we make two more minor advances. First, as we did in the previous section, we clarified the explicit links the SACS Toolkit shares with case-based complexity science and case-based modeling. In this section we make a second minor advance: we integrate the qualitative and mathematical outlines of the assemblage algorithm into a new updated version, representing current practice. Let us explain.

Given the limited space of a book chapter we felt the most useful rendition of the SACS Toolkit would be to lay out the updated assemblage algorithm, threaded with key conceptual and methodological points found in our general reviews of the SACS Toolkit. In this way we aim to provide a holistic, albeit brief, sketching of the toolkit that will hopefully give the reader enough grounding to tackle past treatments and future developments of the SACS Toolkit—interested readers could turn to our book or article for more information. And so, beginning with recasting a topic in complexity terms, through model construction and to the final leveraging of the model to answer the researcher's question(s), we now conduct a brief excursion through the Assemblage process. As a side note, we will use the following examples from our recent research: the first on medical professionalism [6, 7] and community health [4, 5]. We cite them here so we do not have to repeat them below.

31.3.2 Assembling a Complex System: A Basic Overview

The assemblage algorithm involved a series of seven basic steps. They are outlined as follows:

31.3.2.1 Converting One's Topic into a Complex Systems Framework

This first step comes in two parts. To begin, researchers need to conceptualize their topic in complex systems terms. For those relatively new to the study of complex systems, we recommend a starter text that defines the basic concepts for a complex system, such as Mitchell [15]. Conceptualizing a topic in complex systems terms means looking at your topic as a complex system/network and asking yourself such questions as (1) What will be gained by studying my topic as a complex system? (2) Do I really know what it means that my topic is emergent or self-organizing? (3) Can I thinking about my topic as a system or a network, evolving over time? Or (4) can I think of my topic in terms of the interactions amongst variables and parts, rather than the parts themselves? For example, in our study of medical professionalism we turned the concept on its head, realizing that professionalism could be thought of as a complex system, instead of a single entity. Furthermore, as a system, we saw medical professionalism comprised of several competing types (dominant profiles around which cases cluster) of professionalism, a few of which were vying for control over its future trajectory; namely, nostalgic, entrepreneurial, and lifestyle professionalism. This was an extremely novel way of thinking about professionalism. As another example, in our study of community health we examined a Midwestern county in the USA, treating its 20 major communities (cases) as a complex system—Figure 31.1, which we discussed earlier, is the cluster map for these 20 communities. Thinking about this county as a complex system was novel insomuch as it forced us to think of its 20 communities as a network, interrelated, and interdependent.

Once a topic is recast in complexity terms researchers are ready to pose a complex systems research question. This second part is just as challenging as the first, as researchers really need to make sure that the research question is a complexity science one, and not a conventional question wrapped in the new language of complexity science. For example, in our study of medical professionalism we examined how medical professionalism, as a complex system, has evolved over the last decade, including which professional types were the most dominant. This proved to be a very novel question, which only thinking in terms of complex systems could have allowed, as previous research treated

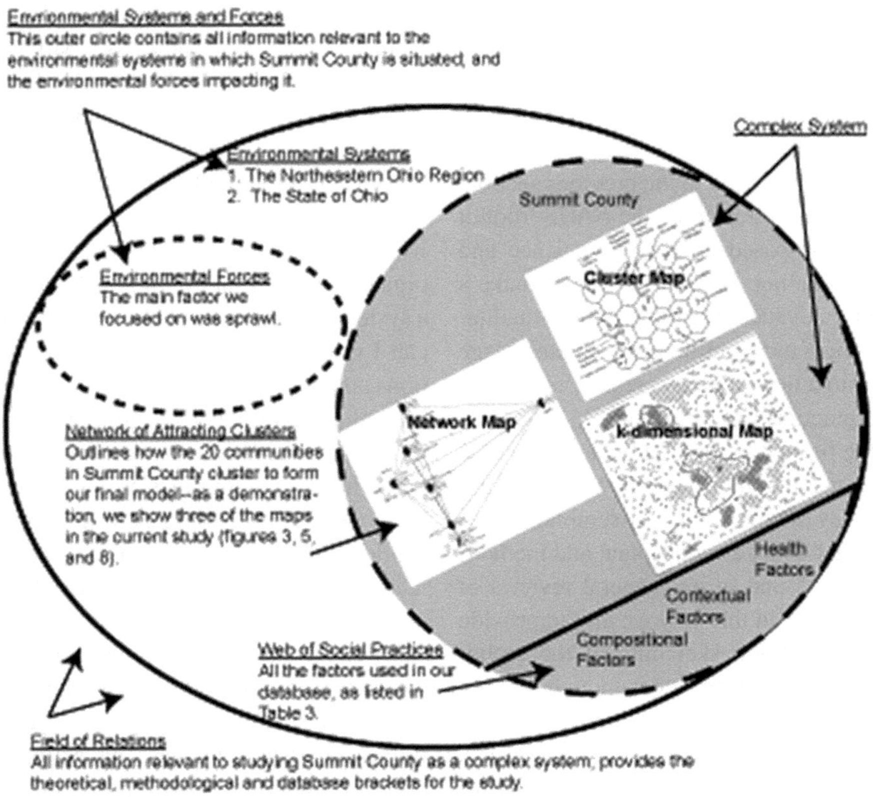

Fig. 31.2 Example of the initial model created for our community health Study

professionalism as a singular profile, and anything else as a deviation from it. As another example, in our community health study, thinking about the county as a network allowed us to ask how the health and wellbeing of the more affluent communities, across time/space, have kept the poor communities caught in a poverty trap, despite all efforts to get out of this poverty—again, a very novel question that required a complex systems viewpoint.

31.3.2.2 Building an Initial Model

Perhaps slightly counter-intuitive, the second step in the assemblage algorithm is to construct the initial model of the complex system of study. This initial model is essential because it forces researchers right from the beginning to see their topic as a complex system and to begin, albeit very fuzzily, to employ the vocabulary of social complexity theory in the building of their model. It is also crucial because it helps to define how the database for the study should be built and developed across time and what sorts of techniques to use to assemble and data-mine the model.

In our study of medical professionalism, for example, we built an early model depicting the fight in the 1990s and 2000s between three major professional types (nostalgic, entrepreneurial, and lifestyle) as they sought control over the future of medical professionalism—see our article for a picture of the model. This was very helpful because it pushed us to see how professionalism could be a system of competing types and how these different types could and were influencing the ethical behavior of physicians in the USA. As another example, as shown in Fig. 31.2, in our study of community health, our initial model was a conceptual map of our county, onto which we projected the various issues we saw contributing to the health disparities between the affluent and poor communities. This was very useful because it got us thinking about our topic as a system, helped us to identify the cases and factors we wanted to study, and helped us build our database.

31.3.2.3 Constructing the Database

Initial model in hand, the third step in the assemblage algorithm is to construct the database. The database—qualitative, numerical, or otherwise—can be usefully thought of (and, in the case of numerical data, actually assembled) as a table. In such a table, the rows are the cases. Each case is composed of a series of factors (aka variables, sets, etc). Each factor is a column. Together, these columns form the k-dimensional profile of the cases. Profiles are comprised of two types of factors: social practices and environmental forces. Social practices refer to all the factors that make up a complex system of study; environmental forces are those impacting the complex system of study; they are all put together to make the profile for the cases [3]. Construction of the database can either proceed from social practices and environmental forces to cases or from cases to social practices. Key to building the database is adopting a "data-mining" perspective: that is, researchers need to realize that modeling complex system is an iterative, evolving, and dynamic process that pushes the database to grow and change as the model is fleshed out and the framework is applied to the topic.

For example, in our study of medical professionalism, the social practices used to construct our case profiles had to do with ten key aspects of medical work. These practices included values such as altruism; skills such as interpersonal communication; economic practices such as entrepreneurialism; and personal beliefs such as an emphasis on lifestyle. In turn, environmental forces included commercialism and government regulation. In our community health study, the practices included individual factors such as household income and community level factors such as quality of school system. The main environmental force we examined was suburban sprawl.

31.3.2.4 Constructing the Field of Relations

Here, in Step 4, is where the actual model building process really gets going: creating the **field of relations**. As the term implies, the SACS Toolkit is all about relationships: the similarities and dissimilarities amongst the cases based on differences in profile; the relationships (ties, connections links) amongst the cases as a network; and the relationships amongst the factors making up the profiles. These are the three main relationships the SACS Toolkit studies, and we will visit them all again in Step 5 as we assemble the network of attracting clusters. The only difference in the next step is that we will seek to condense these relationships from full matrices to simplified maps.

The first set of relationships is referred to as the **proximity matrix**; the second as the **adjacency matrix**; and the third as the **correlation matrix**—all three matrices are algebraic terms to describe what these relationships look like mathematically. However, we use the same terms for qualitative and historical inquiry, primarily for the purposes of consistency, noting however that the construction of such matrices in these latter forms will be more loosely defined and assembled and, perhaps, even metaphorical.

Proximity Matrix: The proximity matrix is the most important of the matrices, as will be seen when map generation is discussed in the next step. Within the proximity matrix each case has a profile which shows how similar or dissimilar the case is from all the other cases. Similarity or dissimilarity is determined holistically, by looking at all of the dimensions (the complete profile) of a case and comparing them to the dimensions of other cases. Qualitative researchers can build such a matrix in any qualitative software package; quantitative studies can build such a matrix in any statistical software package. For example, in our study of medical professionalism, we used all historical and qualitative data and therefore built our profiles and sorted (compared and contrasted) them by considering a variety of cases we found in the academic medicine literature, newspapers, and through first-person qualitative interviews. In our community health study, we used an actual statistical database matrix, so the initial sorting of cases was done statistically and computationally—See Castellani and Rajaram [3] for a complete explanation.

Adjacency Matrix: Secondly there is the adjacency matrix which displays relationships or links amongst cases in the database. Think here of the new science of networks and hubs, strong ties, and small worlds, etc. The focus in this second matrix is identifying what the key relationships amongst the cases are relevant to some question of interest. In other words, whereas the proximity matrix compares all of the dimensions of one case to another to determine their similarity or dissimilarity, the adjacency matrix looks for ties or connections between two cases depending on their values on one of their shared dimensions. For example, in our medical professionalism study, we thought of the seven major professional types as a network of cases, with each case labeled according to its type and the relationships amongst these cases having to do with professional relationships of one type or another.

As our example of medical professionalism suggests, because each case contains many dimensions, it is possible to build multiple adjacency matrices. The key point in building these matrices is to look for meaningful or informative relationships among the cases that will help the researcher model their topic. For example, in our medical professionalism study, we were very interested in, from a network perspective, how students learn their professionalism, through their interactions and relationships with peers and clinical faculty. So, this has been one of the networked relationships we have been exploring.

Correlation Matrix: Finally there is the correlation matrix which consists of relationships among the social practices and environmental forces themselves. In full matrix form this means conceptualizing statistically all pairwise correlations for the factors in a study. In more practical terms, the relationships between factors could be approached with a variety of techniques or perspectives. For example, in our community health study, we were very interested in the link between micro-level residential mobility behaviors (where people moved over time) and community-level health outcomes (how residential migration patterns lead to the segregation of rich and poor communities). Whereas in our medical professionalism study we were interested in the correlation between altruism and commercialism—two of the ten factors on medical work we used to construct our profiles.

31.3.2.5 Constructing the Network of Attracting Clusters

On their own, the three matrices comprising the field of relations provide a fundamental mapping of the relationships in the database but, particularly when there are a large number of cases or dimensions (as the SAC Toolkit was designed for), are less useful for modeling the topic. Their usefulness is limited because of the overwhelming amount of data generated when analyzing all possible relationships in the database, which make identifying key relationships and patterns among the cases difficult. To overcome this limitation the next step in assemblage is to condense and compress the field of relations into a series of maps: the cluster, network, and dimensional maps, collectively known as the **network of attracting clusters**.

Compression involves taking the field of relation's different matrices and condensing all the relationships therein to a smaller set of salient and common patterns of relationships. Each matrix has its own type of map: the proximity matrix is turned into cluster maps; the adjacency matrix is turned into network maps; and the correlation matrix is turned into dimensional maps.

Together, these maps make up the network of attracting clusters, which, as the name implies, is a model that combines maps of the main profiles (clusters around which the cases cluster) in a system and the relationships these clusters and their cases share as a network, as well as the relationships amongst the dimensional factors of which they are comprised.

Cluster Maps: As we have discussed repeatedly in this chapter, the cluster maps seek to identify the most common profiles among cases in the system by grouping or clustering cases with common profiles. As shown in Fig. 31.1, cases are grouped around an attractor point, or centroid around which similar cases cluster. All cases are clustered in this way, compressing the collection of cases to a smaller collection of clusters that each contains a set of cases. A series of clusters thus enables comparison across groups of similar cases, instead of requiring that every case be compared with every other case—which is often impossible in large databases. Clustering can be done using traditional case-comparative method, Ragin's QCA or computationally, as in our usage of k-means cluster analysis and the self-organizing map neural net—see 1 and 3 in references for more information.

It is possible to generate multiple cluster maps by starting off with different types of cases or by clustering at different levels (e.g., a few big clusters or many smaller clusters). The proximity matrix can be compressed using a variety of techniques, depending upon the data used. In the case of numeric data, compression can be done using cluster analysis and the self-organizing map algorithm; for qualitative data basic sorting techniques can be used.

Note: in small databases, clustering may be very simple, amounting to little more than sorting five or seven cases. In such instances, compression may not be necessary; and so the proximity matrix and the cluster map are similar. Most databases, however, require some type of compression.

Network Maps: Network maps are directly informed by the cluster maps. Various techniques commonly used in network analysis and the new science of networks (e.g., hubs, degrees of separation, cliques, etc) are applicable for generating this kind of map. The goal is to go beyond just finding common profiles, as in Ragin's QCA, to understand how these profiles are influenced by the relationships and interactions amongst the cases. For example, in our medical professionalism study, we discussed how clinical faculties transmit their professional type to students through the hidden curriculum. Again, in a small database, the network map and the proximity matrix may be similar.

Dimensional Maps: Dimensional maps explore the relationships either between dimensions within a case or between dimensions in different cases. While the SACS Toolkit is primarily concerned with cases and relationships amongst them, dimensional maps are still useful for understanding dynamics within a case and in some instances between them, which both inform the larger focus of studying cases. Further, studying dimensional relationships across cases can inform network maps, as they suggest interactions or links between cases. Many techniques can be used to study these relationships from qualitative, statistical, agent based, to equation based. For example, in our medical professionalism study, we examined how commercialism (as an environmental force) impacted the importance of altruism in the various professional types.

Stitching the Model Together: Once the above maps have been generated, it is time to bring them together to create an integrated model. Looking over the maps, researchers at this stage should ask themselves: (1) What do these maps tell us about the model as a whole (for one point in time/space)? And (2), what do the maps tell us about the cases? Using these questions as prompts the researcher takes the three maps to create a complete network(s) of attracting clusters or single map of the given time point.

The aim for this stage is to make a map of the **negotiated ordering** of the system for this point in time/space. Negotiated ordering, as the name implies, involves the spatial and conceptual arrangement of the clusters in relation to each other, the dynamics and relationships between the clusters, how their ordering relates to the larger system in which they are part and eventually (as time points are added) negotiated ordering also incorporates the trajectory of the clusters over time. For example, Fig. 31.2 is a rough sketch of how all the maps in our model came together to create a picture of the county we studied and its 20 communities.

Furthermore, if multiple cases or cluster levels are examined or different network linkages are examined, it is also possible to make different network of attracting clusters for each time point. For example, in our medical professionalism study, we examined medical professionalism at three different levels: micro, meso, and macro, examining how our three maps came together at each level and across levels.

Time/Space: After a network(s) of attracting clusters has been completed for one time period this is repeated for other time periods, from the database to the matrices, to the initial maps and a new final network(s) of attracting clusters. This is an important point. Complexity science is ultimately about modeling complex systems across time/space.

The SACS Toolkit follows the same logic (1) Cases are not static they are dynamic and change over time. (2) Therefore, as a case based methodology, the best way to study a system is to study how the cases (or clusters) develop across time/ space.

Deciding how many time-points a topic should examine should be informed by the questions the researcher is attempting to answer as well as the topic that is being studied (e.g., some systems are more stable than others, and different levels of stability may call for more time periods). To model the system across time, the network of attracting clusters is assembled on a timeline as a series of discrete time points. These discrete time points can be connected longitudinally using qualitative, computationally, statistically, or other techniques. In particular the researcher should examine the negotiated ordering of the cases/clusters present in the network(s) of attracting clusters to understand how the system is arranged and changes over time. For example, in our community health study, we examined how the 20 communities in our county evolved over a ten-year period (specifically in terms of sprawl) and the impact this had on community-level health outcomes. In our medical professionalism study, we conceptualized it as a social movement that has been evolving over the last two decades through the negotiated conflict amongst these major professional types, and in response to environmental forces and their interplay with these types.

Validity Checking: Given that just about anything in the health sciences can be seen as complex, researchers need to be careful that their study is nothing more than the same old ideas restated in the fancy language of complexity science. As such, throughout the modeling process it is necessary to do a series of validity checks. Key questions researchers should ask themselves are (1) Does modeling this topic as a system offer any substantively, theoretically, or methodologically meaningful new insights beyond conventional modes of study for this topic?; and (2) Am I forcing my topic to fit the SACS Toolkit framework or does it naturally connect to and develop with the framework? If the model passes these questions, then it is time to transition to conclude the study.

31.3.2.6 Concluding One's Study

At some point it is necessary to end model construction. Like Step 1, this last step is in two parts: drawing a study to close and answering the initial research question.

A strong signal that it is time to advance to the next part of the model or to end model construction overall is reaching the saturation point. The saturation point is the time in which adding new parts to the complex model or generating new/ different maps yield marginal insights or differences from past attempts. Another sign that model construction or the present step should come to a close is when the researcher violates the validity check on forcing the model to fit the topic. If new additions or maps require forced or drawn out explanations, it is likely approaching time to end the given stage of model construction. One thing the researcher should be wary of is iterative looping disorder—the need to keep iterating on the model for fear of missing some details. Saturation or violation of the validity check is an indication that iterations should come to an end.

Finally the last step in assemblage is one that has already begun throughout model construction: answering the research question. The model settled upon as iterations on the working model come to a close become the final model, a series of network(s) of attracting clusters, used to study the topic of interest. One final point is that not all parts of the model may be necessary to answer the research question that motivated the study or some parts of the model may be more salient for the particular question.

For example, in our medical professionalism study, we realized that there is almost no end to the detail we could explore, given our model is conceptualized simultaneously at the macro, meso, and micro level. So, we have built our general model and have realized that we can go back to this model, repeatedly, to data mine it (and, also, add or develop new data) to address specific questions we want to explore, such as how, in medical schools, new students learn their professional type through theirs social networks and the larger informal and hidden curriculum in which they are situated. In our community health study, we only really addressed one key environmental force, sprawl, and its impact on community-level disparities in health. But, there are so many other forces that can be addressed, such as the evolution of the health care systems that care for the people living in our county of study.

References

1. Byrne D, Ragin C, editors. The sage handbook of case-based methods. CA: Sage; 2009.
2. Castellani B, Hafferty F. Sociology and complexity science: a new field of inquiry. Germany: Springer; 2009.
3. Castellani B, Rajaram R. Case-Based Modeling and the SACS Toolkit: A Mathematical Outline. Computational and Organizational Theory; forthcoming.
4. Rajaram R, Castellani B. Modeling complex systems macroscopically: case/agent-basedm modeling, synergetics and the continuity equation. Complexity, 2012. Online version: http://onlinelibrary.wiley.com/doi/10.1002/cplx.21412/abstract.
5. Castellani B, Rajaram R, Buckwalter G, Ball M, Hafferty F. Are communities complex systems? A case study and empirical test. Proceedings of the Center for Complexity in Health; 2011.
6. Hafferty F, Castellani B. The increasing complexities of professionalism. Acad Med, Flexner Centenary Issues. 2010;85(2):288–301.
7. Castellani B, Hafferty F. The complexities of professionalism: a preliminary investigation. In: Delese Wear, Julie Aultman's (Eds) Professionalism in medicine: critical perspectives. New York: Kluwer Academic Publishers, 2006
8. Byrne D. Complexity theory and the social sciences. London: Routledge; 1998.
9. Rihoux B, Ragin C, editors. "Configurational Comparative methods: qualitative comparative ANalsysi (QCA) and related techniques". Applied social research methods Series, vol. 51. CA: Sage; 2009.
10. Ragin C. Redesigning social inquiry: fuzzy sets and beyond. Chicago: University of Chicago Press; 2008.
11. Byrne D. Complexity, configurations and cases. Theory, Culture and Society. 2005;22(5):95–111.
12. Byrne D. Using cluster analysis, qualitative comparative analysis and invivo in relation to the establishment of causal configurations with pre-existing large-N datasets: machining hermeneutics. In: Byrne D, Ragin C, editors. The Sage handbook of case-based methods. CA: Sage; 2009. p. 260–8.
13. Blackman T. Placing health: Neighborhood renewal, health improvement and complexity. UK: The Policy Press; 2006.
14. Uprichard E. Introducing cluster analysis: what it can teach us about the case. In: Byrne and Ragin's Sage handbook of case based methods, Sage Press, CA. 2009; pp 132–147
15. Mitchell M. Complexity: a guided tour. New York, NY: Oxford University Press; 2009.
16. Richardson K, Cillers P. Special editors note: what is complexity science? A view from different directions. Emergence. 2001;3(1):5–22.
17. Klüver J, Klüver K. Social Understanding: On Hermeneutics, geometrical models and artificial intelligence. Springer 2009

The Drug Prescription Process: A Network Medicine Approach

32

Pierpaolo Cavallo, Sergio Pagano, and Luna Carpinelli

32.1 Network Science and Network Medicine

This section presents the definition of Network Science, and its fundamentals and attributes, the main mathematical aspects of the study of networks and the parameters, or quantitative indicators, used to study and analyze a network from a mathematical point of view. It introduces the concept of "Network Medicine," the science that studies biological and medical phenomena from the point of view of networks.

32.1.1 Network Science: Fundamentals and Attributes

A common definition of Network Science is "… a new and emerging scientific discipline that examines the interconnections among diverse physical or engineered networks, information networks, biological networks, cognitive and semantic networks, and social networks" [50]. In simple words, network science is the study of natural and artificial phenomena by representing them as networks, made of nodes and links, and by determining properties, laws, characteristics, and parameters by which the networks are born, evolve, and end or transform their existence. Network Science can be applied to many different fields: physics, biology, psychology, sociology, economics, computer science, etc.

The study of networks has been, almost exclusively, the domain of a branch of discrete mathematics known as graph theory, and the discipline has seen important achievements in some specialized contexts, the most prominent of which are the social sciences. Indeed, the practice of "social networks analysis" started to develop in the 1920s of the twentieth century, and since then has been an important instrument for studying relationships among social entities, for example: the members of a group and their communications; business corporations and their economic trading networks; or nations and their political relationships.

Toward the end of the twentieth century, there was a renewed interest and research in the study of complex networks, with the publication of reference papers on the "small-world" networks [43] and on "scale-free" networks [5]. Complex networks are networks whose structures are irregular, complex, and dynamically evolving in time. Almost any kind of network mentioned at the beginning of the present paragraph (information, biological, cognitive, social, etc.) is a complex network, as reported in another, now classical, review paper [35]. Network Science, initially applied in Social Science Studies, is being adopted by Medicine and Healthcare Organizational studies. A contemporary view is that social perspectives are

P. Cavallo (✉) • S. Pagano • L. Carpinelli
University of Salerno, Via Ponte Don Melillo, 1, 84084 Fisciano (SA), Italy
e-mail: pcavallo@unisa.it

J.P. Sturmberg and C.M. Martin (eds.), *Handbook of Systems and Complexity in Health*,
DOI 10.1007/978-1-4614-4998-0_32,

closely interlinked with healthcare, and that "they remain fundamental for our understanding of health, illness, and care" [36].

32.1.2 Network Study Parameters

A network can be represented as a graph: a graph is a mathematical entity that consists of a set of nodes connected by links. We use the terms "nodes" and "links," but they are also known as "vertices" and "edges," although the two terms, in some notation, are used to indicate different kinds of elements. Figure 32.1 shows a simple example of a network composed of eight nodes and eight links.

In this example, we find a fundamental characteristic of complex networks: the number of links of each node is not constant, as there are nodes with many links (see nodes 2, 3, and 4) and nodes with a small number of links, with the appearance of an insulated individual (see nodes 7 and 8).

Networks are studied with the mathematical instruments of the graph theory, and they have some definite specific features and quantitative parameters. In an undirected graph, the link between two nodes can be considered in either direction; this link is said to be incident, and the two nodes joined are referred to as adjacent or neighboring, while, in a directed graph, the link between two nodes has a specific and irreversible direction. A good example with which to demonstrate the difference between directed and undirected networks is a social one: a parenthood relationship is, for biological and social reasons, a directed one, while a friendship is, commonly, undirected.

Graphs can be unweighted, that is, any link has the same importance, or "weight," or weighted, where each link has a specific "weight": a good example for these parameters could be a road system with and without a toll. In this example, if one has to carry freight on free roads, one would choose the route in terms of shortest distance (unweighted links) on the basis of cost, but if one has to go through toll roads, the route will likely be chosen in terms of lowest cost, summing fuel consumption plus toll costs (weighted links).

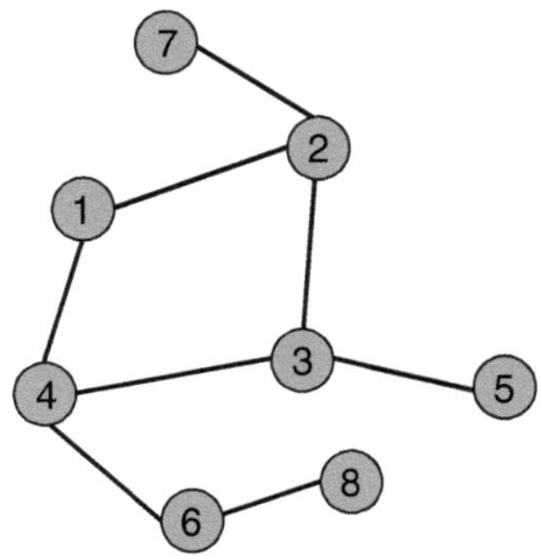

Fig. 32.1 A simple example of a network, composed of 8 nodes and 8 links

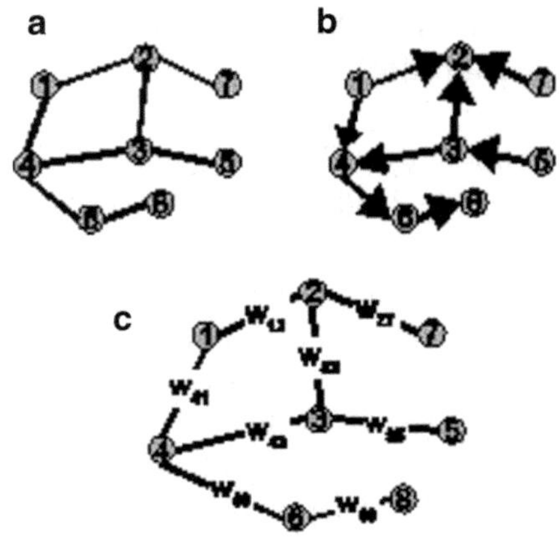

Fig. 32.2 Examples of a simple undirected, directed, and weighted network

Figure 32.2 shows some examples of simple undirected, directed, and weighted networks.

A central concept in graph theory is the possibility of connecting two different nodes of a graph: even if two nodes are not adjacent, they may, nevertheless, be reachable from one to the other, if one jumps from node to node, passing through the available links. This is called a "walk" into the graph, and has a length, that is, defined as the number of edges in the sequence; in graph theory language, a "trail" is a walk in which no link is repeated, while a "path" is a walk in which no node is passed twice. An important parameter of analysis is the "shortest path" or "geodesic distance," that is, the minimal distance between two given nodes, and a graph is said to be "connected"

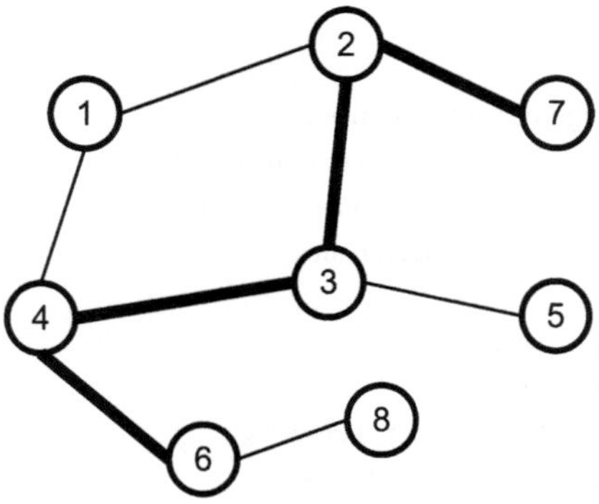

Fig. 32.3 A simple graph with the path between two nodes being highlighted. The distance between the two nodes is four

if, for every pair of distinct nodes, there is a path from one to the other, while if there are nodes present but not reachable from every other node, the graph is said to be "unconnected" or "disconnected." The maximum value of geodesic distance is called the diameter of the graph, as it represents the distance between the two farthest points of the network.

Figure 32.3 shows an example of a simple graph with the path between two nodes highlighted with a thicker line. The distance between the two nodes is four.

Another important parameter for quantitative analysis of a network is the "degree" or "connectivity" of a node. It is the number of links incident with the node; a simple way of explaining this is to compare two airports, a hub with hundreds of flights a day and a small country airport. In terms of degree, measured as the number of airline links, the former will have a degree of hundreds, while the latter will probably have a degree of less than ten.

When we measure the degree of each node in a network, for example, the number of flights on each airport, we can calculate the "degree distribution," that is, the probability that a node chosen at random has a certain degree. For the above example, there are a large number of small airports with a probability of having a small number of flights, while there are a very small number of airports having the probability of a large number of flights and these hubs support most of the activity of the air traffic network.

Table 32.1 describes the most important network parameters, their meanings, and their mathematical symbols.

32.1.3 Types of Networks

Networks can have very different appearances, depending on the number of their nodes and on the average number of the links; we have:

1. Regular networks: usually designed for a certain goal as a chain, a grid, or a regular lattice, e.g., the net of a tennis court.
2. Random networks: networks that are formed by random process: the network we obtain if we scatter a number of buttons across the floor, and tied them in couples at random with thread.
3. Small-world networks: networks that lie somewhere between the extremes of order and randomness, and have the characteristics of short paths and high clustering, i.e., there are many dense groups of nodes, very linked with one another, but single links tie the groups together, making it possible to find a "shortcut" to jump quickly from a group to another.
4. Scale-free networks: networks in which there is a small number of highly connected nodes, the so-called hubs, and a large number of nodes with a low number of links; this architecture has been inspired by the formation of the World Wide Web (WWW). The Barabasi–Albert model of scale-free networks [5] is based on two basic ingredients, growth and preferential attachment, in the sense that a very connected node will have a higher probability to receive many more links than a scarcely connected node.

Figure 32.4 shows examples of regular, random, small-world, and scale-free networks.

Scale-free networks are very important in many fields of scientific research. The scale-free nature of a large number of networks of key scientific interest has been well established [4].

The study of scale-free properties has revealed that the structure and the evolution of networks are inseparable: networks constantly change because new nodes arrive and/or new links are

Table 32.1 Most common terms related to networks

Parameter name	Description
Node or vertex	The objects that are connected together in the network
Edge or link	The connection from one node to another (or the same) in a network
Path	The sequence of links that connect a given node to another one
Shortest path	The smallest number of links between two given nodes
Size	The number of edges in a graph
Diameter	The largest distance in a connected network
Degree	The number of edges connected to a node
Indegree	The number of edges entering a node in a directed network
Outdegree	The number of edges leaving a node in a directed network
Subgraph	A part of a graph containing only part of all vertices, with their links
Complete graph	A graph in which each pair of edges is connected by one link
Weighted graph	A network in which to each link is associated a value, the weight
Strength	An attribute of a node corresponding to the sum of the weights of all the links connected to it
Loop	A link whose endpoints are the same node
Connected graph	It is possible to establish a path from any vertex to any other vertex of the graph
Adjacency matrix	A matrix representation of a network containing for each node all the nodes adjacent to it
Centrality	Measures the relative importance of a vertex within the graph
Closeness centrality	Measures how close, in terms of distance, is a node to all others of the network
Betweenness centrality	Measures the probability of a given node to be in the shortest path between any other randomly chosen nodes in the network
Clustering coefficient	A measure of degree to which nodes in a graph tend to cluster together
Distribution	Arrangement of values taken by one property (typically of nodes or links)
Binning	A way to divide an interval into sections, used to build histograms
Power law	Dependence of a quantity on the power of another, e.g., $P(k) = a\ k^b$
Fat-tail distribution	A distribution of a quantity that does not fall rapidly to zero moving away from the average value

formed and/or older links change, assuming different patterns, and/or assuming different strength. The most prominent example is the WWW that is linked by a very small number of highly connected pages. In the first study of this kind, Barabasi and Albert [5] found that more than 80% of the pages mapped in the WWW had fewer than four links, but a small minority, less than 0.01% of all nodes, had more than 1,000.

In a scale-free network, the distribution of the number of links for each node follows a so-called power law, that is, the probability $P(k)$ for a node to have k connections depends on a power of k, as $P(k) = a\ k^{-b}$ where b is a characteristic exponent.

In other words, in this mathematical relationship, the frequency of an event varies as a negative power of a quantitative attribute of that event, in the above case k is the number of connections of a certain node.

Such networks are characterized by a continuous transition from a large number of nodes having very few connections to few nodes (the hubs) having a very large number of connections.

32.1.4 Network Medicine

Network Medicine can be defined as the science that studies the network effects of biologic and medical occurrences [3]. The studies of Network Medicine range from the network-based understandings of diseases, in terms of

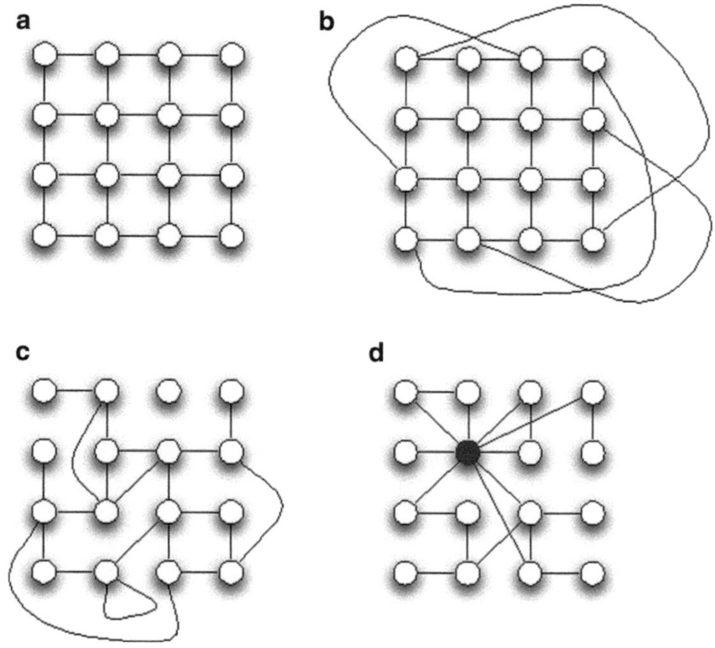

Fig. 32.4 Examples of regular, random, small-world, and scale-free networks

disease classification, to network pharmacology, in order to discover new drugs or to deepen knowledge of existing drugs' mechanisms of action in the "interactome."

To define the interactome, we have to consider that nearly all the cellular components of any living organism exert their functions through interactions with other cellular components; these "partner components" can be located either in the same cell or across cells, and even across different organs.

> In humans, the potential complexity of the resulting network—the human interactome—is daunting: with about 25,000 protein-coding genes, about 1,000 metabolites and an undefined number of distinct proteins and functional RNA molecules, the number of cellular components that serve as the nodes of the interactome easily exceeds 100,000. The number of functionally relevant interactions between the components of this network, representing the links of the interactome, is expected to be much larger [6].

The definition of the interactome, as a network of normal biologic and metabolic processes, is mirrored by the "diseasome," that is, the network of the disease, in which each disease is a node and the links are the common biological features between the diseases.

Considering the interactome from the point of view of the relationships between genes and diseases, a question emerges: are disease genes placed randomly on the interactome, or are there detectable correlations between their location and their network topology? The search for answers has led to a series of hypotheses that tie the interactome to human diseases.

Figure 32.5 shows the relationships between interactome, diseasome, and social networks.

In this scheme, we can see how the interconnections operate: the existence of intricate molecular links between subcellular components implies the possibility that one or more of these components and/or their links might fail. The failure is connected, in turn, to the possibility that a disease, the manifestation of the failure, appears, and there are links into the interactome that make possible to walk through the network, going from node to node. In other words, diseases may not be as independent of each other as medical practitioners currently consider them to be, as it has been demonstrated for the association between obesity and diabetes, in which several genes are associated to both diseases [3], and the presence of obesity elevates the risk of diabetes.

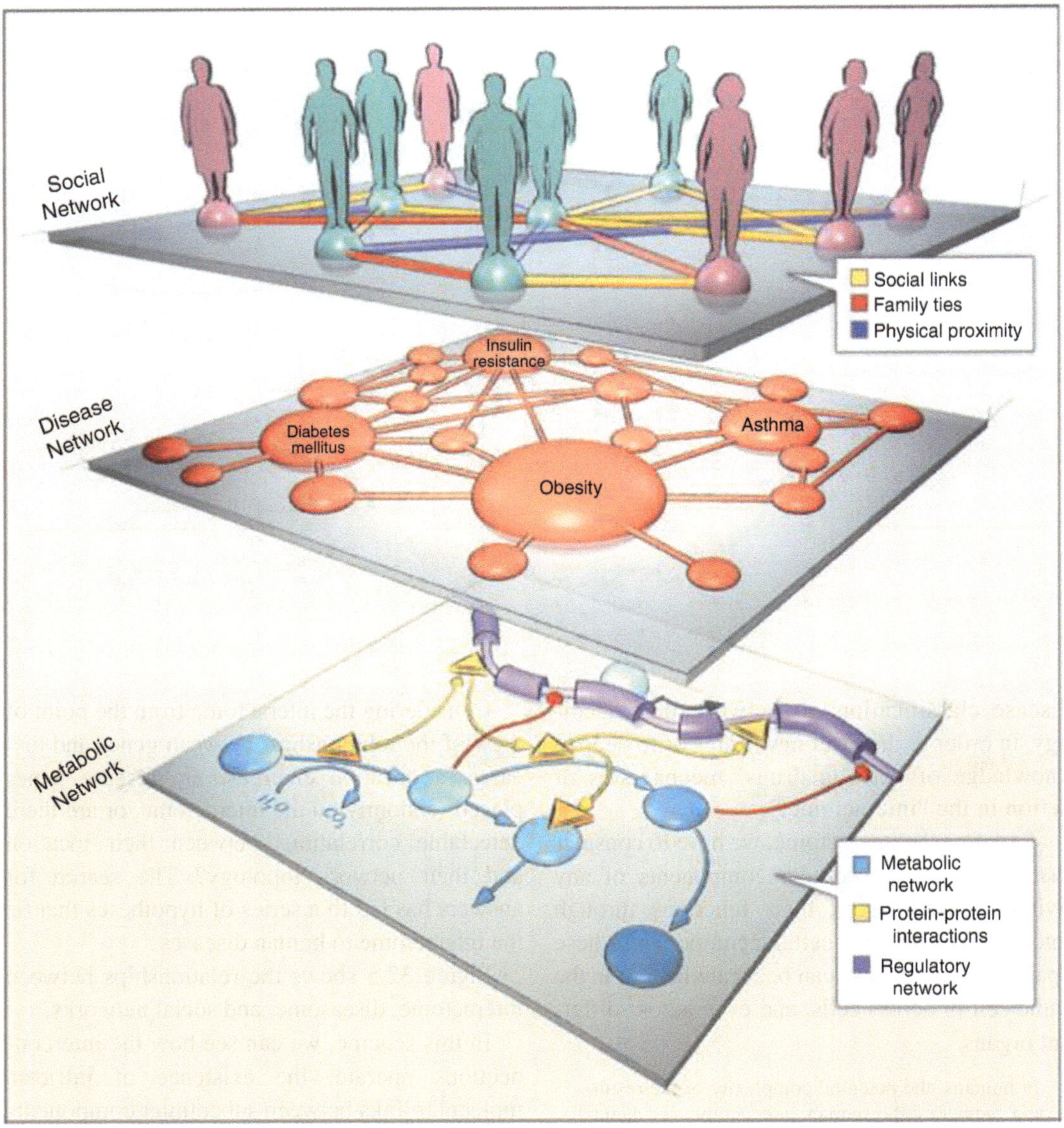

Fig. 32.5 The relationships between interactome, diseasome, and social networks

At the level of our everyday life is the social network. Networks may account for the many environmental and social influences on diseases as well. If we reflect on the numerous and disparate human interactions that encompass social and family links, proximity-based contacts, transportation networks, and the determinants of health, we can appreciate the huge scope for ongoing network research and application in this field.

The Determinants of Health [46] are the factors that combine together to affect the health of individuals and communities, and can be divided into three main groups:

1. The social and economic environment
2. The physical environment
3. The person's individual characteristics and behaviors

These are the main groups, and among them there are modifiable and non-modifiable factors.

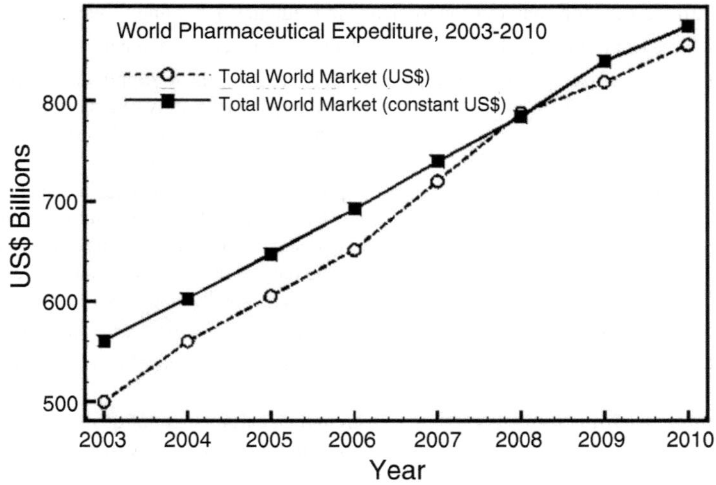

Fig. 32.6 World pharmaceutical expenditure 2003–2010

While one cannot modify one's age or genome, everyone can modify—if willing—his/her behavior, or the environment, at least to some extent.

Our social environment has a deep impact on our behavior. The social networks in which we are embedded and live contain culture, customs, traditions, and beliefs, passed on to us by our family and community, and all influence individual and community health. A clear demonstration of this has been described in a paper on the spread of obesity into a social network [11], in which the results suggest that obesity may spread in social networks in a quantifiable and discernible pattern that depends on the nature of social ties, and on the social distance between two individuals. Among mutual friends—two subjects who both declare each other to be a friend—one becoming obese increases the risk for the other becoming obese by 171%. The social tie between two subjects, each one with his/her own interactome, has acted on the spread of a disease that is largely influenced by behavioral factors.

32.2 Pharmaceutical Expense and Drug Prescription Process

This section is presents a general description of international pharmaceutical expenditure and the process of medical prescribing is summarized in its principal aspects.

32.2.1 Health and Pharmaceutical Expense

Article 25.1 of the Universal Declaration of Human Rights [41] states: "*Everyone has the right to a standard of living adequate for the health of himself and of his family, including food, clothing, housing and medical care and necessary social services.*"

As of year 2011, about 100 countries include health provisions in their constitutions, and access to medications is a fundamental aspect of the right to health. Spending on prescription drugs continues to be an important health care concern, particularly in light of rising pharmaceutical costs, the aging population, and upcoming new "molecular" drugs, which are very promising in terms of higher and wider therapeutic effectiveness and opportunities, but are also very expensive.

The Global Pharmaceutical Market value [21] rose from 500 billion US $ in 2003 to 856 billion US $ in 2010, with a growth forecast rate of 3–6% per year from 2011 to 2015 [20]. Figure 32.6 shows the world pharmaceutical expenditure from 2003 to 2010.

Unfortunately, an amount variably calculated to range from between one-third to half of the US annual health expenditure, i.e., between US $600 and 1,000 billion, is wasted, as it is spent for useless, ineffective, or inappropriate reasons.

Table 32.2 Top sources of inefficiency in medicine, according to World Health Organization (2010)

Source of inefficiency 1	Medicines: underuse of generics and higher than necessary prices for medicines
Common reasons for inefficiency	Inadequate controls on supply-chain agents, prescribers and dispensers; lower perceived efficacy/safety of generic medicines; historical prescribing patterns and inefficient procurement/distribution systems; taxes and duties on medicines; excessive mark-ups.
Ways to address inefficiency	Improve prescribing guidance, information, training and practice. Require, permit or offer incentives for generic substitution. Develop active purchasing based on assessment of costs and benefits of alternatives. Ensure transparency in purchasing and tenders. Remove taxes and duties. Control excessive mark-ups. Monitor and publicize medicine prices.
Source of inefficiency 2	*Medicines: use of substandard and counterfeit medicines*
Common reasons for inefficiency	Inadequate pharmaceutical regulatory structures/mechanisms; weak procurement systems.
Ways to address inefficiency	Strengthen enforcement of quality standards in the manufacture of medicines; carry out product testing; enhance procurement systems with pre-qualification of suppliers.
Source of inefficiency 3	*Medicines: inappropriate and ineffective use*
Common reasons for inefficiency	Inappropriate prescriber incentives and unethical promotion practices; consumer demand/expectations; limited knowledge about therapeutic effects; inadequate regulatory frameworks.
Ways to address inefficiency	Separate prescribing and dispensing functions; regulate promotional activities; improve prescribing guidance, information, training, and practice; disseminate public information.

In addition, it is likely that a large part of the annual world healthcare expenditure of US $5.3 trillion a year follows the same pattern.

Medicines account for 20–30% of global health spending [45], slightly more in low- and middle-income countries, and, therefore, constitute a major part of the budget of whoever is paying for health services. The rise in costs of prescription medicines affects all sectors of the health care industry, including private insurers, public programs, and patients. In recent history, increases in prescription drug costs have outpaced other categories of health care spending. The same report [45] puts the word "medicines" at the very first three places of the "Ten leading sources of inefficiency" in world healthcare. Table 32.2 highlights the detailed description of these first three sources of inefficiency.

The underuse of generics and higher than necessary prices for medicines is connected to a number of factors from market to prescription. This includes factors such as inadequate controls, lower perceived efficacy/safety of generic medicines, and "die-hard" prescribing patterns. There are a number of solutions which can be deployed, such as more prescribing guidance and training, assessment of cost and benefices, and an increase in transparency and monitoring.

The use of substandard and counterfeit medicines, indeed, is connected essentially to markets, being determined by inadequate pharmaceutical regulatory structures/mechanisms, weak procurement systems and, simply, illegal activities, and could be fixed by more control on quality standards. Inappropriate and ineffective use is connected essentially to prescribing, including inappropriate prescriber incentives, unethical promotion practices, induced consumer demands and inadequate regulatory frameworks. This could be addressed by separating the functions between prescribers and pharmacists, and—again, like underuse of generics—by more prescribing guidance and training.

Pharmaceutical expense, thus, is a leading topic in the larger area of healthcare expense, the former being a significant proportion of the latter, but this topic so far has been only barely touched in terms of network theory [52]. Probably a large area of research "is out there," ready to be explored, in which interesting and promising will be to apply complex systems and networks analysis instruments onto pharmacologic, clinical, and pharmacoeconomic data.

32.2.2 Medical Prescription

Medical prescription significance can vary from country to country, but, in general, its definition can be "... *a health-care program implemented by a medical practitioner in the form of instructions that govern the plan of care for an individual patient*" [49].

Medical prescription may regard a diagnostic test and/or a therapeutic intervention, and can show a very variable range of complexity, from a simple diagnostic prescription for a single test or drug to very complex combination of tests and/or therapies, with a wide range of duration, from a single administration to life-long therapy.

The prescription of drugs takes place in a complex environment and involves a number of factors, whose impact may be difficult to unravel, as the habits and behavior of the physician, the physician–patient relationship, the pharmaceutical market and its actions/effects, and the decision-making process itself, may be "*partly unconscious, based on heuristics rather than structured analysis of all relevant information, and partly based on socially less desirable motives*" [10].

Prescribing is becoming increasingly complicated, and there is evidence of poor prescribing by a range of doctors across different settings, whether from errors, under-prescribing, over-prescribing, inappropriate or irrational prescribing [1]. Overall the prescribing process is a complex task which from training to ongoing practice, is difficult to separate into its components of theoretical knowledge and safe quality-based performance [32].

Factors affecting prescribing relate to the prescriber, patient and society, medication, and/or other interventions being performed at the same time, practice environment and organization, available information and other external factors [34].

The role of these factors, excluding the prescriber-related ones, can be summarized as follows:

1. *Patient and societal related.* This group includes the patient's family and medical history, which can be related to single, specific and/or undifferentiated, or multiple illnesses. These factors include also the lifestyle, in an enlarged aspect, including not only lifestyle of patients, but also of the relatives/cohabitants, and the preferences for use of OTC (over-the-counter) medication and natural health products.
2. *Medication related.* This group includes the properties of drugs, in terms of their pharmacology, pharmacokinetics, pharmacodynamics, dosage, formulation, taste, route and ease of administration, side effects and cost.
3. *Practice environment and organization-related.* This group of factors includes the influences from the group of peers, norms, and interaction with specialists and/or opinion leaders.

   The practice environment includes both the technical support (patients and drug data) and the human resource support, which is the various possible interactions with other healthcare professions, such as nurses, pharmacists, educators, dieticians, psychologists, and health informatics experts to improve prescribing. A specific role could be played by organizational factors, such as type of practice (individual or group), number, length and frequency of patient visits, availability of access to specialists and diagnostic facilities, and the transport network, for its role to make possible for the patient to move to the physician or to the center indicated by the physician.
4. *Information and other external factors.* This group includes a huge number of sources of information about drugs, medications and every kind of therapeutic intervention.

   Some sources are directed toward the physician, including the detailing (i.e., the office visits of drug sales representatives), but also a growing activity of "cyberdetailing," also called "e-detailing," that is, a form of web-based detailing, and the provision of drug samples; to the medical world, of course, are also directed the classic marketing instruments, like targeted mailings, websites and call centre calls, plus—last but not least—the specific medical instrument of sponsored scientific conferences.

Some sources are directed to the patients, and are called DTC (or "DTC-ad") that is "direct-to-consumer" advertising; even if this kind of advertising is not permitted in all the countries, or at least is permitted with specific limitations, its role acts everywhere, using Internet, or simply accessing through satellite TV to programs aired by DTC permitting countries.

5. *Other external factors*. This group includes drug reimbursement laws/rules/policies, which can be different between countries with a National Health Systems (NHS) and countries without an NHS.

Even between countries with a national health service, the policies of drug reimbursement, with or without co-payments, vary in relation to the two basic healthcare system designs, the "Bismarck" and the "Beveridge" systems.

The Bismarck system is based primarily on social insurance contributions, while the financing of the Beveridge system comes from tax revenue. These differences also include government policies on physician remuneration, standards of practice from professional organizations, prescribers' concerns about legal liability, regulatory and control measures, and political considerations.

The prescriber-related factors require special attention, starting with the decision-making process [10]. In this process, the physician, rather than the patient, is the key decision maker, but the decision may be subject to a range of influences:

1. Patients can influence prescribing [37], with much greater impact on prescriptions by general practitioners (GPs) than on those by specialists.
2. The physician's prescribing decisions may be influenced by specific demands from relatives, as happens when the prescribing of more drugs is believed to ensure better and quicker improvement [2].
3. Formal and informal interactions with other physicians and medical staff can influence decisions.

In the latter case, the interactions are associated with the type of practice, being stronger in hospital and group practices and weaker for single practitioners and/or for clinical consultants. Hospital pharmacists can greatly influence prescriptions through composition of the hospital formulary to which the prescriber has to adhere.1

Moreover, there is evidence that emotive and cognitive factors play an important role in medical prescribing. In many cases, prescribing depends on ready memory [48]. The treatment decision may not follow a scientific rationale, at least for certain illnesses [29], or in the choice of generic drugs [38]. In these cases, attitudes and motivations of GPs play an important role [31].

In order to understand the nature of medical prescribing, it is thus important to identify psychological, professional, and organizational components of medical prescribing processes, in General Practice and over and above, to using a complex systems approach [27].

32.3 Prescription Behavior and Decision Making

This section presents the conceptual framework of the decision-making process from a psychological point of view, and then the practical implications of this process in the daily activity of drug prescription by physicians. The last part presents the patterns and rules of drug prescription and administration, in the general case from an international perspective, and in the specific setting in which the authors' research has been performed.

32.3.1 The Decision-Making Process and Problem Solving

The decision is the process in which an individual, or a group (decision maker), makes a choice between several alternatives considered (options). The necessary condition to define a *decision* is that the decision maker has before him a number of

[1] Note: hospital formularies, on face value a great way to safe on drug expenditure, have been shown to be more costly, as their protagonists invariably have failed to consider "unintended consequences." For more detail, see John Sterman's talk at the IHI: Systems Methodologies for Solving Real-World Problems: Applications in Public Health. March 22, 2007 (http://videocast.nih.gov/Summary.asp?file=13712).

options: if there is only one choice, in the absence of alternatives, the activity is not a decision.

The first studies on decision-making processes begun in the discipline of economics, but since the early 1950s, psychology has studied this question from a different point of view and identified some models that aim to explain the steps by which people make a decision:

1. *The rational–normative model/absolute rationality* [26]. This model assumes that the human is a perfectly rational decision maker who reaches the best solution satisfying the principle of maximizing results.
2. *The heuristic model or limited rationality* [33]. The representation of the decision maker as an "infallible scientist" was inadequate and was replaced with that of a decision maker who has a limited and intentional rationality.
3. *Theory of the prospectus* [39]. The prospect theory is based on the assumptions that people seem to evaluate each possible outcome of a decision on the basis of a reference point (or status quo) as can be, for example, their situation at the time of the decision. This theory gives great importance to the way it is interpreted the decision problem, since the experimental evidence shows that the same problems, but described differently, giving rise to different decisions [40].

Usually, people decide using a problem solving perspective in contexts where it is not easy to readily interpret all the available information [8]. Often the terms decision making and problem solving are considered equivalent. Yet they are two different and distinct activities: the decision making takes place after the process of problem solving has identified a number of possible guidelines. A problem is considered as a gap or a difference between the desired performance and the real one [28].

It is necessary to use a variety of strategies to find the best solution to a problem, usually following these steps [15]:

1. Identify the problem: when we are aware of the problem, we can take action to resolve it as best we can.
2. Define the objectives: a process that describes precisely what we want to obtain.
3. Collect relevant information: first we need to find adequate information. It may be difficult to decide which data are important and which are not. In this process, the collection of relevant data is generally one of the most difficult stages of the process.
4. Identify alternatives: the decision-making process can take place if we have alternatives of action. There is no way, however, to be sure that the best alternative can be included in those considered. It is possible to try to make sure that all "conventional" solutions have been considered, and then try to suggest innovative proposals.
5. Select the criteria for evaluating the better alternative: logically, we want to choose the best alternative. However, this can only be done if we describe the concept of "better." There should be a criteria or a set of criteria to assess which alternative is the best.
6. Building the model: to establish the relationship between the objective, alternatives, data collection, and evaluation criteria.
7. Estimate the expected results of each alternative: the model built is then used to estimate early the outcome of each alternative.
8. Choose the best alternative with respect to the objective: if all other process steps were done accurately, we can make the choice of the best (which best meets the selection criteria adopted).

Certainly when we take decisions we try to gather as much information in an accurate manner that accounts for the costs and benefits for each of the options available to us. This is done through our "intellectual component" that allows us to reduce the uncertainty margins, and then consequently errors, in everyday situations. However, to make a decision in a "totally rational condition" is only possible if we know all the data and the possible interferences of a situation [9]. Emotions play an important role in the decision-making process, they have a "constructive role in the higher forms of human experience." Numerous studies [12,13] have shown that, generally, people turn to emotions or emotional sensations when situations are complex to evaluate or when time constrains the ability to accurately evaluate of each alternative. In the process of alternative selection there is a continuous interaction between evaluation and emotional regulation [18].

32.3.2 Therapeutic Drug Prescription Behavior

The complexity and sensitive nature of the physician's decision-making behavior appears to be more hybrid and less rational in nature than is often assumed in quantitative, model-based analyses of prescription behavior. The decision-making process is typically complex and influenced by several sorts of factors, each in turn grouping multiple influences [10]:

1. The multiple-party-setting
2. The prescriber's multiple goals
3. The prescriber's multiple sources of information overload
4. The multiple diagnostic and therapeutic uncertainties

The multiple-party setting. The physician rather than the patient is the key decision maker. This observation, however, needs to be put in perspective. First, the patients may still influence prescribing, their impact being markedly stronger on prescriptions by GPs than on that by specialists. Secondly, the physician's prescribing decisions may be influenced by specific demands from relatives, formal and informal interactions with other physicians, and other medical staff [19,25].

Multiple goals. A predominant goal pursued by physicians, is to rationally and exclusively assume their medical responsibility. This hypothesis does not confirm the logical assumption that medical goals generally dominate prescribing decisions, other goals—such as the prescriber's personal financial and socio-psychological goals are also found to be of influence [10].

Multiple sources of information overload. The information processing capacity of physicians is structurally insufficient. They cannot possibly process all the information reaching them, from many different scientific and/or commercial sources, and concerning many different aspects like pathologies, treatments, and pharmacological supply. The typical time pressure plaguing physicians—in combination with the high risk and uncertainty of the prescribing decisions—worsens this structural problem [10].

Multiple diagnostic and therapeutic uncertainties. A major element in the physician's decision-making process is the difficulty of assessing the results of a treatment. There are three groups of structural causes of uncertainty [14,24]:

1. Uncertainty concerning patients, caused by such eventualities like subjective, imperfect reporting by patients, numerous—often unknown—exogenous elements affecting the patient, and also the changing set of patients.
2. Uncertainty concerning the pathology, caused by the fact that there might be multiple explanations for specific complaints, multiple complaints resulting from a single pathology, or multiple pathologies coinciding.
3. Uncertainty concerning the effects of drugs, due to the limited opportunity for experimentation throughout a treatment, the possibly multiple effects of drugs and the possible carry-over effects of drugs.

These elements may stimulate the adoption of risk reducing prescribing strategies, like following opinion leadership or remaining brand loyal. When analyzing decision-making processes, on must distinguish between situations that are routine and non-routine, as the criteria of choice are different [10]. In non-routine situations, involving new products and/or new patients with a complex pathological profile, prescribers typically go through a fairly extensive evaluation, and rely on multiple criteria:

1. Disjunctive or conjunctive rules—often based on main medical effects—reduce the number of alternatives.
2. The remaining options are eliminated in a lexicographic fashion, either on an aspect-by-aspect basis or on an alternative-by-alternative basis (product, product form, brand).
3. Compensatory rules intervene to arrive at an actual choice.

In routine situations, the physician will implicitly and/or explicitly go through a learning process. In fact, physicians apparently have a particularly strong need to remain in control of events, even under a high degree of uncertainty, and—as a result—predominantly acquire information through an active rather than a passive learning process. Pharmaceutical companies' marketing mix instruments that both affect non-routine decisions and reinforce or disrupt established routines. Price, advertising, detailing, samples, and gifts may affect prescription rates of the company's drug products [10].

Price. The price is generally unimportant. However, price may be salient in brand selection for very expensive products, or products prescribed for patients on a very tight budget. Also, GPs and younger physicians seem somewhat more alert to price, and increased government pressure is bound to increase price attention in years to come [19].

Advertising. Traditionally, pharmaceutical companies' promotional efforts almost exclusively concentrated on detailing and free product samples directed to physicians. The impact of *advertising* directed to the physician is assessed by considering drug information published in different media (medical journals, official publications) and originating from different sources (pharmaceutical companies vs. government and professional organizations) [17,25,51].

Detailing. In the highly complex and rapidly evolving drug market, sales representatives have an important information function, both for new and existing products [53].

Samples. Samples are, for instance, thought to create commitment toward sales representatives and their company, and to serve as a reminder of the sales representatives' visit once they have left.

Gifts. In addition to samples, many pharmaceutical companies also offer various gifts (sponsoring of conference participation, travel and lodging, medical education, meals, honoraria, promotional material, and other small gifts such as pens) [44], which mainly aim to enhance the long-term relationship between the company and physicians [14]. Like for samples, the fact that giving gifts has more or less become common practice may be responsible for their diminished effectiveness, but may also imply that not giving these advantages may elicit negative reactions [22].

32.3.3 Drug Prescription and Administration Rules

Drug prescription and administration are performed in many different ways, according to local uses, habits, informal rules, and formal laws. The main behavior patterns that can be found are referred to:

1. The authority to prescribe, limited to the physician or extended to other health professionals (nurses, midwives, etc.) and/or to the pharmacist.
2. The authority to supply/sell the drug, limited to the pharmacies or extended to the physician and/or healthcare institutions.

In other words, we can range from healthcare settings in which the physician only can prescribe and the pharmacist only can sell the drug, or supply it to those who have the right to receive it for free, to settings in which a drug can be prescribed by almost any healthcare professional and it can be sold/supplied by pharmacists, physicians, and other professionals.

National or local legislation regulates who can write a prescription: for example, in the United States, all States and Columbia District allow prescription from Medical Doctors but also, with some limitations, from nurses, midwives, dentists, podiatrists, optometrists, and somewhere from clinical pharmacists, while in Thailand the drugs are supplied directly to outpatients by the prescribing physician into the district hospital [30]. Moreover, there are some classes of drugs that are not subject to medical prescription, the so-called over the counter (OTC) drugs. These are not strictly regulated as a prescription drug, and, in certain healthcare settings, prescribers can write prescriptions for OTC drugs because drug benefit plans may reimburse the patient only if the OTC medication is taken under the direction of a medical practitioner.

32.3.3.1 Drug Prescribing in Italy

According to the Italian rules, only medical doctors, i.e., people with a degree in Medicine and registration to the professional order, can prescribe drugs. Drugs to be supplied on behalf of the Italian National Health Service (INHS) can be prescribed only by a medical doctor employed or in agreement with the INHS, and GPs are the only physicians who can prescribe drugs for the outpatients.

Drug prescriptions are made by GPs using prescription sheets on which a maximum of four different drugs can be prescribed, each with a maximum of two pieces (boxes of other kind of package); however, more than one prescription sheet can be given to a patient as a result of a visit.

Drug prescriptions for outpatients can also be made by physicians operating in an institution, such as a hospital, but usually these physicians limit their prescribing to the discharge report, that the patient takes to the GP, who will carry on all continuing prescribing, and this behavior is usual for referrals, too. In fact, the referred specialist usually recommends therapy and/or diagnostic tests, but the prescription of all is always made by the GP.

According to current Italian law, only a pharmacist can sell a drug or supply it on behalf of the INHS, and drugs are divided into three categories:

1. Type A are supplied to patients by pharmacies free of charge
2. Type B are supplied after a co-payment (so-called ticket)
3. Type C are sold after the full payment, but their price is controlled by the INHS

Also patients are divided into categories, in respect of their right to receive drugs without co-payment: children up to 6 and adults over 65, if their income (family income for the children) is under a certain threshold, do not have to pay anything, the so-called ticket exemption. Ticket exemption is also provided, regardless of age and/or income, for specific diseases carrying patients, and these pathologies are typically chronic and/or rare diseases, such as diabetes, adrenoleukodystrophy, Alzheimer disease, rheumatoid arthritis, etc., officially listed.

The nominal difference between the amount paid by patients and the cost of drugs is covered by the INHS, which in turn expects a tight surveillance on the amount and type of prescriptions made by GPs. This complex situation changes in time, as new drugs and therapies are introduced, ticket exemption rules are modified and socio-economic conditions vary.

32.4 An Application of Network Medicine: The Drug Prescription Network

As an example of a Network Medicine study in a Public Health setting, a set of drug prescription data from 99 GPs, working in Italy and covering a 6-month time period, has been studied and analyzed. The data set, containing a total of 42,965 consultations and 631,232 prescribed drugs, has been transformed into a drug prescription network, where each drug is a node, and different drugs prescribed to the same patient on the same day, are linked together. The resulting networks, describing the entire population or subgroups by patient's age and gender, have been analyzed using the tools of network theory.

32.4.1 Introduction

Different types of drug networks can be built [52]. In our study model, it can be described as consisting of two elements: drugs (nodes) and their contemporary prescription for a given patient (links).

In this network model, the link between two drugs, prescribed at the same time in the same patient, can be also called "co-prescription." As in general Network Science studies, once all the nodes (drugs) and links (co-prescription) are known, one can draw pictures of the network, measure its parameters and properties, and discern every node location within it. Each node, thus, is placed in a "co-prescription" space, analogous to a social space, as mapped by social networks methods, or even to a geographic space, as mapped by photographs or drawings.

The rationale of our research has been to consider the drug prescription process from the point of view of its topology, using single drugs as nodes, and their co-prescription for the same patient at the same moment, as links.

Such an approach could make possible to represent and measure relationships between drugs, as there could possibly be "hub drugs," i.e.,

Table 32.3 Summary of data used in this research

Total	Patients	Prescriptions	Nodes	Links
All ages	42,965	631,232	964	52,915
Age<30	6,882	35,052	494	3,398
30<age<60	20,515	196,787	820	23,775
Age>60	17,177	399,393	830	39,580
	Males			
All ages	19,321	281,435	794	29,864
age<30	3,297	16,078	372	1,709
30<age<60	9,290	89,129	664	12,801
age>60	7,656	176,228	697	22,730
	Females			
All ages	24,832	349,797	896	37,356
Age<30	3,793	18,974	400	2,155
30<age<60	11,978	107,658	743	15,438
Age>60	10,212	223,165	759	28,506

drugs that are often prescribed in association with other drugs, and "isolated drugs," i.e., drugs usually prescribed alone. Moreover, the utilization of hub drugs could change among different groups of patients, as the pathology changes between different genders and age groups. This leads to the need to measure and represent different patterns of drug prescription network in different subgroups.

32.4.2 Materials and Methods

32.4.2.1 Prescription Data Used

The networks we have built are based on the set of prescriptions made by a group of 99 GPs operating in Salerno, a city of 140,000 inhabitants in southern Italy, in the first 6 months of year 2009. The Italian rules for drug prescriptions allow a maximum number of drugs for each prescription sheet, but more than one prescription sheet can be given to a patient as result of a visit. The total number of drug prescriptions collected is 631,232 corresponding to 42,965 patient consultations. The data have been collected from the database of "Consorzio Mega Ellas," a GP medical association based in Salerno, including a total of 150 physicians, and treated in the full respect of current legislation on privacy.

32.4.2.2 Network Construction

The rationale for linking the network nodes is the following: two drugs are connected if they have been prescribed to the same patient during the same medical consultation. The number of times two given drugs have been co-prescribed is recorded as link weight. Similarly the number of times a drug is prescribed, alone and in association, is also computed and associated to each node. These procedures resulted in a network of 964 nodes and 52,915 links. By separating the patients by gender and age (in the ranges 0–30, 30–60, and over 60 years), we have built several sub-networks to investigate possible effects due to patient characteristics.

The main characteristics of the networks obtained are reported in Table 32.3.

The obtained networks can be graphically represented in many different ways, and this processing has been performed using "Pajek" software [7], a free large network analysis and visualization tool.

32.4.2.3 Coding: The Anatomical Therapeutic Chemical Classification System

From the ensemble of prescriptions we have build several networks relating the different drugs. The nodes of the networks are made by the drugs, identified by their common name and ATC code.

Table 32.4 Description of ATC code at anatomical level

ATC code	Anatomical region affected
A	Alimentary tract and metabolism
B	Blood and blood forming organs
C	Cardiovascular system
D	Dermatologicals
G	Genito-urinary system and sex hormones
H	Systemic hormonal preparations, excluding sex hormones and insulins
J	Antiinfectives for systemic use
L	Antineoplastic and immunomodulating agents
M	Musculo-skeletal system
N	Nervous system
P	Antiparasitic products, insecticides and repellents
R	Respiratory system
S	Sensory organs
V	Various

The Anatomical Therapeutic Chemical Classification (ATCC) System [47] is used for the classification of drugs. It is controlled by the WHO Collaborating Centre for Drug Statistics Methodology (WHOCC), and was first published in 1976. In the ATC classification system, the active substances are divided into different groups according to the organ or system on which they act and their therapeutic, pharmacological, and chemical properties.

Drugs are classified in groups at five different levels: at the first level (anatomical) the drugs are divided into 14 main groups, shown in Table 32.4.

The following four levels are the therapeutic subgroup (second level), the pharmacological subgroup (third level), the chemical subgroup (fourth levels), and the chemical substance (fifth level).

Table 32.5 shows an example of the ATC code structure for a common diabetic drug.

In the Appendix, the complete list of ATC codes at level 2 is reported.

32.4.3 Results

The appearance of a network can be very complex even if suggestive: Fig. 32.7 shows drug network at ATC level 2.

Figure 32.7 (left) shows that the network is a complete graph, i.e., all the nodes at ATC level 2 are connected to each other, although the links have different weights, because each drug of each ATC group has been prescribed with each of the other different ATC groups.

If we remove the lower 10% in weight links, that are the less frequent co-prescriptions, we obtain a very different picture of the network, as shown in Fig. 32.7 (right). In this representation, the strong links between drugs belonging to the A (alimentary tract), B (blood), C (cardiovascular), and J (infection) groups are clearly evident.

In order to extract more quantitative information on the way this network is formed, the network analysis tools have been used, and the results are reported in the following.

32.4.3.1 Scale Invariance in Drug Co-Prescription Network

The graphic representation of networks can be very suggestive but, sometimes, not very informative. Important information on the network connectivity is given by the degree distribution,

Table 32.5 Description of the ATC code structure

ATC code	Description
A	Alimentary tract and metabolism (first level, anatomical main group)
A10	Drugs used in diabetes (second level, therapeutic subgroup)
A10B	Blood glucose lowering drugs, excl. insulins (third level, pharmacological subgroup)
A10BA	Biguanides (fourth level, chemical subgroup)
A10BA02	Metformin (fifth level, chemical substance)

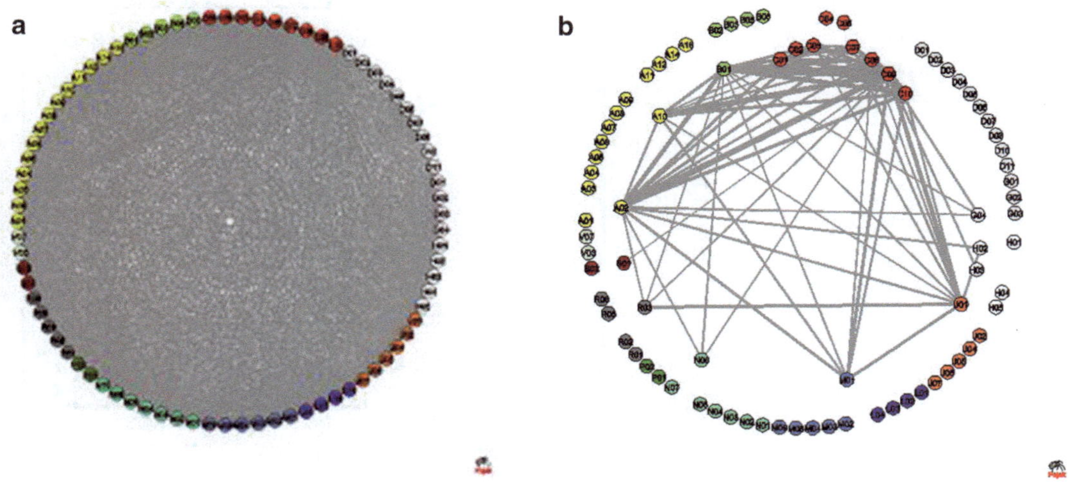

Fig. 32.7 The network is a complete graph (*left*), removing the lower 10% in weight links, shows the strong links between drugs belonging to the A (alimentary tract), B (blood), C (cardiovascular), and J (infection)

Fig. 32.8 The network of drug co-prescription shows a scale-free behavior

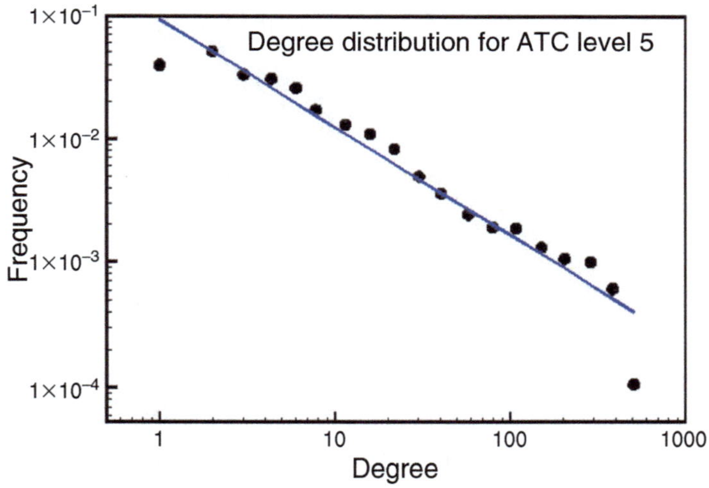

i.e., the relationship between the number of connections (degree; k) of a node and the relative frequency in the network.

Regular networks have a very narrow degree distribution, while random networks show an average degree value, with a given spread around it. Scale-free networks show a characteristic power law dependence of the degree distribution that appears as a negative slope line in a log–log plot.

Surprisingly, the network of drug co-prescription shows a marked scale-free behavior, as shown in Fig. 32.8.

By looking at Fig. 32.8, we note that the degree distribution data points are very well aligned on a line, with the exclusion of the very first and last, a fact due to the finite size of the network. The best fit of the data in log scale gives a slope of -0.88 ± 0.05.

In weighted networks, links do not have the same importance: some links are more important than others, i.e., have a higher weight. In such case, a parameter that could better describe the way the network is connected is the strength, defined as the sum of the connections reaching a node, each multiplied by its weight.

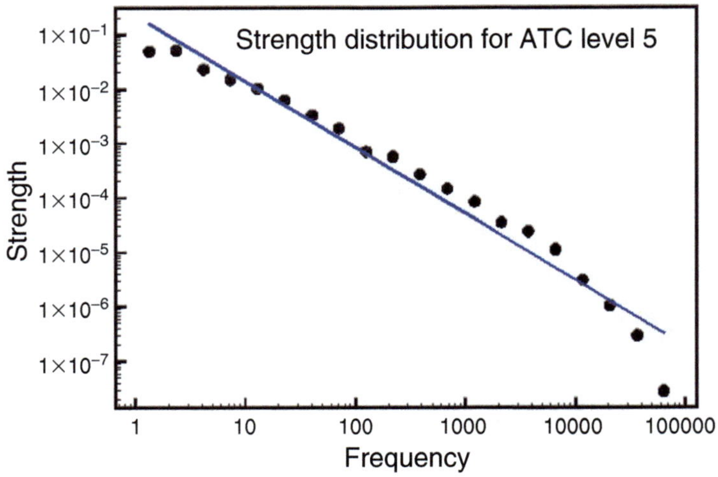

Fig. 32.9 The strength distribution of the drug co-prescription network shows a scale-free behavior

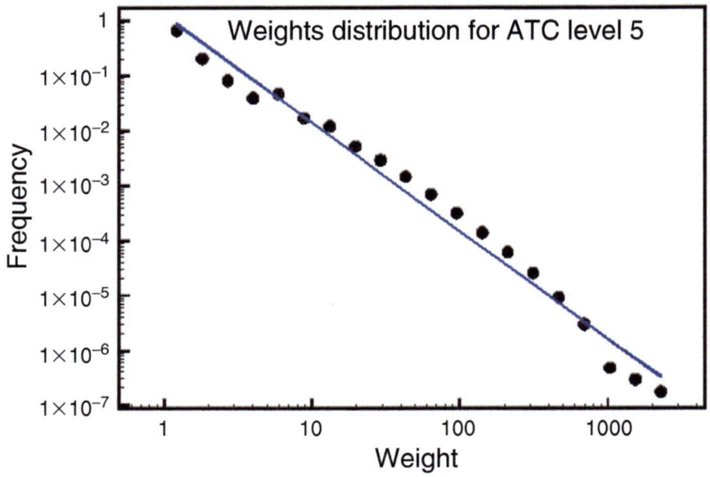

Fig. 32.10 The weight distribution of the drug co-prescription network shows a scale-free behavior

In all networks considered here the links have a weight given by the number of co-prescriptions, therefore an analysis of the strength distribution has also been performed. As a result, the same scale-free behavior, observed for the degree distribution, has been found, as shown in Fig. 32.9, with a characteristic exponent of -1.25 ± 0.06.

Another characteristic often studied in networks is, besides the strength, the weight distribution. Figure 32.10 shows such distribution for the co-prescription network studied.

Also in this case, a clear power law dependence is observed, with a characteristic slope, very close to -2 (-1.97 ± 0.06).

Although all the different graphs in Figs. 32.8–32.10 show a clear linear dependence with a negative slope, for the type of networks considered here, where the links are weighted, the relevant figure for investigating the possible power law dependence of the connections distributions is Fig. 32.9, from which the scale invariance property is maintained over more than six orders of magnitude in the frequency.

By repeating the same analysis in the sub-networks obtained by separating by gender and by age range, we have observed that there is a decisive dependence of the strength distribution exponent (the slopes in Fig. 32.9) on the age range, while there is no significant dependence on gender, as summarized in Table 32.6.

Looking at Table 32.6 it appears that the exponents shows no dependence on the gender and a weak, but significant dependence on the age range of the patients.

Table 32.6 Strength distribution of the networks separated exponents by age and gender

Strength distribution exponent	All	Age<30	30<Age<60	Age>60
All	−1.25±0.06	−1.52±0.06	−1.29±0.07	−1.27±0.06
Males	−1.27±0.06	−1.57±0.06	−1.35±0.08	−1.30±0.07
Females	−1.27±0.06	−1.58±0.07	−1.27±0.06	−1.29±0.06

Table 32.7 The ten most prescribed drugs at the ATC2 level for each gender group

ATC2	Total				Males				Females			
Most prescribed	All ages	<30	30–60	>60	All ages	<30	30–60	>60	All ages	<30	30–60	>60
1	C09	J01	J01	C09	C09	J01	C09	C09	C09	J01	J01	C09
2	J01	R06	C09	B01	J01	R06	J01	B01	J01	R06	C09	B01
3	A02	H02	A02	A02	B01	N03	A02	A02	A02	G03	M01	A02
4	B01	N03	M01	J01	A02	R03	C10	J01	B01	H02	A02	J01
5	M01	R03	C10	C10	C10	H02	M01	G04	M01	A02	N06	A10
6	C10	A02	C07	A10	A10	A02	A10	C10	A10	R03	H03	C10
7	A10	G03	A10	C08	G04	R01	B01	A10	C10	N03	C07	M01
8	C08	R01	B01	M01	C08	M01	C07	C08	C07	B03	G03	C08
9	C07	M01	N06	C07	M01	N06	C08	R03	C08	R01	C10	C07
10	R03	N06	C08	C01	C07	A07	N03	C07	N06	M01	A10	C03

Table 32.8 The ten most co-prescribed drugs at the ATC2 level for each sex/gender group

ATC2	Total				Males				Females			
Most co-prescribed	All ages	<30	30–60	>60	All ages	<30	30–60	>60	all ages	<30	30–60	>60
1	C09	J01	C09	C09	C09	J01	C09	C09	C09	J01	C09	C09
2	B01	H02	J01	B01	B01	H02	C10	B01	B01	H02	J01	B01
3	A02	R03	A02	A02	C10	R03	B01	A02	A02	R03	A02	A02
4	C10	R06	C10	C10	A02	R06	A02	C10	C10	R06	M01	C10
5	A10	R01	B01	A10	A10	R01	C07	C08	A10	R01	C07	A10
6	J01	A02	C07	C08	C08	A02	A10	A10	J01	A02	B01	C08
7	C08	M01	A10	C03	C07	N03	J01	G04	C08	G03	C10	C03
8	C07	N03	M01	C01	J01	M01	C08	C01	C07	M01	A10	C07
9	C03	G03	C08	C07	G04	N06	M01	C03	M01	B03	N06	C01
10	M01	N06	H02	J01	C03	A07	C03	C07	C03	N03	H03	M01

Although a model connecting strength distribution exponents to patient characteristics (age and gender) does not exist to date, Table 32.6 data suggest a possible connection between drug network characteristics and age-related disease epidemiology.

32.4.3.2 Most Prescribed Versus Most Co-Prescribed

The scale-free behavior of the investigated drug networks suggest the existence of "hub drugs," i.e., drugs which are most frequently prescribed with others. To investigate this phenomenon, the drug co-prescription networks represented at ATC level 2 have been analyzed. The results are summarized in Table 32.7, where the top ten most prescribed drugs are listed for all the sub-networks obtained separating the population by age and gender.

Conversely, in Table 32.8 the top ten most co-prescribed drugs are listed for all the sub-networks obtained separating the population by age and gender.

In other words, Table 32.7 shows statistical information on drug prescription, while Table 32.8 reports a connection-related information, typical of network analysis.

In order to further exploit the information from network connections, the "nearest neighbors" of the most co-prescribed drugs, always at ATC level 2, are reported in in Table 32.9.

Table 32.9 Drugs to which are connected the most co-prescribed drug for each gender group

	Total				Males				Females			
ATC2	All ages	<30	30–60	>60	All ages	<30	30–60	>60	All ages	<30	30–60	>60
Most Co-prescribed	C09	J01	C09	C09	C09	J01	C09	C09	C09	J01	C09	C09
To whom is connected in decreasing order	B01	H02	C07	B01	B01	H02	C10	B01	B01	H02	C07	B01
	C10	R03	C10	A02	C10	R03	B01	C10	A02	R03	A10	A02
	A02	R01	B01	C10	C08	R01	C07	C08	C10	R01	C10	C10
	C08		C08	C08	A02		C08	A02	A10		C08	A10
	A10		A10	A10	A10		A10	A10	C08		B01	C08
	C07		A02	C07	C07		A02	G04	C07		A02	C07
			J01					C07			J01	

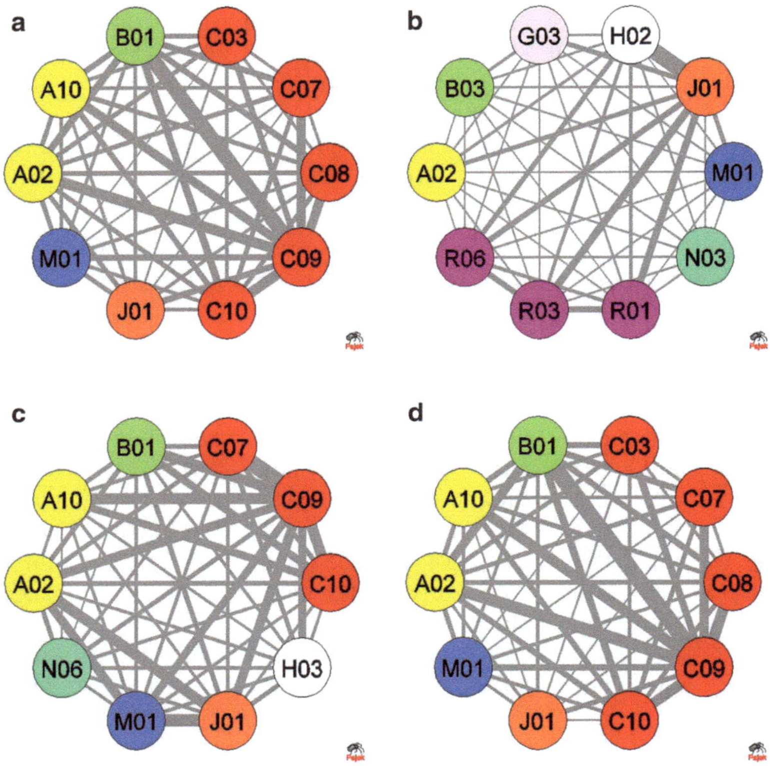

Fig. 32.11 The top ten co-prescribed drugs at ATC level 2, gender by age

For each sub-network analyzed the most co-prescribed drug is presented in the first row of Table 32.9, and, in the corresponding column are listed the drugs with which the head drug is connected, in decreasing order. As an example, for "Males" with age "under 30," the most prescribed drug group, J01 (antibacterials for systemic use) results to be co-prescribed mostly with group H02 (corticosteroids for systemic use), then with R03 (drugs for obstructive airway diseases), and then with R01 (nasal preparations).

This finding is consistent with epidemiological experience that relates the highest demand of drugs for this age/gender group to infections, more likely to happen in the respiratory tract.

In order to better visualize the data of Table 32.9, in Figs. 32.11 and 32.12 a graphic representation of the top ten co-prescribed drugs at ATC level 2 are shown, separately by age and gender.

The circles in Figs. 32.11 and 32.12 represent the nodes, i.e., the most co-prescribed drugs. The color/gray shade indicates the anatomical drug

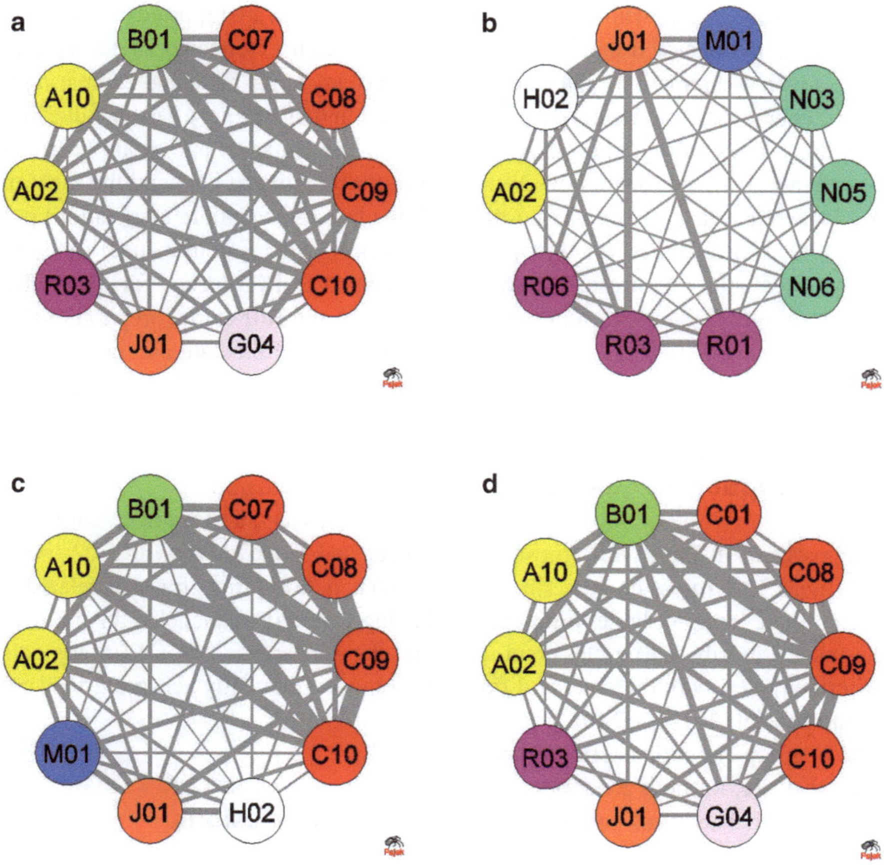

Fig. 32.12 The top ten co-prescribed drugs at ATC level 2, gender by age

group (ATC level 1); while the ATC level 2 code is written in the node. The link thickness is proportional to the number of times the two drugs have been co-prescribed.

This representation shows in a very clear way the difference between age/gender drug co-prescription: in other words, the data commented above, related to the "Males/under 30" column of Table 32.9, are graphically shown in Fig. 32.12, top right.

32.5 Discussion

32.5.1 Significance of Scale Invariance

The main result of this work is the discovery that the network formed by the drugs co-prescription to the patients by GP is scale invariant.

To date there are very few theoretical models that produce scale invariant networks. The most considered is the "Preferential Attachment" network [5,16,23,42].

Within the preferential attachment model, a network is build by adding one node at a time at random in the network, but with a definite probability. In particular the probability that a node is linked to a target node depends linearly on the degree of the target node. The resulting network [5] is a scale free one with a scaling exponent equal to three. We note however that the value of the scaling exponent may depend on the specific choice of relation between node degree and linking probability.

In our case the exponent is always between one and two.

An interpretation of the preferential attachment, in the framework of this work, could be the following: when confronted with the necessity of prescribing a relatively new drug, i.e., a drug for which all adverse interactions may not be known yet, a general practitioner tends to co-prescribe it with it known drugs, i.e., drugs for which there exist a well-known history of adverse interactions. In essence, a new drug is co-prescribed with higher probability together with drugs that have been safely co-prescribed many other times before, rather than with more rarely used ones.

32.5.2 Prescription Versus Co-Prescription Frequency

From the comparison of Table 32.7 (most prescribed) and Table 32.8 (most co-prescribed) it stems out that there are no large differences between the most prescribed drug class and the most co-prescribed ones, with the notably difference of J01 and possibly R01 and M01.

Drugs of J01 group are antibacterials for systemic use, and they are, for both genders:

1. Highly ranked either as most prescribed and most co-prescribed in lower (0–30) and middle (30–60) age groups.
2. Relatively highly ranked as most prescribed, and low ranked as most co-prescribed, in the older age group (over 60).

This finding could be consistent with the fact that in older patients antibacterials are mostly associated with urinary tract infections, as suggested by general practitioners' experience. In such case, the antibacterial prescription is less frequently associated with other drugs. Conversely, in younger patients antibacterials are mostly associated with respiratory tract infections, and the prescriptions are frequently associated with other drugs.

However the purely statistical information on the most prescribed drug class, does not tell with which other drug it has been prescribed. This information is instead easily obtainable from the network analysis, as shown in Table 32.9, and even better visually in Figs. 32.11 and 32.12.

The table reports the drug classes co-prescribed with the most co-prescribed drug (row 1). In each column are reported only the drug classes with the higher frequency of co-prescription.

Looking at the table and comparing the different sex and age groups, few peculiarities pop up.

For the youngest age group (<30) there is no difference between genders: the most prescribed drug class (alone and in association) is J01 (antibacterials for systemic use) and is always associated to R03 (drugs for obstructive airway diseases) and R01 (nasal preparations).

For the intermediate age range (30–60) there is a notable difference in the frequency of co-prescription of B01 (antithrombotic agents) and A10 (drugs used in diabetes), which is reversed for the two genders.

A somewhat similar effect can be seen in the higher age range (>60) with C02 (antihypertensives) and A02 (drugs for acid-related disorders) having inverted frequencies for the two genders, and the appearance of G04 (urologicals) among the most frequently co-prescribed drugs for males.

32.6 Conclusions

Network Medicine is a new concept that can be exploited in different fields—from the network-based study of diseases to network pharmacology. Studies are currently conducted horizontally at one of the three levels, the interactome, the diseasome, and the social network, but the future will see vertical studies too, connecting the three levels.

These studies will offer a greater knowledge not only of the network molecular mechanisms of physiology and pathology, but also of the relationships between molecular and social mechanisms, which are responsible for the social transmissibility of diseases, once considered "non-transmissible," like obesity, diabetes, ischemic heart disease, various cancers, and so on.

A large part of global health spending is accounted for by medicines, and a large part of this expenditure is led by general practitioners' prescribing. Prescribing, not only for drugs but also for diagnostic tests, is a complex process, that takes place in a complex environment, and involves a large number of factors, related to the knowledge, professionalism, and culture of the prescriber and their health system context, but also on patient's personal, social, and cultural features.

Emotive factors can play a central role in this process, but, again, the social network of the physician and the patient can make it easier or more difficult to decide on a correct, effective drug choice, which can lead to better or worse results in terms of health outcomes and healthcare expenditure.

Moreover, different settings and prescribing rules, incentives and constraints, throughout the world can be considered as additional complications, but also as a promising ground of research, in which network science instruments can help to find simple answers to complex questions.

The experimental part of the present chapter has investigated a drug co-prescription network obtained from a database of 99 GPs which contained more than 600,000 drug prescriptions. The most interesting result is that all the drug prescription networks show scale invariance behavior, with characteristic exponents related to the patients age but not to gender; this finding is new, and could be related to a prescribing behavior explainable by a preferential attachment model.

Moreover, by looking at the most frequently prescribed versus the most frequently co-prescribed drugs, specific correlations emerge between drugs belonging to different anatomical and therapeutic groups.

The emerging field of Network Medicine can investigate not only biological/social systems and disease factors, but also behavioral and healthcare planning factors. Drug prescription process has a complex structure that can be evidenced and measured using Network Science instruments.

Therefore, Network Medicine could be a new powerful tool for research in Public Health.

Acknowledgments We gratefully acknowledge Dr. Mario De Santis, General Practitioner and Vice-President of "Consorzio Mega Ellas," for kindly providing the data used in this research.

We also thank Prof. Albert Laszlo Barabasi for his encouragement and for the permission to use his very significant and didactic picture, which shows relationships between the levels of molecular, clinical, and social networks.

32.7 Appendix. ATC Codes (At Anatomical and Therapeutic Level)

A01	Stomatological preparations
A02	Drugs for acid-related disorders
A03	Drugs for functional gastrointestinal disorders
A04	Antiemetics and antinauseants
A05	Bile and liver therapy
A06	Laxatives
A07	Antidiarrheals, intestinal anti-inflammatory/anti-infective agents
A08	Antiobesity preparations, excluding diet products
A09	Digestives, including enzymes
A10	Drugs used in diabetes
A11	Vitamins
A12	Mineral supplements
A13	Tonics
A14	Anabolic agents for systemic use
A15	Appetite stimulants
A16	Other alimentary tract and metabolism products
B01	Antithrombotic agents
B02	Antihemorrhagics
B03	Antianemic preparations
B05	Blood substitutes and perfusion solutions
B06	Other hematological agents
C01	Cardiac therapy
C02	Antihypertensives
C03	Diuretics
C04	Peripheral vasodilators
C05	Vasoprotectives
C07	Beta blocking agents
C08	Calcium channel blockers
C09	Agents acting on the renin-angiotensin system
C10	Lipid modifying agents
D01	Antifungals for dermatological use
D02	Emollients and protectives
D03	Preparations for treatment of wounds and ulcers
D04	Antipruritics, including antihistamines, anesthetics, etc.
D05	Antipsoriatics
D06	Antibiotics and chemotherapeutics for dermatological use
D07	Corticosteroids, dermatological preparations
D08	Antiseptics and disinfectants

D09	Medicated dressings
D10	Anti-acne preparations
D11	Other dermatological preparations
G01	Gynecological antiinfectives and antiseptics
G02	Other gynecologicals
G03	Sex hormones and modulators of the genital system
G04	Urologicals
QG51	Anti-infective and antiseptics for intrauterine use
QG52	Products for teats and udder
H01	Pituitary and hypothalamic hormones and analogues
H02	Corticosteroids for systemic use
H03	Thyroid therapy
H04	Pancreatic hormones
H05	Calcium homeostasis
QI01	Immunologicals for Aves
QI02	Immunologicals for Bovidae
QI03	Immunologicals for Capridae
QI04	Immunologicals for Ovidae
QI05	Immunologicals for Equidae
QI06	Immunologicals for Felidae
QI07	Immunologicals for Canidae
QI08	Immunologicals for Leporidae
QI09	Immunologicals for Suidae
QI10	Immunologicals for Pisces
QI11	Immunologicals for rodents
QI20	Immunologicals for other species
J01	Antibacterials for systemic use
J02	Antimycotics for systemic use
J04	Antimycobacterials
J05	Antivirals for systemic use
J06	Immune sera and immunoglobulins
J07	Vaccines
QJ51	Antibacterials for intramammary use
QJ54	Antimycobacterials for intramammary use
L01	Antineoplastic agents
L02	Endocrine therapy
L03	Immunostimulants
L04	Immunosuppressants
M01	Anti-inflammatory and Antirheumatic products
M02	Topical products for joint and muscular pain
M03	Muscle relaxants
M04	Antigout preparations
M05	Drugs for treatment of bone diseases
M09	Other drugs for disorders of the musculo-skeletal system
N01	Anesthetics
N02	Analgesics
N03	Antiepileptics
N04	Anti-Parkinson drugs
N05	Psycholeptics
N06	Psychoanaleptics
N07	Other nervous system drugs
QN51	Products for animal euthanasia
P01	Antiprotozoals
P02	Anthelmintics
P03	Ectoparasiticides, including scabicides, insecticides, and repellents
QP51	Antiprotozoals
QP52	Anthelmintics
QP53	Ectoparasiticides, including insecticides and repellents
QP54	Endectocides
R01	Nasal preparations
R02	Throat preparations
R03	Drugs for obstructive airway diseases
R05	Cough and cold preparations
R06	Antihistamines for systemic use
R07	Other respiratory system products
S01	Ophthalmologicals
S02	Otologicals
S03	Ophthalmological and otological preparations
V01	Allergens
V03	All other therapeutic products
V04	Diagnostic agents
V06	General nutrients
V07	All other non-therapeutic products
V08	Contrast media
V09	Diagnostic radiopharmaceuticals
V10	Therapeutic radiopharmaceuticals
V20	Surgical dressings

References

1. Aronson JK, Henderson G, et al. A prescription for better prescribing. BMJ. 2006;333(7566):459–60.
2. Awad A, Al-Saffar N. Evaluation of drug use practices at primary healthcare centers of Kuwait. Eur J Clin Pharmacol. 2010;66(12):1247–55.
3. Barabasi AL. Network medicine–from obesity to the "diseasome". N Engl J Med. 2007;357(4):404–7.
4. Barabasi AL. Scale-free networks: a decade and beyond. Science. 2009;325(5939):412–3.
5. Barabasi AL, Albert R. Emergence of scaling in random networks. Science. 1999;286(5439):509–12.
6. Barabasi AL, Gulbahce N, et al. Network medicine: a network-based approach to human disease. Nat Rev Genet. 2011;12(1):56–68.

7. Batagelj V, Mrvar A. Pajek–Program for large network analysis. From http://vlado.fmf.uni-lj.si/pub/networks/pajek/
8. Broadbent DE. Levels, hierarchies, and the locus of control. Quart J Exp Psychol. 1977;29(2):181–201.
9. Buchner A, Funke J, et al. Negative correlations between control performance and verbalizable knowledge: indicators for implicit learning in process control tasks? Quart J Exp Psychol Sect A. 1995;48(1):166–87.
10. Campo K, De Staebel O, et al. (2002) Therapeutic drug prescription behavior: Decision process and marketing mix effects. Antwerp, University of Antwerp, Faculty of Applied Economics, Prinsstraat 13, B-2000 Antwerpen, Web page: http://www.ua.ac.be/tew: 1–44.
11. Christakis NA, Fowler JH. The spread of obesity in a large social network over 32 years. N Engl J Med. 2007;357(4):370–9.
12. D'Zurilla TJ, Goldfried MR. Problem solving and behavior modification. J Abnorm Psychol. 1971;78(1):107–26.
13. D'Zurilla TJ, Sheedy CF. The relation between social problem-solving ability and subsequent level of academic competence in college students. Cogn Ther Res. 1992;16(5):589–99.
14. DeSarbo WS, Degeratu AM, et al. Disaggregate market share response models. Int J Res Market. 2002;19(3):253–66.
15. Dewey J (1910) How we think. D.C. Heath
16. Eisenberg E, Levanon EY. Preferential attachment in the protein network evolution. Phys Rev Lett. 2003;91(13):138701.
17. Freudenheim M (1998). The media business: advertising; influencing doctor's orders. New York Times
18. Goldstein FC, Levin HS. Disorders of reasoning and problem-solving ability. In: Meier M, Benton A, Diller L, editors. Neuropsychological rehabilitation. London: Taylor & Francis Group; 1987. p. 327–54.
19. Hurwitz MA, Caves RA. Persuasion or information? Promotion and the shares of brand name and generic pharmaceuticals. J Law Econ. 1988;31:299–320.
20. IMS (2011a) Total unaudited and audited global pharmaceutical market By Region. from http://www.imshealth.com Top-line Market Data.
21. IMS (2011b) Total unaudited and audited global pharmaceutical market, 2003–2010." From www.imshealth.com Top-line Market Data.
22. Jhon P (2001) Drug company dependent? Medscape Med Students 3
23. Jones JH, Handcock MS. An assessment of preferential attachment as a mechanism for human sexual network formation. Proc Biol Sci. 2003;270(1520):1123–8.
24. Kahn B, Greenleaf E, et al. Examining medical decision making from a marketing perspective. Market Lett. 1997;8(3):361–75.
25. Leffler KB. Persuasion or information? The economics of prescription drug advertising. J Law Econ. 1981;24(1):45–74.
26. March JG, Simon HA. Motivational constraints: the decision to participate. New York: Wiley; 1958.
27. Martin CM, Sturmberg JP. General practice–chaos, complexity and innovation. Med J Aust. 2005;183(2):106–9.
28. Newell A, Simon H. Human problem solving. NJ, Prentice Hall: Englewood Cliffs; 1972.
29. Paredes P, de la Pena M, et al. Factors influencing physicians' prescribing behaviour in the treatment of childhood diarrhoea: knowledge may not be the clue. Soc Sci Med. 1996;42(8):1141–53.
30. Plianbangchang P, Jetiyanon K, et al. Physicians' generic drug prescribing behavior in district hospitals: a case of Phitsanulok, Thailand. Pharm Pract. 2010;8(3):167–72.
31. Rodriguez-Calvillo JA, Lana A, et al. Psychosocial factors associated with the prescription of generic drugs. Health Policy. 2011;101(2):178–84.
32. Ross S, Loke YK. Do educational interventions improve prescribing by medical students and junior doctors? A systematic review. Br J Clin Pharmacol. 2009;67(6):662–70.
33. Simon HA. Motivational and emotional controls of cognition. Psychol Rev. 1967;74(1):29–39.
34. Sketris IS, Langille Ingram EM, et al. Strategic opportunities for effective optimal prescribing and medication management. Can J Clin Pharmacol. 2009;16(1):e103–125.
35. Strogatz SH. Exploring complex networks. Nature. 2001;410(6825):268–76.
36. Sturmberg JP, Martin CM. Complexity and health – yesterday's traditions, tomorrow's future. J Eval Clin Pract. 2009;15(3):543–8.
37. Tan NC, Tay IH, et al. Factors influencing family physicians' drug prescribing behaviour in asthma management in primary care. Singapore Med J. 2009;50(3):312–9.
38. Theodorou M, Tsiantou V, et al. Factors influencing prescribing behaviour of physicians in Greece and Cyprus: results from a questionnaire based survey. BMC Health Serv Res. 2009;9:150.
39. Tversky A, Kahneman D. Judgment under uncertainty: heuristics and biases. Science. 1974;185(4157):1124–31.
40. Tversky A, Kahneman D. The framing of decisions and the psychology of choice. Science. 1981;211(4481):453–8.
41. United Nations A (1948) Universal Declaration of Human Rights. Approved by the General Assembly at its Plenary Meeting on 6 December, 1948, pp. 6. [Lake Success,] 1948.
42. Vazquez A. Growing network with local rules: preferential attachment, clustering hierarchy, and degree correlations. Phys Rev E Stat Nonlin Soft Matter Phys. 2003;67(5 Pt 2):056104.
43. Watts DJ, Strogatz SH. Collective dynamics of 'small-world' networks. Nature. 1998;393(6684):440–2.
44. Wazana A. Physicians and the pharmaceutical industry. JAMA. 2000;283(3):373–80.

45. WHO. The world health report: health systems financing: the path to universal coverage. Geneva: WHO; 2010.
46. WHO (2011) Health Impact Assessment (HIA): the Determinants of Health. WHO programmes and projects. from http://www.who.int/hia/evidence/doh/en/.
47. WHO, Collaborating Centre for Drug Statistics Methodology (2010b) Guidelines for ATC classification and DDD assignment 2011. Oslo, WHO -World Health Organization.
48. Wierenga B, Jong SJ, et al. The decision making process of the family physician in choosing a drug. Ned Tijdschr Geneeskd. 1989;133(3):115–22.
49. Wikipedia.org (2011a). Medical prescription. From http://en.wikipedia.org/wiki/Medical_prescription.
50. Wikipedia.org (2011b) Network science. From http://en.wikipedia.org/wiki/Network_science.
51. Wilkes MS, Bell RA, et al. Direct-to-consumer prescription drug advertising: trends, impact, and implications. Health Affairs. 2000;19(2):110–28.
52. Yildirim MA, Goh KI, et al. Drug-target network. Nat Biotechnol. 2007;25(10):1119–26.
53. Zuger A. Fever pitch: getting doctors to prescribe is big business. New York: New York Times; 1999.

33 Applications of Complex Dynamics: An Approach to Refractory Health Policy Interventions

Renee Crichlow

One person smokes a cigarette, and one person eats potato chips and has a soda; how do such simple actions turn into multibillion dollar health policy burdens? Tobacco abuse and obesity in and of themselves are not the problems; the sequelae of their impact on the health of the individual and the subsequent cost to society are what create the problems.

33.1 Syndemics

These are examples of syndemics.[1] Syndemics are co-arising health problems—tobacco smoking and lung cancer; obesity and diabetes—and are the major focus of this chapter. Syndemics are complex concerns because of the multiple relationships and interactions between multiple agents, including individuals, communities, and industries [1]. These relationships and interactions are necessary for the problems to be created and sustained, and by definition describe a self-organizing system.

For health policy researchers and decision makers, syndemics are ideal examples to study the workings of complex systems. An understanding of human beings and our industries as complex interactive agents may provide a practical approach to addressing complex health policy concerns. Identifying a health policy problem as complex requires an examination of its composition and impact. Complex systems are not linear by nature and are embedded in historical contexts, thus exhibit qualities of sensitivity to initial conditions and path dependence. The interactions within a complex system accumulate overtime and outcomes are not reversible.

Tobacco abuse and obesity offer two examples to discuss public health policy concerns. Developed countries have made significant positive changes in the former, smoking, and are rapidly trying to respond to the latter, obesity, which is a rising concern. Comparing these as cases will further elucidate the underlying mechanisms behind these population-wide health problems and highlighting both approaches to the challenges and various possible strategies for solutions.

33.2 The Emergence of "Big Tobacco" and the Coevolution of "Tobacco Control"

The development of the tobacco industry into "Big Tobacco" prototypically describes a complex health policy concern. "Big Tobacco" is a

[1] Syndemics, as defined by the Center for Disease Control, are two or more afflictions, interacting synergistically, and contributing to excess burden of disease in a population.

R. Crichlow, MD, FAAFP (✉)
Department of Family Medicine, MMC 381, North Memorial Family Medicine Program, University of Minnesota, School of Medicine,420 Delaware Street SE, Minneapolis, MN 55455, USA
e-mail: rcrichlo@umn.edu

J.P. Sturmberg and C.M. Martin (eds.), *Handbook of Systems and Complexity in Health*, DOI 10.1007/978-1-4614-4998-0_33, mega-system where its constituent parts are themselves complex adaptive systems. All are self-organizing systems; they are individual and aggregate components of the mega-system and consist of the multiple stakeholders involved in the production, distribution, and consumption of tobacco products [2]. The self-organization of these complex adaptive systems is based on the exploration, exploitation, stigmergic[2] engagement, and coevolving behaviors with their local environments.

Understanding "Big Tobacco" and Tobacco Control" as complex adaptive systems is essential to identify and address other complex health policy concerns. This is most important because simple interventions to complex problems invariably lead to unintended consequences. A complex health problem may have synergistic impacts on other health factors, e.g., other syndemics that must be addressed simultaneously as they compete for resources in other policy domains like tax revenue. Complex systems solutions that allow for incremental self-reinforcing dynamic systems to develop may work best to address profound health policy challenges.

33.2.1 Historical Context

33.2.1.1 The Emergence of Tobacco

Tobacco, as a product, consumed by humans has ancient roots, originally being cultivated throughout the pre-Colombian Americas. Tobacco has then spread throughout the rest of the world along trade routes by sailors, and tobacco was primarily used in smoking pipes and inhaled snuffs. The centralization of trade under the Spanish Empire then leads to the development of first cigars and subsequently the cigarette. Throughout the Colonial American period, the "cash crop" of tobacco became more prevalent and lucrative. At one point in time, the leaves of the plant were used as currency, nowhere as deeply ingrained as in the culture of Virginia and Maryland, where the tobacco leaves were a part of standard currency for over 150 years. There are early stories of the colonials in America paying over 100 pounds of tobacco for the delivery of a European wife [3].

Throughout 1800s, cigar smoking was on the rise and in the USA in 1901 reportedly four out of five American men smoked at least one cigar per day. In 1901, 43 of the 45 states in the USA had strict anti-cigarette legislation. These early tobacco control laws including city wide cigarette bans which eventually fell by the wayside. With cigarette rolling machine invented in the 1880s, the prevalence of smoking machine rolled cigarettes increased throughout the early nineteen hundreds. A large boost of population wide smoking occurred as a result of the daily cigarette distribution to Allied military personnel, especially during the First and Second World Wars. Hence in the early 1950s, cigarette consumption reached its height, in the USA reaching a prevalence of nearly 60% of American men being daily smokers [3, 4] (Fig. 33.1).

33.2.1.2 The Emergence of "Unintended Consequences"

At this time, the tobacco industry had become well established throughout the world. However, parallel with the increasing prevalence of cigarette smoking began concerns regarding the increasing incidence of lung cancer [5]. In 1912, a monograph linking the two was published by Dr. Isaac Adler. In 1964, after reviewing over 7,000 studies, the Surgeon General of the United States Department of Public Health released the first report of the Surgeon General Advisory Committee on Smoking and Health, conclusively stated that cigarette smoking is a cause of cancer of the lung in men and an important cause of chronic bronchitis. For past 40 years, since the first Surgeon General report on cigarette smoking which resulted in increased regulations, which in turn shifted the user demographics. Yet, even though, according to the Centers for Disease Control, smoking accounts for over 440,000 premature deaths every year in the USA, the Tobacco industry is prospering (Fig. 33.2).

[2] Stigmergy is a mechanism of indirect coordination between agents or actions.

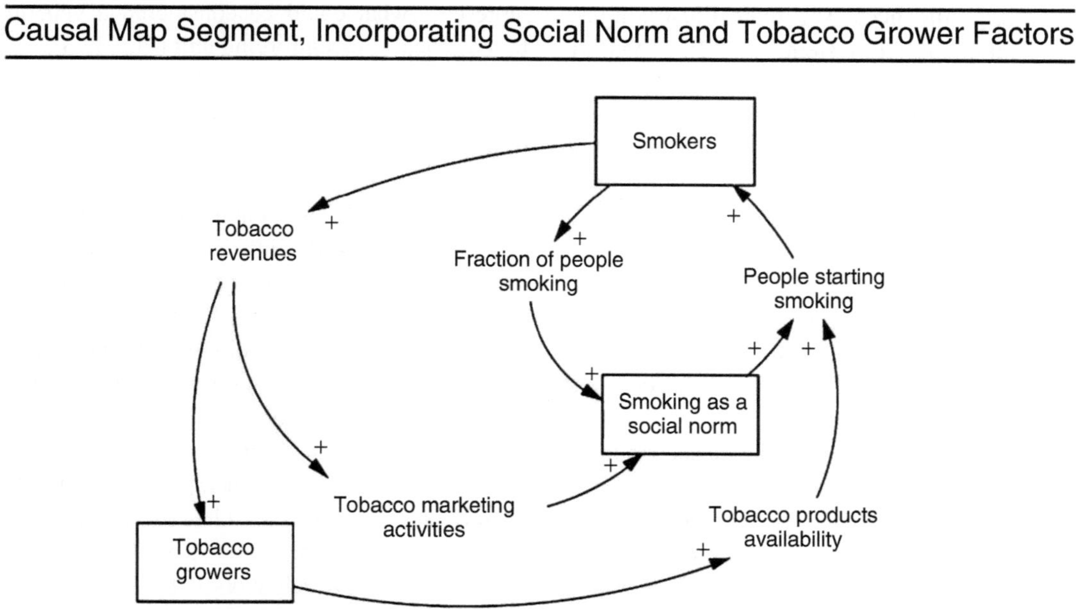

Fig. 33.1 Causal map incorporating social norm and tobacco grower factors. From National Cancer Institute "Greater Than the Sum." Tobacco Control Monograph No. 18

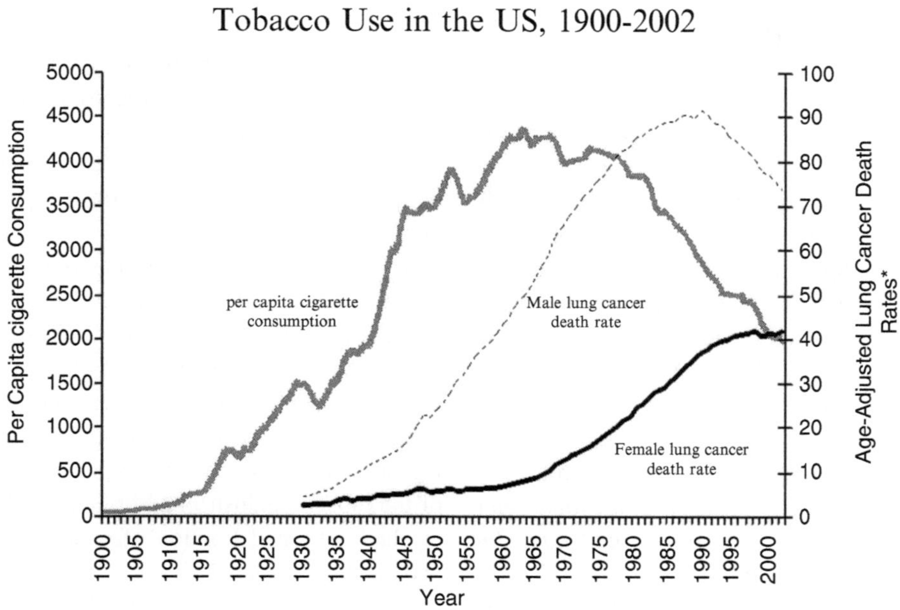

Fig. 33.2 Tobacco use and lung cancer death rates in the USA, 1900–2002

The tobacco industry remains a powerful economic force with profits reportedly growing at 2t per year worldwide, being "*worth*" over $614 billion in 2009.

33.2.1.3 The Coevolution Between the Tobacco Industry and Tobacco Control

Tobacco control as a "priority health policy concern" has ebbed and flowed in response to the prevailing influences over times, illustrating the coevolution of the systems of "Big Tobacco" and "Tobacco Control" in the USA. These fluctuations can be explained by the characteristic behaviors of complex adaptive systems in response to structural changes and feedback from their environment to the system over time [1, 2]. Evaluating this coevolution, we may diagnose areas of influence that have had "success" or "failure" in order to, respectively, sustain, and reproduce or curtail, further investment of resources by either "Big Tobacco" and "Tobacco Control".

The iterative modeling of the tobacco industry and control as a complex health policy concern consisting of coevolving complex adaptive systems will provide a historic model that may help forecast both interventions and conflicts arising from addressing the growing concern of obesity and its syndemics.

33.2.2 The Struggle for Power

In a 1997 article, "Organized Complexity in Human Affairs: The Tobacco Industry" Dr. David Bella wrote quite simply, "*This paper is about power* [2]." In the paper, he describes not only the concept of "power" being an emergent property of a complex adaptive system but clearly illuminates the nature and appearance of the self-sustaining entity that is the Tobacco industry. In contrast, the 2007 National Cancer Institute monograph, "Greater Than the Sum: Systems Thinking in Tobacco Control" the authors clearly state that in dealing with tobacco control, "*The medical model of disease specialization once praised for its utility … is proving inadequate for confronting contemporary challenges*" [1]. Through the monograph, the authors seek to examine a systems approach for "*grappling with complexity interconnectedness, rapid change and uncertainty … to demonstrate the value of systems approach for the entire public health profession.*" Both of these papers present models of systems engaged in feedback with their respective and often interconnected environments. Bella presents a proposal (modified as Fig. 33.1) that represents what could be considered a generic foundational self-reinforcing unit (Fig. 33.3).

This unit is reflected in the systems models of both papers illustrating the development of the tobacco industry and tobacco control systems and their response to both positive reinforcement and threats to their local environments. The basic expression of a self-reinforcing, and thus self-organizing system follows the pattern of augmentation of activities and behaviors that increase the order of the system, providing it with the ability to perpetuate its cycle of sustainability and the suppression of activities and behaviors that otherwise would provide disorder to the self-sustaining system [6]. In this manner, multiple complex adaptive systems, like individual tobacco companies, can respond to similar environmental threats with activities and behaviors that would appear to be conspiracy. The primary drive is actually an adaptive response to facilitate independent sustainability, whether due to positive or negative perturbation, of their system [7].

33.2.2.1 The Emergence of Two Competing Complex Adaptive Systems

As mentioned previously, in the 19th century, there was wide spread use of tobacco, primarily cigars and pipes. There was no technology for high volume manufacture of cigarettes until the end of the the1880s. The early tobacco industry as a complex adaptive system (CAS) was not as robust as today's juggernaut. In the early 1900s, the social climate was quite vociferously anti-cigarette with most states and municipalities having strict anti-cigarette policies [3]. Thus the CAS of the tobacco industry had limited profits, society had a limited fraction of cigarette smokers and the cigarette smoker was seen as the undesirable in society, not as a role model. At that time, the gentleman with the cigar was seen as more socially acceptable. Tobacco still had its political supporters, especially in the States where tobacco was a considerable cash crop for both large and small farms. This influence was demonstrated when at the request of tobacco state congressmen, tobacco was removed from the US pharmacopeia, the list of substances considered drugs,

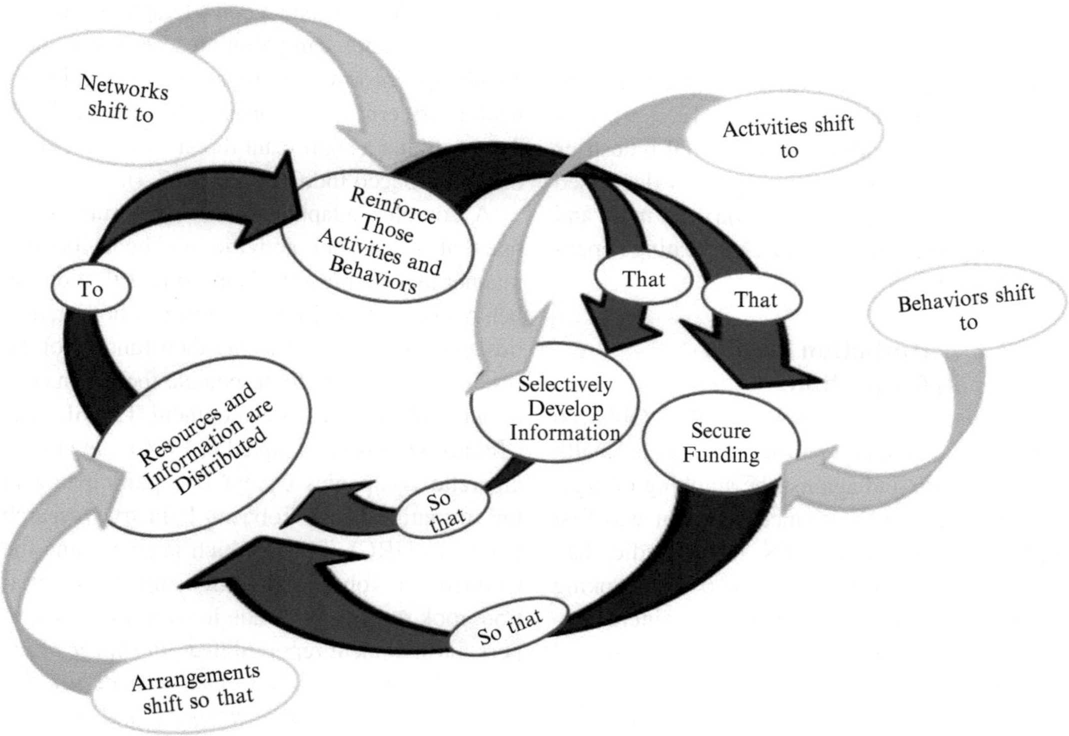

Fig. 33.3 A graphic representation of the feedback forces influencing a generic foundational self-reinforcing unit, modified from D.A. Bella, "Organized Complexity in Human Affairs: The Tobacco Industry"

and this was the compromise seen as necessary for the Food and Drug Act of 1906 to become law [3]. This compromise put tobacco and tobacco products beyond the regulation of the FDA.

The development of high volume cheaper, machine rolled, cigarettes led to an increased need for new smokers. Packaging and advertising became integral to introduce new smokers to their first cigarettes. Early packages of cigarettes were sold with pictures and biographies of famous people as the "card stiffeners" in the cigarette pack, the images of baseball players on these cards invented the genre for these types of sport associated collectibles. Greater political influence and power came to the tobacco industry as a result of the continued increase in these new smokers. Soon anti-cigarette legislation fell by the wayside and remained dormant for almost 70 years [8].

Society's view of cigarette smoking was greatly affected by the images in the media. During the Great Depression, through movie and editorial cartoons, cigars became associated with the oppressive rich robber baron character, whereas the cigarette became the symbol of the gutsy rebel or war hero on the front lines. The social acceptability of cigarette smoking contributed to the tobacco industry's expanding influence and continued to have a synergistic effect on profits and political power, the number of cigarette smokers grew exponential as a result of distribution by the government of packs of cigarettes supplied by the tobacco companies for free to the service men of the armed forces during the two world wars. The post war period in the 1950s the US Tobacco industry had its largest fraction of smokers in society, smoking was seen as a masculine social norm, and almost 6 out of 10 American men were daily smokers. Through advertising tobacco industry money was everywhere supporting magazines, newspapers, radio and the new technology of television. In 1945, the tobacco industry became the second largest source of advertising funds. Thus, the tobacco industry as a CAS became a resource for funding of other industries such as the media and advertising. These industries now had a direct stake in the success of the tobacco industry [2, 3, 9]. This interdependence with other influential industries which are complex adaptive systems decreased tobacco industry's initial isolation even more and spread its power and influence by creating dependent constituencies.

33.2.2.2 Perturbation Results in Adaptation

It was also in the 1950s that the "Tobacco Control" CAS found a partner in science. Health claims both for and against the smoking of cigarettes had been present since the plant was first used. Throughout the 1940s, many studies had been noting the correlation of the rise in smoking prevalence with the rise in lung cancer incidence. Now, however, considerable data on a cause and effect relationship was beginning to emerge from reputable sources. The first publication of major study that definitively linked smoking to lung cancer was published in 1950 [5]. What may be considered a threat to one CAS may be a reinforcing boon to another. Tobacco industry and tobacco control were both affected by this research, and the medical community's response to these findings. Both the "Tobacco Industry" and "Tobacco Control" CAS enlisted the help from physicians. In 1953, leading physicians came out in speeches warning that cigarettes were linked to cancer and the American Medical Association banned cigarette ads in its organizations publications. The tobacco industry filled other publications with advertisements of physicians smoking and claiming healthful benefits from cigarette smoking. Tobacco control advocates at the time reported that many of these cigarette ad laden magazines and newspapers were suppressing adverse health reports related to cigarette consumption [3]. The political influence of the tobacco industry was also still strong. In fact, at that time it was reported that the National Cancer Institute, a government agency, supported by congressional appropriations officially refuted any cause and effect relationship between cigarette smoking and lung cancer. Even though, "Tobacco Control" had a considerable weapon in its anti-cigarette campaign, research was now supplying information regarding a significant health concern, *lung cancer*. This new research demonstrated a significant threat to the resources of the "Tobacco Industry" (Fig. 33.4).

A complex adaptive system that survives is one that adjusts their activities and behaviors in a manner that suppresses disordering activities and behaviors, and positively reinforces both activities and behaviors that secure their funding of the resources necessary to perpetuate their own continued well-ordered development [2, 6]. The "Tobacco Industry" adapted in 1954 all the major cigarette companies except one participated in the formation of the Tobacco Industry Research Council (TIRC). TIRC, which later became the Council for Tobacco Research, upon its initiation, took out full page ads in over 400 newspapers. Its statement reported that "*In charge of the research activities of the Committee will be a scientist of unimpeachable integrity and national repute*" [3]. Through the Tobacco Research Council, the Tobacco industry would fund and publish its own research. In the same year, the longest running most successful cigarette advertising campaign was launched. The Marlboro Man, in 1954, was an ad campaign designed to associate smoking with rugged individualism and masculinity. This was an effort to re-brand the Marlboro cigarette label, which had originally been marketed towards women, since being a filtered cigarette it was considered less manly than unfiltered ones. Within 3 years of the campaigns launch, Marlboro cigarettes increased sales over 300%. The Tobacco industry had continued to expand its market as these types of campaigns brought in significant numbers of first time adolescent smokers. Even in these early encounters, the "Tobacco Industry" and "Tobacco Control" demonstrate the characteristics of coevolving, complex adaptive systems responding to changes in their local environment [1].

Each year the "Tobacco Industry" and "Tobacco Control" CAS adjusted to the realities of their continuously changing environments. In light of increasing health problems being associated with cigarette smoking some organizations,

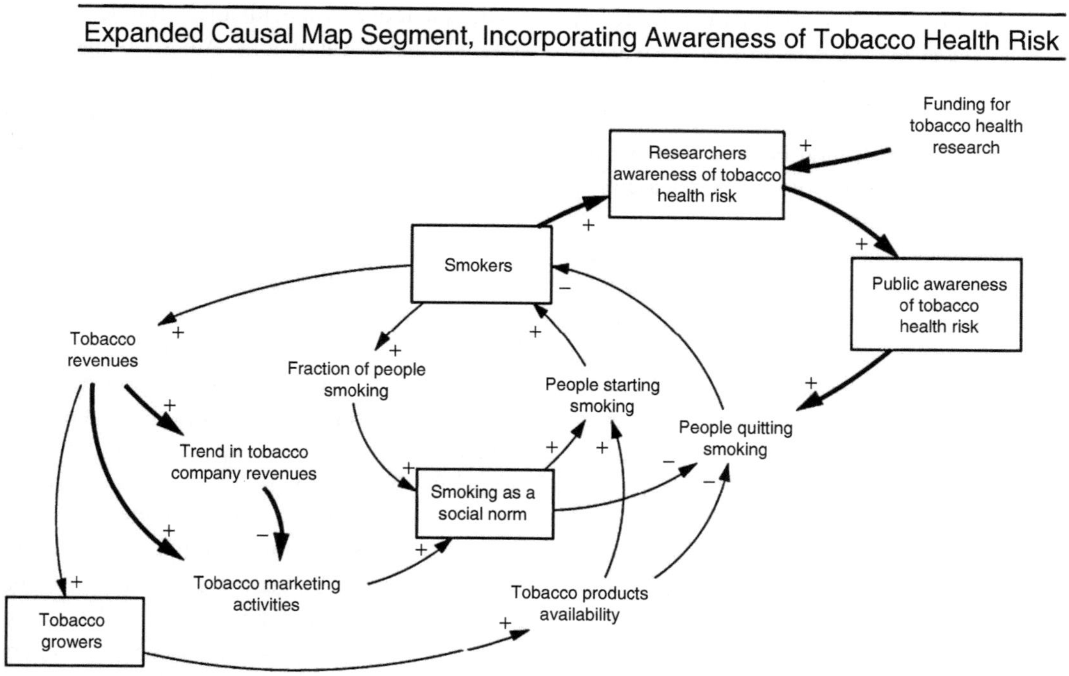

Fig. 33.4 Expanded causal map segment incorporating awareness of tobacco health risk. From National Cancer Institute. Greater Than the Sum. Tobacco Control Monograph 18

particularly some magazines, had stopped taking tobacco industry advertising revenue. More physicians are reported quitting smoking and more conclusive associations were published regarding the adverse health effects of cigarettes. Throughout the later 1950s, claims and counter claims of increased cardiovascular morbidity and mortality, obstetric prematurity and low birth weight babies, and the increased incidence of cancers were the battlegrounds of research, the media, and even civil court claims [8, 9].

33.2.2.3 The "Tobacco Industry" Move

The tobacco industry and its allies addressed every front. Four years after the creation of the Tobacco Research Council, the six major tobacco companies formed a joint lobbying organization. As mentioned prior, *political power* is an emergent property of these complex adaptive systems. The Tobacco industry was preparing to deal with multiple challenges throughout the upcoming decade, with claims of adverse health effects being confirmed, changing the environment and making them vulnerable to possible regulation—thus requiring further adaptation. There were multiple attempts at regulating either the cigarette products or their advertising claims, as well as civil litigation by patients and families with lung cancer. 1960 saw landmark efforts by the unification of tobacco control advocacy. The presidents of the American Cancer Society, the American Heart Association, the National Tuberculosis Association, and the American Public Health Association requested that President Kennedy form a commission to address the "*increasing evidence of the health hazards of smoking*" [3].

33.2.2.4 A Weak Counter Move

In 1962, the Surgeon General announced the formation of the Advisory Committee on Smoking and Health. Attempts by the Federal Trade Commission to provide effective cigarette warning labels were preempted by Congressional action written by the lawyers of the Tobacco industry: "*The Cigarette Labeling act not only gave the industry weak, generalized labels, but preempted litigation by letting the industry argue that the labels had given smokers sufficient warning, and that they undertook smoking at their own knowledgeable risk*" [1, 3]. 1964 was a pivotal year for both of these CASs; the Tobacco industry in the USA was an eight Billion dollar a year business (Ben-David, 1963) and prior to the release of the Surgeon General report they adopted early voluntary advertising policies, "*prohibiting advertising, marketing and sampling directed at young people*". In Tobacco control, the first national anti-smoking coalition is formed, it is called the National Interagency Council on Smoking and Health and was composed of multiple health related medical, nursing and public health agencies both in and outside of the Federal government. The turning point of that fateful year was the publication of The 1964 Surgeon General report "Smoking and Health: Report of the Advisory Committee to the Surgeon General of the Public Health Service". The report included definitive evidence of cigarette smoking as a cause of lung cancer in men. In the following year, Congress passed the Federal Cigarette Labeling and Advertising Act requiring the Surgeon General's Warning on the side of cigarette packs: "*Caution: Cigarette Smoking May Be Hazardous to Your Health*" [3].

33.2.2.5 Changing Focus on "New Markets"

With more organizations and systems clearly expressing serious evidence regarding the dangers of cigarette smoking, we begin to see a shift from individual health to public health approaches. In 1967, the Second Surgeon General's Report: "The Health Consequences of Smoking: A Public Health Service Review," concluded that smoking is the principal cause of lung cancer, and finds evidence linking smoking to cardiovascular disease. The Federal Communications Commission applies televisions "Fairness Doctrine" to cigarette advertising and broadcast stations must donate time for anti-cigarette advertising. In response to decreasing cigarette smoking by its previous consumers, the Tobacco industry sought to find new costumers, increasing its advertising focused toward women and markets outside the USA. In 1968, the Virginia Slims cigarette market campaign sought to capitalize on the women liberation movement with the "You've come a long way baby" campaigns. Ads showing "liberated women" demonstrating their new freedoms by lighting up the slim cigarette made especial to attract the female smoker. Smoking in Latin America increased by 32% over the next decade as it continued to decrease in the USA.

33.2.2.6 The Public Health Concerns Prevail

In the decades that followed the health concerns of the dangers of second hand or "passive smoking" were confirmed. This clearly shifted the individual "vice" of smoking squarely into the arena of public health. In 1970, the World Health Organization (WHO) took a public position against cigarette smoking and its health dangers. In 1971, the 5th Surgeon General report "The Health Consequences of Smoking: A Report of the Surgeon General" demonstrated the dangers of second hand smoke further and recommended a government ban on public smoking. This further isolated the tobacco industry and its positions on cigarette smoking. Over the next two decades, municipalities and industries spread smoking bans in public spaces including airplanes. Public awareness of the health risks of cigarette smoking led to decreased cigarette consumption in the USA, and eventually to the removal of previous government subsidies to tobacco growers. The consequence was fewer farmers being dependent on the crop and subsequently more and more production of tobacco consumed in the cigarettes of the USA was from foreign production.

33.2.2.7 Desperate But Successful "Loop Hole" Strategies

US tobacco companies diversified their investments and many modified their names as they purchased other industries such as food products and liquor companies. Tobacco industry research programs that were evaluating health effects of cigarette smoking were being dismantled as they were discovering untoward effects of their own product [1, 2]. Tobacco control organizations were developing new campaigns and new allies, with one important event being the ban of cigarette commercials from television.

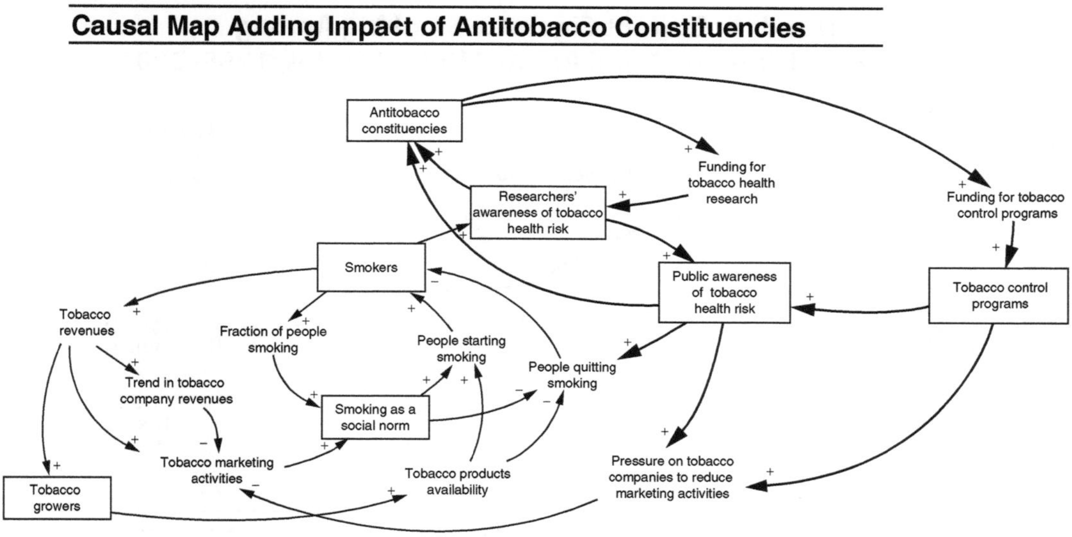

Fig. 33.5 Causal map adding impact of antitobacco constituencies. From National Cancer Institute. Greater Than the Sum. Tobacco Control Monograph No. 18

The tobacco industry had to find creative mechanisms for product placement advertising. One highly successful strategy was "*product placement*" in television shows, movies and "*sponsoring*" of sporting events. Research during that time period demonstrated that the percentage of people smoking in movies was significantly disproportional to the population at large [10], and tobacco control found these strategies very hard to counter, especially in the adolescent populations.

33.2.2.8 Big Tobacco "Forced" to Its Knees, But Not Dead

The largest impact of smoking control policies resulted in the 1997 from the Attorneys General Tobacco Settlement deal, where the tobacco industry in an effort to avoid piecemeal attacks by multiple litigation efforts, settled an agreement with individual states to spend 360 Billion dollars over 25 years, with the money being mainly used for anti-cigarette smoking campaigns. In this settlement, the tobacco industry became the single largest funder of Tobacco control efforts in the USA (Fig. 33.5).

Local and state municipalities then proceeded to continue the attack on cigarette consumption by significantly increasing taxes on individual packs of cigarettes. These efforts, which would likely have met with greater resistance in the periods of the tobacco industry's greatest political power, had now become a key factor in the continued reduction of cigarette smoking. Marginal income smokers, such as the poor and adolescent smoker are more likely to quit, decrease their daily consumption or never start smoking due to the increased price burden. In a last ditch effort, the tobacco industry responded with decreased pricing and/or free packs of cigarettes with purchases. The overall impact of this long campaign of "Tobacco Control" versus "Tobacco Industry" was a decrease in adult smoking rates from 42% in 1965 to 20% in 2008 [4] (Fig. 33.6).

33.2.3 Nicotine Addiction

One factor that we have not yet addressed in this effort is the addictive quality of the product. Cigarettes contain nicotine. Nicotine is a naturally occurring compound in tobacco and has been known to be addictive for decades [8, 9, 11–14]. The tobacco industry also adds known quantities of this addictive compound to cigarettes. They say it is for the consumers,

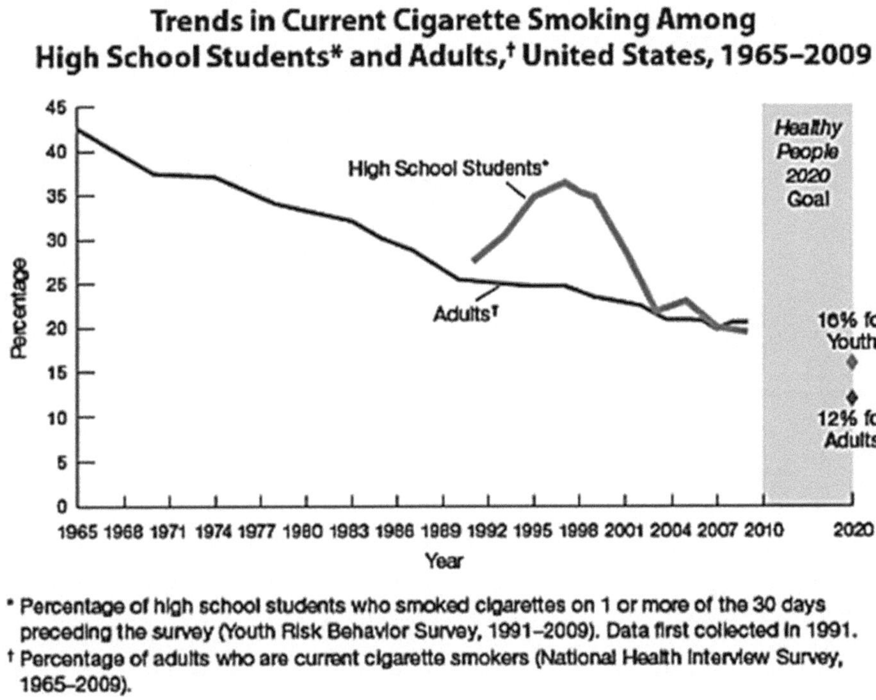

Fig. 33.6 Trends in cigarette smoking 1965–2009. CDC

who by the tobacco industry's reports, feel that the "taste" of the cigarette is affected by the nicotine. The presence of nicotine, an addicting agent in cigarettes clearly contributes to its past and present popularity, but it is only one factor in the interplay in the complex adaptive systems of the tobacco industry and tobacco control. Nicotine affects primarily the individual smoker and its greatest issue comes into play with the programs necessary to facilitate smoking cessation.

The most significant impact in cigarette smoking control is stopping the advent of the new smoker. However, the efforts toward smoking cessation have been greatly aided by the understanding of the addictive qualities of nicotine. The tobacco control campaigns funded by the Tobacco Settlement provide nicotine supplementation, replacement in the form of gums, patches, and lozenges at low or no costs, and have greatly aided in smoking cessation programs at the state level. There is a known quantity of nicotine [8, 11, 12, 14] that is the threshold for addiction. Tobacco control has tried to have the tobacco industry reduce levels below this threshold but has failed to have an impact on this facet of cigarette control. The tobacco industry claims that they are unable to find a consistent way of monitoring the intake of nicotine in the in vivo process of a human smoking. Lower levels of nicotine often mean that "addicted" will inhale more deeply in order to achieve their desired physiologic response. Therefore, work from this particular angle has been a "tow-edged sword" (Fig. 33.7).

33.2.4 Lessons Learnt

The Tobacco industry and Tobacco control system have been engaged and interacting for decades [1, 3]. Following these two coevolving complex adaptive systems overtime demonstrates clearly the organic nature of their interactions to the positive and negative feedbacks and reinforcement from their ever changing environments [2].

Fig. 33.7 The intertwined history of "Big Tobacco" versus "Tobacco Control" (illustration by Joachim Sturmberg)

One system's threat may appear to be the others gain, and yet with adaptation as the key the responses are not always predictable. The loss of access to the television advertising is a good example. The Tobacco industry then found product placement and sporting events as arenas to engage consumers that were not easily regulated by tobacco control and, as an unintended consequence, may have been able to influence even more customers [10]. The tobacco industry's use of foreign tobacco suppliers may have significantly decreased their political power base by removing farmers who had been the constituents of the tobacco industry's potential political allies. Each complex adaptive system seeks to reinforce the activities and behaviors that augment its own sustainability, through exploration and exploitation of the resources it seeks in its environment. Learning from these engagements is critical for planning health policy interventions for complex public health concerns.

33.2.5 The Importance of Logistics

Lessons that can be learned from the historical review of these complex processes are profound. When engaging in these interventions people would often use the vernacular of *war*. They use phrases like *tactics, strategy, engagements, battles and campaigns*. After reviewing the interplay of the "Tobacco Industry" and "Tobacco Control" as complex adaptive systems and evaluating their entanglements overtime, one might find greater understanding in addressing these complex problems by focusing on the logistics.[3] Logistics describes the means of supplying needed resources to maintain the power and influence of an organization. Logistics is how to truly impact a complex public health problem by understanding the

[3] Logistics, are the operations dealing with the procurement of resources necessary to maintain, personnel, facilities, and services of an organization.

sources of a CAS's self-maintenance are and how policy can alter these resources. The tactics will need to evolve based on the long term strategy and local conditions of the time, but complex adaptive systems need resources, and resources are provide by logistics.

The tobacco industry was at its zenith in the USA when it was socially acceptable to smoke cigarettes, the majority of the male population smoked , multiple industries including media and agriculture, were dependent on the tobacco industry's dollars, and there was no conclusive evidence of adverse health effects from cigarette smoking [4]. Logistically, the tobacco industry as a complex adaptive system appeared to be an impregnable juggernaut. It had a self-perpetuating revenue stream with the financial and political power to protect itself and its allies from regulation or encroachment. Logistically, it therefore appeared invulnerable.

Tobacco control slowly grew in its own power to influence the public discourse when adverse health effects started being correlated to the prevalence of cigarette smoking, and began to solidify its position when evidence of the tobacco syndemics of lung cancer and heart disease came into play. Tobacco control was able to strengthen its own logistical support as a result. The evidence of the real impact of cigarette smoking provided a crack in the alliance that had protected the tobacco industry. Smoking became less and less the social norm and eventual has become in many circumstances a socially unacceptable behavior. Tobacco control was able to interrupt the ability of the Tobacco industry to function in self-protective manners by isolating them from other power bases, by promoting the evidence-based threat to public health. As a result, tobacco control continued to be able to gather more adverse health data, and when the tobacco industry's power base had waned further their isolation as a system made them vulnerable to regulation and litigation [2, 3]. Eventually, the combined effects of decreasing the number of smokers and agricultural producers in the USA, and an increased burden on the funding of supply by eliminating subsidies for tobacco production and increasing taxing of tobacco product at the shop door decreased demands by the consumer [1].

Tobacco control, benefitting from an out-of-court settlement with the tobacco industry, suddenly had the required revenue for advertising against cigarette smoking and the promotion of treatment of the adverse health effects of cigarettes, resulting in the eradication or severely decreased use of cigarettes. In response, the tobacco industry sought other markets for its product and now has the greatest yearly financial gains in overseas markets-an issue having great ethical dimensions. This further reinforces the premise that "*complex problem must be solved adaptively with evolving feedback*" [1]. As noted in the accompanying figures from the CDC report "Greater Than the Sum" the complex adaptive systems of the "Tobacco Industry" and "Tobacco Control" can be modeled in many iterations of their components and subsequent feedbacks. This is helpful in understanding the roles of the stakeholders in these systems and how their relationships coevolve in response to the adaption of the prevailing forces in their shared environment.

33.3 Guiding the Obesity Campaign

This iterative modeling of "Tobacco Industry" and "Tobacco Control" as a complex health policy concern consisting of coevolving complex adaptive systems will provide a historic model that may help forecast both interventions and conflicts arising from addressing the growing concern of obesity and its syndemics.

Obesity and its syndemics of diabetes and cardiovascular disease are rapidly becoming the leading public health concern in the developing world. In the year 2000, the Surgeon General Dr. C. Everett Koop reported that in the previous 15 years rates of obesity had increased over 48% and had become the number one preventable cause of death in the USA. Childhood obesity is also on the rise while obesity rates in the 12- to 19-year-old population that were 5% in 1976, had reached were over 18% in the year 2008.

In looking for causal relationships, we have found multifaceted answers. Over the same time period as obesity increased in developed nations:

- The number of people with sedentary lifestyles increased.
- Physical activity in schools decreased.
- Eating out increased, and the cost of eating out decreased.
- Daily caloric intake increased.
- The consumption of whole foods decreased.
- The consumption of packaged foods increased.

In addition, at the point (1985) where the curve of obesity starts to increase its slope, the sweeteners of most sodas and juices switched from sugar to the cheaper alternative, High Fructose Corn Syrup, HFCS.

33.3.1 The Potential Role of High Fructose Corn Syrup

Correlation does not infer causation, but one thing we have learnt from the tobacco engagement is that correlation does merit investigation. The early studies of the adverse effects of cigarette smoking came from examining the increased prevalence of cigarette smoking correlated to the increased incidence of lung cancer (see Fig. 33.2 Tobacco and Lung Cancer 1900–2002). Subsequently, the temporal association of increased HFCS and obesity has spawned investigations [15–21]. Other public health interventions are occurring to address the obesity epidemic that is directed at other possible contributors. Though there is some early data that has raised concerns regarding HFCS and its interaction with the physiology of humans. Like at the beginnings of concerns about tobacco, these questions and concerns have led to similar responses from the producers of HFCS.

33.3.1.1 Emerging Anti-HFCS Evidence

Early research into HFCS has found some interesting conclusions. A Princeton University team of researchers have noted as a result of their studies that "*all sweeteners are not equal when it comes to weight gain*". Their studies noted rats with access to HFCS gained significantly more body weight than the control groups and that the "*... increase in body weight with HFCS was accompanied by an increase in adipose, fat, notably in the abdominal region ...*" [22], they also reported that even with the same number of calories consumed, the rats ingesting HFCS gained more weight than the control group counterparts.

Though they have raised considerable levels of concern, these early finding are preliminary and have not been reported as conclusive evidence of HFCS in the obesity epidemic. Already there is negative feedback from many consumer who are reading labels and when possible avoiding purchasing of products containing HFCS in favor of other sweeteners. The individual consumer is also a complex adaptive system who is adapting to information in their environment. This in turn has caused feedback to the complex adaptive system responsible for the production of the HFCS, the corn sweetener industry.

33.3.1.2 The Corn Sweetener Industry Coevolutionary Moves

The corn sweetener industry is responding to this perceived threat in a manner consistent with Bella's R Attractor diagram of a complex adaptive system. It is encouraging activities and behaviors that increase the information and activities favorable to its own sustainability. In response to this reduced demand, the corn sweetener industry has its own political lobbying and scientific research organizations responding with the dissemination of advertisements and publications disputing the adverse health claims and supporting its product as a natural product. One of the studies funded by a HFCS company has found that their, "*expert panel concluded that HFCS does not appear to contribute to overweight and obesity any differently than do other energy sources*" [21]. Among the conclusions of this industry, sponsored study was decreased cigarette smoking had contributed to the obesity epidemic [21]. In addition to research sponsored by the HFCS industry, there have even been efforts to change the HCFS name to a more wholesome sounding moniker, "corn sugar", as opposed to High Fructose Corn Sugar.

33.3.2 Towards a Whole System Approach

Addressing the obesity epidemic will not be so simple even if HFCS is found to be a significant factor. There are still other complicating factors that have to be dealt with in developed nations such as the USA, like the relationships of food production, distribution and consumption, and Western population's increasingly sedentary lifestyles.

This is a multifaceted complex and complicated public health policy concern that may require fundamental changes in the funding and infrastructure of some of our deeply ingrained beliefs as a society. Who determines what and how you decide to eat does not have a simple answer. Many people are already under the misconception that they are making these choices unaided and any attempt to influence their choice or their children's choice of foods may be misinterpreted as an attempt to infringe on their freedoms. This is where the lessons from the Tobacco systems may be helpful. In looking to address the obesity problem, the approach that steers away from the individual health choice and focuses on the subsequent impact and costs to society may have a better chance to shift the debate and in subsequent iterations get closer to addressing the core problems.

Identifying the current stakeholders is really necessary in addressing this epidemic. If we are able to provide knowledge and resources to create allies instead of belligerents to produce these fundamental shifts, resources will be preserved to address the real problem and not spent in unnecessary antagonisms.

The Federal government subsidizes many of the foods that contribute to our ability to have greater caloric ingestion than previous generations; e.g., we have in the past provided extremely calorie dense, highly processed food in our public schools' lunch and breakfast programs. Whole food and especially vegetable intake has been significantly decreased in these programs. New initiatives such as the "My Plate" design for nutrition advice and improved regulation of the quality and composition of school prepared foods are striving to increase consumption of whole foods including vegetables, but are early in their initiation and have not had time to note effects as yet.

These actions however are consistent with the lessons learned from the Tobacco models. The shifting of resources to multiple complex adaptive systems necessary to support these initiatives provides opportunities to build more robust health-based food initiatives as aggregate, larger complex adaptive systems. This may shift resources and therefore decrease the power and influence of CASs that have benefited from the current structures present during the ongoing obesity epidemic.

Other methods of shifting resources can be restricting access to the product similar to smoke-free cities and college campuses or taxation of the food products in question. These are sources of controversy amongst those dealing with obesity, which has so many contributing factors, but there are some who are looking to the taxation of products they associate with more obesity as part of the solution.

The government's purchasing power and/or power of distribution affected the growth of the tobacco industry by giving cigarettes to the military during the Second World War. Using this power may be an important tool in the fight against obesity. More schools buying and preparing whole foods would have a significant emphasis for greater support for the infrastructure of production and distribution of healthier food alternatives. The standards set by the government require political will to first pass the policy and second support implementation of the policy.

The Child Nutrition Bill of 2011 has its critics, yet many see it as an early step in the right direction. The level of concern regarding health care cost to society and even military preparedness all were factors used to influence passage of the law. There were also various food industry lobbyists whose impact on the final bill is what has provided concern to some critics.

One law is not sufficient to halt this epidemic. This law does help reinforce some behaviors and activities that are reinforcing in the development of a more healthful obesity prevention CAS. It is one component. Our past system of publicly funding health interventions has been linear, myopic, and grossly ineffective in addressing complex problems. As we have seen a complex health policy problem requires complex solutions addressed overtime in multiple iterations. Tobacco control has been successful not in individual efforts, but the reinforcing behaviors emerging from the sum of its parts.

33.3.3 Encouraging "Whole System" Change. An Example: The Somerville Project

"Shape up Somerville",[4] SUS, is an example how a complex problem like childhood obesity can be counteracted by developing a complex adaptive intervention through a Community-based Participatory Research (CBPR) intervention, including stakeholders at every level of the community: parents, school officials, teachers, school nurses, cafeteria staffs, community groups, local "champions", city officials, city government employees, health care providers and professionals, food service providers, restaurants and local media within the city of Somerville. The project was designed in an effort to determine if a "community-based environmental change intervention" could impact the "energy-equation" of at risk elementary students in an ethnically diverse urban community [23–25]. Interventions changed, what was determined to be an "obesigenic" environment, by increasing physical activity and full day access to healthy food in the schools, homes and community at large [23–25].

The SUS intervention targeted the early elementary age group as there has been a threefold increase in childhood obesity over the last few decades. Children are also at particular risk because they have less control over their food choices and activity levels. Previous school-based interventions have had mixed to poor results, perhaps because a child spends less than 50% of their time in school. The SUS intervention utilized an ecologic approach, looking to address multiple aspects of a child's environment including the energy balance between food intake and physical activity levels, food consumption choice in a child's environment at school and in the home, education of children and parents, improving playgrounds, walkways, and cycle ways.

As an intervention for a complex problem, the Shape up Somerville results indicate that a CBPR intervention may provide a model for how to initiate the development of a complex adaptive system designed to alter outcomes refractory to linear, single focus interventions, like a solely school-based food consumption interventions. One challenge, arising from a grant dependant project like SUS, is to maintain the self-reinforcing feedbacks necessary to sustain a complex adaptive change strategy through the multiple iterations required to achieve a "new" self-sustaining system. The "whole of community" approach achieved self-sustainability, once grant funding ceased, as residents have appreciate their improved health and wellness, a large numbers of community residents continue to use the modified

---

[4] The key features of the project are summarized on the Somerville website (http://www.somervillema.gov/departments/health/sus).

Shape Up Somerville is a city wide campaign to increase daily physical activity and healthy eating through programming, physical infrastructure improvements, and policy work. The campaign targets all segments of our community, including schools, city government, civic organizations, community groups, businesses, and other people who live, work, and play in Somerville.

This effort began as a community-based research study at Tufts University targeting 1st through 3rd graders in the Somerville Public Schools. Today there is Coordinator working on active and healthy living programs supported by the Health Department and a Taskforce that is a collaboration of over 11 initiatives and 25 stakeholders involved in working on various interventions across the city, such as:

- School Food Service
- Teachers teaching an-School Curriculum
- After School programs using a new curriculum
- Parent, City Employee and Community Outreach
- Restaurants
- Walkability and Safe Routes to School
- Extension of the Community Path
- School Nurses and Pediatricians
- Policy Initiatives
- Farmers markets and community/school gardens
- Monthly community celebrations of Walk/Ride Days

The major of the City of Somerville describes the implementation of the program at http://www.youtube.com/watch?v=WT7NqVLhteE.

bike/walkways, and restaurants that serve healthy food choices increased their business. These changes all provide ongoing positive feedback to maintain the system change; however, care is needed to identify contravening influences early.

33.4 Summary

As complex systems are by their nature nonlinear, modeling one system cannot predict outcomes any more than one may predict the weather. Modeling one system may, however, provide an ability to forecast under what conditions any specific intervention may have greater impact. We know that from the Tobacco CAS models, when health concerns can be directly associated with a single source product, i.e., cigarette smoking and lung cancer, early correlation leads some consumers to stop using that product.

Research will be supported by both the industry of concern and the health care system; this early information may be contradictory. The CAS that feels threatened by these investigations may seek to react on many fronts, including making a more healthful alternative to its product, e.g., filters on cigarettes. In addition other means will be explored, like initiating counter research [21], starting high profile advertising campaigns to influence the public conversation, lobbying politicians, seeking new markets for their product, or changing the name of the product of concern.

Based on the processes observed in the Tobacco CASs, one might forecast the HFCS industry response to the development of adversarial CASs promoting alternative calorie sources. Specifically, corn industry funding "research counsels" [21], expanding markets for corn products with international markets seeing more HFCS exports, and influencing legislation that would mandate "Flex-Fuel" engines so more cars would consume the corn product ethanol. This latter shift would risk altering agricultural production from food to transportation and raises concerns regarding "food security" and the rising food costs in exchange of support of the motor vehicle industry and in essence fuelling another CAS. Understanding the adaptive and coevolutionary behaviors will be able to guide what systems we could develop to address our leading arising health policy concern, obesity and its multiple syndemics of diabetes, and heart disease.

A final point demanding consideration is the role of government. Through taxation of a product that has unhealthful associations, government puts itself in a compromising position. Governments would be advised to avoid dependence on these revenues generated from known dangerous products, otherwise government may be less motivated to stop those behaviors—an old conundrum.

> "This vice brings in one hundred million francs in taxes every year. I will certainly forbid it at once-as soon as you can name a virtue that brings in as much revenue" Napoleon III, (The Tobacco Almanac, page 310, edited by David B. Moyer, 1998).

This complication can be avoided if the revenue is EXCLUSIVELY earmarked for cessation and treatments of the adverse effects of the product/s in question. In this situation, as consumption of the product decreases, the revenue loss would have less of an overall budget impact and the government would be less complicit in the maintenance of unhealthful consumption. In this manner, maximizing the impact of taxation as a tool to decrease initiation and consumption of cigarettes proved an effective tool and should continue. It has had such a great success with cigarettes that taxation on consumption will likely find its way, in some manner, in the battle to prevent and decrease obesity.

Other efforts would depend on the social will, which also coevolves, and may now be ripe like the reclassification of tobacco as a drug. Another effort might be to find a method to reproducibly monitor nicotine affective blood absorption level consumption in humans so as to regulate that cigarettes could not contain more than the known addictive threshold of nicotine [11–14].

Propositions in this chapter to put into daily practice. Complex Health Policy concerns require

- When possible, identify all stakeholders prior to interventions.
- Identify all levels of influence and impact of the etiology of concern, not the disease but the causes of the disease.
- Identify the prevailing social conditions to determine their readiness for change or recalcitrance.

- Understand single-focused interventions will always have unintended consequences in non-linear systems.
- Repeatedly evaluate outcomes and adjust subsequent iterations as indicated.
- When considering a community or public health intervention develop a counter CAS that changes the environment and addresses the logistics of the CAS you would like to impact. For instance, provide positive alternatives for and decreased incentives for the stakeholders within the CAS whose impact you desire to change.

References

1. National Cancer Institute. Greater Than the Sum: Systems Thinking in Tobacco Control NIH Pub No 06–6085 2007 April 2007.
2. Bella DA. Organized complexity in human affairs: the tobacco industry. J Bus Ethics. 1997;16(10):977–99.
3. Borio G. The tobacco timeline. Accessed Aug 2011. Available at http://www.tobacco.org/History/Tobacco_History.html.
4. Centers for Disease Control and Prevention. Percentage of adults who were current, former, or never smokers, overall and by sex, race, Hispanic origin, age, and education. National Health Interview Surveys, selected years—United States, 1965–2000. Available at http://www.cdc.gov/tobacco/research_data/adults_prev/adstat1print.html.
5. Doll R, Hill AB. Smoking and carcinoma of the lung; preliminary report. Br Med J. 1950;2(4682):739–48.
6. Sterman JD. Business dynamics: systems thinking and modeling for a complex world. New York: McGraw-Hill/Irwin; 2000.
7. Parunak HV. Expert assessment of human-human stigmergy. Ann Arbor, MI: Altarum Institute; 2005.
8. Dani JA, Balfour DFJ. Historical and current perspective on tobacco use and nicotine addiction. Trends Neurosci. 2011;34(7):383–392.
9. Russo P, Nastrucci C, Alzetta G, Szalai C. Tobacco habit: historical, cultural, neurobiological, and genetic features of people's relationship with an addictive drug. Perspect Biol Med. 2011;54(4):557–77.
10. Glantz SA, Kacirk KW, McCulloch C. Back to the future: smoking in movies in 2002 compared with 1950 levels. Am J Public Health. 2004;94(2):261–3.
11. Grenhoff J, Svensson TH. Pharmacology of nicotine. Br J Addict. 1989;84(5):477–92.
12. Ortells MO, Barrantes GE. Tobacco addiction: a biochemical model of nicotine dependence. Med Hypotheses. 2010;74(5):884–94.
13. Shadel WG, Shiffman S, Niaura R, Nichter M, Abrams DB. Current models of nicotine dependence: what is known and what is needed to advance understanding of tobacco etiology among youth. Drug Alcohol Depend. 2000;59 Suppl 1:S9–22.
14. Wayne GF, Carpenter CM. Tobacco industry manipulation of nicotine dosing. Handb Exp Pharmacol. 2009;(192):457–85.
15. Ackroff K, Sclafani A. Rats' preferences for high fructose corn syrup vs. sucrose and sugar mixtures. Physiol Behav. 2011;102(5):548–52.
16. Akgun S, Ertel NH. The effects of sucrose, fructose, and high-fructose corn syrup meals on plasma glucose and insulin in non-insulin-dependent diabetic subjects. Diabetes Care. 1985;8(3):279–83.
17. Bray GA, Nielsen SJ, Popkin BM. Consumption of high-fructose corn syrup in beverages may play a role in the epidemic of obesity. Am J Clin Nutr. 2004;79(4):537–43.
18. Kanarek RB, Marks-Kaufman R. Developmental aspects of sucrose-induced obesity in rats. Physiol Behav. 1979;23(5):881–5.
19. Kanarek RB, Orthen-Gambill N. Differential effects of sucrose, fructose and glucose on carbohydrate-induced obesity in rats. J Nutr. 1982;112(8):1546–54.
20. Moeller SM, Fryhofer SA, Osbahr 3rd AJ, Robinowitz CB, Council on Science and Public Health, American Medical Association. The effects of high fructose syrup. J Am Coll Nutr. 2009;28(6):619–26.
21. Forshee RA, Storey ML, Allison DB, Glinsmann WH, Hein GL, Lineback DR, et al. A critical examination of the evidence relating high fructose corn syrup and weight gain. Crit Rev Food Sci Nutr. 2007;47(6):561–82.
22. Bocarsly ME, Powell ES, Avena NM, Hoebel BG. High-fructose corn syrup causes characteristics of obesity in rats: increased body weight, body fat and triglyceride levels. Pharmacol Biochem Behav. 2010;97(1):101–6.
23. Economos CD, Folta SC, Goldberg J, Hudson D, Collins J, Baker Z, et al. A community-based restaurant initiative to increase availability of healthy menu options in Somerville, Massachusetts: Shape Up Somerville. Prev Chronic Dis. 2009;6(3):A102.
24. Goldberg JP, Collins JJ, Folta SC, McLarney MJ, Kozower C, Kuder J, et al. Retooling food service for early elementary school students in Somerville, Massachusetts: the Shape Up Somerville experience. Prev Chronic Dis. 2009;6(3):A103.
25. Burke NM, Chomitz VR, Rioles NA, Winslow SP, Brukilacchio LB, Baker JC. The path to active living: physical activity through community design in Somerville, Massachusetts. Am J Prev Med. 2009;37(6 Suppl 2):S386–94.

34 Designing for Health Promotion, Social Innovation, and Complexity: The CoNEKTR Model for Wicked Problems

Cameron D. Norman and Andrea L. Yip

34.1 Introduction

Global health care and promotion in the twenty-first century faces challenges posed by aging populations, mass migration, environmental threats, and the rise of new types of infectious diseases; issues that are interconnected, dynamic in nature, and without precedent. These are commonly called *wicked problems*: ones that are complex, ill formulated, "joined-up", confusing, and require multi-party collaboration to address [1]. Wicked or complex problems are resistant to the linear and simple solutions sets that are often employed in health systems that typically come with a push towards models of best practice and evidence-based medicine [2, 3]. While these solution sets demonstrate utility in certain areas pertaining to healthcare, they typically fail when applied to health promotion and population health [4] which commonly deal with multi-causal, contextually bound, and dynamic problems. It is for this type of problem space that complexity science and systems thinking are most appropriate [5]. In terms of action, it is often the place frequented by designers [6].

The absence of a clear strategy for applying evidence in complex settings does not render it useless, but nor does it encourage a harder push towards methods of synthesis that are more appropriate for clinical medicine or doing what the late systems scholar and management professor Russell Ackoff calls "*doing the wrong things righter*". The way forward resides somewhere in between, where evidence can be gathered, synthesized, and applied in a manner that recognizes complexity and novelty, while drawing on the diversity that comes from such contexts and allows innovations to emerge. Design and design thinking is one such approach that marries complexity with action [7].

Design thinking is a means of engaging with content in a manner that supports intentional change in contexts with ambiguous or sometimes nonexistent data and holds an orientation towards addressing problem solving. In the case of health promotion, the primary desired outcome is positive holistic health and reduced or eliminated health inequities. The manner in which the biological and social determinants of health influence wellbeing is highly varied from setting to population and contextually bound making it a complex domain. The inability to control complex environments poses a challenge to public health professionals seeking to facilitate change. Complexity

C.D. Norman, Ph.D. (✉)
Dalla Lana School of Public Health,
University of Toronto, 155 College Street, Room 586,
Toronto, ON M5T3M7, Canada

CENSE Research + Design,
757-155 Dalhousie Street, Toronto, ON M5B2P7, Canada
e-mail: cdnorman@cense.ca

A.L. Yip, M.P.H.
Dalla Lana School of Public Health,
University of Toronto, 155 College Street, Toronto,
ON M5T3M7, Canada
e-mail: andrealyip@gmail.com;

J.P. Sturmberg and C.M. Martin (eds.), *Handbook of Systems and Complexity in Health*,
DOI 10.1007/978-1-4614-4998-0_34, © Springer Science+Business Media New York 2013

does not deny the ability to influence change, rather it provides parameters on expectations for change with an acknowledgement that change is non-linear, resistant to control methods, and problematic in terms of yielding predictive outcomes. A designer in this regard falls to what Herbert Simon suggested as "*one who changes an existing situation into a preferred one*" [8]. This approach to design suggests that the act of creation with purpose is something that distinguishes a designer from others.

In health promotion, the focus of intervention is often social in structure and content, thus requiring methods of population engagement that recognize and deal with these social forces. Compounding this is the scale of intervention, which often oscillates between individuals, groups, and communities through to systems simultaneously. This chapter considers how we can effectively address complexity in health in a meaningful way using design thinking to explore complex phenomena and leverage tools that work with, rather than challenge, these dynamics. Food security is an issue complex in its scope and nature that we profile later to demonstrate use of social media tools and the applications of a design thinking informed model for health promotion we developed called CoNEKTR. We begin by looking at the concept of design thinking, followed by an introduction to social media as a means of bringing together design and systems thinking.

34.2 Design Thinking

Design thinking emerged from the field of industrial design and its popularity is often attributed to the work of industrial design firm IDEO [9] and its leaders David and Tom Kelley, Bill Moggridge, and Tim Brown. Industrial designers, having achieved much success with creating products for public consumption, began turning their attention from hard product development towards engineering human services and programmes in the early 1990s. This was partly due to the increasing use of social science and arts-based approaches to learning through empathy that generated insight into the desires, needs, and interests of humans engaging with designed products [10]. At the same time, design knowledge initially generated through engineering and architecture was expanded by the engagement of professionals from disciplines such as anthropology, psychology, business, and health sciences [11, 12]. As design firms began applying their craft to social problems [13, 14] and institutions such as hospitals and schools, there was growing interest in applying design thinking to addressing problems within the business and human services sectors [11, 15, 16].

It is estimated that 80% of the environmental impact of any product or service is determined at the design stage [17], which has enormous implications for the way we create social systems. Whether it is the layout of a neighbourhood, structure of a school system, or role assignment within a workplace, the design of social systems is likely to have major implications for human behaviour. An examination of most theories of health behaviour will find an inordinately high value placed on adapting people to existing problems, rather than designing opportunities to prevent problems from forming in the first place. The dynamic and overlapping nature of socially anchored problems like chronic disease create a complex adaptive system, rather than a linear, disconnected one [18]. In this complex system behaviours mutually influence one another to manifest problems that differ from context to context and present challenges for the application of theories focused on unidirectional change. These wicked problems are defined as: "a class of social system problems which are ill formulated, where the information is confusing, where there are many clients and decision makers with conflicting values, and where the ramifications in the whole system are thoroughly confusing" [6]. This is health promotion.

Wicked problems have increasingly fell into the domain of design [6, 19] or design thinking, which is described as an approach to leveraging creativity purposefully to find, frame, and produce novel solutions to problems [c.f.,14, 16, 20].

Nobel laureate Herbert Simon [8] argued for design this way:

> Everyone designs who devises courses of action aimed at changing existing situations into preferred ones. The intellectual activity that produces material artifacts is no different fundamentally from the one that prescribes remedies for a sick patient or the one that devises a new sales plan for a company or a social welfare policy for the state (p.111).

The fit between design, complexity science, and health promotion is demonstrated in the focus on these wicked problems. Scientific thought wrestles with challenges posed by complexity in that no directive, reductive, clear, or compartmentalized solution can be generated. Thus, no wicked problem can be wholly solved; this is evident in health promotion where there exists no single solution for issues like poverty, homelessness, or discrimination. While design takes such problems as its primary focus [21], it often lacks the rigour of documentation that scientific inquiry provides.

Design thinking provides a bridge between systems science and health promotion practice, with an ability to manage complexity and emphasize the importance of social interaction. It highlights many aspects of social psychology such as empathy [22–24], visualization [25, 26], teamwork [27], multiplicity [27], and the importance of issue framing [6]. Design also amplifies the potential for collaborative thinking and problem solving in a manner that fits with complex systems. Not only does it support a coordinated and focused effort to engage people around a particular problem space [28], it draws upon the insights of diverse and multi-level actors positioned within the system to inform the design process. Thus, the advancement of design thinking for health benefits from considering systems and psychological theories in combination with health promotion concepts.

Two of the central concepts tied to design thinking have been creativity and innovation [11, 15, 29, 30], which provide useful corollaries to complex adaptive systems. Complex systems become so through the emergence of new patterns that are brought about through the dynamic interchange between diverse agents [31, 32]. The information that is exchanged and the patterns that create opportunities for recombination and formation of new structures; thus any system that encourages diversity and interaction is likely to produce complexity. Social media platforms provide such an environment.

34.3 Social Media

Social media is enabled by what is referred to as Web 2.0 technologies [33, 34], which include a host of platforms and tools that enable anyone with an Internet-enabled connection to create, modify, distribute, and consume content without knowledge of computer programming. These platforms include popular tools such as Facebook, Twitter, and YouTube and are used by billions of people every day for the purposes of entertainment, information seeking, social support, and enterprise development.

Social media and the social networks that they operate through provide an environment that offers the most explicit example of the manner in which complexity intersects with health promotion, design thinking, and systems thinking. Social media is any form of networked electronic communication platform that derives its value from user engagement. Services like Facebook, YouTube, Twitter, and Flickr are functionally useless without content and participation of its users and their interconnections. Social media is also a disruptive force, which is the kind that best fit with complex situations. Thackara (2005) extends this argument: "*design thinking, in combination with Internet-enabled networks, and wireless communications, can reshape whole production processes, even the entire logic and structure of an industry*" (p.18).

Social media opens up opportunities for engagement with health content through the use of diverse media forms in settings that can be adapted to meet the needs of the user. Whereas creation of content online for the early World Wide Web was limited to those with some programming skill, social media make it easy to create, re-work, and distribute content with the most basic Internet skills. The increasing availability and accessibility of mobile technology tools such as phone handsets, tablets, and portable media players have shifted Internet traffic from hardwires to wireless channels [35], creating new spaces for social cohesion [36, 37] and opening up the potential for the public to access health information wherever they are and whenever they need it.

34.4 The CoNEKTR Model

The CoNEKTR (Complex Network Electronic Knowledge Translation Research) Model [38] focuses on social problem solving and issue identification based on the science of systems, design thinking, and behaviour change theories and leveraged through social media tools. Drawing on complexity science, the model also draws flexible, dynamic boundaries around the system being engaged and provides guidance for bringing diverse groups together to address complex problems in practical terms. The CoNEKTR model is a 10-step model that is guided by five processes. Generally, the model involves facilitated face-to-face group interaction followed by an extended knowledge development and exchange process that can take place remotely using electronic communication tools and possible additional face-to-face meetings. In addition to design, systems thinking and health promotion, the model is also founded on a variety of complementary approaches for social learning and exchange:

1. **Community of Practice (CoP).** A CoP is a self-organized, voluntary, and focused collective of agents (people and organizations) who collaboratively work towards understanding on an issue or problem [39, 40]. The CoP approach has been particularly effective within health contexts where there is diverse expertise, expertise is not co-located, and Internet technologies are used for communication [41–43].
2. **Self-organized group facilitation and engagement**. Among the most widely used models of self-organized group learning are Open Space Technology [44], the World Café approach [45], and the Unconference [46, 47]. All of these models seek to foster interaction between multiple stakeholders, who would not likely assort with one another.
3. **Online Communities**. Electronic communities of practice (eCoP) can serve as ways to link health professionals who may not have the means due to physical proximity or time to otherwise collaborate and share experiential practice knowledge [42, 43].
4. **Social Media**. Social media refers to a constellation of shared technologies that derive their value from the participation of users through directly creating original content, modifying existing material, contributing to a community dialogue, and/or integrating various media together to create something unique. It allows the traditional consumer to become a prosumer in their contributing to product generation [26, 48].
5. **Collaborative decision-making and emergent action**. Public engagement for idea generation and refinement has been widely discussed in the popular business literature [48–50] under various names such as "crowdsourcing" or "collective innovation" and reflects a process of collaborative design, decision-making, and social emergence [18, 30]. Through exchange of very little information and simple rules, complex patterns can emerge, producing knowledge or action that is both unanticipated and impossible to create if developed by individuals working independently [51].
6. **Enhancing empathy through social and experiential learning**. Empathy is generated through interaction and engagement with others, which enhance positive attitudes towards them [23, 24] and when connections between groups separated by social power are formed [22]. Empathy is a core concept in design and enables the creators and users of a product or service to learn from each other and connect beyond.
7. **Arts-based research.** Issue exploration, strategy development, and prototyping are enhanced through the use of arts-informed methods and are a hallmark feature of design thinking [14, 25]. Arts-informed methods such as sketching, scrapbooking, participatory theatre, or poetry favour tangible forms of self-expression, supporting the notion of "learning by doing" [14], and can enhance

creativity by opening new spaces for ideas to emerge and interact.

34.4.1 CoNEKTR in Practice

A key feature of the CoNEKTR model is that it is designed to leverage the collaborative potential and communication reach provided by social networks. The model's implementation begins with an initial face-to-face event focused on a topic of importance (focus), which serves as a catalytic probe, a conceptual device that focuses attention and sets the conditions to foster the emergence of new attractors, stimulating creativity by providing a set of boundary conditions that focus cognitive energy. These conditions support creative problem exploration with groups [29].

Attendees to a CoNEKTR-informed event bring not only expertise, but also knowledge networks related to the topic. The introduction of a catalytic probe stimulates activity that bridges structural holes in the interaction space, which expands the number of weak ties in the network. Establishing networks of networks is the key, rather than completely reconfiguring existing agent networks in a manner that could have unintended negative consequences. CoNEKTR works to initially foster collaboration between existing, homophilous agents (groups or individuals) and extending communication and engagement outward to those agents with more heterogeneous characteristics over time [52]. See Fig. 34.1.

The catalytic probe is a problem or topic of interest that serves to focus group activity. Grounding knowledge generation and exchange in an explicitly social process encourage ideas to be prototyped and evolve rather than force proscribed solutions at the outset; an approach is consistent with design thinking [14, 20] and models of health promotion intervention development [53]. Social exchange and prototyping generate collective feedback allowing ideas most likely to achieve wider acceptance (and subsequent adoption) to emerge, get tested and refined by participants, and deployed more widely. Full prototypes may not always be possible within the constraints of a given gathering, however mental models can be developed and tested through thought experiments as participants work possible solutions. Fuller prototyping is likely to occur as interactions extend beyond the initial face-to-face encounter. Implementing the model also initiates an opportunity to integrate knowledge through practice by having groups apply lessons learned to projects identified at the initial meeting and then supported by online engagement afterwards.

34.4.2 Five Processes of CoNEKTR

CoNEKTR's pathways for action occur through a series of channels facilitated by knowledge of complexity, social network theories, behaviour change, and design. There are five key processes involved in the CoNEKTR model. The first two are the most intense and interactive, while the latter provide the momentum for action beyond the initial event. The five processes work in a cyclical manner, building on the work done in successive steps. With each step, the outreach increases along with the potential for new information to be generated.

34.4.2.1 Key Processes

1. **Face-to-face interaction**. Invited participants are put together in an event that alternates between an organized and self-organized state that is closely modelled on the Unconference. The Unconference is similar to other participatory organization models such as Open Space Technologies [44] or the World Café [45], where most organization is performed by the participants. These methods have been used successfully in supporting learning in a variety of contexts [47]. However, as it is likely that participants will have a lack of knowledge of the boundaries and components of the system (e.g. invitation list, participant backgrounds) a purely self-organized model of leadership is impractical. Instead, a convener (individual or group) initiates action and sees things to a point where participants can take over effectively.
2. **Social modulation**. CoNEKTR's facilitated discussion format is designed to provoke discussion and draw connections between topics of relevance between diverse agents in the system.

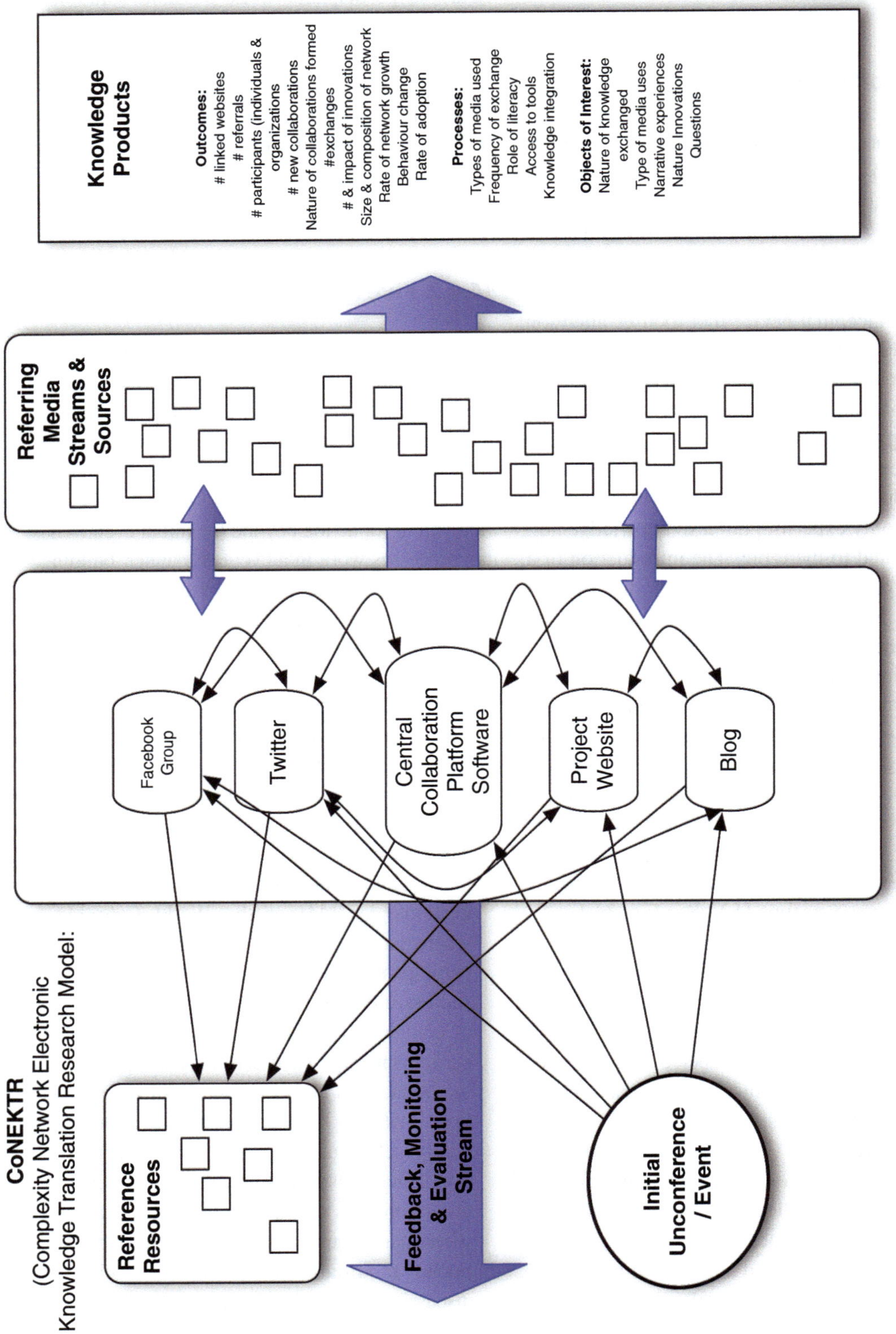

Fig. 34.1 CoNEKTR – Model

Active facilitation specifically creates a space for inclusive dialogue by encouraging actors to convene around a particular topic based on their interest and lived experiences rather than formal or educational "expertise". The emphasis is on a dialogical process in which individuals come together as equals to solve common problems [54] in an environment of respect, hope, and trust, where facilitators "model and promote empathy, reinforce active listening skills, and encourage participatory discussion" [55]. This is consistent with common design practice recommendations [14, 20, 56]

Anyone can be creative if given opportunities for supported expression [30, 57], which is why this is where conveners spend most of their energy. Conveners and their team serve as modulators and encourage self-organization into groups based on an initial process of issue identification. This ideation is performed in a free-form manner with facilitators who create social spaces for creative engagement with each other and the material. The process of collaborative design often involves use of tools and strategies that involve creative visualization, use of arts and crafts materials to aid in prototype development, and methods that involve both physical and mental activity [58].

In the modulation role, the design iteration process is sped up to enable early, preliminary synthesis on areas of loose organization using content analysis to sort through ideas and presented back to the group for approval and refinement before moving on to the next step. The process is done in a transparent manner whereby attendees can see how ideas were generated and how synthesis was achieved. Details of this process are described in earlier works [38].

The CoNEKTR model was designed to suit the constraints of many health professionals and community and business leaders, while providing a highly focused, guided process that begins with a single-day event, followed by a second event held in close proximity (preferably within 3–6 weeks of the first event). This space in between events allows for time for reflection without reducing the momentum and idea generation, and also allows for additional participants with similar vested interests to be identified between events and brought into the discussion.

These lessons and themes are posted to an accessible online social networking platform and used to seed groups or engagement with other common spaces that can enhance discussion [59]. The process continues to generate and deploy catalytic probes generated from discussion themes to invoke further conversation, stimulating attractors and guiding action from the resulting activity generated. In the online networks facilitators seek connections within the community to identify the emergence of new patterns, acting to reinforce existing beneficial and intervening to dampen unhelpful patterns (i.e. those that are not congruent with the group's stated wishes).

3. **Evaluation.** Feedback is a critical component of a complex adaptive system and evaluation is the mechanism by which data can be gathered to form the kind of sensemaking process that allows agents in the system to learn from one another and adapt. Data are used in an iterative manner to inform decision-making, further action, and ongoing evaluation [53]. Evaluation methods can vary with those most likely to capture the effects of social networks, to reflect activity within the system as a whole, and to provide clear, utilization-focused feedback, being preferred. Consistent with complexity science theories and research, a utilization-focused or developmental evaluation approach for using these methods is the preferred option [60, 61].
4. **Feedback.** Drawing on lessons learned from community-based research [62] evaluation data are collected, analysed, and shared in a manner that is transparent to the participants. Feedback also serves as a self-correcting method of dampening attractors that are not productive and supporting those that are. Ongoing waves of data collection with feedback consistent with both a developmental evaluation [60, 61, 63, 64] and action-research approach [65–67].
5. **Integration**: Knowledge integration occurs when what has been learned has been

integrated into normal practice [68–70]. This integration involves taking the innovation and using it to create new ways of working, policies and strategies, and having such innovations sustain themselves over time. Although the steps are akin to the Diffusion of Innovation model [71, 72], the process of knowledge integration describes an active, iterative process of putting knowledge into action. Knowledge integration is not just additive, it is synthetic, influencing the entire system that it becomes a part of where learning is transformative. Such multiple and transformative iterations provide for environments of supportive experimentation and innovation.

34.4.3 Ten Steps of CoNEKTR

The CoNEKTR model is implemented in ten steps to which the five processes are overlaid. However, the number of steps and the degree of implementation can be modified to suit the needs of participants, the scope of the problem, and the resources available. CoNEKTR is designed to be flexible and responsive, not rigidly applied. The steps involved in the model are described in Table 34.1.

34.5 Applications of CoNEKTR: The "Food4Health" Project

To illustrate how the CoNEKTR model can operate in practice, we present a case study from our work with the Youth Voices Research Group, an innovation-focused health promotion research unit based at the Dalla Lana School of Public Health at the University of Toronto in Toronto, Canada. The *Food4Health* project provided a mechanism to test a systemic intervention strategy aimed at fostering cooperative interconnections across the food system as a whole, which included growers, processors, distributors, retailers, marketing leaders, and disposal agents. The initiative included representatives of small and large for-profit and non-profit organizations, educators, researchers, citizen advocates, and food service leaders. The overall goal of the project was to strengthen awareness within the system and overall capacity for collaborative action on food system health issues including: food security, environmental stewardship, and agri-food innovation. Within this mandate was an emphasis on the impact and involvement of youth and young adults in this system.

The CoNEKTR model was applied as part of a 10-week rapid-response, proof-of-concept project that began with an initial face-to-face event entitled *Food4You!* held at the University of Toronto campus. As the first step of the CoNEKTR model, the event brought together 34 youths and 25 adults who were recruited using a combined maximum variation and snowball approach. Particular attention was paid to include participants from areas with typically low representation and high needs on issues of public health including: low-income areas of the community, those from visible minorities, immigrants and refugees, and individuals from rural centres.

The event began with a brief orientation to the project and its objectives. Once consensus was reached on an operating definition of food security to provide a stable point of discussion (step 2), participants engaged in a large-group brainstorm-like ideation session, where ideas were noted on flip-chart paper in full transparent view of others. After the initial idea generation was completed, participants independently voted on the ideas and posted them on the wall using sticky notes (step 3). These ideas were collated and sorted using a rapid content analysis by members of the research team (step 4). Five themes emerged from the data:

1. The food industry (production)
2. Food availability and selection
3. Taste and appeal
4. Cultural appropriateness
5. The ability to access self-sustaining methods for food production (e.g. urban gardening)

These themes formed the basis for discussion groups for the afternoon session (step 5).

Table 34.1 Steps in the CoNEKTR Process

Step	Activities	Description
1.	Convene group	Recruit participants connected to a specific topic from diverse facets of the population(s) connected to the topic. Participant selection is done purposively, relying on both snowball and purposive sampling strategies to capture maximum variation (**enhance diversity**), while also exploring some depth in existing **networks**
2.	Introduce topic	Provide detailed outline of the activities and plan for the day, pointing to areas where established knowledge exists (**path dependency**), while encouraging areas to create novel knowledge around a specific topic for discussion (**catalytic probe**). Set the specific **boundary conditions** for the activity
3.	Collective and independent brain storming	Within the boundaries, provide a set of **minimum specifications (simple rules)** to guide the process aimed at encouraging **diverse perspectives** to be expressed in the generation of ideas within the established boundaries. This assists in provoking **creative responses** to the challenges of the topic
4.	Content analysis and thematic organization of data	**Emergent patterns** within the data are observed and re-organized in a manner that is transparent to the participants (that is, there is a clear, visible method of organizing the data underlying its structure for the group). Content analysis serves to modulate the complexity of the information system by ensuring that information is neither chaotic (lacking organization) nor oversimplified (lacking flexibility)
5.	Create discussion groups	Encourage **self-organization** of groups around the general themes. It is highly recommended to have an assigned person not directly involved in the discussions (e.g. research assistant) serve as the scribe to document the process and capture the **novel information** generated in the group, otherwise there is at least one participant not able to fully participate in the activity or is unable to properly take notes
6.	Summarize key points from each group	An appointed spokesperson or two from each group provides a short summary of the discussion including novel points, areas of contention and agreement, and possible suggestions for future activities. Participants in other groups are encouraged to provide **feedback** on the ideas and add to them
7.	Future planning	Feedback is used to generate new **catalytic probes** to stimulate interest in ongoing discussion and dialogue for the group as they prepare to self-organize online and develop new collaborative opportunities and extend those generated at the face-to-face event
8.	Post notes and ideas online	Further opportunities for **feedback** is provided through the web platform by having the summary notes posted online, preserving the **information** generated during the face-to-face event. This includes the evaluation data collected from the face-to-face meeting
9.	Modulate discussion and identify key connectors and connections	Participants in the discussions continue to **cooperate**, while their ideas **compete** for **attention** and contribution from the community, creating **attractors**. Discussion modulators monitor the groups and encourage interaction within and between them as they form online seeking to **amplify attractors** that are advancing the purpose of the discussion and dampening **attractors** that are not to reduce unnecessary **noise** and unhelpful **variation**
10.	Identify core activities for action	Based on the groups' activity and the **emergence** of patterns of activity, specific **leverage points** for action/change are identified and the necessary resources are coordinated by the modulation team and organized to support action. An evaluation (**feedback generation**) plan is included in this process
Repeat the process as necessary in a series of iterative, evolutionary stages		

Note: Complexity and systems-related processes noted in bold

The five participant-designed breakout sessions generated insights and discussion that led to three broad-based areas of focus (step 6):

(a) The idea that food is too expensive and choice is limited in places where youth spend time, particularly schools, including university campuses.
(b) Food education is not comprehensive enough (both in terms of physical health, such as nutrition, and mental health, such as eating disorders) and that young people often are misled or ill informed about food choices and their consequences for themselves and their communities.
(c) "*We are the food industry*" and thus, through action, have control over what foods we eat, how they are produced, and where they come from.

In the late afternoon, these three broad-based themes were used to focus discussion on potential actionable steps that could inform the next stage of the process (step 7). The next step of the process involved taking the ideas and action steps online to the social media environments organized by the conveners. This included the Youth Voices Research Group's Facebook page, an action page on the youth-focused network TakingITGlobal, a Ning customized Web-based networking platform (http://studentlife-engagement.ning.com/), and the project Twitter feed, @food4health (steps 8 & 9).

One of the products of the discussion was a common interest in developing a sustainable food policy at the University of Toronto, viewing the campus as a model for the larger food system. This provided the foundation for the next phase of the project (step 10). The Youth Voices Research Group served as the convener for the first three face-to-face meetings, while a student-led group emerged as the leader for the final meeting and took responsibility for continuing the project beyond the initial project phase with the goal of achieving longer-term sustainability.

Taken together, the CoNEKTR model applied through Food4Health enabled a mass exchange of ideas, coordinated networking between diverse groups within a complex system that hadn't interacted before, the means to do rapid prototyping of ideas, and a strategy to bring those ideas to bear on a health problem over time. An evaluation of the first wave of project using document review, social media use statistics and monitoring, and qualitative interviews with the staff and participants suggested that the project was successful in creating new connections, promoting knowledge translation, and generating useful innovations. One of the products of this initiative was that a seed project started by one group was able to gain support to develop into a national programme. Other outputs included new funding applications, joint ventures, and strategic partnerships between groups that had the same interests, but no means of connecting. Rather than simply meeting at an event, the CoNEKTR model applied to Food4Health enabled a mechanism for relationships to grow within the context of a project that focused mutual interest and energy.

34.6 Conclusion

The CoNEKTR model brings concepts from complexity science, systems thinking, and design thinking together in a strategic manner to address complex health problems that facilitates action. By amplifying the relationship development processes that facilitates co-design, group collaboration, and opportunities for meaningful participation, CoNEKTR can aid groups working in complex health environments navigate complexity and assist in the transformation of group thinking into meaningful and sustained change. The model emphasizes the opportunity to articulate a diversity of voices, advancing steps towards greater health equity.

Implementation of the CoNEKTR model has limitations that require notation, particularly with respect to the manner by which making connections (networks) between individuals can exacerbate inequalities, rather than to reduce them [73]. Creating more frequent, regularized interactions that are designed to support change over a longer period of time is one way to reduce these potential inequalities. Reducing competition, increasing cooperation, and connecting the project foci to participants' own interests and real-world context is another. It is also why there is an explicit emphasis on transparent communication and rapid feedback through evaluation and providing the results to participants quickly after data are collected to inform decision-making. This provides continued evidence of effectiveness to support organizations (or agents affiliated with organizations) to show the return on investment in time and resources for networking.

Social media scholars have argued for some form of political or coordinating overlay in support of collaborative decision-making on broad, population issues such as health [74, 75], suggesting that complete self-organization may not be practical or useful to advancing social learning and action. The CoNEKTR model aims to balance the benefits of self-organization and self-determined action with a coordinated approach that provides boundaries and stimulates appropriate attractors within the system to provoke change. It is also designed to adapt to the needs of busy people who nonetheless face complex problems and seek opportunities to leverage their assets through networks to address them.

The CoNEKTR model offers those within the health system a vehicle for engaging with complexity and designing potential means for addressing complexity in practical terms. Through engaging diversity within a dynamic system with intention, the application of design thinking provides health practitioners and researchers with some means of influencing changing health events in a manner that is harmonious with complex adaptive systems, rather than conflicts with it. In providing this vehicle for engagement, the promise of achieving population health impacts through complexity may be brought closer to reality.

References

1. Rittel H, Webber M. Dilemmas in a general theory of planning. Policy Sci. 1973;4:155–69.
2. Biswas R, et al. The process of evidence-based medicine and the search for meaning. J Eval Clin Pract. 2007;13(4):529–32.
3. Sturmberg JP, Martin CM. Complexity and health – yesterday's traditions, tomorrow's future. J Eval Clin Pract. 2009;15(3):543–8.
4. Norman CD, Best A. Family-based programmes for preventing smoking by children and adolescents: commentary. Evid Based Child Health. 2009;4(2): 890–2.
5. Norman CD. Health promotion as a systems science and practice. J Eval Clin Pract. 2009;15(5):868–72.
6. Buchanan R. Wicked problems in design thinking. Des Issues. 1992;8(2):5–21.
7. Kolko J. Exposing the magic of design: a practitioner's guide to the methods & theory of synthesis. New York, NY: Oxford University Press; 2011.
8. Simon HA. The sciences of the artificial. 3rd ed. Cambridge, MA: MIT Press; 1996.
9. IDEO. IDEO [cited 2012 February 27, 2012]. Available from http:// www.ideo.com (2012).
10. Brown T, Wyatt J. Design thinking for social innovation. Stanford Soc Innov Rev. 2010; 31–35.
11. Kelley T. The art of innovation: lessons in creativity from IDEO, America's leading design firm. New York, NY: Crowne Publishers; 2001.
12. Mau B, Institute Without boundaries. Massive change. New York, NY: Phaidon Press; 2004.
13. Berger W. Glimmer: how design can transform your life, your business, and maybe even the world. Toronto, ON: Random House Canada; 2009.
14. Brown T. Change by design. New York, NY: Harper Collins; 2009.
15. Kelley T. The ten faces of innovation: IDEO's strategies for defeating the devil's advocate and driving creativity throughout your organization. New York, NY: Doubleday; 2005.
16. Martin R. The design of business: why design thinking is the next competitive advantage. Boston, MA: Harvard University Press; 2009.
17. Thackara J. In the bubble: designing in a complex world. Cambridge, MA: MIT Press; 2005.
18. Sawyer RK. Social emergence: societies as complex systems. Cambridge, UK: Cambridge University Press; 2005.
19. Coyne R. Wicked problems revisited. Des Stud. 2005;26(1):5–17.
20. Ambrose G, Harris P. Basics design: design thinking. West Sussex: Ava Publishing; 2010.
21. Barnwell M. Design, creativity and culture: an orientation to design. London: Black Dog Publishers; 2011.
22. Baston CD, et al. Empathy and attitudes: can feeling for a member of a stigmatized group improve feelings toward the group? J Personal Soc Psychol. 1997;72(1):105–18.
23. Ickes W. Empathic accuracy. J Personal. 1993; 61(4):587–610.
24. Stephan WG, Finlay K. The role of empathy in improving intergroup relations. J Soc Issues. 1999; 55(4):729–43.
25. Buxton W. Sketching user experiences: getting the design right and the right design. San Francisco, CA: Morgan Kaufmann; 2009.
26. Wurman RS. Information anxiety 2. Indianapolis, IN: QUE; 2001.
27. Baty S. Solving complex problems through design. Interactions. 2010;17(5):70–3.
28. Innes JE, Booher DE. Planning with complexity: an introduction to collaborative rationality for public policy. London: Routledge; 2010.

29. Sawyer RK. Explaining creativity: the science of human innovation. Oxford: Oxford University Press; 2006.
30. Sawyer RK. Group genius: the creative power of collaboration. New York, NY: Basic Books; 2008.
31. Coveney P, Highfield R. Frontiers of complexity: the search for order in a chaotic world. New York, NY: Random House; 1995.
32. Miller JH, Page SE. Complex adaptive systems: an introduction to computational models of social life. Princeton, NJ: Princeton University Press; 2007.
33. Deshpande A, Jadad A. Web 2.0: could it help move the health system into the 21st century? J Mens Health Gend. 2006;3:332–6.
34. Seeman N. Web 2.0 and chronic illness: new horizons, new opportunities. Healthc Q. 2008;11(1):104.
35. Cisco Systems. Cisco virtual networking index: global mobile data traffic forecast update 2010–2015. Cisco Systems: San Jose, CA; 2011.
36. Ling R, Stald G. Mobile communities: are we talking about a village, a clan, or a small group? Am Behav Sci. 2010;53(8):1133–47.
37. Ling R. New tech, new ties: how mobile communication is reshaping social cohesion. Cambridge, MA: MIT Press; 2008.
38. Norman CD, et al. Designing health innovation networks using complexity science and systems thinking: the CoNEKTR Model. J Eval Clin Pract. 2010;16(5):1016–23.
39. Wenger E. Communities of practice: learning, meaning and identity. Cambridge, UK: Cambridge University Press; 1998.
40. Wenger E, Snyder W. Communities of practice: the organizational frontier. Harv Bus Rev. 2000;78(1):139–45.
41. Li L, et al. Evolution of Wenger's concept of community of practice. Implement Sci. 2009;4(1):11.
42. Norman CD, Huerta T. Knowledge transfer & exchange through social networks: building foundations for a community of practice within tobacco control. Implement Sci. 2006;1(1):20.
43. Ho K, et al. Electronic communities of practice: guidelines from a project. J Contin Educ Heal Prof. 2010;30(2):139–43.
44. Owen H. Open space technology: a user's guide. New York, NY: Berrett-Koehler; 1997.
45. Brown J. The World Cafe: shaping our futures through conversations that matter. San Francisco, CA: Berrett-Koehler; 2005.
46. Crossett L, Kraus JR, Lawson S. Collaborative tools used to organize a library camp unconference. Collab Librariansh. 2009;1(2):66–9.
47. Greenhill K, Wiebrands C. The unconference: a new model for better professional communication, in LIANZA Conference Papers 2008. Auckland, NZ: Lianza; 2008. p. 1–18.
48. Tapscott D, Williams AD. Wikinomics: how mass collaboration changes everything. Toronto, ON: Portfolio/Penguin; 2007.
49. Howe J. Crowdsourcing: why the future of the crowd is driving the future of business. New York, NY: Crown Business Publishing; 2008.
50. Surowiecki J. The wisdom of crowds. New York, NY: Doubleday; 2004.
51. Gloor PA. Swarm creativity: competitive advantage through collaborative innovation networks. Oxford, UK: Oxford University Press; 2006.
52. Aral S, van Alstyne M (2009) Networks and novel information: the diversity–bandwidth tradeoff, in Workshop on Information in Networks (WIN). New York, NY; 2009.
53. Skinner HA, Maley O, Norman CD. Developing Internet-based eHealth promotion programs: the spiral technology action research (STAR) model. Heal Promot Pract. 2006;7(4):406–17.
54. Smith WA. The meaning of Conscientizacao: the Goal of Paulo Freire's Pedagogy. Amherst, MA: University of Massachusetts Centre for International Education; 1976.
55. Wallerstein N, Sanchez-Merki V. Freirian praxis in health education: research results from an adolescent prevention program. Heal Educ Res. 1994;9(1):105–19.
56. Lawson B. How designers think: the design process demystified. Amsterdam, NL: Architectural Press; 2006.
57. Sanger J. Seven types of creativity: looking for insights in data analysis. Br Educ Res J. 1994;20(2):175–85.
58. Stickdorn M, Schneider J, editors. This is service design thinking. Amsterdam, NL: BIS Publishers; 2011.
59. Surman M, Weshler-Henry D. Common space: beyond virtual community. Toronto, ON: Pearson – Financial Times; 2001.
60. Patton MQ. Utilization-focused evaluation. 4th ed. Los Angeles, CA; 2008.
61. Patton MQ. Developmental evaluation: applying complexity concepts to enhance innovation and use. New York, NY: Guilford Press; 2010.
62. Israel BA, et al. Review of community-based research: assessing partnership approaches to improve public health. Palo Alto, CA, USA. Annu Rev Public Health. 1998;19:173–202.
63. Patton MQ. Developmental evaluation. Eval Pract. 1994;15(3):311–9.
64. Westley F, Zimmerman B, Patton M. Getting to maybe: how the world is changed. Toronto, ON: Random House Canada; 2006.
65. Argyris C, Putnam R, Smith D. Action science: concepts, methods, and skills for research and intervention. 1st ed. San Francisco, CA: Jossey-Bass; 1985.
66. Stringer ET. Action research: a handbook for practitioners. Thousand Oaks, CA: Sage; 1996.
67. Whyte WF. Participatory action research. Newbury Park, CA: Sage; 1991.
68. Best A, Hiatt RA, Norman CD. Knowledge integration: conceptualizing communications in cancer control systems. Patient Educ Couns. 2008;71(3):319–27.
69. Best A, et al. Systems thinking for knowledge integration: new models for policy-research collaboration.

In: McKee L, Ferlie E, Hyde P, editors. Organizing and reorganizing: power and change in health care organizations. Houndsmill: Palgrave McMillan; 2008. p. 154–66.
70. Kerner JF. Knowledge translation versus knowledge integration: a "funders" perspective. J Contin Educ Health Prof. 2006.
71. Oldenburg B, Parcel GS. Diffusion of innovations. In: Rimer BK, Lewis FM, Glanz K, editors. Health behavior and heath education. San Francisco, CA: Jossey-Bass; 2002.
72. Rogers EM. Diffusion of innovations. 4th ed. New York, NY: Free Press; 1995.
73. Small ML. Unanticipated gains: origins of network inequality in everyday life. New York, NY: Oxford University Press; 2009.
74. Shirky C. Here comes everybody: the power of organizing without organizations. New York, NY: Penguin; 2009.
75. Halamka J, et al. Exchanging health information: local distribution, national coordination. Heal Aff. 2005;24(5):1170–9.

The Dynamical Behaviours of Diseases in Africa

35

Winston Garira

35.1 Introduction

More than ever before, there is now great concern about infectious diseases. Africa bears the disproportionate burden of most infectious diseases and the detrimental impact of infectious diseases is currently more strongly felt in Africa. Although chronic diseases, such as cancer and heart disease receive more attention in developed countries, infectious diseases are the most well-known causes of suffering and mortality in Africa and some developing countries. The multiple burdens of infectious diseases represent a demand on health services of Africa far beyond that experienced in developed countries. Infectious diseases, such as malaria, HIV/AIDS and tuberculosis (TB) are a growing health problem in Africa. Resources for addressing health problems of Africa remain disproportionately low when compared with the tremendous disease burden. The United Nations Development Programme (UNDP) report on the Millennium Development Goals (MDGs) [1] cautions that the health goals of the MDGs will not be met by 2015 in the neediest countries, and, in fact warns that the situation in Africa may actually worsen. The variety of intervention programmes that can be implemented to control these infectious diseases and the limited resources available in Africa to combat these infectious diseases in addition to the existence of already strained and weak public health infrastructure results in "infectious diseases" in Africa being a complex system.

The complexity of factors underlying infectious diseases in Africa include:

(a) Healthcare systems which are poorly funded leading to poor access to essential healthcare.
(b) The context in which these health systems operate which include:
    - Violence, war and political unrest.
    - People living crowded without enough nutritious food to eat.
    - The unavailability of clean water to drink and sanitary systems to dispose safely of human waste.
    - The warm, moist climate conditions which make it easy for many disease parasites and bacteria to grow and spread.
    - The deepening poverty across the continent which is creating fertile ground for the spread of infectious diseases.
(c) The growth of the HIV/AIDS epidemic which has infected up to 30% of adults in some African countries and its synergistic interactions with many other infectious diseases prevalent in Africa causing re-emergence of infectious diseases previously thought to be largely under control, such as malaria and tuberculosis and also because its own infectivity and clinical course is altered by other infections [2–7].

W. Garira, Ph.D. (✉)
Modelling Biomedical Systems Research Group,
Department of Mathematics and Applied Mathematics,
University of Venda, P. Bag X5050,
Thohoyandou 0950, South Africa
e-mail: winston.garira@univen.ac.za

J.P. Sturmberg and C.M. Martin (eds.), *Handbook of Systems and Complexity in Health*,
DOI 10.1007/978-1-4614-4998-0_35,

(d) Evolution of drug resistance strains of many prevalent infectious diseases in Africa resulting in diseases such as pneumonia, bacterial dysentery, malaria, sexually transmitted infections, and tuberculosis, once manageable by available therapies, becoming increasingly more difficult and costly to treat [8–12].
(e) The emergence of new infectious diseases, such as Ebola [13].
(f) Lack of capacity for health research. In order to respond effectively to the growing health threats due to infectious diseases and even the transfer of these health risks, Africa needs to develop capacity for health research. For now, the programmatic response to the complex infectious disease pattern in Africa is moving forward without a great deal of scientific investigation to facilitate a comprehensive evidence-based action.

The interactions of these factors affect the control of infectious diseases in several ways. Take for example the interactions between HIV/AIDS and tuberculosis alone. Since the late 1980s, TB control has markedly deteriorated in Africa because of the HIV pandemic [14]. In South Africa alone, two million people co-infected with MTb (Mycobacterium tuberculosis and HIV) [15] were reported by the year 2000. In Zimbabwe, the ministry of health has reported that the number of TB cases increased by 400% between 1990 and 1999. Therefore, because of the HIV infection epidemic, TB mortality in Africa has also increased. In fact, the face of HIV/AIDS in Africa is TB. Many people, perhaps over a third, of all adults in Africa carry a latent TB infection, which is suppressed by a healthy immune system. When HIV weakens the immune system, it can no longer control the latent infection and active TB may develop. The impact of HIV infection on TB is an especially serious problem because TB can be transmitted through casual contact. Consequently, Africans who are not at risk of HIV infection can become TB infected as an offshoot of the HIV/AIDS epidemic. In addition, some people who are TB infected (both HIV positive and HIV negative) receive inadequate drug treatment while others fail to adhere to their treatment. The result is that drug-resistant strains of TB are appearing, making it more expensive, if not impossible, to treat the disease. But, even if the biological basis for antimicrobial resistance and the importance of contributing factors are known, combating the problem poses a difficult challenge in Africa due to limited human and institutional capacities and lack of financial resources and the complexity of a host of other contextual factors underlying antimicrobial resistance in Africa which include lack of access to appropriate drugs; weak regulatory authorities; inadequate national drug policies; sub-optimal curricula regarding antimicrobials in schools of pharmacy and medicine; lack of access to accurate, unbiased, up-to-date drug information; incorrect diagnosis and inappropriate prescribing practices by both public and private sector providers, including drug sellers; sub-standard pharmaceutical products; poor patient compliance with prescribed drug treatment; lack of public awareness of the consequences of inappropriate drug use [16].

Apart from the synergistic interactions of HIV/AIDS and TB, the overlapping endemicity of infection by HIV and many other tropical pathogens in Africa resulting in a high rate of co-infection adds to the complexity of infectious disease challenge in Africa. Co-infections with HIV and other endemic pathogens are common causes of morbidity and mortality in Africa. In addition, progression of HIV infection is accelerated by many co-infections, and the risk, severity, and complications of co-infections often are increased in the presence of HIV infection. Tuberculosis and malaria are examples of co-infections for which previous methods of control and treatment are no longer sufficiently effective. These common co-infections also cripple economic development, perpetuate or exacerbate poverty, create substantial social burdens, and contribute to political instability [17, 18]. For example, recent studies using conventional modelling frameworks by this author and his co-workers which sought to examine the synergistic interactions between HIV/AIDS and other infections afflicting the African continent such as HIV/AIDS and tuberculosis [5, 6], malaria and HIV [4], Schistosomiasis and HIV [7] also highlighted the fact that understanding the complexity of infectious diseases in Africa requires an understanding of complex systems. Although co-infections with endemic pathogens are extremely common in Africa, the complexity of the spectrum of co-infections, co-pathogenesis, and the course and outcomes of co-infections with HIV and endemic pathogens are not being adequately understood. The safety and efficacy of treatment strategies for individuals co-infected with HIV and tropical pathogens compounds the complexity of infectious disease challenge in Africa. A consensus regarding a single universal definition of a complex system does not yet exist. We propose the following definition for complex systems.

Definition 1.1
A complex system is any system featuring a large number of components arranged in structure(s) which can exist on many scales with multiple interactions whose aggregate activity is non-linear and being characterized by openness (so that it may be difficult to determine system boundaries), a history (so that the past helps to shape the present behaviour), emergence (so that patterns emerge from interaction of system components), co-evolution (so that as the environment changes, the system also changes to ensure best fit), and self organizing.

In the context of infectious diseases in Africa, this definition implies that, the 'aggregate', that is, the infectious disease condition we want to treat or prevent, features many aspects to it at various scales and different levels, and brings together biological, epidemiological, clinical, public health issues, and their functional context with multiple interactions. For example, TB cannot be contained in Africa by simply making anti-TB drugs available, neither can resistance to anti-TB drugs be contained by improving adherence to the drugs alone e.g. DOTS (directly observed treatment strategy) because a multiplicity of factors involved in all these issues need to be taken into account as well as the functional context. However, when viewed together (the 'aggregate') the structure and function to prevent, treat and contain TB in Africa represents a complex system whose sum is greater than its parts.

There is therefore an urgent need to develop new strategies to better control the spread of infectious diseases in Africa. To that end principles and techniques in mathematical modelling are increasingly being applied to inform policy interventions for disease management activities.

Definition 1.2
Mathematical modelling is the process of using various mathematical structures, such as graphs, equations, diagrams, scatter plots, tree diagrams to represent Real World situations.

A model provides an abstraction that reduces a problem to its essential characteristics and it allows the user to eliminate the unimportant details so that the user can concentrate on the relevant decision variables that are present in a situation. This increases the opportunity to fully understand the problem and its solution. Once a model has been developed and used to answer questions, it should be critically examined and often modified to obtain a more accurate reflection of the observed reality of that phenomenon. In this way, mathematical modelling is an evolving process; as new insight is gained, the process begins again as additional factors are considered. Generally, the success of a model depends on how easily it can be used and how accurate its predictions are. Mathematical models can help us figure out which decisions will have the largest impacts on outcomes and can provide comprehensive examination of the assumptions that enter into decisions in a way that purely verbal reasoning and debate cannot. Mathematical models can be used to [19–21] (1) systematically compare alternate strategies, (2) determine the key issues in decision making, and (3) identify gaps in current knowledge. These models can be classified according to use (descriptive or optimization), degree of randomness (deterministic or stochastic) and degree of specificity (specific or general).

The purposes of disease modelling are [22]:

1. The model formulation process clarifies assumptions, variables and parameters.
2. The behaviour of precise mathematical models can be analyzed using mathematical methods and computer simulations.
3. Modelling allows explorations of the effect of different assumptions and formulations.
4. Modelling provides concepts such as a threshold, reproduction number, etc.
5. Modelling is an experimental tool for testing theories and assessing quantitative conjectures.
6. Models with appropriate complexity can be constructed to answer specific questions.
7. Modelling can be used to estimate key

parameters by fitting data.

8. Models provide structures for organizing, coalescing and cross-checking diverse pieces of information.
9. Models can be used in comparing diseases of different types or at different times or in different populations.
10. Models can be used to theoretically evaluate, compare or optimize various detection, prevention, therapy and control programs.
11. Models can be used to assess the sensitivity of results to changes in parameter values.
12. Modelling can suggest crucial data which needs to be collected.
13. Modelling can contribute to the design and analysis of epidemiological surveys.
14. Models can be used to identify trends, make general forecasts, or estimate the uncertainty in forecasts.
15. The validity and robustness of modelling results can be assessed using ranges of parameter values in many different models.

In disease modelling two types of modelling frameworks are useful: conventional modelling framework and complex systems modelling framework. In Sect. 35.2, we discuss in detail conventional modelling framework for infectious diseases and in Sect. 35.3 we discuss the complex systems modelling for infectious diseases. We end the chapter with some concluding remarks in Sect. 35.4.

35.2 Conventional Modelling Framework of Infectious Diseases

Conventional modelling of infectious diseases is predominantly built around the use of ordinary differential equations and is also sometimes called deterministic modeling or compartmental modelling of infectious diseases. We call them conventional models because they are the most commonly used models to inform policy on disease programming. They are commonly used because they require less data, are relatively easy to set up, and also relatively easy to analyse. The conventional models for infections are also called compartmental models because they categorize the whole population into different subgroups (compartments). Further, the models specify the transition rates between the compartments.

These models attempt to describe and explain disease processes in populations. The population referred to could be a human population or a population of immune cells depending on whether the model describes the disease at the human population level or at the host immune system–pathogen level. This type of modelling does not emphasise the microscopic entities explicitly, but estimates the behaviour at macroscopic levels. For this reason, this type of modelling is also sometimes called top-down approach [23, 24]. In general, six steps are involved in developing conventional models for infections. The six steps involved are outlined in Table 35.1. Conventional models or deterministic models are based on the premise that if "enough" information on the system is known at a specific time, then its future behaviour can be predicted exactly. So, it does not take into account random variations giving a fixed and precise result and is approximate when large populations are involved. "Enough" here refers to information adequate to describe a particular phenomenon though this is rarely available.

In the deterministic continuous-time framework, the dynamical behaviours of diseases can be modelled by a system of non-linear differential equations in the form:

$$\begin{aligned} \dot{x}_1 &= f_1(x_1, x_2, \ldots, x_n, \lambda_1, \lambda_2, \ldots, \lambda_m) \\ \dot{x}_2 &= f_2(x_1, x_2, \ldots, x_n, \lambda_1, \lambda_2, \ldots, \lambda_m) \\ \vdots\ &\ \vdots \qquad \vdots \\ \dot{x}_n &= f_n(x_1, x_2, \ldots, x_n, \lambda_1, \lambda_2, \ldots, \lambda_m) \end{aligned} \tag{35.1}$$

This can be written in more compact form as

$$\dot{\mathbf{x}} = \mathbf{f}(\mathbf{x}, \lambda) \tag{35.2}$$

where $\mathbf{x}$ and $\mathbf{f}$ are vectors with components f_1, $f_2, \ldots, f_n$ and $x_1, x_2, \ldots, x_n$, respectively, and the parameters $\lambda_1, \lambda_2, \ldots, \lambda_m$. The functions fi are assumed to be smooth enough so that through each point x_0 there passes a unique solution of (35.1). In disease modelling, the variables represent numbers of individuals, or fractions of the whole population, or

Table 35.1 Steps involved in developing a conventional infectious disease model and their description

Step	Description
Step 1	The first step in conventional modelling consists of having a complete and realistic picture of the biology of the disease, for example, the duration of the period of infectivity, incubation period, immune status after infection, the type of immune response (humoral or cell—mediated), the type of target cells, etc.
Step 2	The second step is to collect data on the demographical, epidemiological, and biological characteristics of the infection (transition rates) and the population (birth and death rates). For models at the host immune system–pathogen level, we take data for the rates of interactions and kinetics from human derived experiments, non-human primates, animal data (mouse and rabbit) when no human or primate data are available, and other published experimental data
Step 3	Determine the type of compartments and therefore variables in the model
Step 4	Assume the correlative relationships between the variables. Modelling enforces clarity of thought and assumptions must be explicitly formulated allowing one to see implications or make predictions which may not otherwise be obvious
Step 5	Formulate the ordinary differential equations. This is normally done by setting up a system of first order differential equations which incorporate variables from step 3 and parameters from step 2 using correlate assumptions from step 4
Step 6	Analyse the model to predict some results. The analysis of the conventional models is generally in three parts which are: (1) derivation of the equilibrium points, (2) derivation of stability conditions and threshold expressions, and (3) analysis of transient dynamics through numerical methods

cell populations or population of pathogens, therefore they should be positive or zero for all times $t \geq 0$, and if this fails then the model should be discarded since it violates a basic aspect of the biological reality. The parameters λ_i represent, epidemiological and/or demographic parameters, or clinical parameters depending on whether we are modelling an infectious disease at the human population level or are modelling an infectious disease at the host immune system–pathogen level. The conceptual foundations of these two types of conventional models are discussed in the following two sub-sections.

35.2.1 Conventional Modelling of Disease at Human Population Level

In the conventional modelling of infectious diseases at the human population level, we normally use multi-states and so the x_i, $i=1, 2,\ldots, n$ given in equation system (35.1) represent the various disease states which segment the modelled human population into a set of distinct compartments each exhibiting different characteristics with respect to the infectious disease. The λ_i, $i=1, 2,\ldots, m$ are usually demographic and epidemiological parameters. For infectious disease modelling in the simplest cases, we usually have only five basic states that are modelled which are, passively immune (M), susceptible (S), exposed (E), infective (I), and removed (R), as described in Table 35.2. The existence of these five states, that is, x_i, $i=1, 2,\ldots, 5$ together with links between them are usually defined by demographic and epidemiological parameters λ_i, $i=1, 2, \ldots, m$ which is usually sufficient to provide a broad outline of the particular infectious disease model being used. Based on the use of demographic parameters, a particular model may describe an acute (epidemic over a short time) or a chronic epidemic (epidemic over a long time). Acute epidemic models usually do not have demographic parameters, such as birth rate and death rate while chronic epidemic models do.

Making use of the states outlined in Table 35.2, it is standard to name the model after the states considered in the modelling process and the possible transitions between the various states. For example, one of the simplest models is called the susceptible, infective, and removed (SIR) model, named after the three states described in the modelling. Many other infectious disease models, considering the five states are shown in Table 35.3.

Table 35.2 The states or variables used in conventional models of infectious diseases at human population level and their description

Symbol	Epidemic variable	Description
M	Passively immune	Density of People who have acquired temporary immunity to a particular disease without having ever been infected. An example of this state would be newborn infants with antibodies against the disease passed from their mother. These antibodies eventually disappear from the body at which time the infant moves into the Susceptible state
S	Susceptible	Density of People who are healthy, but who could potentially develop the disease
E	Exposed	Density of People who have been infected and with the disease, but who are still in the latent period (with or without symptoms of the disease) and who cannot transmit the disease to others
I	Infective	Density of People who are infected with the disease (with or without symptoms of the disease) and who are capable of transmitting the infection to others
R	Removed	Density of People who have either died or recovered from the infection thereby acquiring immunity (temporary or permanent) from this infection

Table 35.3 Many other conventional basic infectious disease models formed by considering combinations of the five states given in Table 35.2

Model	Characteristics
SI	Once a susceptible member of the population (*S*) has been infected with the disease, he or she is immediately infective (*I*) and capable of transmitting the infection to others. No recovery from the disease is possible
SIS	Same as the *SI* model, except that recovery from the disease is possible. However, upon recovery, a subject is immediately susceptible (*S*) once again. That is, recovery from the disease does not confer any immunity against future infection
SEI	Same as the *SI* model, except that, following initial exposure (*E*) to the disease leading to infection, there is a latent or incubation period during which the disease can be passed on to others
SEIS	Same as the *SIS* model, except that, following initial exposure (*E*) to the disease leading to infection, there is a latent or incubation period during which the disease cannot be passed on to others
SIR	Same as the *SI* model, except that recovery (*R*) from the disease is possible. Once recovered from the disease, there is lifelong immunity from re-infection
SIRS	Same as the *SIR* model, except that post recovery immunity is only temporary. Following a period of immunity, a subject may become susceptible (*S*) once again
SEIR	Same as the *SIR* model, except that following initial exposure (*E*) to the disease leading to infection, there is a latent or incubation period during which the disease cannot be passed on to others
SEIRS	Same as the *SEIR* model, except that post recovery immunity is only temporary
MSEIR	Same as the *SEIR* model, with the addition of subjects who are passively immune (*M*) from infection when they enter the population
MSEIRS	Same as the *MSEIR* model, except that post recovery immunity is only temporary

An example would be the general basic SIR Model (both the endemic and non-endemic SIR) illustrated in Fig. 35.1. The basic general SIR model divides the population into three classes or compartments that are susceptible $S(t)$, containing individuals who are healthy, but could potentially develop the disease, infective $I(t)$, containing individuals who are capable of infecting susceptible, and recovered $R(t)$, containing individuals who are immune. The model assumes a constant population size so that $N = S(t)+I(t)+R(t)$. A recruitment-death demographic structure is assumed. Recruitment of new susceptible is represented by the function $\lambda_R f_R(S,I,R)$. In most infectious disease models, $f_R(S,I,R)=1$ which represents constant recruitment or $f_R(S,I,R)=N=S(t)+I(t)+R(t)$ which represents variable recruitment. The natural death is assumed to be proportional to the population number in each class, with constant rate $\mu > 0$. Individuals may recover at a rate α and the transfer function αI corresponds to the exponential

Table 35.4 Possible extensions to the conventional basic infectious disease conventional models given in Table 35.3

Number	Extension
1.	Inclusion of several demographic effects such as migration etc.
2.	Inclusion of effects of age structure
3.	Inclusion of effects of sex structure
4.	Inclusion of risk groups
5.	Inclusion of various types of interventions
6.	Inclusion of effects of co-infections
7.	Inclusion of more complex forms of transmission functions
8.	Inclusion of effects of seasonal/time variations
9.	Inclusion of effects of drug resistance
10.	Extensions to complex systems models

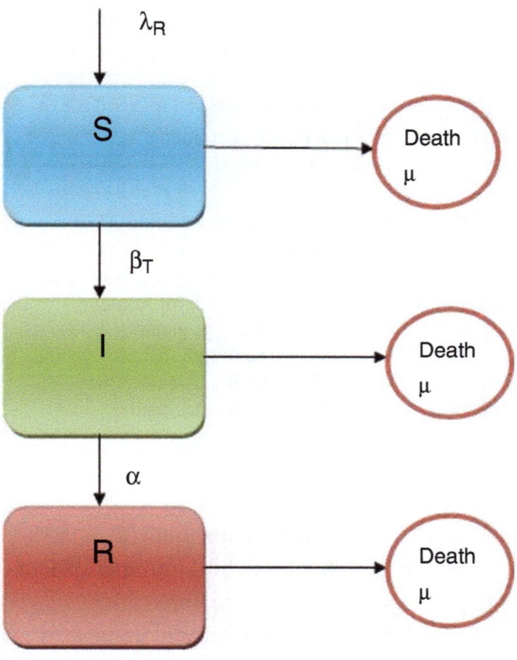

Fig. 35.1 Schematic representation of the general basic SIR model

distribution waiting times in the infectious class, thus we have $p(t) = e^{-\alpha t}$ as the fraction that is still in the infective class at time t units after entering this class and $1/\alpha$ as the mean waiting time. The model assumes a general disease transmission function of the form $\beta_T f_T(S, I, R)$. Transmission functions determine the rise and fall of infectious diseases because the dynamics of an infectious disease to a large extent is determined by how new cases of infections are generated. The most commonly used transmission functions are the mass action transmission function where $\beta_T f_T(S, I, R) = \beta_T SI$ (which assumes that disease transmission is proportional to sizes of S and I compartments and the frequency dependent transmission function where $\beta_T f_T(S, I, R) = \beta_T SI / N$. Various other types of transmission functions can be used (see [25] and references therein).

The assumptions result in the following system of differential equations,

$$\begin{aligned} \dot{S} &= \lambda_R f_R(S, I, R) - \beta_T f_T(S, I, R) - \mu S \\ \dot{I} &= \beta_T f_T(S, I, R) - (\mu + \alpha) I \\ \dot{R} &= \alpha I - \mu R \end{aligned} \quad (35.3)$$

Various authors have analysed some specific cases of this general basic SIR model (see [28] and references therein).

Extensions can be made to various basic infectious diseases models by incorporating more realistic features of infectious disease. A summary of possible extensions are given in Table 35.4.

35.2.2 Conventional Modelling of Disease at Host Immune System: Pathogen Level

The techniques of modern molecular biology have produced much knowledge about pathogens infecting humans, such as viruses, bacteria, protozoa, helminthes and prions and their infection processes, and some fine details about the immune responses mounted by the host to fight the infection. To complement this knowledge, mathemati-

Table 35.5 Summary of uses of conventional models at host immune system–pathogen level and their description

Identifying important immune mechanisms and disease processes	Enhance understanding of the mechanism of a specific disease or its key principles. This will include using mathematical models to understand the dynamic interaction between a specific pathogen and the immune system or its critical component, leading to failure of pathogen elimination prior to the onset of clinical symptoms or to interference with the formation of lasting protective memory (as noted during HIV, tuberculosis or malaria infection, for example)
Elucidating evolution of pathogens	Revealing attributes of the pathogen under study that makes it able to invade the immune system
Suggesting directions of treatment strategies	This entails using mathematical models to assess novel drug or immune therapies by evaluating a therapeutic agent based on its impact on pathogen dynamics. This may include evaluation of a single or combinatorial drug regime or immunological treatment, including therapeutic vaccines, for its ability to suppress pathogen replication as determined by pathogen dynamics
Preclinical evaluation of drugs	Host immune system–pathogen models can be used to aid the design and preclinical evaluation of vaccines and drugs: Such models enable us to more fully understand the cause and treatment of disease by making it possible to explore the actual mechanics of a disease and to test new therapies, all within a virtual environment on the computer, thus without exposing humans to the actual risk of the medication if not proven to be effective
Complements limitations of laboratory instruments	Modelling events that are beyond resolution of current laboratory instruments and enhance biological insight where biological laboratory experiments are limited
Enhance the development of accurate human population level disease specific models	Specific in-host processes affect the spread or control of disease in a population. Host immune system–pathogen models assist in assessing biological in-host characteristics of a disease that affect its spread or control in a population. This involves factoring in of critical in-host disease factors (such as pathogen tropism, incubation period, type of immune response, natural vaccination, pre-existing immunity, latency or the emergence of CTL (cytotoxic T-lymphocytes) escape or drug-resistant mutants) in the design disease-specific epidemic models to more accurately predict the patterns of spread of a disease or its effective control measure in a population. Such measure may include vaccination, antimicrobial usage, isolation etc.

cal techniques have been used to facilitate the analysis of the dynamics of host responses to infectious pathogens. The host immune system–pathogen level mathematical models model the interface between the immune response and the microbial milieu within the human body. As such their applications are concerned with examining human host's innate and adaptive immune responses to a wide variety of pathogens including viruses, bacteria, fungi, and parasites. These disease models at host immune system–pathogen level facilitate new avenues of inquiry that hold potential for advancing significantly the biochemical, pharmacological, immunological and molecular biological understanding of how infectious agents, such as virus, bacteria, protozoa, helminthes or prions and the human body interact through mathematical modelling. Host immune system–pathogen interactions are complex, nonlinear systems that cannot be understood without the help of detailed mathematical analysis. Common uses of models at host immune system–pathogen level are detailed in Table 35.5.

In the conventional modelling framework at host immune system–pathogen level, we assume that microbial pathogenesis gives a fixed and precise reproducible result and a further simplifying assumption is made that statistical variations in microbial pathogenesis are relatively unimportant. Moreover, it is assumed that state transitions occur at each point of time and the only independent variable is time. These assumptions are used in model building resulting in a system of ordinary differential equations (35.1). For conventional mathematical models of disease at host immune system–pathogen level, target cell populations, infecting pathogens, and immune cells are the biological variables contained in the model and are transformed to mathematical variables and these mathematical variables are assumed to carry the biological properties of the biological variables under study in a sound bio-mathematical model. Then these variables are used to construct functions (ordinary differential equations) that describe the interactions and relationships between the different variables. After the mathematical model has been calibrated, it is then analysed or is used in biological investigations from which biological results are obtained analytically or numerically.

Like in modelling disease at human population level most conventional mathematical models at the host immune system–pathogen level are also compartmentalized. The simplest models have only one compartment, representing the whole human body. For more elaborate models, there is a secondary anatomical compartmentalization of the whole human body. Depending on the level of detail required, this anatomical compartmentalization of the model may be based on (1) the site of production of immune cells in the human body, such as the spleen, thymus, lymph nodes or the bone marrow and/or (2) the site of action of immune cells such as the liver and blood for malaria, and the lung for tuberculosis. Within these anatomical compartments there can be further compartmentalization of the model based on cells involved during the infection process. These cells could be the target cells with different cell states, such as healthy susceptible target cell, exposed target cells, infected target cells, or they could be compartmentalized based on different species of the immune cells involved. These cells could typically be either cell-mediated immune cells or humoral immune cells or both. Humoral immunity refers to the component of the immune system involving antibodies. Antibodies are secreted by B cells and circulate as soluble proteins in the plasma and lymph. Antibodies bind to the surface of extracellular infectious particles in order to effect, amongst other functions, signalling and activation of macrophages to engulf and process antibody-bound infectious particles. Cell-mediated immunity is an immune response that does not involve antibodies but rather involves the activation of natural killer cells (NK-cells), the production of antigen-specific cytotoxic T lymphocytes, and the release of various cytokines in response to an antigen.

In order to illustrate conventional disease modelling at host immune system–pathogen level, we introduce a basic general model. This basic general model can be adapted to model any disease caused by an infectious pathogen. In such a model, the basic immune response model contains five variables: the density of uninfected target cells (T_S), the density of exposed target cells (T_E), the density of infected target cells (T_I), the density of the infecting pathogen (P) at the site/place of infection (that could be either in the blood or lung or any other site in the human body depending on the infecting pathogen), and the density of the immune cells (C) that could be either cell-mediated or humoral or both. Uninfected cells are usually assumed to be recruited at a constant rate Λ_S from a source within the body, such as the bone marrow (macrophages in the case of tuberculosis infection or red blood cells in malaria infection) or the thymus (in the case of HIV infection, CD4+ T cells). These variables and their definitions are given in Table 35.6.

Based on the variables in Table 35.6, a general basic model at host immune system–pathogen level is given by a system of differential equations (35.4).

$$
\begin{aligned}
\dot{T}_S &= \Lambda_S - \beta_S f_S(T_S, P) - \mu_S T_S \\
\dot{T}_E &= \beta_S f_S(T_S, P) - \mu_E T_E - \delta T_E \\
\dot{T}_I &= \delta T_E - \eta f_I(T_I, C) - \mu_I T_I \\
\dot{P} &= N\alpha T_I - \gamma f_p(P, C) - \mu_P P \\
\dot{C} &= \Lambda_C + \kappa f_c(T_I, C) - \mu_C C
\end{aligned}
\tag{35.4}
$$

A schematic representation of this basic general model is shown in Fig. 35.2.

The first equation in system (35.4) describes the dynamics of susceptible target cells. The first term Λ_S, represents the rate of supply of new target cells from a source within the human body, such as the bone marrow (macrophages in the case of tuberculosis infection or red blood cells in the case of malaria infection) or the thymus (CD4 T-cells in the case of HIV infection).

Table 35.6 Variables in a general conventional basic model at host immune system–pathogen level and their description

Symbol	Immune variable	Description
T_S	Susceptible Target Cells	Populations of target cell, for a single pathogen. These uninfected cells are normally assumed to be supplied at a particular rate from a source such as the thymus
T_E	Exposed Target Cells	Populations of exposed target cells
T_I	Infected Target Cells	Populations of infected target cells
P	Pathogen	Populations of the infecting pathogens at the site/place of infection that could be either in the blood or lung or any other site in the human body depending on the infecting pathogen. The infecting pathogen could be a virus, bacteria, protozoa, helminthes or prions
C	Immune Cells	Populations of the immune cells that could be either cell-mediated or humoral or both

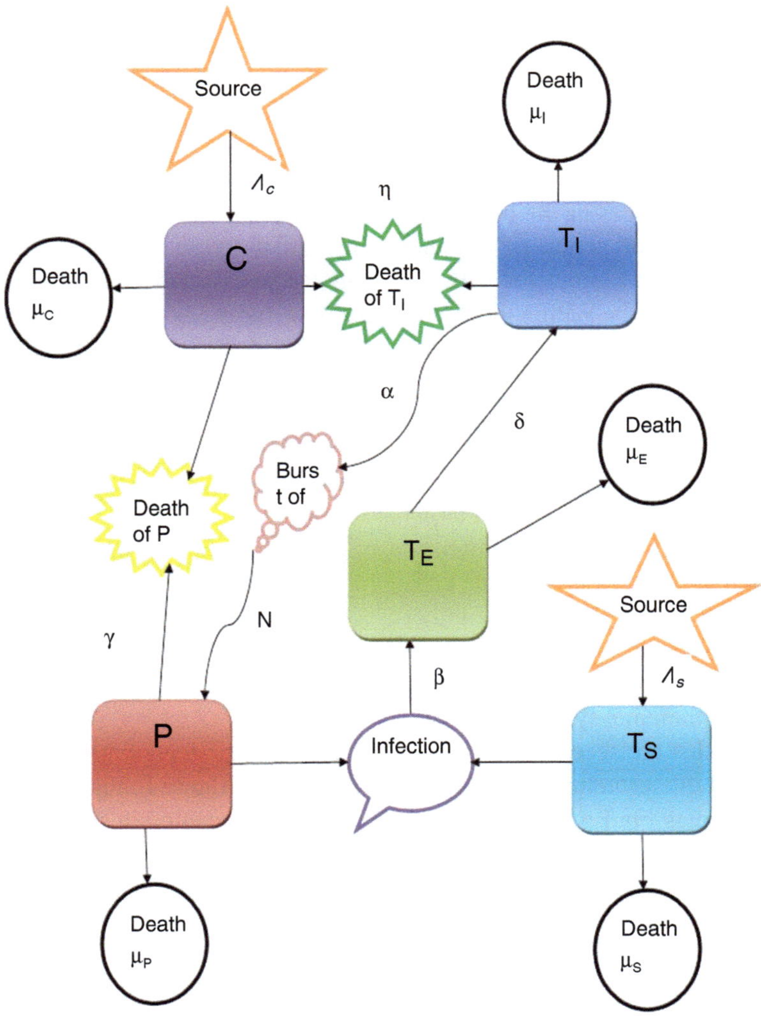

Fig. 35.2 Schematic representation of a general basic conventional model at host immune system–pathogen level

This is followed by a term describing the infection of target cells by the pathogen. Target cells are infected by the pathogen at a rate β_S. Infection is represented by the function $f_S(T_S, P)$, which estimates the rate of contacts between the target cells and the pathogen that result in successful infection of target cells. In most cases, the infection is modelled by a mass action term so that $f_S(T_S, P) = T_S P$. The term $-\mu_S T_S$ represents the natural death of susceptible target cells so that on average their life span is $1/\mu_S$.

The second equation in system (35.4) models the dynamics of exposed target cells (cells which are infected but are not yet producing the pathogen). The first term on the right-hand side of this equation represents a gain term for exposed target cells. The second term represents natural death of exposed target cells due to having a life span that averages $1/\mu_E$. The last term represents the progression of exposed target cells to the productively infected state at a rate δ.

The third equation in system (35.4) models the dynamics of productively infected target cells. In this equation, the first term on the right-hand side represents a gain of productively infected target cells. The second term models direct killing of productively infected target cells. The direct killing is modelled by the function $f_I(T_I, C)$ which estimates the rate of contacts between productively infected target cells and immune cells that result in successful killing of pathogen infected target cells at a rate η. Again in most cases, the direct killing term is modelled by a mass action term so that $f_I(T_I, C) = T_I C$. Infected target cells are also normally lost by cytopathic effects of the pathogen at a rate α and natural death at a rate μ_I so that if we do not take into account the effect of immune response, they have a finite life span that averages $1/(\alpha + \mu_I)$.

The fourth equation in system (35.4) models the dynamics of the pathogen. The first term represents the source of the pathogen within the human body. Pathogens are released by the burst of infected target cells. The second term describes the direct killing of the pathogen by immune cells at a rate γ. Direct killing of the pathogen is represented by the function $f_p(P, C)$, which also estimates the rate of contacts (directly or indirectly) between immune cells and the pathogen. The killing here may be through indirect contact in the sense that death of pathogen may be due to contact between substances produced by immune cells such as cytokines or antibodies. Pathogens are assumed to have an average life span of $1/\mu_P$.

The last equation in system (35.4) models the dynamics of immune cells. The immune cells here could be cell-mediated or humoral or both. The first term on the right-hand of this equation represents the rate of supply of immune cells from a source within the human body to the site of infection and the second term accounts for immune stimulation or proliferation of immune cells which is assumed to occur at a rate κ. Immune cell proliferation is represented by the function $f_c(T_I, C)$ which depends on infected target cells and immune cells. In the simplest forms of the basic model, this function can also be represented by a mass action term for that $f_c(T_I, C) = T_I C$. The last term represents the natural death of immune cells at a rate μ_C. In the case where most of the non-linear functions in the basic model are absent except the infection term, then the basic model will not incorporate any immune response. A summary of the non-linear terms in the basic model and their description is given in Table 35.7.

The presence of non-linearity means that changes in immune system conditions do not occur in a proportional or linear manner. This allows the immune system to fluctuate in response to infectious pathogen challenge and varying environmental demands, to assure survival.

Just like in conventional modelling of infectious diseases at human population level, various types of models or more complicated models at host immune system–pathogen level may be developed from this simple general basic model. A summary of possible extensions are given in Table 35.8. The only difference is that here extensions are only done to one general basic model, while in the case of conventional infectious disease models at human population level extensions are done to different types of basic models as described in Table 35.3.

Table 35.7 Functions modelling immunological processes which are often the origins of nonlinearity in models at host immune system–pathogen level

Function	Process
$f_S(T_S, P)$	Infection of target cells by the pathogen
$f_I(T_I, C)$	Killing of infected target cells or killing of intracellular pathogen by immune cells
$f_P(P, C)$	Killing of extracellular pathogen by immune cells
$f_C(T_I, C)$	Immune stimulation or Proliferation of immune cells

Table 35.8 Possible extensions to the basic model at host immune–pathogen level and their description

Extension	Description
Involvement of various types of immune cells (immune responses)	Assuming that various types of immune cells are involved or that the immune response is polyclonal, so the basic model may be extended to account for the effects of the different types of immune cells involved or to take into account the polyclonal nature of the immune response
Effect of infection process occurring in several anatomical compartments	Assuming that the infection process happens in several anatomical compartments, e.g. for malaria the liver stage infection and the blood stage infection
Complexity of life cycle of infecting pathogen	Taking into account the complexity of the life cycle of the infecting pathogen such as pathogens associated with vector borne diseases where the pathogen has different life forms in both the human host and the vector
Effect of therapeutic interventions	Incorporating the effects of various types of therapeutic interventions
Effect of drug resistance	Assuming the presence of different strains of the infecting pathogen
Effect of co-infection	Assuming co-infection by different types of pathogens
Involvement of various types of target cells	Assuming the involvement of various types of target cells such as the involvement of macrophages and CD4 T cells as target cells in HIV infection
Effect of more complicated nonlinear terms	Including more complicated nonlinear terms in the key processes such as those given in Table 35.4

35.2.3 Modelling Diseases in Africa

Below, we discuss some of the conventional modelling efforts at the human population level with reference to three major diseases affecting African communities: HIV/AIDS, malaria and tuberculosis. Modelling has been used to compare, plan, implement, evaluate and optimize various detection, prevention and therapy programs to control these endemic conditions.

35.2.3.1 HIV/AIDS

To ensure a comprehensive response to HIV/AIDS, many African countries have adopted a number of intervention strategies comprising prevention, treatment and care. The key elements in comprehensive HIV/AIDS prevention, treatment and care in Africa which may be investigated using conventional and/or complex systems modelling frameworks include the following:

1. Behaviour change programmes especially for young people and populations at higher risk of HIV exposure, as well as for people living with HIV/AIDS.
2. Prevent sexual transmission of HIV by promoting the use of male and female condoms as a protective option along with abstinence, fidelity and reducing the number of sexual partners.
3. Prevent mother-to-child transmission (MTCT).
4. Prevent the transmission of HIV through injecting drug use, including harm-reduction measures.
5. Ensure safety of blood supply.
6. Prevent HIV transmission in healthcare settings.

7. Promote greater access to voluntary HIV/AIDS counselling and testing while promoting principles of confidentiality and consent.
8. Integrate HIV/AIDS prevention into HIV/AIDS treatment services.
9. Focus on HIV/AIDS prevention among young people.
10. Provide HIV-related information and education to enable individuals to protect themselves from infection.
11. Confront and mitigate HIV/AIDs-related stigma and discrimination.
12. Prepare for access and use of vaccines and antiretrovirals.
13. Treatment of people coinfected with HIV and other diseases.

HIV/AIDS has many interrelated narratives. A HIV/AIDS model has to take into account many different "storylines" affecting the behaviour of this endemic. Here, we provide key findings from this author's and his collaboratorss modelling work on the effects of antiretroviral treatment [26, 27], prophylactic antiretroviral drugs administered to infected pregnant mothers and the effects of breastfeeding on. The details of the models can be found in the reference papers [26–35] based on mother-to-child transmission (MTCT) [28], vaccination [29], condom use [30], male circumcision [31] public health educational campaigns [32, 33], and voluntary testing and counselling [34]. Factors that have been in-cooperated into our modelling work are listed in Table 35.9.[1]

We build a basic HIV/AIDS model with a single class of infectives that we extended to incorporate stage progression and antiretroviral therapy which allows AIDS patients to undergo amelioration. Data sources are those of the demographic and epidemiological parameters from Zimbabwe.

This model was analysed to assess, the effects of the incubation period and period of infectivity on the transmission dynamics of the disease (Fig. 35.3, left), and the impact of the disease and migration patterns on the population of HIV/AIDS patients in Zimbabwe (Fig. 35.3, right).

The infection dynamics of HIV and the early treatment with antiretrovirals shows a dramatic effect on the disease dynamics (Fig. 35.4), allowing many infected people to have extended life expectancies.

Knowing the basic disease dynamics allowed us to expand the model to integrate people's life style behaviours, especially the effect of sexual and reproductive behaviour, on the effects of the

Table 35.9 Variables and parameters used in the HIV models (in alphabetical order)

AIDS-related death rate
Apoptosis rates
Average incubation period
CTL proliferation rates
CTL supply rates
CTLs specific to mutant strain virus
CTLs specific to wild-type strain virus
Death rate of CD4+ T cells
Effectiveness of the vaccine
Emigration rate
Infant birth rate
Infants and adults AIDS-related death rate
Infants and adults average incubation period
Infants breast feeding infection rate
Mutant strain infected CD4+ T cells
Mutant strain infectious virus
Natural CTL death rates
Natural death rate
Probability of transmission
Proliferation rates of CD4 T cells
Proportion of infected babies born from infected mothers
Proportion of mature individuals that are female
Proportion of susceptible babies born from HIV infected mothers
Proportion of vaccinated recruits
Rates CD4+ T cells are infected by virus
Rates CTLs lyse infected cells
Rates CTLs reduce infectivity
Rates of acquiring new sexual partners
Recruitment rate into sexual maturity
Reproduction number
Stimulation constants
Supply rate of CD4+ T cells
Susceptible vaccination rate
Uninfected CD4+ T cells
Virus burst sizes
Virus clearance rates
Virus cytopathic rates
Virus mutation rate
Wild-type strain infected CD4+ T cells
Wild-type strain infectious virus

[1] For the details of all models refer to the original papers.

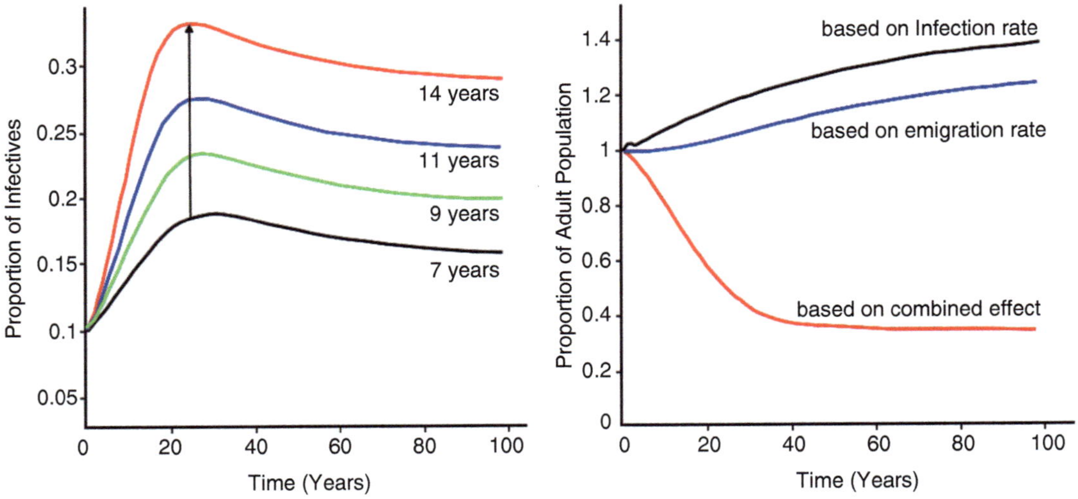

Fig. 35.3 The effect of an increasing incubation period on the proportion of HIV/AIDS infections in the population (*left*) and the demographic impact of the HIV/AIDS epidemic on the Zimbabwean population taking account of infection rate, migration rate and combined effects of infection and migration (*right*) (modified from [30])

rate of susceptibles and the demographic impact of HIV/AIDS in communities using antiretroviral therapy. For a long time, the sexual behaviour of treated AIDS patients did not receive any serious attention because being an AIDS patient appeared to imply a death sentence. In this context, the sex life of AIDS patients seemed a secondary issue making prevention focused on sexual behaviour hard to imagine, thus we analysed the models to gain insight into the effects of sexual and reproductive behaviour of treated AIDS patients.

Sexual Behaviours

An HIV/AIDS model for heterosexual transmission was developed and analysed in [30]. The model divided the population into a two sex structure consisting of females and males. We used the model to focus on the effects of condom use as a single-strategy approach in HIV prevention in the absence of any treatment. Initially, we modelled the use of male condoms and further extended the model to incorporate the use of both female and male condoms. The model included two primary factors in condom use to control HIV that are condom efficacy and compliance. The exposure risk of infection after each intervention was obtained. Basic reproductive numbers for these models were computed and compared to assess the effectiveness of male and female condom use in a community. We noted from the study that in general more females than males are infected by HIV among heterosexuals. The models were numerically analysed to assess the effectiveness of condom use on the transmission dynamics of HIV/AIDS. The study showed that condom use as a preventive measure for HIV/AIDS is more effective in preventing HIV/AIDS transmission if both the female and the male condoms are used, but the associated compliance for effective control or eradication may be too high and perhaps unattainable in Africa which has a high HIV prevalence, suggesting that other control strategies should be adopted to effectively control the epidemic in these settings (Fig. 35.5).

The HIV/AIDS model for heterosexual transmission presented in [31] was extended to model male circumcision as a possible preventive strategy for HIV/AIDS. We incorporated the effects of condom use in the model as another preventive strategy for controlling HIV/AIDS. Basic reproductive numbers for these models were computed and compared to assess the effectiveness of male circumcision and condom use in a community. The models

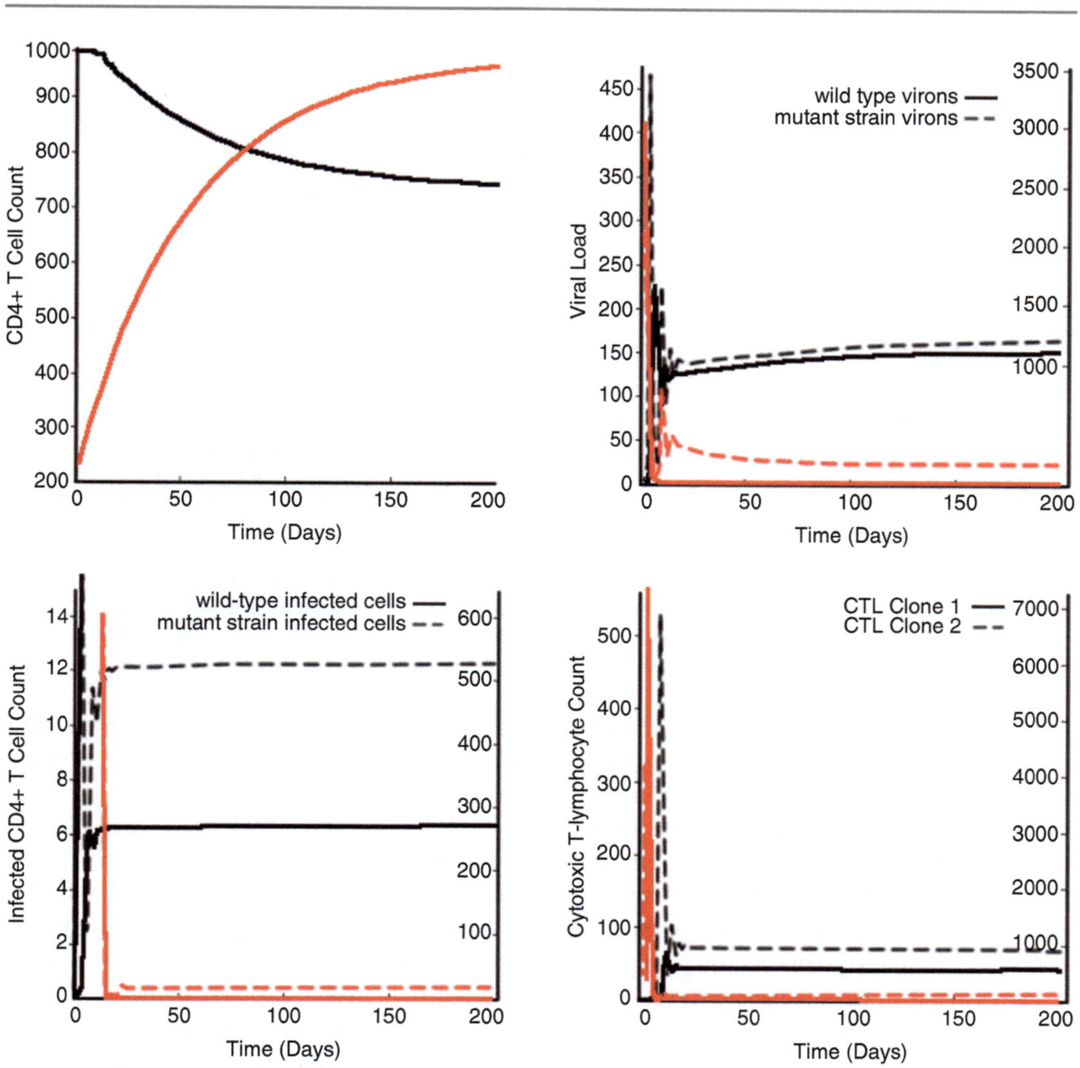

Fig. 35.4 The model kinetics of CD4+ T cell count, viral load, infected CD4+ T cell count and cytotoxic T-lymphocyte count for wild-type and mutant strain HIV. The infection kinetics of HIV infection (*black*) and the effects of treatment of HIV infection (*red*) are superimposed (modified from [36])

were numerically analysed to assess the effects of the two preventive strategies on the transmission dynamics of HIV/AIDS. The study revealed that in the continuing absence of a preventive vaccine or cure for HIV/AIDS, male circumcision is a potential preventive strategy of HIV/AIDS to help communities slow the development of the epidemic if implemented jointly with condom use. The study provided insights into the possible community benefits that male circumcision and condom use as preventive strategies provide in slowing or curtailing the epidemic.

Breast Feeding

An age and sex-structured HIV/AIDS model consisting of two age groups, children and adults, was proposed and the important mathematical features of the model were analysed in [28]. The model considered both MTCT and heterosexual transmission of HIV in a community. MTCT can occur prenatally, at labour and delivery or postnatally through breastfeeding. In the model, we considered the children age group as a one-sex formulation and divided the adult age group into a two-sex structures consisting of females and males. The basic reproductive number $R0$ for the model

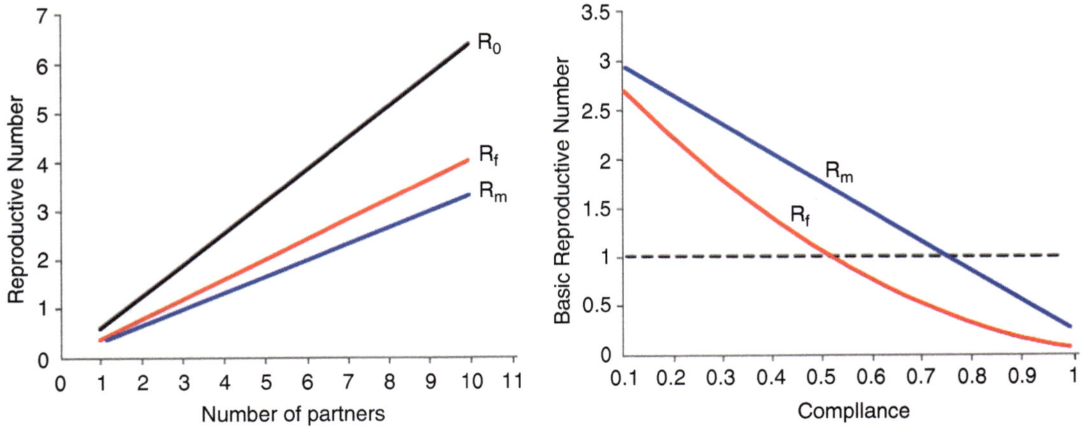

Fig. 35.5 Trend R_0, R_m, R_{fm} of the basic reproductive numbers for varying number of sexual partners per time unit (female = male). Sexual partners are varied from 1 to 10 per unit time (*left*), and the impact of male and female condom use on the basic reproductive numbers with increasing rates of compliance (*right*). The model assumes a condom efficacy of 92% for both males and females. The *dashed line* is $R_m = R_{fm} = 1$. The trends show that for the case of Zimbabwe, a male condom compliance of more than 70% would be required to reduce R_m to values below unity and if both female and male condoms are used, a compliance of more than 50% would be required to reduce R_{fm} to values below unity (modified from [31])

showed that the adult population is responsible for the spread of HIWAIDS epidemic, thus up-to-date developed HIV/AIDS models to assess intervention strategies have focused much on heterosexual transmission by the adult population and the children population has received little attention. We numerically analysed the HIV/AIDS model to assess the community benefits of using prophylactic antiretroviral drugs in reducing MTCT and the effects of breastfeeding in settings with high HIV/AIDS prevalence ratio. The numerical simulations of the model illustrated that the use of antiretroviral drugs to reduce MTCT in communities with high HIV prevalence results in an increase of the population though less is known about their possible risk. Numerical results showed that the increase in HIV infection of infants through breastfeeding will result in a decrease of the population for communities with high HIV prevalence, suggesting that breastfeeding of infants by HIV-infected mothers has detrimental effects in settings with high HIV prevalence. We also showed that the associated risk of morbidity/mortality due to not breastfeeding results in a decrease of the population. Thus, there is no conclusive answer whether HIV-infected mothers should breastfeed or not due to the dilemma of competing risks. We concluded that more research is needed to determine which aspect has a more negative impact on population growth than the other for different settings (Fig. 35.6).

The Impact of Sex Workers

We also considered a sex-structured model for heterosexual transmission of HIV/AIDS in which the population is divided into two classes, those involved in high-risk sexual activities and those involved in low-risk sexual activities [32]. The model considered the movement of individuals from high to low sexual activity groups as a result of public health educational campaigns, thus in this case public health educational campaigns are resulting in the split of the population into risk groups. The model was extended to incorporate commercial sex workers, and their role in the spread of HIV/AIDS in settings with heterosexual transmission was explored. The basic reproductive numbers for the models were obtained and used to assess the possible community benefits of public health educational campaigns. We concluded from the study that the presence of commercial sex workers fuels the epidemic among heterosexuals, and that public health educational campaigns among the high risk heterosexual population can help to slow or eradicate the epidemic.

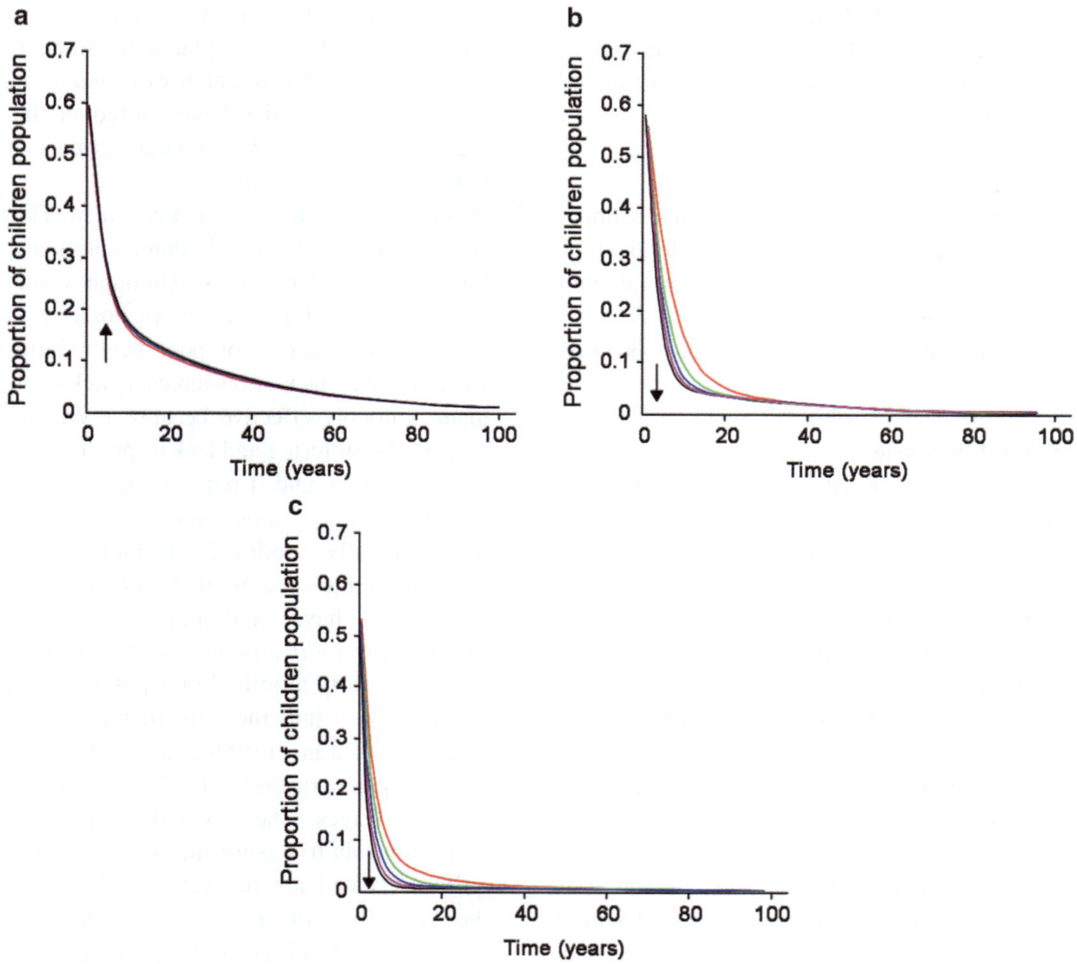

Fig. 35.6 Simulation results of HIV-infection of babies (**a**) Proportion of HIV-negative babies born to HIV infected mothers as a result of anti-retroviral therapy; *arrow* shows direction of increase of the proportion of susceptible babies from 0 to 1 in steps of 0.5. (**b**) Proportion of HIV-infected babies due to breastfeeding; *arrow* shows direction of increase of breastfeeding rate from 0.2 to 1 in steps of 0.2. (**c**) Decrease in childhood population as a result of increasing childhood mortality rates; *arrow* shows direction of increase of infant natural death rate from 0.2 to 1 in steps of 0.2 (modified from [29])

Vaccination and Prevention

A HIV/AIDS vaccination model for sexual transmission with explicit incubation period was presented and the model considered prophylactic vaccination of sexually immature (pro- and early-adolescents) and mature susceptibles in a community [29]. The HIV/AIDS vaccination model was analysed and extended to incorporate condom use based on efficacy and compliance. Further, the HIV/AIDS vaccination model with condom use was extended by incorporating treatment which allows AIDS patients to undergo amelioration. The basic reproductive numbers for the models were computed and compared to assess the possible community benefits of using prophylactic vaccines, condoms and treatment with amelioration of AIDS patients. It was concluded from the study that vaccination and condom use can reduce R_0 to values below unity but treatment with amelioration intended to lengthen the lives of AIDS patients may result in higher numbers of HIV infections and fail to reduce R_0 to values less than unity as intended for disease control.

Summary of HIV/AIDS Modelling

These models considered the individual risk of acquiring HIV through sexual transmission and highlight that:

1. HIV prevalence ratios are increased
   (a) By long incubation periods,
   (b) Treatment of AIDS patients without positive change in sexual behavior, and
   (c) An increase in the recruitment rate into sexual maturity.
2. AIDS-related deaths reduce adult population size.

35.2.3.2 Malaria

The malaria pathogen can be combated either in its human host or mosquito vector and both strategies have received enormous attention over the years. Some of the control strategies for malaria which can be studied on the African continent using conventional and/or complex systems modelling frameworks at human population level include:

1. Larval control: This strategy includes methods such as the destruction of mosquito breeding sites, to reduce the number of mosquitoes.
2. Indoor residual spraying: Spraying reduces mosquito longevity. This strategy is also likely to kill mosquitoes that rest indoors after feeding so it increases the chances of killing infected mosquitoes.
3. Insecticide-treated bed nets (ITNs): Roll Back Malaria RBM has been promoting the use of insecticide treated bed nets in many countries including regions of Africa to reduce the transmission of malaria; and has succeeded in doing so in many regions. As some recent studies have shown [37], ITNs have had a significant impact on disease prevalence and mortality.
4. Intermittent prophylactic treatment: This is a new area of research that involves administering antimalarial drugs at regular intervals, even to those who are not sick, to reduce parasitaemia load. This is essentially similar to the treatment taken by travellers from malaria-free regions when visiting malaria-endemic countries. This form of control would most likely be applied in areas of high transmission where almost everyone has some *Plasmodium* in their blood. Intermittent prophylactic treatment is also carried out for pregnant women and for infants. Initial studies have started in this area and have shown significant effects in reducing infant mortality.
5. Prompt and effective case management: This strategy involves the quick identification and treatment of malaria cases. Although it may seem obvious, it is not always possible in many places because of poor health infrastructure and a lack of resources. Quick treatment is doubly effective because it directly reduces the suffering and lack of productivity due to malaria, and it reduces the transmission of infection to mosquitoes.
6. Transgenically modified mosquitoes: As there are some species of *Anopheles* mosquitoes that have an immune response to kill the *Plasmodium* parasites, there is hope that genetically modified mosquitoes could be introduced into the wild that would be incapable of transmitting malaria. This is a promising area of research, although still in its early stages. There would be need for strict controls to ensure that the new mosquitoes created are not accidentally given the capability of transmitting other diseases, such as influenza or AIDS. As these mosquitoes would be immune to malaria infection, having a population of only transgenically modified mosquitoes would eliminate the transmission of malaria. However, we would expect some wild-type mosquitoes to persist in the population. Li [38, 39] has proposed some population models for the introduction of transgenic mosquitoes.
7. Vaccination: Several approaches to malaria vaccine development target different stages of the parasite's life cycle in humans and mosquitoes. Pre-erythrocytic vaccines are designed to prevent the parasite's infective sporozoite stage from entering or developing within the liver cells of an individual bitten by an infected mosquito. Erythrocytic (asexual blood stage) vaccine prevents the malaria parasite from entering or developing in red blood cells. Transmission blocking (sexual stage) vaccine

is directed against the sexual stage antigens of the life cycle to prevent fertilisation.

We have developed and analysed mathematical models of the epidemiology of malaria. The resulting models were then qualitatively analysed for the existence and stability of their associated equilibria through the use of extensive mathematical techniques. In addition to allowing the determination of various epidemiologic thresholds (such as the reproductive numbers), analysis enabled us to gain deeper insights into malaria control in malaria endemic areas of Africa. Comprehensive numerical techniques were also employed to verify analytic results and illustrate possible efficient methods of controlling malaria (Tables 35.10 and 35.11).

A transmission model of human malaria in a partially immune population with three discrete delays was formulated for variable host and vector populations [40]. These are latent period in the host population, latent period in the vector population and duration of partial immunity. The results of the mathematical analysis indicated that when the average number of mosquitoes infected from both the infectious and partially immune humans is less than a certain value then the model exhibits a backward bifurcation which is the coexistence of a stable disease free equilibrium with a stable endemic equilibrium. We deduced from model analysis that an increase in the period within which partial immunity is lost increases the spread of the disease. In a partially immune population transmission blocking vaccines will be effective in reducing the spread of the infection. Numerically, we deduced that treatment of the partially immune humans assists in reducing the severity of the disease and that transmission blocking vaccines would be effective in a partially immune population (Fig. 35.7).

Malaria Vaccine

A separate study examined a vaccination model for malaria which incorporates the effects of a pre-erythrocytic vaccine on the transmission dynamics of the disease [41]. The model was analysed and extended to incorporate the effects of erythrocytic and transmission blocking vaccines. The reproductive numbers for the models were computed and compared to assess possible community benefits of using the different vaccines. Analysis of the reproductive number showed that effectiveness of the vaccine depends critically on its efficacy, infection blocking, disease modification and transmission blocking properties, the proportion vaccinated and the duration of its effectiveness. The minimum vaccination rate required to eradicate malaria depends on its endemicity in each region. We deduced that, pre-erythrocytic vaccines work faster than transmission blocking vaccines in reducing the number of infectious humans. In our analysis, we were able to quantify the effectiveness of combining

Table 35.10 Variables and parameters used in the malaria models

Human birth rate
Natural death rate of humans
Malaria-induced death rate
HIV-induced death rate
Effective contact rate for HIV infection
Rate of progression to AIDS stage
Probability that a bite from infectious mosquito(s) leads to infection
Contact rate between human and mosquito
Treatment rate
Rate of resistance evolution
Recovery rate of untreated infections
Recovery rate of treated infections
Rate of acquiring immunity
Rate of partial immunity
Factor by which treatment reduces rate of acquiring immunity
Period within which immunity is lost
Period between infection and infectiousness in humans
Rate at which eggs develop into mature adult female mosquitoes
Probability that a bit from an infectious non immune human leads to infection
Probability that a bite from an infectious treated human(s) leads to infection
Probability that a bite from an infectious immune human(s) leads to infection
Period between infection and infectiousness in mosquitoes
Natural birth rate of mosquitoes
Natural death rate of mosquitoes
Excess death rate due to presence of parasite in mosquitoes

Table 35.11 Variables and parameters used in the tuberculosis models

Recruitment rate
Natural mortality rate
Contact rate
TB induced death rate
Probability of being infected
Natural rate of progression to active TB
Natural recovery rate
Relapsing rate
Treatment rate for the latently infected
Treatment rate for the infective ones
AIDS related death rate
Rate at which HIV-positive progress to active TB
Rate at which HIV-positive suffering from TB progress to AIDS
Probability of being infected with HIV
Natural rate of progression to AIDS
Rate at which a HIV-positive infected with M-TB progresses to AIDS
Rate at which an AIDS case infected with M-TB progresses to TB
Treatment rate for AIDS cases

two or more subunit vaccines over using only one subunit vaccine. Our analysis showed that a vaccine that reduces the infectious period of an infected human always reduces the number of secondary infections. Several candidate malaria vaccines are currently in development. This study provided useful tools for assessing the effectiveness and potential population level impact of the vaccines with various properties.

Drug Resistance

A mathematical model for malaria treatment and spread of drug resistance in an endemic population was presented in [8]. The model considered treated humans who remained infectious for some time and partially immune humans who are also infectious to mosquitoes although their infectiousness is always less than their non-immune counterparts. The model was formulated by considering delays in the latent periods in both mosquito and human populations and in the period within which partial immunity is lost. The mathematical properties of the model were explored. Analysis of the reproductive numbers showed that if the treated humans become immediately uninfectious to mosquitoes then treatment will always reduce the number of sensitive infections. If however treated humans remain infectious then for treatment to effectively reduce the number of sensitive infections, the ratio of the infectious period of the treated humans to the infectious period of the untreated humans multiplied by the ratio of the transmission rate from a treated human to the transmission rate of an untreated human should be less than one. Our results showed that the spread of drug resistance with treatment as a control strategy depends on the ratio of the infectious periods of treated and untreated humans and on the transmission rates from infectious humans with resistant and sensitive infections (see Fig. 35.8). Robust numerical analysis was performed to assess the effects of treatment, recruitment rate of susceptible mosquitoes and the mortality rate of mosquitoes on the spread of resistance and infection. The study provided insight into the possible intervention strategies to be employed in malaria endemic populations with resistant parasites by identifying important parameters such as the use of combination therapy and rates of transmissions of sensitive and resistant infections.

Preventive Efforts

We formulated a deterministic model with two latent periods in the non-constant host and vector populations, in order to theoretically assess the potential impact of personal protection, treatment and possible vaccination strategies on the transmission dynamics of malaria [42]. The thresholds and equilibria for the model were determined. The model was analysed qualitatively to determine criteria for control of a malaria epidemic and was used to compute the threshold vaccination and treatment rates necessary for community-wide control of malaria. In addition to having a disease-free equilibrium, which is locally asymptotically stable when the epidemiologic threshold known as the basic reproductive number is less than unity, the model exhibits the phenomenon of backward bifurcation where a stable disease-free equilibrium coexists with a stable endemic equilibrium for a certain range of associated reproductive

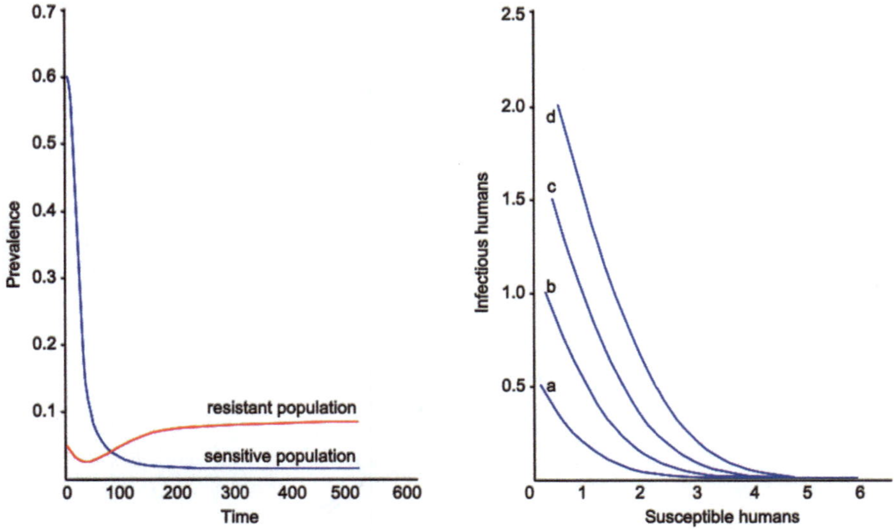

Fig. 35.7 In endemic areas the coexistence of infectious humans with drug sensitive parasites and infectious humans with drug resistant parasites reach an endemic equilibrium between sensitive (*blue*) and resistant infections (*red*) (*left*) (modified from [40]). The phase plane shows the relationship between infective humans against susceptible humans. As the proportion of infectious humans increases, the proportion of susceptible humans decreases. The graph also shows that for certain demographic and epidemiological parameters the result in the reproductive number being less than unity a disease free state emerges. The graph was obtained by varying the initial conditions of S_h (susceptible) and I_h (infectious). (a) $S_h = 0.1$ and $I_h = 0.5$ (b) $S_h = 0.2$ and $I_h = 1.0$. (c) $S_h = 0.3$ and $I_h = 1.5$. (d) $S_h = 0.4$ and $I_h = 2.0$. The fixed model parameters are: birth rate (human and mosquito), natural death rate (human and mosquito), disease related death rate (human and mosquito), rate of oviposition, inoculation rate, incubation period, time to build up immunity, rate of partial immunity, recovery rate and probability to get infected (modified from [40])

number less than unity (Fig. 35.9, left). From the analysis, we deduced that personal protection has a positive impact on disease control, but to eradicate the disease in the absence of any other control measures the efficacy and compliance should be very high. Our results showed that vaccination and personal protection can suppress the transmission rates of the parasite from human to vector and vice-versa (Fig. 35.9, right). If the treated populations are infectious then certain conditions should be satisfied for treatment to reduce the spread of malaria in a community. Among the interesting dynamical behaviours of the model, numerical simulations showed a backward bifurcation which gives a challenge to the designing of effective control measures because assumptions for the control of the disease based on the most often used threshold condition (the reproductive number) will be violated.

Malaria and HIV Co-infection Dynamics

Malaria and HIV are some of the leading causes of morbidity and mortality in Africa. The interactions of these two infections has a bidirectional influence on the transmission, clinical manifestations and treatment outcomes of both diseases largely due to the HIV-associated immunosuppression which contributes to more frequent and more severe malaria and reduced efficacy of antimalarials in pregnant women and adults which in turn is also affected by the endemicity and stability of malaria transmission. Many questions on the interactions between malaria and HIV have not yet been addressed. Some of the key issues about the interactions of malaria and infections in Africa which may be investigated using conventional and/or complex systems modelling frameworks include the following:

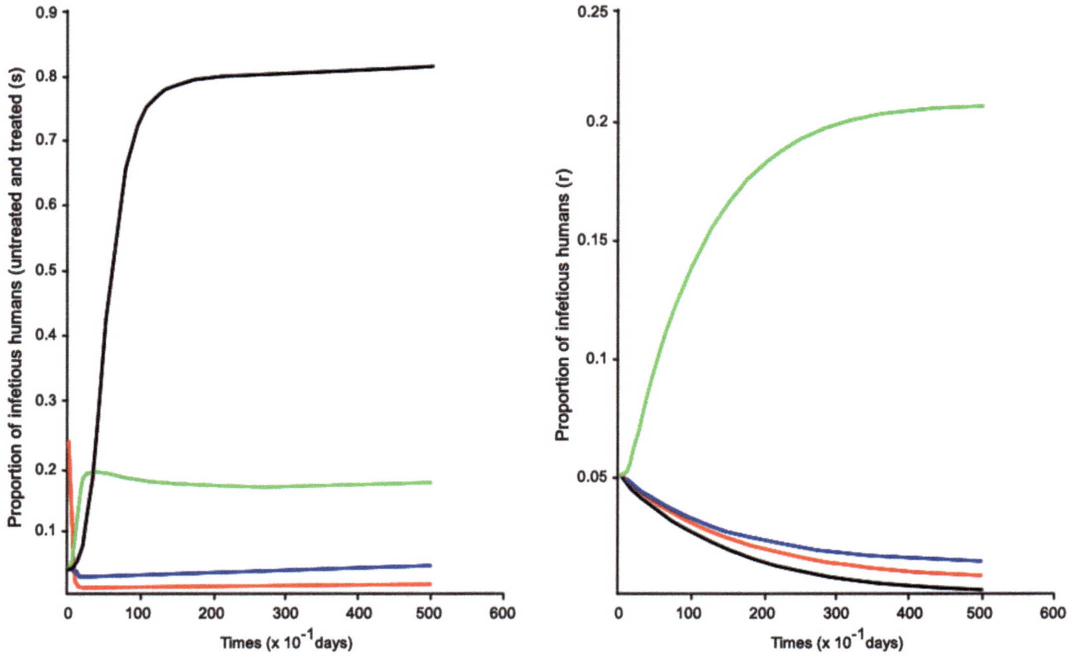

Fig. 35.8 Simulation results showing the dynamics of (*left*) proportion of infectious human population with drug sensitive parasites and (*right*) proportion of infectious human population with drug resistant parasite for corresponding values of reproductive numbers R_0 = 216.0 (*black line*), R_T^S =1.79 (*red line*), R_{01}^S =5.8 (*blue line*) and R_{02}^S = 151.78 (*green line*). The letters (s) and (r) denote drug sensitive and drug resistant parasites respectively. The various reproductive number symbols denote the reproductive number when there is no treatment (R_0), when there is treatment and the treated humans are non-infectious (R_T^S), when there is treatment and the treated humans are infectious and the ratio of the infectious period of the treated human to the infectious period of the untreated human multiplied by the ratio of the transmission rate from a treated human to the transmission rate of an untreated human is less than one (R_{01}^S) and when there is treatment and the treated humans are infectious and the ratio of the infectious period of the treated human to the infectious period of the untreated human multiplied by the ratio of the transmission rate from a treated human to the transmission rate of an untreated human is greater than one (R_{02}^S) (modified from [41])

- Establishing the effect of HIV on malaria in children.
- Measuring the impact of the current HIV epidemic on malaria control programs.
- Investigating whether improved clinical management of malaria in HIV-infected people is having a significant impact in the fight against HIV.
- Investigating whether acute malaria episodes accelerate HIV progression and increase transmission.
- Investigating the interactions and effects of co-administration of antimalarials and antiretrovirals.

We designed and rigorously analysed a deterministic compartmental model for the transmission dynamics and interaction of HIV and malaria with application to communities in Africa [4]. The model considered the epidemiologic synergy between sexually transmitted HIV and malaria. Initially, the HIV-only and malaria-only models were qualitatively examined. The main theoretical results obtained are as follows:

1. The HIV-only model has a globally asymptotically stable disease-free equilibrium whenever a certain epidemiological threshold (R_H) is less than unity and unstable if this threshold exceeds unity.
2. The HIV-only model has a unique endemic equilibrium whenever the aforementioned threshold exceeds unity. For the case where no AIDS-related mortality is considered, this

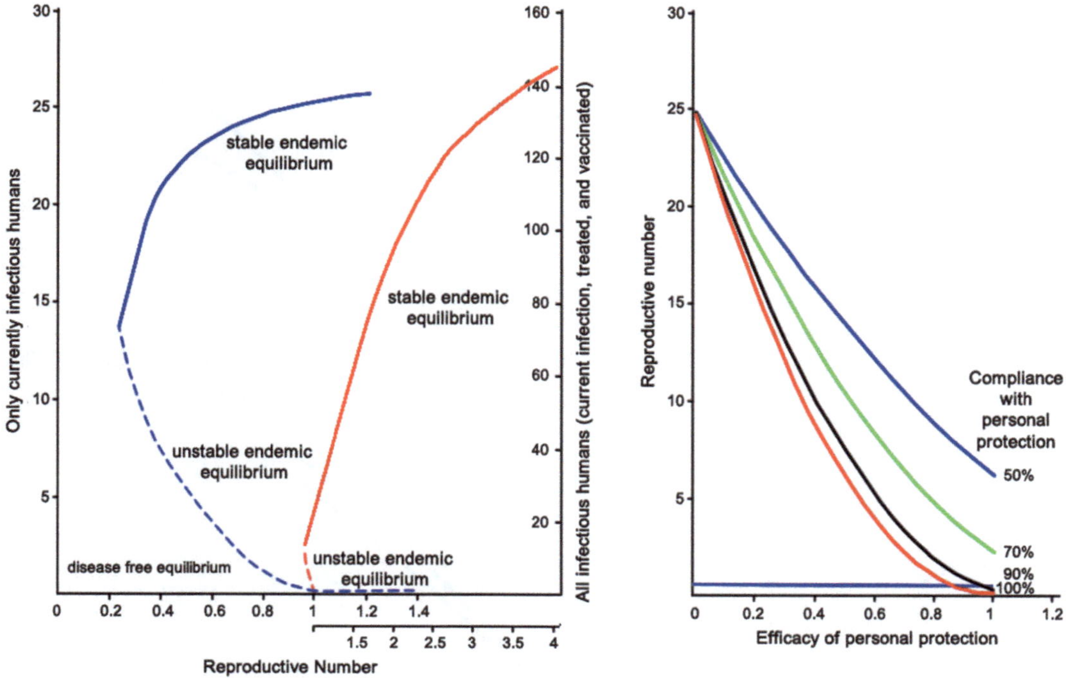

Fig. 35.9 Backward bifurcation diagrams for malaria transmission model for varying transmission probabilities [0.0,1.0] for number of currently infected people, and all people with current active infection, treated and vaccinated (*left*); and the effect of adherence to personal protection at various levels of efficacy of personal protection on malaria spread (modified from [42])

endemic equilibrium is globally asymptotically stable whenever it exists.

3. Unlike the HIV-only model, the malaria-only model undergoes the phenomenon of backward bifurcation, where the associated stable disease-free equilibrium co-exists with a stable endemic equilibrium when the corresponding reproduction number (*R*M) is less than unity.
4. The full HIV-malaria model is shown to have a locally asymptotically stable disease-free equilibrium when its reproductive threshold is less than unity, and unstable if the threshold exceeds unity. It also undergoes the phenomenon of backward bifurcation under certain conditions.

Numerical simulations of the full HIV-malaria model show the following:

1. The two diseases co-exist whenever the reproduction number of each of the two diseases exceed unity (regardless of which number is larger).
2. The number of new cases of malaria at steady state seems to always exceeds that of HIV.
3. The assumed reduction in sexual activity of individuals with malaria symptoms results in decrease in the number of new cases of HIV and the mixed HIV-malaria infection, while increasing the number of new cases of malaria.
4. The HIV-induced increase in susceptibility to malaria infection has marginal effect on the number of new cases of HIV, but significantly increases the number of new cases of the dual HIV-malaria infection.

This study provides the first in-depth analysis of the transmission dynamics of HIV and malaria in a population using conventional modeling framework (for example, in Fig. 35.10 we illustrate some numerical results of the effect of malaria and HIV co-infection on mortality). There are a number of ways this study can be extended, including incorporating preventive and therapeutic strategies for HIV (such as the use of

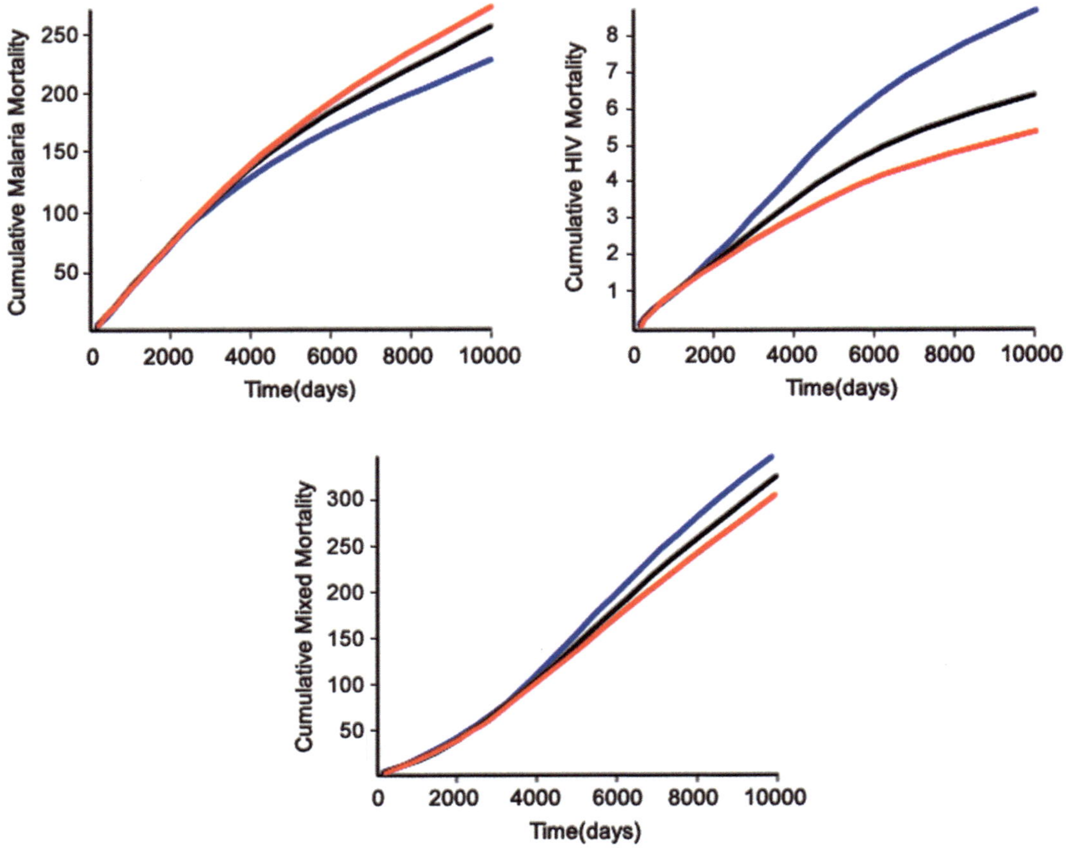

Fig. 35.10 Malaria and HIV co-infection. Cumulative mortality due to (**a**) malaria, (**b**) HIV and (**c**) the mixed infection, as a function of time. Parameters included are: transmission probability for malaria in vectors ($\beta V = 0.9$), biting rate of mosquitoes ($bM = 0.25$), effective contact rate for HIV infection ($\beta H = 0.0007$) and varying values of $\psi > 1$, the parameter which accounts for the assumed amplified HIV–related mortality due to dual infection with malaria for $\psi = 1$ (*blue line*), $\psi = 2$ (*black line*) and $\psi = 3$ (*red line*). With $\psi = 1$, it is assumed there is no amplification of mortality due to co-infection. (**b**) and (**c**) show that while the cumulative number of mortality due to HIV infection and mixed (HIV—malaria) infection increases with decreasing ψ, the cumulative number of new cases of malaria (**a**) increases with ψ increasing from 1 to 3 in steps of 1. Therefore, as mortality due to HIV is amplified as a result of co-infection cumulative mortality due malaria also increases while the cumulative HIV and mixed HIV—malaria mortality also rises (modified from [4])

anti-retroviral therapy, condom use, voluntary HIV testing and screening) and malaria (such as the use of treatment and prophylactic drugs, vector-reduction strategies and personal protection against mosquito bites) and the acquisition of malaria immunity for adults in malaria-endemic settings, following repeated exposure (the latter would be somewhat of a daunting task since both diseases affect the immune system). It would also be interesting to consider the possible consequences of HIV-Malaria co-infection in mother-to-child transmission of HIV.

Summary of Malaria Modelling

In summary, the results obtained from the development and analysis of the mathematical models of the epidemiology of malaria includes the following:

1. A decrease in the latent period of the vector increases the infection in a population.
2. Transmission blocking vaccines are effective in reducing the disease in a population where there are partially immune humans.
3. Secondary infections can be reduced if the ratio of the infectious periods of the vaccinated popu-

lation to that of the unvaccinated population is less than the inverse of the product of the factors by which transmission rates from mosquitoes and from humans is reduced. Erythrocytic vaccines which reduce the infectious period always reduce disease transmission.
4. Spread of antimalaria drug resistance is independent of the intensity of transmission but depends on the infectious periods and transmission rates and for treatment to effectively reduce the number of sensitive infections a certain condition should be met if the treated do not immediately become uninfectious.
5. Effective control or eradication of malaria using personal protection only may be difficult because compliance and personal protection required to achieve this may be too high, or perhaps unattainable, suggesting that multiple control strategies should be adopted in a population. The study also revealed that knowledge of different levels of immune protection that exist in a population assists in evaluating effective intervention strategies in a given area (endemic or non-endemic). Most of the models illustrated backward bifurcation which implies that the standard epidemiological implication of the the reproductive number being less than unity does not always guarantee disease elimination and this provides a challenge to the implementation of intervention strategies.
6. Comment on co-infection with HIV issues.

35.2.3.3 Tuberculosis

Using conventional modelling techniques, we developed mathematical models to assess the effects of various tuberculosis (TB) intervention strategies and incorporated the effects of chemoprophylaxis, treatment, holistic approach (combined chemoprophylaxis and treatment), vaccination, and co-infection of HIV/AIDS and TB in TB control. We additionally analysed the possible benefits of the pre- and post-exposure vaccines currently under development. A TB model with exogeneous re-infection and disease relapse was proposed [43]. Local stability of the disease-free equilibrium and uniqueness of the endemic equilibrium were investigated. The model was further extended to incorporate chemoprophylaxis for the latently infected and treatment for individuals with active TB. From the comparison of the reproduction numbers, it was deduced that chemoprophylaxis of the latently infected is more effective than treatment of infectives only when there is one infective in a fully susceptible population. However when there is a pool of latently infected and active TB cases, treatment of infectives is more necessary than chemoprophylaxis as an intervention strategy as this results in shortening the infectious period and averts possible TB related deaths.

We formulated and qualitatively analysed models incorporating three strains of TB [11]. Local stability of the disease-free equilibrium and the endemic equilibrium were investigated. The centre manifold theory was used in the investigation of the local stability of the endemic equilibrium and to show the existence of backward bifurcation. Using numerical simulations we deduced that case detection, isoniazid preventive therapy and treatment of drug sensitive TB while reducing the spread of drug sensitive mycobacterium tuberculosis (M-TB) results in an increase of multi-drug resistant tuberculosis (MDR-TB) cases supporting the argument that bacterial resistance develops as a result of selective pressure on non-resistant strains due to antibiotic use. The same also applies for the treatment of MDR-TB which results in an increase of extremely drug resistant tuberculosis (XDR-TB) cases. Numerical simulations supported our analytical conclusions.

TB and HIV/AIDS

TB and HIV/AIDS are common co-infections and were analysed in a separate model [6]. The global stability of the disease-free equilibrium of an HIV/AIDS only model and the local stability of the full model (HIV/AIDS and TB co-infection model) were investigated. Effects of HIV/AIDS on TB dynamics were shown with the aid of numerical simulations. From the study, we conclude that treatment of AIDS cases result in a significant reduction of numbers of individuals progressing to active TB. Further, treatment of latent and active forms of TB results in delayed onset of the AIDS stage of HIV infection (Fig. 35.11).

Mathematical models were presented and studied to assess the impact of chemoprophylaxis, treatment of drug sensitive TB, drug resistance and antiretroviral therapy on TB cases in areas with high HIV/AIDS prevalence [43]. Local stability of the disease-free equilibrium and endemic equilibria were investigated. From the analysis of critical treatment rates, it is noted that:

1. Treatment using the first line TB drugs is more effective in controlling TB in settings without or with low levels of drug resistant TB.
2. TB treatment with first line drugs is equally effective in treating drug sensitive TB cases co-infected with HIV/AIDS. Combined HIV/AIDS and TB treatment in individuals co-infected with HIV/AIDS and TB results in a significant reduction of TB cases, but not for HIV/AIDS as AIDS treatment without behaviour change results in an increase of AIDS individuals due to a reduction of AIDS related deaths.

Models to examine the possible benefits of the pre-exposure and post-exposure vaccines currently under development were presented in [44]. It is deduced from the analysis of the last two models that the use of the pre-exposure vaccine currently under development coupled with chemoprophylaxis and treatment is more effective in controlling TB requiring approximately half the time required by the post-exposure vaccine currently under development coupled with treatment of infectives to eradicate the disease. However, in the case where there is a pool of latently infected individuals and a high number of the infective population the use of the post-exposure vaccine currently under development for the latently infected coupled with treatment of active TB is equally effective in controlling TB.

35.3 A Proposal for a Complex Systems Modelling Framework for Infectious Diseases in Africa

In order to explore the dynamic complexity of infectious diseases in Africa, we need to consider the complex systems modelling framework. This is a kind of system-wide modelling framework. This system-wide modelling approach can still focus on specific disease process, as is the case with conventional infectious disease models, while incorporating a deeper understanding of how the surrounding African health system functions (physically, socially, and organizationally). The complexity of infectious disease in Africa and their co-evolving nature means that the prevention and treatment of each infection cannot be considered in isolation of other infectious diseases, the overall health system, and the functional context of this health system, as is the case with conventional modelling. Conventional modelling approaches are based on an intervention/outcome methodology instead of a comprehensive holistic one.

Complex systems models are also called stochastic models because they incorporate probabilistic behaviour with some random events and also rely on inter-individual chance variation in risks of exposure, disease, and other factors. These models have several advantages, one of which is that they allow follow-up of each individual in the population on a chance basis. These models, however, can be laborious to set up and need many simulations to yield useful predictions. A particularly widely used complex system modelling framework or stochastic modelling framework which can also be used for infectious diseases in the agent-based model [18, 24, 45, 46].

Conventional disease-modelling efforts focus on only a limited aspect of the disease process. In order to create more realistic and useful tools, it is important we use the integrated approaches of complex systems modelling and develop more comprehensive systems dynamic models of infectious diseases that integrate disease factors across scales from the molecular to the cellular, organismal, epidemiological, health system as well as its functional context.

Agent-based simulations attempt to model the system under consideration by means of interactions between agents at a local level [18, 24, 46]. Very very little has been done on the African continent in terms of modelling infectious diseases using agent-based models. An agent in an epidemic model would be an individual person, whilst in a model of the immune response an

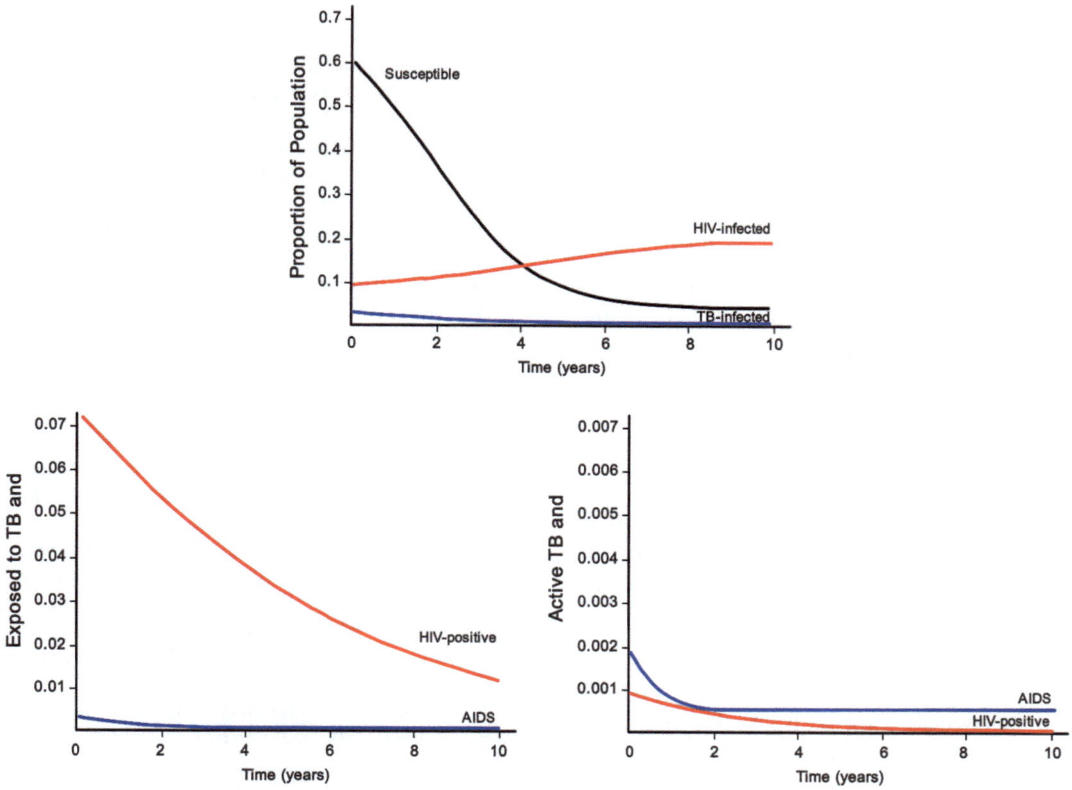

Fig. 35.11 TB and HIV co-infection. *Top panel* shows the population distribution of those susceptible, being HIV-positive and having AIDS. *Bottom left panel* shows the TB exposures to patients being HIV-positive and having AIDS, *bottom right panel* shows the active TB cases amongst patients being HIV-positive and having AIDS. Note the scale change by a factor of 10^{-1} from one panel to the other (modified from [6])

agent would represent an individual cell, whether that cell be a T-cell, B-cell, pathogen infected cell, or any of the other cells that participate in the immune response. The agents then interact with each other based on simple rules. Out of such simple local interactions complex global phenomena can emerge. Hopefully, this will enable us to determine useful control strategies against the spread of the infectious diseases in Africa, as well as giving us a greater understanding of the immune response. The immune response is dynamic and includes growth and replenishment of cells and in-built adaptability, through mutation of its defences to meet new threats. It also includes aspects of cell mobility, which may be captured by means of defining the movement and affinity of cell types in a defined spatial framework.

35.4 Conclusions

The task of reversing the current complex disease patterns in Africa through understanding their dynamical behaviours requires new conceptual modelling approaches different from the current single disease focused conventional modelling approaches. The behaviours of these infectious diseases need to be understood in the broader context of the health system in place. Conventional models are not based on one comprehensive model that captures all essential features of a particular infectious disease, all plausible characteristics of the infection and all available interventions. Development of complex systems models to address the numerous disease programming issues involved in addressing the infectious disease prospects for analyzing and understanding the components and dynamical behaviours of infectious diseases in Africa in a holistic manner. Although conventional modelling tools have been extensively studied for a long time, their analytical approaches have not yet provided us with the urgently needed new solutions to the prevention and treatment of many infectious diseases afflicting the African continent. Therefore, it is clear that the methodology and techniques of complex systems modelling must be used to provide new solutions for preventing and curing the infectious diseases in Africa. For Africa, we propose a joint infectious disease complex systems modelling framework that includes the infectious disease transmission component (a dynamic epidemiological model) and a health system component (an operational systems model) to support the prevention and treatment decisions of infectious diseases. Complex systems modelling goes beyond the reductionist approach of breaking complicated phenomena into simple variables; new properties and behaviours evolve from the interactions between individual components [26, 45].

References

1. UNDP. The Millennium Development Goals Report 2005. New York: United Nations Development Program. Accessed online at http://millenniumindicators.un.org/unsd/mi/mi_dev_report.asp.
2. Corbett EL, Steketee RW, Kuile FO, Latif AS, Kamail A, Hayes RJ. HIV-1/AIDS and the control of other infectious diseases in Africa. AIDS Africa III. 2002;359:2177–87.
3. Magombedze G, Garira W, Mwenje E. In-vivo mathematical study of HIV-1 and Mycobacterium tuberculosis co-infection dynamics. J Biol Syst. 2008;16(3):357–94.
4. Mukandavire Z, Gumel AB, Garira W, Tchuenche JM. Mathematical analysis of a model for HIV-Malaria co-infection. J Math Biosci Eng. 2009;6(2):333–63.
5. Magombedze G, Garira W, Mwenje E. Modelling the TB/HIV-1 co-infection and the effects of treatment. Math Popul Stud. 2010;17(1):12–64.
6. Bhunu CP, Garira W. Modelling HIV/AIDS and tuberculosis coinfection. Bull Math Biol. 2009;71(7):1745–80.
7. Bhunu CP, Tchuenche JM, Garira W, Magombedze G, Mushayabasa S. Modelling the effects of schistosomiasis on the transmission dynamics of HIV/AIDS. J Biol Syst. 2010;18(2):277–97.
8. Chiyaka C, Garira W, Dube S. Effects of treatment and drug resistance on the transmission dynamics of malaria in endemic areas. Theor Popul Biol. 2009;75(1):14–29.
9. Bhunu CP, Mukandavire Z, Garira W. A two strain tuberculosis transmission model with therapy and quarantine. Math Model Anal. 2009;14(3):291–312.
10. Magombedze G, Garira W, Mwenje E, Bhunu CP. Mycobacterium tuberculosis treatment and the emergence of a multi-drug resistant strain in the lungs. In: Tchuenche JM, Chiyaka C, editors. Infectious disease modelling research progress. New York: Nova Publishers; 2009. p. 197–227. ISBN 978-1-60741-347-9.
11. Bhunu CP, Garira W. Mathematical analysis of a two strain HIV/AIDS model with antiretroviral treatment. Acta Biotheor. 2009;57(3):361–81.
12. Magombedze G, Garira W, Mwenje E. Modelling the human immune response to Mycobacterium Tuberculosis in the lungs. J Math Biosci Eng. 2006;3(4):661–82.
13. Bhunu CP, Garira W. Modelling the transmission of multidrug-resistant and extensively drug-resistant TB. In: Tchuenche JM, Mukandavire Z, editors. Advances in disease epidemiology. New York: Nova; 2009. p. 195–220. ISBN 160741452X.
14. Aluwong T, Bello M. Emerging diseases and implications for millennium development goals in Africa by 2015 – an overview. Vet Ital. 2010;46(2):137–43.
15. Currie CS, Williams BG, Cheng RC, Dye C. Tuberculosis epidemics driven by HIV: is prevention better than cure? AIDS. 2003;17:2501–8.
16. North RJ, Jung Y. Immunity in tuberculosis. Annu Rev Immunol. 2004;22:599–623.
17. USAID. Reducing the threat of infectious diseases of major public health importance. USAID's initiative for to prevent and control infectious diseases. 1998. http://transition.usaid.gov/our_work/global_health/id/idstrategy.pdf.
18. Alam SJ, Meyer R, Norling E. A model for HIV spread in a South African village. CPM Report No.: 08-186. 2008. http://cfpm.org/cpmrep186.html.
19. Stillwaggon E. Complexity, cofactors and the failure of AIDS policy in Africa. J Int AIDS Soc. 2009;12:1–9.
20. Daun S, Clermont G. In silico modeling in infectious diseases. Drug Discov Today Dis Models. 2007;4(3): 117–22.
21. Magombedze G, Garira W, Mwenje E. Modelling the immunopathogenesis of HIV-1 infection and the effect of multi-drug therapy: the role of fusion inhibitors in HAART. J Math Biosci Eng. 2008;5(3):485–504.
22. Garira W, Musekwa SD, Shiri T. Optimal control of combined therapy in a single strain HIV-1 model. Electron J Differ Equ. 2005;2005(52):1–22.
23. Hethcote WH. The basic epidemiology models: models, expressions for R_0, parameter estimation, and applications. In: Stefan MA, Xia Y, editors. Mathematical understanding of infectious disease. Singapore: World Scientific; 2008. ISBN 978-981-283-482-9.
24. Li X, Wang Z, Lu T, Che X. Modelling immune system: principles, models, analysis and perspectives. J Bionic Eng. 2009;6:77–85.
25. Galea S, Riddle M, Kaplan GA. Casual thinking and complex systems in epidemiology. Int J Epidemiol. 2010;39:97–106.

26. McCallum H, Barlow N, Hone J. How should pathogen transmission be modeled? Trends Ecol Evol. 2001;16(6):295–300.
27. Mukandavire Z, Garira W. HIV/AIDS model for assessing the effects of prophylactic sterilizing vaccines, condoms and treatment with amelioration. J Biol Syst. 2006;14(3):323–55.
28. Mukandavire Z, Chiyaka C, Garira W, Musuka G. Mathematical analysis of a sex-structured HIV/AIDS model with a discrete time delay. Nonlinear Anal Theory Methods Appl. 2009;71(3–4):1082–93.
29. Mukandavire Z, Garira W. Age and sex structured model for assessing the demographic impact of mother to child transmission of HIV/AIDS. Bull Math Biol. 2007;69(6):2061–92.
30. Mukandavire Z, Garira W, Chiyaka C. Asymptotic properties of an HIV/AIDS model with a time delay. J Math Anal Appl. 2007;330:916–33.
31. Mukandavire Z, Garira W. Sex-structured HIV/AIDS model to analyse the effects of condom use with application to Zimbabwe. J Math Biol. 2007;54(5):669–99.
32. Mukandavire Z, Garira W. Modelling circumcision and condom use as HIV/AIDS preventive control strategies. J Math Comput Model. 2007;46(11–12):1353–72.
33. Mukandavire Z, Garira W. Effect of public health educational campaigns and the role of sex workers in the spread of HIV/AIDS. J Theor Popul Biol. 2007;72(3):346–65.
34. Mukandavire Z, Garira W, Tchuenche JM. Modelling effects of public health educational campaigns on HIV/AIDS transmission dynamics. Appl Math Model. 2009;33(4):2084–95.
35. Bhunu CP, Mushayabasa S, Garira W, Ngarakana-Gwasira E, Tchuenche JM. Is the world doing enough for the poor? A case of HIV/AIDS testing and counselling. World J Model Simul. 2010;6(3):163–76.
36. Shiri T, Garira W, Musekwa SD. A two-strain HIV-1 mathematical model to assess the effects of chemotherapy on disease parameters. Math Biosci Eng. 2005;2(4):811–32.
37. Igwe PC, Ebuehi OM, Inem V, Afolabi BM. Effect of the use of insecticide-treated bed nets on birth outcomes among primigravidae in a peri urban slum settlement in South-East Nigeria. SA Fam Prac. 2007;49(6):1.
38. Li J. Simple mathematical models for interacting wild and transgenic mosquito populations. Math Biosci. 2004;189:39–59.
39. Li J. Heterogeneity in modeling of mosquito populations with transgenic mosquitoes. J Differ Equ Appl. 2005;11(4–5):443–57.
40. Chiyaka C, Garira W, Dube S. Transmission model of endemic human malaria in a partially immune population. J Math Comput Model. 2007;46(5–6):806–22.
41. Chiyaka C, Garira W, Dube S. Mathematical modelling of the impact of vaccination on malaria epidemiology. Int J Qual Theor Differ Equ Appl. 2007;1(1): 28–58.
42. Chiyaka C, Tchuenche JM, Garira W, Dube S. A mathematical analysis of the effects of control strategies on the transmission dynamics of malaria. Appl Math Comput. 2008;195(2):641–62.
43. Bhunu CP, Mukandavire Z, Garira W, Zimba M. Tuberculosis transmission model with chemoprophylaxis and treatment. Bull Math Biol. 2008;70(4):1163–91.
44. Bhunu CP, Garira W, Mukandavire Z, Magombedze G. Modelling the effects of pre-exposure and post-exposure vaccines in tuberculosis control. J Theor Biol. 2008;254(3):633–49.
45. Perrin D, Ruskin HJ, Crane M. An agent – based approach to immune system modelling: priming individual response. World Acad Sci Eng Technol. 2006;17:80–6.
46. Bauer AL, Beauchemin CAA, Perelson AS. Agent – based modelling of host–pathogen systems. Inf Sci (NY). 2009;179(10):1379–89.1 For the details of all models refer to the original papers.

Part VI

Health Organizational Systems

The health organisational system provides the overarching structure for healthcare services delivery. It encompasses a wide range of issues, including the domains covered by the author's of these chapters: the evaluation of system wide policy initiatives, organisational leadership and change management for practitioners, practices and large health service organisations. Other important topics address the broad field of education covering self-awareness and personal growth, coping with uncertainty in a changing world, formal health professional education and self-organising education driven by the "issues of the moment".

Health Transition Funds Initiatives: Implications for Health System Reform and Evaluation

36

Margot Félix-Bortolotti and Carmel M. Martin

36.1 Introduction

In the past decades, countries have reformed or restructured their health systems with different strategies to provide universal and accessible care, to improve the quality of services delivery and their outcomes and most of all to provide health care that is affordable. The main concern was to keep the cost under control while ensuring services effectiveness and efficiency [1–3]. Recently, Canada has adopted different strategies in the form of short term funding or transition funds with several envelopes designed to look at specific issues affecting health systems as instruments for policy reform. These envelopes have generated hundreds and hundreds of grants to study or find innovative ways to improve existing health services in order to meet population need—with the view that will reduce gap in health status without further increasing health system cost [4–6]. All of these initiatives require some form of mandatory evaluation to account for the effectiveness of the projects.

This chapter is not so much on the merit of transition funds as policy reform—but more on the challenges that such multifaceted initiatives pose for health system evaluation. Thus, a brief overview of current strategies will serve as a point of departure. Particular attention will be given to social determinants of health and disparities in the system across different populations—also it will highlight some of the systemic issues that need to be integrated in project and program evaluation. Throughout this chapter, the authors utilise both inductive and deductive arguments in an attempt to examine the complexity[1] of health system reform and the strategies that transform them to ensure their survival. The Canadian experiences with transition funds initiatives to reform health system services delivery was used as an illustration.

This chapter will argue that the prevalent model based on a "logic model"[2] determined in

M. Félix-Bortolotti, R.N., B.Sc. Soc., M.P.A., Ph.D. (✉)
Sociologist Independent Researcher Consultant,
7 Bellfield Street, Ottawa, ON, Canada K2B6K6
e-mail: margotfelix@rogers.com

C.M. Martin, M.B.B.S., M.Sc., Ph.D., M.R.C.G.P., F.R.A.C.G.P.
Department of Public Health and Primary Care, Trinity Innovations Centre, The O'Reilly Institute, and Visiting Academic, Trinity College, Dublin, Ireland
e-mail: martinc7@tcd.ie

[1] Complexity refers to something that is "made of (usually several) closely connected parts" (Oxford Dictionary). Complex Systems such as the health system possess a number of linked agents and properties. At the same time each agent can be seen to be both a cause and effect of the other agents which make it difficult to identify the starting point. For more information see Kurtz, C. & Snowden, D. (2003) The new dynamics of strategy: sense-making in a complex and complicated world. *IBM Systems Journal* 42(3) 462–83. These explain why it is difficult to measure the real impact of the multifaceted initiatives engendered by the transition funds.

[2] See Section 36.5.1 for more details on the logic model.

J.P. Sturmberg and C.M. Martin (eds.), *Handbook of Systems and Complexity in Health*,
DOI 10.1007/978-1-4614-4998-0_36, © Springer Science+Business Media New York 2013

advance by the funders which guides the approach to projects evaluation, limits the application of theoretically based inquiry and the consideration of important issues that could enhance policy decisions. There is a need to construct a theoretical-based framework that has the potential to produce more comprehensive and in depth evaluation—that will offer the possibility to circumvent some of the shortcomings of the prevalent one. In addition, such a theoretical framework will provide not only the mechanism for explanatory evaluation, but also has the capacity to expand the scope and include key missing indicators that are essential for policy reform, designed to address disparity between populations. Similarly, it will give evaluators the much needed scope and operational independence to preserve their objectivity and get the most out of the process.

What will ensue is an overview of health system reform focusing on transition funds policy goals initiatives—and the challenges they pose for evaluation of changes in a complex adaptive system. Similarly, the health needs and issues confronted by the system will be examined to account for social determinants of health. In addition, the prevalent approach to projects evaluation will be assessed in order to identify the strengths and the shortcomings—then, a theoretical framework will be proposed to assist evaluators or others people involved in policy implementation, with their future endeavours, followed by lessons learned from experiences.

36.2 Overview of Health Systems Reform

Industrialised countries spend about 8–16% of their GDP in their respective health systems [7]. The ultimate policy goals of providing universal and accessible care have provided to a great extent global achievements in the improvement of population health in many developed countries. Despite this, health inequities have not diminished, and persistent disparities in access and in health status exist in some segments of the population within and across countries [8]. There are well-known fundamental systemic issues such as uneven distribution of resources, fragmentation of care and services, lack of emphasis on health promotion and prevention, and inadequate attention being paid to some service providers in rural and remote areas [9]. These issues are in flagrant contradiction to the stated policy objectives relating to access equity and health outcomes [3, 10–12]—most notably in high income nation such as Canada, where profound health and social inequities persist between aboriginal[3] and non-aboriginal populations [6, 13, 14]. Therefore, the policy goal of improving health status and reducing health disparity within nations [8, 10, 11] remains difficult and hard to achieve—even for the most advanced health systems where universal health care is the norm.

With the global economic downturn, the social envelop in which health services are an integral part is under tremendous pressure from cost cutting measures. Pressures to achieve better expenditure control and/or greater productivity and efficiency need to be balanced against deeply rooted moral imperatives to maintain universal access to necessary care, and to improve the equity with which services are distributed across social groups [15–17]. Hence health system renewal has become a perennial matter—almost everywhere, reforms were being contemplated, organised or implemented, some in direct contradiction to others [1–3]. Each has claimed to make the system more responsive to user needs; paradoxically, most are really designed to bring its component parts under particular financial control.

For instance, Canada which is the focus of this chapter has adopted the strategy of transition funds with three consecutive rounds [4–6] as mechanism to improve health services without any substantial increase in the percentage of GDP on health care. Canadian health policy aims to provide universal access, portable and affordable

[3] Aboriginal peoples in Canada are three distinct political and cultural entities. They are recognised by the Constitution Act of 1982 as Indians (First Nations), Inuit and Métis. In the last census of 2006 there were 1,172,785 Aboriginal peoples which constitute four percent of the Canadian population (Statistics Canada 2008).

health services to the population. The strategies are directed toward initiatives that could improve the quality of existing health system delivery—at the same time decrease disparity between the population via better coordination, integration or adaptation of services [9]. Ultimately, these efforts aim to continually contribute to the improvement of the overall population health status.

36.2.1 Transition Funds Strategies the Canada Experiences

Transition funds in general are joint policy effort between federal, provincial and territorial governments—they are used as instruments to reform the system of health services delivery. The first round of Health Transition Fund (HTF), which lasted for 4 years, was created to encourage and support evidence-based decision making in health care reform focusing on access, quality, integration, health outcomes, cost-effectiveness and transferability [4]. The HTF-targeted initiatives in four priority areas: home care, pharmaceutical issues, primary health care and integrated services. Other emerging health issues, such as, Aboriginal, rural/telehealth, seniors, children and mental health were also identified as priority [4, 9].

The second round, the Primary Health Care Transition Fund (PHCTF) implemented over a 6-year period (2000–2006) established on a per capita basis by the government of Canada to support the efforts of provinces and territories and other stakeholders to develop and implement transitional primary health care reform initiatives—as part of the overall renewal of Canada Health care systems. The view is that primary health care improvements are essential for the renewal of health services. The goal was to improve access of comprehensive health services to a defined population with an increased emphasis on health promotion, disease and injury prevention—and chronic disease management [5].

The third round Aboriginal Health Transition Fund (AHTF) was rolled out over a 5-year period 2005–2010. The policy strategy was to support cross-jurisdictional integration of health systems[4] via adaptation of current provincial/territorial health services systems for the betterment of all Aboriginal peoples. The aim in long term is to improve access to health services in order to address the disparity that exists in the health status between the Aboriginal peoples and the overall Canadian population [6].

36.3 Contextual Challenges for Policy Initiatives: Implications for Evaluation

The Canadian health system delivery services are under the constitutional jurisdictions of the provincial/territorial governments. The 10 provinces and 3 territories are distinct entities and autonomous in their decisions-making process—on how they establish priorities for health care—even when there is broad national consensus on policy direction. The implementation depends very much on the provinces' own history, and particular socio-political, economic sphere with no interference from the federal government. In addition, the three territories are divided into four self-governing regions and their communities/municipalities are embedded in complex jurisdictional agreements and service provisions between the Aboriginal peoples, the federal and respective provincial governments to form a complex organisational structure of four levels of bureaucracy and government. This has the potential to create decision traps when it is time to address urgent issues to improve the health of the Aboriginal

[4] First Nations and Inuit peoples for their health care must use two different systems. The federal government supplies limited primary health care and health promotion services to First Nations reserves and to Inuit; whereas the provincial and territorial governments dispense universal health care services to all Canadians, including First Nations, Inuit and Métis and those Aboriginal peoples residing off reserve. It is well acknowledged that there is a need for better coordination and integration of the services in order to meet the need of Aboriginal people in Canada (ref. For more information see the Royal Commission on Aboriginal peoples, the Standing Senate Committee on Social Affairs and the Commission on the Future of Health Care in Canada (Romanow Commission), the list is not exhaustive [6].

peoples—and constitute a tremendous challenge for multijurisdictional projects.

Implementation of such health policy goals in diverse political, social, cultural and economic dynamic systems is challenging. The services are by nature decentralised, multifaceted—delivered in various independent regional and local institutions or organisations with different financial and jurisdictional agreements which create a local silo effect, yet they are also interdependent. Moreover, meeting health needs and improving population health status are convoluted and difficult to assess or evaluate even with well accurate indicators. There are many intervening variables and factors that come into play in health status disparity. Many of these are not only related to physical health factors but to a great extent directly associated with social factors. In fact, the fundamental causes are complex, and usually endemic to social, political, historical, economic and environmental factors, built up from generation to generation [8, 10]. The disparity in the overall health status of the Aboriginal peoples as compared to other Canadians is well documented [18–20]. There are endemic issues of substance abuse (alcohol, drug and cigarette) consumption, high incidence of suicide and chronic diseases among this population [21–23]. In this instance, one can trace the distinct influences of social determinants of health

Thus, projects or programs development and implementation for target population should take into account these issues in order to reduce the gap. For that matter, assessment of health services or programs design to improve population health status and reduce disparity should include key indicators of social determinants of health. In the same vein, the integration and adaptation of health systems, services and programs can neither be done in isolation nor can it disregard the interplay between structure and agency as well as the people[5] affected by these.

Information about the living conditions, physical environment and the critical factors or social determinants influencing health are required for effective public health interventions [24]. Another aspect is to examine how these issues affect the population. The evaluation of initiatives to improve health services for population, must consider the context, the location, the risk factors in order to determine the health needs. Arguably, in omitting these, one will miss important information. Some health problem can be symptomatic to a broader issue associated with social determinants. These certainly have implications for evaluation and policy decisions.

36.4 Health Needs and Issues

Health needs and issues cannot be examined without talking about health status nor can it be examined without social determinants and the context and the location. Canada is a vast land, second largest country in the world [25] with a relatively small population of 34,482,779 millions [26]. The geographic distance constitute a challenge to provide services to some segments of the population. For instance, the communities in the vast arctic northern regions are scattered over vast distances between not only themselves but also the major centres further south [27]. For communities where virtually all transport is by air, expenses are much greater than in regions of similar population where road and rail transport are available. This situation in turn creates an uneven development or disparity in the health and health services between Northern and Southern Canada. Therefore, geographic and transportation issues have hindered the socio-economic development that might alleviate health problems associated with socio-economic status. Moreover, because of the small population, the large distances and the lack of road network connecting villages, it is necessary to provide a wide range of basic services in every village—even to the smallest. Having to supply several separate points of service for schools, clinics, transportation infrastructure and other services is extremely expensive on a per capita basis [24, 28]. Hence, make it difficult to benefit from economy of scale as it happens in larger sized population. This creates

[5] People are referred to not only those who are receiving the services, but also those who are providing it.

a vicious circle for people to have to travel far away for specialise services. As one can see providing health services to the vast northern and remote rural areas is a challenging task for the Canadian health system.

36.4.1 Social Determinants of Health

Social environment—to some extent—does have a potent effect on attitudes, values, beliefs, social skills, as well as coping resources and behaviour. Indeed, it influences how individuals respond to life events that can affect their health or well-being. Social environment includes all those external social factors that protects and favours well-being or which makes people vulnerable and susceptible to harm. The setting, institutions, the community and public policy as well as structures that contribute to attain and maintain health on a continuum are also part of the equation [24, 28]. One social factor is of such fundamental impact on health is poverty.

36.4.1.1 Demographic Issues

Demography[6] is a crucial indicator in particular for the northern regions with a rapid growth of population [29] at a time of perennial budget constraint in the health sector where services are already scarce and stretch to the limit is a cause for concern. In view of the fact, social determinants influence children, youth and adults alike, but they tend to exhibit different health issues in each life stage. Social determinants not only have a differential impact on health across a life span, but subsequent health issues may themselves create conditions or chain reactions (i.e. other determinants) that subsequently affect health. For instance, poverty is associated with increased substance use, which in turn decreases money available to spend on food and basic need-thus can lead to stressful family environments and diminished social support, which is linked to, among other things, depression and suicide [24, 28]. All of these have an impact on youth and adult alike and lifestyle and can carry from generation to generation. Many children live in poverty and are affected by several risk factors (family violence, abuse and consumption) that are likely to compromise their developmental health status. Paradoxically, there are few services and prevention programs for this extremely vulnerable clientele [24, 29].

36.4.1.2 Physical Environment and Family Unit

The current settlement structures engendered to a great extent the pervasive outcomes of housing shortages and poor standard of existing homes. Couple to that the extreme weather conditions contribute greatly in wear and tear on homes in the northern region [18, 30]. Lack of suitable/affordable home has created bad living conditions in these communities. Inadequate ventilation and moldy housing have been associated with several health problems including high incidence of severe asthma and allergies among the population [30–32]. Although the impact of crowded housing is harder to measure than the impact on physical health—housing conditions have been associated with stress and family violence in all age groups [22, 31, 32]. For instance in conditions of excess numbers of people in the household, which is very much the case in the North—children often have no place to study or play, while adults have no privacy. These conditions can bring extra stress, which has the potential to increase behavioural and learning problems in children and adolescents including substance use and other social problems. Likewise family violence can affect very much the family unit resulting in negative impact on the overall well-being of the population [24, 27, 28].

36.4.1.3 Health Behaviour and Lifestyle

These are well-recognised health indicators. Lifestyle include the over alcohol consumption, which is directly related to increase incidence of mortality. Excessive smoking is another behaviour affecting health, with the etiological effect of high rates of heart disease and increasing rates of lung cancer. Alcohol consumption and smoking

[6] Demography is concerned with characteristics of a population such as how young or old the population is birth and death rates, and population growth.

during pregnancy increase the prevalence of poor physical, emotional and intellectual development among children. Finally, unhealthy lifestyle such as lack of exercise and poor diet has been associated with Type II Diabetes [24, 28, 33, 34]. These behaviours and lifestyle must be accounted within the socio-political and economic context of the health of the population. It is in this brief contextual overview that the Canadian health system contemplated its renewal with the policy reform or renewal strategies of transition funds. The foregoing has certainly implications for project evaluation and policy recommendations.

36.5 Prevalent Approach to Projects Evaluation

This section will examine specifically the official approach to transition funds evaluation. It will argue that the current standardised model which guides evaluation of the initiatives limits the possibility for evaluators to comprehensively address complex socio-political issues, and bypasses important issues that could enhance policy decisions. It is by no mean an assessment of the different definitions and approaches to project and program evaluation that exist in the literature—nor is it a review of projects evaluated during the transition funds. The chapter takes the view that in the context of health policy initiatives—it is more productive to talk about the purpose of evaluation, rather trying to sort out the myriads descriptions that exist in the field. It is even trickier trying to make sense of the words process, impact and outcome evaluations because different disciplines or participants in their definition of evaluation frameworks make use of these words—they may not necessarily have the same connotations. Here evaluation is viewed as policy instrument conceived and developed to produce knowledge, provide useful feedback that could inform decision makers—and could also be used to improve health system services delivery. Improvement presumes that there will be changes in the system conducive to maximisation [35, 36] of services in order to meet the populations' needs at the same time decrease disparity.

36.5.1 The Logic Model

The logic model is central to Canadian transition fund evaluations and has been used for both formative[7] and summative[8] evaluation of the transition funds [4–6, 9, 16, 17]. It establishes how the initiatives of each funding envelop is understood or intended to give specific results. The model presents four linear components: inputs, activities, outputs and outcomes. These constitute the logical flow: the first is related to resources such as funding, human resources and equipment; the second concerned with work activities, programs or processes; the third constitutes the immediate outputs of the work that are delivered; the fourth are short-term, intermediate and long-term outcomes or results [4–6]. This is by far the approach of predilection for governments or not profit organisations, for that matter the prevalent one in transition funds [9, 16, 17, 37]. A framework based on the model was developed to serve as guideline for evaluation and writing report. This approach accounts to some extent for outcomes and ensure standardisation in projects evaluation and report. At the same time, it has also constraint one's thinking to current activities and programs, as well as limiting valid research questions [38]. The typical: what, where, whom, how queries were clearly identified in the transition fund evaluations. However, the why was conspicuously missing.

This can explain the shortcoming of the evaluation explanatory strength in the initiatives generated by the transition funds [9, 16, 17]. In fairness, an attempt was made to produce a comprehensive framework which has generated useful data, but again the analytical drive was curtailed and reduced to a simple exercise or mechanistic data analysis. Put it simply the underpinning of

[7] Formative evaluations aim at strengthening or improving the thing or object that being evaluated—they assist in the exploratory phase of programs or initiatives. This is usually done before the development and the implementation processes.

[8] Summative evaluations, in contrast examine the effects or outcomes. They help also with the effectiveness and efficiency of the initiatives that have been implemented. The outcomes evaluation can be immediate, intermediate and long-term.

the model acts as a deterrent for evaluators to go beyond the logic model framework—and to link causal effects of social determinants which are crucial for health system services evaluation. The same can be said for the lack of connectedness not only between issues, context and location but also between structures and agency. As a result, conceals crucial understanding of the real issues that could enhance policy decision and services. Thus, the relevance of why query in health system services evaluation.

While acknowledging the merit of the logic model framework, one tends to agree with the criticism that it is an accountability driven model [39]. Some authors have argued that evaluations are becoming more and more part of bureaucrats or policy makers' support systems. With the logic model framework "policies and programs are often treated as black box" [40]. Even with the increasing trends in program theory development or "theory-driven" evaluation which is not in reality even close to the true meaning of the term [41–44], in reference to Merton[9] "middle-range" theories tradition and its relevance in evaluation [42]. The so-call program logic cannot be treated as program theory; it is merely an evaluation practice.

Other issue, transition fund evaluation had short term life with the hope the result yield would benefit in the long term. However, there are no mechanisms or ways to verify or predict the sustainability of perceived outcomes improvement or progresses obtained. The net impacts or effects of programs and interventions are rarely available [16, 40]. Therefore, the proposed framework for evaluation and writing the report of the transition funds initiatives, albeit comprehensive is nothing else than a larger black box. The problem is how to go beyond the framework while respecting required mandate. At this juncture, it will be more beneficial to encompass the model with other approaches in order to accommodate for key missing variables, especially when dealing with complex social issues such as services to improve population health status. If one really wants to avoid the "black box" syndrome—a broad scope is needed to ensure that responsibility to the population and social welfare is maintained.

[9] *Cited in (Page 7) by Leeuw [40].*

36.6 Exploring Other Approaches to Health System Evaluation

The health transition funds are allocated to projects, as either demonstrations or applied translational research. As a case in point, the recently completed Aboriginal Health Transition Fund aimed at health system services amelioration or improvement via evidence based, innovation and adaptation or integration of existing services. The overall thrust was these initiatives is a short term perturbation of current systems that in the long term will lead to a transformation of the health system that will improve the services at the same time decrease disparity between the populations [4–6]. Evidently, no single approach can address the multifaceted initiatives engendered by this overarching policy goal. This chapter takes the view that complex issues cannot be fairly evaluated without the context and the location of projects in question. In the same vein, the particularity of each province/territory, the uneven development of health and health services between the regions, territories, mainstream and remote population warrant a broader assessment to account for determinants of health. Furthermore, neither the integration and adaptation of health systems, services and programs can be done in isolation, nor can it disregard the interplay between the structure, process and agency as well as the people who receiving or providing the services affected by these proposed changes. Hence, integration and adaptation of health systems and services in partnership with others systems is not simply a mechanical activity but a very complex and intense political process deeply entrenched in the respective province, the regional government and their bureaucratic apparatus. In this, tensions and contradictions are accepted to be an inherent part of the process.

Health is a social phenomenon and its systems and services regardless the countries are heavily influenced by politic and economic. The multifaceted of health services coupled with the mandate to integrate or adapt multi-layer cross-jurisdictional sectors—call for a comprehensive theoretical framework to sort out the implementation impacts of transition fund projects, not only on the organisations and services, but also on the people affected by the process and vice versa. Thus, it is necessary to take this approach to guide the analysis for a fair assessment, similarly avoid giving preferences to some issues over others. While keeping in mind, "*there is no such thing as analysis without theory… rational techniques themselves are based on meta-theoretical assumptions about the nature of evidence and logic and every problem definition is laced with political, sociological, and economic theory*" [45].

36.6.1 Theoretical Framework

The theoretical framework will guide the analysis and helps identify gaps—and establish relationship between issues, context and location, as well as directly answer to the transition funds main request—thus ensuring consistency across evaluations and facilitate a national level of analysis to inform policy decisions. The analysis can take several forms which involved an iterative process of moving back and forth between the information obtained in the documents and in empirical material from the interviews, the focus groups and questionnaires survey. This also entailed an examination of the data not only at the conceptual and theoretical level, but also to look at the methodological implications. In this instance, data can be organised under the broad categories proposed by the transition fund guideline for developing a project evaluation and writing the report without affecting the analysis integrity of the evaluation.

Based on the foregoing a variety of methodological procedures could be employed to gather evidence and established relationship, for a comprehensive non linear assessment analysis that will connect the issues not only with the context and the location, but also with the structures and people that will be affected by the transformation. A theoretical framework,[10] linking complexity theories with political economy and sense-making process, connecting to theoretical work on structure and agency open up the possibility to go beyond the mandated box.

Complexity theories support the notion of organisations as dynamic, living and social systems. In this, it is assumed that research observations target patterns of relationships, interactions and processes, and over time are the key to understanding of complex system. A search for patterns implies attention to the flow of behaviour within systems rather than merely describing static behaviour. Consensus is identified as the most appropriate method to synthesise research/evaluation findings in this context [46].

Political economy[11] seeks to locate the movement of society in the forces of change as production and reproduction are transformed. This perspective leads to an understanding of how people or a group or organisations have shaped their own lives or existences, under imposed conditions and what strategies they have used to withstand the changes. Tensions and contradictions within society that produce struggles and resistance to the prevailing order are integral to political economy [47–49]. Moreover, a political economy perspective helps to conceive dialectically "objects and subjects of study taken to be fixed or static and unproblematic in the empiricist and positivist traditions of political science and neoclassical economics" [50]. As a whole, political economy assists us in the understanding of conflicts between structure, process and providers (agents) action; between health and intersectoral practitioners relationship with

[10] The theoretical framework was first tested in the PhD thesis (M. Felix-Bortolotti) 2004 and demonstrated its relevance in the study of complex phenomenon; subsequently was adapted for Virtual Office of Synthesis (C.M. Martin, M. Felix-Bortolotti) 2006; Was recently adapted to guide selected Aboriginal Health Transition Fund projects external evaluation C. Martin, M. Felix-Bortolotti.

[11] This is not a specific theory but a general approach to social analysis that stresses the interconnection of social, political and economic processes in society.

the states[12] and their apparatus; between thoughts and real conditions; between management, policymakers and the grassroots, and along with the trend to look at these features in fragmentation [47, 50, 51].

Sense making is viewed as a process by which people give meaning to experience [52, 53]. The process is mainly focused on understanding the dynamics of the phenomena mentioned above and assists in the formulation of interview questions that identify "what" and "why" a person's perceived situation, gaps, bridges and outcomes. For example, questions that elicit data on a person experiencing a situation include: "What issue were you dealing with and why?" "What led you to confront this issue and why?" and "What did you hope to achieve?" Questions designed to elicit information on gaps include: "What was confusing about the situation and why?" and "What prevented you from better understanding the situation and why?" Questions constructed to elicit bridges include: "What answers helped you better understand the issue?" and "What ideas or conclusions came to your mind and why?" [53].

Structure refers to any recurring pattern of social behaviour; or, more specifically, to the ordered interrelationships between the different elements of a social system or society. The idea of social structure points out the way in which societies, and institutions within them, exhibit predictable patterns of organisation, activity and social interaction. This relative stability of organisation and behaviour provides the quality of predictability that people rely on in everyday social interaction. Social structures are inseparable from cultural norms and values that also shape status and social interaction. Structure is generally agreed to be one of the most important but also most elusive concepts in the social sciences [54]. Agency is concerned with the capacity of individuals to act independently and to make their own choices, although not under conditions of their own choosing or merely as a result of ideas that rise independently to their minds [51]. It is also referred to the ability of people to change the institutions in which they live.

This overall approach has the potential to produce a more comprehensive evaluation and thus assist the different stakeholders and policy makers in their decisions. It offers the possibility to circumvent some of the shortcomings of the accepted framework for projects evaluation. In addition, the theoretical framework provides not only the mechanism for explanatory evaluation, but also expands the scope to include key missing indicators that are essential for policy reform designed to address disparity between populations. Similarly, gives evaluators the much needed tools to safeguard their neutrality and maximise the positive impacts of their work [36].

36.6.1.1 Limitations

Some grant initiatives evaluations are likely to be affected by mandate, limited scope to one or a number aspects of programs that may/or not be essential to address the real issue. In this case, evaluations findings are prone to being dismissed as insufficient to assist in policy decisions regarding the programs [36]. Pressures of time and resource are major constraints. Often project final reports and evaluations are operated concurrently to meet the same deadline—with the risk of creating discrepancy between the final report and the evaluation. Large distances as seen in the Canadian context and constrained financial resources can impede evaluators to go in the fields to obtain affected communities perceptions of the initiatives. Thus, the views of key stakeholders, leaders or community's representatives prevailed to the detriment of grassroots population—these contribute to asymmetrical findings in the qualitative evaluation.

Data gathered from primary official sources and complimented by unofficial findings via interviews and others—while the data could dependable as much as possible, it is impractical to warrant that they are totally accurate for several reasons. These include limitations in finance and infrastructures involved in data collection for

[12] Here is referred to a governed entity or sub- entity such as federal, provincial, regional and municipal government as well as their policy and bureaucracy.

an exhaustive evaluation given the projects short duration. Comparisons are difficult to make and differences difficult to assess when out of context. One also needs to go beyond the written reports and texts to the empirical interviews and other primary research and confront them back and forth. It is important to know about the context and the location in order to see the nuances and meaningfully interpret differences.

The request for outcomes evaluation and measurement of success on the part of the HTF provider is legitimate. However, it is premature to assert with certainty that the initial outcomes can be sustained, because project's success is contingent on multiple variables. The assessment of the impact of a project or the outcomes on the target population is usually completed after such a project has been implemented for some time. In order to actually measure the impact, it will be necessary to talk to the community and to the recipients of services as well as to project personnel. Time, geographic distance and late funding can definitively be a limitation. Knowing these factors can help mitigate some of the limitations outline above.

36.7 Lessons from Experiences

At the minimum transition fund policy projects can demonstrate that is feasible to integrate the activities or services of independent organisations. It is also feasible to adapt the concepts of existing program to meet the needs of specific population as evident in some of the success stories in the summative evaluation report. The difficulties of measuring the real impact and insure the sustainability of success obtained still remains. In some instance, projects have improved the delivery of services by streamlining there organisational process, without any substantial increase in access to services [9, 16, 17]. This is attributable not only to the complexity of the initiatives but also to the nature of transition funds which tend to run for a short time period. Paradoxically, the logic model that underpins the projects evaluation is calling for long term outcomes. Obviously, the incremental or piecemeal approach to system reform aiming at reducing systemic disparity between populations is debatable. It will require an intersectoral approach [55] to operationalise this righteous goal. Although the Canadian health system was used as a case study, this analysis can find its relevance elsewhere, with a caveat that it is essential to always locate the system under study in its historical context and the political economy as well as the organisational structure in which it is embedded.

Provincial/territorial governments have their own history and political economy, sometimes difficult to reconcile. Complex jurisdictional relationships, mandates, regulations, devolution processes—in which the Canadian health care system is embedded, can impede projects implementation—resulting of a further delay in the relatively short timeframe of transition funds. Some adjustment had to take place in order to account for time lost. This will affect to some extent the expected outcomes. Moreover, project final evaluation supposes to be done after the implementation. In reality, pressures of time force both processes to run concurrently at the end. The following part will highlight some of key constraints, barriers and enablers that evaluators need to take into consideration when evaluating specific policy reform initiatives implemented via research, better coordination and integration or adaptation of services.

36.7.1 Constraints and Barriers

Integration/adaptation of services when dealing with existing jurisdictions which are autonomous in their decisions can be cross-cutting in the negotiation process. The potential for competitive programs were being run concurrently by other major partners that share objectives is very high. The duplication of services and a competing demand on overburdened human resources can also affect the effectiveness of partnership, and impede adaptation/integration. Organisations and services carry their own philosophy, value and objectives, which sometimes make it difficult to conciliate. Barriers to adaptation are the cultural challenges of existing ways of living and doing things. Moreover, establishing partnerships with autonomous services or programs is not simply done. It requires time for negotiation to bring about trust and positive outcomes. Changing organisational culture and structure, as well as obtaining community support for integration and sustainability of projects, cannot be done overnight. Health care institutions transformation call for lengthy negotiations.

Similarly, predetermine initiatives can find it difficult to adapt health system services, because in some circumstances adaptation and integration demarcation line can be blurred. In realty, the initiatives or projects may be more an integration of services and activities than an adaptation. It is important to remember that services or programs which are working in affluent part or in southern provincial regions, might not necessarily work in the northern regions or remote rural areas. In this case, focusing on the concepts of the services and adapt them to the cultural, geographic, lifestyle and environment realities of the target population will make more sense. Transition fund projects acted as a vehicle in streamlining certain services and activities, hence bringing about a more concerted approach to address the health status issues of the population

In the case of research project, barriers take the form of the time it takes to deal with multiple levels of government and bureaucracy layers at federal, provincial, territory and community when conducting sensitive research issues that affect the population. Project leaders can find it difficult to streamline the line of communication with multiple stakeholders, especially when the role of each player is ill defined. Also maintaining rigorous research methodology can prove very difficult exercise to achieve in program funding.

36.7.2 Enablers

Partnerships to be effective call for openness in the relationship between the different partners involved in the projects or the programs. The structure must allow enough flexibility to accommodate differences as well as sufficient time to build trust. The partners need to be opened to suggestions to facilitate the integration. A rigid structure will not yield any result.

In cross-jurisdictional grant initiatives, researchers or people involved in health system implementation and evaluation, prior initiating health system projects—must understand how each level of governments and jurisdictions operate to minimise bureaucratic frustrations. Establishing and streamlining the line of communication especially when dealing with multiple stakeholders are primordial. Clearly define the role of each other in each project; specify the importance of pre-assessment of the field as a sine qua non condition especially when dealing with sensitive issues. Pre-assessment will increase not only awareness among partners but also will give the opportunity to build trust and account for cultural and jurisdictional differences. These enablers are keys to both internal and external evaluators. This is especially crucial for the external evaluators, who usually come into the evaluation process at the end.

Initial collaborations among a wide range of other agencies if sustained can yield some improvement in health services; hence improve health status of target population. Long-term integration and sustainability are primordial for lasting outcomes. Some initiatives may require ongoing funding in term of resources to sustain the immediate or intermediate outcomes. Longer-term outcomes depend on structural change and sustainable funding, which is difficult to assess in the context of transition funds projects. The principal outcome is anticipated to be more effective knowledge transfer of information about the population issues that have being investigated. The findings from research projects can assist in the feasibility of increased capacity for integrating federal, provincial/territorial health systems, program and services to the needs of target population for specific issues. The findings have the potential to be a catalyst for improved adaptation of the different health systems, programs and services. Transition funds are excellent vehicle for potential innovation for the promotion of policy, planning and services for the well-being and health needs of population.

The expectation to form a partnership, to be effective, requires a good line of communication without too many intermediaries. Respect for each other's jurisdictions, roles, expertises and domains are primordial. Quality time to build trust facilitates functional partnership. Setting realistic goals as well as assessing the magnitude and the feasibility of a project in a short timeframe is also important. This is crucial especially for cross-jurisdictional projects or programs where provincial/territorial, regional differences come into play. A project which is much more integrated into existing resources with sufficient infrastructure and duration to enhance capacity on the ground would serve the population better. It is important to take stock of the existing milieu and its resources and capacities more thoroughly, before attempting to implement what is a good idea.

When dealing with systems, services or program integrations with the expectations of creating new partnerships and new ways of doing things in a culture of organisational silos, it is important to have the transition fund for a much longer period. This is even more problematic when cross-jurisdiction partnerships come into play. Seeking community participation is also essential for the sustainability of integration or adaptation of programs or services. All these processes require time to materialise.

36.8 Summary and Conclusion

This chapter has examined the health system policy reform in the Canadian context. Specifically, it has focused on the transition funds initiatives as policy instruments to reform health system and the challenges that such multifaceted projects have posed for evaluation. It used also the current strategies as a springboard to demonstrate the complexity of this overarching goal. Particular attention was given to social determinants of health and disparity in the system between the populations—and some of the systemic issues that considered being crucial in projects evaluation dealing with disparities was highlighted. Throughout the document, the authors utilised both inductive and deductive arguments to make the case for non-linear approach to evaluation of health system.

Herein it has argued that the Logic Model which guide the approach to projects evaluation, limit the possibility for evaluators to go beyond the propose framework—thus conceal important issues that could enhance policy decisions. A theoretical framework, linking complexity theories with political economy and sense-making process, connecting to theoretical work on structure and agency was proposed to go beyond the model. This offered the possibility to guide the evaluation analysis to not only answer the required questions but also assist in connecting the different phenomena that are generated or emerged from the process. At the same time give evaluators the much needed tools to preserve their objectivity and get the most out of their work for the betterment of the system.

Health systems, services or organisational adaptation/integration cannot happen overnight, for that matter it is very difficult to establish cross-jurisdictional and organisational partnerships. When dealing with independent services and organisations, one most allow sufficient time to engage, build trust and know what is feasible or not and come to an agreement. There are differences emanated from the political economy of each region, on the different approaches to problem solving, on the mission and philosophy of particular organisation as well as their priority and strategic direction. All that comes into play when negotiating partnerships for projects or alliances which were not part of an established plan. The other factor can be associated with the scope and unrealistic goal and objectives within the set timeframe of the Health Transition Funds.

Moreover, improvement of health system services delivery implies change from the status quo. Consequently, it is crucial for evaluators or people involved in implementation to locate projects or initiatives within the context of the past and larger society in order to promote change. Likewise be aware of the forces that shape the scope and outcomes of evaluations, the health care structures and the agencies—but also the political economy in which they are embedded and the struggles around their development. Contradictions and tensions are also part of the process. For that matter, changes are inherent to system survival, this explain the intrinsic dynamism of health system. Incontestably, when population needs change, the system will be under pressure to transform as well—and reform regardless of how it is labelled or conceived will be part of the equation. This is also true for health system evaluation. In final analysis health system policy evaluation has proved to be as intricate as the system itself, which one sets out to evaluate. It is hoped that this attempt will help evaluators or others people involved in policy initiatives, with their future work.

References

1. OECD. The reform of health care: a comparative analysis of seventeen OECD countries. Paris, Organization for Economic Co-operation and Development. Health Policy Stud. 1994; (5).
2. Buchan J, Hancock C, Rafferty AM. Health sector reform and trends in the United Kingdom Hospital Workforce. Med Care. 1997;35(10):143–50.
3. Docteur J, Howard O. Health-care systems : lessons from the reform experience. OECD Health Working Paper. Paris, France. 2003. Available at http://www.oecd.org/data.oecd/5/53/22364122.pdf (last accessed 11 Nov 2011).
4. Health Canada. Health Care System The health Transition Fund (HTF) 1997–2001. Ottawa, Canada. 1997. Available at http://www.hc-sc.gc/hcs-sss/ehealth-esante/infostructure/finance/…/index-eng.php (last accessed 11 Nov 2011).
5. Health Canada. Primary Health Care Transition Fund (PHCTF) 2000–2006. Ottawa, Canada. 2000. Available at http://www.hc-sc.gc.ca/hcs-sss/prim/phctf-fassp/index-eng.php (last accessed 10 Nov 2011).
6. Health Canada. Aboriginal Health Transition Fund (AHTF) 2004–2005 to 2009–2010. Ottawa, Canada. 2004. Available at http://www.hc-sc.gc.ca/fniah-spnia/services/acces/ahtf-eng.php (last accessed 10 Nov 2011).
7. OECD. OECD Health Data 2011. 2011. Available at http://www.org/document/16/0,2340,en_2649_34631_2085200_1_1_1_1,00.html (last accessed 06 Nov 2011).
8. World Health Organization. Indigenous Health – Australia, Canada, Aotearoa New Zealand and the United States – Laying Claim to a Future that Embraces Health for us All. Background paper, No 33. Health Systems Financing The path to Universal Coverage. 2010. Available at http://www.who.int/healthsystems/topics/financing/healthreport/HNO33.pdf (last accessed 29 Oct 2011).
9. Health Canada. Sharing the Learning: Health Transition Fund: Synthesis Series: Primary Health Care. Prepared by Ann L. Mable and John Marriott. 2002. Available at http://www.hc-sc.gc.ca/hcs-sss/pubs/acces/2002-htf-fass-prim/index-eng.php (last accessed 27 Oct 2011).
10. World Health Organization. Commission on Social Determinants of Health Final Report. Geneva, Switzerland. 2008. Available at http://www.who.int/social_determinants/thecommission/finalreport/en/index.html (last accessed 10 Nov 2011).
11. World Health Organization. World Health Assembly Assembly Resolution 62.14 Reducing health inequities through action on the social determinants of health. Geneva, Switzerland. 2009. Available at http://apps.who.int/gb/ebwha/pdf_files/WHA62-REC1/WHA62_REC1-en-P2.pdf (last accessed 25 Oct 2011).
12. World Health Organization. The World Health Report: Health Systems Financing: the path to universal coverage. Geneva, Switzerland. 2010. Available at http://www.who.int/whr/2010/en/index.html (last accessed 25 Oct 2011).
13. Blueprint on Aboriginal Health: A 10 Year Transformative Plan. Prepared for the meeting of First Ministers and leader or national Aboriginal organizations, 24–25 Nov 2005. A Work in Progress Ottawa, ON. Available at http://www.hc-sc.gc.ca/hcs-sss/alt_formats/hpb-dgps/pdf/pubs/2005-blueprint-plan-abor-auto/plan-eng.pdf (last accessed 10 Nov 2011).
14. Health Council of Canada. Health Care in Canada Clearing the Road to Quality. Annual Report to Canadians, Toronto, Canada. 2005. Avalaible at http://www.healthcouncilcanada.ca/docs/rpts/2006/ExecSumEnglish2006.pdf (last accessed 28 Oct 2011).
15. World Health Organization. European Health Care Reforms Analysis of Current Strategies Summary. Regional Office for Europe Copenhagen. 1996. Available at http://www.euro.who.int/__data/assets/pdf_file/0005/136670/e55787.pdf (last accessed 11 Nov 2011).
16. Health Canada. Primary Health Care Transition Fund Summative Evaluation Final Report. Department Executive Committee on Finance, Evaluation and Accountability (DEC-FEA), Ottawa, Canada. 2008.
17. Health Canada. Summary of Initiatives Final Evaluation, Ottawa, ON. 2007. Available at http://www.hc-sc.gc.ca/hcs-sss/alt_formats/hpb-dgps/pdf/phctf-fassp-initiatives-eng.pdf (last accessed 29 Oct 2011).
18. Royal Commission on Aboriginal Peoples. Report of the Royal Commission on Aboriginal Peoples, Supply and Services Canada, Ottawa, ON. 1996.
19. Waldram J, Herring D, Young K. Aboriginal health in Canada. In: Historical, cultural, and epidemiological perspectives. 2nd ed. Toronto, ON: University of Toronto Press; 2006.
20. Adelson N. The embodiment of inequality: health disparities in aboriginal Canada. Can J Public Health. 2005;96:S45–61.
21. Kirmayer LJ, Boothroyd LJ, Hodgins S. Attempted suicide among youth: psychosocial correlates and implications for prevention. Can J Psychiatry. 1998;43:816–22.

22. Health Canada. First Nations, Inuit and Aboriginal Health. A Statistical Profile on the Health of first Nations in Canada: Determinants of Health, 1999 to 2003. 2009.
23. O'Neil JD. Suicide among Canadian aboriginal peoples. Transcult Psychiatry Res Rev. 1994;31(1):3–58.
24. Hodgins S. Health and what affects it in Nunavik how is the situation changing? Kuujjuaq, QC: Nunavik Regional Board of Health and Social Services. 1997. ISBN 2980335428.
25. United Nations. Statistical Yearbook Fifty Second Issue. New York. 2008. Available at http://unstats.un.org/unsd/syb/syb52.pdf (last accessed 10 Nov 2011).
26. Statistics Canada. Quaterly population estimates and factors of demographic growth, provincial perspective, Ottawa, Canada. 2011. Avalaible at http://www.statcan.gc.ca/daily-quotidien/110622/dq110622a-eng.htm (last accessed 10 Nov 2011).
27. Inuit Tapiriit Kanatami. National Inuit position paper regarding the CCME Canada-Wide strategy for the management of municipal wastewater effluent and environment Canada's proposal regulatory framework for wastwater, Ottawa, ON. 2008. Available at http://www.itk.ca/about-itk (last accessed 23 Oct 2011).
28. Loppie C, Wien F. Health inequalities and social determinants of aboriginal peoples' health. National Collaborating Centre for Aboriginal Health. 2009. Available at http://epub.sub.uni-hamburg.de/epub/volltexte/2009/3060.pdf/NCCAH_Loppie_Wien_Report.pdf (last accessed 24 Oct 2011).
29. Statistics Canada. 2006 Census: Aboriginal Peoples in Canada in 2006: Inuit, Métis and First Nations, 2006 Census: Inuit. 2005. Available at http://www 12.statcan.ca/census-recensement/2006/as-sa/97-558/p6-eng.cfm (last accessed 22 Feb 2011.
30. Canada Mortgage and Housing Corporation. 2001 Census Housing Series Issue 6: Revised Aboriginal Households. Research Highlight, Socio-economic Series 04–036. Ottawa, ON: CMHC, August; 2004.
31. Young TK, Mollins J. The impact of housing on health: an ecological study from the Canadian Artic. Artic Med Res. 1996;55:52–61.
32. Health Canada. Literature review evaluation strategies in aboriginal substance abuse programs: a discussion. pp. 12,13. 1998. Available http://www.hc-sc.gc.ca/fniah-spinia/alt_formats/fnihb-dgspni/pdf/pubs/ads/literary_examen_reviews-eng-pdf (last acessed 11 Oct 2011).
33. Jenkins A, Gyorkos T, Culman K, Ward B, Pekeles G, Mills E. An overview of factors influencing the health of Canadian Inuit infants. Int J Circumpolar Health. 2003;62(1):17–39.
34. Tait H. Aboriginal peoples survey, 2006: Inuit health and social conditions. Analytical paper, Statistics Canada, catalogue no. 89-637-X no.001. 2008. Available at http://www.statcan.gc.ca (last accessed 10 Feb 2011).
35. House ER. Assumptions underlying evaluation models. Educ Res. 1978;7(3):4–12.
36. Slattery D. The political inherency of evaluation the impact of politics on the outcome of 10 years of evaluative scrutiny of Australia's mandatory detention policy. Eval J Australasia. 2010;10(1):17–27.
37. Treasury Board of Canada Secretariat. Performance measurement for the Government On-Line Initiative : Logic Model. Ottawa, ON. 2004. Available at http://www.tbs-sct.gc.ca/si-as/performance/performance07-eng.asp or http://www.tbs-sct.gc.ca/si-as/performance/performance-eng.pdf (last accessed 10 Nov 2011).
38. Millar A, Simeone JT, Carnevale JT. Logic models: a systems tool for performance management. Eval Program Plann. 2001;24:73–81.
39. Blewden M. Developing evaluation capacity and use in the New Zealand philanthropic sector: what can be learnt from the US experience ? Eval J Australasia. 2010;10(1):8–16.
40. Leeuw F. Evaluation: a booming business but is it adding value ? Eval J Australasia. 2010;9(1):3–9.
41. Scriven M. Minimalist theory: the least theory that practice requires. Am J Eval. 1998;19(1):57–70.
42. Merton R. Social theory and social structure. New York: Free Press; 1968.
43. Weiss CH. Theory-based evaluation: past, present, and future. In: Rog DJ, Fournier D, editors. Progress and future directions in evaluation: perspectives on theory, practice and methods: new directions for evaluation, vol. 76. San Francisco: Jossey-Bass; 1997.
44. Weiss CH. Evaluation: methods for studying programs and policies. 2nd ed. Upper Sadle River, NJ: Prentice Hall; 1998.
45. Pal LA. Public policy: an introduction. 2nd ed. Nelson: A Division of Thomson Canada Limited; 1992. p. xiii.
46. Martin CM, Kasperski J. Developing interdisciplinary maternity services policy in Canada. Evaluation of a consensus workshop. J Eval Clin Pract. 2010;16(1):238–45.
47. Armstrong P, Armstrong H. Wasting away the undermining of Canadian Health Care. Canada: Oxford University Press; 1996.
48. Clement W. Introduction: whither the new Canadian political economy. In: Clement W, editor. Understanding Canada. Montreal and Kingston: Mc Gill-Queens University Press; 1997. p. 3–18.
49. Félix-Bortolotti M. Part 1 – unravelling primary health care conceptual predicaments through the lenses of complexity: a position paper for progressive transformation. J Eval Clin Pract. 2009;15:861–7.
50. Clement W, Vosco LF. Introduction. In: Clement W, Vosco LF, editors. Changing Canada political economy as transformation. Montreal & Kingston: McGill-Queen's University Press; 2003. p. xi–xxxii.
51. Armstrong P, Armstrong H, Coburn D. The political economy of health care. In: Armstrong P, Armstrong H, Coburn D, editors. Unhealthy times political economy perspectives on Health Care in Canada. Don Mills: Oxford University Press; 2001. p. vii–x.
52. Weick K. Sensemaking in Organisations. London: Sage; 1995.

53. Dervin B, Foreman-Wernet L, Lauterbach E. Sense-making methodology reader: selected writings of Brenda Dervin. Cresskill, NJ: Hampton Press; 2003.
54. Sewell WH. A theory of structure: duality, agency, and transformation. Am J Sociol. 1992;98(1):1–20.
55. World Health Organization. Primary health care. Report of the International Conference on Primary Health Care, Alma-Ata, USSR, 6–12 Sep 1978. Geneva. (Health for All Series, No 1). 1978.

Putting Complexity to Work: Supporting Practitioners in Health Systems

37

Christine Broenner and Patrick Beautement

37.1 Introduction

This chapter sheds a light on the practical use of insights stemming from Complexity Science for successful transformations of day-to-day practice in the wide field of health systems. A lot of research work has been conducted in the past years regarding how Complexity Science explains complex phenomena that have been observed in the health sector and described in numerous studies, articles and discussion papers. However, little has been proposed to people working in the sector regarding the practical implications of the opportunities offered by the successful exploitation of Complexity Science insights. Practitioners—those who deal with complex realities in their day-to-day work, for example those involved in policy-making, community matters, humanitarian aid and emergency response or financial and spatial planning or organisation management—often lack a systematic appreciation of and approach for how to deal with complexity. These practitioners work in routine situations, but also with the unpredictable and novel events in dynamic environments that we call Complex realities—defined as: *'Real-world situations which co-evolve with humans in an environment and in a dynamic manner which cannot be stopped and which can only be changed through engagement and influence'*. In health systems, a wide variety of specific issues in such complex realities are described, which a few examples regarding the aspects of patient care shall illustrate: Matlow et al. [1] point out that the coordination of multi-disciplinary care is a key ingredient of quality patient care; Leykum et al. [2] pick up on the fact that the identification of effective ways to improve care of patients with chronic disease has been difficult because the non-linear nature of practice has not been recognised and systematically addressed; Browne and Varcoe [3] argue that in relation to Aboriginal health, nurses have to develop greater critical awareness of politics, culture and history, and how these have shaped people's health; Singhal [4] reviews the complex factors which affect the provision of high quality care of nursing homes and points out the importance of the quality of the relationships between nursing home staff members. These papers indicate that Complexity Science can provide useful explanations of individual complex phenomena observed. To work and engage with these dynamic phenomena, however, require appropriate integrative ways-of-working in the diverse areas in or related to the health sector—ways that are immediately usable to practitioners in their particular context as discussed below.

This chapter presents insights gained concerning what practitioners require and provides recommendations of how to engage appropriately with complexity leading to the development of

C. Broenner (✉) • P. Beautement
The abaci Partnership LLP,
10 Beauchamp Road, Tewkesbury, GL20 7TA, UK
e-mail: christine@abacipartners.co.uk; patrick@abacipartners.co.uk

J.P. Sturmberg and C.M. Martin (eds.), *Handbook of Systems and Complexity in Health*,
DOI 10.1007/978-1-4614-4998-0_37, © Springer Science+Business Media New York 2013

such alternative ways-of-working. The analysis is based on the outcomes of a one-day Workshop called 'Putting Complexity to Work—Supporting the Practitioners' held on the 24th September 2009 at Warwick, UK as part of the European Conference on Complex Systems (ECCS '09).[1] The aim of the Workshop was to improve understanding of how the insights coming out of Complexity Science could be harnessed to support practitioners. The Workshop was run as part a series of so-called Satellite Workshops which focussed on particular aspects or phenomena in Complexity Science. All other Satellite Workshops at the Conference dealt with topics on theory and computational methods in Complexity Science. In contrast, this Workshop specifically had the intention to bridge the gap between Science and Practice, as it was recognised that Complexity Science could give more support to Practitioners—but only if their needs and working context were better understood. People in the health sector and in many other interdisciplinary domains do manage complex situations every day—yet insights from Complexity Science are not routinely employed as part of the language, approaches and techniques used by practitioners in their tasks. The Workshop wanted to investigate why not, what was needed by Practitioners from Complexity Science and, as a result, develop suitable approaches and techniques for working with complexity. To this end, it brought practitioners and researchers together to discuss how the 'Complexity Community' can better tailor their insights to provide practical, relevant support in these situations. This chapter illustrates practitioners' perceptions of complexity they expressed during the Workshop and examines the issues they raised when considering how to 'operationalise' insights from Complexity Science. Specifically, the needs for working with complex situations towards successful transformations and change that practitioners have expressed will be addressed. To that end, insights that Complexity Science offers for working in complex situations are analysed to see how they can support health practitioners to put complexity to work in their own context. The chapter will conclude with the presentation of a systematic appreciation that was developed for deriving options for change in a given context. It is based on insights from Complexity Science and meets practitioners' requirements.

37.2 Practitioners' Perceptions of Complexity

In the Workshop, complex realities were viewed from the perspectives of the participants belonging to the various (practitioner and academic) communities as shown in Fig. 37.1. The health sector is represented in all of theses communities, for example in the functions as nurse, medical researcher, policy maker, care service provider, epidemiologist, community volunteer, medical equipment supplier, health and safety official, development aid worker etc.

Using the World Café approach developed by Brown and Isaacs [5], participants were invited to share and explore their working experiences in situations that they perceived as complex and to discover the experiences of others. In small groups, they considered the question: *'What are the challenges you have faced and which insights have you gained from dealing with complex realities in your day-to-day work?'* to which each individual participant contributed a narrative or drawing illustrating their own experiences. The aim here was to gain better understanding of each others' perceptions and challenges, and to identify and share common themes and issues.

Although, as we have seen in Fig. 37.1, the Workshop participants came from different working domains, the *'Perceptions of complexity'*, the patterns experienced and the observations regarding complexity made, were the same across many of the domains. A broad range of issues emerged, among them those perceived as positive ones as well as those which seemed

[1]A White Paper describing in full the conduct of the Workshop and of the material arising is available from: http://www.abaci.net/library/eccs09_pctw_white-paper_v1-1.pdf.

The Workshop was sponsored by The *abaci* Partnership LLP and ASSYST.

Fig. 37.1 A range of practitioners' perspectives on complex realities

challenging. It became apparent that these issues that had been raised fell under the following thematic headings:

37.2.1 What Is Complexity, How Do We Recognise It? How Do We Perceive, Recognise, Understand, Reason About and Visualise Complex Situations?

The way in which people perceived, recognised and understood complex situations appeared to be very diverse and was thought to depend largely on their mindset, background, assumptions about the World and the context of the task in which they were engaged. Their experience was that, in practice, applying common-sense with confidence is a virtue which cannot and does not need to be explained by theory. Schön [6] similarly points out the value and necessity of using practical learning experiences obtained by practitioners against technical rationality.

The workshop facilitators noted that, among the practitioners, there seemed to be a drive to 'translate' the complex realities that they perceived into complexity theory terms, regardless of relevance, often inappropriately and even though Complexity Science describes real-world situations at an abstract level that may not add value to practice—indeed making an inappropriate translation into 'Complexity Science' might impoverish understanding. This conversion was considered unnecessary though, as practitioners understood the complexity issues in their own context and on their own terms very well. The starting point for engaging with a situation for practitioners is the natural context, people's experiences with and perceptions of it. Where Complexity Science might label a situation as complex, for people their experience in practice of the situation might be 'simple'. What practitioners did need to appreciate though is how complexity 'works' in their specific context—they knew what it is as they deal with it every day, so wanted to understand better what they could do about it in a practical way and, in their terms, and how they could engage appropriately with and influence different types of complex phenomena in their working context.

Practitioners in health systems and other domains have looked to Complexity Science for dealing with their day-to-day complex environment. For many, describing individual complex phenomena in Complexity Science terms illustrated the situation at hand in a new light, and was considered helpful to understand and demonstrate previously undetected interactions of components or the dynamics in a context. In the health sector, many examples of this kind of analysis have been published, such as by Jayasinghe [7], who looks at a list of general implications that the incorporation of Complexity Science to the population health discourse would have. Eppel et al. [8] deduce principles for practice for New Zealand's Public Policy from Complexity Science. Booth et al. [9] recognise that taking a complexity perspective to look at organisational change offers a better match with everyday human experience of change compared to the predominant current desire for control and predictability in organisational change. Plsek [10] derives some recommendations from Complexity Science for a more effective adoption of innovation in health care. Whilst establishing such principles, using recommendations and perspectives based on Complexity Science for a particular context is a major shift in practice by itself. Workshop participants voiced their concern that, in spite of such explanations and insights that Complexity Science provides, they still lacked practical, integrative approaches and techniques for embracing those principles and for engaging with the complexity of their own context. Also, on the health research side, the actual utility and benefit of Complexity Science is still being formulated. Griffith [11] gives a short summary of ways of thinking that Complexity Science can or might offer to primary health care research in support of practitioners.

Though many Complexity Science terms have found their way into everyday language, a certain unease with using the terminology in day-to-day practice was observed, as many of the scientific concepts and their implications were difficult to grasp. There were also observations that the Complexity Science terminology was, in trying to adapt it to and interpret it for a specific context or own worldviews, used incorrectly by many; so confusion and misunderstandings follow. A lot of participants felt that demystifying the Complexity Science terminology and explaining it in terms that are oriented at practice or are practice-centred was necessary. This is because complexity scientists use the academic abstractions of scientific terminology (e.g. emergence and co-evolution) for the description of natural complexity (i.e. complexity as it is in the real world) and of complex environments, their components and the phenomena arising in these environments.

It was felt that people largely introduced 'their own complexity'—what in other words can be labelled as self-imposed complicatedness. This was considered to be a side-effect of the various organisations, languages and abstractions and contrivances created, and a result of trying to compensate for natural complexity through adding more rules and processes. Yet despite this one still did not know how to '*make sense*' of it all. Also in these cases it was hoped—wrongly—that Complexity Science could help 'solve' the situation, as the self-created complicatedness was perceived (incorrectly) as being part of natural complexity and its emergent phenomena.

Participants expressed that they lacked pragmatic approaches for large-scale 'whole-system' understanding of their working context—including an appreciation of how to engage with the whole system at macro level in the real world. What forms part of such a view of systems is for example described by Capra [12] who advocates that the properties of the parts of a system can only be understood from the dynamics of the whole, whereby a part in a whole is actually not a part, but merely a pattern in an inseparable web of relationships. In practitioners' terms obtaining such understanding for their own context includes the characterisation of the environment in which the context is set and of its patterns. It was widely recognized in the Workshop that systematic approaches and techniques to capture and describe contexts in a way that grasp the dynamics and interactions in 'open' systems (such as human communities) as compared to 'closed systems' (such as machines) were not available to practitioners. Clearly though, in these two forms of situation, there are different types of phenomena at work that pose a variety of challenges and that require appropriate behaviours and capabilities to be employed, as Beautement and Broenner [13] illustrate in a comparison of suitable approaches to open and closed situations.

37.2.2 What Are the Limits on Analysis, Modelling and Verification of Complexity?

Practitioners asked: Is complexity computable? If not, which techniques and tools are appropriate? How do we quantify and validate models and tools—how do we work out which ones are appropriate to employ in which circumstances? What is the role of prediction in complex situations?

Participants discussed whether real-world complexity was 'computable' and, if not, which techniques and tools would be appropriate to understand complexity and to obtain outcomes. It was recognised that computability for complex environments in practice is constrained by assumptions and other limits such as the variety of boundaries, levels/nestings, scales (e.g. in time and space) that have to be embraced. In complex, open situations boundaries are arbitrary—one cannot 'enclose' such situations within a 'system boundary' to 'understand the problem'. In practice, most of the boundaries are contrivances to aid understanding and do not exist in reality. As such, even the term 'Health System' introduces such boundaries, but only some of these are systems in a system thinking sense and can be bounded and 'managed' in a way described for example by Armson [14]. When using these boundaries and making other limiting assumptions about the real-world in one's thinking or during computation it is inevitable that the wider real-world influences will contribute to or result in so-called unintended consequences. Modellers can assist practitioners through being clearer about the assumptions, constraints and limitations underlying models and by helping to understand the consequences of these constraining factors in practice.

In many environments in which people live, and where one tries to intervene and 'help', the realities are emergent, non-linear and unpredictable. Deterministic thinking in such situations is not appropriate. Many practitioners have accepted that trial and error along the way may be better—though difficult to defend—and that mindset changes by many involved are required accordingly. Most real-world patterns, phenomena and behaviours are unpredictable—beyond the so-called 'prediction-horizon'—and are, by definition, 'unknowable' with certainty.

'Measuring' change and benefits is commonly carried out with the help of indicators—and these indicators should be appropriate to the approaches used for a context. When they are not, such indicators give 'wrong' significance to what is happening. This can also mean that events that arise through novelty—the unexpected effects, opportunities and benefits that nobody had accounted for or considered possible during financial and project planning—are not accounted for with indicators which are not suitable to use in complex and dynamic situations. Musgrove [15] looks at the formulation and input variables for some health indicators in use in complex situations such as 'health sector performance' and makes recommendations regarding underlying assumptions that can cater for the complexity at hand, but does promote numerical indicator measurement and the simplification of variable input into indicator formulation and calculation. Practitioners in the Workshop concluded that they are in need of sets of indicators and metrics which are more appropriate for the various kinds of interventions they may employ/types of complex phenomena they may face.

Moving from linear to non-linear thinking, from using the same key variables for 'predictable' outcomes and the same models independent of the context, was perceived as a challenge by practitioners. It came up again and again that many scientific and engineering methods are not appropriate for complex contexts because they are predicated on limiting assumptions about the world. Real-world realities mean that there are significant constraints on people's ability to observe and know about certain types of phenomena in practice—some are difficult to observe even for experienced professionals. This is not helped by the fact that although scientific concepts or methods may be fine in theory. If the data needed to make the concepts work is not available or can't be collected then, as such, they are not fit-for-purpose in practice. Martens and Thomas [16] recognise the limitations of models for complex situations in their work on modelling the impact of climate change on malaria. However, modelling has its place and, as they point out, can play a role in gaining a greater understanding of what they call *multiple cause-and-effect relationships based on available knowledge and reasoned guesses* (p. 12).

Discussion also took place concerning the quantification and validation of models and tools. It was felt that part of the problem here was the mindset/language used. In many domains, it was usual to talk about 'optimum', 'validate', 'prove', 'targets' etc. on the one hand, but on the other hand it was deemed necessary that the language of 'success and failure' had to change to one which recognised the nature of complex environments. It was acknowledged that the drive for tangible outcomes and to 'quantify at all costs' created distorted perceptions, caused people to look for inappropriate indicators and so misread events.

37.2.3 What Is Involved in Enabling Effective Communication and Collaboration in Complex, Open-Ended and Unpredictable Situations?

What needs to be different, if anything, about communication, language and negotiation in complex contexts?

Participants felt that developing common ground and a rich understanding of each other's values and intents among the communities of interest involved in specific contexts was important. This was in recognition of the fact that neither one person, nor one group alone can assemble the understanding necessary to bring about changes in a complex situation. It was furthermore acknowledged that effective communication and collaboration required good communication skills, appropriate language and techniques for balanced negotiations. What practitioners need are alternatives to standardised taxonomies in order to embrace the diversity of ways of communication, of forms of collaboration and related structures for engagement that are encountered in practice.

Workshop participants were aware of and were employing a range of (culture-dependent) techniques for exploring contexts and carrying out thought experiments in a cross-disciplinary manner with those sharing a real-world context of interest. In literature, O'Daniel and Rosenstein [17] for example endorse collaboration and the establishment of a culture to support communication and team collaboration in health care organisations as this, as they argue, enhances clinical outcomes. McMurtry [18] promotes ways to think about effective interdisciplinary work of health teams that does not rely on the notion of consensus, as to him interdisciplinary teams can work very effectively, even in the absence of such consensus.

It was also recognised that there were many domains of discourse to be accommodated (e.g. directing, ordering, agreeing, influencing) and that communication between people was often influenced by overtones of status, position, power and expertise. In this respect, participants commented on the lack of good facilitators in their working world, individuals who are able to let opinions, ideas and topics emerge, while guiding people effectively through discussions and 'processes'.

37.2.4 How Do We Accommodate the (Necessarily) Diverse Views and Perspectives Which Are Inevitable in Human Endeavours?

What are effective ways of working with a wide range of 'behaviours' across stakeholders and actors? Who are the ones best placed to understand and explain the dynamics of a context?

- It was recognised that, when dealing with complex realities, a diversity of perspectives, views and behaviours across stakeholders and actors must be accepted and employed, and multiple narratives be allowed for, even if conflicting values and interests are underlying. The importance of considering and comparing various viewpoints and perspectives, from different levels and scales, both for the understanding and exploration of a context and for the creation of novelty and emergence of innovation and opportunities, was acknowledged. Practitioners were aware that in some projects or in the introduction of, for example, social or health policies expected benefits have not been delivered. This is not the least because, in neglecting the need for obtaining a global picture and understanding the multiple dimensions of interventions and in scoping the issues to address, the projects or policies are often subject to external, ill-informed narratives. It was realised that outcomes are often best achieved by participation, i.e. that the people of the community for which change is sought, are probably best placed both to understand the dynamics of their environment and to effect the changes required. Participation also did not mean consultation, but actual involvement of people in understanding a context and illustrating actors, influences and interests in that context. This also meant accepting that externally imposed interventions are not the only way to bring about the required change. Capturing multiple visions and balancing these still proved to be a key challenge in many circumstances for practitioners, but it was felt that such social intangibles have to be reflected in appropriate ways. In the development aid context, a lot of experiences have been gained with participatory methods for capturing viewpoints and local knowledge of people in different kinds of situations. Chambers [19] offers a thorough review and appraisal of such methods.

- Trust, emotion and ethics were considered important factors in the work of practitioners and must be factored into any analysis, as does the question of what is 'rightness' in a complex situation. The capability for grasping different viewpoints and for building trust in complex situations from engaging with people who are perhaps not likeminded, could be portrayed as a transdisciplinary one. Transdisciplinarity is described by Nicolescu [20] who argues that reality is not something that exists on only one level, but on many, and that only transdisciplinarity can deal with the dynamics brought about by the action of several levels of reality at once. Max-Neef [21] states that in the transit from discipline to transdicipline getting glimpses from different levels of reality generates reciprocal enrichment that may facilitate the understanding of complexity. Transdisciplinary capabilities are ultimately dependent on the personal qualities of people and practitioners are aware that there cannot be a systematic way for building trust and for allowing for emotions.

37.2.5 In Which Ways Can the Effects of Feedback, Failure and Learning Be Useful?

How do these effects relate to the social phenomena that underpin collaboration and purposeful activity?

- An aspect which was extensively discussed in various contexts was how feedback, failure and learning in organisations came about. These factors are clearly an issue within and between organisations that are or have to be collaborating. Feedback and learning are specific topics for those practitioners who deal with project and programme evaluations and 'measures of performance'. Ackermann [22] presents the challenges and dilemmas experienced in evaluating a nationwide strategy for healthy weight in Switzerland. Practitioners noted that time and again, evaluations did not include the kind of feedback that would help answer the question 'Have we made any difference?' in terms of *'goodness of outcome'* for the actual beneficiaries that were targeted by an intervention, neither were they designed to enable actual learning to be implemented across consecutive assignments. Instead, what is often evaluated is whether the project management has been successful and whether pre-determined outcomes have been achieved—independently of the dynamics of actual events.

- The topic of feedback and learning, the organisation culture and its way to interact with others in a context and with the changing dynamics is related to most of the others previously summarised, covering issues such as fostering trust (as this provided a space for negotiation), leaving space for 'error' (i.e. active learning), engendering and recognising the value and utility of informal (human-scale) interactions and adopting appropriate mindset and language (e.g. of respect, recognition, power, incentives and reward). These transdisciplinary challenges mean that for current 'single-issue' solutions necessary interdependencies with other relevant factors are need to be identified and implemented.

37.2.6 What Are the Dynamics of Structures in Societies and Organisations?

It is perceived that governmental (institutional) structures are inhibiting—what are the consequences? How do we go about creating nurturing institutions and organisations?

- Another item of extensive discussion related to the way that governmental (institutional) structures are inhibiting the human-scale activities of concern to practitioners. Many of the key factors that characterise inhibiting structures and prevent the development of nurturing organisations and institutions were identified at the Workshop. These include that institutional structures are too rigid and clamped, that the views of the world that form the basis for the work are too mechanical, and that many structures inhibit the human-scale interaction and activities within organisations but also the ones of concern especially to many frontline workers. It was felt that defendable alternatives to these inflexible, stifling forms were needed in practice. In reality, it is a big challenge to transition from a feeling of certainty obtained through planning and forecasting, to allow for learning, to encourage risk-taking and to move to a state where continual adaptation to a changing complexity is the norm.
- In the context of dynamics of structures, the participants also discussed the different types of institutional, organisational and social forms in general, how they related and interacted, how to morph them/transition between them and what their lifecycles were. In practice, the most important aspect was how formal structures related to the informal social forms such as community groups or volunteers—which were often (wrongly) dismissed as insignificant. The issues at stake are that on the one hand, there are particular structures in government, institutions and organisations that are assumed to be found in other groups as well—and that is not necessarily the case. On the other hand, community groups are fluctuating phenomena which do not have conventional types of underlying graspable structures at all. This asymmetry between the 'top-down' and the 'bottom-up' as they are often referred to in terms of relationships and organisational dynamics and management, is well illustrated by Conn [23] in her Social Eco-System Dance Model (Fig. 37.2), in which the '*space of possibilities*' between the horizontal relationships of the fluid informal network in communities and community groups and the vertical, hierarchical detailed regulation of other structures is characterised.
- Being able to demonstrate clearly and, even better, provide an alternative to the (potentially damaging) consequences of the mismatch of these kinds of structures, their inability to interact and interoperate, and the utility of different mechanisms to policy-makers and managers was seen as being important. Vergeer et al. [24] look at the consequences of aid mechanisms in health sector early recovery and note that in fragile states the health sector at these stages require both assistance in rebuilding and ongoing delivery of basic services, efforts which need to be coordinated accordingly.

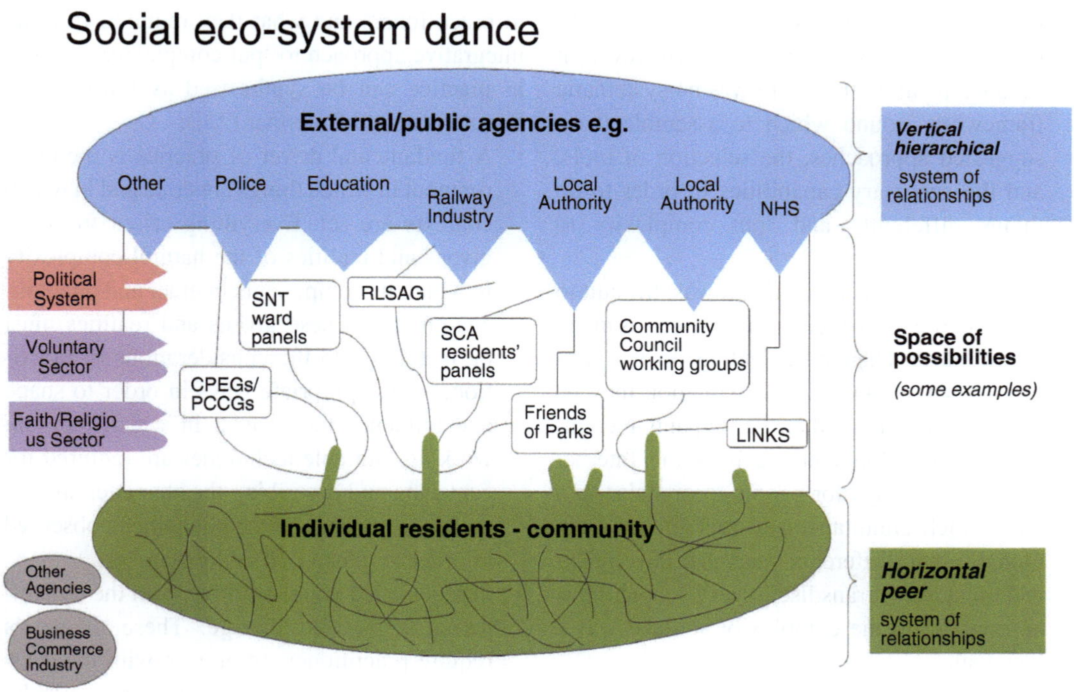

Fig. 37.2 The Social Eco-System Dance Model

37.2.7 Practice: How to Effect Self-Sustaining Change in Ways-of-Working?

How should we deal with complexity and change as a route to achieve desired outcomes? What is involved in putting things into Practice—what are the issues to address, tools and techniques required etc.? How can they be matched appropriately to Practitioners' tasks?

- At the heart of 'putting complexity to work' was how to deal with the complexity and change at work in one's own context. This included achieving a better understanding of how change came about, how to understand it, engage with it and influence it. It also meant to accept the dynamic, 'always-on' and ever-changing nature of the situation which practitioners are *'co-evolving'* with (i.e. a change cannot be made without affecting both the situation itself and one's own place in the changing events). Putting complexity to work also requires the understanding that phenomena, events, actors and objects *'self-organise'*—in other words, structures and patterns will form spontaneously (also in ways we cannot ever completely understand)—regardless of whether we intervene or not. The metaphor of going in a kayak on a fast-flowing river was found helpful as it expressed clearly the need to be able to influence on-the-fly—where there may not be time for extensive 'planning'. A related issue was the question previously raised of who was in the best position to achieve change—that it may not necessarily be the person formally responsible—and so enabling initiative and partnering was important. A request was made that Complexity Science could provide better understanding of the mechanisms underlying change and transition and help practitioners experiment with change through providing appropriate simulations that help them understand what their options are. These should not attempt to be absolutely predictive/ prescriptive models, but instead help inform practitioners' thinking.
- The participants, at various times and places in the Workshop, discussed 'Putting things

into Practice' in terms of approaches, techniques and tools required etc. However, it became apparent that there was no systematic framework around which to assemble these suggested approaches, the selection of tools, and the necessary capabilities in order to do things differently and 'put complexity to work'.

- In addition, it was noted that it was difficult to make hard-and-fast recommendations as many of the needs change depending on the context—what is 'right' in one situation may be inappropriate in another. Participants recognised that developments such as the Internet offered new ways for people to organise in a way which simultaneously recognised commonality and difference, and that developing and fostering transdisciplinary capabilities appropriate to the complexity at stake was a key goal.

37.3 'Operationalising' Complexity: Summary of Practitioners' Issues and Needs

These issues raised in the Workshop session clearly lead to the question of how to operationalise complexity? Which mindsets, approaches, tools and techniques might be useful to the various communities shown in Fig. 37.1? Some of the suitable approaches, changes and transformation challenges to address them were identified by participants, as well as: things which are common across the various practitioner communities, those factors which are crucial and/or controversial; bottle-necks (inhibitors) or, on the contrary, factors that offer novel opportunities (enablers) for working in complex real-world situations in practice. What had become obvious across the domains is the need for a comprehensive integrative, but not standardised, approach to bring about change in complex environments, which is flexible enough to be independent of the context in which a practitioner is working.

Based on the Workshop participants' contributions and on experiences from other practitioners, four complementary strands of issues of concern to practitioners and what they required from an integrative approach to 'put complexity to work' in practice can be synthesised as follows (see Beautement and Broenner [25]):

- A fundamental driver of practice is the environment in which things are done and in which contexts are set. Everything arises from the givens and realities of the natural complexity in which its components interact and phenomena emerge. These givens and realities must be factored into the considerations of practitioners in appropriate ways in order to shape and influence the context in which they are working. Suitable techniques are required for capturing and describing the characteristics of the environment and the phenomena observed in order to gain a rich understanding.
- Practice is all about working with the dynamics of real-world change. These dynamics require practitioners to engage with the ever-changing phenomena, when they are available, if self-sustaining change is to be achieved. It is accepted that such situations and dynamics are generally transient. In consequence that means that they can only be influenced by affecting the mechanisms that cause them to arise, by grasping fleeting opportunities and by trading off tensions that develop, all of which require a thorough understanding of the possibilities of how dynamics can evolve.
- In Practice, continuous reflection about the changing situation is required. An objective perspective on the realities of a context is provided by so-called *'contextual complexity'*, a term also used by Kurtz and Snowden [26].[2] The term *'experienced complexity'* in contrast refers to the way in which people describe the realities in a/the same context from a subjective view. This continuous consideration and comparison of contextual and experienced complexity and the factors surrounding the practicalities of influencing change in the specific context are essential

[2]Kurtz and Snowden also coined the term 'contextual complexity', as part of their Cynefin Framework. The meaning they attach to the term is complementary to the way used here.

and enable the practitioner to judge the appropriateness of options and interventions.

- Effective Practice depends upon and is augmented by capabilities which are appropriate to working with the ever-changing nature of complex situations. Specific types of appropriate, *'complexity-worthy'*[3] capabilities (such as mindsets employed, information available and accessible, level of skills competency, capability and capacity to collaborate etc.) need to be employed and matched dynamically to the context as required in order to explore and reflect on the various *'spaces of possibility'* and to engage effectively with the overall dynamics in a transdisciplinary manner.

37.4 Insights and Principles from Complexity Science Towards Solutions for Practice

Whilst Complexity Science has produced many important insights about complexity and the phenomena that arise in complex situations, summarised for example by Mitleton-Kelly [27], insights are not accessible without some sort of interpretation and synthesis into everyday concepts that practitioners need in order to understand how complexity 'works' in their specific context and how to engage with it successfully. Building on the practitioners' observations about complexity in the previous section, the following paragraphs examine, through the practitioners' lens, what Complexity Science can offer to practice in a form that is directly usable. The inverse, looking at practitioners and practice through the Complexity Science lens, would not provide material of immediate utility. Rather, it would repeat exercises many practitioners have gone through, which left them without pragmatic, accessible approaches for action. Therefore, in this section, insights are interpreted, analysed and synthesised into everyday concepts that, together with practitioners' needs, are the basis for a complexity-inspired technique for bringing about change which is presented at the end of this chapter.

Complexity Science has provided a basis for grasping the underlying complexity of situations in which practitioners work. Fundamentally, four 'axioms of complexity' can be derived that underpin the emergence of complex phenomena:

- A suitable environment exists in which the phenomena can arise and be sustained;
- There are components or entities in the environment—objects or agents, which have suitable properties that enable them to interact with each other at many levels time and scale-free in novel ways;
- Interactions of the components take place in the environment, and;
- Patterns are generated from the interactions which are persistent enough to be detectable as features—i.e. complex phenomena appearing at different times, at different scales and in different modes emerge.

Complexity Sciences describes a set of real-world givens and realities that surround these four axioms that are fundamental to practitioners' work. A list of these givens and realities, relevant to everyday situations (adapted from Beautement and Broenner [28]) follows:

37.4.1 Complexity Is Normal

Complexity is a normal, everyday phenomena arising from the natural world, and dealing with it is in principle an everyday skill of people, one that is powered by common sense, responsibility and initiative. This complexity offers opportunities which people can take and do take advantage of. Complexity, chaos and the randomness of things however are usually being talked about with a negative connotation. Such a mindset, unfortunately, precludes people from seeing the possibilities and opportunities for novelty and innovation arising and available in the world, a fact which is not helped by manufactured and contrived solutions in use that clamp down the emergence of such possibilities. As a consequence for practice there are, as Cohen and

[3] 'Complexity-worthy' capabilities—being fit for purpose and being able to cope with the unexpected, such as a sea-worthy ship in stormy seas.

Stewart explain [29], a lot of 'simplicities'—useful features and phenomena in the world—available 'for free', which health practitioners and members of other communities can look out for and be ready for to turn into advantage by putting the underlying complexity to work.

37.4.2 Change is Continual and Inevitable

Another given is that change comes about continually and inevitably—it occurs everywhere and everybody changes with the world via unavoidable *co-evolution* which affects both individuals and systems and is operational at all levels, scales or domains. It is complexity itself that generates this endless novelty spontaneously through *emergence* as Morowitz [30] clearly illustrates. Accepting this dynamic, the 'always-on' and ever-changing nature of the situation within which practitioners are co-evolving with means appreciating that people cannot make a change without affecting both the situation itself and their own place in the changing events. This is one of the reasons why practitioners need to iteratively reflect on the realities of their working context and on the conduct of practice to make space for considerations looking at what is occurring through co-evolution.

37.4.3 You Can't Turn Back Time

One of the realities is called the 'arrow of time' as Davies [31] describes it. It is hard to accept but provable fact that, even if one could turn the clock back and put everything 'as it was' and start again, the outcomes would be different every time because it is impossible to recreate the *initial conditions*, anywhere, exactly as they were. Following Prigogine [32], the arrow of time results in futures that are characterised by multiple possibilities. As a result, it has to be accepted for practice that because one can never be sure of the initial conditions and, because of non-linearity, many of these patterns, phenomena and behaviours will be unpredictable and hence are, by definition, 'unknowable' in advance with any certainty.

37.4.4 Causality Can Be Influenced

The focus on singular or primary causes is called 'direct causality' and is usually understood in terms of events unfolding along a direct chain of events (see Batten [33]), as if they already existed somewhere, with each effect inexorably determining the next, like a machine. Alternatively, so-called *'systemic causality'* thinks in terms of networks of criss-crossing causal relationships through which consequences propagate and interact—exhibiting emergent behaviours along the way. The reality is that there will always be aspects of evidence that link so-called causal elements together that will remain unknowable—a mystery whose causes are apparently invisible and undetectable. For practitioners, causality in relation to the emergence of patterns is both of interest and feasible (such as the spread of an epidemic), but determining causality concerning specific instances is not because it is not achievable in practice with any useful certainty.

37.4.5 There Are Limits to Prediction

Related to the causality issue is the predictability issue. Many different options being considered may result in similar outcomes and the inverse, that similar options may generate very different outcomes. This is because it is not possible to detect, observe or know all the things that are going on in reality and therefore exact prediction of individual outcomes and their effects is not possible. However, as already indicated, general predictions can be made about dynamic patterns and emergent phenomena arising from the collective behaviour of groups. For example, Reynolds [34] discovered the simple basis underlying complex behaviour, for instance flocking and swarming, which is also relevant to predicting conduct such as the spread of public reactions over health and of disease itself.

37.4.6 There Is No Single Objective Worldview

Complexity Science tells us that capturing a full representation of the factors that generate specific complex phenomena, which is close to being reliable or accurate, is almost impossible as Gell-Mann [35] shows. This is because most of the information required is unknowable or unobservable by humans or devices or too transient in practical terms. So, no single objective world-view exists which can be used to form an all-encompassing model in these dynamic situations. Modelling is limited by the assumptions it has to make about the world and, as Hofstadter [36] demonstrates, is necessarily and provably incomplete in real situations. Therefore it should be employed with caution and the results produced viewed critically and with common sense.

37.4.7 Effects Are Non-linear Phenomena

Complexity Science describes non-linear systems as those with a *lack of proportionality* between cause and effect, between 'input' and 'output'—where small disturbances can lead to large consequences that are very sensitive to the initial conditions. Heylighen [37] furthermore states that when the effects are larger than the causes, there is positive feedback in that initially small perturbations reinforce themselves and become more and more intense. This has been captured in the public imagination by the so-called *'Butterfly Effect'*, where the flap of a butterfly's wings on one side of the world can, in theory, lead to a hurricane on the other (see Lorenz [38]). The concept of non-linearity has practical relevance, because it means that the steps on the route between stimulus and desired change will jump in unpredictable ways across levels and scales and will propagate by a variety of means, in other words: there is no guarantee of there being a 'fully-connected graph'. Evidence from Complexity Science, neatly summarised by Albert Einstein who said *As far as laws refer to reality they are not certain and as far as they are certain they do not refer to reality* indicates that a target-based mindset based on absolute outcomes is usually flawed, because it is predicated on an unrealistic need for the world to be stable and predictable—which is in clear contradiction with the very reality indicated by practitioners earlier in this chapter.

37.4.8 Repertoires Must Be Appropriate to Connectedness

Complexity Science states that the interaction between components generates the networks of connectedness of the real world and that these interactions can be influenced to change outcomes—but only if the interventions can be matched to the structure of the network of interactions. This assertion is based on a key principle, that Complexity Scientists call the *law of requisite variety*, originally a cybernetics principle formulated into a law by Ashby [39]. It states, in effect, that to engage with or affect something, abilities must match with the nature of things being affected. The more options, i.e. the bigger the wiggle-room that can be generated, the better a system is able to respond to or bring about change. In practice, this means that flexible systems which generate a large range of diverse options are better able to generate the requisite variety and cope with or influence change in their environment than those which are tightly optimised around a set of initial conditions (which would inevitably fail if these conditions changed). Here, the capabilities available in the practitioners' world, which earlier on were characterised as 'complexity-worthiness', come into play as their degree of appropriateness determines the range of outcomes of interventions that are feasible.

37.4.9 Effects Propagate Across Scales

An important related aspect revealed by Complexity Science is the way in which influences and changes can propagate and cascade, via many intermediate levels and mechanisms, sometimes in a way that at first-sight might seem counter-intuitive. Barabasi [40] and Watts [41] describe how this 'surprising' propagation of effects that can occur because natural networks are *'scale-free'* and the heterogeneous nodes that make up these networks are interconnected in a manner that facilitates rapid cascades. The nodes that are highly connected with other nodes are called hubs. They play an important role and have disproportionate influence on the whole network. Scale-free networks are very resilient—the whole network will not necessarily be damaged or disturbed by a single intervention. However, interventions can quickly propagate across the network via the highly interconnected nodes to parts of the network far from, for example, the initial well-meaning intervention. Therefore, a practical insight is that effects may continue to reverberate in the environment of their own accord long after the original trigger ceased to exist.

Grasping and understanding these givens and realities described above is essential for practitioners. The insights and analysis provided informs what is appropriate for the actual 'doing' in daily practice in any context in complex situations. The following section describes an approach for operationalising the insights in a pragmatic manner that puts complexity to work.

37.5 Putting Complexity to Work: A Landscape of Change

The preceding sections in this chapter presented practitioners' issues and needs when working with complex situations, and looked at relevant insights from Complexity Science for use in practice. By working, appropriately, with the underlying causes of complex phenomena described above, there are plenty of degrees of freedom and options available for selectively shaping and influencing change in the real world pragmatically and thus for putting complexity to work. The insights from Complexity Science indicate that, for bringing about change, it is necessary that practitioners appreciate and reflect on the realities of and within their ever-changing context in a way that enables them to iteratively gather, express and explore the features present and the kinds of phenomena they are facing. Practitioners also have to reason and reflect about practice to be able to set the conditions for 'doing' by examining mismatches and tensions which fundamentally drive the dynamics of practice and change. In addition, practice depends on their (individual, organisation, community) fitness for effectively and appropriately bringing about self-sustaining change in their working context.

To help understand how appropriateness might be judged, a complexity-inspired 'Landscape of Change' (Fig. 37.3) has been developed (adapted from Beautement and Broenner [42]). It provides a structured way for practitioners to identify, given certain opportunities, what may be driving change in a context, and hence indicates ways to engage with and take advantage of dynamic situations in a purposeful manner. The Landscape's design addresses practitioners' concerns by offering the possibility to identify and map ways of achieving change dynamically in and for a specific context. The design principles of the Landscape of Change are based on a synthesis of the insights previously discussed. Basically, practitioners need to appreciate two things: how change comes about and how it can be influenced. Therefore, the dimensions of the Landscape are as follows:

- The drivers of change, which are based on the four axioms of complexity—environment, the components, their properties, behaviour and interactions—are shown along the top-left axis of Fig. 37.3.
- The ways in which change can be brought about and influenced (discussed in the following paragraph) are along the right-hand axis.

The ways in which change can be brought about, previously mentioned, are the topdown or bottom-up ones which Mintzberg et al. [43] characterise briefly in the light of effecting successful transformations. In many cases, the necessity to employ both ways of change is being promoted when proposing solutions, as for instance Ebi et al. [44] do in their work on

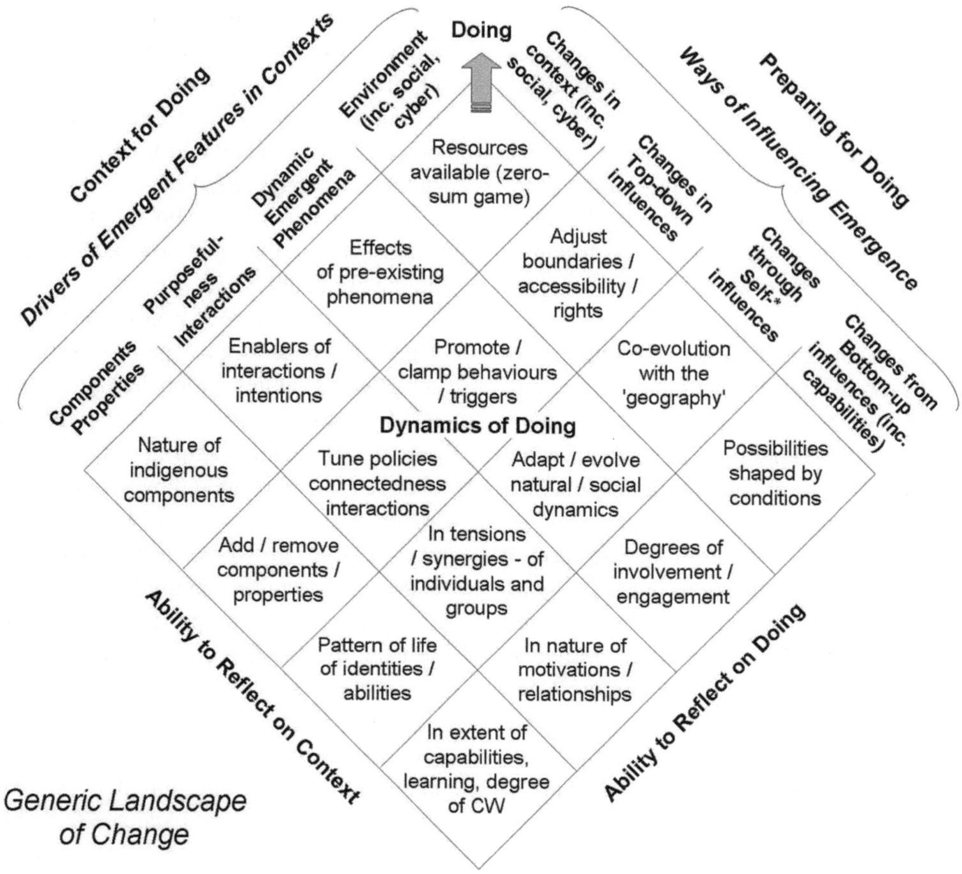

Fig. 37.3 The landscape of change

community-based adaptation to health impacts posed by climate change. Other change theories, especially after the increasing recognition of the insights that Complexity Science provides took hold, promote the 'self-' changes in the broadest definition (for example self-organising, self-regulatory etc.) which occur in the kind of natural 'complex adaptive reflexive systems' explained by Mathieson [45] . In complex situations, these three ways, top-down, bottom-up and 'self-' in dynamic interaction—along with changes to the environment in which the context is set itself—can be the avenues of change which form the right-hand axis in the Landscape of Change which are examined in more detail below:

- Making Changes in the Context/Environment. These cells in the Landscape represent the environment dimensions of change where practitioners would make adjustments to, for example, the physical environment itself. This might be done by: altering the physical geography, through draining malaria prone swamps; or changing the nature of the 'background' phenomena that are active in the environment, such as air pollution levels causing health impacts. Practitioners may need to consider the degree to which the environment is enabling or inhibiting interactions and intentions and whether conditions are stable enough over time to achieve a specific aim, for example, the adaptation of vulnerable communities to climate change. Changes can also be brought about through altering indigenous components in the environment (an example would be the location of health centres in a region), and through 'seeding' the space, so preferred phenomena are more likely to come about. An example would be increasing the density of hand-disinfection dispensers in hospitals to encourage and improve hygiene by visitors and staff. Depending on the time-horizon, greater or lesser efforts may be made to change the environment so that it becomes more suitable to the tasks at hand.

- Changing the Nature of Top-down Influences. This row in the Landscape of Change represents the top-down influences that are active in a context, such as incentives, policies, interventions, rules and permissions. Examples are legislation on stem-cell research, medical interventions by governments following a natural disaster, health policies or disease specific actions and programmes (e.g. regarding polio eradication). It also includes the ways in which accessibility rights or '*boundaries*' might be changed, for instance with geographical and organisational boundaries in health service provision, and how behaviours might be encouraged or clamped (e.g. by offering or withdrawing rewards). Top-down influences can encourage connectedness and facilitate interactions, for example through introducing federated—necessarily different yet interoperable—Health Management Information Systems. They can also add or remove components or change their properties (such as suggesting that actors with certain capabilities should be included in community-based care-taking of elderly persons).
- Changes through Self-organisation/Self-regulation. Changes in this part of the Landscape of Change come about on their own accord as a result of the ongoing interactions in the context or of co-evolution with the environment. There is a threshold where change occurs when the underlying conditions for 'doing' something in the situation achieve a favourable critical mass, the so-called *tipping point*. This point can be stimulated for example by influencing social drivers (manifested in behavioural norms, ethos etc.), or it might arise when synergies or tensions between groups and/or individuals change the dynamics in the context—such as the effects of changes in peer-pressure on young women taking up smoking. The Landscape can help practitioners identify and track the indicators of those underlying factors—though it is only the people themselves who will make the transition—and so the timeline of occurrence cannot be predicted.
- Changing the Nature of Bottom-up Influences. This row in the Landscape of Change represents the dynamics which are driven bottom-up by individuals or by the underlying complexities that impact change. These changes can be triggered by people, individuals or in any organisational form, who think/act locally, but whose action and thinking cause effects with potentially broad impact. Here, the innate properties of components and the availability of appropriate complexity-worthy capabilities are a fundamental driver. Examples of bottom-up influences are workshops for health service and residential care staff or programs initiated by the staff themselves to improve local access to health-care. Certain phenomena that occur from bottom-up are also enablers for change at higher levels, and manifest themselves in cascades of emergent phenomena. Examples here include training and health education of women at local level in developing countries with little health service provisions which leads to the improvement of child health in entire regions.
- A fifth option to bring about change, though not represented in the Landscape, is the 'doing nothing' one. The option can be suitable in all the cells and across the Landscape of Change, when things just follow their own course without active intervention. Such behaviour can be triggered by the idea of opening up a new space of possibilities, but also when there is a trade-off between what practitioners would like to do with what is likely to be possible, given the circumstances, and what is viable given the capabilities and the means available to them.

Summary: Successful Operationalisation of Complexity in the Daily Practice of Practitioners in Health Systems

In order to engage effectively with the on-going dynamics in a situation and to bring about change practitioners need to gain an appropriate appreciation of the underlying complexities which are driving the phenomena and features active in a context.

This chapter has provided a translation of relevant principles from Complexity Science into an integrative technique, the Landscape of Change, with the help of which practitioners who are working in dynamic, complex situations can do the following:

Engage with the dynamics and complexity of the situation as part of practice.

Understand and continuously reflect on their own context.

Trade off tensions that arise in these situations from the underlying complexity.

Consider the possibilities to bring about change by employing and adapting suitable complexity-worthy capabilities.

The Landscape of Change enables practitioners, engaged in endeavours of dynamic health systems, to systematically judge and determine appropriate ways of working required for putting the complexity at hand to work effectively.

Acknowledgments The authors would like to thank the ASSYST Program which co-sponsored the ECCS Satellite Workshop 'Putting Complexity to Work—Supporting the Practitioners'. We would also like to thank all participants of the workshop for their contributions.

References

1. Matlow AG, Wright JG, Zimmerman B, Valente M, Thomson K. How can the principles of complexity science be applied to improve the coordination of care for complex pediatric patients? Qual Saf Health Care. 2006;15(2):85–8.
2. Leykum LK, Pugh J, Lawrence V, Parchman M, Noël PH, Cornell J, McDaniel RR. Organizational interventions employing principles of complexity science have improved outcomes for patients with Type II diabetes. Implement Sci. 2007;2:28.
3. Browne AJ, Varcoe C. Critical cultural perspectives and health care involving aboriginal peoples. Contemp Nurse. 2006;22(2):155–67.
4. Singhal A. Which nursing home would you put your mother IN [Internet]. Plexus Institute, Complexity Series; 2007. Available from http://www.plexusinstitute.org/resource/resmgr/docs/which-nursing-home-mcdaniel.pdf.
5. Brown J, Isaacs D. The World Café. Shaping our Futures through Conversations that matter. San Francisco: Berrett-Koehler Publishers; 2005.
6. Schön DA. The reflective practitioner. How professionals think in action. New York: Basic Books; 1983.
7. Jayasinghe S. Conceptualising population health: from mechanistic thinking to complexity science. Emerg Themes Epidemiol. 2011;8:2.
8. Eppel E, Matheson A, Walton M. Applying complexity theory to New Zealand public policy. Principles for practice. Policy Q. 2011;7(1):48–55.
9. Booth BJ, Zwar N, Harris M. A complexity perspective on health care improvement and reform in general practice and primary health care. Aust J Prim Health. 2010;16(1):29–35.
10. Plsek P. Complexity and the adoption of innovation in health care. accelerating quality improvement in health care [Internet]. Strategies to Speed the Diffusion of Evidence-Based Innovations Conference, Washington, DC. 27–28 Jan 2003. Available from http://www.nihcm.org/pdf/Plsek.pdf.
11. Griffiths F. Complexity science and its relevance for primary health care research. Ann Fam Med. 2007;5:377–8.
12. Capra F. The web of life. London: Harper Collins; 1996.
13. Beautement P, Broenner C. Complexity Demystified, – a guide for practitioners. Axminster: Triarchy Press; 2011. p. 17.
14. Armson R. Growing wings on the way: systems thinking for messy situations. Axminster: Triarchy Press; 2011.
15. Musgrove P. Indicators: how much complexity is desirable? Rev Bras Epidemiol. 2002;5 suppl 1:73–82.
16. Martens P, Thomas C. Climate change and malaria risk: complexity and scaling, chapter 2. In: Takken W, Martens P, Bogers RJ, editors. Environmental change and malaria risk: global and local implications. Wageningen: Wageningen UR Frontis Series; 2005. p. 3–14.
17. O'Daniel M, Rosenstein AH. Professional communication and team collaboration, chapter 33. In: Hughes RG, editor. Patient safety and quality. an evidence-based handbook for nurses. Rockville, MD: Agency for Healthcare Research and Quality (US); 2008.

18. McMurty A. Reinterpreting interdisciplinary health teams from a complexity science perspective. Univ Alberta Health Sci J. 2007;4(1):33–42.
19. Chambers R. Revolutions in development inquiry. London: Earthscan; 2008.
20. Nicolescu B. Manifesto of transdisciplinarity. Albany: State University of New York Press; 2002.
21. Max-Neef MA. Foundations of transdisciplinarity. Ecol Econ. 2005;53:5–16.
22. Ackermann G. Evaluation of complex interventions in social systems [Internet]. IUHPE World Conference in Geneva, 14 Jul 2010. Available from http://iuhpe.gesundheitsfoerderung.ch/downloads/en/Programme/PDFs/Wild_Side/Ackermann—2010-Complexity-and-evaluation-def.pdf.
23. Conn E. Community engagement. A social eco-system dance. [Internet]. European Conference on Complex Systems 2009. Available from http://www.abaci.net/library/eccs03_eileen-conn.pdf.
24. Vergeer P, Canavan A, Rothmann IA. Rethink on the Use of Aid Mechanisms in Health Sector Early Recovery [Internet]. Amsterdam: KIT Development Policy and Practice; 2009. Available from http://www.gsdrc.org/go/display&type=Document&id=3258.
25. Beautement P, Broenner C. Complexity Demystified – a guide for practitioners. Axminster: Triarchy Press; 2011. p. 47.
26. Kurtz CF, Snowden D. The new dynamics of strategy: sense-making in a complex world. IBM Syst J. 2003;42(3):462–83.
27. Mitleton-Kelly E. Ten principles of complexity and enabling infrastructures. In: Complex systems and evolutionary perspectives of organisations: the application of complexity theory to organisations. London: Pergamon; 2003. p. 3–20.
28. Beautement P, Broenner C. Complexity Demystified – a guide for practitioners. Axminster: Triarchy Press; 2011. p. 58.
29. Cohen J, Stewart I. The collapse of Chaos. Discovering simplicity in a complex world. London: Penguin Books; 2000.
30. Morowitz HJ. The emergence of everything: how the world became complex. Oxford: Oxford University Press; 2002.
31. Davies P. The cosmic blueprint. Order and complexity at the edge of chaos. London: Penguin; 1995.
32. Prigogine I. The end of certainty. New York: The Free Press; 1997.
33. Batten DF. Changing our brains: systemic causality in complex human systems. Proceedings of the 18th conference on behavior representation in modeling and simulation, Sundance, UT, 31 March–2 April 2009; p. 9–14.
34. Reynolds CW. Flocks, herds, and schools: a distributed behavioral model. Comput Graph. 1987;21(4): 25–34.
35. Gell-Mann M. Fundamental sources of unpredictability. Santa Fe: Santa Fe Institute. Conference on fundamental sources of unpredictability. Mar 1996; 10.1.1.7.1.: 7–8.
36. Hofstadter DR. Gödel, Escher, Bach: an Eternal Golden Braid. London: Penguin Books; 1983.
37. Heylighen F. Complexity and self-organization. In: Bates MJ, Maack MN, editors. Encyclopedia of library and information sciences. New York: Taylor & Francis; 2008.
38. Lorenz EN. Deterministic nonperiodic flow. J Atmos Sci. 1963;20(2):130–41.
39. Ashby WR. An introduction to cybernetics. London: Chapman & Hall; 1956.
40. Barabasi AL. Linked: how everything is connected to everything else and what it means. London: Plume; 2003.
41. Watts DJ. Six degrees: the science of a connected age. New York: W.W. Norton & Company; 2003.
42. Beautement P, Broenner C. Complexity Demystified – a guide for practitioners. Axminster: Triarchy Press; 2011. p. 134.
43. Mintzberg H, Ahlstrand B, Lampel J. Strategy Safari. Harlow: Pearson Education; 2009.
44. Ebi KL, Semenza JC. Community-based adaptation to the health impacts of climate change. Am J Prev Med. 2008;35(5):501–7.
45. Mathieson G. Complexity and managing to survive it [Internet]. DSTL; 2005. Available from http://ismor.cds.cranfield.ac.uk/collected-papers-of-graham-mathieson/complexity-and-managing-to-survive-it-playing-with-cars/@@download/paper/cars.pdf.

Implementing Adaptive Health Practice: A Complexity-Based Philosophy of Health Care

38

N. Marcus Thygeson

38.1 Introduction

This chapter describes an approach to health care derived from the application of Dr. Ronald Heifetz' Adaptive Leadership model to the practice of medicine and other health professions. A prior publication introduced the theory and evidence supporting the potential value of this model, which I now call Adaptive Health Practice [1]. Here I further elaborate the concept and propose specific, concrete systems changes and behaviors that could be combined in such a way as to pilot this model in actual practice. The objective of this chapter is to describe Adaptive Health Practice (AHP) in sufficient detail that interested healthcare professionals can successfully pilot these concepts in real practice settings.

The goals of AHP are to help patients be healthy, resilient, and maximally autonomous producers of their own health; to increase productivity and reduce community sick-care costs, due to both reduced morbidity and more efficient return to health; and to provide a sustainable, satisfying professional life for providers (less burn out, futility, and sense of ineffectiveness).

Implementing AHP promises to promote the triple aim [2] by addressing unhealthy behavior and unwarranted variation in preference- and supply-sensitive care, and by changing the "simple rules" governing the healthcare system from illness- to health-focused, provider-centered to patient-centered, and reductionist, technical medicine to holistic, adaptive medicine. AHP can also improve organizational performance by teaching providers adaptive leadership skills and by giving them a new set of tools for coping with previously frustrating and demoralizing patient behaviors.

The chapter starts with a summary of the healthcare system problems that motivated the development of the AHP model. This is followed by a brief description of Adaptive Leadership, how it addresses the motivating problems, and how it applies to healthcare practice. I then suggest concrete changes to current practice routines that would be necessary or beneficial to implementing AHP in the actual care of patients. These changes are organized by the conventional "SOAP note" categories of the clinical encounter: subjective, objective, assessment, and plan. I also discuss organizational or policy levels changes that will be necessary or supportive for implementing AHP.

To be explicit, the suggestions in this chapter are often speculative. Where there is evidence to support a suggestion I have cited it. However, many of the ideas in this chapter, while promising and solidly based (in my view) on good theory, have not been shown to "work," and need to be tested with appropriate rigorous evaluation methods that are context-sensitive [3, 4].

N.M. Thygeson, M.D. (✉)
Blue Shield of California,
50 Beale Street, San Francisco, CA 94105, USA
e-mail: mthygeson@gmail.com

J.P. Sturmberg and C.M. Martin (eds.), *Handbook of Systems and Complexity in Health*,
DOI 10.1007/978-1-4614-4998-0_38, © Springer Science+Business Media New York 2013

38.2 Motivation

Modifiable health status is mostly determined by environmental, social, and behavioral factors, but rather than effectively addressing these root causes of poor health, we focus almost all of our healthcare resources on using medical technology to treat illness [5–7]. The leading causes of death in the United States are all related to behavior issues (tobacco use, overeating, insufficient physical activity, inappropriate use of alcohol and psychoactive drugs, and risky behavior) [8]. However, our efforts to help our patients change their behavior have been relatively ineffective. Frustration with this lack of efficacy has engendered a sense of futility about behavior change in many providers [9–11]. This further reinforces our tendency to focus on technical interventions, like medications and surgery, to treat illness.

Unfortunately, medical technology is often a "double-edged sword," in that it has the potential to both heal and harm patients. Frequently that harm presents in a most ironic way—the treatment actually aggravates the condition it was intended to treat. Examples of this are common in health care. For instance, we are all familiar with the fact that narcotics induce addiction and chronic pain problems. In the past decade the United States experienced an epidemic of opioid use after long-acting powerful oral narcotic analgesics entered the market [12]. The pain medication makes the pain problem worse, a classic example of the system dynamics archetype "fixes that fail" [13]. Such observations are not new. It was the famous French physician Armand Trousseau who observed:

> You should treat as many patients as possible with new drugs while they still have the power to cure [14].

Likewise, proton pump inhibitors (PPIs—heartburn medicines) induce heartburn in normal people and perpetuate heartburn symptoms in patients with gastroesophageal reflux [15]. The prevalence of gastroesophageal reflux disease (GERD) has doubled in the United States since PPIs were introduced [16].

Antidepressants sometimes induce suicidality, frequently lead to tachyphylaxis, and often produce a withdrawal syndrome with symptoms of depression that mimic the symptoms that led to the drugs being prescribed in the first place [17, 18]. Antidepressant treatment may be superior to placebo only in patients with very severe baseline symptoms of depression, and no better for patients with mild-to-moderate depression, the vast majority of patients currently treated [19]. Interpersonal and cognitive behavioral therapy, in contrast, may produce persistent benefit, and reduce or prevent recurrence of depression [20].

Long-acting beta agonist asthma medications increase deaths from asthma [21]; spine fusion surgery for back pain increases the risk of more spine surgery, chronic low back pain, and disability [22, 23]; percutaneous coronary interventions are no more effective at preventing heart attacks or cardiac death than medications [24]; intracoronary stents, if not accompanied by continuous treatment with antiplatelet agents (which are associated with bleeding risks), clot off and cause heart attacks; and aggressive treatment of incurable metastatic non-small-cell lung cancer shortens and reduces the quality of life, compared to early palliative care [25].

For each of the conditions mentioned above, the patient's behavior, social system, and environment play important roles in the genesis or management of their illness. Non-pharmacological techniques for managing pain; changing diet and losing weight for GERD; removal of allergens for asthma; and physical rehabilitation for back pain—these are all examples of effective non-technical approaches to helping patients with these conditions. Unlike medical technology, these types of changes generally do not have negative health consequences, and often they improve other health conditions. Thus, helping a patient make the environmental, social, and behavioral changes required to quit smoking, increase physical activity, eat a healthy diet, lose weight, and develop stress-management skills will treat not only their back pain, but will also treat or reduce their risk for all the "plagues" of modern life: diabetes, hypertension, coronary artery disease, arthritis, GERD, depression, and cancer.

Our relative inability to effect changes in the environmental, social, and behavioral determinants of health, and our consequent "overreliance" on technical interventions to manage illness, produces large amounts of avoidable waste and harm. Waste, because instead of helping our patients make the changes they need to be healthy, we are prescribing and administering treatments to them that would often be unnecessary if we were more effective at facilitating behavior change, and because those treatments often induce the demand for further technical interventions. For example, medications often require follow-up lab testing to monitor drug levels and possible side effects. Harm, because those treatments too frequently aggravate the underlying condition, or produce other kinds of morbidity (e.g., side effects of medications, or early and late complications of surgery).

Imagine how health care would be different if we could partner with our patients to effectively address the environmental, social, and behavioral determinants of their (and our) health.

Why aren't we more effective at supporting patient behavior change?[1] For one thing, behavior change, and changing social and environmental systems, is hard work. It requires the patient to make significant changes in the people, places, beliefs, and behaviors to which they are attached. It requires them to learn new things, and give up old, familiar, comfortable things. Helping people make these kinds of changes is also hard work. It requires patience, skill, and lots of time. The working conditions in most medical practices do not provide the time or systems required to support patients in doing the work to be healthy [26]. Both patients and health care providers avoid the work, effort, and time required to effect these kinds of changes.

Behavior is strongly influenced by social factors, like our work, our family, and our social network. However, the evidence base regarding effective social interventions is limited [27, 28]. This reinforces our tendency to neglect these opportunities, and further undermines our ability to help patients adopt healthier lifestyles.

Also, health care workers, for the most part, are not trained to do this kind of work. Physicians are educated and trained about the technical aspects of medicine, but most get little training on behavior change, and even less about managing the social and environmental determinants of health. Public health and medicine are separate disciplines, both professionally and pedagogically. Nurses, family medicine physicians, mental health professionals, and social workers are trained to a greater degree on these issues, but even in those professions the dominant culture and business models in health care reinforce the use of technical interventions and discourage the systematic application of practices required to address the environmental, social, and behavioral determinants of health.

Perhaps our most effective, and in most cases our only, social "intervention," the provider–patient relationship, is systematically undervalued in the fee-for-service system that prevails in the United States. Current health-care financial models pay for procedures and relative value units, not talking to patients. Profits for both providers and manufacturers come from efficiently delivering the maximum number and intensity of technical interventions to patients. A recent article in the New York Times, entitled "Talk Does Not Pay, So Psychiatry Turns to Drug Therapy," illustrates this story vividly [29]. Only 10 % of psychiatrists in the United States still provide psychotherapy. The rest now spend their time with patients in brief medication adjustment visits. "I had to train myself not to get too interested in their problems, and not to get sidetracked trying to be a semi-therapist," one psychiatrist is quoted.

Fundamentally, our preference for technical interventions, and our lack of efficacy in facilitating patient behavior change, arises from an underlying failure to apply the principles of complexity science to healthcare practice.

[1] Rick Botelho provides further suggestions in his chapter.

Table 38.1 Technical versus adaptive challenges

Technical challenges	Adaptive challenges
• Simple or complicated problems	• Complex problems
• An expert somewhere already knows the solution (a puzzle)	• Solution is unknown and must be discovered (a mystery)
• A technical intervention exists or can be constructed to solve the problem	• No expert or technical intervention can solve the problem
• Solution does not require material learning and behavior change by the person(s) experiencing the problem	• Solution requires learning and behavior change by the person(s) experiencing the problem
• The challenge is addressed by identifying and applying the expertise and technical interventions required to resolve the problem (technical work)	• The person(s) with the challenge must work (overcome resistance) to discover (learn) and adopt the new beliefs, attitudes, and behaviors required to resolve the challenge (adaptive work)

38.3 Adaptive Health Practice: A Complex Systems Approach to Health Care

Adaptive Health Practice is a theory of health care resulting from the application of Adaptive Leadership to health care practice. Adaptive Leadership (AL) is an approach to organizational leadership that was introduced in the 1990s by Dr. Ronald Heifetz, a psychiatrist who teaches at Harvard's Kennedy School of Government. As suggested by Dr. Heifetz' first profession, AL is informed by the discipline of behavioral health practice, and in fact one of Heifetz's first published descriptions of AL was of a physician helping a patient and family cope with the patient's terminal illness ([30], p. 73 et seq.) A well-developed and practical exposition of AL as applied to organizational leadership is now available [31].

Adaptive leadership applies to both organizations and human beings because both are complex adaptive systems. As discussed elsewhere in this book,[2] complex adaptive systems have a variety of characteristics, including high levels of interconnectedness, feedback loops, nonlinear behavior, sensitive dependence on initial conditions, emergence, fractal structures, power law dynamics, and learning (the ability to adopt new attitudes, beliefs, and behaviors, and to shed old ones).

The theory of AL starts with the recognition that complex adaptive systems face two kinds of challenges, adaptive and technical (See Table 38.1). Technical challenges are typically simple or complicated problems that can be identified and specified with considerable accuracy, and are amenable to solution by experts using methods and technologies that do not require major changes in the system itself. In medicine, ocular cataract is a technical challenge which is amenable to a technical solution—replacement with a prosthetic intraocular lens.

Adaptive challenges, on the other hand, are complex problems that are often hard to identify or specify completely. Addressing adaptive challenges requires that the system (e.g., the patient) learn new attitudes, beliefs, and behaviors. As such, adaptive challenges cannot be resolved by outside experts or technical interventions. Only the system itself (the patient, or the members of an organization) can do the adaptive work required to overcome adaptive challenges. Technical interventions are top-down; adaptive work is bottom-up. Growing up, coping with death and dying, learning to manage a complex medical regimen, adopting new lifestyle behaviors, and deciding between two or more approximately equivalent treatment options with different side-effect profiles are all examples of health-related adaptive challenges. The work of overcoming these challenges must be done by patients. Using the methods of adaptive leadership, clinicians can facilitate patients' adaptive work, but they cannot do it for them.

Adaptive work, as the word "work" implies, requires effort and is difficult. Behavior change, whether by people or organizations, involves giving up old attitudes, beliefs, and behaviors and adopting new ones. This process invariably produces feelings of intrapsychic conflict and fear and

[2] See Chapter 3.

loss for those involved, and thus adaptive work is generally resisted or avoided. The practice of adaptive leadership is a set of behaviors that individuals (e.g., clinicians) can use to help complex adaptive systems (patients) overcome work avoidance and do the adaptive work required to overcome adaptive challenges. Note that adaptive leadership is not contingent on authority. All healthcare providers, and indeed laypersons including other patients, can exercise adaptive heath leadership.

A fundamental principle of AL is that you cannot solve adaptive challenges with technical interventions. However, because technical interventions often appear to be relatively easy and straightforward, it is quite common for both clinicians and organizational leaders to use technical interventions to address adaptive challenges. Substituting technical interventions for adaptive work is ineffective, wasteful, and potentially harmful. Examples of this phenomenon are legion, as described in detail in the "Motivation" section above, and readily evident to all thoughtful clinicians.

Technical interventions can and should be used to facilitate patient adaptive work. Examples of this include using nonaddictive pain medications to relieve back pain so patients can do physical therapy; video games that promote physical exercise (so-called "exer-gaming"); and use of light therapy, antidepressant medications, or electroconvulsive therapy to help depressed patients get to the point where they can do the adaptive work (psychotherapy, regular exercise, coping with stressors) that is required to achieve and maintain remission.

Also, many health problems are a combination of adaptive and technical challenges. While replacing an ocular cataract is a technical challenge, deciding when to do it and what artificial lens to use are adaptive challenges. Indeed, almost all medical technical interventions involve patient adaptive work including making preference-sensitive decisions in the face of unavoidable ambivalence; coping with pain, suffering, and loss of function; learning self-management skills; and dealing effectively with the healthcare financing and delivery system.

The technique of AL starts with being mindful of the distinctions between adaptive and technical challenges and work, and the importance of not substituting technical interventions for adaptive work. The fundamental purpose of adaptive leadership is facilitating adaptive work, and this requires addressing work avoidance. This in turn involves creating a holding environment (a close, therapeutic relationship), and regulating the "heat" so that the person or persons doing the adaptive work are motivated to do it, but not so overwhelmed by fear, confusion, or exhaustion that they are unable to engage in it. Successful AL requires the ability to constructively manage the conflict that usually arises when dealing with work avoidance. Finally, in addition to helping people overcome work avoidance, AL involves finding ways to facilitate them doing the adaptive work of learning the new attitudes, beliefs, and behaviors required to address their adaptive challenges. These concepts are illustrated graphically in Fig. 38.1.

Adaptive work gets done when the "tension" is maintained in a productive zone. The term "tension" can refer to interpersonal conflict (i.e., between provider and patient, or patient and others in their social network) or intrapersonal conflict within the patient. If there is too much tension, the patient will feel overwhelmed or exhausted and will avoid the work. Work avoidance also occurs if there is not enough conflict. The art of adaptive health practice involves the clinician maintaining the level of tension in the productive range, so that the patient remains engaged in doing the work.[3]

Work avoidance looks the same, whether it is due to too much tension, or not enough, However, the two situations require different responses from the provider. If the patient is overwhelmed, the provider needs to "lower the heat" by doing some combination of the following: (1) setting fewer or less ambitious goals with the patient, (2) bolstering their adaptive capacity, or (3) strengthening the therapeutic relationship. On the other

[3]Here and elsewhere in this chapter, I use language that focuses on the clinician as the agent managing the clinical encounter. I do this because it conforms to the current dominant care delivery mental model and practice norm (attractor), and therefore will be most readily understood by the intended reader. However, this is a simplification of the true circumstances. Both the clinician and the patient have the opportunity to be aware of the levels of tension related to doing the adaptive work, and either can take action to "change the conversation" so as to address work avoidance.

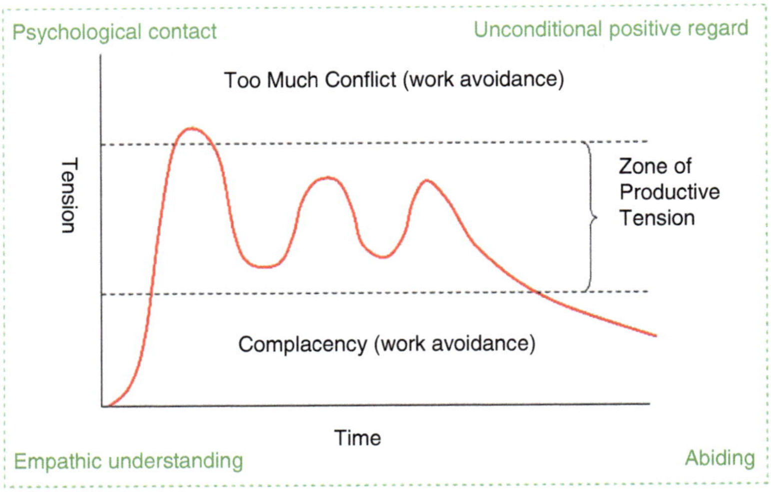

Fig. 38.1 A therapeutic relationship characterized by psychological contact, unconditional positive regard, empathic understanding, and abiding enables a clinician to safely maintain the tension in a productive range that enables a patient to engage in the adaptive work required to be as healthy as possible (Adapted from Heifetz et al., John Scott, and Carl Rogers [30, 65, 77])

hand, if the patient is complacent, the provider needs to "raise the heat" to make the patient more uncomfortable with the status quo.

Therefore, a key element of AHP is learning how to differentiate these two states. This can be done using an iterative process of *observe–interpret–intervene* ([31], p. 32). The provider observes the patient's behavior in the context of the relationship history, and makes an initial interpretation or hypothesis as to whether the patient's work avoidance is due to being overwhelmed or complacent. Then, based on this assessment, the provider takes an action intended to move the tension into the productive range while monitoring the patient's behavior for evidence that may confirm or disprove the initial interpretation. Based on this further cycle of observation and interpretation, the provider then makes another intervention. In this iterated way the provider adjusts the level of tension so as to support the patient in doing their adaptive work and build adaptive capacity while continuing to strengthen the therapeutic relationship. In essence, the provider is using an exploratory sense-making process to determine the state of the patient's system and the appropriate therapeutic response.

At this point, a few words about clinician–patient conflict management are in order. Conflict is not in general an explicit part of the typical clinical encounter, and we usually try hard to minimize both implicit and explicit conflict between ourselves and our patients. Also, the power differential between providers and patients tends to dampen explicit conflict. Moreover, most clinicians see their patients relatively infrequently. Only mental health providers and physical therapists (who almost by definition are trying to facilitate patient adaptive work) typically see their clients on a weekly or more frequent basis. Because we see our medical patients so infrequently, when we do see them we are often focused on reestablishing our relationship with them, and may be disinclined to "raise the heat" to promote adaptive work. Thus, the idea of using conflict in a medical practice as a tool with which to help patients do their adaptive work seems problematic. We will need to develop new structures and processes in care delivery to overcome this limitation.

We and our patients also carry with us conflict management styles from our personal lives that influence how we manage conflict in the clinical encounter. As we strive to adopt an AHP approach, we will need to explore and understand our own approach to conflict management and learn more constructive "best practices" where appropriate.

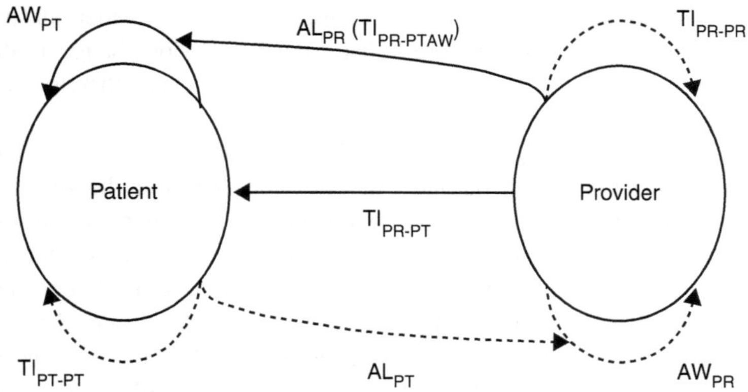

Fig. 38.2 Categorization of adaptive health practice behaviors. *TI* technical interventions, *AW* adaptive work, *AL* adaptive leadership, *PR* provider, *PT* patient. *Solid arrows* indicate primary AHP behaviors. *Dashed arrows* indicate behaviors that are less central to the model

In medicine we typically approach our work in our professional, "expert" role. The expert role works well when implementing a technical intervention, but facilitating adaptive work requires that we show up as nonexperts. Thinking of ourselves as experts gives us an illusion of understanding and control which is counterproductive when trying to help patients do their adaptive work. To quote Heifetz:

> Anybody operating with a theory of leadership that assumes that experts know what is best, and that then the leadership problem is basically a sales problem in persuasion, is in our experience doomed at best to selling partial solutions at high cost. ([29], p. 70)

"Partial solutions at high cost" might be a suitable tagline with which to describe much of modern medical practice.

Adaptive health practice consists of understanding the concepts described above; seeing both ourselves and our patients as complex adaptive systems; correctly diagnosing (distinguishing between) patients' adaptive and technical challenges; facilitating patient adaptive work; and using technical interventions judiciously—in particular as a complement to, not a replacement for, adaptive work.

These concepts and behaviors are illustrated schematically in Fig. 38.2. Providers typically provide technical interventions to patients (TI_{PR-PT} in the diagram); patients do adaptive work (AW_{PT}); and providers facilitate patient adaptive work using the methods of adaptive leadership (AL_{PR}). In addition, providers have their own adaptive work to do changing their philosophy of care and practice routines (AW_{PR}); patients can self-administer technical interventions (TI_{PT-PT}; e.g., over-the-counter medications, "exer-gaming," or smart-phone health promotion applications); and patients can practice adaptive leadership to facilitate provider adaptive work (AL_{PT}). Note that the technical interventions self-administered by patients may be less likely to interfere with or undermine patient adaptive work because they are generally less potent than the technical interventions prescribed by providers. Finally, providers may "self-administer" technical interventions (TI_{PR-PR}) that facilitate AHP (e.g., reconfiguring the electronic health record, changing compensation models, and adding health coaches to the care team). In the figure, the solid arrows indicate the primary components of the model. The dashed arrows indicate second-order behaviors. Note also that recommending technical interventions that facilitate patient adaptive work ($TI_{PR-PTAW}$) is a subset of adaptive leadership behaviors.

A key part of adaptive leadership has to do with the leader (provider) taking care of themselves, so they can continue to function effectively in their role. High levels of emotional exhaustion and burnout occur in health care, especially among nurses and physicians [32]. Adaptive health practice explicitly calls on us to do the adaptive work of changing the culture and practices of health care in order to make it more sustainable for health care workers.

Adaptive leadership also provides us with some clues about how to better care for ourselves in our clinical practice. Heifetz and Linsky call out the importance for leaders of avoiding role-self confusion, a "trap" into which it is all too easy for healthcare practitioners to fall ([33], pp. 187–206). Our role is largely defined by and based on the expectations of others. In healthcare, and especially in medicine, we readily embrace and identify with our role—it is high status and very ego-syntonic. In our work setting, patients and many of our coworkers will generally respond to us in terms of our perceived role. If we confuse our self with our role, we are at risk for internalizing any criticism, negative emotions, or bad clinical outcomes and taking them personally, thereby adding to our stress and also impairing our ability to really connect at the interpersonal level with our patients. It is that genuine interpersonal connection that forms the basis of the healing relationship which is so important for effective adaptive leadership.[4]

Leadership is an activity, not a role or position, and thus all clinicians, regardless of their level of training or status, can practice AL when caring for patients. The principles of AL are as applicable to nursing and other allied and complementary health professions as they are to medicine. In fact, I have found that nurses often grasp these concepts immediately, while physicians, oddly enough, often find them "too abstract."

Adaptive health practice involves changing three key mental models that dominate modern medical practice. First, providers (and often patients) believe that healing is something physicians, as experts, do to patients, not something patients do for themselves. Second, both patients and clinicians have an excessive faith in the ability of medical technology to promote health and cure disease, and a corresponding blind spot when it comes to recognizing the harm that medical technology, especially allopathic technology, causes. Third, many providers and patients believe that patient behavior change is difficult if not impossible to effect and therefore such efforts are largely futile.

The alternative, adaptive health mental models are just the opposite: health is an emergent phenomenon that patients create for themselves; technology is not a solution for adaptive problems, and should be used very judiciously; and patients are capable of doing the adaptive work to change their health behavior, and providers can and must learn the skills required to help them do this. In fact, there is growing behavioral health evidence about the patient's role as healer, and the clinician's role as facilitator [35]. All of us are familiar with patients who have been able to make dramatic changes in their social circumstances and behavior, and consequently in their health status. We should not blame patients for our own ineffectiveness as change facilitators.

In adaptive health practice, technology has three main uses: (1) relief of suffering that overwhelms patient coping or impedes adaptive work; (2) facilitation of adaptive work (e.g., "Wii-habilitation")[5]; and (3) management of symptoms and risk factors, and prevention of disability and death not amenable to, or pending the effect of, adaptive work. A corollary is that every effort should be made to avoid using technology that undermines patient adaptive capacity. In general, the more invasive or impactful a technical intervention, the more disruptive it is to patient physical resilience or psychological adaptive capacity, and the more likely it is to be associated with both short- and long-term harm. Consequently, AHP requires adopting a minimalist philosophy regarding health care technologies. We should follow a "minimum necessary potency" rule. This is what *primum non nocere* really means.

The adaptive leadership framework applies to healthcare practice at two levels. At the condition level, practicing adaptive leadership can help our patients respond to specific adaptive challenges, such as managing a chronic illness. At the person level, we can help our patients build their adaptive

[4] See Mimi Guarnieri's *The Heart Speaks* for a vivid example of how showing up as a person, not in our clinician role, can induce an important therapeutic breakthrough ([34], pp. 80–81).

[5] I first heard this bon mot from Lars Oddsson at the Sister Kenney Research Institute.

capabilities in general, so that they are globally more resilient and autonomous. Indeed, AHP in its fullest expression involves supporting patients in becoming maximally autonomous producers of their own health and sick care.[6] This is perhaps the most transformational characteristic of AHP, in that it goes against the nearly universal tendency of social systems to foster dependency, maintain their own existence, and perpetuate the challenge they were designed to address [36].

38.3.1 Adaptive Health Practice and Complexity Science

Ultimately AHP is about how clinicians and patients work together to facilitate patient adaptive work to change the environmental, social, and behavioral determinants of health or, put another way, the nested hierarchy of complex systems from which emerge suboptimal health states. Certain principles about effecting change in complex systems are readily applicable. The first is that the current state of the system is an emergent phenomenon resulting from repetitive iteration of "simple rules" built into the system structure.[7] A given health condition or state can be seen as an emergent pattern of behavior (an attractor) conditioned on the biopsychosocial structure of the specific individual of interest. To change the state of this system, it is necessary to change the underlying structure and simple rules governing the system's behavior. Putting this in human terms, to help patients adopt healthy behavior, we need to understand their social and environmental context and the health beliefs and attitudes that govern their current behavior. Most patients will have both health-promoting and health-reducing cognitive, social, and environmental resources. AHP involves building on the health-promoting assets and helping patients overcome their health-reducing constraints.[8]

A second principle is that complex systems tend to exhibit homeostatic behavior as a result of multiple levels of balancing feedback loops, and consequently it is often necessary to intervene at multiple levels of the system to effect lasting beneficial change. Thus, we need to partner with patients to help them make changes not only in their behavior but also in their physical environment and social systems. To do this we need to educate ourselves about the science of behavior change, the role of social networks in health, and the influence of social and environmental factors on behavior and health. At a minimum, we need to learn how to fully leverage the most basic, and perhaps the most important, social network intervention we have at our disposal—the therapeutic clinician–patient relationship.

In this regard, a critical but unanswered question is what will happen to the physician–patient relationship, and who will own the "healing relationship," as we increasingly move to a more "industrialized" model of care exemplified by the

[6] The sense of this statement is not materially altered by allowing for the possibility that some of the value in the therapeutic relationship comes from patients not being autonomous, or at least not more autonomous than they want. Rephrasing this, an AHP strives to help patients be as autonomous in their health and sick care as they want to be, with a corresponding commitment to help them grow in their desire to be autonomous to the extent that it is health promoting.

[7] The concept of "simple rules" is grounded in the mathematics of nonlinear dynamical systems. Such systems can be described by iterated self-reflexive functions of the form $X(t+1)=F(Xt)$, where the function F describes or defines a transformation on X at time t to produce the outcome $X(t+1)$. In human and social terms, the function F represents habitual thoughts, beliefs, and behaviors ("simple rules") that are iterated repeatedly. In Aristotle's words: "We are what we repeatedly do. Excellence, then, is not an act, but a habit." Mental models and family and organizational culture are manifestations of iterated simple rules. Simple rules function at the level of our unconscious and conscious thoughts, and our modes of relating with other people and things in our environment. The ability to be aware of and change the self-reflexive function itself (second-order self-reflexivity) is what makes complex adaptive systems adaptive.

[8] I find that often complexity science describes or explains phenomena that our ancestors captured succinctly in "folk wisdom." Thus, "experience is the best teacher" and the Vietnamese expression "a thousand hearings isn't worth one seeing, and a thousand seeings isn't worth one doing" both express the insight that one of the best ways to change the simple rules of a system is to expose the system to new experiences. These new experiences may then induce learning (change in the simple rules). "Act your way to a new way of thinking" is how the Positive Deviance community has expressed this [37].

concepts of Accountable Care Organizations and the Patient-Centered Medical Home, both of which carry expectations that some form of highly standardized and systematic "team-care" will replace the more purely dyadic physician–patient relationship. Patients may develop more than one "healing relationship" in these new care models. Given the infrequency of physician–patient interactions, the resource costs associated with an MD degree, and the fact that physicians are not trained, in general, to support patient adaptive work, it may be better to have someone like a health coach or community health worker be the person who is primarily responsible for facilitating patient adaptive work.

A third "systems thinking" principle for this work comes from the "soft" end of the systems thinking spectrum [38]. Patients and clinicians often, if not usually, mean different things by words such as "health" and "well-being," and we approach our mutual dialog from very different positions of knowledge and power. AHP involves developing a shared understanding with the patient about the meaning of these words, a shared understanding as to the goals of the therapeutic process, and the development of a non-hierarchical relationship based on trust and mutual regard.

This amounts to perceiving the clinical encounter as a "complex responsive process of relating," and working to change the nature of the conversation between clinician and patient [39]. In Ralph Stacey's words,

> The social, in human terms, is a highly sophisticated process of cooperative interaction between people in the medium of symbols in order to undertake joint action. ([39], p. 88)

This describes the clinical encounter and adaptive health practice. The process of cooperative interaction proceeds, for the most part, as conversation. In health care, the primary conversation is between clinician and patient. Changing the process requires us to change the conversation. Furthermore,

> as with interaction between bodies, the social, so with interaction of a body with itself, mind, there is the experience of familiar repetition of habit and the potential of spontaneous change. ([39], p. 89)

Stacey argues that the source of transformation (spontaneous change) arises in the unpredictable response of the "I" to the "me"—the response to perceiving how others see us (self-reflexivity, again). If the clinician's role is to help a person change their current health attractor, an important tool is to reflect back a perception of the patient's circumstances that somehow triggers a novel response of the patient's "I." Another approach is to facilitate change in the patient's social or environmental circumstances such that the patient's complex responsive process of relating changes, either because of change in the contextual factors directly, or because contextual change (new experience) induces a novel response of the "I" to the "me."

At its core, AHP is about changing the current health state pattern (attractor) that emerges from an individual's complex responsive processes of relating to the world around them, including the clinician. We can think of this mysterious process (adaptive work) as learning.[9] Learning (the creation of new knowledge) arises in social contexts from informal, exploratory exchanges between individuals, whether face-to-face or mediated by print or electronic media. It typically involves the reciprocal processes of changing structure to facilitate new behavior, and facilitating new behavior to change the structure ("act your way to a new way of thinking") [37]. Again, the medium of agency here is conversation. Changing the pattern of communication leads to knowledge creation. One person, intentionally changing the pattern of communication, creates new knowledge and changes the complex responsive process of relating.

Stacey's prescription for organizational leaders, which applies I think equally well to clinicians practicing AHP, is to be open to emergent conversational themes of health (as patients experience it) and adaptive work, and to "*seek to articulate them when this seems likely to shift the patterns of conversational life*" in the clinical encounter ([39], p. 235).

[9] Because people and relationships are complex adaptive systems, we expect that often this learning and behavior change will be experienced as a relatively sudden, qualitative "phase change"—an "aha!" moment of insight or rapid change. A goal of adaptive leadership is to induce such salutogenic phase changes.

Stacey identifies several other characteristics of conversations with transformational potential, all of which are relevant to AHP. Potentially transformational conversations typically have the "*dynamics of fluid spontaneity, liveliness, and excitement*," and include participants with diverse points of view. Perhaps as a consequence of the latter, they often produce, or are at risk of producing, misunderstanding. Additionally, because of their transformational potential, they may be perceived as threatening the continuity of identity, current power relations, and dominant ideologies, and therefore are often associated with anxiety and resistance (work avoidance). Changing the conversation, so as to change the complex responsive process of relating between clinician and patient, is challenging, risky, adaptive work ([39], pp. 181–183). Few patients or clinicians would describe their mutual conversation as spontaneous, lively, or exciting. Nor do such encounters typically involve diverse points of view, there being generally only one provider in the room, following a set script, with the patient's point of view too often remaining unexplored or even actively suppressed, especially if it represents a threat to the clinician's "expert" identity, the power differential between clinician and patient, or the dominant (allopathic) ideology.

The injunction not to use technical interventions to solve adaptive challenges is also justifiable on complexity science principles. This is a restatement of Ashby's Law of Requisite Variety: it takes complexity to master complexity [40]. Adaptive challenges are complex. Technical challenges and interventions are simple or complicated. Resolution of adaptive challenges requires a solution that is also complex—adaptive work. Technical interventions will not resolve adaptive challenges.

38.3.2 Relationship of Adaptive Health Practice to Other Views of the Clinician–Patient Encounter

The principles of adaptive health practice are not entirely new in health care. Many techniques have been developed and piloted in health care to help patients do adaptive work. These techniques include motivational interviewing [41], behavior change programs based on the trans-theoretical model [42], shared decision making [43], a wide variety of psychotherapeutic methods [35], and relationship-centered care [44]. However, none of these methods have achieved widespread adoption in clinical practice, except in limited areas like addiction medicine. Systematically incorporating many of these approaches into an existing clinical setting would be an efficient path to implementing an adaptive health practice.

In recent years, there has been a wave of interest, publications, and policy initiatives in support of deploying patient-centered medical homes (PCMH). While much of this work focuses on developing systems to provide more efficient and effective technical care for chronic disease, there is growing attention to the underlying philosophy of care and recognition that in addition to improving the technical aspects of care we need to do a better job of addressing the underlying determinants of health and incorporate "salutogenic" components into our practices [45]. The PCMH literature also is beginning to address the adaptive work that practices need to undertake to successfully adopt the model [46].

The principles of adaptive health practice can also be recognized in some of the "integrative" philosophy of care underlying the use of the set of alternative technical interventions that are characteristic of complementary and alternative medicine [47].

Indeed, AHP has much in common with integrative medicine. One could even argue that integrative medicine, because it is a familiar term, might be a better "name" for this concept. The philosophy of integrative medicine, as defined by those at the forefront of the field, is entirely compatible with adaptive health practice. However, the term "integrative medicine" is associated in the minds of many people (especially allopathic physicians and allied health professionals) with complementary and alternative health-related technologies such as acupuncture, chiropractic, homeopathy, naturopathy, herbal therapy, aroma therapy, energy healing, and other "healing arts" which are viewed by many as nonevidence-based quackery. Moreover, complementary and alternative medical practices, like allopathic practices, are vulnerable to excessive promotion of their own brand of expertise and "technical" interventions. (Although to be fair, integrative medicine providers have a reputation for practicing in a more holistic way, which in some cases at least means facilitating patient adaptive work.) Finally, while integrative medicine fits well with the AHP framework, AHP emphasizes and provides additional specific tools and techniques with which to facilitate patient adaptive work. Because of the connotations associated with "integrative health," and because the important point of distinction is facilitating patient adaptive work, not the specific technology being offered, I have chosen to continue to use the expression "adaptive health practice" to describe this philosophy of care.

Many patients want to receive care that is more "holistic" and less focused on technical interventions, especially the potent interventions of allopathic medicine with their attendant power to harm. The widespread use of complementary and alternative medical services is an indication of this, and survey evidence also supports it [48]. Anecdotally, many of my coworkers make use of these kinds of services on their own, and say they would welcome the opportunity to work with more traditional providers offering a truly integrative, adaptive health practice approach.

Before proceeding with a set of suggestions for implementing AHP, I need to address an important potential criticism of the whole concept, namely, that the current model of sick care and practice is itself "adaptive," or else it would not exist. This argument rests on the idea that the current approach to care in the United States (and most other developed countries) emerged as an adaptive consequence of nearly universal human social conditions. The fact that the model of care is similar throughout the developed world reinforces the impression that the status quo is in fact adaptive, and implies that suggesting an alternative could be "mal-adaptive" and unrealistic.

My counterargument is that just because the current model of care is the result of a self-organizing, path-dependent process rooted in the history of human social development does not mean that it is the only possible care delivery stable state (attractor) available to us in the present day. The well-documented international (and inter-community) variability in healthcare systems is supportive evidence that multiple healthcare delivery attractors exist (albeit tightly clustered, for the most part). We have the capability to explore the health care solution space in hopes of identifying other potentially even more adaptive states like AHP. What follows is an attempt to describe such a new attractor in the care delivery fitness landscape, and a process for transitioning from the current attractor to the new one.

38.4 Implementing Adaptive Health Practice

So what does AHP look like in practice? No one really knows, because to the author's knowledge it has never been specifically piloted.[10] However, a wide variety of programs or processes that support adaptive work have been piloted and studied in clinical settings. What follows is an attempt to outline at a high level a set of behaviors, business processes, and practice routines (some of which are evidence-based) that could instantiate AHP in an outpatient health clinic.

38.4.1 Preliminary Matters

Clinical practice starts when the patient first encounters the clinic itself and decides to seek care there. Many clinics share information with patients about the clinic and the philosophy of care of the providers. In an adaptive health practice, information for patients needs to explicitly state the principles of care described above so that patients can make an informed decision about whether the practice's philosophy of care matches their own. In particular, the patient information should make clear that the practice uses medical technology judiciously so as to minimize the risk of harm, and seeks to support each patient, within the limits of their capacity, to become the source of their own

[10] See chapters by Curt Lindberg and others in this volume.

health. Because most patients will probably not appreciate a theoretical description of AHP and its benefits, information about the practice philosophy should be presented in a way that resonates for patients. There is much to learn about how best to communicate with patients about this philosophy of care, and which patients will be most comfortable with the AHP model. The best approach might be to provide this information in a variety of media and formats, including most importantly the opportunity to have an actual (potentially transformative?) conversation with a representative of the practice (ideally a clinician).

After the patient selects the clinic, the patient and provider meet for the first time. In an adaptive health practice, this initial visit is an important opportunity to collect novel information from the patient and reinforce the key principles of AHP. Because it is a well-established norm, I start with the traditional approach to organizing and conducting the clinical interaction—the "SOAP note" (subjective, objective, assessment, and plan). This begins with subjective history gathering, followed by gathering of objective data obtained from the physical exam and various kinds of diagnostic testing. After the data gathering comes assessment, including diagnosis and prognosis, and then the plan of action. If AHP is to differ from conventional medical practice, we expect that each of the components of the SOAP construct will be significantly different in an adaptive health practice clinic.[11]

Physician–patient communication patterns will also need to be dramatically different in an AHP. Current patterns of physician communication, including the tendency to interrupt, interrogate, and lecture patients, are adaptive behaviors physicians develop to cope with time pressures and rapidly gather information, despite patients' tendency to be circumstantial and tangential, that can be used for illness diagnosis and management. The same conversational patterns are very mal-adaptive when one is trying to help a patient make the changes in their lives they need to be healthy.

38.4.2 Subjective: The History

During this part of the encounter, the clinician works with the patient to identify the nature of the presenting problem. Given patient tendencies to avoid adaptive work, the clinician must focus on whatever problems the patient explicitly identifies, as well as problems that the patient may not recognize or acknowledge. (In this respect AHP is no different from what savvy clinicians have been doing all along.) In addition to the usual components of the history (chief complaint, history of present illness, past medical history, family and social history, and review of physical systems), in an adaptive health practice the clinician will also want to collect information about a broader range of issues, including more detail about the patient's lifestyle, health beliefs and psychological outlook, social support, social networks, and environment. This information will be valuable during the assessment and plan phases of the encounter.

A high-quality health appraisal should be a routine part of history taking in an adaptive health practice. Health appraisals typically include questions about lifestyle and behavior, socioeconomic status (education, income, job status), health-related quality of life, functional status, and productivity. In addition, more detailed information should be obtained about the patient's social network and social support, both given and received. A number of validated questions and surveys are available for assessing these factors [49]. In particular, the absence of someone in whom the patient can confide, or a sparse social network, is clearly associated with worse health outcomes and in themselves should be identified as health issues of concern [27, 28]. A practical social survey tool that provides actionable information needs to be developed, and the impact of its use, and the social interventions arising from it, needs to be studied. A potentially useful approach to collecting social

[11] However, note that the SOAP note format is a cultural artifact that is part of the "sick-care" frame and will tend to conserve the current form and content of the "complex responsive process of relating" that is the clinical encounter. We need a new health and adaptive work oriented conversation between clinicians and patients. The SOAP note construct is a constraint that may inhibit that transformation. I have included it here because it is familiar and there is much that can be accomplished within its constraints; but as AHP evolves we may adopt a different framework for many AHP clinical encounters.

network information in the clinical encounter has been described by Sluzki [28].

Another novel addition to history taking is assessing patient complexity. A variety of methods or concepts have been explored, including the "Vector Model" and the INTERMED project [50–52]. In addition to clinical complexity (severity and comorbidity), these surveys assess the patient's socioeconomic challenges, nature of the relationship with the care delivery system, and stability of home and work life. An older but still not routinely used inventory is the Social Adjustment Rating Scale of Holmes and Rahe [53].

The assessment of psychological factors is typically limited in the usual medical history. However, a patient's health beliefs and psychological outlook strongly influence their ability to undertake adaptive health work. In adaptive health practice, a much broader range of information needs to be collected. This includes questions about a patient's health beliefs and attitudes, searching for a history of childhood trauma, and evidence of depression such as the PHQ-2 or PHQ-9. In addition, it may be useful to assess factors like autonomy orientation [54], learning mind-set ([35], pp. 167–183), patient activation [55], and self-regulatory strength [56]. This information is critical, because to help patients build their adaptive capacity clinicians need to "meet them where they are." If we encourage our patients to undertake adaptive work that is beyond their current capacity, we are likely to overwhelm them, create excessive conflict, and induce a sense of failure or futility that discourages further adaptive work.

38.4.3 Objective: The Physical Exam and Diagnostic Testing

This portion of the clinical encounter may be least altered by the AHP approach. Doing a physical examination and obtaining and reviewing the results of diagnostic tests are still important technical components of the examination. However, in addition to revealing potential evidence (signs) of illness, the physical exam provides important information about patient adaptive capacity, and in particular the level of physical fitness and self-care.

Collecting objective data, however, is not the only goal of the physical exam. In fact, with the increasing use of imaging and laboratory testing to collect biomedical information the importance of the physical exam as a source of objective evidence has declined in modern medical practice. Regardless of the information collected, the physical exam plays an important role in building the clinician–patient relationship. Permission to look at and touch another person's naked body is a unique attribute of the clinical relationship. Only to healthcare providers, and our closest family members (our parents when we are young; our intimate partners when we are sexually mature), do we grant this permission. That is, only in our most trusted relationships do we expose ourselves in this vulnerable way. Thus, the physical exam is an opportunity for the clinician to reinforce (or weaken) that trust, and trust is a critical element in creating the healing relationship which is essential for adaptive leadership and facilitating adaptive health work. A carefully and respectfully performed physical examination not only provides useful objective information about a patient's condition but also strengthens the clinician–patient relationship and in that way increases our ability to help our patients do the work they need to do to be healthy. With this in mind, the current downward trend in the expert use of the physical exam is clearly in the wrong direction [57].

38.4.4 Assessment: Distinguishing Between Adaptive and Technical Challenges, and Evaluating Adaptive Capacity

There are four important tasks that need to be accomplished during the assessment phase of the clinical encounter. First, the clinician needs to consciously review the patient's circumstances from a systems perspective. What are the system characteristics from which the patient's current health challenges are emerging? Next, the clinician needs to explicitly differentiate between the adaptive and technical challenges faced by the patient, so that each category is appropriately addressed. In addition, the clinician needs to assess the patient's adaptive capacity or resilience, and finally the quality of the therapeutic relationship in terms of how well it will support the level of conflict required to support the patient's adaptive work.

38.4.4.1 Evaluating the System from Which the Problems Emerge

Fundamentally, our job as clinicians is to understand our patients as complex adaptive systems, embedded or nested in a hierarchy of other complex adaptive systems, and to see our patients' health problems as emergent phenomena. Without this understanding, we are likely to be inefficient and ineffective, or even harmful, in our care. That this concept is not new (George Engel proposed the "biopsychosocial model" in 1977) does not mean it is passé [58].

One way to conceptualize "health" is as the balance between the complexity of an individual's coping mechanisms (their adaptive capacity) and the complexity of their environment. Ashby's "Law of Requisite Variety" says that the variety of the control system must be as great as the variety of the system to be controlled [40]. That is, a system must be at least as complex as the process it is trying to control.[12] Adaptive health challenges can be framed as arising when the patient's complexity or adaptive capacity is exceeded by environmental complexity, variability, or stress. Separately evaluating the patient's internal and environmental complexity, and finding ways to increase the former while reducing the latter, is a "systems" approach to health promotion that would be consistent with Ashby's Law of Requisite Variety.

As noted above, the complexity of a patient's circumstances might be measurable using the Social Readjustment Scale or one of the emerging patient complexity assessments. We (and especially the patient) need to understand with at least a moderate amount of specificity who the stakeholders are with respect to a patient's health, and their roles and interests. What factors are contributing to the patient's health challenges, and which ones are improving it? What keeps the patient from improving their health? What's working for them?

How best to measure patient complexity is less clear. Presumably it is some function of biophysical factors (physical "fitness" and scale-free physiologic variability), and psychological factors like self-regulatory strength and what Robert Kegan calls "mental complexity" [62], and others would call mindfulness—the ability to maintain a meta-awareness and do second-order learning. In addition to mindfulness or "mental complexity," it seems likely there are other dimensions of adaptive capacity arising from other assets or strengths, including youth, health status, socioeconomic status (up to a point), and social connectedness that should be incorporated into any measure of patient complexity. I discuss the assessment of patient adaptive capacity at greater length below.

38.4.4.2 Distinguishing Between Adaptive and Technical Challenges

Another critical task during the assessment phase of the clinical encounter is to accurately diagnose, and distinguish between, the adaptive and technical challenges that the patient faces. There are at least two approaches to doing this. One approach would be to develop the problem list and explicitly flag each problem as adaptive, technical, or mixed. In the case of "mixed" adaptive/technical problems, it is best to decompose them into their component challenges and list each one separately. In some cases, a particular adaptive challenge may be more general, and relevant to other issues on the problem list, in which case it should be listed as a separate problem in its own right.

For instance, the overweight, middle-aged patient with anxiety, depressed affect, and high blood pressure is a frequent sight in clinical practice. A clinician might list each of these issues (overweight, anxiety, depressed affect, and hypertension) on the problem list. Underlying each of

[12] The work of Goldberg and West on the association between aging and disease and loss of complexity in the variability of physiologic parameters like heart beat, respiration, and gait is consistent with this definition [59–61].

them, however, is likely to be a lack of physical activity, high levels of psychosocial stress, dysfunctional relationships, and an unhealthy diet. These adaptive challenges, because they are common to more than one of the identified problems, should be explicitly identified in the problem list. In fact, they typically represent the "true diagnoses"; the other problems are symptoms of the underlying disorder—an unhealthy lifestyle arising from the interaction between the patient and a particular social and physical environment.

A second approach is to create a template for the assessment portion of the clinical encounter that forces or encourages the clinician to explicitly identify or flag the presence or absence of common adaptive challenges such as substance use, lack of physical activity, poor diet, and low social support. Modifying the electronic health record so that it prompts the clinician to make these diagnoses during the assessment phase would provide systematic support and facilitate the normalization of AHP practices.

38.4.4.3 Evaluating Patient Adaptive Capacity

Evaluating a patient's adaptive capacity is another critical component of the assessment phase of the encounter. "Low adaptive capacity" is an adaptive challenge in its own right, and if identified belongs on the patient's problem list. Tools like the Patient Activation Measure can be used to assess the patient's capacity for adaptive health work [55]. Asking about readiness to change, and using motivational interviewing techniques, can identify the adaptive challenges that are "ripest"—where the patient is ready to start doing adaptive work. Integrating the information obtained during the history part of the exam about psychological and social factors that influence adaptive capacity, such as learning mind-set, autonomy orientation, self-regulatory strength, and sparse or dysfunctional social networks, are also important tasks during the assessment phase.

One simple approach to assessing self-regulatory strength might be to ask patients about the "white coat compliance" phenomenon [63, 64]. Many patients find it difficult to adhere to health promotion regimens. After contact with a clinician, their adherence may be high for some time afterwards, but then it decays. Such patients may become more adherent before and in anticipation of an upcoming clinical encounter. These patients are likely to have lower levels of self-regulatory strength than patients whose adherence does not fluctuate in response to contact with their healthcare providers. Synchronizing the frequency of clinical encounters with the patient's natural "adherence cycle" might create a "resonance" phenomenon that could improve adherence and clinical outcomes overall. By identifying the existence of an "adherence cycle," we also create the opportunity to help a patient do the adaptive work of identifying and addressing the factors that contribute to the cycle, thereby building their adaptive capacity, so they can decrease the amplitude and increase the period of the cycle, and become less dependent on us.

38.4.4.4 Evaluating the Quality of the Clinician–Patient Relationship

Finally, our ability to help our patients do adaptive health work is critically dependent on the strength of our relationship with them. A dysfunctional provider–patient relationship is also an adaptive challenge—for both the patient and the provider. The assessment process should include explicit evaluation of the quality of the clinician–patient relationship. A wide variety of factors need to be considered. Does the patient receive care only or mostly from the one provider, or do they have to see multiple providers and specialists? Is there a pattern of trust, honesty, abiding, and valuing in the relationship? Are power and information shared, or closely held? Tools to evaluate these issues are being developed and tested [65].[13]

Even if such a tool is available, and especially in its absence, clinicians should incorporate meta-conversations about the healing relationship into their patient encounters. Examples of such meta-conversations include asking patients for permission to touch them during the physical exam,[14] acknowledging

[13] See also chapter by John Scott in this book.

[14] I am indebted to Dr. David Moen for this idea.

that sometimes patients are fearful or hesitant to tell their providers the full truth about their condition, fears, or hopes, and encouraging patients to review their medical records to make sure they are accurate. These meta-conversations have the potential to both assess, and improve, the relationship. Thinking about the patient and provider as being members of a team, meta-conversations are a way to promote team reflexivity, which has been shown to improve team performance [66, 67].

Existing patient experience surveys, such as the Consumer Assessment of Healthcare Providers and Systems (CAHPS) set of survey instruments, include items that assess aspects of the clinician–patient relationship that are relevant to adaptive health practice. These surveys are collected and analyzed at the organizational unit or provider level, however, and do not currently provide information about the individual clinician–patient relationship.

38.4.5 Plan: Adaptive Leadership in Action

Once we have, in collaboration with our patient, "diagnosed the system," differentiated between the patient's adaptive and technical health challenges, and assessed the patient's adaptive capacity and the adequacy of the therapeutic relationship, we then need to design a plan of care that properly addresses both the technical and adaptive challenges faced by the patient. Much of the health work to be done is the patient's, and therefore the care plan needs to be codesigned if it is to be as effective as possible.

There are three important elements to the adaptive health practice care plan: (1) following an iterative adaptive leadership framework; (2) codeveloping with the patient an appropriate set of interventions to address the patient's adaptive challenges; and (3) using technical interventions judiciously in order to facilitate patient's adaptive work and minimize harm. All three elements are essential, and none are reliably part of current clinical practice.

38.4.5.1 Adopting an Iterative Approach: Observe, Interpret, and Intervene

The clinician needs to adopt an adaptive leadership framework for discussing the care plan with the patient. This means that the clinician, having observed (the subjective and objective portions of the clinical encounter), and interpreted (the assessment portion of the encounter), now needs to intervene *in an iterative fashion* to develop a plan of care that maintains the patient in a productive range of tension with respect to doing their adaptive work, builds the patients adaptive capacity, and reinforces or strengthens the therapeutic relationship, thereby enabling even more ambitious adaptive work in the future.

The iterative nature of this process is important, and distinctly different from typical current practice. Too often, the Plan portion of the clinical encounter happens with the clinician, hand on the door knob (at least metaphorically), giving a list of recommendations and instructions to the patient, without stopping to adequately assess the patient's reaction. If the patient fails to follow these instructions, as is so often the case, the clinician at the next encounter is likely to label the patient as non-adherent, rather than recognize that they are manifesting work avoidance (assuming they understood the clinician in the first place).

Observing that the patient is struggling with work avoidance, the clinician then needs to interpret—is the work avoidance due to complacency (e.g., do they not understand the health importance of the circumstances), or is it due to overload—are they feeling overwhelmed by the work? This interpretation is critical, as it determines the appropriate next step.[15] If the patient is complacent, having further discussions with them about the importance of the clinical problem, at a level and in a fashion that promotes understanding and increases motivation, is the proper next step. In

[15] One can also describe this phenomenon in terms of motivational interviewing by equating lack of motivation with complacency and low self-efficacy with feeling overwhelmed.

some cases, this is just a matter of better education; in others it means helping them do the adaptive work of adopting more realistic health beliefs. On the other hand, if the patient's work avoidance (non-adherence) is due to being overwhelmed, the next step is to reduce the demands on them by reducing the number of, or setting more modest, short-term goals, or simplifying their therapeutic regimen, while finding ways if possible to bolster their coping skills and adaptive capacity (building self-efficacy).

If you cannot tell whether the patient's work avoidance is due to complacency or overload, you can try an experiment by acting "as if," and then observing the patient's response. For instance, if you act as if the patient has overload and "turn down the heat," and this leads to the patient being more engaged, you have evidence that the patient was in an "overload" state. If on the other hand they remain disengaged, it is more likely that they are in a complacent state, in which case you can "raise the heat" again in the hopes of engaging them.

During the "intervention" phase, the clinician should use certain best practices to engage the patient. These include motivational interviewing, as noted above, and celebrating successes and what's going well, rather than focusing only on what's not. It builds trust and engagement, and the patient's sense of self-efficacy, which is strongly correlated with successful behavior change. There is a growing body of evidence supporting this "positive" approach to therapy and behavior change [68].

Deficits in the clinician–patient relationship may contribute to both kinds of work avoidance. Lack of trust in the provider, or failure to codesign the care plan with the patient, may cause them to either ignore us, or be unwilling to commit the time and energy required for doing adaptive health work. Over time, repeated iteration of this observe–interpret–intervene cycle, if properly done, should support the patient in maintaining their health engagement in a productive zone and progressively strengthens the therapeutic relationship, especially as the patient experiences benefits from the relationship.

38.4.5.2 Codeveloping Interventions that Support Adaptive Work

There are a large number of evidence-based interventions that can support patients in doing their adaptive health work. Referrals for psychotherapy or to a social worker or social welfare agency are standard practice, although perhaps underused, in many existing medical practices. Other resources include behavior change support programs such as 12-Step Programs, Weight Watchers®, smoking cessation programs, and the growing number of other health promotion, self-care, and decision support programs offered by large health care delivery systems, health plans, community organizations, and fitness companies. There are also many resources for patients on the Internet, including patient support groups like PatientsLikeMe.org, high-quality educational content and decision support materials, and helpful videos on YouTube. A growing number of smart-phone "apps" are available to help patients trying to adopt healthier lifestyles. In an adaptive health practice, recommendation of or referral to such resources should be routinely provided and tailored to the patient's identified adaptive challenges. In order to do so, clinicians have to develop processes for becoming, and staying, as well-informed and expert in the application of these interventions as they are with more traditional medical technologies. Ultimately, to provide these recommendations and referrals systematically, they should be built into clinic work flows and the electronic health record.

38.4.5.3 The Judicious Use of Technical Interventions

In addition to promoting tools and resources that support patient adaptive work, clinicians need to adopt, and educate their patients about the wisdom of, a more minimalist approach to using medical technologies like medications and procedures. This can be, as all clinicians know, challenging work. For psychological, cultural, and commercial reasons, both patients and clinicians tend to be overly optimistic about the benefits and harms associated with modern medical care technologies. Not only do we have to become better informed ourselves about these issues, but we also need to find ways to convince our patients that much of what they see on television or read in the print media is misleading or wrong. In modern medicine, more is often not better. As with the treatment of GERD, using less potent medicines, and sometimes no medicines at all, can lead to better outcomes.

Implementing this in clinical practice will involve cultural work—surfacing and challenging the largely unquestioned assumptions about the value of medical technology; becoming more aware of the literature about, and the presence in our patients of, various forms of iatrogenic morbidity; and addressing variation in clinician "propensity to intervene" [69]. In an adaptive health practice, clinicians will welcome the use of decision aids and readily adopt risk communication best practices because they will recognize that both they and their patients need help making evidence-based, unbiased decisions [70, 71]. In an AHP, the use of relative risk when discussing benefits and harms of an intervention should be a "never event."

It will be equally important to address the prevailing business model that currently reinforces the overuse of medical technology. Money is the measure of value in our society. Whatever the various rewards of healthcare practice, the dominant incentives that drive the behavior of all healthcare providers, at least to some extent, are insurance reimbursement arrangements and provider compensation models. These incentive systems still largely reward providers for the productivity of their organizations and themselves, with productivity measured not in units of health produced or quality-adjusted life years saved, but in relative value units, office visits, hospital days, patients seen, or hours worked.

38.4.6 Getting Started

The description of AHP above represents a major departure from the status quo. Given the challenges presented by adoption of the patient-centered medical home model [72], it is hard to see how existing clinics could possibly succeed in converting themselves to the AHP model, which is an even more radical transformation. That said, what follows is an attempt to sketch out a development path that I believe is feasible. The general approach is outlined in Table 38.2.

First, it will be necessary to establish that a sufficient proportion of the staff of an existing clinic share the vision and are willing to do the work to become an adaptive health practice. This assessment should include also an evaluation of the clinic's organizational adaptive reserve, as described by Nutting and colleagues [73]. Staff participating in patient care should be trained on motivational interviewing, the 5A's model [74, 75], and shared decision making. At the beginning, it may be more efficient to fully train only one or two staff members on these methods. These staff can then act in a consulting role as resources for the other staff. An AHP also needs to have compatible psychology and social work professionals available, either in the practice or as consultants, to help with the design of the program and to participate in patient care. Finally, an AHP will probably benefit from incorporating, or forming partnerships, with high-quality complementary and alternative health practitioners. Many of these practitioners are already skilled in supporting patient adaptive work, and the technical interventions they use are generally harmless and can be therapeutic (as a result of the placebo effect, if nothing else).

Once it has been established that the clinic staff are willing, and the culture supports it, the initial steps an existing clinic should take in adopting the AHP model are listed in Table 38.2. Developing ways to communicate the practice's AHP philosophy of care to patients is an early requirement. This can be done even if the operational support systems required to fully implement AHP are not yet in place. In fact, just raising the issue with patients invites them into a dialog with providers from which the providers may well learn a lot about how the patients are already trying to follow adaptive health practices in their own lives, and how providers can better support them in those efforts.

Another early step in developing an AHP is to reconfigure the clinical encounter by adopting some of the easier-to-implement ideas discussed above. In particular, it should be relatively easy to revise the paper or electronic templates used for the subjective, assessment and plan parts of the clinical

Table 38.2 Implementing an adaptive health practice—some practical next steps

Clinic development work:
1. Establish a "critical mass" of support with fellow staff
2. Evaluate the clinic's adaptive reserve, and address deficiencies 3. Develop and train staff on motivational interviewing, the 5A's, and shared decision making
4. Identify psychology and social work partners
Implement patient communication strategy:
1. Develop materials, scripts, etc.
2. Train staff
Modify the clinical encounter template:
1. Add additional questions about lifestyle, self-assessed health status, and social structure and support
2. Explicitly and separately identify adaptive and technical challenges
3. Develop a "pick list" of interventions that support patient adaptive work and can be incorporated into the therapeutic plan
Use shared decision making to ensure the evidence-based and preference-sensitive use of "minimum necessary potency" technical interventions.
Adopt a supportive business model:
1. Shared risk or savings contracts
2. Membership fees
Measure a balanced set of outcomes. Possible examples of potentially useful metrics include:
1. Patients
a. Health—self-assessed health status, lifestyle behaviors, risk factor control
b. Engagement and experience—CG-CAHPS, PAM
2. Providers
a. Maslach Burnout Inventory
b. Discretionary decision making ("propensity to intervene") [69]
3. Clinic
a. Financial metrics
b. Employee engagement
4. Population (in partnership with payers)
a. Total cost of care
b. Utilization (e.g., office visits, medication use, ambulatory care sensitive hospital admissions)

encounter so that they support AHP. With respect to taking a patient's history, an AHP should be able at a minimum to collect information about four key lifestyle issues (physical activity, fruits and vegetables intake, tobacco use, and safe alcohol consumption) and routinely refer patients to programs or other services that support lifestyle change [76]. An AHP should also be able to implement a process for routinely collecting information about the existence, quality, and changes in the key relationships in the patient's life, so that social challenges can be identified, discussed with the patient, and incorporated into the therapeutic regimen.

With respect to the assessment and plan, the templates supporting these phases of the clinical encounter need to be modified so as to explicitly and separately call out and distinguish the technical and adaptive challenges faced by the patient. The adaptive challenges are the patient's to work on, and this needs to be clear to both provider and patient. Lists of resources and interventions supporting adaptive work should be developed and incorporated into the clinical encounter template, so that they can be easily incorporated into the therapeutic plan. All technical interventions, and especially those used to address or manage a symptom related to an adaptive challenge, need to be carefully assessed, in collaboration with the patient, to ensure that the "minimum necessary potency" rule is being followed.

A practice adopting an AHP model must adopt a business model that supports it. Options include

(1) contracting with major payers to assume manageable financial risk for pharmacy, specialty, and facility costs (e.g., a shared savings model based on some combination of total cost of care and quality outcomes); or (2) a membership model where patients pay an affordable monthly or annual membership fee that is sufficient to cover those services provided by the clinic that are not reimbursable by the usual health insurance policies.

An important challenge for an AHP is to maintain the truly professional goal of helping patients be as autonomously healthy and independent of the healthcare system as possible. The typical business model, where providers own their own business and due to market inefficiencies are not fully accountable to their customers, may not provide adequate safeguards to protect against a "reversion to type." Establishing an AHP as a consumer cooperative or, if corporate practice of medicine laws is a barrier, as a professional corporation with an exclusive contractual relationship with a consumer cooperative may be a way to ensure that the AHP maintains this high standard of professionalism.

In approaching all of this work, the practice should follow organizational development methods based on complexity science principles, such as emergent design, minimum specifications, "chunking," appreciative inquiry, and positive deviance [78, 79]. The approach to leadership needs to be facilitative, not prescriptive, which will be a challenge given the traditional medical model to making things happen ("doctor's orders") [46, 80].

Table 38.2 also lists some approaches to measuring the impact and performance of an AHP. A successful implementation of an AHP will be good for patients, providers, the clinic, and the population. Key measures include patient health, provider burnout, and total cost of care. None of these things are routinely measured or reported in conventional medical practices.

38.5 Summary and Conclusions

Adaptive health practice involves distinguishing between the adaptive and technical challenges faced by patients, using the principles of adaptive leadership to support patient adaptive work, and administering technical interventions judiciously according to the principle of "minimum necessary potency." Adaptive health practice is a philosophy of care based on the application of the principles of complexity science and Heifetz' Adaptive Leadership model to clinical healthcare practice. AHP promises to give us a way to address the major challenge facing modern medicine—finding a way to support our patients in making the changes in their lives they need to be healthy. Because of our ineffectiveness in health promotion, and an overemphasis on technical health interventions that are often more harmful than we recognize, we have a healthcare system that generates a lot of iatrogenic disease and "specific counterproductivity," as described many years ago by Ivan Illich [36].

Recognizing that patients are complex adaptive systems, the principles of AHP represent a fundamental change in the simple rules that govern modern medicine. AHP expands the focus in health care from using medical technology for treating and preventing illness to also leveraging patients' self-healing capacity to promote health. AHP expands the role of the clinician from expert prescriber to collaborative health coach. AHP expands the patient's role from passive recipient of care to active, autonomous generator of their own health. Finally AHP gives us a new set of tools with which to address persistent, pervasive clinical challenges like poor provider communication, patient nonadherence, and overuse and misuse of medical technology. AHP supports the systematic implementation of a wide variety of evidence-based approaches to care that are currently underutilized, including shared decision making, motivational interviewing, the 5A's, the transtheoretical model, and the integration of psychological and social support with care delivery. AHP is consistent with other major theories of health and care, including Engel's biopsychosocial model, integrative health, and the socioecological model of health [81].

To some extent AHP and our current approach to practice represent a kind of polarity, with sick care being one pole, and health promotion the other. Both poles are necessary, and an excessive focus on one causes unavoidable dysfunction. The best outcomes occur when the health delivery system has a balanced focus and manages the polarity by doing, and integrating, both things well, while avoiding the pitfalls or limitations that arise from an excessive focus on one or the other [82].

Implementing AHP will be a challenge, but given that it directly addresses critical and persistent issues facing modern medicine, it seems "worth a go." In this chapter I have outlined a preliminary approach that may be a reasonable place to start. The work of implementing AHP is itself adaptive work, involving lots of learning and behavior change. I hope this chapter represents a convincing argument that it is time for us to stop avoiding this work, and to start doing it, in collaboration with our patients and sympathetic colleagues.

Acknowledgements The author wishes to thank Curt Lindberg, Val Ulstad, Larry Morrissey, C. J. Peek, David Moen, and Keith McCandless for their helpful comments and suggestions. Jim Begun, Doug Wholey, George Isham, and Terry Crowson played influential roles in the development of the concept of adaptive health practice. In addition, this work would not have been possible without the financial support of the Archibald Bush Foundation which funded the author's Bush Medical Fellowship.

References

1. Thygeson M, Morrissey L, Ulstad V. Adaptive leadership and the practice of medicine: a complexity-based approach to reframing the doctor-patient relationship. J Eval Clin Pract. 2010;16(5):1009–15.
2. Berwick DM, Nolan TW, Whittington J. The triple aim: care, health, and cost. Health Affairs (Millwood). 2008;27(3):759–69.
3. Berwick DM. The science of improvement. J Am Med Assoc. 2008;299(10):1182–4.
4. Pawson R, Tilley N. Realistic evaluation. London: Sage Publications; 1997.
5. Evans RG, Barer ML, Marmor TR. Why are some people healthy and others not? The determinants of health of populations. New York: Walter de Gruyter; 1994.
6. Schroeder SA. We can do better—improving the health of the American people. N Engl J Med. 2007;357:1221–8.
7. McGinnis JM, Williams-Russo P, Knickman JR. The case for more active policy attention to health promotion. Health Affairs (Millwood). 2002;21(2):78–93.
8. Mokdad AH, Marks JS, Stroup DF, Gerberding JL. Actual causes of death in the United States, 2000. J Am Med Assoc. 2004;291:1238–45.
9. McIlvain HE, Backer EL, Crabtree BF, Lacy N. Physician attitudes and the use of office-based activities for tobacco control. Fam Med. 2002;34(2):114–9.
10. Vogt F, Hall S, Marteau TM. General practitioners' and family physicians' negative beliefs and attitudes towards discussing smoking cessation with patients: a systematic review. Addiction. 2005;100(10):1423–31.
11. Jay M, Kalet A, Ark T, McMacken M, Messito MJ, Richter R, Schlair S, Sherman S, Zabar X, Gillespie C. Physicians' attitudes about obesity and their associations with competency and specialty: a cross-sectional study. BMC Health Serv Res. 2009;9:106. doi:10.1186/1472-6963-9-106.
12. Stannard CF. Opioids for chronic pain: promise and pitfalls. Curr Opin Support Palliat Care. 2011;5(2):150–7.
13. Wolstenholme EF. A core set of archetypal structures in system dynamics. Syst Dyn Rev. 2003;19(1):7–26.
14. Shapiro AK. Contribution to a history of the placebo effect. In: Barber TX, DiCara LV, Kamiya J, Shapiro D, Stoyva J, editors. Biofeedback and self-control. Chicago: Aldine; 1974. p. 222.
15. Reimer C, Søndergaard B, Hilsted L, Bytzer P. Proton-pump inhibitor therapy induces acid-related symptoms in healthy volunteers after withdrawal of therapy. Gastroenterology. 2009;137:80–7.
16. Everhart JE. Gastroesophageal reflux disease. In: Everhart JE, editor. The burden of digestive diseases in the United States. US Department of Health and Human Services, Public Health Service, National Institutes of Health, National Institute of Diabetes and Digestive and Kidney Diseases. Washington, DC: US Government Printing Office; 2008. NIH Publication No. 09-6443. p. 69–72.
17. Fava GA. Can long-term treatment with antidepressant drugs worsen the course of depression? J Clin Psychiatry. 2003;64(2):123–33.
18. Hughes S, Cohen D. A systematic review of long-term studies of drug treated and non-drug treated depression. J Affect Disord. 2009;118(1–3):9–18.
19. Fournier JC, Derubeis RJ, Hollon SD, Dimidjian S, Amsterdam JD, Shelton RC, Fawcett J. Andidepressant drug effects and depression severity: a patient-level meta-analysis. JAMA. 2010;303(1):47–53.
20. Hollon SD, Thase ME, Markowitz JC. Treatment and prevention of depression. Psychol Sci Public Interest. 2002;3(2):39–77.
21. Salpeter SR, Buckley NS, Ormiston TM, Salpeter EE. Meta-analysis: effect of long-acting β-agonists on severe asthma exacerbations and asthma-related deaths. Ann Intern Med. 2006;144:904–12.
22. Deyo RA, Gray DT, Kreuter W, Mirza S, Martin BI. United States trends in lumbar fusion surgery for degenerative conditions. Spine. 2005;30:1441–5.
23. Martin BI, Deyo RA, Mirza SK, Turner JA, Comstock BA, Hollingworth W, Sullivan SD. Expenditures and health status among adults with back and neck problems. JAMA. 2008;299(6):656–64.

24. Bucher HC, Hengstler P, Schindler C, Guyatt GH. Percutaneous transluminal coronary angioplasty versus medical treatment for non-acute coronary heart disease: meta-analysis of randomized controlled trials. BMJ. 2000;321:73–7.
25. Temel JS, Greer JA, Muzikansky A, Gallagher ER, Admane S, Jackson VA, Dahlin CM, Blinderman CD, Jacobsen J, Pirl WF, Billings JA, Lynch TJ. Early palliative care for patients with metastatic non-small-cell lung cancer. N Engl J Med. 2010;363:733–42.
26. Cifuentes M, Fernald DH, Green LA, Niebauer LJ, Crabtree BF, Stange KC, Hassmiller SB. Prescription for health: changing primary care practice to foster healthy behaviors. Ann Fam Med. 2005;3 Suppl 2:S4–12.
27. Smith KP, Christakis NA. Social networks and health. Annu Rev Sociol. 2008;34:405–29.
28. Sluzki CE. Personal social networks and health: conceptual and clinical implications of their reciprocal impact. Fam Syst Health. 2010;28(1):1–18.
29. Harris G. Talk does not pay, so psychiatry turns to drug therapy. New York Times, March 5, 2011
30. Heifetz RA. Leadership without easy answers. Cambridge: Harvard University Press; 1994.
31. Heifetz RA, Grashow A, Linsky M. The practice of adaptive leadership: tools and tactics for changing your organization and the world. Boston: Harvard Business Press; 2009.
32. Shanafelt T, Sloan J, Habermann T. The well-being of physicians. Am J Med. 2003;114(6):513–7.
33. Heifetz RA, Linsky M. Leadership on the line: staying alive through the dangers of leading. Boston: Harvard Business School Press; 2002. p. 187–206.
34. Guarnieri M. The heart speaks: a cardiologist reveals the secret language of healing. New York: Touchstone; 2007.
35. Bohart AC, Tallman K. How clients make therapy work: the process of active self-healing. Washington, DC: American Psychological Association; 1994.
36. Illich I. Limits to medicine. Medical nemesis: the expropriation of health. London: Marion Boyars; 1975/2002.
37. Pascale R, Sternin J, Sternin M. The power of positive deviance: how unlikely innovators solve the world's toughest problems. Boston: Harvard Business Press; 2010.
38. Jackson MC. Systems thinking: creative holism for managers. Chichester: Wiley; 2003.
39. Stacey RD. Complex responsive processes in organizations: learning and knowledge creation. London: Routledge; 2001.
40. Ashby WR. Requisite variety and its implications for the control of complex systems. Cybernetica. 1958;1(2):83–99.
41. Rollnick S, Mason P, Butler C. Health behavior change: a guide for practitioners. Edinburgh: Churchill Livingstone; 1999.
42. Prochaska JO, Redding CA, Evers KE. The transtheoretical model and stages of change. In: Glanz K, Rimer BK, Viswanath K, editors. Health behavior and health education: theory, research and practice. 4th ed. San Francisco: Jossey-Bass; 2008.
43. Makoul G, Clayman ML. An integrative model of shared decision making in medical encounters. Patient Educ Couns. 2006;60:301–12.
44. Beach MC, Inui T, The Relationship-Centered Care Research Network. Relationship-centered care. J Gen Intern Med. 2006;21:S3–8.
45. Fortney L, Rakel D, Rindfleisch JA, Mallory J. Introduction to integrative primary care: the health-oriented clinic. In: Rindfleisch JA, editor. Integrative medicine, Part I: Incorporating complementary/alternative modalities. Prim Care. 2010;37(1):1–12.
46. Nutting PA, Crabtree BF, Miller WL, Stnage KC, Stewart E, Jaèn C. Transforming physician practices to patient-centered medical homes: lessons from the National Demonstration Project. Health Affairs. 2011;30(3):439–45.
47. IOM (Institute of Medicine). Integrative medicine and the health of the public: a summary of the February 2009 Summit. Washington, DC: The National Academies Press; 2009.
48. United Kingdom General Medical Council. Self care. A national view in 2007. London: UK Department of Health; 2008.
49. National Institutes of Health. Demographic and social factors. Cognitive and emotional health project: The healthy brain. [Internet]. http://trans.nih.gov/CEHP/HBPsocialconnectmeasures.htm. Accessed 28 July 2011.
50. Safford MM, Allison JJ, Kiefe CI. Patient complexity: more than comorbidity. The vector model of complexity. J Gen Intern Med. 2007;22 Suppl 3:382–90.
51. Stiefel FC, Huyse FJ, S lner W, Slaets JPJ, Lyons JS, Latour CHM, van der Wal N, de Jonge P. Operationalizing integrated care on a clinical level: the INTERMED project. Med Clin North Am. 2006;90:713–58.
52. de Jonge P, Huyse FJ, Stiefel FC. Case and care complexity in the medically ill. Med Clin North Am. 2006;90:679–92.
53. Holmes TH, Rahe RH. The social readjustment rating scale. J Psychosom Res. 1967;11:213–8.
54. Williams GC, Deci EL, Ryan RM. Building health-care partnerships by supporting autonomy: promoting maintained behavior change and positive health outcomes. In: Suchman AL, Botelho RJ, Hinton-Walker P, editors. Partnerships in healthcare: transforming relational process. Rochester: University of Rochester Press; 1998.
55. Hibbard JH, Mahoney ER, Stock R, Tusler M. Do increases in patient activation result in improved self-management behaviors? Health Serv Res. 2007;42(4):1443–63.
56. Schmeichel BJ, Baumeister RF. Self-regulatory strength. In: Baumeister RF, Vohs KD, editors. Handbook of self-regulation: research, theory, and applications. New York: The Guilford Press; 2004. p. 84–98.

57. Feddock CA. The lost art of clinical skills. Am J Med. 2007;120(4):370–8.
58. Engel GL. The need for a new medical model: a challenge for biomedicine. Science. 1977;196(4286): 129–36.
59. Goldberger AL. Fractal variability versus pathologic periodicity: complexity loss and stereotypy in disease. Perspect Biol Med. 1997;40(4):543–61.
60. Goldberger AL, Giles F. Filley lecture. complex systems. Proc Am Thorac Soc. 2006;3(6):467–71. doi:10.1513/pats.200603-028MS.
61. Goldberger AL, West BJ. Chaos and order in the human body. MD Comput. 1992;9(1):25–34.
62. Kegan R, Lahey LL. Immunity to change: how to overcome it and unlock potential in yourself and your organization. Boston: Harvard Business School Publishing; 2009.
63. Podsadecki TJ, Vrijens BC, Tousset EP, Rode RA, Hanna GJ. "White coat compliance" limits the reliability of therapeutic drug monitoring in HIV-1-infected patients. HIV Clin Trials. 2008;9(4):238–46.
64. Cramer JA, Scheyer RD, Mattson RH. Compliance declines between clinic visits. Arch Intern Med. 1990;150:1509–10.
65. Scott JG, Cohen D, DiCicco-Bloom B, Miller WL, Stange KC, Crabtree BF. Understanding healing relationships in primary care. Ann Fam Med. 2008;6: 315–22.
66. Gittell JH, Fairfield KM, Bierbaum B, Head W, Jackson R, Kelly M, Laskin R, Lipson S, Siliski J, Thornhill T, Zuckerman J. Impact of relational coordination on quality of care, postoperative pain and functioning, and length of stay: a nine-hospital study of surgical patients. Med Care. 2000;38(8):807–19.
67. Gurtner A, Tschan F, Semmer NK, Nägele C. Getting groups to develop good strategies: effects of reflexivity interventions on team process, team performance, and shared mental models. Org Behav Hum Decis Process. 2007;102:127–42.
68. Seligman ME, Rashid T, Parks AC. Positive psychotherapy. Am Psychol. 2006;61(8):774–88. doi:10.1037/0003-066X.61.8.774.
69. Sirovich S, Gallagher PM, Wennberg DE, Fisher ES. Discretionary decision making by primary care physicians and the cost of U.S. healthcare. Health Affairs. 2008;27(3):813–23.
70. Gigerenzer G, Wegwarth O, Feufel M. Misleading communication of risk. BMJ. 2010;341:c4830.
71. Gigerenzer G, Edwards A. Simple tools for understanding risks: from innumeracy to insight. BMJ. 2003;327:741–4.
72. Solberg LI. How can we remodel practices into medical homes without a blueprint or a bank account? J Ambul Care Manage. 2011;34(1):3–9.
73. Nutting PA, Crabtree BF, Miller WL, Stewart EE, Stange KC, Jaén CR. Journey to the patient-centered medical home: a qualitative analysis of the experiences of practices in the National Demonstration Project. Ann Fam Med. 2010;8 Suppl 1:s45–56.
74. Glasgow RE, Davis CL, Funnell MM, Beck A. Implementing practical interventions to support chronic illness self-management in health care settings: lessons learned and recommendations. Jt Comm J Qual Saf. 2003;29(11):563–74.
75. Glasgow RE, Ockene J, Pronk N, Goldstein MG. Translating what we have learned into practice: Principles and hypotheses for addressing multiple behaviors in primary care. Am J Prevent Med. 2004;27(2 Suppl):88–101.
76. Khaw KT, Wareham N, Bingham S, Welch A, Luben R, Day N. Combined impact of health behaviors and mortality in men and women: The EPIC-Norfolk Prospective Population Study. PLoS Med. 2008;5(1):e12. doi:10.1371/journal.Pmed.0050012.
77. Rogers CR. The necessary and sufficient conditions of therapeutic personality change. J Consult Psychol. 1957;21(2):95–103.
78. Zimmerman B, Lindberg C, Plsek P. Edgeware: Lessons from complexity science for health care leaders. Washington, DC: VHA; 1998.
79. Singhal A, Buscell P, Lindberg C. Inviting everyone: healing healthcare through positive deviance. Bordentown, NJ: Plexus; 2010.
80. Begun JW, White KR. The challenge of change: inspiring leadership. In: Lindberg C, Nash S, Lindberg C, editors. On the edge: nursing in the age of complexity. Bordentown, NJ: Plexus; 2008.
81. Stokols D. Establishing and maintaining. healthy environments: toward a social ecology of health promotion. Am Psychol. 1992;42(1):6–22.
82. Johnson B. Polarity management: identifying and managing unsolvable problems. Amherst: HRD Press; 1992.

39 Mayo Clinic: Making Complex Healthcare Simpler

Kent D. Seltman and Leonard L. Berry

"… *Mayo has accomplished in 7 weeks what the outside world could not accomplish in 2 years … What Mayo did the first day that we were here would have taken easily a month or longer at home. And again, [the institutions at home are] renowned medical facilities. It was like test scheduling—you would walk from the exam room to the front desk, and they gave you an itinerary and you already had the test scheduled—that would have taken 2 weeks at home. [At Mayo] it was already set up, and you were having it done in an hour. And then the results of that were quickly sent back to the doctor. Then the doctor was back to you. It was just amazing*" [1]. This account taken from a video posted on the Internet recounts the experience of a seriously ill patient who travelled halfway across the USA for care at Mayo Clinic in Rochester, Minnesota. The patient and her spouse are surprised at how Mayo Clinic dealt with the needs of this patient—so quickly, efficiently, and simply.

For this patient, Mayo Clinic was functioning the way it is designed to function—which happens most of the time, but not always. Mayo Clinic is an exemplary institution, but it is not perfect by any means. Like other organizations, Mayo is on a perpetual journey to improve. The authors of this chapter have studied Mayo Clinic carefully for more than a decade and have seen several significant improvements over this time. We describe here how Mayo Clinic typically serves its patients today. This chapter is based in part on the research we did for our book-length study of Mayo Clinic [2] and also on interviews we have done following the book's publication.

For more than a century, Mayo Clinic has created clinical and interpersonal experiences that pleasantly surprise patients. Often these are service experiences that the patient and family could not even imagine as possible. Patient stories have created a steady demand from more than seven million additional patients who have returned home with their own stories. These stories shared with their family and friends throughout the USA and across the world have made Mayo Clinic the most preferred US provider for complex medical care according to an independent national survey [3].

The story behind Mayo's success is not about a carefully wrought strategy devised a century ago, but rather it is about a simple concept: *the needs of the patient come first*. Mayo Clinic was established by doctors who first articulated this idea shortly before their clinic began growing patient volumes by more than 100% per year for a few years early in the second decade of the twentieth century.

The concept worked then, and it still works a century later. Today, "The Needs of the Patient

K.D. Seltman, Ph.D. (✉)
Mayo Clinic, 802 Fox Pointe Lane, SW Rochester, MN 55902, USA
e-mail: kent@seltman.com; seltman.kent@gmail.com; seltman.kent@mayo.edu

L.L. Berry, Ph.D.
Mays School of Business, Texas A&M University, 201P Wehner Building, Marketing Department, College Station, TX 77843-4112, USA
e-mail: berryle@tamu.edu

J.P. Sturmberg and C.M. Martin (eds.), *Handbook of Systems and Complexity in Health*,
DOI 10.1007/978-1-4614-4998-0_39, © Springer Science+Business Media New York 2013

Come First" is identified as the primary value of Mayo Clinic. But the statement is far more than that, for it serves also as the primary strategy. Certainly, Mayo Clinic does have a detailed strategic plan to guide its direction on many fronts, but the strategic plan constantly looks back to the primary value. In directing both strategy and operations, leaders and care providers think and ask as they go about their work each day: "What would patients need to recover more quickly from surgery?" "Is this equipment purchase actually required to better meet the needs of our patients?" "How will this change affect the experience of the patient?"

With so much focus in management circles on the "new"—new concepts, new theories, new models, and new technologies—it is refreshing, revealing, and inspiring to learn from a world-renowned institution that established its foundation for success in the early 1900s and continues to depend on that foundation in its twenty-first century success. Mayo Clinic illustrates that an organization's basic concept of the business can be so right that it can endure for ages. Mayo Clinic also shows that a not-for-profit organization can operate so well that most other businesses can learn from it. The Clinic showcases the promise of the "modern–traditional" enterprise that aligns strategy with values, innovation with tradition, talent with teamwork, and science with art.

In our analysis of Mayo Clinic we show, first, how the understanding of the simple statement, "the needs of the patient come first," has a living presence inside Mayo and, thus, is relevant to virtually each of the 45,000 employees who currently work at the Clinic. Second, we look at "teamwork," another simple value/strategy that governs how Mayo Clinic goes about meeting the patient's needs. Third, we describe how service delivery systems create the infrastructure enabling efficient, time-compressed delivery of care. Finally, we describe how the melding of outpatient (clinic) and inpatient (hospital) services into a single organization facilitates healthcare delivery by aligning virtually all employees on the best interests of the individual patient being served.

39.1 The Patient-First Legacy

Dr. William J. Mayo and his brother Dr. Charles H. Mayo along with their father Dr. William Worrall Mayo founded Mayo Clinic. As the trio worked together in the family business, they designed new surgical tools and new surgical procedures that brought them wide recognition in their lifetimes. But those achievements have been largely superseded by a century of new medical science. Their most enduring contribution to society is the model of patient-focused care that they developed, along with the institution and culture to deliver patient-focused care.

Dr. Will, the older brother, eloquently articulated the patient-first value in a commencement address to the Rush Medical College in 1910: "*The best interest of the patient is the only interest to be considered, and in order that the sick may have the benefit of advancing knowledge, union of forces is necessary. … It has become necessary to develop medicine as a cooperative science*" [4]. This statement, which identifies two complementary values, has shaped Mayo Clinic for each succeeding generation. Mayo Clinic's contemporary pronouncement of its core value—"the needs of the patient come first"—is obviously derived from this statement. As Dr. Mayo's statement suggests, the needs of the patient can be met only by partnerships and teamwork among the employees providing and supporting patient care.

"The needs of the patient come first" core value has significantly contributed to the long-term success of Mayo Clinic. It is relevant and important to Mayo Clinic's primary customers: patients, their families, and referring physicians. But the value is also relevant and important to each employee of Mayo Clinic—the doctors, nurses, clinical technicians, administrators, and large support staff. As a physician noted, "The maintenance worker who replaces the light bulb in my office sees his work as enabling me to care for my patients." An accreditation inspector asked a custodial staff member what she did; the answer surprised the inspector—"I save lives" was the response of the worker who was cleaning a hospital room for the next patient.

This housekeeping employee's response flowed from the core of her soul—it was not a response that she had been trained or "scripted" to give. She reveals that she did not see her job as just cleaning a room for the next patient; rather she understood her work as an important contribution as a member of the patient care team. Caregivers and service workers in healthcare are typically people who love to help others, so they feel rewarded by giving their best to those they serve. The "needs of the patient" value comes to life anew each day through the human experiences orchestrated by staff in their labour-intensive services delivered to patients and their families.

Many healthcare providers aspire to put the needs of patients first. However, in some organizations the focus on patients' needs is little more than a marketing slogan on a billboard or on a lapel pin on a doctor's lab coat. In other organizations, the value is an admirable and aspirational goal but not yet realized. Few, if any, healthcare organizations succeed with the consistency of Mayo Clinic in keeping the spirit of "the needs of the patient" alive and relevant in daily tasks. After living with this value for more than a century, "the needs of the patient come first" has become integrally woven into the fabric of the institution.

When a value becomes "part of an employee's DNA," it directs not only the way the day-to-day work is performed, but it also gives employees the power and moral authority to act in unique situations. Explicit permission is not needed when an employee sees a patient need that requires action. If the employee's choices are getting back to work on time or taking 10 minutes to get a wheelchair for a patient who seems unsteady, the patient will get a wheelchair. Exceptional service frequently results when employees spontaneously act on the authorization provided by this value.

Matthew McElrath, formerly the Chair of Human Resources for Mayo Clinic Arizona, shares this story of employee empowerment:

> I … ended up as a patient in the ICU at Mayo Clinic Hospital. Dr. Trastek [CEO of Mayo Clinic Arizona at that time] and his wife were returning to Arizona from a trip to Rochester and learned of my being hospitalized and came to the hospital to visit me.
>
> What was remarkable wasn't that they came to visit me—I was deeply touched that they wanted to come and see me, but what made it so remarkable was that … the nurse let me sleep and turned Dr. Trastek away at my door.
>
> When I woke later that day, she said to me, "You had some visitors, but I turned them away. I hope you don't mind—but one in particular I feel a little strange about."
>
> I asked, 'Why?'
>
> And she said, "Dr. Trastek and his wife came by to see you … I told them that you were sleeping and I really wanted you to sleep."
>
> I said, "Thank you very much, it's the best thing, I'll catch up with him later."
>
> And she was like, "Is that OK—you don't mind?"
>
> I said, "Absolutely." And I thought to myself, "Here is a great example where the nurse asked herself what is the best thing to do for the patient and that's what she did." She knew that the best thing for me was to sleep and even if it meant her shooing away the CEO from my door.

Mayo Clinic patients often comment on the extra time given by a doctor when "bad news" is shared. Others see unexpected empathy when they notice through their own tears that the doctor's eyes have tears as well. Occasionally, we hear from patients that a Mayo Clinic doctor shared a home phone number or a personal cell phone number in the event that a critical medical problem occurred. Another patient was thrilled to have a nurse practitioner listen to her 45-min account of her chronic medical problem, and then the surgeon came in and said that he thought he could fix it—and he did. This came after four other well-known physicians at other institutions refused to take the time to listen to this patient. The common theme in these and hundreds of other stories is that special, unexpected support and care were offered in the patient's moment of great need. In each case, the Mayo employee earns some personal satisfaction from helping Mayo Clinic deliver on its promise of making "the needs of the patient come first."

Patient centricity also functions in deliberations in the committees and governing boards. A retired

Mayo Clinic CEO, Dr. Robert R. Waller, recalls that in his years on the Board of Governors he and his colleagues frequently faced complex decisions. When it was difficult to achieve unanimity, someone would invariably ask, "Yes, but what is best for the patient?" Shirley Weis, Chief Administrative Officer of Mayo Clinic, echoes Dr. Waller, "Our value of patient-needs-first helps cut through a lot of chatter in meetings. Just ask, 'Is this right for the patient or not?' And that gets you centred properly on the issue." She illustrates with an example dealing with the electronic medical record; the issue concerned how quickly physicians would need to return to a computer before they would need to sign on again. The sign-on routine is an annoyance, of course. However, when one physician asked what patients would feel best meets their needs for privacy, a decision was quickly reached that favoured patients' needs.

The pervasive presence of the "needs of the patients" in the minds of 45,000 employees provides focus for the entire workforce. It facilitates decision-making, defines appropriate interpersonal behaviours, humanizes a tissue sample in a laboratory, and brings special meaning to the work of staff cleaning patient rooms. The value becomes a touchstone that simplifies and clarifies much that happens inside Mayo Clinic each day.

39.2 Team-Based Medical Care

The Mayo Brothers, Dr. Will and Dr. Charlie, were surgeons in the era just following the development of safe, reliable anaesthesia which for the first time gave surgeons time to perform complex surgeries of the abdomen, chest, neck, and head. Their surgical skills plus their early adoption of antiseptic and, shortly thereafter, aseptic methods resulted in large numbers of successful outcomes. Their clinical results were conveyed across the USA and around the world by the accounts of patients and family members. These patient stories, as well as the brothers' publication of their experiences in medical journals, were the basis of their growing fame which in turn created high demand for their services as early as the mid-1890s. To help meet the demand of patients coming to Rochester, Minnesota, the Mayo Brothers began hiring other doctors to triage those who would benefit from surgery and to follow up with the post-surgery care that patients required.

The Mayo Brothers were members of the early fraternity of surgical stars, but to their credit, they had the humility and insight to recognize that the doctors who were working for them as salaried physicians were developing expertise that the brothers, as surgeons, did not have. One colleague excelled at differential diagnosis through the analysis of blood, urine, and gastric fluids. Another focused on the use of the new X-ray technology in diagnosis; and another doctor specialized in pathology of surgical specimens. As the practice grew, another doctor became a student of kidney disorders. This was in the era when most of the doctors in the USA were general practitioners who both performed surgeries and provided medical treatment and thought that each doctor needed to know everything. Specifically, the specialization of doctors that was unfolding in the Mayo Brothers' practice came before the formal conceptualization of medical and surgical specialties that exist today.

By offering their sincere respect for the expertise developing in their salaried staff members, the brothers came to identify the second important value—team-based care—that underpins the Mayo Clinic model of clinical care. At the beginning of the twentieth century, the Mayo Brothers recognized that medicine was so complex that no single person could know everything. They proceeded to develop their team into what is known today as an integrated, multi-specialty group practice. To expedite diagnosis and treatment of patients, Mayo Clinic has assembled what is today a group of more than 2,800 staff physicians, medical scientists, and clinical and research associates. This team provides the deep expertise and understanding necessary to diagnose and treat the frequently complex and sometimes rare diseases of the more than 500,000 patients who come to Mayo Clinic each year. And importantly, these team members function together as working parts of a single complex organism.

The Mayo Brothers chose not to think of themselves as the "stars" others considered them to be for they knew that their success was the result of team effort. At Mayo Clinic, the team-based culture recognizes only the stars that stand out in a constellation. It is the team that is recognized, not the individuals. This approach simplifies the management of Mayo Clinic because it removes from the organizational culture the corrosive effect of "star-power politics" in identifying priorities and allocating resources. If a would-be star physician or administrator at Mayo Clinic tells leadership his/her demands must be met or otherwise there will be a resignation, there is likely to be a resignation; claimants of stardom do not fit well in the Mayo Clinic culture.

Most importantly, team-based medical care at Mayo Clinic facilitates the services delivered to its patients. An actual patient whom we refer to as "Don" came to Mayo Clinic with a terminal cancer diagnosis from a faculty surgeon at the university medical centre in his hometown. The surgeon had described to Don a horrific surgery that would have compromised his quality of life forever, though the doctor confided to Don's wife that he had less than a year to live with or without the surgery. Don decided quickly that he'd travel a thousand miles from home to see if he might experience one of the "miracles" that he'd heard sometimes happen at Mayo Clinic.

On his first day as a patient at Mayo, Don met first with a surgeon who stated, "I'm taking surgery out of the discussion as there are better ways to treat your cancer." Don was surprised and impressed to hear from a surgeon that he would not do surgery. A few hours later, Don met with a medical oncologist and a radiation oncologist who were experts in treatment of tumours of the head and neck. And before that first day at Mayo Clinic ended, the three physicians gathered in an exam room with Don and his wife to describe their recommendations: a 3-month regimen of chemotherapy and radiation treatments that they described as very uncomfortable and challenging. Don took this advice, and because they were so impressed after his first day at Mayo Clinic, he and his wife chose to live in a hotel room in Rochester for what they knew would be the worst 3 months of their lives. Now, 7 years later, Don is cancer-free and living a full life—one not compromised by the treatment that saved him.

Don's story illustrates very well the concept of team medicine at Mayo Clinic. A treatment team was assembled for Don in a few minutes after the Mayo surgeon knew what needed to be done. The surgeon, as the "quarterback" of Don's care, contacted the two oncology colleagues who became members of Don's three-physician team—a team that still exists for Don's cancer monitoring today. But, of course, there were more than physician players on Don's care team—the nurses who administered his chemotherapy and the radiation technologist who administered the radiation treatments. And in addition to those clinical players, Don also counts Rose as one of his Mayo team members. Rose is a receptionist in radiation oncology; her cheerful face and kind words of encouragement helped buoy Don's spirits in the nadir of his treatment journey.

The Mayo Brothers also included the nurses at Saint Marys Hospital as important members of the care team, and as additional healthcare professions have developed over the last 100 years, they have also joined the doctors and nurses as team members. Many of these workers labour in laboratories or administrative functions away from direct contact with patients—but their distance does not diminish their importance. Some ascertain, for instance, that transfused blood is a safe match for the patient. Others identify the pathogens infecting a patient's body as well as the sensitivity of the pathogens to potential medications.

Teamwork is so important that employment screening of all new employees—doctors through the rest of the workforce—emphasizes the identification of team players. Prospective employees who ask about "how much autonomy" they will have may well be screened out. Over 60% of Mayo's physicians have had some training at Mayo—typically between 1 and 8 years. Those trainees who are invited to join the staff will have demonstrated their internal values during their training with the patient-first teamwork values being of utmost importance. Mayo Clinic has trained many healthcare professionals who have not wanted to join the staff as they feel constrained by teamwork. Many of these have achieved stardom elsewhere. Employees throughout the organization test their comfort with the teamwork atmosphere during their early years of employment. Those who are not comfortable typically resign within the first 3–5 years.

Another contributor to the teamwork atmosphere of Mayo Clinic is the policy of straight-salary employment. This means, for instance, that a surgeon's salary is not impacted either by performing or not performing surgery on a given patient. Accordingly, the surgeon who told Don that he would not perform surgery did not make any personal financial sacrifice. Often a Mayo physician or surgeon will refer a patient to a colleague in the same specialty if he/she feels that the colleague would be able to better serve the needs of the patient.

Since money is used as a motivator in many, if not most, businesses in the USA, those looking in from outside Mayo Clinic wonder how Mayo manages to attain the high levels of productivity needed to survive financially without the use of incentive compensation. Several factors keep Mayo Clinic staff members motivated to perform high volume and high quality work. Typically, Mayo Clinic's physicians and its administrative leaders and the allied health professionals have high levels of intrinsic motivation—they have not been paid by anyone to earn their high grades and academic honours in their training. Additionally, people who choose to work at Mayo Clinic are driven more by altruistic than by financial motives. Mayo Clinic is a not-for-profit organization, so they know from the outset that Mayo Clinic is not focused on creating wealth—corporate wealth or personal wealth. Also important is the peer pressure that individuals feel from highly productive colleagues in the next exam room or office cubicle. Physicians in particular are also motivated by the transparency afforded by the common medical record that collects everything about an individual patient in one document. All Mayo doctors are keenly aware that their clinical notes, the clinical studies they request, the diagnoses they make, and the treatments they offer will be read by world-class experts who work with them on the patient's medical team. No one wants to be perceived as a marginal doctor by his/her colleagues. Work-life becomes uncomfortable for those who feel they may not be measuring up to Mayo standards of productivity and quality by their peers. Mayo Clinic's stellar reputation around the world also sets the internal standards for all Mayo Clinic employees—doctors, receptionists, and custodians—as they don't want to be the one who lets down the Mayo Clinic brand.

39.3 Service Systems

Though invisible to most customers, great service organizations operate effectively because of systems that make it easy and natural for front-line service providers to perform effectively and efficiently. The "amazing" service referenced in the testimonial at the beginning of this chapter did not result from heroic behaviours by the patient's team members. Having the next appointments scheduled by the time a patient moves from the exam room to the checkout desk is embedded in Mayo's operational model. Based on years of experience in other healthcare organizations, the patient in the opening story could not have even imagined that something this simple could ever exist.

Beginning with the hiring of Dr. Henry Plummer in 1901, Mayo Clinic has enjoyed the benefit of industrial engineers and systems designers. Though trained as a physician and hired to be in charge of the laboratories, Plummer had a natural aptitude for engineering and systems thinking. He was not pigeonholed by his assigned tasks in the laboratory, and in 1905, Dr. Plummer created the integrated medical record of each individual patient at Mayo Clinic. Prior to this, Mayo physicians—like most physicians in that day—kept the clinical notes on each patient in a book they maintained in their office. Plummer understood the value of having everyone functioning as part of a single organism focused on the individual patient. With the integrated medical record, everyone's contributions were in one place for all to use in the care of the total patient.

But to realize the benefit of the single record, Plummer and his successors needed to engineer the means to physically deliver the patient's paper medical record to each doctor or laboratory that needed to see it. So systems of lifts and chutes and vacuum tubes were built into Mayo's clinic buildings until the electronic transmission of medical records was implemented beginning in the 1990s.

The centralized system that amazes patients with appointments scheduled so promptly and effortlessly is what provided Don with appointments with three different doctors on his first day at Mayo Clinic. The central appointment system that exists today began to take shape in the early 1950s and today is largely computerized. Initially, the rules for different kinds of appointments were collected in 3-ring binders, and the most common appointment types were memorized by many experienced schedulers. But by the late 1990s that system was beginning to collapse from the weight of its own complexity. (When a grocery cart heaped with 3-ring binders was rolled into a Mayo Clinic committee meeting, the visual exhibit delivered the mandate for change.) At that time, almost all of the millions of physician and laboratory appointments scheduled each year required one Mayo employee to speak to another Mayo employee by phone.

By 2008, scheduling was largely managed by a computer that uses decision rules for over 8,000 different types of appointments to optimize an appointment itinerary for each patient. The computer sequences appointments based on rules—for example, waiting times required between tests after one test uses a contrast solution. The appointment system also considers the travel time between clinic locations and even includes wait times for riding elevators. (Times were determined by industrial engineers who pushed a colleague with a stopwatch in a wheelchair from one place to another on campus.) The appointment system can now quickly identify the best sequence and timing of appointments for the patient based on the complex rules matched against open appointment slots in the calendars of laboratories and physicians. This is a clear example where the needs of the patient helped drive simplicity by rooting out unnecessary complexity.

For this appointment system to work, physicians must surrender personal control of their appointment calendar to a computer that uses the rules they provide. When appointments were scheduled by two employees speaking by phone, the physician's secretary on one end sometimes looked after the best interest of the doctor. But the computerized system provides results that have satisfied sceptics while eliminating millions of dollars of expense incurred by the old labour-intensive system. Moreover, the computerized system better matches Mayo's primary value—the needs of the patient come first.

The appointment system also uses complex algorithms to predict appointment demands. For example, using the past yields of neurology consultations from every 100 new general medical exams, the program can identify the number of neurology consultation appointment slots that will be needed 12 weeks in the future based on the number of general medical appointments already scheduled. The methodology, of course, works for other specialty practices as well. Open slots are built into the calendars of down-stream practices. So the ability of Mayo Clinic to accommodate the request from Don's surgeon for same-day or next-day consultations in medical oncology and radiation oncology is built into the operating procedures.

The system also can detect growth patterns that will identify prospective demand, say, for an additional MRI scanner or an additional gastroenterologist. Anticipating appointment logjams before they occur enables Mayo to maintain a more consistent service experience for its patients. In the past, Mayo has unintentionally created dissatisfied patients by schedule bottlenecks. Lack of slots for an MRI or a physician in a given specialty meant that patients' appointments could not be scheduled within a time frame that matched the expectations and tolerance levels of patients. Today, one of the most important operational indicators tracked is the percent of patients completing their diagnostic itineraries within the week that they begin.

As we write, Mayo Clinic is focused on creating an even more flexible healthcare delivery system that incorporates information technology to deliver monitoring services to patients as well as remote consultations with physicians. Already, Mayo Clinic has in place a program for referring physicians to help determine the need to send a patient to Mayo Clinic. This service joins the referring physician and patient on a secure, live Internet video link to discuss the patient's medical history, interpret test results, and determine appropriate next steps. This technology can, of course, work around the world as long as the appropriate broadband infrastructure is in place.

One of the most humane elements of Mayo Clinic's delivery of clinical care is the rapid turnaround of test results. Radiologists begin reading imaging studies as soon as the images are completed rather than "batching" them for reading at a later time, be it the evening or the next day or two. So, women in the breast clinic with a suspicious lump do not need to endure long delays for results after a mammogram as the report typically comes to their breast physician within an hour or two. In the echocardiography laboratory, the cardiologist posts the results of the exam to the patient's medical record within 5 min after the test is completed. Immediately after drawing a patient's blood in the outpatient lab, the phlebotomist puts the labelled tubes of blood on a conveyor to the lab or passes them through a small window in the wall. Technologists then place most samples into a computerized blood analyser that automatically sends the results to the medical record of the patient usually within 30–60 min.

These examples cover a small sample of the work performed in recent years by the approximately 100 industrial engineers and computer system experts who design and implement new systems and system improvements for Mayo Clinic. The guiding themes that run through these projects are (1) simplification of operations will almost always improve the patient's experience, (2) excellent systems enable employees to deliver consistent service without extraordinary effort, and (3) efficient service delivery reduces the cost of service for patients, Mayo Clinic, and, ultimately, for society.

39.4 One Mayo Clinic

What we have described above is possible largely because Mayo Clinic is one organization. Although Mayo Clinic operates large group practices in Jacksonville, Florida, Scottsdale/Phoenix, Arizona, and Rochester, Minnesota, it is functioning today as a single institution. One Board of Governors guides the overall organization. Each campus has an executive operations team that oversees campus operations. The group practices in Rochester, Florida, and Arizona each operate both outpatient clinics and hospitals on their campuses. The hospitals and the clinics on each campus operate as a single organization. Thus, Mayo Clinic can be accurately described as a vertically integrated healthcare delivery system.

It was not always this way. As founded by the Mayo Brothers, Mayo Clinic was just an outpatient clinic and the laboratories needed to support the outpatient practice. Beginning in the early days of the original Dr. Mayo—the father, William Worrall Mayo—the doctors in the community used hospitals and hotels in Rochester for their patients. But in response to a community need sensed by the Sisters of St. Francis of Assisi, the sisters were determined to build a new hospital, and the three Doctors Mayo agreed they would use the hospital if it were built. The hospital opened in 1889 as a community hospital, but this was an era in which a great deal of anti-Catholic bias raged through the American Midwest. The Mayos stood by their promise to have their patients cared for by the Franciscan Sisters, but other community physicians did not take their lead. Thus, by default rather than design, Saint Marys Hospital became a closed-staff hospital devoted to the care of Mayo Clinic patients. Because Saint Marys Hospital could not accommodate all their patients, the Mayo physicians continued to use another Rochester hospital that predated Saint Marys; that hospital became Rochester Methodist Hospital 1954. Like Saint Marys, Rochester Methodist Hospital was also a closed-staff hospital for the exclusive use of Mayo Clinic physicians.

While Mayo Clinic did not operate either of these hospitals, the working relationships between the clinic and the hospitals were very symbiotic, unlike the typical community hospitals where the voluntary medical staff is structured out of many independent physician groups that compete with one another and sometimes with the hospital itself. In early 1980s, it became clear to the Franciscan Sisters that their order was no longer bringing in sufficient new members for them to continue operation of Saint Marys Hospital, so they offered to deed the hospital to Mayo Clinic in an exchange for an agreement to continue operating the hospital as a Catholic institution. For different reasons, the board of Rochester Methodist Hospital also decided to present that institution to Mayo Clinic. Both transactions were signed on the same day in 1986.

These transactions provided the opportunity to simplify the management of the hospitals. As these institutions were assimilated into a single organization—Mayo Clinic—two significant changes occurred. Where the three entities had each had a board of trustees to oversee their operations, after the merger just one remained—the Mayo Clinic Board of Trustees. Previously, if Mayo Clinic wanted to move a clinical service such as transplantation from one hospital to the other, that move would have an impact on the finances and operations of both hospitals. Getting hospital board approval was time consuming. Further, inefficiencies resulted from duplications when both hospitals, for instance, operated emergency departments.

More significantly, however, the traditional administrative leadership of the hospitals gradually faded away over a period of more than 5 years as one after another of the clinical and administrative groups from the two hospitals were merged with their counterparts at the clinic. For instance, the finance operations of the hospitals were absorbed into Mayo Clinic's finance department. Support services such as housekeeping, maintenance, grounds, and the mailroom were each integrated into respective campus-wide departments that served everything Mayo—the two hospitals and all the clinic and laboratory facilities.

Nursing was one of the last merged and perhaps the most important in the operation of the hospitals. Since Mayo had not centralized management of the dozens of nurses at the clinic, the merger initially involved just the staffs in the two hospitals. A nurse administrator from one of the hospitals was appointed as the Chair of the Nursing Department of Mayo Clinic. This title resulted from a deliberate decision to create an organizational nomenclature analogous to that used with the medical staff. Since physician department chairs had reported to the physician chair of the Board of Governors beginning in the 1920s, the new Chair of the Department of Nursing reported to the Chief Administrative Officer who was the administrative counterpart of the physician chair of the Board of Governors. Ultimately, the nurses working in the clinic's outpatient operations joined their inpatient colleagues in the Nursing Department. Today, surgeons and nurse managers in the hospitals jointly manage the operating rooms and surgical nursing units, and this model is replicated for all hospital nursing units. Thus, a physician from the discipline such as cardiology works with the nurse manager for the unit serving, for instance, cardiology patients. The core clinical services such as radiology and laboratory medicine seamlessly serve both outpatient and inpatient operations.

Ultimately, clinical management of the hospitals was assigned to the Clinical Practice Committee, one of the most important of Mayo's operating committees. The Clinical Practice Committee and its subcommittees oversee all clinical activities on each campus. To accommodate oversight of the hospitals, the Clinical Practice Committee created two sub-committees: the Outpatient Practice Committee and the Hospital Practice Committee. The physician chair of the Hospital Practice Committee is the single individual ultimately responsible for the operation of the hospitals. In a minor variation of the physician/administrator dyad that pervades throughout clinical management at Mayo, the team for the hospitals is a triad—the physician, the chair of nursing and chair of hospital operations. Unlike a typical hospital administrator, this physician focuses primarily on clinical activities and patient care in the hospitals. All the other management functions are wrapped into campus-wide management.

While leaders carefully track utilization and expenses at many levels of campus operations, the hospital and outpatient operations financials are not reported separately. Likewise, financial summaries of the various clinical departments are not reported because each is important only as part of the whole organization. From the outset, Mayo Clinic adopted a "one-bucket" model for managing its revenues—surgeons, neurologists, endocrinologists, and all other doctors are paid with Mayo Clinic money, not just the money derived from their clinical practices. Mayo Clinic's group model demands a comprehensive coverage of specialties, some of which are reimbursed well and others that are not. The one-bucket model enables Mayo's leaders to fund specialties and clinical services that do not fully pay their own way but are important to address comprehensively the many needs of patients. Mayo staff salaries across all disciplines are competitive with other premier academic practices. All Mayo Clinic employees—nurses, medical technologists, administrators, computer technicians, physicians, and the security officers—receive their salary from the same bucket of revenue. This arrangement goes a long way in aligning all employees on the same priorities—particularly the primary value of the "patient's needs."

In typical American communities, administrators trained in business disciplines manage hospitals. Hospital boards are made up largely of the community's business leaders. As a result hospital management can be focused on numbers—financial numbers. In trying to keep the focus on patients, Mayo Clinic is run by doctors. These are actual care-giving doctors—doctors who maintain a clinical practice while overseeing and leading operations. Further, Mayo Clinic is led by highly trained physicians who typically have little academic training in business and finance. As practicing physicians, the doctors leading Mayo Clinic are expected to advocate first for the best interests of the patient rather than the business side of healthcare. While physicians are the undisputed leaders of Mayo Clinic, each physician leader is paired with an administrative leader. At the highest level, the physician CEO is paired with a business-trained Chief Administrative Officer who oversees business operations. This model of the physician/administrator dyad is replicated throughout clinical operations. The physician leader of a clinical department such as ophthalmology is responsible for the medical staff and the clinical standards; the administrator of ophthalmology is responsible for managing the allied health staff supporting the outpatient practice and the clinical laboratories. The administrator supports the physician leader in developing strategy and the budget and monitoring operations, among other roles.

Mayo Clinic's operation as a single organization, as well as its dyad management model that pairs physicians and administrative leaders, dramatically simplifies the politics that confound much of healthcare management in the USA today. Struggles between physicians and administrators for power and control are moot at Mayo Clinic—physicians are in charge. However, since physicians have as much to lose—if not more—than administrators should Mayo Clinic fail to manage well, the physician–administrator pair shares the same goal of success. Though the doctor looks first at the best interest of the patient and the administrator tries to determine how to make that happen, they are both interested in the success of both the patient's best interest and the long-term financial survival/success of Mayo Clinic.

39.5 Conclusion

In the early 1980s, the Mayo Clinic Board of Governors realized they were on the cusp of major change. The conversations with the Sisters of St. Francis of Assisi and Rochester Methodist Hospital had begun; Mayo's leaders knew the hospitals would soon become the responsibility of Mayo Clinic. Mayo Clinic at that time employed about 5,000 employees; the leaders could see that Mayo Clinic's management responsibilities would soon nearly triple in size. The Board was also opening new Mayo Clinic campuses in Florida and Arizona. A major angst within the organization was the possibility that Mayo might become too large to manage. Leaders sought counsel from management consultants and professors. A consensus emerged: 5,000 employees were thought to be the maximum that could be managed effectively. The experts may have been correct in the business environment of that moment, but they had failed to account for the impact of the brewing revolution in information technology—desktop computers, the Internet, real-time video links, email, integrated back-office computer systems, and more to come. Today, with more than 9 times the maximum staff size that was recommended in the early 1980s, Mayo Clinic still operates well and is a single organization.

Mayo Clinic is not paralysed by the complexity that often comes with size because Mayo operates more simply than most other large healthcare organizations. We've discussed above the single organization, but in addition to the large structural simplicity, there are thousands of smaller simplifications. For instance, self-service systems in Human Resources simplify profile updates and information requests. HR files are more organized and complete and maintained more cost-effectively than could have been imagined in 1985. Common finance infrastructure creates more accurate and timely reporting. Moreover, by sticking to its core values and continually reinvesting in its operational infrastructure to manage complexity, Mayo Clinic has sustained its success.

In this chapter, we have focused on Mayo Clinic's mission of clinical care because it shares much in common with other healthcare providers. But Mayo also has two other complementary missions: medical education and medical research. These activities, however, operate within the management model that we have described here.

Our intent has been to discuss the ways that Mayo Clinic has simplified structures and processes in healthcare delivery with the goal of encouraging other organizations to do the same. We have described three major themes:

1. Guiding values—(a) the needs of the patient come first, and (b) teamwork in healthcare delivery. Together these simplify strategic and operational priorities—the "what" and the "how" of Mayo Clinic.
2. The infrastructure designed to create and maintain superb customer service—(a) the single medical record that captures all inpatient and outpatient services delivered to each individual patient, and (b) the integrated, computerized appointment management system. A common medical record is within the grasp of all medical organizations today.
3. Successful operation as one organization that benefits from organizational simplifications: (a) melding of hospital and clinic operations, (b) a "one-bucket" revenue principle where revenue is allocated to serve the best interest of patients, and (c) one governing board with local management teams overseeing operations at each practice site.

In essence, Mayo Clinic institutionalizes simplicity with a focus on a primary value that is part of the fabric of the organization, with insistence on a no-star, teamwork model for clinical care as well as administrative operations, with well-engineered systems that simplify operating a complex organization of 45,000 employees, and with a single-organization model designed both to simplify operations through common information technology and minimize organizational politics.

Mayo Clinic has had 12 CEOs over its history. Each has been a physician who has demonstrated fealty to the values and management and governance models originally created by the Mayo Brothers. No new leader has been hired at Mayo with the task of fixing a broken organization. Mayo has more than a century-plus of history invested in the value of patient needs. If a new leader were to try to change Mayo's primary value, it is likely that his/her leadership career would be brief. A simple value that genuinely benefits those who are served and those who serve can endure for generations as demonstrated by the real-world experience of Mayo Clinic.

Acknowledgements We acknowledge all Mayo Clinic employees whose outstanding patient-focused clinical care and personalized service have created the story that we are privileged to share with others. In addition, we acknowledge and thank the hundreds of Mayo Clinic patients and employees who have given their time during our research so we can fully understand both the organization and their personal stories.

References

1. Al Errato's Mayo Clinic Story. http://www.youtube.com/watch?v=t0U86sBPcxc.
2. Berry LL, Seltman KD. Management lessons from Mayo Clinic. New York: McGraw-Hill; 2008.
3. Joe M. Inguanzo. PRC National Consumer Perception Study. Professional Research Corporation, Omaha, NE (publication pending).
4. William J. Mayo. The necessity of cooperation in medicine, delivered at the Rush Medical College commencement, 15 June 1910, originally published in the Collected Papers by the Staff of Saint Marys Hospital, Mayo Clinic; 1910; vol. 2, p. 557–66; and reprinted verbatim in *Mayo Clinic Proceedings*, vol. 75; 2000. p. 553–6 (The quote used in this chapter appears on p. 554 in the *Mayo Clinic Proceedings* reprinting).

40 Embracing Uncertainty: Complexity-Inspired Innovations at Billings Clinic

Curt Lindberg, Margaret Hatch, Virginia Mohl, Carlos Arce, and Elizabeth Ciemins

What "you" can do … is to become more skillful in participating in the relationships you already participate in, in generating the knowledge you already generate with others, by paying attention in a different way [1].

Stacey in
Complex Responsive Processes in Organizations, pp. 8–9

40.1 We Were Sick of It

Nancy Iversen is a former critical care nurse and one of the leaders in the battle against emerging infections at Billings Clinic. She is also a marathon runner whose practiced physical endurance is applied professionally as well. She leaned against the fence by the track at the Montana State Games, her direct blue gaze fixed on the 1,500 meter runners rounding the third curve in 90° heat, evaluating the runners for heat stroke and reflecting on deteriorating morale in the Infection Control department back in July 2005, when frustrated staff members decided they needed profound change in what they were doing.

"We were *sick* of it. What we were doing—collecting and presenting data, imparting best practices and nagging people to wash their hands—wasn't working. Urging people to follow the established CDC [Centers for Disease Control and Prevention] guidelines and organizational policies wasn't working either," Nancy recalled.

The data showed increasing numbers of patients acquiring serious infections from MRSA, or methicillin-resistant *Staphylococcus aureus*, during their hospitalization at Billings Clinic. Nancy and her team urgently wanted new ideas.

Nancy had started to learn about Complexity Science and she knew Billings Clinic CEO Nicholas Wolter, MD, was a long-time student of Complexity Science and the insights it offers for health care and organizational operations. He had sent her to a Complexity Science workshop in 2005, an inspired decision with unanticipated results. "It was at this conference that I heard stories of how the Positive Deviance approach helped diverse communities solve really difficult, even

C. Lindberg, D.Man, M.H.A (✉)
Partnership for Complex Systems and Healthcare Innovation, Billings Clinic, 801 North 29th Street, 37000, Billings, MT 59107-37000, USA
e-mail: CLindberg@billingsclinic.org

M. Hatch, M.Div.
Pastoral Care, Billings Clinic, 801 North 29th Street, P.O. Box 37000, Billings, MT 59107-37000, USA

V. Mohl, M.D., Ph.D.
Administration, Billings Clinic, 801 North 29th Street, P.O. Box 37000, Billings, MT 59107-37000, USA

C. Arce, M.Ed.
Organizational Development, Billings Clinic, 801 North 29th Street, P.O. Box 37000, Billings, MT 59107-37000, USA
e-mail: carce@billingsclinic.org

E. Ciemins, Ph.D., M.P.H.
Center for Clinical Translational Research, Billings Clinic, 801 North 29th Street, P.O. Box 37000, Billings, MT 59107-37000, USA
e-mail: eciemins@billingsclinic.org

J.P. Sturmberg and C.M. Martin (eds.), *Handbook of Systems and Complexity in Health*,
DOI 10.1007/978-1-4614-4998-0_40,

intractable, problems without ongoing outside help," Nancy said. "I left the meeting with a new sense of hope and excitement, and an attitude of 'Why not try this?'. I also knew Nick was behind me. He suggested I go out and mix things up."

40.1.1 Mix Things Up

In 2006, Nancy and Billings Clinic joined a network of six hospitals committed to a future where patients do not acquire life-changing MRSA infections. As Nancy explained, "We formed a collaborative team and learned to really listen, interact and relate to one another differently. We took action, practiced new ways of doing things and discovered solutions together." In doing so, the team embodied a characteristic of complex systems, that the whole is greater than the sum of its parts. Nancy continued, "We learned that solutions cannot be imposed; rather, they arise from within and through shared experiences." In order to change outcomes for patients, Nancy and her colleagues knew rhythms of the workplace would have to change, and old habits would have to be unlearned. Turbulence was inevitable. Nancy was undaunted; "What did we have to lose?"

What Nancy and the team had to lose, in fact, was the collegiality that permeates a close-knit unit such as the ICU (Intensive Care Unit). Nancy's team was working with innovators who were bringing Positive Deviance into a field where it had never been applied before, and so initially they faced skepticism and minimal support. "They [the ICU staff] were really frustrated with us. We were one of the initial hospital sites using this method, and it took us a while to shake out how to use this new approach to effect change. We only had experience using linear tools to address complex problems," she said. "We didn't have the right language and we were burdened with a lot of technical details. We had to take things on without planning and implementing, and it was hard."

It was so hard that the ICU twice rebelled and asked to withdraw from the project. A frustrated ICU nurse asked during one meeting, "*When* are we going to stop working on this?" Nancy's team was confronted with doctors balling up protective gowns and throwing them on the floor, sarcastic email chains, and committees that took four months to accept the CDC definition of contact isolation. But the data on declining infection rates were on their side, and they knew how to share those data. Upon seeing graphical evidence of declining infections, the ICU staff answered their own question on when they would stop working on this: "*Never*!"

And in the silence that followed that emphatic response, something new emerged.

The ICU dramatically reduced its infection rate and lengthened the time between transmissions. The ICU has experienced only two healthcare-associated MRSA infections in nearly 4 years (January 2008–September 2011), an extraordinary achievement in the stressful, fast-paced environment of intensive care (Fig. 40.1).

"This was the ICU's problem to fix," Nancy concluded. The 1,500 meter runners were now gone, and the sprinters were setting their blocks in the noonday sun. "It takes respect, a history of trusting relationships, informed leadership and agreement from top levels of administration that things are not working before change can happen."

Nick Wolter, Billings Clinic CEO, reflecting on the MRSA initiative, observed that success emerged from changed relationships more than from technical solutions.

Transformational change requires self-discovery, and not everyone has the courage or the patience to tolerate the task. Not surprisingly, Nancy is still a marathoner.

40.2 Ample Bedrock

The leadership commitment, trust and relationship-building that allowed the Positive Deviance innovations to flourish had deep roots.

Billings Clinic CEO Nicholas Wolter's wide-ranging philosophical exploration of Complexity Science began in the late 1990s at a time of unprecedented transformation in U.S. health care.

His response to the challenge was to find and listen to the brightest, most interesting and inquisitive people available. He initiated meetings and invited current thought leaders in

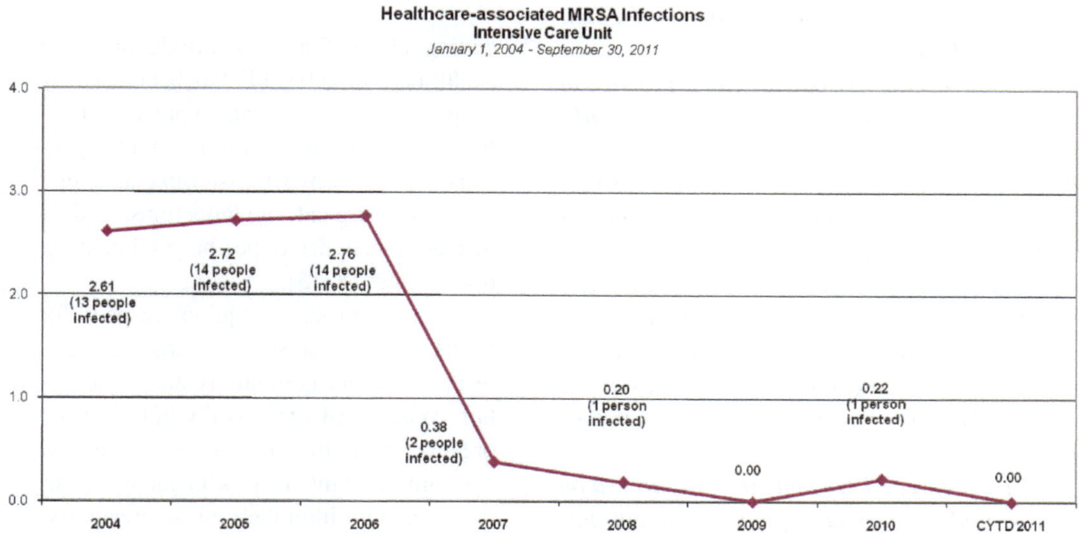

Fig. 40.1 Healthcare-associated MRSA infections per 1,000 patient days

Complexity Science to connect with Billings Clinic staff members. He had conversations and he read. He read a lot, starting with a big box of books and articles supplied by Curt Lindberg, a pioneer in introducing complexity concepts into health care. "It was an invitation to explore, back when Complexity was new. It was a chance to hang out with serious scientists, health care administrators, a wild variety of disciplines, not necessarily mainstream," Nick recalled.

"It was heavy, too," he added, referring to the box. He listed some of the authors: "Brenda Zimmerman, E. O. Wilson, of course." Curt added, "Ralph Stacey, Jeff Goldstein." Nick finished, "Remember Gareth Morgan's book Images of Organizations?" [2]

Nick and Curt were reminiscing about the fledgling relationship between Complexity Science scholars and the new organization created by the integration of the for-profit multispecialty group practice, Billings Clinic, and a not-for-profit hospital, Deaconess Medical Center. The organization is now called Billings Clinic.

The exploration began through the VHA (a national network of health care organizations) Complexity and Health Care Learning Network. "We read, we met, and we got curious about Complexity and how it might apply to health care organizations," Nick said.

The development of Billings Clinic was similar to any other organization. Nick explained, "You need some bedrock, some stability. You need to balance the complexity and chaos associated with a complex organization in a sea of change with some very clear values.

After the merger [between Billings Clinic and Deaconess Medical Center] things were a little too chaotic. We needed to focus on some basic functions like solid administration, physician leadership, and relationships of trust. We had to have strong enough relationships so we could define ourselves. For us, the 1998 mediation session came at an especially critical time when some were thinking of undoing the merger because of deep differences between physician leaders from the Clinic and leaders from the hospital about the nature of effective organizations and approaches to change. At that session we had board members from both organizations, plus a critical mass of physicians. The conversations led to a pragmatic acceptance of the direction we were going and a realization that together we could create something new.

"We hadn't spent much time on vision. As we turned our attention to defining the new Billings Clinic, some basic principles emerged." It was agreed that Billings Clinic should aspire to

become an outstanding medical foundation built upon the following ideas:

- *A multi-specialty physician group practice in which a "community of physicians" works together.*
- *The partnering of physicians, excellent business managers, professional staff, and volunteers creates a team whose synergies drive success.*
- *Not-for-profit, community-owned and governed.*
- *Mission driven decision-making dedicated to a higher purpose in the community and the region.*
- *An obsessive dedication to quality and service.*

These became what are called the Cornerstone Principles.

"Then, between 1998 and 2001, we worked on the essentials. The direction provided enough clarity for us to get the job done. Our team recruited the right kind of physicians and we developed an Internal Board of Physicians to be a part of all decisions. By 2002 we had developed the process and capacity to improve. It took awhile," Nick said, returning from answering an incoming message. His reflections stayed on track despite the interruption. "Staff morale improved. And we realized that we no longer had a we-they culture [between the hospital and the clinic]. People went through a process to become the new Billings Clinic, and we decided we wanted to play on the national health care stage."

About Positive Deviance

Positive Deviance (PD) is a social and behavioral change process that rests on the observation that most communities and organizations contain individuals and groups whose different (deviant) practices yield better (positive) results than peers who have access to the same resources. The process was created by Jerry and Monique Sternin in 1990 to address childhood malnutrition in Vietnam. It has since been used to tackle other health-related challenges in the developing world such as neonatal and maternal mortality, HIV/AIDS prevention, school attendance, and female genital cutting [3–7]. Since its introduction into health care in 2005, PD has been used by hospitals in the Americas to prevent infections caused by multidrug-resistant organisms, to improve hypertension management, to reduce surgical site infections, and to increase adherence to posthospital medication regimes [8–15].

The PD process requires respect for local culture and a belief that true expertise resides in every community and organization. When used successfully in healthcare organizations, the process is guided by front-line staff members who uncover positive deviants within their midst, create new PD practices through abundant interaction, and determine how best to diffuse these effective practices.

About Billings Clinic

Billings Clinic in Billings, Montana, is a physician led, not-for-profit multispecialty medical group practice integrated with a hospital and long-term care facility, serving the communities of Montana, northern Wyoming, and the western Dakotas. Billings Clinic is the largest healthcare organization in the region, with 3,500 employees, including 310 physicians, physician assistants, and nurse practitioners.

Billings Clinic has a main location plus two urban branch clinics and additional physicians located in Red Lodge, Bozeman, Columbus, Colstrip and Miles City, Montana, and Cody, Wyoming. It also manages three small hospitals. Patients from rural communities make up 46% of hospital admissions and 41% of outpatient visits. Annually, Billings Clinic hospital has more than 15,000 hospital admissions accounting for 67,199 inpatient days and experiences over 900,000 patient encounters, including 38,922 visits to the Emergency and Trauma Center.

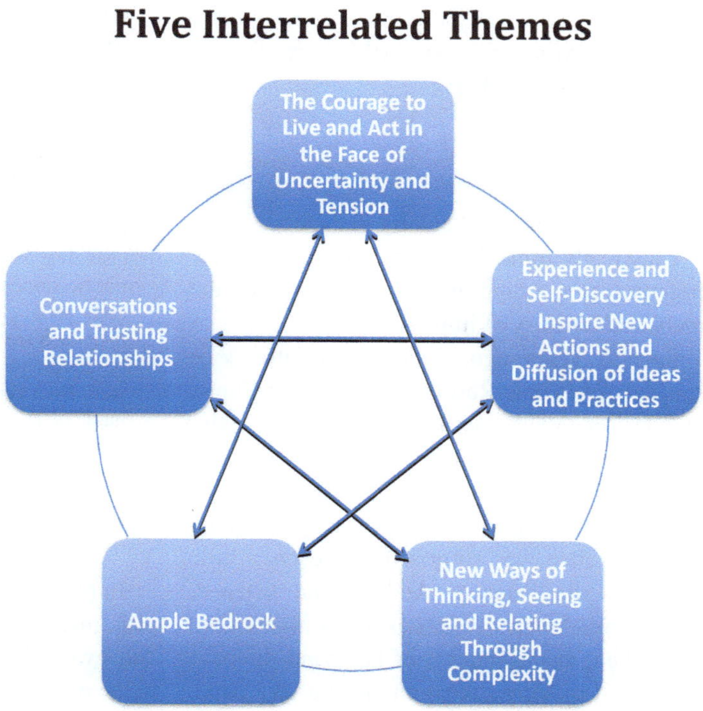

Fig. 40.2 Five interrelated themes

40.3 The Story Writing Process

Before continuing with the story about Billings Clinic's journey with Complexity Science, we would like to describe how we uncovered and made sense of this story. Our first step was to organize a meeting of colleagues who had been involved in Complexity Science-related efforts in health care over the years and invite them to tell stories about their involvement in a Complexity Science learning activity or a project informed by Complexity. After sharing these stories, we asked the participants to identify common themes and general insights.

We then selected stories that seemed to best illustrate the Complexity Science work at the Clinic. Generally, two or three of us interviewed individuals who played central roles in the stories, hoping our different perspectives would help us capture the fullness of the stories. The full set of story drafts was reviewed by the key players with these questions in mind: What stands out for you in the stories? What themes do you see? What should we change to more accurately depict your story and the larger Billings Clinic story?

From this process and our deliberations five themes emerged:

- Ample Bedrock
- Conversations and Trusting Relationships
- New Ways of Thinking, Seeing, and Through Complexity Science
- The Courage to Live and Act in the Face of Uncertainty and Tension
- Experience and Self-Discovery Inspire New Actions and Diffusion of Ideas and Practices

We found strong evidence of many of these themes in each of the stories; they were strongly interconnected (Fig. 40.2).

The five themes along with a narrative that developed from two extended interviews with CEO Nick Wolter provide the basic structure of the chapter. Dr. Wolter's narrative is presented in italics. Each theme is illustrated with a story. To provide a fuller depiction of this undertaking, a timeline of key events and milestones is provided at the end of the chapter (Fig. 40.6).

40.4 Conversations and Trusting Relationships

40.4.1 Jazz

Kellee Fisk did not particularly like jazz.

Classically trained in piano and clarinet, she preferred performing music as written to the uncharted territory of improvisation.

Several years ago, Kellee, currently Vice President of People Resources at Billings Clinic, attended a VHA workshop on Complexity Science in Park City, UT. At the beginning of one session, a jazz band walked into the room, set up their instruments, and started to play together. Though Kellee was surprised and even a little confused, she listened. And as she listened, Kellee's mind meandered back to her high school days when, at the request of her band director, she started playing saxophone in the jazz band. Like many classically trained musicians, she found herself intimidated by those minimalist jazz charts that offered only an occasional chord to guide the music. After years mastering intricate passages of notes on a page of music, the sketchy chord suggestions placed casually on a jazz chart seem like inadequate guidance for a collection of people trying to make music together. "You're forced to let it grow, evolve, move," Kellee reflected. "I thought I was a linear thinker, but this kind of music sparked my creative side."

40.4.1.1 Planning for an Uncertain Future

At the time of the Park City workshop, Kellee was Director of Marketing and Planning. "I liked the planning piece," she shared. "I was intrigued by how we think about planning when the future is uncertain." Like jazz, there may be some "givens," similar to a recommended chord sequence, but how an institution moves into the future looks a lot like how a jazz band spontaneously creates music. The ability to improvise, within some boundaries, is the key to making music.

Kellee flipped through her well-worn, heavily highlighted copy of *Edgeware* to back up her point and read the following passage:

> Being a successful leader these days therefore requires recognition that the world is not the way it used to be, or the way it used to be perceived. And no more so than in health care, where uncertainty and unpredictability can no longer be denied or dismissed…During this time of crisis and ambiguity, a top down approach to management comes up short. Leaders cannot command outcomes—the business environment is too complex, too fast-changing. Faced with the reality that change cannot be stopped, leaders have to change how they think about change. Once, setting a goal and planning each step to achieve that end was regarded to be good change strategy, and sometimes worked well. Now adaptability and flexibility are the hallmark of a robust organization, which calls upon different leadership skills [16].
>
> Zimmerman, Lindberg, Plsek in *Edgeware: Lessons from Complexity Science for Health Care Leaders,* p. 58

In planning for the future of a large and complex institution, the challenge Kellee saw was "how we unleash creative and adaptive behaviors in the midst of chaos."

40.4.1.2 Let It Be Messy

As Vice President of People Resources, Kellee often finds herself facilitating conversations in groups. As a student of Complexity Science (see text box), Kellee has been learning how to use Liberating Structures, and she excels in finding new ways to bring people together for conversation.

An example of her leadership growth is evident in what Kellee calls "our PSE journey." "Personal Service Excellence" sets the standard for what every Billings Clinic employee strives for in their interactions with patients, patients' families and fellow employees. Personal Service Excellence is spelled out in nine simple rules (see Fig. 40.3).

What is striking about these rules is that they emerged from the Employee Board—a group of employees who meet monthly to problem solve issues and challenges concerning service to each other and our patients. With guidance from former Walt Disney executive Scott Lily, PSE was built from the ground up, and Kellee found that she had to be a partner in the process rather than direct it. This came as something of an epiphany to her, "I realized that I couldn't control it all. We just had to stay the course. We had to let it be messy for awhile."

Personal Service Excellence (PSE)
Billings Clinic Service Expectations

Personal Connection
Find ways to create relationships with our patients and guests. Use their names in every interaction. Be respectful of patients and make sure that patient information is kept confidential. Never discuss patients and their care in public areas or with your family and friends.

No Passing Zone
When you pass a patient or guest off to someone else, take personal responsibility to follow through. Anticipate and help patients and guests with their needs. Listen to our patients and guests. Be courteous. Speak clearly and avoid jargon. Educate families about procedures and provide a comfortable atmosphere for waiting.

Make a Point
Escort our patients and guests to their destination, whenever possible, or take them part of the way there. Avoid pointing.

Personal Accountability
Take pride in this organization as if you own it. Pick up litter and dispose of it properly. Contribute to the overall cleanliness of the environment. Accept the responsibilities of your job. Adhere to policies and procedures. Live the values of this organization. Hold each other accountable to follow the Service Expectations. *You are Billings Clinic.*

Always Strive to Exceed
Find ways to say "yes" to our patients and guests. Give options and think of ways you can exceed expectations. Use service recovery as a way to create an advocate. Thank our patients and guests for choosing Billings Clinic.

Internal Service
Serve and care for our coworkers as we would our patients and guests. Treat our colleagues as professionals through courtesy, honesty, and respect. Encourage other people's work; praise whenever possible. Make new staff members feel welcome.

10 Foot Circle
Engage and acknowledge everyone who comes into your circle of influence. Smile and introduce yourself when appropriate. When riding the elevators, initiate conversation and hold the door open for others.

Positive Image
Present a positive image through friendly body language, facial expressions, and appearance. Be clean and professional. Follow dress code policies and wear your identification badge correctly at all times.

Phone Image
Create a welcoming atmosphere by answering the phone in four rings or less with a personal greeting. Identify your department and yourself, then ask, "How may I help you?" Know how to operate the telephones in your area. Provide the correct number before transferring a call. Get the caller's permission before putting them on hold and thank the caller for holding.

Fig. 40.3 Billings Clinic service expectations for Personal Service Excellence (PSE)

The Employee Board at Billings Clinic is an example of a creative complex adaptive system in action. The Employee Board demonstrates what can happen when a group of people from many departments are gathered, invited to raise issues of any kind, and then asked to generate solutions. "Leadership emerges," Kellee observed. People who do not really know each other find ways to work together to solve problems, and very often the people closest to the problems are the ones finding the solutions. This bottom-up approach to problem solving can lead to much quicker and more sustainable change.

The result of "staying the course" and "letting it be messy" on the Employee Board-led PSE journey was a profound and noticeable culture change at Billings Clinic. The results are clearly noticeable as you walk through Billings Clinic. Employees greet you, smile at you, and ask if you need help. It is not uncommon to see an employee walking a patient to their appointment or pick up a piece of trash. Although the Employee Board did not begin their PSE journey with the intention of changing the organizational culture, that was the result (Fig. 40.3).

40.4.1.3 Pushback and Renewal

Despite the noticeable energy among staff, not all managers were ready to embrace newly empowered employees. The theory sounded good, but how it emerged in real life was surprising to those higher up the organizational ladder. "We got some pushback from leaders and doctors when we implemented the YAG cards," Kellee offered as an example. "YAG" stands for "You Are Great!" and employees are encouraged to recognize greatness in other employees by giving YAG cards to each other. Each YAG card lists the PSE Service Expectations, and employees mark which particular service behavior a fellow employee has demonstrated through great service to others. Although these small cards represented what seemed like a small change, those higher up in the organization were surprised by the energy that came with this shift and by the movement's ability to sustain itself.

"This stuff helps people manage their own angst," Kellee observed. It allows leaders to step back and find a new approach that might work better in their current situation than a traditional approach. For instance, rather than writing a protocol or a new policy, a leader might empower employees to write the new protocols or policies themselves and then develop ways to share and disseminate them among staff members.

Working with groups of people within a large institution can be "a little scary"—you never know what issues will be brought up or what solutions might emerge. But, if you can slow down, ask good questions, and trust the process, "you might get a little jazz."

"Jazz was not my favorite," Kellee Fisk admitted. "But now I have more appreciation for it."

Nick Wolter's reflections on PSE and Complexity:

"We were taking on a lot, we had a big appetite for change. We put a lot on the plate and set a big fast pace in the midst of a sea of change. It was, and is, a complex environment. Where do you want to take the risk and put something in motion when you don't know what's going to happen? And no matter what your position is, you are only able to influence about 15% [referring to Gareth Morgan's '15% principle']. That's both liberating and a challenge." His head tips to one side and he smiles. "What if you pick the wrong 15%?"

"Personal Service Excellence (PSE) was one of our first endeavors after the merger. We knew service excellence had to be one of our first priorities. If we took better care of each other we would take better care of our patients. We wanted to be an early adopter of this concept. The work included relationship building. Employees were empowered to create ways to encourage PSE, and then we had to help our leaders who weren't prepared for all these empowered employees." Nick grinned. Kellee had nodded her head at the same remembrance. Both mentioned that employee groups invented their own approaches to PSE and at first there was a lot of skepticism, especially among physicians.

People were saying, 'Is this the next project of the day?' Now people can see the difference. It's a rare person that walks into Billings Clinic and doesn't see it and feel it.

Nick observed, "In a complex organization, you don't work with just one person. Some people need more structure. Some people need more time. You need to be present, sending the message of encouragement and making it really clear you're involved. It isn't really as easy as we'd like it to be but the work is worth it. Recognize that there is not one thing you can do—bam—and it will all be better. Sometimes you just have to keep trying."

"It's the relationships with people like Nancy that allowed others to take the risk of approaching a problem with tools that hadn't yet been applied to health care."

When asked why he stayed interested in Complexity Science even before the MRSA project, even before he had evidence that it could be used effectively to tackle unsolved problems in health care, Nick Wolter said it was mostly because thinking about Complexity had changed him and had changed how he thought about his role and the organization and individuals for whom he was responsible. "I had to change how I thought about my role as a leader and stop using the typical hierarchical frame." He leaned forward a little. "How do you lead, how do you follow, how do you relate to others? It was rewarding even when we weren't sure the science was applicable."

Nick recalled a skeptical inquiry made by a conference attendee about the real utility of Complexity concepts in organizations. Complexity scholar Ralph Stacey had responded, "There is nothing more practical than the way you think."

About Liberating Structures

Liberating Structures (LS) are a collection of 33 easy-to-use methods to help groups have more creative, productive conversations. Each LS is defined by a basic structure and offers guidelines for organizing conversations and interactions. The guidelines are designed to draw out diversity in groups, encourage attentive listening, foster participation by all members, and support creative self-organization. Examples of Liberating Structures are:

- Open Space
- Appreciative Interview
- Discovery and Action Dialogues
- Conversation Café
- Nine Whys
- What, So What, Now What Design Debrief
- Wise Crowds

Liberating Structures have been collected and developed by Keith McCandless and Henri Lipmanowicz. More information on LS can be found at www.liberatingstructures.com.

40.5 New Ways of Thinking, Seeing, and Relating Through Complexity

40.5.1 Everything Is Simple and Neat—Except, of Course, the World

Dr. Bob Merchant chairs the department of pulmonary medicine. He leaned back in a chair in the Intensive Care Unit conference room, contemplating his first exposure to Complexity Science.

A quiet, modest man, Bob described himself as being in the learning phase when it comes to Complexity Science. He then began to recall a series of articles he read over 10 years ago in the journal *Science*. The articles that caught his attention explored signaling pathways in biological systems and dynamics in ocean ecosystems from a complex systems perspective.

> One of the most striking aspects of physics is the simplicity of its laws…The ideas that form the foundation of our worldview are also very simple indeed. The worldview is lawful, and the basic laws hold everywhere. Everything is simple, neat, and expressible in terms of everyday mathematics, either partial differential or ordinary differential equations. Everything is simple and neat—except, of course, the world [17].
>
> Goldenfeld and Kadanoff in *Science*, p. 87

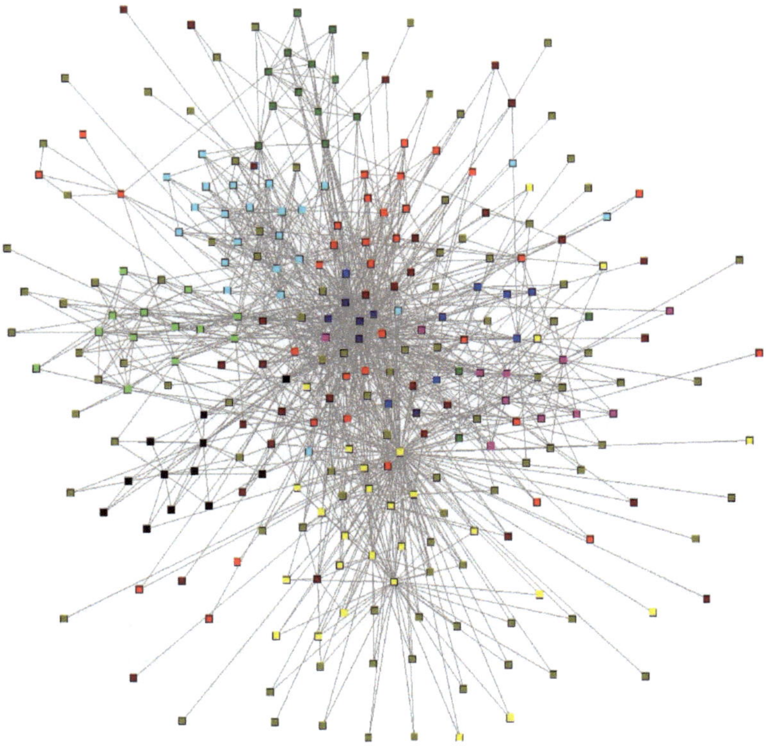

Fig. 40.4 Collaboration network in MRSA prevention

In the medical arena, Bob's first experience with Complexity Science concepts was through the work done at Billings Clinic on prevention of MRSA infections using the Positive Deviance (PD) process. What fascinated him was "how you could actually use insights from Complexity to influence behavior change" in social systems. He described his first impression of Positive Deviance as "sitting around in bean bag chairs holding hands." Then he began to appreciate the scientific principles underneath the process. Through "nodal analysis" (a social network analysis conducted in conjunction with the PD work that displays who worked with whom in infection prevention) he saw how "sitting around in bean bag chairs" led to new conversations and relationships and changed important clinical outcomes (lower MRSA infection rates) (Fig. 40.4). "This was a true scientific endeavor, not just something whimsical, where the mechanism of causality was plausible."

Bob Merchant began to understand other issues more fully by adopting a Complexity perspective. Complexity Science helped validate what he was learning about some medical conditions like insomnia. He realized that using linear thinking to understand and treat the condition was inadequate. He went on by reflecting on his experience working in the intensive care unit. Bob explained that he had learned that you "have to be very collaborative to practice" in the ICU. He now appreciates this even more deeply and believes he engages in collaboration better through an appreciation for all voices in the department, better listening, and an understanding of the importance of relationships.

40.5.2 Variability and Resilience in Social and Physiologic Systems

Billings Clinic is joining an international experimental study using heart rate variability (HRV) analysis to inform clinical decision making on when to remove patients from ventilators. A small study conducted by Dr. Andrew

Seely at the Ottawa Hospital Research Institute suggested that changes in HRV were potentially good indicators of when patients could resume breathing on their own. Bob is serving as the principle investigator at Billings Clinic. When asked about his desire to join the study, he explained that weaning from ventilators is an interesting area of study with a rich body of publications but with few articles that are helpful clinically. He views HRV as an "integrative assessment," a window into physiologic resilience. The trial is "proof of the concept that you can study complex variability and use findings to guide clinical decision making."

Ventilator associated pneumonia (VAP) was another example Bob raised. He noted that VAP "clearly fits the definition of a complex issue." The condition, its prevention and treatment are complex. "I think we've had more success in prevention in the past couple years because we've tapped Complexity Science-inspired practices," he said. There are good national guidelines and protocols but the toughest issue is getting them to happen. "We adopted a more intensive team and network approach," he said. "A dietician recommended that probiotics be used, respiratory therapists came up with innovative strategies to prevent contaminated secretions from entering the trachea and monitoring all ventilated patients for pneumonia, and nurses found ways to keep patients' heads elevated. It wasn't that I, as ICU director, came up with a list of things for staff to implement. That approach would not have had the same success. Our VAP rates now are close to zero."

"Now," he said, "instead of getting pissed and coming down on someone, we recognize VAP as a complex issue and together figure out what can be modified in our practices. Instead of a directive coming down from above, solutions get implemented before I know about them." Bob expressed what this meant to him as a leader. "It's gratifying to know that so many staff members are owning the problem and working together collaboratively and intensively."

When faced with a seemingly intractable problem or impossible situation, Nick describes how he and other organizational leaders proceed: "We gather a group of the smartest, most interesting and committed people we can find, and we start exploring new ways of understanding or new ways of acting."

"In circumstances where there are no obvious answers, I think about connecting the right people, providing time for conversations, and building trusting relationships. I consider what processes should be used, and I stay alert to the direction suggested by the group's interaction."

"For example, we are looking at changing how we deliver primary care, not just to address the primary care shortage but to create better and better coordination of care. We're also looking at how we can advance the use of palliative care. These are examples of where we are currently using Complexity Science ideas to help clinicians become more comfortable with change processes."

40.6 The Courage to Live and Act in the Face of Uncertainty and Tension

40.6.1 Creating a Tapestry

Deborah Agnew, a practicing pediatrician who serves as Division Chief of Primary Care, described her experience with Complexity Science. (Curt Lindberg's observations about the Clinic's evolving approach to primary care development are in italics.)

"My dad was the first trained cardiac surgeon in Montana and the surrounding states. He was the 25th partner at the Billings Clinic; I was about number 200! A lot has changed … but not everything. He loved his work; he was able to bring something to the region that was so valuable and so transformational. Most of the time he just seemed to be working so hard and having so much fun! And he loved his partners; it was a real community or big family back then. We all spent so much time together. Those days have passed. But as I watched those doctors, and particularly him, I saw how much medicine meant to them and brought to the people around them. I guess I came to understand that it was a calling, almost noble at times, which invigorated them. Even so, I didn't think that I wanted to be a doctor, because of how demanding I knew the work was. But the call to have a skill that could improve people's lives and serve the community eventually convinced me. I started my career at a migrant farm worker clinic in eastern Washington. I never thought I'd return to Billings, but I did, attracted in part by the pioneering spirit of the Clinic; I think this is the spirit that attracted my dad too."

40.6.1.1 It's Okay Not to Know

"The first time I heard about Complexity Science was from Nick [Wolter, CEO, Billings Clinic]. It was about five years ago at a retreat soon after the Division Chief role was created. He talked about his hopes, his vision for our new roles. He also spoke about self-organization, adding his view that it is okay if we don't know exactly where we were going [with the new role] and how everything would unfold. I've since learned more about the science through experience and context. The scientific principles from Complexity resonate with me."

"My initial experience with Liberating Structures [see box] was when Nancy Iversen came to a Division Chiefs meeting and conducted a Fishbowl Conversation with colleagues. Members of the group guiding the Clinic's effort to eliminate MRSA infections sat in a circle and reflected on the progress they were seeing in infection prevention practices as well as some remaining challenges. Some of what they said about the behavior of some physicians was disturbing; I was flabbergasted. The fishbowl wasn't rehearsed; it was real, it was emergent. It was so different from the usual PowerPoint presentation on infection rates we'd seen in the past. What they did was courageous. The reaction among the Division Chiefs was fascinating: some tried to minimize the problem; others, including me, wanted to help. The group fighting MRSA left it in our hands, saying, 'We can't do anymore with physicians without your help.' We had to respond."

"When we were having conversations across the organization about what issues to focus on in the next phase of our work with complex systems concepts, I was hoping Primary Care would be selected. I knew we had to try something new."

"2010 was a very hard year for me. I got an email from a very respected physician saying I had lost my credibility, that I didn't understand them [physicians in the Primary Care Division]. It came from a place of kindness. This email was a turning point for me. I realized I had to listen more, to engage and invite my colleagues into building better models for primary care, to regain trust and credibility."

Curt reflected on that same time period: Beginning in fall 2010, I began interacting with Deb Agnew and some of her leadership colleagues about the Clinic's plans to develop new models for primary care. My role was to serve as a resource by helping leaders consider complex systems perspectives.

Over the course of multiple conference calls with Deb and others it was obvious that tension was high around the lack of progress and the pressure to produce. Anxiety reverberated through the phone lines. Talk was of time pressure, the need to 'bring the staff along.' It was mentioned that there was little dialogue with staff about the vision. In one call Deb remarked, "the organization is as tense as it's ever been and some practices are beginning to hold meetings where senior leaders are not welcome."

40.6.1.2 A Social Network Map of Change Was Before Me

Deb remembered that "When we were designing one of our early 2011 Primary Care planning sessions somebody said during a conference call, 'Let's talk about some of the good things we're doing in primary care.' At first I thought, well, there is nothing to talk *about*. But then I thought of Positive Deviance and realized there were actually lots of good things going on in our various primary care practices. We ended up inviting those doing innovative work to share their stories in short roundtable sessions with their primary care colleagues from across our system. For example, one physician had partnered effectively with a physician assistant. Another was pioneering work in palliative care by facilitating conversations around the organization. Another group of physicians had figured out a way to effectively communicate with patients about their primary care 'team.' There was a really good turn out. I saw real energy in the room as nurses and doctors learned from each other! Instead of doing my usual PowerPoint presentation, I moved from one roundtable discussion to the next and heard fascinating stories. It was heartening—all is not lost, I thought. I could now visualize how this group could advance primary care. A social network map of change was before me."

Later in January another conference call was held involving the leadership team. It was lively, lighter and free-flowing. Many insights and next step possibilities were raised. Could we use our new primary care practice, Primary Care Partners, as a place for experimentation? Let's bring the members of this practice together and help them create their own vision. Let's meet the different practices where they are, recognizing that they have different histories and cultures, serve different populations, and have varying capabilities and inclinations when it comes to change. Why not explore the concept of adaptive clinical leadership with the staff [18]. *We should block out time each week in the schedules for staff to meet and work on issues.*

"This got me thinking about how I work as a primary care leader," recalled Deb, "I thought about how primary care leaders work with each other and with our staff. How we lead, how we engage staff really matters. I asked Curt to help organize a session where we could explore these topics."

From these and other interactions with primary care staff, a new framework for primary care development at Billings Clinic emerged. It was built around the idea of a primary care community comprised of various neighborhoods (regions and patients served by different practices) and a variety of homes (the different practices). This framework recognizes that a common purpose, as well as certain systems and values are needed to bind the community together. Those in the community can learn from the different experiences, skills and orientations in the neighborhoods and homes. It was inspired by the understanding that adaptable complex systems are characterized by a paradoxical mix of order and disorder, uniformity and variation (Fig. 40.5).

Fig. 40.5 Primary Care Community

40.6.1.3 It's the Opposite of Being Alone

Deb remembered that "One of our primary care nursing managers, Lisa Preeshl, asked me to help design and facilitate a retreat with the Geriatrics team at a difficult time for the group. This was the first time I had done something like this. I used a variety of Liberating Structures and story sharing to help the group hear themselves articulate what was important, to them; this helped them to create a direction and communicate more fully. Everybody came, nobody left. We didn't accomplish our whole list of objectives, but we accomplished a lot. Hope and genuine engagement were very evident."

"This was a powerful experience for me. I was at the edge of my comfort zone. I don't have clinical expertise in geriatrics, and the practices actually run very differently from my own, but I found it was much easier than usual for me just to listen. I was there as a facilitator and an expert of sorts in 'the process of empowering change'. I feel they saw me as being on their side, as more understanding."

Deb concluded, "There has been a shift in me. I am more comfortable than I used to be when things are unclear, when I don't know. I see things differently; it's not all pre-determined. I imagine larger opportunities now. It's liberating and risky. You have to believe it's all about creating a

Year	Activities
1998	Wolter joins VHA Complexity Science and Health care Learning Network and participates with Billings Clinic colleagues in regular meetings. These included visits to Santa Fe Institute and sessions with prominent complexity scientists like Stuart Kauffman, Ralph Stacey, John Holland, Ary Goldberger, Jeffrey Goldstein, and Edward O. Wilson.
2000	Wolter and Lindberg help form Plexus Institute, an organization devoted to helping people understand and use complex system principles.
2001	Billings Clinic becomes one of the first organizational members of Plexus and administrators and physicians participate in numerous education offerings. Additionally, the publication, *Edgeware: Lessons from Complexity Science for Health Care Leaders*, is distributed and read by members of the senior leadership team. As a result of the CEO's references to and insights from those leaders who attended Plexus gatherings, curiosity and interest grow as the organization seeks opportunities to better understand how to use the "novel" concepts found in Complexity Science.
2004	Plexus Institute and Harvard Center for Health Systems Improvement host workshop on health care quality improvement and complexity. Wolter and Lindberg learn about Positive Deviance and meet Jerry Sternin.
2005	Billings Clinic joins partnership with five other health care organizations, Plexus Institute and CDC to conduct the first significant application of Positive Deviance in health care. Effort was directed at reducing infections caused by the methicillin-resistant *Staphylococcus aureus* (MRSA). Cross-functional partnerships between areas such as infection prevention and organizational development accelerate the involvement of staff members. The dramatic declines in MRSA infection rates attract more curiosity and interest among Billings Clinic staff.
2008	Billings Clinic and Plexus Institute collaborate to advance Complex Systems agenda and engage more staff. PD is used to tackle additional issues like hypertension management. There are now more overt references to concepts such as Positive Deviance as an established improvement method. Because of growing awareness of Complex Systems concepts and the importance of how employees interact, more thought is given to meeting design, and Liberating Structures are used more frequently. The dynamic energy of respected groups like the "Employee Board" reinforces the organization's new learning in Complexity Science.
2010	Billings Clinic joins with six other organizations and leading Complexity Science researchers to form The Quality Commons, a network devoted to improving health care in the US by tapping Complex Systems insights and advancing a Complexity Science-health care research agenda. Internally, Billings Clinic forms a Complexity Science and Health Care Learning Network and expands use of Liberating Structures.
2011	Billings Clinic establishes Partnership for Complex Systems and Healthcare Innovation and names Lindberg as director.

Fig. 40.6 A Timeline of Key Events in Billings Clinic's Complexity Journey

network, a tapestry. When, as a leader, you're no longer expected to know all the answers, it feels like the opposite of being alone."

CEO Nick Wolter's conversation shows self-reflection and joy from the day-to-day interactions with colleagues. He also goes back frequently to his early experience with the VHA Complexity and Health Care Learning Network. "One thing that was pretty apparent to me was that the relationships with great thinkers in this area were personally rewarding. They were thinking about things in a much broader perspective; they were thinking about how systems of all kinds change. Conversations with these scholars added to the enjoyment and satisfaction of my job and enrichment, both organizationally and personally. And what's personally rewarding can become a contagious passion."

"We observe that from the learning opportunities available at Billings Clinic and from experiences with complexity-informed approaches many others are coming to share this passion and approach their work differently."

"A lot of what we were learning was as we were doing. It's only later, on reflection, that you can see how the perspectives and tools we were using worked."

40.7 Experience and Self-Discovery Inspire New Actions and Diffusion of Ideas and Practices

40.7.1 Swimming in the Ocean

Eric Saberhagen is a soft-spoken internal medicine physician at Billings Clinic who reflected on his experience with Complexity Science while sitting in "The Commons," a large open space with huge windows, a café, and comfortable seating. Eric works at one of Billings Clinic's satellite primary care outpatient clinics on the west end of town.

In collaboration with Billings Clinic's Center for Clinical Translational Research, Eric's practice was chosen to participate in a national Learning Collaborative on Hypertension (HTN) Management. Eric was designated as "Physician Champion" for the project.

"I became an internist in the first place because it's like swimming in the ocean," Eric said. "There's a lot you don't know." Eric enjoys the challenge of dealing with that ambiguity. Though after years swimming in that ocean, Eric began to feel that he had the basics of his practice down and that he was ready to start looking at the bigger healthcare system. "You can either sit back and complain or involve yourself in change and try to make the system better," he said. "I'd like to spend my career participating in something bigger than myself." The hypertension project was an opportunity for such involvement.

Initially, the Billings Clinic hypertension team enlisted a standard approach to addressing HTN management: educate patients, develop clinical protocols, and create a registry. Early in the collaborative, a fellow team member suggested trying Positive Deviance (PD), given the effective work being done at Billings Clinic on MRSA prevention. Eric had no direct experience with PD. He knew about it generally from his training in pediatrics but was unfamiliar with PD as a professional quality improvement approach. He was skeptical, but willing.

40.7.1.1 Changing Physician Behavior

According to the World Health Organization, uncontrolled hypertension may be the single greatest contributor to deaths worldwide. HTN is an identifiable and fixable problem. The barriers to addressing it are, to some degree, habits that medical professionals have developed around routine behaviors. "People don't necessarily want to apply diligence to something they think they already know. Everyone thought they already knew how to do this," Eric observed.

Instead of the usual segmented and siloed approach to quality improvement and behavior change, the team began the project by using one of the Liberating Structures called Discovery and Action Dialogues (DADs) (see box). Initially, this approach generated anxiety and resistance. Strictly clinical interactions were common between nurses and physicians, and these were new types of conversations, outside of the traditional clinical setting. Eric took a lead role, together with a Translational Research nurse, in initiating these dialogues. The open-ended yet thought-provoking DAD questions helped shape the subsequent HTN program. Questions like, "In your own practice, how do you know you have an accurate blood pressure?" led to the discovery that physicians had no idea if their nurses knew how to take an accurate blood pressure. In fact, most physicians had *never* seen their nurses take a blood pressure, and nurses were unaware that so many blood pressures were being retaken. These realizations inspired curiosity, concern, and ultimately the desire to discover what others were doing. Nurses and physicians role-played the process by taking each others' blood pressures. As a result, at least one physician ceased retaking the blood pressure of every single one of his patients, adding 30 minutes to his day. Even patients began to act differently. Physicians remarked on patients coming to visits prepared, with sweaters removed and two feet firmly on the floor.

Eric described the process: "We used PD to change group dynamics and thinking for group change." When asked what he learned from the experience, he replied, "I guess we learned how to change physician behavior."

40.7.1.2 Creating a Shared Protocol from the Bottom-Up

As part of the PD process, nurses developed their own protocol for blood pressure measurement.

The final protocol looked very much like the standard national protocol, but it felt different. The nurses *owned* this protocol and therefore used it. Nurses took great pride in what they had developed. In fact, nurses at Billings Clinic continue today to remind physicians when their patients' blood pressures are in the "gray zone" (140–150/90–100 for an uncomplicated case of HTN) or when patients are not meeting other established targets. They remember the targets, because they were an integral part of the process. Eric spoke proudly of these changes, "We were breaking practice patterns, and changes have been sustained. People want to do quality work, and I think that, when people look back over this project, they feel a sense of accomplishment."

Including nurses in the process created a sense of transparency. There were fewer hierarchical interactions, and a "leveled" team worked on HTN management. There was increased focus on communication and nurse–doctor office flow. Eric reflected that "all these things that weren't about hypertension got worked on," and perhaps those unintended "by-products" of this project had inherent value, too.

40.7.1.3 Linking Math to Complexity Science

For Eric, this work reminded him of the mathematics he really enjoyed as an undergraduate student-higher math where uncertainty plays a role and is embraced, where linear methods no longer produce the results you are looking for, where disorder and chaos are welcomed. "In math, you get to a certain level where everything falls apart and gets philosophical again, similar to Complexity Science."

Asked how he has changed because of this process, how he approaches problems differently, Eric said he finds these approaches to be a more inspiring way to run teams. He finds these methods are applicable to a wide variety of issues beyond disease and health care, including staffing and physical space issues. Using a PD approach to address HTN management was more effective than he thought it would be. Despite early doubts, he was pleasantly surprised by the results. Changing conversations proved successful in breaking resistant and ineffective patterns of behavior and producing positive outcomes.

40.8 Closing Observations

Eric Saberhagen's story and all the others in this chapter demonstrate the relevance of Complexity Science to a wide range of important challenges faced by health care organizations, their clinicians, and leaders. Quality improvement, infection prevention, organizational change, service, employee engagement, and leadership development are among the issues at Billings Clinic that benefited from a Complexity Science perspective.

Engaging the minds and interests of physicians in the dynamics of change is another. Once physicians at Billings Clinic appreciated the scientific rigor behind such processes as Positive Deviance and social network mapping, their interest in Complexity was piqued. This interest led to fresh thinking about organizational change and openness to new applications of Complex Systems thinking. The leadership Bob Merchant displayed in the MRSA prevention effort and the clinical study of heart and respiratory rate variability among ICU patients is an example.

This learning among physicians and other professionals at the Clinic provided meaningful opportunities for personal growth. Every story in the larger Billings Clinic story is replete with increased understanding of organizational dynamics, of serious rethinking of assumptions about change, of examinations of how to interact with and engage colleagues, and of acceptance of uncertainty as a constant companion.

> *"There are times when we don't know what to do; acknowledging this provides space for emergence of new ideas and directions."*
>
> *Nicholas Wolter, MD*
> *CEO, Billings Clinic*

Discovery and Action Dialogues (DAD)

One of the primary vehicles created by the hospitals and Positive Deviance consultants for engaging front line staff in the PD process was the Discovery and Action Dialogue (DAD) [10]. These dialogues are facilitated small group conversations. They are relatively short and often impromptu sessions convened at staff convenience. DAD facilitators use the following questions, which can be easily modified to reflect the issue being addressed, to spur creative conversations:

- How do you know if your patient is MRSA positive?
- In your own practice, what do you do to prevent MRSA transmissions—to yourself or patients or other providers?
- What stops you from doing these things all the time?
- Is there anyone who has a way of doing things that helps them overcome these barriers?
- Do you have any ideas about what to do next?
- What can we do now? Any volunteers?
- Who else needs to be involved?"

Acknowledgments The authors would like to express their appreciation to Julie Burton, Public Relations Coordinator, Billings Clinic and Prucia Buscell, Communication Director, Plexus Institute for their fine editorial assistance in the preparation of this chapter.

References

1. Stacey R. Complex responsive processes in organizations: learning and knowledge creation. London: Routledge; 2001.
2. Morgan G. Images of organization. Newbury Park: Sage; 1986.
3. Pascale R, Sternin J. Your company's secret change agents. Harv Bus Rev. 2005;83:2–12.
4. Pascale R, Sternin J, Sternin M. The power of positive deviance: how unlikely innovators solve the world's toughest problems. Boston: Harvard Business Press; 2010.
5. Zeitlin M, Ghassemi H, Mansour M. Positive deviance in child nutrition—with emphasis on psychosocial and behavioral aspects and implications for development. Tokyo: United Nations University Press; 1990.
6. Mackintosh U, Marsh D, Schroeder D. Sustained positive deviant child care practices and their effects on child growth in Viet Nam. Food Nutr Bull. 2002;23:16–25.
7. Marsh DR, Schroeder DG, Dearden KA, et al. The power of positive deviance. BMJ. 2004;329:1177–9.
8. Awad SS, Palacio CH, Subramanian A, et al. Implementation of a methicillin-resistant Staphylococcus aureus (MRSA) prevention bundle results in decreased MRSA surgical site infections. Am J Surg. 2009;198:607–10.
9. Ellingson K, Muder RR, Jain R, et al. Sustained reduction in the clinical incidence of methicillin-resistant *Staphylococcus aureus* colonization or infection associated with a multifaceted infection control intervention. Infect Control Hosp Epidemiol. 2011;32:1–8.
10. Lindberg C, Norstrand P, Munger M, et al. Letting go, gaining control: positive deviance and MRSA prevention. Clin Leader. 2009;2:60–7.
11. Lindberg C, Clancy TR. Positive deviance: an elegant solution to a complex problem. J Nurs Adm. 2010;40:150–3.
12. Lloyd J, Buscell P, Lindberg C. Staff-driven cultural transformation diminishes MRSA. Prevent Strateg. 2008;Spring:10–5.
13. Marra AR, Guastelli LR, deArajo CM, et al. Positive deviance: a new strategy for improving hand hygiene compliance. Infect Control Hosp Epidemiol. 2010;31:12–20.
14. Marra AR, Guastelli LR, deArajo CM, et al. Positive deviance: a program for sustained improvement in hand hygiene compliance. Am J Infect Control. 2011;39:1–5.
15. Singhal A, Buscell P, Lindberg C. Inviting everyone: healing healthcare through positive deviance. Bordentown, NJ: Plexus; 2010.
16. Zimmerman B, Lindberg C, Plsek P. Edgeware: lessons from complexity science for health care leaders. Irving, TX VHA; 2001.
17. Goldenfeld N, Kadanoff L. Simple lessons from complexity. Science. 1999;284:87–9.
18. Thygeson M, Morrissey L, Ulstad V. Adaptive leadership and the practice of medicine: a complexity-based approach to reframing the doctor–patient relationship. J Eval Clin Pract. 2010;16:1009–15.

A Complexity Science Perspective of Organizational Behavior in Clinical Microsystems

41

Holly J. Lanham

41.1 Introduction

Complexity science is increasingly being used to understand organizational behavior in health care organizations. This chapter reviews aspects of organizational behavior that are important for understanding clinical microsystems from a complexity science perspective. I begin by introducing complexity science, briefly reviewing its history in studying health care organizations and defining key terms. I then narrow our focus to the topic of organizational behavior in clinical microsystems. Although this chapter discusses organizational behavior in clinical microsystems from a complexity science perspective, it is important to note that not all of the research referenced in this chapter was approached using complexity science as a theoretical frame. Nonetheless, each of the topics discussed below are consistent with a complexity science perspective and are important to consider when thinking about the human element of health care delivery in clinical microsystems. The chapter concludes with a discussion of quality improvement interventions from a complexity science perspective.

H.J. Lanham, Ph.D., M.B.A. (✉)
The University of Texas Health Science Center at San Antonio,7703 Floyd Curl Drive, San Antonio, TX 78229, USA
e-mail: Holly.Lanham@mccombs.utexas.edu

41.2 Complexity Science

Complexity science is the study of complex systems. Several researchers have discussed the fundamental aspects of complex systems [1–4]. Cilliers' [5] definition, shown in Table 41.1, is both elegant and highly recognized among complexity scholars.

Complex in this chapter is defined as comprising many heterogeneous interdependent parts, where the nature of the interactions among the parts influences the functioning of the system at local and global levels in unpredictable ways. Clinical microsystems are made up of individuals with diverse professional backgrounds (e.g., nurses, physicians, medical assistants, therapists, and dietitians) who bring different expertise, professional values, and worldviews to the context of health care delivery. As patients are cared for, they receive inputs from many heterogeneous individuals who interact in providing their care. The quality of these interactions is important to the overall functioning of a clinical microsystem and to the care that patients receive. Members of clinical microsystems often interact (including interactions with patients) in ways that generate unpredictable outcomes. This unpredictability derives from the nonlinearity in complex systems—thus, even when each variable and its relative importance are known, the specific outcomes of the system can be impossible to predict [6]. Unpredictability in complex systems creates challenges for both understanding and managing clinical microsystems.

J.P. Sturmberg and C.M. Martin (eds.), *Handbook of Systems and Complexity in Health*,
DOI 10.1007/978-1-4614-4998-0_41, © Springer Science+Business Media New York 2013

Table 41.1 Ten criteria for complex systems [5]

1.	Complex systems are comprised of a large number of elements.
2.	The elements in a complex system interact dynamically.
3.	The level of interaction is fairly rich.
4.	Interactions are nonlinear.
5.	The interactions have a fairly short range.
6.	There are loops in the interactions.
7.	Complex systems are open systems.
8.	Complex systems operate under conditions far from equilibrium.
9.	Complex systems have histories.
10.	Individual elements are ignorant of the behavior of the whole system in which they are embedded.

Complexity science is a useful framework for studying clinical microsystems [7–18]. From large integrated hospital settings to small primary care practices to community-based nursing homes, complexity science research has yielded valuable insights into the delivery of health care. Until the introduction of complexity science, the dominant conceptualization of an organization was that of a mechanistic system characterized by predictability and affinities for equilibrium [19]. From a complexity science perspective, clinical microsystems are complex systems made up of heterogeneous agents and are characterized by nonlinear dynamics [1, 5]. Heterogeneity among agents improves the ability of organizations to solve problems and learn [20, 21]. Nonlinearity in complex systems implies that small inputs in one area of a system may have disproportionately large effects in another area and that the relationship between inputs and outputs is often unpredictable [19]. As agents in complex systems interact locally over time, they self-organize, forming stable patterns of organizing [3]; they exhibit emergent properties, properties at one level of a system that cannot be understood by analyzing the same property at another level of the system [22, 23]; and they coevolve with their environments [24–26], making the contexts in which agents and organizations operate an important contributor to the achievement of desirable outcomes.

One particular area of health care that complexity science has impacted is the study of organizational behavior in clinical microsystems. The human aspect of health care delivery creates unique challenges. For instance, in health care delivery the patient is both the product and the consumer (setting aside the issue of third-party payers playing a role in consumption as well). Additionally, the production of health outcomes is generally not as straightforward as the production of concrete, automobiles, or printers. In addition to poorly understanding many of the contributors to health within a given clinical microsystem and the organization in which it exists, numerous social and behavioral factors that are situated outside of the walls of the health care delivery setting influence patients' health outcomes (e.g., the health of one's social network, availability of healthy foods, one's activity level, and one's occupation). This chapter examines several key organizational behavior topics of health care delivery in clinical microsystems. These topics are consistent with a complexity science perspective and have yielded new understandings of health care delivery systems and new insights about their management.

41.3 Organizational Behavior

Organizational behavior is the study of the activities of individuals and groups in organizations. The goal of organizational behavior research is to improve understanding of how individual and group dynamics affect and are affected by the organizations in which they exist. Organizational behavior research draws primarily from psychology and sociology to examine topics such as cognition, decision-making, negotiation, motivation, cooperation, leadership, diversity, power and influence, communication, coordination, interrelating, learning, and sensemaking.

This chapter focuses on group aspects of organizational behavior that are important for understanding and managing clinical microsystems from a complexity science perspective. Activities such as interrelating, reflection, learning, and sensemaking are particularly salient and central to this discussion. I also discuss health care improvement efforts from a complexity science perspective, paying particular attention to the role of adaptive reserve and conversation in improvement interventions.

41.4 Clinical Microsystems

Clinical microsystems are small groups of clinicians and staff working together to provide health care to a population of patients [27]. They are often described as the small systems that are at work where patients and health care delivery meet. Neonatal intensive care teams, nursing home units, primary care practices, and orthopedic surgery teams are examples of clinical microsystems. Clinical microsystems are situated in the middle of the larger system of health care, nested in the community and the organizational systems of care and around the provider–patient system and the self-care system (see Fig. 41.1). Essential components of a clinical microsystem include clinical and business objectives, linked processes, shared information, and the production of services and care that can be measured as performance outcomes. Clinical microsystems are not static entities; rather, they evolve over time as they and the systems in which they operate interact.

The *organization* has been the traditional level of analysis for examining health care delivery. Thinking about health care delivery at the clinical microsystem level of analysis is valuable because it is at this level that health care is physically delivered to patients. Clinical microsystems are a critical level for patient safety interventions. It is at the clinical microsystem that errors in direct patient care processes are detected or missed; that patients decide to ask or withhold questions of their physician; and that nurses and physicians respect or disrespect each other. It is at this level of analysis that patient-centered care is delivered, that medication reconciliation occurs, and that signs of a patient's increased risk for hospital readmission appear. It is at the level of the clinical microsystem that patients are appropriately cared for or unintentionally but fatally harmed. Well-functioning clinical microsystems can interrupt, avert, or diminish the impact of medical errors on patient harm. For these reasons, understanding organizational behavior in clinical microsystems is an important part of understanding health care delivery and improving the quality of care delivered to patients.

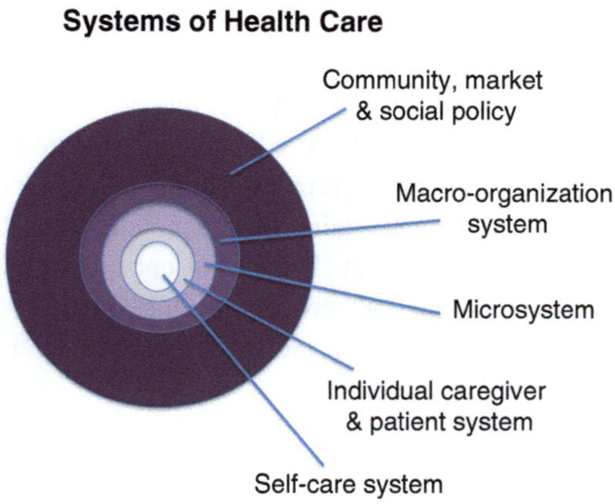

Fig. 41.1 Clinical Microsystem situated in Systems of Health Care Model (adapted from Godfrey et al, 2004)

41.5 Implications of Complexity Science for Thinking About Organizational Behavior in Clinical Microsystems

What does complexity science offer to the understanding of organizational behavior in clinical microsystems? In the USA and other nations, health care leaders are looking to complexity science to address a diversity of clinical and management challenges in health care. Clinical applications of complexity science include advances in diagnostic and treatment tools for a wide range of diseases. Management applications of complexity science in clinical microsystems include developments in areas such as relationship-centered models of practice redesign, emergent leadership, and complexity science-based efforts to reduce hospital-acquired infection rates.

An important notion that complexity science contributes to the study of organizational behavior in clinical microsystems is that of irreducible uncertainty. Uncertainty has traditionally been conceptualized as reducible with more information or better information processing. Traditional efforts to manage uncertainty in organizations focus on information gathering activities and information processing capacity [28–31]. Complexity science recognizes the presence of reducible uncertainty as well as irreducible uncertainty. Irreducible uncertainty is a type of uncertainty that cannot be reduced with more information. Irreducible uncertainty is present in patient care processes, illness and disease trajectories, and in the interactions between the two [2, 32].

The difficulty in managing irreducible uncertainty is that it arises from the nonlinear dynamics in complex systems. In complex systems, even when the variables and the relationships among the variables are known, predicting system behavior precisely over time turns out to be impossible [6]. Even sophisticated experimentation and simulation techniques cannot resolve irreducible uncertainty. For this reason, outcomes in complex systems are typically discussed in terms of probabilities and ranges of possibilities as opposed to precise predictions [19]. Irreducible uncertainty is difficult to manage; however, strategies for managing uncertainty that emphasize uncertainty absorption have been shown to be effective [33–35]. Managing the interrelationships among individuals, reflection, learning, and sensemaking are additional strategies for absorbing uncertainty in complex systems. Each of these strategies is discussed in the following section.

41.5.1 Interrelating in Clinical Microsystems

The ways that people in clinical microsystems interrelate are important. Relationships between clinicians and patients, between physicians and nurses, and between clinicians and non-clinicians are all important to consider in clinical microsystems.

Clinicians often have an instinctive understanding of how their relationships with patients influence healing. The relationship between providers and patients has been well studied [36–38]. In investigating relationships between clinicians and their patients, Scott et al. [38] identified three processes that foster *healing relationships* between physicians and their patients: (1) valuing and creating a nonjudgmental emotional bond, (2) appreciating power and consciously managing clinician power in ways that benefit patients, and (3) abiding, or displaying a commitment to caring for patients over time. From these processes, trust, hope, and a sense-of-being-known emerge in patients. Finally, clinicians who are self-confident, mindful, capable of managing their own emotions, and knowledgeable are more likely to foster healing relationships with their patients than clinicians who lack these skills [38].

In addition to the relationships between clinicians and patients, the relationships among members of clinical microsystems are important [13, 15, 39, 40]. Complexity science-based research in primary care settings demonstrates the key role that practice relationships play in understanding practice change efforts. Work conducted by Miller et al. [15] provides the insight that practices that function like a jazz ensemble may have an advantage over practices that lack this jazz-like quality. This work also makes the point that variation in physician behavior can be a source of strength in practices and that quality interrelationships among all members of primary care practices are critical for creating good jazz. The following quote from Miller et al. [15] illustrates how a complexity science perspective can inform thinking about interrelating in clinical microsystems:

> The traditional, largely unsubstantiated, view is that the best way to improve care is to eliminate variation. A view of family practice informed by complexity science suggests otherwise. In complex adaptive systems, agents in the practices create responses to changing circumstances—they improvise, or play practice jazz. Jazz players are often seen as role models of sensemaking and improvisational behavior. They know a general musical structure, and within that they create jazz. Bad jazz occurs when one person plays what the others cannot make sense of and build on. All the players have an interdependent responsibility to create good jazz. When good jazz players hear something unexpected, they make sense of it and improvise. Dealing with the uncertain nature of complex adaptive systems involves thinking in terms of making sense of what is emerging. How can I improvise to use whatever happens to further the

Table 41.2 Seven characteristics of work relationships [13]

Trust
The willingness of an individual to be vulnerable to another individual.
Diversity
Differences in individual perspectives of the world. Diversity is important for problem-solving and learning.
Mindfulness
Openness to new ideas; new ways of doing things; fully engaged presence; rich discriminating awareness to detail; seeking novelty even in routine situations.
Heedful interrelation
Interaction in which individuals pay attention to the task at hand (their job) and at the same time are sensitive to the way their actions affect the group.
Respectful interaction
Honest, self-confident, and appreciative interaction among individuals that creates new meaning.
Social and task relatedness
• Social relatedness is characterized by nonwork-related conversations and activities. • Task relatedness is characterized by work-related conversations and activities.
Rich and lean communication
The use of an effective mixture of communication channels for transferring messages. Face-to-face conversation is a form of rich communication and is effective when information being transferred is highly uncertain or ambiguous. Impersonal documents are lean channels of communication and are most effective when information being transferred is clear and nonthreatening.

system's development? It involves building on emergent characteristics of the complex adaptive system to develop patterns of social interaction among agents that give them confidence in each other, lead to small wins, and enhance the capacity to learn from unpredicted events.

Beyond the jazz metaphor of practice relationships, several characteristics of practice relationships have been identified as important for practice performance. A model of practice relationships developed by Lanham et al. [13] identified seven characteristics of relationships as important in distinguishing practices based on delivery of preventive services. The characteristics are mindfulness, heedful interrelating, respectful interaction, trust, diversity, social/task relatedness, and rich and lean communication (see Table 41.2 for definitions). Through comprehensive analyses of qualitative and quantitative data from over 160 primary care practices operating across the USA, Lanham et al. observed these seven characteristics in practices where preventive care was consistently provided at a high level and were absent, or present at low levels, in practices where preventive care was provided less consistently.

In addition to being associated with the delivery of preventive care, patterns of work relationships have been linked to practice level patterns of electronic health record use in ambulatory care settings [41]. In this study, practices with fragmented relationships displayed heterogeneous electronic health record use and practices with cohesive relationships displayed homogeneous use of this technology. Moreover, the practices that displayed electronic health record use that was uniformly high across all physician–nurse teams were the only practices to exhibit high *mindfulness* and high *respectful interaction*, suggesting that these two relationship characteristics may be particularly salient as clinical microsystems work to effectively incorporate health information technology into their work practices. Both of these studies were informed by a complexity science perspective focusing on the nature of interrelationships in clinical microsystems.

With a focus on coordinating interdependent work tasks, *relational coordination theory* is another authoritative framework for understanding interrelating in clinical microsystems [39]. In relational coordination theory, attributes of relationships and communication are mutually reinforcing and are considered important to clinical microsystem performance. Attributes of relationships include: shared goals, shared knowledge, and mutual respect; attributes of communication include communication with others that is frequent, timely, accurate, and problem-solving in nature [39]. Relational coordination theory focuses on task-based relationships as opposed to personal relationships. This framework has successfully distinguished high from low performing clinical microsystems in a variety of different health care delivery settings including both inpatient [42] and outpatient care settings [43].

Regardless of the specific framework used, complexity science enables an appreciation of the fundamental role that interrelating plays in the work of clinical microsystems. Interrelationships are important on their own, but they are also important because they form the foundation on which other key organizational behaviors occur. The remainder of this chapter discusses three organizational behaviors, reflection, learning, and sensemaking, from a complexity science perspective.

41.5.2 Reflection in Clinical Microsystems

Because events that take place in clinical microsystems are often unpredictable, reflection is important. Reflection is the ongoing critical and developmental examination and evaluation of action. Reflection can occur during action and after action. These different modes of reflection are often described as reflection-*in*-action and reflection-*on*-action. Schon's influential work on *reflection in practice* argues the need for professionals to critically reflect on their everyday practice in order to stimulate much needed learning from experience [44]. This type of reflection is thinking about what one is doing while one is doing it. Reflection-in-action is an important skill for professionals to have and is tantamount to being able to think on one's feet and apply, or not apply, previous experience to new circumstances.

Reflection-on-action is also important in clinical microsystems. Allocating time and space for members of clinical microsystems to think together and collectively debrief about past events (e.g., mistakes and errors resulting in patient harm, near misses, unexpected critical events, and failed/successful improvement attempts) is difficult from a resource perspective, but can result in improved learning, sensemaking, and improvisational capacity. *After action reviews* (AAR), initially developed by the US Army, are an example of a reflection-on-action technique. AAR are structured debriefing processes whereby participants discuss what happened, why it happened, and how it could have occurred differently. Another mechanism that facilitates reflection-on-action is a *reflective adaptive process* [45]. Reflective adaptive process (RAP) sessions are designed to improve work relationships in clinical microsystems, and then based on these relationships, encourage members of clinical microsystems to regularly meet to discuss clinical and nonclinical matters of importance. Reflection in clinical microsystems allows the integration of multiple diverse perspectives on the issues at hand, and ultimately enables learning and sensemaking in clinical microsystems.

41.5.3 Learning in Clinical Microsystems

Clinical microsystems are complex systems made up of individuals that interact and learn as new information is exchanged. Learning is essential in complex systems because events rarely repeat themselves exactly as they occurred in the past [19]. For this reason, activities such as learning from samples of one [46], learning from rare events [47], and learning from mistakes [48] are key strategies for learning in clinical microsystems that are consistent with a complexity science perspective.

Learning from a complexity science perspective emphasizes the dynamic nature of knowledge. Learning in clinical microsystems should be ongoing, or continual, in order for it to positively shape action. Learning from a complexity science perspective is less focused on what individuals can learn on their own through, for example, continuing education courses and online training modules, and is more focused on learning that occurs through interactions of individuals, their exchanges of information, and co-creation of new meaning. Learning sessions, also called learning collaboratives, are consistent with a complexity science perspective. In these sessions, diverse individuals who may or may not know each other, but who are facing the same set of challenges come together to share experiences, exchange ideas and collectively create new solutions. Agendas for learning sessions tend to be broad and not overly structured allowing time for unplanned conversations on topics that arise during the session.

Several characteristics of learning in clinical microsystems are important to consider. For effective learning to occur, individuals should feel secure in speaking up and contributing to the learning of the group. Edmondson [49] demonstrated the critical role that *psychological safety* plays in learning in clinical microsystems. Psychological safety has been identified as a key antecedent to speaking up and learning in a variety of different clinical microsystems including neonatal intensive care units and cardiac surgery teams. Research has also shown the need for *mindful learning* in clinical microsystems [50]. Mindful learning is learning that is open to new explanations for ordinary events. Mindful learning is being present rather than operating on autopilot. It is paying attention to the nuances of the patient who is being operated on right now. Mindful learning seeks novelty, even in routine situations. It is picking up on something new, something not quite right about a well-known, established patient and following through on that *rich point* [51] that was almost overlooked. Mindful learning includes input from others—not for justice, equality or fairness, but because it results in better outcomes. Finally, *reciprocal learning* plays a role in clinical microsystems. Reciprocal learning occurs when individuals who are working together learn together by sharing knowledge and insights with each other in back-and-forth, building, and iterative processes [52]. With this type of learning, people frequently learn from others and also frequently teach others within the same clinical microsystem. Reciprocal learning has been associated with primary care practices' readiness for delivering comprehensive care to patients with chronic disease [52].

In summary, clinical microsystems wanting to facilitate effective learning should create an environment where it is safe to speak up without fear of negative repercussions, where people are open to new explanations for repeated observations, and where people treat learning as an ongoing and iterative activity.

41.5.4 Sensemaking in Clinical Microsystems

Sensemaking is a powerful tool for managing uncertainty in clinical microsystems. "Sensemaking is a diagnostic process directed at constructing plausible interpretations of ambiguous cues that are sufficient to sustain action" [53]. The link between sensemaking and action is important. Sensemaking and action are interdependent—sensemaking informs action and action informs sensemaking. The sense that is made at any one point in time depends on many things [54], and the nature of the interdependencies among people in a group will affect the quality of sensemaking that occurs in that group. For example, a clinical microsystem consisting of people who are willing to be vulnerable to each other will assign different meaning to an experience than the meaning assigned by a clinical microsystem consisting of people who are unwilling to display vulnerability. Similarly, clinical microsystems that are open to learning new ways of thinking about problems will make sense of the same situation differently than those that are set in routine ways of thinking about problems.

Because clinical microsystems are dynamic and unpredictable, they need the capacity to continually make sense of their environment. Leykum et al. [55] found that the ability of physician inpatient care teams to make sense of an ever-shifting clinical context was significantly associated with length of hospital stay and patient complication rates. This study also found that care team relationships characterized by trust, mindfulness,

Table 41.3 Examples of interrelating, reflection, learning, and sensemaking in practice

Organizational behavior	Practice A	Practice B
Interrelating	People in this clinical microsystem: ◦ Listen with empathy and build on others' contributions, even those that are unexpected ◦ Understand how their actions affect the ability of others to complete their tasks ◦ Respect their ideas and the ideas of others ◦ Are willing to speak up even when their opinion may be unpopular ◦ Interact timely and frequently so that problems can be solved	People in this clinical microsystem: ◦ Do not build on the contributions of others ◦ Lack the ability to see how their work fits into the work of the clinical microsystem ◦ Overvalue their own ideas and undervalue the ideas of others ◦ Keep their opinions to themselves ◦ Avoid interacting with others, even when they have ideas that they believe can contribute to solutions
Reflection	This clinical microsystem regularly allocates time and space for members to talk about previous action—even when time is scarce. Evaluation of action a high priority.	This clinical microsystem lets the demands of its day-to-day work take precedence over talking about previous action. Evaluation of action is rare.
Learning	People in this clinical microsystem: ◦ Feel safe expressing their ideas without negative side effects ◦ Are open to new ways of doing things ◦ Look for moments in time that tell a different story than the one usually told ◦ Teach each other new things ◦ Learn from each other	People in this clinical microsystem: ◦ Feel that sharing their ideas could result in punitive action ◦ Look to established routines as a guide for action ◦ Keep their ideas to themselves, including things that could help others in their work
Sensemaking	When faced with ambiguity, people in this clinical microsystem seek a variety of plausible understandings	When faced with ambiguity, people in this clinical microsystem seek high likelihood explanations

Practices A and B are both fictional; Practice A representing an idealized example of a practice with effective interrelating, reflection, learning, and sensemaking and Practice B representing a practice with less effective interrelating, reflection, learning, and sensemaking heedful interrelation, and respectful interaction displayed more effective sensemaking strategies than teams with lower levels of these relationship characteristics. Sensemaking is an important but often overlooked aspect of health care delivery in clinical microsystems. Table 41.3 provides examples of how each of these organizational behaviors might appear in practice.

41.6 Clinical Microsystem Interventions from a Complexity Science Perspective

Behavior change in clinical microsystems is difficult [56] and diffusion of knowledge can be unpredictable [34, 57, 58]. Clinical microsystems struggle to successfully scale up and spread effective practices across diverse settings. Improvement intervention failures are often attributed to a lack of resources, limited understanding of local contexts, and competing demands. I argue that many intervention failures can be attributed to the notion that intervention designs frequently lack an understanding of clinical microsystems as complex systems.

Research has demonstrated an increased effectiveness of interventions using design elements consistent with a complexity science perspective [59, 60]. Across a sample of 46 interventions, those that took interdependencies, self-organization, coevolution, and nonlinearity into account achieved significantly higher success rates than interventions that did not consider these elements. At the same time, quality has been discussed as an emergent property—meaning that organizational quality of care arises from the local interactions of individuals [13]. This complexity science-based conceptualization of quality suggests that intervention designs should incorporate methods to assess and improve (if necessary) the relationship infrastructure of clinical microsystems prior to initiating change.

Additional aspects of improvement interventions that should be considered here include adaptive reserve and conversation. An evaluation of the National Demonstration Project of the patient-centered medical home found that primary care practices that displayed high levels of adaptive reserve (the ability to keep up with rapid developments and environmental changes) successfully implemented more elements of the patient-centered medical home model than practices that displayed lower levels of adaptive reserve [61]. Adaptive reserve is made up of several sub-constructs including relationships, learning culture, work environment, communication, teamwork, and leadership. Many of these sub-constructs are consistent with a complexity science perspective. Conversation also plays an important role in clinical microsystem improvement interventions [12]. Fostering high-quality conversation is not as easy as it might sound. Conversation that is collaborative, meaning-making, and improvisational is more likely to enable effective sensemaking and learning during interventions than conversation that is unidirectional, didactic, and scripted [12]. Using complexity science to frame improvement interventions in clinical microsystems brings into focus organizational behavior aspects such as interdependencies, self-organization, coevolution, nonlinearity, adaptive reserve, and conversation that might be overlooked in traditionally framed interventions.

41.7 Conclusion

Clinical microsystems are complex systems characterized by unpredictability. Using this perspective, interrelating, reflection, learning, and sensemaking are discussed as key organizational behaviors that are important in understanding and managing clinical microsystems. Organizational behavior aspects of improvement interventions are also discussed from a complexity science perspective. Interventions that recognize interdependencies, self-organization, coevolution, and nonlinearity and that foster adaptive reserve and conversation are more likely to be effective than interventions that overlook these aspects of clinical microsystems undergoing change. Using complexity science to frame our thinking about human behavior in the delivery of health care to patients can be a powerful tool in the pursuit of high-quality health care.

References

1. Anderson P. Complexity theory and organization science. Org Sci. 1999;10(3):216–32.
2. Bar-yam Y. Making things work: solving complex problems in a complex world. New England: Complex Systems Institute, Knowledge Press; 2004.
3. Kauffman S. At home in the universe: the search for laws of self-organization and complexity New York. NY: Oxford University Press; 1995.
4. Waldrop MM. Complexity: the emerging science at the edge of order and chaos. New York: Touchstone by Simon and Schuster; 1992.
5. Cilliers P. Complexity and postmodernism: understanding complex systems. New York, NY: Routledge; 1998.
6. Liebovitch LS. Fractals and chaos simplified for the life sciences. Oxford: Oxford University Press; 1988.
7. Anderson RA, Issel MI, McDaniel Jr RR. Nursing homes as complex adaptive systems: relationship between management practice and resident outcomes. Nurs Res. 2003;52:12–21.
8. Arndt M, Bigelow B. Commentary: The potential of chaos theory and complexity theory for health services management. Health Care Manage Rev. 2000;25(1):35–8.
9. Bazzolli GJ, Shortell SM, Dubbs N, Chan C, Kralovec P. A taxonomy of health networks and systems: bringing order out of chaos. Health Serv Res. 1999;33(6):1683–717.
10. Begun JW. Chaos and complexity: frontiers of organizational science. J Manage Inquiry. 1994;3(4):329–35.
11. Crabtree BF, Nutting PA, Miller WL, McDaniel Jr RR, Stange KC, Jaén CR, et al. Primary care practice transformation is hard work: insights from a 15-year developmental program of research. Med Care. 2010;49(Suppl):S28–35.
12. Jordan ME, Lanham HJ, Crabtree BF, et al. The role of conversation in health care interventions: enabling sensemaking and learning. Implement Sci. 2009;4:1–13.
13. Lanham HJ, McDaniel Jr RR, Crabtree BF, et al. How improving practice relationships among clinicians and nonclinicians can improve quality in primary care. Jt Comm J Qual Patient Saf. 2009;35:457–66.
14. McDaniel Jr RR, Driebe DJ. Complexity science and health care management. In: Blair JD, Fottler MD, Savage GT, editors. Advances in health care management, Vol 2. Stamford, CN: JAI Press; 2001. p. 11–36.
15. Miller WL, et al. Practice jazz: Understanding variation in family practice using complexity science. J Fam Pract. 2001;50:872–8.
16. Plesk P. Redesigning health care with insights from the science of complex adaptive systems. In: Crossing the quality chasm: a new health system for the 21st century. Washington, D.C.: National Academy Press, Institute of Medicine; 2001. p. 322–35.
17. Nutting PA, Crabtree BF, Miller WL, Stange KC, Stewart EE, Jaen CR. Transforming physician practices to patient-centered medical homes: lessons from the National Demonstration Project. Health Aff. 2011;30(3):439–45.
18. Zimmerman B, Lindberg C, Plesk P. Edgeware: insights from complexity science for health care leaders. Irving, TX: VHA; 1998.
19. Sornette D. Critical phenomenon in natural sciences. Berlin: Springer; 2006.
20. Jehn KA, Northcraft GB, Neale MA. Why differences make a difference: a field study of diversity, conflict, and performance in workgroups. Admin Sci Q. 1999;44(4):741–63.
21. Page SE. Diversity and complexity. Princeton, NJ: Princeton University Press; 2011.
22. Epstein JM, Axtell R. Growing artificial societies: social science from the bottom up. Washington, DC: The Brookings Institutes Press; 1996.
23. Holland JH. Emergence: from chaos to order. Reading, MA: Addison-Wesley; 1998.
24. Allen PM, Strathern M, Baldwin JS. The evolutionary complexity of social economic systems: the inevitability of uncertainty and surprise. In: McDaniel Jr RR, Driebe DJ, editors. Uncertainty and surprise in complex systems. Heidelberg: Springer; 2005. p. 31–50.
25. Chu D, Strand R, Fjelland R. Theories of complexity: common denominators of complex systems. Complexity. 2003;8(3):19–30.
26. Vidgen R, Wang X. Coevolving systems and the organization of agile software development. Inform Syst Res. 2009;20:355–76.
27. Mohr JJ, Batalden P, Barach P. Integrating patient safety into the clinical microsystem. Qual Saf Health Care. 2004;13(Suppl II):ii34–8.
28. Daft RL. Organization theory and design. St. Paul, MN: West Publishing; 1989.
29. Galbraith JR. Designing complex organizations. Reading, MA: Addison-Wesley; 1973.
30. Shannon CE. A mathematical theory of communication. Bell Syst Techn J. 1948; 27(4):379–423, 623–56.
31. Tushman ML, Nadler DA. Information processing as an integrating concept in organizational design. Acad Manage Rev. 1978;3(3):613–24.
32. West BJ. Where medicine went wrong: rediscovering the path to complexity. Singapore: World Scientific Publishing; 2006.
33. Ashmos DP, Duchon D, Hauge FE, McDaniel Jr RR. Internal complexity and environmental sensitivity in hospitals. Hosp Health Serv Admin. 1996;41(4):535–53.
34. Boisot M, Child J. Organizations as adaptive systems in complex environments: the case of China. Org Sci. 1999;20(3):237–52.
35. March JG, Simon HA, Guetzkow HS. Organizations. Cambridge, MA: Blackwell; 1958.
36. Hjortdahl P, Borchgrevink C. Continuity of care—influence of general practitioners' knowledge about

their patients on use of resources in consultations. BMJ. 1991;303:1181–4.
37. Sturmberg J. Comprehensiveness of care in a small group practice and its relationship to patient utilization patterns. Melbourne: Monash University; 1998.
38. Scott JG, Cohen D, DiCicco-Bloom B, MillerWL SKC, Crabtree BF. Understanding healing relationships in primary care. Ann Fam Med. 2008;6(4):315–22.
39. Gittell JH, Seidner R, Wimbush J. A relational model of how high performance work systems work. Org Sci. 2010;21(2):490–506.
40. Safran DG, Miller WL, Beckman H. Organizational dimensions of relationship-centered care: theory, evidence, and practice. J Gen Intern Med. 2006;21:S9–15.
41. Lanham HJ, Leykum LK, McDaniel Jr RR. Same organization, same electronic health records (EHR) system, different use: Exploring the linkage between practice member communication patterns and EHR use patterns in an ambulatory care setting. J Am Med Inform Assoc. 2011;19(3):382–91.
42. Gittell JH, Fairfield K, et al. Impact of relational coordination on quality of care, post-operative pain and functioning, and length of stay: a nine hospital study of surgical patients. Med Care. 2000;38(8):807–19.
43. Noel PH, Lanham HJ, Palmer RF, Arar NH, Leykum LK, Bowers K, Parchman ML. Relational coordination and reciprocal learning: Importance of interdependencies among primary care teams for chronic illness care. Health Care Manage Rev; 2012 (Epub ahead of print).
44. Schön DA. The reflective practitioner. New York: Basic Books; 1983.
45. Stroebel CK, McDaniel Jr RR, Crabtree BF, et al. How complexity science can inform a reflective process for improvement in primary care practices. Jt Comm J Qual Patient Saf. 2005;31(8):438–46.
46. March JG, Sproull LS, Tasmuz M. Learning from samples of one or fewer. Org Sci. 1991;2(1):1–13.
47. Christianson MK, Farkas MT, Sutcliffe KM, Weick KE. Learning through rare events: significant interruptions at the Baltimore & Ohio Railroad Museum. Org Sci. 2009;20(5):846–60.
48. Edmondson AC. Speaking up in the operating room: how team leaders promote learning in interdisciplinary action teams. J Manage Stud. 2003;40:1419–52.
49. Edmondson A. Psychological safety and learning behavior in work teams. Admin Sci Q. 1999;44:350–83.
50. Langer EJ. The power of mindful learning. Cambridge, MA: Perseus Books; 1997.
51. Agar M. The professional stranger: an informal introduction to ethnography. 2nd ed. New York: Academic Press; 1996.
52. Leykum LK, Palmer R, Lanham HJ, Jordan ME, Noel PH, Parchman ML. Reciprocal learning and chronic care model implementation: results from a new scale of learning in primary care settings. BMC Health Serv Res. 2011;11:44.
53. Weick KE. Managing the unexpected: complexity as distributed sensemaking. In: McDaniel Jr RR, Driebe DJ, editors. Uncertainty and surprise in complex systems: questions on working with the unexpected. Berlin: Springer; 2005. p. 51–65.
54. Weick KE. Sense-making in organizations. Thousand Oaks, CA: Sage Publications; 1995.
55. Leykum LK, Kumar P, Parchman ML, McDaniel Jr RR, Lanham HJ, Agar M. Use of an agent-based model to understand clinical systems. J Artif Soc Soc Simul. 2012;15(3):2.
56. Grol R, Grimshaw J. From best evidence to best practice: effective implementation of change in patients' care. Lancet. 2003;362(9391):1225–30.
57. Berwick DM. Disseminating innovations in health care. JAMA. 2003;289(15):1969–75.
58. Rogers EM. Diffusion of innovations. 5th ed. New York: Free Press; 2003.
59. Leykum LK, Pugh J, Lawrence V, Parchman M, Hoel PH, Cornell J, McDaniel Jr RR. Organizational interventions employing principles of complexity science have improved outcomes for patients with type II diabetes. Implement Sci. 2007;2:28.
60. Leykum LK, Parchman M, Pugh J, et al. The importance of organizational characteristics for improving outcomes in patients with chronic disease: a systematic review of congestive heart failure. Implement Sci. 2010;5:1–34.
61. Miller WL, Crabtree BF, Stange KC, Nutting PA, Jaén CR. Primary care practice development: a relationship-centered approach. Ann Fam Med. 2010;8 Suppl 1:S68–79.
62. Holland JH. Hidden order: how adaptation builds complexity. New York: Addison Wesley; 1995

42 Reverse the Epidemic Tides of Unhealthy Habits: Personal, Professional, and Leadership Challenges

Rick Botelho

Politics, economics, and health are interdependent. Inequalities in the economic domain have profound negative impacts on the health domain [1]. This chapter challenges the prevailing worldview and values inherent in the power and control structures within our political, financial, and disease care systems. Political philosophies and public policies have mal-distributed money, power, and resources at global and national levels that favor the 1% economic elite of the population.

> *An imbalance between rich and poor is the oldest and most fatal ailment of all republics.*
> Plutarc 64–123 AD

Socioeconomic inequalities[1] have increased over the past 35 years [2]. Global inequities (within and between countries) in health care access and outcomes are getting worse, including the USA. One billion people do not have access to basic affordable health care [3]. The poor have higher rates of unhealthy habits and chronic diseases [4]. The negative consequences of the widening gaps between the 1% economic elite club and the 99% of humanity are the risk factors for civil unrest, terrorism, wars, and revolutions.

R. Botelho (✉)
Faculty & Student Development (2009–2010), Florida International University, Miami, FL, USA

Family Medicine (1985–2009), University of Rochester, Rochester, NY, USA

85 Eastland Ave, Rochester, NY, USA
e-mail: rbotelho@me.com

[1] *Branko Milanovic* (lead economist at the World Bank's research division) *The Haves and Have Nots*. This event was recorded on 8 February 2011 in Old Theatre, Old Building http://www.youtube.com/user/lsewebsite#p/c/0C1E7F2DC649DD98/94/OS2vuLCXVPU

Inequality is a surprisingly slippery issue, involving not just straightforward comparisons of individuals, but also comparisons of price and consumption differences around the world—and over time. In this lecture, Branko Milanovic, the lead economist at the World Bank's research division, will approach the issue in a new and innovative way, focusing on inequality in income and wealth in different time periods and contexts: from inequality in Roman times (and how it compared with inequality today), to depictions of wealth inequality in literature (Pride and Prejudice and Anna Karenina), to inequality across generations of a single family (the three generations of Obamas illustrating this theme). As for global inequality today, the talk will examine its main cause (differences in average incomes between countries), the role China and India might play, and, perhaps most importantly, whether global inequality matters at all, and if does, what can we do to reduce it. Branko Milanovic is one of the world's leading experts on inequality. He is lead economist at the World Bank's research division in Washington DC, a visiting fellow at All Souls College, Oxford, and the author of The Haves and Have Nots: A Brief and Idiosyncratic History of Global Inequality.

J.P. Sturmberg and C.M. Martin (eds.), *Handbook of Systems and Complexity in Health*,
DOI 10.1007/978-1-4614-4998-0_42,

42.1 From Gross Inequities to Fairness in Wealth and Health

In what ways and to what extent are our advocacy efforts for universal justice and health equity hollow rhetoric in our public policies and laws [5]?

A WHO Commission (2009) reported that the social determinants of health are mostly responsible for health inequities: the unfair and avoidable differences in health status seen within and between countries [6]. These "*inequities are killing people on a grand scale*," even greater than the tobacco pandemic that is estimated to kill one billion people in the twenty-first century [7, 8]. At the macro-level, our toxic environments and diseases-producing cultures create downward spiral of negative social influences to produce higher rates of unhealthy habits and chronic diseases, especially among the poor. This Commission regards the reduction of these health inequities as an ethical imperative.

The World Economic Report (2011) reported that non-communicable diseases (cardiovascular diseases, chronic respiratory diseases, cancer, diabetes, and mental health) will create output loss of US$47 trillion over the next two decades. This saving could eradicate two-dollar-a-day poverty among the 2.5 billion people in that state for more than half a century [9]. What can we learn from these political inconsistencies, economic disparities, and ethical hypocrisies to erase our rhetoric-action gaps in our public policies?

To illustrate with historical examples, the Declaration of Independence (4 July 1776) stated all men are created equal and excluded women. In 1865, the 13th Amendment abolished slavery in the USA (1870), the 15th Amendment provided male slaves with the right to vote. Hegemonic masculinity prevailed [10]. It was not until 1920 that the 19th Amendment gave women the right to vote. It took 144 years before the true spirit of the Declaration of Independence was implemented. Despite this progress, gender inequities still remain a major global determinant of health disparities [11–13].

The Declaration of Human Rights asserts that health care is a human right in 1948 [14]. Yet this is not the case in many third world countries and in the USA. How many more years will it take these countries to assure universal access to even basic health care, regardless of the ability to pay?

Our scientific worldviews about disease have profound impacts on how we address health and health care issues, including the pandemics of unhealthy habits and chronic diseases. The ethnocentric medical worldview of health care is the absence of disease and psychopathology. However, health goes beyond the physical malfunctioning of the mind–body machine, or a broken cog in a wheel. It is much more than preventing and treating diseases.

In 1948, the constitution of the World Health Organization (WHO) defined health as "*a state of complete physical, mental and social well-being, and not merely the absence of disease or infirmity*" [15]. This individualistic construct of health oversimplifies the complex dynamics of health, because it inadequately addresses the ethical, spiritual, political, economic, educational, environmental, and cultural factors affecting the development of health. Therefore, the etiology of health is much more socio-ecological than individualistic one. Consequently, this static, convergent, prescribed, and narrow-minded definition of health as the ultimate individualistic outcome has many limitations, because it does not address the myriad of aetiological factors affecting the dynamics of health. The development of optimal health is contingent on individual and social learning processes. This lifelong learning process from cradle to grave includes healthy death and dying.

This raises questions about how health care and health should be conceptualized and defined [16, 17]. Health care is a human right. But the multidimensional concept of health is not a right, because health care plays only a minor role in the development of health. Health is a shared responsibility that involves individuals, families, and communities cultivating healthy environments, healthy habits, and well-being. This social conceptualization of health also includes individual and group happiness.

Health is an individual, community, and cultural journey of spiritual and psychological development toward higher levels of consciousness, well-being, happiness, virtues, and functioning, based on understanding the neurophysiological development of our learning processes within our brain, beginning in early childhood [18]. Chronic toxic stress in early childhood impedes brain development with long-term psychosocial complications and is a risk factor for heart disease and diabetes [19]. The National Child Traumatic Stress Network aims to change the course of children's lives by changing the course of their care [20]. Conversely, happiness enhances the capacities and capabilities of our brain learning centers and the prospects of achievements and success. Achievement and success do not guarantee happiness.[2]

A divergent, dynamic, evolving, generative, and ethical model of health incorporates a blend of scientific worldviews and multiple constructs to address the ecological complexities of facilitating behavior change at micro-, meso-, and macro-levels. At the micro-level, the psychosocial-spiritual worldview of health includes concepts, such as well-being (PERMA[3] positive emotions, engagement, relationship, meaning, and achievement), social coherence, and the development of resilience,[4] character strengths, and virtues [21–24]. Individuals can improve their health and self-care of diseases by cultivating the development of any combination of the following skills:

- Experiences of contentment, despite negative circumstances
- Autonomy [25], resilience [26], and social coherence [21, 22]
- Activation, vitality, intrinsic motivation, and self-efficacy [27–33]
- Inspirational pursuit of virtues, such as open-mindedness, equanimity, and inclusiveness [24]
- Self-regulation of health, behavior, emotions, and controlling behaviors [34–37]
- Self-control of desires and destructive impulses [38–40]
- Polarity management between individual and community values[41]
- Amelioration of flaws and closed-mindedness

The worldviews of complex adaptive systems and ontology are incorporated into this ecological and dynamic model of health. Individuals, communities, and cultures can use this dynamic model to develop their own operational definitions of health in ways that strive for equal opportunities in justice, wealth, and health.

At the meso-level, the systems and organizational worldview of health is based on improving the socio-behavioral dynamics for health and wellness by synergistically cultivating transformational leadership development and health-promoting cultures, environmental stewardship, and the protection of the community and individual rights against abuses.

At the macro-level, the ethical and sociopolitical worldview of health is based on a vision of fostering socioeconomic fairness and implementing the values of universal justice, equal opportunities, and transparent accountability. This vision becomes an attractor and alignment factor for galvanizing efforts to reverse the worst inequalities in wealth and health in the history of humankind.

Within this chapter, Socratic questions are posed to evoke critical reflections that aim to deepen your cognitive, emotional, perceptual, and spiritual understanding about the complexity of health. This reflective learning process may help you appreciate the need for continuing personal, professional, and leadership development (CPPLD) programs in order to facilitate

[2] In his humorous TED talk, Shawn Achor explains how happiness activates the brain learning centers in ways that enhance the prospects of achievements and success. http://www.youtube.com/watch?v=GXy__kBVq1M.

[3] Martin Seligman gives a brief introduction to positive psychology and flourishing using PERMA

http://www.amareway.org/holisticliving/01/martin-seligman-on-flourishing-us-zeitgeist-2010/.

[4] In his RSA talk, Martin Seligman addresses the benefits of fostering resilience to prevent Post-traumatic Stress Disorder in the military. http://www.thersa.org/events/video/vision-videos/martin-seligman.

behavioral changes at multiple levels. This chapter will explore the implications of the following questions. How can we create and use CPPLD frameworks at the:

Micro-level to

- Foster greater open-mindedness and ongoing dialogues about healthful change?
- Learn in adaptive and transformative ways?
- Experience the process of developing personal evidence about deep change?
- Improve our health habits as we coach others to do the same?

Meso-level to

- Disseminate health coaching programs and related programs?
- Create learning organizations and communities about health promotion [42, 43]?
- Develop social networks of transformational leadership development?
- Develop professional and social movements?

Macro-level to

- Align, merge, and converge our diverse worldviews about health and disease?
- Build common ground for the common good?
- Adopt participatory democracies that truly represent the interests of all people?
- Decentralize our hierarchal power structures within our health care and political systems?
- Reverse the epidemic tides of unhealthy habits and diminish the mounting burdens of chronic diseases?

42.2 How Can We Overcome Our Short-sighted, Closed-mindedness?

Almost 50% of preventable deaths are related to unhealthy habits [44]. These behavioral epidemics are getting worse, along with their escalating costs of chronic diseases [9, 45–48]. Our disease-producing cultures and toxic environments have precipitated these global catastrophes [49]. These cultures and environments prevail over the values of health promotion and disease prevention. Furthermore, we are addicted to the quick fixes of curing diseases, such as medications to cure obesity and the genomic treatments of diseases, such as personalized medicine. In effect, we have disease care systems, not health care systems. Our short-sighted, closed-minded, and outside-in (positivistic) worldviews about disease prevail over the long-sighted, open-minded, and inside-out (ontological) worldviews about health.

42.3 Transform from Disease Care to Health Care Systems

A decentralized shift in power from our representational democracies and our hierarchical systems of complicated disease care toward participatory democracies and complex systems of adaptive health care will help to facilitate a transformation from:

- Disease-producing cultures and toxic environments that generate negative social influences and encourage unhealthy decision-making into
- Health-promoting cultures and environments that create positive social influences and foster healthy decision-making

Without such transformations, our behavior change programs for health promotion, disease prevention, and disease management will remain inadequately developed and poorly integrated into our "yet to be developed" health care systems.

Transformational leadership is essential for effective change [50]. To integrate behavior change learning innovations into all sectors of society, we need to deliver longitudinal, lifelong health literacy that helps people develop appropriate health and disease self-management skills at all educational levels and life stages [51–55]. To foster this lifelong learning process, we can journal and document the outcome of our learning processes in our computerized personalized health record (PHR) that can interface our electronic medical records. These curricula combined with the proactive use of the PHR can foster participatory health uprisings that activate the development of "*4p-patients*": prepared, proactive, proficient, and partnering. The *Health Buddy Coaching* (HBC) program is an example of such a curriculum that aims to enhance the performance

Table 42.1 Contrasting scientific worldviews

Positivistic worldview	Complexity worldview	Ontological worldview
Metaphor: microscope	Metaphor: telescope	Metaphor: mirror
Science of "It"	Science of "Its and Networks"	Science of "I and We"
Objectivity	Interactivity	Subjectivity
Reductionism	Holism	Authentic self
Mechanistic processes	Organic processes	Development of virtues
Controlled	Dynamic	Evolutionary
Experiments	Naturalism or contextualism	Consciousness
Linear causality	Nonlinear causality	Phenomenonology
Predictability	Emergence	Being and becoming

of our health promotion, disease prevention, and disease management programs, within and beyond our health care delivery systems.

HBC learning experiences can help you understand why evidence-based guidelines inadequately address the complexities of behavioral change. Researchers generate objective data to predict the impact of a particular intervention on the average patient.[5] This (outside-in) scientific evidence cannot provide any specific guidance on what will work for any particular individual. However, experiential learning (inside-out) methods can overcome the limitations of brief evidence-based interventions that address predominantly surface change (thoughts and feelings).

The HBC learning process begins with you by developing your own personal evidence about deep change. Becoming the principal investigator, researcher, and author of your own behavioral change, you discover what works for you, in particular, by collaborating with others who are using the same learning process. Experiential learning methods greatly improve the impact of evidence-based guidelines in reversing these epidemics and reducing the burden of chronic diseases.

To promote healthy habits at individual and population-based levels, we need to adopt holistic and ecological perspectives that integrate the worldviews of ontology and complex adaptive systems (CAS). In short, the whole and the parts are interacting internally and externally, and changing each other dynamically over time, in linear and nonlinear ways, and for better and for worse [56]. How can we:

- Evolve in upward spirals of positive social influences toward healthy habits, happiness, and well-being?
- Avoid regressing in downward spirals of negative social influences toward unhealthy habits and chronic diseases?

The comparison table (Table 42.1) outlines broad distinctions in the worldviews of positivism, complex adaptive systems, and ontology that shape and change how we respond to issues related to health and disease.

We can blend worldviews to develop more sophisticated, dynamic, and diversified ways of evolving, implementing, evaluating, and scaling up process innovations that synergistically facilitate positive changes at multiple levels.

The metaphor for understanding the scientific worldview of positivism is the microscope. The science of "it" focuses on objectivity. Expert researchers conduct randomized control studies to produce evidence-based guidelines. These guidelines help practitioners work episodically with patients on addressing the surface aspects of behavior change.

Unfortunately, current public policies, reimbursement methods, and clinical models do not support the systematic implementation of evidence-based interventions for behavior change in health care settings [57–62]. Consequently, brief interventions are of marginal impact at population-based levels. Furthermore, evidence-based interventions developed solely on positivism are too static and simplistic for addressing the ecological, nonlinear dynamics of behavioral epidemics [56, 63–66].

[5] See also: Chap. 4—Sturmberg & Miles: The Complex Nature of Knowledge.

To reverse these epidemics, we need to expand our modernist worldview to include the postmodernist worldview or lens. Metaphorically, this represents a shift from the microscopic worldview of reductionism, specialism, and technology to include the telescopic worldview of holism, generalism, and humanism [67].

The metaphor for understanding the scientific worldview of complex adaptive systems is the telescope. The science of networks and "its" (such as environmental factors and public policies) helps to understand what shapes our health behaviors. For example, social networks influence the development of (un)healthy behaviors [68–71]. Socio-environmental influences lie outside of the health sector and play significant roles in shaping health behaviors [72–81]. Health in All Policies (HiAP) is essential for addressing these social–environmental determinants of health [82, 83]. Such policies have been successfully implemented in Finland, beginning with the North Karelia Project in 1972 [84–88]. The Finnish HiAP operationalized the vision of Ottawa Charter (1986): make health an easy choice [83, 89].

We also need sociopsychological innovations within and beyond the health sectors to further promote health and healthy behaviors [90]. This learning process involves discovering how to influence the sociopsychological dynamics of health for the better [91]. This involves understanding and using nonlinear complexity concepts: such as attractors, self-organizing processes, evolutionary processes, network dynamics, and catastrophes [92]. For example, the attractor for promoting healthy habits can arise from generating meaningful experiences based on catalytic learning innovations. Long-term educational investments are needed to develop, test, refine, and evaluate dynamic, complex process innovations.

Catalytic innovations create social change through replication and scaling up of needed services to underserved (or not served at all) populations that are convenient, inexpensive, easy-to-use, and beneficial [93]. Positive HBC experiences can cultivate meaningful camaraderie and create multiple hives of crowdsourcing in ways that galvanize our social networks to promote healthy habits. The effective adoption of catalytic innovations will also benefit from building on, and developing synergy with preexisting and new models of care, such as the community-oriented primary care and the chronic care model [94–100]. Participatory research methodologies and systemic realist reviews provide ways to refine catalytic innovations iteratively over time and add rigor in evaluating dynamic, complex process innovations for cultivating healthy habits [101–115].

Ideally, such social and health improvement initiatives involve community capacity-building that are sustained by learning opportunities for skills and leadership development, asset-based approaches, linkages to networks, and partnerships with multi-communication channels, participatory decision-making to create a sense of community, and developmental pathways. These improvements and pathways are enhanced by having a shared vision with clear goals, community need assessments, process and outcome monitoring processes, and commitment to action together with dissemination and sustainability methods [106]. Our health care systems have yet to learn how to leverage community engagement and mobilization to accelerate intersectoral progress toward health equity by coordinating political, advocacy, public policy, economic, educational, social, and health improvement initiatives [116].

A neo-modernist worldview adopts a synthetic and adaptive approach that optimally blends worldviews to reverse the epidemics of unhealthy habits and reduce the burdens of chronic diseases. Metaphorically, this approach includes the ability to zoom back and forth between the microscope and telescope, as well as the ability to observe the observer. We need to expand our scientific worldviews to include ontology.

The metaphor for understanding the scientific worldview of ontology is the mirror. This involves reflecting about ourselves and working with others who act as mirrors by shining light on our blind spots. Such reflective learning experiences can enhance our self-awareness about our blind spots. We can learn how to rectify shortcomings in our worldviews, discrepancies in our value systems, our distorted perceptions, and flaws in ways that cultivate virtues. This learning process can also help us reflect about ourselves in ways that we can develop our own personal evidence about behavioral change.

As practitioners, staff, and/or patients alike, we can use HBC programs to learn, collaborate, and coach each other longitudinally over time in how to improve our health habits. With this paradigm shift from scientific to personal evidence, we become the principal investigator, researcher, and author of our own behavior change. We make deep change and discover our own solutions for change [117].

42.4 How Can You Facilitate Your Continuing Personal, Professional and Leadership Development?

You can use the Continuing Personal, Professional, and Leadership Development (CPPLD) frameworks to explore and reflect more deeply about the ecological complexities of facilitating behavioral change, from individual to cultural levels. This learning process will help you understand more about your developmental needs at the:

Personal level:

- Embark on a journey of lifelong learning about behavioral change
- Address the flaws of closed-mindedness worldviews
- Cultivate the virtues of open-minded worldviews
- Experience adaptive and transformative learning methods
- Develop personal evidence about deep change
- Improve your health habits as you help others do the same

Professional level:

- Implement HBC pilot studies within health care and community settings
- Support practitioners and staff to improve their health habits
- Help patients and their families improve their health habits
- Adopt a motivational role when the "fix-it" health advisor role does not work
- Develop coaching and motivational skills to facilitate health behavior change

Leadership level:

- Manage polarities to prevent dysfunctional polarizations
- Align, merge, and converge diverse worldviews
- Enhance transformational leadership skills
- Disseminate HBC programs within health care and community settings
- Create learning organizations and communities
- Build professional and social movement
- Develop transformational health care policies
- Facilitate sociopolitical transformation
- Cultivate health-promoting cultures

42.4.1 Identify Deficiencies in Your Education

As health practitioners and lay health coaches, you can use your communication and counseling skills to help your patients, peers, and family members change their unhealthy habits. Evidence-based guidelines designated for busy health care settings typically describe brief interventions for health behavioral change. These interventions predominantly address surface change, such as practitioners giving information and advice to patients, eliciting patients' good intentions and negotiating action plans with them.

Not surprisingly, low-complexity interventions have only a limited impact on helping individuals change. These guidelines do not address how to work with patients who do not respond to evidence-based interventions. What is missing from these guidelines for you and your patients? And what is missing in your education?

The internal change process for patients remains implicit. In other words, patients are often unaware of how you are working with them to help them change. Furthermore, professional training does not adequately prepare practitioners in how to facilitate the internal process of change, particularly when patients resist change, such as die-hard smokers. This is like the blind leading the blind.

Even when patients are ambivalent and willing to consider change, they are often unaware, to varying degrees, of their emotional resistance to change [118–121]. Patients seldom get explicit guidance from their practitioners on how to identify, address, and lower their resistance. Furthermore, evidence-based guidelines provide no assistance on how to work dynamically with populations longitudinally over decades. Health behavior change is a longitudinal and complex learning process across generations.[6]

Risk behaviors are the greatest threat to patient safety. In effect, individuals with risk behaviors are like airplanes without black boxes. They are on flight paths toward accidents waiting to happen. Without access to the black boxes, families cannot understand why these airplanes crashed. Despite their risky flight paths, many individuals have no resistance at all, because they are enjoy their unhealthy habits: for example, "I love smoking cigarettes." Practitioners have no or minimal training in how to engage "so-called" resistant patients in dialogues about change and therefore do not waste their time with them.

Evidence-based guidelines adopt silo approaches to addressing the different risk behaviors: in other words, separate programs for each risk behavior. This is an inefficient way of learning about behavior change. An integrated approach can avoid these duplications of efforts with separate programs [122]. Instead of having separate programs for each risk behavior, practitioners can use the same motivational process to help ambivalent and resistant patients change any unhealthy habit [122, 123].

42.4.2 Address Deficiencies in Your Education

To overcome the limitations of evidence-based guidelines, you can use experiential learning methods to help patients learn how to develop their own personal evidence about deep change, such as changing their perceptions about risks, benefits, and harms of unhealthy habits, and lowering their resistance. You also can help patients decide whether to change their value system, or at least work within and from their value systems to effect change.

Experiential learning methods can work synergistically with evidence-based guidelines. Evidence-based interventions can raise the floor of your organizational performance, but only personal evidence can help your organization break through the ceiling. The process of individuals developing their own personal evidence about change provides the greatest quality improvement opportunities for enhancing organizational performance. Together, they can both improve the performance of your behavior change, disease prevention, and disease management programs.

The epidemics of unhealthy habits totally overwhelm the capacities of the health care system. There will never be enough trained practitioners to coach patients working one-on-one. To address this challenge, the health care system must develop alternative approaches to bridge the chasm between the needs for individualized learning opportunities about behavior change and the professional supply of such services.

Most individuals change their unhealthy behaviors without professional help [124]. Collaborative learning programs about behavior change can accelerate this self-recovery process. The challenge is how do we develop and provide low-cost, high-impact, and high-reach learning programs to diverse populations.

Self-reflective learning exercises can set up collaborative learning opportunities (offline, online, or both) for yourself, colleagues, patients, and their family members. Such blended learning methods help learners make sense of their learning experiences and coach each other to improve their health behaviors. Group collaboration and commitment in completing these learning exercises will help you learn more about how to combat these epidemics.

---

[6] Mechai Viravaidya: How Mr. Condom made Thailand a better place. This video provides a longitudinal perspective about how to do the right thing and how to do things right.

http://www.ted.com/talks/mechai_viravaidya_how_mr_condom_made_thailand_a_better_place.html.

At TEDxChange, Thailand's "Mr. Condom," Mechai Viravaidya, walks us through the country's bold plan to raise its standard of living, starting in the 1970s. First step: population control. And that means a lot of frank, funny—and very effective—talk about condoms.

42.5 How Can You Use the Science of "I and We" for Your CPPLD

The science of "I and i" addresses how to enhance your self-awareness and learning about oneself. Using a combination of learning, communication, and coaching methods, you can explore how your worldviews affect your behavior. Instead of complying with preset, teacher-centered learning objectives, you determine the breadth and depth of your learning outcomes, in accordance with your preferences and developmental needs. This self-focused approach does not presuppose what you need to learn about, in terms of developing your own personal evidence about deep change. You explore and discover what you specifically need to understand about yourself in order to change. What works for you emerges from your learning experiences, in collaboration with others.

To explore the ontological science of "I and i" (being and becoming), you use experiential learning methods to address the complexity of our personal, professional, and leadership development [125]. This science of "i" focuses on the psychological aspects of heart and mind: such as passion, worldview, ego state, individual identity, character strengths, and flaws such as obsessional self-centeredness ("what's in it for me"). The science of the "I" focuses on the soul: the spiritual experience of transcendent Oneness in order to facilitate the development of virtues (Table 42.2).

Collaborative learning methods can help us explore the science of "We and we." This "science of we" focuses on the sociopsychological aspects of ourselves: such as tribes, political parties, religious affiliations, professional groups, and sports teams. These groups form coalitions and alliances to compete for hierarchal positioning and limited resources, with "Win–Lose" consequences.

The "science of We" focuses on our consciousness on our shared humanity. For example, do we identify more with our individual tribe or with being part of the human race? Are we divided by our conflicting tribes or unified by our universal humanity? These questions pose challenges to our identities about "We and we" and our relationships between We and we. Healthy co-opetition occurs when "We" *co-op*erate to foster constructive com*petition* for mutual gains and create "Win–win" outcomes. Co-opetition is a word derived from cooperation and competition.

To move beyond the limitations of the "science of i and we," transcendental experiences of "I and We" can help us expand our worldviews, so that we can build common ground for the common good. The science of "We" focuses on the spiritual experiences of universal transcendence and consciousness, such as interfaith coalitions and universal peace movements. The science of "We" focuses on what is in our best global interest, such as working for common good and environmental stewardship. The transformational leadership capability of shifting between the psychosocial sciences of "i and we" and the spiritual sciences "I and We" involves cultivating self-awareness and equanimity.

Table 42.2 Virtues and character strengths [24]

Science of "I and i"
Virtue 1: Wisdom and Knowledge
Character strengths—creativity (originality, ingenuity), curiosity (interest, novelty-seeking, openness to new experience), open-mindedness (judgment, critical thinking), love of learning, perspective
Virtue 2: Courage
Character strengths—bravery, persistence (perseverance and industriousness), integrity (authenticity and honesty) and vitality (zest, enthusiasm, vigor, and energy)
Virtue 3: Humanity
Character strengths—love, kindness (generosity, nurturance, care, compassion, altruistic love, and "niceness"), social intelligence (emotional intelligence, personal intelligence)
Virtue 4: Justice
Character strengths—citizenship (social responsibility, loyalty, and teamwork), fairness, and leadership
Virtue 5: Temperance
Character strengths—forgiveness and mercy, humility and modesty, prudence, and self-regulation
Virtue 6: Transcendence
Character strengths—appreciation of beauty and excellence (awe, wonder, elation), hope (optimism, future-mindedness, future orientation), humor (playfulness), and spirituality (religiousness, faith, and purpose)

A neurophysiological understanding of our consciousness can enhance our self-awareness about what contributes toward greater open-mindedness or closed-mindedness. In this higher state of evolutionary consciousness, we deactivate the right posterior superior lobe of our parietal brains to induce the transcendent experiences of universal Oneness (I/We) [126–128]. Activation of our frontal neocortex enhances our capabilities to exercise higher cognitive functions and moral judgments. In this holistic state of nonattachment to ego, we are more likely to experience equanimity in ways that foster greater open-mindedness. When the frontal neocortex is activated and the right parietal lobe of the brain is deactivated, we are more likely to:

- Reflect and act in more virtuous ways
- Control inappropriate impulses that arise from our reptilian brains
- Modulate the downsides of egoism
- Address our flaws more effectively

A neurophysiological understanding of human consciousness enhances our self-awareness about what contributes toward greater closed-mindedness. Fear, anger, and anxiety from any perceived threats can induce regression toward close-mindedness. The over-activated reptilian brain (fight, flight, or freeze) deactivates the frontal neocortex and suppresses the higher cognitive function and moral judgments, and activates the right parietal lobe to restore a personal sense of self. In extreme states of negative emotional reactivity such as self-righteous rage, the unbridled, aggressive ego runs amock. Might is right. The amoral ego mocks and attacks opponents with ruthless impunity, and commits transgressions and violent atrocities[7] without regret. In its most evil form, we can regress to a reptilian level of consciousness: a "life-and-death" survival of the fittest with mindless killing [129].

[7] Daniel Pinker asserts that acts of violence are on a decline despite social pessimism to the contrary.

http://www.thersa.org/events/video/vision-videos/the-better-angels-of-our-nature.

42.6 Enhance Self-Awareness and Self-Focused Learning

The following questions may evoke reflections about scientific methods for exploring "I and i." This exploration of self-awareness provides a way to discover how you may hinder or enhance the process of facilitating change. To what extent are you aware of your:

- Worldview, assumptions, and biases?
- Cognitive, emotional, perceptual, and spiritual aspects of self?
- Self with the capital "I" (nonattached ego state of universal oneness) versus a small "i" (ego state)?
- Self in relationship to facilitating change at individual, family, organizational, community, and cultural levels?

These questions prepare you to explore your continuing personal, professional, and leadership development. The science of "I and i" can help you see your blind spots. Self-reflective and collaborative learning processes provide a way to address the experiential complexities of learning about oneself. Collaborative learning can shine light on your blind spots, so that you can get glimpses of your worldview or the lens that shapes the interpretation of your experiences. However, this is not a straightforward process.

> *We learn from experience that men* (and women) *do not learn from experience.*
> George Bernard Shaw

Self-reflective learning exercises may help you learn more effectively about change when you collaborate with others. When you share your learning experiences together, you can cultivate wisdom about the change process.[8] Learning

[8] This audio book excerpt (http://www.randomhouse.com/features/wisdomofcrowds/Wisdom2.mp3) from the Wisdom of Crowds (http://www.randomhouse.com/features/wisdomofcrowds/audio.html) by James Siroweicki highlights the benefits of such group learning, but he also identifies the detractors who question the value of this phenomenon.

Derek Powazek applies those ideas to the web, concentrating on how to design websites that empower people to work together to create something truly awesome (http://www.youtube.com/watch?v=RX-7xwPPY8I).

how to improve your own health habits along with your peers (before helping your patients or clients) are prerequisite learning experiences for students and practitioners, before they work with patients and their families.

> *Everyone thinks of changing the world but no one thinks about changing himself (or herself)*
> Leo Tolstoy

The science of "I and i" is an inside-out, learning process of becoming the principal investigator, researcher, and author of your own behavioral change. You can engage your peers and/or family members as coinvestigators or coaches. This mutual learning process involves taking turns and reversing the roles in the coach–coachee relationship.

The first CPPLD framework outlines an introspective process for personal discovery. It provides some guidance in how to explore narrative health. Exploring the story of your own health involves cultivating curiosity about self, with the goal of developing your own personal evidence about deep change.[9] The quote of T.S. Eliot may evoke greater curiosity about the science of "I and i."

> We must never cease from exploration. And the end of all our exploring will be to arrive where we began and to know the place for the first time.

It takes courage to explore deep change. Maya Angelou claims that "*of all the virtues, courage is the most important. Without courage, it is impossible to practice the other virtues.*" Real courage involves moving beyond the comfort zone of your own worldview to face challenges, uncertainties, and vulnerabilities, and endure the pain and suffering of change.

Self-reflective learning exercises can help you decide where you can focus your attention, in terms of making deep change. Mindfulness practice provides different perspectives on how you can focus your attention and enhance your self-awareness: focusing on rest, focusing in, focusing out, focusing on positive emotions, and focusing on change.[10] Mindful practice can also enhance emotional self-regulation to address the challenges of making deep change.

Changing your unhealthy habits and helping others to do the same involves enhancing your self-awareness about the internal process of deep change.[11] In addition to curiosity about self, courage, and focused attention, it also takes self-awareness to become the lead author of your own behavioral change. The discovery process involves using your imagination and creativity to make sense of your learning experiences, so that you can foster greater open-mindedness, expand your worldview, and make deep change (Fig. 42.1).

As you gain experience in helping yourself and others (colleagues, family members, or friends) improving your health habits, you will enhance your coaching skills over time. Beginning within your organization, you can invite trainers, colleagues, staff, and their families to experience this coaching process and then replicate the same process for patients/clients and their families. By joining a wisdom learning community, you can further enhance your leadership development skills.

Transformational leaders address the questions of "*why, how, and what,*" in *that order*. This inside-out approach helps to develop effective

[9] Noble laureate Carol Greider emphasizes the critical role of curiosity in the pursuit of scientific inquiry and discovery (http://www.youtube.com/watch?v=sVRPneReK4o&feature=related). Although her work focused on the biomedical genome, what would it take for you to develop a fascination to learn about unraveling the psychosocial genome of unhealthy habits? What is the telomerase equivalent for addressing unhealthy habits? What are the catalysts for healthy change? Important prerequisites for change include courage and self-awareness.

[10] Shinzen Young is known for his innovative "interactive, algorithmic approach" to mindfulness, a system specifically designed for use in pain management, recovery support, and as an adjunct to psychotherapy. In this video clip he describes five forms of mindfulness practice (http://www.youtube.com/watch?v=2SE5O9tjqMo&feature=related). Reflecting on what, how, and why you accept and reject different forms of mindfulness practice are self-awareness exercises in themselves, in terms of understanding your worldview, values, and assumptions.

[11] To further explore the spiritual meaning of self-awareness, you can watch Antony de Mello, a Jesuit priest and psychotherapist widely known for his books on spirituality, who challenges your "deeply held" assumptions (http://www.youtube.com/watch?v=4Y3Q7H2urto).

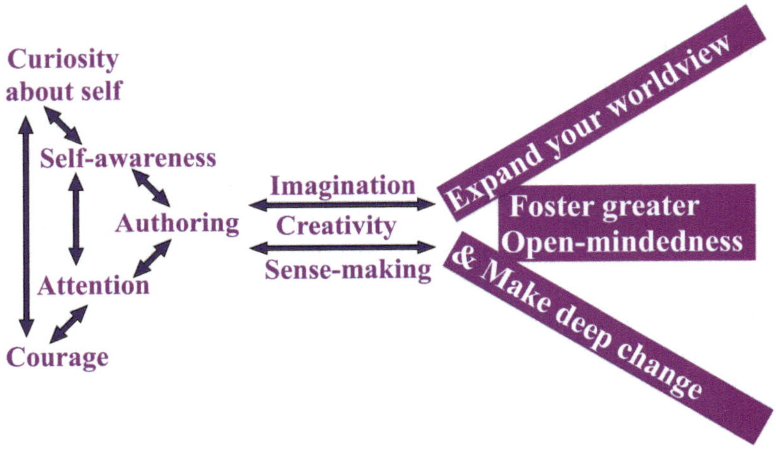

Fig. 42.1 CPPLD Framework 1: Foster greater open-mindedness

change leadership teams. Effective leaders inspire a shared sense of purpose, urgency, and ownership about the need for change.[12] They are the most potent catalysts for change when they galvanize collective action toward a shared vision. To improve health care outcomes at population-based levels, high-performing leaders use both experiential learning processes and evidence-based guidelines synergistically to promote healthy habits, and self-care of chronic diseases.

42.7 How Can You Understand Differences in Worldviews?

Worldview is the lens through which you perceive, evaluate, interpret, and make sense of your learning experiences. Genetic, family, social, cultural, and environmental influences can shape your worldview about health. Your worldview shapes the life story of your own health. How can you discover what influences shaped your worldview and your health story? Narrative health is the process of telling, deconstructing, and reconstructing your story in ways that improve your health behaviors and self-care of diseases. How will you become the principal investigator, researcher, and lead author of your own behavior change?

42.7.1 Cultivate Open-mindedness and Diversity

The CPPLD frameworks will help to:

- Discover how worldviews limit and enhance our capacities and capabilities to reverse the epidemics of unhealthy habits
- Explore the extent to which behaviors are individually and socially determined
- Understand more about the complexities of behavior change, from individual to cultural levels

Frameworks 2 and 3 provide ways to reflect about how open-minded and closed-mindedness worldviews enhance and hinder our continuing personal, professional, and leadership development. All individuals, groups, and communities have combinations of virtues and flaws. The struggles between virtues and flaws operate at intrapersonal and interpersonal levels within organizations and communities. To what extent do virtues triumph over flaws, or flaws prevail over virtues?

Transformational leaders inspire individuals to take on these struggles, work for the greater good, and bring out the best in others. Such transformation begins with oneself. Personal development is at the heart, mind, and soul of this learning

[12] Simon Sinek, a leadership expert, identifies how great leaders focus on the inside-out question of "why" to order to galvanize teamwork for positive action, inspire mass movements, and enhance human welfare (http://www.youtube.com/watch?v=u4ZoJKF_VuA).

Table 42.3 CPPLD Framework 5: Reflecting on the nature and valence of your interpretations

Interpretation	Exclusive	Inclusive	Inclusive discernment
Options	Either/or	Both/and	Either/or and Both/and
Bipolarity	Positive or Negative	Positive and Negative	Degrees of Positive and/or Negative
Multiple polarity	Positive, Neutral or Negative	Positive, Neutral, and Negative	Degrees of Positive, Neutral, and/or Negative

process. After all, how can we change others if we cannot change ourselves? How can we foster greater open-mindedness and work more effectively with a diversity of worldviews?

The CPPLD frameworks of open-mindedness and close-mindedness provide contrasting perspectives on different levels of spiritual consciousness. Spirituality is about our purpose and meaning in life, our values, our connectedness to others (including higher powers), and the cultivation of virtues (see Table 42.3), as opposed to flaws, such as exclusion and intolerance.

Flawed and virtuous aspects of our spiritualities shape our memes, metaphors, and worldviews, and vice versa. Memes are the cultural equivalents to genes [130]. They unconsciously predetermine our disposition to behave in certain ways, for better or worse. Memes are cultural ideas (such as individual freedom over tyranny), symbols (such as the swastika sign[13]), rituals, and practices that are transferred from individual to individual and from generation to generation. How can we create healthy memes that promote health and well-being?

Metaphors shape our worldviews about disease care and health care. Metaphors are often implicit and embedded in our everyday language. Do health care practitioners operate more as mechanics fixing train wrecks, or as organic farmers growing healthy food? To what extent do the machine and garden metaphors shape our disease care and health care systems?

[13] Many cultures and religions have adopted variations of the Swastik sign. This sign is popular in the Hindu religion. Swastika is derived from the Sanskrit word "Svastika," which means well-being.

42.7.2 Virtuous and Flawed Spirituality

Virtuous spirituality (see CPPLD Framework 2) uses healthy memes, such as the Hindu svastika symbolizing well-being, and holistic metaphors, such as health is an organic garden. The effective use of memes and metaphors fosters cohesion, inclusion, healing, trust, and equanimity in ways that cultivate open-minded worldviews. In turn, our worldview helps us develop healthy identities, resilience, and beliefs in ways that foster positive assumptions, perceptions, biases, thoughts, and feelings (Fig. 42.2).

Flawed spirituality (see Fig. 42.3) generates negative memes (such as the Nazi swastika) and inappropriate metaphors (health is a machine) that foster separation, fragmentation, suffering, alienation, fear, and emotional reactivity, such as fear, anger, and anxiety. Our emotional reactivity fosters close-mindedness and inhibits our capacities to elicit and receive needed feedback for our ongoing development. In turn, closed-minded worldviews create negative assumptions, thoughts and feelings, distorted perceptions, and prejudices that create flawed belief systems and unhealthy identities.

A dynamic polarity between open- and closed-mindedness operates within all of us, and within our group dynamics within organizations and communities, to varying degrees. This battle between open-minded and closed-mindedness worldviews can:

- Descend destructively into downward spirals of negative social influences.
- Create stalemate between negative and positive social influences.
- Ascend constructively into upwards into spirals of positive social influences.

When we move toward greater open-mindedness, we enhance our abilities to be more

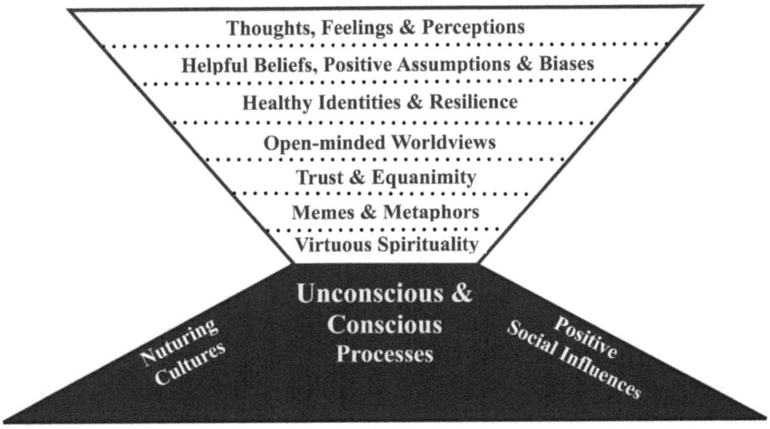

Fig. 42.2 CPPLD Framework 2: Reflecting about open-minded worldviews

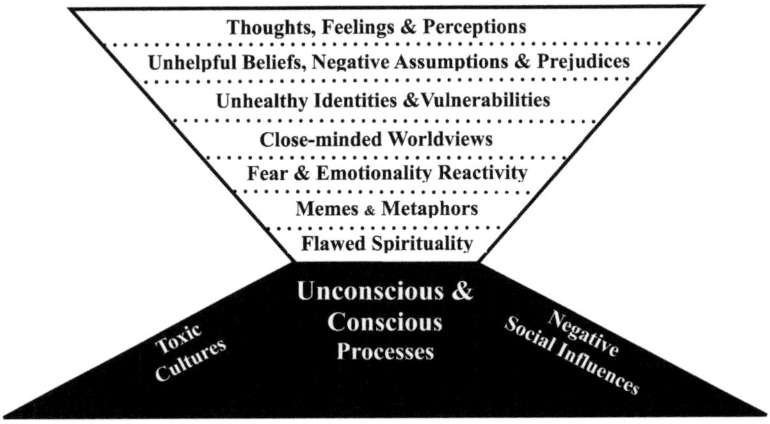

Fig. 42.3 CPPLD Framework 3: Reflecting about closed-mindedness worldviews

inclusive, understand more about our difference in worldviews, and engage into more meaningful dialogues about the change process.

Adaptive open-mindedness is a flexible learning disposition for addressing complex issues that do not have predetermined or predictable answers. Complex problems call for collaborating with others who have different worldviews into order to develop more sophisticated and elegant solutions to manage complexity and polarities in worldviews. Diversity competence is the ability to work with diverse worldviews in ways that cultivate shared wisdom to address complexities.

The CPPLD frameworks can help you move beyond the comfort zone of linear causality, certainty, and predictability to embrace nonlinearity, vulnerabilities, and emergence with greater courage, intention, and purpose.[14] These frameworks can help you explore and understand more about the interactional benefits of being, doing, and performing, as part of individual and collaborative learning processes. This humanistic approach facilitates evolving ways of "*relating, becoming*, and *evolving*," in terms of facilitating your own personal, professional, and leadership growth (see CPPLD Framework 4).

[14] Mel Schwartz, a psychotherapist, describes how you can cultivate your growth by changing the relationships between your comfort and discomfort zone (http://www.youtube.com/user/schwartzbwu?feature=mhum%22%20%5Cl%20%22p/u/21/FVzMiu8KoW0).

42.7.3 Ways of Relating, Becoming, and Evolving (Fig. 42.4)

Polarities provide ways to understand better your worldview and assumptions: what shapes your lens in making interpretations. When you take an observer role to your own behavior, you are adopting the meta-position, a reflective state of non-ego attachment. In this state, you are likely to

- Become more aware of what shapes your lens or worldview
- Understand better what influences how you make interpretations

The CPPLD Framework 5 provides a way to reflect about what affects the nature and valence of your interpretations, thoughts, and feeling, in terms of positive and/or negative reactions: for example, either/or, both/and, or a continuum between both/and, including the neutral position (Table 42.4).

The initial step is to become more aware of, and acknowledge the extent to which your mind is making these reflex interpretations all the time. To what extent are you aware of:

Fig. 42.4 CPPLD Framework 4: Ways of relating, becoming, and evolving

- What shapes and changes whether you respond in dichotomous (such as good or evil) or relativistic ways (a blend of thoughts and feelings with varying degrees of good and evil)?
- How your interpretations help and hinder the process of making sense of your learning experiences?
- How you can enhance your capabilities to manage complexity and polarities by taking different perspectives to interpret your experiences?

A CPPLD framework of polarities across five domains may help you reflect about what shapes the lens of your worldview. The exploratory process may help you expand your worldview, so that you can understand values and worldviews that are contrary to your natural inclination. This learning process begins by becoming more aware of, and understanding your own worldview. How can you understand differences in worldviews if you do not understand your own worldview, assumptions, and interpretations?

These polarity frameworks will help you reflect on the diversity of worldviews that help and hinder the prospects of galvanizing and converging our diverse worldviews for the common good. The following questions aim to evoke reflections about polarities and the challenge of understanding different worldviews.

1. Ethical and Spiritual Domains

- Individual versus community values
- Individual freedom versus universal justice
- Flaws versus virtues
- Unaccountability versus transparency
- Religious fundamentalism versus spiritual relativism
- Western and Eastern philosophies
- Reductionism versus holism

Table 42.4 CPPLD Framework 5: Reflecting on Interpretative Lens

Interpretative Lens	Exclusive either/or	Inclusive both/and	Inclusive discernment Either/or & Both/and
Bipolarity	Positive or Negative	Positive and Negative	Degrees of Positive and/or Negative
Multiple polarities	Positive, Neutral or Negative	Positive, Neutral and Negative	Degrees of Positive, Neutral and/or Negative

Examples of Self-focused Inquiries
To what extent
- Are you naturally inclined toward individual versus community values?
- Do unaccountable individual values violate community values?
- Do overbearing community values violate individual values?

2. Political and Philosophical Domains
- Right-wing versus left-wing politics
- Capitalistic versus social entrepreneurship
- Materialism versus meaningful-ism
- Hyper-consumerism versus sustainable development
- Rugged individualism versus robust interdependence
- Shareholders for short-term profits versus stakeholders for the long-term good

Examples of Self-focused Inquiries
To what extent
- Are you naturally inclined toward right-wing versus left-wing political philosophies?
- Can you expand your worldview to understand political philosophies that are contrary to your natural inclination, so that you can address contrasting assumptions of these political philosophies?
- Are you naturally inclined toward working for corporations based on for-profit motives that serve shareholders versus public organizations based on community and stakeholder needs that serve the common good?

3. Authority, Power, and Control Domains
- Taking control versus taking charge collaboratively
- Positional authority (alpha behaviors) versus influential guidance (beta behaviors)
- Ego attachment versus ego-less non-attachment
- Arrogant hubris versus bold humility
- Domineering versus empowering behaviors
Examples of Self-focused Inquiries
- Under what circumstances are you more inclined to take control, act in authoritarian and ego-driven ways, act with arrogant hubris, strive for predictability, and domineer others?
- To what extent can you work collaboratively, act as an influential and ego-less guide, act with bold humility, embrace emergence, and empower others?

4. Thinking and Feeling Domains
- Rigid dichotomous thinking versus flexible relativistic thinking
- Analytical rationality versus intuitive creativity
- Task-driven orientation versus empathic relational processes
- Emotional reactivity versus equanimity
- Linear predictability versus nonlinear emergence
Examples of Self-focused Inquiries
- Under what circumstances are you more inclined toward dichotomous thinking, analytical rationality, task orientation, emotional reactivity, and linear predictability?
- To what extent do you have the capabilities of relativistic thinking, intuitive creativity, empathy, emotional self-regulation, and emergence?

5. Doing and Being Domains
- Competitiveness versus cooperativeness
- Diversity intolerance versus diversity competence
- Narcissism versus altruism
- Individually focused versus collaborative work
- Striving for certainty versus embracing uncertainty
- Risk aversion versus risk-taking
- Dysfunctional polarizations versus polarity management

Examples of Self-focused Inquiries
- Under what circumstances are you more inclined toward
  - Competitiveness and diversity intolerance?
  - Self-interest and individually focused work?
  - Risk aversion and striving for certainty?
- To what extent do you
  - Have the capabilities of cooperativeness and diversity competence?
  - Foster altruism, and collaborative work?
  - Embrace risk and uncertainty?
  - Have polarity management skills to address dysfunctional polarizations?

42.8 How Do We Effectively Manage Polarities in Diverse Worldviews?

Adopting health-promoting memes and metaphors can expand our worldviews. Metaphors that reflect our values and worldviews are embedded in our everyday language. According to Aristotle, changing our metaphors can change our worldviews.

The dominant metaphor and worldview in medical care has been the machine. The mechanistic worldview of diseases can be traced back to Rene Descartes, the seventeenth century philosopher and protagonist for mind–body dualism. In turn, Descartes influenced Sir Isaac Newton who adopted the scientific worldview rooted in cause–effect linearity, certainty, and predictability.[15] This mechanistic worldview is inadequate for addressing the complex cultural, sociopolitical, and scientific challenges for the twenty-first century [63].

To expand our worldviews about science, disease, and health, we must go beyond:

- Our linear, reductionist, and predictive worldview of disease to include
- Holistic, ecological, and emergent perspectives about narrative health based on experiential learning processes that address the psychosocial-spiritual challenges of change [131–134]

Our constricted and conflicting worldviews jeopardize the F-PHC vision and mission of eliminating and reversing the epidemics of unhealthy habits. Our mal-aligned worldviews are our rate-limiting step that constrains our capabilities and capacities to develop professional and social movements that promote health.

These limitations call for overcoming shortsighted closed-mindedness, so that we can mount increasingly more sophisticated responses to these epidemics: in other words, expand, align, merge, and converge our diverse worldviews to promote health. Our greatest untapped human resource is our abilities to foster greater open-minded adaptability and reduce closed-minded rigidity. "*Think big, act small*" will help us address the following questions more effectively.

How do we:

- Build forms to foster open-minded adaptability?
- Create leadership development programs to enhance our capabilities of managing complexity and diverse polarities in worldviews?
- Evolve increasingly more sophisticated ways of reversing the epidemics of unhealthy habits?
- Escalate the dissemination of effective, low-cost, high-impact, and high-reach behavior change programs?

Online learning platforms and programs along with social media provide highly scalable ways to expand our worldviews about health and health care in ways that build professional and social movements [135, 136]. We can use such information technology in synergistic ways to enable our health care systems to work on the audacious vision of eliminating the epidemics of unhealthy habits. Such a journey is without an end. The art of inspirational leadership is to create urgency and buy-in around a shared vision so that it becomes a beacon for generating self-sustaining, long-term commitments for healthy change [137]. This learning process involves cultivating the bright side of open-minded leadership in ways that illuminate the dark side of closed-minded leadership.

The polarity of closed-mindedness versus open-mindedness represents a never-ending continuum of evolving development. The process of discovering that you are more closed-minded or less open-minded than you think that you are involves becoming aware of your distorted perceptions about oneself and others. Distorted perceptions about oneself can create positive biases by thinking better of oneself. The most flagrant distortion is to view a flaw as a virtue. The courageous act of redressing distorted perceptions is a genuine virtue. This learning process is best done in collaboration within a safe and trusting learning organization or community, because others can act as mirrors to your blind spots and provide needed feedback for change and improvement.

A task for transformational leaders at all levels within organizations and communities is to share their learning experiences about fostering virtues and addressing flaws. Such acts of courage involve sharing our strengths and vulnerabilities. Effective leaders manage the polarity dynamics between flaws and virtues in ways that:

[15] Here Mel Schwartz describes the dehumanizing effects of the machine, and highlights the limitations of this reductionist and fragmenting Worldview. Understanding how this Worldview informs your thoughts and beliefs is key to affecting change (http://www.youtube.com/user/schwartzbwu#p/u/21/FVzMiu8KoW0).

- Create upward spirals of positive social influences that cultivate virtues
- Prevent downward spirals of negative social influences that exacerbate flaws

Understanding polarity dynamics between open-mindedness and closed-mindedness, in relationship to developing virtues and exacerbating flaws, is critically important to developing authentic transformational leaders.

42.8.1 Understanding Dysfunctional Polarizations Between Value Systems

Opposing political worldviews create conflicts and impasses that hinder our capabilities and diminish our capacities to cultivate healthy habits and foster self-care of chronic diseases at population-based levels. Rigid polarizations between right-wing and left-wing political philosophies based on individual and community values, respectively (e.g., individual freedom from community and government regulations versus transparent accountability for universal justice), result in avoidant or attack behaviors, with incompatible political philosophies that are incapable of dialog.

The word freedom has multiple meanings, depending on political philosophies affecting the context of the situations: consider the distinctions between "individual and group freedoms from," such as oppressions, discrimination, and prejudice, versus the "individual freedom to" free speech and the pursuit of wealth. The Arab uprisings have become a rally call for a "freedom from" oppressions arising from fascist tyrannies and "ruling aristocracies" of corrupt wealth. At the other extreme is the Ayn Rand's "freedom to": a self-righteous form of ethical self-interest. This individualistic philosophy is an extremist counterpoint to her escape from her middle-class childhood experiences living in the communist Russia. She lived under the oppressive tyranny of collectivism.

Ayn Rand's philosophy repudiates collectivism and ethical altruism. She glorifies the just accumulation of individual wealth and power, without government infringements. Her influence ricochets in US politics today, with rally calls for freedom from Big government, regulations, and taxes in order to justify a freedom to self-centered, self-interests without regard to social consequences.

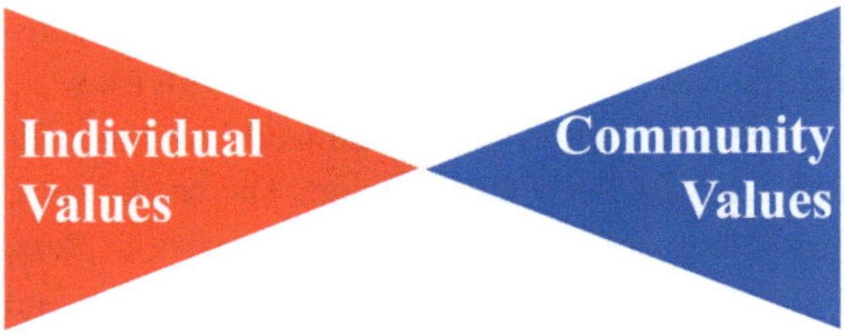

Fig. 42.5 CPPLD Framework 6: Avoidant worldviews

The globalization of corporations and financial institutions, together with greater freedom from national governance, has given rise to a covert form of tyranny: the inequalities of wealth and economic power over universal justice. This has been compounded by government deregulations and amoral greed in the financial markets. Financial institutions acted in their short-term interests: profits-for-profit's sake without regard to the long-term consequences. This corruption led to the inevitable economic meltdown in 2008.

42.8.1.1 Avoidant Worldviews: Looking in Opposite Directions

The religious abuse of power led to the separation of church and state and the development of democratic governments with partisan politics and secular societies. Given this history, democracies have tried to keep political and religious worldviews apart. However, it is impossible to keep religion and spirituality out of politics entirely. They influence politics in implicit and explicit ways (Fig. 42.5).

42.8.1.2 Confrontational Worldviews: Staring Down Each Other

Democratic elections are vivid flash points of confrontational worldviews**,** followed by periods of departing worldviews when the political majority works on its political agenda (Fig. 42.6).

Destructive behaviors arise from uncompromising stances that are fixated on short-sighted, closed-minded, political worldviews. When leaders, politicians, and policy-makers adopt such closed-minded worldviews, they make biased attributions about the predominant causal determinants of unhealthy habits. They develop sim-

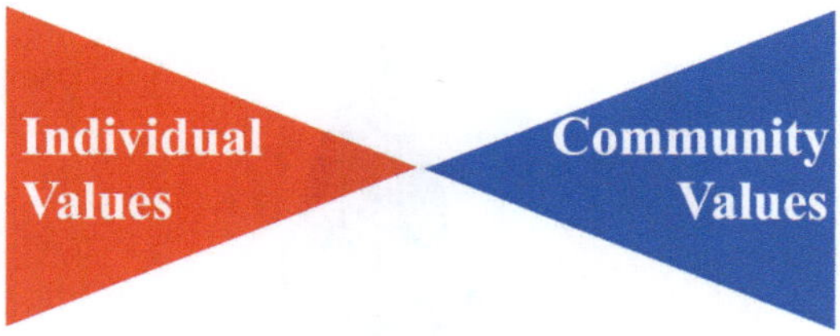

Fig. 42.6 CPPLD Framework 7: Confrontational worldviews

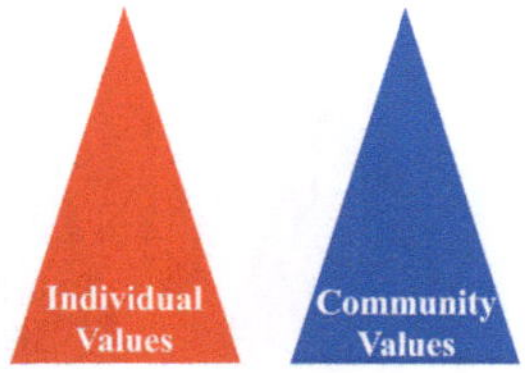

Fig. 42.7 CPPLD Framework 8: Aligning contrasting worldviews

plistic public policies for changing unhealthy habits that are based on either individually focused or socially determined approaches, in accordance with their respective right-wing and left-wing political philosophies.

With either-or, winner-takes-all thinking, politicians can create oppositional practices and even counterproductive approaches that fail to address the ecological complexities of facilitating behavioral change at multiple levels. These warring factions work against the common good and destroy the possibility of using polarities (individual-community values) in win/win ways. Consequently, competing political worldviews, un-informed by evidence-based and experiential learning approaches, create flip-flopping, contradictory public policies that are incapable of addressing the socio-behavioral dynamics of healthful change. These policy failures exacerbate the widening health outcome disparities between the rich and the poor.

In summary, our short-sighted, divergent, and mal-aligned worldviews have created toxic environments and cultures that mass produced diseases on unprecedented scales in the history of humankind. These toxic environments and disease-producing cultures prevent us from mounting effective countermeasures against the negative social influences that created these epidemics of unhealthy habits.

So why are we closed-minded to understanding the limitations of our worldviews? Domineering behaviors, quest for absolute certainty, and negative emotional reactivity (fear, anxiety, and anger) can induce closed-mindedness, diversity intolerance, and dysfunctional polarizations. Closed-mindedness can reinforce rigid dichotomous thinking, foster uncritical thinking, impair analytical rationality, and inhibit intuitive creativity. In other words, our flaws exploit differences to make our differences worse: in other words, increase dysfunctional polarizations. These negative consequences drain and kill organizational and community vitality that is essential for developing effective teamwork for health promotion.

42.8.2 Align, Merge, and Converge Diverse Worldviews

How can we:

- Move beyond closed-minded rigidity?
- Stop making our flaws worse?
- Avoid cultivating rigid extremism at both ends of the polarities?
- Stop working against each other in destructive and counterproductive ways?
- Reverse dysfunctional polarizations and downward spirals of negative social influences that produce diseases?
- Effectively manage polarities and create upward spirals of positive social influences that promote healthy habits?

42.8.2.1 Aligning Contrasting Worldviews

The process of aligning worldviews based on an advocacy for both individual and community values means that we must agree to travel in the same direction with some common vision and purpose. The alignment of our diverse worldviews can avoid the futile, dysfunctional polarizations that flip-flop between avoidance and confrontation (Fig. 42.7).

42.8.2.2 Mingling Worldviews

The process of mingling worldviews involves advocates for individual and community values becoming more familiar with the other, without trying to change anything (Fig. 42.8).

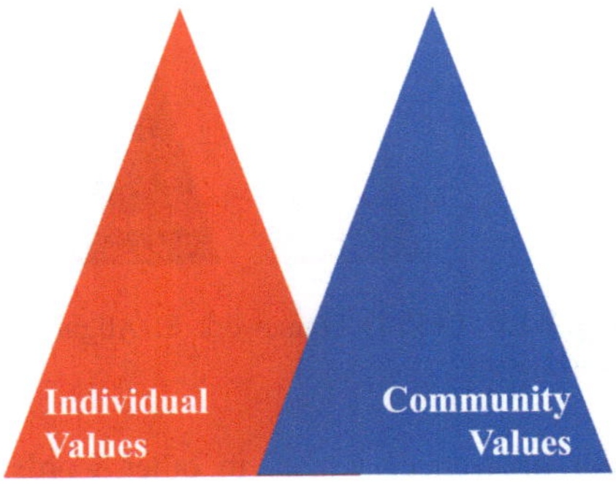

Fig. 42.8 CPPLD Framework 9: Mingling of worldviews

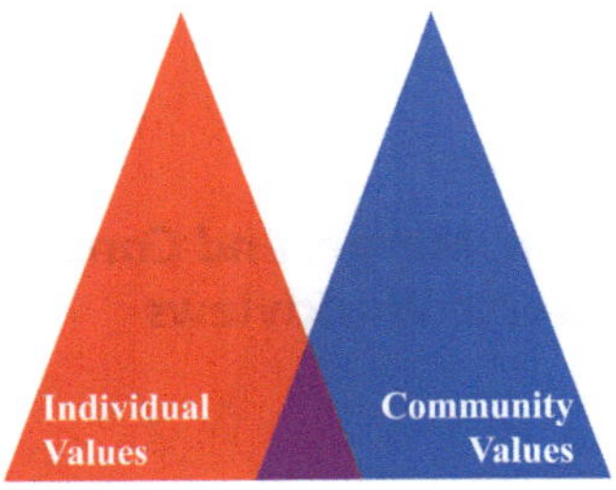

Fig. 42.9 CPPLD Framework 10: Merging short-sighted worldviews

42.8.2.3 Merging Short-sighted Worldviews

The process of merging worldviews involves advocates for individual and community values discovering common ground for possible collaboration and working on a shared goal. The purple triangle (a combination of blue and red) represents common ground (Fig. 42.9).

42.8.2.4 Merging Long-sighted Worldviews

The process of converging worldviews involves advocates for individual and community values developing a longitudinal perspective about the need to collaborate together. This learning process helps potential partners in developing greater common ground for collaboration (Fig. 42.10).

42.8.2.5 Building Common Ground

Advocates and leaders of individual and community values become less strident, more flexible, and accommodating, so they further increase the common ground between these contrasting values systems (Fig. 42.11).

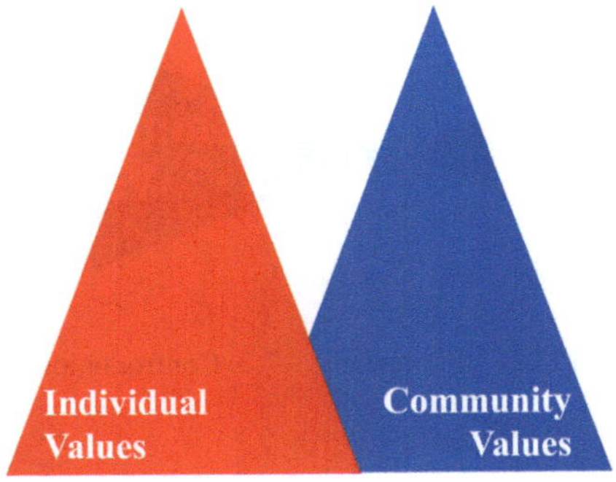

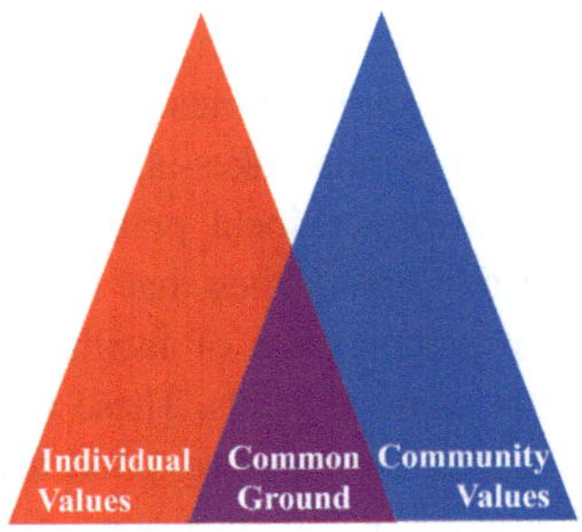

Fig. 42.10 CPPLD Framework 11: Converging worldviews

Fig. 42.11 CPPLD Framework 12: Building common ground

42.8.2.6 Building Common Ground for the Common Good

Advocates and leaders of individual and community values become invested in each other's futures. They build a foundation of common ground so that both parties can work synergistically together toward the common good. The upper purple triangle represents the spearhead for the common good (Fig. 42.12).

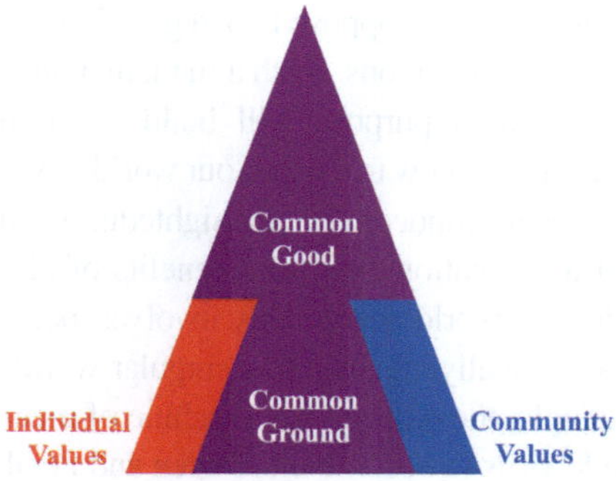

Fig. 42.12 CPPLD Framework 13: Building common ground for the common good

42.8.2.7 Virtuous Guidance of Sociopolitical Polarities

We can imbue symbols and logos with profound meaning and use them for sacred purpose. Used appropriately, they inspire ethical and virtuous behaviors: do the right thing and do things right. To avoid repeating the same mistakes across generations, we can use these symbols to rally effective leaders to generate ethical and sociopolitical transformations needed to cultivate health-promoting environments and cultures. With effective stewardship of this transformation, we can create upward spirals of positive social influences that prevail over the downward spirals of negative influences (Table 42.5).

The CPPLD Framework 14 is an upward-pointing symbol for virtuous leadership and stewardship, with specific meanings attached to these colors (Fig. 42.13).

The indigo-green triangle symbolizes the cultivation of equanimity leadership and stewardship needed to create self-organizing, sustainable development cycles. Such equanimity is essential for navigating through diverse worldviews and contrasting polarities in win/win ways that align and converge worldviews to build common ground for the common good.

Using a tricycle as a metaphor, the lower purple triangle represents the seat, and the upper purple triangle is the large front wheel, with red and blue trailer wheels. The ethical role is to steer the purple wheel forward in the right direction. The red and blue wheels may vary in speed and in size over time. This leadership challenge

Table 42.5 Color code meaning of logo

Indigo	Ethical leadership and stewardship
Red	Individual values
Blue	Community values
Purple	Common ground for the common good
Green	Sustainable development

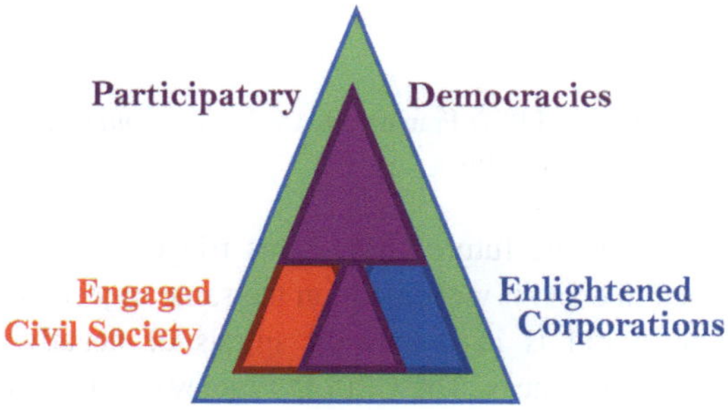

Fig. 42.13 CPPLD Framework 14: Virtuous guidance of sociopolitical polarities

is to keep the tricycle moving in the right direction, so it can facilitate individual, organizational, community, and corporate development in addressing this perennial question. Are we doing the right thing and doing things right for the long haul?

Corporations, financial institutions, and consumer society have individual values as their largest wheel leading the tricycle forward. Civil society is the smallest wheel (purple) and is the underpowered, social force in politics. Representational democracy is also a small wheel (blue).

This tricycle is driving around faster and faster in ever smaller circles. This inward spiral of nonsustainable development, combined with the increasing socioeconomic inequities in wealth, has worsened the social determinations of health, caused environmental degradations, and reduced biodiversity.

42.8.2.8 Flawed Guidance of Sociopolitical Polarities

With increasing overpopulation and finite resources, we will face a myriad of increasing challenges: fossil fuel pollution and shortages, global warming, water supply shortages and contamination, increasing food prices, rising waters and flooding, and reduced biodiversity: an

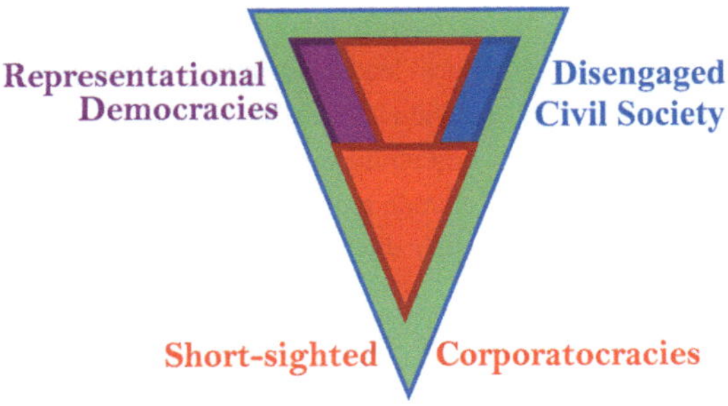

Fig. 42.14 CPPLD Framework 15: Flawed guidance of sociopolitical polarities

unsustainable future [138]. This tricycle is destined to crash if we remain on the same trajectory. Our world is in downward spirals of negative social influences. The CPPLD Framework 15 is a downward-pointing symbol for flawed leadership and stewardship, driven by short-sighted corporatocracies and amoral financial institutions (Fig. 42.14).

42.8.2.9 Political Stewardship of Participatory Democracies

How can we:

- Move beyond our short-sighted, close-minded, and dysfunctional political polarizations that prevent us from creating healthy cultures and environments?
- Build learning organizations and communities in ways that create positive social influences that develop virtues and foster healthy decision-making?

To repair this broken tricycle, we must create social networks that help individuals, organizations, communities, and corporations learn how to manage the polarities between individual and community values. This learning process can help us make better policies and decisions to promote health. We can use our differences in worldviews to create common ground for the common good in ways that create upward spirals of positive social influences. These positive learning experiences have the power to reverse the epidemics of healthy habits, restore the environment, and promote sustainable development.

First, we must align our diverse worldviews toward the same ideal vision with a sense of urgency. We also need buy-in to merge our worldviews in the same direction, as opposed to regressing to dysfunctional polarizations. Such a fundamental agreement of shared purpose will build commitment toward learning how to expand our worldviews with greater open-mindedness, far-sightedness, and a greater appreciation about the benefits of plurality and diverse worldviews. This involves becoming less categorically attached to a singular worldview: for example, the disease care machine of medicine. This allows us to become more open and flexible to the sociopolitical transformation needed to build adaptive health care systems: ideally, cultivating gardens of health, happiness, and well-being.

Expanding beyond the worldview of the machine metaphor to the garden metaphor is an essential for cultivating adaptive health care systems. We need large-scale transformation in our worldviews, so that we can address these monumental challenges, especially given our current political power structures and limitations of representational democracies. Corporatocracies domineer over representational democracies. Amoral and asocial corporatocracies do not act in the best long-term interests of civic society, the environment, and democracies. Such corporations (for example, the tobacco companies) create toxic environments that mass produce diseases and environmental degradation. They put their own profit and survival ahead of the overall common good. This situation calls on civil society to develop social movements to shift the power base toward building participatory democracies that work with enlightened enterprises[16]: in other words, ethical and socially responsible corporations. This shift will help participatory democracies, civil society, and enlightened corporations to develop transparent accountability in working more effectively for the common good. We need such political and sociocultural transformations, so that we can reverse the epidemics of unhealthy habits (Fig. 42.15).

[16] In his 2011 annual lecture about enlightened enterprises, RSA chief executive Matthew Taylor explores how business can combine strategies for competitive success with commitment to social good.
http://www.thersa.org/events/audio-and-past-events/2011/enlightened-enterprise.

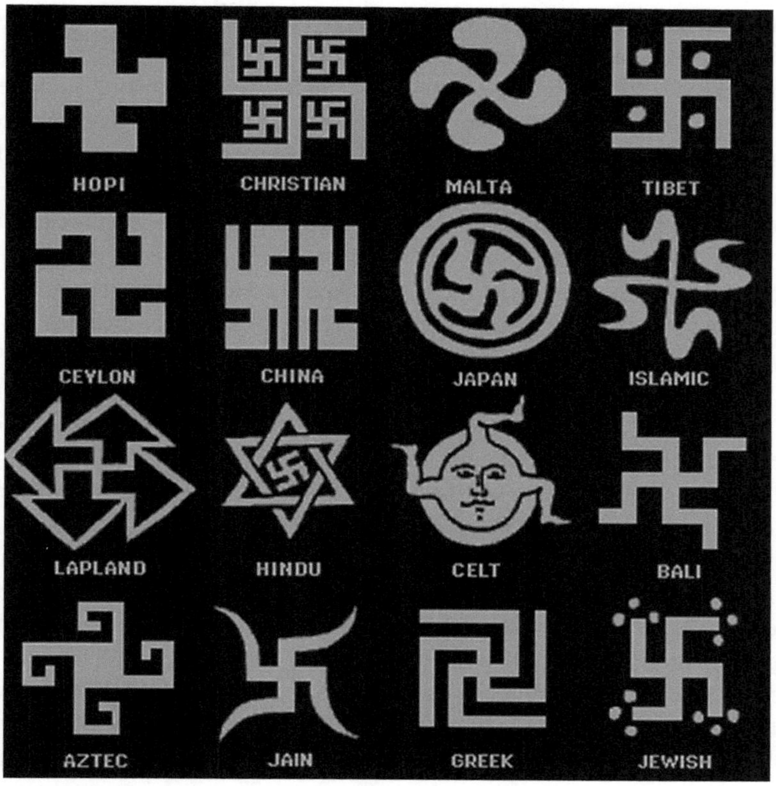

Fig. 42.15 CPPLD Framework 16: Political stewardship of participatory democracies

Civil society could become a major political force in collaborating with participatory democracies and enlightened corporations to:

- Advocate for universal justice and equity in health care
- Develop long-range visions of mutual interests
- Align, merge, and converge our divergent worldviews about health and disease
- Create ample common ground for galvanizing our efforts to reverse the epidemics of unhealthy habits
- Empower practitioners and patients to build professional and social movements for health

To work toward these goals, we need to enhance our empowerment and leadership skills to foster greater open-mindedness, equanimity, and diversity competence in ways that cultivate healthy behaviors. We can use our virtues and our differences in worldviews to make a difference. Working in complementary and synergistic ways, we can evolve toward higher levels of global consciousness to create positive social influences that foster ethical behaviors and virtues. In this way, we create healthy environments and cultures that support the development of healthy habits. Such adaptive learning is essential for fostering transformational leadership and effective interdisciplinary teamwork. Such teamwork enables more effective dissemination of HBC programs first to students, practitioners, and staff in health care settings and then to patients and their families in the community.

42.8.3 Inspire Transformational Leadership Development

To develop transformational leadership skills, we can use the synthetic frameworks of CPPLD to foster equity, universal justice, inclusiveness of divergent worldviews, transparent accountability, and transformative learning in ways that facilitate the development of virtuous behaviors at multiple levels. What does "synthetic" mean?

At the most elemental form, synthetic approaches involve using varying degrees of both–and perspective (for example, relational

process and task orientation) as opposed to either–or approaches (relational process versus task orientation). This worldview involves managing polarities in complementary ways to address dilemmas and questions, such as:
How do we manage the polarities between

- Relational process and task orientation in ways that foster effective partnerships and teamwork that improve the process and outcomes of health care?
- Community values and individual values in ways that reduce inequities in heath care outcomes?

This synthetic approach does not attempt to integrate irreconcilable differences between these polarities. Instead, this approach uses our differences within these polarities to find some common ground for working in ways that make a positive difference. This is more about creating collaborative partnerships working on a peace mission than competitive enemies going to war. We can manage these polarities in ways that peace prevails over war and in ways that unite rather than divide our common humanity.

For example, how can social responsibilities foster individual responsibilities and vice versa in synergistic ways? This win/win approach avoids the contentious shortcomings of either–or judgments: right–wrong, good–evil, or love–hate. These divisive judgments create dysfunctional polarizations between individuals and groups that impair our capabilities of working in collaborative win/win ways. Our political, religious, spiritual, community, and organization leaders and policy-makers can use these CPPLD frameworks to help students, practitioners, patients, and their families develop greater open-mindedness and use appropriately dualistic, multiplistic, and relativistic thinking as needed [139]. These frameworks can help to foster dynamic, adaptive, and emergent learning that works toward the common good.

42.8.3.1 Call for Galvanized Actions

Transformational leadership, based on virtue-based ethics, is essential for aligning, merging, and converging our diverse worldviews in order to build common ground for the common good. Such preparation will enable us to transform from:

- Our disease-producing cultures and toxic environments that generate negative social influences and encourage unhealthy decision-making into
- Health-promoting cultures and environments that create positive social influences and foster healthy decision-making

These macro-level changes are essential for synergistically leveraging meso-level and micro-level interventions (and vice versa) in ways that blend ontological, positivistic, and complexity worldviews about health and diseases.

The vision of reversing and eliminating the epidemics of unhealthy habits can help us keep our eyes on the horizon. Dreaming of healthy futures, we can inspire ourselves toward 100% health to sustain our focus and maintain efforts to work on this vision.

To work on the HBC mission, we must keep our feet on the ground and lead by example: improve our health habits and help others to do the same. This mission involves implementing HBC programs for our faculty, students, practitioners, staff working within our educational and organizational settings, cultivating learning organizations to effectively deliver these programs, and creating learning communities to disseminate these programs to patients and families, within and beyond the health sector.

Reversing the epidemic tides of unhealthy habits and decreasing the burdens of chronic diseases are lifelong learning processes. To contribute toward this vision, you can join a self-organizing, learning community: go to www.HealthBuddyCoaching.com, access free resources, and sample the HBC learning process.

Acknowledgements I would like to thank Dr. Joachim Sturmberg for his editorial support on this chapter.

Health Buddy Coaching Program: Changing Ourselves Our Families, Organizations, Communities and Cultures (2013). This book is about developing transformational leadership and disseminating HBC programs. The behaviour change guidebook, Health Buddy Coaching: Discover Your Path to Health Habits (2013) is for practitioners, staff, patients and their family members.

References

1. Wilkinson R. The spirit level: why more equal societies almost always do better. New York: Bloomsbury Press; 2009.
2. Milanovic B. The haves and the have-nots: a brief and idiosyncratic history of global inequality. New York: Basic Books; 2010.
3. Carr D. Improving the health of the world's poorest people. Washington, DC: Population Reference Bureau; 2004.
4. Manning WG, Keeler EB, Newhouse JP, Sloss EM, Wasserman J. The costs of poor health habits. Cambridge, MA: Harvard University Press; 1991.
5. Harrington J, Stuttaford M. Global health and human rights: legal and philosophical perspectives. London; New York: Routledge; 2010.
6. World Health Organization, Commission on Social Determinants of Health. Closing the gap in a generation: health equity through action on the social determinants of health. Geneva: WHO; 2008. p. 1–256.
7. Wipfli H, Samet JM. Global economic and health benefits of tobacco control: Part 1. Clin Pharmacol Ther. 2009;86(3):263–71.
8. World Health Organization. WHO report on the global tobacco epidemic, 2008: The MPOWER package. Geneva: World Health Organization; 2008.
9. Bloom DE, Cafiero ET, Jané-Llopis E, Abrahams-Gessel S, Bloom LR, Fathima S, Feigl AB, Gaziano T, Mowafi M, Pandya A, Prettner K, Rosenberg L, Seligman B, Stein A, Weinstien C. The global economic burden of non-communicable diseases. Geneva: World Economic Forum; 2011.
10. Connell R, Messerschmidt JW. Hegemonic masculinity: rethinking the concept. Gend Soc. 2005;2005:19.
11. Hausmann R, Tyson L, Zahidi S. The global gender gap report 2011. Geneva: World Economic Forum; 2011.
12. Sen G, Ostlin P, George A. Unequal, unfair, ineffective and inefficient. Gender inequity in health: why it exists and how we can change it. Final Report to the WHO Commission on Social Determinants of Health. Geneva: WHO; 2007.
13. Evans T, Mafubelu D, AbouZahr C, de Zoysa I, Moreno GC. Women and health: today's evidence tomorrow's agenda. Geneva: World Health Organization; 2009.
14. United Nations. United National Universal Declaration of Human Rights. New York: United Nations; 1948.
15. WHO. Preamble to the Constitution of the World Health Organization as adopted by the International Health Conference, New York, 19 June–22 July 1946; signed on 22 July 1946 by the representatives of 61 States (Official Records of the World Health Organization, no. 2, p. 100) 1948, 7 April.
16. Huber M, Knottnerus JA, Green L, van der Horst H, Jadad AR, Kromhout D, et al. How should we define health? BMJ. 2011;343:d4163.
17. Saracci R. The World Health Organisation needs to reconsider its definition of health. BMJ. 1997; 314(7091):1409–10.
18. Harvard University Center on the Developing Child. http://developingchild.harvard.edu/.
19. National Scientific Council on the Developing Child and the National Forum on Early Childhood Policy and Programs. Working Papers on Childhood Development No. 1–11. 2004–2011. Available from: http://www.developingchild.harvard.edu.
20. NCTSN. National Child Traumatic Stress Network: Changing the course of children's lives by changing the course of their care. Charlotte, NC: NCTSN; 2011.
21. Antonovsky A. Health, stress, and coping: new perspectives on mental and physical well-being. San Francisco, CA: Jossey-Bass; 1979.
22. Antonovsky A. Unraveling the mystery of health: how people manage stress and stay well. San Francisco: Jossey-Bass; 1987.
23. Seligman ME. Flourish: a visionary new understanding of happiness and well-being. New York, NY: Free Press, Simon & Schuster; 2011.
24. Peterson C, Seligman M. Character strengths and virtues: a handbook and classification. New York: Oxford University Press; 2004.
25. Deci EL, Eghrari H, Patrick BC, Leone DR. Facilitating internalization: the self-determination theory perspective. J Pers. 1994;62(1):119–42.
26. American Psychological Association. The Road to Resilience http://www.apa.org/helpcenter/road-resilience.aspx.
27. Niemiec CP, Ryan RM, Patrick H, Deci EL, Williams GC. The energization of health-behavior change: examining the associations among autonomous self-regulation, subjective vitality, depressive symptoms, and tobacco abstinence. J Posit Psychol. 2010;5(2): 122–38.
28. Ryan RM, Frederick C. On energy, personality, and health: subjective vitality as a dynamic reflection of well-being. J Pers. 1997;65(3):529–65.
29. Ryan RM, Deci EL. A self-determination theory approach to psychotherapy: the motivational basis for effective change. Can Psychol. 2008;49(3):186–93.
30. Greene J, Hibbard JH. Why does patient activation matter? An examination of the relationships between patient activation and health-related outcomes. J Gen Intern Med. 2011;27(5):520–6.
31. Hibbard JH, Mahoney E. Toward a theory of patient and consumer activation. Patient Educ Couns. 2010;78(3):377–81.
32. Bandura A. Health promotion by social cognitive means. Health Educ Behav. 2004;31(2):143–64.
33. Bandura A. Self-efficacy—toward a unifying theory of behavioral change. Psychol Rev. 1977;84(2): 191–215.
34. Salovey P, Rothman AJ, Detweiler JB, Steward WT. Emotional states and physical health. Am Psychol. 2000;55(1):110–21.
35. Cameron LD, Leventhal H. The self-regulation of health and illness behaviour. New York, NY: Routledge; 2003.
36. Vohs KD, Baumeiste RF (2010) Handbook of self-regulation. Research, theory, and applications. 2nd ed. England: Guilford; 2010.

37. de Ridder D, Bertha J, de Wit F. Self-regulation in health behavior. Chichester, England: Wiley; 2006.
38. Hofmann W, Baumeister RF, Forster G, Vohs KD. Everyday temptations: an experience sampling study of desire, conflict, and self-control. J Pers Soc Psychol. 2012;102(6):1318–35.
39. Hofmann W, Friese M, Strack F. Impulse and self-control from a dual-systems perspective. Perspect Psychol Sci. 2009;4(2):162–76.
40. de Ridder DTD, Lensvelt-Mulders G, Finkenauer C, Stok FM, Baumeister RF. Taking stock of self-control: a meta-analysis of how trait self-control relates to a wide range of behaviors. Pers Soc Psychol Rev. 2012;16:76–99.
41. Johnson BA. Managing polarities: identifying and managing unsolvable problems. Amherst, MA: HRD Press; 1992.
42. Senge PM. The fifth discipline: the art and practice of the learning organization. New York: Random House; 2006.
43. Block P. Community: The Structure of Belonging. San Francisco, CA: Berrett-Koehler Publishers, Inc.; 2008.
44. Mokdad AH, Marks JS, Stroup DF, Gerberding JL. Actual causes of death in the United States, 2000. JAMA. 2004;291(10):1238–45.
45. Raine KD. Obesity epidemics: inevitable outcome of globalization or preventable public health challenge? Int J Public Health. 2012;57(1):35–6.
46. Hipple B, Lando H, Klein J, Winickoff J. Global teens and tobacco: a review of the globalization of the tobacco epidemic [Review]. Curr Probl Pediatr Adolesc Health Care. 2011;41(8):216–30.
47. Lando HA, Wilson K. Combating the global tobacco epidemic [Editorial]. Prev Med. 2010;50(1–2):11–2.
48. Choi BC, McQueen DV, Puska P, Douglas KA, Ackland M, Campostrini S, et al. Enhancing global capacity in the surveillance, prevention, and control of chronic diseases: seven themes to consider and build upon. J Epidemiol Community Health. 2008;62(5):391–7.
49. Magnusson RS. Rethinking global health challenges: towards a 'global compact' for reducing the burden of chronic disease. Public Health. 2009;123(3):265–74.
50. Bass BM, Riggio RE. Transformational leadership. Sussex: Psychology Press; 2005.
51. Mogford E, Gould L, Devoght A. Teaching critical health literacy in the US as a means to action on the social determinants of health. Health Promot Int. 2011;26(1):4–13.
52. Lindstrom B, Eriksson M. From health education to healthy learning: implementing salutogenesis in educational science. Scand J Public Health. 2011;39(6 Suppl):85–92.
53. Lorig KR, Ritter PL, Dost A, Plant K, Laurent DD, McNeil I. The Expert Patients Programme online, a 1-year study of an Internet-based self-management programme for people with long-term conditions. Chronic Illn. 2008;4(4):247–56.
54. Holman H, Lorig K. Patient self-management: a key to effectiveness and efficiency in care of chronic disease. Public Health Rep. 2004;119(3):239–43.
55. Reed RL, Battersby M, Osborne RH, Bond MJ, Howard SL, Roeger L. Protocol for a randomised controlled trial of chronic disease self-management support for older Australians with multiple chronic diseases. Contemp Clin Trials. 2011;32(6):946–52.
56. Van Beurden EK, Kia AM, Zask A, Dietrich U, Rose L. Making sense in a complex landscape: how the Cynefin framework from complex adaptive systems theory can inform health promotion practice. Health Promot Int. 2011 [Epub ahead of print].
57. Green LA. Prescription for health: round 1 initial results. Ann Fam Med. 2005;3 Suppl 2:S2–3.
58. Flocke SA, Kelly R, Highland J. Initiation of health behavior discussions during primary care outpatient visits. Patient Educ Couns. 2009;75(2):214–9.
59. Simons VA, Flynn SP, Flocke SA. Practical behavior change counseling in primary care. Prim Care. 2007;34(3):611–22, vii.
60. Flocke SA, Crabtree BF, Stange KC. Clinician reflections on promotion of healthy behaviors in primary care practice. Health Policy. 2007;84(2–3):277–83.
61. Green LA, Cifuentes M, Glasgow RE, Stange KC. Redesigning primary care practice to incorporate health behavior change: prescription for health round-2 results. Am J Prev Med. 2008;35(5 Suppl): S347–S9.
62. Dodoo MS, Krist AH, Cifuentes M, Green LA. Start-up and incremental practice expenses for behavior change interventions in primary care. Am J Prev Med. 2008;35(5 Suppl):S423–30.
63. Jayasinghe S. Conceptualising population health: from mechanistic thinking to complexity science. Emerg Themes Epidemiol. 2011;8(1):2.
64. Sturmberg JP, Martin CM. Knowing–in medicine [Review]. J Eval Clin Pract. 2008;14(5):767–70.
65. Ellis B, Herbert SI. Complex adaptive systems (CAS): an overview of key elements, characteristics and application to management theory. Inform Prim Care. 2011;19(1):33–7.
66. Paina L, Peters DH. Understanding pathways for scaling up health services through the lens of complex adaptive systems. Health Policy Plan. 2012;27(5): 365–73.
67. Kneebone R. Total internal reflection: an essay on paradigms. Med Educ. 2002;36(6):514–8.
68. Rosenquist JN, Murabito J, Fowler JH, Christakis NA. The spread of alcohol consumption behavior in a large social network. Ann Intern Med. 2010;152(7): 426–33, W141.
69. Christakis NA, Fowler JH. The collective dynamics of smoking in a large social network. N Engl J Med. 2008;358(21):2249–58.
70. Fowler JH, Christakis NA. Dynamic spread of happiness in a large social network: longitudinal analysis over 20 years in the Framingham Heart Study. BMJ. 2008;337:a2338.
71. Christakis NA, Fowler JH. The spread of obesity in a large social network over 32 years. N Engl J Med. 2007;357(4):370–9.
72. Kapinos KA, Yakusheva O. Environmental influences on young adult weight gain: evidence from a natural

experiment [Comparative Study]. J Adolesc Health. 2011;48(1):52–8.
73. Fisher A, van Jaarsveld CH, Llewellyn CH, Wardle J. Environmental influences on children's physical activity: quantitative estimates using a twin design. PLoS One. 2010;5(4):e10110.
74. Ershow AG. Environmental influences on development of type 2 diabetes and obesity: challenges in personalizing prevention and management [Review]. J Diabetes Sci Technol. 2009;3(4):727–34.
75. Ramos R, Ferreira-Pinto JB, Brouwer KC, Ramos ME, Lozada RM, Firestone-Cruz M, et al. A tale of two cities: social and environmental influences shaping risk factors and protective behaviors in two Mexico-US border cities [Comparative Study]. Health Place. 2009;15(4):999–1005.
76. Cummings KM, Fong GT, Borland R. Environmental influences on tobacco use: evidence from societal and community influences on tobacco use and dependence [Review]. Annu Rev Clin Psychol. 2009;5:433–58.
77. Chow CK, Lock K, Teo K, Subramanian SV, McKee M, Yusuf S. Environmental and societal influences acting on cardiovascular risk factors and disease at a population level: a review. Int J Epidemiol. 2009;38(6): 1580–94.
78. Vrazel J, Saunders RP, Wilcox S. An overview and proposed framework of social-environmental influences on the physical-activity behavior of women [Review]. Am J Health Promot. 2008;23(1):2–12.
79. Richards R, Smith C. Environmental, parental, and personal influences on food choice, access, and overweight status among homeless children. Soc Sci Med. 2007;65(8):1572–83.
80. Ball K, Timperio AF, Crawford DA. Understanding environmental influences on nutrition and physical activity behaviors: where should we look and what should we count? [Editorial]. Int J Behav Nutr Phys Act. 2006;3:33.
81. McCrady BS, Zucker RA, Molina BS, Ammon L, Ames GM, Longabaugh R. Social environmental influences on the development and resolution of alcohol problems. Alcohol Clin Exp Res. 2006;30(4):688–99.
82. Puska P, Stahl T. Health in all policies-the Finnish initiative: background, principles, and current issues [Review]. Annu Rev Public Health. 2010;31:315–28.
83. Hancock T. The Ottawa Charter at 25. Can J Public Health. 2011;102(6):404–6.
84. Puska P. The North Karelia Project: nearly 20 years of successful prevention of CVD in Finland. Hygie. 1992;11(1):33–5.
85. Puska P, Tuomilehto J, Nissinen A, Salonen JT, Vartiainen E, Pietinen P, et al. The North Karelia project: 15 years of community-based prevention of coronary heart disease. Ann Med. 1989;21(3):169–73.
86. Puska P, Nissinen A, Salonen JT, Toumilehto J. Ten years of the North Karelia Project: results with community-based prevention of coronary heart disease. Scand J Soc Med. 1983;11(3):65–8.
87. Puska P, Tuomilehto J, Salonen J, Neittaanmaki L, Maki J, Virtamo J, et al. Changes in coronary risk factors during comprehensive five-year community programme to control cardiovascular diseases (North Karelia project). Br Med J. 1979;2(6199):1173–8.
88. Puska P. The North Karelia project: an attempt at community prevention of cardiovascular disease. WHO Chronicle. 1973;27(2):55–8.
89. WHO. Ottawa charter for health promotion. Geneva: World Health Organization; 1986.
90. Christensen CM, Grossman JH, Hwang J. The innovator's prescription: a disruptive solution for health care. New York: McGrall-Hill; 2009.
91. Stacey RD. Strategic management and organisational dynamics: the challenge of complexity. 5th ed. Upper Saddle, NJ: Prentice Hall; 2009.
92. Mennin S. Complexity and health professions education: a basic glossary. J Eval Clin Pract. 2010;16(4): 838–40.
93. Christensen CM, Baumann H, Ruggles R, Sadtler TM. Disruptive innovation for social change. Harv Bus Rev. 2006;84(12):94–101, 63.
94. Downs TJ, Larson HJ. Achieving millennium development goals for health: building understanding, trust and capacity to respond. Health Policy. 2007; 83(2–3):144–61.
95. Gabbay RA, Bailit MH, Mauger DT, Wagner EH, Siminerio L. Multipayer patient-centered medical home implementation guided by the chronic care model. Jt Comm J Qual Patient Saf. 2011;37(6): 265–73.
96. Goroff M, Reich MR. Partnerships to provide care and medicine for chronic diseases: a model for emerging markets. Health Aff. 2010;29(12):2206–13.
97. Oprea L, Braunack-Mayer A, Rogers WA, Stocks N. An ethical justification for the Chronic Care Model (CCM). Health Expect. 2010;13(1):55–64.
98. Gofin J, Foz G. Training and application of community-oriented primary care (COPC) through family medicine in Catalonia, Spain. Fam Med. 2008;40(3):196–202.
99. Art B, De Roo L, De Maeseneer J. Towards unity for health utilising community-oriented primary care in education and practice. Educ Health (Abingdon). 2007;20(2):74.
100. Iliffe S, Lenihan P. Integrating primary care and public health: learning from the community-oriented primary care model [Review]. Int J Health Serv. 2003;33(1):85–98.
101. Macaulay AC, Jagosh J, Seller R, Henderson J, Cargo M, Greenhalgh T, et al. Assessing the benefits of participatory research: a rationale for a realist review. Glob Health Promot. 2011;18(2):45–8.
102. Jagosh J, Pluye P, Macaulay AC, Salsberg J, Henderson J, Sirett E, et al. Assessing the outcomes of participatory research: protocol for identifying, selecting, appraising and synthesizing the literature for realist review. Implement Sci. 2011;6:24.
103. Wong G, Greenhalgh T, Westhorp G, Pawson R. Realist methods in medical education research: what are they and what can they contribute? Med Educ. 2012;46(1):89–96.

104. Greenhalgh T, Wong G, Westhorp G, Pawson R. Protocol–realist and meta-narrative evidence synthesis: evolving standards (RAMESES). BMC Med Res Methodol. 2011;11:115.
105. Pawson R, Greenhalgh T, Harvey G, Walshe K. Realist review–a new method of systematic review designed for complex policy interventions [Review]. J Health Serv Res Policy. 2005;10 Suppl 1:21–34.
106. Liberato SC, Brimblecombe J, Ritchie J, Ferguson M, Coveney J. Measuring capacity building in communities: a review of the literature. BMC Public Health. 2011;11:850.
107. Downey LH, Anyaegbunam C, Scutchfield FD. Dialogue to deliberation: expanding the empowerment education model. Am J Health Behav. 2009; 33(1):26–36.
108. Horowitz CR, Robinson M, Seifer S. Community-based participatory research from the margin to the mainstream: are researchers prepared? Circulation. 2009;119(19):2633–42.
109. Wells K, Jones L. "Research" in community-partnered, participatory research. JAMA. 2009;302(3):320–1.
110. Bogart LM, Uyeda K. Community-based participatory research: partnering with communities for effective and sustainable behavioral health interventions. Health Psychol. 2009;28(4):391–3.
111. Goh YY, Bogart LM, Sipple-Asher BK, Uyeda K, Hawes-Dawson J, Olarita-Dhungana J, et al. Using community-based participatory research to identify potential interventions to overcome barriers to adolescents' healthy eating and physical activity. J Behav Med. 2009;32(5):491–502.
112. Kennedy C, Vogel A, Goldberg-Freeman C, Kass N, Farfel M. Faculty perspectives on community-based research: "I see this still as a journey". J Empir Res Hum Res Ethics. 2009;4(2):3–16.
113. Jernigan VB. Community-based participatory research with native american communities: the chronic disease self-management program. Health Promot Pract. 2009;11(6):888–99.
114. Nastasi BK, Hitchcock J. Challenges of evaluating multilevel interventions. Am J Community Psychol. 2009;43(3–4):360–76.
115. Hebert JR, Brandt HM, Armstead CA, Adams SA, Steck SE. Interdisciplinary, translational, and community-based participatory research: finding a common language to improve cancer research. Cancer Epidemiol Biomarkers Prev. 2009;18(4):1213–7.
116. WHO. Health equity through intersectoral action: an analysis of 18 country case studies. World Health Organization, Geneva: Her Majesty the Queen in Right of Canada, represented by the Minister of Health; 2008.
117. Botelho RJ. Health Buddy Coaching: Discover Your Path to Healthy Habits. Rochester, NY: MHH Publications; 2013.
118. Anderson C, Stewart S. Resistance. In: Anderson C, Stewart S (eds). Mastering resistance. New York: Guilford; 1995. p. 1–38.
119. Bar-Levav R. Thinking in the shadow of feelings: a new understanding of the hidden forces that shape individuals in society. New York: Simon and Schuster; 1988.
120. Butler C, Rollnick S, Stott N. The practitioner, the patient and resistance to change: recent ideas on compliance. Can Med Assoc J. 1996;154(9):1357–62.
121. Shaffer HJ, Simoneau G. Reducing resistance and denial by exercising ambivalence during the treatment of addiction. J Subst Abuse Treat. 2001;20(1):99–105.
122. Botelho RJ. Motivational practice: promote healthy habits and self-care of chronic diseases. 2nd ed. Rochester, NY: MHH Publications; 2004.
123. Miller WR, Stephen R, Conforti K. Motivational interviewing: preparing people for change. 2nd ed. New York: Guilford Press; 2002.
124. Klingemann HK. Coping and maintenance strategies of spontaneous remitters from problem use of alcohol and heroin in Switzerland. Int J Addict. 1992; 27(12):1359–88.
125. Kolb DA. Experiential learning. Chicago: Prentice Hall; 1984.
126. Horgan J. Rational mysticism: dispatches from the border between science and spirituality. New York, NY: First Mariner Books; 2004.
127. Urgesi C, Aglioti SM, Skrap M, Fabbro F. The spiritual brain: selective cortical lesions modulate human self-transcendence. Neuron. 2010;65(3):309–19.
128. Johnstone B, Glass BA. Support for a neuropsychological model of spirituality in persons with traumatic brain injury. Zygon. 2008;43:861–74.
129. Peters KE. Understanding and responding to human evil: a multicausal approach. Zygon. 2008;43: 681–704.
130. Dawkins R. The selfish gene. New York: Oxford University Press; 2006.
131. Vader JP. Spiritual health: the next frontier. Eur J Public Health. 2006;16(5):457.
132. Hawks SR, Hull ML, Thalman RL, Richins PM. Review of spiritual health: definition, role, and intervention strategies in health promotion. Am J Health Promot. 1995;9(5):371–8.
133. Engel GL. The clinical application of the biopsychosocial model. Am J Psychiatry. 1980;137(5):535–44.
134. Green LW, Kreuter MW. Health program planning: an educational and ecological approach. 4th ed. New York: McGraw-Hill; 2005.
135. Tapscott D, Williams A. Wikinomics: How mass collaboration changes everything. 2006.
136. Simon P. The age of the platform: how Amazon, Apple, Facebook and Google redefined business. Les Vegas, NE: Motion Publishing; 2011.
137. Kotter JP. Leading change. Boston, MA: Harvard Business School Press; 1996.
138. Rees W. Whats blocking sustainability? Human nature, cognition, and denial. Sustain Sci Practice Policy. 2010;6(2):13–25.
139. Perry Jr WG. Forms of intellectual and ethical development in the college years: a scheme. New York: Holt, Rinehart, and Winston; 1970.

Health Professions Education: Complexity, Teaching, and Learning

43

Stewart P. Mennin

43.1 Introduction

Health professions education is about learning to care for and about fellow human beings and the environment we coinhabit. Health, as a resource for daily life and well being, emerges from a constellation of interdependent conditions embedded in webs of social, cultural, political, economic, and geographic events that influence access to food, water, shelter, employment, education, safety, peace, a stable ecosystem, sustainable resources, social justice, and equity [3]. Becoming a health professional is necessarily relational and dynamical. Relationships are about the quality of the exchanges we have, i.e., communication, and the extent to which they are adaptive for shared understanding, health, and well being [4]. Health professions in this chapter refer to all persons who have earned the privilege to care for those in need and to those who promote and optimize healthy lives and healthy families. How we think about and understand what we do as teachers and practitioners to help prepare future health professionals is guided by cultural and historical frameworks that evolve through periodic paradigm shifts [5–12].

Whenever new concepts, methods, and tools become available, there is a period of flux as the landscape of understanding rearranges itself into a new configuration [7, 13, 14]. We are at a turning point [8], a time and place of self-organized criticality [7], on the cusp of a paradigm shift [13] in health professions education. Complexity science is a relatively new field of study that has only recently entered the lexicon of health professions education. It is a different paradigm [15], one that deals with experiences familiar to educators and health practitioners and uses different concepts and terminology that, at first, seem strange [16–18]. It was not that long ago that words and concepts such as formative assessment, constructivism, and modified Angoff were strange as well. In time, we explored them, and others, learned them and made them part of our professional language and practice. Common usage of the word complex refers to things that are interconnected in ways that are not obvious or easy to understand. It is impossible to define complexity with precision because, "*Definitions require stability, the very element complexity neither has nor aspires to have. Instead complexity*

The past is not what it used to be.
JA Goldstein [1]

To think complexly is to adopt a relational, a system(s) view. That is to look at any event or entity in terms, not of itself, but of its relations.
WE Doll Jr and D Trueit (p. 845) [2]

S.P. Mennin (✉)
Professor Emeritus, Department of Cell Biology and Physiology; Assistant Dean Emeritus, Educational Development and Research, School of Medicine, University of New Mexico, Albuquerque, NM, 87131, USA

Mennin Consulting & Associates, Inc. and Mennin Consultoria em Saude, Ltda., Rua Inhambu 1708, Sao Paulo, Brazil
e-mail: menninsp@gmail.com

J.P. Sturmberg and C.M. Martin (eds.), *Handbook of Systems and Complexity in Health*,
DOI 10.1007/978-1-4614-4998-0_43, © Springer Science+Business Media New York 2013

asks us to see, to deal with a world in continual flux; but a world that does have patterns to it, patterns that bind and structure through their interplay" (p. 841) [2]. Some of the basic characteristics and properties of self-organization in complex systems that have been described earlier in this book [18, 19] and elsewhere [9, 18, 20–22] are highlighted with some examples in Table 43.1.

Historically, the dominant perspective in Western science and most health professions education institutions has been a biomechanical derivative of Cartesian and Newtonian concepts [6, 8, 12]. It is a linear cause and effect paradigm in which numerical and analytic thinking favoring prediction and control [24] enjoys supremacy over narrative explanation [25]. Complexity science is an outgrowth of traditional science that is complementary with it and extends our ability to study problems and situations that, until recently, have been intractable to traditional research methods unable to deal with so many interacting variables at the same time [9]. It moves us from thinking about a framework of individual elements to one that is relational where patterns and forms emerge from dynamic interactions and interdependencies. It offers a perspective in which the relationships among things are more important than the things themselves [26]. It sees the world, nature, and life as open systems continuously exchanging energy across permeable (fuzzy) boundaries, a world that organizes itself in ways that are not obvious or predictable from knowing the parts and where the whole is different from the parts [20]. A system is taken to mean an assemblage of interacting components "whose *essential properties arise from the relationships between its parts*" (p. 27) [9]. It is the study of a world in which novelty and familiarity, cooperation and competition are dialectic not dualistic [27] and it has immediate relevance for economics, biology, sociology, psychology, health, health care, healthcare management and leadership, and health professions education [8, 9, 18, 26].

The growth and acceleration of science in all its forms has contributed to spectacular discoveries, inventions, and research that have changed the world and improved the ways we live and work. The scientific method was founded in physics and works best in situations that have few measurable interacting variables [6, 7, 28]. Most problems in the natural world like nature and the nervous system, for example, are continuously adaptive and nonlinear. They do not lend themselves to linear mechanical solutions and explanations [7, 22, 29–31]. Life is complex. It seems reasonable to approach the study of health, health care, and the education of future health professionals by embracing both complicated and complex events [16, 21, 32–34]. Reductionist, biomechanical models based on linear causality will, in the near future, come to occupy a more limited sphere of influence as the need to understand the dynamics of complex interactions among multiple agents and actors contributes to our practice and understanding in ways that the study of agents and actors as isolates cannot [35–38]. Instead of seeing everything as linear and reduced to its smallest parts, it becomes important to see how the dynamics of interactions at the local level give rise to the properties of the whole, how the various systems, organs, cells, and molecules interact, how social conditions influence individuals and vice versa [9, 20].

What is needed now are ways to move health professionals and educators in this direction to help recognize when it is more appropriate to incorporate concepts of complexity into their work. The biomechanical view continues to be important and necessary. The challenge is to determine the extent to which situations are ordered (biomechanical—simple or complicated) or disordered (complex or chaotic) and acting accordingly [16, 19, 39–42]. It is not that one size fits all. Rather, it is that complex systems require considerations of both time and space in ways that cannot be understood with linear reductionist thinking [33].

The goal of this chapter is to make more accessible concepts of complexity science relevant to health professions education in general and learning and teaching in particular. Future health professionals will practice and live in an increasingly complex world in which reductionist explanations cannot sufficiently illuminate complex health situations.

Table 43.1 Conditions for self-organization, with health professions education, clinical and biological examples (From Mennin [23])

Conditions for self-organization	Health professions education examples	Clinical and biological examples
Open to the outside, far from equilibrium	Students and teachers are free to come and go; students encounter new information and new experiences daily; patients come and go in clinics and hospitals; agents are changing constantly	Biological systems are open and exchange energy with the outside world (e.g., the respiratory system, the gastrointestinal system). Global epidemics like H1N1, SARS, and HIV/AIDS arise from the openness, fluidity, flow, and mobility of people and infectious agents
Large number of interacting elements, multiple short feedback loops	Many students, teachers, patients; continuous exchange of information among different agents	Many patients, workers, and staff in the hospital or community clinic change as a result of their interactions
Fuzzy boundaries	Boundaries that are ill defined and permeable hold agents together to promote interactions; context can serve as a fuzzy boundary; boundaries can be physical, biological, or conceptual; examples include ground rules for small-group discussions, curriculum general learning outcomes, rules of ethics and professionalism	Semi-permeable cell membranes frame cellular interactions; groups of health workers may have varying roles depending on conditions; an emergency room has defined space yet extends to other units in a hospital
Agents change through multiple nonlinear local interactions	Feedback and formative assessment; self, peer, and group; the exchange of ideas in groups affects outcomes and data; a patient's history influences the doctor's approach, sequence of questions and actions which, in turn, affect the patient, who in turn, affects the doctor	Small changes in clotting factors make big differences in hemodynamics; large expenditures in health care do not necessarily result in greater health indices or greater equity

43.2 Health Professions Education Is Relational and Complex

It is encouraging to see an increase in the incidence of experiential, interactive, and relational learning, especially in small groups, in health professions education over the past 30–40 years, more so in some parts of the world than others [43]. Small-group, problem-based, and team-based learning, once viewed as innovative and experimental, is now commonplace in medicine, nursing, and allied health professions curricula. Outcome and competency-based learning have been added to the lexicon of curriculum planning and evaluation [44]. Assessment, once almost exclusively focused on psychometrics, is now viewed more as a question of education than as one of measurement, and educators discuss assessment as being coembedded with learning and teaching [45]. The essential nature of effective communication (low tech–high touch) for health and the practice of health professionals are undisputed in an age of technology and information. Professionalism and ethics, especially in research, was not on anyone's radar screen 35 years ago [43, 46–48]. Clinical skills and patient contact, once relegated exclusively to the last years of medical education now are present more often in the early years of study and the experience is a valued integrative part of curricula [49–53]. Clinical clerkships (rotations) in remote and rural sites have been shown to be equally, if not more, effective educational experiences compared to those based in tertiary care centers [52]. Multidisciplinary, multiprofessional, and interprofessional learning is gaining traction. Authentic workplace learning and assessment are natural extensions of early clinical and community learning experiences. The impact of information technology, computers, tele-health, tele-consulting, distance learning, and online resources cannot be underestimated, nor can the challenges they pose to relationship-centered care and relationship-centered teaching and learning [4, 5, 54]. The availability of educational online resources (MedEdWorld http://www.mededworld.org/ and MedEdPortal https://www.mededportal.org), access to prescription drug data on small hand-held electronic devices and increased attention to electronic medical records are pushing the applications of technology to new frontiers as the Internet reaches more and more corners of the globe. The basic sciences, rooted in experimental measurement and observation, long the foundation of medicine are more important than ever. Reframing them as the sciences basic to medicine that serve to explain and amplify experience rather than functioning primarily as prerequisites to clinical experience is an important next level in the continued pursuit of integration. Such views promote relevance and integration that has been elusive ever since Flexner proposed a strong foundation in the basic sciences should precede clinical practice [55].[1]

Now, in the year of the 100th anniversary of the Flexner report on medical education, it is time to move on to earlier access to learning in authentic settings where practical experience creates a need to know that is explained by theory and the sciences basic to medicine [5, 25, 56]. The foundation of experiential learning in an era of rapid change and exchange of information requires attention to technology in the service of learning and health in ways that strengthen curricula and pedagogies that are rich, recursive, relational, and robust [6, 23, 57, 58]. Leadership and management skills appropriate for complex situations are needed to successfully keep healthcare systems and academic health education institutions and hospitals functioning and relevant as competition for declining resources increases [21]. It is naive to think that a reductionist, linear causal model of thinking and reasoning is sufficient to sustain health professions education and to understand the social dynamics fundamental to explanations of learning [42, 59–61]. Education for the health professions is about relationships, interactions, and interdependencies that contribute to complexity. The world is becoming more complex and that includes health professions education. Alternative perspectives based on complexity thinking are relevant to health professions education.

43.2.1 Learning Is Complex and Embodied

> What do you think you understand with? With your head? Bah!
> *Nikos Kazantzakis, Zorba the Greek*

Kazantzakis created Zorba as a sensuous and passionate man who lived life simply and deeply. He was fully engaged with the all his senses, his whole being. Learning, understood as complex, is something that emerges from the whole person, body, and mind. Cognition has been described as embodied [62–68].

> By using the term embodied we need to highlight two points: first, that cognition depends upon the kinds of experience that come from having a body with various sensorimotor capacities, and second, that these individual sensorimotor capacities are themselves embedded in a more encompassing biological, psychological, and cultural context. By using the term action we need to emphasize once again that sensory and motor processes, perception and action, are fundamentally inseparable in lived cognition. Indeed, the two are not merely contingently linked in individuals; they have also evolved together [65] (p. 173).

Lakoff extended embodied mind to abstract thought.

> … the structures used to put together our conceptual systems grow out of bodily experience and make sense in terms of it; moreover, the core of our

[1] It is worth noting that Flexner very soon after the implementation of his reforms became aware and concerned about the "unintended consequences" of the emerging "sole focus" on the basic sciences in medical education and the plea to not loose the virtues of the humanities. He wrote: *… the very intensity with which scientific medicine is cultivated threatens to cost us at time the mellow judgment and broad culture of the older generation at its best. Osler, Janeway, and Halsted have not been replaced.* (Flexner A. (1930) Universities, American, English and German (p. 95)).

> conceptual system is directly grounded in perception, body movement, and experience of a physical and social character. Thought is imaginative, in that those concepts which are not directly grounded in experience employ metaphor, metonymy, and mental imagery-all of which go beyond the literal mirroring, or representation, of external reality. It is this imaginative capacity that allows for "abstract" thought and takes the mind beyond what we can see and feel. The imaginative capacity is also embodied—indirectly—since the metaphors, metonymies, and images are based on experience, often bodily experience. [62] (p. iv)

The perspective of embodied thought supports and argues for organizing and integrating learning activities, simulations, and authentic learning environments in curriculum planning and implementation. The concept of embodied cognition fits well with a trend toward increased interactive and transactive learning, i.e., small-group activities, early clinical and community experiences, and integration of clinical skills with the "basic sciences" [53, 57, 69–71].

Complexity is transdisciplinary. Per Bak's Nobel prize winning work on punctuated equilibrium and self-organized criticality began with studies on the complex behavior of sand piles [7]. Learning, like avalanches in sand piles and earthquakes, is an adaptive, self-organized change the fundamental processes of which are relevant to how we think about, understand, and organize learning.

> To make this less abstract, consider the scenario of a child at the beach letting sand trickle down to form a pile. … In the beginning, the pile is flat, and the individual grains remain close to where they land. Their motion can be understood in terms of their physical properties. As the process continues, the pile becomes steeper, and there will be little sand slides. As time goes on, the sand slides become bigger and bigger. Eventually, some of the sand slides may even span all or most of the pile. At that point, the system is far out of balance, and its behavior can no longer be understood in terms of the behavior of the individual grains. The avalanches form a dynamic of their own, which can be understood only from a holistic description of the properties of the entire pile rather than from a reductionist description of individual grains: the sand pile is a complex system. … The behavior of the sand pile mimics several phenomena observed across many sciences, which are associated with complexity. [7] (pp. 2–3)

The sand pile reaches a state, which Bak called self-organized criticality, that is highly sensitive to disturbance, poised and ready, and just one grain of sand away from an avalanche. There is no omniscient hand arranging the grains of sand to maximize the chances of an avalanche. It happens spontaneously based on the local interactions and conditions, i.e., there is no self in self-organization [72]. Thousands of neuronal synapses on dendrites, like grains of falling sand, influence the conditions of a postsynaptic cell enhancing and inhibiting the probability of an action potential. The state of the postsynaptic cell just prior to an action potential is one of self-organized criticality, like a sand pile that is one grain of sand away from an avalanche [7, 29, 30]. A similar process occurs on a larger scale among groups of neurons, and among groups of groups of neurons. The dynamical nature of neural oscillations is interdependent interactions of the nervous system with itself, the body in which it exists, and the environment in general. What emerges are multiple levels of self-organized criticality for optimal sensitivity and responsiveness to the environment [29, 73, 74]. In a similar way, the avalanche of political unrest that erupted in Tunisia in Spring 2011 reached a point of self-organized criticality that changed the world. Systems achieve such critical states episodically rather than existing in them continuously. The episodic nature of change is referred to as punctuated equilibrium. The fossil record of life on earth, for example, shows relatively short periods of catastrophe, extinction, and rapid speciation followed by long periods of quiescence and quasi-stability. Similarly, behavior is observed for earthquakes. This concept of change in complex systems is different than a slow incremental process advocated by Darwin's theory of evolution and by the traditional literature on management, change and leadership [75–77]. Punctuated equilibrium is visible in local and global economies, social movements, biological cycles, oscillations, learning, and curriculum change. Rogers observed the average length of time between awareness of an innovation and its adoption was about 3.5 years, and it did not matter very much what the innovation was [78].

Change in complex systems cannot be explained by linear cause and effect. Learning is a dynamical complex process that depends on multiple diverse agents interacting in ways that are recursive (loop back on themselves in nonlinear ways), variably iterative (variations due to different initial conditions), reflective (a process that promotes self-organization), and interdependent (agents that participate in exchanges are each changed and changes affect the entire organism) [79]. From this perspective, the experience in experiential learning functions as triggers of change. It does not direct the nature of the change itself, like the falling grains of sand trigger an avalanche but do not determine the size or the frequency of the avalanche because it is the collective state of the sand pile, i.e., its history, that determines what happens and it is the learner's history and structure that determines what is learned [41, 42, 80].

> Learning, then, is a matter of transformations in the learner that are simultaneously physical and behavioral—which is to say, in biological terms, structural. Learning is certainly conditioned by particular experiences, but it is "due to" the learner's own complex biological-and-experiential structure, not an external stimulus. [80] (p. 13)

The learner's structure, their body, is a living expression of their life history. Structure can be thought of as rigid and fixed, as in architecture, buildings, preplanned procedural steps, and inflexible organizational charts. This perspective gives rise to concepts such as foundations, platforms, scaffolds, fixed outcomes, inflexible contexts, and hierarchical top-down organizations. In addition, structure can also be understood from the perspective of ecosystems and living organisms in which the history of the system and organism is embodied [65, 68, 81] and is itself the instrumentality and identity by which and through which it engages the world. The organism is coembedded in, and interdependent with, other organisms and systems on multiple levels in a dynamic that is dialectic rather than dualistic. Structure, in this sense, is best characterized in global and general terms. Learning, then, comes to be understood as dynamical, a "*becoming*" rather than *an acquisition* or fixed state and, like health, requires continuous interrelationships and the sociopolitical and cultural dynamics that inform and form life as it adapts to changing circumstances in an interdependent environment. If learning is interactive and dynamical, i.e., in motion, then perturbation and ambiguity are essential characteristics to be welcomed rather than diminished, excluded and factored out. The ability of teachers, students, and patients to question, disagree, and choose is fundamental to disturbing the status quo and to the movement of a complex adaptive system toward a critical self-organized state optimally responsive to contingency and, what Stuart Kauffman called, "*the adjacent possible*" [7, 72].

Two practical examples are illustrative. Arrow and Henry [82] studying 16 operating room teams learning minimally invasive cardiac surgery found dramatic differences in the ability of the teams to master the technique. They observed that some teams learned effectively. Others failed to improve with experience. Three factors predicted success: (1) the ease with which team members spoke up during operations; (2) the amount of boundary spanning communication with other hospital groups; and (3) the extent to which the surgeons acted as coaches for their teams. These three factors boosted the flow of energy and information, i.e., the quality of the exchanges within the group, moving it into a more productive dynamic state. Highly successful groups operate in a "hot" complex adaptive state characterized by positive feedback cycles that boost member energy and channel it into high group productivity. A second example from Arrow and Henry is the study of the complex dynamics of several mental health treatment teams with both permanent staff and with short-term staff, i.e., residents and interns, at a large psychiatric facility [82]. They found that tension between the training mission and the primary task of developing, monitoring, and adjusting therapeutic plans for patients was a source of problems. Team dynamics were often marred either by vehement interdisciplinary arguments or by boring meetings that put people to sleep. These patterns suggested groups marooned in "frozen" fixed state dynamics characterized by negative feedback cycles, and low energy and productivity. Constant turnover, inexperienced leaders (the psychiatric interns), and unclear group boundaries made it hard for the groups to sustain and deploy member energy and ideas for group productivity and learning. Arrow and Henry, examining group dynamics from a complexity perspective, proposed that in the healthcare arena, where multidisciplinary groups are common and tasks, patients, and dynamic events ensure high variety and regular perturbations, effective learning and performance is most likely when teams are able to operate in a dynamic state in which the complexity of the variables in play must match or exceed the complexity of responses to the situation. The above examples illustrate that complex activities and events require equally complex approaches in which the quality of the exchanges are the most important characteristic [4, 57, 69, 82, 83]. We are social beings who for most of our educational lives were taught and examined as individuals, where independent investigation and self-directed learning were encouraged. For the sake of the patients, the importance of collaborative learning and collective work needs to be emphasized much more in contemporary health professions education. Similarly, continuing professional education traditionally has focused primarily on the single practitioner in spite of the fact that the collective health team, more than the individual practitioner, is the key to improving healthcare practices and outcomes. Permanent education is a collective orientation to group problem solving and learning in the workplace developed by the Pan American Health Organization [84]. It emphasizes collaborative learning in ways that can be understood as a complex adaptive activity [82, 83, 85].

43.2.2 Complexity and Teaching

Historically, teaching was primarily about the transmission of content. It made sense in an era when access to printed material and images was limited and the few people that could read and had books taught others by telling what was in them [6, 42, 86]. Today we have too much information,[2] too much content. Content without action in context is inert and context is structurally determined by history and contingency. Without context, words and actions have no meaning. Teachers, now more than ever, are challenged to deal with the role of content in understanding. Some teachers, and many students, still mistake telling about and describing the parts as a substitute for understanding the whole. All teachers want students to learn and it seems natural to want to simplify and reduce complex events to make them more accessible to learners, to have a formula, a best practice, and a correct answer [41, 87]. However, this approach is limiting and misleading because complex situations are not compressible and do not have well-defined best practices. Expertise may not be helpful in complex situations [16, 39, 41, 88]. Teaching and learning in and about complex situations requires patience, reflection, and a tolerance for ambiguity; a willingness to try something to see what happens before formulating a response [41]. One goal for teachers is to help students learn how to be adaptive in a situation rather than to look for the right answer, especially when teaching in authentic settings and the workplace where skills like reflection, feedback, communication, ethics, and cultural adaptation are essential [5, 25, 89–91].

Teaching and learning are disturbing activities. A teacher's role is to disturb the status quo among learners and to create conditions that promote and trigger learning (self-organization). Teaching is about shaping connections, exchanges, and relationships, weaving contexts, using narrative, posing questions, and sharing interests and passions for curiosity, exploration, discovery, and understanding ([4, 92], Royson HA. Personal Communication. Placitas, New Mexico, 2000). Meaningful exchange is the precursor of adaptive learning. Teaching also needs to be a caring profession, especially for health

[2] Note the distinctions between information, knowledge, and wisdom. As T. S. Elliot put it in his poem *The Rock*: Where is the Life we have lost in living? Where is the wisdom we have lost in knowledge? Where is the knowledge we have lost in information?

professions educators. How we treat students and the way we are together as human beings makes a difference in teaching and learning [5, 90, 93]. Suchman, and the group at the University of Rochester, have pioneered relationship-centered patient care as a model for humanism and health care [54]. Caring has long been central to the mission of nursing and nursing education [12]. The locus of control and power between teachers, students, and patients has begun to change from teacher-centered to student-centered. Hopefully, it will keep going to become relationship-centered. Increased attention to ethics, professionalism, and humanism is contributing to a trajectory that embraces relationship-centered teaching [5, 58, 90, 94, 95].

Questions perturb the status quo, redirect the flow, intrigue, stimulate curiosity, and call attention to what teachers want students to notice [96]. Well-formulated questions can move discussions toward a divergence of understanding and the exploration of the frontiers of understanding [97]. Abraham [98] refers to questions as a *control parameter* in the language of dynamic systems theory [73, 79].

> A control parameter is a variable outside the system to which the system is sensitive and that moves the system through different states. Teachers use control parameters to perturb the attractor states of the students. In pedagogy, a control parameter is the teaching method. Some teaching methods are more successful than others [98] (p. 301).

Abraham proposed a ranking of pedagogical approaches accorded to the frequency of perturbations important for learning: small-group learning (problem-based learning) > social interaction models > inductive models > deductive models lectures [98]. A recent study reported student-to-student interactions to be highest in problem-based learning, followed in descending order by team-based learning and then lectures [99]. Diversity and multiple interactions with short loop recursive feedback are features of learning in complex adaptive situations. Teachers need a simultaneous appreciation of place, action, and time. Content itself may not change much but the context in which it is relevant is in continuous motion, has permeable boundaries and is far from being in equilibrium [9].

Teaching and learning have to do with boundaries. Boundaries make it possible to tell things apart, to recognize something that has defining characteristics, even to question if a problem or object belongs in a particular category. Boundaries make it possible to have ideas, concepts, and to know how things fit together under particular conditions. Teachers explore boundaries by pointing to things and questioning. Boundaries are paradoxically constraining and liberating.

> We often fall into the trap of thinking of a boundary as something that separates one thing from another. We should rather think of a boundary as something that constitutes that which is bounded. This shift will help us to see the boundary as something enabling, rather than as confining. ... All social systems, and thus all living systems, create, maintain, and degrade their own boundaries. These boundaries do not separate but intimately connect the system with its environment. They do not have to be just physical or topological, but are primarily functional, behavioral, and communicational. They are not "perimeters" but functional constitutive components of a given system. As an example of this logic, think of the eardrum. It forms the boundary between the outer and the middle ear, but at the same time it exists in order to let the sound waves through. As a matter of fact, if it was not there, the sound waves would not be able to get through at all! If the boundary is seen as an interface participating in constituting the system, we will be more concerned with the margins of the system, and perhaps less with what appears to be the evident or "central" components of the system. ... Social systems are obviously not limited in the same way. Parts of the system may exist in totally different spatial locations. The connections between different components could be seen as virtual, and therefore the system itself may exist in a virtual space. This much should be self-evident to most inhabitants of the global village, but two important implications should be kept in mind. The first is that noncontiguous subsystems could be part of many different systems simultaneously. This would mean that different systems interpenetrate each other, that they share internal organs. How does one talk of the boundary of the system under these conditions? A second implication of letting go of a spatial understanding of boundaries would be that in a critically organized system we are never far away from the boundary. If the components of the system are richly interconnected, there will always be a short route from any component to the "outside" of the system. There is thus no safe "inside" of the system, the boundary is folded in, or perhaps, the system consists of boundaries only. Everything is

> always interacting and interfacing with others and with the environment; the notions of "inside" and "outside" are never simple or uncontested. [100]

Teachers cannot transmit or manage knowledge because knowledge is an action in a flexible bounded context in time and place not a fixed object to be given or managed [39, 41].

43.3 Summary

Health professions education is a complex continuous process in which students learn to care for fellow human beings across a wide range of cultural, political, and economic conditions. Education for a complex world shifts the focus from linear causality among individual elements to nonlinear recursive relationships, interactions, and interdependencies among multiple diverse elements over time. Learning comes to be understood as embodied and holistic rather than fragmented and separated into silos of disciplines or mental models and representations. Open dynamical systems reach a state of maximal sensitivity to perturbation, called self-organized criticality, and new structures and changes emerge episodically over time as punctuated equilibrium. The same process is discussed in relation to the emergence of learning and the continuous and episodic reorganization of categories, ideas, and understandings. The role of experience is described as a trigger instead of a direct cause for adaptive change. The history and structure of the learner together with the immediate local conditions at the time determine what is learned.

Teaching and learning involve disturbing the status quo. Questions and questioning serve to clarify differences and help us to recognize the frontiers of understanding. The metaphor of an edge and frontier of understanding involves indefinite boundaries. Teachers work with boundaries to highlight and clarify differences. It is important to remember that teaching is a caring profession. The longstanding focus of nursing on caring and the newer model of relationship-centered patient care have relevance to teaching and learning.

Addressing these issues and challenges is not an incremental problem. It is paradigmatic and requires a different understanding of the nature of meaning and relating [39, 41, 87, 101]. Complexity thinking and dynamical systems theory offer a way to unify teaching and learning.

References

1. Goldstein JA. Classic Paper: Emergence and identity. E:CO. 2007;9(3):75–81.
2. Doll Jr WE, Trueit D. Complexity and the healthcare professions. J Eval Clin Pract. 2010;16:841–8.
3. Westin S, Johnsen R. List-based systems and gatekeeping on the international agenda [Editorial]. Eur J Gen Pract. 1998;4:53–4.
4. Gergen KJ. Relational being: beyond self and community. Oxford: Oxford University press; 2009.
5. Bleakley A, Bligh J, Browne J. In: Hamstra SJ, editor. Medical education for the future: identity, power and location. Dordrecht: Springer; 2011.
6. Doll Jr WE. A post-modern perspective on curriculum. New York: Teachers College press; 1993.
7. Bak P. How nature works: the science of self-organized criticality. New York: Springer; 1996.
8. Capra F. The turning point: science, society, and the rising culture. Toronto: Bantam Books; 1982.
9. Capra F. The web of life: a new scientific understanding of living systems. New York: Anchor books; 1996.
10. Engestrom Y. From teams to knots: activity-theoretical studies of collaboration and learning at work. Cambridge: Cambridge University Press; 2008.
11. Engel C. Medical education in the 21st century: the need for a capability approach. Capability. 1995;1(4): 23–30.
12. Davidson AW, Ray MA, Turkel MC, editors. Nursing, caring, and complexity science: for human-environment well-being. New York: Springer; 2011.
13. Kuhn TS. The structure of scientific revolutions. Chicago: University of Chicago press; 1962.
14. Goldstein JA. The unschackled organization: facing the challenge of unpredictability through spontaneous reorganization. Portland: Productivity Press; 1994.
15. Kuhn L. Complexity and educational research: a critical reflection. In: Mason M, editor. Complexity theory in the philosophy of education. Oxford: Wiley; 2008. p. 169–80.
16. Zimmerman B, Lindberg C, Plsek P. Edgeware: insights from complexity science for health care leaders. Irving, TX: VHA; 2001.
17. Mennin S. Complexity and health professions education: a basic glossary. J Eval Clin Pract. 2010;16:838–40.
18. Cilliers F. Understanding complex systems. In: Sturmberg JP, Martin CM, editors. Handbook of systems and complexity in health. New York: Springer; 2012 p. 39–62.

19. Sturmberg JP. The complex nature of knowledge. In: Sturmberg JP, Martin CM, editors. Handbook of systems and complexity in Health. New York: Springer; 2012.
20. Cilliers F. Complexity and postmodernism: understanding complex systems. London: Routledge; 1998.
21. McDaniel RR, Driebe DJ. Complexity science and health care management. In: Fottler MD, Savage GT, Blair JD, Payne GT, editors. Advances in health care management. Amsterdam: Elsevier Science; 2001. p. 11–36.
22. Kauffman S. At home the in universe: the search for laws of self-organization and complexity. New York: Oxford University press; 1995.
23. Mennin S. Self organisation, integration and curriculum in the complex world of medical education. Med Educ. 2010;44:20–30.
24. Radford M. Prediction, control and the challenge to complexity. Oxf Rev Educ. 2008;34(5):505–20.
25. Sullivan WM, Rosin MW. A new agenda for higher education: shaping a life of the mind for practice. San Francisco: Jossey-Bass; 2008.
26. Capra F. The hidden connections: integrating the biological, cognitive, and social dimensions of life into a science of sustainability. New York: Doubleday; 2002.
27. Kelso SJA, Engestrom DA. The complementary nature. Cambridge, MA: MIT Press; 2006.
28. Glouberman S, Zimmerman B. Complicated and complex systems: what would successful reform of medicare look like? *Discussion paper No. 8,* 2002. Available from: https://docs.google.com/viewer?url=http://www.plexusinstitute.org/resource/collection/6528ED29-9907-4BC7-8D00-8DC907679FED/ComplicatedAndComplexSystems-Zimmerman Report_Medicare_reform.pdf.
29. Edelman G. Second nature: brain science and human knowledge. New Haven: Yale University Press; 2006.
30. Freeman WJ. How brains make up thier minds. New York: Columbia University Press; 2000.
31. Freeman WJ, Nunez R. Restoring to cognition the forgotten primacy of action, intention and emotion. J Conscious Stud. 1999;6(11–12):ix–xix.
32. Glouberman S, Zimmerman B. Complicated and complex systems: what would successful reform of medicare look like? Comm Future Health Care Can. 2002;8:1–54.
33. Plsek P, editor. Redesigning health-care with insights from the science of complex adaptive systems. Crossing the quality chasm: a new health care system for the 21st century. Washington DC: National Academy press; 2001.
34. Borrell-Carrio F, Suchman AL, Epstein RM. The biopsychosocial model 25 years later: principles, practice, and scientific inquiry. Ann Fam Med. 2004;2(6):576–82.
35. West BJ. Homeostasis and Gauss statistics: barriers to understanding natural variability. J Eval Clin Pract. 2010;16:403–8.
36. West BJ, editor. Where medicine went wrong: rediscovering the path to complexity. New Jersey: World scientific; 2006.
37. Ahn AC, Tweari M, Poon C-S, Phillips RS. The limits of reductionism in medicine: Could systems biology offer an alternative? PLoS Med. 2006;3(6):e208.
38. Federoff HO, Gostin LO. Evolving from reductionism to holism: is there a future for systems medicine? JAMA. 2009;302(9):994–6.
39. Stacy RD. Complex responsive processes in organizations: learning and knowledge creation. Stacy RD, Griffin D, Shaw P, editors. London: Routledge; 2001.
40. Stacy RD. Strategic management and orgnaizational dynamics. London: Pitman Publishing; 1996.
41. Snowden D. Complex acts of knowing: paradox and descriptive self-awareness. J Knowl Manage. 2002;6(2):100–11.
42. Davis B, Sumara D, Luce-Kapler R. Engaging minds: changing teaching in complex times. 2nd ed. New York: Routledge; 2008.
43. Ludmerer KM. Time to heal: American medical education from the turn of the century to the era of managed care. Oxford: Oxford University press; 1999.
44. Smith SR. Outcome-based curriculum. In: Dent JA, Harden RM, editors. A practical guide for medical teachers. 3rd ed. Edinburgh: Churchill Livingstone, Elsevier; 2009. p. 161–7.
45. Schuwirth LWT, van der Vleuten CPM. Changing education, changing assessment, changing research? Med Educ. 2004;38:805–12.
46. Groopman J. How doctors think. Boston: Houghton Mifflin; 2007.
47. Stern DT, editor. Measuring medical professionalism. Oxford: Oxford University press; 2006.
48. Cooke M, Irby DM, Sullivan W, Ludmerer KM. American medical education 100 years after the Flexner report. N Engl J Med. 2006;355:1339–44.
49. Dornan T, Littlewood S, Margolis SA, Scherpbier A, Spencer J, Ypinazar V. How can experience in clinical and community settings contribute to early medical education? A BEME systematic review. Med Teach. 2006;28(1):3–18.
50. Mennin S, Kaufman A, Urbina C, McGrew M. Community-based medical education: toward the health of the public. Med Educ. 2000;34:503–4.
51. Mennin S, Petroni-Mennin R. Community-based medical education. Clin Teach. 2006;3:90–6.
52. Worley P, Silagy C, Prideaux D, Newble D, Jones AH. The parallel rural community curriculum: an integrated clinical curriculum-based in rural general practice. Med Educ. 2000;34:558–65.
53. Duban S, Kaufman A. Clinical skills: enhancing basic science learning. In: Kaufman A, editor. Implementing problem-based medical education: lessons from successful Innovations. New York: Springer; 1985. p. 89–105.
54. Suchman AL. A new theoretical foundation for relationship-centered care: complex responsive process of relating. J Gen Intern Med. 2006;21: S40–S4.

55. Flexner A. Medical education in the United States and Canada. A report to the Carnegie Foundation for the Advancement of Teaching 1910. http://www.carnegiefoundation.org/sites/default/files/elibrary/Carnegie_Flexner_Report.pdf.
56. Dornan T, Mann K, Scherpbier A, JS, editors. Medical education: theory and practice. Edinburgh: Churchill Livingstone, Elsevier; 2011.
57. Mennin S. Small-group problem-based learning as a complex adaptive system. Teach Teach Educ. 2007; 23:303–13.
58. Mennin S. Teaching, learning, complexity and health professions education. JIAMSE. 2010; 20(2s):162–5.
59. Norman G. Chaos, complexity and complicatedness this: lessons from rocket science. Med Educ. 2011;45:549–59.
60. Bleakley A, Brice J, Bligh J. Thinking the post-colonial in medical education. Med Educ. 2008;42:266–70.
61. Friedenberg J. Dynamical psychology: complexity, self-organization and mind. Litchfield Park, AZ: ISCE Publishing; 2009.
62. Lakoff G. Women, fire, and dangerous things: what categories reveal about the mind. Chicago: University of Chicago press; 1987.
63. Johnson M. The meaning of the body: aesthetics of human understanding. Chicago: University of Chicago press; 1999.
64. Gallagher S. How the body shapes the mind. Oxford: Clarendon press; 2005.
65. Varela FJ, Thompson E, Rosch E. The embodied mind: cognitive science and human experience. Cambridge, MA: The MIT Press; 1991.
66. Bateson G. Mind and nature: a necessary unity. Cresskill, NJ: Hampton Press; 2002.
67. Clark A. Being there: putting brain, body, and world together again. Cambridge, MA: The MIT Press; 1997.
68. Gibbs Jr RW. Embodiment and cognitive science. Cambridge: Cambridge University press; 2005.
69. Arrow H, McGrath JE, Berdahl JL. Small groups as complex adaptive systems: formation, coordination, development, and adaptation. Thousand oaks: Sage; 2000.
70. Engestrom DA. From teams to knots. Cambridge: Cambridge University press; 2008.
71. Group WS. Increasing the relevance of education for health professionals: Report of WHO study group on problem-solving education for the health professions. Geneva: Group WS; 1993.
72. Kauffman S. The origins of order: self-organization and selection in evolution. Oxford: New York; 1993.
73. Guastello S, Koopmans M, Pincus D, editors. Chaos and complexity in psychology: the theory of nonlinear dynamical systems. Cambridge: Cambridge University Press; 2009.
74. Hollis G, Kloos H, Van Orden GC. Origins of order in cognitive activity. In: Guastello SJ, Koopmans M, Pincus D, editors. Chaos and complexity in psychology: the theory of nonlinear dynamical systems. Cambridge: Cambridge University press; 2009. p. 206–41.
75. Wheatley M. Leadership and the new science: discovering order in a chaotic world. 2nd ed. San Francisco: Berrett-Koehler; 1999.
76. Hazy JK, Goldstein JA, Lichtenstein BB, editors. Complex systems leadership theory: new perspectives from complexity science on social and organizational effectiveness. Boston, MA: ISCE Publishing; 2007.
77. Goldstein J. Riding the waves of emergence: leadership innovations on complex systems. In: Business AU-SoMa, editor. Plexus Institute: Garden City; 2010.
78. Rogers EM. Diffusion of innovations. 5th ed. New York: Free Press; 2003.
79. Kelso SJA. Dynamic patterns: the self-organization of brain and behavior. Cambridge: MIT Press; 1995.
80. Davis B, Sumara D. Complexity and education: inquiries into learning, teaching, and research. Mahwah: Lawrence Erlbaum Associates; 2006.
81. Johnson M. The body in the mind: the bodily basis of meaning, imagination, and reasoning. Chicago: University of Chicago Press; 1990.
82. Arrow H, Henry KB. Using complexity to promote group learning in healthcare. J Eval Clin Pract. 2010;16:861–6.
83. Stacy RD. Complexity and group processes: a radically social understanding of individuals. London: Routledge; 2003.
84. Brito PQ, Roschke MA, Ribeiro EC. Educacion permanente, proceso de trabajo y calidad de servicios de salud. In: Haddad J, Roschke MA, Davini MC, editors. Educacion permanente de personal de salud. Washington, DC: OPS/OMS; 1994. p. 34–60.
85. Otero Ribeiro EC, Mennin S. Continuing medical education: Guide supplement 35.2 - Viewpoint. Med Teach. 2010;32:172–3.
86. Davis B, Sumara D. 'If things were simple…': complexity in education. J Eval Clin Pract. 2010; 16:856–60.
87. Snowden D. The Cynefin Framework; 2010. http://www.youtube.com/watch?v=N7oz366X0-8&list=PL5CB210886A3E51AE&index=1&feature=plpp_video. Accessed 3 Jan 2012.
88. Griffin D. In: Stacy RD, Griffin D, Shaw P, editors. The emergence of leadership: linking self-organization and ethics. London: Routledge; 2002.
89. Bleakley A. Blunting Occam's razor: aligning medical education with studies of complexity. J Eval Clin Pract. 2010;16:849–55.
90. Bleakley A. Social comparison, peer learning and democracy in medical education. Med Teach. 2010;32(32):878–9.
91. Bruffee KA. Collaborative learning: higher education, interdependence, and authority of knowledge, Second edition ed. Baltimore: The Johns Hopkins University press; 1999.

92. Christensen CR, Garvin DA, Sweet A, editors. Education for judgement: the artistry of discussion leadership. Boston, MA: Harvard Business School Press; 1991.
93. Haidet P, Stein HF. The role of the student-teacher relationship in the formation of physicians: the hidden curricuum as process. J Gen Intern Med. 2006;21:S16–20.
94. Bates CK, Babbot S, Williams BC, Stern DT, Bowen JL. Are we making progress in medical education? J Gen Intern Med. 2006;21(5):542-3.
95. Stern DT, Papadakis M. The developing physician: becoming a professional. New Engl J Med. 2006;355(17):1794–9.
96. Christensen CR. The discussion teacher in action: questioning, listening, and response. In: Christensen CR, Garvin DA, Sweet A, editors. Education for judgement: the artistry of discussion leadership. Boston: Harvard Business School Press; 1991. p. 153–74.
97. Cole M, John-Steiner V, Scribner S, Souberman E, editors. Mind in society: the development of higher psychological processes. Cambridge, MA: Harvard University Press; 1978.
98. Abraham JL. Dynamical systems theory: application to pedagogy. In: Tschacher W, Dauwalder J-P, editors. The dynamical systems approach to cognition: concepts and empirical paradigms based on self-organization, embodiment, and coordination dynamics. Singapore: World scientific; 2003. p. 295–307.
99. Kelly PA, Haidet P, Schneider V, Searle N, Seidel CL, Boyd F. A comparison of In-class learner engagement across lecture, problem-based learning, and team learning using the STROBE classroom observation too. Teach Learn Med. 2005;17(2):112–8.
100. Cilliers P. Complexity. In: Sturmberg JP, editor. Complexity and health; 2012.
101. Morin E. Seven complex lessons in education for the future. Paris: UNESCO; 2001.

44 Understanding Clinical Complexity Through Conversational Learning in Medical Social Networks: Implementing User-Driven Health Care

Amy Price, Shivika Chandra, Kaustav Bera, Tamoghna Biswas, Pranab Chatterjee, Ralph Wittenberg, Neil Mehta, and Rakesh Biswas

44.1 Introduction

With the increasing pressures to keep up with the most current literature, skills such as the ability to be empathic with patients, emotional intelligence, the ability to collaborate, and well-rounded decision making may be set aside in the race for information mastery. This can affect patient care. It is well documented that a significant part of a patient's recovery can be attributed to the dynamics between the patient and those they come to for medical care. In fact even placebo intervention is more successful when the patient trusts in and can form a therapeutic alliance with the treating professional [1].

Our chapter outlines how the use of social media and dynamic networking and interaction can improve medical student learning and contribute to patient care. Scenarios which revolve around conversational learning for improved medical care will be explored as a proposed model for future innovation. Students, patients, and physicians share conversations and interventional outcomes through conversational learning. In this way, you the reader can explore the strengths, weaknesses, and viability of learning through collaboration. This model of learning transcends geographical and cultural limits by taking advantage of social networking and collaboration sites like Facebook, Twitter, and Wikipedia.

We explore the power of empathy and shared decision making in unraveling of layers of human complexity, and providing fulfillment in doctor patient relationships. This "forest" of complexity is made sense of using the tools of social networking, peer mentoring, and dynamic cross-cultural interaction. Students, physicians, and patients will share insights about the complex dynamics of human nature and simple solutions in the seeds of interactive communication. We present this exploratory platform as a seed with which all can engage. In every seed the answer to growth is contained from the materials that are within it. For the seed to reach its full potential,

A. Price
Oxford University, Oxford, UK

Edmonton UK, N182Ha, Miami, FL, USA
e-mail: dr.amyprice@gmail.com

S. Chandra
Kasturba Medical College, Mangalore, India

K. Bera • T. Biswas
Medical College Kolkata, Kolkata, India

P. Chatterjee
University College of Medical Sciences and Guru Teg Bahadur Hospital, New Delhi, , India
e-mail: mail@pranab.in

R. Wittenberg
Georgetown University,
Washington, DC, USA
e-mail: r.e.wittenberg@gmail.com

N. Mehta
Stanford University, Stanford, CA, USA
e-mail: mehtan1@yahoo.com

R. Biswas, M.D. (✉)
Department of Medicine, People's College of Medical Sciences and Research Center,
Bhanpur, Bypass Road, Bhopal 462037,
Madhya Pradesh, India
e-mail: rakesh7biswas@gmail.com

J.P. Sturmberg and C.M. Martin (eds.), *Handbook of Systems and Complexity in Health*,
DOI 10.1007/978-1-4614-4998-0_44, © Springer Science+Business Media New York 2013

and to create a complex forest it must first die and be reborn as a new beginning that will take root and grow with the ability to reproduce after its own kind. We now present to you the journey of our planted seed, and our pathways to transcend complexity in order to meet human need as the journey have evolved for us. We share our story in the hope that this evolving innovation will inspire you to build on our emerging foundation.

As you read, defining ideas will spring up within you. We exhort you to act on these ideas so that together we can overcome the challenges of complexity with the simplicity of clear communication, peer support, and shared empathy. Above all our message is that we as healers choose to communicate openly with respect and compassion with all the stakeholders in user-driven health care (UDHC).

The word "complexity" in itself encompasses its definition. For example, Bonavita and Simone [2] define clinical complexity as:

> Clinical complexity encompasses multiple levels, including all the disorders and conditions experienced by a person along cross-sectional and longitudinal contexts, the diversity of severity levels and courses of clinical conditions, but also the plurality of values of people experiencing health problems and seeking help for them.

To put it simply; "*The evidence never makes a decision, people do*" [3]. Choices are based on the understanding of evidence and how it is intertwined with needs, financial constraints, policy, and culture. Maskrey shares with us that the problem is complex rather than complicated and helpfully points out that complicated problems can be distilled into stages where each element can be tackled whereas a complex problem is infiltrated by external cultural or other influences. "*Attempting to fix complex problems with an approach suited to complicated problems leads to frustration*" [4]. Complex problems are compared to the art of bringing up children and how the intervention used for one child can produce different results in another yet learning can continue to take place, suggesting adaptive mechanisms can improve complex problems over time. Education and communication combined with tools to build decision-making skills are suggested as first line approaches [5].

We tend to deal with complexity with a stepwise mentality better reserved for complicated rather than complex interactions. In real-life, ambivalence, human preference and cognitive dissonance may compete with the medical solution on offer. Often we are hampered by limited resources, a lack of training and the pressure of time. This is where the dynamics of human synergy and shared solutions can contribute.

In order to demystify the definition for our basic understanding and the context of this chapter, we can break down this definition into sub-components. The "multiple levels" denote that clinical complexity is the layered manner in which variables in the health system interplay with each other to form a clinical picture. These variables can alter an otherwise "defined" picture and contribute to a change in the manner it presents, its outcome, or even the way it is viewed by the participant or the observer.

To illustrate an example, imagine you are an attending physician at an outpatient clinic. A 50-year-old lady comes to you with a complaint of headache for the past month and seeks your opinion on the same. As a physician, you would have several differential diagnosis lined up as soon as the patient mentions the term "headache." However, several other factors now come into play, such as:

1. What social background does the lady hail from, would she be compliant with the treatment plan you are considering?
2. Would she be able to afford further testing, is she under a particular insurance plan?
3. Does she have a family history of a particular disease (brain cancer/migraine) that you/she are not aware of?
4. Are there members in her locality/family who have recently had similar symptoms?
5. What other comorbidities are present that might contribute to the causation/exacerbation of her current symptoms?
6. Does the patient have any genetic variations which might influence the way the prescribed drugs act?
7. Are there any cultural practices which may influence your treatment plan?
8. Does your patient take alternative medicine remedies which may influence the course of the disease pathology?

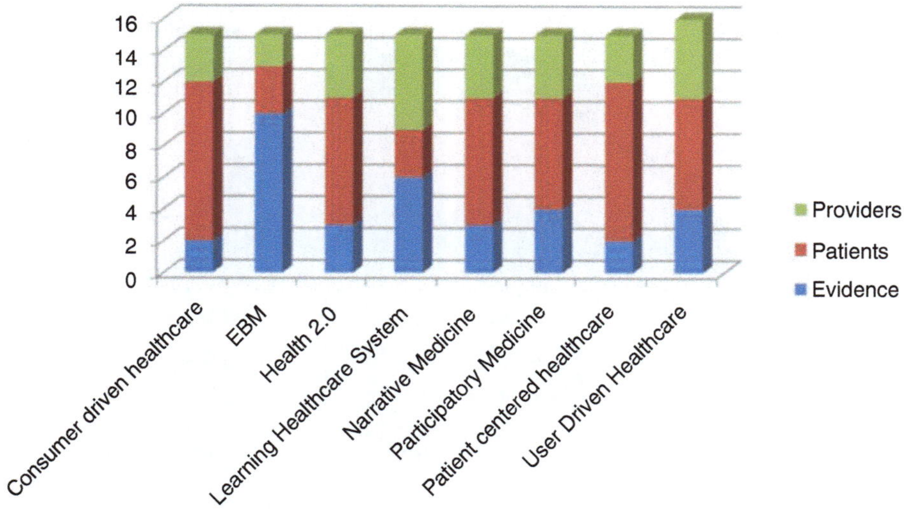

Fig. 44.1 Relative emphasis on three healthcare drivers: providers, patients, and evidence among various current concepts in health care. Reproduced with permission from [7] Ross, S (2011), A Lexicon for User-Driven Healthcare, *International Journal of User-Driven Healthcare,* 1(1), 50–54

Therefore, a simple "headache" could be the outcome of fatigue or the harbinger of a malignant process, or any number of other possibilities. In addition to the diagnostic uncertainty, the clinician needs to consider socio-demographic profiles, disease pathology, comorbidities, epidemiologic trends, genetic variants, socioeconomic resources, and various other factors. The fluidity of variables necessitates that the improvised solution is as dynamic as the problem. The outcome of the disease process depends not only upon the diagnostic accuracy, but also on whether the clinician addresses these contributory variables.

For instance, the patient can be negatively or positively affected by remarks or observations made by the staff we employ or by ourselves using language that is not empathic or even may be pejorative. In the above scenario, the patient may be asked what her "complaint" is which can cause her to say nothing as she does not want to be labeled as a complainer or to become defensive and seek to justify her position. Although it is accepted medical language, is it wise communication? In the effort to conserve time we often ask a patient, "What is your problem? What have you been doing to yourself lately" Patients may answer but others may take this as the physician inferring that blame and responsibility is laid on them. Another way to phrase this could be "Thank you for coming, what brings you here today?" or simply "Tell me what your concerns are?" The patient is now free to communicate unencumbered by blame. A simple language change to get the same information may facilitate better outcomes. In interactive groups we can discuss and practice the effectiveness of changing conversational strategies to gain maximum effectiveness with minimal input.

In order to deal with the evolving complexity that pervades health systems, the "user-driven healthcare model" is slowly evolving in shared online platforms between patients and clinicians alike. UDHC brings improvements in healthcare services by incorporating a concerted approach where clinical problems are solved through asynchronous discussions between "users" in a collaborative interface. This may entail a complex Web 2.0 interface or even a basic group discussion [6]. The "user" in such a case may be a medical student, a health care practitioner, an advocate, a patient, or a caregiver.

What sets UDHC apart from evidence-based health care (EBHC) is that while EBHC mainly concentrates on preexisting data or literature to reach a conclusion, UDHC tries to propagate both preexisting evidence as well as individual inputs from doctors or people who may have prior experience in contextually dealing with the same topics (Fig. 44.1).

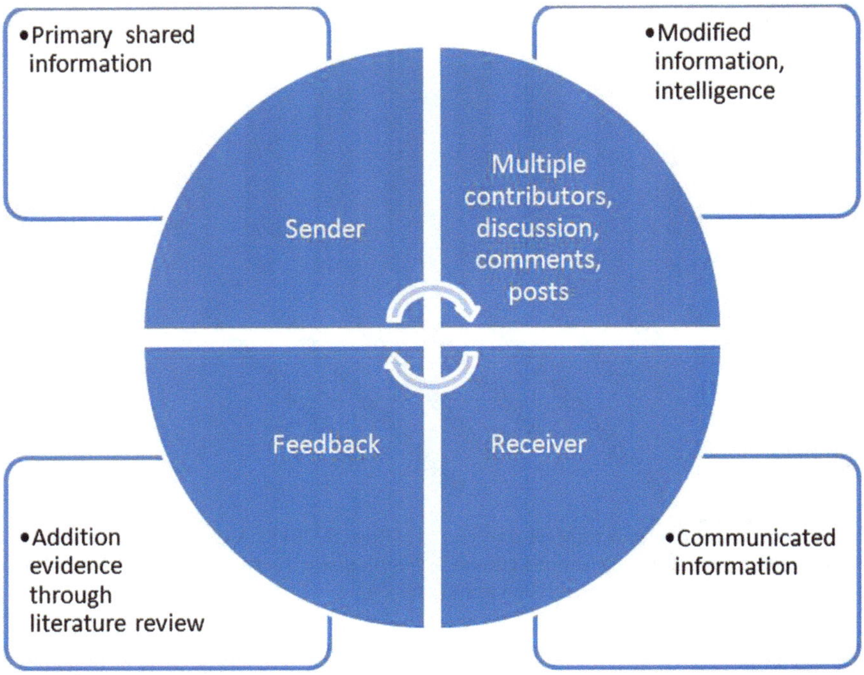

Fig. 44.2 Diagrammatic representation of the flow of information in a user-driven network

The UDHC hypothesis is that interactive conversational learning between multiple stakeholders in health care (such as patients and health profession students) who present a topic related to a specific uncertainty or a general uncertainty around treatment decisions can create stimulating secondary learning resources in medicine [8].

The flow of information in UDHC is free from the restrictions that exist if one were starting from a traditional clinical perspective. Instead, it asks "users" to chart their own path in solving a clinically complex problem with the information flowing in multiple trajectories. However, the beauty of this approach is not only does it enable free-flowing conversation which might explore otherwise neglected areas in the traditional structured way, at the same time all the multiple trajectories ultimately converge at a clinically significant conclusion (Fig. 44.2).

In the information to follow, we discuss sequentially our individual experiences with clinical complexity as users in the form of narratives and the solutions we have developed in practice. The common themes of the solutions we propose have been developed in response to the "felt need" of the users.

To enumerate:

- The solutions discussed are *simple, easy to implement, and avail of existing resources* which require little or no financial investment.
- The interventions are *targeted at various levels of users* ranging from medical students and healthcare practitioners to healthcare consumers and advocates.
- They promote *collaborative discussion and problem-based learning* among the various subgroups of the healthcare system and thereby maintain a dynamic discussion base.
- The user maintains the momentum of the knowledge translation
- Brainstorming among users in a *moderated and supervised environment* helps the group to *ideate, innovate, and implement new solutions* which are patient-centric due to a *shared pooling of resources* such as knowledge, expertise, and experience.
- There is no solution which is "absolutely right." Clinical complexity does not have a "unified theory" and therefore, the *solutions*

are as versatile as the problem at hand. Individual users can modify them as per their requirements.

- The key to dealing with complexity is each individual's willingness and *ability to evolve*. The user is provided with the tools to adapt the flow of information to his/her needs.

44.2 Individual Experiences with Clinical Complexity

In the study of EBHC we find evidence gaps in the bridge between knowing what and knowing how. This is coupled with the patient's expectation that the medical provider will know what to do and how to accomplish healing. A patient seldom understands the gap between knowledge, experience, and relationship. The physician may be expected to be responsible for establishing a relationship with a patient within the confines of the 10 minute appointment segments commonly reimbursed by funding bodies. There seems to be no universal measure to which patients can refer to for a standard of care Patients are left to practice survival in his/her own way.

44.2.1 Personal Experience Narrative

Dr. Amy Price

44.2.1.1 The Patient as a Dynamic Rather than an Object Problem

I do remember after my hand was run over I noticed molds in the office that could be used to lengthen the tendons and straighten inflicted limbs. While waiting for the nurse (a long time) I tried them on and found ones that fit. The nurse came in and said *you can't have them they are consultant X's and he saves them for special patients.* I said *I am special, please give him my thanks.* I further explained, *I need them more than he does and they are not his, they are paid for by the taxpayers. Please, if he wants these few out of this whole drawer let him come and get them.* By this time there was a man standing in the door who quickly disappeared (I found out later it was consultant X). At my next visit he was there and asked how my independently prescribed treatment was working out for me. I thanked him.

Sensory contact beyond the clinical space opens new cueing areas and demonstrates that a patient is a dynamic being rather than an object problem to be solved. I always read the posters, play with the models and visit the lab if I have to go for medical care and as a clinician I provide people with interesting educational tools while they are waiting to see me.

In the training for psychology, there are multiple papers, minimal lab work and little patient contact. I felt such joy when I held a human brain in my hands and later saw brain slides in three dimensional spaces. My classmates generally hated the lab and were content to explore rat brains or write papers about the morals of human experimentation but for me a real brain was like a world into tangible knowledge. Patients and providers have varied interest and learning curves and this makes collaboration and dynamic networking a valuable addition for medical care.

In my first clinical experience I was thrust without supervision into a mental health ward. Treating mentally ill institutionalized patients' in real time was so different from reading about them in books. I would have welcomed a security net of conversational learning, some mentor to novice interaction or even peer-to-peer support but at that time in history it was not available.

44.2.1.2 Information Bulking Due to Clinical Complexity

In this world of complexity, medical professionals deal with far more than relationship, treatment, and diagnosis. They wrestle with constantly changing reimbursement codes, new procedures, changing patient expectations, rules of engagement, and legal complications. In personal injury litigation cases this is even more pronounced. Many physicians require full separate offices just to deal with the paperwork and complications that these cases produce. Industry lawyers often load up the case with extraneous paperwork if it is likely to go to litigation. This lessens the likelihood of anyone looking at the case in depth and reduces the pool of lawyers that would be willing to engage to assist the patient. The doctor is frustrated because patient care is threatened along with physician reimbursement. Additionally, treatments are questioned by those who are administrators without the full qualifications to practice medicine and the unrelenting tension can increase stress to potentially slow the patient's rate of healing.

I would suggest that a major overhaul is in order to avoid excess charges and the squandering of precious professional time. It would be good to investigate how expenses could be standardized while insisting that bulking up data reports be penalized and managed because in the end it does seem to be about managing knowledge. This is best accomplished in a collaborative, dynamic setting where stakeholders can declare their interests and propose solutions that could then be ratified by all parties.

For instance, in my case, with over one million US dollars in medical expenses, the defense insurance companies managed to bulk this up with five filing cabinets with five drawers each of reports/depositions and other non-medically necessary paperwork. In reality this could be condensed to about 40 pages of real information. This is where students fortified through Cochrane-based journal clubs and evidence-based medicine strategies can excel by cutting through bad science and extraneous information to extract for their patients the best available intervention in the shortest possible time.

Recently, I was invited to do an appraisal of a critical review on Traumatic Brain Injury for a Students Cochrane Journal Club (Informer). This was a wonderful experience. I met dedicated medical students who were committed to evidence-based learning. I found their insight as students surpassed that of many practicing clinicians. After the review to my surprise several physicians faced with an unexpected traumatic brain injury case took the time to e-mail me and ask for my input on best available evidence.

44.2.1.3 Dynamic Solutions for Clinical Complexity

It seems there are misunderstandings and unspoken assumptions with present day diagnostic interventions and expectations. This erodes patient trust and physician confidence and competence. As a patient and as a clinician, I would prefer to support open access for publications and peer-to-peer support groups where difficult cases could be discussed and multiple options captured by collaborating with others. In this way doctors and patients can escape the narrow thinking that comes from imperfect diagnostics, litigation fears, and bureaucratic roadblocks. I would prefer my physician was emotionally supported by colleagues as they navigate difficult decisions with limited options.

I experienced a troublesome incident with my granddaughter that resulted from inexperience and rigid dogma. She presented with double pneumonia but her parents were sent home with only cough syrup and X-rays were denied. They refused treatment saying the condition was routine and not called for. The parents called me and the child was clearly close to respiratory arrest. I took her back to the hospital immediately and demanded a specialist consultant. The child was found to have serious pneumonia bilaterally; she required oxygen for several days, drainage, and extensive treatment. In a less medically observant family atmosphere she would have faced death.

There are also times when diagnostic treatment is a needless extra expense and contributes nothing to treatment. In my own experience two low-resolution MRIs were ordered for my cervical spine. They were of such low quality that the surgeon wanted a third. I refused consent since no useful purpose for the first two had been established and they were costly. After looking at the medical bills I also eliminated other unnecessary provider visits like follow up and to receive test results. My solution: mail the test results and if I need an appointment I will call. I treated follow up the same way since the visit cost me several hundred dollars each time on a limited income. I read the instruction sheet, and informed all that if anything happens I would check in.

We would be well advised to consider the maxim adopted in the UK for considering patient input in the process of decision making "Nothing about me without me!" [9].

44.2.2 Clinical Complexity and Listening

Ralph Wittenberg and Amy Price

When people notice that they are not listening, the question is "What is getting in the way?" In this context, listening has two meanings, which happens frequently between parents and children. Listening in this context actually means obeying. A lecture on the benefits of listening is not nearly as effective as discovering what the attitudes in the listener are that cause them some kind of discomfort. This is exactly what Freud discovered over a century ago.

In formal psychoanalysis, patients were given only one instruction and that was to say out loud whatever was on their mind at the moment. Of course, no patient could do that. Early on, in his frustration, Freud squeezed his patient's heads like a tube of toothpaste. But then he realized that there was an unconscious element getting in the way (what is referred to as resistance). At that point he would ask, "What was making it difficult to talk?"

An infinite number of answers would come out, all of which were exactly right for a particular patient at a particular time. This began the process of accessing the unconscious. For example, patients often would say that they were embarrassed to not want to obey the doctor and feared that he would get mad if they revealed it. This was perfectly reasonable, but even better; it might be the gateway for the patient to become aware of even greater mistrust of the authority figure.

There are two main forms of authority. One is the "father knows best" version and the other comes from a person's depth of knowledge and wisdom, which the recipient is then free to question or even accept. If someone believes they know what's best for the "great unwashed," he or she has to be willing to give up the comforting illusions of omnipotence or omniscience in order to change. I do not think the vast majority of experts are aware of their feelings.

This is a widespread problem in health care. The higher you are in the pecking order the greater the temptation to know it all. This is manifest in the idea that doctors believe that they should be able to keep all of their patients from dying. They act like they know just when it will or will not happen. The problem arises in the fact that patients die anyway, in fact all of them. This makes many doctors feel guilty or inadequate when it happens. I think this is a major factor in "burnout" (Wittenberg [10], Personal correspondence).

44.2.2.1 Environmental Complexity Shapes Decision Making

Our understanding of patients can change because of the context in which we see them [11]. Most of us have experienced recognition difficulties when we find a familiar landmark changed or moved to an unspecified location. Likewise, medical professionals can be influenced by the environment or setting in which the patient is seen. For instance, when a patient presents with significant chest pain in an emergency room he/she is more likely to receive a diagnosis of myocardial infarct than the during a routine office visit. Likewise, awareness of typhoid fever pathology may not be as sharp in a northern climate unaccustomed to international travelers and it is often not picked up because the physicians have never been exposed to an actual typhoid fever patient. In a treatment scenario physicians may treat snakebite cases noting high side effects and limited efficacy and blame the antivenom when perhaps the quality of the antidote was compromised in production or stored improperly. In scenarios like these, interactive shared decision making across social media can make a difference and really help patients and medical professionals alike.

There is impressive medical information for low and high resource countries alike made available through shared collaboration and decision making in the format of medical Wikipedia. James Heilman shares how available information and collaborative effort allowed him and his peers to write about an unfamiliar topic to meet medical information needs. "*Yes it is a little strange that I who has never seen a case of dengue fever during my 10 years of practice have contributed extensively to an article on the topic and one that gets over a 100,000 page view a month. This is due to the platform on which I write (Wikipedia).*" Heilman has orchestrated meetings with medical schools and universities on multiple continents to share his vision of providing free medical information access for all through the efforts of Wikipedia [12].

44.2.3 Generating Positive Behavioral Change in Cultural Complexity[1]

Ralph Wittenberg

Dr. Ralph Wittenberg allowed us to share the post he contributed in HIFA 2015 as he responded to behavioral change for the problem of genital mutilation. All agreed change was needed but were conflicted about ways to move forward in order to enact change.

Wittenberg shares that the simple answer is we don't change anyone. They have to change themselves just as we do. The real question is how you create the conditions under which people will want to do things differently. The "Fathers Now" program in New Jersey is an intervention where men are taught by example and mentoring to be the fathers they never had. As proof of concept it is relevant that every man in the "Fathers Now" program was recently released from prison yet with this intervention the expected recidivism rate dropped from 60% to 3% [13] (http://fathersnow.com).

People have been trying to make criminals behave from time immemorial. This is done by imposing external standards and punishments, for example. Since we have two million people incarcerated in the United States at this time it is a system that clearly doesn't work. But that does n't keep people from doing the same old stuff again, and again, and again.

I'm a psychiatrist who's been in practice for fifty one year's whose job it is, supposedly, to change people. One of the things I must say out loud at least every couple of days is, "People are not rational." By that I mean they do things for reasons that they are unaware of and aren't influenced by whether what they do actually works or not. This is certainly true for the way people raise children. Most often it's based on a system of fear and punishment, or should I say terror. "You have to put the fear of God into children," "Spare the rod and spoil the child," and finish up by saying, "That'll teach you!" Yell at them saying things that are so horrible you would never say them to a grown up. Most often you get children who may be outwardly compliant but inwardly defiant. With little children this defiance is often expressed in wet or soiled diapers or a variety of physical symptoms from skin rashes, to digestive problems, to asthma. When they grow up they are plagued with constant unconscious guilt which corrupts their sense of self-esteem.

If you pay attention to long-term trends there is a lot to be optimistic about. In the Middle Ages half of the male population died in battle. As recently as World War I, one million men were killed in one battle in France. In World War II there were almost 20,000,000 military casualties alone. Fast forward to the Vietnam War. In 7 years the US military only suffered 60,000 casualties. In the 10 years of the Afghan war there are less than 10,000. There are also far fewer wars, even though it seems like there's war all the time. In the last 150 years there have been dramatic changes in the status of women. They did not get the vote in the USA until 1919. As recently as the 1970s, in the state of Virginia, wives were considered chattels.

When I went to medical school there were only three women enrolled in a class of 75. Presently, women comprise over 50% of medical school classes. "Women's Lib," which established women's personal freedom, especially sexual, did not begin until the 1970s.

These are trends that have taken a long time to develop and have required passionate people willing to promote them at their own personal risk. As much as people say they want change it is always scary. The status quo is known. To change it means to face the unknown, which is very hard for most people to do. The desire for change is internal. Just look at a baby learning to stand and walk. There is no stopping him or her. As many times as they fall they get up again. There is nothing like the look of triumph on their faces when they succeed. They do not question

[1] This section was edited by Amy Price.

their ability or the danger. This is true of the vast majority of children, but there are some who have already had traumatic experiences that inhibit this wonderful mastery of the world.

There are many negative influences in the world; the people who were supposed to fix them do not, and often they are, in fact, the ones creating the problems.

In the field of medicine the greatest advances have been the results of people whose passion to change things included risking their own life or reputation. Sir Edward Jenner who inoculated himself with cowpox, Walter Reed who exposed himself to yellow fever by allowing mosquitoes to bite him, Marie Curie, who died of radiation poisoning from this miraculous substance, radium and it is X-rays which had tremendous potential for the diagnosis and treatment of disease. Amazingly enough, then, one person, like you, can change the world. This runs contrary to the notion that you have to have huge enterprises to bring about change.

44.2.4 Answers for Complexity in Dynamic Networking

Amy Price

Negative experiences shared in the chapter are in sharp contrast to the caring concern of a group of practitioners who meet in an interactive forum to discuss patient care. Sometimes these doctors will share test results, medical films, and treatment to date with their patient's full consent. At times the patients themselves are videoed or speak with other physicians through Skype.

Recently, in a forum of these medical practitioners I was asked for my input about cognitive interventions and mental health. The conversation was shared with the patient's consent. I was afforded a window into how challenging the conditions were for hardworking physicians in low resource settings. One question was asked about how to keep insulin cold when the village had no refrigeration available. The answer was that commonly the insulin was kept underground clay urns. The doctors pursued saying this was not good enough and as a team they worked out solutions to invest in refrigeration for the village by initiating microenterprise ventures and through the innovative use of solar panel technology.

Another situation was not as productive. Here a young woman died despite the best efforts of Internet consultancy; sadly it was not through a lack of medical expertise but rather because of the lack of resources. Even I with no medical school training was searching for a solution to help the patient fight for life. Learning and fighting together forms an invisible bond created by the need for a solution and the possibility of imminent failure. With collaboration none of us had to shoulder the grief alone, there was comfort in the realization that although the outcome was a failure we as a team gave our very best shared resources. The grief was palpable in the group and although not much was said there was an implicit support and the absence of the blame culture so prevalent in many fields.

On another occasion I was impressed as these dedicated professionals discussed how to solve the clinical problem of a young mentally disabled woman who took off her garments publically because of the irritation of menstruation and a lack of understanding for the social consequences. The suggestions were practical, affordable, and doable. More solutions were discovered in cooperative synergy than any one medical professional could hope to access alone.

I was later invited to explore and contribute to an online group called Tabula Rasa. Doctors, consultants, and medical students in training offer vignettes of real patient conditions complete with lab results, imaging, and medical history. Everyone works together to solve the medical complexities and to suggest best ways forward. In one example, I was astounded to see the brain image of a man where a glioblastoma had taken over roughly half of his brain. This patient was asymptomatic until just the day before when he fell off a ladder and was brought to the hospital. Another patient carried the tragic story of being offered a donor organ only to find after the transplant that the donor organ was infected and it cost this man his life. Real people where things have gone wrong in medicine are best met first on an online platform replete with mentors and peer support so that when the student sets off in

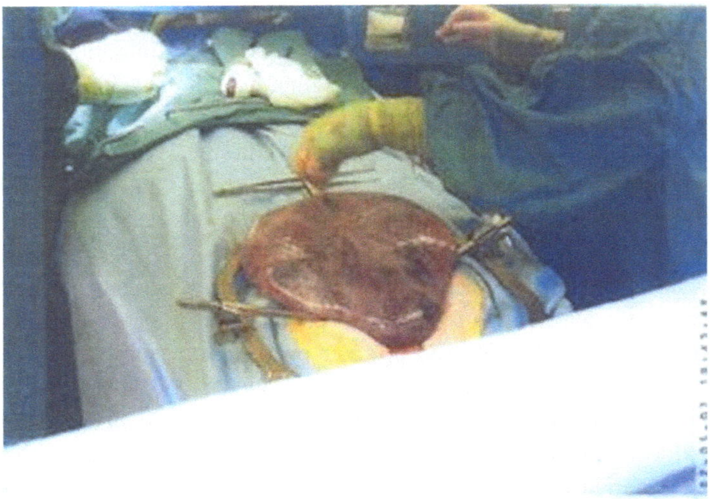

Fig. 44.3 Peg's surgery: The tumor was a sex cord stromal tumor which was confined to the ovary (Stage one). During the surgery, the surgeon evacuated 6.5 l of fluid and a tumor which when decompressed measured 30–35 cm across (Personal photograph used with permission from Peg Ford)

medical practice they will have lived through the experiences no textbook can teach with the supportive supervision of those that mentor them.

Another group of physicians battled for the life of a 13-year-old lupus patient. After exhausting all known forms of evidence and after seeking advice from professional contacts they turned to the grueling task of a 4 h search through electronic medical records to find patterns that could point to successful outcomes for this condition. Their young patient recovered. In expectation of the criticism to come for using what is considered lower level evidence their response is profound.

> Did we make the correct decision for our patient? We may never really know but in the light of experience as guided by intelligence and in the practice of medicine, one cannot do better than that. [14].

44.2.5 Dealing with Complexity: How Patients Can Help

Amy Price and Shivika Chandra

Peg Ford is an ovarian cancer survivor who was fortunate to be diagnosed at an early stage (Stage 1: See Peg's tumor, Fig. 44.3) and was referred to a gynecologic oncologist who performed a life-altering surgery on her. Now almost completing her fifth year of NED (No Evidence of Disease), Peg has transformed herself from a patient survivor to a patient research advocate, championing the cause of early detection and cure of ovarian cancer for *ALL* women.

Her efforts to raise awareness have led to the implementation of the "Survivors Teaching Students: Saving Women's Lives" [15] program of the Ovarian Cancer National Alliance at the UCSD School of Medicine. In this teaching module, patient survivors share their personal stories with third-year medical students to facilitate experiential learning and supplement it with evidence-based decision making, thereby sensitizing young doctors to the topic. Peg also has teamed up with a local gynecologic oncologist, Dr. Afshin Bahador to participate in his CME (continuing medical education) grand rounds in major hospitals.

As a patient advocate, Peg and others like her help bridge the gap between doctors and patients by providing a collaborative channel of communication. The involvement of informed patients in the education of healthcare professionals helps to mold the young doctors into more empathetic physicians who can then in turn empower their patients to be more proactive in healthcare-related decisions. This is a simple yet powerful way to decrease the development of complexity at the grass root level itself.

44.2.6 Anecdotes from the Wards: A Medical Student's Introduction to Clinical Complexity

Shivika Chandra

Clinical rotations have always been my favorite part of medical school. While mastering the basic sciences and performing procedures in the laboratory are an integral part of our medical curriculum, the first time I felt aware of my responsibility as a "doctor" was when I entered the wards and understood my role in the healthcare system.

During one of my Pediatric clinics, a 6-year-old child with features of moderate dehydration secondary to acute gastroenteritis was brought in by his distraught mother. As my attending examined the child, he asked us to evaluate what form of rehydration we would like to consider. Suddenly, all eyes were on the child as we tried to incorporate the weight–height formulae, the percentage of various oral rehydration solutions (ORS) and to assess the degree of dehydration. The irritable sick child needed reassurance and overwhelmed by this unsolicited attention, he began to cry. He was promptly whisked away to the wards for further treatment. The attending instructed us to go read about rehydration and come back the following day with the answer, the catch being we had to look for an answer "beyond the textbooks."

Caught in a quandary, many of us visited the library to seek answers in dusty journals which were archived in the library back volumes. Having done a research project sometime back, I decided to use PubMed to seek the answer and was rewarded by a number of studies which addressed the issue. I finally chose a Cochrane review titled "Oral versus intravenous rehydration for treating dehydration due to gastroenteritis in children." Since this was a systematic review, it collated the data from a number of studies and the text of the entire paper was freely accessible in India (under the patronage of the Indian council of Medical Research). The review showed no obvious difference in the oral versus intravenous routes of administration of fluids, but noted that 1 in 25 cases treated orally would fail and requires intravenous fluids.

The next morning as I presented the findings in clinic, not only did I impress my attending, but I also realized how research was not an entity onto itself. Incorporating research into current clinical practice, as a physician enabled me to provide the highest standard of care to my patient.

After my presentation, I was given the privilege of being the acting intern for the child. Over the course of 3 days, he received oral fluids under the watchful eye of his mother and was closely monitored for signs of dehydration. He quickly recovered, and on day 4, as he chased me around the ward during our morning rounds, my attending made the decision to discharge him. I was given the task of educating the patient's family and with the help of evidence, WHO guidelines on Oral Rehydration Solutions and Charts, I explained to the patient's mother about how she could take simple preventive measures against gastroenteritis and administer ORS to the child at home.

The experience helped me learn the following points:

1. Clinical guidelines for even the most common problems are constantly updated in the face of new evidence, as a physician our challenge is to keep abreast with the current trends and at the same time make decisions individualized to each patient.
2. Accessibility to research is a universal right for healthcare providers.
3. Academic centers of excellence which focus equally on medical training, education, and research are the need of the hour.
4. Access to research is not enough; students need to learn how to utilize the resources available.
5. By involving students in clinical practice from an early stage, their knowledge-seeking behavior is encouraged. Patients provide a better stimulus to learn than textbooks and didactic lectures.
6. Evidence needs to be critiqued before being applied in patient care. Statistics and research-based outcomes do not always translate into patient-relevant outcomes.
7. Supervised research projects, journal club discussions, and clinical discussions during hospital rounds can improve the students' capacity to search and gauge the data available, analyze its quality, and apply it to the clinical problem at hand.

44.2.7 Snapshot of Clinical Complexity in Primary Care

Rakesh Biswas and Shivika Chandra

The following case vignette (Fig. 44.4) was shared by Physician A. among the members of a

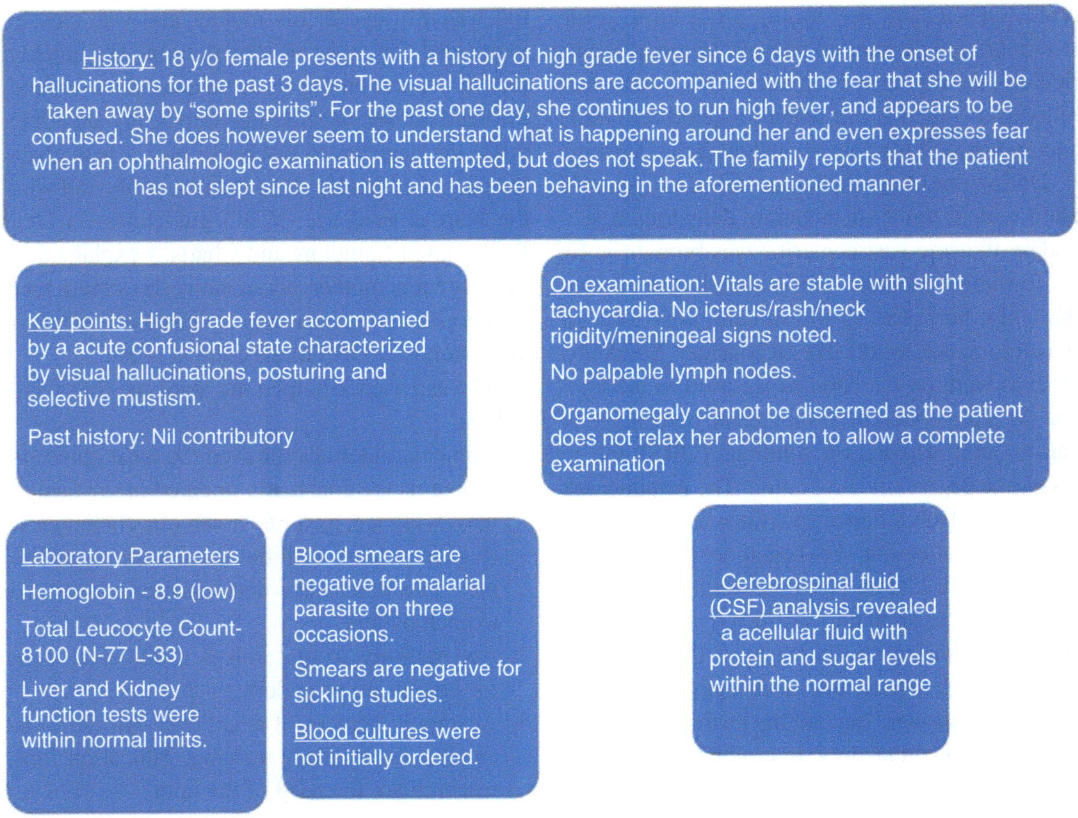

Fig. 44.4 Case vignette from Indian rural primary care

clinical discussion forum. Dr. A. is a practicing clinician, initially trained as a pediatrician, who now practices in a rural setting in the Ganiyari region in India as a primary care physician.

Based on the above presentation, Dr. A. made a provisional diagnosis of an *acute confusional state with fever* and started the patient on an antibiotic (ceftriaxone) suspecting enteric fever. Another diagnosis he considered was of an *oculogyric crisis* but since there was no history of antiemetic use and the patient failed to respond to the standard treatment for this condition (50 mg of promethazine), this diagnosis was discarded.

Dr. A. asked the other physicians on the network to aid him in solving this diagnostic dilemma, i.e., "*how should one approach this patient?*"

Dr. B., an intensivist who works in an urban hospital setting in the USA, responded that while searching for evidence which was best tailored to answer the current diagnostic dilemma, he came across a paper titled "Acute febrile encephalopathy in adults from Northwest India" [16].

Based on the causes listed on the study, Dr. B. then suggested the following differentials and plan:

1. Ideally, brain imaging studies (CT or MRI) to rule out a brain abscess.
2. Among viral encephalitis—the only treatable condition is herpes simplex virus encephalitis (HSVE) which is managed with intravenous acyclovir and symptomatic treatment with anti-epileptics as required. Dr. B. had come across HSVE in his practice and had noted that 5–10% of HSVE can have normal CSF initially, serial CSF analysis may be positive [17].
3. He also reasoned that cerebral malaria (CM) could present with a negative peripheral smear.
4. He ruled out the differential diagnosis of leptospirosis as no hepatosplenomegaly was present and no lymphadenopathy and/or rash were noted.

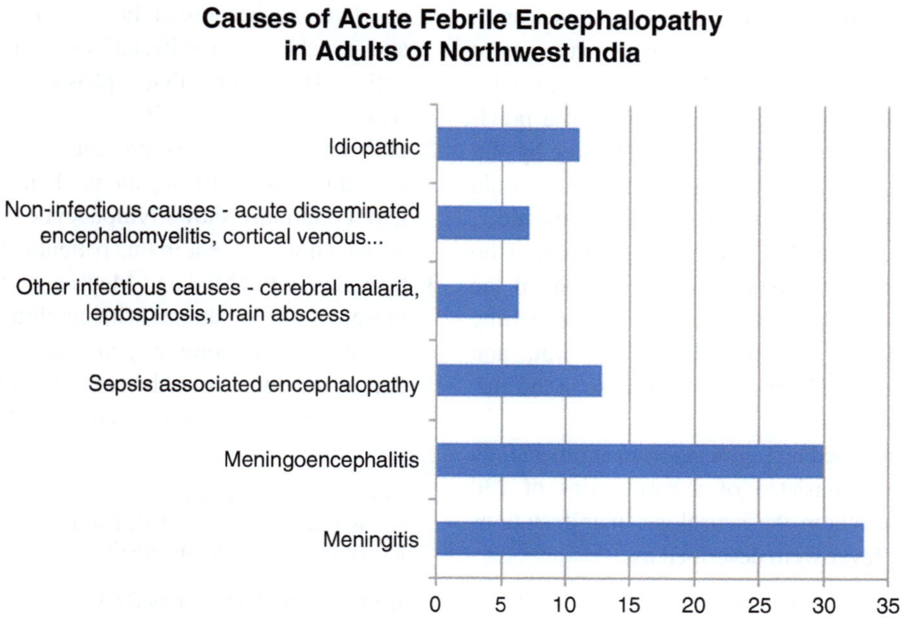

Fig. 44.5 Diagrammatic representation of the causes of acute febrile encephalopathy based on the study by Bhalla et al. [16]

5. Keeping in mind the normal CSF study, he also ruled out the possibility of tuberculosis leading to this presentation (Fig. 44.5).

In a nutshell, based on the evidence at hand, Dr B. suggested the following course of action:

1. Empirically treat the patient for HSVE and CM and anti-epileptics.
2. If patient does not get better despite treatment, consider brain imaging studies.

In response, Dr A. clarified the following:

1. In his practice, patients with severe malaria with cerebral manifestations rarely presented with a negative peripheral blood smear. Also, in this case, the smear was checked thrice for malarial parasites and was negative consistently.
2. The patient was in an acute confusional state and there were no seizures or loss of consciousness reported to date.

Course of action: He agreed that there was a strong possibility that this could be a presentation of HSVE and would initiate treatment with acyclovir and follow up on the imaging studies. He narrated his experience in treating a symptomatic patient with a negative CSF study with acyclovir, and the patient dramatically improved within 24 h of starting the treatment. Dr A. also shared that in a rural setting, the cost issues had to be considered as acyclovir is an expensive drug.

1. After analyzing the risk:benefit ratio of administering an antimalarial, despite the negative peripheral blood smear, he was in favor of starting treatment with artesunate (antimalarial) to rule out the possibility of cerebral malaria, no matter how rare.
2. As Dr A. stated, the strongest likelihood was of enteric fever and a strongly positive Widal test seemed to suggest as much. Despite the test being strongly positive, Dr A. mentioned that he was unsure of the diagnostic utility of the Widal test and asked for the group's inputs on the same. His opinion was that the need of the hour was a systematic review which addressed this question.

Dr C. contributed to the discussion with some points at this junction. Dr C. is a practicing primary care physician in an urban hospital setting in India.

1. He highlighted the fact that though the Widal's diagnostic utility was in doubt, there was no alternative way to diagnose enteric fever (blood cultures in the best of centers yield 50–70% positivity).

2. He suggested that continuous patterns seen in fever charts for the first few days may suggest enteric sepsis (or any other bacterial sepsis for that matter). He gave a reference of a previously published study on fever charting by him and his colleagues [18] and suggested it could be adopted as a supplemental diagnostic aid.
3. According to Dr C., very often patients who presented in a manner similar to that of the lady in the case vignette did end up getting treated for multiple (bacterial, viral, and malarial) etiologies as was the case in the current case.

Dr B. then pulled up segments from another study [19] which consisted of a case series of 230 patients in whom the neurologic manifestations of enteric fever were described as:

> Neurological manifestations were seen in 63 (27.1%) patients. Of these, 27 (42.8%) patients had typhoid delirium state and 36 (57.2%) had specific neurological complications. Among specific neurological complications, encephalitis (25%), psychiatric manifestations (19.44%), cerebellar ataxia (19.44%), and meningitis (13.89%) were the dominant features.

Dr A. also gave his inputs regarding the diagnostic dilemma:

1. He felt that despite the massive levels of gastrointestinal infections (including invasive diarrheas) he came across in the rural setting, enteric fever was rare of late. He attributed this to the fact that rural India lacks a sewage system and in most cases of enteric fever, the organism thrives in the sewage system and it is when the drinking water supply is contaminated with sewage, humans become infected.
2. Dr A. also mentioned that his rural practice was equipped with a functional microbiological lab which made it easier to run diagnostic tests on fecal samples, etc.

Dr A. then asked the group *if altered sensorium (whether stupor/confusional state/delirium) were seen with severe malaria or HSVE.*

Dr B. responded with the following learning points:

1. After following the group's discussion and reading about enteric encephalopathy, he concurred that the diagnosis enteric fever was most suitable because of the normal CSF study and psychiatric manifestations seen in this patient. He reflected that "aphasia or mutism" was seen in 3% cases [20].
2. HSV encephalitis could present with psychiatric symptoms in 70% of patients [17]. However, CSF commonly shows lymphocytic pleocytosis which was absent in this patient [21].
3. Searches in PubMed and Medscape regarding presentations in patients with cerebral malaria revealed that neurologic symptoms like obtundation and coma could be seen. No acute psychiatric manifestations were noted in the spectrum.

Dr B. summarized as follows:

1. Working diagnosis: Enteric Encephalopathy
2. Differential diagnosis: HSVE

Progress Notes as Shared by Dr A.

"I started *olanzapine* on 19th morning (Day 4) 5 mg per day via nasogastric route. Got an MRI done the same day, and found a normal scan. Repeat CSF study was also normal.

After procuring Acyclovir yesterday afternoon (Day 3), she was started on it intravenously with 10 mg/kg q8h. After I stopped dexamethasone on Day 2, the patient had high fever which lasted from Day 2 evening till today morning (Day 6). After receiving four doses of acyclovir, today evening she has started talking. She asked her mother where she has been brought to

Is this acyclovir that is working? Or olanzapine? I believe it is the former, since the fever has also come down."

Post Follow up notes

1. Dr A. decided to arrange for funds such that the patient could undergo a HSV PCR to confirm the diagnosis.
2. He queried if acylovir is effective if started so late in the course of the disease (>48 h) as seen in this patient.

Dr C. reflected that it was difficult to say if this was the effect of acyclovir or a delayed effect of ceftriaxone. He cited a study titled "Herpes simplex encephalitis treated with acyclovir: diagnosis and long term outcome," [22, 23] to widen the discussion [23].

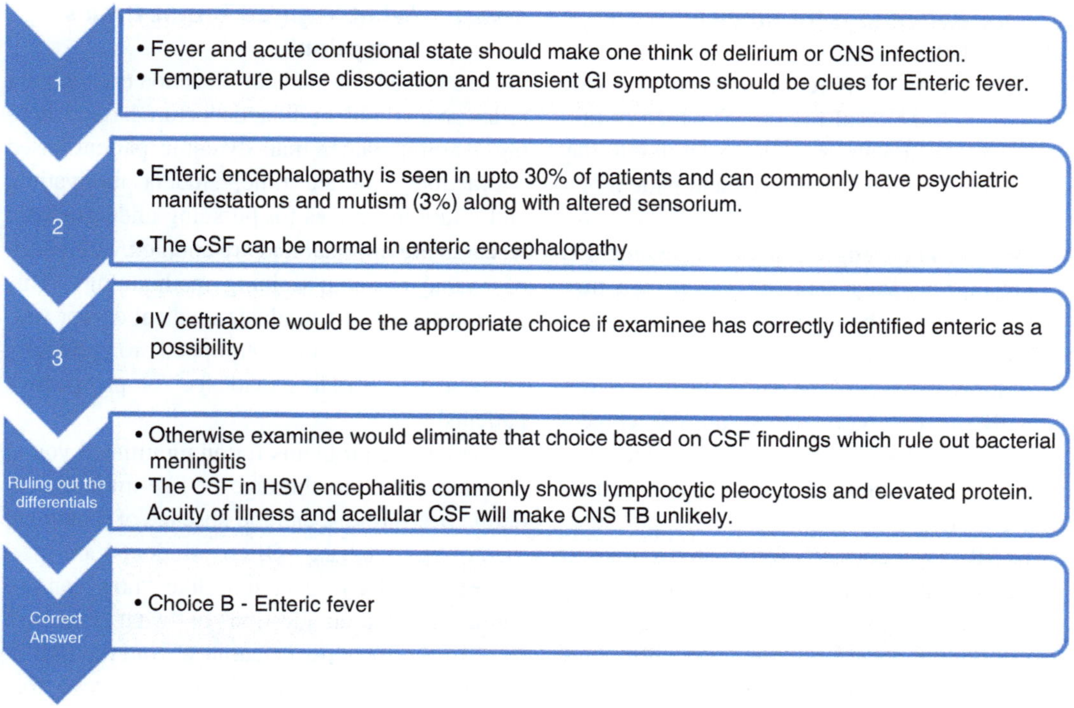

Fig. 44.6 Stepwise manner in deducing the answer to the clinical question

44.2.7.1 Helping Students to Learn from Our Learning

Based on the discussion, Dr B. created *a learning resource* for undergraduate medical students in the form of a clinical question which requires them to arrive at a diagnosis based on the laboratory and clinical findings discussed in the vignette**.**

44.2.7.1.1 Question

18-year-old previously healthy female presented with high grade fever for 6 days. Over the past 3 days she is experiencing visual hallucinations and fear that she will be taken away by some spirits. Since yesterday, she is having confusion and mutism. In the beginning of this illness she had loose stools for a few days that have resolved now. Her review of systems is negative other than these symptoms. On examination, she is inattentive and constantly looks around but is not agitated. She does not speak but appears to understand. Her temperature is 38.9 °C (102 °F), her heart rate 100/min, and her blood pressure 110/70. She has no neck rigidity, no rash, no lymphadenopathy or splenomegaly. Labs show WBC count of 8,100 with 77% polymorphonuclear leukocytes (neutrophils). Liver and kidney function tests are normal. CSF shows WBC 2 per HPF, glucose 3.9 mmol/l (70 mg/dl), protein 40 g/l. Peripheral smear is negative for malarial parasites. Blood cultures are pending. Contrast enhanced CT-scan of the head does not show any space occupying lesions or intracerebral hemorrhage. Which of the following is the most appropriate initial treatment for this patient with fever and acute confusional state?

(a) Artesunate
(b) Ceftriaxone
(c) Acyclovir
(d) Anti-tubercular therapy

Figure 44.6 highlights the stepwise approach the student might follow in deducing the answer to the multiple choice question listed above. Though the discussion itself offered various differentials and the option of starting multiple treatments at a given time, the question has only a single correct answer and therefore students have to analyze the data in the clinical stem comprehensively before arriving at the answer.

44.2.7.2 Reflections for Medical Educators

1. There is a stark difference between "textbook cases" and the "real-life cases" which healthcare professionals see in their day-to-day practice. Diagnostic accuracy and patient outcomes depend upon a multitude of factors—physician knowledge, experience, availability of resources (drugs, laboratory facilities), time of presentation, being a few.
2. **Clinical complexity** has been represented in the above conversations where the patient was finally treated with almost all the choices listed in the formulated multiple choice question (MCQ) meant for undergraduate medical education.
3. The MCQ was correctly formulated as a single response answer for *initial* therapy for the given case scenario. However, this approach to designing a medical education curriculum with necessary "straight answers" may delay the development of an ability in the medical student toward confronting medical pluralism (and its attendant uncertainty and clinical complexity)?
4. Many cases which present in the clinics lack well-defined symptoms and students should be trained to follow an evidence informed manner to workup the case and decide upon the right course of action that navigates around clinical complexities.

44.3 Local Solutions for Global Impact

This section deals with how interventions aimed at supplementing routine medical education can be used to handle clinical complexity.

44.3.1 Tabula Rasa: A Student-Led Conversational Learning Platform, Exploring the Nuances of *"Clinical Complexity"* in Medicine and Medical Education

Kaustav Bera and Tamoghna Biswas

44.3.1.1 What Inspired Tabula Rasa's Inception?

The undergraduate medical education traditionally relies on textbook or didactic lecture based teaching–learning rather than dynamic patient-based interaction. Allotted time, student motivation, and future incentives for pursuing undergraduate research are minimal and are coupled with a lack of formal research training methodology. This can translate into a graduate medical doctor who may be insufficiently equipped to critically appraise the available evidence for prospective patients.

Two major problems for an uninitiated young researcher are the shortage of like-minded academic peer groups, and the absence of a national mentoring database. Often, students are perplexed with basic questions like "how can we frame our research question" or "what is the protocol to get the ethical clearance from the IRB"?

The founders of Tabula Rasa believe that medicine, in its true sense, is something that reaches beyond the drudgeries of classrooms and textbooks and becomes a way of life. Learning in this dynamic and exponentially growing realm of science can be achieved with peer-to-peer interaction between learners, and with minimum communication barriers between mentors and students.

44.3.1.2 How Did Tabula Rasa Happen?

Although the concept of Tabula Rasa was brought in the confines of Medical College, Kolkata, first as a journal club which would foster interesting discussions among fellow medical students, which would give them insight, outside the formal curriculum, soon it became more than just another journal club as it amalgamated the inputs from students, doctors, and allied medical professionals from all over the world through an asynchronous system of discussion that could be spontaneous without the need of any rigid framework.

Tabula Rasa is an initiative which invites participation from everyone and anyone who has a Facebook profile.

With the omnipotence and omnipresence of Web 2.0, the membership increased rapidly from

30 students to over 300 members. The members presently hail from diverse areas of the world, from multiple institutions and they are at various stages of their professional development.

The bulk of members are undergraduate students, and Facebook group moderation is done mostly by them, however, the list of our members includes postgraduate trainees, residents, and faculty staff of various institutions.

Tabula Rasa has inculcated a way of conversational learning in the group, with the knowledge and information sharing platform being multiuser, asynchronous, and interactive. These include (but are not limited to) online clinical problem solving, reading real-life stories of virtual patients, and UDHC. Tabula Rasa engages in "self-teaching" and conceptualization through discussions. The initial purpose of the group, i.e., forming and maintaining a collaborative learning interface, is thus mostly fulfilled with the Web 2.0 forum.

44.3.1.3 Impact

Tabula Rasa was started with the motto: "*Per aspera ad astra*" (*Through difficulties to the stars*). Although the journey is just beginning and invites exploration into an alternative system designed for medical learning, in the short span of time Tabula Rasa has achieved much of its objective when it comes to giving students a fresh view on learning. By inculcating discussions as a method of learning and by also providing a platform where they get to interact with people who are at the helm of their respective fields, it has made them think out of the box for solutions to real world scenarios. Not only are they trained for future clinical encounters but they also come to appreciate the experiences shared by experts which could very well be faced by them, 5 years down the road. This mode of learning without any structured framework or guidelines promotes lateral thinking and "out of the box" approaches which are handy tools to possess when faced with the constant challenges of a practicing physician.

While the benefits of conversational learning are harvested, concerns about its potential loopholes remain. These include human bias in conversations which might adversely affect scientific communication (but can be possibly reduced through "sense-making" of the narratives).

Two issues of concern are: a lack of grading the reliability of the information supplied and the concern of potentially breaching patient confidentiality in a Web 2.0 platform like Facebook. This latter concern is managed through strict moderation of the group and adherence of all participants to good ethical practices.

44.3.2 Cochrane Students' Journal Club: Teaching the Basics of Evidence-Based Medicine Using Accessible, Affordable, and High-Quality Educational Tools

Shivika Chandra

44.3.2.1 Background

Name of Project: The Cochrane Students' Journal Club (CSJC) [24]
Country and area of interest our work focuses on: South Asia region with emphasis on India, Pakistan, Bangladesh, and Sri Lanka
The team: A group of recent medical graduates, medical students, healthcare professionals, epidemiologists, research methodology, and literature searching experts
The Story Behind it: In 2010, a few members of the team from India rotated at a medical university in the USA as visiting medical students. During their electives, they experienced firsthand how journal clubs were an established component of the medical curriculum in the USA. Not only did the School of Medicine host regular journal clubs, but individual academic departments as well made it a priority that one of the faculty members discussed a research paper with students (medical students, residents, fellows on that rotation) at least once a week. These exercises helped stimulate the students to understand the research available, dissect its quality and grade, and accordingly apply it to the clinical problem at hand.

After returning home to India, the team realized that if medical students are taught to think analytically and approach clinical problems systematically beginning in the undergraduate years, then a new generation of physicians could be introduced to the practice of evidenced-informed health care.

The Idea: To create an online learning resource for medical students of South Asia that is accessible, affordable, and focused on the dissemination of evidence-informed health care in a peer-interactive and mentor-supervised manner [25].

44.3.2.2 How We Went About It

44.3.2.2.1 Identify the Target Audience and Their Needs

We realized that such an interactive mentor-guided learning exercise was something relatively unexplored in our region and yet something which students would benefit immensely by.

It was rare to find faculty who would allocate time in their hectic schedules for such a learning exercise which was viewed as "extracurricular." Physically setting up and running such meetings in medical colleges in our region would be something that was difficult on a large scale—especially in colleges where students faced administrative and infrastructure hurdles. In order to facilitate the setting up of local journal clubs we needed a paradigm shift in the way research was viewed in the SA region.

44.3.2.2.2 Address the Needs with Available Resources

1. In order to bypass the difficulties encountered in getting an actual physical space and resources for the journal club, we decided to host it on an online forum.
2. A group of young medical graduates with a passion for the cause volunteered to moderate the club as well as promote mentor–novice interactions.
3. Universal access to the research paper: While some medical colleges in the region have great libraries and access to the online medical journals, there are also an equal number of colleges who do not. The Cochrane database of systematic reviews is available free of cost in India (Courtesy of Indian Council of Medical Research), Sri Lanka, and Bangladesh and therefore it was decided to base the journal club's exercises on Cochrane systematic reviews [30].
4. In order to make the journal club a sustainable exercise, it was decided to host it on a monthly basis with the discussion spaced over the course of a month.

44.3.2.2.3 Method

A student moderator (usually a recent medical graduate) is placed in charge of each monthly topic's discussion and as a part of his/her responsibilities has to:

1. Select a clinical case and modify it as per the discussion at hand.
2. In conjunction with the subject expert for the month, frame a suitable set of clinical/methodological questions to stimulate the students participating in the discussion.
3. Work with the literature search expert to make sure that a search strategy is posted online after giving students some time to explore evidence on their own.
4. Work with the subject expert/mentor for the month to facilitate an appraisal which addresses the clinical questions linked to the problem as well as demystifies the review for the students.

44.3.2.2.4 Impact

The journal club is a continuously evolving dynamic intellectual enterprise which is responsive to the needs of its users. The efficacy of any intervention is based on its effect on the outcome. In a similar manner, the impact of the CSJC can be evaluated by the feedback received from the participants. We are in the process of setting up a formal survey to gauge the efficacy and some of the individual feedback we have received on the initial stages from students and faculty alike has been very encouraging.

Website URL: http://csjconline.informer.org.in/

44.3.2.2.5 What Makes Our Idea Unique?

- First online journal club in existence in the South Asian region.
- Provides an interactive environment and overcomes geographic boundaries which would otherwise limit attendance.
- "Necessity is the mother of invention"—This idea was created by a group of young medical

graduates based on their experiences as medical students and is relevant to the needs of the region.

- The online social networking format (blog) adapted for use makes it technologically easy for the moderators to conduct the discussions and for participants to navigate.
- The monthly discussions are automatically archived online and so are available for quick reference as and when needed.
- Collaboration of the journal club with organizations such as INFORMER (The Forum for Medical Students' Research) and the South Asian Cochrane Network (SACN) helps to foster symbiotic partnerships between student and professional bodies at a organizational level and harness resources (personnel and technological).
- In association with the various representative groups of the Cochrane Collaboration (healthcare consumers, professionals, and advocates) the CSJC team hopes to build upon an integrated learning platform for all user-driven learning.

44.4 Harnessing Social Media Toward Sustainable Solutions to Tackle Clinical Complexity

Neil Mehta, Pranab Chatterjee, and KaustavBera

In a medical education scenario where information exchange is becoming a necessity, asynchronous learning platforms using cloud-based tools are rapidly gaining popularity. With the expressed benefits of access to information from any place, reduction in physical data collection, and rapid dissemination, the use of social media in medical education also comes with the benefit of connecting globally. In an era when the buzzword is to think global and act local, this is an ideal way out. The problem remains in harnessing the powers of social media. If improperly managed, this may seem like a chaotic torrent of data flowing too fast and furious to rein in. Like early man who discovered fire, we have in our hands a faithful servant and a destructive master. Yet, standing at the crossroads of an epoch of information, it would be an opprobrious act to turn away just by being intimidated by the risks ahead.

The main purpose of this section is to chart out ways in which the apparently wild torrent of information coming in through the sluices of social media can be categorized and tamed. The risks of using social media for didactic purposes have been debated ad nauseam, so instead of focusing on that aspect, it is more productive to see the way out of the quandary.

44.4.1 Coordinating Problem Solving by Using Dynamic Networking Concepts Prevalent in the New Age Web 2.0[2]

44.4.1.1 Using Google Reader with Facebook to Coordinate Problem Solving

The challenge at hand was to create an automated stream of key journal articles that would be posted in an online forum stimulating discussion among the members of a residency program. The strategy behind the concept was to integrate this with the social media activities of the program members so that they could fit it in their social schedule without having to take out specific time for the discussion. With more medical students and residents using Facebook [26], it was a prudent step to integrate the system in such a manner that it would be accessible only through Facebook. Dr. Neil Mehta has proposed a system which works from within Facebook and encourages member participation as well (Fig. 44.7).

Steps involved

- Faculty member/chief residents set up Google Reader accounts.
- Subscribe to RSS feeds for some selected journals in their specialty.
- Periodically (at least weekly) review the feeds and share some of the most relevant articles in Google Reader.
- Find the RSS feed for their shared items list.
- Create a Group in Facebook (possibly a closed group so only members can view and comment).

[2] See glossary at the end of the chapter.

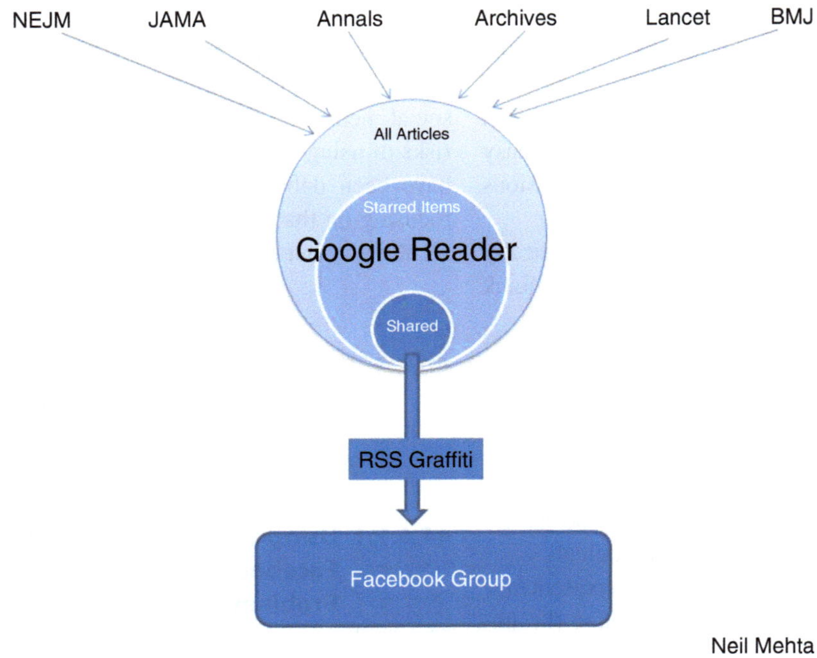

Fig. 44.7 Google reader platform

- Authorize the Facebook Application RSS Graffiti for this Group (only group admins can do this).
- Create a feed in RSS Graffiti with the RSS feed of the shared items list.
- Ensure that specific settings are modified and test for the posts to show up within about 30 min.
- RSS Graffiti for your Facebook account needs to be authorized by you before it will work on your Facebook Group.
- You must be the admin for that Facebook Group.
- Edit the feed and under the Filter tab, make sure you change the date to when there are some items to show. By default it is set to the time of creation of the feed in RSS Graffiti.
- Edit the feed and under the More tab, make sure you are posting as yourself and not as the Group.
- Invite/add members to the Group.

44.4.1.2 The Google+ Video Journal Club Concept

Working on a similar premise based on the concept of Dr. Mehta, a more focused discussion can be set up through video-conferencing, through the Hangout feature in Google's new social media platform Google + which allows free video conferencing facilities. Currently, the system is limited to ten simultaneous participants but that number will increase soon.

44.4.2 Blogging as a Tool of Medical Education

44.4.2.1 Keeping Up with Research

Most journals today have an active blogging section. The BMJ has undoubtedly been a leader in this aspect, using BMJ Blogs, blog comments, and Rapid Response as an effective means to connect with its audience, which would be instantaneous. Moreover, in the generation of Web 2.0 there has been a steady stream of dynamic repositories for topicwise blog posts where science posts on blogs from all over the web are collected and at the same time vetted through peer posting.

Furthermore, this has also led to the advent of blogging networks on topics as diverse as the whole of Science to covering extremely specific fields, which have successfully brought under their aegis, some exquisite scientists and science communicators who chew down and digest research articles and breakthroughs in their fields, thus making it easy for the general viewer to grasp the material.

44.4.3 E-mail Groups

44.4.3.1 Thinking Global, Acting Local: HIFA2015

Perhaps you have not considered the power of an e-mail discussion group to act as a policy changing tool. The members of HIFA 2015 (Health Information for All by 2015) raised a storm when publishers retracted over 2,500 titles from HINARI [27] the free/low cost article repository supported by the WHO. Ultimately all the retracted titles were restored when academics from all parts of the world raised their voices online in protest [28]. Without HIFA 2015 the cry of medical professionals in low resource countries may have gone unnoticed and patients would die for lack of freely available knowledge.

44.4.4 Clinical Problem Solving: Caregiver–Careseeker Closed, Moderated E-Mail Discussion

In developing countries, there still remain massive pockets of areas which still have poor healthcare delivery and healthcare professionals having access to poor healthcare management and information. In such scenarios, a rudimentary distance-diagnostic service, using e-mail based platforms to share patient information with specialists can come in very handy.

The key in this exercise is protection of the patient's privacy and proper information delivery. Though the professionals serving in these areas with poor access to specialists like radio-diagnosticians, toxicologists, critical care specialists, etc. are more concerned with the delivery of the best possible care, they also need to be aware of protecting the rights of the patient. This is especially significant in a setting where the patient may be unaware of his/her own rights in this regard.

The role of moderated e-mail lists has been demonstrated by the Caregiver–Careseeker peer-to-peer discussion group run by Dr. Rakesh Biswas. The moderator removes patient identifiers and forwards diagnostic or clinical queries by the group members for specialists to opine on. This has, in real time, and real life, proved to be an invaluable, often life-saving tool, while serving as a wonderful educational platform for the other members as well.

44.4.5 Bypassing Pay Walls: Guerilla Open Access

Aaron Swartz, Greg Maxwell and others have made social media their weapon to wage the crusade on behalf of the open access supporters. While Swartz has been charged with federal offenses for hacking into JSTOR and downloading 4.8 million articles, Maxwell has followed suit by uploading 32 gigabytes of published articles that he himself acquired by legal means to a prominent torrent downloading site.

While these two debonair demagogues have come forward and given the guerilla warfare a face, there remain whispers of other insurgents raising their heads. One can hear the tales of such outlaws. There are supposedly, information sharing groups on the Internet for those who seek references, on Twitter, and even on Google groups for e-mail-based seekers of medical articles. While the existence of these shadow lurkers remains subtle, there is no doubt about their ability to deliver access and to bring justice and knowledge to the ones left behind.

44.4.6 Twitter as a Tool for Medical Education

44.4.6.1 Networking

If one believed 140 characters was too little to go by, one could ask the systems engineer who got fired for a rant against his employer. Twitter is an innocuous looking tool with an immense power. In an age when the attention span of people is getting shorter than ever, Twitter summarizes the first law of networking.

Hit early, hit hard and keep it short and simple.

With traditional journals and academic bodies of professionals taking to twitter, the power of this simple idea to demonstrate the "six degrees of separation" concept is now beyond a shadow of doubt.

44.4.6.2 Peer-to-Peer Communication

One of the most striking examples of the use of social media as a post publication peer review filter was when a paper in Science claiming the discovery of key genes in predicting longevity was thoroughly taken apart on social media sites, and especially on Twitter by specialists who claimed to have had encountered technical issues using the described methodology. The paper was ultimately retracted on grounds of technical errors [29].

44.5 Assessing Clinical Competency Amidst Clinical Complexity

Rakesh Biswas

Developing clinical competency amidst clinical complexity may be analogous to enabling the blind men of Indostan (analogous to medical students/teachers as described in the Elephant poem, http://en.wikisource.org/wiki/The_Blindmen_and_the_Elephant) to develop a vision for the patient (elephant) in its entirety to provide the best possible form of care.

If basic sciences represent its "trunk" that probes/searches for its food (e.g., precursors toward revolutionizing therapy such as dendritic cells) then clinical sciences perhaps represent its "oropharynx" that gulps down whatever "help" basic sciences have to offer.

A competent student helps facilitate the information flow between the basic and clinical sciences to develop a smooth workflow for the patient beneficiary of his/her education.

A student may be considered competent if s/he has the necessary competence to see the pattern emerging from the given "information" surrounding the patient in the form of major "micro" themes around his "macro" problem of weakness, weight loss, and shortness of breath.

If s/he can identify the anatomical area giving rise to the problem using all the clinical modalities that supply the information (also asking for more information where necessary), the change in the patients microanatomy that gave birth to the problem, identify the external "microorganisms" that incited a change in the patient's "microanatomy" and finally facilitate a plan for tackling the problem utilizing available evidence-based resources, the patient beneficiary would have derived the necessary benefit from our "medical education" system.

Clinical complexity challenges the notion of "*competency*," which can be extremely variable and we may not always require all students to have the same competencies while they strain to center their efforts on the elephant/patient.

The authors are reminded of this everyday on following the interactions of Dr A.'s, Dr B.'s, and Dr C.'s network of "online clinical problem solving" as well as the medical students interacting on "tabula rasa" where each participant has some newer insight to offer around a given patient's problem and many of these just evolve spontaneously often in a bid to help the real patient.

Just going through these interactions makes one aware of the diversity of the group's individual competencies that go toward generating insights around the patient.

Many of these students participating in real patient-centered problem-solving exercises revel in the spirit of teamwork that these exercises bring out.

A retrospective qualitative analysis of these "patient problem" centered conversations may help to identify which "individual competency-based sharing" ultimately benefits the best decision on the patient.

44.6 From Complexity to Point-of-Care: A Case in Point

Shivika Chandra

With India, for example, ranking second among the most-populated countries in the world, the morbidity and mortality patterns reflect demographic patterns. Extrapolating results and findings from studies conducted outside India often lead to erroneous outcomes. Socioeconomic norms, religious practices, availability of basic resources (clean water, food, and sanitation), overcrowding, and the lack of education are parameters which tend to skew the interpretation and translation of data.

44.6.1 Points to Ponder

1. Complexity is an innate constituent of any healthcare system. Though the variables which govern complexity can be identified, the manner in which they interplay needs to be assessed and addressed on an individual basis.
2. An evidence-based paradigm adopted early in clinical training helps professionals to identify, appraise, select, and apply the best available resources towards patient care.
3. Interactive discussions between various users in user-driven healthcare platforms help both patients and providers to align therapeutic goals as well as discuss prevalent socioeconomic–cultural factors which may influence a given health issue.
4. Web-based social media platforms for problem-based learning and knowledge translation are accessible and affordable health education tools, which require little to no infrastructural investment and are ideal for use in low-to-middle income countries.

With a low doctor-to-patient ratio and limited healthcare resources in low to middle income countries, often the newly initiated resident physician finds himself in the outpatient department with a case load of fifty plus patients (maybe as high one hundred patients per day). In order to accommodate all the patients within the time constraints of the clinic, each patient is scheduled for 10–15 min with the physician. In this short interval, the physician has to take a focused history, do a brief physical examination, address any queries the patient may have, and review the treatment plan. A point to be noted here is that these clinics are not continuity clinics. Residents are posted under various units for training and therefore in each rotation, they attend the outpatient department (OPD) of that particular unit. This means that patient and doctor have only 15 minto establish rapport and forge a therapeutic alliance.

The following narrative by a young trainee doctor can be used as a starting point for us to identify complexity and also assess how the solutions described in this chapter can be applied towards solving the cause of the same. The text-boxes interspersed in between the narrative highlight the possible solutions.

As a new resident in a psychiatry unit, on my first outpatient clinic, I met Mr. Anand, a 23-year-old male, who was diagnosed with schizophrenia 3 years ago. Over the past year, it was noted that his symptoms were not improving. During his past two visits, the resident-in-charge had noted this and mentioned "Review antipsychotics and consider change of treatment." A beginner to the field of psychiatry, I quickly reviewed the treatment plan and found that the recommendation of care were appropriate to current treatment guidelines for the condition. On questioning the family, I found them supportive of his condition and Anand himself was compliant with the treatment. In order to gauge his willingness to participate in his care, I asked him to show me to take out the tablets he had with him and show me how many of them he would take everyday.

Points to Keep in Mind in Those 15 min

Patients should

- Communicate with their providers
- Be informed
- Be proactive in decisions involving their health
- Understand the limits of the health system and explore alternate viable options to circumvent them

HealthCare Providers should

- Listen to the patient
- Be informed
- Make evidence-based decisions which are best applicable to their patient
- Make the patient an active participant in his/her care
- Base therapeutic goals on patient-centered outcomes

I was not very surprised to see that the dosage of the drugs he was taking was inaccurate. The reason was yet to be ascertained. On further questioning the family, I came to understand that Anand was the most educated of his family having studied up to fourth grade. The rest of the family, like a significant proportion of the population which seeks care at government hospitals, was illiterate.

The prescription that the doctors wrote for them was issued on a "free drug slip" which provided them the drugs free of cost under a government sponsored scheme for low income groups. While issuing the drugs, the pharmacy retained the slip as proof of delivery. Therefore, the patient was left with the drugs but no documentation of the dosage required. Rather than making their way back into the crowded OPD, they relied on their memory to remember dosage and scheduling of the medication. Being a family of laborers, it was difficult for a single person to get a leave of absence to regularly accompany Anand for his follow-up visits. Therefore, among the family of six, there were different opinions as to what the dosing schedule was.

It is interesting to note that complexities do not necessarily have to originate from the disease pathology or comorbidities alone. Poor knowledge, miscommunication, socioeconomic factors—all of these can lead to an interplay of factors which despite being "nonmedical" affect the clinical outcome of the patient. A good clinician should be able to identify the problem areas and recapitulate didactic knowledge and merge it with current research to make evidence-informed decisions which are individualized to his/her patient keeping a comprehensive view of the complexities involved.

The course of action I chose based on hearing this was, instead of writing out the standard slip alone, I also made another slip for the family to keep. Replacing numbers and abbreviations with crosses and other symbols, I showed them alternative ways to visualize the dosage scheduling. In order to further simplify the process, I asked them to purchase a simple ruled notebook and bring it with them on each visit. This would allow the resident-in-charge to maintain the follow-up notes and prescription in one record. While the hospital record and even the prescription slip would be retained with the hospital, the notebook could serve as their reference point. I asked them to request the pharmacist to show them the appropriate drugs against the list and place them in separate covers. Even if they were unable to read the names on the labels, they could identify the medication on the basis of the cover/symbols as explained to them by the pharmacist.

This entire exercise took me around 20–25 min with Anand. I was among the last to finish with my cases, but the overall experience was rewarding. I made sure that I checked with my other patients if they had a similar problem. By taking a minute more and writing out a separate drug slip, I gained a better understanding of the complexity in my health care system and took a step towards addressing my patients' concerns.

Dealing with complexity does not require expertise. A blend of common sense and experience is often sufficient to tackle it. What is required for any individual is to realize the fact that the healthcare system is fraught with complexity and to be aware of the methods to tackle them.

Key Messages

For healthcare providers

1. Listen to the patient—Miscommunication or non-communication contributes to clinical complexity.
2. The focus of care should be the formation and maintenance of a therapeutic alliance between the patient and the provider and not therapy-based alone.
3. In resource-poor nations especially, it is vital that physicians adopt an evidence-based approach to manage cases such as to best utilize available resources.
4. Often a thorough clinical history and a focused physical examination can reveal far more than a battery of diagnostic tests.

For patients

1. Patient survivors as health advocates and teachers have a valuable role in shaping the younger generation of healthcare providers into more empathetic caregivers [31]. Peg Ford a cancer survivor and patient advocate says it this way, "*I envision Patient Advocates as the bridge*

between the patient community and the scientific/ medical world, working hand-in-hand to advance research and impact changes to better serve both communities".

2. Patients can be empowered to participate and contribute to their healthcare decisions along with their physician.
3. Patients need to make an attempt to keep themselves abreast with current developments in health care via consumer blogs, user-driven healthcare forums, social media platforms etc. An informed patient can play a more proactive role in deciding what clinical outcomes are relevant for assessment in routine practice as compared to the outcomes delineated by research studies.

For students

1. As a student it is essential to realize that medicine is one subject which has to be "learnt" and cannot be "taught."
2. Clinical complexity is a part and parcel of health care and students should not expect a clearly demarcated set of symptoms which allude to a predefined disease. Variables influence the course of any disease pathology and therefore the best teachers of medicine are not the textbooks but the patients themselves.
3. The use of social media as educational tools to disseminate concepts of evidence-based medicine in a user-driven learning environment help the students to conceive and develop a sound foundation of core clinical skills and knowledge upon which they can build.
4. Reiterative problem-based learning in a mentor-supervised, mentee-initiated atmosphere such as online journal clubs, clinical discussion forums help the students to constantly recapitulate and update their knowledge by developing the skills needs to appraise and apply current contextual evidence.

For policy makers and healthcare advocates

1. A collaborative approach to deal with clinical complexity would aid policy makers to frame an appropriate infrastructural framework to implement appropriate solutions.
2. The nature of the complexity sometimes requires a multilevel intervention and by formulating a common action plan with health care providers and seekers alike, the policymakers can address health concerns in a more comprehensive manner.

44.7 Conclusion

The key learnings from this chapter are obtained from different author perspectives toward clinical complexity. They offer a shared understanding of different solutions to optimize healthcare outcomes through promotion of "user-driven health care" where clinical problems are solved through asynchronous discussions between "users" in a collaborative interface. What sets UDHC apart from EBHC is that while EBHC mainly concentrates on preexisting data or literature to reach a conclusion, UDHC tries to propagate both preexisting evidence as well as individual inputs from doctors or people who may have prior experience in contextually dealing with the same topics.

Development of "clinical competency" has a "complex–nonlinear trajectory." One way to look at this "complex road map" to a student's "competency" would be to observe all the "real" patient decisions the student has actively engaged in while utilizing his/her understanding of the basic sciences and ensure that all these are explored within a connected team of learners and facilitators. This may go toward providing a judicious formative assessment of the student. Even in those that volunteer to mentor and guide there is a growing recognition that knowledge is dynamic, effective and practical as it grows in the synergy provided through conversational learning and mutual support.

Our goal is to further develop (through suggestions, scenarios, and inspirational insights contained in this chapter) other dynamic and meaning based medical learning opportunities. We see social networking as a growing force where excellence in lecture and textbook knowledge can be merged with experiential conversational learning to bring the best of both worlds consisting of great academic potential united with applied learning to promote EBHC in meaningful practice. Our desire is that in the near future medical student interaction; mentoring and online collaboration with the use of social media will be a resource accessible from every student's 360° learning portfolio.

Conflict of Interests:

The authors have no financial conflict of interests but the solutions represented in the chapter are colored by their passion in this area and these solutions still require further experimentation by neutral bodies before they can be accepted as concrete evidence.

Acknowledgements Priyank Jain and Yogesh Jain who are two of the physicians described in the illustrative example of "clinical complexity" in this chapter thank

The team at the Cochrane Students' Journal Club, The Forum for Medical Students' Research, India (INFORMER) and the South Asian Cochrane Network and Centre, Vellore.

The team of students and faculty members from Tabula Rasa.

Peg Ford—patient advocate extraordinaire for sharing her story and inputs with us.

Vasumathi Sriganesh, CEO at QMed Knowledge Foundation India for her help in citation finding.

Glossary of Social Tools

Blogs Blogs are content management systems that can host text, images, audios or video in an interactive format where users can place comments and create a discussion. Free platforms are available on Google's Blogspot (http://blogger.com) or the open access Wordpress (http://wordpress.com) services.

E-mail Groups Are moderated lists which allow sharing of information, video, audio, or other attachables and hence act as asynchronous discussion platforms. Free e-mail groups can be created on Google (http://googlegroups.com) or Yahoo (http://groups.yahoo.com) or Posterous (http://posterous.com).

Facebook A social networking site that allows peer-to-peer interaction, build groups with different levels of privacy (from open to secret), build fan pages to showcase content, and create albums to handle images. Also allows video and text chats.

Google+ A Google powered social network that has similar functionality as Facebook with a special video chat system called "Hangouts."

Google Reader A tool to read updates from sites that can generate dynamic "feeds" also known as RSS.

RSS "Really Simple Syndication" is a system which generates dynamic feeds whenever new updates or additions are made to a site. These can be pulled in by tools like Google Reader to read the updates.

Twitter Microblogging site which allows for sharing of images, video, or audio, along with links. Contents of a single post are to be limited to 140 characters.

YouTube Free video hosting site run by Google that can host a number of different formats of varying lengths.

References

1. Goldacre B. Bad science. London: Fourth Estate; 2008.
2. Bonavita V, De Simone R. Towards a definition of comorbidity in the light of clinical complexity. Neurol Sci. 2008;29 Suppl 1:S99–102.
3. Maskrey N. https://www.jiscmail.ac.uk/cgi-bin/webadmin?A0=EVIDENCE-BASED-HEALTH. Accessed 29 June 2011.
4. Maskrey N, Underhill J, Hutchinson A, Shaughnessy A, Slawson D. Getting a better grip on research: a simple system that works. InnovAiT. 2009;2(12):739–49. doi:10.1093/innovait/inp106.
5. Price A. Let the evidence speak. Editorial. The International Society for Evidence-Based Health Care Newsletter. 4 July 2011. http://isehc.blogspot.com/p/newsletters.html
6. Biswas R, Martin C, Sturmberg J, Shankar R, Umakanth S, Shanker K, Kasthuri AS. User driven health care—answering multidimensional information needs in individual patients utilizing post EBM approaches: a conceptual model. J Eval Clin Pract. 2008;14:742–9.
7. Ross S. A lexicon for user-driven healthcare. Int J User Driven Healthcare. 2011;1(1):50–4.
8. Biswas T, Bera K, Biswas R (2011) Creating secondary learning resources from the BMJ Case Reports through social-media based discussion groups. BMJ Case Reports Blog. http://blogs.bmj.com/case-reports/2011/10/28/creating-secondary-learning-resources/. Accessed 6 Nov.
9. Rees B. An introduction to the NHS & NHS reforms revolution or evolution? European M Health Alliance Summer Symposium 2011.

10. Wittenberg R. HIFA 2015 Email Forum 'Genital mutilation'; 2011. http://www.hifa2015.org/hifa2015-forum/. Accessed 10 Oct 2011.
11. Heneghan C. Diagnostic tests; 2010. http://www.cebm.net/index.aspx?o=1023. Accessed 08 Oct 2011.
12. Heilman J. Wikipedia and medical knowledge. Personal correspondence; 2011.
13. Fathers Now. http://fathersnow.com. Accessed 12 Oct 2011.
14. Frankovich J, Longhurst C, Sutherland S. Evidence-based medicine in the EMR era. N Engl J Med. 2011;365(19):1758–9.
15. Survivors teaching students: saving women's lives by the Ovarian Cancer National Alliance. http://www.ovariancancer.org/about-us/survivors-teaching-students/. Accessed 12 Dec 2011.
16. Bhalla A, Suri V, Varma S, Sharma N, Mahi S, Singh P, Khandelwal NK. Acute febrile encephalopathy in adults from Northwest India. J Emerg Trauma Shock. 2010;3(3):220–4.
17. Anderson W. Herpes Simplex Encephalitis Workup. Medscape Reference. http://emedicine.medscape.com/article/1165183-workup#aw2aab6b5b1aa. Accessed 7 Nov 2011.
18. Biswas R, Dhakal B, Das RN, Shetty KJ. Resolving diagnostic uncertainty in initially poorly localizable fevers: a prospective study. Int J Clin Pract. 2004;58(1):26–8.
19. Lakhotia M, Gehlot RS, Jain P, Sharma S, Bhargava A. Neurological manifestations of enteric fever. JIACM. 2003;4:196–9.
20. Scragg J, Rubidge C, Wallace HL. Typhoid fever in African and Indian children in Durban. Arch Dis Child. 1969;44(233):18–28.
21. Koskiniemi M, Vaheri A, Taskinen E. Cerebrospinal fluid alterations in herpes simplex virus encephalitis. Rev Infect Dis. 1984;6(5):608–18.
22. Herpes simplex encephalitis treated with acyclovir: diagnosis and long term outcome, http://www.ncbi.nlm.nih.gov/pmc/articles/PMC2169720/. Accessed 29 June 2011.
23. McGrath N, Anderson NE, Croxson MC, Powell KF. Herpes simplex encephalitis treated with acyclovir: diagnosis and long term outcome. J Neurol Neurosurg Psychiatry. 1997;63(3):321.
24. Cochrane Students' Journal Club. http://csjconline.informer.org.in/. Accessed 12 Nov 2011.
25. Chandra S, Shah KN, Sriganesh V. The Cochrane Students' Journal Club - creating a secondary learning resource for gathering and appraising evidence: An example of rational use of medicine to prevent malaria relapse. Int J User Driven Health Care. 2011;1(4):31–41.
26. Gray K, Annabell L, Kennedy G. Medical students' use of Facebook to support learning: insights from four case studies. Med Teach. 2010;32:971–6.
27. Koehlmoos T, Smith R. Big publishers cut access to journals in poor countries. Lancet. doi: 10.1016/S0140-6736(11)60067-6.
28. Kmietowicz Z. Publishers withdraw 2500 journals from free access scheme in Bangladesh. BMJ. 2011. doi:10.1136/bmj.d196.
29. Garner J, O'Sullivan H. Facebook and the professional behaviours of undergraduate medical students. ClinTeacher.2010;7:112–5.doi:10.1111/j.1743-498X.2010.00356.x.
30. The Cochrane Collaboration. http://www.cochrane.org/. Accessed 12 Nov 2011.
31. Salzburg Global Seminar. The greatest untapped resource in healthcare? Informing and involving patients in decisions about their medical care, 12–17 Dec 2010 (Session 477). http://www.SalzburgGlobal.org/go/477. Accessed 10 Oct 2011.

Part VII

The Future of Healthcare

The final section of this *Handbook* explores aspects of the future developments in health care and the changing health systems. There are many unexpected outcomes and unintended consequences in health care, health services improvement exercises and healthcare reforms. A systems and complexity perspective provides unique insights into the dynamics and potential consequences of systems change, and stimulation of opportunities for innovation. Not surprisingly some of the issues and solutions postulated by the authors of the chapters in this section may appear controversial; some can be understood on the basis of *initial condition* of current systems, and some are based on *emerging* discoveries.

Systems cannot be predicted, but understanding the principal behaviours of systems allows us to analyse the potentials and the likely pitfalls of current and proposed developments. These chapters, no doubt, will provoke ongoing innovation, discovery and emergence of health.

45 Making Sense: From Complex Systems Theories, Models, and Analytics to Adapting Actions and Practices in Health and Health Care

Carmel M. Martin and Joachim P. Sturmberg

It is the theory that decides what we can observe.

Albert Einstein [1]

Theories rarely arise as patient inferences forced by accumulated facts. Theories are mental constructs potentiated by complex external prods (including, in idealized cases, a commanding push from empirical reality). But the prods often include dreams, quirks, and errors—just as we may obtain crucial bursts of energy from foodstuffs or pharmaceuticals of no objective or enduring value. Great truth can emerge from small error. Evolution is thrilling, liberating, and correct.

By Stephen Jay Gould

I think a major act of leadership right now, call it a radical act, is to create the places and processes so people can actually learn together, using our experiences.

Margaret J. Wheatley[1]

Common to complex systems are two fundamental themes—the universal interconnectedness and interdependence of all phenomena, and the intrinsically dynamic nature of reality [2]. "*At each level of complexity we encounter systems that are integrated, self-organizing wholes consisting of smaller parts and, at the same time, acting as parts of larger wholes*" (Capra [3]).

Many of the original health complexity leaders, like Plesek, Dooley, Berwick, and Lindberg, have pioneered the approaches that led to many diverse pathways to health improvement. But have these ideas been sufficiently adopted by decision makers to make a difference [4]? We stand at crossroads in relation to health systems evidence, funding and organization, particularly in the USA. Many cherished reductionist ideas about the nature of disease and illness and how systems should be organized to improve health outcomes have not delivered their promises, e.g. "disease management carve outs," and are found to be costly, fragmenting, and ineffective [5]. In addition, comparative metrics have revealed many unintended and unwelcome consequences such as widening health inequities, overtreatment, and poor access to care to name a few [6, 7]. However, the opportunity to influence policy and practice has never been greater, as major shifts in

[1] Writer and management consultant who studies organizational behavior. Her approach includes systems thinking, theories of change, chaos theory, leadership, and the learning organization: particularly its capacity to self-organize.

C.M. Martin (✉)
Department of Public Health and Primary Care, Trinity College, College Green, Dublin D2, Ireland
e-mail: martinc7@tcd.ie

J.P. Sturmberg
Department of General Practice, Monash University, Melbourne, VIC, Australia

The Newcastle University, Newcastle, PO Box 3010,Wamberal, NSW 2260, Australia
e-mail: jp.sturmberg@gmail.com

J.P. Sturmberg and C.M. Martin (eds.), *Handbook of Systems and Complexity in Health*, DOI 10.1007/978-1-4614-4998-0_45, © Springer Science+Business Media New York 2013

healthcare provision are being forced by the international financial crises [8].

Futile medicine or *care* is a relatively new term to describe a medical procedure or treatment that cannot achieve its stated goals or produce its expected benefits with an acceptable level of probability regardless of repetition and duration of treatment [9]. For example, disease management, based on "evidence-based" guidelines and protocols, has failed to produce the desired outcomes, as person-centred care was "written out of the script" [10]. Simultaneously, the costs of health care are growing exponentially and have reached prohibitively high levels internationally, and in the USA in particular. *Futile* or *unhelpful care processes* have been identified as the major source, illustrating the *unintended consequences* of a *large-scale systems change* [11].

John Sterman [12] and other systems dynamics thinkers have demonstrated that the failure of all major health system reforms has been *foreseeable* as the most basic consequences of a change on the neighboring agents have not been taken note of. Despite these repeated and overwhelming failures, the health sector has been very slow to adopt complex adaptive systems theory and science. What has attracted considerable interest, as it largely focuses on cost control, has been lean management approaches, developed for manufacturing by Toyota [13]. "Health reform," not having its own theoretical underpinning, made the all too common mistake of implementing solutions decontextualized from its environment.

45.1 Perturbing Conversations About Complex Health Systems

Making health systems truly patient-centric, adaptable and responsive requires acknowledgement of their complex adaptive nature. We require ongoing conversations and several ways of sense-making to understand and respond to the dynamics arising from the systems self-organizing properties which requires ongoing learning.

Three themes underlying the workings of health systems are explored: the notion of innocent information; the continual process of making sense of the values and assumptions in health care; and the ever increasing content and changing context of health systems knowledge.

45.1.1 Prevailing Assumptions

Current mainstream evidence is seen through a particular lens that views problems and solutions as simple or complicated, and for which evidence-based solutions are available, and that there is a good business model for all the activities in health care for investors and governments. This view assumes that every problem is reducible if only enough research is conducted and the evidence is appropriately assembled. Recently, and prefaced with an "of course," proponents of this worldview argued that this linear and static evidence should be contextualized by human ethics and values [14]. On the other hand, with health care being increasingly delivered by industrial business or government conglomerates, these values are increasing dollar bottom lines [15].

However, no healthcare system is static, linear, and appropriately assembled. Polarities, contradictions, and tensions in values arise from the "specific perspectives" associated with the "special expertise" of the various "workers" in the healthcare system. These phenomena result in nonlinear dynamics, the constituent attribute of systems containing feedback loops; this is the norm in all "socially determined" systems, including healthcare systems.

Though there is no doubt that simple and complicated approaches to certain health service problems, like the time-out procedure in the operating room to ensure the right patient is receiving the right operation, have been highly successful. The fundamental of simple and complicated approaches is the checklist, hailed as a major disruptive innovation in health care, and has been copied from "the cockpit environment of planes" and "the pit stop" in motor car racing circuits [16]. Implemented as the "model for practice," particularly in the operating room and emergency department environments, it has dramatically improved health-outcomes, reduced complications and saved lives. As in "the cockpit" checklist approaches to ensure the "proper predetermined" procedures focus individuals and teams to always perform the "correct simple and/or complicated steps" in a timely fashion.

However, the push to implement checklists to all healthcare environments has not proved particularly successful outside of these types of complicated instrumental activities. In particular, the attempt to transform primary care into "a sum of checklists," such that if boxes are ticked payments follow,[2] has not improved outcomes beyond existing trends such as the UK Quality and Outcomes Framework [17–19]. What works well in some domains of health care such as operating theatres does not work well in many others such as primary care practices. Checking checklists, as pushed by some, is *reduction ad absurdum.*

45.1.2 Complexity Assumptions

Many *technical* as well as *value questions* will never be easily answered in their dynamic and ever changing challenging environments. We should reflect on our ancestors' approaches that used many different skills in observing and responding to *patterns* in nature in order to live and survive by listening, feeling, and sensing. These are particularly useful strategies where phenomena are unpredictable, but need to be managed.

Making sense of options and dilemmas in a realistic time frame is essential to inform decision making in a complex healthcare system environment [3]. How do we determine *how to make sense* of different approaches in and across different healthcare sectors? We suggest that learning to engage with sense making dialogues and knowledge developments that aid decision making in complex adaptive clinical practice and health systems management. This is the way forward to address the pressing challenges as healthcare systems continue to struggle to adapt to changing internal and external constraints [20].

Complexity science is a disciplined approach to studying complex adaptive systems. Complexity science provides a sophisticated approach to studying the complex adaptive systems composed of numerous, varied, simultaneously interacting parts or "agents," from molecules or bacteria in a biological process, to individuals or businesses in an economy. Informatics leverages computing power and sophisticated software tools to significantly enhance decision making by applying analytical methodologies to massive amounts of data. Together, Complexity Science and Informatics, have impacted every industry from manufacturing to biotechnology, and can be leveraged for healthcare delivery and health system reform.

45.2 Sensemaking or Sense-making

Sensemaking [21] or Sense-Making [20],[3] is essentially the process of people giving meaning to their experience, based on identification of *patterns*—either internal or external [22]. *A pattern is a particular arrangement of elements that constitute a model to be used or emulated. Patterns are patterns if they can do something, if they can cause something to occur with some regularity. ... a pattern* [is] *an arrangement that expresses a reproducible and meaningful relationship between relatively independent components* [23]. Although studied for centuries, and an essential part of biological

[2] In 2004 a new contract between Primary Care Trusts (PCTs) and General Practitioner (GP) practices was negotiated. The new contract's centerpiece, the Quality and Outcomes Framework (QOF), included 146 check list targets in four domains (clinical, organizational, patient experience, and other services), which are revised periodically. The cost of QOF, around £600 million in the first year, and around £1 billion thereafter, formed part of the planned increased investment in primary medical care services. "To date, there is no evidence that the high expenditure on QOF can be linked to improvements in health outcomes. The high expenditure on the program makes it critical to be sure that the performance improvement is not achieved at the expense of other more valuable initiatives, services, or nonmeasurable aspects of patient care." [14]

[3] The term "sensemaking" has primarily marked three distinct but related research areas since the 1970s: Sensemaking was introduced to human–computer interaction by PARC researchers Russell, Stefik, Pirolli, and Card in 1993, to information science by Brenda Dervin, and organizational studies by Karl Weick. In information science the term is most often written as "sense-making." In both cases, the concept has been used to bring together insights drawn from philosophy, sociology, and cognitive science (especially social psychology) [22].

survival mechanisms, "sensemaking" has developed in distinct but related research areas since the 1970s.

Distinct but related research areas relevant to understanding health include the disciplines of mathematics, biological modeling, philosophy, sociology and cognitive sciences, communication studies, complexity sciences, informatics, and knowledge engineering. Of particular importance to making sense of patterns in different domains applicable to human health are:

- Communications—Brenda Dervin [20],
- Intelligence and multi-ontology sense-making: the Cyenfin framework—Dave Snowden [24],
- Organizational studies—Karl Weick [25],
- Mathematical modeling—Bruce West [26], Stephen Guastello [27], David Katerndahl [28], and others, and
- Artificial intelligence and informatics pattern recognition—Eric Horwitz [29], data-mining—David Riaño [30], and computational linguistics—Carl Vogel [31]

Sensemaking has emerged as an area of interdisciplinary study in response to practical challenges of knowledge management in health systems. It has three major streams pertaining to: the individual and real-time communications; organizational sense making; and intelligence as in strategy, policy, and business [32]. In addition, mathematics, statistics, and computational systems are producing prediction with feedback models to help explain and understand emerging patterns in real world systems, especially in the fields of bioinformatics, health informatics, and health system dynamics. Sense making is about identifying patterns and their meanings, so that best individual or collective responses can be made.

45.2.1 Patient-Centered Care Case Study: Applying Sense-Making

In order to demonstrate sense-making implications, a real clinical case study is provided as an example. An extract from a care worker report January 27, 2012 describes brief narratives of "health and wellness" from the perspective of a chronically ill couple. It challenges the prevailing notion of protocol-based disease management.

Eileen Murphy (a 68-year-old lady who has suffered a bipolar type syndrome since her stroke and has multiple physical complaints)—*Doing better today apart from a pain in her right eye, she thought it was the start of a migraine. Got advice from her pharmacist and has taken x2 paracetamol for same (had only taken them a few minutes before I rang) so she was hoping that would help. She is wearing a corset for her back. She was just discharged from hospital yesterday following 2 days in hospital for her "collapse." Eileen may be going into respite next Monday for a week, home for a break on the weekend; and then back in again for 2 weeks. She was to be assessed between today/tomorrow re same at her house. If she deteriorates, she is to contact her PCP. She rates her health today as 8/10.*

Mick Murphy (Eileen's husband and caregiver under stress with pressures of caregiving and his own medical problems include alcohol dependency)—*In great form today, went to Alcoholics Anonymous meeting last night and really enjoyed it, (it always seems to give Mick a boost when he attends). He was a little stiff first thing this morning when he woke up but was fine after an hour and no pain or soreness from fall yesterday. Is hoping wife Eileen gets into respite for a while this will give him a little break at home (although he will still be running errands for her while she is in respite). Granddaughter is doing fine so this is a weight off Mick's mind. Feels a hundred per cent today 10/10.*

45.2.2 Sense-Making: Dervin's Concept of "Innocent Information"

Dervin is a leading theorist in the field of communications. Her sense-making theory and research have arisen to address the gaps, tensions, and even contradictions among different ways of knowing and different levels of information in complex systems. The conceptual framework for sense-making [20] encompasses polarities that disrupt decision making and how conceptualizations of information impede its development and use (Fig. 45.1).

"*Innocent information*" is a deliberate misnomer for idealized information that is true,

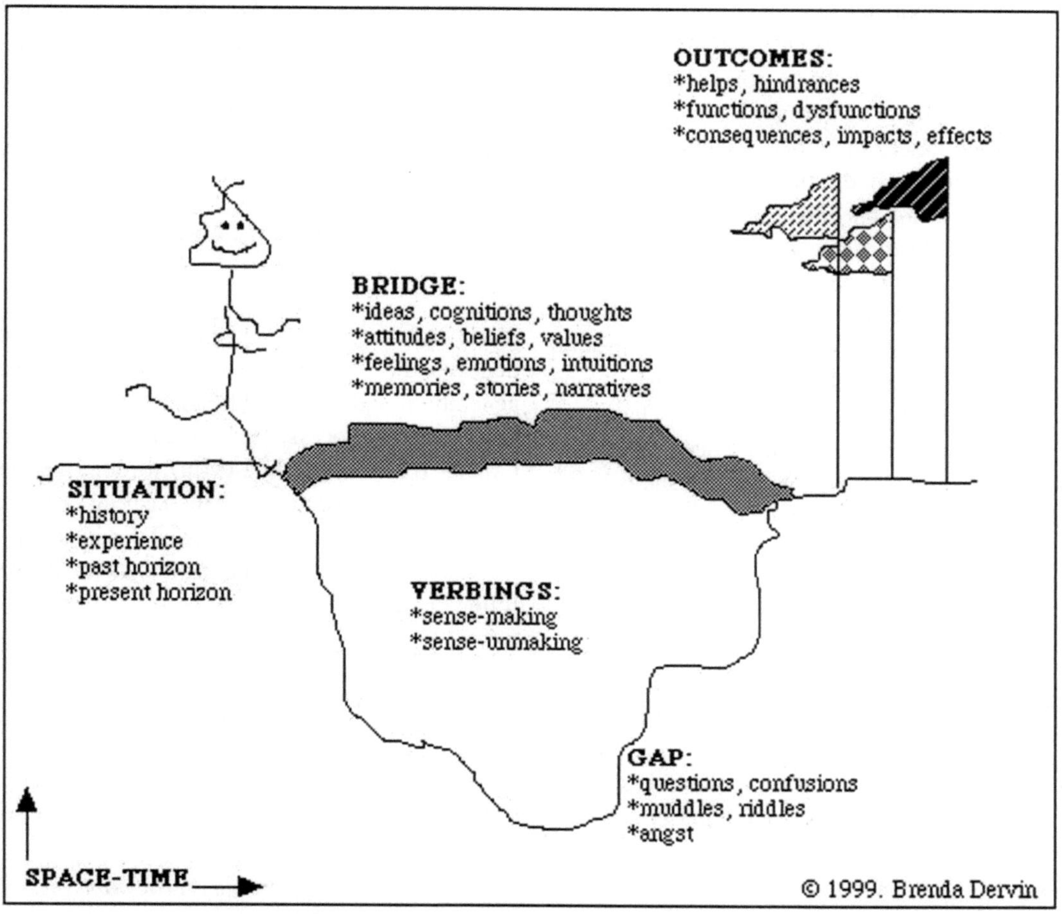

Fig. 45.1 Dr. Dervins original artwork of the Sense-Making Metaphor (reproduced under Wikimedia Commons)

value-free, and objective. Yet, in the real world, information is always socially constructed and subject to assumptions, bias, and distortions. The pervasive polarities of our times are shaped in particular directions by dominant professional, political, economic, and media discourses.

Dervin categorizes the main polarities as clear versus fuzzy, complicated versus complex and nonlinear, objective versus subjective, and research versus practice. These polarities are highly pertinent to the assumptions we make in our everyday discourses in medicine and health care. Dervin's polarities can be modified for our purposes as:

- Order versus chaos,
- Explicit (quantitative) versus implicit (qualitative and sensing) pattern recognition,
- Information (tacit) versus evidence (explicit),
- Autonomy versus collaboration,
- Cognitive versus emotional,
- Rule bound versus intuitive,
- Mind versus body, and
- Community versus individual.

45.2.2.1 The Case Study: Personal Journeys Through Dervin's Lens

The journey. The trajectory of Mick and Eileen's processes to cope with Eileen's psychological and physical state (*body* and *mind*) encompasses Mick's drinking in the past and Eileen's numerous hospital admissions. Although not drinking now, Mick is still vulnerable and he is recovering from a recent fall. Eileen has limited motivation or capacity for self-managing her health.

Bridges and Gaps. Mick's narratives—Eileen is in and out of hospital and respite care, and he is not happy about that. He feels that his wife is getting progressively worse mentally despite all her counseling and hospital admissions. Mick is going out walking as normal. Eileen's narratives range from describing totally excellent health to major excursions in erratic behavior. "*while talking to <Mick> on the phone, Eileen picked up the extension and was listening in and was shouting down the phone that he was a liar and a cheat and then hung up.*" Eileen fluctuates in her self-care and smokes and does not take her medication properly and has had hospital admissions and admissions to respite care on a monthly basis.

Sense-making. Mick wants Eileen to go into permanent care. He feels well when she is being cared for by others, as she can become abusive and very demanding. Eileen wants to sell the house but still return to the community. Her current trajectory despite government subsidies is very expensive on the family finances.

Outcomes for the individuals. What makes this couple feel well today? The outcome that they both desired has been put in place albeit only in the short term—a hospital admission for Eileen.

The situation is at least temporarily resolved with brief institutional placement. There are fuzzy areas of interpersonal relationships and mutual dependence and frustration. The communication of need is not *innocent information*, as the couple must express their wants for the caring burden to shift in such terms that the health system can diagnose and implement an appropriate action.

Health, illness, and quality of life have been viewed through reductive lenses, similar to disease approaches. Yet, the situation, meaning and process of construction of the individual experience is highly personal and contextual. There is social and emotional and physical relief for both Eileen and Mick, and for a short time at least, both rate their health as very good, even though they have a poor outlook.

"Innocent information" about Eileen's medical and psychiatric condition exists in the medical records mapping and defining her clinical care. However, it is not innocent information that is presented to the clinical providers, but information constructed through the eyes of major actors and across many polarities. Information is an ongoing flow and dynamic. It is only by taking charge and configuring information that we can design truly responsive and efficient care.

45.2.3 Sense-Making: The Cynefin Framework

Kurtz and Snowden [33], based on a dynamic understanding of knowledge, view knowledge generation and knowledge management as a *sense-making* process. They used the Welsh term *Cynefin* (pronounced kun-ev'in) to describe the nature of knowledge simultaneously as a thing and a flow—knowledge being in constant flux. They explained the meaning of Cynefin in the following terms:

> *It is more properly understood as the place of our multiple affiliations, the sense that we all, individually and collectively, have many roots, cultural, religious, geographic, tribal, and so forth. We can never be fully aware of the nature of those affiliations, but they profoundly influence what we are. The name seeks to remind us that* ***all human interactions are strongly influenced and frequently determined by the patterns of our multiple experiences, both through the direct influence of personal experience and through collective experience expressed as stories*** (emphasis is ours) [33].

A loose translation of the term is "*place of belonging*," which is reflected in the Cynefin framework's five domains—the known where cause and effect relationships are generally linear, empirical, and not disputed; the knowable where cause and effect relationships exist but may not be fully known or only known by experts, gained through systematic methodologies, and relies on trust; the complex or emergent, where cause and effect relationships can be perceived but not clearly defined or predicted and events are fully understood only in retrospect; the chaotic or random where no apparent cause and effect relationships are evident; and the central space of disorder where conflicting views reside resulting from different perspectives on the same issue.

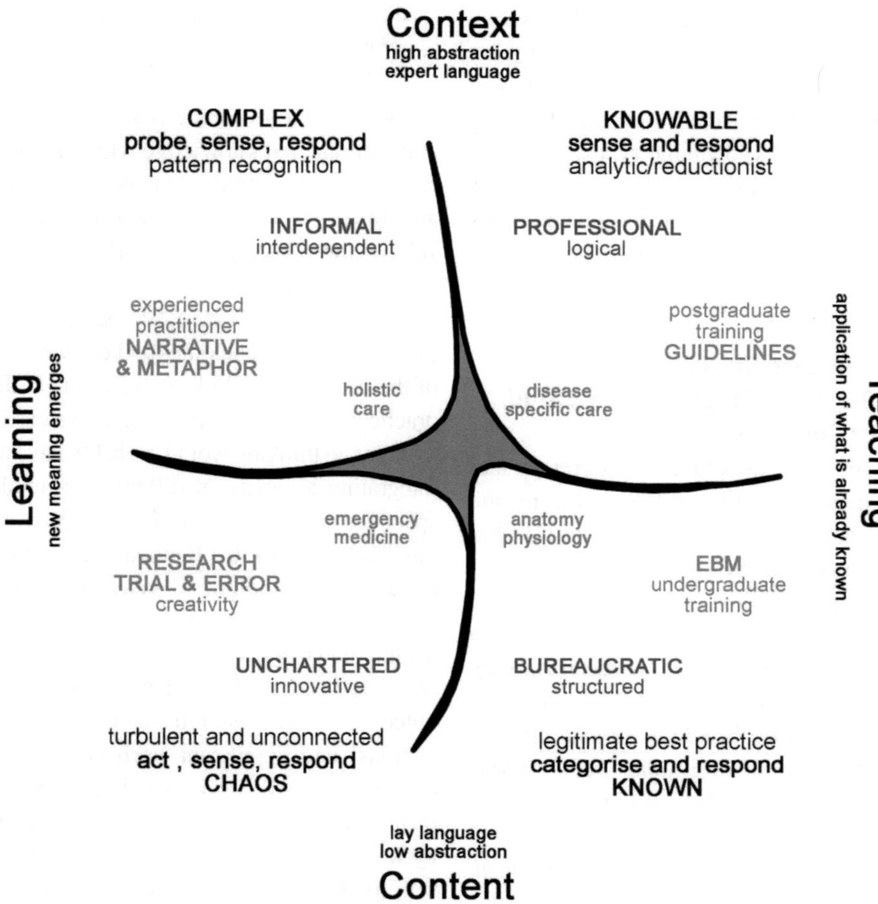

Fig. 45.2 The Cynefin model of knowledge in medicine

"Medical knowledge is inherently uncertain, and we require a context-driven flexible approach to knowledge discovery and application, in clinical practice as well as in health service planning" [34]. Thus, it is of utmost importance to understand "*knowledge" as a* personal construct *achieved through* sense making. *Specific knowledge aspects temporarily reside in either one of four domains—the known, knowable, complex, or chaotic, but* new knowledge *can only be created by challenging the* known *by moving it in and looping it through the other domains*" [34].

The knowledge distribution in medicine is depicted in Figure 45.2. Different disciplines predominantly practice and operate within one specific domain, utilizing specific knowledge dynamics. Standardization is achieved by developing and adhering to protocols and checklists, guidelines provide some discretion in decision making, pattern recognition relies to a large extent on sensing change before objective features are clearly evident, and "educated interventions" are the prerogative for chaotic situations. It is, however, important to realize that "[t]*hough in clinical care we may operate predominately in one knowledge domain, we also will operate some of the time in the others*" [34].

Complex adaptive systems science views knowledge simultaneously as a thing and a flow, as constructed, as well as constantly changing. Health knowledge is simultaneously explicit and implicit with certain aspects already well known and easily transferable, and others that are not yet fully known and must still be learned. At the same time, certain knowledge aspects are predominantly concerned with content, whereas others deal with context. While some fields like primary care act predominantly in a complex person-centered environment and surgery acts in a simple and complicated environment, all require the whole gamut of knowledge types. The skill and art of sense making in clinical care and in healthcare systems is to identify different domains and knowledge flows to facilitate appropriate practices.

45.2.3.1 The Case Study—Clinical Practice: Simple, Complicated, or Complex

The following day, the situation progresses and Mick becomes ill as Eileen leaves for respite care. The strain of caregiving takes its toll, but also there is an element of guilt and loss.

Mick Murphy—*Looking forward to having a break from Eileen for a few days. Exhausted today and has a bit of a cold, has been running around all morning trying to get Eileen ready to go into respite. Thinks he will be fine in a day or 2. Mick will be collecting some prescriptions from GP later in the week so if not feeling well then he will have a check-up with GP.*

Eileen Murphy—*ok as per Mick's assessment. Went into respite today and will be there until Friday afternoon, they don't have a bed over the weekend so Eileen will be home and will return to respite the following Monday for x2 weeks.*

45.2.3.2 The Importance of Narratives

According to Charon et al. [35], there are three fundamental tensions upon which medicine finds itself (when coming from a medical evidence paradigm)—how to manage health when there is known and unknown lack of evidence, how to particularize the universal or average case from statistical analyses to the individual journey, and how to personalize or embody the evidence about the body for an individual experience and of the patient journey. Their formulation of "Narrative-Based Medicine" complements "Evidence-Based Medicine" in that:

> Clinical evidence examines the known and unknown. Clinical circumstances integrate the universal and particular. Patients' values speak to both body and self. By virtue of its capacity to recognize the tensions fully, narrative medicine can lend to evidence-based medicine the methods of respecting its three circles of attention [35].

What is the evidence base that is the simple and complicated knowledge with which the clinical care system should manage the care of Eileen and Mick? Is it as straightforward as the right side of the Cynefin framework being integrated with the left side?

Can we identify evidence-based interventions that will improve outcomes and can we make sense of the narratives of Mike and Eileen? Is there best practice or even better practice in this case?

Cynefin thinking would probably have preferred integrating a variety of different stakeholder perspectives and values, before unilaterally establishing what is best [36]. Reducing the known to the reductionist paradigm of science as "evidence" curtails a broader understanding of what is known and unknown about the nature of health. For example, interventions have demonstrated that better outcomes of care—significant decrease in hospitalizations, and significant decrease in emergency department visits, were associated with programs that had complex adaptive system characteristics—agents who learn, interconnections between agents, self-organization, i.e., order is created without explicit hierarchical direction; and coevolution, i.e., the patient, the healthcare providers and health system, and the environment influence each other's development [34].

Perhaps, we will find more useful information if we reflect on Bury's types of patient narrative:

- *Contingent narratives*—belief about the origins of disease, proximate causes in the illness journey, and the immediate impact of illness on physical health. Contingent narratives are most closely aligned with medical care and are personal and explanatory. Thus, health and illness are intensely personal matters [37].
- *Moral narratives*—The second type of narrative identified by Bury [38] are moral narratives which explore the relation between the person, his/her illness, and his/her social identity. This implies a threat to the individual's inner and social being by the biological components of their own body. An individual experiences loss during falling ill. The worst biographical events are ones that bring about

the loss of part of the individual's and affect both function and survival.

- *Core narratives*—The third type of narrative is a "core narrative" [38] that evokes the lay persons' connection with the prevailing cultural meanings pertaining to illness and suffering. Here, narratives can be seen to be "heroic, tragic, ironic, comic, or regressive-progressive."

In relation to the content of narratives, particularly the contingent, Martin [39] has pragmatically identified three narrative streams appropriate to clinical care:

1. Narratives of the person, their life stages, and, if applicable, their caregiver.
2. Narratives of the body and the disease and illness that describe the trajectory through all phases of the illness.
3. Narratives of the treatment journey—from prevention to palliation such narratives include the personal narratives of the physicians, nurses, and all healthcare team members involved in patients' healthcare journeys.

45.2.3.3 Returning to the Case Study

Eileen's narrative expresses her vulnerability, while Mick has a stoic and heroic narrative. Yet, when she goes into respite care, he becomes ill and "runs around looking for antibiotics" to keep him well. Both their bodies and minds are failing in different ways, due to the intense pressures of Eileen's mental condition. It is a story of deterioration; yet hope that support through respite care will bring relief. There is no narrative about disease on the devastation that disabling conditions have wrought—what is the root cause of what is making them ill? Thus, clinical care is challenged to make sense of how to "heal" people with complex states of health and illness, without protocols and standards to follow.

The polarities between the objective biomedical and the subjective well being as the dominant discourses between health professionals and patients are evident. It would be futile to expect major changes in health outcomes for this couple, yet major improvements in improving quality of life and alleviating illness can be achieved by respite and supportive care in the case of this couple. The tensions and contradictions of *wanting to be apart and yet wanting to stay together* are delicately balanced.

45.2.4 Weick and Sensemaking

Weick, a leading theorist in organizations sought to uncover simple patterns underlying what appears to be complex organizational behavior [21]. The central approach to his ideas and theory building was to find patterns that edit particulars into a more compact summary that allows people (including theorists) to anticipate and thread their way through the complexities of everyday social life. It is important to know that the project has been based on three central assumptions regarding communication practice (a) that it is possible to design and implement communication systems and practices that are responsive to human needs; (b) that it is possible for humans to enlarge their communication repertoires to pursue this vision; and (c) that achieving these outcomes requires the development of communication-based methodological approaches [40]. Environments coexist and to a large extent are created by the organization—the organization does not play a passive observational role but an active role in sense making and adaptation. Organizational systems theory sees healthcare organizations as a living adaptive "organism" [41]. Fluctuations or contingencies from the environment are adjusted to by organizational change. The nature of change can be strategic—in health systems (e.g., offering new ways of care delivery), tactical (e.g., developing closer relationships between key players and incentives) or cultural (e.g., offering staff training and support for change) [18] (Fig. 45.3).

45.2.4.1 The Case Study: Changing Personal Journeys Applying Weick's Lens

The original healthcare organization—the hospital—cared for the ill, the vulnerable, and the poor. As the modern Medical Industrial Complex (MIC) emerged due to ecological changes in technologies, culture, and expectation, hospitals and healthcare organizations changed. Standardized

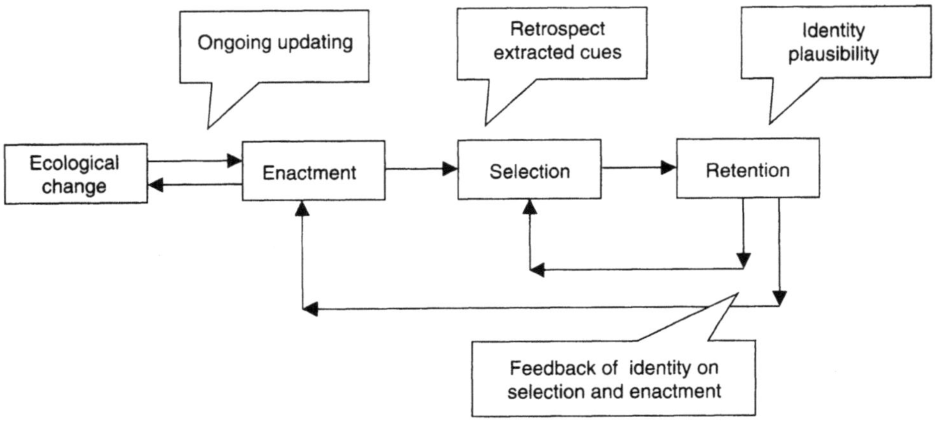

Fig. 45.3 The relationship among enactment, organizing and sensemaking (Source: Jennings and Greenwood 2002 [40] who adapted from Weick 1979;139 [41])

work processes and practice became the norm for mass delivery of health care. This process ignores the enactment, organizing, and sensemaking feedback loops identified by Jennings [41]. Yet like the earlier QOF example, results in high expenditures on standardized activities at the expense of other more valuable initiatives, services, or nonmeasurable aspects of patient care. In this case, there are no ways to deal with the care of Eileen and Mick, who fall outside the metrics, yet utilize significant public and private resources. How can we sensitize health systems to the real ecology of real patient journeys, rather than to the profit margins of large and small investors? The Medicare Innovation Fund is seeking to transform American medicine and health care to do just that, but it needs to make progress on sensitizing care incentives to respond to more complex cases [42, 43].

45.2.4.2 Implications for Healthcare Practice

Taking on leadership in complex health organizations, Thygeson, Morrissey, and Ulstad [44] using organizational complexity models, espoused initially by Heifetz [45], translate organizational learning into adaptive leadership in the doctor–patient relationship. These approaches seek to make sense of emerging identities, enactments, and dynamic feedback between the external and internal health system ecology. Adaptive leadership by definition straddles the polarities, tensions and contradictions of the clinical encounter, and the higher levels of clinical care and healthcare organizations. It will be interesting to see how the work of the clinician, demonstrated by Katerndahl et al. [46], evolves with more emerging frameworks and innovative leadership models.

45.2.4.3 The Case Study: Applying the New Lens

In relation to the case study, the organizations that treat and manage the care of Eileen and Mick need to recognize the real nature of their journey, rather that deliver standardized "disease management," standardized "behavioral health," and standardized chronic illness "self-management," which have no impact on their deterioration.

The ecology of health care will continue to change rapidly. The democratization of social media and networking and the pervasive pressure from internet technologies, which increasingly shifts the process of personal meaning making, to online and collective sensemaking. However, will this participation be truly innocent or will the loudest voices rather than the voices of the disempowered, chronically ill, and less educated be swamped by others more articulate and well resourced.

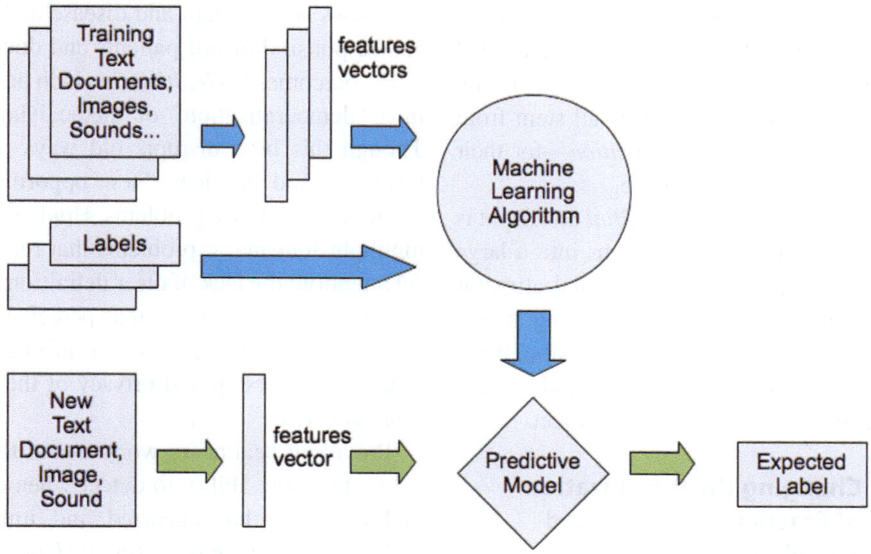

Fig. 45.4 Example of Machine Learning Concepts (See Chap. 27 Table 27.1 for more information)

45.2.5 Another Way of Sensemaking: Informatics, Computations and Modeling

Alan Turing, Benoit Mandelbrot, and John von Neumann among others pioneered the mathematical and computational analyses of complex systems. An important aspect of their work is the appreciation of the *initial condition* of the system before entering new information; even very small difference at the start can result in very major differences after even few iterations. Adding new information to a system impacts on its feedback processes which in turn shape the behavior of complex interacting systems.

Some systems with many interactions among highly differentiated parts produce surprisingly simple, predictable behavior (such as a programmable mechanical routine or process), while others generate behavior that may be impossible to predict, even though these systems feature simple laws and few actors or agents (e.g., the mathematic equation resulting in the Mandelbrot set) [47].

This early work has been taken further by others and forms the basis for modern information systems in health care [48]. Being able to handle a vast amount of clinical data has the potential to bring about large improvements in clinical, administrative, and financial healthcare delivery. Analyzing large datasets, particularly where the data have been merged from several sources, introduces problems of data integrity and noise. The challenge is to transform these data into information that then can be turned into new knowledge.

Machine learning and predictive modeling are techniques used in data mining[4] and have the potential to facilitate the examination of the dynamics of disease and the improvement of clinical quality. By its nature the field of complex adaptive systems deals with environments that change because of the interactions that have occurred in the past. Health informatics and bioinformatics already successfully use machine learning methods to stratify patients who need intensive support to keep them from having multiple expensive hospitalizations [49] yet such analytics can only go so far in understanding the complexities of the human journey and relationships (Fig. 45.4).

45.2.5.1 The Case Study: Applying this Lens

In our case study, in order to understand how Mick and Eileen arrived at their current health destinations, it is important to understand the *initial*

[4] For example, cluster analysis, principle component analysis, decision tree learning, Bayesian network models, artificial neural networks, and genetic programming.

conditions of their situation. Eileen was involved in a car accident, with Mick driving, which resulted in her injured spine. Their complex relationship, his drinking, her bipolar diagnosis all stem from the car accident—the *initial condition*—for their current "organization" of their life.

A healthcare system whose *initial condition* is a small business model that morphs into a large business model, has as its key internal attractor "financial gain", which makes adapting to the "external health needs of people" a tension, if not a contradiction. Service strategies and patient need and expectation are unlikely to meet.

45.2.5.2 Changing the Organization of Practice, Research, and Knowledge

The introduction of clinical changes is a cyclic task, meaning that the processes under examination operate in an environment that is not static. Traditional linear methods of analysis cannot address the nonlinearity and recursive feedback loops, that complexity and chaos theory and artificial intelligence and machine learning are designed to achieve [50].

Continuous flow of information is now possible and increasingly essential. The time it takes to cycle and feedback information must be as short as possible—and real time in many situations. On the other hand, one needs time for systems to reach some level of stability in order to function well and this takes time. Reflection is needed, even in systems like healthcare delivery that are fast paced.

Although these technologies greatly enhance information flow and information quality, this does not automatically equate to providing better knowledge that would result in wise decisions. Cilliers pointed to the importance of *a certain slowness* for the effective and efficient functioning of a complex adaptive system [51]. Sturmberg and Cilliers highlighted the implications of *such slowness* to medical practice—time is required to humanize medical care [52].

Health 2.0, the Democratization of Health Care

The rapid developments within the communications and electronics industries have challenged our views about health and disease and the roles and responsibilities of patients and doctors. One of the outcomes is *Health 2.0*, which has resulted in a "democratization" of medical knowledge. Though this both disrupts old ways of "doing business" and provides "new opportunities," it has its very real own problems. Hughes et al. [53] highlight four major problems that require careful attention: the lack of clear definitions; the loss of control over information as perceived by doctors; the safety and accuracy of information; and issues of ownership and privacy of the information collected.

Providing healthcare workers and their clinical systems the ability to detect when an abnormal condition has occurred and immediately address the problem is central to *Health 2.0*. This enables operations to build-in quality at each process and to separate men and machines for more efficient work (Fig. 45.5).

Continuous Quality Improvement in Health Care

Many of the continuous quality improvement strategies in healthcare have been adopted from the Toyota Production System (see Appendix). Jidoka, along with just-in-time, are the two pillars of this system. Jidoka is sometimes called autonomation, meaning automation with human intelligence. It is a quality control process that involves four principles: (1) detect the abnormality; (2) stop; (3) fix or correct the immediate condition; and (4) investigate the root cause and install a countermeasure. It is a system of production that makes and delivers just what is needed, just when it is needed, and just in the amount needed. Continuous improvement of an entire value stream or an individual process can create more value with less waste. There are two levels of improvement: (1) system or flow improvements that focus on the overall value stream and (2) process improvements that focus on individual processes. These approaches have, e.g., been applied in clinical delivery of better timed insulin administration systems [54], as well as quality improvement in organizational behavior [55].

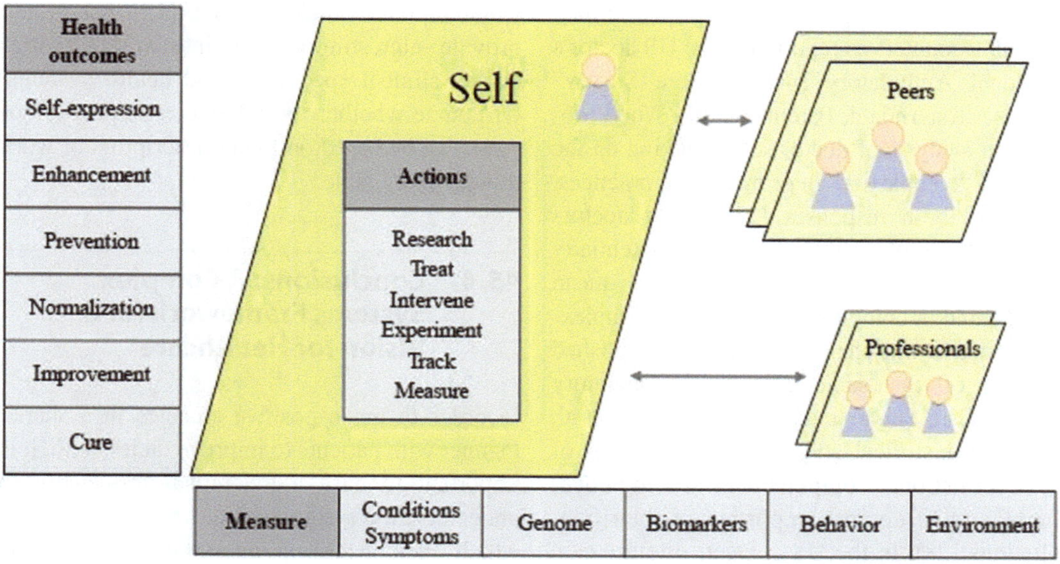

Fig. 45.5 A model of Health2.0 (reproduced under Wikimedia Commons)

Solving "Wicked" Problems

The Complex Network Electronic Knowledge Translation Research Model[5] [56] represents an attempt to address complex problems, like health promotion, using a social organizing approach based on complexity science, social learning theories, design thinking, and models of knowledge exchange, translation, and integration. The model, developed by Norman and his colleagues, approaches sense making across polarities of knowledge and values by mobilizing the diverse strengths and knowledge of multiple key stakeholders. The "new" knowledge about complex problems provides the basis for creating viable response options within the constraints of existing public health and healthcare practices.

45.3 Discussion

Sense-making emerges as a transdisciplinary process. The seamless exchange of knowledge and research approaches across traditional disciplinary silos provides new ways of conceptualizing and addressing systemic questions and solutions [21].

Key components that contribute to sense-making include [22] the following:

- Identifying polarities, contradictions and tensions in health systems knowledge—including nonlinear linear patterns in quantitative research, diverse values and perspectives,
- Identifying the predominant appropriate patterns for different health related activities—from checklists to probing and sensing, and to creative problem solving,
- Developing the intellectual foundation for multiple ways of knowing with a neutral space for negotiation, discourse and conflict resolution; with the development of
- A system of values that is open, participatory, respectful and focused on the "real world" of individuals, and
- Principles for communication and leadership.

"Innocent information" that is shaped by the pervasive polarities of our times and the directions by dominant professional, political, economic, and media discourses, must be continually challenged by democratizing processes of sense-making and decision making. Increasingly, this should be accomplished by engaging patient as central to the process.

Around a time of major health system reform, primary care doctors in the USA are reflecting on

[5] For more details, see Norman and Yip., Chap. 34.

the increased "complexity" in their everyday clinical encounters as reported in the US doctor's "National Ambulatory Medical Care Survey" database. Katerndahl, Parchman, and Wood provided an analysis of doctor sense-making on the nature of "complexity" in primary care practices, based on these responses [28]. What doctors mean by "complexity" is fuzzy and potentially not objective or accurate, and whether they mean complicated or complex in the sense of complexity science is not clear. In a recently published analysis of the same National Ambulatory Medical Care Survey database, Katerndahl et al. propose a historical and innovative analysis of the (informational, computational, and cognitive) complexity of doctors' reporting of their consultations [28]. In the US context, primary care doctors are perceived by decision makers to operate a lower order of cognitive sophistication compared to other medical specialties such as cardiology or psychiatry. However, when shifting the perspective from relative work value based on the severity of illness or disturbance of a constrained bodily system such as in cardiology or psychiatry with a limited range of problems and therapies to analyzing diversity and variability of clinical processes [28], family medicine patterns demonstrated a greater diversity of diagnostic and knowledge inputs with more treatments and other outputs and less time to process these cognitive steps.

They represent another form of sense making with nonlinear modeling using information theory and computational complexity. A series of polarities and questions emerge for further research and analysis, and ultimately decision and policy making.

Are these data accurate or inaccurate, clear or unclear, complex or complicated? Does increasing diagnostic and therapeutic diversity, and broadening scope of practice from cradle to grave in rapidly changing social contexts, amount to increasing complexity and thus increased work value for family medicine? Further in-depth analysis of the actual work of decision making and sense-making in clinical practice by doctors, patients and others, linked to current research on clinical decision support, would produce valuable outputs [13]. This would provide interesting insights into work activities across clinical specialties and health systems. Of course whether the relative "complexity" of work can be based on linear descriptors of work though is arguable.

45.4 Conclusions: A Complex Systems Framework for a Vision for Healthcare

In order to make positive changes in a shared manner with patients to improve their health, it is important to have a theoretical framework to understand the system and system level states, in which their health journey takes place. The framework of the "complex adaptive system" defines the scope, boundaries, and limits; it locates the current state in relation to shared vision, values, and goals [57].

The vision and goals of a system provide *leverage points* where specific interventions are most likely to be effective for a "whole of systems" change, and thus a greatest potential for improvement of system performance [58]. The key unit of knowledge and understanding is the pattern that cuts across linear and nonlinear representations of the real world in the form of evidence, narrative and stories, tacit and explicit information and experiences.

A framework for understanding and intervening in the system of an individual patient narrative would contain four types of leverage points around pattern identification at different health system levels:

1. Values/beliefs leverage points
   - Related to the intrinsic philosophy that is fundamental to the individual's experience of care.
2. Goals leverage points
   - Related to the expectations and intended outcomes of interventions change.
3. Information leverage points
   - Related to the availability of narrative feedback to health professionals and system stakeholders.

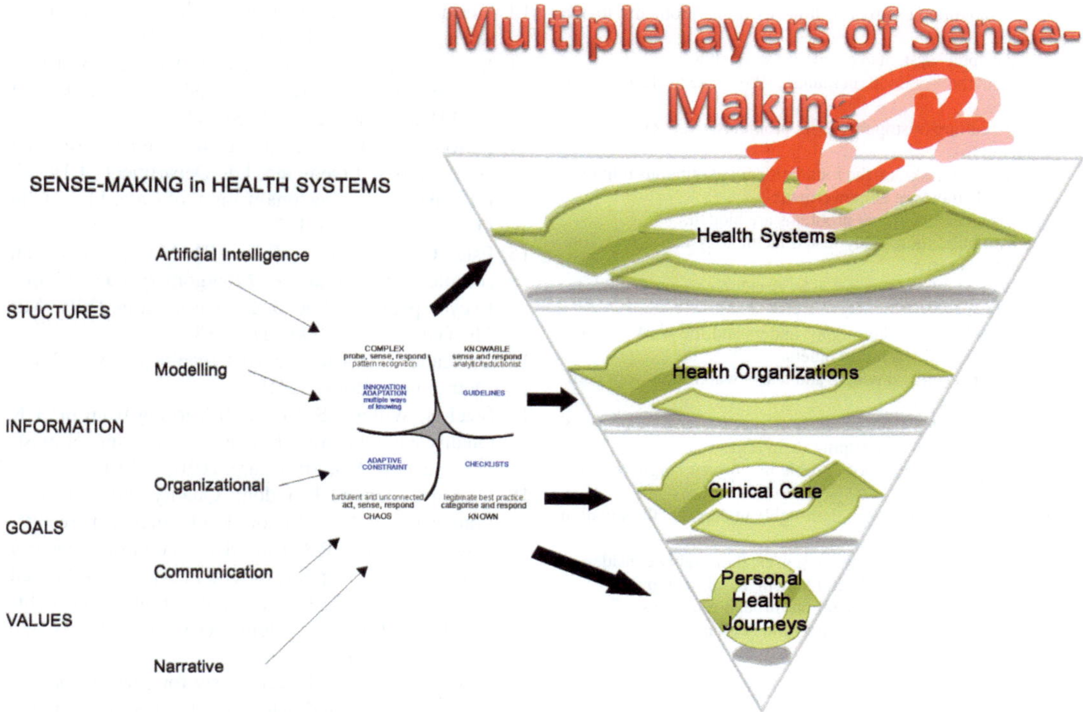

Fig. 45.6 Sense making with different approaches at multiple interconnected health system layers within a complex adaptive systems framework

4. Structures leverage points
   - Related to specified roles, responsibilities, and authorities that define the boundaries of the patient journey and enable a healthcare provider to perform their functions.

Sensemaking about different levers in health systems range from the micro- to the macro-levels, all of which are highly interconnected (Fig. 45.6).

In conclusion, there are narratives and a range of emerging typologies of narratives. In order to make sense of these narratives, it is important to identify a clear frame of reference with the aim of locating the narratives under consideration in their appropriate theoretical, conceptual and operational frameworks. Only then, one can identify and make sense of the rich patterns that emerge from our patients' stories in everyday practice, and link them to the increasingly complex knowledge of interdependencies of human biology and environments. Clinical care and larger health ecosystems inevitably shape and should be shaped by their patients' narratives and experiences.

45.5 Appendix

The underlying principles, called the Toyota Way, have been outlined by Toyota as follows:

Continuous improvement

- ***Challenge*** *(We form a long-term vision, meeting challenges with courage and creativity to realize our dreams.)*
- ***Kaizen*** *(We improve our business operations continuously, always driving for innovation and evolution.)*
- ***Genchi Genbutsu*** *(Go to the source to find the facts to make correct decisions.)*

Respect for people

- ***Respect*** *(We respect others, make every effort to understand each other, take responsibility and do our best to build mutual trust.)*
- ***Teamwork*** *(We stimulate personal and professional growth, share the opportunities of development and maximize individual and team performance.)*

External observers have summarized the principles of the Toyota Way as:

- Long-term philosophy
- Base your management decisions on a long-term philosophy, even at the expense of short-term financial goals.

The right process will produce the right results

- Create continuous process flow to bring problems to the surface.
- Use the "pull" system to avoid overproduction.
- Level out the workload (heijunka). (Work like the tortoise, not the hare.)
- Build a culture of stopping to fix problems, to get quality right from the first.
- Standardized tasks are the foundation for continuous improvement and employee empowerment.
- Use visual control so no problems are hidden.

Use only reliable, thoroughly tested technology that serves your people and processes.

- Add value to the organization by developing your people and partners
- Grow leaders who thoroughly understand the work, live the philosophy, and teach it to others.
- Develop exceptional people and teams who follow your company's philosophy.
- Respect your extended network of partners and suppliers by challenging them and helping them improve.

Continuously solving root problems drives organizational learning

- Go and see for yourself to thoroughly understand the situation (Genchi Genbutsu).
- Make decisions slowly by consensus, thoroughly considering all options (Nemawashi); implement decisions rapidly.
- Become a learning organization through relentless reflection (Hansei) and continuous improvement (Kaizen).

References

1. Heisenberg W. Physics and beyond. New York: Harper & Row; 1971. p. 63.
2. Capra F. The web of life. London: HarperCollins; 1996.
3. Capra F. The systems view of life, in the turning point: science, society, and the rising culture. New York: Simon and Schuster; 1982.
4. Castellani B. Sociology and complexity science: a new field of inquiry. Understanding complex systems. 1st ed. New York: Springer; 2009.
5. Coleman K, et al. Untangling practice redesign from disease management: how do we best care for the chronically ill? Annu Rev Public Health. 2009;30:385–408.
6. McDonald R, Roland M. Pay for performance in primary care in England and California: comparison of unintended consequences. Ann Fam Med. 2009;7(2):121–7.
7. Bardach NS, Cabana MD. The unintended consequences of quality improvement. Curr Opin Pediatr. 2009;21(6):777–82.
8. Reschovsky JD, et al. Following the money: factors associated with the cost of treating high-cost Medicare beneficiaries. Health Serv Res. 2011;46(4):997–1021.
9. Segal. Available at http://www.selegal.org/glossary.htm.
10. Rula EY, Pope JE, Stone RE. A review of healthways' medicare health support program and final results for two cohorts. Popul Health Manage. 2011;14(S1):S3–S10.
11. Averill RF, et al. Achieving cost control, care coordination, and quality improvement through incremental payment system reform. J Ambul Care Manage. 2010;33(1):2–23.
12. Sterman JD. Learning from evidence in a complex world. Am J Public Health. 2006;96(3):505–14.
13. Joosten T, Bongers I, Janssen R. Application of lean thinking to health care: issues and observations. Int J Qual Health Care. 2009;21(5):341–7.
14. Fulford KWM. Bringing together values-based and evidence-based medicine: UK Department of Health Initiatives in the 'Personalization' of Care. J Eval Clin Pract. 2011;17(2):341–3.
15. Estes C, Harrigton C, Pellow D. Medical industrial complex. In: Borgatta E, Montgomery RJV, editors. Encyclopedia of Sociology. Farmington Hills, MI: The Gale Group. 2000;1818–1832.
16. Gawande AA. The checklist manifesto. New York: Metropolitan Books; 2010.
17. Steel N, Willems S. Research learning from the UK Quality and Outcomes Framework: a review of existing research. Qual Prim Care. 2010;18(2):117–25.
18. Cashin C. United Kingdom: Quality and Outcomes Framework (QOF). Major developments in results-based financing (RBF) in OECD countries: country summaries and mapping of RBF Programs 2011 [cited 2012 February 26th]. Available at http://www.rbf-health.org/rbfhealth/system/files/Case%20study%20UK%20QOF.pdf.
19. Serumaga B, et al. Effect of pay for performance on the management and outcomes of hypertension in the United Kingdom: interrupted time series study. BMJ. 2011;342:d108.
20. Dervin B, Foreman-Wernet L. The sense-making methodology reader: selected writings of Brenda Dervin. Cresskill, NJ: Hampton Press; 2003.
21. Weick K, Sutcliffe KM, Obstfeld D. Organizing and the process of sensemaking. Org Sci. 2005;16(4):409–21.
22. Wikipedia. Sensemaking. Available at http://en.wikipedia.org/wiki/Sensemaking. Accessed 27 Feb 2012.
23. Cohen IR. Tending Adam's Garden: evolving the cognitive immune self. London: Academic Press; 2000.
24. Snowden DJ. Multi-ontology sense making: a new simplicity in decision making. Inform Prim Care. 2005;13(1):45–54.
25. Weick K. Sensemaking in organisations. In: Whetten D, editor. Foundations for organizational science. London: Sage; 1995. p. 1–231.
26. West BJ, Grigolini P. Complex Webs: anticipating the improbable. New York: Cambridge University Press; 2011. p. 375.
27. Guastello SJ. Chaos as a psychological construct: historical roots, principal findings, and current growth directions. Nonlinear Dyn Psychol Life Sci. 2009;13(3):289–310.
28. Katerndahl DA, Wood R, Jaen CR. A method for estimating relative complexity of ambulatory care. Ann Fam Med. 2010;8(4):341–7.
29. Horvitz E. From data to predictions and decisions: enabling evidence-based healthcare. Series on Data Analytics, Computing Community Consortium; 2010.

30. Riaño D, European Society for Artificial Intelligence in Medicine. Knowledge representation for healthcare: data, processes and guidelines: AIME 2009 workshop KR4HC 2009, Verona, Italy, July 19, 2009: revised selected papers. Lecture notes in artificial intelligence, subseries of lecture notes in computer science. New York: Springer; 2010. p. 194.
31. Vogel C. N-gram distributions in texts as proxy for textual fingerprints. In: Esposito A et al., editors. The fundamentals of verbal and non-verbal communication and the biometrical issue. Amsterdam: IOS Press; 2007. p. 189–94.
32. Martin CM. Making sense of polarities in health organizations for policy and leadership. J Eval Clin Pract. 2010;16(5):990–3.
33. Kurtz C, Snowden D. The new dynamics of strategy: sense-making in a complex and complicated world. IBM Syst J. 2003;42(3):462–83.
34. Sturmberg JP, Martin CM. Knowing–in medicine. J Eval Clin Pract. 2008;14(5):767–70.
35. Charon R, Wyer P, NEBM Working Group. Perspectives, narrative evidence based medicine. Lancet. 2008;371(9609):296–7.
36. Simone JD. Reductionist inference-based medicine, i.e. EBM. J Eval Clin Pract. 2006;12(4):445–9.
37. Bury M. Health and illness in a changing society. London: Routledge; 1997.
38. Bury M. Illness narratives: fact or fiction? Sociol Health Illn. 2001;23(3):263–85.
39. Martin CM. The care of chronic illness in general practice. PhD Thesis, National Centre for Epidemiology and Population Health, Australian National University: Canberra; 1998.
40. Jennings P, Greenwood R. Enactment as an Institutional Change Mechanism: A Counterpoint to Weick. In: Westwood R, Clegg S, editors. Paradigm Wars. London: Blackwell Publishing; 2002.
41. Weick KE. The Social Psychology of Organizing, second edition. New York: McGraw Hill. 1979:139.
42. Charland K. Pay for performance comes to Medicare in 2009. Healthc Financ Manage. 2007;61(9):60–4.
43. McClellan M, et al. A national strategy to put accountable care into practice. Health Aff. 2010; 29:982–90.
44. Thygeson M, Morrissey L, Ulstad V. Adaptive leadership and the practice of medicine: a complexity-based approach to reframing the doctor–patient relationship. J Eval Clin Pract. 2010;16(5):1009–15.
45. Heifetz RA. Leadership without easy answers. Cambridge: Harvard University Press; 1994.
46. Katerndahl D, Parchman ML. Dynamical difference in patients encounters involving uncontrolled diabetes: an orbital decomposition analysis. J Eval Clin Pract. 2010;16(1):211–9.
47. Tan J, Wen HJ, Awad N. Health care and services delivery systems as complex adaptive systems. Commun ACM. 2005;48(5):36–44.
48. Westbrook JI, et al. Multimethod evaluation of information and communication technologies in health in the context of wicked problems and sociotechnical theory. J Am Med Inform Assoc. 2007;14(6): 746–55.
49. Zmiri D, Shahar Y, Taieb-Maimon M. Classification of patients by severity grades during triage in the emergency department using data mining methods. J Eval Clin Pract. 2012;18(2):378–88.
50. Marsland S, Buchan I. Clinical quality needs complex adaptive systems and machine learning. Stud Health Technol Inform. 2004;107(Pt 1):644–7.
51. Cilliers P. On the importance of a certain slowness. Stability, memory and hysteresis in complex systems. Emerg Complex Org. 2006;8(3):106–13.
52. Sturmberg JP, Cilliers P. Time and the consultation – an arguement for a "certain slowness". J Eval Clin Pract. 2009;15(5):881–5.
53. Hughes B, Joshi I, Wareham J. Health 2.0 and Medicine 2.0: tensions and controversies in the field. J Med Internet Res. 2008;10(2):e23.
54. Yamamoto JJ, et al. Facilitating process changes in meal delivery and radiological testing to improve inpatient insulin timing using six sigma method. Qual Manag Health Care. 2010;19(3):189–200.
55. Juarez A, et al. Barriers to emergency departments' adherence to four medication safety-related Joint Commission National Patient Safety Goals. Jt Comm J Qual Patient Saf. 2009;35(1):49–59.
56. Norman CD, et al. Designing health innovation networks using complexity science and systems thinking: the CoNEKTR model. J Eval Clin Pract. 2010;16(5):1016–23.
57. Martin CM, Kaufman T. Addressing health inequities: a case for implementing primary health care. Canadian Family Physician Médecin de Famille Canadien 2008;54(11):1515–7, http://www.biomed-experts.com/Experts/Abstract.aspx?pid=19005107.
58. Martin CM, Kaufman T. Addressing unfinished business in primary health care (PHC): a shared framework across Canada is needed to implement PAHO/WHO 'new orientations' for adaptive PHC-based health systems. In: A Commissioned Report for the Canadian Alliance of Community Health Centre Associations; Primary Health CareTransition Fund (PHCTF) National Evaluation Strategy. Northern Ontario School of Medicine: Ottawa; 2007.

Modeling the Paradox of Primary Care

46

Johnie Rose, Rick Riolo, Peter Hovmand, Sarah Cherng, Robert Ferrer, David A. Katerndahl, Carlos R. Jaén, Timothy Hower, Mary C. Ruhe, Heide Aungst, Ana Diez Roux, and Kurt C. Stange

A paradox exists in the outcomes of primary care: despite delivering apparently poorer quality disease care compared to that delivered by specialists, primary care is associated with better population health, lower inequality, and lower cost. Understanding the dynamics that give rise to this paradox could lead to better-informed interventions to promote more patient-centered, holistic, equitable, and cost-effective models of care. In this chapter, we articulate the paradox and how complexity science principles can make sense of its contradictions. We suggest a novel approach to advancing understanding through a participatory group modeling process to build and conduct experiments with an agent-based computational model.

J. Rose, M.D., Ph.D. (✉) • M.C. Ruhe • H. Aungst • K.C. Stange, M.D., Ph.D.
Department of Family Medicine, Division of Research, Case Western Reserve University School of Medicine, 11000 Cedar Ave, Ste. 402, Cleveland, OH 44106-7136, USA
e-mail: jxr109@case.edu; kcs@case.edu

R. Ferrer, M.D., M.P.H. • D. Katerndahl, M.D. • C.R. Jaén, M.D., Ph.D.
Department of Family & Community Medicine, University of Texas Health Science Center at San Antonio, San Antonio, TX 78229-3900, USA
e-mail: ferrerr@uthscsa.edu; katerndahl@uthscsa.edu; jaen@uthscsa.edu

R. Riolo
Center for the Study of Complex Systems, University of Michigan Ann Arbor, MI, USA
e-mail: rlriolo@umich.edu

P. Hovmand
Director, Social System Design Lab, Brown School, USA
e-mail: phovmand@wustl.edu

S. Cherng
University of Michigan, Ann Arbor, Michigan, USA
e-mail: scherng@umich.edu

T. Hower
Associate Director, Social System Design Lab, Brown School, USA
e-mail: thower@gwbmail.wustl.edu

A.D. Roux
Center for Social Epidermiology and Population Health, University of Michigan, School of Public Health, USA
e-mail: adiezrou@umich.edu

46.1 The Paradox of Primary Care

The US healthcare system provides an illustrative example and a cautionary tale for the world. Its strikingly poor performance (ranked 37th by the World Health Organization [1] despite spending far more than any other nation on health care [2]) often is attributed to fragmentation [3] and the undervaluation of primary care [4–7].

Despite its apparent importance for a high-value healthcare system, there is a paradox about primary care [8, 9]. Many studies show that primary care clinicians provide disease care that is less concordant with evidence-based guidelines than care provided by specialists in those diseases [10–12]. Yet, research at the level of whole people, communities, and countries finds that primary care is associated with better quality [13, 14], lower cost [15–17], less inequality [18–23], and better health for whole people and populations [16, 17, 24, 25].

J.P. Sturmberg and C.M. Martin (eds.), *Handbook of Systems and Complexity in Health*, DOI 10.1007/978-1-4614-4998-0_46, © Springer Science+Business Media New York 2013

As primary care is reinvented in a new era [26–32], an understanding of the dynamics that give rise to the paradox of primary care (PPC) can be used to assure that patient-centered aspects of primary care that provide value are not unintentionally *de*valued by interventions that may improve disease care but fail to integrate, personalize, and prioritize care in ways that maximize its patient-centered and population outcomes [3, 33, 34].

46.2 Health Care as a Complex Adaptive System

The fundamental tenets of primary care—first contact accessibility, a comprehensive, whole-person approach, integration and coordination of care, and a focus on ongoing relationships with people, families, and communities [25, 35, 36]—bring added value beyond the reactive treatment of disease. This is true at both the individual and population levels. While the importance of primary care in an effective healthcare system may seem obvious, far less obvious are the *mechanisms* by which these tenets of primary care interact to produce better outcomes than would be predicted based on an aggregation of individual disease-specific processes and outcomes. Accordingly, understanding these mechanisms will not likely come through traditional reductionist approaches which attempt to explain phenomena in terms of a set of isolatable, linear cause–effect relationships. Applying the framework and tools of complexity science can allow movement beyond detection and description of a set of independent associations to a data-driven depiction of processes [37].

A number of features of healthcare delivery place it in the realm of *complex adaptive systems*. A complex adaptive system (CAS) can be defined as

> a collection of individual agents with freedom to act in ways that are not always totally predictable, and whose actions are interconnected so that one agent's actions change the context for other agents' [38].

Specific CAS features which are observable in healthcare delivery systems include:

- *Tight interconnections between some individuals* [39, 40]—In any system involving multiple autonomous agents, the actions of one person can, in turn, affect the actions of others. This can occur through competition for resources, family connections and other social influences, vicarious learning, or other means of altering the environment in which decisions are made or probabilistic events occur. Examples abound in the context of health systems. At a strictly biological level, transmission of infectious diseases is dependent on interactions between infectious and susceptible individuals. At the behavioral level, health practices [41]—both positive and negative—and health literacy [42] may spread across groups through learned behavior and knowledge acquisition. Indeed, it may be more difficult to conjure scenarios where the actions of an individual have no effect on others.
- *Interaction between individuals and their environments*—The actions of agents in complex adaptive systems affect their environments. In turn, these environments affect the health and behaviors of agents. Examples include the toxic effects of parental smoking on children, or the beneficial effects of increased fresh food availability brought about by increased demand at the neighborhood level. In both instances, the environment essentially serves to mediate the effects of individuals on other individuals. Further, healthcare system features (such as insurance differences or cultural and language barriers) can alter access for some groups [43–45]. The implications of this feature of complex systems are that location and membership matter and that attempts to understand complex processes may be hindered by a failure to account for the impact of place and relationships.
- *Numerous feedback loops*—In complex systems, the outcomes of a process frequently also function as determinants of that process. The resulting balancing (negative feedback) and reinforcing (positive feedback) behavior may serve to either mitigate or amplify the effects of perturbations to the system [39, 46]. In addition to the numerous balancing feedback loops involved in maintaining physiological homeostasis, an example of balancing

loops at a higher level can be seen in long-term patterns of physician specialty choice. Shortages in a specialty may bid up salaries and draw medical graduates toward a specialty with current or impending personnel shortages. To the extent that these higher salaries successfully lure young physicians into that specialty, the growth rate of salaries may begin to level off, ratcheting down the financial incentive to enter the specialty [47]. The result, over the long term, is a mix of practitioners whose practice specialties reflect the demands (if not the needs) of payers and patients [48].

A common example of a reinforcing feedback loop, on the other hand, is the tendency for inactivity to beget inactivity. As individuals become less active, they may gain weight. This weight gain renders physical activity more difficult and more likely to cause injury, leading to greater weight gain in a vicious cycle [49]. As another example consider the so-called "death spiral" seen in the pricing of health insurance plans which offer richer benefits. An initial small price differential between plan A (a restrictive health insurance plan with significant co-pays) and plan B (a higher cost, but more comprehensive low out-of-pocket cost plan with significant patient choice of providers) may grow exponentially larger over time as the lowest utilizing individuals in Plan B each year make a rational economic decision to switch to Plan A—leaving behind an ever-shrinking pool of patients with ever-higher expected levels of utilization in Plan B [50].

- *Temporal lags and tradeoffs* [46]—Distinct downstream consequences of an individual's action or of a system perturbation often occur on different timescales so that near-term and long-term responses may differ significantly. Negative health behaviors which lead to short-term pleasure but cause chronic disease provide classic examples at the individual level. In terms of health systems, Sterman [46] distinguishes *low-leverage policies* from *high-leverage policies.* Low-leverage policies require a significant effort for a net benefit which, in the long term, is ultimately quite small. High-leverage policies, on the other hand, can generate significant gains from small efforts. A characteristic of many complex systems such as health systems is that the policies favoring short-term proximal improvements tend to be low leverage, leading to negative unintended consequences in the longer term which offset initial improvements. In contrast, high-leverage policies tend to create "worse before better" scenarios, yielding benefits over a longer time period [51, 52]. Failure to understand or consider downstream consequences can yield low-leverage policies that ultimately exacerbate problems. Consider, for example, a practice driven by rising costs and stagnant growth in reimbursement to shorten appointment times and expand their panel of patients in an attempt to increase revenue. In order to accommodate these shorter visits, the providers in the practice (consciously or otherwise) decrease the amount of time spent on chronic disease prevention and patient education. This reduces costs in the short term. In an environment of capitations, the resulting increased revenue may ultimately be dwarfed by the additional costs associated with more frequent chronic disease exacerbations. This ultimately increases the costs in the long term. An effective high-leverage solution might therefore aim to increase chronic disease prevention efforts, actually increasing short-term costs.
- *Adaptability*—In complex adaptive systems, individuals are not bound to behave according to the same rules over time. Experience (either personal or vicarious) can modify their rules of behavior; that is to say that they may *learn.* At an individual level, this could manifest in improved health behaviors following education by a provider or, more dramatically, following a myocardial infarction. Adaptation may also occur at higher, subsystem levels—for instance, at the level of the institution. Frequently, subsystems *coevolve* in that an adaptive response by one changes the environment so that the other subsystem must in turn adapt [40]. Consider, for instance, the coevolution that occurred throughout the 1990s as large-scale consolidation of insurers was

answered by subsequent consolidation of hospitals and providers in many large healthcare markets [53]. Coevolution occurs among different agents across multiple levels within healthcare systems [54]. For example, primary care practices adapt their operations in response to healthcare system and payment structure shifts [55, 56], and the resulting changes in practices and their capacity to meet population health needs result in system reorganization and policy-level changes [40, 57]. The cascading changes that arise through coevolution lead to *perpetual novelty*—versus convergence to equilibria—that is characteristic of complex adaptive systems [58].

- *Threshold behavior*—Many phenomena which play a role in health and healthcare delivery are characterized by relationships where large effects manifest only once some critical threshold in inputs is surpassed. These relationships are seen across numerous scales, from the accumulation of stimuli which induce the firing of a single neuron to societal "tipping points" [59, 60]. The very act of seeking care abides by this type of pattern. Individuals tend not to seek a tiny amount of health care on a day when they feel slightly less well than usual. Instead, there is some threshold of illness or of concern beyond which individuals will make the conscious decision to seek care.

Phenomena associated with the fundamental CAS features described above can coalesce to give rise to the following properties:

- *Nonlinear relationships*—Often, in complex adaptive systems, effects are not proportional to inputs over a broad range. Feedback loops, adaptation, and thresholds, in particular, may conspire over time to produce systems which either are remarkably stable or could exhibit dramatically different system behavior depending on small differences in inputs [61, 62].
- *Emergence*—Though nonlinear, the general form of many complicated relationships may be predictable, even if the specifics are not. For instance, despite the many parameters and subsystems involved, we understand the nonlinear relationship between declining blood pressure and onset of fatal dysrhythmias within fairly narrow bounds. Similarly, we are able to predict approximate homeostasis in concentrations of blood electrolytes in a healthy person despite fluctuating water and salt intake [63]. A hallmark of complex adaptive systems, however, is the appearance of patterns which might not have been predictable a priori even with a very high-resolution understanding of all subsystems. Interdependent processes among interconnected individuals embedded in a responsive environment sometimes produce unpredictable results. Emergent phenomena are not predictable based on the aggregated expected outcomes from artificially isolated subsystems [64]. This is a key difference between systems which are merely complicated and those which are complex [65].

46.3 Why a Systems Modeling Approach

> The capacity of the human mind for formulating and solving complex problems is very small compared with the size of the problem whose solution is required for objectively rational behavior in the real world… (HA Simon [66]).

This principle of *bounded rationality* proffered by Herbert Simon [67] suggests the perils of linear solutions to complex problems. A human tendency exists to seek simplified causal relationships where effect is predictable and follows cause in close succession. As discussed in the previous section, complex systems often do not exhibit this type of behavior.

A staple of causal analysis is multivariable regression, a type of statistical modeling in which "empirical regularities" [37] are sought by attempting to isolate independent associations between effectors and outcomes [40]. This, by definition, is a reductionist approach. Such uncoupling of processes, which interdigitate in the real world, results in a loss of complexity. The inclusion of interaction terms in regression analyses may account in a limited way for effect modification. But, in doing so, one is still bound to certain distributional assumptions and to deterministic predictability.

Systems modeling approaches, on the other hand, attempt to model processes in a way that preserves complexity. One such systems modeling approach is *system dynamics* (SD) modeling [68]. System dynamics models consist of sets of nonlinear ordinary differential equations which describe flows and accumulations of people, money, knowledge, or virtually any other entity of interest over time. The equations are initially based on a conceptual or mental model of what are considered the relevant relationships with respect to the focal problem being studied, and then tested and revised using a variety of qualitative and quantitative data sources including key informant interviews, focus groups, direct observation, literature, and administrative data [68]. These simulations focus on feedback mechanisms that are hypothesized to generate a dynamic behavior (e.g., S-shaped growth, oscillation, equilibrium), reflected in one or more graphs over time. As such, SD models allow extensive experimentation in ways that would not be feasible or ethical in the real world, and in an environment where all assumptions are known and changeable. In system dynamics models, emergent behavior arises from the interaction of nonlinear balancing and reinforcing feedback mechanisms over time. Furthermore, because they can be expressed and analyzed as explicit equations, these models and insights about the emergent behavior from the system of feedback loops are communicated readily using causal loop diagrams (CLD) or stock and flow diagrams (SFD) of a model.

Because of their ability to preserve many of the features of complex systems, system dynamics models have been used extensively to address questions in health policy [68, 69]. They are less helpful, however, for representing the heterogeneity of individuals. System dynamics models are aggregate models and focus on representing the feedback relationships between aggregated variables (e.g., means, sums, frequencies). In system dynamics, capturing heterogeneity of actors requires disaggregating a model to represent subgroups. However, this approach is practically limited to two or three dimensions.

An alternative systems modeling technique is *agent-based modeling (ABM)* [39, 70–72]. Like system dynamics modeling, ABM takes a mechanistic approach, attempting to recreate the dynamics of an actual system as opposed to simple description. A fundamental difference between SD models and ABMs is that the former directly model changes over time of important observable, aggregate system features, while the latter model behaviors of individual agents over time [73]. Agent-based models operate most fundamentally at the *individual* (as opposed to the aggregate) level. In these models, a system of individual agents with some initial but changeable set of decision-making rules interact over time in a defined environment [37, 72]. These interactions give rise to emergent system-level outcomes, i.e., macrolevel patterns and structures. An appeal of these simulations is the transparency with which one can examine the experiences of single agents as well as the distributions of traits and outcomes in a hypothetical population. In effect, the product of an ABM run is a longitudinal epidemiological database for a hypothetical population.

ABMs lend themselves well to simultaneous nesting at multiple levels and across multiple dimensions (patients/households/neighborhoods, patients/physician practices, patients/ethnic groups, etc.). This feature allows the modeling of individuals within the context of overlapping hierarchies. In addition, interactions of heterogeneous agents in the real world can differ based on location relative to one another and based on features of the environment. Care seeking, and health in general, are intimately tied to place [74]. Capturing the dynamics that emerge from spatial relationships can help to elucidate the mechanisms driving the paradox of primary care. ABMs also allow the modeling of agent interactions within complex topologies, e.g., social networks [73].

The specification of model rules by small groups of modelers leaves models vulnerable to engrained biases arising from a relatively narrow range of experiences and viewpoints. In addition, model building by experts who understand the phenomena from the point of view of high-level aggregate data may not adequately represent the on-the-ground experience of those living the phenomena depicted by the model. In the section that follows, we discuss an approach for avoiding these pitfalls by engaging a range of stakeholders—real-world "agents"—in the actual model-building process.

46.4 Participatory Group Model Building

46.4.1 Group Model Building

Group model building (GMB) is a method for involving stakeholders and content- and modeling experts in the process of developing simulation models. Advantages of GMB with broad stakeholder input include more rigorous critique of assumptions and parameters underlying the model; greater "buy in" to high-leverage policy recommendations; and increased likelihood that the model will be used by others for research, teaching, and policy analysis [75]. To date, this participatory approach has primarily been applied to system dynamics models which fit well with the stock and flow diagrams typically used to specify SD models [76–78]. However, the process of GMB also fits with certain features of ABMs. People tend to think in terms of individuals (agents) and their worldviews, characteristics, and behaviors. With prompting, most people can be encouraged to begin to specify characteristics of environments and to consider how the interaction of multiple factors over time may lead to emergent phenomena, such as the PPC.

There are a number of different approaches to GMB. In this chapter, we focus on structured GMB, which involves leading participants through a series of structured small group exercises [79]. Each exercise is guided by a written protocol or GMB "script" [77]. Each script defines a set of inputs from previous scripts, the nature of expected outputs, and steps that are needed to successfully produce the intended outputs [80]. Early exercises focus on progressively solidifying and exploring participants' conceptions about the salient properties of agents (e.g., important individual traits, decision rules, and adaptive behavior of physicians and patients), rules for interaction and communication between agents, and rules for interaction between agents and their environments. As a model is developed based on stakeholder input, sessions increasingly focus on exploring the model in real time, assessing face validity of outputs, and determining priorities for enhancements to the model.

In GMB, some type of consistent visual framework is needed to help participants understand and interact with the model being developed. In SD, for instance, participants can interact with the model through the use of diagrams (e.g., CLDs and SFDs) representing the underlying mathematical structure of the system [81]. Extending the approach to ABMs, a standard framework can be used to help to define and track all agent types, the environment, and interaction rules in a succinct, comprehensible manner. This framework can encompass not only numeric and narrative summaries of model scenarios or output, but also neighborhood-level graphical representations—both static and animated. In addition to providing clarity during the model development process, this framework can guide the ultimate conveyance of methods and results to a scientific audience.

Like any participatory research approach [82, 83], GMB can involve varied levels of stakeholder involvement. Involvement can range from interviews of key informants, to stakeholder involvement in selecting or conceptualizing the problem, interacting with model, to making decisions about modeling process, formulations, data, analysis, and implementation [79].

A group modeling process offers numerous theoretical advantages over traditional expert-informed approaches but is not without unique pitfalls. The most obvious advantage is the incorporation of diverse inputinput that is tempered by a range of ideologies, philosophies, priorities, and experiences. The hope is that such circumspection yields a final product with greater resemblance to reality, greater face validity, and, ultimately, better "buy-in" from decision makers. Soliciting this stakeholder input, however, does not guarantee these outcomes. Throughout the process, a careful balance on the part of investigators is required between leading and being led by group participants. Too much of the former imposes investigators' hypotheses of system mechanics into the model, while too much of the latter may result in a model which either does not capture the key drivers of outcomes or, alternatively, attempts to reproduce irrelevant detail. The "sweet spot" for group facilitators lies in conveying a basic understanding of modeling, clearly defining goals of the endeavor, provoking deep thought about questions posed, knowing when to dwell on an aspect of the process versus when to forge ahead, and objectively bringing vague ideas to the point of operationalization.

46.4.2 Establishing and Increasing Validity

A number of approaches can be used to enhance confidence in, and understanding of, the PPC model [69, 84], the results it generates, and thus the inferences drawn from it. These approaches include engaging outside experts as auditors, calibration with healthcare system and population data, and sensitivity analysis [85–87]. Model calibration is a process for estimating certain unknown parameters by running a model repeatedly as these unknown parameters are systematically varied. Best fitting parameter sets are selected based on minimizing the discrepancy between observed and model-predicted outcomes. The "observed" data used for calibration of models of the PPC may come from sources such as the Medical Expenditure Panel Survey (MEPS), National Ambulatory Care Survey (NAMCS) [88, 89], Dartmouth Institute [90], comparative effectiveness research by Starfield [17, 23, 24, 35, 91–94], Medicare data, and other sources. Sensitivity analysis involves varying constants, parameters, and initial conditions to determine to what extent results are sensitive to uncertainties in particular parameters [87]. This step is important since, given the complexity of the system involved, it is unlikely that all parameters will have firm estimates from data or literature. Ongoing engagement of GMB stakeholders, and periodic involvement of outside experts, relevant data and theory, calibration, and sensitivity analysis can advance model robustness and trustworthiness as the model evolves.

Both the use of GMB and data-driven calibration represent forms of methodological "triangulation" in which different information sources and viewpoints are used to challenge, corroborate, and ultimately to refine and deepen understanding [95–97].

46.5 Application to the Paradox of Primary Care

Figure 46.1 illustrates application of the GMB process. This participatory process involves (1) defining the problem, (2) identifying the relevant stakeholders for the problem and engaging them in developing a concept model around the mechanisms involved in the problem, (3) using the concept model to program an agent-based computer model, and (4) iteratively refining the model based on further input from stakeholders, and eventually parameterizing the model based on relevant data.

For the PPC problem, we would envision the core stakeholders as patients, caregivers, and primary care practice members. These stakeholders would be gathered together to discuss their experience with the paradox, and to begin to explore together their concepts of how the paradox might emerge. If they were working to set up a system dynamics model, they would draw out the relationships between the variables in stock and flow diagrams. However, to enable programming an agent-based model, participants will be asked to consider the relevant agents, and what characteristics of the agents' approach to the world are pertinent. Participants will specify healthcare system and environmental characteristics relevant to health care and health, and how these might be related to agent characteristics.

Using the resulting stakeholder-developed concept model, computer scientist members of the team can then program an initial agent-based model. The resulting process and model outputs (disease care guideline–concordant quality, whole person functional health, population health, cost, and equity) will be shared with the stakeholders. Surprises will be discussed, details added, and—over time—relevant data used to parameterize the model to reflect what already is known about the paradox. Parameters may be estimated directly from published data or from analysis of large

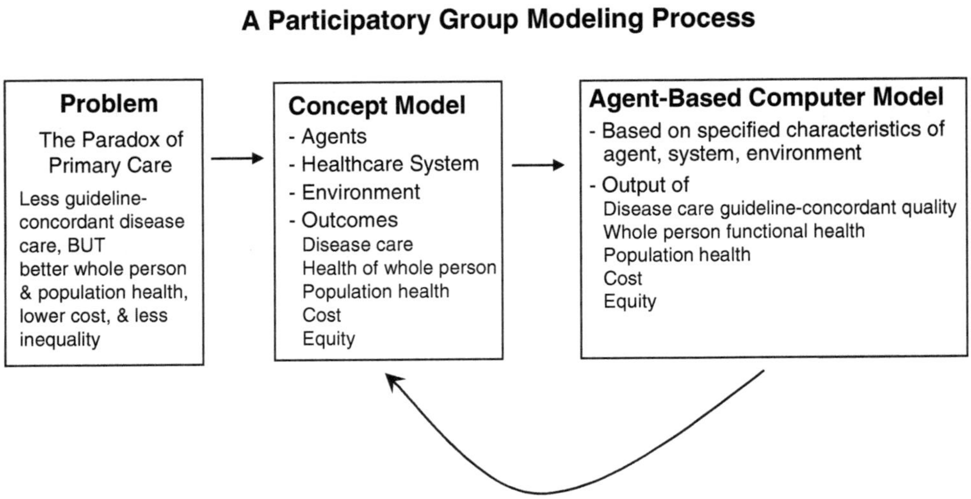

Fig. 46.1 Group model building to elucidate the paradox of primary care (1) define the problem, (2) identify the relevant stakeholders for the problem and engage them in developing a concept model around the mechanisms involved in the problem, (3) use the concept model to program an agent-based computer model, and (4) iteratively refine the model based on further input from stakeholders. With successive iterations of the cycle of concept-model-revision and agent-based-model-building, real world data will increasingly used to estimate parameter values—both directly and through calibration

datasets (e.g., distribution of the number of chronic conditions for different age groups) or may be estimated through a process of calibration to real-world data (e.g., the reduction in treatment side effects afforded by "whole person" care as opposed to solely disease-focused care).

Eventually, additional stakeholders, such as other healthcare providers, policymakers, payers, or healthcare administrators may be involved to bring in needed additional perspectives, and additional data brought in to reflect relevant phenomena.

46.6 Implications and Uses of a Model of the Paradox of Primary Care

A participatory-built ABM could be used by stakeholders including patients, clinicians, and policymakers to conduct exploratory modeling exercises [98] in order to test hypotheses about the mechanisms influencing the paradox of primary care. Shedding light on the mechanisms of the PPC will improve our understanding of which aspects of a healthcare system, and primary care in particular, promote patient-centered care and improve whole-person health—itself an emergent property which cannot be fully understood as the sum of organ system function or the sum of diseases. In a time of accelerating health system change, this understanding is vital in order to avoid unintended consequences which would destroy the emergent properties of primary care that result in improved population and personal health, sustainability, and equity.

A robust model of healthcare seeking and delivery may be applied to any number of questions relevant to patients, practitioners, healthcare system managers, and policymakers. The strategy described here of using GMB to develop an agent-based model offers an alternative to the traditional reductionist approach of association-seeking and cause-isolation. Instead, it seeks to capture the dynamics of a complex adaptive system by building it from the ground up *in silico* and then experimenting with it with the help of multiple stakeholders.

An agent-based model of the paradox of primary care which is informed by a GMB process has the potential to overcome the unintended negative consequences of linear, reductionist conceptualizations of health care, and to advance our ability to improve the health of people and populations.

References

1. World Health Organization. Press Release WHO/44: World Health Organization assesses the world's health systems. Geneva: World Health Organization; 2000. http://www.who.int/inf-pr-2000/en/pr2000-44.html. Accessed 22 July 2011.
2. Organization for Economic Co-Operation and Development. OECD Health Data 2009–Frequently Requested Data; 2009: http://www.oecd.org/document/16/0,3343,en_2649_34631_2085200_1_1_1_1,00.html.
3. Stange KC. The problem of fragmentation and the need for integrative solutions. Ann Fam Med. 2009;7(2):100–3.
4. Sandy LG, Bodenheimer T, Pawlson LG, Starfield B. The political economy of U.S. primary care. Health Aff. 2009;28(4):1136–45.
5. Starfield B. Is US health really the best in the world? JAMA. 2000;284(4):483–5.
6. Phillips Jr RL, Bazemore AW. Primary care and why it matters for US health system reform. Health Aff. 2010;29(5):806–10.
7. Rohde J, Cousens S, Chopra M, Tangcharoensathien V, Black R, Bhutta ZA, et al. 30 Years after Alma-Ata: has primary health care worked in countries? Lancet. 2008;372(9642):950–61.
8. Stange KC. The paradox of the parts and the whole in understanding and improving general practice. Int J Qual Health Care. 2002;14(4):267–8.
9. Stange KC, Ferrer RL. The paradox of primary care. Ann Fam Med. 2009;7(4):293–9.
10. Harrold LR, Field TS, Gurwitz JH. Knowledge, patterns of care, and outcomes of care for generalists and specialists. J Gen Intern Med. 1999;14(8):499–511.
11. Solomon DH, Bates DW, Panush RS, Katz JN. Costs, outcomes, and patient satisfaction by provider type for patients with rheumatic and musculoskeletal conditions: a critical review of the literature and proposed methodologic standards. Ann Intern Med. 1997;127(1):52–60.
12. Smetana GW, Landon BE, Bindman AB, Burstin H, Davis RB, Tjia J, et al. A comparison of outcomes resulting from generalist vs specialist care for a single discrete medical condition: a systematic review and methodologic critique. Arch Intern Med. 2007;167(1):10–20.
13. Baicker K, Chandra A. Medicare spending, the physician workforce, and beneficiaries' quality of care. Health Aff. 2004; Suppl Web Exclusives:W184–97.
14. Roetzheim RG, Gonzalez EC, Ramirez A, Campbell R, van Durme DJ. Primary care physician supply and colorectal cancer. J Fam Pract. 2001;50(12):1027–31.
15. Greenfield S, Nelson EC, Zubkoff M, Manning W, Rogers W, Kravitz RL, et al. Variations in resource utilization among medical specialties and systems of care. Results from the medical outcomes study. JAMA. 1992;267(12):1624–30.
16. Friedberg MW, Hussey PS, Schneider EC. Primary care: a critical review of the evidence on quality and costs of health care. Health Aff. 2010;29(5):766–72.
17. Starfield B. Primary care: balancing health needs, services, and technology. New York, NY: Oxford University Press; 1998.
18. Shi L, Starfield B. The effect of primary care physician supply and income inequality on mortality among blacks and whites in US metropolitan areas. Am J Public health. 2001;91(8):1246–50.
19. Shi L, Macinko J, Starfield B, Wulu J, Regan J, Politzer R. The relationship between primary care, income inequality, and mortality in US States, 1980-1995. J Am Board Fam Pract. 2003;16(5):412–22.
20. Shi L, Macinko J, Starfield B, Politzer R, Wulu J, Xu J. Primary care, social inequalities and all-cause, heart disease and cancer mortality in US counties: a comparison between urban and non-urban areas. Public Health. 2005;119(8):699–710.
21. Shi L, Macinko J, Starfield B, Politzer R, Xu J. Primary care, race, and mortality in US states. Soc Sci Med. 2005;61(1):65–75.
22. Shi L, Macinko J, Starfield B, Politzer R, Wulu J, Xu J. Primary care, social inequalities, and all-cause, heart disease, and cancer mortality in US counties, 1990. Am J Public Health. 2005;95(4):674–80.
23. Starfield B. Primary care and equity in health: the importance to effectiveness and equity of responsiveness to people's needs. Hum Soc. 2009;33(1/2):56–73.
24. Starfield B, Shi LY, Macinko J. Contribution of primary care to health systems and health. Milbank Q. 2005;83(3):457–502.
25. Donaldson MS, Yordy KD, Lohr KN, Vanselow NA, editors. Primary care: America's health in a new era. Washington, DC: National Academy Press; 1996.
26. Dentzer S. Reinventing primary care: a task that is far 'too important to fail'. Health Aff. 2010;29(5):757.
27. Howell JD. Reflections on the past and future of primary care. Health Aff. 2010;29(5):760–5.
28. Margolius D, Bodenheimer T. Transforming primary care: from past practice to the practice of the future. Health Aff. 2010;29(5):779–84.
29. Chan M. Return to Alma-Ata. Lancet. 2008;372(9642):865–6.
30. Lawn JE, Rohde J, Rifkin S, Were M, Paul VK, Chopra M. Alma-Ata 30 years on: revolutionary, relevant, and time to revitalise. Lancet. 2008;372(9642):917–27.
31. van Weel C, De Maeseneer J, Roberts R. Integration of personal and community health care. Lancet. 2008;372(9642):871–2.
32. Chan M. Primary health care as a route to health security. Lancet. 2009;373(9675):1586–7.
33. Stange KC. A science of connectedness. Ann Fam Med. 2009;7(5):387–95.

34. Stange KC. The generalist approach. Ann Fam Med. 2009;7(3):198–203.
35. Starfield B. Primary care: concept, evaluation, and policy. New York, NY: Oxford University Press; 1992.
36. McWhinney IR, Freeman T. Textbook of family medicine. 3rd ed. New York: Oxford University Press; 2009.
37. Auchincloss AH, Diez Roux AV. A new tool for epidemiology: the usefulness of dynamic-agent models in understanding place effects on health. Am J Epidemiol. 2008;168(1):1–8.
38. Plsek PE, Greenhalgh T. Complexity science: the challenge of complexity in health care. BMJ. 2001;323(7313):625–8.
39. Diez Roux AV. Complex systems thinking and current impasses in health disparities research. Am J Public Health. 2011;101(9):1627–34.
40. Litaker D, Tomolo A, Liberatore V, Stange KC, Aron DC. Using complexity theory to build interventions that improve health care delivery in primary care. J Gen Intern Med. 2006;21 Suppl 2:S30–S4.
41. Collins C, Phields ME, Duncan T. An agency capacity model to facilitate implementation of evidence-based behavioral interventions by community-based organizations. J Public Health Manag Pract. 2007; Suppl:S16–23.
42. Allen K, Zoellner J, Motley M, Estabrooks PA. Understanding the internal and external validity of health literacy interventions: a systematic literature review using the RE-AIM framework. J Health Commun. 2011;16 Suppl 3:55–72.
43. Fiscella K, Franks P, Gold MR, Clancy CM. Inequality in quality: addressing socioeconomic, racial, and ethnic disparities in health care. JAMA. 2000; 283(19):2579–84.
44. Starfield B. The hidden inequity in health care. Int J Equity Health. 2011;10(1):15.
45. Braveman PA, Egerter SA, Mockenhaupt RE. Broadening the focus: the need to address the social determinants of health. Am J Prev Med. 2011;40(1 Suppl 1):S4–18.
46. Sterman JD. Learning from evidence in a complex world. Am J Public Health. 2006;96(3):505–14.
47. Berenson RA, Rich EC. US approaches to physician payment: the deconstruction of primary care. J Gen Intern Med. 2010;25(6):613–8.
48. Goroll AH. The future of primary care: reforming physician payment. N Engl J Med. 2008;359(20): 2087–90.
49. Davis EM, Stange KC, Horwitz RI. Childbearing, stress and obesity disparities in women: a public health perspective. Matern Child Health J. 2010; 16(1):109–18.
50. Pauly MV, Zeng Y. Adverse selection and the challenges to stand-alone prescription drug insurance. Front Health Policy Res. 2004;7:55–74.
51. Nutting PA, Miller WL, Crabtree BF, Jaén CR, Stewart EE, Stange KC. Initial lessons from the first national demonstration project on practice transformation to a patient-centered medical home. Ann Fam Med. 2009;7(3):254–60.
52. Stange KC, Nutting PA, Miller WL, Jaén CR, Crabtree BF, Flocke SA, et al. Defining and measuring the patient-centered medical home. J Gen Intern Med. 2010;25(6):601–12.
53. Weil TP, Pearl GM. Merger mania: physicians beware. Physician Exec. 1998;24(2):6–11.
54. Stange KC, Breslau ES, Dietrich AJ, Glasgow RE. State-of-the-art and future directions in multilevel interventions across the cancer control continuum. J Natl Cancer Inst Monogr. 2012;2012(44):20–31.
55. Miller WL, Crabtree BF, McDaniel R, Stange KC. Understanding change in primary care practice using complexity theory. J Fam Pract. 1998;46(5):369–76.
56. Miller WL, McDaniel Jr RR, Crabtree BF, Stange KC. Practice jazz: understanding variation in family practices using complexity science. J Fam Pract. 2001;50(10):872–8.
57. Crabtree BF, Nutting PA, Miller WL, Stange KC, Stewart EE, Jaén CR. Summary of the National Demonstration Project and recommendations for the Patient-Centered Medical Home. Ann Fam Med. 2010;8 Suppl 1:S80–90.
58. Holland JH. Hidden order: how adaptation builds complexity. Cambridge, MA: The MIT Press; 1995.
59. Gladwell M. The tipping point: how little things can make a big difference. New York, NY: Little, Brown and Company; 2002.
60. Mabry PL, Olster DH, Morgan GD, Abrams DB. Interdisciplinarity and systems science to improve population health: a view from the NIH Office of Behavioral and Social Sciences Research. Am J Prev Med. 2008;35(2 Suppl):S211–24.
61. Goldberger AL. Non-linear dynamics for clinicians: chaos theory, fractals, and complexity at the bedside. Lancet. 1996;347(9011):1312–4.
62. Katerndahl DA. Lessons from Jurassic Park: patients as complex adaptive systems. J Eval Clin Pract. 2009;15(4):755–60.
63. Holt TA, editor. Complexity for clinicians. San Francisco: Radcliffe; 2004.
64. Stange KC. Ways of knowing, learning, and developing. Ann Fam Med. 2010;8(1):4–10.
65. Sturmberg JP. Systems and complexity thinking in general practice. Part 2: application in primary care research. Aust Fam Physician. 2007;36(4):273–5.
66. Simon HA. Administrative behavior; a study of decision-making processes in administrative organization. 2nd ed. New York: Macmillan; 1957.
67. Simon HA. Models of bounded rationality. Cambridge, MA: MIT Press; 1982.
68. Homer JB, Hirsch GB. System dynamics modeling for public health: background and opportunities. Am J Public Health. 2006;96(3):452–8.
69. Forrester JW, Senge PM. Tests for building model confidence in system dynamics models. In: Richardson GP, editor. Modelling for management. Vol II:

Simulation in support of systems thinking. Brookfield, VT: Dartmouth; 1980. p. 413–32.
70. Bankes SC. Agent-based modeling: a revolution? Proc Natl Acad Sci USA. 2002;99 Suppl 3:7199–200.
71. Epstein J. Generative social science: studies in agent-based computational modeling. Princeton, NJ: Princeton University Press; 2007.
72. Gilbert N. Agent-based models. London: Sage; 2007.
73. Parunak HVD, Savit R, Riolo RL. Agent-based modeling vs. equation-based modeling: a case study and users' guide. Proceedings of Multi-agent Systems and Agent-based Simulation [serial on the Internet]. 1998. http://www.citeulike.org/user/lavinp/article/7681604.
74. Berry W. Health is membership. In: Wirzba N, editor. The art of the commonplace: the agrarian essays of Wendell Berry. Berkeley: Counterpoint; 2002. p. 144–58.
75. Rouwette E, Vennix JAM, Mullekom TV. Group model building effectiveness: a review of assessment studies. Syst Dyn Rev. 2006;18(1):5–45.
76. Vennix J. Group model building. New York: Wiley; 1996.
77. Andersen DF, Richardson GP. Scripts for group model building. Syst Dyn Rev. 1997;13(2):107–29.
78. Richardson GP, Andersen DF. Teamwork in group model building. Syst Dyn Rev. 1995;11(2):113–37.
79. Hovmand PS, Brennan L, Chalise N, editors. Whose model is it anyway? The 29th International Conference of the System Dynamics Society, Washington, DC, 25–29 July 2011.
80. Hovmand PS, Etiënne AJA, Rouwette E, Andersen DF, Richardson GP, Calhoun A, et al. Scriptapedia. http://www.systemdynamics.org/other_resources.htm/scriptapedia. Accessed 3 Apr 2011.
81. Lane DC. Diagramming conventions in system dynamics. J Oper Res Soc. 2000;51(2):241–5.
82. Macaulay AC, Commanda LE, Freeman WL, Gibson N, McCabe ML, Robbins CM, et al. Participatory research maximises community and lay involvement. North American Primary Care Research Group. BMJ. 1999;319(7212):774–8.
83. Minkler M, Wallerstein N. Community-based participatory research for health: from process to outcomes. San Francisco, CA: Jossey-Bass; 2008.
84. Sterman JD. Business dynamics: Systems thinking and modeling for a complex world. Irwin: McGraw-Hill; 2000.
85. Marino S, Hogue IB, Ray CJ, Kirschner DE. A methodology for performing global uncertainty and sensitivity analysis in systems biology. J Theor Biol. 2008;254(1):178–96.
86. Helton JC, Johnson JD, Sallaberry CJ, Storlie CB. Survey of sample-based methods for uncertainty and sensitivity analysis. Reliab Eng Syst Saf. 2006;91(10–11):1175–209.
87. Bankes SC. Tools and techniques for developing policies for complex and uncertain systems. Proc Natl Acad Sci USA. 2002;99 Suppl 3:7263–6.
88. Gilchrist VJ, Stange KC, McCord G, Bourget CC, Flocke SA. A comparison of the National Ambulatory Medical Care Survey (NAMCS) measurement approach with direct observation of outpatient visits. Med Care. 2004;42(3):276–80.
89. Valderas JM, Starfield B, Forrest CB, Sibbald B, Roland M. Ambulatory care provided by office-based specialists in the United States. Ann Fam Med. 2009;7(2):104–11.
90. Baicker K, Chandra A. Medicare spending, the physician workforce, and beneficiaries' quality of care. Health Aff. 2004; W4.
91. Starfield B, Lemke KW, Bernhardt T, Foldes SS, Forrest CB, Weiner JP. Comorbidity: implications for the importance of primary care in 'case' management. Ann Fam Med. 2003;1(1):8–14.
92. Starfield B, Shi L. The medical home, access to care, and insurance: a review of evidence. Pediatrics. 2004;113(5 Suppl):1493–8.
93. Starfield B, Lemke KW, Herbert R, Pavlovich WD, Anderson G. Comorbidity and the use of primary care and specialist care in the elderly. Ann Fam Med. 2005;3(3):215–22.
94. Starfield B, Shi L, Grover A, Macinko J. The effects of specialist supply on populations' health: assessing the evidence. Health Aff. 2005; Suppl Web Exclusives:W5-97–107.
95. Duffy ME. Methodological triangulation: a vehicle for merging quantitative and qualitative research methods. Image J Nurs Sch. 1987;19(3):130–3.
96. Mercer SL, DeVinney BJ, Fine LJ, Green LW, Dougherty D. Study designs for effectiveness and translation research: identifying trade-offs. Am J Prev Med. 2007;33(2):139–54.
97. Cohen DJ, Crabtree BF. Evaluative criteria for qualitative research in health care: controversies and recommendations. Ann Fam Med. 2008;6(4):331–9.
98. Bankes S. Exploratory modeling for policy analysis. Oper Res. 1993;41(3):435–49.

47 Healthcare Reform: The Need for a Complex Adaptive Systems Approach

Joachim P. Sturmberg, Di M. O'Halloran, and Carmel M. Martin

All meaningful and lasting change starts first in your imagination and then works its way out. Imagination is more important than knowledge.

Albert Einstein

The notion of health care, and in particular primary health care, being a public good was strongly expressed in the Declaration of Alma-Ata in 1978 [1]. If this position is accepted, the provision of health care should be participatory, assist individuals to remain or become socially and economically productive, promote social justice, and also involve the food, education, employment, and social sectors. It follows that it is fundamentally Government's responsibility to provide the policies and ensure the necessary resources to establish comprehensive national health systems and health-promoting environments regardless of which jurisdictions, sectors, or organisations are providing health services (Table 47.1).

If governments around the world had embraced these declarations, then our health systems should by now be strongly person-focused to address individual need, flexible in structure, and with a sufficiently qualified health workforce to provide the right services when and where needed. Or at least be closer to the WHO vision. Our health systems would be locally interconnected to create health-promoting environments for the whole community, especially across social, occupational, and wider built and natural environments. Reform—which will always be needed in such settings—would more likely equate to "tweaking" in light of changing patient and community needs and/or emerging new knowledge and technologies.

However, as we are all well aware, the WHO's vision remains futuristic and largely unrealised. One rightly might ask: How could so many countries fail so dismally.[1] In this chapter we expand on the *common good* notion entailed in the WHO charter, before outlining the prevailing healthcare discourse at the beginning of the second decade

J.P. Sturmberg (✉)
Department of General Practice,
Monash University, Melbourne, VIC, Australia,

The Newcastle University, Newcastle,
PO Box 3010, Wamberal, NSW 2260, Australia
e-mail: jp.sturmberg@gmail.com

D.M. O'Halloran
Department of General Practice,
140 Stevens Rd, Glenorie, NSW 2157, Australia
e-mail: dohalloran01@bigpond.com

C.M. Martin
Department of Public Health and Primary Care,
Trinity College, Dublin,
College Green, Dublin D2, Ireland
e-mail: martinc7@tcd.ie

[1] Bhutan seems to be the only exception; the country follows the Buddhist ideals that development of human society should benefit from the side-by-side development of material and spiritual growth, and that both aspects reinforce each other. Development is measured in "growth in national happiness" based on the principles of promotion of sustainable development, preservation and promotion of cultural values, conservation of the natural environment, and the establishment of good governance.

J.P. Sturmberg and C.M. Martin (eds.), *Handbook of Systems and Complexity in Health*,
DOI 10.1007/978-1-4614-4998-0_47, © Springer Science+Business Media New York 2013

Table 47.1 Excerpts from the WHO declaration at Alma-Ata 1978

Article IV The people have the right and duty to participate individually and collectively in the planning and implementation of their health care.
Article V Governments have a responsibility for the health of their people which can be fulfilled only by the provision of adequate health and social measures. A main social target of governments … should be the attainment … of a level of health that will permit them to lead a socially and economically productive life. Primary health care is the key to attaining this target as part of development in the spirit of social justice.
Article VII, Section 4 in addition to the health sector, all related sectors and aspects of national and community development, in particular agriculture, animal husbandry, food, industry, education, housing, public works, communications and other sectors; and demands the coordinated efforts of all those sectors.
Article VIII All governments should formulate national policies, strategies and plans of action to launch and sustain primary health care as part of a comprehensive national health system and in coordination with other sectors…
Article X An acceptable level of health for all the people … can be attained through a fuller and better use of the world's resources, a considerable part of which is now spent on armaments and military conflicts.…
It [The International Conference on Primary Health Care] urges governments, WHO and UNICEF, and other international organizations, as well as multilateral and bilateral agencies, nongovernmental organizations, funding agencies, all health workers and the whole world community to support national and international commitment to primary health care and to channel increased technical and financial support to it, particularly in developing countries.

of the twenty-first century, and exploring how this discourse hinders, rather than helps, the evolution of the health system envisaged by the WHO. The final section describes a pragmatic model of a *patient-centred healthcare model* based on *complex adaptive systems* (CAS) principles and advocates the *case for change*.

47.1 The *Common Good* Notion for Health and Healthcare

The WHO Charter defines health care in terms of a *common good*.

The concept of the *common good* goes back as far as Plato and Aristotle who argued that government's purpose is to pursue "good" for the whole of the community. Philosophical discourse since then has often pushed the notion of utilitarianism (Bentham and Mill): that one should fundamentally strive toward the "greatest happiness for the greatest number" of people. This notion tacitly accepts that some people will be left behind, and in the medical context, this is usually those most marginalised and in greatest need. Modern philosophers such as Anscombe have returned to the *altruism* of the virtue ethics of Aristotle, pushing for a balance between positions, while in Eastern philosophy, the balance notion is expressed in the Confucian view of striving for harmony with other people.

The *common good* notion has been "reconstructed" by economists in an attempt to "count" and "account" for its value to the economy. In economic philosophy common goods are distinguished from public goods. Common goods (like renewable resources) can be used freely, but overuse or domination by one/some will result in depletion, and minimise accessibility for others, whereas public goods (like police or fire brigade) are always freely available to all. Economic philosophy defines as social goods, a distinct type of public goods, which include societal activities like social services, education, and health care. These goods are produced largely based on social policy principles aiming to benefit all human welfare. Whether social goods should be provided by private enterprise, Government, or both remains a complex and often contentious issue. In any case Government always remains responsible for the provision of social goods as their provision is ultimately government's responsibility

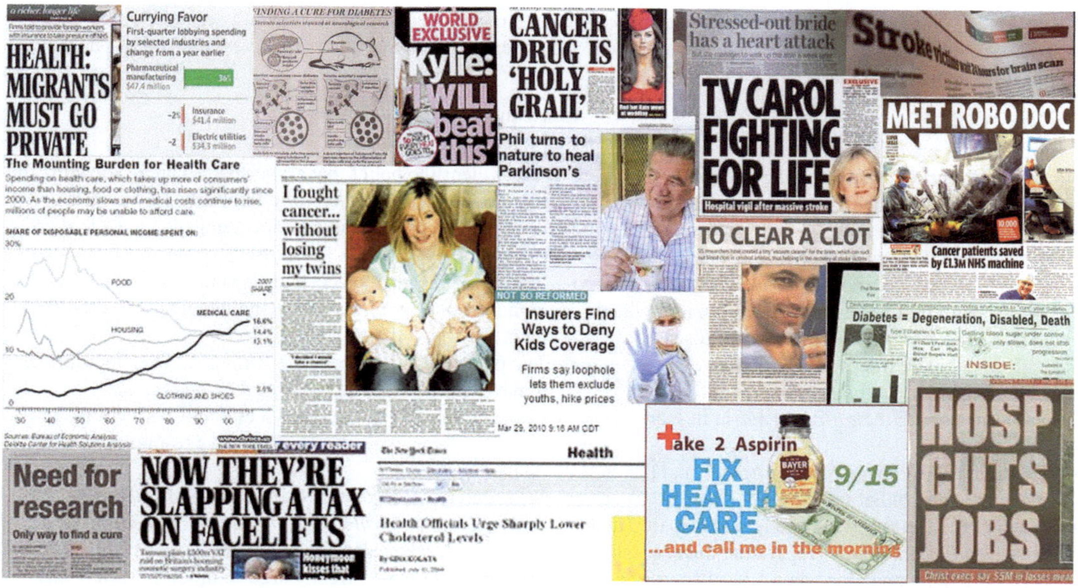

Fig. 47.1 Health stories in the media

and a community priority, or because some social goods, based on "economic rationale", would otherwise not be provided at all.[2]

Regardless of personal preferences in terms of public versus private provision of health care, the driving force behind healthcare provision should be the *betterment of an individual's health*. Hence the Aristotelian definition of a *common good* applies to healthcare services. They should be freely accessible, and at the same time carefully guarded against overuse/inappropriate use. Despite the apparent logic of this position, prevailing health professional attitudes and public discourse unknowingly demonstrate the difficulty of achieving this *common good* outcome.

47.2 The Prevailing Healthcare Discourse

The public discourse about health care over recent decades appears to have two main themes characterised metaphorically as either "fighting and/or winning the war against disease" or as "health care is a market place" (Fig. 47.1). And while "how can we best work together to achieve desired health experiences and health outcomes for our patients" has become an increasingly loud discourse in the USA and UK, largely missing is the discourse about "what constitutes health".[3] The prevailing notion of health is being driven by the medical industrial complex including clinicians, the health workforce, and the pharmaceutical industry.

These commonly found metaphors provide insight into the perceptions, thinking, and acting of much of the current political, health financing, and health professional leadership [2]. This tendency is fuelled by the well-recognised reality that media outlets thrive on drama and conflict, and will heighten these elements, or even go so far as to create drama where little or none exists. The implications of these prevailing war and economic metaphors have been outlined by Annas [3] and his insights are summarised in Table 47.2.

[2]George Annas re-emphasised this: The "economic perspective" *ignores the inability of the market to distribute goods and services whose supply and demand are unrelated to price.*

[3]For a detailed discussion of these notions, see Chaps. 14–17.

Table 47.2 Annas explored the military and market metaphors and their impacts on health service policies and the practice of medicine in 1995, some 18 years ago, and to a large extent these metaphors still prevail in today's healthcare and health system reform discourse

Military metaphor	Consequences
• Medicine is a *battle* against death. • We are almost constantly engaged in *wars* on various diseases, such as cancer and AIDS. • Diseases attack the body, patients fight the disease, and doctors *intervene* or counter attack. • Doctors are mostly specialists and backed by allied health professionals all of whom are trained to be *aggressive, fight* these *invading* diseases with weapons designed to *knock them out.* • Doctors give orders in the *trenches* and on the front lines, using their *armamentaria* in search of breakthroughs. • Treatments are *conventional* or *heroic*, and the brave patients soldier on. • We engage in triage in the emergency department, *invasive procedures* in the operating theatre, and even defensive medicine when a legal enemy is suspected.	• Ignore *costs.* • Strengthens the belief that all problems can be solved with more *sophisticated technology* and scientific advances, prompting hospitals and doctors to engage in *medical arms races.* • War analogies lead to acceptance as inevitable that organizations are *hierarchical* and largely dominated by men. • The patient's body becomes a *battlefield*, thus appropriate to have *short-term, single-minded tactical goals.* • Concentrates on the physical, sees control as central, and encourages the expenditure of massive resources to *achieve dominance.* • We failed to assert that medicine, like war, *should be financed and controlled only by the government.* • The metaphor has also become mythic. As a historian of war, John Keegan, correctly argues, modern warfare has become so horrible that 'it is scarcely possible anywhere in the world today to raise a body of reasoned support for the opinion that war is a justifiable activity.'
Market metaphor	**Consequences**

• Health plans and hospitals *market products* to *consumers*, who *purchase them on the basis of price.*
• Medical care is a business that necessarily involves *marketing* through advertising and competition among suppliers who are *primarily motivated by profit.*
• Health care becomes *managed care.*
• *Mergers* and *acquisitions* become core activities.
• *Chains* are developed, *vertical integration* is pursued, and antitrust worries proliferate.
• *Consumer choice* becomes the central theme of the market metaphor.
• In the language of insurance, consumers become *'covered lives'* or even *'money-generating biological structures'.*
• Economists become *health-financing gurus.*
• The role of doctors is radically altered as they are *instructed by managers* that they can no longer be patient advocates (but instead must advocate for the entire group of covered lives in the health plan).
• The goal of medicine becomes a *healthy bottom line* instead of a healthy population.
	• Emphasis is placed on
	  o *Efficiency*
	  o *Profit maximisation*
	  o *Customer satisfaction*
	  o *The ability to pay*
	  o *Planning*
	  o *Entrepreneurship*
	  o *Competitive models*
	• The ideology of medicine is displaced by the *ideology of the marketplace.*
	• Trust is replaced by *caveat emptor.*
	• There is *no place for the poor and uninsured* in the market metaphor.
	• *Business ethics* supplant medical ethics as the practice of medicine becomes corporate.
	• *Non-profit medical organizations may tend to be corrupted* by adopting the values of their for-profit competitors.
	• A *management degree* becomes at least as important as a medical degree.
	• *Public institutions*, which by definition cannot compete in the for-profit arena, *risk demise, second-class status, or simply privatisation.*
	• *Patients, as consumers*, are to make decisions that are governed by *corporate entities.*
	• The market metaphor *conceals the inherent imperfections of the market* and *ignores the public nature of many aspects of medicine.*
	• It *ignores the inability of the market* to distribute goods and services whose supply and demand are unrelated to price.

(continued)

Table 47.2 (continued)

Market metaphor	Consequences
	• The metaphor *pretends that there is such a thing as a free market in health insurance* plans and that purchasers can and should be content with their choices when unexpected injuries or illnesses strike them or their family members. • The reality is that American markets are o Highly regulated, o Major industries enjoy large public subsidies, o Industrial organizations tend towards oligopoly, and o Require strong laws that protect consumers and offer them recourse through product-liability suits to prevent profits from being too ruthlessly pursued.

Compiled from Annas: Reframing the debate on health care reform by replacing our metaphors [3]; italics emphasise pertinent concepts from the paper

We do not have to agree with Annas's every comment to appreciate the accuracy of his observations. Besides succinctly indicating the problems common to healthcare systems around the world, this analysis also offers a way to develop a more meaningful discourse to lead us towards a more effective and efficient healthcare system.

These metaphors have become even more interwoven in public discussion and have continued to shape research agendas, policy settings, and public expectations; for example, research funding for specific diseases has far outperformed research funding for public and environmental health concerns. Therapeutic interventions and treatments are too often promulgated by industry interests by funding vested interest groups [4–6] who largely fail to indicate such "conflicts of interest" [7, 8] and which in part are supported by the medical profession [9], and/or through direct-to-consumer advertising in the media [10].

However, these metaphors completely fail to take account of the fast changing epidemiology of morbidity: most patients suffer with complex conditions and social circumstances that fit neither *war* nor *market* metaphors. Patients with complex needs simply cannot be managed within the confines of standardised (military) management (and frequently profit oriented) protocols, or siloed specialised care delivery such as smoking cessation, weight loss, acute mental health care, or chronic disease management.

47.3 The *Healthcare Vortex* as a Metaphor for Patient-Centred Health Care

To change the nature of this professional and public discourse and enable a meaningful healthcare system to emerge, one has to firstly contemplate three basic questions: *"what is health"*, *"what is health care intended to achieve"*, and given the answer to this question, *"how can effective health care best be designed and achieved"*.

47.3.1 What Is Health?

A number of chapters in this Handbook have been dedicated to understanding health. In short *health* is a personal state reflecting the balance between physical, emotional, social, and cognitive/sense-making experiences. Physical integrity, functional performance, and subjective experience produce an entropic state most consistent with viability.

47.3.2 What Is Health Care Intended to Achieve?

As health resides within the individual as his/her own personal experience, health care must primarily aim to improve each patient's own unique and ever changing experience of their own health in light of their particular needs and circumstances. This understanding of health care implies that to be effective, health care must be broader than the prevailing narrow focus on discrete disease entities and their specific biomedical management.

In addition to consideration of the organic, psychological, and spiritual dimensions, improving a person's health experience more often than not also involves attention to their local environment, such as access to nutritious food, safe housing, open spaces, quality education, a secure workplace, and accessible social infrastructure. In consequence, every individual's health experience is continually evolving as individuals influence, and are influenced by, their complex and changing environments.

47.3.3 How Can Effective Healthcare Best be Designed and Achieved?

If health itself is a complex adaptive state which arises in a complex adaptive environment, it follows that health systems must also be organised—and operate—as complex adaptive systems. Equally, it follows that the core focus of a complex adaptive health system should be—first and foremost—the patient and his/her health experience.

The Mayo Clinic exemplifies this by ensuring that its core value statement, *"The needs of the patient come first"*,[4] is reflected in every aspect of the organisation's objectives, rules (or operating principles), structures, decision-making processes, and delivery of care. The Mayo's dedication to meeting the patient's needs has become, and is widely recognised to be, a valid and proven driver to the simultaneous achievement of effective and efficient care for each person.

47.3.4 Healthcare Systems Should be "Driven" by Core Values and Guided by "Simple Rules"

We are what we repeatedly do.
Excellence, then, is not an act but a habit.

Aristotle

Self-organisation is the core characteristic of complex adaptive systems. Two features govern self-organisation in such systems and organisations: first, *core values* which define the system's focus or mission and thus 'drive' the fundamental approaches underpinning the system/organisation, and second, *simple rules* (or operating principles) which provide its operational framework. Self-organisation then arises spontaneously due to local interactions between neighbouring agents and has system-wide repercussions. Acting together and applied continuously, *core values* and *simple rules* provide the freedom as well as the necessary constraints for agents to interact in ways that allow the system to let its *inherent properties emerge*; "colloquially we refer to this as achieving the system's goals" [11, 12].

47.3.4.1 The Importance of Values and Ethics

Core values are those that remain unchanged in a changing world: they do not change in response to market, financial, or administrative changes,[5] and sustain the organisation in times of challenge. *Values and ethics* are the "soul" of what an organisation stands for, how it conducts itself, and how it guides the behaviour of its members. They underlie *the work an organisation does*, individually and collectively, and provide the *moral compass* that guides *how* the necessary work gets done—the strategies employed—to achieve the organisation's goals. In other words, they define the "governing principles" of the organisation and uniformly apply to all of its organisational units.

Core values are not descriptions of the work an organisation does or the strategies it employs to accomplish its mission, i.e. they are not

[4] See Chap. 39 by Seltman and Berry.

[5] For more detail see, e.g. the training guides of the National Park Service (http://www.nps.gov), especially.

What are core values? http://www.nps.gov/training/uc/whcv.htm.

How Will Core Values be Used? http://www.nps.gov/training/uc/hwcvbu.htm.

National Leadership Council on Core Values. http://www.nps.gov/training/uc/tcv.htm.

- Operating practices,
- Business strategies,
- Cultural norms, or
- Competencies.

This may sound trivial, however, as the Appendices highlight, with a few exceptions the value statements of most health organisations (Health Departments, Medical Institutions, Professional Associations and Clinical Institutions) are focused inwards on operating practices, business strategies, cultural norms, and competencies without defining what the organisation really stands for, and in what manner the organisation's work will be done (and surprisingly, most refer only vaguely to the health and the care of patients).

47.3.4.2 Understanding Simple Rules

In a complex adaptive system, complex behaviour results from observance of "simple rules" or operating principles. Simple rules are generated when people central to the success of a particular endeavour consider the relationship between the organisation's goals and its core values, enabling the crafting of "simple rules" (operating principles) which best express the integration of these two elements, and describe the manner in which the organisation's work would be undertaken.

If an organisation does not take the time to reflect and incorporate core values into the organisation's culture and behaviours, it is likely to be compromised. Simple rules should be "how to" statements that provide concrete direction. *Consistently* applied, they enable continuing *self-organisation* and *emergence* of system structures and functions[6] [13]. Usually a "handful" of simple rules are all that is needed; too few do not provide a sufficient boundary, and too many constrict creativity.

An obvious example is the Mayo Clinic's core value: *the needs of the patient come first.* The development of effective strategies to achieve this outcome requires ongoing evaluation of the approaches to patient-centred care in order to develop appropriate operating principles to guide subsequent decision making and system function.

Health systems are no different, with the agents in complex adaptive healthcare systems requiring an agreed *core value* or *shared vision* on which to focus, and by which they are continuously, individually, and collectively influenced, 'driven', and in combination with the rules/operating principles, constrained. The suggested core *value* for health care was earlier described in terms of patients' *health experiences,*[7] i.e. optimising the experience of their own state of health.

This notion takes account of both the multiplicity of influences on health and the fact that not all illness and disease may be reversible to the pre-illness or pre-disease state. *Health care* thus should be the means of helping patients to *make sense of their illness and disease experiences,* allowing for the emergence of a *new state* of *good health experience*.

47.3.4.3 The Vortex as a Complex Adaptive System Metaphor

The characteristics and functions of complex adaptive systems can be appreciated through a very practical metaphor—the bathtub vortex [14]. The attractor in that system is the combined effect of gravity and the plug hole: remove the plug and the water forms a vortex—and as we all recall from our childhood, disturbing the vortex always results in the restoration of the vortex very close to its pre-disturbance state—illustrating the

[6] These insights are not new as such, e.g. Dent & Holt described the application of complex adaptive systems thinking in the context of the US Air Force in the following terms: *Its most powerful "attractors of meaning" are its core values: "integrity first," "service before self," and "excellence in all we do." …*

Successful organizations of the future will maximize its members' understanding of the mission and vision, optimize the connections inside and outside the organization, and then seek to remove barriers to formal structural change. They will be identifiable by their core values, beliefs, and culture. An organization's values will be the source of its self-referential stability and order. They will provide the organization with a strong anchor that will enable self-renewal without experiencing chaos. The emergence of informal structures within the organization will be encouraged and supported but not prescribed. Senior leadership will not insulate organizational members from the realities of the environment. Organizational change will be governed only by a few guiding principles or core values that are central to the identity of the organization [13].

[7] See Chap. 15 by Sturmberg for more details.

self-organising function based on its attractor, or system driver.

The effects of disturbing the vortex at different levels will lead to changes at its apex with varying degrees of time delays. The effects, and therefore the amount of interactions within the vortex, increase exponentially towards its apex, and disturbances closer to the apex result in more dramatic changes to the vortex than at its top, which in fact may ultimately not result in any changes at all.

47.3.4.4 The *Healthcare Vortex*

Building on this metaphor, the *healthcare vortex* can be seen as a means of visualising the different agents and their interactions in the healthcare system. The healthcare vortex takes account of the various levels within the healthcare system, traditionally referred to as the macro-, meso-, micro-, and nano-level. Each level has its own agents and interaction; however, as the vortex model highlights, all agents are interconnected within and across all other levels, and are constantly and dynamically focused on its attractor [15].

Since health care is ordered towards people retaining or regaining their experience of good health, a *good personal health experience* [16–19] should be the attractor (or shared vision) for the whole health system. In operational terms, this would translate to a person's particular needs shaping the configuration of all relevant agents and their interactions within the health system. This ideal requires all system levels to respond *adaptively* (rather than prescriptively) to a challenge which allows the emergence of best solutions for individuals and communities (Fig. 47.2).

In reality, current health system drivers and rules in many countries generally relate to performance and/or financial factors, rather than the patient's health experience, resulting in continuing distortion of the "health vortex" (system) function. This dysfunction of the system, i.e. the inability of current health systems to respond effectively to actual and predicted increases in changing morbidities, workload, and costs, has created an environment where real health system reform may yet be possible. The next section will outline some of the key issues to inform real health system reform.

47.4 Planning for Change

The preceding section outlined the principles, structures, and functions of complex adaptive systems. Applying these principles to health system reform is not easy, given the prevailing patterns of thinking about disease rather than health, and the associated reliance on ingrained structures and procedurally driven functions in most healthcare organisations.

To better appreciate the problems of the current healthcare system, this section explores some of the foundational problems with current healthcare systems.

47.4.1 Health and Illness in the Community

Contrary to popular belief most people are *healthy most of the time*. Epidemiological studies have repeatedly shown that most people in the community feel healthy, and most do not perceive a need to seek help from any health professional. Figure 47.3 shows the health and illness experience reported by Kerr White et al. [20] in 1961. This health and illness curve follows a Pareto distribution (also known as the 80:20 split): the "normal" nonlinear distribution typically found in biological/ecological systems.

Kerr's study provides the following insight: at any point in time 80% of people in the community have no *perceived* healthcare needs, 80% of those seeking health care only require care from a primary care physician (i.e. 16% of the community)—which preferably should be provided by their familiar and trusted general practitioner/family physician, and the need of 80% of the remaining 20% is met by secondary care (i.e. 3.2% of the community), leaving a mere 0.8% of the community requiring tertiary care.

Larry Green et al. [21] repeated this epidemiological study in 2001 and found that the community epidemiology of health and illness had not changed. The persisting bottom line is: only 4% of the community had serious disease conditions

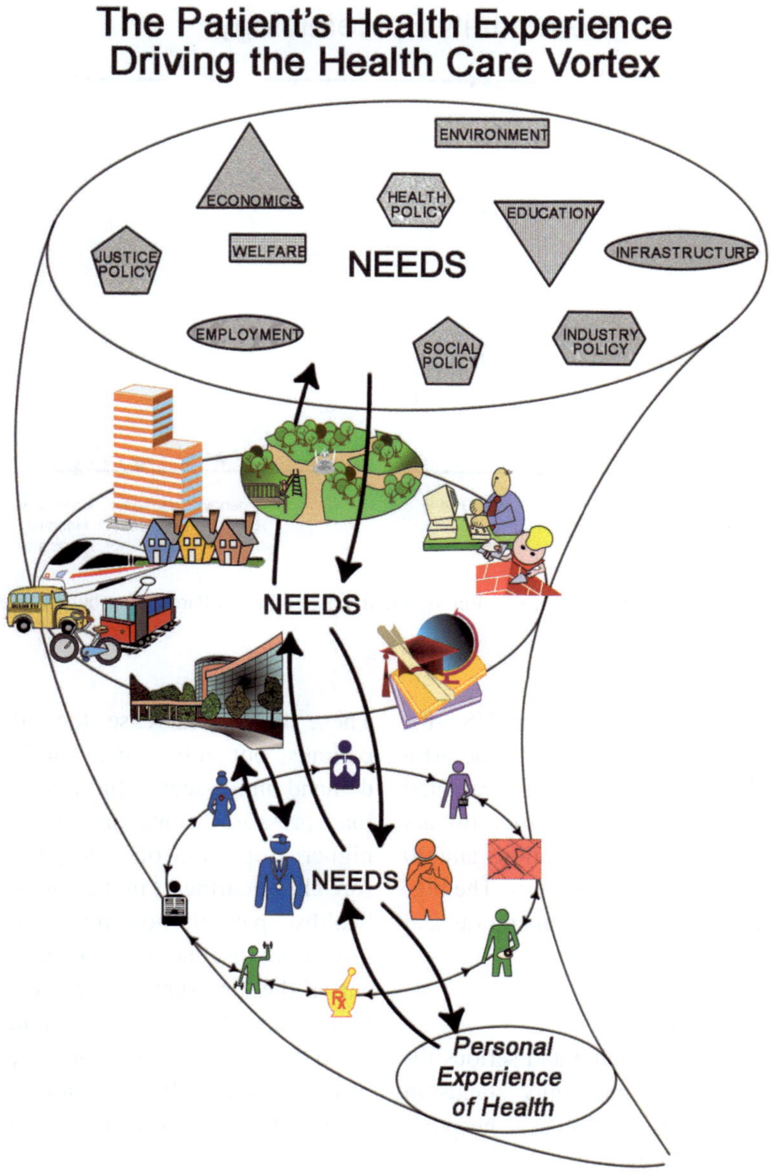

Fig. 47.2 The healthcare vortex

for which 3.2% required secondary care, and 0.7% tertiary care.

Comparing the two sets of data indicates an unchanging level of serious conditions and the need for tertiary care, combined with a rising level of health anxiety and an increase in the number of patients reporting symptoms and concerns (750 to now 800/1,000).

Somewhat surprisingly, seeking health care at a physician's office has decreased from 250 to 217/1,000 [20, 21]. The difference may have been compensated for by other forms of community-based care (home care, hospital-based outpatient care). It is, however, worth noting that the community appears to have lost trust in the medical profession as the sole source of health care, with a substantial number of patients now also seeking help from alternative medical care providers. Overall, it appears that the population's need for hospitalisation has, if anything, slightly

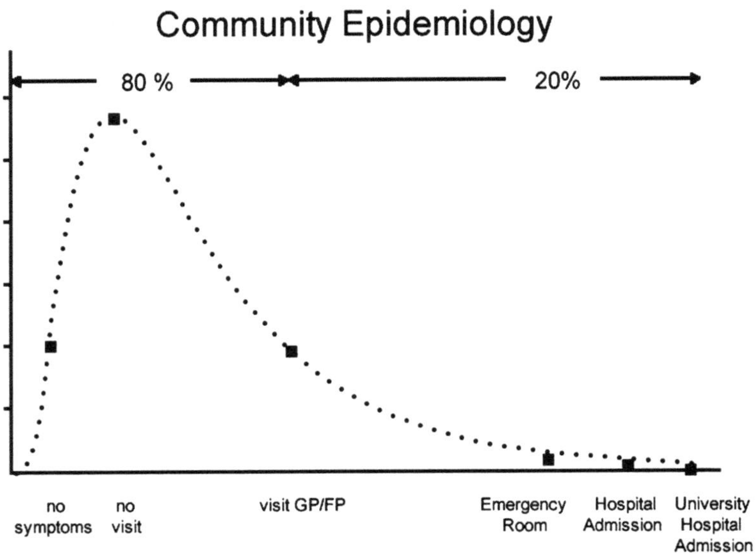

Fig. 47.3 Community epidemiology of health, illness, and disease follows a Pareto distribution (plotted from White et al.'s data [21])

decreased (Fig. 47.4).[8] However, in the USA for example, 10% of the population consume 80% of the health budget, and 50% of those are related to hospitalisation. Every year 48 billion dollars are wasted in avoidable hospitalizations and 20 billion dollars in avoidable admissions. The biggest costs, however, are in administration and fraud [22].

47.4.1.1 Implications

This reassuring pattern of health and serious illness distribution in the community stands in sharp contrast to the disease-focused, hospital-based, tertiary care priorities of policy makers, funders, educators, and the media.

Though the population in most western societies is ageing, older people generally remain healthy in terms of their health experience and day-to-day function. Much of the care provided in the secondary hospital sector can now be provided in day-only or short-term admissions due to advancing technologies. These factors increase the "throughput" of patients, but only somewhat increase overall demand on the sector. Increased hospital workload pressures come predominantly from the higher proportion of aged people (and therefore chronic illness) in the community. As "less healthy" patients now remain in the community or are discharged back into the community early, the pressure on the community care sector has also increased rapidly. Without a rapid increase in community-based primary healthcare capability and capacity, the sustainability of the whole system is threatened.

47.4.2 Healthcare System Orientation

The "healthcare world" is divided by two perspectives; the first is that health care should largely focus on the primary healthcare sector as the means to enhance safe and healthy living environments and to prevent chronic, potentially disabling diseases. The alternative perspective is that health care should largely be a "repair service" providing any available, largely technology-based intervention to fix diseased parts of the human body.

[8] With a growing total population, a steady proportion of the population requiring hospitalisation will still result in an increase in the absolute number requiring hospitalisation, significantly impacting on workloads and costs.

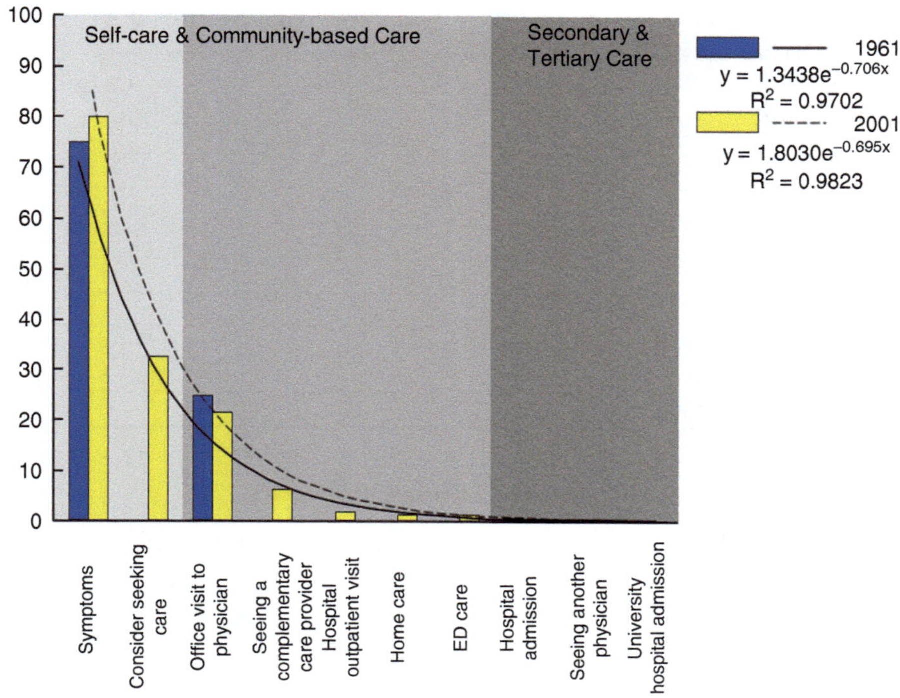

Fig. 47.4 Changes in community epidemiology between 1961 and 2001. Note the increase in illness anxiety, without any increase in disease requiring secondary or tertiary care (plotted from data by White et al. [21] and Green et al.'s [22])

In reality, a well-functioning health system will need to provide both type of services; however, the evidence demonstrates that the *predominant* orientation towards either primary care or secondary/tertiary care has well-defined consequences in terms of health system effectiveness and efficiency. The difference between these two orientations affects far more than government health system policy, strategy, and funding processes. The healthcare workforce: professional education, roles, attitudes, and the focus with which patients are approached, as well as the ratio between primary and secondary care providers, all are profoundly affected. This in turn has implications for overall healthcare costs and health outcomes.

47.4.2.1 Defining Health Systems' Orientation

The importance of a health system's orientation towards primary medical or secondary/tertiary care has been emphasised by Starfield and her colleagues. They demonstrated that a healthcare system's orientation can be described firstly by its professional practice at the delivery level:

1. First contact,
2. Longitudinality,
3. Comprehensiveness,
4. Coordination,
5. Family-centredness, and
6. Community orientation

and secondly its general focus at the health policy level:

1. Type of system,
2. Financing,
3. Type of primary care practitioner,
4. Percent active physicians who are specialists,
5. Professional earnings of primary care physicians relative to specialists,
6. Cost sharing for primary care,
7. Patient list,
8. Requirements for 24 h coverage, and
9. Strength of academic departments of family medicine (Fig. 47.5) [23].

The following discussions explore the impact of health system orientation on patient outcomes, costs, and health workforce ratios, structure, and practice.

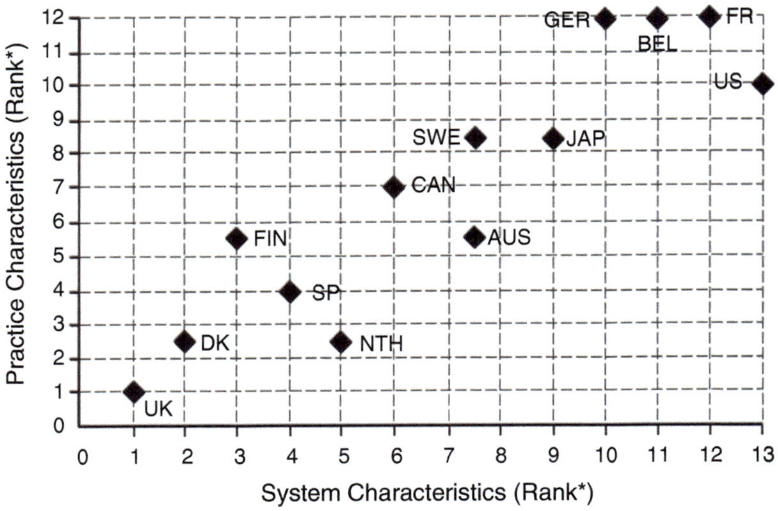

Fig. 47.5 System and practice characteristics facilitating primary care, early-mid 1990s, 1 = highest, 13 = lowest (with permission [24])

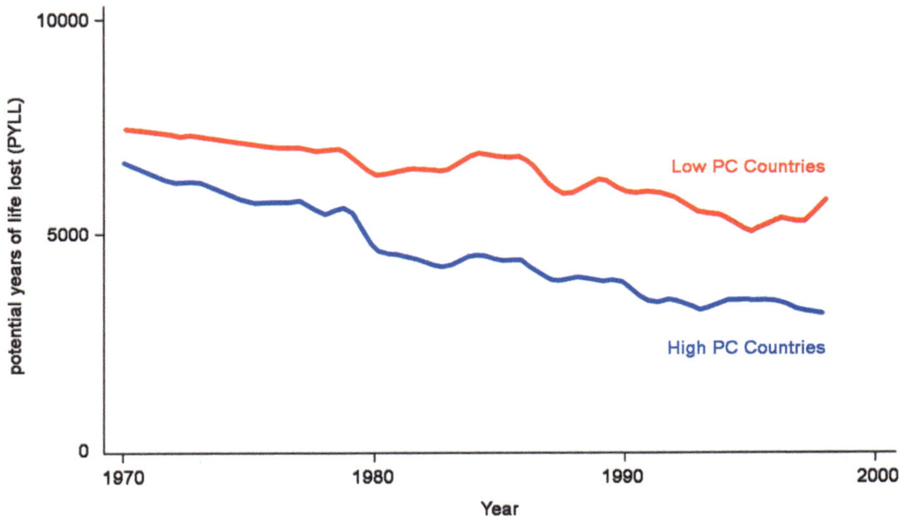

Fig. 47.6 Potential Years of Life Lost (PYLL) for countries with a low and high primary care orientation. (Modified from Starfield, based on data from Macinko et al. [25])

47.4.2.2 Health System Orientation and Mortality

The healthcare system's orientation has a major impact on mortality. *Primary care orientation* has a significant positive impact on all-cause mortality, all-cause premature mortality, as well as cause-specific mortality from asthma and bronchitis, emphysema and pneumonia, and cardiovascular disease and heart disease. The magnitude of these differences is shown in Fig. 47.6 in terms of potential years of life lost (PYLL). In 1998 the risk of losing one life year was almost double in low compared to high primary care oriented countries [24].

47.4.2.3 Health System Orientation and Healthcare Expenditure

The healthcare system's orientation also affects overall healthcare expenditure. Figure 47.7 shows that in general terms a *higher primary care*

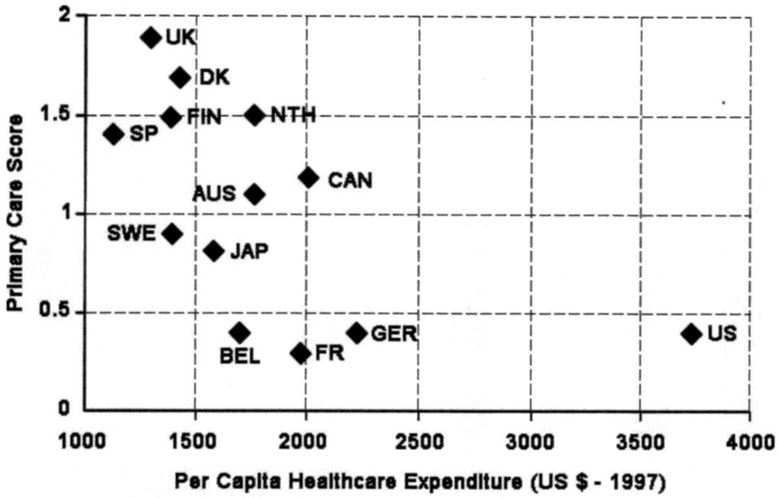

Fig. 47.7 Primary care score vs. healthcare expenditures, 1997; 0 = low, 2 = high (with permission [24])

orientation is associated with lower per capita spending on health and that countries with a low primary care orientation have a higher per capita expenditure on health care.

International comparisons also show that spending more on health care is not a guarantee of achieving better objective or subjective health outcomes. The most obvious exemplar is the USA which spends the highest proportion of its GDP on health care, yet does not cover the whole community and has amongst the worst health status indicators of all OECD countries [25].

47.4.2.4 Health System Orientation and Workforce Structure

The health system's orientation is also reflected in the ratio of primary care general practitioners/family physicians to secondary care providers (specialists/sub-specialists) which has a direct impact on morbidity and mortality.

A *greater primary care orientation* shows a positive correlation between primary care physician supply and population health outcomes in relation to all-cause mortality, cancer, heart disease, stroke, infant mortality, low birth weight, life expectancy, and self-rated health [26]. In the USA an increase of 1 additional primary care physician per 10,000 population improves these outcomes on average by 4% [26]. The figures are similar for the UK where an increase of 1 additional primary care physician per 10,000 population decreases all-cause mortality by 3–10% [27].

Other studies have shown the same relationships in regard to disease-specific outcomes. An increase of primary care physician decreases, whereas an increase in specialists increases, the risk of late-stage diagnosis of colorectal cancer [28]: an increase in primary care physician increases the likelihood of early-stage diagnosis of breast cancer [29]; an increase in family physicians of 1 in 10,000 population decreases the incidence of and mortality from cervical cancer [30] and increases the early diagnosis of melanoma [31].

47.4.2.5 The Link Between Workforce Structure and Healthcare Expenditure

Skinner and colleagues [32] showed the interdependence of health system orientation, workforce structure and density, and workforce performance, with mortality (effectiveness) and expenditure (efficiency). They found significant inverse relationships between consistently providing "good care" (left panel in Fig. 47.8) as well as the "number of specialists" involved in the care of a patient with a myocardial infarction (right panel in Fig. 47.8) on post-myocardial mortality and costs (bottom panel in Fig. 47.8).

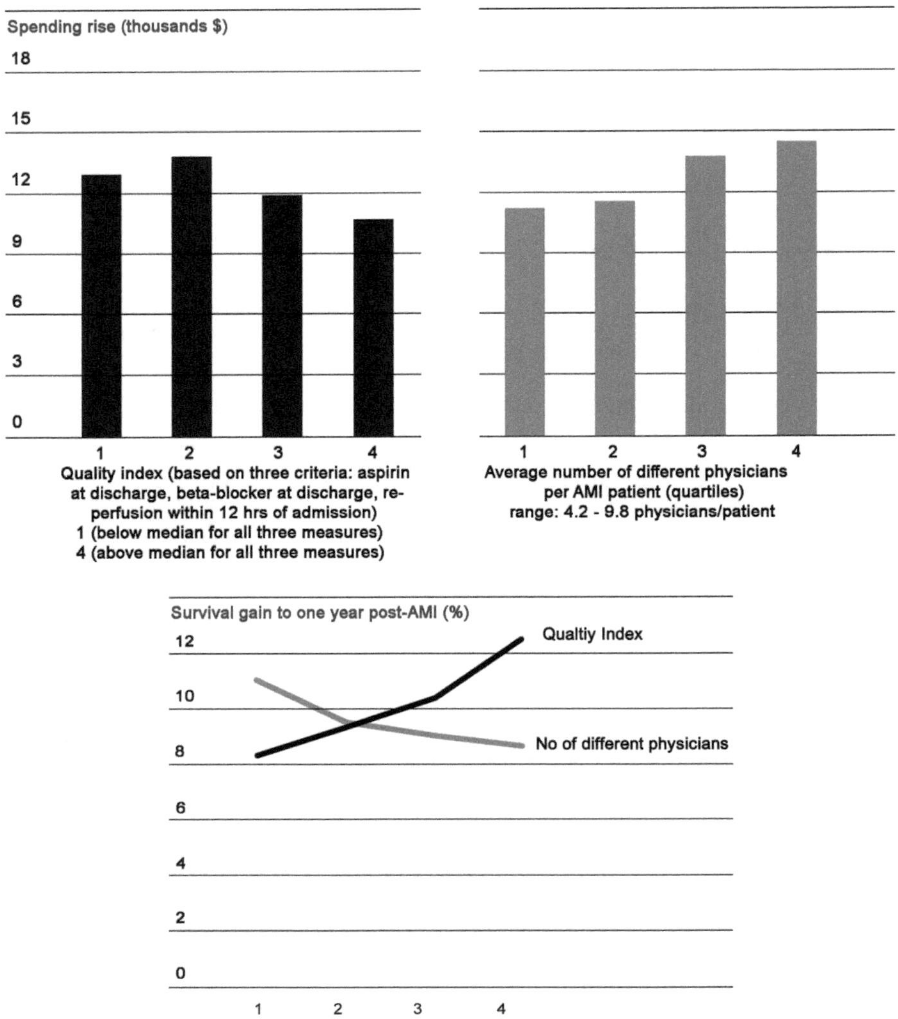

Fig. 47.8 Association of regional quality of care for acute myocardial infarction (AMI) and average number of physicians per AMI patient (quartiles) with changes in survival and spending, 1986–2002 (adapted from [34])

These findings can be summarised as doing a *few proven things right all the time* with the *minimally necessary care team* is better for the patient and the system as a whole, and is probably more rewarding at the personal and interpersonal level for patients and providers alike.

47.4.2.6 Implications

The structure of the healthcare system is reflected in the system's general orientation which has multiple interconnected consequences affecting its effectiveness and efficiency. Given the positive outcomes associated with *primary care orientation,* healthcare reform efforts should logically focus on building the strong, integrated primary healthcare foundations essential to health system effectiveness and efficiency.

Key strategies would include strengthening the standing and resourcing of primary health care and providing greater incentives for healthcare professionals to enter and remain in the primary healthcare workforce. Incentives need to address quality, equity, and most importantly the provision of patient-centred, comprehensive, coordinated, and continuing care in the community.

47.4.3 The Importance of a Patient Focus

People experience health and illness in their own unique ways [16, 17]: as Kant put it: *Every man has his particular way of being in good health.*[9] Nevertheless, health and illness care have long been seen as less important to achieving "good" health outcomes than disease-specific care [34], generally delivered based on clinical protocols and/or "evidence-based" guidelines.

Specific conditions or diseases are labels [35] to categorise certain typically biomedical patterns of complaints. However, such categorisation provides only a narrow view of the individual's health and illness experience, just as it fragments the understanding of multiple diseases in the context of the patient in front of us.

From a system's perspective, raising the distinctions between patient and disease-focused healthcare orientation is not a dichotomous question of either–or. The two are interconnected; however, the foreground orientation of patient-centredness or disease-centredness has major impacts on the health workforce's thinking, workings, actions, and the outcomes of those actions, as discussed below.

47.4.3.1 What We Mean by Person-Centred Care

A prerequisite for person(patient)-centred care, particularly for those with complex conditions and/or circumstances, is as McWhinney stressed [36], the development and maintenance of a *therapeutic relationship*[10] which itself requires a significant level of provider continuity [37] to build the necessary familiarity and trust [38, 39]. The beneficial effects of such a relationship are well documented in the medical literature [40–45], and its absence has been shown to leave patients vulnerable [46] while increasing risks and costs.

Though an ongoing relationship is more likely to result in patient-centred care, patient-centredness itself is a *provider attribute*, namely valuing the patient and her/his concerns foremost, over and above any specific disease. Reflecting a way of thinking and appreciation, patient-centred care can be provided by any individual provider and provider team in any healthcare setting.

Implicit in the concept of person (patient)-centred care is an acceptance of the importance of *sense-making* as a fundamental aspect of health and health care. As has been discussed previously, most people are experientially, rather than condition-specifically, ill. Hence most of the work in person (patient)-centred care will focus on achieving a mutual understanding of the nature of the person's illness experience,[11] [12] which can then guide therapeutic approaches.

Importantly, understanding health and illness as somato-psycho-socio-semiotic in nature requires as detailed an understanding of human physiology, patho-physiology, and the effects

[9] For more details, see Sturmberg, Chap. 15.

[10] The *therapeutic relationship*—sometimes also referred to as the 'doctor as a drug'—as a source of healing has been downgraded as a placebo effect, e.g. see Benson [33]. However the field of psychoneuroimmunology has provided the "scientific" explanations of the workings of the *therapeutic relationship*: the dampening of the pituitary–adrenal axis.

[11] For somato-psycho-socio-semiotic nature of health, refer to Sturmberg, Chap. 15.

[12] The importance of the patient experience, and the impact of harnessing this experience, has been demonstrated by Ben Heywood and colleagues from *PatientsLikeMe* (http://www.patientslikeme.com).

PatientsLikeMe was inspired by the personal experiences with an ultimately fatal condition in the co-founders family. PatientsLikeMe *conceptualized and built a health data-sharing platform that ... can transform the way patients manage their own conditions, change the way industry conducts research and improve patient care.*

PatientsLikeMe follows *four core values: putting patients first, promoting transparency ("no surprises"), fostering openness and creating "wow." We're guided by these values as we continually enhance our platform, where patients can share and learn from real-world, outcome-based health data.*

For a detailed discussion on PatientsLikeMe, see e.g. TEDxCambridge—Ben Heywood tells the story of PatientsLikeMe (http://www.youtube.com/watch?v=n3NVG-pVDIs)

For its impact, see e.g. Paul Wicks, Timothy E Vaughan, Michael P Massagli & James Heywood. Accelerated clinical discovery using self-reported patient data collected online and a patient-matching algorithm. Nat Biotechnol 2011;29(5):411–14.

of "therapeutic" interventions as it does an understanding of the psychological and ecological factors impacting on health, illness, and diseases. Put differently, health, illness, and diseases are the result of the interactions within the patient's broader networks of his/her genetics, family, society, and environment: they cannot be separated from each other.

47.4.3.2 The Status Quo of Disease-Centred Medicine

Disease-centredness, in the form of looking at one specific condition at a time and in isolation (explicitly excluding co-morbidity considerations), characterises much current medical practice and clinical research. This way of thinking is part of the fragmentation (also known as specialisation) of medicine, and fragmentation, in turn, perpetuates this way of thinking. Fragmentation, it is argued, is necessary as "knowledge is increasing exponentially": an argument that demonstrates the error of equating "information" to knowledge.[13]

These confusions are fuelled by the belief in the randomised controlled trial (RCT) as the "gold standard" of gaining "knowledge". The "single-mindedness" underpinning the RCT is also its greatest problem [47, 48]. As Fortin highlights, many RCTs exclude patients with multiple co-morbidities to ensure internal validity. However, this approach diminishes the external validity, i.e. the generalizability, of results. Fortin showed that looking at a primary care population (i.e. 96% of the community), many patients with common conditions, like hypertension, have co-morbidities, and most would have been excluded from RCTs based on reported inclusion/exclusion criteria [48].

What do we know about the outcomes of disease-centred care?

Disease-centredness, as typically exercised in specialty/sub-speciality practice, shows that patients seeing a specialist have a 41% higher hospitalisation rate and a 12% higher prescription rate compared to those being cared for in a more primary care-focused health maintenance organisation (HMO) [49]. Long-term health outcomes for patients with hypertension and NIDDM, for example, showed no differences regardless of the type of care system or physician type [50].

Differences occur even within different disease-focused settings, for example the marked differences in the approach to patients with ischaemic heart disease between Texas and New York. Despite patients presenting in different settings symptom severity at the time of hospitalisation did not differ amongst patients, however, patients in Texas had a coronary angiography in 45% of cases compared to 30% in New York. Angiography did not show any difference in three-vessel or left main stem disease, the pathologies associated with a high risk of sudden death. Equally Texans had significantly more coronary angioplasties (15% vs. 7%) and a significant trend of higher coronary artery bypass surgery (15% vs. 13%). The long-term outcome, despite much lower intervention rates, showed lower mortality, lower morbidity, and better functional and subjective health for patients in New York[14] [51].

Examples of hospital and/or specialised based practice variations such as these abound, and are now the subject of countless strategies in developed countries to reduce variation and its associated costs. Variations in healthcare spending across the USA, Europe, and Australasia are well documented. The National Health Expenditure data show total per capita healthcare spending ranging from $4,000 in Utah to $6,700 in Massachusetts. Spending variations across smaller geographic units have also been documented using Medicare data. County-by-county analyses by the National Center for Policy Analysis show Medicare per capita spending in 2008 varied from just over $5,000 in Nobles County, Minn., to $8,500 in Rice County, Kan.

[13] For more details on this, refer to Chap. 4

[14] The likelihood of death for patients treated in New York than those treated in Texas showed a hazard ratio of 0.87 (CI 0.78–0.98), the likelihood of suffering from angina amongst patients in Texas than New York was significantly higher (OR 1.41; CI 1.13–1.76), and the ability to NOT perform instrumental activities of daily living requiring ≥ 5 METs was significantly greater (OR 1.62; CI 1.26–2.07)

Similarly, researchers with the Dartmouth Atlas Project found that among 306 hospital referral regions, Medicare spending per patient ranged from more than $16,000 in some areas to less than $6,000 in others. Lower costs were often associated with better care [52–54]. Value-based payments linked to outcomes in capitated systems is the major international strategy.

As AHRQ states: *'The concept of value-based health care purchasing/payment is that buyers/ funders should hold providers of health care accountable for both cost and quality of care. Value-based purchasing brings together information on the quality of health care, including patient outcomes and health status, with data on the dollar outlays going towards health. It focuses on managing the use of the health care system to reduce inappropriate care and to identify and reward the best-performing providers. The key elements of value-based purchasing include: Contracts spelling out the responsibilities of employers as purchasers with selected insurance, managed care, and hospital and physician groups as suppliers; Information to support the management of purchasing activities; Quality management to drive continuous improvements in the process of health care purchasing and in the delivery of health care services; Incentives to encourage and reward desired practices by providers and consumers; Education to help employees become better heath care consumers'*[55].

The reality of Value-Based Purchasing is that it is experimental and still developmental with need for ongoing improvements [56]. For example the UK Quality and outcomes framework achieved process change without demonstrable changes in health outcomes [57], confirming earlier findings across time and context settings [58].

47.4.3.3 Evaluating Patient-Centred Care

Quality of care between generalists and specialists has been debated extensively, with the conclusions usually being that specialists provide better *disease-specific process of care* compared to generalists, but that generalists achieve *better overall health outcomes* without increased morbidity or mortality, an observation known as the "paradox of primary care" [34]. This difference is most pronounced in the presence of multimorbidity.

Patient-based outcomes require measures other than "hard" bio-medical indicators. As patients usually evaluate health from a personal perspective, measures of different aspects of functional health and overall well-being are much more useful, e.g. SF-36 or WONCA-COOP charts. Figure 47.9 shows how different common specific conditions affect patients rather differently; note that up to 70% of all specific conditions occurred in the presence of one or more co-morbidity [47].

More important is the finding that a person's disease and level of disease control shows little relationship to their overall health experience, as shown in patients with Type II diabetes. Neither overall blood sugar control nor change in diabetic control showed any relationship with patients' general health experience (depicted in Fig. 47.10), or other components of the SF-36 scales. These findings were independent for five covariates: insulin use, number of diabetic complications, duration of diabetes, education, and number of hyper-, or hypoglycaemic episodes during the preceding month [59].

A caveat is needed at this point—not all outcomes of medical care, especially negative ones, relate to the medical treatments provided. The patient is an important additional "variable". A study looking at predictors for complications in diabetes showed that hypertension and smoking habits are associated with a dramatic increase in complications in insulin-dependent diabetics, whereas poor compliance with visits was strongly associated with poor outcomes in non-insulin-dependent diabetics. In addition, the risk of poor outcomes increased with "no longer modifiable" patient characteristics of age (older patients), level of education (<7 years), and occupation (low skilled workers) [60].

47.4.4 Implications

Person-centredness means genuinely putting the patient's *perception* of need first: person-

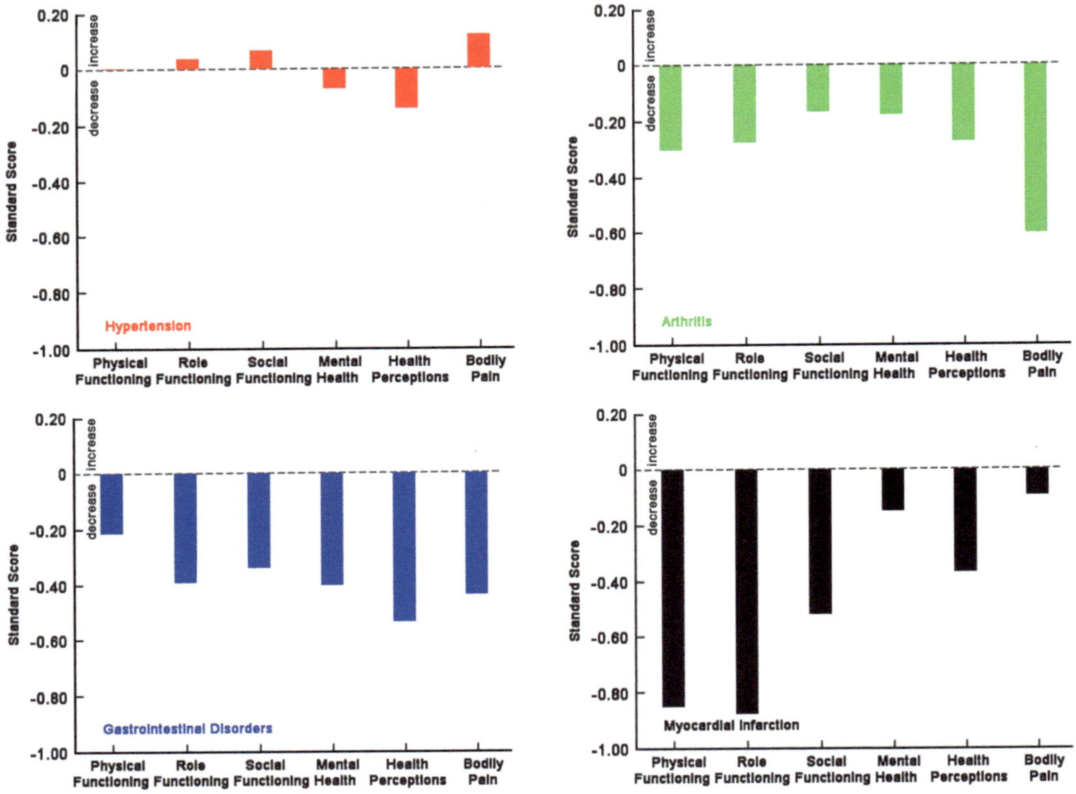

Fig. 47.9 Change of the functional health profiles of patients with four common conditions in a general community setting. The *dotted line* indicates the health of patients without chronic conditions. Each condition affects various aspects of functional health in a different way. Hypertension has the least impact on functional health, and gastrointestinal disorders and myocardial infarction the greatest. Note: within the group, variance is high, and the functional health of patients with more than one chronic condition is substantially worse (adapted from [50])

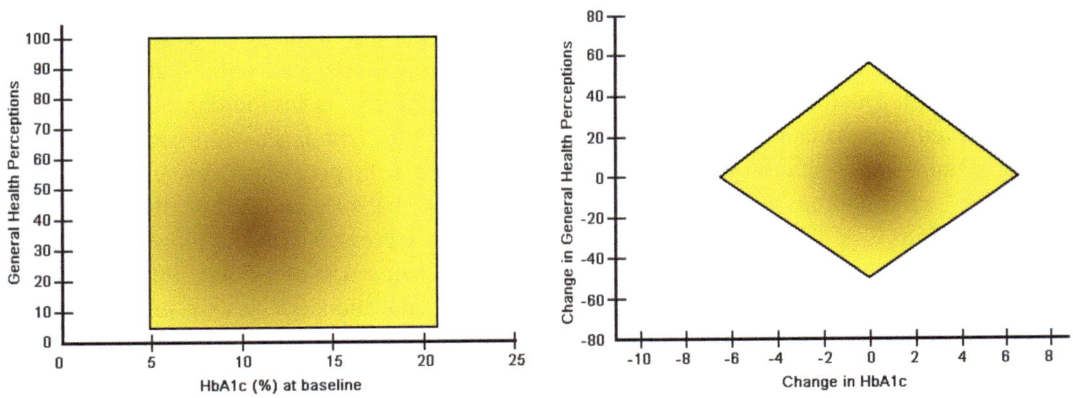

Fig. 47.10 Effect of diabetes control on general health perception. The left-hand panel shows the probability density plot of health perception for a given HbA1c level; the right-hand panel shows the effect of change in HbA1c level and change in health perception after 12 months (the darker the colour the greater the density, adapted from [61])

centredness is the antithesis of disease-focused, protocol-driven care.

Patient-centred health professionals have a profound impact on the healthcare system. They benefit the patient in ways that are independent of the process of disease-specific care processes, and by being more discerning in the application of healthcare resources, their care is more cost-effective.

Breaking the current cycle of disease-focused care requires not only a greater focus on primary care orientation of the system at large, but it also requires an understanding of the doctor characteristics that lead and perpetuate disease-centred care approaches. Policy makers and funders have three fundamental challenges.

- The first challenge is to come to terms with the fact that most patients experience good health most of the time, and thus would benefit from health-promoting support across all aspects of their life and their community.
- The second is addressing the structure of the healthcare workforce. This is a particularly difficult challenge as it has two interrelated components. Current practices have firmly shaped the way one "sees and acts", which in turn reinforces one's personal perceptions or worldview which provide levels of comfort and certainty. Being "guided" and having a "protocol" to go by lessen anxieties associated with the inherent uncertainties of patient presentations. These insights have profound implications for future student selection (requiring a focus on resilience when confronted with uncertainty) and education (requiring different approaches and role models[15]), and for the need to increase the ratio of primary care generalists to specialists (or 'partialists').
- The third challenge influencing the media's obsession with acute events in tertiary settings—combined with the "publicity machine" of pharmaceutical, technology, and research industries which continue to raises hopes of "cures for everything".

[15] For a more detailed consideration of this aspect, see Stewart Mennin, Chap. 43

47.5 A Pathway for Change

Medicine is a social science, and politics is nothing but medicine in a broader perspective.

Rudolf Virchow, 3 November 1848 [61]

Fundamentally health and health care are *common goods* in the sense of Aristotelian virtue ethics: an understanding that has progressively lost ground over the past century to rationalist and economic perspectives. As the latter approaches are clearly failing, there is an urgent need for a broad discourse exploring and subsequently defining the purposes and core values for a health system that is effective, efficient, equitable, sustainable, *and* makes sense to all.

The findings outlined in this chapter were intended to highlight a fundamental insight, namely that the health system is *a complex adaptive system*, and that to move forward on healthcare reform thinking therefore requires a shift to embrace the very different nature of complex adaptive systems, their nonlinear responses to potential interventions and innovations, and the far-reaching repercussions across the whole of the system and society at large.

Our discussion has attempted to demonstrate that:

- Most people experience "*good enough*" health and do not seek healthcare services.
- Most people experiencing the need for health care only require primary healthcare services.
- The subjective health experience, and therefore the perceived need for health care, is not related to the severity of a disease state nor to a change in disease state.
- Only 3.2% of the community requires secondary, and <1% tertiary care.
- Primary care achieves better overall health outcomes, measured as self-reported health, morbidity, and mortality, for people with common chronic conditions than "single disease-focused" services.
- Primary care achieves these outcomes at a substantially lower cost, and

- Good access to primary care produces the greatest health gains for those suffering the greatest disadvantage.

Given this, more strenuous efforts are required to re-orient our health systems away from its current predominant disease and hospital focus, and towards patient-centred health care grounded in strong, integrated primary health care as the foundation of effective and efficient healthcare systems.

47.5.1 Value Change Equals Health System Change

Ultimately healthcare reform requires a change in the system's *core value* which can then guide the reform process and establish an anchor for its ongoing activities. Our discussions suggested that: *patients' needs are at the centre of the health system,* and that this would be a suitable summation of a *core value* statement for a patient-centred healthcare system.

47.5.2 A New Set of *Simple* ("How To") *Rules*

As discussed previously, "How to" rules (or operating principles) apply to all agents in the healthcare system, regardless of their level of operation. To arrive at a set of *simple rules* requires deliberation on *how* core values and knowledge can be combined into clear statements to guide decision making, or more specifically, deliberation on *how* statements—given an understanding of complex systems' dynamic behaviours—need to integrate core values with the related knowledge base to produce simple rules which guide[16] health professionals at service delivery levels to best *meet the needs of the patient.*

Suggested "*simple rules*" might be:

- Understand the patient's needs.
- Develop ongoing relationships of trust between patients and their key providers.
- Consider and understand the patient's context before delving into detail.
- Explore the effects of your intended actions on other agents in the system.
- Consider time delays between actions and outcomes.

47.5.3 Operationalise at Different Health System Levels

A complex adaptive health system will necessarily be a *decentralised* system that allows *different* but *mutually agreeable* solutions to emerge that best meet the patients' and communities' needs. Such a system continually learns from feedback at the *local* level,[17] allowing local level needs and global objectives to co-evolve—in simple terms, the principle of appropriate subsidiarity is in operation: everyone works hand in hand across all domains, decisions are made at the right "local" level, and there is no loss of sight of either necessary detail or the whole picture.

Adhering to the perpetual application of *simple rules* should allow the emergence of operational policies, strategies, and guidelines ordered towards shared objectives such as the progressive improvement of:

- Patients' health experience.
- Access, equity, respect, and empowerment.
- Capacity in health professional teams to provide *person-centred, continuing*, and *coordinated care* for individuals and groups.
- Appropriate community health services for minority and at risk populations.
- Coordinated cross-portfolio preventive services (e.g. smoking, obesity, motor vehicle accidents).

[16] Suggested points 4 and 5 relate to the system dynamic behaviours, the one-to-many relationship of each agent, and the time delay common to inputs into a system.

[17] A key lesson, summarised by Peter Allen, Prof of Management at the Cranfield University School of Management, is: *No learning if agents cannot read the feedback!*, *Centralised organizations will stop Local learning*, and the *Need to design Organizations that allow learning*.

47.5.4 Potential Consequences

Adhering to core values would keep the health system clearly focused, and shared objectives combined with a set of simple rules would enable agents in all parts of the health system to achieve "best possible" solutions through self-organisation within the opportunities and constraints of their local environment.

In this environment, the role of government becomes more clearly to focus on providing policy directions, allocating resources appropriately, oversighting the overall system performance against its key objectives, and adjusting policy levers and resources in light of both system outcomes and feedback received.

Key national policy development would necessarily include:

- The development of an appropriate national policy framework, strengthening the role, objectives, standing, and resourcing of primary health care.
- Definition of greater incentives for healthcare professionals to enter and remain in the primary healthcare workforce.
- Definition of incentives to address quality, efficiency, equity, and most importantly, the provision of patient-centred, comprehensive, coordinated, and continuing care in the community.

Potential local strategy solutions could include:

- Co-location of medical and community services within and beyond the health sector.
- Building flexible practice-based primary care teams to include nurses, allied health professionals, and community outreach services.
- Expansion of e-solutions including patient-controlled personal health records, with for example patient information stored on swipe cards to improve critical information flow between care settings.
- The ongoing engagement of primary healthcare services and their communities in local service planning and community development and enablement with a focus on promoting healthy community environments.

Ultimately, each healthcare provider's central responsibility remains to identify every individual patient's needs, and to achieve appropriate person-centred, coordinated, and integrated care. The challenge for those of us considering system redesign is how best to reorient and reconfigure our existing healthcare systems such that this responsibility is both clear and achievable.

And finally, one cannot avoid facing the difficult issue of healthcare economics. Embracing the altruistic virtue philosophy of "health care as a *common good*" and aiming to achieve effective, efficient, and equitable care necessarily lead to consideration of the most appropriate government financing strategies, and within this, the strengths and weakness of existing business models in delivering care.

What are the strengths and weaknesses—not just of various blended payment options, but of the business models employed in public, private, and not-for-profit sectors, and which models are most/least compatible with the values, policy goals, and system operations we have described? Following on from this, what is government's role in terms of regulating the financing of health care?

Should the future directions outlined in this chapter come to pass, it may follow that there are clinical service, business, and corporate models which could be seen as incompatible with society's broader healthcare needs? The pressures on our existing health system dictate that health system reform efforts will inevitably continue. In this environment, it will be critical to ponder and debate questions such as these.

47.6 Appendix 1: Core Value Statements of Government Health Departments

47.6.1 Secretary of State for Health, UK, July 2010[18]

Our values

1. It is our privilege to be custodians of the NHS, its values and principles. We believe that the NHS is an integral part of a Big Society, reflecting the social solidarity of shared access to collective healthcare, and a shared responsibility to use resources effectively to deliver better health.

[18] http://www.dh.gov.uk/prod_consum_dh/groups/dh_digitalassets/@dh/@en/@ps/documents/digitalasset/dh_117794.pdf

2. We are committed to an NHS that is available to all, free at the point of use, and based on need, not the ability to pay. We will increase health spending in real terms in each year of this Parliament.
3. The NHS is about fairness for everyone in our society. It is about this country doing the right thing for those who need help. We are committed to promoting equality and will implement the ban on age discrimination in NHS services and social care to take effect from 2012. The NHS Commissioning Board will have an explicit duty to address inequalities in outcomes from healthcare services.
4. We will uphold the NHS Constitution, the development of which enjoyed cross-party support. By 2012, the Government will publish the first statement of how well organisations are living by its letter and spirit. The NHS Constitution codifies NHS principles and values, and the rights and responsibilities of patients and staff. It is about mutuality; and our proposals in Chap. 2 for shared decision-making by patients, their carers, and clinicians will give better effect to this principle. It is also about NHS-funded organisations being good employers; and our plans in Chap. 4 will give organisations and professionals greater freedoms, leading to better staff engagement and better patient care.
5. Current statutory arrangements allow the Secretary of State a large amount of discretion to micromanage parts of the NHS. We will be clear about what the NHS should achieve; we will not prescribe how it should be achieved. We will legislate to establish more autonomous NHS institutions, with greater freedoms, clear duties, and transparency in their responsibilities to patients and their accountabilities. We will use our powers in order to devolve them.

47.6.2 Health Canada, Canada, October 2011[19]

Core Values—Our Values in Action

The Dialogue on Values and Ethics resulted in the identification and description of the following core values for Health Canada. In pursuing and fulfilling our vision and mission, we achieve personal, organizational and public good by:

Taking Pride in What We Do

We are motivated and guided by our personal integrity.

(a) We recognise our potential
(b) We take the initiative to improve ourselves and the way we do things
(c) We act with sound judgment

Building a Workplace Community

We respect each other and work together in a healthy environment.

(a) We embrace diversity and nurture empowering relationships
(b) We communicate honestly and effectively
(c) We create an environment that promotes learning and innovation

Caring for the People of Canada

We advance the public good with purpose and passion while honouring democratic values.

(a) We provide credible information, reliable advice and quality services
(b) We establish and maintain good working relationships with our stakeholders
(c) We responsibly and wisely manage resources entrusted to us

By being accountable for our values and their integration in our work, we lay the foundation for excellence at Health Canada.

47.6.3 Department of Health and Ageing, Annual Report 2010–2011, Australia[20]

Our Values

We value:

- An apolitical, impartial and professional environment.
- The importance of achieving results for the Government and the community.
- Delivering services to the public fairly, effectively and impartially.
- Transparency, accountability and responsiveness.
- A workplace that is fair and free of discrimination.
- Diversity and equity in employment.
- The highest ethical standards.
- Innovation.
- Respect.

These principles reflect the Australian Public Service Values in the Public Service Act 1999 (Section 10) and the department's People Strategy.

47.6.4 Office of Public Health and Science, 2011, USA[21]

Values

The OPHS has identified and defined five core values, which are listed below.

Put People First

[19] http://www.hc-sc.gc.ca/ahc-asc/activit/about-apropos/index-eng.php#val

[20] http://www.health.gov.au/internet/main/publishing.nsf/Content/annual-report2010-11

[21] http://www.hhs.gov/about/budget/fy2011/2011cj.pdf

- Honor the public's trust and confidence
- Respect for colleagues and the public health professions
- Recognise the invaluable contributions of OPHS staff

Integrity
- Adhere to the highest ethical standards
- Ensure products and services are truthful, accurate, and comprehensive
- Assure health research conforms to scientific norms
- Recognise that privacy and safety of human participants is paramount

Excellence
- Conduct programs and activities guided by science and driven by results
- Delineate clear and enforce consistent accountability for program outcomes
- Design programs and activities so that rigorous program evaluations can and will be performed
- Promote public health that is effective, efficient, and community-delivered

Diversity
- Embrace the richness of OPHS' diversity and seek to strengthen it
- Value the diversity of our Nation and the perspectives brought by differences in race
- Ethnicity, gender, age, and socio-economic status
- Believe that all Americans should benefit from advances in health promotion

Leadership through collaboration
- Commit to disease prevention and health promotion
- Believe that collaboration and coordination builds effective, efficient, responsive
- Sustainable public health systems
- Foster input from all relevant partners and stakeholders in program operations

47.7 Appendix 2: Core Value Statements of Health Institutions

47.7.1 Institute of Health Improvement, USA[22]

Our Values

These operating values are core principles for work in IHI. They guide the behaviour and choices of all staff, faculty, and the Board of Directors.

Without Boundaries: The people of the IHI compose a single organization, with common systems, common knowledge, and unconditional teamwork.

Speed and Agility: We change our own work and respond as quickly as the health care systems we serve need us to. Our past work need not ever be our future work. We are always willing to change.

Focus on Subject Matter: Our concerns are health and health care; we are not wedded to specific methodologies. We remain always open to new approaches to the continual improvement of care. Results for patients and communities define our success.

Valuing Volunteers: We network together people who have expertise and knowledge, so that they can teach each other, help others, and improve the work of IHI. These people are our "faculty." Their work is the lifeblood of the IHI. We will make their experience with IHI the most satisfying of their professional lives.

Customer Focus: To achieve our mission, we must serve and delight those who shape and deliver health care. Their satisfaction—100% satisfaction—is our uncompromising aim, in everything that we do.

Honesty: To achieve our mission, we must earn and preserve the trust of those we attempt to help. To do so, we must tell the absolute truth about ourselves and our work, reporting both failures and successes with equal discipline, and seeking the views and opinions of people outside our organization.

Transparency: We are an institute without walls. Those who work with us, no matter where or when, should feel informed and welcomed. We work always in daylight.

Orderliness: Disorder is waste, which neither we nor health care can afford. We will be lean in our work, and continually reduce waste and disorder. We practice what we teach.

Celebration and Thankfulness: Our mission is long, and our work is not easy. We take time to look back, as well as forward, to thank each other, and to take pride in what we do.

47.7.2 Centre for Disease Control, Atlanta, USA[23]

CDC Core Values

Accountability—As diligent stewards of public trust and public funds, we act decisively and compassionately in service to the people's health. We ensure that our research and our services are based on sound science and meet real public needs to achieve our public health goals.

Respect—We respect and understand our interdependence with all people, both inside the agency and throughout the world, treating them and their contributions with dignity and valuing individual and cultural diversity. We are committed to achieving a diverse workforce at all levels of the organization.

Integrity—We are honest and ethical in all we do. We will do what we say. We prize scientific integrity and professional excellence.

[22] http://www.ihi.org/about/Pages/IHIVisionandValues.aspx

[23] http://www.cdc.gov/about/organization/mission.htm

47.7.3 Cancer Council Western Australia[24]

Our Values

- We will serve the people of Western Australia to reduce the impact of cancer.
- We will improve our services to the people of Western Australia based on the best available evidence and practice.
- We will respect and value the community, volunteers and our staff.
- We will work toward our vision and mission through teamwork and collaboration.
- We will act with integrity and honesty.
- We will strive to achieve equity in the provision of programs and services.
- We will manage and apply funds entrusted to us by the community in the most effective way.

47.8 Appendix 3: Core Value Statements of Professional Associations

47.8.1 American Medical Association, USA[25]

Core values

1. Leadership
2. Excellence
3. Integrity and ethical behaviour

47.8.2 Canadian Medical Association, Canada[26]

CMA Values

Our values describe what is important to us and outline the behaviours required to make us successful.

- *Excellence:* We strive to be the best in everything we do for physicians and their patients and we are committed to continuous learning and improvement.
- *Professionalism:* We strive to unite physicians around the fundamental tenets of high quality standards, patient safety, professional autonomy, accountability, responsiveness and physician health and well-being.
- *Integrity:* We uncompromisingly adhere to the highest ethical standards and honesty in representing our members and conducting our business.
- *Compassion:* We foster an individual and corporate culture of caring for physicians, their patients and each other.
- *Rassembleur:* We strive to bring together diverse communities of interest in the pursuit of common goals.

47.8.3 Royal College of General Practitioners (RCGP), UK[27]

Our Values

The RCGP is the heart and voice of General Practice and as such:

We protect the principle of holistic generalist care which is integrated around the needs of and partnership with patients.

We are committed to equitable access to, and delivery of, high quality and effective primary healthcare for all.

We are committed to the theoretical and practical development of general practice.

47.8.4 American Association of Family Physicians (AAFP), USA[28]

Values

AAFP and its Members are Committed to Care that is…

- Equitable for all people
- Centered on the whole person within the context of family and community
- Based on science, technology, and best available evidence
- Supported by lifelong professional learning
- Grounded in respect and compassion for the individual

47.8.5 Royal Australian College of General Practitioners (RACGP), Australia[29]

RACGP values statement

"we serve with integrity, strive for excellence, foster GP unity, advocate for health equity and embrace the diversity of our profession"

[24] http://www.cancerwa.asn.au/aboutus/mission/

[25] http://www.ama-assn.org/ama/pub/about-ama/our-mission.page

[26] http://www.cma.ca/aboutcma/history-mission-vision-values

[27] http://www.rcgp.org.uk/about_us/vision,_purpose_and_priority.aspx

[28] http://www.aafp.org/online/en/home/aboutus/theaafp/strategicplan.html

[29] http://www.racgp.org.au/nsc/education

47.9 Appendix 4: Core Value Statements of Clinical Institutions

47.9.1 Mayo Clinic, Rochester, USA[30]

Primary value
The needs of the patient come first.
Value statements
These values, which guide Mayo Clinic's mission to this day, are an expression of the vision and intent of our founders, the original Mayo physicians and the Sisters of Saint Francis.
Respect
Treat everyone in our diverse community, including patients, their families and colleagues, with dignity.
Compassion
Provide the best care, treating patients and family members with sensitivity and empathy.
Integrity
Adhere to the highest standards of professionalism, ethics and personal responsibility, worthy of the trust our patients place in us.
Healing
Inspire hope and nurture the well-being of the whole person, respecting physical, emotional and spiritual needs.
Teamwork
Value the contributions of all, blending the skills of individual staff members in unsurpassed collaboration.
Excellence
Deliver the best outcomes and highest quality service through the dedicated effort of every team member.
Innovation
Infuse and energise the organization, enhancing the lives of those we serve, through the creative ideas and unique talents of each employee.
Stewardship
Sustain and reinvest in our mission and extended communities by wisely managing our human, natural and material resources.

47.9.2 Central Coast Health, July 2008, Australia[31]

Values
Integrity × *Teamwork* × *Best Practice*
Accountability × Social Justice

[30] http://www.mayoclinic.org/about/missionvalues.html

[31] http://www.nscchealth.nsw.gov.au/clinicalservicesplan/ServicesPlan/Chapters/003812873.pdf

47.9.3 Brigham and Women's Hospital, Boston, USA[32]

Values
- *Quality Patient Care:* Delivering quality patient care is the center of everything we do.
- *Teaching Excellence:* We seek to uphold the highest standards in training health care professionals.
- *Research Leadership:* We continuously seek new ways to demonstrate our leadership role in research.
- *Customer Focus:* Our focus is to serve our customers.
- *Respect for the Individual:* We recognise and value the contributions of every individual.
- *Teamwork:* We work toward a unified approach to developing health care solutions.
- *Embracing Change:* Embracing change will help us to be successful.
- *Operational Efficiency:* We strive for efficient and effective delivery of services.

References

1. WHO. Declaration of Alma-Ata. International Conference on Primary Health Care, Alma-Ata, USSR, 6-12 September 1978. Geneva: World Health Organisation1978.
2. Lakoff G, Johnsen M. Metaphors we live by. London: The University of Chicago Press; 2003.
3. Annas GJ. Reframing the debate on health care reform by replacing our metaphors. N Engl J Med. 1995; 332(11):745–8.
4. Moynihan R. Key opinion leaders: independent experts or drug representatives in disguise? Br Med J. 2008;336(7658):1402–3.
5. Astra Zeneca. Patient groups. Astra Zeneca International; [cited 2012 05-Jan-2012].
6. Mintzes B. Should patient groups accept money from drug companies? No. Br Med J. 2007;334(7600):935.
7. Herxheimer A. Relationships between the pharmaceutical industry and patients' organisations. Br Med J. 2003;326(7400):1208–10.
8. Ball D, Tisocki K, Herxheimer A. Advertising and disclosure of funding on patient organisation websites: a cross-sectional survey. BMC Public Health. 2006;6(1):201.
9. Kent A. Should patient groups accept money from drug companies? Yes. Br Med J. 2007;334(7600):934.
10. Donohue JM, Cevasco M, Rosenthal MB. A decade of direct-to-consumer advertising of prescription drugs. N Engl J Med. 2007;357(7):673–81.
11. Heylighen F. Complexity and Self-organization. In: Bates MJ, Maack MN, editors. Encyclopedia of

[32] http://www.brighamandwomens.org/about_bwh/mission.aspx

Library and Information Sciences. New York: Taylor & Francis; 2008.
12. Gell-Mann M. Complex adaptive systems. In: Cowan GA, Pines D, Meltzer D, editors. Complexity: metaphors, models, and reality, Santa Fe Institute studies in the sciences of complexity, Proc Vol XIX. Reading, MA: Addison-Wesley; 1994. p. 17–45.
13. Eisenhardt KM, Sull DN. Strategy as simple rules. Harv Bus Rev. 2001;79(1):107–16.
14. Capra F. The web of life. London: HarperCollins Publishers; 1996.
15. Dooley KJ. A complex adaptive systems model of organization change. Nonlinear Dynam Psych Life Sci. 1997;1(1):69–97.
16. Sturmberg JP. The personal nature of health. J Eval Clin Pract. 2009;15(4):766–9.
17. Sturmberg JP. The Foundations of Primary Care. Daring to be Different. Oxford San Francisco: Radcliffe Medical Press; 2007.
18. Sturmberg JP, Martin CM, Moes M. Health at the centre of health systems reform—how philosophy can inform policy. Perspect Biol Med. 2010; 53(3):341–56.
19. Batalden P, Ogrinc G, Batalden M. From one to many. J Interprof Care. 2006;20(5):549–51.
20. White K, Williams F, Greenberg B. The ecology of medical care. N Engl J Med. 1961;265(18):885–92.
21. Green L, Fryer G, Yawn B, Lanier D, Dovey S. The ecology of medical care revisited. N Engl J Med. 2001;344(26):2021–5.
22. Institute of Medicine. The Healthcare Imperative: Lowering Costs and Improving Outcomes—Workshop Series Summary2011: Available from: http://iom.edu/Reports/2011/The-Healthcare-Imperative-Lowering-Costs-and-Improving-Outcomes.aspx.
23. Starfield B, Shi L. Policy relevant determinants of health: an international perspective. Health Policy. 2002;60(3):201–18.
24. Macinko J, Starfield B, Shi L. The Contribution of Primary Care Systems to Health Outcomes within Organization for Economic Cooperation and Development (OECD) Countries, 1970–1998. Health Serv Res. 2003;38(3):831–65.
25. Starfield B, Is US. Health really the best in the world? J Am Med Assoc. 2000;284(4):483–5.
26. Macinko J, Starfield B, Shi L. Quantifying the health benefits of primary care physician supply in the United States. Int J Health Serv. [10.2190/3431-G6T7-37M8-P224]. 2007;37(1):111–26.
27. Gulliford MC. Availability of primary care doctors and population health in England: is there an association? J Public Health. 2002;24(4):252–4.
28. Roetzheim RG, Pal N, Gonzalez EC, Ferrante JM, Van Durme DJ, Ayanian JZ, et al. The effects of physician supply on the early detection of colorectal cancer. J Fam Pract. 1999;48(11):850.
29. Ferrante JMMD, Gonzalez ECMD, Pal NMPH, Roetzheim RGMDM. Effects of physician supply on early detection of breast cancer. J Am Board Fam Pract. 2000;13(6):408–14.
30. Campbell RJ, Ramirez AM, Perez K, Roetzheim RG. Cervical cancer rates and the supply of primary care physicians in Florida. Fam Med. 2003;35(1):60–4.
31. Roetzheim RG, Pal N, Van Durme DJ, Wathington D, Ferrante JM, Gonzalez EC, et al. Increasing supplies of dermatologists and family physicians are associated with earlier stage of melanoma detection. J Am Acad Dermatol. 2000;43(2):211–8.
32. Skinner JS, Staiger DO, Fisher ES. Is technological change in medicine always worth it? The case of acute myocardial infarction. Health Aff (Millwood). 2006;25(2):w34–47.
33. Benson H, Epstein M. The Placebo Effect. A Neglected Asset in the Care of Patients. JAMA. 1975;232:1225–1227.
34. Stange KC, Ferrer RL. The paradox of primary care. Ann Fam Med. 2009;7(4):293–9.
35. Rosenberg CE. The tyranny of diagnosis: specific entities and individual experience. Milbank Q. 2002;80(2):237–60.
36. McWhinney IR. The Importance of being different. William Pickles Lecture 1996. Br J Gen Pract. 1996; 46(7):433–6.
37. Hjortdahl P. Continuity of care: general practitioners' knowledge about, and sense of responsibility toward their patients. Fam Pract. 1992;9(1):3–8.
38. Fugelli P. Trust—in general practice. Br J Gen Pract. 2001;51(468):575–9.
39. Freeman G, Olesen F, Hjortdahl P. Continuity of care: an essential element of modern general practice? Fam Pract. 2003;20(6):623–7.
40. Hjortdahl P, Lærum E. Continuity of care in general practice: effect on patient satisfaction. Br Med J. 1992;304:1287–90.
41. Hjortdahl P. The influence of general practitioners' knowledge about their patients on the clinical decision-making process. Scand J Prim Health Care. 1992;10:290–4.
42. Sturmberg JP, Schattner P. Personal doctoring. Its impact on continuity of care as measured by the comprehensiveness of care score. Aust Fam Physician. 2001;30(5):513–8.
43. Saultz JW, Lochner J. Interpersonal continuity of care and care outcomes: a critical review. Ann Fam Med. 2005;3(2):159–66.
44. Guthrie B, Saultz JW, Freeman GK, Haggerty JL. Continuity of care matters. Br Med J. 2008; 337(aug07_1):a867.
45. Freeman G, Hughes J. Continuity of care and the patient experience. London: The Kings's Fund; 2010.
46. Sweeney K, Gray D. Patients who do not receive continuity of care from their general practitioner—are they a vulnerable group? Br J Gen Pract. 1995;45:133–5.
47. Stewart AL, Greenfield S, Hays RD, Wells K, Rogers WH, Berry SD, et al. Functional status and well-being of patients with chronic conditions. Results from the Medical Outcomes Study. J Am Med Assoc. 1989; 262(7):907–13.

48. Fortin M, Dionne J, Pinho G, Gignac J, Almirall J, Lapointe L. Randomized controlled trials: do they have external validity for patients with multiple comorbidities? Ann Fam Med. 2006;4(2):104–8.
49. Greenfield S, Nelson E, Zubkoff M, Manning W, Rogers W, Kravitz R, et al. Variations in resource utilization among medical specialties and systems of care. J Am Med Assoc. 1992;267(12):1624–30.
50. Greenfield S, Rogers W, Mangotich M, Carney M, Tarlov A. Outcomes of patients with hypertension and non-insulin-dependent diabetes mellitus treated by different systems and specialties. J Am Med Assoc. 1995;274(18):1436–44.
51. Guadagnoli E, Hauptman P, Ayanian J, Pashos C, McNeil B, Cleary P. Variation in the use of cardiac procedures after acute myocardial infarction. N Engl J Med. 1995;333(9):573–8.
52. The Dartmouth Institute for Health Policy and Clinical Practice. Tracking the care of patients with severe chronic illness: Copyright 2008 The Trustees of Dartmouth College; 2008.
53. Tuijn SM, Robben PBM, Janssens FJG, van den Bergh H. Evaluating instruments for regulation of health care in the Netherlands. J Evaluat Clin Pract. 2011:no-no.
54. Doran T, Campbell S, Fullwood C, Kontopantelis E, Roland M. Performance of small general practices under the UK's quality and outcomes framework. Br J Gen Pract. 2010;60(578):e335–44.
55. AHRQ. Value-Based Payments2012: Available from: http://www.ahrq.gov/qual/meyerrpt.htm#head3.
56. Jarousse LA. Value-based purchasing and bundled payments. Hosp Health Netw. 2011;85(5):6. following 32, 2.
57. Cashin C. United Kingdom: Quality and Outcomes Framework (QOF). 2011 [cited 2012 February 26th]; Available from: http://www.rbfhealth.org/rbfhealth/system/files/Case%20study%20UK%20QOF.pdf.
58. Kjellstrand CM, Kovithavongs C, Szabo E. On the success, cost and efficiency of modern medicine: an international comparison. J Intern Med. 1998; 243(1): 3–14.
59. Weinberger M, Kirkman M, Samsa G, Cowper P, Shortliffe E, Simel D, et al. The relationship between glycemic control and health-related quality of life in patients with non-insulin-dependent diabetes mellitus. Med Care. 1994;32(12):1173–81.
60. Nicolucci A, Carinci F, Ciampi A. On behalf of the SID-AMD Italian Study Group for the Implementation of the St. Vincent Declaration. Stratifying patients at risk of diabetes complications: an integrated look at clinical, socioeconomic, and care-related factors. Diabetes Care. 1998;21(9):1439–44.
61. Bauer A. "Die Medicin ist eine sociale Wissenschaft"—Rudolf Virchow (1821–1902) als Pathologe, Politiker und Publizist. Medicine—Bibliothek—Information. 2005;5(1):16–20.

A Complexity Science Approach to Healthcare Costs and Quality

48

Yaneer Bar-Yam, Shlomiya Bar-Yam, Karla Z. Bertrand, Nancy Cohen, Alexander S. Gard-Murray, Helen P. Harte, and Luci Leykum

48.1 Introduction

The US healthcare system suffers from high costs and low quality compared to healthcare systems internationally [1–3], as measured by reported life expectancy [4] and infant mortality [5]. High rates of nosocomial infection (infections acquired in healthcare settings) as well as adverse drug effects (errors in the administration of medication) manifest the need for improvement in the system of care. At a cost of $2.5 trillion annually [6] the system is not delivering affordable, effective care.

The paradox of higher costs and lower quality makes clear the existence of a systemic problem. How can we fix it? Complex systems science provides tools to address this question directly. In this chapter we provide eight scientifically based steps toward reducing costs and improving quality. Our suggestions arise from an analysis of the US healthcare system in particular, but they are broadly applicable when adapted appropriately. The eight steps are:

1. *Separate simple care* from complex care.
2. *Empower workgroup competition* as an incentive and avoid regulating costs or quality.
3. *Create superdoctor teams* to rapidly diagnose and treat highly complex conditions.
4. *Accelerate intake routing* to rapidly identify the right provider.
5. *Add redundancy to improve communication* to prevent prescription errors.
6. *Create disinfection gateways* at spatial boundaries to reduce hospital-based infections.
7. *Use e-records for research* to supplement clinical studies.
8. *Promote "First Day" celebrations* to encourage healthy behavior.

Additional reading is available in the references provided at the end of the chapter.

48.2 Separate Simple Care

Healthcare work may be divided into two types: simple care, which is the same for many people, and complex care, which is different for each individual.

Simple care includes preventive services, such as health screenings, vaccinations, and healthy-habits counseling sessions. Complex care includes the individualized diagnosis and design of treatment.

Physicians are specially trained to diagnose and treat complex medical conditions. Nonetheless, one finds that physicians and their offices are typically responsible for simple, standard care in addition to complex, individualized care.

That poses a problem. Asking the same organizational structure to provide mass-applicable

Y. Bar-Yam (✉) • S. Bar-Yam • K.Z. Bertrand •
N. Cohen • A. S. Gard-Murray • H. P. Harte • L. Leykum
New England Complex Systems Institute,
238 Main Street Suite 319, Cambridge, MA 02142, USA
e-mail: yaneer@necsi.edu

J.P. Sturmberg and C.M. Martin (eds.), *Handbook of Systems and Complexity in Health*,
DOI 10.1007/978-1-4614-4998-0_48, © Springer Science+Business Media New York 2013

preventive care and complex individual care is like asking an expert violin craftsman to provide all the chairs for a new concert hall. The mismatch between the organization and the task leads to ineffectiveness and inefficiency.

Ironically, instead of streamlining the delivery of high-volume simple services, most cost-reduction efforts to date have tried to make complex tasks simpler and faster. Industrial-style efficiency is poorly applicable to doctors' diagnoses and treatment of individual patients, however. Trying to speed and simplify doctors' work assembly-line style reduces doctors' time to make complex decisions, which is not a good idea if we want doctors to be careful and make the best decisions possible. At the same time, many healthy patients are receiving insufficient preventive care, since doctors are being asked to provide many of these services. The volume of preventive care needed is too great for the current system to handle it effectively.

What can be done? The solution is to separate the tasks. Let doctors perform the complex tasks that they do well and delegate preventive-care tasks such as vaccinations to an organization suited for simple, repetitive tasks.

In many hospitals and doctors' offices, simpler tasks such as drawing blood and taking X-rays are performed by professionals trained for these specific, frequent tasks. This idea can be applied much more broadly.

We can improve the healthcare system dramatically by separating the simple services that many healthy people need even further, delegating them not just to different individuals but to different organizations.

We are beginning to see this concept in programs that make flu shots available in supermarkets and airports, and in the growing number of "retail clinics.[1] "

Since retail outlets at malls and supermarkets serve many people with similar needs, "retail clinics" make sense; they can readily provide routine and preventive care such as health screenings, vaccinations, and dissemination of public health information. These clinics have the additional advantage of locating preventive services where healthy people frequently go, rather than requiring them to make less convenient trips to their physician's office.

CVS is installing MinuteClinics in its pharmacies. Walmart has such clinics at over 50 locations and is planning thousands. These clinics, originally developed to provide routine treatment for minor problems such as strep infections, now also offer preventive services, including vaccinations, cholesterol and other tests, and school physicals.

What is the payback in widespread retail clinics? To be sure, a retail setting offers convenience and efficiency in implementing preventive care via large volume, simplicity, and a focus on healthy people. Whenever a large number of similar tasks are to be performed, the medical system is well served by moving such care from physicians' offices or hospitals to the retail setting.

Besides easing the burden on doctors—freeing them for complex tasks for which their time is now too limited—the separating-out of mass care from individual care would streamline high volume processes. This would address the excess costs that arise when the performer and the task don't match.

More people would gain access to routine care due to the convenience of location and avoidance of the need to make appointments, travel to the doctor, and wait.

What is clear is that making preventive healthcare more accessible can reduce illness, further easing the burden on the medical system. It is time for the medical and insurance communities to embrace this solution.

Scientific Principle

For an organization to perform tasks effectively, it must be organized so that it can match both the scale (rate of repetition) and complexity (variety) of those tasks. When large-scale tasks are performed by an organization designed for complex tasks, the result is inefficiency. When high complexity tasks are assigned to an organization designed for large-scale tasks the result is nonoptimal (wrong) acts, i.e., errors. Separating tasks by

[1] For a detailed discussion of retail clinics see Chap. 49.

scale and complexity enables simple, rote, mass-applicable to be performed by individuals and organizations (retail clinics in this case) well suited to those tasks and complex tasks to be performed by individuals and organizations (physician practices) designed for those tasks.

48.3 Empower Workgroup Competition

The current use of "managed care" to improve medical performance and reduce costs is inherently flawed. To understand where the flaws lie, it's useful to ask: What roles should government, management and practitioners play in healthcare and in healthcare decision-making? Today, they often serve in the wrong roles. That's because they serve within the wrong system structures. Neither traditional centralized management nor free market competition scenarios work for healthcare and yet those settings are perceived as the only two options.

Whether by government or by insurer, efforts to control healthcare costs are typically deployed within a setting of centralized management to dictate how to allocate limited funds. Yet healthcare is a highly complex system, and, as we see in the failure of the USSR and other centralized economies, centralized control doesn't work for complex systems.

At the same time, free-market competition, which may result in rapid improvements for, say, the electronics industry, is ill-suited for managing healthcare. After all, patients generally can't shop around for the best hospital.

Are these the only two options?

There is a third: *empowering workgroup competition*. Groups of care providers who together can be responsible for medical outcomes and other performance metrics compete against other groups in the same hospital and between hospitals. Workgroups must become teams in a peer competition to improve care. This team competition approach combines the best aspects of both the free market and centralized management; it allows the spirit of competition to spur advances and improve performance, while still allowing management to set objectives.

In free-market competition, the goal is financial gain. In workgroup competition, the objective is to be a top performer according to carefully designed metrics that measure both cost and quality. This kind of competition works for sports teams, students competing for grades, and in other competitions where the goal isn't just to make money.

In order for medical care to improve, the people engaged in providing that care, who know the most about what to do, must be the ones who have control over care decisions, and must be the ones with responsibility for outcomes. However, performance should not be measured at the level of individual doctors and nurses, because outcomes often rely on an entire workgroup's performance—e.g., how nurses or physicians communicate information across a shift change has a huge impact on outcomes, and communication relies on how people work together.

How does a system empower the people who work together, and who can take on such responsibilities—and not tell them what to do? Through workgroup competition. Fostering workgroup competition means first identifying and solidifying groups that can be responsible for outcomes.

Nurses at a nursing station responsible for the care of patients in a specific part of the hospital could be a workgroup, taking responsibility for improvement in areas like infection rates and patient satisfaction.

In many hospitals, nurses, technicians, and an anesthesiologist are assigned to a surgery based on who happens to be available. In some, however, each surgeon has his or her own team of nurses and techs, and an anesthesiologist with whom he or she works. This team-based approach allows workgroup members to get to know each others' styles and to work smoothly together. A team approach also allows them to improve their outcomes as a group.

One emergency room, in the Washington Hospital Center of Washington, DC, takes just such a team approach, dividing the emergency room staff into teams who care for a patient from entry to release or admission to the hospital.

Workgroups must be of the right size and function to be able to improve based on the measures evaluated. Some efforts have attempted to produce competition between entire hospital systems, but these units are too large and unwieldy. They focus responsibility on the hospital system management. Since these efforts do not directly measure workgroup performance, they do not enable practitioners to take responsibility for their outcomes or work together to improve their performance. Having provider workgroups rather than management assume the responsibility for healthcare performance in a competitive environment is key.

Once workgroups are created, the metrics of competition must be designed. Healthcare administrators should determine workgroup-performance measures that cover health outcomes, costs, lengths of hospital stays, and patient satisfaction. Remember—what is measured is what will be improved: performance metrics must be designed carefully and should be revisited periodically for updates and improvements.

Finally, workgroups' performance on each of the measures should be publicized to all the groups on a regular basis—say, monthly. The workgroups would compare their results with those of other workgroups providing similar types of care, like other surgical teams or other nursing stations, and then be responsible for their own improvement.

Care must also be taken to set an appropriate tone for the competition. Just as with sports, rules must be established which will encourage good sportsmanship. Also, as teams improve their performance, swapping team members among them causes better strategies to be shared and improve the performance of all teams.

When workgroups' scores are in focus, both the scores and the performance they measure improve radically. Members of a workgroup will work together to improve their performance. They will innovate, and they will emulate improvements that they see others adopting.

Some hospital systems publicize their performance measures so that consumers can compare hospitals or systems. This won't improve performance, however, because patients can't always change providers, and because hospital systems don't have much control over their practitioners' performance.

Arranging hospital operations such that groups consistently work together, and then identifying these workgroups and entering them into a spirited, friendly competition based on well-designed performance metrics will lead to dramatic improvements in results.

Besides improving performance on the measures evaluated, workgroup competition will change the way efforts to control healthcare costs are directed. We will move from a limited focus on cutting healthcare costs to a broader focus on improving the healthcare system at lower cost.

48.3.1 Why Centralized Management Doesn't Work in Healthcare

- Decisions about care should be made by the people who understand the need for care, the healthcare providers.
- The decisions must be made on a case-by-case basis. Abstractions according to generalized rules degrade the effectiveness of care.

48.3.2 Why Free Market Competition Doesn't Work in Healthcare

- The consumer has limited knowledge of how well each hospital or doctor's practice works.
- The ability to choose is often limited by emergency circumstances, capacity of the provider, insurance policies, or restrictions on changing providers.

48.3.3 Why Team Competition Works

- Professionals care about comparisons with their peers.
- They have the most knowledge and ability to improve outcomes.
- Team competition, like competition in biological evolution, is the natural way to improve performance.

Scientific Principle
Highly complex systems arise through evolutionary change that involves replication, variation, and selection. We can understand this in social learning as mimicry (replication), creativity (variation), and competition based on comparative evaluation, i.e., performance feedback (selection). Highly complex tasks cannot be performed by a single individual (they exceed an individual's complexity). It is not even possible for an individual to manage or design a system to meet the challenge of high complexity. Instead evolutionary processes that provide for feedback and learning of the group performing the task enable progressive improvement. Just as team competition drives improvement in sports, competition between teams of care providers enables improvement of outcome measures.

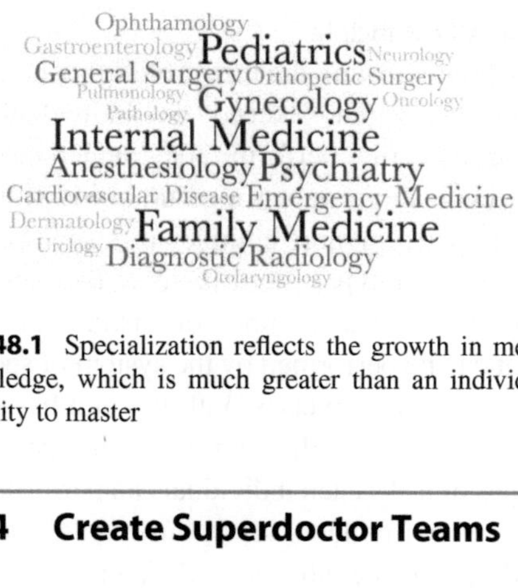

Fig. 48.1 Specialization reflects the growth in medical knowledge, which is much greater than an individual's capacity to master

48.4 Create Superdoctor Teams

A hundred years ago, physicians were generalists, treating most medical conditions. Humanity didn't have nearly as much medical knowledge and knowhow back then so that for the most part a single doctor could master what was known. That has changed.

Medical knowledge now far exceeds a single expert's ability to master it. Medical students receive a general training and then they specialize, seeking to learn just one small piece of what we know about medicine (Fig. 48.1).

Specialists have become essential because of the complexity of care. The more we learn, the more kinds of specialists are needed. Increasingly, however, it is necessary to have patients see multiple specialists for a single problem, which causes fragmentation and delays the care. Furthermore—and critically—the interplay between multiple causes of a single condition, or multiple aspects of its treatment, makes it difficult for the separated specialists to address such complex problems.

What is the solution?

A human being is a single working system and specialists must be able to work together as an integrated unit for diagnosis and treatment. Specially constituted teams of physicians and other care providers who work together on a regular basis should address the more complex problems. The cost of having such a team in place might seem high, but for complex cases such a team will prove to be more effective and less costly than the alternative—the difficulties, delays, and costs inherent in multiple appointments. The challenge is making sure the teams can work together smoothly and efficiently, and with better results than specialists working separately.

A well-integrated team of specialist physicians can be thought of as a "superdoctor." In order for medical teams to be superdoctors, they must get to know each other's strengths and styles and act together seamlessly. Well-integrated teams have the combined specialized knowledge of each member and more: they have the ability to relate these different domains of knowledge and combine them in new ways. Moreover, they can act rapidly with this combined knowledge. They can be an important part of the solution to the problems of fragmentation.

Such teams have become standard practice in cancer care, where specialists in imaging, surgery, radiation therapy, and chemotherapy often meet and work together to treat patients. The wide diversity of cancers and of individual responses to treatment make the team approach necessary for effective care. These teams generally also include nonphysician practitioners. While the team approach is most widely used for cancer, some medical centers, recognizing the problem of fragmentation in care, are using the team approach for other conditions.

To be most effective, superdoctor teams need to work together on a regular basis. If you were to throw together several sports players—even professional athletes—to play as a team without training together, they would not play as well as they would with team members who they were used to. Similarly, medical teams must "practice" together to fully leverage their collective ability.

We can take clues for the formation of superdoctor teams from the types of cases that currently require many specialist appointments—teams should be formed that can handle these cases together more effectively than the specialists could working separately. The advantage of the team is not just the ability to do what the individual specialists would do separately; it's the ability to treat a wide range of conditions effectively, to make very subtle distinctions that are important for effective care, to solve the cases that are the most difficult due to the interplay of multiple causes or complications.

The capabilities of superdoctor teams will grow through being challenged, and they will learn from experience. Innovation in their composition and testing their abilities is key. We can only discover their effectiveness through observing how they respond to challenges. Measuring their effectiveness brings us back to Step II.

Superdoctor teams can assume the dynamics we described in Step II: Empower Workgroup Competition, competing against one another and continuously pushing the boundaries to improve care and reduce costs. By measuring their performance, we can learn how to build more and more effective teams, both in terms of choosing types of specialists to be on a team and in their specific interactions.

It is important to note that, in trying to stem the cost of specialist care, alternative cost-cutting approaches have been tried but have not been successful.

Some have proposed having primary care physicians treat more cases, to reduce the number of specialists that patients see as a way of reducing healthcare costs. This approach, though at times politically popular, is ineffective. Family physicians can treat a certain set of conditions, but they do not have the specific knowledge to treat many complex, more specialized conditions.

Of course, we will still need primary care physicians—many problems are best treated by a single person knowledgeable about a wide range of conditions. We also will continue to benefit from individual specialists, or from specialists, who don't normally work together collaborating for particular patients. This works fine for problems of intermediate complexity.

However, for the increasing number of highly complex cases, superdoctor teams are necessary for comprehensive, integrated, cost-effective, quality care.

We must take steps to form innovative specialist teams that can treat the most complex cases successfully and cost-effectively. Introducing such teams is essential if we are to put our vast medical knowledge to effective use.

Scientific Principle

We continue with the theme of highly complex tasks that exceed an individual's capacity to perform or understand. Specialization enables a group of individuals to perform more complex tasks by routing certain types of task to a particular individual. The number of distinct tasks that can be performed by the system of specialists grows linearly with the number of individuals (it is the sum of the number of types of tasks each individual can perform). However, a collaborative team enables each individual to contribute a different dimension to the task performed by the group, so that the number of types of tasks can be as high as the product of the number of tasks each individual can perform. As there are increasing number of specializations, and conditions that require multiple specialists to address, this implies that we have reached the point in complexity of care that teams are increasingly necessary to address complexity of medical care.

48.5 Accelerate Intake Routing

One of the strengths of our current healthcare system is that the critical task of routing patients to appropriate specialists is performed by primary care physicians. They are also often on call to respond to inquiries about urgent care or referral to an emergency room.

How patients are routed to appropriate care is a crucial aspect of running an efficient medical system. As we consider changes to our healthcare system, it is vital to note the importance of accurate and timely routing, to focus resources on keeping this process reliable, and to improve upon it where possible. Routing is known in medical circles as triage. Unlike the triage in disaster or wartime, triage in a modern medical setting simply refers to the act of directing patients to the appropriate care.

Today, most routing occurs in the primary care physician's office or through the physician's after hours call-in system. This process often works smoothly, with knowledgeable family physicians, internists or pediatricians assessing patients in a timely manner and routing them appropriately. They may treat a patient directly or refer to an appropriate specialist or clinic.

When care is required after-hours and the PCP is not available, patients call a service to reach the doctor on call. Some medical offices provide extended-hour urgent visits. Some practices and insurance agencies also provide 24-h phone access to medical professionals through a "nurse line." These call-in services can provide feedback to the patient as to whether he or she should go to the emergency room, or whether their condition can wait.

However, the medical routing system doesn't always work efficiently.

Individuals who are uninsured, or who do not have a primary care provider, often resort to the emergency room for nonemergency care. Without a PCP to act as a router, and without access to a 24-h phone support, these individuals' options are limited. If their medical problem occurs at night, or if they can't make it to a walk-in clinic during the clinic's hours because of work, childcare, or other responsibilities, the emergency room becomes the only option. This puts a huge strain on the emergency care system, since a large portion of its resources must be devoted to treating or routing these cases rather than on the truly urgent ones.

Even for people with insurance and a PCP, the intake system often doesn't work well. The wait for an initial appointment with a family physician averages 20 days in the United States as a whole. According to a 2009 survey, the average wait time to see a family practitioner in Washington, D.C., was 30 days; in Los Angeles, it was 59 days; and in Boston, 63 days.

This delay in the initial care and routing process indicates that something is amiss with the intake system. The influx of patients seeking initial evaluation and referral to specialists overtaxes the primary care system. What's more, delays in diagnosis can have serious health consequences.

The initial triage decision often falls to the receptionist, who is generally unprepared to properly make such decisions. When the doctor is too busy to see everyone who wishes to be seen, the receptionist is put in the position of deciding which patients must be seen urgently and which can wait for an appointment—an appointment that may be several weeks away.

No matter how good the care is once a patient gets to the right place, delays mean that the healthcare system isn't working well, not for quality of care, where delays may compromise the patient's health, and not for costs.

Even where the current routing system works well, it can be improved using advances in technology. The advent of email and other forms of electronic information and communication have changed our expectations of response time. The medical routing system should make use of these advances to build on the existing structure, speed response time, and improve outcomes.

Perhaps the best way to think about an effective rapid-response system is "Triage on Steroids" … or "Super-911."

A routing service should be available 24 h a day, seven days a week, and be performed by a knowledgeable and capable medical professional. This is the gold standard of medical routing. The

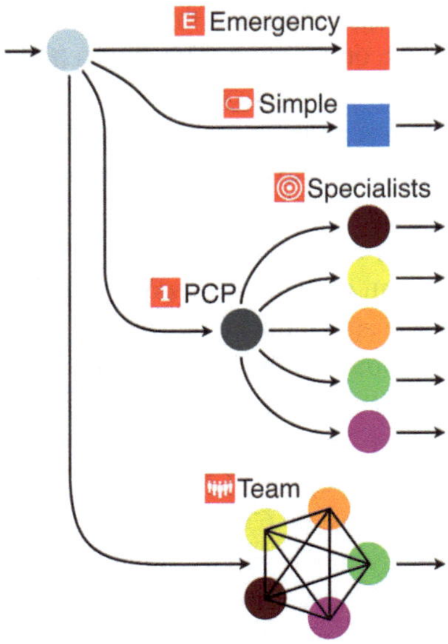

Fig. 48.2 No caption

system should be accessible via phone, Internet, and in person. Most of the traffic should be handled by phone or electronically; for in-person routing, the staffed routing site might be near to—but separate from—an emergency room, or perhaps at a pharmacy.

The first task of a routing service is to serve as a "Super-911," to identify emergencies and reassure if urgent care is not needed.

Second, if it is not an emergency, the intake specialist could determine the complexity of the response needed. Patients requiring a simple, standard response, especially preventive care such as flu shots, could be directed to a preventive care clinic (see Step I). Patients with more complex problems should be directed to their primary care provider or, where the determination can be readily made and the primary care provider is otherwise overburdened, an appropriate specialist. Finally, patients with especially complex problems might be referred to a superdoctor team (see Step III) (Fig. 48.2).

To accelerate the appointments for an initial evaluation, some family practices are adopting a system known as "open access," designed to facilitate routing and expedite care. In an open access setting, no appointments (or only a limited number) are made significantly in advance. Instead, patients call into the office when they require care and are given an appointment that same day. This system provides an opportunity for a very rapid initial evaluation, allowing for routing decisions to be made literally hours after symptoms manifest.

There is another difficulty that could be addressed with creative use of new technology. Often, the best person to determine whether a particular specialist should be seen is the specialist himself or herself. But a patient moving from specialist to specialist to find out who should provide treatment is not a good strategy. It is inefficient and potentially costly in health consequences.

One approach to solving this problem is to use information routing rather than patient routing. The key is information-gathering and communication. Most of what happens at an initial medical visit to a clinic or primary care physician is a gathering of key information that will serve to determine which specialist should be seen. In information routing, after the initial visit, the information, not the patient, would be forwarded to a number of specialists.

The specialists could rapidly evaluate whether, based upon this limited information, they should be seeing the patient. Or, a specialist might provide a question—if the patient has such and so a symptom or such and so a test result, then they should be seeing the patient, e.g., "If the patient's ears hurt while the other symptoms occur, she should see me. If not, I'm not the right specialist for this case."

This information-based routing system, on the specialist level, serves patients better, and costs less, than patients being sent around to several specialist appointments in order to route them correctly.

An "everywhere and always-on" routing system could be made available instead of the more usual answering services, by primary care providers, provider systems, or insurers. Such a system would relieve emergency rooms of having to perform the routing of nonurgent cases, freeing them to focus on the urgent and emergency care they are supposed to be providing. This type of routing system would also relieve some of the burden of PCPs, and shorten patients' wait times for routing and treatment significantly.

It is clear that an accessible and reliable 24/7 accelerated triage mechanism—staffed by intake specialists and augmented by an information-transfer system—will dramatically improve our medical system's cost-efficiency and ability to serve patients well. Wait time for care will be dramatically reduced, emergency rooms will be put to their proper use, and the burden on primary care providers will be lightened. Augmenting and improving our existing routing system is crucial to improving healthcare quality for all.

Scientific Principle

The rate of response of a system is a key aspect of its ability to perform time-sensitive tasks (such as medical care). A system only responds as well as its rate-limiting step, i.e., the step that takes the longest time to perform. The current design of the US healthcare system involves the use of "gatekeepers" who serve to route individuals to the care they need. Evidence suggests that this task has become the rate-limiting step. The gatekeepers become overloaded, wait times to see them are extended and the routing process itself often involves many time-delayed steps. Improving the system behavior involves: explicitly recognizing response time as a critical dimension of care, identifying the time of first contact rather than the time of intake as the beginning of the medical response process, expanding the set of gatekeepers to allow adequate parallel intake channels, and improving the routing function to better utilize available channels of care that work in parallel. These interventions also should reduce the complexity of the initial intake process.

48.6 Improve Communication

Ten years ago, the Institute of Medicine's report on the extent of serious medical errors brought the issue to the attention of medical professionals and the public. According to the Institute and the FDA, medication-related errors cause over one million harmful drug events each year. Even one case of medical error may result in tragedy for those directly affected and may be traumatic for the professionals involved.

How can this problem be solved?

First, it must be said that the often-suggested electronic prescription system is not the solution to medication errors—unless the system is well designed. Research shows that different electronic systems affect errors quite differently, ranging from eliminating 99 % of them to increasing the error rate and all possibilities in between. Moreover, these systems can cause a variety of unanticipated side effects that compromise patient safety. This paradox can be understood once the real sources of medical errors are understood.

For many errors, the solution lies in adding redundancy. What does this mean?

To explain, we can turn to another context where the prevention of errors is important: writing checks.

Where money is involved, we are careful to make sure the information is conveyed clearly. To this end, we write the amount twice, in both words and numerals. This is done, purely and simply, to prevent errors. Electronic check-writing systems also make sure that critical information is "double-checked." Another example is the double entry of e-mail addresses or passwords when one registers for online accounts. Why enter the same information twice? To make sure it is correctly received.

The same principle of redundancy can and should be applied to writing prescriptions.

Why isn't this done already? The system we use today for writing prescriptions was developed when there were far fewer medications. As the number of possibilities increases we have to be increasingly careful to make sure that enough information is communicated so that the right prescription is delivered.

Thus, whether written or electronic, what matters is how well the system is designed.

The caveat is that every critical piece of a prescription must be written twice, to ensure that few if any errors occur. There are five critical pieces of information on a standard prescription: Patient, Drug, Dose, Route (oral, intravenous, etc.), and Time. Each of these must be written in two ways or double-checked after electronic entry.

For example, the patient could be designated by both name and ID number. The drug could be doubly specified by writing both the medication and the indication (the condition for which it is prescribed), or both the generic and trade names. Dosage, route of drug delivery, and time of administration could be written out fully and abbreviated, rather than given only in abbreviated form.

For electronic systems, auto-completion and simple check boxes should be avoided. These items are more prone to error precisely because they are quick and easy. Instead, it is important to have the prescriber provide all key information longhand and verify it. Writing something twice admittedly takes more time but the prevention of errors, as in writing checks, must be considered of primary importance (Fig. 48.3).

It is possible to write less when there is less potential for misunderstanding. For example, if the route is already determined by the medication, then the route can just be indicated as "Standard." For now, however, we should be conservative in shortcuts; once medication errors are dramatically reduced, we can carefully study which efficiencies can be implemented without errors being introduced.

Electronic systems also should be carefully designed to avoid distraction and disruption. The difference between a well-designed intuitive way of entering prescriptions, with appropriate redundancy, and a poorly designed system is the difference between success and failure.

Communication is not only central to prescription errors, but is also central to other forms of medical errors. Errors generally arise not because of an individual's action, but because of the way individuals work together. Improvement of communication and coordination is often the solution. The development and competition of workgroup teams recommended in Step II are key to reducing errors throughout healthcare, because such groups can improve local team coordination and communication.

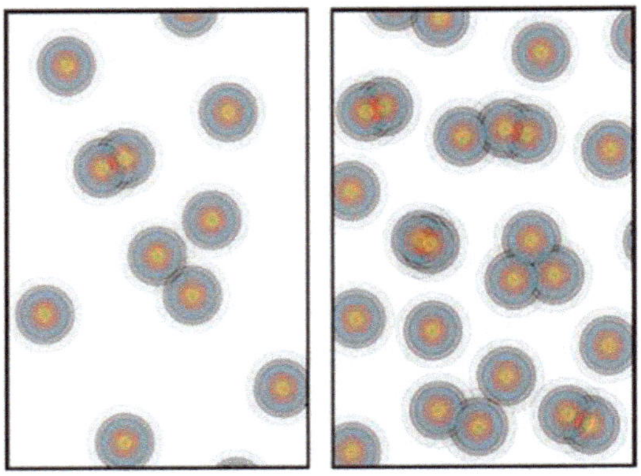

Fig. 48.3 Hitting the right target is harder when there are more targets. The more medications and treatments there are, the more accurate the system has to be to avoid errors that shift from one of the possible medications to another

In looking for ways to solve the problem of medical errors, improving upon the analysis of the source of medical errors is important too. Too often, the process of examination only looks at a specific error—what went wrong in this particular case—and a particular practice is blamed, and a particular solution is offered to that practice. Instead, we should abstract from the level of individual errors and find the patterns among effective and ineffective cases.

Focusing on just the individual error is as ineffective as a tennis player only practicing the one last shot he or she missed, over and over. In most cases, it makes more sense to work on improving speed, agility, and the player's ability to respond to a large set of possible shots. The next challenging shot will not be the same as the last one.

The same holds true for medical errors: understanding the many possible ways errors can occur rather than just the last one, and the way things work correctly, will allow us to recognize the weaknesses and improve the strengths of the system.

The Institute of Medicine originally reported up to 100,000 deaths per year due to medical errors, and up to $29 billion in additional costs incurred. More recent reports have found these numbers to be even higher. The price in human and financial terms is too great. We can and must fix the problem.

Scientific Principle
According to information theory, adding redundancy to a message dramatically reduces the error rate when it is transmitted through noisy channels. Human-to-human communication can be analyzed as occurring through a communication channel, whether oral, hand written, or through electronic means. Information theory uniquely identifies the origin of errors and how to alleviate them. Errors in the medical system that result from miscommunication can be analyzed using this approach and the introduction of the necessary redundancy can be used to drastically reduce medical errors. Prescription miscommunication has been documented as a major cause of loss of life and other adverse outcomes and should be addressed in this way. The common misconception that automation (electronic prescription systems) reduce error rates is incorrect, unless they introduce the necessary redundancy specified by information theory.

48.7 Create Disinfection Gateways

Infections acquired in hospitals, known as HAIs or nosocomial infections, are often resistant to antibiotics and thus particularly dangerous. Each year, the estimated 1.7 million infections cause nearly 100,000 deaths in the United States. Many patients in hospitals, nursing homes, and clinics become sicker from these infections than they were before they sought care. These infections also play a significant role in costs—HAI hospital costs alone were recently estimated at between $30 and $45 billion.

Current recommendations for reducing hospital-acquired infections target the patient's immediate environment and interactions with care providers. Hand washing by care providers before and after patient contact is a key part of protocols in patient-focused transmission prevention. The wide variety of other measures include identifying patients who enter the hospital with infections for additional isolation, extra care to avoid catheter associated infections, and augmented surface sanitation—of bed rails and controls, light switches, partition screens, faucet handles, and the like.

Collectively, recommended protocols have been shown to reduce transmission, are cost effective, and could be more widely adopted. Still, the attention and effort involved are significant and progress in eliminating infections is slow.

Underlying the widespread prevalence and difficulty in addressing these infections is the large number of contacts between care providers and patients. Because there are so many contacts, the effort involved in making every contact safe is huge and this effort burdens already busy care providers.

How can we speed up progress?

We need to expand our view beyond the point of contact between patients and providers to think in terms of the overall process of transmission within a hospital and between care facilities.

Each transit across a boundary between domains should be considered as a potential "transmission" of pathogens that will infect a unit, ward, floor, building, or care facility. At these boundaries, protocols of disinfection should be designed to reduce pathogen transfers from one domain to another. The boundaries between domains should be like airlocks, disinfecting people and objects that pass through them.

What protocols should these boundaries have? Since there are relatively few such crossings as compared to the number of patient contacts overall, we can consider more extensive decontamination procedures than just hand washing, such as clothing sanitation and the cleaning of cell phones and other personal effects. There is evidence that lab coats, PDAs, cell phones, and the like act as repositories for pathogens and can be responsible for HAI transmission. The protocol should still be efficient, and it can be. Staging such intensive interventions at the gateways could significantly reduce the flow of pathogens between patients (Fig. 48.4).

We have to consider how pathogens are transferred: from one patient to the surfaces and fabrics near that patient to the care providers and their clothing, cell phones, and pagers. From there the

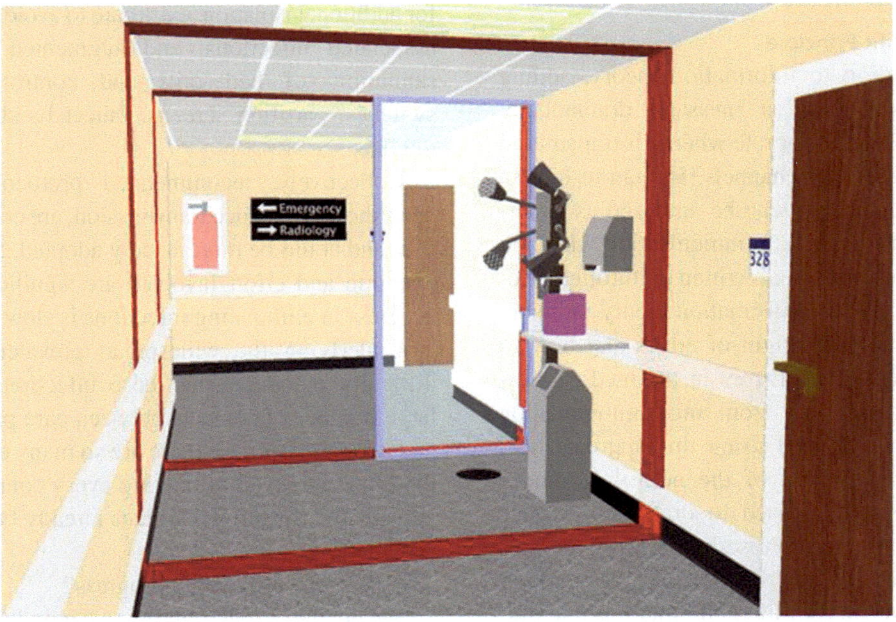

Fig. 48.4 A mockup of a disinfection gateway for use in inhibiting the spread of infections

pathogens are transferred either directly to another patient or to the fabrics and surfaces in common areas or around that patient from which they eventually reach that patient at a different contact opportunity.

Most of the possible transmission events happen because of the large number of contacts within a local ward between patients and doctors, nurses, medical technicians, food service people, and cleaning staff. Each of these contacts has the potential to transfer pathogens between patients, and to contaminate objects in shared spaces, such as computer keyboards.

If there were no virulent pathogen in the ward in the first place, none of those possible transmission events could actually transfer virulent pathogens. Using boundary protocols to reduce transmission between wards would eliminate a large number of potential transmission events among the individuals within each ward.

With the use of boundary protocols, there would be a reduction not only in the transmission of existing pathogens but also in the emergence of new resistant strains. The high number of physical contacts makes medical care facilities a uniquely fertile environment for pathogens to evolve into more virulent strains. By blocking the spread of infection between areas, we can cut down on the appearance of virulent pathogens as well as their prevalence.

Would everyone have to go through disinfection at these airlocks? Visitors and patients entering a hospital for an appointment don't present the same level of risk (though they might be tested for infection themselves). Unlike care providers who go from patient to patient to patient, they don't act as agents for transmission. Accordingly, the same protocols need not apply. Similarly, a caregiver who is only interacting with a single patient need not undergo this process. Furthermore, these protocols could be overridden for the sake of speed in the event of an emergency—when protocols are generally observed, a single contact is unlikely to transmit pathogens.

The same principles of containment are behind biological membranes that prevent transmissions between parts of the body and are the reason why the immune system is concentrated in the high speed transport system of the body—the blood. It is the reason we have regulations about plant and animal products crossing national borders. Conversely, the absence of such boundary protections in an increasingly interconnected world has promoted the rise of highly virulent new strains of pathogens and the risks of global pandemics.

Reducing the probability of transmission at each provider-to-patient contact by hand washing and other protocols is still a good idea. At the same time, the flow of pathogens through a hospital and overall transmission between sites can be dramatically reduced. This can be done by creating additional levels of transmission prevention at key internal boundaries in the care facility and between care facilities.

The cost of hospital-based infections is high and using high-leverage methods to eliminate them is the way to go. By instituting protocols at geographic domain boundaries, at low cost, we can dramatically reduce their transmission.

Scientific Principle

In many systems the difference between a geographically partitioned and a well-mixed one is significant. In a well-connected system, the large number of pairwise interactions makes highly likely the spreading of any transmittable condition. Indeed, analysis shows that infections of high virulence and transmissibility only survive in highly connected systems. Understanding the flow of contagion through a system involves mapping out the set of contact points and the network of transmissions that result. Today the focus on reducing infection transmission in hospitals is on reducing the likelihood of transmission through each of the many individual contacts. However, because the system is highly connected, the probability of transmission is high even when there is a low individual contact transmission probability. Spatial, and more generally, hierarchical partitions are a powerful approach that reduces the overall connectivity of the system and thus dramatically decreases the sustainability of infections, inherently making the system not conducive to highly virulent and transmissible strains.

48.8 Use e-Records for Research

Electronic records, which have become increasingly prevalent in recent years, represent a valuable repository of medical data. There are over 300 million people in the United States, most of whom are receiving some sort of medical care. If anonymized medical records were made available to researchers, these e records could be leveraged to improve care at low cost.

In today's quest to answer questions about medicine and human health, the large-scale, controlled clinical trial is central. New drugs, surgical techniques, non surgical interventions, and medical devices are typically tested in such studies, which require the creation of control and test groups, controlling for confounding factors such as age and lifestyle, and the tracking of patients.

While these studies are essential for testing new drugs and interventions, not every medical question can or should be tested using a clinical trial, given the human and monetary resources that are required to conduct such studies.

Physicians and researchers already have other accepted ways of advancing medical knowledge.

Observational studies and chart reviews, typically analyze groups of patients based on the condition displayed or the intervention used, and are often performed when large randomized studies are infeasible. They can yield important results even without the controls needed for clinical trials.

Physicians author case reports as a means of sharing their experiences with especially instructive cases. These reports are regarded as valuable parts of the medical literature and are a standard part of peer-reviewed medical journals. The medical field recognizes and accepts knowledge gained through observational studies, chart reviews and case reports in addition to controlled clinical studies.

With the use of e-records, the scope of chart reviews and the sharing of case reports are dramatically expanded by many orders of magnitude. This represents a unique opportunity to leverage vast amounts of newly available data to explore many medical questions.

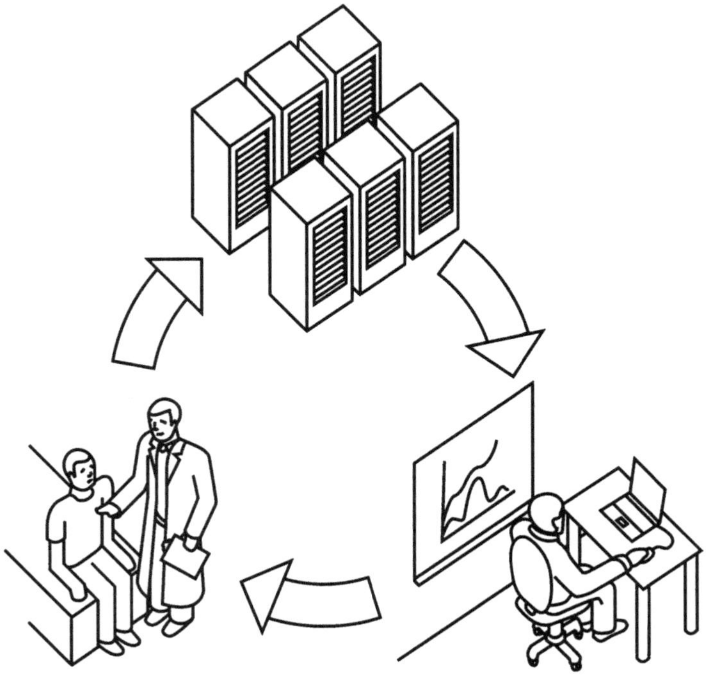

Fig. 48.5 Research based on medical care data can help improve healthcare

Recently, the drug Vioxx®[2] was recalled due to side effects causing higher rates of heart attack than comparable drugs. The recall took place one month after results of a study using medical records from 1.4 million people were reported. This recall is one example of how collecting and using data from actual patients can lead to medical advances.

Another example of data gathering that has led to significant recent advances is the ongoing Framingham Heart Study. This longitudinal study has tracked over 10,000 individuals from three consecutive generations, monitoring their physical health and lifestyle choices, in order to learn about cardiovascular disease. That information has been used by researchers to make many advances, ranging from genetics to the role of social networks.

Many other studies have been based upon survey and medical reporting data collected by the CDC. The data that could be made available dwarfs current resources.

As previously noted, medical care data is logged for many millions of people in the United States. We can leverage the sheer volume of available data, combined with new pattern-recognition methods and theoretical advances in data analysis, to increase our knowledge in ways beyond the practical reach of other methods (Fig. 48.5).

At the very least, analyzing these data could reveal previously unknown connections–for example, between a particular medication, a bit of medical history, and a seemingly unrelated disease—that would provide clues as to what questions should be pursued in formal, controlled studies.

Also, since each person's medical records may cover many years, we can learn about long-term effects much more easily and cost-effectively by analyzing these available data than by conducting longitudinal studies on a particular therapy. Thus, we can use these data to discover long-term effects that may otherwise not be detected at all.

Leveraging the availability of care data to increase our knowledge can't and shouldn't replace controlled studies or physician experience. But it can be a powerful and cost-effective tool, allowing us to utilize huge amounts of information and new methods of analysis to increase our medical knowledge, improving our ability to treat patients and take care of ourselves.

[2] Rofecoxib.

Data—be they from study results or individual experiences—are the raw material that we use to build our medical knowledge. Gaining access to such a huge volume of new information about human health is like inventing a microscope that can see objects that are much smaller, or a telescope that can see much farther away. This new data can lead to many new discoveries.

Analyses of these data would be an important addition to the medical research toolkit, augmenting traditional research methods.

The governmental agencies that oversee various aspects of healthcare practice and research should work with medical organizations to make electronic medical records available in an anonymous, analyzable form. They should encourage use of this vast, important resource to propel our knowledge of medicine forward.

Scientific Principle
Our increasingly complex world yields massive quantities of data, and we now have the scientific knowledge to perform pattern recognition on the data. Scientists are utilizing such "big data" methods in areas as diverse as genomics, finance, and crime prevention. If made available, the vast corpus of medical records should result in the discovery of opportunities for advancement in medicine. This approach complements the more traditional and more controlled framework of specially designed clinical trials.

48.9 Promote "First Day" Celebrations

Any discussion about improving our healthcare system must acknowledge the important role that is played by people caring for themselves and their loved ones. The most important step we can take to improve the healthcare system is to support and inspire an informed and widespread level of personal care.

Major health issues are related to behavior—smoking, alcohol consumption, diet, exercise, even safe driving. Other health issues must be addressed partly through behavior, including remembering to take medications.

Addressing public health problems such as obesity is at times viewed as the responsibility of government, medical professionals, or fast food chains. But these problems should also be addressed by individuals working to change their own behavior.

Yet, when we do turn to individuals to improve their own health habits, we often overlook the real potential in ensuring their success via support groups of friends and coworkers, and via support mechanisms such as community institutions and social traditions.

Al Gore's message calling for us all, as individuals and collectively, to be responsible for our planet, resonates in this instance. We can all take responsibility to safeguard our health. We need a culture of healthy people in a healthy world.

How do we realize this vision?

Building on the tradition of setting aside a time for New Year's resolutions, we can promote lifestyle change with the use of "First Day" celebrations, which will convey health information and will draw forth personal commitments to healthier living.

The fundamental purpose of these celebrations, resonating with "today is the first day of the rest of your life," is to celebrate healthy lifestyles for the new year. This will promote and reinforce our existing societal traditions and our recognition of the natural yearly cycle as one of renewal and improvement.

Health is serious business, but people should take care of their health in a positive way, mindful of new opportunities rather than focusing only on dangerous risks.

"First Day" also builds on "First Night," the popular New Year's Eve festivals full of arts, family activities, and cultural entertainment. Started in 1976, First Night built upon people's natural tendency to celebrate the New Year and channeled that impulse toward constructive cultural activities and fun.

First Day should not be driven solely by individuals—companies, communities, towns, cities, and states can all play a role. The Centers for Disease Control and Prevention (CDC) articulates a vision of health as pervading all aspects of life. We can leverage personal and community participation to improve public health.

Perhaps surprisingly, Walmart has led the way. In 2007, Walmart launched a program in which employees design and carry out "personal sustainability projects" including anything from recycling at home to quitting smoking to getting more exercise. Originally focused on the environment, participants naturally included personal health projects. Indeed, health for oneself, one's family, community, country, and world are all linked—both in effect and in desire and commitment for a better life.

Through this program, Walmart provides the framework for employees to exercise their capabilities. Working alongside others to accomplish goals has a positive effect on what people can accomplish. Employees self-monitor their progress for several weeks and are encouraged to make the improvements long term. Coworkers encourage one another to meet their goals. Walmart's popular program has been a great success, helping many employees improve their lives.

This idea can be made into a national or global activity of personal and collective improvement. Aligning it with New Year's celebrations is a natural thing to do.

The preceding week, employers and government agencies can provide information and events. Organizations of different types—companies, religious organizations, schools, towns, states—can set up programs that encourage people to take responsibility for their own health and lifestyle, and they can provide supportive communities toward that end. The organizations themselves can undertake new commitments to improve social health and community well-being.

Some people may want their goals and commitments to be private or to share them with friends; others may be pleased to share them publicly. The key is for familiar institutions and networks to support each person's desire to improve his or her life and each person's journey toward better health.

Social network follow-up interactions can be planned. Internet-based and mobile device apps with calendars, reminders, and checklists can be developed to support people in reaching their goals.

We can dramatically improve health by inspiring individual responsibility and action. When people embrace their health as a personal opportunity and are also given community support, they reveal tremendous power to make lasting improvements in their own lives and each other's.

Scientific Principle

Behavioral change can propagate through social network links. The importance of individual behavioral choices to major public health problems and health more generally is well known. Individual responsibility for health can be socially reinforced. The power of social influence can be engaged to encourage individual healthy lifestyle choices. The most effective way to achieve a large response from a system is to engage its existing natural modes of activity. A yearly "First Day" celebration leverages the existing culture of "New Year's resolutions" and the natural yearly cycle of renewal.

48.10 Executive Summary of Real World Implementation

- *Separate simple care*, especially simple preventive services required for large numbers of people. Nurses or other appropriately trained individuals, in easily accessible settings like retail clinics, can provide prevention to healthy individuals.
- *Empower workgroup competition.* Apply performance measures at the level of the performing group and make the measurements visible to competing groups. Trust this as an incentive and avoid regulating costs or quality.
- *Create superdoctor teams.* Innovate in teams of physicians and professional providers to

rapidly diagnose and treat highly complex conditions. Just as string quartets can play more complex music than soloists, teams can provide highly complex care.

- *Accelerate intake rout*ing to rapidly identify the right provider. Separate emergency, simple and complex cases.
- *Add redundancy to improve communication* to prevent prescription errors in paper and electronic systems. Electronic systems vary widely in effectiveness; increasing redundancy and reducing distraction are the keys to preventing errors.
- *Create disinfection gateways* at internal boundaries between sections of hospitals and between care facilities to dramatically reduce hospital-based infections. This type of boundary-based prevention is a high-leverage strategy.
- *Use e-records for research*. The large volume of data represents a unique resource for advances in medical knowledge and effective care.
- *Promote "First Day" celebrations*. Personal and community engagement in renewal is the most important force for improved health.

Acknowledgements This work was supported in part by the Centers for Disease Control and Prevention and by an anonymous donation to the New England Complex Systems Institute.

References

1. Kozhaya N. Two myths about the American health care system. Montreal Economic Institute; 2005.
2. World Health Statistics 2009. World Health Organization; 2009.
3. Roehr B. Health care in US ranks lowest among developed countries. Br Med J. 2008;337:a889.
4. Kulkarni SC, Levin-Rector A, Ezzati M, Murray CJL. Falling behind: life expectancy in US counties from 2000 to 2007 in an international context. Popul Health Metrics. 2011;9:16.
5. Oestergaard MZ, Inoue M, Yoshida S, Mahanani WR, Gore FM, Cousens S, et al. Neonatal mortality levels for 193 countries in 2009 with trends since 1990: A systematic analysis of progress, projections, and priorities. PLoS Med. 2011;8(8):e1001080.
6. Centers for Medicare and Medicaid Services, National Health Statistics Group. National health care cxpenditures data. 2012.

Further Readings

Section 1: Separate Simple Care

Bar-Yam Y. Improving the effectiveness of health care and public health: a multi-scale complex systems analysis. Am J Public Health. 2006;96:459–66.

Bar-Yam Y. Making things work: solving complex problems in a complex world. Cambridge, MA: NECSI Knowledge Press; 2005. p. 239.

Bar-Yam Y. Multiscale variety in complex systems. Complexity. 2004;9:37–45.

Baskas H. More airports adding flu shots for fliers [Internet]. USA Today 2008 Oct 15. Available from http://www.usatoday.com/travel/columnist/baskas/2008-10-14-flu-shots_N.htm.

Bohmer R. The rise of in-store clinics—threat or opportunity? N J Engl Med. 2007;356:765–8.

Groopman J. How doctors think. New York: Houghton Mifflin; 2007.

Keckley PH, Underwood HR, Gandhi M. Retail clinics: facts, trends, and implications; update and implications. Washington DC: Deloitte Center for Health Solutions; 2008.

Lambreaw J. A Wellness Trust to prioritize disease prevention. Brookings Institution. Report number: 2007-04, 2007.

Maciosek M, Coffield A, Edwards N, Flottemesch T, Goodman M, Solberg L. Priorities among effective clinical preventive services: results of a systematic review and analysis. Am J Prev Med. 2006;31:52–61.

Mehrotra A, Liu H, Adams JL, Wang MC, Lave JR, Thygeson NM, Solberg LI, McGlynn EA. Comparing costs and quality of care at retail clinics with that of other medical settings for 3 common illnesses. Ann Intern Med. 2009;151:321–8.

Mehrotra A, Wang MC, Lave JR, Adams JL, McGlynn EA. Retail clinics, primary care physicians, and emergency departments: a comparison of patients' visits. Health Aff. 2008;27:1272–82.

Promoting healthy lifestyles. Chicago: American Medical Association; 2008.

Rudavsky R, Pollack CE, Mehrotra A. The geographic distribution, ownership, prices, and scope of practice at retail clinics. Ann Intern Med. 2009;151:315–20.

Scott MK. Health care in the express lane: the emergence of retail clinics. Oakland: California Healthcare Foundation; 2006.

Thygeson M, Van Vorst KA, Maciosek MV, Solberg L. Use and costs of care in retail clinics versus traditional care sites. Health Aff. 2008;27:1283–92.

U.S. Preventive Services Task Force Guide to Clinical Preventive Services. U.S. Department of Health and Human Services, Agency for Healthcare Research and Quality. Report number: 07-05100, 2007.

Yarnall K, Pollak K, Ostbye T, Krause K, Michener J. Primary care: Is there enough time for prevention? Am J Public Health. 2003;93:635–41.

Section 2: Empower Workgroup Competition

Arrow KJ. Uncertainty and the welfare economics of medical care. Am Econ Rev. 1963;53:941–73.

Bar-Yam Y. Making things work. Cambridge: NECSI Knowledge Press; 2005. See page 239.

Berwick M, James B, Coye MJ. Connections between quality measurement and improvement. Med Care. 2003;41:30–8.

Carrol B, Tomas S. Team competition spurs continuous improvement at Motorola. Natl Prod Rev. 1963;14:1–9.

Forsyth D. Group dynamics. Belmont, CA: Wadsworth, Cengage Learning; 2006.

Johnson DW, Maruyama G, Johnson R, Nelson D, Skon L. Effects of cooperative, competitive, and individualist goal structures on achievement: A meta-analysis. Psychol Bull. 1981;89:47–62.

Krugman P. Why markets can't cure healthcare. New York Times [Internet]. 2009 Jul 25. Available from: http://krugman.blogs.nytimes.com/2009/07/25/why-markets-cant-cure-healthcare/.

Levi D. Group dynamics for teams. Thousand Oaks: Sage Publications; 2007.

McGlynn EA. Selecting common measures of quality and system performance. Med Care. 2003;41:39–47.

Nadler DA. The effects of feedback on task group behavior: A review of the experimental research. Organ Behav Hum Perform. 1979;23:309–38.

Plsek P. Redesigning health care with insights from the science of complex adaptive systems. In: Institute of Medicine. Crossing the quality chasm: A new health system for the 21st century. Washington DC: National Academy Press; 2011.

Porter ME, Teisberg EO. Redefining health care: Creating value-based competition on results. Boston: Harvard Business School Press; 2006.

Schon D, Argyris C. Organizational learning II: Theory, method and practice. Reading, MA: Addison-Wesley; 1996.

Schon D, Argyris C. Organizational learning: A theory of action perspective. Reading, MA: Addison-Wesley; 1978.

Senge PM. The fifth discipline: The art and practice of the learning organization. New York: Doubleday; 1990.

Szarka FE, Grant KP, Flannery WP. Encouraging organizational learning through team competition. Eng Manag J. 2004;16:21–31.

Werner M, Asch DA. The unintended consequences of publicly reporting quality information. J Am Med Assoc. 2008;293:1239–44.

Section 3: Create Superdoctor Teams

Burke D, Herrman H, Evans M, Cockram A, Trauer T. Educational aims and objectives for working in multidisciplinary teams. Australas Psychiatry. 2000;8:336–9.

Calman K, Hine D. A policy framework for commissioning cancer services: A report by the expert advisory group on cancer to the chief medical officers of England and Wales. London: Department of Health; 1995.

Chang JH, Vines E, Bertsch H, Fraker DL, Czerniecki BJ, Rosato EF, Lawton T, Conant EF, Orel SG, Schuchter L, Fox KR, Zieber N, Glick JH, Solin LJ. The impact of a multidisciplinary breast cancer center on recommendations for patient management. Cancer. 2001;91:1231–7.

Denvir MA, Pell JP, Lee AJ, Rysdale J, Prescott RJ, Eteiba H, Walker A, Mankad P, Starkey IR. Variations in clinical decision-making between cardiologists and cardiac surgeons; a case for management by multidisciplinary teams? J Cardiothorac Surg. 2006;1:2.

Fendrick A, Hirth R, Chernew M. Differences between generalist and specialist physicians regarding Helicobacter pylori and peptic ulcer disease. Am J Gastroenterol. 1996;91:1544–8.

Fleissig A, Jenkins V, Catt S, Fallowfield L. Multidisciplinary teams in cancer care: are they effective in the UK? Lancet Oncol. 2006;7:935–43.

Forsyth D. Group dynamics. Belmont CA: Wadsworth, Cengage Learning; 2006.

Gottlieb K, Sylvester I, Eby D. Transforming your practice: what matters most. Fam Pract Manag. 2008;15:32–8.

Harrold LR, Field TS, Gurwitz JH. Knowledge, patterns of care, and outcomes of care for generalists and specialists. J Gen Intern Med. 1999;14:499–511.

Hayward RA. The Calman–Hine report: a personal retrospective on the UK's first comprehensive policy on cancer services. Lancet Oncol. 2006;7:336–46.

Hearn J, Higginson IJ. Do specialist palliative care teams improve outcomes for cancer patients? A systematic literature review. Palliat Med. 1998;12:317–32.

Levi D. Group dynamics for teams. Thousand Oaks, CA: Sage Publications, Inc.; 2007.

Medical Univerity of South Carolina, Hollings Cancer Center. Multidisciplinary team members [Internet]. 2010 [cited 2012 Jan 9]; Available from: http://www.muschealth.com/cancer/headandneckcancer/members.htm.

Senge PM. The fifth discipline: The art and practice of the learning organization. New York: Doubleday; 1990.

Starfield B, Shi L, Macinko J. Contribution of primary care to health systems and health. Milbank Q. 2005;83:457–502.

Stone VE, Mansourati FF, Poses RM, Mayer KH. Relation of physician specialty and HIV/AIDS experience to choice of guideline-recommended antiretroviral therapy. J Gen Intern Med. 2001;16:360–8.

Section 4: Accelerate Intake Routing

Arnst C. A new practice: the doctor will see you today. Kaiser Health News [Internet]. 2010 Jul 14. Available from: http://www.kaiserhealthnews.org/Daily-Reports/2010/July/14/1khnstory.aspx?print=1.

Bodenheimer T, Lo B, Casalino L. Primary care physicians should be coordinators, not gatekeepers. J Am Med Assoc. 1999;281:2045–9.

Forrest CB. Primary care gatekeeping and referrals: effective filter or failed experiment? Br Med J. 2003;326:692–5.

Gatley S, Grace A, Lopes V. E-referral and e-triage as mechanisms for enhancing and monitoring patient care across the primary-secondary provider interface. J Telemed Telecare. 2003;9:350–3.

Grumbach K, Keane D, Bindman A. Primary care and public emergency department overcrowding. Am J Public Health. 1993;83:372–8.

Grumbach K, Selby JV, Damberg C, Bindman AB, Quesenberry C, Truman A, Uratsu C. Resolving the gatekeeper conundrum: what patients value in primary care and referrals to specialists. J Am Med Assoc. 1999;282:261–6.

Lattimer V, George A, Thompson F, Thomas E, Mullee M, Turnbull J, Smith H, Moore M, Bond H, Glasper A. Safety and effectiveness of nurse telephone consultation in out of hours primary care: randomised controlled trial. Br Med J. 1998;317:1054–9.

Leibowitz R, Day S, Dunt S. A systematic review of the effect of different models of after-hours primary medical care services on clinical outcome, medical workload, and patient and GP satisfaction. Fam Pract. 2003;20:311–7.

Merritt Hawkins & Associates. 2009 survey of physician appointment wait times. Texas: MH&A; 2009.

Moyer CA, Stern DT, Dobias DS, Cox DT, Katz SJ. Bridging the electronic divide: patient and provider perspectives on e-mail communication in primary care. Am J Manag Care. 2002;8:427–33.

Murray M, Tantau C. Redefining open access to primary care. Manag Care Q. 1999;7:45–55.

Poole SR, Schmitt BD, Carruth T, Peterson-Smith A, Slusarski M. After- hours telephone coverage: the application of an area-wide telephone triage and advice system for pediatric practices. Pediatrics. 1993;92:670–9.

Richards R, Navarro ML, Derlet RW. Survey of directors of emergency departments in California on overcrowding. West J Med. 2000;172:385–8.

Section 5: Improve Communication

Alvey L. FDA and ISMP launch campaign to reduce medication mistakes caused by unclear medical abbreviations [press release]. Silver Spring, MD: Food and Drug Administration; 2006 Jun 14.

Ammenwerth E, Schnell-Inderst P, Machan C, Siebert U. The effect of electronic prescribing on medication errors and adverse drug events: a systematic review. J Am Med Inform Assoc. 2008;15:585–600.

Ash JS, Sittig DF, Poon EG, Guappone K, Campbell E, Dykstra RH. The extent and importance of unintended consequences related to computerized provider order entry. J Am Med Inform Assoc. 2007;14:415.

Bar-Yam Y, Smith M, Wachman A, Topolski S. Prescription form with redundancy, version 0.5. New England Complex Systems Institute. 2010 Aug 26. Available from: http://necsi.edu/research/management/health/prescription.html.

Bar-Yam Y. Making things work. Cambridge MA: NECSI Knowledge Press; 2004. See chapter 11.

Bar-Yam Y. System care: multiscale analysis of medical errors—eliminating errors and improving organizational capabilities. New England Complex Systems Institute. Report number 09-01-2004, 2004.

Bonnabry P, Despont-Gros C, Grauser D, Casez P, Despond M, Pugin D, Rivara-Mangeat C, Koch M, Vial M, Iten A, Lovis C. A risk analysis method to evaluate the impact of a computerized provider order entry system on patient safety. J Am Med Inform Assoc. 2008;15:453–60.

Borden S. Computer entry a leading cause of medication errors in U.S. health systems: percentage of reported errors steadily increased from 1999 to 2003 [press release]. Rockville MD: United States Pharmacopeia; 2004 Dec 20.

Chaudhry B, Wang J, Wu S, Maglione M, Mojica W, Roth E, Morton SC, Shekelle PG. Systematic review: Impact of health information technology on quality, efficiency, and costs of medical care. Ann Intern Med. 2006;144:742–52.

Christian CK, Gustafson ML, Roth EM, Sheridan TB, Gandhi TK, Dwyer K, Zinner MJ, Dierks MM. A prospective study of patient safety in the operating room. Surgery. 2006;139:159–73.

Colpaert K, Claus B, Somers A, Vandewoude K, Robays H, Decruyenaere J. Impact of computerized physician order entry on medication prescription errors in the intensive care unit: a controlled cross-sectional trial. Crit Care. 2006;10:R21.

Experts to probe factors behind overdose error. Edmonton Journal [Internet]. 2007 May 8. Available from: http://www.canada.com/edmontonjournal/news/story.html?id=adc17250-f77b-4b8f-b97f-d6657fe12e8d.

Food and Drug Administration. Medication errors [Internet]. [Cited 2012 Jan 9]; Available from: http://www.fda.gov/drugs/drugsafety/medicationerrors/default.htm.

Goldstein A. Overdose kills girl at Children's Hospital. Washington Post [Internet]. 2001 Apr 20. Available from: http://www.highbeam.com/doc/1P2-439883.html.

Graban M. Statistics on healthcare quality and patient safety [Weblog]. Leanblog. 2009 Aug 9 [cited 2012 Jan 9]; Available from: http://www.leanblog.org/2009/08/statistics-on-healthcare-quality-and/.

Hamming RW. Error detecting and error correcting codes. Bell Syst Tech J. 1950;29:147–60.

Han YY, Carcillo JA, Venkataraman ST, Clark RSB, Watson RS, Nguyen TC, Bayir H, Orr RA. Unexpected increased mortality after implementation of a commercially sold computerized physician order entry system. Pediatrics. 2005;116:1506–12.

Holdsworth T, Fichtl RE, Raisch DW, Hewryk A, Behta M, Mendez-Rico E, Wong CL, Cohen J, Bostwick S, Greenwald BM. Impact of computerized prescriber order entry on the incidence of adverse drug events in pediatric inpatients. Pediatrics. 2007;120:1058–66.

Institute of Medicine. Preventing medication errors: quality chasm series. Washington DC: National Academies Press; 2007.

Institute of Medicine. To err is human: building a safer health system. Washington DC: The National Academies Press; 2000.

Kennedy AG, Littenberg B. A modified outpatient prescription form to reduce prescription errors. Jt Comm J Qual Patient Saf. 2004;30:480–7.

Koppel R, Metlay JP, Cohen A, Abaluck B, Localio AR, Kimmel SE, Strom BL. Role of computerized physician order entry systems in facilitating medication errors. J Am Med Assoc. 2005;293:1197–203.

Leape LL, et al. Systems analysis of adverse drug events. J Am Med Assoc. 1995;274:35–43.

Leape LL. Errors in medicine. J Am Med Assoc. 1994;272:1851–7.

Liebovitz D. Health care information technology: A cloud around the silver lining? Arch Intern Med. 2009;169:924–6.

Loughran S. In-hospital deaths from medical errors at 195,000 per year, HealthGrades study finds [press release]. Denver: HealthGrades; 2004.

MacWilliams FJ, Sloane NJA. The theory of error-correcting codes. Amsterdam: Elsevier BV; 1977.

Perrow C. Normal accidents: living with high risk technologies. Princeton, NJ: Princeton University Press; 1999.

Quality Interagency Coordination Task Force. Doing what counts for patient safety: federal actions to reduce medical errors and their impact. Report to the president on medical errors. 2000 Feb.

Quality Interagency Coordination Task Force. National summit on medical errors and patient safety research, summary [Internet]. 2006 Nov 30 [cited 2012 Jan 9]; Available from: http://www.quic.gov/summit/summary1.htm.

Reason J. Human error: models and management. Br Med J. 2000;320:768–70.

Robeznieks A. Data entry is a top cause of medication errors: training and design are seen as keys to reducing electronic prescribing errors. American Medical News [Internet]. 2005 Jan 24. Available from: http://www.ama-assn.org/amednews/2005/01/24/prsa0124.htm.

Shannon CE. A mathematical theory of communication. Bell Syst Tech J. 1948;27:379–423. 623–656.

Shapiro R. Preventable medical malpractice: revisiting the Dennis Quaid medication/hospital error case [Weblog]. The Injury Board Blog Network. 2010 Aug 9 [cited 2012 Jan 9]; Available from: http://virginiabeach.injuryboard.com/medical-malpractice/preventable-medical-malpractice-revisiting-the-dennis-quaid-medicationhospital-error-case.aspx.

Streeter CL. Redundancy in organizational systems. Soc Serv Rev. 1992;66:97–111.

Teich JM, Merchia PR, Schmiz JL, Kuperman GJ, Spurr CD, Bates DW. Effects of computerized physician order entry on prescribing practices. Arch Intern Med. 2000;160:2741–7.

US Dept. of Health and Human Services, Agency for Healthcare Research and Quality. Computerized provider order entry [Internet]. [Cited 2012 Jan 9]; Available from: http://psnet.ahrq.gov/primer.aspx?primerID=6.

Walsh KE, Adams WG, Bauchner H, Vinci RJ, Chessare JB, Cooper MR, Hebert PM, Schainker EG, Landrigan CP. Medication errors related to computerized order entry for children. Pediatrics. 2006;118:1872–9.

Wears RL, Berg M. Computer technology and clinical work: still waiting for Godot. J Am Med Assoc. 2005;293:1261–3.

Weiner SJ, Schwartz A, Weaver F, Goldberg J, Yudkowsky R, Sharma G, Binns-Calvey A, Preyss B, Schapira MM, Persell SD, Jacobs E, Abrams RI. Contextual errors and failures in individualizing patient care: a multicenter study. Ann Intern Med. 2010;153:69–75.

Section 6: Create Disinfection Gateways

Agency for Healthcare Research and Quality. Health care-associated infections [Internet]. [Cited 2012 Jan 9]; Available from: http://psnet.ahrq.gov/primer.aspx?primerID=7.

McCaughey B. Unnecessary deaths: the human and financial costs of hospital infections, 3rd edition. New York: Committee to Reduce Infection Deaths; 2006.

Backman C, Zoutman DE, Marck PB. An integrative review of the current evidence on the relationship between hand hygiene interventions and the incidence of health care-associated infections. Am J Infect Control. 2008;36:333–48.

Bascetta CA, Edwards N, Brown D, Peterson E, Richardson AE, Legeer SS, et al. Health-care-associated infections in hospitals: an overview of state reporting programs and individual hospital initiatives to reduce certain infections. U. S. Government Accountability Office, Report number: GAO-08-808, 2008.

Boyce JM, Pittet D. Guideline for hand hygiene in health-care settings: recommendations of the healthcare infection control practices advisory committee and the HICPAC/SHEA/APIC/IDSA Hand Hygiene Task Force. Infect Control Hosp Epidemiol. 2002;23:S3–40.

Boyce JM. Environmental contamination makes an important contribution to hospital infection. J Hosp Infect. 2007;65:S2,50–54.

Braddy CM, Blair JE. Colonization of personal digital assistants used in a health care setting. Am J Infect Control. 2005;33:230–2.

Brady RRW, Fraser SF, Dunlop MG, Paterson-Brown S, Gibb AP. Bacterial contamination of mobile communication devices in the operative environment. J Hosp Infect. 2007;66:397–8.

Brady RRW, Verran J, Damani NN, Gibb AP. Review of mobile communication devices as potential reservoirs of nosocomial pathogens. J Hosp Infect. 2009;71:295–300.

Bures S, Fishbain JT, Uyehara CFT, Parker JM, Berg BW. Computer keyboards and faucet handles as reservoirs of nosocomial pathogens in the intensive care unit. Am J Infect Control. 2000;28:465–71.

Center for Disease Control. Monitoring hospital-acquired infections to promote patient safety—United States, 1990–1999. Morb Mortal Wkly Rep. 2000;49:149–53.

Centers for Disease Control and Prevention. Healthcare-associated infections (HAIs) [Internet]. [Cited 2012 Jan 9]; Available from: http://www.cdc.gov/hai/.

Donker T, Wallinga J, Grundmann H. Patient referral patterns and the spread of hospital-acquired infections through national health care networks. PLoS Comput Biol. 2010;6:e1000715.

Duckro N, Blom DW, Lyle EA, Weinstein RA, Hayden MK. Transfer of vancomycin-resistance enterococci via health care worker hands. Arch Intern Med. 2005;165:302–7.

Eber MR, Laxminarayan R, Perencevich EN, Malani A. Clinical and economic outcomes attributable to health care-associated sepsis and pneumonia. Arch Intern Med. 2010;170:347–53.

Eckmanns T, Schwab F, Bessert J, Wettstein R, Behnke M, Grundmann H, Ruden H, Gastmeir P. Hand rub consumption and hand hygiene compliance are not indicators of pathogen transmission in intensive care units. J Hosp Infect. 2006;63:406–11.

Eckstein BC, Adams DA, Eckstein EC, Rao A, Sethi AK, Yadavalli GK, Donskey CJ. Reduction of Clostridium difficile and vancomycin- resistant Enterococcus contamination of environmental surfaces after an intervention to improve cleaning methods. BMC Infect Dis. 2007;7:61.

Emori TG, Culver DH, Horan TC, Jarvis WR, White JW, Olson DR, Banerjee S, Edwards JR, Martone WJ, Gaynes RP. National nosocomial infections surveillance system (NNIS): description of surveillance methods. Am J Infect Control. 1991;19:19–35.

Fridkin S, Welbel S, Weinstein R. Magnitude and prevention of nosocomial infections in the intensive care unit. Infect Dis Clin North Am. 1997;11:479–96.

Institute for Healthcare Improvement. Reducing healthcare-associated infections [Internet]. 2011 [cited 2012 Jan 9]; Available from: http://www.ihi.org/explore/HAI/Pages/default.aspx.

Klevens RM, Edwards JR, Richards Jr. CL, Horan TC, Gaynes RP, Pollock DA, et al. Estimating health care-associated infections and deaths in U.S. hospitals, 2002. National Center for Infectious Diseases, Center for Disease Control and Prevention. Report number: 122, 2007. See pp 160–166.

Kramer A, Schwebke I, Kampf G. How long do nosocomial pathogens persist on inanimate surfaces? A systematic review. BMC Infect Dis. 2006;6:130.

Leykum L, Bar Yam Y. The rational for system level strategies of infection control. New England Complex Systems Institute. Report number 08-17-2010, 2010.

Lipsitch M, Bergstrom CT, Levin BR. The epidemiology of antibiotic resistance in hospitals: paradoxes and prescriptions. Proc Natl Acad Sci. 2000;97:1938–43.

Macartney KK, Gorelick MH, Manning ML, Hodinka RL, Bell LM. Nosocomial respiratory syncytial virus infections: the cost-effectiveness and cost-benefit of infection control. Pediatrics. 2000;106:520–6.

McCaughey B. Hospital scrubs are a germy, deadly mess: bacteria on doctor uniforms can kill you. Wall Street Journal [Internet]. 2009 Jan 8. Available from: http://online.wsj.com/article/SB123137245971962641.html.

McDonald LC, Jarvis WR. Linking antimicrobial use to nosocomial infections: the role of a combined laboratory-epidemiology approach. Ann Intern Med. 1998;129:245–7.

Muto CA, Jernigan JA, Ostrowsky BE, Richet HM, Jarvis WR, Boyce JM, Farr BM. SHEA Guideline for Preventing Nosocomial Transmission of Multidrug-Resistant Strains of Staphylococcus aureus and Enterococcus. Infect Control Hosp Epidemiol. 2003;24:362–86.

O'Grady NP, Alexander M, Dellinger EP, Gerberding JL, Heard SO, Maki DG, Masur H, McCormick RD, Mermel LA, Pearson ML, Raad II, Randolph A, Weinstein RA. Guidelines for the prevention of intravascular catheter–related infections. Infect Control Hosp Epidemiol. 2002;23:759–69.

Pittet D, Hugonnet S, Harbarth S, Mourouga P, Sauvan V, Touveneau S, Perneger TV. Effectiveness of a hospital-wide programme to improve compliance with hand hygiene. Lancet. 2000;356:1307–12.

Rauch EM, Bar-Yam Y. Long-range interactions and evolutionary stability in a predator-prey system. Phys Rev E. 2006;73:020903.

Scott II RD. The direct medical costs of healthcare-associated infections in U.S. hospitals and the benefits of prevention. Centers for Disease Control and Prevention. 2009.

Sébille V, Chevret S, Valleron A. Modeling the spread of resistant nosocomial pathogens in an intensive care unit. Infect Control Hosp Epidemiol. 1997;18:84–92.

Sébille V, Valleron A. A computer simulation model for the spread of nosocomial infections caused by multidrug-resistant pathogens. Comput Biomed Res. 1997;30:307–22.

Silvestri L, Petros AJ, Sarginson RE, de la Cal MA, Murray AE, van Saene HKF. Handwashing in the

intensive care unit: a big measure with modest effects. J Hosp Infect. 2005;59:172–9.
Steere C, Mallison GF. Handwashing practices for the prevention of nosocomial infections. Ann Intern Med. 1975;83:683–90.
The Department of Health and Human Services. HHS Action plan to prevent healthcare-associated infections [Internet]. [Cited 2012 Jan 9]; Available from http://www.hhs.gov/ash/initiatives/hai/actionplan/.
Treakle AM, Thorn KA, Furuno JP, Strauss SM, Harris AD, Perencevich EN. Bacterial contamination of health care workers' white coats. Am J Infect Control. 2009;37:101–5.
Wenzel R. The economics of nosocomial infections. J Hosp Infect. 1995;31:79–87.
Williams WW. CDC guideline for infection control in hospital personnel. Infect Control. 1983;4:S326–349.
World Health Organization. Prevention of hospital-acquired infections: a practical guide. 2nd ed. Malta: WHO; 2002.
Yokoe DS, Mermel LA, Anderson DJ, et al. A compendium of strategies to prevent healthcare-associated infections in acute care hospitals. Supplement article: executive summary. Infect Control Hosp Epidemiol. 2008;29:S12.

Section 7: Use E-Records for Research

Allison JJ, Wall TC, Spettell CM, Calhoun J, Fargason CA, Kobylinski R, Farmer R, Kiefe CI. The art and science of chart review. Joint Comm J Qual Improv. 2000;26:115–36.
Altman DG, Schulz KF, Moher D, Egger M, Davidoff F, Elbourne D, Gøtzsche PC, Lang T. The revised CONSORT statement for reporting randomized trials: explanation and elaboration. Ann Intern Med. 2001;134:663–94.
Benson K, Hartz AJ. A comparison of observational studies and randomized, controlled trials. New Engl J Med. 2000;342:1878–86.
Black N. Why we need observational studies to evaluate the effectiveness of health care. Br Med J. 1996;312:1215.
Centers for Disease Control and Prevention. Data & Statistics [Internet]. [Cited 2012 Jan 9]; Available from: http://www.cdc.gov/datastatistics/.
Centers for Disease Control and Prevention. National Center for Health Statistics [Internet]. [Cited 2012 Jan 9]; Available from: http://www.cdc.gov/nchs/.
Centers for Disease Control and Prevention. WONDER online databases [Internet]. [Cited 2012 Jan 9]; Available from: http://wonder.cdc.gov/.
Christakis NA, Fowler JH. The spread of obesity in a large social network over 32 years. New Engl J Med. 2007;357:370–9.
Des Jarlais DC, Lyles C, Crepaz N, TREND Group. Improving the reporting quality of nonrandomized evaluations of behavioral and public health interventions: the TREND statement. Am J Pub Health. 2004;94:361–6.
Framingham Heart Study: A Project of National Heart, Lung, and Blood Institute and Boston University (1948-present) [Internet]. 2012 Jan 3 [Cited 2012 Jan 9]; Available from: http://www.framinghamheartstudy.org/.
Graham DJ, Campen D, Hui R, Spence M, Cheetham C, Levy G, Shoor S, Ray WA. Risk of acute myocardial infarction and sudden cardiac death in patients treated with cyclo-oxygenase 2 selective and non-selective non- steroidal anti-inflammatory drugs: nested case-control study. Lancet. 2005;365:475–81.
Haley RW, Schaberg DR, Mcclish DK, Quade D, Crossley KB, Culver DH, Morgan WM, Mcgowan Jr JE, Shachtman RH. The accuracy of retrospective chart review in measuring nosocomial infection rates: results of validation studies in pilot hospitals. Am J Epidemiol. 1980;111:516–33.
Jenicek M. Clinical case reporting in evidence-based medicine. 2nd ed. London: Hodder Arnold Publication; 2001.
Kidd M, Hubbard C. Introducing journal of medical case reports. J Med Case Rep. 2007;1:1.
Levy D, DeStefano AL, Larson MG, O'Donnell CJ, Lifton RP, Gavras H, Cupples LA, Myers RH. Evidence for a gene influencing blood pressure on chromosome 17: genome scan linkage results for longitudinal blood pressure phenotypes in subjects from the Framingham Heart Study. Hypertension. 2000;36:477.
Narcisi G. Clinical data repositories: boosting patient care & research for a data-intensive future. CMIO. 2010:8–9.
O'Donnell CJ, Lindpaintner K, Larson MG, Rao VS, Ordovas JM, Schaefer EJ, Myers RH, Levy D. Evidence for association and genetic linkage of the angiotensin-converting enzyme locus with hypertension and blood pressure in men but not women in the Framingham Heart Study. Circulation. 1998;97:1766–72.
Richardson ML, Chew FS. Radiology Case Reports: a new peer-reviewed, open-access journal specializing in case reports. Radiol Case Rep. 2006;1:1–3.
Rubin R. How did Vioxx debacle happen? USA Today [Internet]. 2004 Oct 12. Available from: http://www.usatoday.com/news/health/2004-10-12-vioxx-cover_x.htm.
Schoenberg R, Safran C. Internet based repository of medical records that retains patient confidentiality. Br Med J. 2000;321:1199–203.
Schwartz A, Pappas C, Sandlow LJ. Data repositories for medical education research: issues and recommendations. Acad Med. 2010;85:837–43.
University of Virginia School of Medicine. Clinical Data Repository (CDR) [Internet]. 2010 Jun 2 [Cited 2012 Jan 9]; Available from: http://www.medicine.virginia.edu/clinical/departments/phs/informatics/cdr-page.
Victora CG, Habicht J, Bryce J. Evidence-based public health: moving beyond randomized trials. Am J Pub Health. 2004;94:400–5.

Wake Forest University School of Medicine, Clinical Data Repository (CDR).

Weitzman ER, Kacil L, Mandl KD. Sharing medical data for health research: the early personal health record experience. J Med Internet Res. 2010;12:e14.

Section 8: Promote "First Day" Celebrations

Barbaro M. At Wal-Mart, lessons in self-help. New York Times [Internet]. 2007 Apr 5. Available from: http://www.nytimes.com/2007/04/05/business/05improve.html?pagewanted=all.

Centers for Disease Control and Prevention. CDC's healthy communities program. 2011 Sep 23 [cited 2012 Jan 9]; Available from: http://www.cdc.gov/healthycommunitiesprogram/.

Minkler M. Personal responsibility for health? A review of the arguments and the evidence at century's end. Health Educ Behav. 1999;26:121–41.

Norcross JC, Mrykalo MS, Blagys MD. Auld lang syne: success predictors, change processes, and self-reported outcomes of New Year's resolvers and nonresolvers. J Clin Psychol. 2002;58:397–405.

Perez AP, Phillips MM, Cornell CE, Mays G, Adams B. Promoting dietary change among state health employees in Arkansas through a worksite wellness program: the healthy employee lifestyle program (HELP). Prev Chronic Dis: Pub Health Res, Pract, and. Policy. 2009;6:A123.

The National Association of Community-Based New Year's Eve Arts Festivals. Welcome to First Night USA [Internet]. [Cited 2012 Jan 9]; Available from: http://firstnightusa.org/.

Turner G. Peer support and young people's health. J Adolesc. 1999;22:567–72.

Wal-Mart announces expansion of associate-driven personal sustainability projects [press release]. Bentonville, AR: Wal-Mart; 2007 Apr 5.

Wal-Mart. 2009 global sustainability report: associates' personal sustainability projects [Internet]. [Cited 2012 Jan 9]; Available from: http://walmartstores.com/sites/sustainabilityreport/2009/s_ao_psp.html.

Wal-Mart. The personal sustainability project [Internet]. [Cited 2012 Jan 9]; Available from: http://walmartstores.com/sites/sustainabilityreport/2007/associates-Personal.html.

Zimmerman RS, Connor C. Health promotion in context: the effects of significant others on health behavior change. Health Educ Behav. 1989;16:57–75.

Opportunities in Delivery of Preventive Services in Retail Settings

49

Yaneer Bar-Yam, Dion Harmon, Keith Nesbitt, May Lim, Suzanne Smith, and Bradley A. Perkins

Y. Bar-Yam (✉)
New England Complex Systems Institute,
238 Main St. Suite 319, Cambridge, MA 02142, USA
e-mail: yaneer@necsi.edu

D. Harmon
New England Complex Systems Institute,
238 Main St. Suite 319, Cambridge, MA 02142, USA

Two Sigma Investments,
LLC, 379 West Broadway, 5th Floor, New York, NY 10012, USA

K. Nesbitt
New England Complex Systems Institute,
238 Main St. Suite 319, Cambridge, MA 02142, USA

School of Design, Communication & Information Technology, University of Newcastle,
ICT Building—ICT3.07, University Drive, Callaghan, NSW 2308, Australia

M. Lim
New England Complex Systems Institute,
238 Main St. Suite 319, Cambridge, MA 02142, USA

University of the Philippines,
Diliman, Quezon City 1101, Philippines

S. Smith
New England Complex Systems Institute,
238 Main St. Suite 319, Cambridge, MA 02142, USA

Speech Technology and Applied Research Corp.,
54 Middlesex Tpke, Bedford, MA 01730, USA

B.A. Perkins
Centers for Disease Control and Prevention,
1600 Clifton Rd., Atlanta, GA 30333, USA

Vanguard Health Systems, 20 Burton Hills Blvd, Suite 100, Nashville, TN 37215, USA

49.1 Introduction

Improving the US healthcare system requires not only providing care to the uninsured but also addressing a set of linked organizational and motivational issues, including enhancing the role of wellness and prevention. It is widely acknowledged that system organization is key to healthcare improvement [1]. Previous studies [2, 3] have pointed to separating wellness and prevention services from acute care as a central component of healthcare transformation. Even while chronic and preventable diseases have become the dominant cause of loss of life, incentives and perceptions limit the delivery of services that could prevent them [2]. A complex systems analysis implies that the optimal organizational structure depends critically on the scale of repetition and complexity of tasks to be performed. Applied to health care, the separation of acute care from selected preventive services should enable improved matching of organizational structure to function [3–5].

The distinction between simple repetitive and complex tasks is apparent in the proper and improper application of efficiency. Complex tasks—including most medical diagnosis and treatment—require extensive training and careful decision making to determine which one of many possible actions should be performed. Efficiency is detrimental to complex tasks, as shorter times and streamlining through standardization curtails the necessary decision making. In contrast,

J.P. Sturmberg and C.M. Martin (eds.), *Handbook of Systems and Complexity in Health*,
DOI 10.1007/978-1-4614-4998-0_49, © Springer Science+Business Media New York 2013

simple repetitive tasks are amenable to rapid execution by streamlined processes. Without such efficiency the necessary repetitions may not be achieved due to insufficient manpower and other resources. Conserving limited resources in simple repetitive tasks enables those resources to be utilized for high complexity tasks. The distinction between simple and complex is manifest in the allocation of tasks within a hospital, ranging from laundry to diagnosis. It is also apparent in a great success of public health—smallpox eradication—that relied upon two processes: mass immunizations and "surveillance and control" [6]. The former, a simple repetitive task, was made highly efficient to enable 100 million vaccinations over 5 years. The latter, a high-complexity task, was performed by specially trained teams to identify individual cases and vaccinate their households and close contacts.

A focus of public health prevention efforts is achieving higher levels of delivery of clinical preventive services [7–9]. Hundred thousand additional lives could be saved annually by increasing five services to 90% delivery [10]: tobacco cessation counselling, aspirin chemoprophylaxis, influenza immunizations, colorectal cancer screening, and breast cancer screening. However, these long-standing recommendations [11] have not achieved their full potential impact and therefore we consider the obstacles to and opportunities for improvement based upon an analysis of organizational structure.

Retail clinics have been introduced in the USA [12] as a response to the need for convenient and affordable care, and are gaining popularity as a mechanism of simple care delivery. Retail clinics, also known as convenient care clinics, are medical institutions established within a retail setting, such as a shopping mall. They are generally staffed by nurse practitioners or physician's assistants—individuals who have medical training and are able to write prescriptions. Distinct from "urgent care" centers, they follow a retail model—they provide only relatively simple services that can be delivered rapidly by nonphysician providers and allow payment by insurance as well as directly by consumers [13]. Retail clinics diagnose and treat a variety of common ailments: the common cold, the flu, ear infections, allergies, injuries, rashes, etc. They also provide some preventive services. They may be open for extended hours and 7 days a week. Retail clinics join pharmacies and opticians in using retail locations for a medical purpose.

As of 2011 there were over 1,200 retail clinics in the USA [14]. Some are independently operated, while others are managed as part of pharmacy or retail chains, including CVS, Walgreens, and Target, within their store locations. It is estimated that one in six Americans have already visited a retail clinic, while almost half are receptive to the idea; young, healthy individuals are particularly receptive to the use of retail clinics [15]. The accepted purpose of retail clinics is to provide more convenient locations and times, and lower prices than traditional physician practices.

As an innovation in the healthcare system, retail clinics have provoked concerns reflected in prominent recommendations for investigation and legislation [16]. The debate has not been informed by a framework adequate to analyze the potential contributions of various service delivery models. In this chapter we propose that the low-complexity, large-scale nature of a retail clinic's practice allows for the advantages of efficiency to be realized, thus providing much needed services at lower cost, and relieving the burden of overcrowding in traditional medical practices. These advantages are particularly relevant to simple preventive services that are needed by a large fraction of the population on a frequent basis.

We analyzed preventive services to identify the organizational structure that would be most effective in delivering them and determine whether they would benefit from high-efficiency processes that are characteristic of retail service organizations. For each service we evaluated the number of delivery repetitions that are required for the target population; the time, number of personnel and cost required to deliver the services in physician-based practices; the level of training required for performance of preventive services; and what reductions in time and cost would be obtained through efficiency in a retail context.

49.2 Methods

We estimated the required number of repetitions (from target population and frequency), time, and cost in the traditional healthcare setting for 35 of the services recommended by the United States Preventive Services Task Force (USPSTF) [4], the Advisory Committee on Immunization Practices (ACIP) [17], and the Centers for Medicare and Medicaid Services (CMS) [18]. Where recommendations specify "regularly," e.g., for short informational counselling sessions, we assume annual frequency. Medication prescriptions were incorporated into the frequency of the related contact service. The time to provide the service includes the contact and related administrative time estimated using 2007 CMS Physician Fee Schedule (PFS) [19] in Relative Value Units (RVUs), which can be converted consistently to time based on one RVU as equivalent to 0.5 h of physician time, as stated in military policy and studies (15.4 RVUs for 8 h in 2003) [20]. For cases where recommended and estimated contact time are both available, we found them to be similar, with one exception: diabetes self-management is recommended to be 30 contact minutes, while the RVU-based estimate is 22.5 min. For this case, we use the recommended contact time.

Unless otherwise specified, costs are from CMS specifications [19, 21, 22]. We did not include the allocation for malpractice insurance, which for each service is under 5% of total costs, or the "budget neutrality factor." Varying costs are those expected to decline when services are subject to process streamlining. This includes all contact and administrative times. Fixed costs include chemoprophylaxis medications and immunization vaccines because medication manufacture and delivery is already a combination of patent protection and industrial efficiency. Multiple service options or medication types were averaged by usage and cost data [22, 23]. Laboratory costs were treated as varying because rates [21] do not use existing lower cost options (e.g., over-the-counter immunologic fecal occult blood tests) or have remained fixed over time (e.g., laboratory rates for serum potassium and creatinine, 2002–2007). Current total cost estimates use available coverage data [7]. Where necessary, costs were inflated to 2007 using 3% annual increases.

We evaluated the potential efficiency gains from process improvement including shorter times for task execution and lower costs. The empirical "learning curve" [24–26] quantifies increasing efficiency for tasks performed repetitively by individuals and for entire industries. The relationship between the time t_1 required to perform the first repetition and the time required for the nth repetition, t_n, is

$$t_n = t_1 n^{-z}.$$

The learning-curve parameter z varies between industries, clustering around 0.15 and generally in the range (0.07, 0.4). We used $z=0.15$ for our estimates and conservatively considered only individual provider repetitions, not industry-wide repetitions, to estimate efficiency gains. The conclusions are robust to varying these assumptions. We did not directly consider specific mechanisms for efficiency, e.g., group informational sessions. Costs follow a similar behavior. However, we considered the base (first repetition) cost to be half the physician provider costs, because lower salaried employees can provide high-repetition, low-complexity services.

The "learning curve" neglects idle time—the true cost and time for a retail employee depends on the demand. Retail demand depends on the level of promotion by retailers themselves. We made calculations for an individual provider that performs $m=20$ distinct services; increasing m reduces the efficiency gains, because it reduces the number of repetitions of each service. However, an increased number of services increases the likelihood of sufficient demand to utilize employee time. A change of m by 10% would change the efficiency gain by 2%. The efficiency gains are affected by time estimates through the number of repetitions possible for each provider per year. Where time is estimated roughly due to lack of direct specification (e.g., short counseling sessions) the efficiency factors are robust as even an increase of 100% or decrease of 50% (a factor of 2) yields a change in efficiency of 12%.

To illustrate the effect of efficiency gains, we compared costs and personnel requirements for full coverage by current providers with retail organizations in the third year of operation (from level budgeting in each of 3 years). The number of services provided by an individual was obtained using 1,824 hours per year [27]. Our RVU-derived times underestimate personnel due to the use of physician cost equivalents for administrative time. For tobacco cessation we also calculated the impact of quits on the target population reducing the costs and personnel. We developed a simple model consistent with available data [28] using an exponentially decreasing quit rate saturating at a total of 23% quits.

We also estimate the reduction of acute care costs due to full preventive coverage based upon reviews of the National Commission on Prevention Priorities, which take into account efficacy and adherence [29]. Tobacco cessation healthcare cost reductions were estimated by an exponential decay with a conservative three-year time constant starting in the year following a quit.

We also evaluated which services would be appropriate based upon their apparent complexity, reflected in protocols, diversity of actions, or required certification. Explicit protocols for services and levels of training for providers are available in the relevant literature.

We identified other features of physician offices, retail settings, and dedicated healthcare providers that would be important for the patient or for the physician's ability to provide preventive services to the target population. These include population served, location relative to target population, and core competencies.

49.3 Results

The number of repetitions needed for full coverage of the target population, the time to deliver the service, and the variable and fixed costs are summarized in Fig. 49.1. Across the analyzed services, the number of repetitions ranges from 380,000 abdominal aortic aneurysm ultrasounds to 560 million obesity-information counseling sessions. Repetitions over 100 million also include information sessions for tobacco cessation, hyperlipidemia diet, problem drinking, aspirin chemoprophylaxis, calcium chemoprophylaxis, diabetes nutrition, and screening for hypertension and cholesterol. We excluded depression screening because recommendations state [7] that this service should be provided by a facility that can also provide treatment.

Very few of the preventive services require particular levels of training. The risks associated with preventive services are generally much lower than those of acute care. Mammography is distinct because high levels of training are required to interpret the images, but not to acquire them; the latter is currently performed in some retail settings [30]. There are also tasks that require ongoing organizational mechanisms, such as the timing and record keeping of childhood immunizations, but these were not considered to preclude retail delivery.

Consider tobacco cessation information sessions: The USPSTF [7] recommends one session per year, but only 28% of smokers [10] receive such sessions even though over 70% of smokers indicate they want to quit [31]. CMS authorizes up to eight sessions per year [18]. The benefits have been studied [28] extensively: The estimated quit rate from a single session is 2.5%, and 5% with cessation aids. Multiple sessions raise the quit rate to about 23%. On average, each smoker who quits gains about 6 years of life expectancy and reduces per annum adult healthcare costs by $1,100 (from $10,200 to $9,100, i.e., 12.5%), with most of the benefits accruing after 3 years. An information session has been specified by a simple protocol [32] composed of questions such as: "Does patient now use tobacco? Is patient now willing to quit? …"

Our estimates suggest that providing tobacco cessation information through retail organizations would reduce the cost and effort of providing the service. The per-person cost of the session ($11.40) and medications ($209 per year for the 16% who use them) is $34. If the standards were increased universally to eight times per year (as

Table I: Key properties for retail implementation of preventive services.*

Retail appro-priate(√)	Preventive service	Repetitions for target popuation per year (millions)	Delivery time† (min)	Varying cost ($)	Fixed cost ($)	Current coverage¥	Varying costs	Fixed costs	Total costs	Time	Varying cost	Total cost	Personnel required (reference)	Personnel required (retail 3rd year)	Cost (reference)	Cost (retail 3rd year)	Acute care savings from full delivery‡
		Individual service properties					Current costs			Efficiency factors in year 3			Full delivery comparisons of personnel and cost				
	Abdominal Aortic Aneurysm Ultrasound	0.4	71	$ 89.06	$ -					-	-	-	240	-	$ 34	$ -	
√	Aspirin Counseling	121.0	6	$ 7.88	$ 16.00	36%	$ 343	$ 697	$ 1,040	3.8	7.5	1.4	6,900	1,830	$ 2,889	$ 2,060	$ 12,200
√	Breast Cancer Screening Mammogram^	35.0	75	$ 95.50	$ 34.86	67%	$ 2,239	$ 817	$ 3,057	2.4	4.9	2.4	24,130	9,930	$ 4,563	$ 1,910	$ 223
√	Calcium Counseling	100.0	6	$ 7.88	$ 29.20					3.8	7.5	1.2	5,700	1,510	$ 3,708	$ 3,020	$ 9,020
√	Cervical Cancer Screening	56.0	32	$ 87.35	$ -	79%	$ 3,865	$ -	$ 3,865	2.8	5.7	5.7	16,270	5,750	$ 4,892	$ 860	$ 950
√	Childhood immunization	36.6	25	$ 51.90	$ 63.74	76%	$ 1,446	$ 1,775	$ 3,221	3.0	5.9	1.6	8,340	2,820	$ 4,232	$ 2,650	$ 5,670
√	Chlamydia Screening	15.8	4	$ 63.10	$ -	40%	$ 399	$ -	$ 399	4.1	8.2	8.2	580	140	$ 997	$ 120	$ 200
	Cholesterol Screening high risk	9.6	6	$ 2.53	$ 183.31					-	-	-	530	-	$ 1,784	$ -	$ 180
√	Cholesterol Screening	76.8	6	$ 2.53	$ 492.23	31%	$ 60	$ 11,378	$ 11,438	3.8	7.6	1.0	4,210	1,110	$ 37,998	$ 37,830	$ 4,060
	Colon Cancer Screening: Colonoscopy	3.7	297	$ 374.80	$ -	33%	$ 461	$ -	$ 461	-	-	-	10,110	-	$ 1,398	$ -	
√	Colon Cancer Screening: FOBT	42.4	4	$ 10.23	$ -	33%	$ 143	$ -	$ 143	4.1	8.2	8.2	1,550	380	$ 434	$ 50	$ 2,778
	Colon Cancer Screening: Sigmoidoscopy	1.5	99	$ 124.68	$ -	33%	$ 62	$ -	$ 62	-	-	-	1,370	-	$ 189	$ -	
	Depression Screening	223.0	4	$ 5.00	$ -					-	-	-	8,070	-	$ 1,115	$ -	
	Depression Treatment	55.8	76	$ 95.50	$ 92.94					-	-	-	38,550	-	$ 10,515	$ -	
√	Diabetes Management Counseling	84.0	30	$ 29.18	$ -					2.9	5.7	5.7	23,030	8,050	$ 2,451	$ 430	
√	Diabetes Nutrition Counseling	168.0	23	$ 28.42	$ -					3.0	6.0	6.0	34,540	11,480	$ 4,775	$ 790	
√	Diabetes Screening	37.0	6	$ 14.88	$ -					3.8	7.6	7.6	2,030	530	$ 551	$ 70	
√	Diet Counseling	296.0	23	$ 28.42	$ -					3.0	6.0	6.0	60,860	20,230	$ 8,412	$ 1,400	
√	Folic Acid Counseling	55.8	6	$ 7.88	$ 48.00	33%	$ 145	$ 884	$ 1,029	3.8	7.5	1.1	3,180	840	$ 3,118	$ 2,740	$ 59
√	Hearing Screening	37.0	2	$ 2.53	$ -	12%	$ 11	$ -	$ 11	4.6	9.2	9.2	680	150	$ 94	$ 10	
√	Hypertension Screening (except lab)	190.3	2	$ 2.53	$ 99.46	33%	$ 159	$ 6,246	$ 6,405	4.6	9.2	1.0	3,480	750	$ 19,409	$ 18,980	$ 10,900
√	Hypertension lab	27.1	6	$ 20.87	$ -					3.8	7.6	7.6	1,490	390	$ 566	$ 70	
√	Influenza Immunization	87.0	16	$ 19.13	$ 12.25	43%	$ 716	$ 458	$ 1,174	3.2	6.4	2.1	12,400	3,860	$ 2,730	$ 1,330	$ 1,755
	Initial Preventive Physical Exam	2.0	93	$ 117.85	$ -					-	-	-	1,710	-	$ 236	$ -	
√	Injury Prevention Counseling	20.3	31	$ 39.41	$ -					2.8	5.7	5.7	5,790	2,040	$ 800	$ 140	
√	Obesity Counseling	539.3	18	$ 23.11	$ -					3.1	6.2	6.2	90,170	28,900	$ 12,462	$ 2,000	$ 1,930
	Osteoporosis Screening Imaging	10.5	109	$ 138.70	$ -	-				-	-	-	10,460	-	$ 1,456	$ -	$ 312
	Pneumococcal Immunization	2.0	8	$ 10.61	$ 27.03	65%	$ 14	$ 35	$ 49	-	-	-	150	-	$ 75	$ -	$ 230
√	Problem Drinking Counseling	42.6	18	$ 24.25	$ -					3.1	6.2	6.2	7,120	2,280	$ 1,033	$ 170	
√	Problem Drinking Screening	215.0	4	$ 5.00	$ -	13%	$ 140	$ -	$ 140	4.1	8.2	8.2	7,780	1,900	$ 1,075	$ 130	$ 1,825
√	Prostate Cancer Screening	40.0	16	$ 20.46	$ -					3.2	6.4	6.4	5,920	1,860	$ 818	$ 130	
√	Tetanus-diptheria Booster	22.3	16	$ 19.71	$ 18.87	2%	$ 9	$ 8	$ 17	3.2	6.4	1.8	3,180	990	$ 860	$ 490	
√	Tobacco Cessation Counseling	371.2	10	$ 12.51	$ 4.26	4%¥	$ 163	$ 54	$ 216	3.5	7.0	2.8	33,580	9,660	$ 6,225	$ 2,250	$ 9,650
√	Vision Screening for Adults	37.0	2	$ 2.27	$ -	17%	$ 14	$ -	$ 14	4.7	9.4	9.4	610	130	$ 84	$ 10	$ 745
√	Vision Screening for Children	52.8	6	$ 7.58	$ -	36%	$ 144	$ -	$ 144	3.8	7.6	7.6	2,890	760	$ 400	$ 50	
	TOTALS						$ 10,533	$ 22,353	$ 32,885				437,600		$ 142,377	-	$ 62,687
	TOTAL OF RETAIL APPROPRIATE (As indicated by first column)						$ 10,457	$ 22,318	$ 32,774				366,650	118,270	$ 125,609	$ 79,690	$ 61,965

Notes:
* Costs are in millions except for individual service properties.
√ Greater than 15 million required per year and limited complexity.
^ The acquisition of mammograms for breast cancer screening is included but their interpretation cost is considered fixed and personnel does not include the necessary 8,830 physicians.
† Contact time and physician equivalent of administrative time
¥ Estimates of current coverage (where available) [7], tobacco cessation standards usually reported are one per year for 28% coverage, we use eight per year for 4% coverage
‡ Acute care savings from per-capita estimates.[30] Colon cancer screening acute care savings are combined.

Fig. 49.1 Key properties for retail implementation of preventive services suggested by CMS)—370 million sessions annually—the cost in the current system would be $6.2 billion and would require 34,000 full-time personnel for that service alone. By contrast, full coverage may be achieved in a retail setting in the third year at a cost of $1.7 billion per year and 7,900 personnel (if we do not include the reduction of the target population due to quits, the values would be $2.25 billion and 9,700 personnel). By the end of the third year 21% of smokers would be expected to quit, constituting 30% of the individuals who say they want to quit, and 4% of the population as a whole. We estimate $1.57, $3.77, and $5.83 billion in acute care savings in years 2, 3, and 4, respectively. Differences in healthcare costs and life expectancy [33, 34] imply these quits eventually save $8.9 billion per year in acute care costs ($9.7 billion for a budget of $2.25 billion per year due to improved coverage in the first 2 years), and result in a combined life expectancy gain of 59 million years.

Contrast the case of colonoscopies for colorectal cancer screening: USPSTF recommends one every 10 years [7, 35] with alternatives of sigmoidoscopy or fecal occult blood tests. The number of colonoscopies required is relatively small (low scale): only two million per year. The procedure itself and the common use of anesthesia are significantly more complex than tobacco cessation counseling. Thus, high service complexity and low scale make colonoscopies not well suited for a retail setting.

Our calculations show that we can achieve a fivefold reduction in variable costs of simple large-scale services if they are delivered by efficiency-oriented organizations (Fig. 49.1). Such efficiency gains can be achieved with 500 repetitions of a service per provider per year (a factor of 2.5 in time reduction and factor of 2 reduction in base costs of employees). Where medication costs are a significant fraction of total costs, efficiency gains for the total service are smaller.

A significant barrier to full implementation of clinical preventive services in the existing healthcare system is the actual workload burden. While recognized in publications [36] this realization has yet to influence policy, perhaps due to a lack of identified alternatives. We estimate full delivery in the current system would require over 400,000 full-time personnel for these services alone (Figs. 49.1 and 49.2). A retail context would reduce the required number of personnel through efficiency gains, reduce the level of training required, and place the services in an industry that can more readily deploy the necessary workforce.

49.4 Discussion

Performing repetitive tasks using a suitable process results in dramatic efficiency gains. While such an approach does not apply to most medical care, it can apply to flu shots and other simple tasks. Historically, industrial efficiency gains resulted in widespread availability of key goods and services that were not available previously. The industrial revolution was in large part driven by mass production. Model-T Ford cars were introduced in 1909 at $850, a significantly lower price than competitors and further declined in price to $260 in the 1920s [37, 38], consistent with the empirical "learning curve" [24, 25]. The time required to assemble a single car declined from 12.5 h in earlier production systems to 93 min [37, 38]. The total production of Model-Ts was 15 million, a fraction of the annual need for many preventive services today. While the Model-T production line example is not directly applicable to health services, the approach of high efficiency is today similarly embedded in retail services.

Our analysis suggests that widespread retail availability could dramatically improve the delivery of preventive services to underserved populations in the USA.

We caution that misapplication of efficiency to complex tasks that require careful decision making by highly-trained individuals would lead to ineffective execution. Because the individual actions are distinct, repetition should not be expected to lead to efficiency gains. With continued pressure to reduce healthcare costs, it is imperative to distinguish those services to which efficiency can be applied and those to which it cannot. Without such differentiation, efforts to

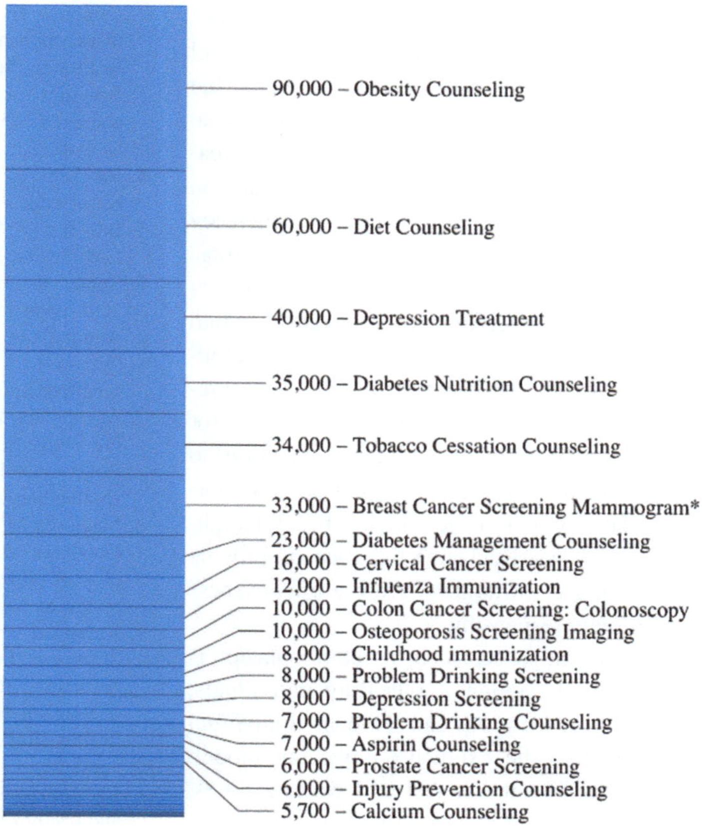

Fig. 49.2 Personnel requirements for fully delivering clinical preventive services in the current healthcare provider setting. (*) Mammography includes both acquisition and interpretation

reduce costs would lead to less effective services, rather than increased efficiency.

The largest scale preventive services are most appropriate for retail organizations. Setting the threshold at services corresponding to 500 per 10,000 population (or 250 for each of the 59,000 US pharmacies [39]) and avoiding complex services, we obtained the set marked as retail-appropriate in Fig. 49.1. Lifestyle and chemoprophylaxis counselling is the largest set of high benefit services for retail implementation.

The total cost of the identified retail-appropriate services that would be transferred from physician to retail delivery by this model, excluding costs that are not transferred, such as medications, is approximately $14 billion (allowing delivery of services for which no estimates are available to be at 20%), corresponding to 3.2% of national expenditures on physician and clinical services [40]. Why should physicians endorse the transfer of healthcare tasks to others, even such a small percentage? Because unlike other services, physicians are not necessary to guarantee quality delivery; because of the resulting dramatic increase in delivery of needed prevention services; because of the reduction of pressures to deliver these services; to enable physicians to focus on tasks requiring their higher training; and because improving the efficiency of the healthcare system allows more resources to be devoted to tasks that need them, and increases the ability of the system to meet demands beyond the current capacity of providers and resources. Everyone, including traditional providers, would benefit from improved healthcare system function, especially if reimbursement is made appropriate by recognizing the distinct needs for simple and complex tasks.

Using a retail context for a restricted set of services is consistent with the existing US healthcare system. Prescription pharmacies provide a service for which physician offices are no longer considered practical. We offer four factors that characterize why physician offices do not (and should not be expected to) provide highly efficient services, including preventive services: The population served, the location of service (travel distance), the number of locations at which the service can be obtained, and the frequency at which the target population is present at the location of service for other reasons. The population served by a physician office is limited. Efficiency arises from repetition and with fewer repetitions, the efficiency gain is smaller. Moreover, efficiency gains in a practice setting cannot be translated into more individuals served. The exclusive relationship between the patient and the physician precludes patient choice. Thus, the drivers for efficiency improvement and cost reduction to compete for patients are not present. Most importantly, the physician office is designed for high-complexity tasks. Replacing this focus to provide efficient delivery would reduce the effectiveness of its primary responsibility.

For preventive services, a retail concept is particularly convenient for healthy individuals, since multiple locations can serve any individual as opposed to the exclusive physician model. This results in lower travel and waiting times. The importance of sufficient demand motivates retailers to promote their preventive services, which would benefit population health. The non-urgent nature of these services reduces their priority among both traditional providers and patients. Wrapping preventive care in the nationally popular focus on health and fitness, rather than in a medical context, may provide new opportunities for wider adoption. Given the empirical evidence that the existing US healthcare system cannot improve upon its current performance efforts, delivering preventive services through retail operations should be supported by policy.

Acknowledgements This work was supported in part by the Centers for Disease Control and Prevention.

References

1. Institute of Medicine (IOM), Committee on Quality of Health Care in America. Crossing the quality chasm: A new health system for the 21st century. Washington, DC: National Academy Press; 2001.
2. Lambrew J. Who has the cure?: Hamilton Project ideas on health care. Washington, DC: Brookings Institution Press; 2008. p. 227.
3. Bar-Yam Y. Improving the effectiveness of health care and public health: A multi-scale complex systems analysis. Am J Public Health. 2006;96:459–66.
4. Bar-Yam Y. Making things work. Cambridge, MA: Knowledge Press; 2005.
5. Bar-Yam Y. Multiscale variety in complex systems. Complexity. 2004;9:37–45.
6. Centers for Disease Control and Prevention. Smallpox: 30th anniversary of global eradication [Internet]. Atlanta, GA: 2008 [cited 2008 May 29]. Available from http://www.cdc.gov/Features/SmallpoxEradication.
7. Agency for Healthcare Research and Quality. U.S. Preventive Services Task Force guide to clinical preventive services. U.S. Department of Health and Human Services. Report number: 07-05100, 2007.
8. Maciosek M, Coffield A, Edwards N, Flottemesch T, Goodman M, Solberg L. Priorities among effective clinical preventive services: Results of a systematic review and analysis. Am J Prev Med. 2006;31:52–61.
9. American Medical Association. Promoting healthy lifestyles [Internet]. 2008 [cited 2008 May 30]. Available from: http://www.ama-assn.org/ama/pub/physician-resources/public-health/promoting-healthy-lifestyles.page.
10. National Commission on Prevention Priorities. Preventive care: a national profile on use, disparities, and health benefits. Washington, DC: Partnership for Prevention; 2007.
11. Agency for Healthcare Research and Quality. Guide to clinical preventive services. 2nd ed. Rockville, MD: U.S. Department of Health and Human Services; 1996.
12. Scott MK. Health care in the express lane: the emergence of retail clinics [Internet]. Oakland (CA): California Healthcare Foundation; 2006 [cited 2008 May 25]. Available from: http://www.chcf.org/topics/view.cfm?itemID=123218.
13. Bohmer R. The rise of in-store clinics—threat or opportunity? N Engl J Med. 2007;356:765–8.
14. Marcus MB. Medical clinics in retail settings are booming [Internet]. USA Today. 2011 Aug 28 [cited 2012 Mar 8]. Available from: http://yourlife.usatoday.com/health/healthcare/story/2011-08-28/Medical-clinics-in-retail-settings-are-booming/50168796/1.
15. Deloitte Center for Health Solutions. Survey of health care consumers. Washington, DC: Deloitte Development LLC; 2008.
16. American Medical Association. AMA calls for investigation of store-based health clinics [Internet]. 2007 [cited 2008 May 29]. Available from: http://www.medicalnewstoday.com/articles/75308.php.
17. CDC Advisory Committee on Immunization Practices. Immunization schedules [Internet]. Atlanta (GA): Centers for Disease Control and Prevention. 2008 [cited 2008 May 28]. Available from: http://www.cdc.gov/vaccines/recs/schedules.

18. Centers for Medicare & Medicaid Services. Medicare preventive services quick reference information [Internet]. 2008 [cited 2008 May 28]. Available from: http://www.cms.hhs.gov/MLNProducts/downloads/MPS_QuickReferenceChart_1.pdf.
19. Centers for Medicare & Medicaid Services. Physician fee schedule relative value files [Internet]. 2007 [cited 2008 May 31]. Available from: http://www.cms.hhs.gov/PhysicianFeeSched/.
20. St. Andrews J. A study of the Relative Value Unit as a practice management tool for provider productivity [Internet]. Miami (FL): Army Medical Department; 2003 [cited 2008 May 28]. Available from: http://www.dtic.mil/cgi-bin/GetTRDocLocation=U2&doc=GetTRDoc.pdf&AD=ADA421282.
21. Centers for Medicare & Medicaid Services. Clinical laboratory fee schedule [Internet]. 2007 [cited 2008 May 31]. Available from: http://www.cms.hhs.gov/ClinicalLabFeeSched/.
22. Centers for Medicare & Medicaid Services. Medicare Part B drug average sales price [Internet]. 2007 [cited 2008 May 31]. Available from: http://www.cms.hhs.gov/McrPartBDrugAvgSalesPrice/.
23. Verispan VONA. Top 200 drugs in 2006 [Internet]. 2006 [cited 2008 May 29]. Available from: http://drugtopics.modernmedicine.com/drugtopics/data/articlestandard//drugtopics/092007/407652/article.pdf.
24. Henderson B. The experience curve reviewed, parts I through IV. Perspectives. 1973;124:25.
25. Yelle LE. The learning curve: historical review and comprehensive survey. Decis Sci. 1979;10:302–28.
26. Newell A, Rosenbloom PS. Mechanisms of skill acquisition and the law of practice. In: Anderson JR, editor. Cognitive skills and their acquisition. Hillsdale, NJ: Lawrence Erlbaum Associates; 1981. p. 1–51.
27. Mishel L, Bernstein J, Allegretto S. The state of working in America 2006/2007. Ithaca, NY: ILR Press; 2007.
28. Solberg LJ, Maciosek MV, Edwards NM, Khanchandani HS, Butani AL, Rickey DA, Goodman MJ. Tobacco use screening and counseling: technical report prepared for the National Commission on Prevention Priorities. Minneapolis, MN: HealthPartners Research Foundation; 2006.
29. Partnership for Prevention. Rankings of preventive services for the U.S. population [Internet]. 2008 [cited 2008 May 29]. Available from: http://www.prevent.org/National-Commission-on-Prevention-Priorities/Rankings-of-Preventive-Services-for-the-US-Population.aspx.
30. See e.g. Baptist Memorial Health Care and Goldsmith's open Tennessee's first retail-based mammography center [Internet]. 2008 [cited 2008 May 28]. Available from: http://www.bmhcc.org/media/news/archivecontent.asp?category=2000&article=News_Item.2004-11-12.7860892798.
31. Centers for Disease Control and Prevention. Cigarette smoking among adults – US 2004. Morb Mortal Wkly Rep. 2005;54:44.
32. U.S. Department of Health & Human Services. Screen for tobacco use status [Internet]. 2007 [cited 2008 May 29]. Available from: http://www.surgeongeneral.gov/tobacco/screen.htm.
33. Centers for Disease Control and Prevention. Annual smoking-attributable mortality, years of potential life lost, and productivity losses—U.S., 1997-2001. Morb Mortal Wkly Rep. 2005;54:625–8.
34. Centers for Disease Control and Prevention. Cigarette smoking attributable morbidity—U.S., 2000. Morb and Mortal Wkly Rep. 2003;52:842–4.
35. Maciosek MV, Solberg LJ, Coffield AB, Edwards NM, Khanchandani HS, Butani AL, McGree DA, Goodman MJ. Colorectal cancer screening. Technical report prepared for the National Commission on Prevention Priorities, Partnership for Prevention [Internet]. 2006 [cited 2008 May 29]. Available from: http://www.prevent.org/data/files/initiatives/colorectalcancerscreening.pdf.
36. Yarnall K, Pollak K, Ostbye T, Krause K, Michener J. Primary care: Is there enough time for prevention? Am J Public Health. 2003;93:635–41.
37. Model T Facts [Internet]. Detroit, MI: Ford Motor Company; 2008 [cited 2008 May 27]. Available from: http://media.ford.com/article_display.cfm?article_id=858.
38. Assembly line [Internet]. Wikipedia. 2001 [cited 2008 May 27]. Available from: http://en.wikipedia.org/wiki/Assembly_line.
39. Pharmaceutical Care Management Association. Consumer access to pharmacies in the United States [Internet]. 2007 [cited 2008 May 28]. Available from: http://pcma-upgrade.webdrivenhq.com/assets/2008-03-25_Research_SKA%20Research%20Consumer%20Access%20to%20Pharmacies%202007%205_1_07.pdf.
40. Dept. of Health & Human Services. National health expenditure accounts, 2006 highlights [Internet]. 2008 [cited 2008 May 28]. Available from www.cms.hhs.gov/NationalHealthExpendData/downloads/highlights.pdf.

Complexity Science in the Future of Behavioral Medicine

50

David Pincus and Stephen J. Guastello

50.1 Introduction

Complexity science offers a new, broader paradigm emerging from the traditional biomedical model of medicine. This new paradigm will inform research and intervention, particularly for the most complex medical conditions such as type-II diabetes (DT2), heart disease, pain, and anxiety-depression spectrum (ADS) disorders. Traditional medical interventions, including those from behavioral medicine, utilize the framework of *disease* to understand etiology and treatment. The disease framework is based on the idea that some exogenous agent, such as a germ, intrudes upon an otherwise healthy body and causes illness. Etiological concerns for health care providers are then logically aimed at identifying these disease agents as simple material causes, and treatment is aimed at protecting against their intrusion, mitigating their harmful effects, or removing them from bodily systems where they may cause harm.

Traditional medical research utilizing this disease perspective logically relies upon the randomized clinical trial (RCT), for understanding treatment and comparable linear regression models for understanding etiology when experimental control is not possible. Indeed, such linear cause-and-effect models such as the RCT are ideal for answering simple questions where the disease model is applicable. In etiology, for example, the majority of studies use linear models to answer simple questions such as: *What is the impact of high glycemic foods on weight gain? What is the impact of emotional dysregulation on blood glucose levels in DT2? What personality factors contribute to the development of chronic pain conditions?* Or *What are the genetic and epigenetic contributions to ADS disorders?* Similarly, common linear treatment questions that are well handled by RCTs include: *How effective is motivational interviewing for increasing adherence to lifestyle interventions for overweight? How effective is a lifestyle fitness intervention for DT2 compared to common medications like metformin? How effective is relaxation imagery training for reducing frequency, intensity, and functional impairment for migraine headaches?* Or *How effective is cognitive-behavioral therapy compared to psychodynamic therapy for mild to moderate depression?* Such simple, independent, and proportional conceptualizations of cause and effect in disease and intervention are indeed necessary first steps in treatment and research contexts. If potent linear causes exist and prove useful, then one may conceptualize disease processes in simple terms that lead to efficient and cost-effective interventions.

D. Pincus
Chapman University, Crean School of Health and Life Sciences, One University Drive, Orange, CA 92866, USA
e-mail: pincus@chapman.edu

S.J. Guastello (✉)
Marquette University, P. O. Box 1881, Milwaukee, WI 53201-1881, USA
e-mail: stephen.guastello@marquette.edu

J.P. Sturmberg and C.M. Martin (eds.), *Handbook of Systems and Complexity in Health*,
DOI 10.1007/978-1-4614-4998-0_50, © Springer Science+Business Media New York 2013

The present chapter will aim to broaden and build upon this simple framework of disease within behavioral medicine utilizing concepts, models, and select research methodologies from complexity science. Complexity science includes theoretical concepts, models, and interventions that are grounded in nonlinear dynamical systems (NDS) theory, which includes many more specific and often overlapping topics (e.g., self-organization, emergence, deterministic chaos, and fractals). For the purposes of this chapter, *nonlinear* may be considered to include a focus on disease processes that are caused by multiple and interactive factors coming from within the organism and from the quasi-independent nesting context in which the organism is situated. Importantly, timing is crucial in such *dynamical* systems, which exhibit disproportional and often complex patterns of change over time. As a result, etiological and treatment effects depend sensitively on timing.

This chapter will lay out a roadmap to the future of behavioral medicine developed within the framework of complexity science. First, complex conditions will be discussed and a case will be made for using NDS to add precision to existing biopsychosocial models (e.g., diathesis-stress) to guide better research and clearer applications. Next, the chapter will build a bridge for practitioners and researchers to use in moving from a traditional and linear understanding of behavioral medicine to a broader nonlinear viewpoint using some specific empirical and conceptual innovations from complexity science. Finally, these suggestions for innovation will be applied to the use of mobile health to allow NDS equipped smart-phone applications to analyze and intervene in the complex behavioral patterns underlying DT2.

Complexity science concepts may appear highly technical and abstract; therefore, it is important to keep in mind that all good clinicians use concepts from complexity science in their day-to-day work with patients. For example, a clinician is considering nonlinear timing effects and the cultural nesting environment when working with a DT2 patient toward relapse prevention at Christmas time (e.g., high exposure to cookies and other holiday treats). Similarly, pain clinicians may frequently encounter the multifaceted, interactive, and dynamic factors of the patient's mood, identity, relationships, and sense of life purpose in designing a course of treatment.

50.2 Complex Conditions

"Complex conditions" are defined by three overlapping features. First, they are *overdetermined*, meaning that they are caused by a range of interacting causes across the biological, psychological, and social spectra. Such overdetermination causes problems for clinical research aimed at the identification of simple causes and for practitioners who practice within strict disciplinary boundaries and specializations.

Second, complex conditions are chronic, as overdetermined biopsychosocial causal factors prove more than sufficient to maintain the disease state over time, and as momentum effects lead to cyclical phenomena whereby the disease processes become self-sustaining over time. Importantly, chronicity is often defined by disease recurrence, rather than stability as is often assumed. The patterns underlying such recurrences can become self-reinforcing, for example as demoralization and stigmatization set in. Furthermore, patterns of chronicity may contain important etiological clues, informing the timing for interventions (e.g., stages of change). Strict practice parameters, specialization, and single point in time models of intervention tend to provide an ill fit for such chronic conditions. For example, even in the best treatment settings for DT2, ADS, chronic pain, or heart disease, the patient typically receives this or that intervention at some regular time interval in a healthcare setting rather than in the day-to-day home, community, or work environments where the disease processes are actually occurring. Most patients would agree that the timing of the phenomenology of these conditions does not conform to their office visits. Furthermore, it is exceedingly rare that a single practitioner has the training or even that a single clinical setting has the staffing or logistics to address the frequent co-occurrence and interactions among these complex and chronic conditions.

Third, complex health conditions are highly unique in the ways they manifest and maintain themselves across individuals, despite the fact that their causal processes are quite similar, including: demoralization, stigma, avoidance, and symptom generalization. Even with the same symptoms presenting, different individuals with the same condition will typically have a unique constellation of causal processes that maintain their disease processes over time. This is a hallmark of systemic cause, understood at least since the 1960s under the term *equifinality*, connoting that unique causes can lead to the same outcomes. By contrast, complex conditions may also exhibit *multifinality*, whereby equivalent causes may lead to the different outcomes [1]. On a clinical level, these concepts provide a foundation for the clinician's instinct to design individually tailored intervention, particularly for complex conditions where one is concerned with longer term outcomes. Once again, the prevailing approach to clinical research and intervention involves the search for a "one size fits all" approach to both investigation and treatment. Diminishing marginal returns apply to treatment standardization efforts for complex conditions, whereby the more specific the models of etiology and treatment become, the less able these models are to account for individual differences. Even within the best RCTs for complex conditions, the treatment components are listed in broad strokes and are highly individualized in actual care (i.e., "lifestyle" interventions). The diminishing marginal utility of treatment standardization is a function of the illnesses themselves, and hypothetically is insurmountable, as idiographic factors, complex cause, equifinality, and multifinality appear to be the rule rather than the exception in the phenomenology of complex conditions.

If complexity science is going to improve upon the trajectories for innovation from traditional science and medicine, it will need to better address this marginal utility problem. Indeed, an approach that embraces alternate notions of multiple, nonlinear, and cyclical cause over time does potentially remove the contrast between standardization and individual differences. As such, complexity science must build upon the best that the state of linear science has to offer with respect to the technology of habit change combined with integrative theory for understanding the dynamics underlying complex biopsychosocial conditions such as DT2, ADS, pain, and heart disease. By elevating the science to understanding the *processes* of disease in addition to the *contents* of disease, one may obtain a more general level of abstraction that will assist in understanding the common processes of etiology, resilience, and recovery that are common across specific syndromes. The new questions may move from: *What causes disease?* And: *How does a treatment work?* To: *How does a self-regulating system break down?* And: *What are the necessary conditions for self-repair?*

The new paradigm in lifestyle change interventions based on complexity science should aim to initiate the simplest possible behavioral changes that will also have the greatest impact on the complex and dynamical biopsychosocial etiology of each individual patient. The behavioral interventions of the future should aim to target the transformative moments during each individual patient's daily routine where a meaningful change is most likely to occur. Like an avalanche effect, the accumulation of well-timed and well-situated behavioral changes has maximum potential for transformative effects in behavioral patterns, personality structure, and social dynamics contributing to complex illness.

An important premise here is that what is most necessary is an opening within the patient to new flows of biopsychosocial information, which will then spread to other areas in the patient's day-to-day life and then to interactions between the patient his or her social context. These are the critical antecedents to transformative moments in lifestyle interventions. The theoretical material to follow in the next section will review relevant concepts from behavior modification that have been shown in numerous clinical trials to be the most effective package of intervention strategies for DT2, nearly twice as effective than the leading pharmaceutical interventions.

DT2 was chosen as the primary syndrome for the remainder of this chapter because it provides a clear example of a medical condition that is poorly understood or treated through the classical notions of disease. By contrast, DT2 is the "poster-child" for the application of complexity science and behavioral medicine to stave off a growing and potentially catastrophic pandemic. It is important to keep in mind, however, that the same complexity concepts that are applied to intervention in DT2 should be equally applicable to the range of frequently co-occurring complex conditions including heart disease, pain, and ADS.

50.3 Complexity Theory and Innovation in Understanding Chronic Conditions

The standard approaches to research and practice are not likely to be sufficient to stave off a growing healthcare crisis in the United States and worldwide [2], because the costliest conditions are frequently the most complex. As detailed previously, these "complex" conditions tend to be chronic and are typically overdetermined through the dynamical interactions of biological, psychological, and social factors that vary across individuals, calling for costly interdisciplinary and individualized treatments. Some of the more prominent of these conditions are: (a) Heart disease—In 2010, an estimated 785,000 first and 470,000 recurrent coronary attacks will occur, contributing to the $316.4 billion to be spent for that year alone [3]. (b) Pain is the most common complaint in primary medicine worldwide, and chronic pain afflicts roughly one in five people worldwide costing $61.2 billion each year in lost productivity in the United States alone [4, 5]. (c) 20.9 million Americans suffer from depression, costing $58 billion annually [6, 7], and 40 million Americans suffer from anxiety, costing well over $42 billion annually [6, 8], while the lifetime prevalence of generalized anxiety is around 40%, making it the most common psychiatric condition in the United States.

Nevertheless, barriers within treatment delivery systems, community stigma, and within patients themselves (e.g., anxiety about anxiety treatment) result in less than 25% of those who need treatment ever seeking help. (d) Type-II diabetes—22 million Americans currently have type-II diabetes (DT2), costing $340 billion a year, and by 2031 this is expected to increase to 32 million people and $1.6 trillion [9]. (e) Overweight serves as a condition in its own right and as a precursor for each of the former conditions. As such could be considered to have costs over and above each of the prior conditions combined. Indeed the tremendous increase in the burden of overweight is reflected in its spread around the globe, and in three to fourfold increases in countries such as the United States over the past 30 years, from roughly 5–10% of the child population to over 30%, and from roughly 20% of adults to more than 60% currently, with half of those impacted technically qualifying as obese in both the pediatric and adult populations [10].

Among these conditions, DT2 appears to have the greatest potential for destruction in both human and economic terms [11]. Yet, it is important to bear in mind that each of the complex conditions overlaps significantly with the other, with mental and physical health problems interacting over time in complex ways, suggesting that the boundaries between "mental" and "physical" health are fuzzy at best. In the case of DT2, each of the other complex physical (cardiovascular disease, chronic pain) and mental illnesses (depression, anxiety, and relationship distress) is common sequelae of the original diabetic illness that further contributes to severity and chronicity of the ongoing diabetic symptoms over time.

Behavior modification delivered within a multimodal, interdisciplinary "lifestyle" approach has been shown to be effective in the most rigorous of clinical trials involving prevention of diabetes onset in at-risk individuals [12] and even more dramatically in reversing DT2 status in people already diagnosed without the use of medication [13]. In addition to statistically significant effects, these results were found to be highly clinically significant, with each study demonstrating a 58% increase in rate of prevention and reversal, respectively, compared to placebo control in the incidence of diabetes among at-risk individuals. Furthermore, this level of response was significantly larger, nearly twice as large; as the 31% reversal obtained from standard medical treatment for non-insulin-dependent individuals (i.e., orally administered metformin).

Given the evidence, one would expect that such lifestyle-oriented interventions would be the dominant mode of care for the prevention and reversal of DT2. Unfortunately this is not the case. Despite the enormous and increasing costs in both human and economic terms, there is an ongoing gap between science and medical practice. The section to follow will introduce some of the key barriers that block widespread access to such treatment approaches.

50.3.1 The Science–Practice Gap

The roadmap to the future of behavioral medicine derives from a much broader shift in perspective from the thinking inherent in even the most progressive biopsychosocial theories [14] in behavioral medicine. Rather than simple acknowledgement that biological, psychological, and social variables can influence one another, the new perspective goes further and is grounded within nonlinear dynamical systems (NDS) theory. NDS is an approach to science and clinical practice that allows for the disproportionate (i.e., nonlinear) and time sensitive (dynamical) interactions of systems, which are comprised of holistic assemblies of interactive variables. The prevailing biomedical model has, by most accounts, been successful in extending and enhancing the quality of life for millions of people suffering from numerous illnesses. Yet, an over-reliance upon statistical procedures grounded in the general linear model and null-hypothesis significance testing places a cap on the ability for research to identify, for example, the common structural features of healthy and unhealthy systems [15]. If the goal of a treatment outcome study includes effectiveness (external validity), efficiency, and processes underlying change then a clinical trial is too myopic. To understand complex and interactive causes, the models and methods used must have the potential to capture such complexity.

Despite the escalating imperative to use "evidence-based practices," a wide gap between available evidence and clinical practice continues [16]. The contribution of scientific limitations to this gap is most obvious perhaps in psychotherapy research. A seminal empirical review by Lambert [17] attributed only 15% of the variance in treatment outcome to treatment factors and 30% to relationship factors between therapist and patient. More recent empirical studies have confirmed this conclusion consistently. For example, Carryer and Greenberg [18] found that the therapeutic alliance and emotional experience within sessions accounted for a combined 30% of variance on reductions in symptoms of depression, whereas meta-analytic evidence has suggested that the actual difference in effect sizes among plausible treatments actually may be zero [19]. It is becoming increasingly apparent that the "evidence" behind the use of "evidence-based treatments" is not telling us very much about why treatment works [20]. "Different" approaches to treatment in psychotherapy often amount to little more than different brands or guilds in the majority of clinical contexts, and quite obviously do not meet the criteria to be considered as distinct scientific theories [21, 22].

These general psychotherapy results extend to the more specific area of behavioral medicine. Furthermore, it appears that most research strategies tend to ignore issues of timing and fit; the one-size-fits-all mentality is inherent within the "active ingredients" approach to designing interventions. Rather, common sense suggests that even the most powerful medicine is useless if a patient cannot swallow it, not to mention the failure of linear biomedical models to account for self-healing mechanisms that are unique to the individual. The placebo effect is often cited as an example of this nagging evidence that something is missing from the biomedical model of illness and treatment, particularly within the context of complex conditions. For example, a recent outcome trial for irritable bowel syndrome [23] demonstrated that placebo pills produced significant improvements even when patients were informed that the pills they were taking were inert and the pill bottles were labeled "Placebo." Despite the great success of modern medicine, the medicines themselves often appear to work in nonmaterial and top-down ways. The manner in which such top down, formal cause operates remains an open topic that is well suited to complexity science.

An additional problem across healthcare that contributes to science–practice gaps can be traced to the complex nature of patients' life conditions that clinicians are trying to navigate. Automated decisions rely on textbook or prototypical examples, but when real-world events deviate in some important way, the deviation carries through a series of decisions that become increasingly off course [24]. When the automated decisions are working well for an extended period of time, the operator can become inattentive to deviations that could trigger a problem. This process may be analogous to behavioral interventions where best practices have been known to produce harmful results for some individuals [25, 26].

This conflict between effectiveness and fit is likely to reach a critical point in the near future of behavioral medicine. The field would continue to be driven both by economic and humanistic considerations, most notably the rising costs and declining access to quality care. At the same time, technological advances like smart phones combined with emerging disciplines like behavioral informatics [27] are expected to create niches for innovation under relatively new health initiatives such as "mobile health" [28]. These innovations create the potential for the collection and analysis of vast quantities of ecologically valid and multifaceted data, nested in real time, and with high potential for interactivity within patients, across the patients' social networks, and among patients' health care providers and third party payers.

The section to follow will describe the standard of care in behavioral modification programs for type-II diabetes toward the development of specific ideas about which NDS models and methods would be most likely to assist in the future of behavioral informatics for complex conditions.

50.3.2 Improving Standards of Care for with Complexity Science: The Example of DT2

Each of the current gold standards in behavior modification in management of DT2 may find more parsimonious theoretical and practical grounding using complexity, rather than traditional science. The most prominent of the current ADA [29] recommendations are summarized as follows: (a) Behavior modification should be delivered within the context of "lifestyle change." Lifestyle-oriented interventions include some focus on building long-term resilience and managing relapse, although the concepts "resilience" and "relapse" are typically not well defined even the best programs or investigations [30]. In addition, lifestyle change typically involves the following ancillary procedures: ongoing monitoring for motivation enhancement, the initial establishment of baselines from which to begin habit change, goal setting, use of reinforcement, enlisting of social support with ongoing access to social information pertaining to nutrition and fitness, and building of additional skills as necessary (e.g., self-concept, interpersonal, problem solving, and mood management). (b) Programs should involve physician-led integration of mental health work within the context of interdisciplinary care. (c) The patient should be an active member of the treatment team. (d) Goals and treatment plans should be reasonable, flexible and should be tailored to fit the circumstances of each individual patient, including approaches to be used for direct glycemic control. A1C levels set to the goal of $<7\%$ in most patients, with the primary focus on increasing activity levels (150 min/week minimum e.g., walking with heart rate at 50–70% maximum and resistance training $3\times$per week) and weight loss (7% or more). Attention should also be paid to motivation and relapse management throughout. (e) Self-management education with ongoing assessment of self-management skills and quality of life should occur throughout the program. Psychosocial interventions should be focused on emotional well-being. (f) Finally, the ADA [29] recommends solid "psychosocial care" described as:

> *Psychosocial screening and follow-up should include, but is not limited to, attitudes about the illness, expectations for medical management and outcomes, affect/mood, general and diabetes-related quality of life, resources (financial, social, and emotional), and psychiatric history … Screen for psychosocial problems such as depression and diabetes-related distress, anxiety, eating disorders, and cognitive impairment when self management is poor* (p. 528).

The primary problem with such an ideal list lies in its implementation. Traditional approaches to medicine and empirical investigation are restricted to delivering or investigating each of these various components and parameters independently. Within the context of treatment, the expense of such a program becomes unbearable, as does the investment of time and energy on the part of the patient. As such, idealistic lists such as these remain an idealization, while most actual patients receive only metformin along with biannual check-ups from their physicians.

An approach to lifestyle change utilizing complexity science would include each of the recommended intervention components and parameters along with several more. The target for innovation above and beyond standard approaches will rest in the way in which these program components are integrated and disseminated seamlessly within each patient's individual daily routines. Within the published literature, most descriptions of how each of the listed treatment components is to be actually carried out, how it is proposed to work, and how the components fit together remains quite vague. The latest report of the ADA [29] is no exception. Yet, from every practical or empirical standpoint the unique fit with each individual patient and a clearer understanding of how change occurs is central. One may assume that the emphasis on pulling treatment components apart and making one-size fits all recommendations comes from reductionism and norm-based reasoning, respectively, ill-fitting aspects of traditional research for the problem of complex conditions.

Nevertheless, traditional science has done an excellent job in identifying the important components of treatment and demonstrating their average rates of success [12, 13]. The goal of complexity science and behavioral informatics must be to build upon and deepen this current knowledge base. As such, complexity science must aim to improve the efficiency of such gold standards to intervention to improve upon current levels of dissemination, while at the same time improving upon empirical markers of effectiveness in reversal and prevention of DT2.

One should expect the clinical trial to remain as the foundation of quality evaluation in the future. Additional, richer sources of data will likely come in the form of multivariate time series, allowing for individually tuned measures of structure and complexity among variables, and within variables over time. For example, an ideal measurement in DT2 would involve the monitoring of blood glucose on a meal-by-meal basis in order to obtain short-term (smallest scale of change possible) time series data as well as long-term follow-ups using more reliable measures of blood glucose and insulin response. Such time series data will allow for causal inferences pertaining to processes underlying change, particularly with the inclusion of additional patient-specific dynamical variables such as nutritional balance, motivation, mood, confidence, personality states and traits, resilience and psychopathology, each measured over time. Furthermore, the utilization of such dynamical and structurally oriented variables within complexity science opens the door to innovation in additional variables that may be of interest. For example, it may be interesting to track *hedonic balance*, defined as the relative flexibility among pleasure obtained among an individual's life domains. It is likely that one structural factor that impacts overweight in general and DT2 in particular is the over-reliance on food as a primary source of pleasure. Increasing an individual's ability to obtain pleasure from a variety of sources (e.g., family, friends, work, and leisure) may decrease the tendency toward restrictive, rigid, and repetitive over-reliance on the cacophony of neuropeptides that are activated through eating calorie dense foods [31, 32].

Multivariate time series designs would also serve to inform the ongoing improvement of interventions via repeated clinical trials by utilizing sophisticated NDS analytic methods. Such toggling between traditional clinical trials and complexity informed process research is well suited to emerging disciplines such as (a) behavioral informatics—the use of computational science and new technologies to improve upon behavioral health [27] and (b) mobile health—a market niche identified by mobile phone carriers for aggressive product research and development [33]. These types of emerging areas of research and development in behavioral medicine will benefit particularly from methods from complexity science that are best suited to handle the large and complex time series that consumers are already producing through their day-to-day interactions with software applications on handheld devices.

This next section elaborates how psychological process research, when combined with NDS' capabilities for modeling temporal dynamics, can produce a new understanding of healthy and unhealthy biopsychosocial processes which may guide the development of more effective and efficient approaches to behavioral intervention. These developments are intended to break new methodological, empirical, and theoretical ground for primary prevention and intervention across a range of complex behavioral health conditions, beginning with DT2.

50.3.3 Biopsychosocial Structure and Change

Evolving beyond the simple biomedical model, modern medicine has increasingly come to realize that most disease processes involve multiple and complex causes that may influence resilience over time (e.g., simple stress-diathesis models). Furthermore, advances in behavioral medicine have left little doubt that even the simplest disease processes (e.g., viral infections) are influenced by habits, larger scale personality factors, and various facets of people's social networks, collectively known as biopsychosocial models [34]. Despite early enthusiasm toward these models, methodological and practical barriers have imposed strict limits upon their application in research and practice. In practice, for example, specialization, programs of education, and economic disincentives limit the availability of interdisciplinary and transdisciplinary treatment options.

In the scientific domain, early general systems models lacked specificity regarding how the biopsychosocial components really interact, and how they unfold over time. The new research paradigm proposed here requires a fine-grained recording of events in time series and analyses that are capable of extracting patterns from both qualitative and also metric data. NDS offers a rich lexicon of constructs such as attractors, bifurcations, saddles, chaos, fractals, self-organization, and emergence [35]. Some of NDS' distinguishing contrasts with conventional empirical thinking are that: (a) Events are understood as having evolved over time. (b) Changes in variance or variability patterns are at least as important as differences in means. (c) The statistical distributions of dynamic events, which affect statistical analyses and interpretations of events, are more likely to be exponential or power law distributions rather than Gaussian (normal) distributions. (d) Outcomes are not proportional to inputs; critical points, discontinuities, and other nonlinear functions are abundant. (e) There are emergent phenomena that cannot be decomposed into simpler elements that add together; the whole is greater than the sum of its parts. (f) Homeostatic equilibrium is *not* an ideal condition in most cases; complex adaptive systems (CAS) are healthiest when they exhibit the level of variability that allows them to adapt to environmental or internal stresses. (g) Complex adaptive systems are, unfortunately, prone to "revenge effects" [36] whereby attempts to make a positive change in the system produce negative results. Revenge effects can occur when the system automatically responds to the intrusions of the change agent to minimize its impact, or when the target of change is closely connected to other, less visible, functions of the system that are inadvertently changed. Revenge effects as a distinct class of iatrogenic phenomena are distinct from the well-known class of "side effects" that are associated with many medications; side effects originate with *properties of the intervention* whereas revenge effects originate from the *structure and complexity of the target system.*

Complexity science has the potential to inform a new area of empirically grounded and integrative theory regarding the structural and dynamical processes underlying resilience and health. Specifically, the material to follow aims: (a) to improve the understanding of the role of flexibility and integration versus rigidity and disintegration in etiology and treatment of conditions across the biological, psychological, and social spectra; (b) to outline methodologies for measuring and understanding these processes; and (c) to advance scientific models that are more relevant to the situations encountered by practitioners. In the broadest sense, the complexity-based models to follow should lead to the emergence of common themes to inform a healthy lifestyle, cutting across the somewhat arbitrary boundaries of physical, psychological, and social health. Furthermore, innovation within the complexity paradigm should lead to an increased focus in professional and public education upon the common dynamical and structural problems underlying each of the simple causes of illness and disease. We will have moved beyond the notion of a "balanced lifestyle" euphemism toward an empirically accessible definition of resilience and health.

50.3.4 Conceptual Innovations

Two conceptual innovations from complexity science may be particularly applicable to the future of behavioral medicine. The first conceptual objective was to create grounded NDS theory to understanding health, illness, resilience, and intervention in biopsychosocial dynamics, which was accomplished in two stages. In the first stage, Pincus [37] proposed that when considered from an NDS theoretical perspective the primary approaches to psychotherapy are equivalent in their aims and mechanisms of change. This equivalency may be extended from traditional therapy to behavioral medicine as well. Beginning at the level of interpersonal dynamics, and integrating family, group, and interpersonal approaches to therapy, various theories of healthy and unhealthy interaction patterns may be combined within an NDS grounded model involving three interactive parameters: control, closeness, and conflict (*3C*'s) [38, 39]. Essentially, flexible interactions are associated with relative balance between interpersonal control and closeness among individuals, while conflict is associated with short-term interpersonal rigidity in the service of longer term systemic adaptation to higher order flexibility. Individuals and relationships evolve through conflict resolution, which leads to renewed balance in closeness and control, which may be observed empirically through NDS indices reflecting more flexible and balanced social interaction patterns over time.

In a second stage, this model was extended to the level of individual therapy to integrate the various common factors of individual psychotherapy (e.g., empathy, therapeutic alliance, and emotional intensity) with the various empirically validated techniques (e.g., behavioral therapies, cognitive restructuring, mindfulness, and emotion-focused techniques [37]). Just as conjoint therapies involve a search for more flexible modes of exchanging information, individual psychotherapy is a search for key information as well, within the individual. Such bound up information may come in the form of connecting larger aspects of the self-system to emotion related to key developmental experiences (e.g., psychodynamic therapies), in the form of more flexible behavior patterns (i.e., behavior therapy), or from more open and flexible cognitive processes (i.e., cognitive therapies). The common theoretical concept concerns rigidity within the impacted biopsychosocial systems. Recent empirical work in experiential therapies has suggested, for example, that dysfunctional emotions such as shame and fear involve tight patterns of recurrence in emotional expression within therapy sessions, whereas their more functional counterparts of assertive anger or grief do not [40]. Recent behavioral therapy accounts have come to similar conclusions, suggesting that processes of experiential avoidance underlie nearly all psychological conditions, and that "psychological flexibility" is the hallmark of resilience in mental and behavioral

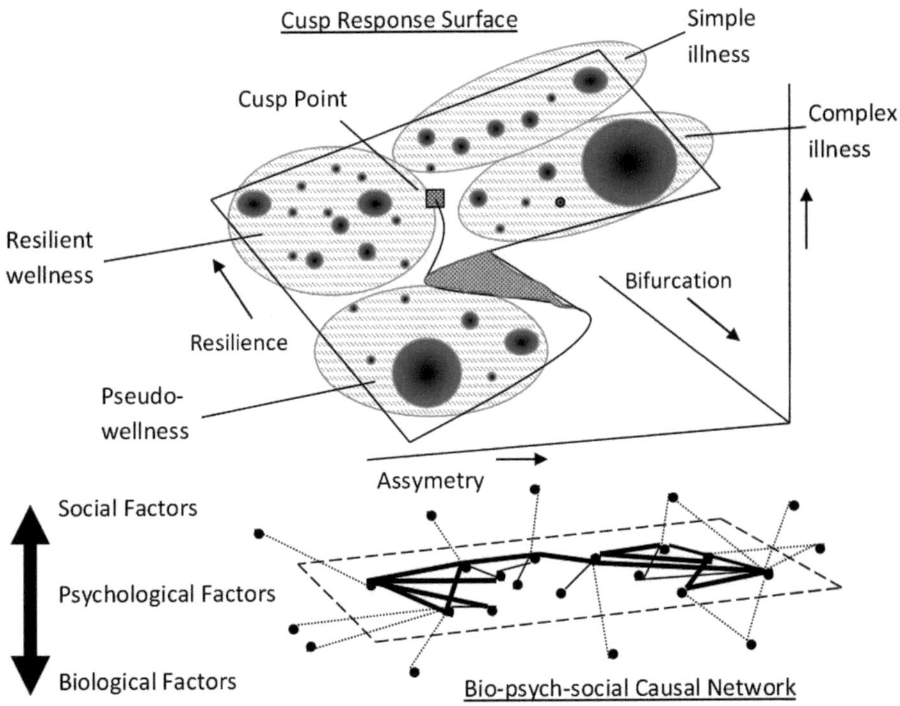

Fig. 50.1 Biopsychosocial causal network and resilience (reprinted with permission from Pincus and Metten [30])

health [41]. The time has arrived to ground these diverse theoretical notions of recursion, constriction, or flexibility in NDS concepts and to measure them using NDS methods [37].

Such theoretical and empirical grounding within NDS allows for the development of potentially more parsimonious models. For example, they may inform the search for patient-specific dynamical keys for unraveling constricted flows of information allowing for the emergence of more open and flexible habits. A dynamical key is a central point in a network of elements that comprise a complex system. It resembles the keystone of an arch; remove it and the arch falls [42]. Furthermore, successful therapy has been shown to be moderated by decreasing entropy within the therapeutic alliance, creating a more coherent environment in which novel information may then be introduced [43]. The identification of dynamical keys and coherence in treatment planning each tend to fall outside of the usual evidence-based therapy knowledge base. Yet each is central to the ubiquitous clinical task of case conceptualization and treatment planning [44] (Fig. 50.1).

The second stage of conceptual innovation extends these ideas of interpersonal and psychological complexity to the NDS conceptualization and measurement of resilience in general medicine [30]. The concept originated with studies of physical and cognitive workloads, and is analogous to the buckling of an elastic beam [45–48]. Imagine a beam made of stiff material that is pin-jointed at both ends, and loads of increasing weight is placed on the beam. The beam will show little response until the load reaches a critical point and the beam buckles and snaps. If the beam were made of more elastic material, it would waffle as the load increased, but would not snap. The discontinuities and elasticity effects can be modeled as a cusp catastrophe (see the cusp response surface located toward the top of Fig. 50.1).

An amplified model [30] (Fig. 50.1) depicts simple illness, complex illness, resilient wellness,

Table 50.1 Categories of illness and wellness predicted by a cusp catastrophe response surface with pathogen exposure as the normal parameter and structural resilience as bifurcation parameter

RESILIENCE	PATHOGEN EXPOSURE: Low	PATHOGEN EXPOSURE: High
High	**Resilient Wellness** Healthy and flexible across biopsychosocial systems with optimal structural connectivity and information flows across systems	**Resilient Illness** In a diseased state while maintaining flexibility across biopsychosocial systems and system connectivity. A good candidate for allopathic intervention alone.
Low	**Pseudo-Wellness** Apparently healthy, but at risk for chronic disease. Significant disintegration across biopsychosocial systems.	**Complex Illness** Disease state is nested within a context of significant disintegration across biopsychosocial systems. Allopathic treatment alone likely to produce temporary or paradoxical effects.

and pseudo-wellness as locations on the response surface of a cusp catastrophe model. Each of these four regions reflects the combination of resilience and pathogen exposure (i.e., combined impact of exogenous disease etiologies such as viruses) as latent parameters (see Table 50.1). These regions are locally stable, with sudden flips in and out of illness as one approaches the critical transition point.

The cusp catastrophe response surface in Fig. 50.1 shows a range of behaviors that are organized around two stable states and a separation between them. A multi-stable attractor diagram (the circular areas in Fig. 50.1) is superimposed on the cusp manifold. This attractor diagram supplements is consistent with the rubberized nature of the cusp model, and it also represents the rigidity or stickiness that is typically observed in the time series of relevant variables when an individual is in a pseudo-well (i.e., at risk for illness and nonresilient) or complex illness (i.e., sick and nonresilient) condition. The dark region on the surface signifies the location where points (representing people and their conditions) are unlikely to fall.

The plane below the surface is known as the bifurcation set, which is a two-dimensional rendering of the manifold of the cusp surface and the positioning of critical points where the behavior of the system changes suddenly. Here it is used to show figurative networks of biological, psychological, and social factors that underlie resilience. The most unstable point on the surface is where the individual does not have a stable network of variables in place, where a structural "disintegration" has occurred. This area is located just below the cusp point and may be empirically verified through nonlinear methods ranging from time series analysis on a single embedded systemic variable to a network analysis of the range of relevant biopsychosocial variables.

This amplified model may be understood as a far more specific account of diathesis-stress processes than one can find in the literature on biopsychosocial models [14, 34]. Stress would constitute the various pathogens that contribute to the *asymmetry* parameter, while diatheses may be modeled as the *bifurcation* parameter, the multi-stable topology of the cusp response surface, or the underlying structural features of the biopsychosocial network—depending upon one's modeling strategy. Essentially, stressors to the system are primarily exogenous; their relationship to health outcomes is linear when diatheses are low, but there is a sudden breakpoint relationship with health outcomes when diatheses are high. Diatheses are essentially structural and primarily responsible for the elasticity that characterizes resilience or its opposite, susceptibility. Diatheses act as trigger variables; the pole of the variable that denotes elasticity contributes to the high bifurcation side of the response surface where the discontinuous changes occur.

Chronicity occurs when both the asymmetry and bifurcation variables are high, such that the individual is locked into the attractors on the response surface for either health or illness. Cusp models also show *hysteresis*, which is a vacillation between the two states. Hysteresis would be observed as an irregular pattern of illness, recovery, and relapse.

In clinical settings it is often the case that the patient was not resilient enough to prevent the need for treatment. As such, treatment aimed at recovery should be aided by concerted attempts at reconnection among critical disintegrations within the patient's biopsychosocial networks. One can also make a case that too much elasticity is not good either. Not many people would feel comfortable driving a car over a rubber bridge. A closer look at the cusp model indicates that resilient wellness, which is desirable, is *not* a stable state. Rather, movement is relatively free between resilient illness and wellness. Thus, an intervention that has a long-lasting effect for resilient patients would explore stiffening processes among biopsychosocial nodes to prevent them from easily slipping back to illness.

Clinically, the measurable goal of treatment for those with complex conditions is for a patient to be able to move between illness and health in a more flexible manner, as is the case along the top edge of the response surface in Fig. 50.1. Once in the area of resilient wellness, increasing the number and variety of attractors (e.g., flexible habits) in this multi-stable region will likely reflect effective relapse prevention (from either of the illness regions).

By contrast, attempts at direct cure for complex illnesses would be expected to backfire for many individuals (e.g., overuse of medications or pathological psychosocial patterns of avoidance). Newer behavioral approaches to therapy have incorporated this long-held wisdom involving pathological avoidance, for example, using the phrase: *if you can't have it, you've got it* [21]. This phrase captures well the often-paradoxical effects of preoccupation with optimal health [41]. Rather, true resilience likely involves a balance between flexibility and integration in the connective fabric among biopsychosocial processes [30].

50.3.5 Methodological Innovations

Three NDS methodologies fit well with the conceptual innovations just described. The first involves a symbolic dynamics technique known as orbital decomposition (OD; cf. 49, 50), to extract patterns in behavioral habits over time, and characterize the level of complexity of those patterns. The second involves the measurement of self-complexity based upon the structural properties underlying reaction time distributions to items on the MMPI-2 personality inventory. The third involves the use of structural equations to assess the structure of the cusp response surface of an individual patient or specific patient subgroups. These three techniques represent conceptual and measurement breakthroughs, which when combined, provide an ideal application of complexity science to the future of behavioral informatics.

50.3.5.1 Symbolic Dynamics

Orbital decomposition is a symbolic dynamics analysis for identifying patterns of events that contribute to a complex series of nominal system states. The methods rely upon the principle that chaotic series result from coupled oscillators, and thus the algorithm searches for the contributing oscillators, although chaos is not necessary nor is it assumed. Based on the mathematical approach of Lathrop and Kostelich [49], this method may be used to quantify the complexity in categorical time series and is the first of its kind in the social or behavioral sciences. Once the algorithm has identified the optimal pattern length, it produces a measurement of topological entropy that can be converted to a largest Lyapunov exponent and then into a fractal dimension. Guastello et al. [50]. introduced some statistical modifications that allowed for comparisons of the final sequences to chance levels of expectation and comparisons with Shannon's entropy statistic, which is popular as a complexity metric in other types of research.

The central equation is:

$$H_T = \lim_{C \to \infty} (1/C) \log_2 \left(\mathrm{tr} M^c \right) \qquad (50.1)$$

where topological entropy (H_T) is the limit as string length, C, goes to infinity of $1/C$ times the base-2 log of the trace of the matrix $\mathbf{M}^c$. $\mathbf{M}^c$ is a matrix showing whether any string of length C is followed by any other string of length C; the trace targets strings that immediately follow themselves.

The earliest applications [50, 51] involved conversations within creative problem-solving groups. Each utterance is coded for the presence of elements from a predetermined scheme that has some theoretical relevance to the researcher. Measures of complexity do depend on the coding scheme that one applies to the data [51]. The procedure is flexible with regard to the number and configuration of simultaneous sets of state variables for which one wants to code. Uses of the technique then spread to clinical applications such as family dynamics [52], group therapy [38], experimental tests of interpersonal conflict [39], sociopolitical violence [53], and family violence [54]. In each case, the patterns that were isolated in the analysis were valuable for interpreting the dynamics of the situation under study.

Furthermore, in studies involving conversations patterned recurrences have been found to be distributed as power laws, with fractal dimensions between 1.0 and 2.0, supporting the dynamical and self-organizing nature of the interpersonal systems under study. A related finding has been the reflection of interpersonal closeness, control, and conflict within structured conversation patterns, accounting for 48% of the variance in those patterns [38], and the association between rigidity and experimentally induced internal conflict within group members [39]. Together, these results suggest that conflict, in the form of rigidity, may spread from individual to group levels and back, and may be observed through a reduction in the fractal dimension of group dynamics. Further methodologies have been developed for identifying critical moments during sessions where changes to conversation structure occur, as well as models testing individual to group contributions to structure or fractal complexity within conversations [55]. Such procedures may be adapted easily to identify key behaviors to target for transformational change, fragile points within behavioral sequences, and to track the structural outcomes of various interventions. Simply, the analytic strategy is ideal for informing the types of small strategic habit changes that could be tracked through interactive relationships with handheld devices.

Orbital decomposition is simple enough to carry out by hand, with a spreadsheet analysis to calculate frequencies of various patterns and a bit of patience. Software is now available, however, that will facilitate studies where numerous conversations may be analyzed and compared [56].

50.3.5.2 Self-Complexity

The second methodology addresses an issue that is as much psychological as it is computational. The psychological concern is the development of a concept of "complexity of the self," which carries notions of rigidity and flexibility. The most common approach [57, 58] has used Q-sort experiments and summaries of participants' self-complexity using a measure equivalent to Shannon entropy. Nearly two decades later, the empirical results became increasingly unclear

[59], with numerous criticisms aimed at the measurement approach [60]. Grounding the concept of self-complexity in "complexity" as it is understood within NDS, and using methods grounded in this theoretical perspective, Pincus et al. [61]. recently discovered that a large amount of self-complexity information is inherent in the response-times to MMPI-2 items. In a forensic sample of 100 men, each person's set of response times is distributed as a power law. The shape parameter of the distribution denotes the complexity of the distribution and a fractal dimension. The shape parameter was significantly correlated with 27 out of 35 clinically relevant MMPI-2 scales, all in the predicted direction. Low self-complexity (rigidity) appears to be associated with broadband psychopathology, across somatic, psychodiagnostic, and social categories of functioning. The newest of the three NDS methods outlined here, the response time method also appears to have great potential as a measure of structural resilience and flexibility in health.

50.3.5.3 Cusp Catastrophe Model for Therapy Evaluation

A third NDS approach to assessment and process research in DT2 and other complex conditions involves the use of catastrophe theory [62], which is a general systems theory for describing and predicting discontinuous changes of events. The cusp model (Fig. 50.1), which is one of the simpler and yet very useful models in the series, consists of two stable states and a bifurcation manifold. Movement along the response surface is governed by two control parameters:

$$z_2 - z_1 = \beta_0 + \beta_1 z_1^3 + \beta_2 z_1^2 + \beta_3 b z_1 + \beta_4 a \quad (50.2)$$

where z_i are measures of the dependent measure at two points in time. Asymmetry variables (a) bring the system closer to the point where a sudden change occurs. Bifurcation (b) variables govern the degree of change and, if strong enough, divide the system between the two stable states. Bifurcation variables are also responsible for *hysteresis*—movements up and down the surface between stable states. β_i are regression weights.

The final analysis for the type of research outlined here involves testing such a model for change in health states. Importantly, one of the first applications of catastrophe theory was for training, therapy, and program evaluation [42, 45, 63, 64]. Not only is one capable of identifying change in the desired direction, but the model will also identify conditions where some people change dramatically and from one stable state to another, while others are less stable or do not change much at all, a much finer parsing of outcomes than is capable using a simple clinical trial design. The cusp is also ideal for modeling relapse patterns. By extending the model to time beyond the end of therapy it will be possible to detect which patients experienced stable outcomes (e.g., resilient wellness) and which ones relapsed (e.g., pseudo-wellness). The cusp effect would also be stronger on longer term follow-up than a linear model based on the same variables. Furthermore, the model also captures recidivism, where patients' progress is only temporary; the backsliding process during treatment, or toggling back and forth between wellness and illness, a manifestation of hysteresis across the two primary thresholds of the cusp response surface (see Fig. 50.1).

Dependent measures for the model would be relatively standard therapeutic outcomes—changes in symptoms and overall functioning as measured by levels of blood sugar, health-related habits, psychiatric symptoms, or personality inventories. Asymmetry variables would correspond to the amount of load on the individual—conflict levels and the severity of external events that instigated treatment seeking. Bifurcation variables would correspond to indicators of rigidity in behavioral habits, such as those identified using OD described previously, as well as the complexity of the self derived from the MMPI-2 response time analysis. For some empirical analyses of cusp models using two different computational procedures, see the chapter by Guastello in this *Handbook*.

The conditions that are identified in OD that are precursors to structural changes in social dynamics would be included as potential bifurcation variables also; the cusp model strongly suggests that the suddenness of such changes results from a bifurcation or triggering effect when the background conditions are favorable (e.g., emotionally corrective experiences in psychodynamic therapies or successful exposure experiences in behavioral therapies). Other themes drawn from qualitative patterns identified through OD could function as asymmetry variables that set the background conditions for illness.

50.4 A Possible Future for Behavioral Medicine

Using the broadest theoretical lens, behavioral medicine views the technology of modifying health habits through the *transtheoretical model* of behavior change [65]. The transtheoretical model, also known colloquially as the *stages of change* model, is perhaps the most widely used theoretical model in high quality health care settings charged with long-term management of diabetes and other behavioral health conditions (e.g., Kaiser Permanente). The transtheoretical model proposes that there are five stages of change that operate in a recursive spiral over time, from pre-change processes to post-change maintenance. This model is another clear example of a complexity model that has not yet been identified as such.

The earliest stage of change is known as *pre-contemplation* and is characterized by minimal levels of insight and little or no motivation for change. The three stages that follow in order are *contemplation*, *preparation*, and *action*, each of which is characterized by increasing levels of insight, motivation, and change related behaviors. The final stage is *maintenance*, where the focus of intervention turns to relapse prevention strategies as well as the consolidation of patient agency within lifestyle processes that maintain and improve long-term biopsychosocial resilience. A key process in this regard is preparation for *bouncing back* from minor *slip-ups*, which are to be expected in complex behavioral health conditions.

One related treatment concern is the tailoring of interventions to fit a particular patient's stage of change. Specifically, patients who arrive at intervention at an earlier stage of change (i.e., from pre-contemplation to early preparation) show greater benefit from insight and motivation enhancement strategies. In the case of complex conditions, these individuals may be considered to be engaging in a process of biopsychosocial repair despite the fact that they are still well within the illness region of the cusp surface. Those at later stages of change (i.e., late preparation to action stages) are better able to make use of active behavior modification procedures (e.g., the use of contingent reinforcements, goal setting, sensitization, and desensitization procedures). These individuals may be considered to be further along in repair and moving toward resilient wellness. Consistent with the model of resilience presented previously, different patients who are at different levels of change will experience varying degrees of nonlinearity and discontinuity in their progression toward maintenance as well as into and out of relapse.

Finally, the upward spiraling and cyclical aspects of the model owe to the idea that behavior modification is an ongoing process akin to spreading activation, positive feedback, or even perhaps to dialectical models. Specifically, once a new behavioral habit has entered maintenance, opportunities for new targets of intervention emerge. Depending upon those targets, a new cycle may ensue, beginning again at pre-contemplation and moving forward in turn to a new plateau involving maintenance. For example, a DT2 patient, once successful at maintaining healthy levels of blood sugar over a matter of days through changes to nutritional intake and physical activity may begin anew in a cycle of change toward even better body mass index (BMI) balance, or related to some quasi-tangential goal pertaining to athletic or interpersonal achievement. Counter to intervention based on materialism, reductionism, and efficient cause (i.e., medicine), such a cyclical process of healing is more consistent with the systemic model of biopsychosocial resilience outlined in Fig. 50.1.

The transtheoretical model can be grounded in a deeper and less descriptive manner when viewed through the lens of Complex Adaptive Systems

(CAS) theory. A CAS is a system that is able to tune its own parameters (via self-organization) in response to some threat to systemic integrity. In general terms, these threats may be considered to be some form of perturbation in the flows of information within the system or between the open system and its nesting environment(s). The response of the CAS to perturbation will typically be a shift toward relative coherence. The resulting decrease in flexibility and creativity of the system is offset temporarily by an adaptive increase in resilience. The human stress response via the hypothalamic-pituitary-adrenal axis (HPA) is an ideal example of such a mechanism in biopsychosocial dynamics.[1] When viewed structurally, the stress response quickly moves biological, psychological, and social processes in the direction of increased coherence, which becomes dysfunctional and rigid only when it is sustained for too long in time [30].

When viewed within the context of CAS theory, the transtheoretical model is simply a description of the relative openness of a biopsychosocial system to novel information flows. When a person is in the contemplation or pre-contemplation stages, he or she is probably experiencing a prolonged stress reaction that is in turn probably related to the illness. Movement toward the action stage would be predicated upon increasing the receptivity of the intrinsic biopsychosocial dynamics of the individual to new information. Paradoxically, a precondition for promoting this receptivity is likely to be an increase in the structure and consistency in the immediate treatment environment (e.g., a strong therapeutic alliance), and also greater coupling within the self-systems of the individual (e.g., self-awareness and acceptance). Such increases in coherence may allow for a reduction in perturbation to information flows within the individual and also less reactivity in the HPA. As such, the CAS experiences the necessary conditions to begin to open up to novel flows of information that can potentially increase the healthy complexity of the individual's relationships, behavioral habits, cognition, and emotional experiences.

[1]For more detail on this aspect, see Bennett et al. Chap. 19.

Fitting a behavioral intervention to the specifics of that individual's stage of change, and the dynamics of his or her impacted CAS (regions involving rigidity and disintegration in Fig. 50.1) is a particular challenge for the future of behavioral medicine science and intervention. Even when one considers the relatively basic processes of operant conditioning, it is critical to avoid reinforcing existing behaviors or new behaviors once established due to high probability of eroding intrinsic motivation and other naturally occurring reinforcements which have successfully emerged and self-organized. Indeed, it is ideal to use consistent schedules of reinforcement during early stages of behavioral acquisition only, fading such reinforcers as quickly as possible to variable schedules and then to contextual factors alone.

The use of CAS theory and related concepts may assist in improving client fit for both widely disseminated electronic media delivered interventions and for research models investigating those interventions. Because the level of explanation for change in a CAS is structural and formal in nature (i.e., pertaining to patterns of structural relations among systemic components and information exchange among those components), the impetus is not on this or that factor (e.g., emotional regulation versus increased exercise), or such and such intervention (e.g., cognitive versus behavioral intervention). Rather, patients can be directed to assess for areas of particular restriction from previous levels of openness, and specific interventions may be selected from a menu of possibilities depending upon relative fit to the individual's structural and dynamical configurations.

For example, if a patient were emotionally blunted, emotional expression and regulation would be the ideal target. For a patient who is primarily avoiding physical exertion, diversity of aerobic activity would be the target. For another patient, diversity or hedonic balance (e.g., diversity in sources of pleasure) related to food intake might be the key. Or, the rigidity could lie within the interpersonal scale, such as the relative shutdown in his or her relationship about the specific diagnosis. Similarly, scientific inquiry may focus on the ability of an intervention to provide rehabilitative improvement to a client's "range of motion" in any of these content-specific domains, rather than addressing simpler questions from RCTs about treatment categories and the variety of clients to which they might pertain.

In addition to the common theoretical umbrella of the transtheoretical model, complexity theory can assist in the integration and enhancement of other modern theories of motivation and habit change including self-efficacy theory [66–68], the theory of planned behavior [69], and newer contextual approaches to behavior therapy [41]. A full integration of these approaches is beyond the scope of the current document [37]. Nevertheless, when interpreted from the perspective of complexity science, these approaches have in common a few key principles: (1) motivation follows behavior in most circumstances, not vice versa; (2) openness and flexibility to self-relevant experience allows for greater flexibility, greater intentionality, and greater adaptation; (3) increases to insight and motivation promote momentum effects, with trajectories toward self-repair and improved resilience; and (4) problems that lend themselves to the theory of planned behavior can be studied more effectively with cusp catastrophe models [70, 71].

This first notion, that motivation and insight follows behavior may be traced back to the earliest modern theories of motivation and emotion (e.g., Canon-Bard and James-Lang theories from the early twentieth century). Complexity science enhances such processes again by leading with, and making central, a patient's use of small self-directed substitutions within his or her habits of daily living. For example, the future of behavioral informatics will likely involve smart phone applications equipped with nonlinear time series analysis algorithms (e.g., orbital decomposition combined with catastrophe analysis) that can identify critical links in an individual's habit chains. Once identified, such singular points may be targeted for key behavioral changes to yield maximum changes to parameters of interest (e.g., introduction of a novel eating pattern that will improve blood glucose levels in a DT2 patient), and yet to not interfere with a patient's normal flows of activity, motivation, sense of agency, or healthy identity.

The second set of processes, which have become central in modern behavior therapies [72] are mindful awareness and self-acceptance, which are thought to lead to greater psychological and behavioral flexibility. Mindfulness is originally a concept from eastern spiritual traditions, associated with an open and nonjudgmental state of consciousness which can be expanded through practice in meditation [34]. Such a metaphysical approach to mindfulness aimed at enlightenment may require a lifetime or even multiple lifetimes to achieve when viewed within the context of eastern traditions. The pragmatism of American behaviorism has already come a long way toward inventing a form of *strategic mindfulness*, and the use of complexity theory in mobile health can carry this effort even further.

One example of this principle of applied mindfulness will involve the use of smart phone technology informed by complexity models to activate key processes of self-repair and behavioral flexibility for DT2 patients through the seamless shifting of patient attention to ongoing habit changes and resulting improvements in his or her meal-to-meal, day-to-day, and week-to-week self and life satisfaction. The heart of mindfulness is the intersection of intention with attention. Simply, when one purposefully focuses attention on a situation, one is better able to counteract automatic action tendencies and to behave in more self-directed ways. Furthermore, when patients: (a) feel supported, (b) notice the positive changes they have made, and (c) experience a sense of ownership over their health decisions they will be most likely to maintain long-lasting and resilient habit changes. Behavioral informatics via smart phone applications can supply such recursive information, for example, through the use of multimedia professional content via smartphone screens, real-time feedback from selected sections of the patient's social network, or the use of personalized avatars. Finally, the future of behavioral medicine may use behavioral informatics to increase momentum effects and to monitor health trajectories by presenting summary statistics in user-friendly formats, such as life-expectancy projections based on choices made in the past day, week, or month, or resilience diagrams using attractor diagrams, catastrophe manifolds, or network topologies (i.e., Fig. 50.1). Vital read-outs like these, akin to the dashboard of a car, can serve to augment limitations in the cognitive functions of human beings which adapt slowly; for example limited cognitive capacity to process nutritional information quickly in the face of the calorie-rich environments of modern life. In addition, such summary information could be designed to be helpful to health care providers, third party payers, or policy makers to provide better fits between these functions and the needs of individual patients.

The use of nonlinear modeling procedures underlying such advances in behavioral informatics could involve, for example, a categorical time series method like OD could identify a highly repetitive and predictable pattern pertaining to binging on carbs, with a sequence involving: Friday evening spousal conflict—shame and withdrawal—weekend binging. Once identified as highly recurrent and predictable, a smart phone application could suggest various interventions and track their impact on increasing the flexibility of this pattern over time. For example, the phone could "push" the patient to be aware of possible conflict in the early afternoon on Friday and suggest a number of assertive strategies toward the aim of conflict resolution; the patient could be pushed to identify and work through the shame and counteract the withdrawal tendency; or the patient could be directed to select from a list of alternate pleasurable behaviors in which to engage as a substitution for the carb binge.

Meanwhile, the "brain" of the application could model results of these various pushes on the catastrophe or other multi-stable attractor surfaces to identify critical points (i.e., cusp points) or thresholds that would need to be reached in order to obtain momentum effects over time. For example, these models may identify the critical threshold(s) in marital satisfaction that will lead to changing momentum toward or away from binge eating.

It is within the realm of near-term development that smart phones could be equipped with highly engaging and potentially empathetic avatars selected and designed by individual consumers. Such an application would be well situated to collect high levels of rich and reliable data pertaining to nutrition, activity, mood, interpersonal relations, heart rate and blood pressure, and myriad other biopsychosocial variables nested in real time. Complexity science in general, and NDS theory in particular is ideally situated to make sense of such data, particularly on a patient-by-patient bases, and within the context of implementing strategic "pushes" of information back to the consumer, increasing both strategic mindfulness along with behavior change.

For example, such a smart avatar-driven application would potentially be in a position to intervene at critical moments within a patients minute-to-minute, hour-to-hour, or day-to-day patterns of living. For instance, one's avatar could warn its patient of a high potential for a migraine, eating binge, or mood episode. If the avatar had learned that notifying a particular loved one, rather than the patient directly, was likely to be a more efficient intervention, then this would be possible as well simply by allowing for a network of avatars to emerge by way of existing social network technologies.

Likewise, the avatar could provide some more active form of intervention. For example, cognitive bias management (CBM, 74–75) appears to be a hallmark example of the future of mobile medicine and the intervention branch that will emerge from behavioral informatics. CBM aims to level out cognitive biases that likely exist across the complex conditions, by presenting patients with competing images and having them actively engage with neutral images while ignoring biasing images as each is presented side by side on a computer or mobile touch screen. For example, one CBM intervention for alcoholics directed patients to attend to and engage in tasks via joystick pertaining to neutral images, which were paired with images related to alcohol (e.g., a glass of whiskey). This simple intervention appeared to significantly decrease the addictive behavioral tendencies of alcoholics. Furthermore, four simple 15-min interventions of CBM prior to traditional inpatient treatment led to an additional 13% increase in sobriety at 1 year follow-up over inpatient treatment alone [73]. Comparable effectiveness has been measured in pilot interventions of CBM for anxiety as well [72]. These early results are encouraging, particularly when one is focused on efficiency and dissemination of intervention for complex conditions. Positive results in such different learning contexts of alcoholism and anxiety suggest broad applicability of this simple automated intervention, including food choices, mood disorders, and the various impulse control conditions.

The expected downside of all these new technological directions is as large as the expected upside, however. Perhaps critics of technological advance are partially correct: Our reliance on technology may increase our dependence and undermine our ability to manage our own survival [74]. Professional psychology has traditionally tried to improve self-reliance, self-confidence, self-regulation, and self-efficacy. Studies are already showing that supportive family members can become a means for outsourcing self-regulation activities [75–77]. And do we really want health insurers and the government telling us what to eat or when to take a laxative? With these caveats in mind, it will be impendent upon the designers of such technology to ensure that patients remain in charge of their own interventions, for example designing intervention programs to be as customizable as possible.

Similarly, to avoid the inherent risks of the standard approaches to data analysis like patient uniqueness, information overload, "garbage-in-garbage out" conclusions, complexity models and methods can play an important role in deriving the algorithms that will drive these coming interventions. From the models of empathy that will drive the questions asked by smart-phone avatars, to the timing and specificity of interventions using CBM, nonlinear models and methods have the unique potential to assess the structural aspects of an individual's current resilience, both in time and also across the quasi-spatial dimensions underlying biopsychosocial resilience and wellness.

References

1. Von Bertalanffy L. General system theory: Foundations, development, applications. New York: Braziller; 1968.
2. Sturmberg JP, Martin CM. The dynamics of health care reform: learning from a complex adaptive systems theoretical perspective. Nonlinear Dynamics Psychol Life Sci. 2010;14:525–40.
3. Lloyd-Jones D, Adams RJ, Brown TM, Carnethon M, Dai S, De Simone G et al. On behalf of the American Heart Association Statistics Committee and Stroke Statistics Subcommittee. Heart disease and stroke statistics—2010 update: a report from the American Heart Association. Circulation: AHA; 2010.169 p. doi: 10.1161/CIRCULATIONAHA.109.192667.
4. Toth C, Lander J, Wiebe S. The prevalence and impact of chronic pain with neuropathic pain symptoms in the general population. Pain Med. 2009;10:918–29.
5. Stewart WF, Ricci JA, Chee E, Morganstein D, Lipton R. Lost productive time and cost due to common pain conditions in the US workforce. J Am Med Assoc. 2003;290:2443–54.
6. Kessler RC, Chiu WT, Demier O, Walters EE. Prevalence, severity, and comorbidity of twelve-month disorders in the National Comorbidity Survey Replication (NCS-R). Arch Gen Psychiatry. 2005;62: 617–27.
7. Wang PS, Simon G, Kessler RC. The economic burden of depression and the cost-effectiveness of treatment. Int J Methods Psychiatr Res. 2003;12:22–33.
8. Greenberg PE, Sisitsky T, Kessler RC, Finkelstein SN, Berndt ER, Davidson JR, et al. The economic burden of anxiety disorders in the 1990s. J Clin Psychiatry. 1999;60:427–35.
9. Fitch K, Iwasaki K, Pyenson B. Improved Management Can Help Reduce the Economic Burden of Type 2 Diabetes: A 20-Year Actuarial Projection. Milliman, Inc; 2010 April 28. 25 p. Retrieved from http://publications.milliman.com/publications/health-published/pdfs/improved-management-can-help.pdf.
10. Hedley AA, Ogden CL, Johnson CL, Carroll MD, Curtin LR, Flegal KM. Prevalence of overweight and obesity among US children, adolescents, and adults, 1999–2002. J Am Med Assoc. 2004;291:2847–50.
11. American Diabetes Association. Type 2 diabetes in children and adolescents. Pediatrics. 2000;105(3 Pt 1):671–80.
12. Toumilehto J, et al. Prevention of type 2 diabetes mellitus by changes in lifestyle among subjects with impaired glucose tolerance. N Engl J Med. 2001;344: 1343–50.
13. Diabetes Prevention Program Research Group. Reduction in the incidence of type 2 diabetes with lifestyle intervention or metformin. N Engl J Med. 2002;346:393–403.
14. Gatchel RJ. Comorbidity of chronic pain and mental health disorders: The biopsychosocial perspective. Am Psychol. 2004;59(8):795–805.

15. Rodgers JL. The epistemology of mathematical and statistical modeling: A quiet methodological revolution. Am Psychol. 2010;61(1):1–12.
16. McHugh RK, Barlow DH. The dissemination and implementation of evidence-based psychological treatments: a review of current efforts. Am Psychol. 2010;65:73–84.
17. Lambert MJ. Implications of outcome research for psychotherapy integration. In: Norcross JC, Goldstein MR, editors. Handbook of psychotherapy integration. New York, NY: Basic Books; 1991.
18. Carryer JR, Greenberg LS. Optimal levels of emotional arousal in experiential therapy of depression. J Consult Clin Psychol. 2010;78:190–9.
19. Wampold BE, Mondin GW, Moody M, Stich F, Benson K, Ahn H. A meta-analysis of outcome studies comparing bona fide psychotherapies: empirically, all must have prizes. Psychol Bull. 1997;122(3): 203–15.
20. Wampold BE. The great psychotherapy debate: models, methods, and findings. New Jersey: Lawrence Erlbaum Associates Publishers; 2001.
21. Frank JD, Frank JB. Persuasion & healing: a comparative study of psychotherapy. 3rd ed. Baltimore, MD: Johns Hopkins University Press; 1991.
22. Popper KR. The logic of scientific discovery. 2nd ed. New York: Routledge; 2002.
23. Kaptchuk TJ, Friedlander E, Kelley JM, Sanchez MN, Kokkotou E, Singer JP, Kowalczykowski M, Miller FG, Kirsch I, Lembo AJ. Placebos without deception: A randomized controlled trial in irritable bowel syndrome. PLoS One. 2010;5(12):1–7.
24. Woods DD, Patterson ES. How unexpected events produce an escalation of cognitive and coordinative demands. In: Hancock PA, Desmond PA, editors. Stress, workload, and fatigue. Mahwah, NJ: Lawrence Erlbaum Associates; 2001.
25. Barlow DH. Negative effects from psychological treatments: a perspective. Am Psychol. 2010;65:13–20.
26. Dimidjian S, Hollon SD. How would we know if psychotherapy were harmful? Am Psychol. 2010;65: 21–33.
27. http://www.behavioralinformatics.org/
28. Riley WT, Rivera DE, Atienza AA, Nilsen W, Allison SM, Memelstein R. Health behavior models in the age of mobile interventions: are our theories up to the task? Translational Behav Med. 2011;1:53–71.
29. American Diabetes Association. Standards of medical care in diabetes—2010. Diabetes Care. 2010;33(supplement 1):1–61.
30. Pincus D, Metten AM. Nonlinear dynamics in biopsychosocial resilience. Nonlinear Dynamics Psychol Life Sci. 2010;14:353–80.
31. Hokfelt T. Neuropeptides in perspective: the last ten years. Neuron. 1991;7:867–79.
32. Hokfelt T, Broberger C, Xu ZQ, Sergeyev V, Ubink R, Diez M. Neuropeptides: an overview. Neuropharmacology. 2000;39:1337–56.
33. Cohen, B. Telecoms pushing for mobile medicine payoff: The future of healthcare could bring a big payoff. May 2, 2011, On-line: http://www.investorplace.com/39330/telecoms-pushing-for-mobile-medicine-payoff/.
34. Pincus D, Sheikh AA. Imagery for pain relief: a scientifically grounded guidebook for clinicians. New York: Routledge; 2009.
35. Guastello SJ, Liebovitch LS. Introduction to nonlinear dynamics and complexity. In: Guastello SJ, Koopmans M, Pincus D, editors. Chaos and complexity in psychology: The theory of nonlinear dynamical systems. Cambridge, MA: Cambridge University Press; 2009.
36. Tenner E. Why things bite back: Technology and the revenge of unintended consequences. New York: Knopf; 1996.
37. Pincus D. Coherence, complexity, and information flow: self-organizing processes in psychotherapy. In: Guastello SJ, Koopmans M, Pincus D, editors. Chaos and complexity in psychology: Theory of nonlinear dynamical systems. Cambridge MA: Cambridge University Press; 2009.
38. Pincus D, Guastello SJ. Nonlinear dynamics and interpersonal correlates of verbal turn-taking patterns in group therapy. Small Gr Res. 2005;36:635–77.
39. Pincus D, Fox KM, Perez KA, Turner JS, McGee AR. Nonlinear dynamics of individual and interpersonal conflict in an experimental group. Small Gr Res. 2008;39:150–78.
40. Pasqual-Leone A, Greenberg LS. Emotional processing in experiential therapy: Why 'the only way out is through. J Consult Clin Psychol. 2007;75(6):875–87.
41. Hayes SC, Strosahl KD, Wilson KG. Acceptance and commitment therapy: an experiential approach to behavior change. New York: Guilford Press; 1999.
42. Guastello SJ. Managing emergent phenomena: nonlinear dynamics in work organizations. Mahwah, NJ: Lawrence Erlbaum Associates; 2002.
43. Tschacher W, Scheier C, Grawe K. Order and pattern formation in psychotherapy. Nonlinear Dynamics Psychol Life Sci. 1998;2:195–216.
44. Eells TD. Handbook of psychotherapy case formulation. 2nd ed. New York: Guilford; 2007.
45. Guastello SJ. Chaos, catastrophe, and human affairs: applications of nonlinear dynamics to work, organizations, and social evolution. Mahwah, NJ: Lawrence Erlbaum; 1995.
46. Guastello SJ. Nonlinear dynamics, complex systems, and occupational accidents. Int J Hum Factor Man. 2003;13:293–304.
47. Guastello SJ, Boeh H, Shumaker C, Schimmels M. Catastrophe models for cognitive workload and fatigue. Theor Issues Ergon Sci, 2011, 12, "ahead of print".
48. Thompson HL. The stress effect: why smart leaders make dumb decisions—and what to do about it. San Francisco: Jossey-Bass; 2010.
49. Lathrop DP, Kostelich EJ. Characterization of an experimental strange attractor by periodic orbits. Phys Rev. 1989;40:4028–31.
50. Guastello SJ, Hyde T, Odak M. Symbolic dynamic patterns of verbal exchange in a creative problem

solving group. Nonlinear Dynamics Psychol Life Sci. 1998;2:35–58.
51. Guastello SJ. Symbolic dynamic patterns of written exchange: Hierarchical structures in an electronic problem solving group. Nonlinear Dynamics Psychol Life Sci. 2000;4:169–88.
52. Pincus D. A framework and methodology for the study of non-linear, self-organizing family dynamics. Nonlinear Dynamics Psychol Life Sci. 2001;5:139–74.
53. Spohn M. Violent societies: an application of orbital decomposition to the problem of human violence. Nonlinear Dynamics Psychol Life Sci. 2008;12:87–115.
54. Katerndahl D, Ferrer R, Burge S, Becho J, Wood R. Recurrent patterns of daily intimate partner violence and environment. Nonlinear Dynamics Psychol Life Sci. 2010;14:511–24.
55. Pincus D, Ortega DL, Metten AM. Orbital decomposition for multiple time series comparisons. In: Guastello SJ, Gregson RAM, editors. Nonlinear dynamical systems analysis for the behavioral sciences using real data. Boca Raton, FL: CRC Press/ Taylor & Francis; 2010.
56. Peressini AF, Guastello SJ. Orbital Decomposition: a short user's guide to ORBDE v1.2 [Computer program]. Retrieved 20 October, 2010 from http://www.societyforchaostheory.org/tutorials.
57. Linville PW. Self-complexity and affective extremity: don't put all your eggs in one cognitive basket. Soc Cogn. 1985;3:94–120.
58. Linville PW. Self-complexity as a cognitive buffer against stress-related illness and depression. J Pers Soc Psychol. 1987;52:663–76.
59. Raefelli-Mor E, Steinberg J. Self-complexity and well-being: a review and research synthesis. Pers Soc Psychol Rev. 2002;6:31–58.
60. Luo W, Watkins D, Lam RYH. Measuring self-complexity: a critical analysis of Linville's H statistic. J Appl Meas. 2008;9(4):357–73.
61. Pincus D, Cadsky O, Kavajian A, Chung C, Horst K, & Asuncion C."Fractal Dimension in MMPI-2 Reaction Times as a Correlate of Psychopathology." Paper presentation at the annual meeting for the Society for Chaos Theory in Psychology and Life Sciences; 2009; Marquette University, Milwaukee, WI, USA.
62. Thom R. Structural stability and morphogenesis. New York: Benjamin-Addison-Wesley; 1975.
63. Guastello SJ. Moderator regression and the cusp catastrophe: Application of two-stage personnel selection, training, therapy, and program evaluation. Behav Sci. 1982;30:201–12.
64. Guastello SJ. Self-organization and leadership emergence in emergency response teams. Nonlinear Dynamics Psychol Life Sci. 2010;14:179–204.
65. Prochaska JO, DiClemente CC, Norcross JC. In search of how people change: applications to addictive behaviors. Am Psychol. 1992;47(9):1102–14.
66. Bandura A. Self-efficacy: toward a unifying theory of behavioral change. Psychol Rev. 1977;84:191–215.
67. Bandura A. Social foundations of thought and action. New York: Prentice-Hall; 1986.
68. Maddux JE. Personal Efficacy. In: Derlega VJ, Winstead BA, Jones WH, editors. Personality: contemporary theory and research. Chicago, IL: Nelson-Hall; 1991.
69. Ajzen I, Manstead ASR. Changing health-related behaviors: an approach based on the theory of planned behavior. In: Hewstone M, Schut HAW, De Wit JBF, Van Den Bos K, Stroebe MS, editors. The Scope of Social Psychology: theory and applications. New York: Psychology Press; 2007.
70. Jacobsen JJ, Guastello SJ. Nonlinear models for the adoption and diffusion of innovations for industrial energy conservation. Nonlinear Dynam Psychol Life Sci. 2007;11:499–520.
71. Smerz KE, Guastello SJ. Cusp catastrophe model for binge drinking in a college population. Nonlinear Dynam Psychol Life Sci. 2008;12:205–22.
72. Hayes SC, Follette VM, Linehan MM. Mindfulness and acceptance: expanding the cognitive-behavioral tradition. New York: Guilford; 2004.
73. Wiers RW, Eberl C, Rinck M, Becker ES, Lindenmeyer J. Retraining automatic action tendencies changes alcoholic patients' approach bias for alcohol and improves treatment outcome. Psychol Sci. 2011;22:490–7.
74. Amir N, Beard C, Burns M, Bomyea J. Attention modification program in individuals with generalized anxiety disorder. J Abnorm Psychol. 2009;118: 28–33.
75. Klosterman C. Eating the dinosaur. New York: Scribner; 2009.
76. Pincus D, Ortega DL, Metten AM. Orbital decomposition for multiple time series comparisons. In: Guastello SJ, Gregson RAM, editors. Nonlinear dynamical systems analysis for the behavioral sciences using real data. Boca Raton, FL: CRC Press/ Taylor & Francis; 2011.
77. Fitzsimons GM, Finkel EJ. Outsourcing self-regulation. Psychol Sci. 2011;22:369–75.

Systems Medicine: A New Model for Health Care

51

Linda MacArthur, Timothy R. Mhyre, Elenora Connors, Sona Vasudevan, Elliott Crooke, and Howard J. Federoff

51.1 The Potential of Systems Medicine

In 2003, Dr. Elias Zerhouni created the National Institutes of Health (NIH) roadmap and articulated an agenda to aggressively pursue a more integrated approach to use research discoveries to impact human health. Working groups focused on three major themes: New Pathways to Discovery, Research Teams of the Future, and Reengineering the Clinical Research Enterprise. The findings illustrated the need to develop science to decipher biological networks, and the need for broad-scale application of bioinformatics and computational methods to biological systems [1]. In March 2007, the Department of Health and Human Services launched the Personalized Health Care Initiative (PHCI) with the aim to accelerate the development of personalized treatment strategies. The program focuses on high-throughput technologies and developing an infrastructure to promote electronic medical records [2].

Systems Medicine has evolved from Systems Biology, an interdisciplinary field that focuses on complex interactions in biological systems, and how these interactions give rise to functions of the organism. Systems Medicine applies these concepts to the development of disease. The premise of Systems Medicine is networks of genes and proteins and their interactions can be measured quantitatively, modeled, and properties characterized that exist only within that network. The transition from wellness to disease results in a measurable perturbation to the network. Moreover, therapeutics development might focus on this clinically important network alteration, and thus we envisage the prescription of network-based treatments intended to drive the system backwards to the well state (Fig. 51.1).

The individual components of networks—the genes, proteins, and metabolites—are called *elements*. The measurements of genes, proteins, and metabolites, such as their levels, their allelic form, or their modifications, are called the *network nodes*. The interactions occurring between nodes are called *edges*. Elements (genes, proteins, and metabolites) and their edges (interactions or influences on each other) are affected by their context, their environment. These interactions and the environment give rise to the systems *emergent properties*. The emergent properties of the network are dynamic and further influenced by aging. When sufficiently robust, perturbations of a network may be manifested by symptoms or

L. MacArthur, PhD • T.R. Mhyre, PhD • S. Vasudevan, PhD • E. Crooke, PhD • H.J. Federoff, MD, PhD (✉)
The Georgetown University Medical Center, Washington, DC, USA
e-mail: macarthl@georgetown.edu; trm36@georgetown.edu; sv67@georgetown.edu; crooke@georgetown.edu; hjf8@georgetown.edu

E. Connors, JD, MPH
O'Neill Institute for National and Global Health Law, Georgetown University, Washington, DC, USA
e-mail: eeconnors@gmail.com

J.P. Sturmberg and C.M. Martin (eds.), *Handbook of Systems and Complexity in Health*, DOI 10.1007/978-1-4614-4998-0_51, © Springer Science+Business Media New York 2013

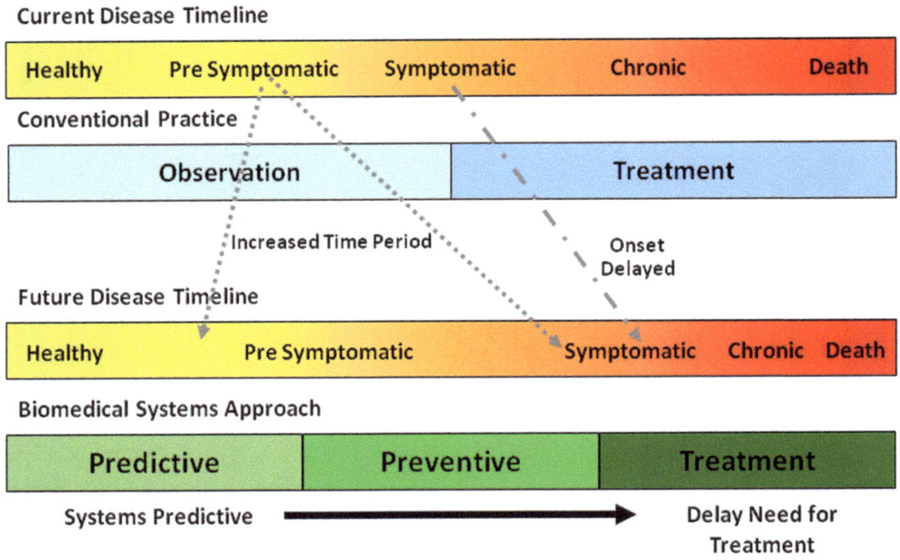

Fig. 51.1 *The promise of systems medicine*. The transition from wellness to disease can be modeled and risk factors stratified. The presymptomatic detection of disease increases the opportunities for halting and reversing disease progression. The goal is to delay disease onset by expanding surveillance and management of the presymptomatic period

signs of disease. Thus, disease ultimately arises from discrete changes of state, a disruption of homeostasis [3].

A number of approaches are under way to model the healthy and diseased state to identify disease-specific risk factors. A strategy pioneered by Dr. Leroy Hood [4] seeks to develop organ-specific fingerprints to reveal healthy and diseased tissues. A screening test of 50 organ-specific probes from 50 target tissues would be used to screen individuals twice a year for early changes in biomarkers that signal diseases before the onset of symptoms. An analogous test is the prostate-specific antigen (PSA) test used to monitor changes in prostate function and to assess the risk of prostate cancer in individuals over time. Although PSA screening is widely employed, the clinical value of a singular measure for predicting prostate cancer is still not clear [5]. A Systems Medicine approach expands this concept to many biomarkers and leverages their interactions and networks with an advantage—a priori knowledge about disease pathways or genetic and environmental risk factors is not required. However, this approach requires baseline data and serial testing to differentiate normal from a disease state. Further, it is not yet clear how the information obtained will be used to stratify risk or to treat the individual.

Other approaches seek to develop markers of disease-specific networks, such as changes occurring during the development of diabetes and Alzheimer's disease. These studies aim to identify early changes in networks that are diagnostic of disease before the onset of symptoms (i.e., in a pre-manifest or clinically silent phase) and are often complemented by data obtained from animal models. Analogous approaches are the blood cholesterol profile used to assess the risk of developing heart disease and biomarker profiles to assess the response of glioblastoma to radiotherapy [6]. The advantage of these approaches is that like the lipid profile (total cholesterol, fractions LDL and HDL, and triglycerides), the results rely on absolute values (compared to a well-established normative range) to assess risk; thus, baseline data are not required. Further, information obtained about disease-specific networks will likely be useful in understanding mechanisms of disease processes and providing additional drug targets. However, this approach is not organ specific, and the data may display greater individual variability and noise, and therefore less sensitivity.

Genome-based approaches use high-throughput DNA sequencing to screen patient DNA for disease-associated alleles and to assess genetic risk. Genome-wide association studies (GWAS) identify single nucleotide polymorphisms (SNPs) in persons with a disease compared with SNPs in healthy controls, and those SNPs that appear significantly more often in the disease group are scored on the strength of their association with the disease. The advantage of this approach is that it is hypothesis free and GWAS studies have identified risk alleles for breast cancer [7], diabetes [8], and Alzheimer's disease [9], among others. The disadvantage is that the individual contribution of SNPs to many diseases may be very small (~1.5%). As proof of principle, whole genome sequencing (WGS) was recently used to interrogate the sequence of 14-year-old fraternal twins diagnosed with dopa responsive dystonia, a disease characterized by dystonic gait, ataxia, and Juvenile Parkinsonism. The twins were being treated with L-dopa, and had some symptom relief. However, WGS identified a third mutation in a gene involving both dopamine and serotonin synthesis and additional supplementation with 5-hydroxytryptophan, a serotonin precursor, led to clinical improvement in both twins [10].

51.2 The Limitations of Systems Medicine

The real potential of Systems Medicine lies in its promise to disentangle the complexity of disease through a biological framework. It seeks through the examination of alterations in the human genome and the measurement of changes to the transcriptome, proteome, or metabolome, to identify risk factors, and to link these to the effects of the environment. These changes must then be measured over time, across the life span. It requires the construction and validation of models to organize data in hierarchies, and mechanisms to incorporate new data to refine existing models. Currently, the state of the science is nascent and unable to address this significant challenge.

One way to imagine the transition of Systems Medicine from the laboratory setting to the clinical setting is to distinguish between three components of the process of network element discovery and application: (1) the use of high-throughput technologies for the *research and discovery* of new risk factors and diagnostic tools; (2) the *application to a clinical setting*; and (3) the *refinement of clinical models and risk stratification* by incorporating new data from patients into existing models. All three steps depend on technologies in existence or under development, such as high-throughput technologies to sample and quantify biological molecules, computational models to understand element (genes, proteins, metabolites) interactions, algorithms to mine information, online databases and repositories to share data and models, and strategies to integrate separate databases to drive towards clinical application.

51.2.1 Research and Discovery

The majority of human studies to date describe the identification of risk alleles in populations. Both whole genome sequencing (WGS) and exome sequencing,[1] which survey exon sequences only, offer the benefit of definitive diagnosis and risk assessment [11]. The rapid development of high-throughput DNA sequencing technologies, such as next generation sequencing, has decreased DNA sequencing costs from over $5000/megabase, or $95,263,072 per genome in 2001, to $0.12/megabase or $10,497 per genome in 2011 [12]. Next generation sequencing of targeted disease panels and exome sequencing have substantially reduced both the cost and complexity of biomarker analysis [13].

To understand how network perturbations manifest in disease, and to identify future drug

[1] Exomes are the protein coding regions of DNA which constitute about 1% of the whole genome. Exome sequencing is used to identify variants found in the coding region of genes which affect protein function. It does not identify structural or regulatory variants associated with diseases in non-coding DNA, which can be identified by whole genome sequencing (WGS).

targets, proteomics is used to sample individual proteomes. Current efforts focus on *connectivity mapping* to find theoretical or functional connections between proteins, genes, and small molecules sharing a mechanism of action, a physiological process, a disease, or drug targets [14]. Starting with a research objective, such as finding novel protein interactions in a metabolic pathway, investigators construct networks of genes and proteins to identify novel members of the networks. The interaction databases are re-queried multiple times, limiting the search to interactions empirically validated versus those that are computationally predicted, to add additional genes to the networks for further study. From this, functionally active and computational networks can be described and novel disease targets discovered. Connectivity maps, such as those developed by the Broad Institute at MIT and Harvard,[2] link gene patterns associated with disease to corresponding patterns produced by drug candidates. The goal is to allow researchers to screen compounds against genome-wide disease signatures rather than a preselected set of target genes [15].

Connectivity maps and other networks must then be validated and provided in platforms that are accessible by biomedical research. Crowd sourcing has emerged as a strategy for data validation. It is organized around challenges in which interested communities compete to verify methodologies around chosen benchmarks. Academic efforts such as DREAM[3] (The Dialogue on Reverse Engineering Assessments and Methods) used crowd sourcing to assess model predictions and pathway algorithms in a Systems Biology database [16] and crowd sourcing may be an effective strategy for constructing protein signaling networks as well [17]. Annotated biological databases such as Biobase (http://www.biobase-international.com) and Ingenuity (http://www.ingenuity.com) improve access to information by curating the peer-reviewed literature and providing guidance for interpreting the relative impact of biomarker discoveries. Access is further improved by the advent of cloud computing, where computation and storage resources are accessed by the Internet at a much lower cost than dedicated on-site resources [18]. Cloud computing has greatly increased the capacity to store, mine, and exchange large amounts of data. User friendly interfaces, such as Galaxy,[4] have been developed to make complex analysis available to researchers with a Web browser [19]. Galaxy Cloud then allows biomedical researchers to move their analysis to local resources or other cloud providers and is currently provided on the Amazon Web Services (AWS) cloud.

51.2.2 Application to a Clinical Setting

For a biomarker to be useful in a clinical setting, it has to be reliable, measurable, specific, sensitive, and predicative. It has to stratify risk. Many clinical laboratories have begun to target 10–100 genes simultaneously. As a proof of concept for its utility in a clinical setting, a retrospective assessment was performed using targeted sequencing for 448 recessive childhood diseases in 104 samples. It reported 95% sensitivity and 100% specificity for detecting and genotyping substitution, insertion and deletion, splicing, and gross-deletion mutations, when custom baits were included to capture non-exonic mutations and boundaries of gross deletions, insertions, and rearrangements (which are difficult to detect with next generation sequencing [13]).

Hurdles for implementation into the clinic include a lack of a clinical-grade general database of disease mutations and biomarkers. Plans to develop this are currently being explored by the National Center for Biotechnology Information,[5] the Human Genome Variation Society,[6] and the Human Variome Project[7] [20]. New business models will be needed to facilitate the transition from clinical research to early

[2] http://www.broadinstitute.org/cmap/

[3] http://www.The-dream-project.org/

[4] http://www.cloudgalaxy.com

[5] http://www.ncbi.nlm.nih.gov

[6] http://www.hgvs.org

[7] http://www.humanvariomeproject.org

adoption by clinical practice. In addition, healthcare workforce training for physicians, advanced practice nurses, and pharmacists, among others, will need to incorporate this paradigm into educational programs. In addition, the acceptance of Systems Medicine by professional bodies that certify practitioners will drive the establishment of systems knowledge criteria that will become part of the certification process. Finally, new programs of continuing medical education (CME) which introduce and build knowledge of Systems Medicine will be required to update practicing physicians.

51.2.3 Refinement of Clinical Models and Risk Stratification

Preemptive genotyping has emerged as a way to identify genetic differences (SNPs) associated with variable drug responses such as reduced drug efficacy or increased toxicity. An example is the widely used antiplatelet drug clopidogrel (Plavix®). Clopidogrel must be converted to an active metabolite, a step partly dependent on the cytochrome P450 2C19 enzyme [21]. Loss of function of the allele is relatively common, with poor metabolizers displaying an increase in cardiovascular events [22–24]. Further, poor metabolizers represent 2–3% of whites and blacks and up to 15% of Chinese [25]. The link between genotypes and variable drug response prompted the US Food and Drug Administration to begin to incorporate pharmacogenetic information in drug labels [26] and to add warnings to over 115 drugs based on biomarker data. However, genetic variations are not absolutes but rather altered probabilities of particular outcomes, and physicians are often uncertain how to incorporate genetic information into their decision making with individual patients [25].

Efforts are under way to link genetic information and electronic health records to alert physicians of patients genetics to help guide decision making. Vanderbilt launched PREDICT (Pharmacogenomic Resource for Enhanced Decisions in Care and Treatment) [27] to measure Plavix responses in patients undergoing heart catheterization. Of the 3,000 patients tested, over 600 were reported to be poor metabolizers of Plavix. Vanderbilt has recently started testing for gene variants associated with muscle toxicity in patients treated with the cholesterol-lowering drug simvastatin. Doctors will receive electronic prompts if a patient carries alleles associated with adverse drug reactions.

Thus, short-term strategies for the application of Systems Medicine to the clinic will likely introduce new tests to survey a small set of genomic, proteomic, and metabolomic profiles in high-risk populations. This will require a hierarchical validation of data that is likely to involve consortia to organize strategies for standardization and interpretation of data. Further, issues regarding who gets tested must be decided within the context of what disorders will most benefit with respect to improved outcomes. The latter is not simply a clinical issue but also one with ethical implications. There will need to be mechanisms to cross-validate data across independent cohorts and methods to archive data will need to be standardized [28].

Implementation in the clinic will require the concurrent development of annotated databases for reporting results from tests and their association with clinical outcomes over time. Strategies for interpretation of high-throughput data must be developed. An example of this is Informatics for Integrating Biology and the Bedside[8] (i2b2), an NIH-sponsored initiative at Roadmap National Centers for Biomedical Computing [29] with a main goal of providing clinical investigators with software tools necessary to collect and manage project-related clinical research data. The combination of i2b2 with Pentaho,[9] a business intelligence software package, is intended to provide an intuitive Web-based interface for nontechnical users and allows clinicians to visualize data and perform additional data mining (Fig. 51.2).

---

[8] https://www.I2b2.org

[9] http://www.pentaho.com

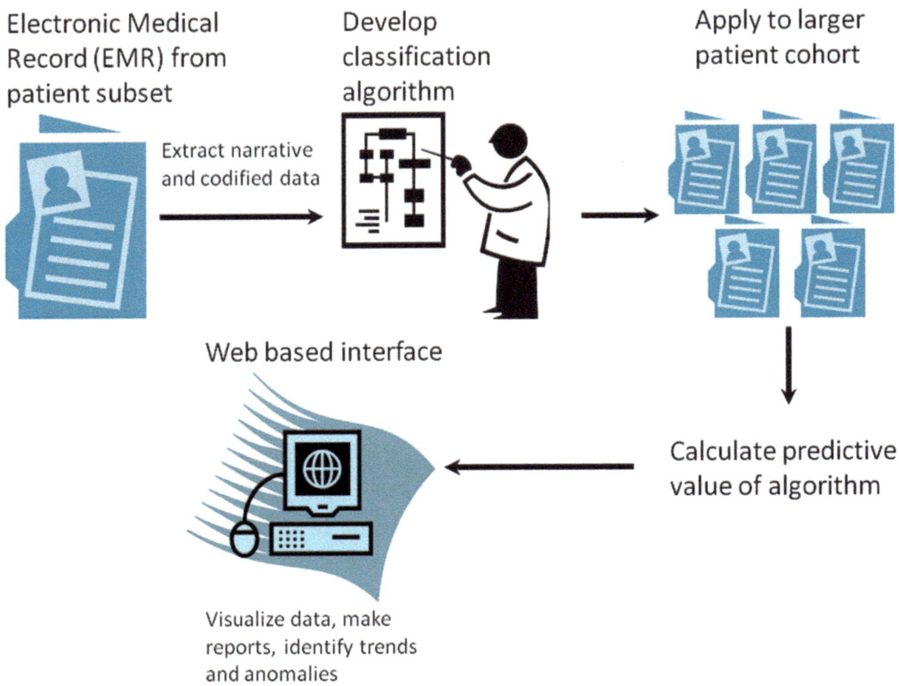

Fig. 51.2 *Integrating diagnostic algorithms into clinical practice*. Software programs have been developed to mine electronic medical records from patients for disease-specific narrative data and codified clinical data. The data are used to develop classification algorithms which are tested on larger patient cohorts. Different algorithms can be compared for their positive predictive values. A Web-based interface allows users to visualize and further analyze the data

51.3 Ethical, Legal, and Policy Implications

The ethical and regulatory issues of implementing Systems Medicine are numerous and complex. The examples discussed here are not meant to be all encompassing but rather to provide a preview of issues that are likely to arise as Systems Medicine is implemented. A myriad of questions arise, including ethical considerations, the costs to beneficiaries and payors, privacy of medical information, and civil liberties.

51.3.1 Ethical Considerations

Systems Medicine holds the potential for remarkable gains in the diagnosis of disease and the delivery of health care. Notwithstanding these gains, ethical issues arise such as when is the appropriate time to intervene when a high-risk factor is identified? What are the implications for screening patients for diseases that have no cure? If a patient is diagnosed as being in very early stages of Alzheimer's disease, but there is no protocol to halt its progression, then what are the legal and ethical responsibilities and consequences for the provider, the insurance company, and the family of the individual? Does an early-state diagnosis entitle a patient to be entered into clinical trials, and if so, how are the study protocols funded? If intervention or treatment is available, would heightened risk factors warrant the public or private insurance coverage of an early-state diagnosis?

Furthermore, what are the implications for preconception carrier testing and neonatal screening? Prenatal diagnosis is fraught with ethical concerns such as when to disclose minor genetic variations to the patient. The number of recognized genetic disorders increases each year and for some of the most severe, prenatal screening may dictate termination of the pregnancy. With more advanced testing, the implications for common spectrum scale disorders are profound. If risk factors for more diseases can be detected prenatally, is there potential for large-scale termination of pregnancy based on a less severely disabling disorder [30], or would the opposite occur with the creation of a new market for prenatal specialists that work with infants determined to have high-risk factors for a disease such as autism, which affects an estimated 1 in 110 children in the USA [31].

51.3.2 Cost and Payment

The Affordable Care Act passed on March 23, 2010, requires newly issued insurance plans, qualified plans participating in state Health Insurance Exchanges, and Medicare to cover certain preventive care services at no cost to the patient (The Patient Protection and Affordable Care Act-Public Law 111-148 and the Health Care and Education Reconciliation Act of 2010 Public Law 111-152). Furthermore, it provides incentives to Medicare and Medicaid beneficiaries to complete programs that work to modify unhealthy behaviors. Substantial delivery system reforms in the Affordable Care Act encompass principles that diverge from healthcare treatment in isolation, and move towards more holistic care. This includes payment policies that focus on health outcomes a new focus on coordinating care for the most high-cost consumers, and a collaborative network of providers.

Insurers are beginning to provide coverage for, more personalized treatments, but the high costs of tests and treatments could lead to increased short-term costs prior to the realization of long-term savings. Given that coverage determinations can impact the availability and use of new technologies and innovations, how viable is a paradigm shift that may not recognize savings until years to come? Could Systems Medicine, with its promise of early screening, detection, and risk stratification, be considered "preventive care" that has low or no cost to beneficiaries, or one that should be paid by Medicare? Payments for health services are complex, multifaceted, and political in nature. In a Systems Medicine approach, who would decide what health benefits were covered by insurance versus those being the responsibility of the patient? With National Health Expenditures (NHE) projected to grow to $4.6 trillion in 2020 (almost 20% of Gross Domestic Product), who will pay? If the political discussion continues to focus on cuts now, not later, how does a paradigm shift occur for a system that may not realize healthcare savings for years to come?

51.3.3 Data Ownership, Access, Privacy and Oversight

In 2009, Google launched Google Health with the aim of enabling individuals to store personal genomic and health information online so that relevant information can be retrieved through a Google search in the future. The advent of consumer-directed health care (CDH) has been further fostered by the recent availability of affordable genomic tests for consumers. Genetic testing is now available to the public through direct-to-consumer marketing [32] and costs of exome sequencing have recently fallen to approximately $1000 per exome [33]. As with other laboratory tests, the US Government has some regulatory oversight of genetic testing. However, tests conducted by clinical laboratories themselves with "home-brew" reagents are not regulated in the same way as genetic tests manufactured by industry and sold for clinical diagnostic use. Therefore, the clinical validity of a large portion of genetic tests is not currently regulated and should be thoughtfully assessed. How do we ensure the reliability and predictive value of consumer-requested tests? Moreover, how do we ensure data are protected?

Patient-driven databases raise questions such as who owns the information, who has the right to maintain it or destroy it, and who has access to the information—the individual and/or a medical professional? Would physicians be required to report certain diseases? If Systems Medicine is about risk and preventive medicine, then who orders and pays for the tests if they are consumer initiated and what oversight is necessary? Databases could contain highly sensitive personal information and a breach could have profound consequences due to the amount and type of information available. The Genetic Information Non-Discrimination Act (GINA, Public Law 110-233) prohibits discrimination based on genetic information for health insurance and employment. What additional safeguards need to be considered to impose additional protections against profiling based on disease-specific risk factors, and to protect consumers from inaccurate tests and profit-motivated advertising?

51.3.4 Civil Liberties, Liability, and Access

Once the ability to pre-symptomatically detect disease is realized, what role, if any, would the government play in implementing strategies to reduce disease and promote positive behaviors in high-risk populations? Numerous disease management strategies and behavioral programs in both the public and private sectors have proved successful in reducing rates and complications from certain diseases, such as Type II Diabetes. To what degree would interventions prove successful or be appropriate on a national scale? To what extent would patients be responsible if they refused to participate? Current policies, such as incentives that reward healthy habits (gym memberships) and disincentives that punish unhealthy habits (cigarette tax), are well recognized, but what role would physicians, the government, or a private sector insurance company be expected to play?

What if an individual is treated to prevent a disease, but the predictive diagnosis was wrong? How do you ensure that no harm is done if treating a healthy person? Who is responsible if harm is done? The link between socioeconomic status and health status is well documented, and the poor and minorities have historically had unequal access to new technologies, particularly those that are expensive. Current disparities in health outcomes remain. Would the implementation of a Systems Medicine approach further increase healthcare disparities? Will it be possible to implement these practices and technologies in remote locations or in areas with a high number of patients at a lower socioeconomic status? How can we guarantee there is equal access to the diagnostic techniques and therapies when already millions of Americans lack access to adequate primary care?

51.4 Proposal for a Systems Curriculum for Health Professional Education

"All the forces in the world are not as powerful as an idea whose time has come" said Victor Hugo, and this thought is applicable to the practice of Systems Medicine. Along with the explosion of data, comes the realization that new educational programs are needed to train healthcare providers to use new tools and resources to make the practice of Systems Medicine a reality.

A patient could conceivably arrive at a clinic with terabytes of molecular data and an assumption that the clinician will use this information for treatment. This change, involving the use and analysis of enormous quantities and variety of data, will require a new type of physician—one with a grasp of modern computational sciences, the "-omic" technologies, including genomics, transcriptomics, proteomics, metabolomics, and lipidomics, and a systems approach to medicine (Table 51.1). Such tools are multidisciplinary and are difficult to capture in traditional medical school curricula. A Systems Medicine curriculum would need to cover these core concepts, as well as statistics, bioinformatics, and courses geared toward developing analytical thinking.

A number of medical schools, including Emory University School of Medicine, Duke University School of Medicine, Dartmouth Medical School, and The Johns Hopkins School of Medicine have established computational medicine and bioinformatics departments that have the goal of combining expertise in biology, computer science, statistics, and imaging to train scientists who will partner with clinicians to use informatics to treat disease [34]. While these pro-

Table 51.1 High-throughput technologies

• *Genomics*—DNA sequence variation
• *Proteomics*—protein and peptide measurements via two-dimensional gel-electrophoresis, mass spectrometry, or HPLC with mass spectrometry. Includes detection of modifications such as phosphorylation and glycosylation.
• *Translational bioinformatics*—using computational tools and techniques for the analysis of large biological and clinical databases.
• *Systems biology*—understanding gene and protein networks in biological systems.
• *Metabolomics*—measurement of all small-molecules known as metabolites.
• *Pharmacogenomics*—identifying genetic differences associated with variable drug response.
• *Epigenomics*—Cell-specific modifications that influence gene expression, and that can be heritable, but are independent of changes in DNA sequence such as DNA methylation and histone acetylation.
• *Biomedical informatics*—acquisition, analysis, and application of biomedical data and knowledge.
• *Glycomics*—measurement of carbohydrates.
• *Interferomics*—transcript correcting factors (i.e., RNA interference).
• *Lipidomics*—measurement of lipids.
• *Transcriptomics*—gene expression measurements by DNA microarrays or serial analysis of gene expression (SAGE).

grams are still evolving, a number of suggested courses and their applications are listed below:

1. *Bioinformatics* (*Genomics and Proteomics*): This course should introduce bioinformatics concepts and methods, such as genomic sequence analysis, comparative genomics, molecular evolution, protein sequence analysis, and structural and functional analysis. Students and clinicians should be trained to use databases for literature (PubMed), genetic phenotypes and diseases (MIM, Orphanet, GeneCards, dbSNP), protein information (NCBI RefSeq, UniProtKB), three-dimensional structures and their visualization (PDB, PyMol, Chimera), pathways (KEGG, REACTOME), drugs (DrugBank), DNA/protein sequence searches and alignments (BLAST, Clustal), genome browsing (UCSC), and phylogenetic analysis (MEGA).
2. *Translational Bioinformatics*: A natural extension of the bioinformatics course described above is the application of the tools and resources to large-scale clinical data, with a goal to develop predictive tools and models. This course should help students and clinicians understand current technologies used in translational research and identify tools and resources to perform specific bioinformatics analysis, with an emphasis on assessment of large-scale multi-omic data. A focus placed on current and emerging technologies such as next generation sequencing, gene expression analysis, cytogenetics, and network and pathway analyses would assist in interpretation of published literature. Students should be able to interrogate existing informatics platforms dedicated to translational research, such as Georgetown University's Database of Cancer (G-DOC), which was developed with this goal in mind [35].
3. *Systems Biology*: This approach can help one understand how the interaction of genes and environment influences the development of disease and may facilitate the development of new therapies. For example, type II diabetes, cardiovascular diseases, colon cancer, postmenopausal breast cancer, dementia, and depression were shown to constitute a diseaseome (network of related diseases) of physical activity [36]. This course should introduce microarray analysis, proteomic informatics, and regulatory network and pathway analyses, and students should learn how "omics" data can be analyzed to understand the biology of normal and diseased states.
4. *Metabolomics*: This is the study of global metabolite profiles in a system (cell, tissue, or organism) under a given set of conditions and has been successful in identifying potential biomarkers associated with diseases such as hepatocellular carcinoma [37] and schizophrenia [38]. This course should provide students and clinicians with a basic understanding of qualitative and quantitative mass spectrometry-based metabolomics as applied to a clinical setting, including for drug metabolism and toxicity, disease progression, and the effect of

therapies (drugs, radiation) on the overall metabolic state of the patient.

5. *Pharmacogenomics*: This is the study of how an individual's genome affects the body's response to drugs. For example, variable responses to anticoagulants, such as warfarin, are attributed to SNPs in VKORC1 (vitamin K epoxide reductase complex subunit 1) [39] and CYP2C9 (cytochrome P450 2C9) [21], leading to increased risk of bleeding complications. This course should include training in the use of resources, such as dbSNP (a tool to mine SNP data), PharmGKB (a comprehensive resource for pharmacogenomics), and strategies for incorporating patient genotyping data into their treatment plan.
6. *Epigenomics/Epigenetics*: Epigenetics examines heritable modifications that influence gene expression but that are independent of changes in DNA sequence. Epigenomics uses high-throughput methods for measuring genome modifications including methylation, acetylation, and phosphorylation. This course should expose students to appropriate technologies and datasets, such as NCBI's epigenome viewer, expression omnibus, and atlas, and provide students a mechanistic understanding of how genome-wide modifications regulate gene activity and contribute to human diseases.
7. *Biomedical Informatics*: A course in biomedical informatics should train students and clinicians in the application of computers to the practice of medicine through the use of Electronic Health Records, standards and terminology, health information exchange, clinical decision support, public and consumer health informatics, telemedicine, and imaging informatics. The American Medical Informatics Association (AMIA) currently leads this effort with its popular 10×10 courses, the goal of which is to strengthen the breadth and depth of the biomedical and health informatics workforce.

Medical education in the USA generally adheres to a model involving 2 preclinical years, with a full range of basic and applied sciences, and 2 years of clinical experience with exposures to specialties. In an effort to meet the new demand for physicians with informatics expertise, several universities now combine medical school training with bioinformatics graduate training. Columbia University offers an MD/PhD program in Biomedical Informatics and, as is standard for this model, students enter the doctoral program after 2 years of medical school, complete their doctoral dissertation in Bioinformatics within 3–4 years, and then return to medical school. Georgetown University's MD/MS program offers a 5-year combined program with two preclinical years followed by a 1-year Masters concentration in Systems Medicine. The program incorporates a Capstone project in which students develop research projects that teach them how to apply their knowledge to discover clinical biomarkers, such as those associated with cancer relapse in patients or the progression of diabetes.

For practicing physicians, Continuing Medical Education (CME) credits may be an avenue for training physicians to incorporate bioinformatics into their clinical practice. While still emerging, Health Information Technology, Neurocritical Care Bioinformatics, and Advanced Bioinformatics are available for CME credits through accredited medical education providers. As the bioinformatics field evolves concurrently with CDH, it will become necessary to integrate bioinformatics training into the standard medical curriculum.

51.5 Building a New Model: A Hypothetical Future for Applying Systems Medicine to Prevent the Onset of Type 2 Diabetes

The greatest success of high-throughput technologies has been with diseases conferred by a small number of loci where heritability itself is the risk factor. In contrast, the majority of human diseases are caused by the interaction of inherited vulnerability with environmental factors and are genetically complex, with multiple individual genes in the network contributing a small amount of risk to the development of disease. Thus, genotyping

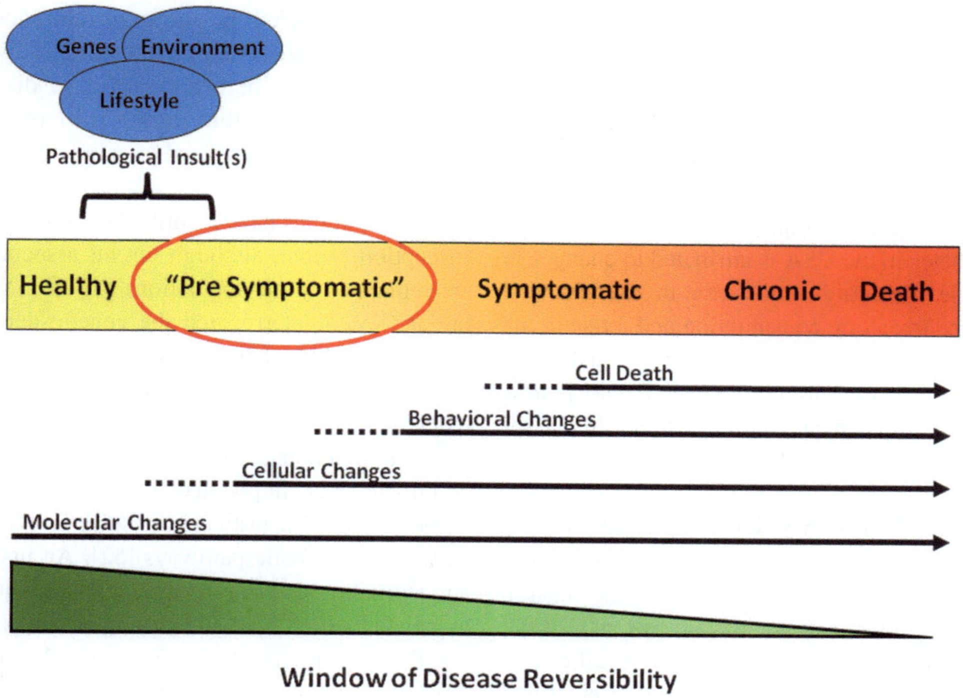

Fig. 51.3 *Halting the progression of diabetes*. A Systems Medicine approach can help elucidate genetic, environmental, and lifestyle risk factors for the onset and progression of diabetes, sometimes years before symptoms become apparent. It models emergent properties unique to individual systems that ultimately lead to healthy or diseased states

to stratify risk is often not more informative than simple observations of family history, current health status, and diet and exercise habits. Is there a future for the application of Systems Medicine for these types of diseases? To address this question, we examine the feasibility of applying Systems Medicine to Type 2 diabetes mellitus (Fig. 51.3).

51.5.1 Type 2 Diabetes Mellitus and Its Impact

Type 2 diabetes mellitus (T2DM) accounts for 90–95% of all diagnosed cases of diabetes. It typically begins as insulin resistance and progresses to loss of insulin production. Risk factors for T2DM include a family history of diabetes, obesity, physical inactivity, impaired glucose metabolism, and a history of gestational diabetes [40]. As of this writing, the incidence of diabetes in the USA is estimated at 25.8 million children and adults, comprising 8.3% of the population, with 18.8 million confirmed cases and an estimated 7 million undiagnosed cases. These estimates are based on fasting glucose and glycated hemoglobin (hemoglobin A1c) levels obtained from samples collected in 2005–2008. From these values, it was estimated 35% of US adults aged 20 years or older, and 50% of adults aged 65 years or older, had prediabetes. Further, an astonishing 1.9 million new cases of diabetes were diagnosed in people aged 20 years and older in 2010 [41].

Adults with diabetes have substantially increased morbidity and mortality: heart disease deaths and stroke are 2–4 times greater than in adults without diabetes. Diabetes is the leading cause of kidney failure, non-traumatic lower-limb amputation, and new cases of blindness among adults in the USA. An estimated 60–70% of diabetics have mild-to-severe peripheral nervous system damage and T2DM is a risk factor for developing Alzheimer's disease [42]. The total

US nationwide costs associated with diagnosed diabetes in 2007 were estimated at $174 billion–$116 billion for direct medical costs and $58 billion for indirect costs including disability, work loss, and premature mortality. The average medical expenditures for individuals with diabetes are 2–3-fold higher than those without. The increase in diabetes in the USA is attributed to changes in lifestyle including a decrease in exercise and increase in caloric consumption and a rise in the incidence of obesity. While it previously presented most often in middle age, T2DM is now appearing in younger people due to its close association with obesity.

51.5.1.1 Achieving Risk Stratification

Diabetes is often present for years before becoming clinically apparent. By the time insulin deficiency manifests as a diagnosis of T2DM, considerable pancreatic beta cell insufficiency has already occurred [43]. Current clinical and laboratory predictors, such as body mass index and fasting blood glucose, often reflect extant disease [44]. There is now convincing evidence, however, that glycated hemoglobin (A1c), which provides a measure of glycemic control over the past 2–3 months, can stratify the risk of diabetes-induced cardiovascular disease. In a study involving 11,000 participants screened with both the glucose and A1c tests, higher A1c levels were associated with up to 16.5-fold higher risk of diabetes and twofold higher risk of heart disease [45].

High-throughput profiles of metabolic status (metabolomics) recently uncovered a significant association between concentrations of five branched-chain and aromatic amino acids in blood and the predisposition to diabetes [46]. Samples were obtained from participants in the Framingham Offspring Study who developed diabetes during a 12-year follow-up. Interestingly, adding molecular profile data only slightly improved disease prediction when applied to a random diabetes-free sample from the study. However, when controls were age matched for known diabetes risk factors such as age, body mass index, and fasting glucose, there were highly significant changes in the amino acids that predicted the future development of diabetes in otherwise normoglycemic individuals up to 12 years before the onset of disease. These data suggest that amino acid profiling may have greater value in high-risk individuals.

GWAS studies have identified over 40 T2DM susceptibility loci, although not all associations are replicated in all populations [47–49]. While initial GWAS interrogated the genetic determinants of T2DM as a dichotomous phenotype (disease vs. no disease), follow-up studies used meta-analysis of all GWAS data for glycemic traits in nondiabetic participants and revealed additional loci implicated in fasting glucose homeostasis [50], fasting proinsulin levels [51], and other metabolic pathways [52]. An updated age-stratified analysis using metabolic alleles associated with T2DM in Framingham improved the ability to predict the onset of T2DM in participants younger than 50 years [53]. The problem remains that the use of GWAS data to stratify risk adds only a minor advantage over conventional risk factors such as BMI, fat mass, and insulin resistance [46, 54]. Among the general population, obesity remains the highest predictor for the development of type 2 diabetes. However, the studies suggest that molecular measures have the greatest value for determining which obese individuals are at highest risk for developing T2DM.

A recent study combined GWAS data with metabolomics to identify genetically determined metabotypes (GDMs) [55]. The authors identified 37 GDMs to indicate key genetic loci related to differences in human metabolism. Of these, large effects (10–60%) were seen for 25 loci, in contrast to GWAS alone. One locus for the glucokinase (hexokinase 4) gene is a risk factor for the development of T2DM [50]. The authors reported that fasting levels of mannose (a substrate of the enzyme), which were lower in individuals that carried the risk allele, had a stronger association than elevated glucose with the risk allele. These data suggest an integrated approach, combining molecular profiles from genomics and metabolomics, will be more powerful in predicting who is at risk for developing T2DM.

51.5.1.2 Environmental Risk Factors

The discovery of risk alleles for T2DM has provided a biological context with which to understand the contribution of environmental factors to the development of the disease. An example of this is the link between increased consumption of high-fructose corn syrup and the development of T2DM. The consumption of sucrose and high-fructose corn syrup in the USA increased from 64 g per day in 1970 to 81 g per day in 1997 [56]. While dietary fructose improves glucose tolerance in adults with T2DM [57], large amounts of dietary fructose are associated with increased free fatty acids in blood, a risk factor for coronary heart disease. The lower insulin release after fructose compared to glucose reduces lipase activation in adipose tissue and impairs free fatty acid clearance [58].

The increase in free fatty acids affects the regulation of genes associated with T2DM. A genome-wide promoter analysis of DNA methylation sites was performed in skeletal muscles from patients with T2DM. It identified hypermethylation of a promoter for a gene involved in mitochondrial function (PGC-1α). Hypermethylation was increased acutely in myotubes by free fatty acids but not insulin or glucose and was associated with reduced mitochondrial content in diabetic subjects [59]. As skeletal muscle plays an important role in the development of insulin resistance, these data suggest a mechanistic link between high dietary fructose consumption and the establishment of T2DM.

Large prospective studies are currently under way to understand genetic and environmental interactions and their contributions to disease. Examples include the UK Biobank study,[10] a major medical research initiative in the United Kingdom. The study aims to recruit 500,000 people aged 40–69 years with a focus on collecting longitudinal data relevant to cancer, heart disease, arthritis, and dementia. The American Cancer Society Prevention Study-3[11] aims to enroll 500,000 people aged 30–65 years with no prior history of cancer to understand the lifestyle, behavior, environmental, and genetic factors that cause or prevent cancer. The Kaiser Permanente Research Program on Genes, Environment, and Health (RPGEH)[12] is a longitudinal clinical study that incorporates environmental exposure data from the Geographic Information System (GIS) with genetic, biomarkers, and environmental data derived from the biospecimens (New Models for Large Prospective Studies Symposium, Bethesda MD 2010) (Fig. 51.4).

51.5.1.3 Managing the Ethical Issues

Entry into a well care system where the risks for disease will be stratified has the potential to discourage alignment between patients and providers. Many of the issues are described in Sect. 51.3. The codification of policy through rule making will be essential for the at-risk patient, particularly one who might have lifelong T2DM. Among the most relevant issues are abrogating any form of discrimination, ensuring that individuals have ownership of their data and health decisions and that payors provide appropriate incentives to align wellness and behavior, and ensuring long-term costs of management of chronic illness with potential comorbidities are endorsed as a societal responsibility. Under such conditions, a provider will be empowered to engage in the preventive and well care for patients at risk for T2DM, and upon presentation of overt diabetes, seamlessly move to optimally deploy resources for their disease management.

Current paradigm

A healthy child and his parents visit their pediatrician for a typical well visit. A careful review indicates a strong family history for major diseases, including diabetes, heart disease, and stroke. However, the risks cannot be stratified. The generic recommendations for getting plenty of exercise, eating a healthy diet, and avoiding risky behaviors are discussed.

Future paradigm

A healthy child and his parents visit their pediatrician for a typical well visit. A careful review indicates a strong family history for major diseases, including diabetes, heart disease, and stroke. A blood sample is drawn for molecular and genetic profiling. The results reveal a higher than normal hemoglobin A1c level and mannose/glucose ratio for his age group and evidence for early preclinical insulin resistance. Further, these profiles are high in relatives currently symptomatic for

[10] http://www.ukbiobank.ac.uk

[11] http://www.cancer.org/Research

[12] http://www.rpgeh.kaiser.org

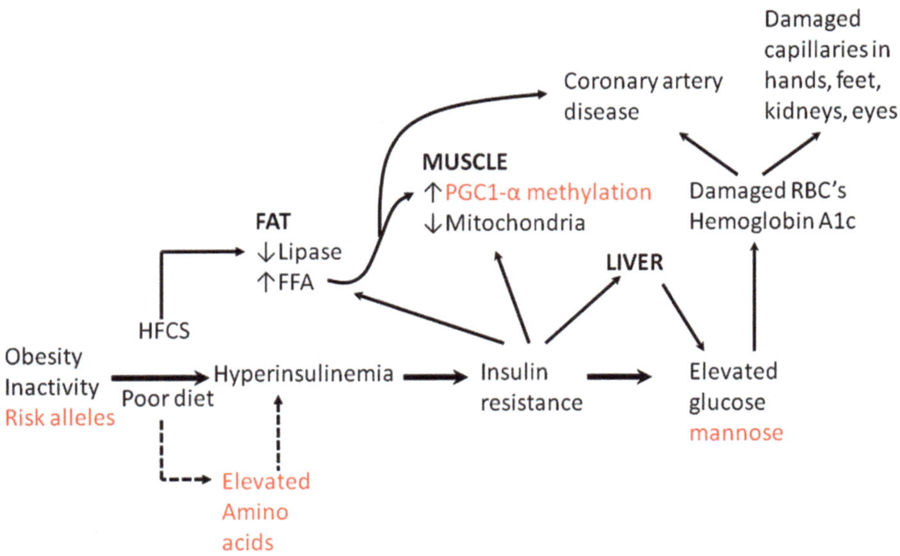

Fig. 51.4 *A network of some of the risk factors associated with the development of T2DM*. Risk factors discovered by "omics" studies are shown in red. Insulin resistance directly affects the uptake of glucose by fat, muscle, and liver. Liver serves as an additional source of glucose. *HFCS* elevated high-fructose corn syrup. *FFA* free fatty acids

diabetes. The risk is sufficiently high, given his family history and profile, to modify the child's diet to reduce insulin secretion, to increase daily exercise, and to warrant observation of the child every 6 months for a further change in biomarkers. A discussion ensues between the physician and parents regarding whether and when to treat the child with medications if the modification in diet and exercise alone are not effective.

51.6 Concluding Thoughts

Just as lipoprotein and cholesterol screens have led to dramatic changes in how we attempt to predict and prevent the onset of cardiovascular diseases, the advent of Systems Medicine proposes to use high-throughput screening of clinical populations to identify inherited vulnerabilities through multimodal data integration, to stratify risk and assess environment–gene interactions, and to initiate a management plan to mitigate the risk of disease (Fig. 51.5). It is a shift from disease management to secondary and subsequently primary prevention. Like many advances, it arrives ahead of detailed regulation or a complete understanding of its utility and applications, and without a roadmap for navigating the ethical implications. It challenges the status quo of physician-directed health care with the advent of consumer-driven health care, and requires a rethinking of who owns patient data, who should pay to collect it, and how should it be protected. It creates issues as to how to incorporate and validate information obtained from patients, and if successful, it will be a paradigm shift from symptom management and treatment of disease to the risk assessment, prediction, and prevention of diseases before the symptoms emerge. Its ultimate success will hinge on demonstration of clinical utility and overall cost effectiveness. Systems Medicine adoption by the clinical community will require a well-orchestrated evolution from current practice guidelines using the most robust data-driven process. A realignment of reimbursement to promote wellness and interventions to delay manifest disease are required to create the financial incentives to restructure clinical practices. Systems Medicine adoption by the patient community mandates effective educational programming, financial incentives to promote compliance, and policy changes to make the requisite resources available to ensure broad

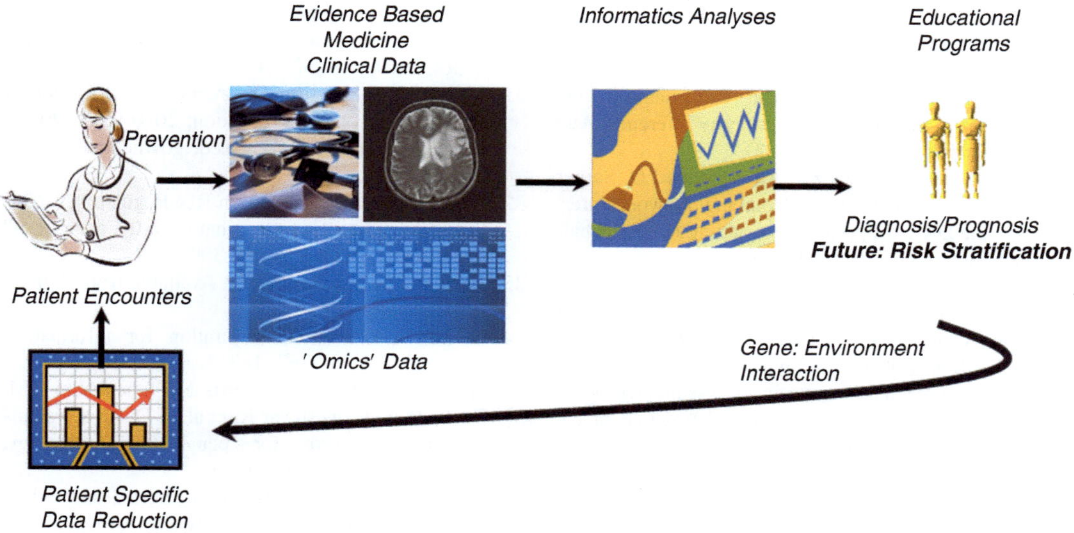

Fig. 51.5 *Building a new model.* A shift in health care from symptom and disease management to the use of bioinformatics to detect disease will require integration of clinical and bioinformatics data and changes in the way physicians are educated and provide medical care to patients

access. The challenges are substantial, but the rewards heralded by Systems Medicine comport with expectations for a healthier society. In short, we must collaborate across the public, private, and governmental sectors to secure the mandate to effect such change.

Acknowledgements The authors are grateful to Dr Edmund Pellegrino for his comments on the manuscript. The authors recognize support from grant W81XWH-09-1-0107 from the Telemedicine and Advanced Technology Research Center, USAMRMC.

References

1. Zerhouni E. Medicine. The NIH Roadmap. Science. 2003;302(5642):63–72.
2. Services HaH. Personalized Health Care for Informed and Effective Choices. [cited 2011]; Available from: http://www.hhs.gov/myhealthcare/goals/index.html.
3. Sauer U, Heinemann M, Zamboni N. Genetics. Getting closer to the whole picture. Science. 2007;316(5824):550–1.
4. Hood L, Heath JR, Phelps ME, Lin B. Systems biology and new technologies enable predictive and preventative medicine. Science. 2004;306(5696):640–3.
5. Mitka M. Experts debate utility of PSA velocity as a criterion for prostate biopsy. J Am Med Assoc. 2011;305(15):1522.
6. Rivera AL, Pelloski CE, Gilbert MR, Colman H, De La Cruz C, Sulman EP, et al. MGMT promoter methylation is predictive of response to radiotherapy and prognostic in the absence of adjuvant alkylating chemotherapy for glioblastoma. Neuro Oncol. 2010;12(2):116–21.
7. Jostins L, Barrett JC. Genetic risk prediction in complex disease. Hum Mol Genet. 2011;20(R2):R182–8.
8. Imamura M, Maeda S. Genetics of type 2 diabetes: the GWAS era and future perspectives [Review]. Endocr J. 2011;58(9):723–39.
9. Sleegers K, Lambert JC, Bertram L, Cruts M, Amouyel P, Van Broeckhoven C. The pursuit of susceptibility genes for Alzheimer's disease: progress and prospects. Trends Genet. 2010;26(2):84–93.
10. Bainbridge MN, Wiszniewski W, Murdock DR, Friedman J, Gonzaga-Jauregui C, Newsham I, et al. Whole-genome sequencing for optimized patient management. Sci Transl Med. 2011;3(87):87re3.
11. Kingsmore SF, Saunders CJ. Deep sequencing of patient genomes for disease diagnosis: when will it become routine? Sci Transl Med. 2011;3(87):87ps23.
12. Wetterstrand K. DNA Sequencing Costs: Data from the NHGRI Large-Scale Genome Sequencing Program Available at: http://www.genome.gov/sequencing-costs. Accessed 11-27-2011.
13. Bell CJ, Dinwiddie DL, Miller NA, Hateley SL, Ganusova EE, Mudge J, et al. Carrier testing for severe childhood recessive diseases by next-generation sequencing. Sci Transl Med. 2011;3(65):65ra4.
14. Lamb J, Crawford ED, Peck D, Modell JW, Blat IC, Wrobel MJ, et al. The Connectivity Map: using gene-expression signatures to connect small molecules, genes, and disease. Science. 2006;313(5795):1929–35.

15. Lamb J. The Connectivity Map: a new tool for biomedical research. Nat Rev Cancer. 2007;7(1):54–60.
16. Stolovitzky G, Monroe D, Califano A. Dialogue on reverse-engineering assessment and methods: the DREAM of high-throughput pathway inference. Ann N Y Acad Sci. 2007;1115:1–22.
17. Prill RJ, Saez-Rodriguez J, Alexopoulos LG, Sorger PK, Stolovitzky G. Crowdsourcing network inference: the DREAM predictive signaling network challenge. Sci Signal. 2011;4(189):mr7.
18. Afgan E, Baker D, Coraor N, Goto H, Paul IM, Makova KD, et al. Harnessing cloud computing with Galaxy Cloud. Nat Biotechnol. 2011;29(11):972–4.
19. Goecks J, Nekrutenko A, Taylor J. Galaxy: a comprehensive approach for supporting accessible, reproducible, and transparent computational research in the life sciences. Genome Biol. 2010;11(8):R86.
20. Kohonen-Corish M, Weber TK, Lindblom A, Macrae F. Report of the combined meeting of the International Society for Gastrointestinal Hereditary Tumours, the Human Variome Project and the National Cancer Institute Colon Cancer Family Registry, Duesseldorf, Germany, 24 June 2009. Fam Cancer. 2010;9(4):705–11.
21. Hulot JS, Bura A, Villard E, Azizi M, Remones V, Goyenvalle C, et al. Cytochrome P450 2C19 loss-of-function polymorphism is a major determinant of clopidogrel responsiveness in healthy subjects. Blood. 2006;108(7):2244–7.
22. Mega JL, Close SL, Wiviott SD, Shen L, Hockett RD, Brandt JT, et al. Cytochrome p-450 polymorphisms and response to clopidogrel. N Engl J Med. 2009;360(4):354–62.
23. Simon T, Verstuyft C, Mary-Krause M, Quteineh L, Drouet E, Meneveau N, et al. Genetic determinants of response to clopidogrel and cardiovascular events. N Engl J Med. 2009;360(4):363–75.
24. Collet JP, Hulot JS, Pena A, Villard E, Esteve JB, Silvain J, et al. Cytochrome P450 2C19 polymorphism in young patients treated with clopidogrel after myocardial infarction: a cohort study. Lancet. 2009;373(9660):309–17.
25. Roden DM, Shuldiner AR. Responding to the clopidogrel warning by the US food and drug administration: real life is complicated. Circulation. 2010;122(5):445–8.
26. Lesko LJ, Zineh I. DNA, drugs and chariots: on a decade of pharmacogenomics at the US FDA. Pharmacogenomics. 2010;11(4):507–12.
27. Dolgin E. Preemptive genotyping trialed to prevent adverse drug reactions. Nat Med. 2011;17(11):1323.
28. Meyer P, Alexopoulos LG, Bonk T, Califano A, Cho CR, de la Fuente A, et al. Verification of systems biology research in the age of collaborative competition. Nat Biotechnol. 2011;29(9):811–5.
29. NIH TNIoH. NIH Roadmap-National Centers for Biomedical Computing. 2011 [updated Nov 22, 2011; cited 2011 Nov 29]; Available from: http://www.bisti.nih.gov/ncbc/.
30. Walsh P. Asperger syndrome and the supposed obligation not to bring disabled lives into the world. J Med Ethics. 2010;36(9):521–4.
31. CDC. Autism Spectrum Disorders (ASDs). Center for Disease Control and Prevention; 2010 [cited 2011]; Available from: http://www.cdc.gov/ncbddd/autism/data.html.
32. Wright CF, Hall A, Zimmern RL. Regulating direct-to-consumer genetic tests: what is all the fuss about? Genet Med. 2011;13(4):295–300.
33. 23andme. 2011 [cited 2011]; Available from: https://www.23andme.com/.
34. Gewin V. Education: Inspiration for informatics. Nature. 2011;478(7367):143–5.
35. Madhavan S, Gusev Y, Harris M, Tanenbaum DM, Gauba R, Bhuvaneshwar K, et al. G-DOC: a systems medicine platform for personalized oncology. Neoplasia. 2011;13(9):771–83.
36. Pedersen BK. The diseasome of physical inactivity–and the role of myokines in muscle–fat cross talk. J Physiol. 2009;587(Pt 23):5559–68.
37. Chen T, Xie G, Wang X, Fan J, Qiu Y, Zheng X, et al. Serum and urine metabolite profiling reveals potential biomarkers of human hepatocellular carcinoma. Mol Cell Proteomics. 2011;10(7):M110 004945.
38. Xuan J, Pan G, Qiu Y, Yang L, Su M, Liu Y, et al. Metabolomic profiling to identify potential serum biomarkers for schizophrenia and risperidone action. J Proteome Res. 2011;10(12):5433–43.
39. Rieder MJ, Reiner AP, Gage BF, Nickerson DA, Eby CS, McLeod HL, et al. Effect of VKORC1 haplotypes on transcriptional regulation and warfarin dose. N Engl J Med. 2005;352(22):2285–93.
40. Herder C, Karakas M, Koenig W. Biomarkers for the prediction of type 2 diabetes and cardiovascular disease. Clin Pharmacol Ther. 2011;90(1):52–66.
41. CDC. National diabetes fact sheet: national estimates and general information on diabetes and prediabetes in the United States, 2011. Atlanta, GA: U.S. Department of Health and Human Services, Centers for Disease Control and Prevention; 2011.
42. Neumann KF, Rojo L, Navarrete LP, Farias G, Reyes P, Maccioni RB. Insulin resistance and Alzheimer's disease: molecular links & clinical implications. Curr Alzheimer Res. 2008;5(5):438–47.
43. Tabak AG, Jokela M, Akbaraly TN, Brunner EJ, Kivimaki M, Witte DR. Trajectories of glycaemia, insulin sensitivity, and insulin secretion before diagnosis of type 2 diabetes: an analysis from the Whitehall II study. Lancet. 2009;373(9682):2215–21.
44. Wilson PW, Meigs JB, Sullivan L, Fox CS, Nathan DM, D'Agostino Sr RB. Prediction of incident diabetes mellitus in middle-aged adults: the Framingham Offspring Study. Arch Intern Med. 2007;167(10):1068–74.
45. Selvin E, Ning Y, Steffes MW, Bash LD, Klein R, Wong TY, et al. Glycated hemoglobin and the risk of kidney disease and retinopathy in adults with and without diabetes. Diabetes. 2011;60(1):298–305.

46. Wang TJ, Larson MG, Vasan RS, Cheng S, Rhee EP, McCabe E, et al. Metabolite profiles and the risk of developing diabetes. Nat Med. 2011;17(4):448–53.
47. Billings LK, Florez JC. The genetics of type 2 diabetes: what have we learned from GWAS? Ann N Y Acad Sci. 2010;1212:59–77.
48. Kooner JS, Saleheen D, Sim X, Sehmi J, Zhang W, Frossard P, et al. Genome-wide association study in individuals of South Asian ancestry identifies six new type 2 diabetes susceptibility loci. Nat Genet. 2011;43(10):984–9.
49. Imamura M, Iwata M, Maegawa H, Watada H, Hirose H, Tanaka Y, et al. Genetic variants at CDC123/CAMK1D and SPRY2 are associated with susceptibility to type 2 diabetes in the Japanese population. Diabetologia. 2011;54(12):3071–7.
50. Dupuis J, Langenberg C, Prokopenko I, Saxena R, Soranzo N, Jackson AU, et al. New genetic loci implicated in fasting glucose homeostasis and their impact on type 2 diabetes risk. Nat Genet. 2010;42(2):105–16.
51. Strawbridge RJ, Dupuis J, Prokopenko I, Barker A, Ahlqvist E, Rybin D, et al. Genome-wide association identifies nine common variants associated with fasting proinsulin levels and provides new insights into the pathophysiology of type 2 diabetes. Diabetes. 2011;60(10):2624–34.
52. Kim YJ, Go MJ, Hu C, Hong CB, Kim YK, Lee JY, et al. Large-scale genome-wide association studies in east Asians identify new genetic loci influencing metabolic traits. Nat Genet. 2011;43(10):990–5.
53. de Miguel-Yanes JM, Shrader P, Pencina MJ, Fox CS, Manning AK, Grant RW, et al. Genetic risk reclassification for type 2 diabetes by age below or above 50 years using 40 type 2 diabetes risk single nucleotide polymorphisms. Diabetes Care. 2011;34(1):121–5.
54. Janipalli CS, Kumar MV, Vinay DG, Sandeep MN, Bhaskar S, Kulkarni SR, et al. Analysis of 32 common susceptibility genetic variants and their combined effect in predicting risk of Type 2 diabetes and related traits in Indians. Diabet Med 2012;29(1):121-127.
55. Suhre K, Shin SY, Petersen AK, Mohney RP, Meredith D, Wagele B, et al. Human metabolic individuality in biomedical and pharmaceutical research. Nature. 2011;477(7362):54–60.
56. Elliott SS, Keim NL, Stern JS, Teff K, Havel PJ. Fructose, weight gain, and the insulin resistance syndrome. Am J Clin Nutr. 2002;76(5):911–22.
57. Moore MC, Davis SN, Mann SL, Cherrington AD. Acute fructose administration improves oral glucose tolerance in adults with type 2 diabetes. Diabetes Care. 2001;24(11):1882–7.
58. Chong MF, Fielding BA, Frayn KN. Mechanisms for the acute effect of fructose on postprandial lipemia. Am J Clin Nutr. 2007;85(6):1511–20.
59. Barres R, Osler ME, Yan J, Rune A, Fritz T, Caidahl K, et al. Non-CpG methylation of the PGC-1alpha promoter through DNMT3B controls mitochondrial density. Cell Metab. 2009;10(3):189–98.

Epilogue

I would rather be vaguely right than precisely wrong.

John Maynard Keynes

Keynes summarises the aspirations, intelligences and knowledge shared in this *Handbook* in one insightful sentence. Systems and complexity thinking opens up our ability to comprehend the structure and function of dynamic systems and to appreciate the effects of system perturbations and shifts over time. But it never attempts to reduce a system's function to precise or fixed states. Depending on the current state or *initial condition* of any system under study, many *different but mutually agreeable* outcomes are possible and—as we all know from "real life experience"—will occur. This awareness provides a much clearer understanding of the fallacies perpetuated by the still prevailing reductionist and deterministic framework of research and practice.

Neurophysiology has shown why, indeed, it is difficult for one to see the forest for the trees. Unsurprisingly then, many cannot see the failings of the prevailing paradigm despite the repeating negative results of implementing highly controlled experimental findings in real-world dynamic systems.

Obviously even a book of this size will not be able to cover all there is to know about the application of systems and complexity in the health-care domains. And not all who have done so much outstanding work in the area in their specific fields of interest were able to contribute to this Handbook. We especially want to refer readers to the works of Ary Goldberger [1–5], Lewis Lipsitz [6–8], Reuben McDaniel [9–12], Will Miller and Ben Crabtree [13–15], James Koopmans [16–20], Soo Downe [21, 22], Christina Economos [23–26], Tony Suchman [27–29], and Alan Sheill and Penny Hawe [30].

We are hopeful that this *Handbook* may become the catalyst to start a broad network of systems and complexity researchers and practitioners in the health professions domain, and we encourage our readers to make contact with the editors to facilitate initial connections.

Information is "a difference which makes a difference"

Gregory Bateson

In closing we refer interested readers to the *Forum of Systems and Complexity in Health,* currently hosted by the *Journal of Evaluation in Clinical Practice*, as a place for ongoing exchange of systems and complexity-related discourses across the health system.

J.P. Sturmberg and C.M. Martin (eds.), *Handbook of Systems and Complexity in Health*,
DOI 10.1007/978-1-4614-4998-0, © Springer Science+Business Media New York 2013

References

1. Goldberger AL, Bhargava V, West BJ, Mandell AJ. On a mechanism of cardiac electrical stability. The fractal hypothesis. Biophys J. 1985;48(3):525–8.
2. Goldberger AL, West BJ. Applications of nonlinear dynamics to clinical cardiology. Ann NY Acad Sci. 1987;504(1):195–213.
3. Goldberger AL. Fractal mechanisms in the electrophysiology of the heart. IEEE Eng Med Biol Mag. 1992;11(2):47–52.
4. Goldberger A. Non-linear dynamics for clinicians: chaos theory, fractals, and complexity at the bedside. Lancet. 1996;347:1312–4.
5. Goldberger AL, Amaral LAN, Hausdorff JM, Ivanov PC, Peng C-K, Stanley HE. Fractal dynamics in physiology: alterations with disease and aging. Proc Natl Acad Sci USA. 2002;99 Suppl 1:2466–72.
6. Lipsitz LAMD, Mietus J, Moody GB, Goldberger ALMD. Spectral characteristics of heart rate variability before and during postural tilt: relations to aging and risk of syncope. Circulation. 1990;81(6): 1803–10.
7. Lipsitz L, Goldberger A. Loss of 'complexity' and aging. Potential applications of fractals and chaos theory to senescence. JAMA. 1992;267(13):1806–9.
8. Iyengar N, Peng CK, Morin R, Goldberger AL, Lipsitz LA. Age-related alterations in the fractal scaling of cardiac interbeat interval dynamics. Am J Physiol Regul Integr Comp Physiol. 1996;271(4):R1078–84.
9. McDaniel R. Strategic leadership: a view from quantum and chaos theories. Health Care Manage Rev. 1997;22(1):21–37.
10. Stroebel CK, McDaniel RRJ, Crabtree BF, Miller WL, Nutting PA, Stange KC. How complexity science can inform a reflective process for improvement in primary care practices. Jt Comm J Qual Patient Saf. 2005;31(8):438–46.
11. McDaniel RRJ, Lanham HJ, Anderson RA. Implications of complex adaptive systems theory for the design of research on health care organizations. Health Care Manage Rev. 2009;34(2):191–9. doi:10.1097/HMR.0b013e31819c8b38.
12. Jordon M, Lanham HJ, Anderson RA, McDaniel Jr RR. Implications of complex adaptive systems theory for interpreting research about health care organizations. J Eval Clin Pract. 2010;16(1):228–31.
13. Miller W, Crabtree B, McDaniel R, Stange K. Understanding change in primary care practice using complexity theory. J Fam Pract. 1998;46(5):369–76.
14. Crabtree BF, Miller WL, Aita VA, Flocke SA, Stange KC. Primary care practice organization and preventive services delivery: a qualitative analysis. J Fam Pract. 1998;46(5):403–9.
15. Miller WL, McDaniel Jr RR, Crabtree BF, Stange KC. Practice Jazz: understanding variation in family practices using complexity science. J Fam Pract. 2001; 50(10):872–8.
16. Koopman JS, Longini IMJ. The ecological effects of individual exposures and nonlinear disease dynamics in populations. Am J Public Health. 1994;84(5): 836–42.
17. Koopman J, Jacquez G, Chick S. New data and tools for integrating discrete and continuous population modeling strategies. Ann NY Acad Sci. 2001; 954: 268–94.
18. Koopman JS. Infection transmission science and models. Jpn J Infect Dis. 2005;58(6):S3–8.
19. Li S, Eisenberg JNS, Spicknall IH, Koopman JS. Dynamics and control of infections transmitted from person to person through the environment. Am J Epidemiol. 2009;170(2):257–65.
20. Kim J-H, Riolo RL, Koopman JS. HIV transmission by stage of infection and pattern of sexual partnerships. Epidemiology. 2010;21(5):676–84.
21. Downe S. Beyond evidence-based medicine: complexity and stories of maternity care. J Eval Clin Pract. 2010;16(1):232–7.
22. Downe S, McKeown M, Johnson E, Comensus Community Involvement T, Comensus Advisory G, Koloczek L, et al. The UCLan community engagement and service user support (Comensus) project: valuing authenticity, making space for emergence. Health Expect. 2007;10(4):392–406.

J.P. Sturmberg and C.M. Martin (eds.), *Handbook of Systems and Complexity in Health*, DOI 10.1007/978-1-4614-4998-0, © Springer Science+Business Media New York 2013

23. Economos CD, Hyatt RR, Goldberg JP, Must A, Naumova EN, Collins JJ, et al. A community intervention reduces BMI z-score in children: shape up Somerville first year results[ast]. Obesity. 2007;15(5):1325–36.
24. Goldberg JP, Collins JJ, Folta SC, McLarney MJ, Kozower C, Kuder J, et al. Retooling food service for early elementary school students in Somerville, Massachusetts: the shape up Somerville experience. Prevent Chronic Dis. 2009;6(3):A103.
25. Economos C, Folta S, Goldberg J, Hudson D, Collins J, Baker Z, et al. A community-based restaurant initiative to increase availability of healthy menu options in Somerville, Massachusetts: shape up Somerville. Prevent Chronic Dis. 2009;6(3):A102.
26. Economos CD, Curtatone JA. Shaping up Somerville: a community initiative in Massachusetts. Prev Med. 2010;50:S97–8.
27. Matthews D, Suchman A, Branch WJ. Making "Connexions": enhancing the therapeutic potential of patient-clinician relationships. Ann Intern Med. 1993;118(12):973–7.
28. Borrell-Carrió F, Suchman A, Epstein R. The biopsychosocial model 25 years later: principles, practice, and scientific inquiry. Ann Fam Med. 2004;2(6):576–82.
29. Suchman AL. A new theoretical foundation for relationship-centered care. Complex responsive processes of relating. J Gen Intern Med. 2006;21 Suppl 1:S40–4.
30. Shiell A, Hawe P, Gold L. Complex interventions or complex systems? Implications for health economic evaluation. Br Med J. 2008;336(7656):1281–3.

Index

J.P. Sturmberg and C.M. Martin (eds.), *Handbook of Systems and Complexity in Health*, DOI 10.1007/978-1-4614-4998-0, © Springer Science+Business Media New York 2013

E

F

N

R

S

MIX
Papier aus verantwortungsvollen Quellen
Paper from responsible sources
FSC® C105338

If you have any concerns about our products,
you can contact us on
ProductSafety@springernature.com

In case Publisher is established outside the EU,
the EU authorized representative is:
Springer Nature Customer Service Center GmbH
Europaplatz 3, 69115 Heidelberg, Germany

Printed by Libri Plureos GmbH
in Hamburg, Germany